Environmental Forces on Engineering Structures

Proceedings of the First International Conference
held at Imperial College, London, July 1979.

[International Conference on Environmental
Forces on Engineering structures 1st 1979]

Edited by

C. A. Brebbia, *University of Southampton*
P. L. Gould, *Washington University*
J. Munro, *Imperial College, London*

A HALSTED PRESS BOOK

JOHN WILEY & SONS
New York — — Toronto

Published in the U.S.A. and Canada
by Halsted Press, a Division of
John Wiley & Sons, Inc., New York.

© The several contributors named in
the list of contents, 1979

Library of Congress Cataloging—in Publication Data
Main entry under title:

Environmental forces on engineering structures.

    "A Halsted Press book."
    "Edited version of the papers presented at the
First International Conference on Environmental
Forces on Engineering Structures, held at Imperial
College, London, in July 1979."
    Includes non-conference papers.
    1. Structural dynamics--Congresses. I. Brebbia,
C. A. II. Gould, Phillip L. III. Munro, J.
IV. International Conference on Environmental Forces
on Engineering Structures, 1st, Imperial College,
1979.
TA654.E58   624'.176   79—16733
ISBN 0—470—26820—4

Printed in Great Britain by
Redwood Burn Limited
Trowbridge & Esher

Preface

This book contains an edited version of the papers presented at the First International Conference on Environmental Forces on Engineering Structures, held at Imperial College, London in July 1979. The topic of the Conference was engineering structures subjected to natural environmental forces as the main design requirement. This includes wind, ocean environment and earthquake effects on structures as well as such other topics as temperature effects, fatigue, vehicle loading etc. A special session was devoted to the dynamic behaviour of structures.

The increase in the size and complexity of engineering structures has necessitated new and more accurate methods of analysis. This book provides a unique state of the art report for the methods as well as their applications to different problems.

The topic of wind on structures is discussed in detail, its dynamic effects and comparison of analytical with experimental results. Applications are presented for several different large structural shapes such as cooling and offshore towers, buildings, bridges etc.

Large and complex structural shapes are frequently used for offshore structures, where a knowledge of the dynamic behaviour is important. Methods of analysis have recently been developed for the structures which have much in common with those employed to study the effects of winds. The problems are somewhat more complex as in addition to the wave loads, the structure may be subjected to currents, temperature gradients, slamming and impact forces, etc. The structural system itself may complicate the problem, as tethered, moored or fixed structures will behave very differently.

In the past earthquake effects on structures have been analysed much more carefully than wind or wave effects. Much of the analytical work now used to predict the behaviour of structures under environmental loads has originated with earthquake engineering and it was important to include this topic in the Conference in order to evaluate some of the new developments in this field.

A series of papers on other environmental loading topics and vibrations are also included. They give an idea of the variety of subjects covered by environmental loadings, their importance in engineering and the need to understand better the dynamics — and in some cases the dynamics and buckling of large structural systems.

The Editors

CONTENTS

SECTION I   WIND AND EARTHQUAKE EFFECTS

SECTION I

WIND AND EARTHQUAKE EFFECTS

ENVIRONMENTAL LOADINGS ON CONCRETE COOLING TOWERS - TYPES,
LIKELIHOOD, EFFECTS AND CONSEQUENCES
Phillip L. Gould

Professor and Chairman, Department of Civil Engineering
Washington University, St. Louis, MO 63130, U.S.A.

INTRODUCTION

The major loadings for which large cooling towers must be designed are environmental as opposed to mechanical in origin.
Among the more familiar are temperature variations, wind
pressure and earthquake motion. There is also the possibility
of icing in cold climates, missiles driven by extreme wind conditions and base disturbances due to poor soil, blasting or
mining.

The objectives of this paper are (1) to describe some
commonly encountered environmental loading conditions (thermal,
wind and seismic); (2) to examine the likelihood of each
occurring; (3) to characterize the predominant structural
effects; and (4) to point out some consequences for design.

In an attempt to accomplish these objectives, it is realized
that the state-of-the-art in each case is progressing. Therefore, it is expected that the procedures described in this
paper will be modified and refined as the characterization
and the data base for the various loading types are expanded.

DESCRIPTION OF ENVIRONMENTAL LOADS

Thermal Loading
Thermal effects on hyperbolic natural draft cooling towers
have been studied by Larrabee, Billington and Abel (1974).
They considered two conditions: (1) Rising hot air in an operating tower which heats the inside of the shell; and (2)
Solar heating of one side of a tower. They studied the magnitude of the induced displacements due to solar heating and
the comparison of measured to calculated displacements.

Since the stresses and displacements calculated from the
method of Larrabee, et.al. are very dependent on the assumed

thermal gradients, it is important to have realistic data. Krätzig, Peters and Zerna (1978) mention a 35°C gradient for the uniform operating conditions and a 25°C gradient for the solar heating condition.

A deficiency in the constant (with height) thermal gradient assumption is the failure to recognize that the temperature difference must dissipate at the interface between the inner and outer regions of the shell at the underside of the lintel (Fig. 1).

Figure 1. Lower lintel region of shell

Since the gradient in this vicinity has a major effect on the
induced loads in the columns, a more refined and realistic
local description would be desirable.

The action of natural wind on hyperbolic cooling towers has
been studied extensively since this is the prevailing design
condition for most tall towers. It is generally
assumed that the wind storm is larger in breadth than the
tower diameter and that the tower is an isolated structure,
unaffected by surrounding buildings or towers. Normally, the
dynamic wind loading is separated into mean and fluctuating
components. The mean component is essentially static and the
fluctuating component is largely quasistatic since resonance
effects are usually minor, Niemann (1977).

Commonly, vertical wind profiles based on the terrain roughness and a designated return period is used in accordance
with applicable standards for tall structures.

With respect to the circumferential wind distribution, the
circular cross-section of the tower results in flow separation and a rapidly varying external normal pressure loading
as shown in Fig. 2.

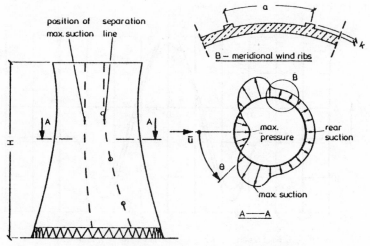

Figure 2. Wind loading on shell
[from Niemann(1977)]

Such pressure distributions have been obtained by a combination of wind tunnel and full scale measurements. At $\theta = 0°$
incident to the wind, the maximum external pressure is the
stagnation pressure. Other key values are the angle of separation and the difference between the maximum suction and the
rear suction, Δp, as shown in Fig. 3.

6

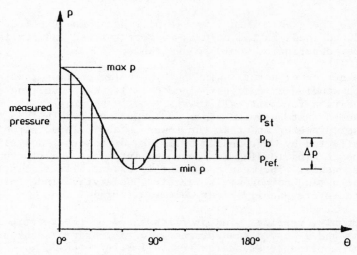

Figure 3. Definition of pressure data in
full-scale measurements
[from Niemann(1977)]

The circumferential variation of the external pressure is conveniently defined by normalized pressure coefficients c_p as
shown in Fig. 4.

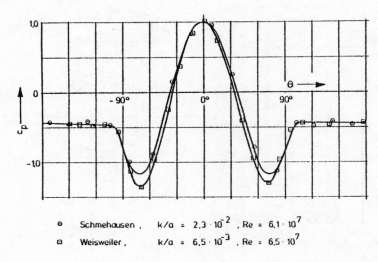

| ⊙ | Schmehausen , | $k/a = 2.3 \cdot 10^{-2}$, | $Re = 6.1 \cdot 10^7$ |
| ⊡ | Weisweiler , | $k/a = 6.5 \cdot 10^{-3}$, | $Re = 6.5 \cdot 10^7$ |

Figure 4. Full-scale pressure distributions
[from Niemann(1977)]

Particularly significant is the pressure rise coefficient Δc_p which corresponds to Δp and which has been shown to be a function of the surface roughness k/a (Fig. 2), as shown in Fig. 5.

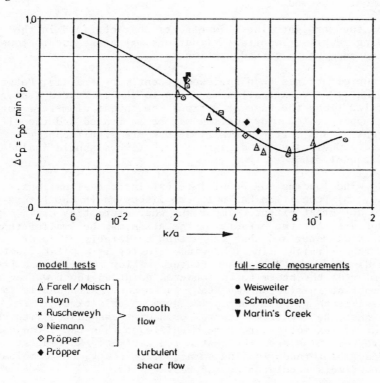

modell tests

△ Farell / Maisch
⊡ Hayn
× Ruscheweyh
⊙ Niemann
◇ Pröpper

} smooth flow

◆ Pröpper turbulent shear flow

full - scale measurements

● Weisweiler
■ Schmehausen
▼ Martin's Creek

Figure 5. Influence of surface roughness on the pressure rise coefficient [from Niemann(1977)]

The effect of surface roughness is reflected in the design pressure curves presented by Niemann (1977).

	k/a	I	II	III
K 1,0	$3 \cdot 10^{-2}$	$1 - 2.0(\sin\frac{90}{70}\varphi)^{2.267}$	$-1.0 + 0.5\{\sin[\frac{90}{21}(\varphi-70)]\}^{2.395}$	-0.5
K 1,1	$2 \cdot 10^{-2}$	$1 - 2.1(\sin\frac{90}{71}\varphi)^{2.239}$	$-1.1 + 0.6\{\sin[\frac{90}{22}(\varphi-71)]\}^{2.395}$	-0.5
K 1,2	$1.2 \cdot 10^{-2}$	$1 - 2.2(\sin\frac{90}{72}\varphi)^{2.205}$	$-1.2 + 0.7\{\sin[\frac{90}{23}(\varphi-72)]\}^{2.395}$	-0.5
K 1,3	$8 \cdot 10^{-3}$	$1 - 2.3(\sin\frac{90}{73}\varphi)^{2.166}$	$-1.3 + 0.8\{\sin[\frac{90}{24}(\varphi-73)]\}^{2.395}$	-0.5

Figure 6. Pressure distributions for design purpose [from Niemann(1977)]

Variations with height are not included in the preceding curves.

In addition to the external pressure loading, an internal suction is usually present and a constant value of $c_{ps} = 0.5$ is commonly assumed.

Besides the straight line wind effects characterized in the preceding remarks, important situations may arise due to tower grouping and due to tornado action.

The group effect has been reviewed recently by Krätzig, Peters and Zerna (1978). Attention is drawn to shells spaced closer than 1-1/2 times the base or 2 times the throat diameter. For such groupings, the side suction may be increased and the circumferential and vertical distributions are significantly affected. Obviously the design of such towers should include a wind tunnel study.

Tornado wind loading may be of interest in some situations. To date, this condition has not been discussed in the literature to any great extent. Important characteristics of tornados are that the breadth of the storm may be smaller than the tower diameter and that the motion is largely circular rather than straight line. The wind velocity is a point function of the distance to the tornado center, making the distance from the center of the storm to an impacted point on the tower at a given instant in time an important parameter. Although it is not possible to present quantative data at this time, it is likely that the negative external pressures associated with the circular vortex motion are more significant than the aerodynamic pressure resulting from the obstructed airflow. To realistically describe tornado loading on a cooling tower, the combination of both effects should be considered.

Another aspect of tornado wind loading which differs from the large scale windstorm usually considered in design is the vertical profile. Indications are that the usual power law and logarithmic distributions based on surface roughness are not particularly applicable to tornados.

Seismic Loading
The effect of an earthquake acting on a cooling tower has been studied extensively with the assumption that the entire shell is subject to a ground motion which imparts a uniform translation, horizontal or vertical, to the base of the shell. The resulting accelerations introduced into the structure are dependent on the physical properties of the shell, primarily the mass, natural frequencies and damping, and the characteristics of the ground motion, primarily the intensity, fluctuation and duration of the shock.

The mass of the shell may be readily calculated, and the

damping is generally taken as 2-5% of critical in accordance with common practice for concrete structures. Natural
frequencies in the symmetric (j=0) and antisymmetric or beam
(j=1) modes may be calculated by analytical means with fairly
good confidence, particularly if the columns at the base of
the shell are realistically modelled, Gould, Sen and Suryoutomo (1974). Typical values for the lowest frequencies of
tall towers are 5-7 Hz for j = 0 and 2-3 Hz for j = 1.

The design earthquake is ideally based on the local seis-
micity and is generally established from technical and economical considerations. It may be presented in the form of a
complete or partial time history or a response spectrum, as
shown in Fig. 7, which reflects only maximum effects.

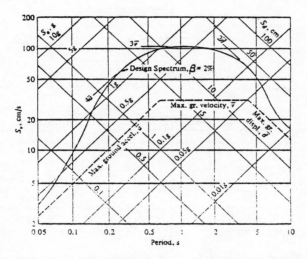

Figure 7. Design response spectrum constructed
           from maximum ground-motion characteristics
           [from Clough and Penzien(1975)]

With the dynamic properties of the shell determined and the
appropriate response spectrum specified, the loading on the
shell due to a design earthquake may be established. At
least two and preferably the three lowest frequencies should
be used, especially for the j = 1 case, since the higher
modes affect the forces introduced into the columns in
particular.

To more realistically characterize the influence of an earthquake on a cooling tower, the idealization of uniform base
motion may be generalized to examine the seismic wave effect
and the interaction of the supporting soil with the tower
foundation.

10

The seismic wave effect, as characterized by Scanlan (1976), accounts for the horizontal travel time of the seismic wave across the base of the structure. This time is not insignificant if the structure is broad, such as a cooling tower. It is likely that the resulting effects of an earthquake on such a structure may be less severe than those calculated assuming uniform ground motion because of interference.

Soil-structure interaction is related to the seismic wave effect but can also be considered separately. Essentially, the soil may act as an energy absorbing medium and thereby reduce the seismic forces introduced into the shell. A model of a cooling tower with a shallow foundation and the accompanying soil mass is shown in Fig. 8.

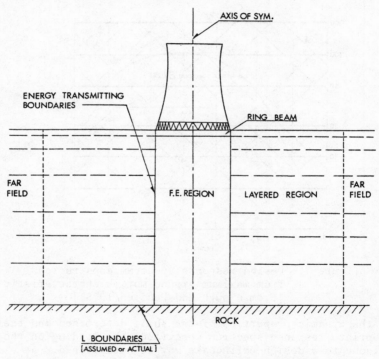

Figure 8.   Soil-structure interaction model
for cooling tower

LIKELIHOOD OF ENVIRONMENTAL LOADS

In specifying and combining various loading conditions for design, it is of interest to consider the expected lifetime of the structure, the probability of a particular condition occurring during this period and the consequences of a failure. Certainly the self-load and equipment loading will

be ever present while the possibility of a tower being hit by
an intense tornado seems to be remote. Within the bounds of
certainty and extreme unlikelihood fall the three conditions
described in the preceding section.

The thermal loading, particularly under operating conditions,
is the most likely to occur and should be considered as such
in design. The frequency of occurrence of major wind storms
is fairly well established through meteorological data and
50 to 100 year return periods are commonly taken for tower
design. On the other hand, the return period for strong
motion earthquakes at a particular site may be very long and
there is a sparsity of data for most regions of the world.
Except for a few locations, it is probably more likely that a
given structure will be subjected to an extreme wind load
than to a strong motion earthquake during its lifetime. This
suggests that with respect to earthquake loading in particular,
the assumption of more risk through the use of smaller design
loadings and the acceptance of the accompanying damage should
a larger event occur might merit consideration. This is, of
course, largely an economic question and is becoming more important since the seismic provisions for cooling towers have
been quite severe in some recent design specifications. A
framework for this approach is the safety concept advanced by
Krätzig, et. al. (1976) in which the probability of failure
is explicitly included in the load factors for the lateral
load.

STRUCTURAL EFFECTS

Method of Analysis
In the studies referred to in this section, the structural
analyses have generally been carried out using finite
difference or finite element approaches based on the bending
theory of thin shells. For the most part, homogeneous and isotropic properties have been assumed but there is some work in
the literature considering orthotropic properties which preports
to represent shell cracking. This consideration may have a
marked influence on the results but is not examined herein.

Thermal Loading
Considering the operating conditions it was found by Larrabee,
et. al. (1974) that overall uniform heating would cause no
stresses except at restrained boundaries as shown in Fig. 9. A
thermal gradient was found to introduce substantial moments
because of the assumed clamping of the base of the shell as
shown in Fig. 10.

12

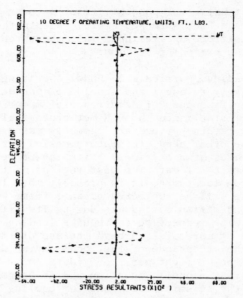

Figure 9. Membrane stress resultants due to operating
temperature, Martins Creek cooling tower
[from Larrabee, et. al. (1974)]

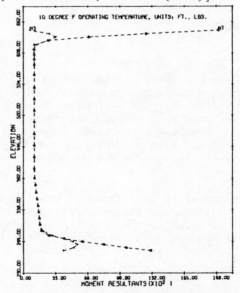

Figure 10. Moments due to operating temperature, Martins
Creek cooling tower
[from Larrabee, et. al. (1974)]

It appears that these findings are somewhat conservative in that the base of an actual shell is far from fully restrained (Fig. 1) and, also, the thermal gradient would be expected to dissipate near the base of the shell as discussed earlier.

Considering the solar heating of one side of the tower, the thermal gradient may be assumed to vary in a convenient harmonic fashion around the shell. Substantial bending stresses were calculated by Larrabee et. al. but, again, the previously mentioned conservative assumptions regarding the base restraint and the thermal gradient were used. An additional effect due to the nonsymmetric loading is a radial displacement which causes the structure to lean,as indicated by the displacements shown in Fig. 11.

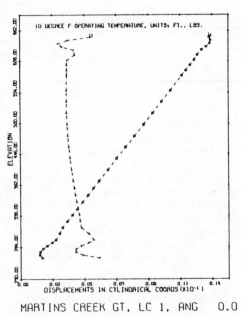

MARTINS CREEK GT. LC 1. ANG   0.0

Figure 11.   Displacements due to operating temperature,
             Martins Creek cooling tower
             [from Larrabee, et. al. (1974)]

Wind Loading
The wind loading, as previously mentioned, may be considered to have steady and fluctuating components. Hashish and Abu-Sitta (1974) have provided an interesting representation for the meridional stress resultant N_ϕ in the lower part of the shell as a function of the mean velocity $\bar{V}$ as shown in Fig. 12.

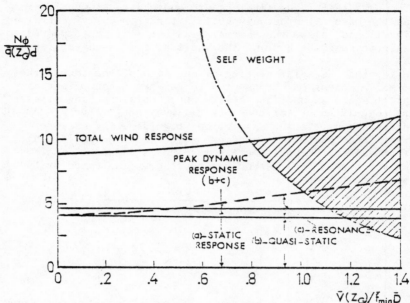

Figure 12. Variation of meridional stress resultant
with wind velocity
[Adapted from Hashish and Abu-Sitta (1974)]

This graph indicates that the peak dynamic response is basically static for the steady component and quasistatic for the
fluctuating component until the velocities become high,where-
upon resonance has some effect. In view of some further
studies, Niemann (1977) and Gould and Basu (1977), an
amplification of perhaps 10% on the static plus quasistatic
value of N_ϕ might be in order for an extreme wind condition.

Also shown in Fig. 12 is the self-weight of the shell plotted
such that the shaded regions correspond to wind velocities for
which the tension due to wind load is greater than the compression due to dead load at the windward meridian.

For a given design wind velocity, a similar plot over the
height of the shell indicates those regions for which net
tension will be present and therefore where reinforcing steel
to resist wind induced tension will be required. A typical
plot is shown on Fig. 13. A very important point with respect to Figure 13 is that *factored* loads should be used as
the basis for the design of the reinforcement. It may be
unsafe to simply take the difference between the dead and
wind load stress resultants and proportion the steel
accordingly, particularly if they are close in magnitude.
In such a case, a modest increase in the wind loading could

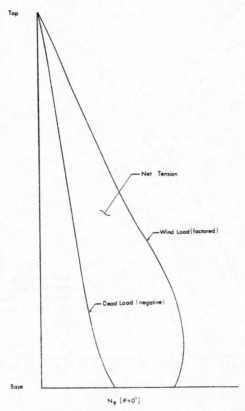

Figure 13. Variation of meridional stress resultant
with height

produce a tremendous increase in the net tension with dis-
astrous consequences.

Similar comparisons should be made for other stress resultants,
especially the hoop stresses.

Earthquake Loading
The effects of lateral earthquake loading are very similar to
those of wind with respect to the shell proper and vertical
earthquake loading has little effect on the shell.

However, the region at the base of the shell in the vicinity
of the shell-column connection is greatly influenced by
earthquake loading, along with the columns themselves. As
shown on Fig. 14, a large relative displacement between the
top and bottom of the columns is introduced resulting in
large column moments and shears. This effect is exacerbated
by the second longitudinal mode response, in particular.

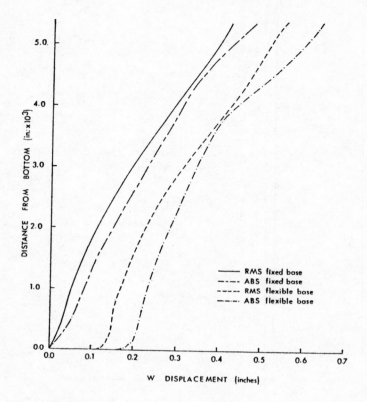

Figure 14.   Maximum normal displacement at θ = 0°
             due to earthquake
             [from Gould, Sen and Suryoutomo (1974)]

Note that an idealized fixed base condition does reveal
this important effect.

Also, the combination of meridional loads resulting from the
overturning moment of the earthquake and horizontal loads
coming from the base shear produces axial along with other
forces and moments in the inclined columns. To evaluate
these forces and to select the most critical case for design,
it is suggested that the displacements and rotations at the
ends of each column be introduced into the stiffness equations
of the column system, Basu and Gould (1978).

Earthquake loading can also produce severe effects on the
foundation system. It has sometimes been casually stated
that "wind controls" or "earthquake controls" for a particular design. In reality, both may be critical with the wind
effects being predominant in the shell and the earthquake

effects being more severe in the columns and foundations.

DESIGN CONSEQUENCES

The major effect of thermal loadings on shell structures is to cause cracking in regions of tensile stress. The cracking, in turn, reduces the stiffness of the cross section so that the resulting forces are generally less than those calculated based on isotropic properties. One may say that the stresses are self-relieving and the initially calculated design forces may be meaningless.

The major problem is then to prevent such cracking from widening to the point where corrosion and other deterioration will be accelerated and the overall integrity of the structure reduced. To this end, placing the reinforcement in two layers, both meridionally and circumferentially, is thought to be beneficial for cooling tower shells.

With respect to the stresses in the shell introduced by wind and horizontal earthquake motion, the necessity of considering factored loads in computing the net tension has been discussed earlier. In addition, it may be necessary to increase the reinforcement due to local bending effects, particularly near the base of the shell and perhaps near the throat and top. Additionally, stability of the tower under wind loading should be considered and this may be the determining factor for the minimum shell thickness.

The cooling tower columns present several design problems. First of all, under lateral loads the design forces may be magnified by the drift of the columns (P-Δ effect) and this should be included. Also, it should be recognized that some columns may be under net tension and biaxial bending and the reinforcement details should reflect this severe condition, particularly the anchorages and splices. It is felt that welded splices are probably the best possibility in this regard. For seismic conditions, confinement steel in the form of hoops or spirals should be provided in accordance with prevailing design codes.

CONCLUSIONS

A discussion of the major types of environmental loadings which may act on tall cooling towers is presented. The emphasis is directed toward those aspects which are important in the design of such structures. In each case the present design representations of the various loading conditions are explored and needs for more realistic representations are identified if possible.

REFERENCES

Basu, P. K. and Gould, P. L. (1979) Finite Element Discretization of Open-Type Axisymmetric Elements, Int. J.
Num. Meth. Engrg.

Clough, R. W. and Penzien, J. (1975) Dynamics of Structures.
McGraw-Hill Co., New York: 542

Gould, P. L., Sen, S. K. and Suryoutomo, H. (1974) Dynamic
Analysis of Column-Supported Hyperboloidal Shells, J. Earthq.
Engrg. Struct. Dyn., 2: 269-279

Hashish, M. G. and Abu-Sitta, S.H. (1974) Response of Hyperbolic Cooling Towers to Turbulent Wind, J. Struct. Div. ASCE,
100, ST5:1037-1051

Krätzig, W. B., Peters, H. L. and Zerna, W., (1978) Naturzug-
kühltürme aus Stahlbeton-Derzeitiger Stand und Entwicklungs-
möglichkeiten, Beton-Und-Stahlbetonbau, 2:37-42; 3:66-72

Larrabee, R. D., Billington, D.P. and Abel, J. F. (1974) Thermal
Loading of Thin-Shell Concrete Cooling Towers, J. Struct. Div.
ASCE, 100, ST12:2367-2383

Niemann, H. T. (1977) Static and Dynamic Wind Effects on
Cooling Tower Shells. Preprint 3031 ASCE Fall Convention
and Exhibit

Scanlan, R. (1976) Seismic Wave Effects on Soil-Structure
Interaction, J. Earthq. Engrg. Struc. Dyn., 4:379-388

NUMERICAL SIMULATION OF THREE-DIMENSIONAL SEPARATED FLOW: A FIRST STEP

Robert L. Street   Peter Y. Ko

Department of Civil Engineering   Department of Civil, Mechanical
Stanford University   and Environmental Engineering
U.S.A.   George Washington University
                                    U.S.A.

ABSTRACT

The forces which arise as the wind acts on a structure are
significantly affected by flow separation. To date only
relatively crude models of the actual separated flow exist,
having been developed from correlations from laboratory and
field data, two-dimensional flows and three-dimensional wake
flow calculations. Techniques developed by the authors for
three-dimensional fully cavitating flow and employing the
finite element method (FEM) and free streamline theory are
shown to be applicable as a first step toward the development
of an accurate numerical simulation of three-dimensional
separated flow about buildings. An algorithm which establishes
the location of the surface of separation by iteration and,
hence, allows rearrangement of the FEM mesh to optimize the
accuracy of the flow solution is described. Tests show that
the errors inherent in this algorithm for locating the separation surface are negligible compared to the discretization
error in the FEM itself. Typical results for the location of
the separation surface behind and forces on ellipsoidal plates
are presented. The logic and means for extending the method
(a) to establish a FEM mesh which varies to obtain high
accuracy in areas of concentrated vorticity (e.g., the separation surface) and (b) to imbed the scheme in a three-dimensional simulation model for turbulent flow past structures are
outlined.

INTRODUCTION

The nature of the problem

An overview of the effect of wind on structures and the
influence of flow separation on wind forces and on the flow
circulation behind structures is given by Sachs (1978,

Chapter 7). He points out that the wind has two main effects.
First, it exerts forces and moments on a building. Second,
the wind distributes air in and around the building. Of
particular interest to this paper is the fact that many buildings interact with the wind to produce a separated flow with
a zone of high vorticity and shear near the separation surface
which demarks the streaming flow and the wake. Plate (1971,
pp. 161 ff) discusses shelter belts or, in particular, two-
dimensional flows in which a barrier (for example, a building)
interacts with the boundary layer flow over the surface of the
earth. Plate discusses the complexity of these flows finding
no less than seven flow zones of different aerodynamic behavior
which can be distinguished.

Review of the Literature
The following review of some of the relevant literature establishes several points. First, a majority of the useful information about wind loading and flow patterns has been obtained
from laboratory model experimentation. Second, the available
numerical calculations are crude. Third, irrotational free
streamline theory has been used to attack this very complex
problem.

Baines (1965) measured pressure distributions on models of
walls and rectangular block structures in a wind tunnel in
which an artificially produced velocity gradient was used to
simulate natural conditions and a constant velocity field was
used for comparison with standard procedures. He deduced
several useful rules for predicting pressure distributions and
wind loads with a specialized wind field and was able to
identify some distinct flow phenomena. Castro and Robins (1977)
conducted an experimental investigation of the flow about
surface-mounted cubes in both uniform and sheared turbulent
flows. The shear flow was an attempt to simulate atmospheric
boundary layer effects. They related the size of the cavity
zone or reversed flow region directly behind their cubes to
flow conditions. They also identified strong vortex effects
in cases of cubes set at a large angle to the flow. They
concluded that theoretical and numerical methods, which could
ultimately be quite helpful in determining the effect of
various parameters on the flow, are only in the preliminary
stages of development. Castro and Robins felt that there was
a great deal yet to be done in theory, in the area of numerical
analysis, and in the area of additional definitive experiments.

The dispersion and mixing phenomena in the vicinity of bluff
bodies is another area of focus in the recent literature.
Robins and Castro (1977a; 1977b) examined the velocity and
turbulence fields associated with both the approach flow and
a building wake in a wind tunnel. Their first paper examined
the flow field, while their second paper focused on the concentration field of a plume of pollutant. They investigated

the flow about a surface-mounted cube in a simulated atmosphere boundary layer, the cube height being 1/10 of the boundary layer height. Their experimental investigation yielded descriptions of the pressure forces on and flow around the cube and qualitative conclusions regarding a number of important effects of body orientation and approach flow characteristics. The pollutant concentration field downstream of the pollutant source was significantly modified by the presence of bluff bodies.

Vincent (1977) used wind tunnel experiments with rectangular blocks in uniform smooth and turbulent air flow to simulate transport of pollutants near buildings. From his experiments he derived an expression for the mean concentration level inside the flow reversal or cavity zone behind a building and argued that two dimensionless parameters could account for the complex physical properties in the wake transport zone. Sheih, et al. (1978) focused on a simpler problem of two-dimensional flow and attempted to construct a simple mathematical expression capable of predicting pollutant flux from the free stream across a mixing zone and into the cavity or separation bubble behind their two-dimensional fence. These last two papers represent initial attempts at constructing predictive equations from experimental data.

Finally, Halitsky (1977) made field measurements of the wind properties downwind of a reactor containment structure. He made simultaneous measurements of concentrations of a tracer released in the lee of the reactor structure and found that these measured concentrations were in agreement with a mathematical wake dispersion model. His analysis provided insights into the nature and properties of building wakes in the atmosphere and their influence on the dispersion of released material.

On the computational side, Hirt and Cook (1972) demonstrated a computing technique for viscous fluid flow in which calculation of three-dimensional flows in the vicinity of one or more block-type structures was possible. The full Navier-Stokes equations were solved with a finite difference scheme and the effects of thermal buoyancy were included. Flow convection was traceable through marker particles and diffusive effects were crudely represented. This early simulation did not, however, include turbulence modeling.

From a purely theoretical point of view Lin and Landweber (1975) considered the flow past a bluff body to be composed of a wake flow and an outer potential flow. For a two-dimensional fence they provided an irrotational free streamline flow solution with a finite separation or flow reversal zone.

CONCEPTUALIZATION

The flow past buildings is turbulent and normally separates.
The flow boundaries are complex, both at solid boundaries and
in the internal separation zones. At least three flow regions
can be defined near structures. First is the "free" turbulent
flow in the boundary layer and over the forward walls of the
building. Second is the high-shear separation zone, together
with a developing mixing region as one moves downstream from
the structure. The third region is the flow reversal bubble
or cavity zone immediately behind the structure. Far downstream (beyond the cavity) a wake flow is observed. Accordingly, one needs to build into a single computational scheme
(i) a numerical method which models flow turbulence, (ii) the
ability to accurately represent the high shear and mixing
zones and (iii) the ability to clearly define the cavity or
bubble region. Our experience suggests that such a computation
scheme can be developed via a combination of three distinct
components and within the present state-of-the-art.

Numerical models of turbulent flow
Reynolds (1976) gave a comprehensive review of the computation
of turbulent flows. Of particular interest is his description
of large-eddy simulations, an approach which Reynolds says is
just beginning to bear fruit. The fundamental idea is to
devise a numerical technique in which large-scale motions and
large-scale turbulence is simulated explicitly; the smaller
scales, which cannot be computed in any real flow in any case,
are modeled. In addition, Ferziger (1977) presented arguments
in favor of large-eddy simulation, described the basis for the
method and presented some typical results. Findikakis and
Street (1979) developed an algebraic model for the small-scale
turbulence, which is not simulated but must be modeled. They
demonstrated that such a turbulent model produces reasonable
physical effects and can account for the effects of buoyancy
on turbulence, i.e., augmentation or suppression depending on
whether the flow is stably or unstably stratified.

Deardorff (1970) demonstrated that the three-dimensional
(primitive) equations of motion could be integrated numerically
in time for the case of turbulent, plane Poiseuille flow at
very large Reynolds numbers. He concluded from his numerical
study of the channel flow that the numerical approach to the
problem of turbulence at large Reynolds numbers was already
profitable with increased accuracy to be expected with modest
increase of numerical resolution. Since 1970 advances in the
computer state-of-the-art have made such calculations within
the realm of a routine, albeit expensive, effort.

Finite element models
The finite element method (FEM) provides the freedom needed to
align a numerical-calculation grid system and to adjust node

point spacing, so as to accurately map regions of concentrated
vorticity, rapid velocity change or rapid concentration change,
etc. Because the separated flow region behind a building has
a close analogy in the free surface jet flow from orifices or
in cavitation resulting from high speed underwater flow, it is
useful to note certain achievements in the free surface flow
area.

Larock and Taylor (1976) demonstrated that the finite element
technique could be used for the analysis of high speed, three-
dimensional free surface jet flows within the context of
potential flow theory. An important feature of their work
and that of Street and Ko (1977) was that the unknown free
surface was located by iteration and the finite element mesh
was adjusted between each iteration to place the free surface
or separation surface in a more accurate or better position.
Street and Ko (1977) [see also Street (1977)] applied the FEM
to three-dimensional fully cavitating flows which are directly
analogous to the separated building flows that are under consideration here.

Findikakis, et al. (1978) combined the FEM with the large-eddy
simulation mentioned above to simulate stratified turbulent
flows in enclosed water bodies. Thus, they demonstrated that
the important features of the FEM could be combined with
numerical models of turbulent flow to produce physically
realistic results.

The surface of separation: an algorithm for coping
We know now that via numerical large-eddy simulation, the FEM
and free surface flow analyses, we should be able to handle a
three-dimensional, turbulent, separated flow about a structure.
However, we need a demonstration that separation surfaces
which normally emanate from the sharp corners of a building
can be accurately delineated. While the details are given in
the next section, we outline here the key mathematical concept.

The important characteristic of separated surfaces is that
they usually begin at a known fixed location and (for the
purposes of establishing the location of an FEM mesh) can be
considered to be coincident with a stream surface which is
tangent to the flow velocity at each point. Such a stream
surface is defined in a Cartesian coordinate system as

$$F(x,y,z) = 0 \tag{1}$$

or, e.g., by

$$F = X(y,z) - x = 0 \tag{2}$$

From differentiation of Equation (2), one obtains

$$dx = X_y dy + X_z dz \tag{3}$$

where the subscript denotes partial differentiation with respect to that variable. However, for the separation surface which is tangent to the velocity field, one can write

$$\left.\begin{array}{rcl} \dfrac{dx}{dt} &=& u, \\[2ex] \dfrac{dy}{dt} &=& v, \\[2ex] \text{and } \dfrac{dx}{dt} &=& w \end{array}\right\} \tag{4}$$

From Equations (3) and (4), one finds

$$vX_y + wX_z = u \tag{5}$$

as the equation of the free surface. It can be shown (Street, 1973, Chapter 9) Equation (5) is a first-order, quasi-linear partial differential equation which has characteristics given by

$$\frac{dx}{u} = \frac{dy}{v} = \frac{dz}{w} \tag{6}$$

and has a unique solution determined by the passage of the characteristics through a single noncharacteristic curve on the surface.

Equation (6) is exactly the same as the normal streamline condition equations; therefore the separation surface is a characteristic surface and has the basic features of a stream surface. Furthermore, the required noncharacteristic curve is precisely the curve which describes the separation points at the edge of the building. Thus, the solution of our problem hinges, first, on an accurate integration of Equation (6) and, second, on an accurate representation of the velocity field, in particular, the conditions at the separation curve. The first condition is easy to satisfy; the second is not. If errors creep in at separation, the exact characteristics through each node on the separation curve uniquely represent different flow conditions (see below).

THE FIRST STEP

The first step in the final development of a three-dimensional, turbulent, separated flow computer code is the development and demonstration of an ability to locate the separation surface accurately and to adjust the FEM mesh for a set of relatively simple three-dimensional separated flows. As a by-product we achieve some useful data on the size of the separation zone and the forces on the structural shapes used. In the present case we have examined the three-dimensional, irrotational, separated flow past plane, ellipsoidal shapes in a rectangular wind tunnel.

Review of the separated flow code

To illustrate the use of the FEM and the algorithm for shifting the separation surface we outline the problem formation and apply it to the case of steady flow past an elliptic plate in a rectangular water tunnel. We parallel Lin and Landweber (1975) here and describe a model which is for the outer potential flow past bluff bodies and which has a closed separation pocket or cavity. The pressure within the cavity is assumed to be constant, i.e., the inertial effects of the trapped fluid are neglected.

The flow is assumed to be incompressible, steady, and irrotational. Hence, it is governed by a velocity potential $\phi(x,y,z)$ which satisfies the Laplace Equation $\nabla^2\phi = 0$ within the fluid and generally

$$\vec{V} = -\nabla\phi \tag{7}$$

so, in the usual notation and Cartesian coordinate system,

$$u = -\partial\phi/\partial x; \quad v = -\partial\phi/\partial y; \quad w = -\partial\phi/\partial z \tag{8a}$$

and, if $q^2 = u^2 + v^2 + w^2$,

$$p + \frac{1}{2}q^2 = \text{constant} \tag{8b}$$

in the absence of gravity influence. The streamline conditions on the separation surface are

$$\frac{dy}{dx} = \frac{v}{u} ; \quad \frac{dz}{dx} = \frac{w}{u} \tag{9}$$

where

$$u_c^2 + v_c^2 + w_c^2 = q_c^2 = \text{constant} \tag{10}$$

As a consequence, in terms of a coordinate system oriented along the streamline, the change of potential on the separation surface is given by

$$-\frac{\partial\phi}{\partial s} = q_c \tag{11}$$

In the FEM we represent a problem solution as a continuous function in terms of values at the edge of finite volumes in the problem domain and with a specified variation locally within such volumes as functions of the edge value. Partial differential equations are replaced by integrals which are subsequently replaced by a set of algebraic equations which can be solved directly and exactly.

There are three parts to our illustrative FEM formulation:

(a) The variational principle (applicable for potential flows).

Functional:

$$\chi = \int_V \frac{1}{2}\left[\left(\frac{\partial \phi}{\partial x}\right)^2 + \left(\frac{\partial \phi}{\partial y}\right)^2 + \left(\frac{\partial \phi}{\partial z}\right)^2\right]dV \qquad (12)$$

in which the functional χ is minimized with respect to the velocity potential values specified at finite element nodes ϕ_i.

(b) The functional representation within an element.

Within element:

$$\phi = [N_1, N_2, ---N_{20}]\begin{Bmatrix}\phi_1 \\ \phi_2 \\ \vdots \\ \phi_{20}\end{Bmatrix} = [N]\{\phi\}^e \qquad (13)$$

We employ a quadratic variation within the hexahedral 20-node element. In Equation (13) the ϕ_i are the values of the velocity potential at the corner and mid-side nodes of the element, while the N_i are shape functions.

(c) The isoparametric element. We have chosen, as noted above, a 20-node hexahedron element. Figure 1 shows a schematic of this isoparametric element in the physical Cartesian space and in its transformed space.

The functional is minimized with respect to the nodal values θ_i within each element (Huebner, 1975); thus,

$$\frac{\partial \chi^e}{\partial \phi_i^e} = \int_V \left\{\frac{\partial N_i}{\partial x}\frac{\partial N_j}{\partial x} + \frac{\partial N_i}{\partial y}\frac{\partial N_j}{\partial y} + \frac{\partial N_i}{\partial z}\frac{\partial N_j}{\partial z}\right\}\{\phi\}^e dV = 0 \qquad (14)$$

or for each element

$$\frac{\partial \chi^e}{\partial\{\phi\}} = [h]\{\phi\}^e = 0 \qquad (15)$$

The global sum over all elements produces a linear algebraic equation set

$$[H]\{\phi\} = 0 \qquad (16)$$

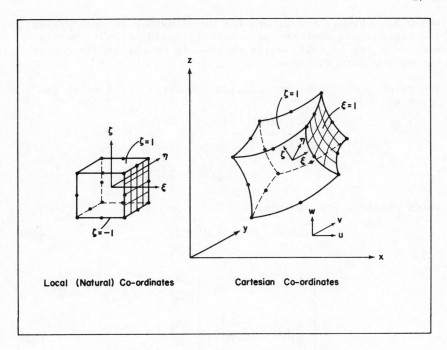

Figure 1. The Isoparametric Element

A unique solution is guaranteed by specification of the ϕ_i or ϕ derivatives along the flow boundaries.

For the isoparametric element of Figure 1 a transform from the physical space to a set of natural local coordinates (ξ,η,ζ) exists such that each element occupies the cube bounded by $|\xi,\eta,\zeta| \le 1$. Generally then (Huebner, 1975)

$$
\left.
\begin{aligned}
x(\xi,\eta,\zeta) &= \sum_{i=1}^{20} N_i(\xi,\eta,\zeta)x_i \\[2ex]
y(\xi,\eta,\zeta) &= \sum_{i=1}^{20} N_i(\xi,\eta,\zeta)y_i \\[2ex]
z(\xi,\eta,\zeta) &= \sum_{i=1}^{20} N_i(\xi,\eta,\zeta)z_i \\[2ex]
\phi(\xi,\eta,\zeta) &= \sum_{i=1}^{20} N_i(\xi,\eta,\zeta)\phi_i
\end{aligned}
\right\} \qquad (17)
$$

where x_i, y_i, z_i, and ϕ_i are the nodal values in each element. The key idea is that the mathematical formulation is easily made on a set of cubes while physically we can handle complex curved boundary surfaces.

The essence of the transformation [Equation (17)] is in the Jacobian

$$[J] = \begin{bmatrix} \partial x/\partial \xi & \partial y/\partial \xi & \partial z/\partial \xi \\ \partial x/\partial \eta & \partial y/\partial \eta & \partial z/\partial \eta \\ \partial x/\partial \zeta & \partial y/\partial \zeta & \partial z/\partial \zeta \end{bmatrix} \tag{18}$$

which relates, for example,

$$dxdydz = |J| \, d\xi d\eta d\zeta \tag{19}$$

and

$$\begin{bmatrix} \dfrac{\partial N_i}{\partial \xi} \\[2mm] \dfrac{\partial N_i}{\partial \eta} \\[2mm] \dfrac{\partial N_i}{\partial \zeta} \end{bmatrix} = [J] \cdot \begin{bmatrix} \dfrac{\partial N_i}{\partial x} \\[2mm] \dfrac{\partial N_i}{\partial y} \\[2mm] \dfrac{\partial N_i}{\partial z} \end{bmatrix} \tag{20}$$

Figure 2 is a schematic of the physical problem. We model the flow past a semi-elliptic plate of semi-axes P and Q in a rectangular tunnel of depth W and width 2D. The plate is assumed to be supported on the floor or roof of the tunnel. The flow velocity is prescribed upstream while a Riabouchinsky image model is used; the flow is assumed to close on an image plate and is symmetric about the cavity midline. Thus, only the forward half of the cavity is shown.

The boundary conditions in a tunnel are straight forward, except that the location and length of the separation surface is generally unknown. With a Riabouchinsky model, the downstream boundary is an equipotential line. If the potential downstream is taken to be zero, then when a separated surface shape is known (assumed or calculated) the potential can be computed on the free surface by Equation (11). The remainder of the flow boundaries are solid or no flow boundaries, except upstream from the plate. Typical boundary conditions are illustrated in Figure 2.

If F, W, D, P, Q, and L are given, there exists a unique value of the pressure coefficient

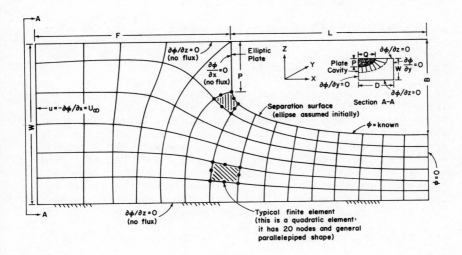

Figure 2. Centerline Schematic of Separated Flow Problem
(Pure drag case: only 1/2 flow needed).

$$\sigma = \left(\frac{q_c}{U_\infty}\right)^2 - 1 = \frac{p_\infty - p_c}{\frac{1}{2} \rho U_\infty^2} \qquad (21)$$

in which q_c is the magnitude of the velocity on the separation surface, U_∞ is the prescribed upstream u-component of velocity, p_∞ is the upstream pressure, p_c is the cavity pressure, and ρ is the fluid density. In our case we elected to prescribe U_∞ and to determine q_c and σ as part of the solution process.

With the given boundary conditions (cf., Figure 2), an assumed separation surface location, and specified geometry of plate and tunnel, the FEM can be used to find the nodal values ϕ_i of the velocity potential and to determine the fluid velocities u, v and w. We next move the free stream surface to a better location.

The total process can be summarized as follows:

(a) Establish the geometry. Input all data including locations of all corner nodes, assumed form of separation surface, assumed U_∞, etc. The computer program then locates all the mid-nodes.

(b) Generate "stiffness" matrix H (Equation 16). Account for various boundary conditions where ϕ_i or ϕ derivatives are specified and for isoparametric mappings.

(c) Solve for ϕ_i by Gaussian elimination and back substitution.

(d) Calculate fluid velocities.

(e) Move the separation surface. Movement is accomplished by integration (cf., Larock and Taylor, 1976, or Street, 1977) along free streamlines beginning at element corner nodes on the edge of the plate. This establishes a set of lines which are tangent to the velocity in accordance with the streamline equations, Equation (9). For pure drag cases, periodic cubic spline interpolation is employed in planes x = constant to locate the element transverse mid-nodes which are missed by the integration process. More complex interpolation is needed for plates set at an angle to the flow.

(f) Establish new values of ϕ on the free surface.

(g) Modify the "stiffness" matrix H. Only elements near the separation surface are changed by the move so only a small part of H is changed.

(h) Return to Step c unless the separation surface movement is less than an arbitrary, preset, small amount. If this happens the solution is complete.

The solution program has been implemented on the IBM 360/91 and 370/168 Triplex System at the Stanford Linear Accelerator Center. With 1325 nodes and 224 elements in a typical case, one iteration (Steps a through h) takes about 45 seconds and solution is achieved in about 15 iterations.

For these simple flows, equality of the pressure coefficients σ [Equation (21)] is an indication of the dynamic similarity of two geometrically similar flows. The force on a plate is given by

$$F = \int_A (P_p - P_c)\, dA \tag{22}$$

where the subscript p denotes the plate and A is its area. From Equations (10) and (22), it can be shown that

$$F = \frac{\rho q_c^2 A}{2} - \frac{\rho}{2} \int_A q_p^2\, dA \tag{23}$$

For a pure drag problem F is the drag. For a plate at an

angle of attack θ relative to the upstream flow the drag D is given by

$$D = F \sin \theta$$

and the transverse force L is given by

$$L = F \cos \theta. \qquad (24)$$

The drag and transverse force coefficients, C_D and C_L respectively, are given as follows:

$$C_D = \frac{D}{\frac{1}{2} \rho U_\infty^2 S} \qquad (25)$$

and

$$C_L = \frac{L}{\frac{1}{2} \rho U_\infty^2 S} \qquad (26)$$

where S is defined as the largest possible projected area of the plate. It is often convenient to normalize C_L and C_D on the velocity q_c on the separation surface. Then,

$$C_D^* = C_D / (1+\sigma) = \frac{D}{\frac{1}{2} \rho q_c^2 S} \qquad (27)$$

and

$$C_L^* = C_L / (1+\sigma) = \frac{L}{\frac{1}{2} \rho q_c^2 S} \qquad (28)$$

where the relationship between σ, q_c and U_∞ is given by Equation (21)

A test: the Rankine oval

In order to test the performance of the selected algorithms, the present procedure has been tested for fully-cavitating flows against the best data available. The results were quite satisfactory. For the present purposes a test of the separation surface shift algorithm was made for a fully three-dimensional flow.

The analytic solution for a uniform flow past a three-dimensional Rankine oval is known so that for an oval of known length and width, one knows the exact surface geometry of the oval and the entire velocity field surrounding it. This information was used to test the present algorithm for moving the free surface. We began by generating a regular, three-dimensional geometric mesh for a typical separated flow past a semi-circular plate in a wind tunnel as shown by the solid lines in Figure 3. Then, the three-dimensional Rankine oval

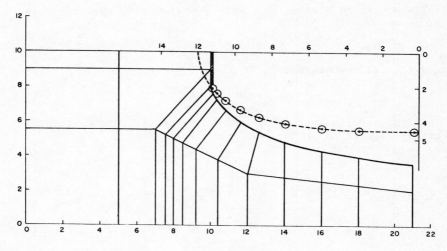

Legend

‒‒‒‒‒   Shape (analytic solution) of cross-section of three-
        dimensional Rankine oval.

  •   Numerical solution from FEM.

_____   Starting position of "free-stream" line, i.e., of
        curve which is to be tangent to velocity field.

Figure 3.   Demonstration of Convergence of Free-Surface
            Shifting Algorithm.

was placed in the separation zone behind the disk such that
the edge of the disk coincided with the analytic boundary of
the oval at some convenient point. At each iteration the
velocity field is computed from the analytic solution for the
Rankine oval and the separation surface is moved from its
original position in accordance with the analytic velocities
for the oval and the separation-surface-moving algorithm. The
moved separation surface converges to the analytic shape of
the Rankine oval. The accuracy obtained with a reasonable
number of iterations is of the order of four to six significant figures. This test shows that if the velocity field is
accurately represented, our iterative procedure for moving the
separation surface is rapidly and stably convergent to any
desired level of accuracy.

Numerical results for irrotational flows
A number of flow configurations have been examined. Figure 4
gives a general schematic diagram of a typical fence or plate
placed on the floor of a tunnel. For pure drag cases the flow
is symmetric about the vertical centerline so only the region
CAOEF of the plate and separation zone and the corresponding
portion of the tunnel are involved in the computation (cf.,

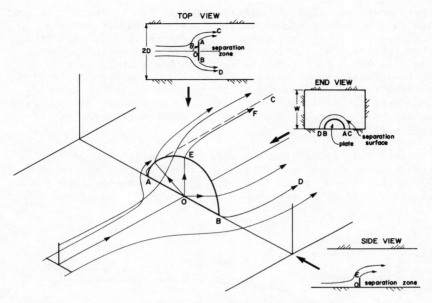

Figure 4. General schematics of flow past a fence
in a wind tunnel.

the equivalent, although inverted, flow in Figure 2). For
pure drag, $\theta = 90°$ (see the top view in Figure 4), while for a
fence at an angle to the flow, $\theta < 90°$. One illustrative case
for $\theta = 60°$ is given here to picture how the iterative shifting of the separation surface converges.

Table 1 presents a set of summary results for various configurations. Total computer output includes flow velocities at

TABLE 1.
Representative Results for Various Configurations

| Fence Type (Size) | Tunnel | | θ (deg) | Half-length of Sep. Bubble (units) | Drag Coeff. C_D^* | Transverse Force Coeff. C_L^* | Remarks |
	Height W (units)	Breadth 2D (units)					
Semi-circle (2 unit radius)	10.0	20	90	10	0.83	0	
"	7.5	15	90	10	0.85	0	
"	5.0	10	90	10	0.87	0	
"	5.0	10	90	5	0.87	0	
Semi-ellipse (P=3.25;Q=7.75)	20.0	80	90	20	0.79	0	See Fig. 2: height of fence = 3.25; width = 15.5 units.
"	10.0	40	90	20	0.83	0	
Semi-circle (2 unit radius)	7.5	15	60	5 (from center of fence)	0.63	0.36	

every node, the pressure distribution on the fence or barrier, and the three-dimensional shape of the separation surface. The drag and transverse force coefficients are computed from the pressure distribution. From Table 1 one sees reinforced the well-known fact that the force on a structure in a tunnel is influenced by the walls. In this case the drag coefficient is seen to increase as the ratio of fence "size" to tunnel cross-section area increases. Such effects must, of course, be accounted for when model results are transformed for proto-type predictions.

Figure 5 illustrates how the separation surface moves from iteration to iteration. This plot corresponds (at iteration 7) to the last entry in Table 1. In this case the flow compu-tation converged to reasonable accuracy in 7 iterations.

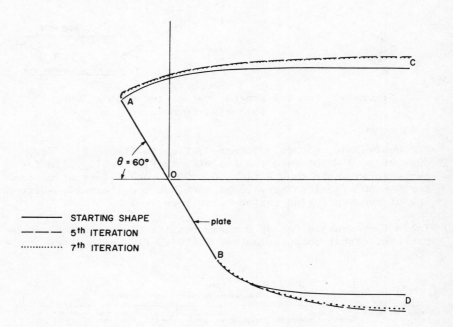

Figure 5. Top view of circular fence ($\theta = 60°$) in tunnel with separation surface lines in plane of tunnel floor (cf., Figure 4).

Figure 6 illustrates the shell of elements lying on the sepa-ration surface. In tests of pure drag, semi-circular-barrier cases, up to 25 iterations were used. Generally, no measura-ble changes in the solution occurred in the last 10 or so iterations; the number of iterations required for reasonable convergence depends on the "quality" of the initial, guessed separation surface location.

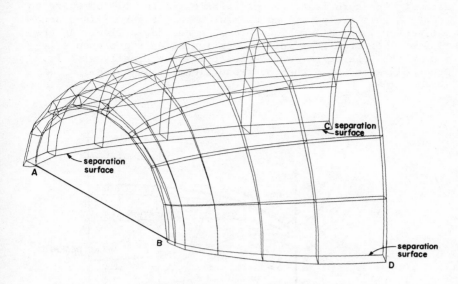

Figure 6.   Shell of elements lying on the
separation surface.

One important feature of the fully three-dimensional flow is
illustrated clearly in Figure 5, viz., the separation surface
curvature is much more severe at one edge of the fence (see
point A).   This requires that mesh or element refinements be
made near the separation curve on the fence to preserve consistent accuracy in the results.   Because our iteration procedure is adjusted to force tangent separation from the fence, a
lack of consistent discretization near separation leads to the
pressure being constant along each streamline in the separation
surface, but different on each streamline.   This is, of course,
inconsistent with the constant pressure assumption of the
analysis.   In the pure drag, semi-circular cases, the pressure
difference across the surface is negligible because the surface
is virtually axisymmetric.   In the elliptic plate cases, where
no effort was made to adjust the element sizes for this problem,
the pressure differences across the surface (between lines AC
and BD for example in Figure 4) is about 7 percent.   For the
"$\theta = 60°$" flow cited, the difference is as much as 25 percent.
This is intolerable and the element sizes must be adjusted.

SUBSEQUENT STEPS

As discussed above the components needed for computation of a
turbulent, three-dimensional, separated flow past a structure
exist.   While the location of the separation surface plays a
crucial dynamic role in the irrotational flow described in the
previous section, the role of the surface-of-separation algo-

rithm in the turbulent flow calculation is to tell us where to
locate a relatively refined element mesh. Figure 7 is a schematic of a vertical section through a separated flow. As the
flow evolves from this initial configuration the elements lying on the separation surface would move vertically with it.
Other elements would then be proportionally adjusted.

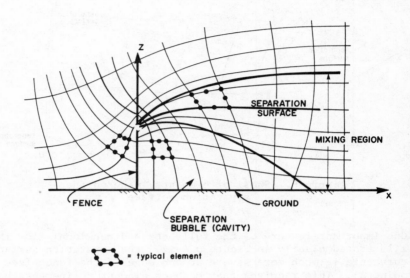

= typical element

Figure 7. The flow configuration behind a fence (after Sheih,
           et al., 1978) with FEM grid overlay to show typical
           variations to accommodate mixing region.

In summary, the actual components required are

(a) a three-dimensional large-eddy simulation computation
scheme based on, e.g., quadratic hexahedron finite elements
(as illustrated in Figure 1).

(b) a separation-surface locating algorithm which locates the
surface and then regenerates the finite element mesh to place
smaller elements always in the regions of high shear and
vorticity.

Both components exist. Only the synthesis is required to
accomplish the desired simulation.

REFERENCES

Baines, W. D. (1965) Effects of velocity distribution on wind loads and flow patterns on buildings. Wind Effects on Buildings and Structures, I, Proc. NPL Conf., London, 198-225.

Castro, I. P., and A. G. Robins (1977) The flow around a surface-mounted cube in uniform and turbulent streams. J. Fluid Mech., 79, 2:307-335.

Deardorff, J. W. (1970) A numerical study of three-dimensional turbulent channel flow at large Reynolds numbers. J. Fluid Mech., 41, 2:453-480.

Ferziger, J. H. (1977) Large eddy numerical simulations of turbulent flows. AIAA Journal, 15, 9:1261-1267.

Findikakis, A. N., J. B. Franzini and R. L. Street (1978) Simulation of stratified turbulent flows in closed water bodies using the finite element method. Finite Elements in Water Resources, Pentech Press, London, 3.23-3.44.

Findikakis, A. N., and R. L. Street (1979) An algebraic model for sub-grid-scale turbulence in stratified flows. Submitted for publication to J. Atmosph. Sci.

Halitsky, J. (1977) Wake and dispersion models for the EBR-II building complex. Atmospheric Environment, 11, 577-596.

Hirt, C. W., and J. L. Cook (1972) Calculating three-dimensional flows around structures and over rough terrain. J. Comp. Physics, 10, 324-340.

Huebner, K. H. (1975) The Finite Element Method for Engineers, Wiley-Interscience, New York.

Larock, B. E., and C. Taylor (1976) Computing three-dimensional free surface flows. Int'l. J. Numer. Meth. Engin., 10, 1143-1152.

Lin, A., and L. Landweber (1975) A free stream-line model for two-dimensional near-wake behind a normal plate or wedges. Proc. 2nd. U.S. Nat'l. Conf. on Wind Engin., IV-25-1-IV-25-2.

Plate, E. J. (1971) Aerodynamic Characteristics of Atmospheric Boundary Layers, U.S. Atomic E. Com., NTIS, Virginia.

Reynolds, W. C. (1976) Computation of turbulent flows. Ann. Rev. Fluid Mech., 8, 183-208.

Robins, A. G., and I. P. Castro (1977a) A wind tunnel investigation of plume dispersion in the vicinity of a surface mounted cube - I. The flow field. Atmospheric Environment, 11, 291-297.

Robins, A. G., and I. P. Castro (1977b) A wind tunnel investigation of plume dispersion in the vicinity of a surface mounted cube - II. The concentration field. Atmospheric Environment, 11, 299-311.

Sachs, P. (1978) Wind Forces in Engineering, 2nd Ed., Pergamon, Oxford.

Sheih, C. M., P. J. Mulhearn, E. F. Bradley and J. J. Finnigan (1978). Pollutant transfer across the cavity region behind a two-dimensional fence. Atmospheric Environment, 12, 2301-2307.

Street, R. L. (1973) The Analysis and Solution of Partial Differential Equations, Brooks-Cole Pub., Monterey.

Street, R. L. (1977) A review of numerical methods for solution of three-dimensional cavity flow problems. 2nd Int'l. Symp. Numer. Ship Hydrodyn., Berkeley, Calif., 237-249.

Street, R. L., and P. Y. Ko (1977) Numerical methods applied to fully cavitating flows, with emphasis on the finite element method. Symp. Hydrodyn. Ship and Offshore Propul. Sys., Det Norske Veritas, Oslo.

Vincent, J. H. (1977) Model experiments on the nature of air pollution transport near buildings. Atmospheric Environment, 11, 765-774.

WIND LOADING ON STRUCTURES

Abdin M. A. Salih

Civil Engineering Department, Khartoum University, Sudan

INTRODUCTION

Studies on wind loading are, at present, receiving an
appreciable attention by scientists in the advanced countries.
In contrast, however, less attention is directed towards these
studies by scientists in developing countries.

In the Sudan and similar countries in the same belt in the
African Continent, many structures (such as bridges,
relatively tall buildings, mast towers, cables, cooling
towers, radar and communication aerials, etc.) are increasingly erected, with little available knowledge on the prevailing
wind loads. In the absence of a local Code of Practice the
design of these structures is normally carried out by adopting the practice abroad where different wind speeds may
persist. Many engineers were often unhappy with that
situation, but it was for Delsi (1968) who openly described
the use of exposure D of the British Standard Code of
Practice CP3 Chapter V, for the Sudan, as unwarranted.

At the moment, however, the Meteorological Department of the
Sudan has established more than 20 gust recording stations
well distributed over the whole area of the country (Fig. 3).
Fifteen of these stations have a recording period ranging
between 27 and 40 years. These measurements were taken
using "Dines Pressure Tube Anemograph MK 11" with response
frequency ranging from one to ten seconds, and were reduced
to a common height of 50 feet.

That meteorological data was analyzed by Salih (1978), using
an extreme value technique as suggested by Sachs (1978),
resulting in few working graphs and tables. From these
graphs and tables the maximum design speed, for different
structure's life-time and at different margins of risk, can
easily be obtained. An updated and simplified picture of

these findings will be shown in this note.

ANALYSIS

It is generally accepted that the wind force (F) can be related to the maximum wind speed (V) by a relation of the form:

$$F = CV^2 \tag{1}$$

where C is a coefficient of proportionality called the shape factor.

The structure's lifetime (T) can also be related to the probability (P) that the maximum velocity will not exceed the extreme value V_{max} , by the relation,

$$T = \frac{1}{1-P} \ . \tag{2}$$

For example if the required lifetime is 50 years, a maximum annual velocity V_{max} which has not occurred for 98% of the annual records is stipulated. Sometimes the recording time may be too short for the required probability. In such cases extrapolation is possible if the available readings can be reduced to a certain graph. This method of extrapolation can be clarified by taking Khartoum's record as an example. In this technique the maximum annual velocities (V_{max}) , after being checked for homogenity and consistency, are plotted against the reduced variate (Y), as shown in Fig. 1.

Table 1 shows Khartoum's annual maximum gust speeds between 1938 to 1977, arranged in an ascending order (Col. 2) with a corresponding rank m as shown in column 3. The probability of their non-recurrence, P (Col. 4), is calculated from the relation:

$$P = \frac{m}{N+1} \tag{3}$$

where N is the number of years of observation. Finally Column 5 indicates the reduced variate, Y , calculated from:

$$Y = - \ln(-\ln P). \tag{4}$$

The required extreme value, V_{max} , is hence obtained from the linear relation, resulting from the best straight line through the field points in Fig. 1, which can be expressed as:

$$V_{max} = AY + B \tag{5}$$

where A , B are constants unique to a specific site. Numerical values for these constants can be found from the relations:

$$A = \frac{\sigma}{\sigma_N} \quad \text{and} \quad B = V_{max} - AY_n$$

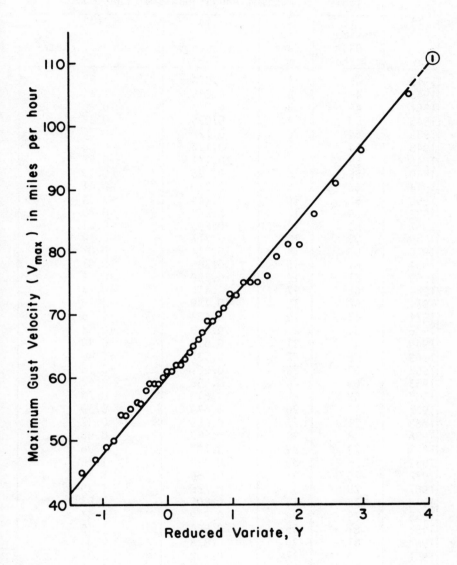

Fig. 1   Extreme Probability Graph of Annual Maximum Wind
Speed at Khartoum.

Table 1. Maximum Annual Gusts at Khartoum.

(1) Year	(2) Maximum Gust mile/hour	(3) Rank m	(4) Frequency P	(5) Reduced Variate Y
1942	45	1	0.024	-1.32
1943	47	2	0.048	-1.11
1961	49	3	0.073	-0.96
1945	50	4	0.098	-0.84
1939	54	5	0.122	-0.74
1948	54	6	0.146	-0.65
1946	55	7	0.171	-0.57
1940	56	8	0.195	-0.49
1944	56	9	0.220	-0.41
1950	58	10	0.244	-0.34
1938	59	11	0.268	-0.28
1947	59	12	0.293	-0.21
1959	59	13	0.317	-0.14
1970	60	14	0.341	-0.07
1954	61	15	0.366	-0.01
1960	61	16	0.390	+0.06
1964	62	17	0.415	+0.13
1974	62	18	0.439	+0.19
1971	63	19	0.463	+0.26
1976	64	20	0.488	+0.33
1949	65	21	0.512	+0.40
1969	66	22	0.537	+0.48
1966	67	23	0.561	+0.55
1953	69	24	0.585	0.62
1965	69	25	0.610	0.70
1977	70	26	0.634	0.79
1968	71	27	0.659	0.87
1956	73	28	0.683	0.96
1973	73	29	0.707	1.06
1952	75	30	0.732	1.16
1972	75	31	0.756	1.27
1967	75	32	0.780	1.39
1975	76	33	0.805	1.53
1955	79	34	0.829	1.67
1958	81	35	0.854	1.85
1962	81	36	0.878	2.04
1951	86	37	0.902	2.27
1941	92	38	0.927	2.58
1963	96	39	0.951	2.99
1957	105	40	0.976	3.72

where σ_N and Y_n are correction factors for a particular sample size, and σ is the standard deviation. For a sample size of 40 (as for Khartoum), Gumbel (1958) suggests the following values;

$$\sigma_N = 1.14132; \qquad Y_n = 0.54362$$

Calculating σ as 14 m.p.h., the best straight line for Khartoum can be obtained as:

$$V_{max} = 60.3 + 12.3 \ Y \ . \tag{6}$$

Equation 6 is plotted in Fig. 1 and seems to fit through the field data in an acceptable manner. Extrapolation from that equation may be used for return periods beyond the present record.

The above technique is similarly followed for the 14 other sites at Wad Medani, Geneina, Damazin, Kosti, El Obied, Malakal, El Fasher, Kassala, Wau, Karima, Juba, Atbara, Port Sudan, and W. Halfa. Similar graphs (Fig. 2) to that of Fig. 1 are obtained and correspondingly similar equations to that of equation 6 are prepared as shown in Table 2. It must be noted, however, that the numbers in Table 2 are corresponding to those in Fig. 2.

Table 2. Best Fitting Lines for Fifteen Stations in Sudan.

No.	Station	Representative Equation
1	Khartoum	$V_{max} = 60.3 + 12.3 \ Y$
2	Wad Medani	$V_{max} = 63.5 + 6.9 \ Y$
3	Geneina	$V_{max} = 61.2 + 9.0 \ Y$
4	Damazin	$V_{max} = 60.9 + 6.1 \ Y$
5	Kosti	$V_{max} = 57.6 + 8.3 \ Y$
6	El Obeid	$V_{max} = 56.3 + 10.6 \ Y$
7	Malakal	$V_{max} = 54.2 + 8.9 \ Y$
8	El Fasher	$V_{max} = 52.7 + 10.0 \ Y$
9	Kassala	$V_{max} = 52.7 + 8.0 \ Y$
10	Wau	$V_{max} = 52.6 + 6.3 \ Y$
11	Karima	$V_{max} = 52.2 + 7.1 \ Y$
12	Juba	$V_{max} = 50.5 + 8.3 \ Y$
13	Atbara	$V_{max} = 48.1 + 9.4 \ Y$
14	Port Sudan	$V_{max} = 47.3 + 10.6 \ Y$
15	Wadi Halfa	$V_{max} = 42.1 + 7.3 \ Y$

Corrections for the maximum speed

Obviously the maximum velocity (V_{max}) for a return period of T years is in fact a statistical average, based on the mean value of several T-year periods and has a 63% possibility of being reached in these years. Thus if the return period T is increased, the percentage risk in selecting any design

44

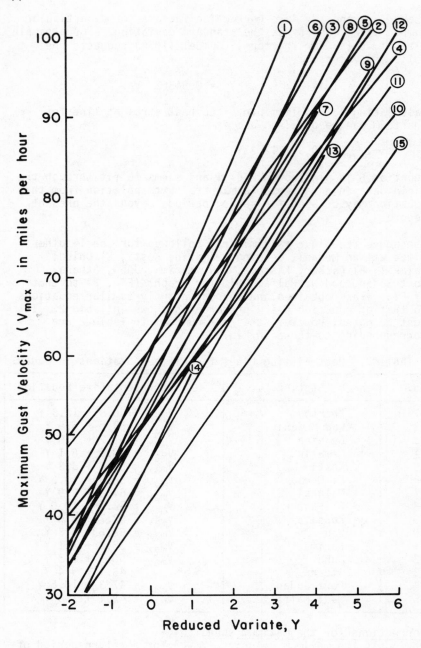

Fig. 2 Extreme Probability Graphs of Annual Maximum Wind
        Speed at 15 Stations in Sudan.

speed is decreased. Taking the example of Khartoum, Table 3
demonstrates the increase in the design speed at different
calculated risks and at different structures desired lifes.
For example a factor of safety not less than 2.16 times the
average maximum velocity ($\overline{V}_{max}$) is badly needed if a
structure of 100 years lifetime is to be constructed with a
marginal risk not exceeding 10%.

Table 3. Maximum Wind Spread at Khartoum ($\overline{V}_{max}$ = 67 m.p.h.)
at Different Desired Lifes and Different Percentages
of Risk.

Desired (Yr.)	20			50			100		
Calcu-lated Risk	0.632	0.300	0.100	0.632	0.300	0.100	0.632	0.300	0.100
T (yr)	20	56	190	50	140	475	100	280	949
V_{max} (m.p.h)	98	110	124	109	122	137	117	130	144.8
$\dfrac{V_{max}}{\overline{V}_{max}}$	1.46	1.46	1.85	1.63	1.82	2.05	1.75	1.94	2.16

Two other corrections are necessary, for height and for gust
period. From the characteristics of the MK II. anemograph,
currently in use in the Sudan, the period of gust measurement can easily be estimated. This estimate may well be
beyond the natural period of the structure, in which case
a correction factor must be added.

Wind speeds, on the other hand, can empirically be correlated
at different heights in stormy, non-convective condition as:

$$\frac{V}{V_o} = \left(\frac{z}{z_o}\right)^p \tag{7}$$

where V = velocity at a height z above ground leve and
V_o = velocity at a reference height z_o (taken here
50 feet).
The value of P can be estimated from prepared tables according to the aerodynamic roughness of the surface. For Khartoum
vicinity, Delis (1968) has suggested a value of 0.085 for P.

Display of results
Some authorities in the world display the maximum wind speed
(V_{max}), at a different return period and heights, in a

national map so as to draw isovents joining sites with similar
velocities. An example is shown in Fig. 3 which demonstrates
the highest gust speeds, of 30 years return period, drawn for
Sudan. It is obvious that while Khartoum area suffers from
the highest gust, yet it represents the largest center for
sophisticated structures of the type mentioned before.

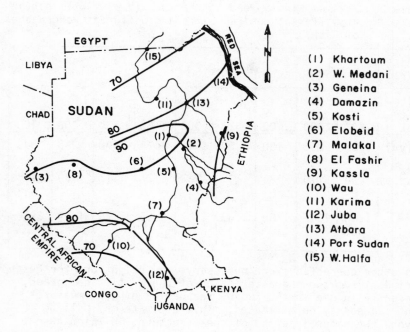

Fig. 3 Highest Gust Speed (m.p.h.), at 50 Feet and of 30
Year Return Period, in Sudan.

CONCLUSION

As a preliminary contribution to the unknown wind load on
structures in the Sudan, the meteorological data from fifteen
gust stations is collected and analyzed. From that analysis
working graphs and tables, for finding the maximum design
speed, are consequently prepared.

Further development to this work is contemplated to follow
soon after the present experiments (with precise pressure
transducers of response up to 0.1 of a second) is completed.

REFERENCES

1. Delsi, M. (1968) Wind and Wind Loading in the Semi-Arid Belt of Central Africa. R.S.C.A., Lettre No. 10-10/1968, Khartoum P.O. Box 1720, Sudan.

2. Gumbel, E. G. (1968) Statistic of Extremes. Columbia University Press, New York, 228.

3. Sachs, P. (1978) Wind Forces in Engineering. Pergamon Press, Oxford, 37-45.

4. Salih, A. M. A. (1978) Wind Forces on Structures. Sudan Engineering Society Journal, No. 23, January, 23-33.

48

SAFETY CONSIDERATIONS FOR NATURAL DRAUGHT COOLING TOWERS UNDER WIND LOAD

H.-J. Niemann

Institut für Konstruktiven Ingenieurbau
Ruhr - Universität Bochum

1. INTRODUCTION

Safety considerations for hyperbolic, natural draught cooling towers have to deal with three design situations mainly:

i   the calculation of meridional and circumferential reinforcement of the shell,
ii the calculation of a proper buckling safety of the shell determining the shell thickness and
iii the design of columns and foundations of the supporting framework.

Each situation may be looked at separately and different safety conditions are relevant.

The actions producing stresses in the structure are
- g : dead load,
- w : external wind load,
- p_i : internal pressure due to wind,
- t : temperature,
- f : foundation settlement, and eventually
- s : earthquake forces.

In a probabilistic safety approach, these actions are the basic loading variables of the limit state function defining the condition for survival or failure of the structure. For a complete calculation of the probability of failure, their statistical properties are required including the cross- correlations. Since such a set of data is not available, the structural safety may in general not be calculated consistently in terms of the probabilistic theory. Therefore, different load combinations are investigated based on experience. The combination rules are recommended in the codes of practice, applying characteristic values of the loads. Such characteristic values are e. g. the 95% - fractile of the specific weight for dead load g, or for the wind load, a wind velocity which will not be exceeded with a probability of .98. A typical load combination for cooling towers would be

$$g + \frac{1}{3} w + t + s \qquad (1)$$

for the investigation of earthquake effects. In equ. (1), the wind load may be reduced since the simultaneous occurrence of wind and earthquake has a low probability. The resistance of the structure is calculated from characteristic values as well and a margin of safety is maintained, applying in general one or two global safety factors. Establishing the safety on the level of the membrane forces, the German code DIN 1o45 results in the case of equ. (1) in

$$1.75 \cdot n\,(g_k,\; \tfrac{1}{3}\,w_k,\; s_k) + 1.00 \cdot n(t_k) \leq n_u \qquad (2)$$

with global safety factor of 1.75 and 1.oo for failure of the reinforcement. In equ. (2), n_u denotes the load bearing capacity and subscript k: characteristic value. In many cases, this "classical" procedure is quite satisfactory and a reasonable level of structural reliability is obtained.

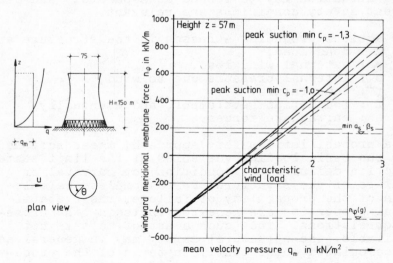

Figure 1: Tensile meridional membrane force due to wind and dead load

But, considering the load combination dead load g and wind w, figure 1 shows a different situation. It is a plot of the design condition (i) mentioned above:

$$n(g, w) = n(w) - |n(g)| \leq \beta_s \cdot a_s \qquad (3)$$

The tensile membrane force, induced by wind action is diminished by the compressive membrane force due to dead load. Their difference has to be carried by the reinforcement of area a_s with yield strength β_s. A design according to

$$1.75 \cdot n(g_k, w_k) \leq \beta_s \cdot a_s \qquad (4)$$

will not produce sufficient reliability, since the membrane forces due to w_k and g_k are of the same order of magnitude and, hence, their difference is small. Therefore, only a limited exceedance of w_k (working state) is allowed for, with too high a probability of occurence. The stresses from dead load are part of the load bearing capacity as far as the reinforcement is concerned. Therefore, it is a straightforward approach to use instead of equ. (4)

$$1.75 \cdot n(w_k) \leq \beta_s \cdot a_s + |n(g_k)|, \qquad (5)$$

or to investigate the load combination

$$g_k + 1.75 \cdot w_k , \qquad (6)$$

which is referred to as "ultimate state". It is implied in equ. (3) that failure occurs when the yield strength of the steel is exceeded. Furthermore, equ. (5) and (6) are not strictly equivalent, since the response of the shell to wind load is not linear due to resonance effects, see figure 1. Resonance may be included in a resonance factor φ. Then, equ. (5) is rewritten as:

$$1.75 \cdot \varphi \cdot n(w_k) \leq \beta_s \cdot a_s + |n(g_k)| , \qquad (7)$$

where $n(w_k)$ is the static and quasi – static response to wind action. This simple extrapolation from the current code of practice is somewhat rough. It is worthwhile to investigate the ultimate – wind – load – factor in equation (7) in more detail since its influence on the amount of reinforcement is rather pronounced. Currently, factors of 1.5o and 1.6o are being used for cooling towers in Germany.

2. RESISTANCE

The condition **ensuring** survival of the cooling
tower against tensile failure from wind load is

$$\varphi \cdot S = \varphi \cdot n(w) \leq \beta_s \cdot a_s + |n(g)| = R \qquad (8)$$

The stresses induced by the action of dead load
increase the resistance R on the right side both
in meridional and circumferential direction since
they are compressive. The parts of equ. (8) are
varying quantities with respect to time or with
respect to several comparable structures. Their
statistical properties are needed for the safety
consideration.

Dead load $n(g)$: The membrane forces $n(g)$ are proportional to the specific weight of reinforced
concrete γ_c and to the shell thickness t. Additionally there is an influence of the imperfections of the middle surface of the shell. The following data (see Krätzig 1978) are characteristic
for cooling towers constructed under rigorous control. The probability distribution of specific
weight is Gaussian with mean value $m_\gamma = 24.5$ kN/m^3
and standard deviation $\sigma_\gamma = 0.2$ kN/m^3. If the 99%-
- fractile is used as the characteristic value,
$\gamma_k = 25$ kN/m , then $\gamma_k / m_\gamma = 1.02$. For a nominal shell thickness of $t_c = 16.0$ cm (characteristic
value), a mean of $m_t = 16.7$ cm with standard deviation $\sigma_t = 0.4$ cm was found. Hence, the nominal
thickness may be interpreted as the 5% - fractile.
A ratio $t_k / m_t = 1.04$ is obtained. The effect of
imperfection is negligible assuming a standard
deviation of the imperfections of t/5.
The coefficient of variation of $n(g)$ is then

$$V_g = \frac{\sigma_g}{m_g} = 2.5\% , \qquad (9)$$

and the characteristic value is related to the mean
by

$$\frac{n(g_k, t_k)}{m_g} = 0.98 . \qquad (11)$$

It may be assumed here, that errors due to the
transformation of loads into stresses are excluded.

Tensile steel force $\beta_s \cdot a_s$: The coefficients of
variation applied usually are $V_\beta = 6\%$ for the
yield strength and $V_a = 5\%$ for the area of reinforcement. Then, the variation of the tensile steel
force is

$$V_{st} = 7.8\% \qquad\qquad (12)$$

and the mean value $m_{st} = m_a \cdot m_\beta$.
As characteristic values, 5% - fractiles are used.
Therefore, the ratio to the mean values is

$$\frac{\beta_{sk} \cdot a_{sk}}{m_\beta \cdot m_a} = 0.83 \ , \qquad\qquad (13)$$

since the probability distributions are Gaussian.
The resulting coefficient of variation and the
relation of characteristic to mean value of the
total resistance R are weighed means of the above
values depending on the relation of m_{st} to m_g. In
the range of $m_{st} \ / \ m_g = 0.7 \ \ldots \ 1.3$, the most unfavourable values are:

$$V_R = 4.5\% \qquad\qquad (14)$$

$$\frac{R_k}{m_R} = 0.94 \qquad\qquad (15)$$

3. CONCEPTS OF WIND LOADING

3.1. Equivalent static load

The simplest concept of wind loading is to represent the wind action, mean and turbulent, by an
equivalent static load which is based on an appropriate gust velocity. A more realistic approach is
to determine an equivalent static load using a
gust response factor B_Y. The design pressure P_D on
some point of the tower surface is then

$$P_D = B_Y \cdot c_p(z, \Theta) \cdot \overline{q}(z) \ , \qquad\qquad (16)$$

where c_p : pressure coefficient,
        z : height above ground,
        Θ : circumferential angle
        $\overline{q}$: mean velocity pressure, averaged over
            600 sec. to 3600 sec.
The gust response factor is related to the action
of mean wind:

$$B_Y = \frac{\overline{Y} + g \cdot \sigma_Y}{\overline{Y}} \quad , \tag{17}$$

where $\overline{Y}$: mean response,
        σ_Y : rms - response to wind turbulence,
        g : peak factor

For convenience, the contribution of the resonant response to turbulence may be separated from the quasi - static by a resonance factor φ:

$$q_D = \varphi \cdot B_{YQ} \cdot \overline{q} \quad , \tag{18}$$

where subscript Q : quasi - static, and

$$\varphi = \frac{\overline{Y} + g \cdot \sigma_Y}{\overline{Y} + g \cdot \sigma_{YQ}} \tag{19}$$

The gust response factor has been calculated for various tower shapes (Abu - Sitta and Hashish,1973; Niemann, 1977) considering the correlations of pressure fluctuations and taking into account the relevant vibrational modes. The computational re- sults were checked by model and full - scale re- sponse measurements (Niemann and Ruhwedel, 1978). The result is that static and quasi - static re- sponse are proportional to the mean wind load for linear behaviour of the structure and that reso- nance increases overlinearly. The quasi - static response factor is in the order of 2 for the maxi- mum tensile membrane force, meridional direction, of a 15o m cooling tower and the resonance factor φ is usually smaller than 1.1 . The design wind pressure may be converted into a corresponding gust velocity and it turns out that a 5 - sec. - gust will produce a good representation of the static and quasi - static response. If the design velocity is based on the fastest - mile of wind, a conver- sion of this basic wind speed into a 5 - sec. - - gust can be readily achieved. A fastest - mile of e. g. 8o mph corresponds to an average velocity over 45 sec. The 5 - sec. - gust speed is greater by a factor of 1.14: a gust factor of 1.3o is needed for the design pressures. Summarizing, the magnitude of the wind load in terms of a design velocity pressure may be established in a reliable manner. The gust response factors for the maxima of other response quantities are of similar size. On the other hand, this approach may be somewhat mis- leading concerning the simultaneous occurrence of several response quantities, e. g. of meridional and hoop force. Therefore, the possible errors are greater when the buckling safety is determined based on an equivalent static wind load.

3.2. Pressure distribution

Apart from the design velocity pressure q_D, the reliability of the predicted wind response depends upon the accuracy of the pressure coefficients c_p. The influence of the pressure distribution on the maximum tensile meridional force is considerable, see figure 1. The extrapolation of wind tunnel data to full – scale was difficult because of the sensitivity of the rounded shape to Reynolds number. Based on additional full – scale measurements this question seems to be clarified. At least it may be said that the accuracy of predicting the pressure distribution is much higher for cooling towers than for many other buildings. Nevertheless, errors may arise from two effects:

i Frequently, several towers are arranged in a group. This will alter the pressure distribution even if a minimum distance of the tower axis of, say, twice the throat diameter is observed.

ii In practical design, the pressure coefficients are assumed independent of height z. This is not true in a profiled flow, and the assumption is not conservative.

Both aspects are not very relevant for the meridional forces but they have some bearing for the hoop force. In a critical situation, wind tunnel tests are recommendable. Anyway, the theoretic and experimental knowledge achieved today makes it possible to predict the effect of mean and turbulent wind with nearly any accuracy desired.

3.3. Exteme wind velocity

The probability $F_u(u)$ of not exceeding an extreme wind velocity u in a given year may be described by a type-I-extreme-value or Gumbel distribution:

$$F_u(u) = \exp\left\{- \exp\left[- U \cdot a \left(\frac{u}{U} - 1\right)\right]\right\} \quad (2o)$$

with U : mode and a : measure of the standard deviation. Predicting extreme velocities with low probability of occurrence, $1 - F_u(u) < 1o^{-3}$, mainly two difficulties arise:

i The meteorological data comprise some 5o years. From such data, the range of small probabilities is extrapolated. This extrapolation is open to doubt. Obviously, there is some upper boundary of physically reasonable velocities which is not

encluded in equ. (2o).
ii The prediction is sensitive to the parameter
   U · a. From wind observations, a range of
   6 < U · a < 12 is obtained. The probability of
   exceeding a velocity u = 2 · U is $2.5 \cdot 10^{-3}$ in
   the first case and $6.1 \cdot 10^{-6}$ in the second. If
   failure occurs at such a velocity, the structural reliability would not be satisfactory in
   the first and overdone in the second.
The mode U has a reasonably clear geographical
pattern and may be used to map the wind intensity,
whereas the scatter of U · a is distributed in a
statistical manner. Nevertheless, the approach of
extreme value distribution seems to be the best
available at the moment.

4. STRUCTURAL SAFETY

4.1. Probability of failure
With resistance R and action S from equ. (8) with
probability density f_R and cumulative probability
function F_S the probability of failure P_f related
to one year reference time may be expressed as

$$P_f = P(S>R) = 1 - \int_{r=0}^{\infty} f_R(r) \cdot F_S(s=r) \cdot dr \qquad (21)$$

The load distribution is obtained from equ (2o),
relating the action s to the wind velocity u by a
linear factor p, $s = p \cdot u^2$. The mean resistance
is connected with the mode U introducing some safety factor γ_u,

$$m_R = p \cdot \gamma_u \cdot U^2 \quad . \qquad (22)$$

Then, $u / U = \sqrt{\gamma_u} \cdot \sqrt{s / m_R}$, and

$$F_S(s) = \exp\left\{ -\exp[-U \cdot a(\sqrt{\gamma_u} \cdot \sqrt{s/m_R} - 1)] \right\} \qquad (23)$$

In equ. (23), the probability of the load s not
exceeding the mean resistance, $s = m_R$, is determined by γ_u. Equ. (21) was evaluated with different
values of γ_u in equ. (23) and different V_R in $f_R(r)$.
The result is plotted in figure 2 assuming U· a=1o.
For a given probability, that the load exceeds the
mean resistance, $1 - F_S(m_R)$, the probability of
failure increases with increasing variation of
the resistance. For small V_R, R is practically
deterministic and the probability of failure is
equal to the probability that $s > m_R$ or that the
corresponding velocity u* is exceeded. Then, the

concept of return periods applies, stating that
$P_f \approx 1 / n$, where n : return period in years.

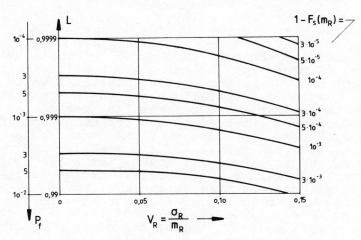

Figure 2: Probability of failure

4.2. Ultimate load factor

The design situation may be expressed in terms of
the ultimate load S*:

$$\varphi \cdot S^* = m_R \quad , \tag{24}$$

or with characteristic values

$$\varphi \cdot \gamma \cdot S_k = R_k \quad , \tag{25}$$

where

$$\gamma = \frac{S^*}{S_k} \cdot \frac{R_k}{m_R} \tag{26}$$

is the ultimate load factor. From equ. (22) and
(25), $S^* = \gamma_u \cdot p \cdot U^2$. As well, the characteristic load S_k is expressed by the mode U :
$S_k = \gamma_k \cdot p \cdot U^2$. Then

$$\gamma = \frac{\gamma_u}{\gamma_k} \cdot \frac{R_k}{m_R} \quad . \tag{27}$$

For a required probability of failure P_f , the load
probability $F_s(m_R)$ is read from figure 2 with consideration of V_R , and γ_u obtained from equ. (23).
If the characteristic wind load has a probability
of o.o2 of being exceeded, $\gamma_k = 1.94$, and the fol-

lowing ultimate load factors are obtained:

$$P_f = 10^{-3} \qquad 2 \cdot 10^{-4} \qquad 10^{-4}$$

$$\gamma = 1.41 \qquad 1.75 \qquad 1.81$$

The reliability achieved by $\gamma = 1.75$ seems to be quite satifactory for the calculation of the reinforcement and the supports of the shell. A value of $P_f = 10^{-6}$ which is generally recommended for sudden failure is certainly too conservative because of some assumptions included in the foregoing considerations. First, the local yielding of the reinforcement will not result in collapse. A redistribution of the stresses will enable the structure to withstand even higher wind load, i. e. the definition of failure used here is conservative to an extent unknown up to now. Second, the prediction of high wind speeds by the Gumbel distribution is conservative particularly in that range which is dominant in the evaluation of the failure integral. Both aspects indicate that an increase of the operational probability of failure is reasonable and that the real reliability will be somewhat greater.

4.3. Buckling safety

Concerning the buckling safety, both the actions of wind and dead load produce compressive stresses and hence reference may be made to characteristic values of the loads in the usual manner. On the other hand, it is much more complicated to describe the statistic properties of the resistance because the imperfections of the middle surface and the cracks affect the resisting critical forces considerably. If n_φ and n_Θ are meridional and circumferential forces, and g is some interaction function, then the design condition may be written as

$$g \left[\gamma_B \cdot \left(\frac{n_\varphi(w_k) + n_\varphi(g_k)}{n_{\varphi crit}} + \frac{n_\Theta(w_k) + n_\Theta(g_k)}{n_{\Theta crit}} \right) \right] \leq 1$$

In a global manner, a safety factor of $\gamma_B = 5$ is recommended in Germany.

REFERENCES

Abu - Sitta, S. H., and M. G. Hashish (1973) Dynamic Wind Stresses in Hyperbolic Cooling Towers. ASCE - Journal of the Struct. Div., pp 1823 - 1835.

Krätzig, W. B. (1978) General Safety Concept of Reinforced Concrete Cooling Towers. Proc. Int.

Symp. on very Tall Reinforced Concrete Cooling Towers (IASS), part 1.

Niemann, H. - J., and J. Ruhwedel (1978) Full -
- scale and Model Tests on Wind - Induced, Static and Dynamic Stresses in Cooling Tower Shells. Proc. Int. Symp. on very Tall Reinforced Concrete Cooling Towers (IASS), part. 2.

Niemann, H. - J. (1977) Gust Response Factors and and Equivalent Static Wind Loads for Natural Draught Cooling Towers. Proc. 2nd Int. Conf. on Struct. Safety and Reliability, Werner - Verlag Düsseldorf.

60

DYNAMIC EFFECTS OF WIND ON OFFSHORE TOWERS

M. Shears and M.H. Bell

Ove Arup & Partners, London, U.K.

INTRODUCTION

The predominant environmental loadings for the structural design of fixed offshore platforms used in the North Sea for oil or gas production purposes arise through wave and current action, and the effects of wind are less important. This is not the case for certain important ancillary structures, such as flare towers, flare booms and communications masts, for which wind and ice provide the main environmental loadings to be considered in the design. Although wind is dealt with in existing design regulations for offshore structures, the subject is not well covered for structures which are sensitive to wind gust fluctuations.

The design offshore towers is often based on static gust loadings only but it is now generally recognised that stochastic dynamic analysis methods provide a more rational means of assessing the response characteristics of tall tower structures to the action of wind gusts. The dynamic analysis uses the hourly mean wind speed and requires a measure of the effective surface roughness of the sea appropriate to the design storm to define the variation of wind speed with height and the turbulence intensity of the wind.

The present state of knowledge regarding wind structure over the sea is very limited and it is not possible to state categoric values for sea surface roughness or for the various wind turbulence parameters needed for the design.

The current rules dealing with wind loadings for offshore structures are based on gross extrapolations of wind data obtained from land based measurements. Very little data is available from wind measurements taken over the sea and published results are of doubtful validity, especially when extrapolated to extreme design storm conditions. Even on land, where measurements have been taken for many years, there is still dispute over certain aspects of the structure of strong winds. Deaves and Harris (1) have indicated the difficulties in obtaining wind measurements of sufficient quality for consistent atmospheric modelling and have set down minimum requirements for useful records. Over the sea, the problem is further complicated by the practical difficulties of taking measurements and by the variable nature of the sea surface.

There is a clear need for better wind data than currently exists for offshore locations and this has been recognised by the Department of Energy, which has recently set up a research project, involving the measurement of wind structure over the sea, to try to provide an improved understanding of the wind characteristics. Initial results from this project may be available in the near future but a universal model of wind over the sea is unlikely to be available for some time. In the meantime the designer must consider a range of wind parameters that is likely to include the design storm conditions.

In this paper, the basic parameters needed to define wind gust loadings for the design of offshore towers are summarised and an outline is given of the possible character of wind waves. The roughness of the sea surface and the corresponding wind structure are discussed in the light of known data. The sensitivity of the dynamic response behaviour of typical offshore tower structures is examined by comparison of results obtained for different sea surface roughnesses.

For simplicity no attempt is made here to investigate the very important dynamic effects which may be introduced by the support conditions of a tower on a platform, or wind turbulence and stream flow effects caused by the platform.

WIND CHARACTERISTICS FOR DESIGN

The design of wind sensitive structures requires an assessment of the static and dynamic components of the wind loading over the structure. In particular the designer requires information on:

(a)   the hourly mean wind speed and the variation of the mean wind speed with height
(b)   the root mean square fluctuating component of longitudinal gust speed
(c)   the wind gust spectrum and coherence functions defining the spatial structure of the turbulence.

A mathematical model describing the variation of these wind characteristics has been developed for land conditions (1) by reference to the fundamental parameters of roughness length of the surface terrain, z_0, the friction velocity, u_*, which defines the surface shear stress and for rough terrains, the displacement height, d, defining the general level of surface obstructions met by the wind. The model is valid for the conditions of strong, steady winds, neutral atmospheric stability, level terrain with no significant topographical irregularities, and sufficient fetch of uniform roughness to ensure equilibrium flow.

It is felt that, since the wind characteristics are not much influenced by the precise form of the surface but by the energy loss and rate of transfer of momentum due to surface friction, the relationships established for the wind over land should also be applicable over the sea, except in the wave height zone. In any event, there is at this stage no alternative, definitive method established for sea conditions.

In the wind model a log-linear velocity profile is developed for the mean wind at height z metres above mean sea level,

$$\overline{V}(z) = \frac{u_*}{k} \left[\ln \frac{z}{z_0} + 5.75 \frac{z}{h} \right] \tag{1}$$

where k is von Karman's constant ($=0.4$) and h is the gradient height,

$$h = u_* / 6f \qquad \text{(f is Coriolis parameter} \cong 10^{-4}) \tag{2}$$

This relationship can be approximated in the height range of interest, above the wave height zone and up to about 200 metres, by the equivalent power low profile,

$$\overline{V}(z) = \overline{V}(10) (z/10) \tag{3}$$

where the exponent, $\propto$, is dependent on both u_* and z_0 and is given by,

$$\propto = \left[1.05 + (0.18/u_*) \right] / \ln (68/z_0) \tag{4}$$

so that, whereas a single z_0 may be applicable to a given site on land, there is not a single power law exponent. Over the sea, the situation is likely to be more variable since z_0 may be expected to change for different sea surface conditions.

Where the wind is obstructed by surface irregularities on land an allowance can be made for the general level of the obstructions by the introduction of a displacement height in the wind profile relationship. Over a sea surface an allowance should also possibly be made for the effect of large wave topography on the local wind profile. Generally, this effect is not taken into consideration,

64

although it is acknowledged that the wave profile interferes with the wind profile to a level of three or four significant wave heights above mean sea level. Most offshore tower structures will be outside this zone of direct influence but the effect of wave profile on the upper regions of the wind profile is not known. In order to determine the nature of this effect, suitable wind measurements are required over a wide range of conditions and to date no published data of this form is available.

The fluctuating longitudinal component of the wind, defined by the rms value, $\sigma_1(z)$, is also related to z_0 and u_* , such that,

$$\sigma_1(z)/u_* = 2.63 \, \eta \left[0.2 \, \ln \, (z/z_0) \right] \, \tfrac{1}{2}\eta^{16} \tag{5}$$

where $\eta = (1 - z/h)$ \hfill (6)

Current regulations for offshore structures (2,3) relate the 3 second gust and the hourly mean wind speeds by means of a gust ratio. For measurements taken with a standard UK anemograph, the 3 second gust speed is given by,

$$\hat{V} (z) = \overline{V} (z) + 3.7 \, \sigma_1 (z) \tag{7}$$

so that the gust ratio, $\gamma(z)$, at any level is,

$$\gamma(z) = \hat{V} (z) \, / \, \overline{V} (z) = 1 + 3.7 \, I_1 (z) \tag{8}$$

where $I_1 (z)$ is the longitudinal intensity of turbulence, and,

$$I_1 (z) = \sigma_1(z) \, / \, \overline{V} (z) \tag{9}$$

Longitudinal wind gust spectra and coherance functions needed by the designer to define the frequency content and spatial distribution of the turbulence are well described elsewhere (1, 4).

It can be seen that the all important parameters to be identified at any site are the surface roughness length, z_0, the friction velocity or wind shear velocity, u_* and the displacement height, d to take account of the variation in base level for the wind profile. For an offshore site account needs to be taken of possible changes in surface roughness, caused by varying sea and weather conditions.

The current rules for the design of offshore structures proposed by the Department of Energy (2) and Det Norske Veritas (3) give details of wind speed profiles and gust ratios for offshore locations, see Table 1. Details of mean hourly and 3 second gust speeds for the offshore areas of the British Isles are given by the Department of

Energy. There is no specific requirement by the Department of Energy for wind sensitive structures to be checked for dynamic wind effects. Det Norske Veritas, however, do suggest a dynamic analysis for appropriate structures and wind stress coefficients of 0.002 and 0.0015 are proposed for rough and moderate sea conditions respectively. The Harris form of the wind gust spectrum is suggested. However, it is noted that the values of wind stress coefficient, gust ratio and wind speed profile exponent are not consistently related on the basis of the wind characteristics described earlier.

Table 1   Code Requirements

Code	Analysis Method	C_{10}	Exponent α		Gust ratio γ	
			Mean hourly	15 sec gust	3 sec gust	15 sec gust
DnV (3)	Dynamic	0.0015 0.0020	0.15	0.106	1.33	1.26
D.O.E. (2)	Static	–	–	0.09	1.37	1.27

MEASUREMENT OF WIND PARAMETERS OVER THE SEA

Many experimenters have attempted to determine the structure of wind over water both in laboratory experiments and in oceanic and lake tests.

The experimental methods generally used in the full scale tests to determine z_0 and u_* are eddy correlation and profile measurements. Other methods using spectral dissipation techniques and wave slope measurements have been used but are less common.

Eddy correlation requires the measurement of the longitudinal rms gust component, $\sigma_1(z)$, and the vertical rms component, $\sigma_3(z)$. The friction velocity is then given by,

$$u_* = \overline{(-\sigma_1 \sigma_3)}^{\frac{1}{2}} \tag{10}$$

and the roughness length, z_0, is determined from Equation 1 using the log term only. Most of the data obtained in this manner, however, have been derived from measurements at a single level, usually at a height of 10 m or less on a beach or a wave buoy, and as a result must be regarded with some suspicion.

Profile measurements have been used to solve for z_0 and u_* from the log term of Equation 1. In such cases measurements have

generally been taken at three of four levels to define the profile but
for most published data all measurement levels are within the first
10 m. This can lead to bias for sea collected data due to the
variation of wave height, while shore collected results may not
reflect oceanic conditions. Krugermeyer et al (11) have shown that
the value of z_O is dependent upon the levels at which profile
measurements are taken.

Much of the data collected experimentally is presented in the form of
the windstress coefficient or surface roughness coefficient.

$$C_Z = (u_* / \overline{V} (z))^2 \tag{11}$$

Table 2 gives a range of windstress coefficients obtained from
different measurements over water. The results appear to be in two
groups, those with a fixed coefficient corresponding to a saturation
drag condition and those with coefficients which vary linearly with
wind speed in the range of measurements. Where the windstress
coefficient has been found to vary, values have been obtained by
extrapolation for a mean wind speed of 40 m/s and entered in Table 2.
Such extrapolation may not be justified, though Whitaker et al (10)
indicate that their formula is applicable up to 40 m/s.

Hsu (12) has collated various experimental data and established a
relationship between wave height and z_O or u_* for wind speeds up to
20 m/s. This relationship indicates z_O values of approximately
0.06m for developing and fully developed sea states.

It is noted, however, that apart from most of the reported
measurements being taken below 10 m height, the wind speeds
recorded were generally very low, often being below 10 m/s and
rarely reaching 15-20 m/s. Taking these and other factors into
account, it is felt that the reported data is of doubtful validity when
extrapolated to the extreme storm conditions appropriate to the
design of offshore structures in the North Sea.

At the present time, therefore, it would appear that there is no
established method of predicting the wind characteristics over the
sea suitable for the design of offshore structures that is supported
by adequate and comprehensive measurements. Consequently it is
necessary for design purposes to consider a range of parameters to
ensure that reasonable upper and lower bounds are taken into account.
In assessing the extent of this range, however, it is useful to
consider the mechanism of wind wave generation and the likely nature
of the wind waves.

Table 2   Full Scale Measurements of Wind Stress Coefficient
C_Z at z metres above mean sea level.

Ref.	Windspeed Range (m/s)	Location, Method of Measurement and Windstress Coefficient. C_Z
5	0-30	Collation of numerous investigations oceanic and laboratory. No eddy correlation data. $\bar{V}_{10} \leqslant 15\text{m/s} \quad C_{10} = \quad 0.5 \; \bar{V}_{10} \; \times \; 10^{-3}$ $\bar{V}_{10} > 15\text{m/s} \quad C_{10} = \quad 2.6 \; \times \; 10^{-3}$
6	4-17	Ocean station(ship) anemometer at 21m dissipation method. $C_{10} = \quad (1.63 \pm 0.28) \, 10^{-3}$ Authors suggest linear increase with $\bar{V}$ not proven.
7	3.5-10	Collation of three data sets on or near shores. Profile measurement method up to 10m. (ν - dynamic viscosity) $C_6 = (1.09 \log (u_* z_0 / \nu) + 0.81) \, 10^{-3}$ in aerodynamically rough regime when $u_* = 0.045 \times \bar{V}_6$, $z_0 = 0.07 \, u_*^2 /g$ at $\bar{V}_{10} = 40$ m/s $\quad C_{10} = 4.4 \times 10^{-3}$
8	3-22	Sand spit. Nova Scotia, at 10m breakers in fetch zone. Two anemometres eddy correlation method. $C_{10} = \quad (0.63 + 0.66 \, V_{10} \pm 0.23) \times 10^{-3}$ at $\bar{V}_{10} = 40$ m/s $\quad C_{10} = 3.3 \times 10^{-3}$
9	3-16	Lough Neagh profile measurement to 16m $C_{10} = \quad (0.36 + 0.1 \, \bar{V}_{10}) \times 10^{-3}$ at $\bar{V} = 40$ m/s $\quad C_{10} = 4.36 \times 10^{-3}$
10	In a Hurricane	Lake Okeechobee. Florida surface slope method. $C_{10} = (2.28 \pm (1-7/V_{10}) \, 2.63) \, 10^{-3}$ Suggested range of applicability $\bar{V}_{10} = \quad 20 - 40$ m/s at $\bar{V}_{10} = 40$ m/s $\quad C_{10} = 4.1 \times 10^{-3}$
11	4-12	From Jonswap data, profile within zone to 8.4m $\quad C_{10} = \quad 1.3 \times 10^{-3} \pm 20\%$

WIND WAVES

Wind flow over a wave boundary differs from that over a fixed
boundary because the wave boundary can deform and translate
horizontally, experiencing changes in its energy and momentum
levels. The rate of transfer of momentum from the wind to the sea
is a key factor determining the wind wave characteristics.

The mechanisms of wave generation and the transfer of energy from
wind to waves are not sufficiently understood to allow analytical
treatment of the mechanisms involved. They have been described
qualitatively (13) as resonance, shear flow, sheltering and wave
breaking, which can occur separately or in combination at various
stages along the fetch.

The sea surface undulations consists of many waves of differing
height and length. Each component is built up to a maximum
steepness until breaking is induced, some energy is thus dissipated
and some fed into longer waves, near whose crests the breaking
occurs. The development of longer wavelengths along the fetch is
illustrated in Figure 1 in which the wave spectrum is shown for
various points along the fetch. The successive curves indicate the
transfer of energy from the shorter period smaller waves to the
longer period higher ones along the fetch, until a fully developed
state is reached. This is the stage at which energy supplied by the
wind is balanced by energy dissipated by the waves.

There is still considerable disagreement on the extent of the transfer
of energy from the wind along a fetch. Wu (5) has proposed that the
main transfer of wind energy to the sea takes place at the high
frequency end of the spectrum, that the longer waves are developed
by energy transfer from shorter waves and a saturation roughness
develops as soon as the wind speed exceeds wave celerity. Other
authorities feel that sea surface roughness increases along the fetch
and provides greater transfer of energy from the wind until a steady
state is reached (13).

In any event, storms can cover hundreds of square miles and the
developed waves along the fetch may travel great distances.
Theories of wave generation and energy transfer are by necessity
based on simplified models, which even so can produce complex
wave conditions. The introduction of additional complications,
such as weather patterns, new storm centres and adverse winds
occurring in regions of developed wave systems, and tidal and
current effects, cannot be dealt with at present.

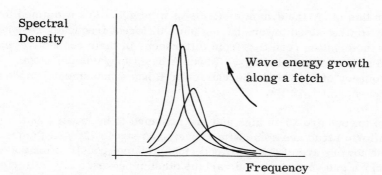

Spectral
Density

Wave energy growth
along a fetch

Frequency

Figure 1   Wave Energy Spectra
showing the concentration of energy into
lower frequency components along a fetch.

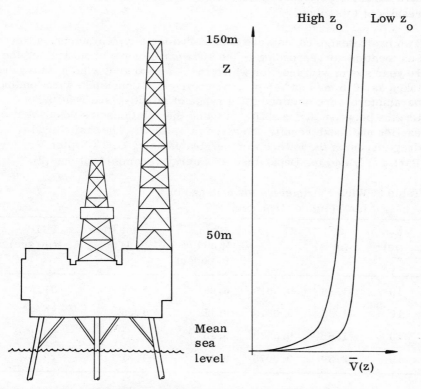

Figure 2   Outline of an offshore platform and tower with mean
hourly profiles corresponding to high and low surface
roughness for the same reference gust velocity.

DYNAMIC RESPONSE OF OFFSHORE TOWERS TO WIND GUSTS

For this study the dynamic effects of wind gusts are examined for two similar sized towers having quite different frequency response characteristics resulting from differences in their respective mass and stiffness distributions. Tower A has a fundamental mode frequency of 0.66 Hz, whereas tower B has a fundamental mode frequency of 1.12 Hz.

The towers are 100m high and are assumed to be located on a platform structure some 50m above mean sea level, see Figure 2. Both towers are of open latticed, tubular steel construction and carry a gas vent riser and various other pipework and ancillaries. For the purposes of this study, the towers have been given identical static wind resistance properties. The effects of vortex shedding, cross-wind turbulence and dynamic interaction with supporting structures will be ignored for simplicity, since they do not alter the results of this study.

Two basic design cases were considered for a typical northern North Sea location corresponding to the 100 year storm without ice and the 10 year storm with ice, for which the 3 second gust wind speeds were taken as 56.5 m/s and 48 m/s respectively. Consistent wind loading parameters were obtained for a range of effective sea roughness lengths between 10^{-4} and 10^{-1} m using the relationships described earlier and these results are given in Table 3. The static wind drag forces on the towers were evaluated using CP3 Chapter V Part 2 (14) and the Department of Energy recommendations (2).

Table 3 Wind Parameters for a Range of z_o
for $\widehat{V}(10) = 48$m/sec.

z_o (m)	u_* (m/s)	C_z	Mean Hourly Exponent	Gust Ratio at 10m	$\overline{V}(10)$ (m/sec)
10^{-4}	1.16	0.0013	0.09	1.52	31.6
10^{-3}	1.33	0.0019	0.11	1.55	30.9
10^{-2}	1.68	0.0033	0.13	1.65	29.1
10^{-1}	2.30	0.0076	0.17	1.81	26.6

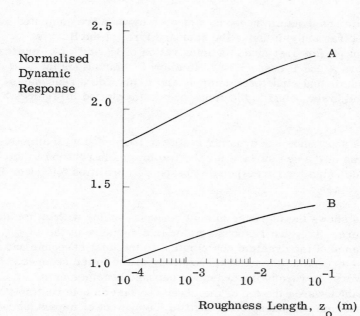

Figure 3. Variation of Dynamic Response with z_o

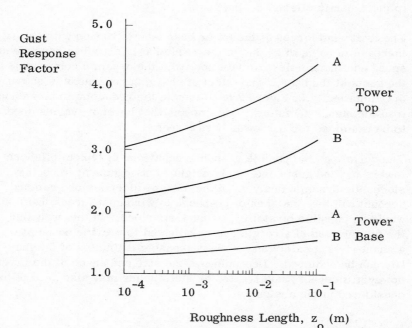

Figure 4. Variation of Gust Response Factor with z_o

The static and dynamic responses of each tower were computed by gust spectral analysis using the standard form of the Harris spectrum (4) for first mode damping ratios of 1% and 2% of critical for towers A and B respectively, to allow for variations in aerodynamic and structural damping and to provide a reasonable upper and lower bound to the range of results plotted in Figures 3 and 4.

Figure 3 shows how the dynamic response increases with effective roughness of the sea surface for the two towers in relation to the reference value for corresponding response quantities for tower B at a z_O value of 10^{-4}.

Figure 4 shows the increase in gust response factor with z_O for the two towers. The gust response factor is a measure of the significance of the dynamic contribution to the total response and is seen to increase as the natural frequency of the tower reduces. For both towers, however, the dynamic response becomes more significant towards the top of the tower, so that an underestimation of sea surface roughness can result in the neglect of important dynamic response contributions in the upper parts of the towers. It is seen that for both towers the base forces are fairly insensitive to the dynamic effects of wind gusts.

The total wind forces at the tower base tend to reduce with an increase in value of z_O due to the reduction in the mean hourly wind speed and the smaller contribution from the dynamic components of the gusts at the base. The effect of changing roughness is shown diagramatically in Figure 2 to illustrate the reduction in mean hourly load for higher z_O values. This means that lower z_O values need to be examined for the tower base forces.

The variation in dynamic response behaviour of typical offshore towers to wind gusts over their height is an argument for using stochastic dynamic analysis in the design, since a more rational account can then be taken of response to wind gust forces than would be possible by static gust analysis alone. In this way also the distribution of strength can be adjusted to provide economy without risking safety and more reasonable estimates of fatigue life can be achieved. Depending on the size and shape of the tower, however, discrete gusts treated as static loads may also need to be considered in the design.

CONCLUSIONS

It has been shown that inadequate or insufficient data exists at present for the extreme storm wind characteristics over the sea to be predicted with any confidence. There is a clear need for better measurements of wind parameters over the sea. It is possible that over a long period of time an offshore tower may experience wind storms in conjunction with a range of effective sea surface roughness. It is unlikely that the most severe storms with the roughest sea conditions would be encountered during a short period of measurement, so a reliable method estimating the effects of extreme storms would still be required for design purposes.

It has been demonstrated that typical offshore towers can be sensitive to the dynamic action of wind gusts and that significant changes in dynamic response can occur for different assumptions regarding sea surface roughness and the corresponding wind characteristics. The dynamic analysis results obtained indicate that adoption of a single value of roughness length or a single value of wind gust factor for a particular design condition , as implied by current regulations for offshore structures, could lead to an underestimate of the dynamic stresses in the tower, if the value chosen was too low.

In the meantime, therefore, until a universal model is available for strong winds over the open sea, it is proposed that the design of offshore towers be carried out for a range of surface roughness lengths with consistent wind structure characteristics. The design should preferably be checked using stochastic dynamic analysis with a reasonable range of damping. The effects of vortex shedding, discrete gusts, modifications to the wind loadings due to flow over the platform and dynamic interaction with supporting structures should also be taken into account.

REFERENCES

1. Deaves D. M. , Harris, R. F. (1978) A Mathematical Model of the Structure of Strong Winds. CIRIA Report No. 76.
2. Department of Energy (1978), Offshore Installations: Guidance on Design and Construction.
3. Det Norske Veritas (1977) Rules for the Design, Construction and Inspection of Offshore Structures.
4. Harris R. I. (1971) The Nature of Wind: The Modern Design of Wind Sensitive Structures. CIRIA, London.
5. Wu, J. (1969) Wind Stress and Surface Roughness at Air Sea Interface. Journal of Geophysical Research, Vol. 74, No. 2 pp 444-445.
6. Denman, K. L. , Miyake, M. (1973) Behaviour of the Mean Wind, the Drag Coefficient and the Wave Field in the Open Ocean. Journal of Geophysical Research, Vol. 78, No. 12.

7.   Sethuraman S., Raynor, G.S. (1975) Surface Drag
     Coefficient Dependence of the Aerodynamic Roughness of the
     Sea. Journal of Geophysical Research, Vol. 80, No. 36.

8.   Smith, S.P., Banke, E.G. (1975) Variation of Sea Surface
     Drag with Wind Speed. Quart. J.R. Met. Soc. 101, pp
     665-673.

9.   Sheppard, P.A. et al (1972) Studies of Turbulence in the
     Surface Layer Over Water (Lough Neagh) Part I
     Instrumentation programme profiles. Quart J.R. Met. Soc.
     98, pp 627-641.

10. Whitaker et al (1975) U.S. Army Corps of Engineers, Coastal
     Eng. Research Centre, Techn. Mem. No. 56.

11. Krugermeyer et al (1978) The Influence of Sea Waves on the
     Wind Profile. Boundary Layer Meteorology 14, pp 403-414.

12. Hsu S.A. (1974) A Dynamic Roughness Equation and its
     application to Wind Stress Determination at the Air Sea
     Interface. Journal of Phy. Ocean., Vol. 4, pp 116-120.

13. Silvester, R. (1974) Coastal Engineering. Elsevier Scientific
     Publishing Co.

14. British Standards Institution (1972) CP3: Chapter V: Part 2:
     1972.

EFFECTS OF SOIL-STRUCTURE INTERACTION ON THE DYNAMIC ALONG-
WIND RESPONSE OF STRUCTURES
G. Solari & D. Stura

Ist. Scienza delle Costruzioni, University of Genoa, Italy

INTRODUCTION

The Authors studied and presented in a previous work (Solari
and Stura, 1979) the problem of the dynamic along-wind response of structures characterized by large flexibility and high
slenderness, including soil-structure interaction phenomenon.
The results showed that, because of the effect of soil flexibi
lity and soil damping, a more realistic representation of the
restraint conditions of the structure is in many cases essential for a correct evaluation of the response.
The conclusion was that the key parameter governing the phenomenon is the ratio between the energy dissipated in the soil
by both radiation and hysteretic damping and the energy dissipated internally by the structure, which is assumed to be linear hysteretic.
When the former energy is prevailing on the latter energy, as
for small values of the structural damping, the soil damping
becomes the most important parameter. It's effect is a deam-
plification of the dynamic response.
In the opposite case the total displacement of the structure
may be greatly increased by the contribution of small rotations
at the base due to the softness of the soil. Moreover the natural frequencies of the system are modified by taking into
account the soil flexibility, and get closer to the dominant
frequencies of the gusty wind. These effects prevail on the
attenuation effects due to the damping in the soil.
Thus the softer is the soil, the more is the amplification of
the dynamic response.
In the present paper the theoretical formulation of the problem
is briefly recalled and the errors associated with certain hy-

potheses and simplifications of the employed analytical models
are discussed and evaluated.
Starting from this general formulation some simple and approxi
mate expressions-which allow the introduction of the concept
of "equivalent clamped prismatic structure"- are derived.
Then in order to systematically investigate and quantify the
reliability and the range of applicability of such simplified
formulas, three kinds of structures have been studied on a wide
range of soil characterized by different properties.

FORMULATION OF THE PROBLEM

The structure has been idealized as an elastic, linear,homoge-
neous beam, in general not prismatic, while the foundation is
considered as a rigid circular disk of infinitesimal thickness
lying on the surface of an elastic half-space.

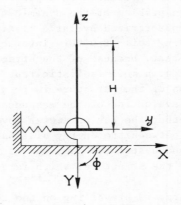

Fig. 1

Refering to fig. 1 the symbols are

$EJ(z), \gamma(z)$	bending stiffness modulus and linear den sity of the beam (function of z)
H	height of structure
R_0, m_f, I_f	foundation radius, mass and rotatory i- nertia
$\omega_j, \psi_j(z)$	j-th natural circular frequency and nor- mal mode of the soil-foundation-structu- re system
G, ν, c_s	shear modulus, Poisson coefficient and shear wave velocity of the soil

Natural frequencies and modes

Natural frequencies and normal modes are obtained by the solution of the eigenvalue problem given by the differential equation

$$\left\{ EJ(z)\,\psi''(z) \right\}'' - \gamma(z)\,\omega^2\,\psi(z) = 0 \tag{1}$$

together with the boundary conditions

$$m_f\,\omega^2\,\psi(0) = \left(EJ\psi''\right)'_{z=0} + K_{11}\,\psi(0) + K_{12}\,\psi'(0) \tag{2}$$

$$I_f\,\omega^2\,\psi'(0) = -\left(EJ\psi''\right)_{z=0} + K_{21}\,\psi(0) + K_{22}\,\psi'(0) \tag{3}$$

$$\psi''(H) = 0 \tag{4}$$

$$\left(EJ\psi''\right)'_{z=H} = 0 \tag{5}$$

When energy dissipation and hysteretic damping of soil are neglected, the dynamic stiffness matrix $[K]$ assumes the form

$$[K] = \begin{bmatrix} k_{11}K_{xo} & k_{21}K_{xo}R_o \\ k_{21}K_{xo}R_o & k_{22}K_{\phi o} \end{bmatrix} \tag{6}$$

in which k_{mn} (m,n=1,2) are dimensionless coefficients depending on a_o and ν (Veletsos and Wei, 1971), $a_o = \omega R_o/c_s$ being the dimensionless frequency parameter;

$$K_{xo} = \frac{8GR_o}{2-\nu} \qquad\qquad K_{\phi o} = \frac{8GR_o^3}{3(1-\nu)} \tag{7}$$

are the coefficients of the static stiffness.

In order to maintain the higher generality, r concentrated masses M_j may be applied to the beam at the z_j (j=1,2,...,r) sections. This is indeed, a more appropriate model for structures such as arc lamps, antennas for radio and metereologic devices, concrete chimneys bearing internal pipes.

The eigenvalue problem of Eq. (1) has been solved applying the variational direct method due to Rayleigh-Ritz, and the following expression of the Rayleigh quotient has been derived

$$R[w] = \frac{\int_0^H EJ\,w''^2\,dz + K_{11}w^2(0) + K_{22}w'^2(0) + 2K_{12}w(0)w'(0)}{\int_0^H \gamma w^2\,dz + \sum_{1j}^z M_j\,w^2(z_j) + m_f\,w^2(0) + I_f\,w'^2(0)} \tag{8}$$

In the class of the allowable functions w(z) only the eigenfunctions $\psi(z)$ have the property of making the Rayleigh quotient $R[w]$ stationary. Then the first N eigenfunctions may be approximated by the expression

$$\psi_j = \sum_{1\ell}^{N} \alpha_\ell^{(j)} \varphi_\ell \qquad (j = 1, 2, \ldots N) \qquad (9)$$

The values of the functional $R[w]$, relative to the eigenfunctions are coincident with the eigenvalues ω^2 .

The accuracy of this method is strictly related to a good choice of the functions φ_ℓ : it would be convenient for these functions to satisfy both dynamic and geometric boundary conditions. In order to meet all these requirements the following method has been used.

Starting from the eigenfunction $\psi_\ell^P(z)$ for the clamped-free prismatic beam

$$\psi_\ell^P(z) = \cos\lambda_\ell z - \cosh\lambda_\ell z + D_\ell(\sin\lambda_\ell z - \sinh\lambda_\ell z) \qquad (10)$$

where λ_ℓ and D_ℓ have the well known values, let us set

$$\varphi_\ell(z) = \psi_\ell^P(z) + A_\ell z + B_\ell \qquad (11)$$

Being $A_\ell z + B_\ell$ linear, the functions $\varphi_\ell(z)$ satisfy the conditions (4), (5) at the free end of the beam for any A_1, B_1. Moreover the arbitrarity of A_1 and B_1 can be used to satisfy the boundary conditions (2), (3).

It follows

$$A_\ell = -2EJ(0)\lambda_\ell^2 \frac{K_{11} - m_\ell\omega^2 + \{D_\ell\lambda_\ell + J'(0)/J(0)\}K_{12}}{(K_{22} - I_\ell\omega^2)(K_{11} - m_\ell\omega^2) - K_{12}^2} \qquad (12)$$

$$B_\ell = 2EJ(0)\lambda_\ell^2 \frac{\{D_\ell\lambda_\ell + J'(0)/J(0)\}(K_{22} - I_\ell\omega^2) + K_{12}}{(K_{22} - I_\ell\omega^2)(K_{11} - m_\ell\omega^2) - K_{12}^2} \qquad (13)$$

This simple and rigorous choice of the functions φ_ℓ , allows a very good determination of first natural frequency and normal mode, but gives some numeric inconveniences in the determination of high order modes.

For this property, Eq. (11) is particularly suitable to define simplified expressions for the determination of the dynamic along-wind response. It is well known that in such problems the contribution of the second and higher vibration modes are reasonably negligible.

In a other work the Authors (Solari and Stura, 1979) largely dealt with the problem of the choice of the allowable functions, with special reference to the numerical aspects of their use.

Dynamic along-wind response

The along-wind displacement $Y(z)$ and the acceleration $a(z)$ may be written as

$$Y(z) = G\varphi(z)\bar{Y}(z) \qquad (14)$$

or

$$Y(z) = \overline{Y}(z) + g_y(z)\,\sigma_y(z) \tag{15}$$

and

$$a(z) = g_a(z)\,\sigma_a(z) \tag{16}$$

where $\overline{Y}(z)$ is the mean displacement, and $Gf(z)$, $g_y'(z)$, $g_a(z)$ are the gust response factor, the displacement peak factor and the acceleration peak factor, respectively.

$\sigma_y(z)$ and $\sigma_a(z)$ are the root mean square of the fluctuating displacement and along-wind acceleration, expressed in the form

$$\sigma_y^2(z) = \int_0^\infty S_y(z\,;n)\,dn \tag{17}$$

$$\sigma_a^2(z) = 16\,\pi^4 \int_0^\infty n^4\,S_y(z\,;n)\,dn \tag{18}$$

n being the frequency and S_y being the spectral density of the response.

The Authors (Solari and Stura, 1979) proposed the following more general expression of S_y, also including the effect of soil-structure interaction.

$$S_y(z;\omega) = \sum_j \sum_k \psi_j(z)\psi_k(z) \int_A \int_A \{ \{ S_p^c(M,M';\omega) + i\,S_p^\Omega(M,M';\omega) \} \cdot$$

$$\{ [\xi_k(\tau;\omega)\,\xi_j(\tau';\omega) + \eta_k(\tau;\omega)\,\eta_j(\tau';\omega)] + i[\xi_j(\tau';\omega)\,\eta_k(\tau;\omega) -$$

$$- \xi_k(\tau;\omega)\,\eta_j(\tau';\omega)] \}\,dA\,dA' \tag{19}$$

which, considering only the real part, may be rewritten as

$$S_y(z;\omega) = \sum_j \sum_k \psi_j(z)\psi_k(z) \int_A \int_A \{ \{ S_p^c(M,M';\omega) [\xi_k(\tau;\omega)\,\xi_j(\tau';\omega) +$$

$$+ \eta_k(\tau;\omega)\,\eta_j(\tau';\omega)] + S_p^\Omega(M,M';\omega) [\xi_k(\tau;\omega)\,\eta_j(\tau';\omega) -$$

$$- \xi_j(\tau';\omega)\,\eta_k(\tau;\omega)]\,dA\,dA' \tag{20}$$

where A is the total exterior area of the structure; M and M' are the centers of the elemental area dA and dA' of coordinates r and r'; S_p^c and S_p^Ω are the co-spectrum and the quadrature spectrum of the loads.

ξ_j and η_j appear in the expression

$$q_j = P(\xi_j + i\,\eta_j)\,e^{i\omega t} \tag{21}$$

of the lagrangian coordinates, which are the solutions of the system

$$(\omega_j^2 - \omega^2) q_j + 2i\,\omega_j \Sigma_\kappa \,\omega_\kappa D_{\kappa j} q_\kappa = \frac{P}{m_j}\, e^{i\omega t}\, \psi_j(\delta) \tag{22}$$

obtained from the study of steady-state vibrations induced by the load $P\delta(z-\delta)\, e^{i\omega t}$.

In Eq. (22) m_j is the j-th modal mass defined as

$$m_j = \int_0^H \gamma(z)\,\psi_j^2(z)\,dz + \overset{\tau}{\Sigma}_\kappa M_\kappa \psi_j^2(z_\kappa) + m_f \psi_j^2(0) + I_f \psi_j'^2(0) \tag{23}$$

and

$$D_{\kappa j} = D_b\, d_{\kappa j} + \frac{1}{2}\, b_{\kappa j} \tag{24}$$

where D_b is the hysteretic damping ratio for the beam, and

$$d_{\kappa j} = \delta_{\kappa j} - (m_j \omega_j \omega_\kappa)^{-1} \{ K_{11} \psi_\kappa(0)\,\psi_j(0) + K_{22}\psi_\kappa'(0)\,\psi_j'(0) + $$
$$+ K_{12}\Big[\psi_\kappa(0)\,\psi_j'(0) + \psi_\kappa'(0)\,\psi_j(0)\Big]\} \tag{25}$$

$$b_{\kappa j} = (m_j \omega_j \omega_\kappa)^{-1}\{ C_{11}\psi_\kappa(0)\,\psi_j(0) + C_{22}\psi_\kappa'(0)\,\psi_j'(0) + $$
$$+ C_{12}\Big[\psi_\kappa(0)\,\psi_j'(0) + \psi_\kappa'(0)\,\psi_j(0)\Big]\} \tag{26}$$

$\delta_{\kappa j}$ being the Kronecker symbol.

In Eqs. (25),(26) C_{mn} (m,n=1,2) are the terms of the damping matrix.

$$[C] = \begin{bmatrix} (2D_\delta k_{11} + \alpha_0 c_{11}) K_{xo} & (2D_\delta k_{12} + \alpha_0 c_{12}) K_{xo} R_0 \\ (2D_\delta k_{12} + \alpha_0 c_{12}) K_{xo} R_0 & (2D_\delta k_{22} + \alpha_0 c_{22}) K_{\phi o} \end{bmatrix} \tag{27}$$

where D_s is the hysteretic damping of the soil, and c_{mn} (m,n= 1,2) are dimensionless coefficients depending on a_0 and ν (Veletsos and Wei, 1971).

Let us observe that in the hypothesis of clamped structure

$$D_{\kappa j} = \delta_{\kappa j}\, D_b \tag{28}$$

so that $[D]$ becomes a diagonal matrix and Eqs (22) are consequently independent. In this case, assuming that in homogeneous turbulence the quadrature spectrum vanishes (Teunissen, 1970), Eq. (20) assumes the well known form

$$S_y(z;\omega) = \Sigma_j \Sigma_\kappa \frac{\psi_j(z)\,\psi_\kappa(z)}{m_j\, m_\kappa \omega_j^2 \omega_\kappa^2} \frac{\left(1 - \frac{\omega^2}{\omega_\kappa^2}\right)\left(1 - \frac{\omega^2}{\omega_j^2}\right) + 4 D_b^2}{\left[\left(1 - \frac{\omega^2}{\omega_\kappa^2}\right)^2 + 4 D_b^2\right]\left[\left(1 - \frac{\omega^2}{\omega_j^2}\right)^2 + 4 D_b^2\right]} \cdot$$

$$\cdot \underset{AA}{\int\int} S_P^c(M,M';\omega)\,\psi_j(\tau)\,\psi_\kappa(\tau')\,dA\,dA' \tag{29}$$

On the contrary, if soil-structure interaction phenomenon is considered, but the damping is small and the resonant peaks are separated, off diagonal terms of the matrix $[D]$ may be neglected, and the following expression, formally analogous to Eq. (29) is obtained

$$S_y(z;\omega) = \sum_j \sum_K \frac{\psi_j(z)\,\psi_K(z)}{m_j m_K \omega_j^2 \omega_K^2} \frac{\left(1-\frac{\omega^2}{\omega_K^2}\right)\left(1-\frac{\omega^2}{\omega_j^2}\right) + 4 D_{jj} D_{KK}}{\left[\left(1-\frac{\omega^2}{\omega_K^2}\right)^2 + 4 D_{KK}^2\right]\left[\left(1-\frac{\omega^2}{\omega_j^2}\right)^2 + 4 D_{jj}^2\right]} \cdot$$

$$\cdot \iint_{A\,A} S_p^c(M,M';\omega)\,\psi_j(\tau)\,\psi_K(\tau')\,dA\,dA' \tag{30}$$

In this case the expression (24) of $D_{kj}(\omega)$ replaces D_b, while m_j , ω_j and ψ_j are the modal masses, the circular frequencies and the normal modes of the whole soil-foundation-structure system.

Although this procedure is widely applied it is convenient to establish some criteria about the approximations of these assumptions as studied by several Authors (Warburton and Soni, 1977; Corsanego and Solari, 1979).

However, in this case, the terms of the matrix $[D]$, are very complicate and make these techniques extremely heavy in the applications.

Then in order to evaluate the reliability of such procedure, a wide range of structures and soils has been analized.

An analogous method has been used for the evaluation of the reliability reached by neglecting first the cross terms in Eq. (29) and then also the contribution of the higher vibration modes to the response.

Simiu (1976) made already use of this method for clamped structures.

EQUIVALENT PRISMATIC CLAMPED STRUCTURE

Let us consider the approximate expression of the first eigen function

$$\psi_1(z) = \psi_1^p(z) + A_1 z + B_1 \tag{31}$$

and the corresponding value of the first natural circular frequency

$$\omega_1^2 = \frac{\int_0^H EJ \psi_1''^2 \, dz + K_{11}\psi_1^2(0) + K_{22}\psi_1'^2(0) + 2K_{12}\psi_1(0)\psi_1'(0)}{\int_0^H \gamma \psi_1^2 \, dz + \sum_{1 j}^{r} M_j \psi_1^2(z_j) + m_f \psi_1^2(0) + I_f \psi_1'^2(0)} \tag{32}$$

82

In order to determine only the fundamental frequency, the following simplified hypotheses can be introduced: 1) the contributions of mass and rotatory inertia are neglected ($m_f=0$; $I_f=0$);
2) the off-diagonal terms of the stiffness matrix $[K]$ are neglected ($K_{12}=K_{21}=0$);3)the in - diagonal terms of $[K]$ are considered to be the static stiffness coefficients ($K_{11}=K_{xo}$; $K_{22}=K_{\phi o}$).
Moreover the beam is considered prismatic and no concentrated
mass is applied to it. From Eq.(8) it follows

$$\omega_1^2 = \frac{\tilde{\omega}_1^2 \, \gamma \int_0^H \psi_1^{p^2} dz + K_{xo}\psi_1^2(0) + K_{\phi o}\psi_1^{'2}(0)}{\gamma \int_0^H \psi_1^2 dz} \tag{33}$$

where $\tilde{\omega}_1$ is the fundamental circular frequency of the clamped
beam.
From Eq.(31) and Eqs. (12),(13) it follows:

$$\psi_1(0) = B_1 = \frac{2EJ\lambda_1^3 D_1}{K_{xo}} \tag{34}$$

$$\psi_1'(0) = A_1 = -\frac{2EJ\lambda_1^2}{K_{\phi o}} \tag{35}$$

If Eqs(34),(35) are substituted in Eq.(33) and an explicit form
is given to λ_1 and D_1 , the following expression is obtained

$$\omega_1^2 = \tilde{\omega}_1^2 \, \frac{\gamma H}{m_1} \left\{ 1 + \frac{4EJ}{H}\left(\frac{1.895}{H^2 K_{xo}} + \frac{1}{K_{\phi o}} \right) \right\} \tag{36}$$

where

$$m_1 = \gamma H \left\{ 1 + 16.483 \, \frac{(EJ)^2}{H^2 K_{\phi o}^2} + 93.695 \, \frac{(EJ)^2}{H^6 K_{xo}^2} + 8 \, \frac{EJ}{HK_{\phi o}} + \right. \tag{37}$$

$$\left. + 15.158 \, \frac{EJ}{H^3 K_{xo}} + 68.067 \, \frac{(EJ)^2}{H^4 K_{\phi o} K_{xo}} \right\}$$

As Roesset (1973) observed, for the range of ω of interest and
typical values of D_s, it is reasonable to assume that the horizontal spring is associated with a viscous dashpot, whereas
the rocking spring has a hysteretic dissipation of energy.
According to a development analogous to the previous one, to
Eqs.(25) and (26) can be given a form:

$$d_{11} = 1 - \frac{(EJ)^2}{m_1 \omega_1^2 H^4}\left(\frac{93.695}{H^2 K_{xo}} + \frac{49.449}{K_{\phi o}} \right) \tag{38}$$

$$b_{11} = \frac{(EJ)^2}{m_1 \omega_1^2 H^4}\left(a_0 c_{11} \frac{93.695}{H^2 K_{xo}} + \frac{98.898}{K_{\phi o}} D_s \right) \tag{39}$$

Calling ξ_{eq} the equivalent damping:

$$\xi_{eq} = D_{11}(\omega_1) \tag{40}$$

and substituting Eq.(38) and Eq (39) in Eq.(24), ξ_{eq} may be written in the form

$$\xi_{eq} = \frac{(EJ)^2}{m_1 \omega_1^2 H^4} \left\{ \frac{m_1 \omega_1^2 H^4}{(EJ)^2} D_b + \frac{93.695}{H^2 K_{xo}} \left(\frac{\omega_1 R_o C_{11}(\omega_1)}{2 c_s} - D_b \right) + \right.$$

$$\left. + \frac{49.449}{K_{\phi o}} (D_s - D_b) \right\} \tag{41}$$

If the approximate expressions (36), (37) and (41) are applied to tall and slender structures, the dimensionless ratios R_o/H and $EJ/HK_{\phi o}$ are small and their squares are negligible. When such hypotheses are introduced, Eqs(36),(37) and (41) may be conveniently replaced by the following conclusive expressions. Setting

$$\mathcal{R} = 4 \frac{EJ}{HK_{\phi o}} \tag{42}$$

one gets:

$$\omega_1 = \tilde{\omega}_1 \frac{1}{\sqrt{1 + \mathcal{R}}} \tag{43}$$

$$\xi_b = \frac{1 + \mathcal{R}}{1 + 2\mathcal{R}} D_b \tag{44}$$

$$\xi_r = \frac{26.24}{(1 + \mathcal{R})^2} \frac{R_o}{\omega_1 c_s} \frac{(EJ)^2}{\gamma H^7 K_{xo}} \tag{45}$$

$$\xi_s = \frac{\mathcal{R}}{1 + 2\mathcal{R}} D_s \tag{46}$$

$$\xi_{eq} = \xi_b + \xi_r + \xi_s \tag{47}$$

It is interesting to observe that:
1) ξ_b, ξ_r and ξ_s represent the contribution to the total damping due to the structural damping and to the mechanisms of radiation and hysteretic dampings in the soil.
2) Among these terms, ξ_r is the term characterized by the largest per cent error. However, it is possible to verify that in the range of tall and slender structures, ξ_r is small in comparison with ξ_b and ξ_s, so that Eq. (47) may be replaced by the following expression

$$\xi_{eq} = \xi_b + \xi_s \tag{48}$$

3) If Eq. (48) is assumed, the static stiffness coefficient K_{xo} does not appear in the simplified expressions of ω_1 and ξ_{eq}. This fact confirms that in the first mode the rocking motion is the most important one as it was observed by many Authors (Novak, 1974; Warburton, 1978).

Starting from the above considerations, the concept of "equivalent prismatic clamped structure" may be introduced.

1) Let us define as "equivalent prismatic structure", a prismatic clamped structure which has the same fundamental circular frequency $\widetilde{\omega}_1$ and a displacement pattern which is in some way analogous to the one of the real structure, clamped at its base. For instance, good results are obtained when the top displacements are equated for a unit uniform horizontal load.

   In this case the moment of inertia J_{eq} of the equivalent structure can be obtained and consequently the value $\gamma_{eq} = 12.36 EI_{eq}/\widetilde{\omega}_1^2 H^4$ follows.

2) If the damping is small, the dynamic response of the structure may be expressed as the sum of two terms, usually referred to as the background part and the resonant part of the response, respectively. The first one is almost independent on the dynamic properties of the structure, while the second term is strongly influenced by the values of the fundamental frequency and by the damping ratio.

   It seems therefore that the best "equivalent clamped structure" is a structure which has the same fundamental frequency and damping ratio of the real structure.

3) In conclusion a good "equivalent prismatic clamped structure" may be defined first by setting $\widetilde{\omega}_1$, J_{eq}, γ_{eq} as specified at the point 1) and then by evaluating ω_1 and ξ_{eq} by Eq.(42) and (48).

It is interesting to point out that in order to introduce such equivalent structure instead of using Eq (31), the first normal mode may be more simply expressed in the applications as $\psi_1(z) = \psi_1^p(z)$. Simiu (1976) demonstred and this paper confirms that the response is not significantly affected by the modal shape.

Computer programs (Simiu and Lozier, 1976) and simplified procedures for calculating the along-wind response (Davenport, 1967; Vellozzi and Cohen, 1968, Vickery, 1970; Simiu, 1976) are both applicable to prismatic, or almost prismatic, clamped structures. In order to evaluate a good approximation of the response, the introduction of the proposed technique generalizes the reliability range of the above procedures.

Moreover, the hypothesis of circular foundation on elastic half-space may also be removed. Many Authors studied and proposed modified expressions of $K_{\phi o}$ in order to account for layered media, embedment conditions and rectangularity of the foundation.

NUMERICAL EXAMPLES

The numerical analysis has been carried out by using a computer program briefly mentioned in an other paper (Solari and Stura, 1979).
Table 1 summarizes the most important characteristics of the following tall and slender structures, selected as case studies:
- a tapered, unlined, welded steel stack (structure A)
- a steel stack as the one above, but gunite lined (B)
- a concrete chimney bearing internal pipes (C).
(R_b and R_t are the base and the top radius respectively).

Structure	H (m)	R_b (m)	R_t (m)	R_o (m)	J_{eq} (m)	γ_{eq} (kg/m)
A	100.	3.50	1.90	7.00	0.72	785.
B	100.	3.50	1.90	7.00	0.72	853.
C	210.	7.80	4.95	19.00	409.26	25000.

Table 1   Description of structures selected as case studies

All these structures are supposed to be placed in suburban area; the mean velocity profile has been taken according to a surface roughness length $z_o = 0.30$ m and a shear velocity $\overline{V}_* = 2.46$m/s. In order to evaluate the approximations and the errors associated with the most important simplified hypotheses introduced in the theoretical formulation of the problem, the calculations have been developed by taking into account:
- the first vibration mode only (scheme 1)
- the first mode, but including the off-diagonal terms of the matrix $[D]$ (2)
- the contribution of the second mode (3)
- the contribution of the second mode and of the cross-terms associated with it (4).
Table 2 describes the percent contributions to the top displacement and acceleration of higher modes in schemes 2,3,4 with reference to scheme 1, for two different restraint conditions at the base of the structure.
Table 3 quantifies the errors in the estimation of the along-wind response for the selected structures on a wide range of soils characterized by different values of the shear modulus G(da N/ cm^2) when the "equivalent prismatic clamped structure" is utili-

	Structure A		Structure B		Structure C	
	displ.	acc.	displ.	acc.	displ.	acc.
Clamped beam						
scheme 2	0.00	0.00	0.00	0.00	0.00	0.00
scheme 3	−0.43	5.82	−0.57	5.78	−0.73	8.36
scheme 4	−0.73	5.82	−1.10	5.79	−1.32	8.42
$G = 1000 \ daN/cm^2$						
scheme 2	0.77	2.10	0.09	0.43	2.99	−0.55
scheme 3	−0.53	2.33	−0.61	3.77	−0.74	6.80
scheme 4	−0.96	2.34	−1.17	3.93	1.59	6.12

Table 2 Percent contribution of higher vibration modes

zed.

In it the true value of n_1 and D_{1f} (n_1) have been compared with n_{leq} and ξ_{eq} calculated by Eq.(43) and Eq.(48).

Moreover the percent errors given by the "equivalent" scheme on the quantities $\overline{Y}$, σ_y , σ_a , q_y and q_a are tabulated.

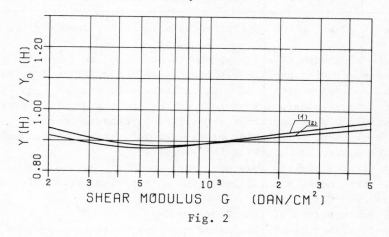

Fig. 2

In conclusion the figures 2,3 and 4 illustrate the ratios bet-ween the maximum displacement Y(H) of the structural systems A,B,C including soil-flexibility and the same quantity $Y_o(H)$ of the clamped structures. Curve (1) corresponds to the real structure while curve (2) is related to the "equivalent pri-smatic clamped structure".

	clamped beam	G 5000	G 2000	G 1000	G 500	G 200
Structure A						
n_1	0.768	0.764	0.759	0.750	0.732	0.684
n_{1eq}	0.768	0.765	0.760	0.752	0.736	0.695
ξ	0.0020	0.0024	0.0032	0.0043	0.0067	0.0134
ξ_{eq}	0.0020	0.0024	0.0030	0.0039	0.0056	0.0093
$(\bar{Y}_{eq} - \bar{Y})/\bar{Y}$	−0.32	−0.31	−0.33	−0.34	−0.40	−0.66
$(\sigma_{yeq} - \sigma_y)/\sigma_y$	−3.42	−2.70	−1.39	−0.01	1.64	4.00
$(\sigma_{\alpha eq} - \sigma_\alpha)/\sigma_\alpha$	−4.81	−3.89	−1.74	1.05	6.42	19.89
$(g_{yeq} - g_y)/g_y$	−0.09	−0.08	−0.03	0.05	0.24	0.79
$(g_{\alpha eq} - g_\alpha)/g_\alpha$	0.03	0.04	0.07	0.10	0.14	0.15
Structure B						
n_1	0.735	0.732	0.727	0.718	0.701	0.656
n_{1eq}	0.735	0.732	0.727	0.720	0.705	0.666
ξ	0.0080	0.0084	0.0090	0.0100	0.0121	0.0179
ξ_{eq}	0.0080	0.0084	0.0090	0.0097	0.0111	0.0144
$(\bar{Y}_{eq} - \bar{Y})/\bar{Y}$	−0.40	−0.37	−0.27	−0.48	−0.47	−0.79
$(\sigma_{yeq} - \sigma_y)/\sigma_y$	−2.16	−1.96	−1.70	−1.29	−0.31	0.01
$(\sigma_{\alpha eq} - \sigma_\alpha)/\sigma_\alpha$	−4.86	−4.43	−4.06	−2.13	1.42	11.39
$(g_{yeq} - g_y)/g_y$	−0.18	−0.16	−0.15	−0.08	0.08	0.51
$(g_{\alpha eq} - g_\alpha)/g_\alpha$	0.11	0.13	0.17	0.21	0.24	0.31
Structure C						
n_1	0.279	0.276	0.272	0.266	0.254	0.226
n_{1eq}	0.279	0.276	0.273	0.268	0.258	0.233
ξ	0.0250	0.0255	0.0262	0.0274	0.0297	0.0355
ξ_{eq}	0.0250	0.0254	0.0260	0.0268	0.0282	0.0308
$(\bar{Y}_{eq} - \bar{Y})/\bar{Y}$	7.32	7.58	7.21	6.80	6.67	6.84
$(\sigma_{yeq} - \sigma_y)/\sigma_y$	6.80	7.18	6.92	6.76	7.14	8.65
$(\sigma_{\alpha eq} - \sigma_\alpha)/\sigma_\alpha$	5.70	6.38	7.09	8.63	11.83	21.14
$(g_{yeq} - g_y)/g_y$	−0.08	−0.05	−0.01	0.08	0.21	0.58
$(g_{\alpha eq} - g_\alpha)/g_\alpha$	0.00	0.00	0.01	0.03	0.03	0.03

Table 3   Comparison between rigorous and approximate analysis

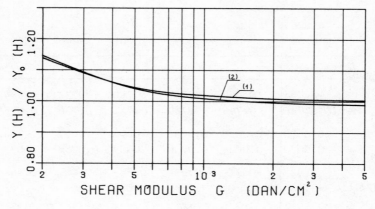

Fig. 3

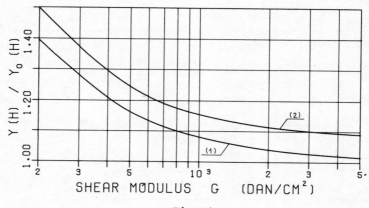

Fig. 4

These diagrams confirm the general criteria summarized in the introduction and point out the good reliability of the proposed simplified technique.

Let us finally observe that when the structural stiffness increases (see structure C), also the errors associated with the utilization of such method increase but the approximation is again wholly acceptable.

References

Corsanego,A. & Solari,G. (1978) Valutazione dell'accoppiamento modale in strutture con piccolo smorzamento. Pubbl. Ist.Sc.Costr. Univ. Genova, 7.

Davenport, A.G. (1967) Gust Loading Factors. J. Struct. Div., ASCE, 93, 3:11-34.

Novak,M. (1974) Effect of Soil on Structural Response to Wind and Eartquake. Earthq. Engng.Struct.Dyn., 3:79-96.

Roessett, J.M & Whitman, R.V. & Dobry,R. (1973) Modal Analysis for Structures with Foundation Interaction. J. Struct. Div., ASCE, 99:399-416.

Simiu, E. (1976) Equivalent Static Wind Loads for Tall Building Design. J. Struct. Div., ASCE, 102, 4: 719-737.

Simiu,E. & Lozier,D.W. (1975) The Buffetting of Tall Structures by Strong Winds. Building Science Series 74, National Bureau of Standards, Washington, D.C.

Solari,G. & Stura, D. (1979) Dynamic Alongwind Response of a Struc tural System Including Soil Flexibility. To appear in Proc. Fifth Int. Conf. on Wind Engng., Fort Collins, Colorado.

Solari, G. & Stura, D. (1979) Methods of Evaluation of Vibration Modes of Structural Systems Including Soil Flexibility. Pubbl. Ist.Sc. Costr.Univ. Genova (in press).

Teunissen,H.W. (1970) Characteristics of the Mean Wind and Tur-bulence in the Planetary Boundary Layer. Review No. 32, Institu-te for Aerospace Studies, University of Toronto.

Vellozzi,J. & Coken,E. (1968) Gust Response Factors. J. Struct. Div., ASCE, 94, 6:1295-1313.

Veletsos,A. & Wei,Y. (1971) Lateral and Rocking Vibration of Foo-tings. J. Soil Mech.Found.Div.,ASCE, 97:1227-1248.

Vickery,B.J. (1970) On the Reliability of Gust Loading Factors. in Proc. Techn. Meet. Conc. Wind Loads on Buildings and Structures, Building Science Series 30, National Bureau of Sandards,Washington, D.C.

Warburton, G.B. (1978) Soil-Structure Interaction for Tower Struc-tures. Earthq. Engng. Struct. Dyn., 6 (in press).

Warburton, G.B. & Soni, S.R.(1977) Errors in Response Calculations for Non-Classically Damped Structures. Earthq.Engng.Struct. Dyn., 5 : 365-376.

THE WIND INDUCED DYNAMIC RESPONSE OF THE WYE BRIDGE

Iain J Smith

Higher Scientific Officer
Bridge Design Division, Structures Department
Transport and Road Research Laboratory

ABSTRACT

As part of the research effort to improve the understanding of the effect of wind on bridges, the Transport and Road Research Laboratory has measured the wind induced dynamic response of the 235 m span cable stayed box girder bridge that carries the M4 motorway over the River Wye. Measurements are presented, which were made during winter months of 1977 and 1978, using automatic recording equipment developed at TRRL.

The main wind induced vibration frequency measured was at 0.46 Hz, which corresponds to the first bending mode. The response occurs mainly when the wind speed is in the range 7-8 m/s and the direction is close to normal to the bridge. This agrees with wind tunnel section tests but the maximum amplitudes predicted by such work have not been attained to date. The strong dependence of vibration amplitude on wind speed, showing a peak in the range 7-8 m/s is characteristic of vortex shedding excitation. The problem of predicting the likelihood of the bridge response from wind speed data is discussed as the oscillation of the bridge in question is a fairly rare occurrence. Finally, the effect on the fatigue life of the structure is discussed. It is shown that it is most unlikely that wind induced oscillations will contribute to a significant shortening of the fatigue life of the bridge.

INTRODUCTION

The work described in this paper was carried out as part of a continuing program of research into the effect of wind on bridges. This work was aimed at obtaining measurements of the response of an actual bridge so that comparison could be made with wind tunnel sectional model tests. Such comparisons have two distinct aspects. Firstly, the sectional model response will approximate to the actual bridge response for the same

aerodynamic excitation. Secondly, the wind tunnel airflow will
also approximate to the actual airflow experienced by the
bridge. The total response of the bridge is related to the
response of the wind tunnel model by a combination of these
two aspects. Sectional model tests can predict the critical
wind speeds for various aerodynamic effects but they do not
always give a reliable guide to the expected amplitudes of
bridge movement during excitation.

As sectional model wind tunnel tests had already been carried
out for the Wye Bridge (Walshe, 1970) it was decided in 1976,
to start a series of observations of wind and bridge movement
at this site. The Wye Bridge carries the M4 motorway across
the Wye close to the river's confluence with the Severn. The
bridge is somewhat dwarfed by its larger neighbour, the Severn
suspension bridge. The Wye Bridge is a cable stayed steel box
girder bridge. Its main span is 235 m between the towers.
Figure 1 shows the main details of the structure. The results
of the wind tunnel tests predicted an instability at relatively low wind speeds which had the characteristics of vortex
shedding excitation. Some measurements made, soon after the
bridge had been completed, by the resident Engineer, showed
that there was in fact an oscillation of the structure in
approximately the predicted wind conditions. The oscillation,
which corresponds to the first bending mode being excited by
the shedding of vortices, was the target of the present
investigation. In addition, it was hoped that information on
gust response in stronger winds would be forthcoming. This
was not so, and the reasons are discussed later. The
characterising of the vortex-shedding response and the comparison with the sectional model tests would give information
about the accuracy of modelling and assist in predicting the
future response of this and similar structures.

Field measurements of a similar nature have been carried out
in US (Gerhart et al 1976), Canada, Norway (Jensen and Hjorth-
Hansen, 1978) and Japan as well as in the UK. The Long's
Creek bridge, Canada (Wardlaw, 1971) and the Onomichi bridge,
Japan (Okuber and Enami, 1971) were both predicted to oscillate by sectional model wind tunnel tests. Long's Creek bridge
gave good agreement with the wind tunnel tests and had to be
altered structurally to reduce the large vibrations which
occurred. On the other hand, oscillation of the Onomichi
bridge was considerably below that predicted by the laboratory
tests. Likewise, the Cleddau bridge, in the UK, showed
a similar result to the Onomichi bridge in that, over a
relatively short period of observation, the measured vibration
amplitudes at the critical wind speeds fell considerably below
those predicted (Eyre and Smith, 1977). The usual explanation
of this is that the Cleddau and Onomichi bridges experienced
turbulent wind conditions due to the surrounding terrain and
that this disruption of the wind lessens the strength of the
stronger winds are mostly from the south west and are thus
normal to the structure. They approach the bridge over the

waters of the Bristol Channel and turbulence will then be less intense than in winds from other directions. Some response should therefore be expected.

DATA COLLECTION AND ANALYSIS

Three propeller anemometers (Gill, 1975), mounted orthogonally, were used to measure the three components of the wind vector and a sensitive accelerometer was used to measure bridge motion. The signals from these sensors were recorded on analogue magnetic tapes by a 4 channel recorder which was under electronic control. The automatic control unit, developed at TRRL by the author (Eyre and Smith 1976), is constantly informed of the direction and magnitude of the wind vector. When the direction and magnitude are within preset limits, a 10 minute tape record is taken of the sensors' output. In this work the direction was required to be from the SW within +40 degrees to the normal to the bridge and the wind speed greater than 5.8 m/s. The bridge normal was taken to be 212° true ie pointing roughly S.W. The anenometers were mounted at the end of a 7 metre beam erected horizontally outwards and normal to the bridge ie pointing S.W. (Figure 2). No data were collected when the wind blew from the north-east as the flow reaching the anemometers would then be affected by the bridge.

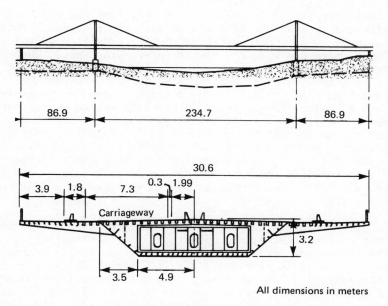

Fig.1 WYE BRIDGE

Fig 2

Each 10 minute tape record was arranged in a suitable format
so that it could be replayed into the computer at TRRL for
automatic analysis. In the first minute, time and data
signals and zero and calibration levels were recorded. This
is followed by an accurately timed 8 minute section containing
sensor outputs and then a 1 minute inter-record gap. The
different sections were noted by shifting the zero level of
the z-axis anemometer at appropriate times. This has the
disadvantage that the signal is recorded at reduced sensitivity and therefore larger error. This inclination signal was
deemed to be of least significance and thus its channel was
chosen to 'double up' as control channel. Computer programs
were written to carry out the analogue to digital conversion,
decoding of time and date signals and scaling and zeroing.
They also combined the data signals to produce values of wind
speed, wind direction and inclination and turbulence factors
over a suitable averaging period.

The accelerometer signal was treated differently. It had been
found from the first year's data that sustained oscillation
of the bridge was rare. The accelerometer naturally emphasised the amplitude of higher frequency vibrations so that
drastic filtering was required to extract the oscillation of
interest at 0.46 Hz. Rather than do this numerically in the
computer it was done after the computer analysis of the wind
signals. The few tape records which actually contained a
significant sustained oscillation were replayed via electronic

filters and recorded by an UV oscillograph. These signals
were measured and scaled in displacement units by assuming
that the motion was effectively simple harmonic. Computer
programs analysing the wind data stored the smoothed, scaled
results on a disc data base. Displacement values were then
entered into this data base. Other programs were produced
which accessed this data base and plotted the histograms and
graphs required for the analysis (see Figures 3 to 11). The
results presented in this paper are mainly based on the
measurements made in the winter of 1977/78. Earlier results
were reported by Eyre and Smith in 1976 (Eyre and Smith,
1976).

RESULTS

The experiment was started on the 19th December 1977 and
continued until the 19th May 1978. During this time recordings
were made on 24 magnetic tapes. On a total of 35 days the
equipment was switched off for operational reasons connected
with the bridge. The observations therefore covered a total
of 116 days. The recording equipment was active, ie waiting
for suitable conditions or recording them, for approximately
1139 hours during this period. After analysis, the tapes
yielded 6488 datum points, each corresponding to a 1 minute
mean value so that on approximately 14% of the period of
observation meterological conditions were favourable for the
bridge to show a response ie the equipment was active.

Frequency of oscillation
On at least five separate records the oscillation was sustained
throughout the 8 minute section. In each case 222 ± 1 clear
oscillations were discernible giving a value for the frequency
of 0.462 ± 0.003 Hz. Confirmation of this was obtained by
playing some tape records into a spectrum analyser where a
distinct peak at 0.46 Hz was found.

The mode shapes and frequencies of this bridge have recently
been investigated by analysing the oscillations caused by
normal trafficking. Results show that the wind induced
oscillation at 0.46 Hz is in fact the first bending mode. The
mode shapes are close to simple sine waves. The frequency
measurements are presented in Table 1.

Mode	Frequency Hz
1st bending	0.46
2nd "	0.74
3rd "	1.14
4th "	1.51
5th "	2.12
6th "	2.93

Table 1. The Wye Bridge Natural Frequencies

Range of wind conditions

The measurements of the wind conditions experienced during these observations are presented in terms of the mean values over 1 minute. Wind speed is obtained by combining the outputs of the three anemometers thus:

$$\text{Wind speed} = (X^2 \times Y^2 + Z^2)^{\frac{1}{2}} \tag{1}$$

Where X, Y and Z are the components along the x, y and z axes. Wind direction is the angle of the projected vector in the x-y plane converted to degrees true (or degrees from north).

$$\text{Wind direction} = \arctan(Y/X) + 212^{\circ} \tag{2}$$

Wind inclination is the angle of the projected vector in the x-z plane.

$$\text{Wind inclination} = \arctan(Z/X) \tag{3}$$

The sign of the value for wind inclination is such as to agree with the method of presenting wind direction data ie it expresses the direction the wind comes from. In this convention, wind blowing towards the ground from above is positive and vice versa. Turbulence factor is taken to be

$$\text{Turbulence factor} = \sigma(u)/U \tag{4}$$

where $\sigma(u)$ is the standard deviation of the wind speed data and U is the mean wind speed. The range of conditions observed and the relative distribution of values are shown in Figures 3 to 7.

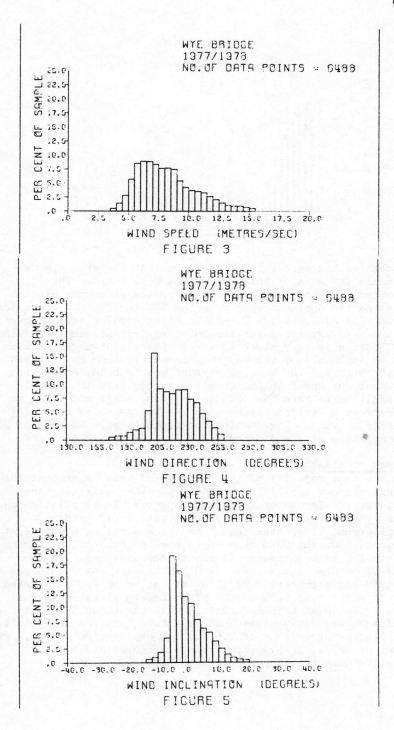

FIGURE 3

FIGURE 4

FIGURE 5

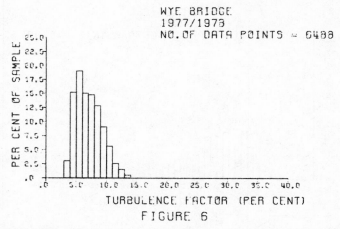

WYE BRIDGE
1977/1978
NO. OF DATA POINTS = 6488

TURBULENCE FACTOR (PER CENT)

FIGURE 6

From these histograms it can be seen that the wind speeds
straddle the area of interest around 7 to 8 m/s. The
turbulence is low as is expected from the geography of the
site and the dominant wind from the SSW is close to the bridge
normal (at 212° true). The inclination results have a lower
accuracy as explained earlier but they indicate that the wind
inclination is very close to zero. The slight negative bias
can be explained by the way the anemometer boom is attached
to the bridge. The boom is slender and is held in position
by steel guy ropes which, when tightened, lift it slightly.
In addition to this the boom is not very stiff and tends to
vibrate in a vertical plane when the bridge moves due to
traffic or wind. The combination of these effects show up as
the negative bias and large scale scatter in Figure 5.

Bridge oscillation
For the vast majority of the duration of this experiment the
bridge did not oscillate appreciably. When examining the
filtered accelerometer records an appreciable oscillation was
taken to be one which had a peak to peak (p/p) bridge
displacement of greater than 5 mm and a duration of at least
1 minute. On the data base there were a total of 6488 datum
points for each parameter (1 minute mean values). In the
case of bridge displacement 6210 points were not considered to
contain any appreciable oscillation ie the bridge oscillated
for approximately 4.3% of the time the experiment gathered
data. Combining this information with the known time the
equipment was active (see RESULTS) shows that the bridge only
oscillated appreciably on 0.6% of the observation period. The
range of bridge displacement values measured and their
relative distribution are shown in Figure 7.

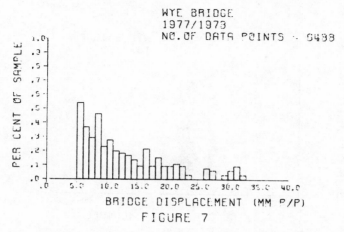

WYE BRIDGE
1977/1978
NO. OF DATA POINTS - 6438

BRIDGE DISPLACEMENT (MM P/P)

FIGURE 7

The relationship between wind speed and bridge displacement is shown in Figure 8 (to avoid overcrowding the graph, displacements less than 5 mm (p/p) were not plotted). This is typical of vortex shedding excitation which occurs when the frequency of vortex shedding from the surfaces of the bridge is close to one of its natural frequencies.

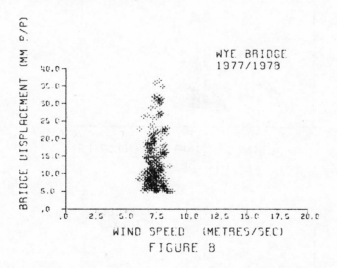

WYE BRIDGE
1977/1978

WIND SPEED (METRES/SEC)

FIGURE 8

From the wind tunnel tests the maximum response is predicted to be between 7.3 m/s and 8.1 m/s, when account is taken of the slightly different frequency assumed in the wind tunnel tests (Walshe, 1970). When the displacement values in Figure 8 are plotted against wind direction, inclination and turbulence Figures 9, 10 and 11 are produced. Taking into account the distribution of these parameters (Figures 4, 5 and 6) no pronounced dependence is shown.

100

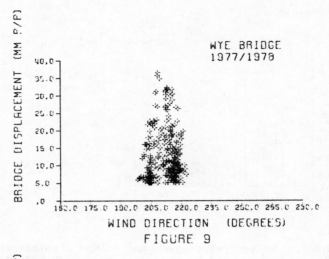

FIGURE 9

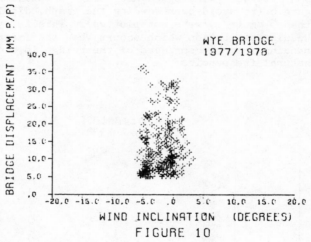

FIGURE 10

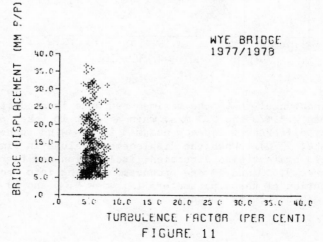

FIGURE 11

Damping

The damping value for the bridge will limit the amplitudes
reached. It is not easy to measure the structural daming of
a bridge under controlled conditions without closing the
structure to traffic and applying some form of artificial
excitation (Tilly, 1977). Estimates can however, be made from
vibration records using autocorrelation techniques. Some such
estimates were made from parts of the records obtained in this
work containing a mixture of wind and traffic vibrations
(mainly traffic). The measurements taken to obtain the
frequencies and mode shapes were also analysed to obtain
damping values. Finally, one of the UV oscillograph records
showed a steady oscillation finishing in a decay to virtually
zero displacement. The rate of decay was evaluated and
expressed in terms of logarithmetic decrement δ. The damping
values are given in Table 2.

Excitation	δ Mean
Wind and traffic	0.023
Traffic	0.019
Decay of displacement signal	0.022

Table 2. Damping estimates

These damping values are approximate and are averages of
measurements having quite high scatter ($\pm$50%) but give an
indication of the level exhibited.

For a welded steel bridge the damping would be expected to be
in the range of 0.02 to 0.08 (Eyre and Tilly, 1977). When
the amplitude of oscillation increases, the cables play an
increasing role in the damping and, as the damping effect of
the cables is non-linear (Wyatt, 1977) the amplitude is
quickly limited.

Gust response

It was hoped that, in addition to the vortex shedding
excitation, response to gust excitation in stronger winds
could be measured. Unfortunately no information was obtained
because high winds did not occur while the equipment was
active.

DISCUSSION OF RESULTS

Comparison of the results in Figure 8 with the wind tunnel
predictions (critical wind speed 7.3-8.1 m/s) shows that the
wind tunnel sectional models predict satisfactorily the

critical wind speed for this aerodynamic effect though the
predictions of amplitude are less so. The wind tunnel
results suggest that a peak to peak (p/p) displacement of 100
to 200 mm, depending on damping and inclination, would be
experienced. The maximum response, however, was 36.6 mm
(p/p) in this work. This is not unexpected. In the wind
tunnel test the airflow was smooth and steady. It is also
the same over all the model and therefore there will be a
high degree of coherence between the vortices shed at
different parts of the structure. This does not necessarily
occur for the real structure. In practice the wind has short
term variations both in space and in time. This reduces the
driving force from the vortex shedding. Coupled with this,
the damping is amplitude dependent so it is likely that the
oscillations experienced in practice will be below the wind
tunnel response.

Excitation of higher modes

Using one accelerometer at mid-span this experiment would
only provide information on odd modes of vibration. The wind
speeds necessary to excite the higher modes can be calculated
from equation 5. The Strouhal number S is a function only of
the geometry of the cross section so that for any wind speed

$$S = NB/U \qquad\qquad\qquad (5)$$

Where N is the frequency of vortex shedding, U is the wind
speed and B is a typical cross section dimension. Using
equation 5 and the data in Table 1 the wind speeds to excite
other modes are calculated and given in Table 3.

Mode (bending)	Wind speed (m/s)
1st	7.7 (measured)
2nd	12.4 (calculated)
3rd	19.1 "
4th	25.3 "
5th	35.5 "
6th	49.0 "

Table 3. Critical wind speeds

It is conceivable for a response at the 2nd mode frequency to
have occurred but as mean wind speeds did not exceed 16 m/s
no higher modes would have been excited by vortex-shedding
during this period. For 2nd mode bending, which is
antisymetrical, the cables would play a much larger part in

damping as the antinodes would be close to the cable
anchorages. This should result in a much reduced response
when compared with 1st mode bending amplitudes.

Prediction of future response

In an earlier section it was estimated that on 14% of the
observation period the wind conditions were such as to
trigger the equipment into operation. The bridge actually
oscillated on 4.3% of the times favourable which gives rise
to a figure of 0.6% for the incidence of oscillation in the
observation period. Figure 7 shows the relative distribution
of these oscillations in terms of displacement value. Using
data presented by Caton (1975) on the distribution of hourly
mean wind speeds in the UK, it is estimated that the
triggering level of 5.8 m/s will, on average, be exceeded
about 48% of the time at this site. The observation period
in question would be expected to show a higher figure than
this as it covers the winter months but taking a 'worst
case' condition a figure of 29% is produced for the likelihood that, given the right wind speed, all other conditions
will be favourable for a response. These figures predict a
total of 8.8×10^4 cycles of oscillation at 0.46 Hz in one
year, distributed as in figure 7. If these calculations are
repeated for a hypothetical experiment where the wind
threshold trigger was set around 18 m/s a much lower figure
is produced. The proportion of time the wind speed exceeds
this trigger value is only 0.2%. The assumption is made
that 29% is again the likelihood that both the wind speed
and all the other conditions will be right and that 4.3% is
the incidence of oscillation. This results in a prediction
of 390 cycles of oscillation at 1.14 Hz per year.

Fatigue effect of aerodynamic response

The modes shapes measured were, to a first approximation,
sine waves and therefore the stresses in the bottom of the
box can be calculated. For 1st mode bending this results in
a relationship between displacement (d) and stress range (S)
given by

$$S = 5.4 \times 10^{-2} . d(mm)MN/m^2 \tag{6}$$

and for 3rd mode bending the relationship is

$$S = 9.4 \times 10^{-1} . d(mm)MN/m^2 \tag{7}$$

Thus for the maximum response found during this observation
period the cyclic stress range would be 2 MN/m². This is far
below typical stress 'cut off' values which are usually
around 30 MN/m². If the 'worst case' 3rd mode bending example
was taken and the highly pessimistic assumption that all of
the 390 cycles per year at 1.14 Hz had a displacement of,
say, 36 mm, the cyclic stress range would be 34 MN/m². For

the lowest class of weld (G) which could be expected in a structure this stress range would give a fatigue life of 7×10^6 cycles. Bridges are assessed for a fatigue life of 120 years and the 'worst case' estimate of the number of cycles due to vortex shedding at this stress range is only 4.7×10^4. The excitation of the structure by vortex shedding thus does not significantly reduce the fatigue life.

CONCLUSIONS

The first bending mode of the Wye bridge has a frequency of 0.462 Hz and responses at this frequency of up to 36 mm peak movement are excited by winds in the range 7.4 to 8.0 m/s.

Wind tunnel tests using sectional models predicted accurately the critical wind speeds for vortex shedding but overestimated the amplitude.

Vortex-shedding excitation is very unlikely to cause large enough stresses or sufficient number of cycles for it to be considered as a possible course of fatigue damage to the Wye Bridge.

ACKNOWLEDGEMENTS

Mr K Yokoyama of the Public Works Research Institute, Ministry of Construction, Japan is thanked for the valuable advice and encouragement he gave while a voluntary worker at TRRL for a year. The co-operation of Avon County Council and its staff during the experiment is also acknowledged.

The spectrum analysis and damping estimates from the recorded data were carried out by Loughborough Consultants Ltd and the measurements of frequencies, mode shapes and damping by Structural Monitoring Ltd, Glasgow.

The work described in this paper forms part of the programme of the Transport and Road Research Laboratory and the paper is published by permission of the Director.

REFERENCES

Caton, P. G. F. (1975) "Standardised maps of hourly wind speeds over the UK". Proc. Int. Conf. on wind effects on buildings and structures, Heathrow.

Eyre, R. and Smith, I. J. (1977) "Wind and traffic induced dynamic behaviour of some steel box girder bridges". Paper 5, Symposium on Dunamic Behaviour of Bridges. Department of the Environment, TRRL Report SR 275, Crowthorne.

Gerhardt, C. L., Nelson D. D. and Greene, D. M. (1976).
"Selected wind and bridge motion data recorded at long-span
bridge sites." Federal Highway Administration, Report FHWA-
RD-76-180. Washington.

Gill, G. C. (1975) "Development and use of the Gill UVW
anemometer". Boundary-Layer Meteorology, 8.

Jenson, N. O. and Hjorth-Hansen, E. (1978) "Dynamic
excitation of structures by wind-turbulence and response
measurements at Sotra Bridge". SINTEF Report No STF 71
A78003. Trondheim.

Okubo, T. and Enami, Y. (1971) "Studies on vortex excited
oscillation of Onomichi Bridge". Proc. Int. Conf. on wind
effects on buildings and structures, Tokyo.

Tilly, G. P. (1977) "Damping of highway bridges: a review".
Paper 1, Symposium on Dynamic Behaviour of Bridges.
Department of the Environment TRRL Report SR 275, Crowthorne.

Walshe, D. E. (1970) "A comparison between full-scale wind-
induced instability of the Wye bridge and that predicted from
sectional-model tests". Ministry of Technology National
Physical Laboratory Aero Report 1318, Teddington.

Wardlow, R L. (1971) "Some approaches for improving the
aerodynamic stability of bridge road decks". Proc. Int. Conf.
on Wind Effects on Buildings and Structures. Tokyo.

Wyatt, T.A. (1977) "Mechanisms of damping". Paper 2.
Symposium on Dynamic Behaviour of Bridges. Department of
the Environment TRRL Report SR 275, Crowthorne.

INVESTIGATION ON THE VIBRATIONAL BEHAVIOUR OF A CABLE-STAYED BRIDGE UNDER WIND LOADS

J.W.G. van Nunen and A.J. Persoon

National Aerospace Laboratory (NLR)
Anthony Fokkerweg 2, 1059 CM AMSTERDAM, The Netherlands

INTRODUCTION

Since the dramatic accident with the Tacoma Narrows bridge it has become normal practice to pay attention to the vibrational behaviour of slender civil structures under wind loads. Over the years a large number of investigations has been conducted on bridge type structures, which has shown that two categories of vibrations with quite different characteristics can be discerned.

In the first category the vibrations of the structure generate unsteady aerodynamic loads, which on their turn influence the dynamic stability of the system. Under certain conditions the system may become unstable which means, that a once started vibration increases its amplitude with every cycle of oscillation until the structure collapses. This phenomenon is called flutter and actually caused the destruction of the Tacoma Narrows bridge. Essentially flutter only occurs when a defined critical wind speed is exceeded. This critical wind speed is to a large extent dependent on the mass and stiffness properties of the structure.

The vibrations of the second category are governed by unsteady aerodynamic loads, which are determined by the outer shape, more specifically the cross section, of the structure and in principle not by the vibrational motion of the structure. In this case vortices are shed periodically from the structure and as a result periodic unsteady aerodynamic loads are generated. The periodicity is directly proportional to the wind speed, thus at a certain wind speed the frequency of the unsteady aerodynamic load will coincide with one of the natural frequencies of the structure. This leads to a response of the structure in its resonance mode with a limited amplitude of oscillation. This amplitude is controlled by the magnitude of the unsteady airloads and the response characteristics of the structure for that specific mode.

In the present paper an investigation is described, which has been performed during the design stage of the new traffic

Figure 1 Picture of the traffic bridge across the river Waal near Ewijk

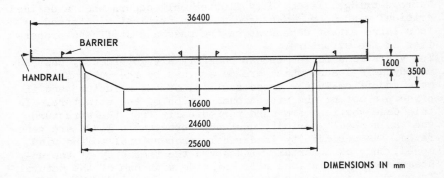

DIMENSIONS IN mm

Figure 2 Sketch showing basic shape of bridge deck

bridge, which crosses the river Waal near Ewijk in the Netherlands. The bridge is of the cable-stayed type and its cross
section consists of a steel box girder which is supported from
two pylons by cables (Figures 1 and 2). The main span amounts
270 m. Because of this rather long length in combination with
the low damping characteristics of steel, it was decided to
study the vibrational behaviour of the bridge under wind loads.
This study in which the two aforementioned categories of vibration were considered, consisted of the following parts:

flutter stability: based upon detailed drawings of the bridge
the vibrational behaviour of the structure
was computed. Using this result flutter
calculations were performed to establish
the flutter stability of the bridge with
respect to wind. The required unsteady aerodynamic loads were derived from "thin plate"
theory. Further, to increase confidence in
these calculations, flutter measurements
were performed on a simple wind-tunnel
model;

vortex-shedding : model experiments were carried out also to
obtain information on the magnitude of the
unsteady aerodynamic loads due to vortex-
shedding from the structure.
The test results were applied in calculations to determine the resonance conditions
for the actual bridge in terms of wind
speed and amplitude of vibration;

full-scale tests : during the erection of the actual bridge vibration measurements were carried out to en-
able a comparison between calculated and
experimental natural frequencies and to obtain information on the damping of the full-
scale structure.
Finally the wind-induced vibrational behaviour of the actual bridge was observed
during a long period. The aim of this experiment was to establish the degree of correlation between calculated and experimental data.

In the following chapters the various parts of the investigation
will be discussed in some more detail.

2 FLUTTER STABILITY

2.1 Vibration computations

To calculate the vibration modes and corresponding natural
frequencies of the bridge the structure was divided into a number of elements of which the mass and stiffness proporties were
derived from detailed drawings. This information was used as
an input into a finite element calculation procedure (FEM). In
Figure 3 a survey is presented of the four lowest vibration

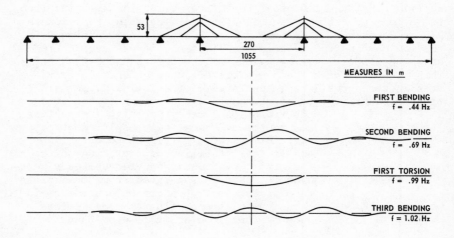

Figure 3 Calculated vibration modes for complete bridge

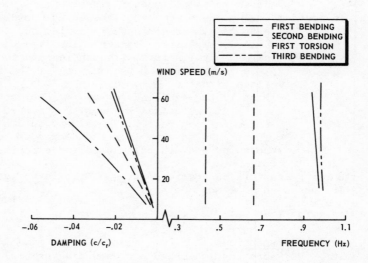

Figure 4 Calculated flutter diagram for actual bridge

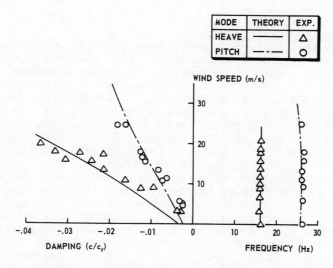

Figure 5 Calculated and measured flutter characteristics for model of bridge deck

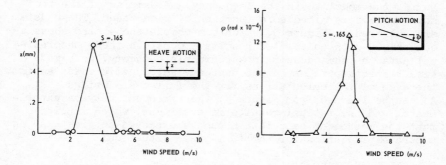

Figure 6 Resonance vibrations observed on model

modes and natural frequencies as determined for the complete bridge.

2.2 Flutter investigation

The computed vibration modes and natural frequencies were used as inputs for flutter calculations. From previous experiences it was known, that for the required unsteady aerodynamic loads use could be made of theoretical results for oscillating, infinitely thin plates (AGARD Manual, 1968). To apply these aerodynamic loads the structure is divided into a number of spanwise strips, which are treated as if they form part of an infinitely long bridge. The aerodynamic interference between the various strips is neglected. This procedure is justified because of the slenderness of the structure.

The results of the computations are presented in a so-called flutter diagram (Figure 4). This diagram indicates how the various natural frequencies and corresponding damping values develop with wind speed. A negative damping value means that damping has to be extracted from the system to sustain a harmonical oscillation or in other words: under these conditions the system is stable. From Figure 4 it can be concluded that the bridge structure is safe with respect to flutter within the range of normally occuring wind speeds.

The calculation results were verified with the help of model tests in a wind tunnel. For this purpose a two-dimensional rigid model of the bridge deck was used, which was supported at both extremities by a set of springs. Through these springs the model was free to oscillate in pitch and heave. The ratio of the frequencies was representative for the ratio of the fundamental frequencies of the full-scale bridge.

During the model tests the frequency and damping of the two modes were determined as a function of wind speed. These data are gathered in Figure 5, which for completeness also shows a comparison with calculated results. To obtain these theoretical data the same assumptions (thin plate theory) were made with respect to the required unsteady aerodynamic loads as in the case of the calculations for the full-scale bridge.

As can be learned from Figure 5, the agreement between experimental and theoretical results is excellent, which implies that the method of calculation can be expected to give also an accurate prediction for the full-scale situation.

3 VORTEX-SHEDDING

3.1 Model experiment

The same model as applied in the flutter investigation, was used to determine the unsteady aerodynamic loads due to vortex-shedding. These model tests were carried out in a smooth air flow, since in this flow condition in general more severe vortex-induced airloads are encountered than in a turbulent airstream. Since in actual practice always some turbulence will be present, the unsteady airloads on the full-scale structure

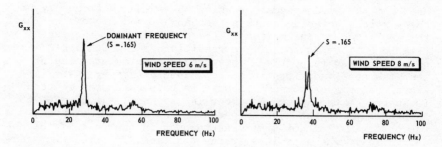

Figure 7 Power spectral densities of hot-wire signal indicating existence of vortex shedding

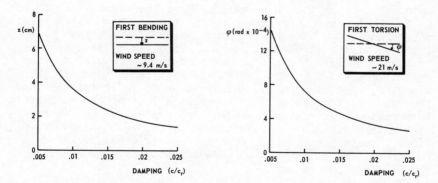

Figure 8 Calculated resonance vibrations for full-scale bridge

thus will very probably be smaller than as predicted from the
model tests.
During this particular model experiment the wind speed was
gradually increased and the response of the model was measured. At the wind speeds for which the shedding frequency of
the vortices coincided with a resonance frequency of the model,
the amplitude of the motion of the model clearly increased
(Figure 6). Comparison of the conditions for which the maximum
of the heaving respectively pitching motion occurs, learns
that the observed vibrations are indeed caused by vortex-induced unsteady aerodynamic loads; in both cases the reduced
frequency or the Strouhal number is the same, viz. S = 0.165.
A further prove that vortex-shedding indeed was the cause,
was obtained from hot-wire measurements in the wake of the model while it was kept in a steady position. The frequency content of this signal exhibited a clear peak (Figure 7) at the
earlier mentioned Strouhal frequency.
From the measured model vibrations the unsteady aerodynamic
loads were determined. This was performed by calculating what
the unsteady load had to be to generate the measured amplitude
of the resonance vibration of the model with its known vibrational characteristics.

3.2 Full-scale predictions

After proper scaling, the unsteady airloads were applied in
calculations for the full-scale bridge. The results are shown
in Figure 8, which presents the amplitude of the resonance
vibration in the middle of the main span of the bridge as a
function of the structural damping. This way of presentation
was chosen, since the actual value of the structural damping
was still unknown in the design stage of the bridge. It should
be noted, that in both the calculations to determine the load
from the model response and the calculations of the full-scale
responses the aerodynamic damping was neglected. As can be observed from Figure 8, the amplitude of the resonance vibrations remains pretty small for this specific bridge and as a
consequence no action was considered to modify the external
shape of the bridge deck.

4 FULL-SCALE TESTS

4.1 Vibrational characteristics

The situation in which one half of the bridge was completed
while work was progressing at the other half, offered an excellent opportunity to carry out some vibration measurements.
In performing vibration measurements on a large and heavy
structure it usually becomes a problem to excite the structure
properly. In the present case this problem was solved by letting a heavy truck (total mass about 10 tons) drive back and
forth the bridge. Through the normal irregularities of the
road deck a random excitation was generated to which the bridge
structure responded in a number of fundamental modes.

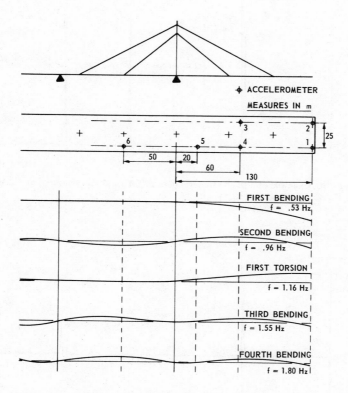

Figure 9 Location of accelerometers and calculated vibration modes for bridge at
half length

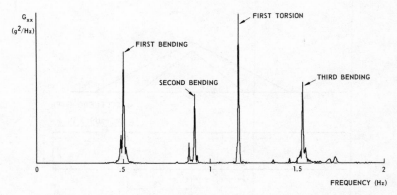

Figure 10 Example of a power spectral density for vibration measured at accelerometer position no. 1

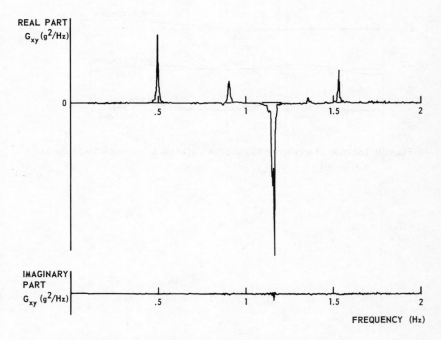

Figure 11 Example of cross power spectral density for vibrations measured at
accelerometer no. 1 and 2

Incidentally the road imperfections were increased by placing wooden beams across the road deck.

The response of the bridge was measured at a number of locations by means of accelerometers. These accelerometers were distributed in such a way, that a reasonable number of vibration modes should be discernable. The actual positions are indicated in Figure 9, in which also some calculated vibration modes are presented for one half of the bridge.

Because of the random excitation the measured responses also were of a stochastic nature. As a consequence the analysis towards natural frequencies and corresponding damping values could only be performed by applying power spectral techniques. An example of the results obtained in this manner is presented in Figure 10. In this plot natural frequencies show up as discrete peaks, while the corresponding damping value can be determined from the width of the peak at half its height. An indication about the vibration modes belonging to each of the natural frequencies can be obtained by determining a cross power spectral density plot of the signals of two accelerometers. Such a plot again shows the natural frequencies, but also the relative phase between the two accelerometers is determined. A typical example of such a cross power spectral density is shown in Figure 11, in which the two accelerometers no. 1 and no. 2 at the extremity of the bridge are considered. This plot learns, that at the third natural frequency the two signals are in counterphase, which corresponds to a torsion mode. All other natural frequencies apparently belong to bending modes.

As a whole it appeared relatively simple to determine the five lowest vibration modes. A comparison between calculated and measured data for these five modes is presented in the following table.

vibration mode	natural frequency (Hz)		damping (c/c_{cr})
	calculated	measured	
First bending	0.53	0.5	0.008
Second bending	0.96	0.91	0.0045
First torsion	1.16	1.16	0.0025
Third bending	1.55	1.54	0.003
Fourth bending	1.80	1.72	0.0035

As can be seen from this table, the agreement between calculated and measured results is very good. As usual, a slight deterioration appears when the vibration modes become more complicated. Further, the extremely low damping values for all vibration modes should be noted. Partly these low values can be ascribed to the fact, that at the moment of execution of the measurements no asphalt was present at the road deck. However, experiences on other bridges have shown also that, even with the pavement added, the damping for steel bridges stays

118

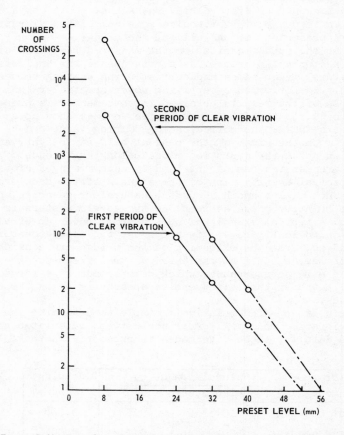

Figure 12 Number of crossings versus preset level as measured on full-scale bridge

mostly well below the values suggested in existing building codes.

4.2 Wind-induced resonance vibrations of full-scale bridge

The availability of the half span bridge was also used to obtain information on the occurrence of wind-induced resonance vibrations on the full-scale structure. For that purpose an accelerometer was positioned at the middle of the road deck near the extreme end of the bridge. In this manner the measured signal was mainly due to oscillations in the fundamental bending mode. The accelerometer signal was fed into an instrument with which crossings of certain preset levels of acceleration could be measured and counted. Together with these data information was gathered concerning speed and direction of the wind at the site. The measurements were continued during a period of about two and a half months. In this period such a variety of weather conditions was encountered, that the results of the measurements could be considered as being sufficiently representative. The results of the countings are gathered in Figure 12.

The first observation that can be made from the test results is, that the bridge deck indeed could encounter resonance vibrations. In fact clear vibrations of the bridge structure were observed during two short measuring periods. As expected, these vibrations only occurred under specified weather conditions: wind blowing across the span of the bridge at a speed of about 11 m/s. Together with the frequency at which this vibration took place $(f = 0.5$ Hz$)$, a Strouhal number of $S = 0.16$ is found, which agrees very well with the value of the model tests in the wind tunnel $(S = 0.165)$. Also good agreement was obtained for the maximum response of the bridge; from the model tests a maximum amplitude of 68 mm was predicted at the position of the accelerometer, while the actual measurements indicate that incidentally an amplitude of about 58 mm was reached.

5 CONCLUDING REMARKS

The present study has revealed that at least for bridges of the box girder type the flutter behaviour can be predicted reasonably well by calculations, in which the aerodynamic loads according the "thin plate" theory are used.

The problem of resonance vibrations due to vortex-shedding has to be studied with the help of model tests in a wind tunnel. Vibration amplitudes as predicted from these model tests appear to agree very well with full-scale values.

6 ACKNOWLEDGEMENT

The authors are indebted to Rijkswaterstaat Directie Bruggen for their permission to present this survey on the investigation for the Ewijk-bridge.

7 REFERENCES

AGARD Manual on Aeroelasticity, Vol.VI; 1968

A STUDY OF THE MEASURED AND PREDICTED BEHAVIOUR
OF A 46-STOREY BUILDING

A P Jeary and B R Ellis

SUMMARY

The response of a 46-storey, 190 m, tall building to both controlled forcing of oscillation and to the wind, has been
monitored.

Measurements of the building's characteristics, together with
the measured wind characteristics, at the time of the tests,
have been used in three design guides. It has been found
that none of these guides perform well in this case. Three
possible reasons contributing to the discrepancies have been
put forward, and areas of research necessary to overcome the
problems have been suggested.

INTRODUCTION

As part of the current BRE investigation of the dynamic
properties of tall structures, tests have recently been conducted on a 46-storey tall building. The purpose of the
investigation is to provide new data for design purposes and
to use these data in order to evaluate the adequacy of codes
of practice and design guides.

The study reported here involved both the controlled forcing
of oscillation of a tall building and the monitoring of its
response to the wind. The controlled forcing allowed the
estimation of the dynamic structural properties. Data about
the wind conditions were obtained from the London weather
centre and from an anemometer, on a crane, about 10 m above
the building. This information was then used in three of the
most up-to-date design guides and the actual response of the
building was compared with the predicted responses.

Severe shortcomings are apparent in all of the guides and
these are discussed.

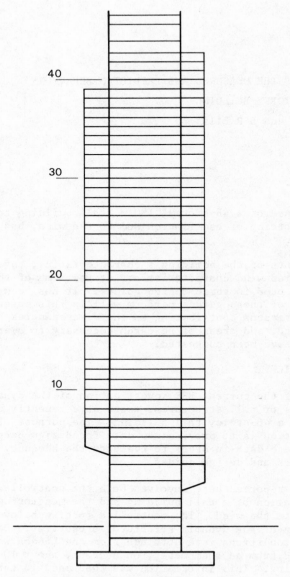

Figure 1 Elevational cross-section of the tower

DESCRIPTION OF BUILDING

Figs 1 and 2 show the basic layout of the structure. The building is a 190 m tall, 46-storey, reinforced concrete office block. It consists of a massive heavily reinforced core, with three 'leaves' attached and supported by large concrete corbels at their bases. Service ducts and lifts are provided inside the core, and the leaves are mainly for office usage. The three leaves are each of different height. Leaf 'A' extends from the 1st to the 43rd levels, leaf 'B' from the 3rd to the 41st, and leaf 'C' from the 5th to the 39th. The exterior of the leaves consists of steel columns, onto which are attached the cladding panels. The concrete floors are cantilevered from the core, and are attached to the steel columns at the periphery. Expansion joints are provided at the interfaces between leaves.

The top of the core is approximately 187 m above ground level and at the time of the forced oscillation tests the building was structurally complete, with the sole exception of the 46th floor. Additionally, the upper three floors were open to the atmosphere, and several minor items at or near ground level were not complete.

TESTS PERFORMED ON THE BUILDING

A vibrator was installed at the 43rd floor level in leaf 'A' (see Fig 2). This vibrator was capable of delivering a uni-directional sinusoidal force of 1070 Newtons at 1 Hz, and this force was generated by contra-rotating masses(1).

Motion of the structure was monitored by the use of servo-type accelerometers. These are very sensitive devices and, when used in conjunction with filters, amplifiers and a spectral analyser, can be used to resolve an acceleration of 10^{-6}g.

The vibrator was used to apply a known force to the structure. In BRE tests of full-scale structures the resonance frequencies and the directions of the fundamental modes of a building are usually established in the following way:

1   Arbitrarily orientate the vibrator, increment the frequency sequentially and produce a response/frequency curve.
2   Select a frequency of resonance and set the vibrator to operate at this frequency.
3   At this resonance frequency, monitor the directional response, by incrementally rotating an accelerometer. Establish the direction of the minimum response.
4   Set the vibrator to a direction which is orthogonal to this minimum response.

124

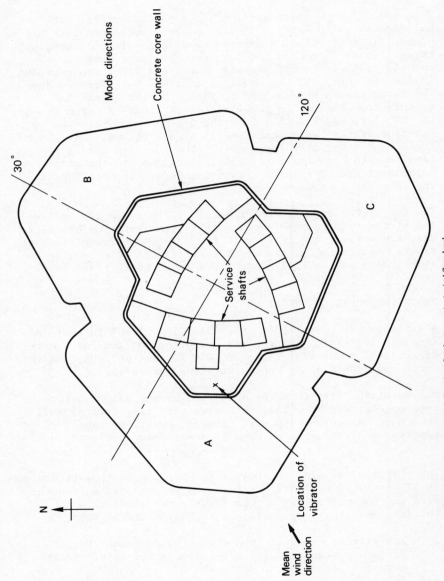

Figure 2 Plan view and modal directions for the tower at a typical floor level

5   Increment the frequency of operation of the vibrator and produce a response/frequency curve.

6   Rotate the vibrator, to a position 90° from the original position, and produce a response/frequency curve.

In this case the directions were established using second translational modes as the frequency separation between these orthogonal modes was greater, and the interference from the wind was less.

The forced oscillation tests were concluded by setting the vibrator to the frequency of each resonance, in turn, and assessing the deflected form of the structure ('mode shape') and the associated damping ratio.

The mode shapes were assessed by taking an accelerometer to various locations throughout the structure and, at each position, comparing the output with the output from a reference accelerometer, located on the 43rd floor.

The damping was assessed by suddenly stopping the exciter and monitoring the ensuing decay of vibration. The form of this decay was then used to estimate the modal damping ratio.

The on-site testing was completed by monitoring, and recording, the response of the structure to wind excitation.

RESULTS OBTAINED FROM THE ON-SITE TESTING

Using the methods outlined in the previous section the five lowest frequency modes, for translation and torsion, as measured, have been derived and are depicted in Fig 3. The natural frequencies, associated damping ratios and modal stiffness 'K' values for the first five modes are presented in Table 1. Modal stiffness values have been obtained using the method outlined in Ref 2.

Figs 4 and 5 show spectra obtained from the recorded response of the building to wind excitation. These show the natural frequencies to be the same as noted in the forced testing. The magnitude of the response in the fundamental translation modes was two orders of magnitude larger than that caused by torsion. This demonstrates that for this building the total response can be characterised, to a very good approximation, by the response in the fundamental orthogonal translational modes.

Autocorrelation plots were used to obtain damping measurements from the response to wind excitation(3), for the two fundamental modes, and Fig 6 shows a typical plot. The values of damping were in close agreement with those obtained in the forced tests.

TABLE 1

STRUCTURAL PROPERTIES

Mass (measured)		$72 \cdot 6.10^6$ kg	$\pm\, 10.10^6$ kg
	Freq	Stiffness	Damping (% of critical)
$E-W_1$	$0 \cdot 440$	$185 \cdot 0$ MN/m	$0 \cdot 5$
$N-S_1$	$0 \cdot 440$	$122 \cdot 0$ MN/m	$0 \cdot 5$
θ	$1 \cdot 553$	$9 \cdot 56.10^{11}$ Nm/rad	$0 \cdot 7$
$E-W_2$	$1 \cdot 670$	$799 \cdot 3$ MN/m	$2 \cdot 8$
$N-S_2$	$1 \cdot 690$	$818 \cdot 6$ MN/m	$2 \cdot 1$

Fundamental Time Constant $\qquad \tau_1 = \dfrac{1}{\zeta_1 W_1}$

$$\tau_1 = \left(\frac{0 \cdot 5}{100} \cdot 2.\pi. \; 0 \cdot 44 \right)^{-1} = 72 \text{ secs}$$

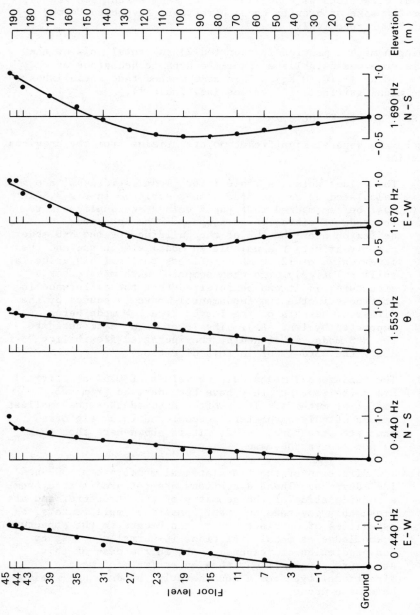

Figure 3 Normalised mode shapes for the fundamental translation, torsion and second translation modes

The incremental rotation of the accelerometer established the modal directions as $+ 30^{\circ}$ and $+ 120^{\circ}$ from north (see Fig 2). For the sake of brevity these directions have been termed N-S and E-W respectively.

Using a method previously reported(2) the total mass of the structure was established from the dynamic behaviour as $72 \cdot 6.10^{6}Kg$ ($\pm 10.10^{6}Kg$). This same method then establishes the modal stiffness 'K' values (see Table 1).

DISCUSSION OF RESULTS

There are several significant points arising from the previous section:

(1) The values noted in Table 1 for 'modal stiffness' are calculated using the mode shapes depicted in Fig 3. It must be remembered that these values are applicable to the mode of vibration as a whole. The enhanced deflexion near the top of the building for the N-S mode greatly affects the calculated modal stiffness, and the lower value of 122.0 MN/m reflects this. (This value is still relatively high when compared with many other structures.) It can be inferred that the difference in stiffness in the two fundamental modes is caused by the different heights of the leaves (the E-W mode being supported by leaf 'A'). The lower value attributed to the N-S mode, is caused by the increased flexibility above the 43rd floor in this direction.

(2) The fundamental modes damping values of 0.5% of critical are quite low, but they have been derived from two different methods. It is often assumed that the smallest values of this parameter, encountered in a reinforced concrete structure, is 1%. It is shown here that this is not always the case.

(3) The direction of the translational modes are at 30° and 120° degrees. These directions are not predictable from a consideration of the geometry of the structure, and are presumably governed by other, possibly small factors, eg asperities of manufacture or attachments to the structure. A knowledge of modal directions is normally necessary for prediction of response, but, in the case of this structure, it is likely that an estimate of modal directionality, from a consideration of drawings would have been wrong.

(4) The response of tower to wind excitation was estimated from a consideration of spectra (Figs 4 and 5 for example). However, the wind data just fail a test for stationarity (the stationarity hypothesis fails at

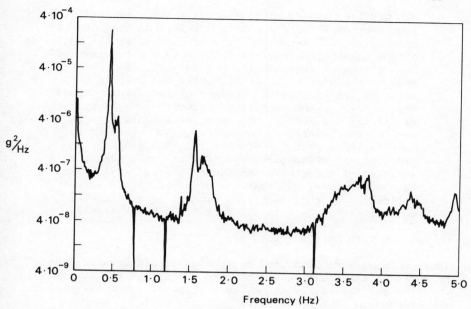

Figure 4 Enhanced uncoherent output spectrum taken from the N–S accelerometer

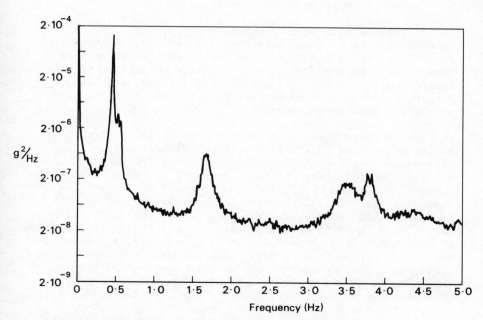

Figure 5 Enhanced unchoherent output spectrum taken from the E–W accelerometer

$\alpha = 0.99)(4)$ and so these spectra must be treated with some caution. There are some conclusions that can be drawn with confidence and this is because of the magnitude of the quantities involved. The response to wind has been estimated from the spectra to be 0.028 mm peak to peak at the 43rd floor level, while the average wind velocity (referenced to a height of 10 m above ground) was 11 m/sec. The confidence interval for the spectra(4) is 11% and a wider interval must be assumed because of the non-stationarity. Considering that the data only just failed a test for stationarity and judging from previous experience with spectra, it would seem that the addition of a further confidence interval of 10% might be conservative, and that the probable steady state excitation of the tower by a wind of 11 m/sec, would be most likely to be within 20% of 0.028 mm peak-peak.

COMPARISON WITH DESIGN GUIDES AND CODES

Three methods have been used to produce predictions for the response of the tower under the wind conditions prevailing at the time the recordings of motion were made. These methods are propounded in the following:

1   Engineering sciences Data Unit item 76001(5)
2   National building code of Canada(6)
3   A method proposed by Dr T Wyatt(7).

These three methods were chosen as being amongst the most up-to-date. Each method requires an input of data about wind conditions (speed and direction) and structural parameters (frequency, damping, mass and stiffness for each normal mode considered). In this case the wind conditions have been obtained from on-site anemometer records and the structural parameters have been measured.

In each of these methods the in-wind response is calculated, but no importance is ascribed to modal directionality. The cross-wind response can be calculated using the Canadian code. All three methods have, at least in part, a common history in the work of Davenport(8), and it is perhaps not surprising that each 'calculates' the dynamic in-wind response to be a similar value. However, this value, in the region of 0.5% of g (for a 'once in 50 years' wind), represents a displacement of approximately 1 mm peak for an 11 m/sec wind, and is almost two orders of magnitude greater than the measured response resolved to include the motions in both fundamental translation modes (0.014 mm peak). If the measured response is multiplied by a peak factor (4.0 for these conditions), to account for the fact that the documents predict maximum excursion of acceleration, then it can be seen that the predictions are each slightly more than an order of magnitude too large.

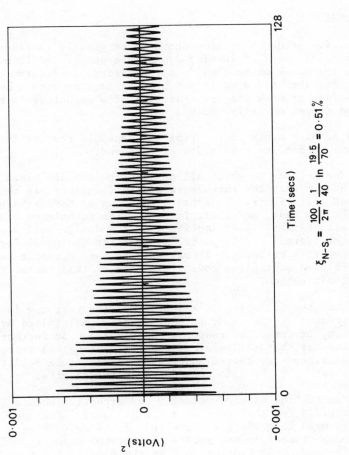

$$\xi_{N-S_1} = \frac{100}{2\pi} \times \frac{1}{40} \ln \frac{19 \cdot 5}{70} = 0 \cdot 51\%$$

Figure 6 Autocorrelation taken from N–S accelerometer

The Canadian code can also be used to calculate the dynamic cross-wind response. This is calculated as being approximately ten times larger than the in-wind response. The measured response of the tower was approximately the same for both orthogonal fundamental translational modes and so this can be seen to be a much larger error. However, it is probable that the cross-wind response of the tower would increase disproportionately as the wind velocity increased, and so the magnitude of the error may decrease.

DISCUSSION

The design guides have been shown to be grossly in error in their predictions of in-wind dynamic response. Additionally there are cases which they do not consider, for instance the situation when the wind is not blowing in line with a mode of vibration, or when the cross-section of a structure is not one of the shapes normally quoted.

It is perhaps worth considering the possible reasons for the various guides overprediction of response.

(1)   The methods on which all of these guides are based(8) assume that the turbulence of the atmosphere is the same at all heights. This has been shown by Harris(9) not to be the case, and that, in fact, the turbulence decreases with height. This implies that the design guides, considered in this paper, overpredict the modal forces for this building. It may well be that a simple height related multiplier could take care of this factor in codification.

(2)   The spectral approach assumes that the response is the product of the input forcing function multiplied by the complex frequency response function(8). It further assumes that all spectral quantities involved represent stationary processes and that a steady state condition exists for the time considered by the spectra.

Fluctuation of wind velocity and direction occurs quite rapidly (of the order of a few seconds). Further, it is assumed that the force exerted on a structure is proportional to the square of the wind velocity. It follows, that fluctuations in the forcing function occurs rapidly. However, the time constant (T) of the building is 72 seconds. If the building were excited with a sinusoidal force of constant phase and magnitude, it would take 4T seconds for the structure to reach 95% of its steady state response and after 72 seconds it would only have reached 37% of this value. Therefore, if spectra were taken under these circumstances, at T and 4T seconds after the 'switch on' of the forcing function, the measured responses would be very different. It is contended that this represents an analogy to the situation which occurs in the natural wind environment, and that buildings with long time constants are not able to reach a steady state response. This situation will always lead to a smaller response than would be expected under steady state conditions.

(3) At the time of testing the top floors of the structure were open to the atmosphere and it is possible that there was some aerodynamic 'spoiling'. This could have reduced the measured response.

The comparisons, between theory and practice, have been made at amplitudes which are much smaller than the design limit. It is probable that the response of the structure could become non-linear at greater amplitudes. Non-linearities have been measured in several cases(10,11,12,13) whilst at normal service loads they may not be significant(14). Since the non-linearities would reduce response this would tend to make the 'once in 50-year wind' predictions, on which the design guides are based, even more in error.

It is perhaps lucky that the predictions fall on the pessimistic side, although such large errors do not offer any reassurance to designers, and it is evident that there are several areas where research is urgently required. Some of these are:

1 Measurements of actual wind forces on real structures.

2 Effects of angle of attack from the wind.

3 Measurement of actual dynamic structural parameters.

4 The theoretical validation of the spectral method for large structures in the natural wind environment.

ACKNOWLEDGEMENTS

The help and expertise provided, to perform the full-scale testing, by G Ellison, J Parry and C Hargis of the University of Bristol, and by C Williams of Plymouth Polytechnic, at a time when most people were enjoying Christmas festivities, is gratefully acknowledged. Also the help and facilities provided by the staff of J Mowlem and Company at the site is gratefully acknowledged.

The authors are indebted to J Mayne and N Cook for their helpful comments on wind structure and predictions.

The work described has been carried out as part of the research programme of the Building Research Establishment of the Department of the Environment and this paper is published by permission of the Director.

134

REFERENCES

1   Severn, R T, Jeary, A P, Ellis, B R, and Dungar, R.
     (1979). Prototype dynamic studies on a rockfill dam
     and on a buttress dam. Conference on Large Dams. ICOLD
     New Delhi, October.

2   Jeary, A P and Sparks, P R. (1977). Some observations
     on the dynamic sway characteristics of concrete
     structures. ACI Symposium on Vibration of Concrete
     Structures, New Orleans, October, and BRE CP 7/78

3   Jeary, A P and Winney, P E. (1972). Determination of
     structural damping of a large multi-flue chimney from
     the response to wind excitation. Proc Inst Civil
     Engineers, Pt 2, 53, Dec 569-572

4   Bendat, J S and Piersol, A G. (1971). Random Data.
     Analysis and Measurement Procedures, John Wiley and
     Sons Inc, New York

5   Engineering Sciences Data Unit. (1976). Item 76001.
     ESOU 251-259 Regent Street, London, September, Amended
     June 1977

6   National Building Code of Canada, (1975) and commentaries
     on Part 4. National Research Council of Canada, Ottawa

7   Wyatt, T A. (1971). The calculation of structural
     response. Paper 6. Proc Seminar on the Modern Design
     of Wind Sensitive Structures. Pub CIRIA

8   Davenport, A G. (1967). The dependance of wind loads
     on meteorological parameters. Proc Int Conference
     on Wind Effects on Buildings and Structures, Ottawa,
     September, Vol 1, pp 19-82

9   Deaves, D M and Harris, R I. (1978). A mathematical
     model of the structure of strong winds. CIRIA, Report
     76, May

10   Whitbread, R E. Private communication.

11   Nielsen, N N. (1966). Vibration tests of a nine-storey
     steel frame building, ASCE Inst of Eng Mech Div, EMI,
     February

12   Rea, D, Bowkamp, J G, Clough, R W. (1968). The
     dynamic properties of McKinley School Building, Earth-
     quake Engineering Research Center, Report 68-4,
     California

13   Foutch, D A. (1978).   The vibrational characteristics
     of a twelve-storey steel frame building.   Earthquake Eng
     and Struct Dynamics, Vol 6

14   Burrough, H L, Jeary A P and Wilson, J M. (1972).
     Structural dynamics of multi-flue chimneys.   Proc 4th
     Int Conference on Wind Effects on Buildings and Struc
     tures, p 497-514.   Cambridge University Press.

136

EXPERIMENTAL STUDY OF WIND FORCES AND WIND-INDUCED
VIBRATIONS AT A STEEL LIGHTHOUSE TOWER

W. Schnabel and E.J. Plate

Institut Wasserbau III
University of Karlsruhe
Karlsruhe, Germany

INTRODUCTION

To evaluate the reaction of tall and slender structures to the
natural wind quasisteady approaches which allow for the effects
of gusts have been used in the past. Recently, increased knowledge of material properties and their statistical distribution has enabled engineers to build higher and more elastic
buildings. This has led to the development of more sophisticated approaches for evaluating the dynamic response of slender
structures to turbulent shear flow (3). Using statistical concepts applied to stationary time series these theoretical models take into account the turbulent structure of the atmospheric wind and especially the coherency of the pressure fluctuations acting on the buildings. Although many wind tunnel
experiments have been carried out to verify the theoretical
model assumptions, limited full scale experimental data are
available. To obtain some information about the structure of
wind loading on a circular shaped tower and the tower's reaction to the wind, measurements were made at a lighthouse in
the North Sea. Special attention was given to the turbulence of
the wind above the sea, to the correlation of the pressure at
different points of the structure and to the correlation of
the reaction to the wind and pressure fluctuations.

EXPERIMENTAL CONFIGURATION

Theories to evaluate the response of slender structures must
take into account the relationship between the characteristics
of the approach flow, the pressure acting on the surface of
the building and the dynamic properties of the structure.
Hence, the experiments at the "Großer Vogelsand" Lighthouse
were concentrated on measuring the

      1. mean wind and the turbulent structure of the
         wind

138

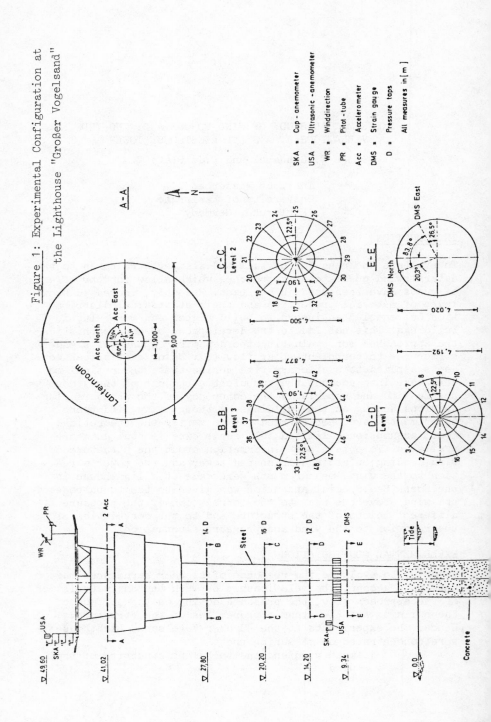

Figure 1: Experimental Configuration at the Lighthouse "Großer Vogelsand"

SKA = Cup - anemometer
USA = Ultrasonic - anemometer
WR = Winddirection
PR = Pitot - tube
Acc = Accelerometer
DMS = Strain gauge
D = Pressure taps
All measures in [m]

2. mean pressure distribution and the
dynamic pressures acting on the tower
3. vibration of the tower.

In comparison to wind tunnel tests, full-scale experiments have
some disadvantages. These should be considered in choosing the
site of an experiment. The dependency of the properties of the
wind on the topography and surface roughness favours a site
that is surrounded by uniform terrain. If, in addition, the
structure itself is symmetrical about a vertical axis of rotation then the profile of the mean wind, the wind structure, the
displacement of the building,and their functional relationship
become independent of the direction of the approach flow.

The experiments were therefore carried out on a circular lighthouse tower (Figure 1), situated about 20 km offshore in the
North Sea. The tower consists of a cylindrical steel shaft
which has a concentrated mass at the top. The windspeeds were
measured on a platform 50 m above water level using a cup anemometer for the longitudinal component of the wind and an ultrasonic anemometer for the lateral and vertical components. Wind
tunnel studies of the optimal design of the top of the tower
(4) have shown that the anemometers are well clear of the disturbed windfield.

Whereas for rectangular structures the position of the zones
of separation are usually determined by the edges, the pressure
distribution around circular cylinders is much more complex.
Although researched extensively at this time it is still not
fully understood in all details. The position of the line of
separation in natural wind depends strongly on the variation
of the direction of the wind which influences the lateral vibration as well as the drag coefficient. Nevertheless, by assuming that on -line vibration is primarily connected with the
gust loading of the structure, dynamic analysis can concentrate
on the zone of positive pressure on the cylinder. This positive
pressure zone is quite independent of the parameters of the
flow.

As the second author was involved in the design stage of the
tower he was able to have installed 45 pressure taps during
construction. These are uniformly distributed at 3 levels
around the steel shaft with an angular horizontal spacing of
22,5 degrees. The taps were connected by pressure tubes to difference pressure transducers or difference pressure transducers
mounted directly on the inside of the shaft. Simultaneous pressure measurements could therefore be obtained at different
points on the outside of the tower. Finally, the dynamic behaviour of the lighthouse was measured using pairs of accelerometer and strain gauges mounted at both the top and the base of
the tower. All signals including the direction of the wind were
recorded in analog form on a 7 channel tape recorder and were
later digitized using a frequency of 20 Hz.

140

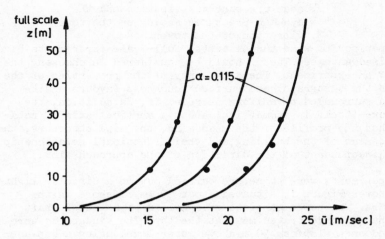

Figure 2: Mean Velocity Profile

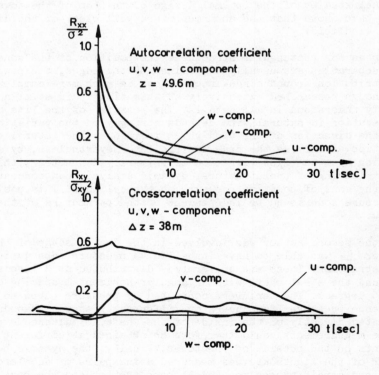

Figure 3: Correlation of Velocity Components

MEAN WIND PROFILE AND WIND STRUCTURE

Ignoring the virtual mass loading, the horizontal loading of a slender towerlike structure by the wind is usually expressed by the overall load p at level z

$$P_{(z,t)} = C_{D(z)} \cdot D_{(z)} \cdot \frac{1}{2} \cdot \varsigma \cdot u^2_{(z,t)} \tag{1}$$

where $u_{(z,t)}$ denotes the velocity of the wind, $D_{(z)}$ the diameter of the tower and ς the density of the air. The longitudinal velocity fluctuations are small compared to the mean velocity $\bar{u}$ and second order terms can be neglected. Hence, the pressures $p_i(t)$ acting at a point of the structure can be divided into a mean part $\bar{p}_i(z)$ and a fluctuating part $p'_i(t,z)$ which shows a linear relation to the variation of the wind

$$\bar{P}_{i(z)} = C_{P_i(z)} \cdot \frac{1}{2} \cdot \varsigma \cdot \bar{U}^2_{(z)} \tag{2}$$

$$P'_{i(t,z)} = C'_{P_i(z)} \cdot \varsigma \cdot \bar{U}_{(z)} \cdot U'_{(t,z)} \tag{3}$$

The drag coefficient $C_{D(z)}$ results from integration of the pressure coefficient $C_p(z)$ around the tower

$$C_{D(z)} = \int_0^{\pi} C_{P(z)} \cdot \cos \varphi \, d\varphi \tag{4}$$

and is assumed to be independent of the turbulence of the on-coming flow.

Along the stagnation line the pressure coefficient $C_p(z)$ is unity and in the absence of additional wind measurements the vertical profile of the wind can be derived using Equation 2. The results are shown in Figure 2. The good agreement between the exponential relationship

$$\frac{\bar{U}}{\bar{U}_R} = \left(\frac{z}{z_R}\right)^{\alpha} \tag{5}$$

and the data supports the use of Equation 2 as was expected from the results of (1) and (5).

To give some indication of the spatial configuration of the eddies affecting the structure the wind measurements in u,v and w direction were analyzed in terms of the autocorrelation function

$$\varsigma_{xx}(\tau) = \frac{\frac{1}{T} \cdot \int_0^T x_{(t)} \cdot x_{(t+\tau)} \, dt}{6^2_{xx}} \tag{6}$$

where x can be replaced by the velocity components u, v or w, and τ is the time lag between the two instants. Using Taylor's hypothesis for the transport of the turbulent structure over short distances by the mean flow $\bar{u}$, scales of turbulence L

142

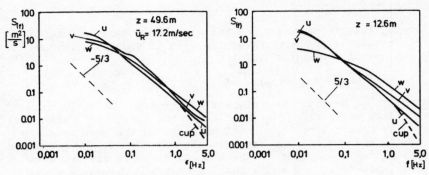

Figure 4: Spectra of Atmospheric Turbulence

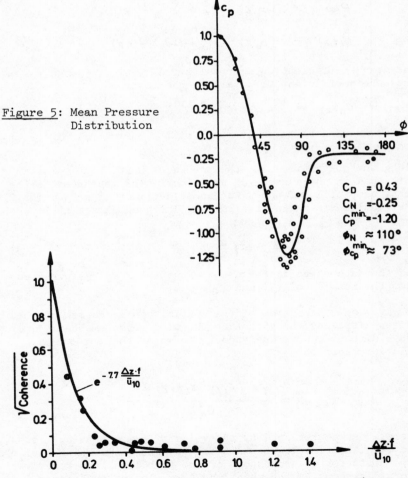

Figure 5: Mean Pressure
Distribution

C_D = 0.43
C_N = -0.25
C_p^{min} = -1.20
$\phi_N \approx 110°$
$\phi_{c_p}^{min} \approx 73°$

$e^{-7.7 \frac{\Delta z \cdot f}{\bar{u}_{10}}}$

Figure 6: Coherence as Function of Vertical Separation to
Wavelength Ratio (after (2))

between 40 and 80 m could be observed. The vertical cross-correlation function between two points 38 m apart still shows significant correlation in the longitudinal component u' of the wind, whereas the cross - correlation of the lateral and vertical components showed no correlation at all. This confirms that the dimensions of the eddies in longitudinal direction are much greater than in lateral or vertical directions.

The energy converted into the eddies is reflected in the power density spectra of Figure 4. The longitudinal component concentrates the energy at a lower frequency than the lateral or vertical component. In addition, Figure 4 shows the energy spectra for the cup anemometer signal. The divergency of this spectra from that measured using the sonic anemometer suggests the use of cup anemometers above a frequency range of 2 Hz is questionable.

THE MEAN PRESSURE DISTRIBUTION AND DRAG COEFFICIENT

The pressure distribution around the shaft of the tower (Figure 5) and from Equation 4 the instantaneous drag coefficient both vary with time. This is due to changes in the direction of the wind, asymmetry in the pressure distribution due to regular vortex shedding and the variation of the wind about its mean (5). The observed range of the drag coefficient was $0,26 < c_D < 0,64$ with the mean value being $\bar{c}_D = 0,43$.

An explanation for the very low values was the low pressure coefficient in the wake of the tower which dropped as low as $c_p = -0,25$.

EVALUATION OF THE WIND PRESSURE

The correlation of the longitudinal velocity component over the frequency range with increasing separation Δz (Figure 6) is expressed by the coherency and was examined in (5). It was found that the decay of the coherency with increasing distance can be approximated by the exponential curve

$$\left| R_{12}\left(\frac{\Delta z \cdot f}{\bar{u}_{10}}\right) \right| = e^{-7,7 \cdot \frac{\Delta z \cdot f}{\bar{u}_{10}}} \tag{7}$$

where $\bar{u}_{10}$ is the mean velocity at height z = 10 m.

Therefore and because of Equation 3 coherency of wind pressures in the stagnation area should also drop exponentially with increasing separation. In the time domain this can be expressed by the peak value of the cross correlation function, that is defined analog to Equation 6. In Figure 6 the cross correlation between the reference velocity $u_R(t)$ and pressure points within the stagnation area are shown. Curve (1) gives the correlation of the wind speed and a pressure signal with no vertical separation, curve (2) and (3) show the correlation of $u_R(t)$ with 2

144

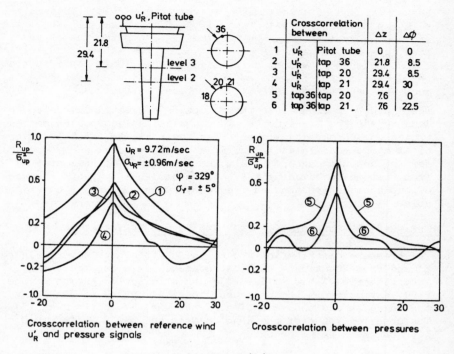

Figure 7: Cross-Correlation Coefficients

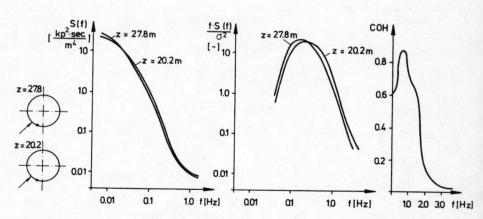

Figure 8: Spectra and Coherence Pressures with the Stagnation Area

pressure points $\Delta\phi=8,5°$ off the stagnation point in 21,8 and 29,4m separation resp..Even a pressure point 30° off the stag- nation point shows some correlation with the reference wind (curve 4), whereas the pressure points in the side lobes or the wake of the pressure distribution have no significant correla- tion with the oncoming flow, which indicates little contribu- tion of those areas of the surface to the buffeting of the structures by gusts.

The same applies for the cross-correlation of pressures at points at the same level: whereas within the stagnation area the correlation is good, in the side lobes and the wake it is rather poor.

Some more information regarding the relationship in the fre- quency range of two pressures is given in Figure 8. There is satisfactory coherency in vertical direction up to about 2 Hz, but above this frequency the pressure fluctuations affect a smaller area than the $\Delta z = 7,6$ m. A similar effect can be ob- served in the coherency of two pressures at points at the same level. Within the side lobes or the wake the coherency tends to be zero over the whole frequency range.

THEORETICAL MODEL FOR THE SIMULATION OF THE DYNAMIC PRESSURE

For separation $\Delta z = 0$ the relationship between wind speed and pressure is given by Equation 2 and Equation 3. However, for increasing distance Δz the pressure fluctuation cannot be fully reproduced using this deterministic approach as the maximum of the cross-correlation function becomes less than unity.

Following the assumption that the pressure and its fluctuation at one point of a surface always depend linearly on the wind and its fluctuation, the pressure may be approximated by a ge- neral model

$$P'_i(t) = \int_0^t u'_R(\tau) \cdot h_i(t-\tau) \, d\tau + P'_R(t) \qquad (8)$$

where u'_R denotes a reference velocity and $p'_i(t)$ the pressure fluctuation at a point i of the structure. The decay of the peak values of the cross-correlation functions with increasing separation of the measured data, as shown in Figure 7, together with Figure 6 imply that it is impossible to fully simulate the pressure at increasing distances from the reference wind speed point. Hence, $p'_R(t)$ is a random time function to compensate for this lack of knowledge. The function $h_i(t)$ is the transfer function that takes into account the physical and geometric properties of the relation, i.e. the mean velocity, the density of the air and the distance of separation.

To describe the general form of the obtained transfer function an expression of the form

$$h_{(t,\Delta z)} = A \cdot e^{-\alpha t} \qquad (9)$$

146

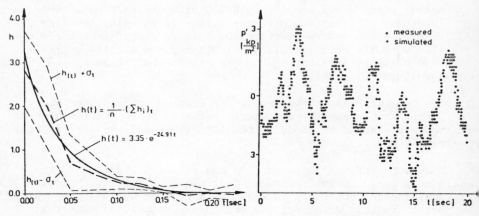

Figure 9: Simulation of the Relationship Wind → Pressure

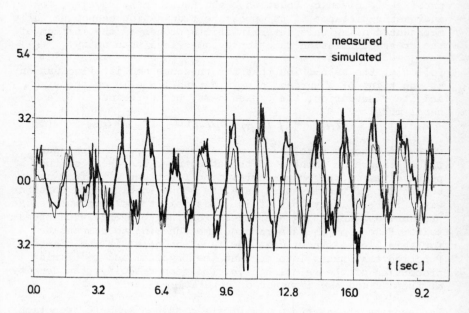

Figure 10: Measured and Simulated Strain Signal

has been chosen. To reduce the effect of the random nature of the time series, a number of transfer functions were calculated and the parameters A, α and determined by least squares fit.

Comparison of measured and calculated pressure series are shown in the computer plots of Figure 9, where satisfying similarity could be obtained. Hence, the forces acting on a unit height of the structure can be expressed by

$$F'_{(z,t)} = D_{(z)} \cdot \int_0^t u'_R(\tau) \cdot h_P(z,t-\tau)\, d\tau + F'_R(z,t) \qquad (10)$$

with the reference velocity $u'_{R(t)}$ and the diameter D of the tower at height z. The transfer function $h_p(z,t)$ includes the correlative effect of the structure of the turbulence on the tower.

Knowing $h_p(z,t)$ and $F'_R(z,t)$ for all z it is possible to compute the total load

$$F'_{(t)} = \int_0^h F'_{(z,t)}\, dz \qquad (11)$$

or the forces $F'_j(t)$, integrated over a section of height Δz_j, and acting on a node j within the section.

$$F'_j(t) = \int_{z_{j-1}}^{z_j} F'_{(z,t)}\, dz \qquad (12)$$

With the time series $F'_{j(t)}$ as load input into one of the common computer programs (i.e. SAP IV(6)) the time deflection relationship can be computed. $F'_{R(t)}$ in Equation 10 is a random force time function that results from the random nature of $p_{r'}(t)$ in Equation 8 and scatter of the parameters in Equation 9. From the analysis of the data there still is some uncertainty on the functional form of the integrated transfer function $h_p(z,t)$ which leads to the consequence of $F'_R(t)$ being too big compared to $F_R(t)$. Therefore, results on Equation 10 still cannot be presented.

Nevertheless, to give an indication on the validity of the model the order of integration in Equation 10 and Equation 11 may be changed, so that the force acting on the tower

$$F'_{(t)} = \int_0^t u'_R(t) \cdot \left[\int_0^h D_{(z)} \cdot h_P(z,t-\tau)\, dz + \int_0^h F'_R(z,t)\, dz \right] d\tau \qquad (13)$$

just becomes dependent from $u'_{R(t)}$ and a function of time integrated over the height of the tower. Due to the random nature of $F'_{R(t)}$ the integral over the height should approach zero.

Together with the parameters of the structure the first integral within the square brackets of Equation 13 can become a transfer function for the response of the tower

$$h^*_{(t)} = f\left(\int_0^h D_{(z)} \cdot h_P(z,t-\tau)\, dz, m, \omega, \delta \right) \qquad (14)$$

where m, ω and δ denote mass, eigenfrequency and damping. Figure 10 shows a measured and computed time-history relationship of the strain at the base, where the transfer function $h^*(t)$ was developed from the reference wind speed $u_R^1(t)$ and the stress $\varepsilon_{(t)}$. The good agreement indicates that the described model is capable of simulating the deflection by means of a measured quantity like a reference wind speed at the top.

CONCLUSIONS

Wind, force, and response measurements at the lighthouse "Großer Vogelsand" are described and analyzed.

The mean wind profile is derived from pressure measurements at the stagnation points together with direct wind measurements. The exponent of the wind profile above the sea is seen to be $\alpha = 0,11$. The measured values of the drag coefficients of the shaft of the tower vary and concentrate around $\overline{C}_D = 0,43$.

The turbulence of the atmosphere is discussed and it is shown, that the spatial structure of the turbulent velocity is reflected in the pressure measurements within the stagnation area. The correlation of the pressures within the stagnation area is seen to be good but diminishing with increasing separation both in vertical and horizontal direction.

A model adopting linear system theory is presented to simulate the pressure fluctuations at points and the overall load at the tower from a single measured reference velocity.

ACKNOWLEDGMENT

The work described herein is supported by the German Science Foundation (Deutsche Forschungsgemeinschaft) within the research program "Buildings Aerodynamics".

REFERENCES

(1) Baines, D. (1963) Effects of Velocity Distribution on Wind Loads and Flow Pattern on Buildings. Proc.Conf. on Wind Effects on Buildings and Structures, Teddington, 198 - 225
(2) Davenport, A.G. (1961) The Spectrum of Horizontal Gustiness near the Ground in High Winds. Quart.Journ.Roy. Met. Soc., 87
(3) Davenport, A.G. (1967) Gust Loading Factors. Journ. of the Structural Division. ASCE, 93
(4) Plate, E.J. and Vogt, H.D. (1975) Wind Tunnel Tests for Aerodynamic Optimization of Helicopter Landing Platforms on Lighthouses. IX Int. Conf. on Lighthouses and Other Aids of Navigation, Ottawa, 1.14, 1 - 18
(5) Huscheweyh,H. (1974) Beitrag zur Windbelastung hoher kreiszylinderähnlicher Bauwerke im natürlichen Wind bei Reynoldszahlen bis Re = 1,4 · 10^7, Dissertation, Inst.f.Leichtbau, RWTH Aachen

(6) Wilson, E.L., Bathe, K.J. and Petersen, F.E. (1973) SAP IV, A
    Structural Analysis Program for Static and Dynamic Response of
    Linear Systems, Univ. Calif. Berkeley, Rep. No. EERC 73 - 11

SIMPLIFIED MODELS OF LARGE-WIND-TURBINE OPEN-TRUSS TOWER

Sankar C. Das

Associate Professor of Civil Engineering
Tulane University, New Orleans, Louisiana, U.S.A.

INTRODUCTION

Recent interest in wind power has resulted in the
initiation of several large wind turbine projects
by NASA-Lewis Research Center as part of the ERDA
wind energy program (ref. 6). As part of this program,
Lewis has designed and constructed a wind turbine
large enough to assess the technology requirements
and the associated operational problems of a large
wind turbine. The 100-kilowatt wind turbine has
been constructed at the NASA-Lewis Research Center's
Plum Brook facility near Sandusky, Ohio. The wind
turbine consists of a 93-foot tall square open truss
steel tower; a bed plate that sits on top of the
tower; a nacelle that houses the alternator, the
gearbox, and the low-speed drive shaft; and two 62.5
foot long aluminum blades. The blades are downwind
of the tower and inclined at an angle of 7° from the
vertical plane. The wind turbine is designed to
produce 100 kilowatts of electric power in an 18-mph
wind at a rotor speed of 40 rpm.

The purpose of this investigation is to develop a
method for quickly modeling a wind turbine tower
structure and the combined system as a system of
masses and springs to study the dynamic characteristics
of the tower structure alone and the combined system
involving nacelle and rotor blades. In this idealized
mathematical model of the tower structure and the
combined system, the mass or mass moment of inertia
of the system are separated from the elastic properties
and equivalent concentrated masses are placed at
the chosen node points to represent the inertia forces
in the direction of the assumed element degrees of
freedom. The expressions for the flexural and torsional

flexibility and or stiffness co-efficients have been
developed in terms of the segmental dimensions and the
elastic properties of the tower structure. All frequencies and mode shapes of the models have been obtained
by direct input of the mass or mass moment of inertia
and stiffness/flexibility matrices in the problem oriented
computer programs such as STRUDL II or NASTRAN.

The frequencies from the idealized model compare closely
with those determined by experiment and those predicted
by using the detailed finite element NASTRAN **representation**. The simplified models developed can also be used
to compute tower fundamental frequencies with the aid
of a pocket-type calculator and could be used to aid
in the preliminary design and sizing of tower members
to meet frequency placement requirements. The simplified
dynamic models can thus be used to find gross errors
in the detailed model, to differentiate between important
modes and those caused by local member vibration, and
to identify couple mode shapes.

MATHEMATICAL MODEL DESCRIPTION

The wind turbine tower structure is composed of four
main columns with diagonal and horizontal bracing, as
shown in fig. 1. Because the tower height is much
greater than its average width, the tower is assumed
to behave like a cantilever beam.

In the idealized mathematical model, the mass properties
of the system are separated from the elastic properties
and equivalent concentrated masses are placed at node
points. A typical node point is located at the intersection of a horizontal member, a diagonal brace, and
the vertical leg, as shown in fig. 1. The four nodal
masses at each tower level are added together, resulting
in a single value for each of the six bays or levels
of the tower. The center of each lumped mass is located
on the vertical centerline of the tower and at the
same elevation of each of the horizontal structural
members.

The elastic properties for the tower structure are
calculated at various levels along the tower height.
The elastic properties are determined by calculating
the area moment of inertia of the structure at each
tower level selected and by using the material mechanical
properties.

Nominal number of masses are used to simulate the structure.
However it is essential that the mathematical model must
faithfully represent all essential characteristics of

the distribution of mass and of stiffness in the structure.
When the structure is represented by a one-dimensional
model, it makes no sense to use very large number of
masses for the preliminary study purpose. Furthermore,
the model should be kept uncomplicated to permit ready
interpretation of the results and thus to yield insight
into the significant aspects of the structural behavior.
The tower structure is thereby simplified to a system model
composed of six lumped masses and six elastic beams.
Fig. 2 shows the tower structural model that is used to
calculate the natural frequencies of the tower in bending.
The model used to calculate the natural frequencies of
the tower in torsion is shown in fig. 3.

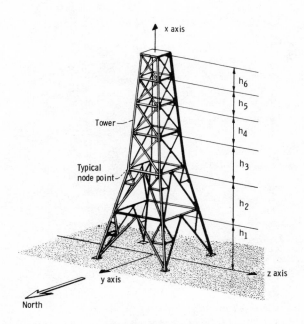

Figure 1. - Typical tower structure.

· EQUATION OF MOTION

Referring to figs.1, 2 and 3, uncoupled system of differential equations in the matrix form involving coordinates y, z and θ are written as

$$\left[M \right] \{\ddot{y}\} + \left[K_y \right] \{y\} = 0 \quad \ldots \ldots \ldots \ldots \ldots \ldots (1)$$

$$\left[M \right] \{\ddot{z}\} + \left[K_z \right] \{z\} = 0 \qquad \ldots \ldots \ldots \ldots \ldots (2)$$

$$\left[J_\theta \right] \{\ddot{\theta}\} + \left[K_\theta \right] \{\theta\} = 0 \qquad \ldots \ldots \ldots \ldots \ldots (3)$$

The mathematical model for the tower structure is symmetric for bending in the y and z directions. As a result, the frequency solutions in the y and z directions will be identical. Since equations (1) and (2) yield identical results, only equation (2) is used in the following analysis.

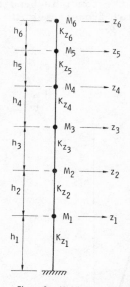

Figure 2. - Model of tower structure for flexural vibration - six degrees of freedom.

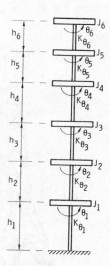

Figure 3. - Model of tower structure for torsional vibration - six degrees of freedom.

Free flexural vibrations in z-direction

In equation (2), the term $\left[K_z \right]$ is the stiffness coefficient matrix for the structure in the z-coordinate direction. Often it is useful to generate flexibility coefficients for the structure. Equation (2) may be rewritten, using the flexibility coefficient matrix $\left[a_z \right]$ rather than the stiffness coefficient matrix $\left[K_z \right]$, as

$$\left[a_z \right] \left[M \right] \{\ddot{z}\} + \{z\} = 0 \qquad \ldots \ldots \ldots \ldots (4)$$

where

$$\left[a_z \right] = \left[K_z \right]^{-1}$$

Equation (4) forms a set of linear second-order differential equations whose solution is given by (ref. 4)

$$\{\ddot{z}\} = -\omega^2 \{z\} \quad \ldots \ldots \ldots \ldots \ldots \ldots \quad (5)$$

From equations (4) and (5),

$$\omega^2 \left[a_z\right] \left[M\right] \{z\} = \{z\} \quad \ldots \ldots \ldots \ldots \ldots \quad (6)$$

or, expanding,

$$\omega^2 \left[a_z\right] \left[M\right] \begin{Bmatrix} z_1 \\ z_2 \\ z_3 \\ z_4 \\ z_5 \\ z_6 \end{Bmatrix} = \begin{Bmatrix} z_1 \\ z_2 \\ z_3 \\ z_4 \\ z_5 \\ z_6 \end{Bmatrix} \quad \ldots \ldots \ldots \quad (7)$$

Dividing equation (6) by ω^2 yields

$$\left(\left[a_z\right] \left[M\right] - \frac{1}{\omega^2}\left[I\right]\right) \{z\} = \{0\} \quad \ldots \ldots \ldots \quad (8)$$

Equation (8) forms a set of homogeneous algebraic equations in z. The solution of these equations is usually called the eigenvalue or characteristics value problem. A non-trivial solution of equation (8) can exist only if the determinant

$$\begin{vmatrix} \left(a_{z_{11}}M_1 - \frac{1}{\omega^2}\right) & a_{z_{12}}M_2 & \cdots \cdots & a_{z_{1n}}M_n \\ a_{z_{21}}M_1 & \left(a_{z_{22}}M_2 - \frac{1}{\omega^2}\right) & \cdots \cdots & a_{z_{2n}}M_n \\ \cdot & & \cdot & \\ \cdot & & \cdot & \\ \cdot & & \cdot & \\ a_{z_{n1}}M_1 & \cdots \cdots \cdots & & \left(a_{z_{nn}}M_n - \frac{1}{\omega^2}\right) \end{vmatrix} = 0 \quad \ldots \quad (9)$$

where the mass matrix is a diagonal matrix,

$$[M] = \begin{bmatrix} M_1 & & & & \\ & M_2 & & & 0 \\ & & \cdot & & \\ & & & \cdot & \\ 0 & & & & \cdot \\ & & & & M_n \end{bmatrix}$$

and the flexibility matrix in the z-coordinate direction is

$$[a_z] = \begin{bmatrix} a_{z_{11}} & \cdots\cdots\cdots & a_{z_{1n}} \\ a_{z_{21}} & a_{z_{22}} & \cdots\cdots & a_{z_{2n}} \\ \cdot & & & \cdot \\ \cdot & & & \cdot \\ \cdot & & & \cdot \\ a_{z_{n1}} & \cdots\cdots\cdots & a_{z_{nn}} \end{bmatrix} \quad \cdots \quad (10)$$

Assume the roots of the frequency equation (9) are $1/\omega_{z_1}$, $1/\omega_{z_2}$, $1/\omega_{z_3}$, . . . , $1/\omega_{z_n}$. Expanding equation (9) with the assumed roots leads to the following equation from reference 7:

$$\frac{1}{\omega_{z_1}^2} + \frac{1}{\omega_{z_2}^2} + \frac{1}{\omega_{z_3}^2} + \ldots + \frac{1}{\omega_{z_n}^2}$$

$$= a_{z_{11}} M_1 + a_{z_{22}} M_3 + \ldots + a_{z_{nn}} M_n \quad \cdot \cdot \quad (11)$$

Natural frequencies of the second and higher modes are often considerably greater than the fundamental frequency. If this holds, all terms on the left side of equation (11), except the first, may be omitted for the approximate determination of the fundamental frequency. It can be written as

$$\frac{1}{\omega_{z_1}^2} \simeq a_{z_{11}} M_1 + a_{z_{22}} M_2 + a_{z_{33}} M_3 + \ldots + a_{z_{nn}} M_n \quad \cdot \cdot \quad (12)$$

Equation (12) is known as Dunkerley's equation (ref. 7) and allows the fundamental frequency to be determined with reasonable accuracy by using longhand calculations.

By using the area moment principle suggested in ref. 1, the flexibility coefficients can be written as

$$a_{z_{ji}} = a_{z_{ij}} = \frac{S_i}{E} \sum_{k=1}^{k=n} \frac{M_i^k x_j^k}{I_i^k} \qquad \ldots \ldots \ldots \qquad (13)$$

For the case of six degrees of freedom (i = 1, 2, . . . , 6 and j = 1, 2, . . . , 6),

$$S_i = \frac{h_i}{n} \qquad \ldots \ldots \ldots \ldots \ldots \ldots \ldots \ldots \ldots \qquad (14)$$

where h_i is the vertical distance between the tower node points and n is an arbitrary number of sections selected between node points. At each selected section, between node points, the area moment of inertia I_i^k is calculated for the tower about the y or z-axis. The bending moment M_i^k is calculated between two particular tower node-point levels (i = 1, 2, . . . , 6) and at a particular tower section (k = 1, 2, . . . , n). The magnitude of the bending moment M_i^k is determined by multiplying a unit horizontal force applied at a node point by the vertical distance x_j^k. The quantity x_j^k is the distance between a section of the tower (k = 1, 2, . . . , n) where the bending moment intensity M_i^k occurs and the unit force applied at a node point (j = 1, 2, . . . , 6). The quantity E is Young's modulus of elasticity for the particular tower structural material.

Free torsional vibration

In equation (3) the term $[K_\theta]$ is the torsional stiffness coefficient matrix for the tower structure. Often it is useful to generate torsional flexibility coefficients for the structure. Equations (3) may be rewritten, by using the torsional flexibility coefficient matrix $[b_\theta]$ rather than the stiffness matrix $[K_\theta]$, as

$$[b_\theta] [J] \{\ddot{\theta}\} + \{\theta\} = 0 \ldots \ldots \ldots \ldots \ldots \ldots \qquad (15)$$

where

$$[b_\theta] [K_\theta] = [I] \qquad \ldots \ldots \ldots \ldots \ldots \ldots \ldots \qquad (16)$$

is the identity matrix.

The solution of equation (15) is similar to that of equation (4). Therefore, the fundamental torsional frequency can be written, by using Dunkerley's equation, as

$$\frac{1}{\omega_{\theta_1}^2} \simeq b_{\theta_{11}} J_1 + b_{\theta_{22}} J_2 + \ldots b_{\theta_{nn}} J_n \quad \ldots \ldots \ldots \quad (17)$$

The torsional spring constant for the tower structure can be developed from figs. 5, 6 and 7 as

$$K_{\theta_i} = \frac{8(A_D)_i E}{h_i} w_i^2 \cos^2 \phi_i \sin \phi_i + 4(\frac{12EI_i}{h_i^3} R_i^2) + 4(\frac{GR_{EQ_i}}{h_i}) \ . \quad (18)$$

where

A_D cross-sectional area of a diagonal brace member

E modulus of elasticity

h_i vertical tower height between node points

w_i half of width of a square horizontal frame

ϕ_i angle of diagonal brace member projected on z-y plane (figs. 5 and 6)

I_i area moment of inertia of a tubular leg taken about centroid of tube

R_i radial distance from tower x-axis to center of a tubular leg (fig. 5)

G shear modulus

R_{EQ_i} equivalent torsional rigidity of tubular tower legs and bracing (ref. 1)

The total torsional spring constant K_θ is composed of three terms in equation (18). The first term accounts for the change in length of the diagonal crossmember bracing as located on the sides of the tower. A torsional moment vector applied along the x-axis of the tower results in tension and compression loads in these crossmembers. The second term in equation (18) accounts for beamwise bending of each of the four main tubular legs of the tower. Again, a torsional moment vector applied along the x-axis of the tower tends to rotate the square frame structure within a horizontal y-z plane. Rotation of each frame causes bending in the tubular tower legs.

Rotation of each frame structure in the y-z plane also causes torsion in each tubular tower leg and in the diagonal cross bracing. The third term in equation (18) accounts for the torsional stiffness of these members. For a typical wind turbine tower structure, the magnitudes of the last two terms on the right in equation (18) were found to be small as compared with the first term. As a result, the last two terms were neglected when the torsional stiffnesses for the tower structure were calculated.

EXAMPLE, RESULTS AND DISCUSSION

The tower structure weight is estimated at 44000 pounds. The total tower height is 93 ft. Multiplying the ratio of bay height to total tower height times the total mass of the tower structure yields the estimated mass, M_i of each tower bay. Each bay of the tower having a height h_i was divided into n = 4 sections (fig. 4). At each of 24 sections of the tower, the area moment of inertia was calculated. The bending moment M_i^j due to a unit horizontal force (fig. 4) sequentially applied to each of the six node points, was computed at each section. Then equation (13) was used to calculate the flexibility co-efficients $a_{z_{ij}}$ The components of the wind turbine are shown in reference 5, with their associated weights. The stair weight (12000 lb.) is divided into six equal weights and lumped at each node point. The nacelle and rotor blade weight (34000 lb.) is lumped at node 6 (fig. 2).

Polar mass moment of inertia J_i (fig. 3) were calculated for each bay of the tower from previously calculated lumped masses M_i and the distances d_i (fig.7). Polar mass moments of inertia for Nacelle mounted on tower with rotor blades oriented vertically and horizontally were also calculated (ref. 3). Torsional flexibility co-efficients of the tower structure $b_{\theta_{ii}}$ were calculated from torsional stiffness co-effient K_{θ_i} as shown in equation (18).

Calculated and experimental tower frequencies are compared in tables 1 and 2. Table 1 presents a summary of tower natural fundamental frequencies without the nacelle and rotor blades. The test data were taken from ref. 5. The values calculated by using finite element NASTRAN model (fig. 8) were taken from ref. 2. The approximate values in table 1 were calculated from the Dunkerley's equations (12) and (18).

Table 2 summarizes tower natural frequencies with the nacelle and rotor blades mounted on top of the tower. The approximate values in table 2 were calculated by using equations(12)

160

and (18).

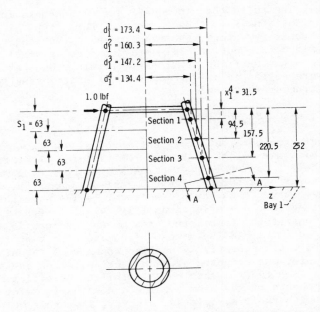

Section A-A: 8.0-inch-diameter extra-heavy pipe; cross-sectional area, 12.76 square inches; area moment of inertia, 105.7 inches[4]

Figure 4. – Location of sections within tower bay 1. (Dimensions are in inches.)

All frequencies of the tower model with and without the nacelle and rotor blades were also obtained by direct input of the mass or mass moment of inertia and stiffness/flexibility matrices in the eigenvalue solution of computer programs such as STRUDL II and NASTRAN. Table 3 presents all six flexural and torsional frequencies of the tower model without the nacelle and rotor blades. Table 4 presents six flexural and torsional frequencies of the tower model with the nacelle and rotor blades mounted on the top of the tower (fig. 8).

The tower first bending and torsional frequencies are reduced by the addition of the stairway, the nacelle and the rotor blades as shown in tables 1 and 2. As a result of this analysis, an error was found in refs. 2 and 5. The first torsional tower frequency with the nacelle and rotor blades (horizontal) was reported as 9.8 hertz in ref. 5 and 9.56 hertz in ref. 2. This approximate analysis predicts the first torsional natural frequency as 3.78 hertz. Ref. 5 reports that a 4.4 hertz mode "appears" to be torsion, with first-mode bending present in the north and south direction. This approximate analysis verifies that the

4.4-hertz mode is in fact, the first torsional mode and that the 9.8 hertz mode is clearly not the first torsional mode. For finite element structural models (ref. 2) that have a large number of degrees of freedom, it is very difficult to determine certain fundamental modes and frequencies.

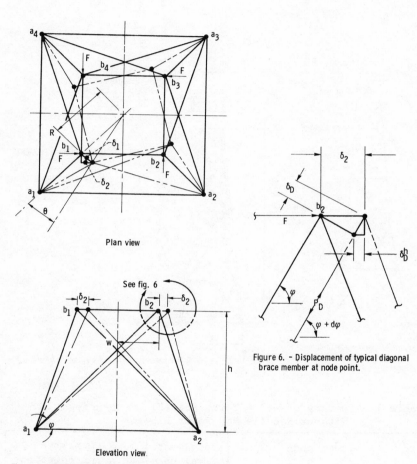

Plan view

See fig. 6

Elevation view

Figure 5. – Displacement of typical tower bay due to torsion.

Figure 6. – Displacement of typical diagonal brace member at node point.

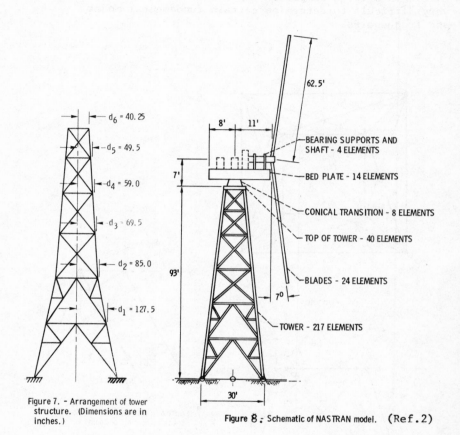

Figure 7. - Arrangement of tower structure. (Dimensions are in inches.)

Figure 8. - Schematic of NASTRAN model. (Ref.2)

Table 1 - Summary of Mod-O Tower Natural Frequencies
Without Nacelle and Rotor Blades

Tower mode	Approximate values	Test data[a]	NASTRAN predictions[a]
	Tower frequency, Hz		
First bending in north-south direction	5.04	4.7	4.76
First Torsional	8.59	10.5	10.1

[a]From reference 5.

Table 2 - Summary of Mod-0 Tower Natural Frequencies
With Nacelle and Rotor Blades on Tower in
Horizontal Position

Tower mode	Approximate values	Test data[a]	NASTRAN predictions[a]
	Tower frequency, Hz		
First bending in north-south direction	2.45	2.1	2.15
First torsional	3.78	4.4[b]	(c)

[a]From reference 5.

[b]Erroneously reported as 9.8 Hz in reference 5.

[c]Erroneously reported as 9.56 Hz in references 2 and 5.

Table 3 - Tower Natural Frequencies Without Nacelle and
Rotor Blades (Direct Input of Mass/Mass Moment
of Inertia and Stiffness/Flexibility
in STRUDL II)

Tower mode	Bending	Torsion
	Tower frequency, Hz	
1	5.19	10.55
2	23.31	22.28
3	58.49	30.14
4	109.11	45.01
5	177.51	59.27
6	252.26	70.11

Table 4 - Tower Natural Frequencies With Nacelle and Rotor
Blades on Tower in Horizontal Position (Direct Input
of Mass/Mass Moment of Inertia and Stiffness/
Flexibility in STRUDL II)

Tower mode	Bending	Torsion
	Tower frequency, Hz	
1	2.29	3.67
2	19.56	14.79
3	54.14	25.95
4	106.29	38.92
5	176.48	55.45
6	252.20	68.83

CONCLUSIONS

The frequencies from the idealized simple model compare
closely with those determined by experiment and those
predicted by using the detailed finite element NASTRAN
model. The simplified models can thus be used to compute
tower fundamental frequencies with the aid of a pocket-
type calculator and could be used to aid in the preliminary
design. These models can also be used to find gross errors
in the detailed finite element models, to differentiate
between important modes and those caused by local member
vibration, and to identify couple mode shapes.

ACKNOWLEDGEMENTS

The results presented in this paper were obtained from
the work of author as Summer Faculty Fellow at the NASA-
Lewis Research Center in 1976. The author wishes to thank
Lewis Research Center for permission to publish the work.
The author also acknowledges B.S. Linscott of Lewis
Research Center for his suggestion of the problem and
help in writing the report of ref. 3.

REFERENCES

1. Blodgett, O.W. (1966) Design of Welded Structures.
    James F. Lincoln Arc Welding Foundation.
2. Chamis, C.C. and Sullivan, T.L. (1976) Free Vibrations
    of the ERDA-NASA 100-KW Wind Turbine. NASA TMX-71879.
3. Das, S.C. and Linscott, B.S. (1977) Approximate Method
    for Calculating Free Vibrations of a Large-Wind-Turbine
    Tower Structure. NASA TM-73754.
4. Hurtz, W.C. and Rubinstein, M.F. (1964) Dynamics of
    Structure. Prentice Hall, Inc.
5. Linscott, B.S. and others (1976) Tower and Rotor Blade
    Vibration Test Results for a 100-kilowatt Wind Turbine.
    NASA TMX-3426.
6. Thomas, R.L. (1976) Large Experimental Wind Turbines -
    Where We Are Now. NASA TMX-71890.
7. Thomson, W.T. (1953) Mechanical Vibrations. Prentice
    Hall, Inc.

WIND LOADING AND MISSILE GENERATION IN TORNADO

T. Theodore Fujita

The University of Chicago
Chicago, Illinois U.S.A.

ABSTRACT

Destructive force in tornado is a result of both aerodynamic pressure induced by the wind and the reduction of pressure due to the vortex itself. The extent of the structural
damage and tornado-induced flying objects depends on the maximum windspeed as well as the duration of the peak wind, which
may last only one to two seconds. Acceleration of objects in
tornado is discussed by introducing new parameters, "ballistic
length" and "half-speed period."

1. INTRODUCTION

Tornado is a violent column of rotating air accompanied by
damaging winds on or near the ground. Such winds over water
are called waterspouts.

Although tornadoes are reported from all over the world,
except in extremely high latitudes, there are specific
locations where damaging tornadoes are likely to develop.
Fujita (1973) revealed that most of the violent tornadoes on
earth occur in the United States, followed by Bangladesh
where killer tornadoes have been reported rather frequently.

While strong tornadoes often occur in Japan, New Zealand, and
Australia, European tornadoes cause strong to severe damage
(see Figure 1.1 for tornado distribution around the world).

Historically, five schemes for tornado classification were
proposed by various authors in an attempt to categorize storms
by the havoc they wreak (see Table 1.1.). Of these schemes,
the Fujita scale (F scale) has been widely used in the United
States, thus classifying all reported tornadoes during the
63-year period, 1916 through 1978. Table 1.2 which excludes
1978 tornadoes shows that F 5 tornadoes were only 0.5% and
F 4, 2.8% of all tornadoes. F 1 tornadoes were the largest
(34.5%) in frequencies followed by F2 storms (28.9%).

166

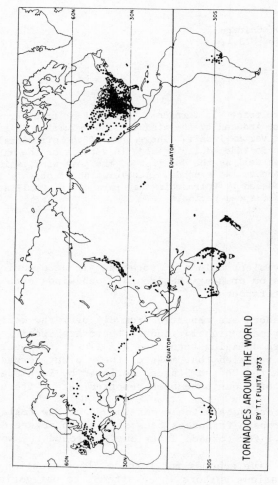

TORNADOES AROUND THE WORLD
BY T.T. FUJITA 1973

Figure 1.1 Tornadoes around the World by Fujita (1973). Tornadoes are rare in high latitudes and over the equator. Most of the northern hemisphere tornadoes rotate counterclockwise, those in the other hemisphere in the opposite direction.

2. WIND FIELD INSIDE TORNADOES

When air spirals in toward the tornado center, its speed increases while pressure decreases. In other words, the air gains
its velocity head while the pressure head is lost.

The linear acceleration along so-called streamline is provided by the resultant of the pressure gradient and the centrifugal forces shown in Figure 2.1. One opposing force is the
drag force, which results in the loss of pressure, P_L . We may
now write

$$P + \frac{1}{2}\rho S^2 + P_L = P_\infty \qquad (2.1)$$

where P is the pressure in vortex; ρ, the air density; S,
the storm-relative velocity; and P_∞ , the pressure at infinity.

VORTEX WITHOUT MASS ACCUMULATION

Equation (1), when re-arranged into

$$\frac{1}{2}\rho S^2 = (P_\infty - P_L) - P, \qquad (2.2)$$

reveals that the linear acceleration continues until P, the
air pressure, reaches zero or the vacuum. The windspeed then
is expressed by

$$S^2 = 2 (P_\infty - P_L) \rho^{-1}. \qquad (2.3)$$

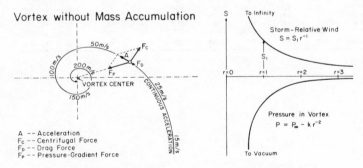

Figure 2.1 Spiral airflow into a vortex without mass accumulation
around its center. In this type of vortex windspeed increases continuously until pressure decreases to the vacuum.

The maximum speed expected to occur when $P_L = 0$ is

$$S = \sqrt{\frac{2 P_\infty}{\rho}} \qquad (2.4)$$

When $\rho = 0.0012$ g/cm^3 and $P_\infty = 1000$ mb we obtain
$S = 408$ m/sec $= 913$ mph which corresponds to Mach 1.19 at 20°C air
temperature.

Years Proposed	Authors	Country	Scales	Classification Schemes
1890	Hazen	U.S.A.	1 through 3+	Five scales ranging from the least destructive to the most severe tornadoes. No descriptions for classification provided.
1945	Seelye	N.Z.	0, 3, and 5	Class 0, no damage; 3, damage to roofs and outbuildings; 5, well-constructed buildings demolished.
1971	Fujita	U.S.A.	F0 through F5	Pegged to windspeeds connecting Beaufort scale with Mach number. Damage specifications and reference photos provided. 24,930 U.S. tornadoes (1916-78) classfied.
1975	Dames & Moore	U.S.A.	1 throuth 6	Damage descriptions correspond to scales with overlapping windspeeds. Assessment scheme relies heavily on roof damages. Generalization to other types of damage not provided.
1976	Meaden	U.K.	T0 through T12	An extension of the Beaufort force. Estimated maximum scale in U.K. was T8, T9 for France and T11 for the U.S.

Table 1.2 **Frequencies** of tornadoes in the United States in each decade since 1916. Based on the DAPPLE Tornado Tape, The University of Chicago.

Scale	1910s (5 yr)	1920s (10 yr)	1930s (10 yr)	1940s (10 yr)	1950s (10 yr)	1960s (10 yr)	1970s (7 yr)	1916-77 (62 yr)	
F 5	13	28	18	23	10	18	17	127	(0.5%)
F 4	24	54	84	134	100	146	128	670	(2.8)
F 3	69	310	253	323	554	610	525	2644	(10.9)
F 2	151	590	723	674	1372	1791	1684	6985	(28.9)
F 1	102	336	441	320	1813	2395	2927	8334	(34.5)
F 0	32	73	273	174	1070	1957	1809	5388	(22.8)
Total	391	1391	1792	1648	4919	6917	7090	24,148	(100%)

It is theoretically possible that a tornado could have a supersonic windspeed near its center. This is why the Fujita scale in Table 1.1 was devised by connecting Beaufort 12 with Mach 1, the speed of sound in the free atmosphere.

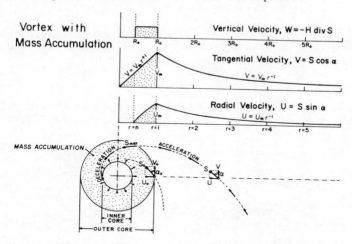

Figure 2.2 Mass accumulation into the outer core induces an opposing pressure to slow down the inflow speed. Both radial and tangential velocities reach maxima at the outer-core boundary.

VORTEX WITH MASS ACCUMULATION

Because the atmosphere cannot store the mass of the continuous inflow, a vortex will have to remove the huge surplus of mass which piles up around its center.

In an actual vortex or in a tornado, the linear acceleration is opposed violently by this surplus mass. As shown in Figure 2.2, the accumulated mass results in the pressure increase all around the vortex center, thus forming a doughnut-shaped outer core in which the inflow decelerates. The inner core which is free from inflow air rotates like a solid disc, providing the necessary centrifugal force to push out the outer core.

Three component velocities, U, V, and W, are expressed schematically in Figure 2.2. It should be noted that the radial velocity decreases to zero inside the outer core, resulting in the vertical velocity by which the inflow air is deflected upward to prevent the mass accumulation.

3. OBJECTS IN TORNADO WINDS

Wind induced by a tornado is the result of the "swirling motion" of the air around the vortex center and the "translational motion" of the vortex itself. Thus, we express l, the wind impinging upon an object in tornado by

$$I = S + T,\qquad\qquad(3.1)$$

where S denotes the swirling velocity relative to the vortex center and T, the translational velocity.

Due to the friction between the air and the surface, 1 should theoretically be zero at the surface. Windspeed increases upward inside the inflow layer as expressed by

$$I = I_m h^{\frac{1}{6}}\qquad\qquad(3.2)$$

where I_m is the maximum windspeed reached at the normalized height $h = 1$ at the top of the inflow layer (see Figure 3.1).

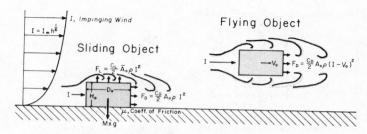

Figure 3.1 Vertical distribution of wind in tornado and various forces acting upon an object prior to its sliding and while flying horizontally.

TEN HYPOTHETICAL OBJECTS

Ten hypothetical objects listed in Table 3.1 and Figure 3.2 are brought into a tornado. Their heights, H_0, are measured vertically, widths, W_0, across the wind, and the depths, D_0, in the direction of the wind.

The windspeed at $\frac{1}{2}H_0$ of each object varies with the overall windspeed and also with H_0. In this paper, the overall windspeed is measured at 5m above the ground level (AGL). The windspeed ratio in Table 3.2 is defined by

Impinging windspeed at $\frac{1}{2}H_0$ / Overall windspeed at 5-m AGL

which decreases, of course, with decreasing height of the object.

Dust particles on the surface, for example, are affected by only 14.7% of the windspeed at 5-m AGL. Once these particles become airborne, they fly with winds at various heights.

THRESHOLD WINDSPEEDS TO SLIDE OBJECTS

The first step in producing an airborne object is to slide the

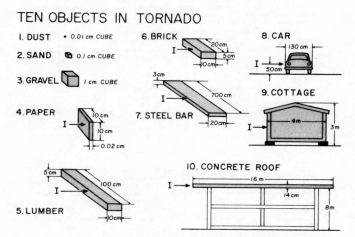

Figure 3.2 Ten objects in tornado, ranging from a particle of dust to a concrete roof section weighing 143 tons.

object off its resting position. Tornado wind, in fact, induces F_D, the drag force and the lift force shown in Figure 3.1. The threshold windspeed, I_T, is obtained by solving

$$\frac{1}{2} C_D A_0 \rho I_T^2 = \mu(Mg - \frac{1}{2} C_L \bar{A}_0 \rho I_T^2) \qquad (3.3)$$

into

$$I_T^2 = \frac{2\mu Mg}{\rho(C_D A_0 + \mu C_L \bar{A}_0)} \qquad (3.4)$$

where $A_0 = H_0 W_0$ is the area of the object facing the impinging wind and $\bar{A}_0 = W_0 D_0$, that of the horizontal surface under the influence of the lift force.

A simple assumption of $\mu = 0.3$, $C_D = C_L = 2$ and $g = 980$ cm/sec^2 for estimation purposes permits us to reduce Eq. (3.4) into

$$I_T = 4.95 \sqrt{\frac{M}{A_0 + 0.3\bar{A}_0}} \qquad (3.5)$$

Table 3.1 Dimensions and masses of 10 objects ranging from dust to concrete roof. Height(H_0), width(W_0) and depth(D_0) are mesured in relation to the impinging wind direction. Windspeed ratio denotes the speed at $\frac{1}{2}H_0$ divided by that at 5-m AGL. "*" indicates the object with elevated base.

Objects	Dimensions height x width x depth	Mass	$\frac{1}{2}H_0$(cm)	Windspeed ratio
1. Dust	0.01 cm Cube	0.025 mg	0.005	0.147
2. Sand	0.1 cm Cube	2.5 mg	0.05	0.215
3. Gravel	1 cm Cube	30 g	0.5	0.316
4. Paper	10x10x0.02 cm	1.4 g	5.0	0.464
5. Lumber	5x100x10 cm	3.5 kg	2.5	0.413
6. Brick	5x20x10 cm	2 kg	2.5	0.413
7. Steel bar	3x700x20 cm	330 kg	1.5	0.380
8. Car	1x3x1.3 m	1.2 ton	*50	0.681
9. Cottage	3x5x4 m	6 ton	150	0.818
10. Concrete roof	0.14x20x16 m	143 ton	*800	1.081

Table 3.2 Threshold windspeed to slide objects off their resting positions. A_0 denotes the area facing the impinging wind and $\bar{A}_0$, that of the top surface affected by the lift force.

Objects	Area A_0+ 0.3 $\bar{A}_0$	Threshold wind speed at $\frac{1}{2}H_0$	at 5m AGL in mph	
1. Dust	0.00013 cm²	0.13 m/sec	0.9 m/sec	2 mph
2. Sand	0.013 cm²	2.2	10.2	23
3. Gravel	1.3 cm²	23.8	75.3	168
4. Paper	101 cm²	0.6	1.3	3
5. Lumber	800 cm²	10.4	25.2	56
6. Brick	160 cm²	17.5	42.4	95
7. Steel bar	0.63 m²	35.8	94.2	211
8. Car	4.2 m²	26.5	38.9	87
9. Cottage	21 m²	26.5	32.4	72
10. Concrete roof	99 m²	59.5	55.0	123

Table 3.3 Ballistic lengths and half-speed periods of objects. Density ratio is the object density, ρ_0 divided by that of air, ρ . Half-speed periods were computed for three windspeeds, 50, 100, and 150 m/sec at 5-m AGL.

Objects	Density ratio	Ballistic length(m)	Half-speed period (sec) 50m/s(112mph)	100(224)	150(336)
1. Dust	2,000	0.2 m	0.027 sec	0.014 sec	0.009 sec
2. Sand	2,000	2	0.19	0.09	0.06
3. Gravel	2,500	25	1.6	0.8	0.5
4. Paper	600	0.12	0.005	0.003	0.002
5. Lumber	600	60	2.9	1.5	1.0
6. Brick	1,700	170	8.2	4.1	2.7
7. Steel bar	6,500	1,300	68	34	23
8. Car	250	325	7.3	3.7	2.4
9. Cottage	80	320	7.8	3.9	2.6
10. Concrete roof	2,700	43,000	796	398	265

Table 4.1   Three methods for estimating tornado windspeeds. The highest windspeeds estimated are limitted to reliable values only. For meteorological anlyses refer to Golden(1976) and engineering, to Mehta(1976).

Analysis Methods	Subjects of Investigation	Highest windspeeds obtained
METEOROLOGICAL ANALYSES	Anemometer records	151 mph in F2 damage area in Michigan
	Shape of funnel clouds	230 mph in Texas and in North Dakota
	Photogrammetric analyses	170 mph without suction vortex, 255 with S.V.
	Ground marks	484 mph, single-object assumption and 180 mph with multiple suction vortices
ENGINEERING ANALYSES	Physical damage to structures	190 mph, railroad bridge; 188 mph, walls
	Displacement of objects	159 mph, overturned stone monument
	Impact force of objects	reliable estimate not available
INTERACTIVE ANALYSES	Duration of peak winds	one second or less in tight-core tornadoes
	Acceleration of objects	2 to 3 times "g" with estimated 300 mph wind

which indicates the threshold windspeed at $\frac{1}{2}H_0$ required to slide the object off its resting position.

Corresponding windspeeds at 5-m AGL can be computed by dividing I_T by the windspeed ratio in Table 3.1. The results obtained in Table 3.2 reveal that only 2- to 3-mph winds will pick up dust and paper while 168 mph wind is required to move the gravel on the surface.

An automobile can be moved by an 87 mph wind. This is why the National Weather Service in the United States recommends not to stay in cars in a tornado. Cottages and mobile homes must be tied down, otherwise, they slide off the foundation in only a 72 mph wind.

HORIZONTAL ACCELERATION OF AIRBORNE OBJECTS

An object, once it becomes airborne, is accelerated by the impinging wind which induces the drag force expressed by

$$F_D = \tfrac{1}{2} C_D A_0 \, \rho \, (I - V_0)^2 \tag{3.6}$$

where V_0 denote the horizontal speed of an object. The acceleration of the object is

$$\frac{d V_0}{d t} = \frac{F_D}{M} = \tfrac{1}{2} C_D \frac{\rho}{\rho_0} D_0^{-1} (I - V_0)^2 \tag{3.7}$$

where ρ_0 is the object density and

$$B_0 = \frac{2}{C_D} \frac{\rho_0}{\rho} D_0 \tag{3.8}$$

is called the "ballistic length" in this paper. Equation (3.7) is solved into

$$V_0 = [I - (I + I \, B_0^{-1} t)^{-1}] \, I \tag{3.9}$$

which shows that the speed of a flying object increases to that of the impinging wind as time, t, increases to infinity, provided that the object does not fall to the ground (see Figure 3.3 and Table 3.3)

The ballistic lengths of paper and dust are only 0.12 m and 0.2 m, respectively, while those of heavy objects are extremely large.

HALF-SPEED PERIOD

The half-speed period, the time required to accelerate an object to one-half of the impinging windspeed, is useful in expressing the acceleration characteristic of an object in tornado.

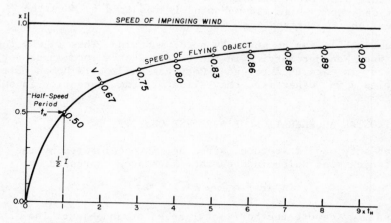

Figure 3.3 Speed of a flying object being accelerated horizontally by an impinging tornado wind.

The half-speed period can be obtained by putting $V_0 = 0.5 \, I$ into Eq. (3.9).
Thus, we have

$$t_H = B_0 \, / \, I. \qquad (3.10)$$

It is evident that the half-speed period is proportional to B_0, the ballistic length, and inversely proportional to 1, the speed of the impinging wind.

Computed values in Table 3.3 reveal that half-speed periods of light objects in tornado winds are less than "one second" while those of large, heavy objects are 10s to 100s of seconds, indicating that they do not move easily with wind.

4. BASIC PROBLEMS OF WINDSPEED ESTIMATES

Meteorological methods for estimating tornado windspeed have
long been dependent upon the "educated guess" which often resulted in a supersonic windspeed. But now, no meteorologist
believes in such an extreme windspeed.

Funnel cloud in tornado is induced by the field of low pressure
in which water vapor condenses into tiny water droplets. Winds
of 230 mph were estimated, based on the shape of funnel clouds.

Circular or cycloidal ground marks left behind tornadoes were
investigated first by Van Tassel (1955) who obtained a 484 mph
windspeed, assuming that marks were drawn by a single object
rotating around the tornado center. Fujita (1970) found that
cycloidal marks are generated by "several" small vortices,
called "suction vortices", which circle around the tornado
center. This new concept reduced the windspeed to about 180 mph.

Photogrammetric analyses of tornado movies provide us with
accurate values of tornado windspeeds. Less than 170 mph windspeeds were estimated in tornadoes without suction vortices
while 255 mph was obtained for those including suction vortices.

The most reliable measurement occured on April 11, 1965 in
Tecumseh, Michigan where 151 mph (scaled off) peak wind was
recorded. The author's aerial survey revealed that the wind
tower (10-m AGL) was located inside the F 2 area.

Engineering estimates are mostly based on physical damage to
structures. Estimated windspeeds, so far, are less than 190 mph.
The pressure forces acting upon a structure are

$$P_A + P_V = P \qquad (4.1)$$

where P_A is the aerodynamic pressure; P_V, the vortex pressure;
and P, the total pressure acting upon structural surfaces.

It is rather difficult to map these pressures separately because windspeeds in tornado change very rapidly. On a large
cooling tower, however, P_A and P_V change gradually as a tornado passes by the tower. Hypothetical pressure fields are
presented in Figure 4.1.

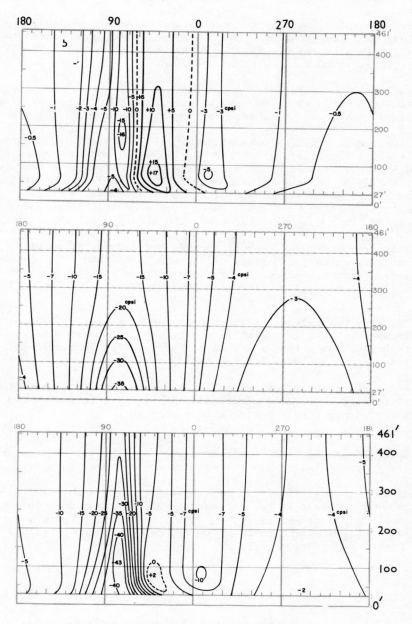

Figure 4.1 Fields of pressure acting upon a hypothetical cooling tower in a tornado vortex. Aerodynamic pressure, P_A, induced by impinging winds (top); vortex pressure, P_V (middle); and the total pressure, P (bottom). Isobars in centi-psi, cpsi.

INTERACTIVE ANALYSIS

Tornado wind and structure form an interactive system leading to the structural failure. Similar to Einstein's "matter tells space how to curve and space tells matter how to move", wind-structure interaction can be expressed by

"wind tells structure how to deform and
structure tells wind how to blow around".

It is evident that a large, strong structure weakens tornado winds before its ultimate failure. Namely, the windspeed estimated based on engineered structure is, in general, lower than that of undisturbed tornado.

Especially when the diameter of tornado is small in relation to its translational velocity, the duration of the damaging wind is very short, say only one to two seconds. Structural failure due to such a short-duration wind requires a considerably larger peak gust speed than those of long-lasting winds.

A tight-core tornado on December 3, 1978 at Bossier City, Louisiana, U.S.A. blew three sections of concrete roof off the Best Western Motel. The dimensions of the roof were that of object 10 in Figure 3.2 and Table 3.1. The three sections of roof, each weighing 143 tons, were blown off and landed "upside down". The maximum airborne distance was 55 m.

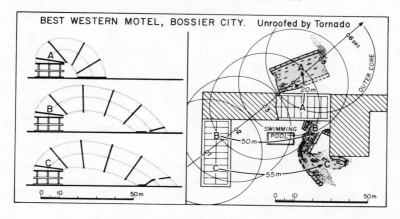

Figure 4.2 Three sections of the roof of Best Western motel at Bossier City, Louisiana which were blown off by the December 3, 1978 tornado. The peak wind lasted only for about one second.

The diameter of the tornado core was 60 m while its center moved across the roof at 20/sec. This means that each roof section was under the influence of the peak wind for only about "one second" when it became airborne.

As shown in Table 3.2, the roof can be slid off with threshold windspeed of 123 mph. A much higher windspeed is required for the roof section to be accelerated by a one-second peak wind and to be rotated into an upside-down position while flying clear across the swimming pool. The wind estimates, based on the motel roof sections, are being performed by the author along with

Robert F. Abbey, Jr. Nuclear Regulatory Commision

and

James R. McDonald, Texas Tech University

Results will be reported at later dates.

5. CONCLUSIONS

Investigation of tornado-structure interaction is extremely important in the United States where over 75% of world's strongest tornadoes have been occuring. Three major aspects of tornadoes being investigated by the author and his collaborators are:-

A. Risk assessment based on historical frequencies and F-scale distribution of tornadoes by geographic locations, months, and hours of the day. Refer to Abbey (1976)

B. Estimate of tornado windspeeds through meteorological, engineering and interactive analyses.

C. Computation of time-dependent pressure field acting upon structures such as cooling towers, mobile homes, etc.

ACKNOWLEDGEMENT

Research on tornadoes reported in this paper sponsored by Nuclear Regulatory Commission under Contract No. 04-74-239.

REFERENCES

Abbey, Robert F., Jr.(1976): Risk Probabilities Associated with Tornado Windspeed. Symposium on Tornadoes, p 177-236.

Dames & Moore (1975): Unpublished document

Fujita, T.T., D.L. Bradbury and C.F. Van Thullenar (1970):Palm Sunday Tornadoes of April 11, 1965. Monthly Weather Review, Vol.98, p. 29-69.

Fujita, T.T. (1971): Proposed Characteristics of Tornadoes and Hurricanes by Area and Intensity. SMRP Res. Paper No. 91, University of Chicago.

Fujita T.T. (1973): Tornadoes Around The World. Weatherwise, 26, p. 56-83.

Golden, J.H. (1976): An Assessment of Windspeed in Tornadoes. Symposium on Tornadoes, p. 5-42.

Hazen, H.A. (1890): Tornadoes, A Prize Essay. American Meteor. Journal, Vol.7, p. 205-229.

Meadon, G.T. (1976): Tornadoes in Britain: The Intensities and Distribution in Space and Time. J. Meteor., Vol.1, p. 242-251.

Mehta, K.C. (1976): Windspeed Estimates: Engineering Analyses. Symposium on Tornadoes, p. 89-103.

Seelye, C.J. (1945): Tornadoes in New Zealand. New Zealand J. of Science and Technology, Vol.27, p. 166-174.

Van Tassel, E.L. (1955): The North Platte Valley Tornado Outbreak of June 27, 1955. Monthly Weather Review, Vol.83, p. 255-264.

Symposium on Tornadoes: Proceedings (696 pages) of the Symposium on Tornadoes, Assessment of Knowledge and Implication for Man. June 22-24, 1976 at Texas Tech University.

EARTHQUAKE FLOOR RESPONSE AND FATIGUE OF EQUIPMENT IN MULTI-
STOREY STRUCTURES

J. C. Wilson

Research Engineer, Department of Civil Engineering and Engineering Mechanics, McMaster University, Hamilton, Canada.

INTRODUCTION

The design of multi-storey structures for seismic areas recognizes the need for adequate protection of structural members
from the dynamic loads imposed during an earthquake. For many
types of structures such as office buildings and commercial
centres, it is sufficient to consider mainly the safety and
serviceability of the primary structural system. Recent earthquakes such as 1971 San Fernando however, have drawn attention
to the behaviour of equipment services within multi-storey
structures, most notably as a result of the failure of some of
these installations during moderate seismic activity. Of major
concern are telephone switching centres, electrical switchgear
stations, hospitals and other similar utilities and public services. In the aftermath of a damaging earthquake these life-
lines and services are vital to a community's emergency plans
but they may be rendered useless unless certain electrical and
mechanical components retain their operational capabilities.

This paper examines the seismic response of multi-storey structures and their equipment installations and attempts to characterize some parameters of equipment response which prove useful in laboratory dynamic testing. Mathematical models of
several typical structures are subjected to a 30-second earthquake excitation and the seismic response at the top floor of
each building is evaluated using the modal superposition technique. This response is then used as an input to a single-
degree-of-freedom (SDOF) representation of an equipment component located on the top floor, and its quasi-harmonic response
and low-cycle fatigue are examined. The paper closes with a
discussion of test methods, which can incorporate characteristics of the previous analysis to test the functional capability
(seismic qualification) of electrical and mechanical equipment
exposed to an earthquake environment.

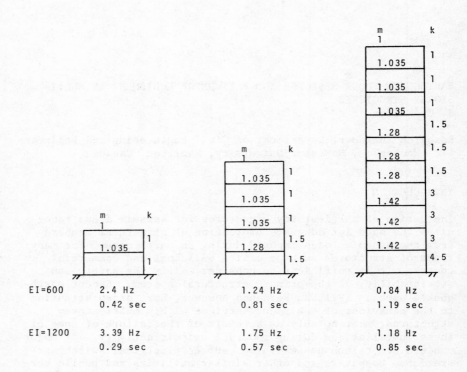

FIGURE 1   STRUCTURAL MODELS

STRUCTURAL MODELS AND SEISMIC GROUND MOTION

Structural Models

The three basic structures selected for study were 2-stories, 5-stories and 10-stories in height. Shown in Figure 1, they are representative of many North American multi-storey buildings designed for dead and live loading and lateral earthquake forces as specified by codes such as the National Building Code of Canada (NBCC 1977) or the Uniform Building Code (UBC 1970). The mass and stiffness coefficients, m_i and k_i respectively,

vary in an approximate linear fashion from top to botton and the dynamic modelling has assumed the buildings to be of a shear-type responding in a linearly elastic manner. This distribution of mass and stiffness is similar to the structural models used by several other researchers (Liu et al, 1977; Bertero and Kamil, 1975; Clough and Benuska, 1967). Since these coefficients serve only to describe the distribution of mass and stiffness within the structure, they must be referenced to specific values (ie., reference EI values) for use in a dynamic analysis. In this study two reference EI values were used. One group of low, medium and high rise structures were modelled using EI=600 and a second, more rigid group of structures used EI=1200. This total of six structural models provided fundamental frequencies covering a range from 0.84 Hz to 3.4 Hz (period range: 1.19 sec to 0.29 sec) and heights of 2, 5 and 10 stories. The fundamental frequencies and periods are shown below each structure in Figure 1. The foundation was assumed rigid in all cases to avoid attenuation of ground accelerations by soil-structure interaction. The structural damping was at 5%, representative of steel and reinforced concrete structures.

Seismic Ground Motion

Each of the structures described above was subjected to the first 30 seconds of the El Centro (May 18, 1940) N-S component as the source of seismic excitation. This event registered a magnitude of 6.3 on the Richter scale, a maximum intensity of X (Modified Mercalli) and a peak ground acceleration of 0.348g. The original Caltech record was digitized at 0.02 second intervals but to retain accuracy for the higher modes in the dynamic analysis, additional points were interpolated by assuming a linear variation of acceleration during each 0.02 second interval. The final interpolated record therefore, contained 3,000 points at increments of 0.01 second and is shown in Figure 2.

DYNAMIC ANALYSIS AND FATIGUE EVALUATION

The mathematical analysis of the seismic response of structure-equipment systems can be conveniently divided into two major sections; (1) evaluation of the seismic environment at the equipment location in the structure, (2) evaluation of equipment response. The results from (1) in the form of a time-history of the top floor (first floor below the roof level) response to the earthquake excitation are used as an input to the base of the equipment in (2). Subsequently, a dynamic analysis of equipment responses allows comparisons to be made amongst the six structural-equipment systems.

Seismic Environment

A dynamic analysis of each of the six structures was performed using the time-history modal superposition technique.

Modal Analysis For each structure the modal frequencies $\{\omega\}$ and mode shapes $[\Phi]$ were calculated using a matrix eigenvalue routine to solve the equation

$$([K] - \{\omega^2\} [M]) \{y\} = 0 \tag{1}$$

where $[K]$, $[M]$, are stiffness and mass matrices and $\{y\}$ is a vector of displacements. Modal participation factors,

$$\Gamma_m = \sum_i M_i \phi_{mi} / \sum_i M_i \phi_{mi}^2 \tag{2}$$

described the contribution made by each mode m, to the overall structural response. The results at this stage produced a set of uncoupled differential equations for the time dependent response of each of the six structures.

Floor Response The seismic floor response at the top floor level in each structure was determined by numerical evaluation of the Duhamel integral D(t), and its time derivatives (Dempsey and Irvine, 1978)

$$D(t) = \frac{1}{\omega} \int_0^t \ddot{x}_g(\tau) h\ (t-\tau) d\tau \tag{3}$$

where $h(t-\tau) = \exp[-\zeta\omega(t-\tau)] \sin [\omega(t-\tau)]$ (4)
and where ζ is damping and $\ddot{x}_g(\tau)$ is the earthquake ground acceleration. The final top floor time-history response, $\ddot{y}_T(t)$ is expressible as,

$$\ddot{y}_T(t) = \sum_{m=1}^{J} \phi_m \Gamma_m \ddot{D}(t)_m \tag{5}$$

with the summation being over J modes.

Spectral Analysis Each of the responses $\ddot{y}_T(t)$ from equation (5) were subjected to a computerized spectral analysis to extract the dominant floor frequency for use in the equipment calculations.

Equipment Response

Modeling an equipment component as a SDOF system provides a convenient means of characterizing floor motions in terms of a few basic parameters which can then be used in a much wider range of generalized applications, such as performance requirements or test specifications. The SDOF equipment representation was assigned the same natural frequency as the predominant frequency in the floor motion calculated by the spectral analysis. This condition can be expected to produce maximum response and exposure to the severest conditions of fatigue. Conservatively, 1% damping was used. The floor motion time-history was input to the equipment and its response through the time domain evaluated by the same Duhamel integral approach.

Fatigue Seismic responses are expected to show rather random amplitude fluctuations and it is convenient to obtain a measure of the number of cycles of a uniform amplitude motion which would produce the same fatigue effect as the entire ensemble of random seismic fluctuations. The method used here essentially follows that of Fischer and Wolff (1973) by determining the number of total equivalent fatigue cycles.

Assuming equipment response is linear elastic, strain may be taken as being proportional to acceleration. If the peak equipment response is R_p, then the fraction of the total fatigue life at peak response used up by any one cycle of lesser amplitude is given as,

$$\frac{N_p}{N} = (\frac{R}{R_p})^\beta \tag{6}$$

where N is the number of cycles to failure at a response amplitude R, and similarly for N_p and R_p. The exponent β is a material constant related to the slope of the log S - log N fatigue-life relationship for a material. For typical materials such as carbon and high alloy steels, β lies in a range of 3 to 5 and as lower β values are more critical, $\beta = 3$ has been used (Duff and Heidebrecht, 1979). The ratio in equation (6) can then be summed for all the peaks in the response to give the total number of equivalent fatigue cycles,

$$N_{eq} = \frac{1}{2} \sum_{i=1}^{n} (\frac{R_i}{R_p})^\beta \tag{7}$$

The 1/2 term is included because the log S - log N relationship considers strain amplitude as a peak-to-peak cycle whereas the sum has included all positive and negative peaks. This value of N_{eq} is dependent upon R_p, the peak response within the record under consideration. To normalize the N_{eq} values obtained from several different structures, equation (6) can be applied again to each N_{eq} giving,

$$\overline{N}_{eq} = (\frac{R_p}{R_f})^\beta N_{eq} \tag{8}$$

where R_p is as before and R_f is the reference value, usually taken as the maximum of all R_p values.

In the next section, results of the application of the above dynamic analysis procedure on each of the six structural-equip-

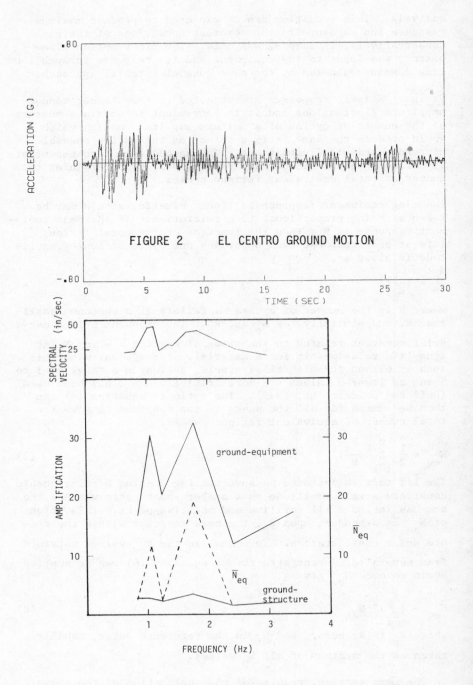

FIGURE 2   EL CENTRO GROUND MOTION

FIGURE 3   AMPLIFICATIONS AND EQUIVALENT FATIGUE CYCLES

ment systems are presented, accompanied by several observations and discussions.

RESULTS AND DISCUSSIONS

The computer calculations of the seismic response of each structure-equipment system are shown in Table 1 in terms of fundamental frequency, peak acceleration, dynamic amplification and fatigue equivalence. The peak accelerations and amplifications were evaluated directly from the time-histories. The normalized equivalent fatigue cycles $\overline{N}_{eq}$, were referenced in each case to 11.34 g, the largest value of equipment-response for all six structures.

TABLE 1 SUMMARY OF DYNAMIC AND FATIGUE ANALYSES

	Structure					
	EI = 600			EI = 1200		
Stories	2	5	10	2	5	10
Fund. Struct. frequency (Hz)	2.40	1.24	0.84	3.39	1.75	1.18
Peak Floor response (g)	.52	.76	.84	.71	1.28	.86
Peak Equip. response (g)	4.17	7.15	6.01	6.19	11.34	10.53
G-S amplif.	1.5	2.2	2.4	2.1	3.7	2.5
G-E amplif.	12.0	20.5	17.3	17.8	32.6	30.3
N_{eq}	29.5	9.4	13.3	12.7	19.0	14.7
$\overline{N}_{eq}$	1.5	2.3	2.0	2.1	19.0	11.7

The amplifications and fatigue cycles in Table 1 are plotted in Figure 3 against the fundamental structural frequencies. In order that the frequency content of the earthquake record may be easily compared with the response, the 2% spectral velocity curve of El Centro has been superimposed at the top of Figure 3.

Computerized time-history plots of the floor and equipment responses in each system are shown in Figures 4 and 5. It should be noted that various vertical axis scales have been used.

From the equipment responses shown in Figure 4 and 5 it is apparent that some time-histories have a considerable level of response at the 30-second cut-off point and it seems that several cycles contribution to $\overline{N}_{eq}$ could be made by allowing equipment response to continue in free-vibration for some time thereafter. Three reasons led to a decision not to consider a continued phase of free-vibration decay. Firstly, since most earthquakes have their significant energy release and strong ground motion phase during the first 15 seconds or so, a 30-second record should adequately represent at least as severe a fatigue condition as a 15-second excitation followed by 15 seconds decay time. Secondly, the contribution to fatigue by free vibration response is directly dependent upon the equipment damping and for 1% damping it can be shown that the amplitude after 15 seconds will only be 39% of the initial amplitude. One percent is quite a conservative estimate and so cyclical contributions at the higher damping levels likely to be found in most real equipment systems would be appreciably smaller (ie., the amplitude of a 5% damped system after 15 seconds is only 0.9%). Third, since fatigue evaluation was only used for relative comparisons of levels of equipment response in various structures, it was convenient to work with one standard time base of continuous excitation. It was felt that introducing a free vibration decay after 15 seconds could sel- ectively favour (give lower $\overline{N}_{eq}$ values) responses which hap- pened to have a low amplitude region at the cut-off time. This approach was felt to adequately account for the significant duration of the earthquake record and would be capable of pro- ducing a set of results which would all be on a consistent and conservative basis.

Structural Response and Floor Motion

The ground-structure amplification was found to vary between 1.5 and 3.7 with four of the structures having amplifications in the range 2.1 to 2.5, and the largest amplification of 3.7 being associated with the more rigid (EI=1200) 5-storey struc- ture. For the four structures with similar amplifications, the amplifications did not demonstrate a significant dependence upon the height of the structure as illustrated by the factor of 2.2 for the 5-story and 2.4 for the 10-storey building (EI=600). Of more influential nature was the matching of a predominant ground motion frequency with a structural funda- mental frequency. In this case, the 1.75 Hz 5-storey structure had a fundamental frequency almost coincident with a major velocity response spectrum peak and showed an amplification of 3.7.

The spectral analysis indicated in each case a predominance of the fundamental structural frequency in the motion of the top floor. This observation which is consistent with other docu- mented reports, often finds a useful application in the devel- opment of 2-degree-of-freedom models of various multi-storey configurations. Modelling of a multi-degree-of-freedom system in this manner must be done with some care however, as mislead-

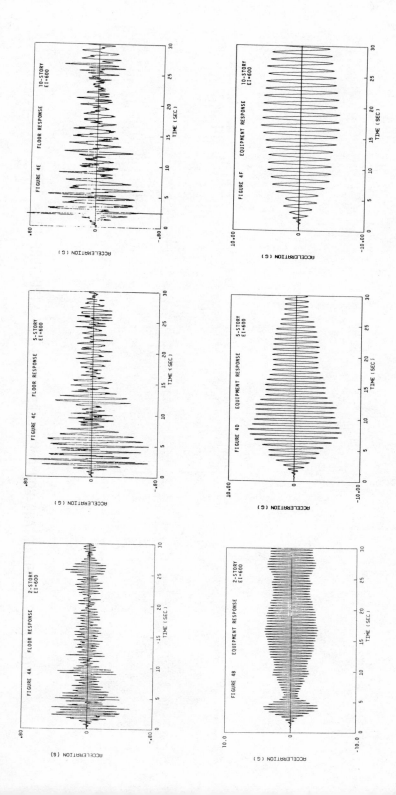

FIGURE 4   FLOOR AND EQUIPMENT RESPONSES FOR STRUCTURE EI = 600

190

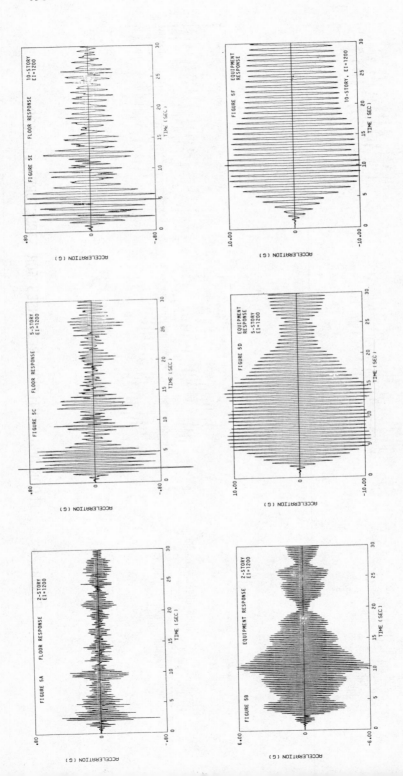

FIGURE 5   FLOOR AND EQUIPMENT RESPONSES FOR STRUCTURE EI = 1200

ing results may arise if there are closely-spaced modal frequencies in the real structure, or if the equipment mass is a significant percentage of the structural mass. These points are mentioned here only in passing and could be usefully investigated in further research into the dynamic modelling of such systems (Aziz and Duff, 1978).

Differences in the waveform response of the various structures are quite evident in Figures 4 and 5. The top floor responses in the 5 and 10-storey buildings generally show an initial 5 to 8 cycles of very strong response followed by a remaining 20 to 25 seconds of motion with peaks at about one-half the amplitude of the initial strong phase. In the low, 2-storey structures the amplitude patterns of floor response (Figures 4a, 5a) are not greatly different from the random ground motion pattern shown in Figure 2. The low buildings act as fairly rigid structures when subjected to the lower frequency inputs of the earthquake. Another variation in response is shown in Figures 5a and 5c where the floor responses contain quite a well-defined sine beat component. This appears after the initial 5 to 10 second phase as the structural response alters from the initial shock response stage to a quasi-harmonic motion.

Equipment Response and Low-Cycle Fatigue

By selecting the SDOF equipment model to have a natural frequency coincident with the dominant floor frequency, and using low (1%) damping, large amplitude, near-harmonic equipment response was observed in at least two cases. The previous section has noted the initial large amplitude cycles in the floor responses of the 1.75 Hz and 1.18 Hz structures. The equipment responses in these structures are likewise similar (see Figures 5d and 5f). An initial 5 second build-up region is followed by about 10 seconds of quite uniform peak response which in turn is followed by a decaying amplitude portion in the remaining 15 seconds or so. The ground-to-equipment amplification is approximately 30 in both cases which represents a response level at least 1.5 times greater than any of the other structures subjected to the same ground excitation.

Normalization of the fatigue equivalence using the 11.34g maximum equipment response emphasized the dominant fatigue effects occurring in the 1.75 Hz and 1.18 Hz systems. Values of $\overline{N}_{eq}$ of 19 and 11.7 respectively, for these systems were the largest calculated and are equivalent to approximately 10 seconds of full amplitude motion at the respective natural frequencies. The remaining four structures had negligible levels of cyclical fatigue with N_{eq} values of only 1.5 to 2.5, representing equivalent full amplitude durations of less than 2.5 seconds.

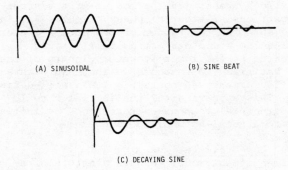

(A) SINUSOIDAL

(B) SINE BEAT

(C) DECAYING SINE

FIGURE 6   SINGLE FREQUENCY TESTS

SEISMIC TESTING

The preceding sections have provided some characteristics on floor motions and equipment responses in multi-storey structures. This final section briefly examines three single frequency shake table tests which are appropriate for testing the operational capabilities of an equipment component in a simulated seismic environment. The test procedure involves mounting the equipment test specimen on the shake table which is then set in motion by an analog wave signal sent to a hydraulic or an electromagnetic actuator system. The single frequency inputs can be selected at closely spaced intervals (1/2 to 1/3 octave intervals) to provide adequate spectrum coverage throughout the entire frequency range. Additional inputs at the specimen's resonant frequencies will serve as fatigue tests for the severest exposure conditions.

Sinusoidal Test

This test consists of an application of a duration and amplitude regulated sinusoidal motion to the test specimen. A typical waveform is illustrated in Figure 6a. By selecting an appropriate duration of table motion, the test can be correlated with the floor response and fatigue effects in a multi-storey seismic environment. For example purposes, suppose that (1) the duration of each test run is to be 7 full cycles of table motion, (2) the multi-storey seismic environment is estimated to produce (in a 1% damped system) a peak SDOF equipment response of 5g and (3) there is an estimated fatigue equivalence of N_{eq} = 12 cycles (referenced to 5g). Using basic vibration theory, 7 cycles of sinusoidal motion will produce an amplification of 17.8. Thus, to simulate the peak response characteristic at the test frequency, a table acceleration of $5/17.8 = 0.28g$ is required. By considering the response build-up phase during excitation and the free vibration decay after excitation, N_{eq} is evaluated from equation (7) for one 7-cycle test run as N_{eq} = 3.7.

Therefore, to simulate the 12 equivalent fatigue cycles of a possible real earthquake event, a minimum of 4 runs (12/3.7) of the 7-cycle, 0.28g test are required. It should be noted that in short duration test applications, free vibration decay will contribute significantly to N_{eq} and hence must be taken into account.

Sine Beat Test

The input waveform for a sine beat test is shown in Figure 6b. An advantage of this method over using a pure sine wave input is that amplification levels are significantly lower thereby permitting a greater degree of control in the operation of the shake table. The high amplifications caused by a sine test can require table displacements to be so small as to present physical control problems for certain equipment configurations and loads. Current test specifications recommend the use of 5 cycles per beat although any number is usually possible. A larger number of cycles per beat increases the amplification factors as well as providing larger fatigue effects. Following the same procedure as for sinusoidal testing, the input levels at each equipment resonant frequency are selected to achieve a predetermined SDOF response (ie., a given response spectrum). A series of sine beats can be used to produce the total number of fatigue cycles, allowing a pause between each beat phase to avoid superposition of amplifications from one beat cycle to the next.

Decaying Sine Test
The application of this test is similar to the sine beat test. By enveloping a sine wave with an exponential decay function (see Figure 6c) it is possible to achieve relatively low amplification factors simultaneously with extended durations of excitation. The decay rate of the enveloping function can be adjusted to control the response with smaller decay rates producing larger amplifications and a greater number of significant fatigue cycles.

Each of the three methods briefly outlined above are acknowledged in current recommended practices for the seismic qualification of safety related equipment for nuclear power plants (CSA, 1977; IEEE, 1975). As well, they should find wide use in the implementation of standard seismic testing procedures for communications equipment, electrical distribution facilities, and other non-nuclear applications.

SUMMARY

Consideration has been given to electrical and mechanical equipment exposed to seismic shaking in multi-story structures. Typical buildings of various heights were dynamically analyzed, floor responses calculated at the top levels, and equipment responses to the floor motions were examined. A method of evaluating effects of cumulative low-cycle fatigue was used to characterize equipment responses and to obtain comparisons of intensities of response for the structures when all were exposed to the same earthquake. Three single frequency shake table tests were discussed, each having time domain characteristics similar to actual structural responses and each capable of producing fatigue effects equivalent to those of a real earthquake.

The following points are noteworthy of mention.
(1) Although structural models used in the analysis were representative of typical low, medium and high-rise buildings, further characterization of multi-story seismic environments in benchmark structures would be useful. In Canada, committees

of the National Building Code (NBCC, 1977) currently have a benchmark system under study for various applications in earthquake and dynamic analyses.

(2) A statistical evaluation of structure-equipment response using several earthquake records could extend the research described in this paper. However, the use of a single event has demonstrated significant levels of response observed when the natural frequency of the structure closely matched the dominant frequency component of the ground motion.

(3) Specialized structures such as nuclear power plants warrant more detailed seismic study because of the stringent safety requirements and the complexity involved with mathematically modelling a reactor structure. In this case, several different models may be examined, and from the ensemble of results, upper bounds on floor response levels can be developed.

(4) Test requirements often specify only spectrum test levels without full consideration to time domain aspects of a seismic event. Accompanying the spectrum with a fatigue equivalence parameter could provide a convenient means of incorporating an earthquake's time and intensity characteristics.

(5) Single frequency laboratory tests can closely simulate the quasi-harmonic floor responses observed in multi-story structures. The three tests outlined in this paper are relatively easy to perform, provide a significant degree of operator control over test specimen response, permit an identification of dynamic properties of the test specimen and are suitable for incorporating into test specifications and requirements for both nuclear and non-nuclear applications.

ACKNOWLEDGEMENTS

The author would like to acknowledge the assistance provided by A. C. Heidebrecht and W. K. Tso in connection with this research.

REFERENCES

Aziz, T.S., and Duff, C.G. (1978) "Decoupling Criteria for Seismic Analysis of Nuclear Power Plant Systems", presented at ASME/CSME Pressure Vessels and Piping Conference, Montreal, Canada.

Bertero, V.V., and Kamil, H. (1975) "Nonlinear Seismic Design of Multistory Frames", Second Canadian Conference on Earthquake Engineering, McMaster University, Hamilton, Canada.

CSA (1977), Standard N289.4 "Testing Procedures for Seismic Qualification of CANDU Nuclear Power Plants", (draft), Sept. 1977.

Clough, R.W., and Benuska, K.L., (1967), "Nonlinear Earthquake Behaviour of Tall Buildings", ASCE. J. Eng. Mech. Div., Vol. 93, EM3, pp 129-146.

Dempsey, K.M., and Irvine, H.M., (1978), "A Note on the Numerical Evaluation of Duhamel's Integral", Earthquake Eng. and Struct. Dynamics, Vol. 6, pp 511-515.

Duff, C.G., and Heidebrecht, A.C., (1979), "Earthquake Fatigue Effects on CANDU Nuclear Power Plant Equipment", Third Canadian Conf. on Earthquake Engineering, Montreal, Canada.

Fisher, E.G., and Wolff, F.H. (1973), "Comparison of Fatigue Effects in Simulated and Actual Earthquakes", Experimental Mechanics, Vol. 12, No. 12, Dec. 1973, pp 531-538.

IEEE (1975), "IEEE Recommended Practices for Seismic Qualification of Class 1E Equipment for Nuclear Power Generating Stations", IEEE Standard 344-1975, The Institute of Electrical and Electronics Engineers, Inc.

Liu, S.C., Fagal, L.W., and Dougherty, M.R., (1977), "Earthquake-Induced In-Building Motion Criteria", ASCE, J. Struct. Div., Vol. 103, No. ST1, January 1977, pp 133-152.

NBCC (1977), "National Building Code of Canada", National Research Council of Canada, Ottawa.

UBC (1970), "Uniform Building Code", Vol. I, International Conference of Building Officials, Pasadena, California.

SECTION II

OFFSHORE

AN INTEGRATED PROCEDURE TO COMPUTE
WAVE LOADS ON HYBRID GRAVITY PLATFORMS

M.Berta, A.Blandino, A.Paruzzolo

TECNOMARE S.p.A., Venice

ABSTRACT

The present paper deals with the major aspects to be taken into account for a reliable wave loading analysis of hy brid structures.

The perturbation effect of the large body components on the surrounding field will be illustrated in terms of diffracted velocities and crest elevation. In addition the influence of this disturbance on the evaluation of the loads acting on the small tubular elements as well as on the large bodies themsel ves will be discussed.

Finally an integrated computer procedure which performs a non linear load analysis accounting for the above aspects will be presented.

INTRODUCTION

The offshore activities in the North Sea have led to the design of many offshore structures to which the usual two-dimensional analysis is no longer applicable.

The concrete gravity platform or the concrete oil storage tank may be mentioned as examples of these large volume three-dimen sional forms. For such structures neither the Morison-type e-quation nor the strip theory can produce any reliable result when evaluating hydrodynamic wave forces and moments.

Adequate numerical schemes based on three-dimensional sink-source technique for potential definition by the use of Green's functions have been provided in order to compute the wave loads on fixed or floating large volume structures, as well as the added mass and damping coefficients and the motions in six

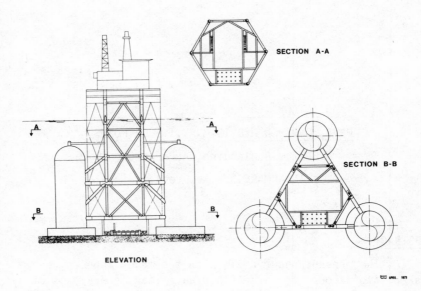

FIGURE No. 1

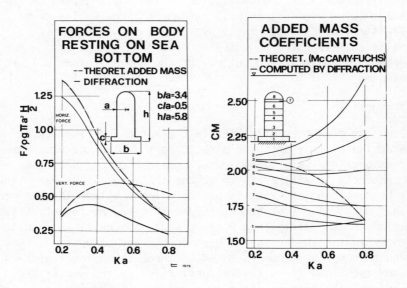

FIGURE No. 2   FIGURE No. 3

degrees of freedom.

However, a common premise of the above methods is that any term higher than the first order of magnitude in wave ampli tude can be neglected for the evaluation of the local loads. Such an approach cannot be applied when designing a hybrid structure mainly composed of large bodies connected by a fram ed structure.

As a matter of fact, for this type of structure the linear inertia loads acting on large bodies and the higher order term force and moments on the framed structure may be equally effective on the total loads and moments. In particular the fram ed structure loads could be significantly affected by the hydrodynamic field perturbation due to the large body presence. Then three different aspects are to be accounted for:
- the diffraction effect when evaluating the total force and moments on the large bodies,
- the field modification around these when evaluating loads on the framed structure,
- the non linear effects on the structure basically related to the use of a non linear wave theory and to the force integration up to the actual free surface elevation.

FORCE EVALUATION ON LARGE BODIES

Before discussing this argument in detail it may be helpful to be more specific about the nature of the structure concerned.

The term hybrid used above refers in this context to a range of structure types which contain at least one component very much larger in diameter or equivalent section dimension and sometimes rather complex in shape compared with the normal tu bular elements of steel jacket. Such structure types can be very diverse in configuration ranging from various combinations of jacket-type steel structure on monolithic base to con crete and steel gravity platform.

The following considerations refer to a particular type of gravity platform, Tecnomare patented, whose general configurat ion is illustrated in Fig.1. It consists mainly of a super-structure, a framed tower and three storage/stabilizing cylin ders and foundation bases.

The particular geometry of the structure's large dimen sion components calls for a wide utilization of the diffract-ion theory method to compute wave loads mainly because of the fluid-body interaction as well as the mutual interaction bet-

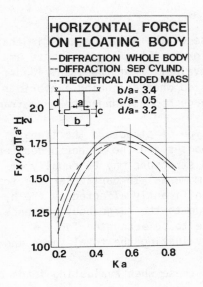

HORIZONTAL FORCE
ON FLOATING BODY
— DIFFRACTION WHOLE BODY
-- DIFFRACTION SEP. CYLIND.
--- THEORETICAL ADDED MASS
b/a= 3.4
c/a= 0.5
d/a= 3.2
d
a
c
b
Fx/ρgπa² H/2
2.0
1.75
1.50
1.25
1.00
0.2
0.4
0.6
0.8
Ka

FIGURE No. 4

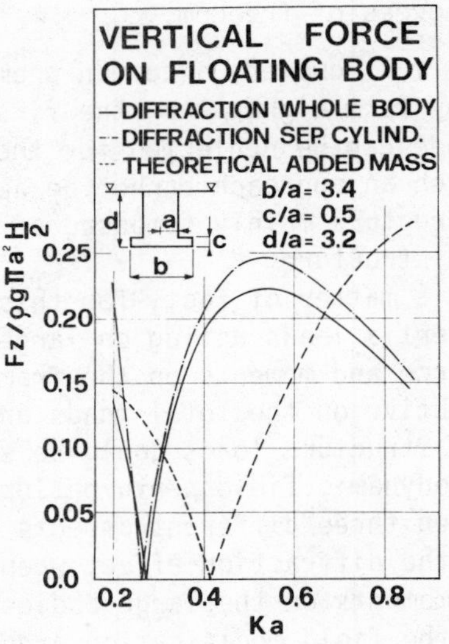

VERTICAL FORCE
ON FLOATING BODY
— DIFFRACTION WHOLE BODY
-- DIFFRACTION SEP. CYLIND.
--- THEORETICAL ADDED MASS
b/a= 3.4
c/a= 0.5
d/a= 3.2
d
a
c
b
Fz/ρgπa² H/2
0.25
0.20
0.15
0.10
0.05
0.0
0.2
0.4
0.6
0.8
Ka

FIGURE No. 5

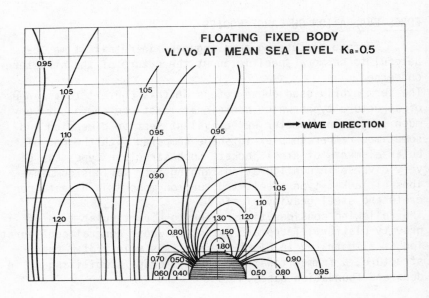

FLOATING FIXED BODY
VL/Vo AT MEAN SEA LEVEL Ka=0.5
WAVE DIRECTION
0.95
1.05
1.05
1.10
0.95
0.95
0.90
1.05
1.10
1.20
1.30
1.20
1.50
0.80
1.80
0.70
0.50
0.90
0.60
0.40
0.50
0.80
0.95

FIGURE No. 6

ween columns placed on top of each other.
These effects are generally great since the general dimensions
of the body are comparable to the usual wave lengths and the
column diameters rather similar.

The acting forces can be derived by integrating the actual pressure over the immersed surface of the body. For this, with reference to the linear diffraction theory, a total velocity potential is defined:

$$\emptyset = (\emptyset_o + \emptyset_D) \; e^{-i\omega t} \qquad (1)$$

where $\emptyset_o$ is the potential relevant to the incident wave with circular frequency ω, and $\emptyset_D$ is associated with the scattering effect.
Finally according to the linearized Bernoulli's equation the hydrodynamic pressure is obtained by

$$p = -\rho \frac{\partial \emptyset}{\partial t} \qquad (2)$$

Fig.2 shows the total horizontal and vertical loads on a single base and cylinder resting on the sea bed. The forces are shown in a non-dimensional form versus the characteristic parameter ka, 'k' being the wave number and 'a' the typical dimension of the upper column.
The figure compares the total load trends computed by the diffraction theory program with the inertia loads calculated by means of the usual Morison's equation and the theoretical added mass coefficient for cylindrical members, flat plates and spherical cupolas.

Fig.3 illustrates the actual added mass coefficients CM as function of ka for different sections along the body vertical axis. They are calculated as mean value over each wave period of the ratio between the horizontal diffracted force and the relevant Froude-Krylov force.
The CM values are compared with the corresponding theoretical added mass coefficient calculated by the McCamy and Fuchs theory for indefinite cylinder piercing the free surface.
Fig.2 clearly demonstrates that the approach through Morison's equation and the theoretical added mass coefficients is not applicable.
In fact, as Fig.3 shows, the computed CM values are a function of the elevation of the body sections considered, and each trend is rather different from the theoretical one.
The main reason for this is that the present body is not at

all amenable to an indefinite cylinder basically because of
the mutual influence of its differently shaped components.
As a result the comparison between the total horizontal load
trends as reported in Fig.2 proves that the approach by means
of Morison's formula and the McCamy and Fuchs CM leads to an
over-estimation of the global horizontal loads almost anywhere within the wave period range.
Such a difference is especially remarkable for the lowest values of ka where actual CM values are nearly all lower than
the theoretical one.
Indeed, as ka increases, the reduced contribution of the sections near the bottom and a sort of balance among the actual
CM values bring computation by the Morison's formula closer
to the diffraction values.
There are even more significant differences for the total ver
tical loads; as for the horizontal ones these occur because
the above mentioned interaction effects greatly modify the
mean values of the usual added mass coefficients adopted for
flat plates and spherical cupolas.

The interaction and diffraction effects become more
and more definitive for the body in floating conditions when
the foundation base is not so far from the mean sea level and
the cylinder pierces the free surface.
In Fig.4 the dashed line represents a horizontal force computation through the theoretical CM which disregardes
both the real fluid body interaction and the mutual interact
ion between its components.
The dotted line represents the horizontal load acting on the
whole body as obtained by a diffraction theory computation on
the separate columns. It does not take into account only the
mutual interaction between columns and so approximates much
better the actual computation (continuous line).
The differences between the same computation methods are even
more significant for the vertical forces reported in Fig.5.
Here the line branches which arise from the discontinuity
point correspond to a phase opposition for the same values of
amplitude force ratio. It can be noticed that this condition
presents a significant shift to greater ka when computing for
ces with Morison's equation.

THE FIELD MODIFICATION AROUND THE LARGE BODIES

So far the large bodies experience forces relevant to
any different interaction effect.
Furthermore the surrounding hydrodynamic field itself reflects

the results due to the presence of large bodies. Therefore, lo
cal loads on trusses lying within an area surrounding the large body, can be evaluated by means of Morison's equation provided that the modified field should be taken into account.
Then its kinematic characteristics must account for the previous procedure effect. For instance, according to Eq.(1), the
x-component wave velocity will be

$$v_x = \partial\emptyset/\partial x = Re\ \{(\partial\emptyset_o/\partial x + \partial\emptyset_D/\partial x)\ e^{-i\omega t}\} \tag{3}$$

where $\partial/\partial x$ denotes partial derivative in x-direction and Re
means the real part.

Fig.6 and the following show the modifications of the
hydrodynamic field due to a single base and cylinder in float
ing condition in terms of horizontal wave velocity at mean
sea level.
They illustrate the equal-value lines of the ratio between
the actual velocity amplitude and the horizontal incident wave velocity.
Fig.6 refers to the diffracted component in line with the incident wave considering the floating body fixed.
From the figure it can be observed that the main effect of the
incident wave is to make the body itself generate waves. Dif
fraction effects occur and the equal-value lines show a distortion in the wave approach direction yielding an asymmetric
al condition in the pattern in front of the body and behind
it. In addition the linear extension of the perturbation zone
has not an equal radial definition along different directions.
Such a distortion becomes much less apparent as the wave
length increases; as a matter of fact a very long wave with
respect to the body dimension can be considered as a uniform
current.
Moreover, in the upstream zone, it is possible to identify an
alternative varying regime of the amplitude ratio as a function of the distance from the body.
Along the incident wave direction the above values are demonstrated to be as function of the ka both for their amplitude
and for the space scale according to which they take place.
In general they are less pronounced and much further away
from each other according to greater wave lengths.

Fig.7 illustrates the diffracted velocity component
transverse with respect to the undisturbed wave heading, in
terms of amplitude ratio over the horizontal incident velocity.

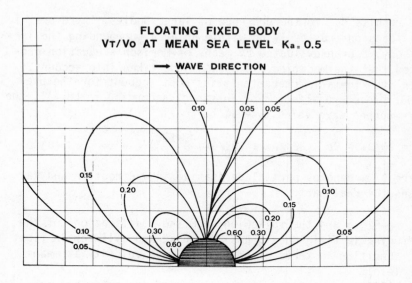

FIGURE No. 7

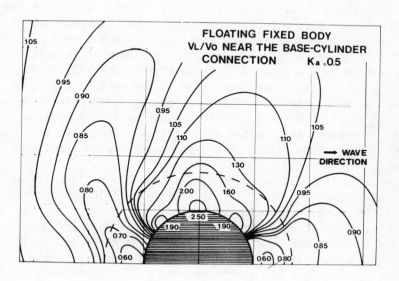

FIGURE No. 8

Here a phenomenon similar to that experienced by a body in a
uniform current can be noted: two separate zones, opposite in
direction, arise upstream and downstream. But as for the above longitudinal diffracted wave velocity, the transversal velocity isoline pattern also shows an asymmetrical condition
with a zero line which is distorted in the wave approaching
direction. The same considerations as in the previous case
can be applied.

According to theoretical considerations, for those ver
tical sections more similar to an indefinite cylinder, the e-
qual-value lines trend is demonstrated to be nearly the same
along the body vertical axis even if the perturbation effect
concerns a decreasing hydrodynamic field.
However this is not the case when a significant variation of
the body geometry occurs.

Fig.8 shows again the amplitude ratio between longitudinal diffracted velocity over incident wave velocity but in
this case near the connection of the two body columns.
The figure clearly demonstrates how the presence of the base
increases the longitudinal velocities of the diffracted waves.
The transversal velocities experience similar effects. This is
mainly due to the base interaction with the vertical component
of the incoming wave velocity.
In particular on the base top this component must satisfy the
boundary condition and so a new velocity field distribution
arises. Outside the base extension this effect is no longer
present and the real isoline pattern is more like the previous
one (Fig.6).

All the above considerations are related to a fixed bo
dy. If a body motion occurs a significant modification on the
wave velocity pattern will take place. Figs 9 and 10 illustra
te this effect taking into account a reliable body motion.
In comparison with Figs 6 and 7, the longitudinal and transversal velocity isolines show lower values and a much less ex
tended zone of perturbation.
The main reason for this is the contribution of the body motion generated waves to the total diffracted field. In terms
of incident field energy distribution, it means that a quanti
ty of the total is devoted to make the body move rather than
to perturbate the surrounding area.
A general conclusion from this is eventually that the actual
body motions considerably affect the total diffracted field
ranging from a maximum perturbation when the body is fixed to

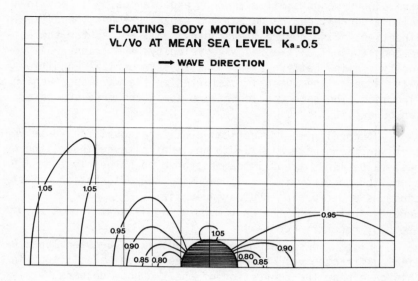

FIGURE No. 9

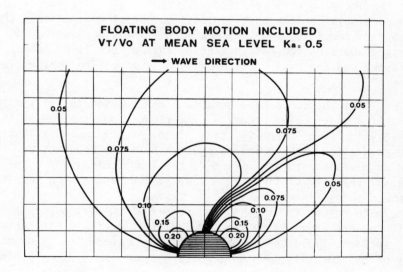

FIGURE No. 10

a more negligible scattering effect as the body is freely
floating.
Anyway, in different directions around the cylinder, it will
be possible to determine the extension of the disturbed zone
as function of ka.
Such a determination is a useful and reliable item for the eva
luation of possible interaction effects among the large bodies
as well as for an estimate of their influence on the framed
structure design.

NON LINEARITY OF THE WAVES AND THE DIFFRACTION EFFECT ON WAVE ELEVATION

Some more aspects are to be taken into account for a
suitable wave force analysis of hybrid structures such as tho
se dealt with in the present paper.
Non linear wave characteristics, particularly near the free
surface, are possibly very significant when designing them.
These aspects are mainly related to the jacket-type structure
although during towing the cylinders also pierce the surface.
In the latter case, however, it is a well established method
to compute the prevailing inertia loads on big bodies in terms
of flow potential theory, linear in wave elevation, over the
wetted body surface cut off at mean water level.
On the contrary, a higher order theory is needed for local
loads on drag-dominated members. The forces on the jacket-type
structure are usually predicted using the Morison's formula
with wave velocity and acceleration given by higher order Sto
kes or stream function theory.
It has been demonstrated that linear and fifth order Stokes
velocity profile differ by 20 percent or more at the free sur
face with relevant differences in drag of over 30 percent.
Moreover, it is necessary to integrate the forces up to the
actual free surface rather than to mean water level. If the
wave is distorted by the diffraction, the error on the total
load may be increased.
This effect may not be disregarded for the present type of hy
brid structure, where a framed structure pierces the surface
very close to the submerged cylinders. As a result, taking
the actual diffracted surface elevation rather than mean immersed depth has a significant effect on drag.
So far the diffraction theory computer program can provide the
designer with the linearized diffracted wave elevation at any
point.

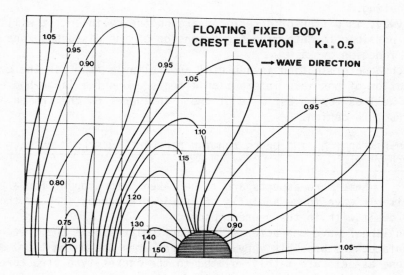

FIGURE No. 11

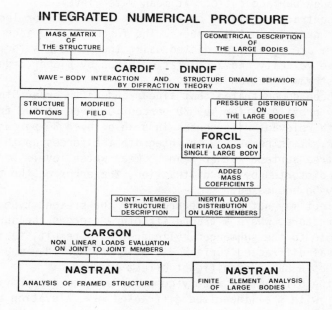

FIGURE No. 12

Fig.11 shows the equal value lines of the ratio of the diffracted pressure over the incident dynamic pressure amplitude which supplies the variation of the disturbed crest elevation, at any point.
This wave pattern is similar to that of the velocity ratio isolines, the maxima and minima being 180 degrees out of phase with respect to those.
The same considerations as in the case of Fig.6 may be made and in particular a reduction of the perturbation effect is demonstrated to occur when the body moves.

THE INTEGRATED NUMERICAL PROCEDURE TO PERFORM THE WAVE LOADING ANALYSIS

As the present type of platform consists both of small and large cylindrical shapes it was necessary to develop an integrated computer procedure which may take into account the various requirements for a complete wave force analysis. It is based on a number of Tecnomare especially developed computer programs; they are actually integrated in order to supply the input data for the structural analysis computer program.

Fig.12 illustrates the flow chart of the integrated computer procedure.
The first calculation step is mainly based on a diffraction theory program. It calculates the wave inertia loads on the large members, as well as the disturbance on the surrounding hydrodynamic field including the structure motion effects.

Owing to the necessity of taking into account the major non linearities of waves and the actual surface elevation, a linear approach is demonstrated to be no longer applicable. In particular this is the case for the global and local load and moment evaluation on the jacket type structure, as well as for the higher-order forces on the large cylindrical bodies. Then one has to rely on a higher order-wave theory and to account for the modification of the hydrodynamic field in terms of wave velocities and accelerations and actual wave elevation.

Thus the problem is to integrate a linear wave theory and a non linear one. The first to compute wave inertia load on large bodies and the relevant scattering effect, the second to determine drag and inertia forces on tubular frames as well as the big bodies drag forces which in some cases cannot be disregarded.
The non linear theory adopted to this purpose is the well-known and recommended Stokes V°.

From the point of view of wave energy, it can be easily recogn
ized that a large amount of the total is related to the wave
first harmonic. This may suggest that, when evaluating the
perturbation of the hydrodynamic field, it is adequate to account only for the modification of the first order wave profi
le, velocity and acceleration.
As a matter of fact such an approach is fully justified where
the monolithic type structures are concerned. This type of
structures is generally amenable to a single and well establi
shed characteristic dimension and thus interferes with the in
cident field according to it.
On the contrary the gravity platform type presented here cannot be described by a single characteristic dimension but
leads to the consideration of several typical dimensions such
as the diameters and heights of the two columns placed on top
of each other, and the interaxis between the storage/stabili-
zing cylinders. Then, each one of the above characteristic di
mensions may present its maximum interaction effect according
to a higher order wave frequency.
As a result the interaction between structure and incident
field cannot be regarded as function of only one wave frequen
cy, but at least the first and the second components of the
Stokes V° wave are to be considered when accounting for the
scattered field modification. It is intended that the higher-
order terms can be regarded as unmodified due to their low
energy.
This is the reason for which the integrated procedure here il
lustrated accounts for the modification of the total wave pro
file, velocity and acceleration computing the actual perturba
tion of the Stokes V° first two harmonics.
This is obtained by running the diffraction computer program
twice in order to evaluate the structure scattering effect re
lated to the fundamental and second order frequency of the in
coming wave. The total modified field around the large bodies
is then derived by the superposition of these effects to the
incident potential. The modified hydrodynamic field is then
totally described by interpolation between a regular mesh of
off-body points for which dynamic velocities, accelerations
and pressures have been evaluated.

     The usual simplified theory based on the Morison equa-
tion is then applied to calculate all wave loads acting on
the whole structure except the inertia wave forces on large
bodies already accounted for by the diffraction theory program.
This is the second step of the integrated procedure.

The drag forces are calculated according to the usual techniques taking into account the local modified velocity and the shape coefficient as function of the Reynolds number.

The inertia forces for tubular elements and flat plates of framed structure are calculated regarding the local modified acceleration through theoretical added mass coefficients.

In addition the computer program can determine the current effect on the drag forces taking into account the non-linear effect of the wave and current velocity superposition; as well as the wind forces acting on the whole structure and the local load due to the member's weight and to the hydrostatic and hydrodynamic pressure.

In a time domain analysis over a wave period all the above forces are actually integrated up to or from the real surface elevation comprehensive of the large body disturbance.

This type of analysis provides directly the member loads input for structural analysis.

AN EXAMPLE OF APPLICATION

The application range of the procedure mentioned covers all the offshore design cases in which a significant perturbation of the hydrodynamic field arises.

This is the case of an integrated tripod-type storage platform as that presented in Fig.1.

As a matter of fact the perturbation zone around its large storage cylinders is so extended as to influence a significant part of the framed structure, and even the cylinders themselves.

The following Table presents the principal results obtained for a single base and cylinder when considered as part of the whole triangular structure.

Three different wave headings are considered; the bow quartering approach is along the line connecting the fore cylinder to an aft one (see Fig.1 , section B-B). The data, relevant to the platform at the towing draft, are expressed in a non-dimensional form for a 0.5 ka value. They are to be compared with the corresponding value produced in Fig.4 for the global horizontal load of the single body (diffraction theory result; continuous line).

TABLE 1

	Fore Cylinder		Left Aft Cylinder		Right Aft Cylinder	
	F_o	$\theta(°)$	F_o	$\theta(°)$	F_o	$\theta(°)$
HEAD SEA in line	1.86	1.0	1.58	168.0	1.58	168.0
transverse	--	--	0.23	119.5	0.23	- 60.5
FOLLOWING SEA in line	1.63	51.5	1.72	-127.5	1.72	-127.5
transverse	--	--	0.12	-145.5	0.12	34.5
BOW QUART.SEA in line	1.93	-177.5	1.75	-143.5	1.73	7.5
transverse	0.36	142.0	0.18	64.0	0.21	-166.5

The above results fully demonstrate the influence of the large bodies on each other.
The Following Sea is demonstrated to be the less significant for the mutual interaction; the global horizontal loads on the three cylinders are rather similar among themselves and to the corresponding value for the single body.
For the Head Sea,as well as for the Bow Quartering Sea, the influence becomes more evident; in particular for the Bow Quartering Sea a total transversal component arises due to the non-symmetrical perturbation with respect to the wave approach direction.

On the contrary, in the present case the field perturbation has been demonstrated to influence the framed structure design in a less significant manner. In fact, the local environmental load on the small members are much less important than those induced by the big bodies and by the deck payload. As a result, even variations up to 100 percent of the local environmental loads obtained by direct calculations which take into account the wave kinematic modifications have scarce influence on the framed structure design.

CONCLUSIONS

A procedure has been presented to improve the analysis of the different wave loading conditions of hybrid structures composed of large members and tubular frames.
The main reasons why this type of structure is not amenable to the usual methods of computing wave loads have been dealt with. Numerical calculations of wave loads on a particularly shaped body as well as of its diffraction effect on the surrounding fluid have been reported in order to prove the interaction between the various elements of a composite large member and the interaction of large members between each other.
The necessity of a non-linear wave theory together with the best knowledge of modified field for evaluating wave loads leads to an integration between computing methods based on dif ferent theoretical approaches.
On this basis the integrated procedure presented has proved to be a powerful and adequate tool in designing large volume struc tures and in particular the gravity platform presented here.

NOMENCLATURE

a	typical body dimension
CM	added mass coefficients
F	amplitude of the force
F_o	$F/\rho g \pi a^2 \dfrac{H}{2}$
g	gravitational constant
H	wave height
$k = \dfrac{2\pi}{\lambda}$	wave number
v_x	x-component wave velocity
V_o	undisturbed wave velocity amplitude
V_L	diffracted wave velocity amplitude in line with wave approach
V_T	diffracted wave velocity amplitude transverse the wave approach
θ	phase
λ	wave length
ρ	fluid density
$\emptyset(w,y,z,t)$	total linear velocity potential
$\emptyset_o$	complex potential amplitude of incident wave
$\emptyset_D$	complex amplitude of diffraction potential

ω   circular frequency of wave

∂|∂x   partial derivative in x-parameter

∂|∂t   partial derivative in time

REFERENCES

1. Morison, J.R., O'Brien, M.P., Johnson, J.W. and Schaff,S.A. (1950) "The force exerted by surface waves on piles" Petro leum Trans., 189, TP2846, p.149-154

2. McCamy, R.C. and Fuchs, R.A. (1954) "Wave forces on a pi- le: a diffraction theory", Tech. Memo. 69, U.S.Army Corps of Engineers Beach Erosion Board, Washington, D.C.

3. Skjelbreia, L. and Hendricksen, J. (1960) "Fifth-order gra vity wave theory", Proc. Coastal Engr. Conf., p. 184-196

4. Wehausen, J.V. and Laitone, E.V. "Surface Waves", Encyclo- pedia of Physics, Vol.9, Springer-Verlag, Berlin, p.446-778

5. Monacella, V.J. (1966) "The disturbance due to a slender ship oscillating in waves in a fluid of finite depth", Journal of Ship Research, 10, No.4, p.242-252

6. Faltinsen, O.M. and Michelsen, F.C. (1974) "Motion of lar ge structures in waves at zero Froude number", The Dyna- mics of Marine Vehicles and Structures in Waves, Paper no.11, The Institution of Mechanical Engineers

7. Garrison, C.J., Rao, V.S. and Snider, R.H. (1970) "Wave in teraction with large submerged objects", Proc. Offshore Technology Conference, Paper OTC 1278

8. Garrison, C.J. and Rao, V.S. (1971) "Interaction of waves with submerged objects", J.Waterways, Harbors & Coastal Engr. Div. Proc., ASCE 97, p.259-277, No. WW2

9. Garrison, C.J. (Jan.1974) "Hydrodynamics of large objects in the sea - Part I: Hydrodynamic analysis", Journal of Hydronautics, AIAA, Vol.8, No.1, p.5-12

10. Garret, C.J.R. "Wave forces on a circular dock", Journal of Fluid Mechanics, Vol.46, Pt.1, p.129-139

11. Løken, A.E. and Olsen, O.A. (1976) "Diffraction theory and statistical methods to predict wave induced motions and loads for large structures", Paper OTC 2502

12. Raman, H. and Venkalanarasaiah, P. (1976) "Forces due to nonlinear waves on vertical cylinders", J. Waterways, Har

bors & Coastal Engr. Div. Proc., ASCE, Vol.102, No. WW3

13. Garrison, C.J. (1976) "Consistent second-order diffraction theory", presented at the Fifteenth Coastal Engineering Conference, Honolulu, Hawaii

14. Lee, C.M. (1968) "The second-order theory of heaving cylinders in a free surface", Journal of Ship Research, 313-327

15. Potash, R.L. (1970) "Second-order theory of oscillatory cylinders", Dissertation, University of California, Berkeley Ca. 157

16. Apelt, C.J. and Macknight, A. (1976) "Wave action on large offshore structures", Proceedings Fifteenth Coastal Engineering Conference, Honolulu, Hawaii

17. Garrison, C.J.,Field, J.B. and May, M.D. (1976) "Drag and inertia coefficients in oscillatory flow about cylinders", in Press, Journal of Waterways, Port & Coastal and Ocean Division, ASCE, 103, No. WW3

18. Hess, J.L. and Smith, O.M.A. (1962) "Calculation of non-lifting potential flow about arbitrary three-dimensional bodies", Rept. No.E.S.40622 (Douglas Aircraft Division, Long Beach, Ca.) also Journal of Ship Research, 1964, 8

19. Garrison, C.J., Tørum, A., Iversen, C., Leivseth, S. and Ebbesmeyer, C.C. (1974) "Wave forces on large volume structures - A comparison between theory and model tests", Paper OTC 2137, Offshore Technology Conference, Houston, Texas

WAVE FORCES ON ELLIPTICAL CYLINDERS BY FINITE ELEMENT METHOD

Suphat Vongvissessomjai and Md. Hanif

Asian Institute of Technology, Bangkok, Thailand

ABSTRACT

This paper presents the application of a diffraction theory
and a finite element technique to analyse the wave force on a
two dimensional object of elliptic section locating both at
the free surface and the bottom. The convergence tests are
successfully performed on both cases.
    Besides the forces on the objects, all the other relevant
parameters, pressure distribution around the object, inertia
coefficient, reflection and transmission coefficients are
also computed. It is found that the horizontal wave force on
the object has unique relation with the reflection coefficient. Free surface and bottom have prominent effects in
increasing the forces on the objects.

INTRODUCTION

Much work has been carried out in recent years on wave forces
exerted on submerged structures by Vongvisessomjai (1973),
Chakrabrai and Naftzger(1974), Yamamoto et al (1974), Nath
and Yamamoto (1974), Tsuchiya and Yamaguchi (1974), Sarpkaya
(1975), Garrison et al (1975), and Vongvisessomjai and Silvester (1976). Work prior to 1974 has been summarised by Hogben
(1974). The bodies studied have included cylinders, spheres,
blocks and other symmetrical shapes. The more complex units
being employed as gravity structures in the oil exploration
industry, McPhee (1975), are being equated to these simple
forms and also tested individually in flumes, Hogben and
Standing (1974). There is a dire need to further this work
and rationalize the whole procedure, Hogben (1974).
    The original achievements in this topic were made by
Morison et al (1950) whose equation generally accepted is a
valid approximation for the calculation of wave forces acting
on objects being considered small when compared with the
incident wave length. However, as the object size is large,

scattering occurs and the incident wave is substantially
disturbed to invalidate the Morison's equation. In addition,
the effect of the free surface and/or the bottom disregarded
in Morison's equation is of considerable importance.

THEORETICAL CONSIDERATION

Boundary Value Problem
The general problem of two dimensional scattering of water
waves in an ocean is formulated under the following usual
assumptions:

    (1) potential flow;
    (2) small wave amplitude;
    (3) simple harmonic incident wave;
    (4) constant depth away from the floating or the sub-
        merged bodies; and
    (5) rigid and perfectly reflective boundaries

The diffraction of water waves due to the obstruction of
a floating and submerged horizontal elliptic objects is
governed by Laplace's equation with a mixed type free surface
boundary condition, a homogeneous Neumann condition on the
objects and the bed, and a radiation condition at infinity.
Assuming sinusoidal motion, the time variation can be suppressed and the formulation becomes

Laplace's equation $\quad \nabla^2 \Phi \quad = 0$ $\hfill (1)$

Free surface B.C. $\quad \partial \Phi_d / \partial y - k_o \Phi_d = 0 \quad$ at $y = 0$ $\quad (2)$

Body B.C. $\quad \partial \Phi_d / \partial n = -\partial \Phi_i / \partial n = n \cdot V$ $\hfill (3)$

Bed B.C. $\quad \partial \Phi_d / \partial n = 0 \quad\quad\quad$ at $y = -d$ $\quad (4)$

Radiation Condition $(\partial \Phi_d / \partial x \mp ik\Phi_d) = 0 \quad$ at $x \to \pm \infty$ $\quad (5)$

where Φ_d is the diffracted potential
    Φ_i is the potential of incident wave with unit amplitude

$$= \frac{g \cosh k(y+d) \exp(ikx)}{\sigma \cosh kd} \hfill (6)$$

σ is the circular frequency
k and k_o are wave numbers defined by the dispersion
    relation

$$\sigma^2 / g = k_o = k \tanh kd \hfill (7)$$

Once the diffracted wave potential is solved, the wave
profile is then given by

$$\eta = -\frac{1}{g} \frac{\partial \Phi}{\partial t} \quad\quad\quad at\ y = 0 \hfill (8)$$

where Φ is the total potential

$$= \Phi_i + \Phi_d \hfill (9)$$

Finite Element Method

The finite element method has been successfully applied to
many water wave problems. The procedures used to solve this
wave diffraction problem follow the computer program for
potential flows with free surface which was developed by Bai
(1972). Only a brief description will be given here.

The potential Φ_d can be decomposed into cosine and sine
mode potentials

$$\Phi_d(x,y,t) = \phi^C(x,y)\cos\sigma t + \phi^S(x,y)\sin\sigma t \qquad (10)$$

The radiation condition (Equation 5) can be modified to

$$\frac{\partial\phi^C}{\partial n} = - k\phi^S \qquad (10a)$$

$$\frac{\partial\phi^S}{\partial n} = k\phi^C \qquad (10b)$$

The radiation boundary is truncated to a distance of 4 times
of the water depth from the elliptic object where local disturbance due to the elliptic object has decayed exponentially
to a negligible magnitude.

Consider a general field problem which is governed by
Laplace's equation

$$\nabla^2\phi = 0 \qquad \text{in } \Omega \text{ (domain)} \qquad (11)$$

and a general mixed type boundary condition

$$\phi_n + \alpha\phi + \beta = 0 \qquad \text{on } \partial\Omega \text{ (boundary)} \qquad (12)$$

where $\alpha(x,y)$ and $\beta(x,y)$ are known functions.
From the variational principle, the solution of the problem
is a function ϕ that minimizes a functional

$$F(\phi) = \iint_\Omega \tfrac{1}{2}|\nabla\phi|^2 \, dxdy + \int_{\partial\Omega} (\tfrac{1}{2}\alpha\phi^2 + \beta\phi)ds \qquad (13)$$

Introducing the decomposition of the potential (Equation 10),
the cosine and sine mode functionals are given as

$$F^C(\phi^C) = \iint_\Omega \tfrac{1}{2}|\nabla\phi^C|^2 \, dxdy - \int_F \tfrac{1}{2}\sigma^2/g(\phi^C)^2 \, dx$$

$$- \int_B (n.v^C)\phi^C \, ds + \int_R k\phi^C\phi^S \, dy \qquad (14a)$$

$$F^S(\phi^S) = \iint_\Omega \tfrac{1}{2}|\nabla\phi^S|^2 \, dxdy - \int_F \tfrac{1}{2}\sigma^2/g(\phi^S)^2 \, dx$$

$$- \int_B (n.v^S)\phi^S \, ds - \int_R k\phi^C\phi^S \, dy \qquad (14b)$$

The fluid domain is subdivided into 8-node quadrilateral
elements within which the function ϕ is supposed to be continuous and bounded. A set of interpolation functions associated with each element is introduced such that

$$\phi = [N]\{\bar{\phi}\} \qquad (15)$$

222

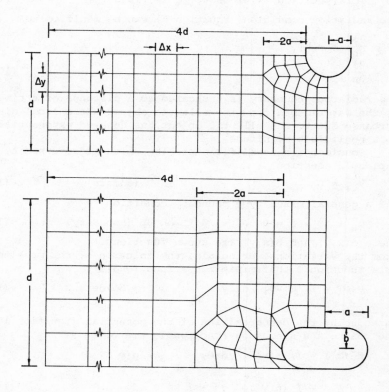

Figure 1   Finite Element Discretization for Objects at
Surface and Bottom

where $\{\bar{\phi}\}$ is the nodal values of ϕ

[N] is the interpolation functions

Substituting Equation 15 into the two functionals (Equations 14a and 14b) which are then minimized to give a system of linear equations

$$[A]\{\bar{\phi}\} = [B] \tag{16}$$

Because the coupled radiation condition ϕ^C and ϕ^S have to be arranged in the matrix $\{\bar{\phi}\}$ as two degrees of freedom at each node, the coefficient matrix [A] is then not symmetric but still banded.

In order to reduce the bandwidth, nodal points should be numbered such that the difference between the highest and lowest nodal numbers are kept as small as possible in an element. In this problem, the numbering can be done consecutively along vertical sections starting from the seaward side to rear end of the object. Numbering should be also in the same direction throughout the domain, either clockwise or counter-clockwise within an element, so that only the one line element would contribute to the surface integral term in the functional.

The area integral was approximated numerically by Gaussian Quadrature while the line integrals were determined exactly.

Convergence Test

The finite element solutions are then tested with known solutions obtained by Dean and Ursell (1959) and Garrison and Chow (1972), respectively for objects at the free surface and the bottom. The final discretizations are shown in Figure 1 and the results shown in Figures 2 and 3 agree perfectly well when the numbers of the element are 258 and 266 for surface and bottom objects respectively and the element length (ΔX) is less than or equal to the one-tenth of the incident wave length (i.e. $\Delta X/L \leq 0.1$). Similar criteria was also suggested by Bai (1972) but Patarapanich (1978) and Sakai (1974) found that $X/L \leq 0.05$ for four node quadrilateral element and triangular element.

OBJECT AT SURFACE

All the results of the analysis are plotted as functions of the relative size ka in the abscissa as usually done by others. However, it is found from this study that the relative depth kd is more important than the relative size ka. Therefore the values of ka should be interpreted as kd that kd = 6ka for d/a = 6.

The model was applied to compute the horizontal force acting on the three elliptic objects as shown in Figure 4. In the range of intermediate depth ($\pi/10 \leq kd \leq \pi$)the relative sizes of the object ka cover the ranges of $\pi/60$ to $\pi/6$ for d/a = 6. The normalized maximum horizontal force increase

224

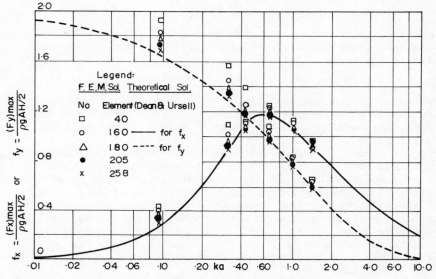

Figure 2 Convergence Test for Force Components on Surface Object

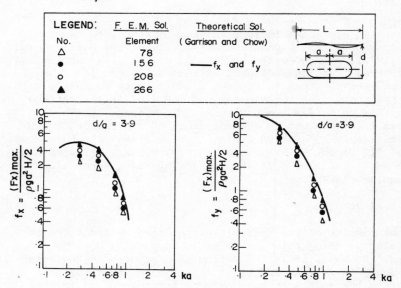

Figure 3 Convergence Test for Force Components on Bottom Object

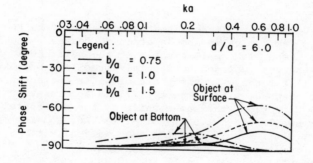

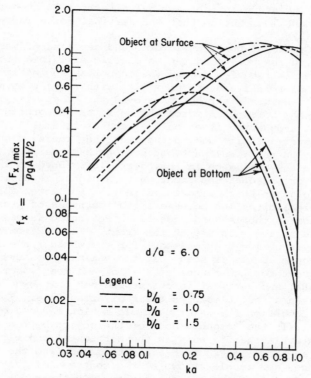

Figure 4 Normalised Max. Horizontal force, f_x v.s. ka

for relative depth kd increases at some value of kd the force attains its maximum value and later on decreases. The more streamline shape with smaller ratio of the half beams of the object in the vertical and horizontal directions b/a experience smaller force. As the phase of water surface profile is 90^0 out of phase with respect to fluid acceleration in horizontal direction, the phase shift of the horizontal force is 90^0 for small relative depth of water kd, and increases to maximum, similar to f_x, at the deep water limit.

The normalized maximum vertical force in Figure 5 is maximum at small relative depth kd and decreases as kd decreases. The water surface profile and fluid acceleration in the vertical direction is in phase, therefore, the phase shift is zero for small kd and increases as kd increases. The more streamline shape in the vertical direction is opposite to the horizontal direction that the vertical force is smaller for higher b/a.

The increase of horizontal force and decrease of vertical force with respect to kd can be clearly explained from the pressure distribution around the object shown in Figure 6a. For small relative depth of water with respect to wave length, the pressure distribution around the object is almost uniform and the horizontal force which is the integration of the pressure around the lower half of object is small. The vertical force is finite due to the fact that there is only force acting on the lower part but not on the upper one resulting in the unbalance in this direction. When kd increases the pressure at the front part is higher resulting in higher force in the horizontal direction but lower force in the vertical direction because of the decrease of pressure at the bottom on the leeward side of object being faster than the increase of the pressure at the front. The maximum pressure at the front ($\theta= 0$ = constant) for the whole range is shown in Figure 7. It is seen that the maximum pressures are almost constant for the ranges of shallow and deep waters; the horizontal and vertical forces depend not only upon the maximum pressures but also upon their distributions. For deep water condition the water particle motion induced by the wave is small in the neighbourhood of the object bottom.

The intertia coefficients in the horizontal and vertical directions as shown in Figure 8 are determined from the respective forces and accelerations at the centres of the objects computed from linear wave theory. In the range of shallow and intermediate depths, C_{Mx} is almost constant and decreases as it approaches the deep water condition. The constant $C_{Mx} \doteqdot 1 + b/a$ = potential value. C_{My} is high for small kd due to the unbalance force explained above.

Coefficient of reflection is small when kd is small (Figure 9), since most of the wave energy flux can penetrate below the object but when kd is large most of the energy flux will be reflected, resulting in C_R increases. C_R varies in

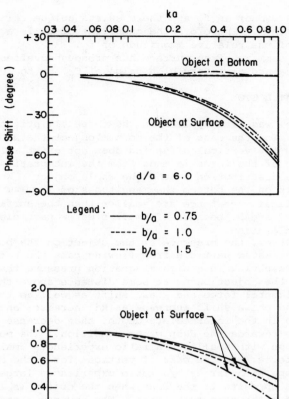

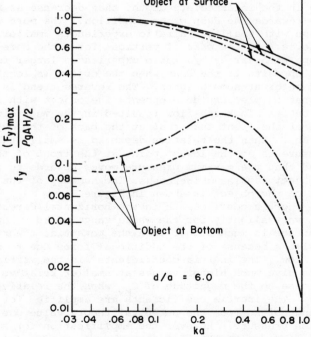

Figure 5 Normalized Max. Vertical Force, f_y v.s. ka

the same manner as f_x and there exists unique relationship between them. Therefore the magnitude of C_R is a good indication of the relative magnitude of f_x.

In summary, free surface has pronounced effect on vertical force and give very high value of C_R (up to 100%).

OBJECT AT BOTTOM

The model was also applied to the three elliptic objects at bottom. For the case of the bottom object the unbalance force in the vertical direction does not exist. The determination of the force is made from the integration of the pressure distribution around the whole object. When the object is on the bottom the magnitudes of maximum pressure and vertical wave force are smaller than the surface object, Figures 7 and 5, because of smaller water particle velocity and acceleration.

However, the presence of the object on the bottom prevents the water particle from flowing pass the bottom of the object resulting in a high stagnation pressure than in the case of the object being at some distance above the bottom. The horizontal force and phase shift as well as the vertical force and phase shift increase as kd increase; and reach maximums in the intermediate depth, then decrease as kd increase towards the deep water condition. The more streamline shape with smaller b/a ratio experiences smaller horizontal force. In the case of vertical force the more streamline shape with larger b/a ratio experiences larger force which is opposite to the case when the object is located on the surface or at some heights. The reverse trend is due to the stagnation pressure development; the object with larger b/a can better block the flow resulted in the much higher pressure at the front than that at the back of the object (Figure 6b). When the relative depth kd is small, the maximum pressure is at the lowest point at the front of the object ($\theta=0$) and that the maximum pressure moves upwards to the top of the object ($\theta=\pi$) as kd increases. The summary of the maximum pressure on the object is shown in Figure 7.

The inertia coefficients in the horizontal direction, Figure 8, vary slightly for the whole range of kd. However, the magnitude is much higher than the potential theory value of $C_{Mx} = 1+b/a$ because of the additional force due to stagnation of flow. The inertia coefficients in the vertical direction give much higher values at smaller relative depth and decrease to the magnitude of C_{Mx} when the relative depth increase. The inertia coefficients are amplified for the whole range of kd from the potential theory value due to the stagnation pressure. However the amplification for small kd is greater due to the fact that the stagnation pressure is of some rather finite magnitude for the whole range of kd while the hydrodynamic force without stagnation pressure is smaller for smaller kd. The more streamline shapes with

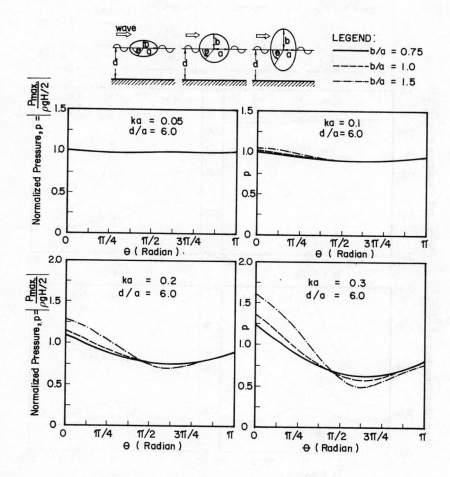

Figure 6 a   Normalized Pressure Distribution on the
Object at the Free Surface

230

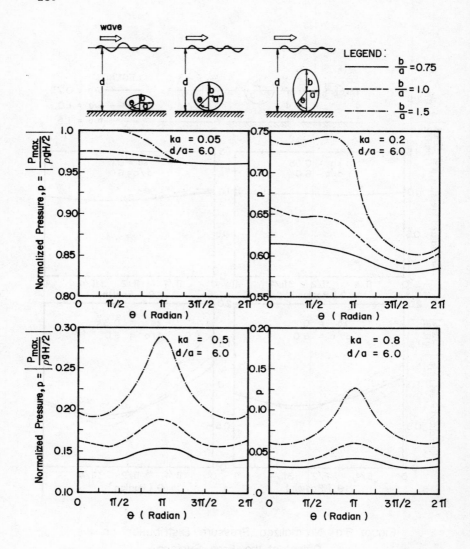

Figure 6 b. Normalized Pressure Distribution around the Object at the Bottom

larger b/a experience smaller values of the inertia coefficient in the vertical direction.

The coefficient of reflection, Figure 9, varies in the same manner as f_x as well as f_y in this case, and that it attains the maximum value in the intermediate depth zone. The magnitude is smaller than the case of surface object.

The existence of bottom, when object is located on it, yields stagnation pressure acting at the lower part of the object. The magnitude of stagnation pressure depends upon the degree of blocking. In the case of small blocking like semi-circular cylinder or rectangular block located on the bottom, Vongvisessomjai and Silvester (1976) showed that inertia coefficients are of the same orders of magnitude as that of the potential theory values. In the case of large blocking like elliptic and circular cylinders, stagnation effect is pronounced for both vertical and horizontal directions.

CONCLUSION

1. The applications of finite element method to the wave force analysis on two dimensional objects of elliptic sections locating at free surface and bottom are first tested by comparing the results with the existing theoretical solutions.
2. Unique relationship exists between simple parameter of the coefficient of reflection and the horizontal wave force on object and that such parameter, can be used to assess the order of magnitude of the force.
3. The pressure distribution around the object is the best parameter in explaining the order of magnitude of wave force on an object.
4. The free surface has dominant effect only on the determination of the unbalance vertical wave force on the object by the integration of pressure distribution around the lower half of the object.
5. The bottom has pronounced effect on the determination of both vertical and horizontal wave forces on an object due to the development of the stagnation pressure underneath the object.
6. The relative water depth is found to be the more important parameter than the relative size in determining the wave forces on objects, coefficient of reflection, effects of surface and bottom, etc.

REFERENCES

Bai, K.J. (1972) A Variational Method in Potential Flows with a Free Surface. Report No. NA72-2, College of Engineering, University of California, Berkeley.

Chakrabrati, S.K. and Naftzger, R.A.(1974), Non-Linear Wave Forces on Halfcylinder and Hemisphere. Proc. ASCE, 100, WW3: 189-204.

232

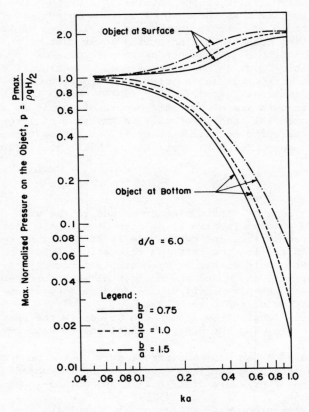

Figure. 7   Max. Normalized Pressure Distribution on the Object

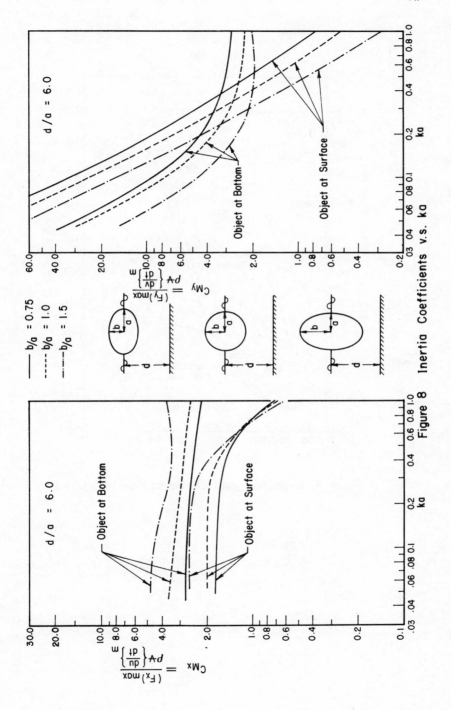

Figure 8 Inertia Coefficients v.s. ka

234

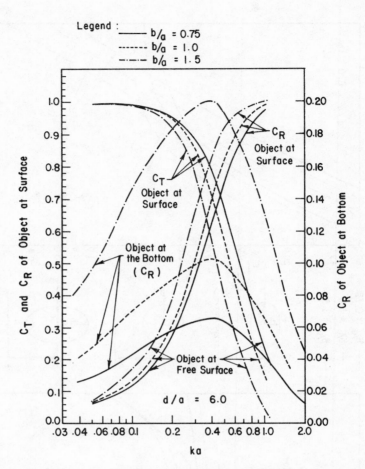

Figure 9   Reflection and Transmission Coefficient v.s. ka

Dean, R.G. and Ursell, F. (1959) Interaction of a Fixed Semi-immerged Circular Cylinder with a Train of Surface Waves. M.I.T. Hyd. Lab., Tech. Report 37.

Garrison, C.J. and Chow, P.Y. (1972) Wave Forces on Submerged Bodies. Proc. ASCE, 98, WW3: 375-392.

Garrison, C.J., Gehrman, F.H. and Perkinson, B.T. (1975) Wave Forces on Bottom-mounted Large Diameter Cylinder. Proc. ASCE, 101, WW4: 343-356.

Hogben, N. (1974) Fluid Loading of Offshore Structures, A State of the Art Appraisal: Wave Load. J. Roy Inst. Naval Arch.

Hogben, N. and Standing, R.G. (1974) Wave Loads on Large Bodies. Proc. Inter. Symp. Dyn. Mar. Verhicles and Struct., 258-277.

Morison, J.R., O'Brien, M.P., Johnson, J.W. and Schaaf, S.A. (1950) The Wave Force Exerted by Surface Waves on Piles. Petr. Trans., Amer. Inst. Min. Metal. Eng., 189,: 149-154

McPhee, W.S. (1975) Drilling and Production Platforms for the Oil Industry. Proc. I.C.E. Conf. Offshore Structures,: 189-196.

Nath, J.H. and Yamamoto, T. (1974) Force from Fluid Flow around Objects, Proc.14th Conf. Coastal Eng., 3, : 1808-1827.

Patarapanich, M. (1978) Wave Reflection from a Fixed Horizontal Plate. Proc. Conf. Water Resources Eng., Asian Institute of Technology, Bangkok, I,: 427-445.

Sarpkaya, T. (1975) Forces on Cylinders and Spheres in a Sinusoidally Oscillating Fluid, ASME, J. App. Mech., Paper 75-APMW-27.

Sakai, F. and Kawai, S. (1974) Application of the Finite Element Method to Surface Wave Analysis. Coastal Eng. Japan, 17,:13-22.

Tsuchiya, Y. and Yamaguchi, M. (1974) Total Wave Forces on a Vertical Circular Cylindrical Pile. Proc. 14th Conf. Coastal Eng., 3,: 1789-1807.

Vongvisessomjai, S.(1973) Wave Force on Submerged Structures. Proc. 1st Australian Conf. Coastal Eng., : 174-181.

Vongvisessomjai, S. and Silvester, R. (1976) Wave Force on Submerged Objects. Proc. 15th Conf. Coastal Eng., 3, : 2387-2412.

DYNAMIC ANALYSIS OF MARINE RISERS INVOLVING FATIGUE AND CORROSION

Sankar C. Das

Assoc. Prof., Dept. of Civ. Engrg., Tulane Univ., New Orleans, LA., U.S.A.

INTRODUCTION

Marine risers (Fig.1) are used to provide a return flow path between a well bore and drill vessel and to guide a drill string to a series of valves called the blowout preventer stack, or BOP on ocean floor. This is also used to support choke, kill and other control lines, to guide tools into the well and as a running string for the BOP stack. The lower ball joint is a moment free connection and usually allows a free rotation up to a maximum of 10 degrees. However, during drilling operations the ball joint angles are limited to about 3 to 4 degrees from the consideration of damages to the ball joint due to rotating drill string inside the riser. The telescopic joint and the upper ball joint (Fig.1) decouple the vessel heave and rotational motions of roll, pitch and yaw effects. The tensioners provide the desired top tension to the riser.

For deep water risers (greater than 1500 ft) tension at the bottom of the riser becomes inadequate and some buoyant materials are used to reduce the effective weight of the risers. The riser is made up of sections of pipe, commonly 50 feet long and of 16 to 24 inch in diameter with 3/8 to 3/4 inch walls.

The riser system is acted on by both static and dynamic forces, and static and dynamic displacements due to its own weight, applied axial tension, winds, currents, waves, external and internal pressure (Ref.4). Determining the design waves and currents is a difficult task. Methods of wave forecasting are available (Refs. 10,12, and 16). However, for our purpose it is assumed that the design winds, waves and currents are obtained from oceanographic and meteorological consultants.

The top tension is controlled by the tensioners for the connected riser configuration (Fig.1). This is probably the most important parameter to be determined from analysis for different environmental conditions. Usual mean values of top tension are 1 to 2 times the suspended weight of the riser above ball joint.

The dynamic analysis of a riser system has been developed to demonstrate how stress analysis is combined with material fatigue behavior and sea water corrosion to arrive at a valid safety factor for the riser design.

MATHEMATICAL MODEL

The tensioned-beam co-planar partial differential equation is used to model the riser structure. The hydrodynamic force terms due to ocean loadings are represented by the modified form of Morrison's equation (Refs.11 and 14).

The differential equation is:

$$M_R \frac{\partial^2 y}{\partial t^2} + C(x) \frac{\partial y}{\partial t} + \frac{\partial^2}{\partial x^2} \{ EI(x) \frac{\partial^2 y}{\partial x^2} \} -$$

$$\frac{\partial}{\partial x} \{ T(x,t) \frac{\partial y}{\partial x} \} = \tfrac{1}{2}\rho C_D D | U_c + U_w - \frac{\partial y}{\partial t} |$$

$$\{ U_c + U_w - \frac{\partial y}{\partial t} \} + \frac{\pi}{4} \rho C_M D^2 A_w - \frac{\pi}{4} \rho (C_M - 1)$$

$$D^2 \frac{\partial^2 y}{\partial t^2} \qquad . \qquad . \qquad . \qquad . \qquad . \qquad . \qquad (1)$$

The right-hand side of the equation is modified Morrison's force term. The symbols used in Eq.1 are defined:

y = y(x,t) is lateral displacement of the riser from the lower ball joint (Fig.1)
x = vertical distance along the riser from the lower ball joint
t = time
E = modulus of elasticity of the riser beam (can vary with x)
I = area moment of inertia of the riser pipe (can vary with x)

ρ = mass density of the sea water
D = riser effective hydrodynamic diameter (can vary with x)
C_D = co-efficient of drag (can vary with x and the time-varying Reynolds number).
C_M= co-efficient of inertia (mass) (can vary with x).
U_c = velocity of the steady current (can vary with x).
U_w = water particle velocity due to wave (varies with x,t, and the horizontal position y(x,t) of the riser).
A_w = water particle acceleration due to wave (varies with x,t, and the horizontal position y(x,t) of the riser)
M_R = mass of the riser per unit length. It includes riser pipe, flow lines, connector, buoyancy materials if any, mud in riser (varies with x)
C(x) = damping function
T = T(x,t) is the effective axial tension in the riser.

The time-history solution of the nonlinear partial differential equation 1 may be obtained using the finite difference numerical method of solution. In the in-house developed ODECO's riser program, finite differences were used to express Eq.1 in an algebraic representation. The algebraic representation of Eq. 1 was done for each nodal point in terms of the deflections of the riser at adjacent points along the riser as well as at other time levels. The damping force expression in Eq.1 was not considered important and thus neglected. The velocities,U_w and the accelerations, A_w of the water particle of the wave were obtained using Stokes fifth order wave theory in ODECO's riser program.

Vessel motion is very important in determining riser behavior. The displacement excitations caused by the vessel motions are best determined by physical modal tests of the vessel concerned (Ref.4). The static offset (excursion) of the top end of the riser is obtained from mooring design and analysis (Ref.4). The vessel motion ties directly to the wave and requires specification of the characteristics of the vessel's station-keeping system and the vessel's response amplitude operators in surge or sway (Ref.4).

Thus vessel motion model used in the program is given by (Refs.4 and 14),

$$S(t) = S_o + S_s \text{ Sin } \left(\frac{2\pi t}{T_s} - \alpha \right) \quad . \quad \quad . \quad \quad . \quad (2)$$

where,

$S(t)$ = location of vessel at any time

S_o = mean vessel static offset (excursion)

S_s = single amplitude of vessel drift (surge or sway)

T_s = period of the vessel drift

α = a phase angle between drift motion and wave.

In actual sea, the phase angle between the response and the wave is random. If this randomness is not properly simulated in the model, a conservative assumption such as to maximize the relative water particle velocity is recommended. Usually the mooring system is so designed (Ref.3) that the horizontal excursion during operation and standby condition are within 3 and 10 percent of water depth respectively. Here the operational condition refers to the maximum weather condition, which may occur several times a year, to which normal drilling operation can be continued. The stand-by condition refers to the condition beyond which the riser is disconnected. As a guideline a 1 to 5 year maximum wind, wave and current may by used.

The program calculates the riser deflection, slope, moment, shear and tension loads for the various time stations; the axial, bending and combined stresses for all nodes along the riser; maximum and minimum deflections; maximum bending and combined stresses for all nodes. In addition this program will handle three different riser configurations-the connected riser, the self-standing riser, and the hanging riser (Ref.14).

EXAMPLE, RESULTS AND DISCUSSION

Studies here are being made to determine if riser fatigue is an important consideration because of the top tension variations and bending stress variations. The high load concentration, aggravated by the action of corrosion, fatigue may be a real structural problem particularly at the connections.

As an example, the dynamic solution which follows was carried out with the following parameters.

Water depth = 1500 ft.

Riser length = 1520 ft.

Riser outside dia. = 16 in.

Wall thickness = 5/8 in.

Mud weight = 90 lbs/ft^3 and 125 lbs/ft^3 (2 cases)

Tension variation = $\pm$ 5% of top tension

Tension ratio = varied from 1.06 to 1.83

Riser hydrodynamic diameter = 26 in.

Load cases
The following two load cases were considered.

Normal maximum drilling condition   Stokes fifth order wave of height equal to 20 ft. and period 13 sec.
Current equal to 0.5 knot at surface, linearly varying to 0 at mudline. Static offset (excursion) equal
to 45 ft. (3% of water depth). Vessel surge is equal
to 4 ft. (double amplitude). Surge phases are 75°
and 180° (2 cases) with respect to wave crest. This
case was used for fatigue analysis.

Maximum standby condition   Stokes fifth order wave
of height equal to 50 ft. and period 13 sec. Current equal to 2.0 knot at surface decreasing linearly to 0.5 knot at 500 ft. depth and then decreasing
linearly to 0 at mudline. Static offset (excursion)
equal to 120 ft. (8% of water depth). Vessel surge
= 25 ft. (double amplitude). Surge phase with respect to wave crest equal to 180°. This case was
used for stress analysis at the most critical sections and were examined against allowable stresses
of API Code (Ref.2) and the AISC Code (Ref.1).

Assumptions
A number of simplifying assumptions were made in
this example.
   1) The riser was assumed to be a continuous pipe
      of constant dimension.
   2) C_D = 0.7 was assumed constant over the riser
      length and with time. C_M = 1.5 was assumed
      constant (Refs. 3 and 11).
   3) Hydrodynamic effects such as Von Karman Vortex
      Trails **were** neglected.
   4) The vessel heave and rotational motions (roll,
      pitch and yaw) were assumed to be decoupled
      from the riser pipe.
   5) A zero bending moment was assumed at each end
      of the riser (ball joints).
   6) The current, wave and the riser displacements
      were all assumed coplanar, thus reducing the
      problem to two dimensions.

Fig.2 shows the riser maximum dynamic bending stress
as a function of water depth for various top tension
ratios. It should be noted that the bending stress
(particularly the maximum value at about 100 ft.
above the bottom) is a function of the ratio of top
tension to the total riser submerged weight. The
upper peak is caused by the velocities of current

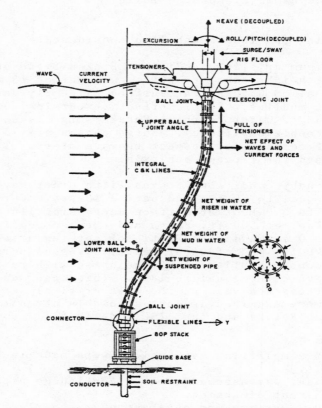

FIG. I MARINE RISER COMPONENTS & PRINCIPAL EXCITATIONS

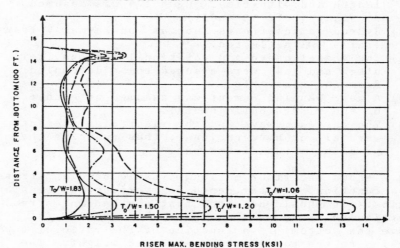

FIG. 2 RISER STRESS VS DEPTH FOR VARIOUS T_0/W
MUD WEIGHT = 125 LBS./CFT.
(NORMAL MAXIMUM DRILLING CONDITION)

and wave. Every material point in the riser is subjected to fluctuating bending stress. Fig.3 shows
the riser maximum and minimum combined stress as a
function of water depth for various top tension ratio. Fig.4 shows the riser bottom ball joint rotation. Since the relatively stiff drill string must
be bent through the same bottom angle inside the
riser system, excessive rubbing and wear usually
takes place at the angle above 4°. This is one of
the most important design criteria that must be met.

High cycle-low stress fatigue

The alternating stress arises from two sources,name-
ly fluctuating bending stress and the fluctuating
tensile stress due to top tension variation of $\pm$ 5%.
The stress will be seperated into effective mean
stress, σ_m and the alternating stress components,
σ_a.

$$\sigma_{Mean} = \frac{\sigma_{max.} + \sigma_{min.}}{2} \qquad . \qquad . \qquad . \qquad . \qquad . \qquad (3)$$

$$\sigma_a = \frac{\sigma_{max.} - \sigma_{min}}{2} \qquad . \qquad . \qquad . \qquad . \qquad . \qquad (4)$$

$$\sigma_H = \frac{pr}{t} \qquad . \qquad . \qquad . \qquad . \qquad . \qquad . \qquad (5)$$

$$\sigma_m = \sqrt{\sigma_{Mean}^2 + \sigma_H^2 - \sigma_{Mean}\sigma_H} \qquad . \qquad . \qquad . \qquad (6)$$

where,
σ_a = alternative stress or stress amplitude
σ_H = hoop stress
p = differential external and internal radial pressure at a pipe section (Fig.1)
r = average radius of the pipe
t = wall thickness of the riser pipe
σ_m = effective mean stress
σ_{Mean} = mean stress
Fig. 5 shows the riser effective mean stress, σ_m as
a function of water depth. Large tension ratios
cause high effective mean stresses at the top. Effective mean stress and alternating stress components
were then introduced into the Goodman diagram (Fig.6)
to predict a factor of safety against a low stress-
high cycle fatigue failure, where the effect of mean
stress is significant on fatigue life (Refs. 5 and
6). Fig. 6 presents an infinite life Goodman diagram for the low stress-high cycle fatigue (usually
the case) for a typical riser steel: Yield stress,
Sy = 50 ksi and ultimate stress, S_u = 70 ksi. Endurance limit of standard steel specimen,
$S_e^{'}$ = 0.5 S_u at 10^6 cycles $\qquad . \qquad . \qquad . \qquad . \qquad (7)$

244

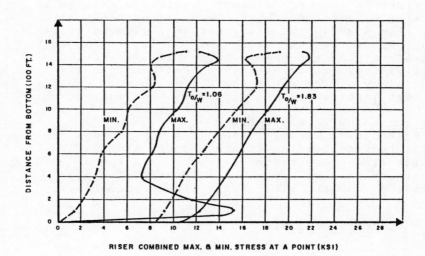

FIG. 3   RISER COMBINED STRESS (TENSILE & BENDING) VS DEPTH FOR VARIOUS T_o/W
UNIT MUD WEIGHT = 125 LBS./CFT
(NORMAL MAXIMUM DRILLING CONDITION)

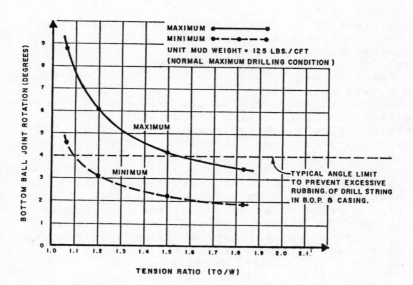

FIG. 4   BOTTOM BALL JOINT ROTATION VS TENSION RATIO, TO/W

Endurance limit of actual part (Ref.5),

$$S_e = (K_a K_b K_c K_d K_d) \; S_e' \qquad . \qquad . \qquad . \qquad . \qquad . \qquad (8)$$

where,

K_a = surface factor; K_b = size factor; K_c = reliability factor; K_d = temperature effect; K_e = stress concentration factor.

In the Goodman diagram (Fig.6) three situations were considered:

1) Stress in the riser pipe itself without stress concentration and corrosion.
2) Stress in the butt weld joint between the riser pipe and its connector with stress concentration but no corrosion.
3) Stress in the butt weld joint between the riser pipe and its connector with stress concentration and corrosion.

Total alternating stress, σa were plotted for the entire riser beam in Figs. 7 and 8. From Eq. 8 and the Goodman diagram (Fig.6) the limiting alternating stress for safety factor, S.F. = 1,

$$\text{Limiting } \sigma_a = S_e \; \left(1 - \frac{\sigma_m}{S_u}\right) \qquad . \qquad . \qquad . \qquad . \qquad (9)$$

Using the appropriate endurance strength, the lines for limiting σ_a were constructed for limits #1, #2 and #3 for the entire riser beam as shown in Figs. 7 and 8. Whenever the total alternating stress exceeds the limiting σ_a, eventual failure is to be expected (limited life). In Fig. 7, due to low top tension ratio and high bending stress, the failure will occur in a corroding weld (limit #3) and safety factors of 1.39 and 1.61 against fatigue failure in a noncorroding weld (limit #2) and in the pipe itself (limit #1) exist respectively near the bottom of the riser. In Fig.8 however due to high tension ratio, a safety factor of 1.40 against fatigue failure will exist in a corroding weld near the top of the riser.

For this riser beam, the maximum combined stresses were also calculated at the various nodal points due to maximum standby condition for two top tension ratios of 1.06 and 1.83 respectively. The total maximum combined stress were 35.7 ksi (0.71 Sy) and 33.6 ksi (0.67Sy) for the top tension ratios of 1.06 and 1.83 respectively. Both these values are higher than the normal design allowable (Refs. 1 and 2) of 0.66 Sy but within the extreme condition increased allowable stress of 0.88 Sy. At the low range of tension ratios the maximum bending stress

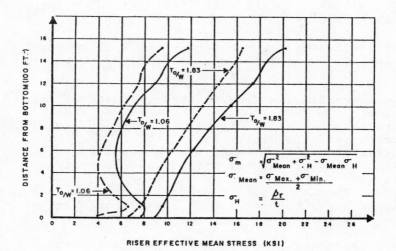

$$\sigma_m = \sqrt{\sigma^2_{Mean} + \sigma^2_H - \sigma_{Mean}\sigma_H}$$

$$\sigma_{Mean} = \frac{\sigma_{Max.} + \sigma_{Min.}}{2}$$

$$\sigma_H = \frac{p_r}{t}$$

FIG. 5 RISER EFFECTIVE MEAN STRESS VS DEPTH FOR VARIOUS T_o/W

MUD WEIGHT = 125 LBS./CFT ————

MUD WEIGHT = 90 LBS./CFT —·—·—·—

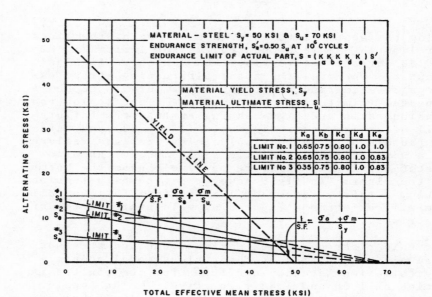

FIG. 6 GOODMAN DIAGRAM SHOWING LIMITING VALUES FOR σ_a & σ_m FOR THE RISER PIPE

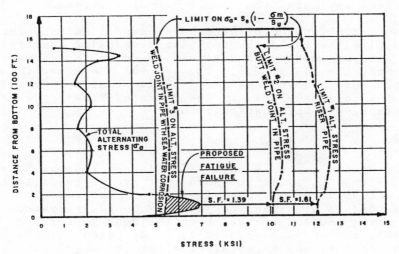

$T_o/W = 1.06$; MUD WEIGHT 125 LBS./ CFT.

FIG. 7

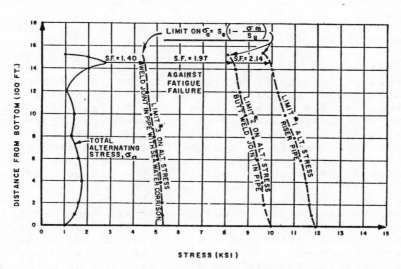

$T_o/W = 1.83$; MUD WEIGHT 125 LBS./CFT

FIG. 8

decreases at a faster rate reducing the combined stress as the tension increases. After the tension ratio reaches an optimum value, the increase in direct stress due to additional tension more than offsets any gain from reduced bending stress.

Cumulative fatigue effects have been calculated for the fixed-bottom deepwater platforms using spectral or discrete method of analysis based on the Palmgren -Miner rule (Refs. 7,8,9,13, and 15). However fatigue analysis and design of marine risers is still in the formative stage. The Palmgren-Miner rule may be expressed mathematically as:

$$\sum_{i=1}^{Total} \frac{n(\sigma_i)}{N(\sigma_i)} \leq 1 \qquad \cdot \qquad \cdot \qquad \cdot \qquad \cdot \qquad \cdot \qquad (10)$$

where $n(\sigma_i)$ is the number of cycles at a stress level σ_i and $N(\sigma_i)$ is the number of cycles to failure at stress level σ_i. Solving for the percent of life expended in a particular joint of a riser system requires knowledge about both the denominators, $N(\sigma_i)$, and numerators, $n(\sigma_i)$, of Eq. (10). The denominators may be obtained after the developments of S-N curves for the various riser connectors involving stress concentration and sea water corrosion. The numerators may be obtained from the developments of cumulative stress-history curves of the connectors (Refs. 8 and 9). However, there are inherent shortcomings in this approach. The first is that the order of loadings is not taken into account. The second shortcoming is that the Palmgren-Miner rule does not take the geometry of the joint into account.

CONCLUSIONS

Riser system fatigue problems are most likely to develop in areas of stress concentration with possible sea water corrosion and stress fluctuations. Therefore low stress - high cycle fatigue may be an important consideration in a deeper water riser system because of the generally higher tensioning levels with variations and the fluctuating bending stress. High stress - low cycle fatigue and the crack propagation situations in the unusual cases were not discussed here.

The proposed analysis design approach were carried over into the riser connectors. These connectors must however withstand the same high tensile and bending stress, must be able to be assembled and disassembled rapidly under the most trying conditions.

In a connector design, the unavoidable stress concentrations should be minimum and the self induced internal stresses are to be reduced as far as possible. Stress concentration will lower the endurance strength of the part and thus lower the allowable σ_a for a given size connector. Selfinduced internal stresses (clamping, bending etc.) add to the effective mean stresses and thus also reduce the allowable σ_a.

From the consideration of both fatigue and strength, the top tension ratio is the most critical factor for safe performance of marine risers. The analysis presented here could be used to determine the optimum tension ratios for a given riser for operations at different water depths and varied environmental loads.

ACKNOWLEDGMENT

Author wishes to express his sincere gratitude to ODECO Engineers,Inc., New Orleans for the company's co-operation. Particular acknowledgement should be given to S. Sengupta for his help in the computer program applications and program modifications.

REFERENCES

1. AISC. (1973) Manual of Steel Construction. 7th. Ed. N.Y.
2. API RP2A.(1977) Planning, Designing, and Construction Fixed Offshore Platforms. API, Dallas, 10,21-23.
3. Childers, M.A.(1976) Environmental Factors Control Station Keeping Methods and Spread Mooring in Ultradeep Waters. Offshore Drilling and Production Technology, Petroleum Engineering Pub. Co., 3-15.
4. Das,S.C. and Sengupta,S. (1978) Dynamic Analysis for Design of Marine Risers. A Short Course on Design and Construction of Offshore Structures, Tulane Univ.,New Orleans, LA.
5. Finnie,I. (1978) Fatigue. A Short Course on Deep-Sea Oil Production Structures, Univ. of California, Berkeley.
6. Hauser,F.E.(1978) Design of Marine Riser System. A Short Course on Deep-Sea Oil Production Structures, Univ. of California, Berkeley.
7. Kallaby, J. and Others (1976) Evaluation of Fatigue Considerations in the Design of Framed Offshore Structures. Paper No. OTC 2609, Offshore Technology Conference, 907-924.

250

8. Maddox, N.R.(1974) Fatigue Analysis for Deepwater Fixed-Bottom Platforms. Paper No. OTC 2051. Offshore Technology Conference.

9. Maddox, N.R. and Wildenstein,A.W. (1975) A Spectral Fatigue Analysis for Offshore Structures. Paper No. OTC 2261, Offshore Technology Conference, 185-198.

10. Mitchel,W.H. (1968) Sea Spectra Simplified. Marine Technology, 17-30.

11. Morgan,G.W. and Peret, J.W.(1976) Applied Mechanics of Marine Riser System. Petroleum Engineering Pub. Co.

12. Myers, J.J. and Others (1969) Handbook of Ocean and Underwater Engineering. McGraw-Hill Book Co, N.Y.

13. Nolte, K.G. and Hansford, J.E. (1976) Closed-Form Expressions for Determining the Fatigue Damage of Structures Due to Ocean Waves. Paper No. OTC 2606, Offshore Technology Conference, 861-872.

14. Sexton,R.M. and Agbezuge, L.K. (1976) Random Wave and Vessel Motion Effects on Drilling Riser Dynamics. Paper No. OTC 2650, Offshore Technology Conference, 391-404.

15. Vughts, J.H. and Kinra, R.K. (1976) Probabilistic Fatigue Analysis of Fixed Offshore Structures. Paper No. OTC 2608, Offshore Technology Conference, 889-906.

16. Wiegel, R.L. (1964) Oceanographical Engineering. Prentice Hall Book Co, N.J.

SPECTRAL DYNAMIC FATIGUE ANALYSIS OF THE ANDOC DUNLIN A PLAT-
FORM
IR D. ZIJP

ANDOC DESIGN/ADRIAAN VOLKER CIVIL ENGINEERING

CONTENTS

252

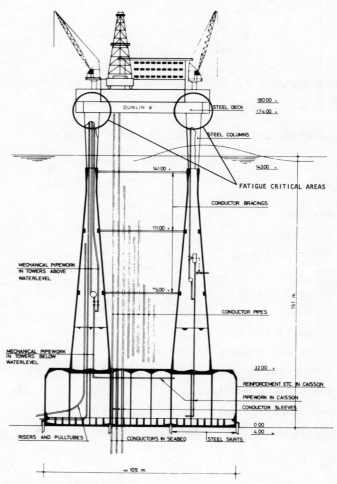

fig. 1   CROSS SECTION DUNLIN A

1. ABSTRACT

Spectral 'static' analysis has proven to be a very valuable tool in the design of offshore platforms.

But for the fatigue analysis of gravity type platforms with their relatively large dynamic response to small waves it is necessary to take the dynamic behaviour of the structure in account.

Therefore in the fatigue analysis of the Dunlin A platform the spectral 'dynamic' analysis technique has been introduced.

This did improve the reliability of the fatigue analysis drasticly, so that less conservative requirements could be used.

The necessary computer programmes have been developed by the IBM computer centre in Zoetermeer and have been extensively tested by the ANDOC design team, using the Dunlin A geometry.

A comparison is made between a deterministic and a spectral analysis as well as between the harmonic solution and time step integration.

2. INTRODUCTION

The platform Dunlin A is an oil-production platform destined for the Dunlin Field in the North Sea and built by ANDOC (Anglo Dutch Offshore Concrete).

The platform (see fig. 1) is a so called gravity structure i.e. the foundation occurs by means of contact pressure with the bottom caused by the self weight of the structure.

A concrete caisson of 100 x 100 x 32 m provides this foundation requirement, but also serves as an oil storage reservoir and as flotation body in several building and transport stages Steel skirts of 4 m height underneath the caisson will prevent scouring and bring the foundation forces to deeper soil layers. On the caisson 4 towers are build of 142 m height, from which the first 111 m are out of concrete and the last 31 m of steel. These towers carry the steel deck, composed out of boxgirders and intermediate floors.

Towers and deck together form a portal frame founded on the concrete caisson. Through the caisson, between two towers, the 48 production wells are guided by means of conductors from the bottom to the deck. The conductors in turn are guided by three conductor bracings.

The deck carries all the equipment like: production equipment, drilling equipment, energy supply, water and fuel tanks, living quarters, heli deck, flare boom, cranes, safety equipment etc.

This equipment is partly on the floors between the boxgirders and inside the boxgirders. (see fig. 2).

Because of the connections between all the equipment, there are numerous holes necessary in the boxgirders. These holes are reinforced by ringstiffeners and sometimes thicker wallplate material.

3. DYNAMIC FATIGUE

Because of these holes the normal stress pattern is changed and high peaks can be expected. Since the structure is a portal frame, the highest moments and thus stresses due to varying wave forces can be expected near the junctions between deck and columns. The large number of waves will give the structure a fatigue loading.

The structure is a dynamic system with the first natural period in the wave period range for small waves. Large dynamic amplification in this range can be expected and therefore 'dynamic' fatigue analysis is an important subject.

4. THE SEA

The sea is a rough, irregular, poorly predictable phenomenon. A sea condition or sea state consists of many waves with different wave heights, periods and directions which occur simultaneously.

One can try to count the number of waves, their wave heights, and periods, but since no two registrations will be the same it is more practical to derive statistical information. To this end a wave spectrum is used which describes the wave energy distribution over the frequency axis. Although it is not completely true it is assumed that each sea state has a uniform shape. Often the Pierson-Moscowitz spectrum shape is used which can be characterised by two parameters H_s and T_z.

$$S_{\eta\eta}(\omega) = \left(\frac{H_s}{2}\right) \frac{692}{\omega(1.09T_z\omega)^4} \exp\left(\frac{-692}{(1.09T_z\omega)^4}\right)$$

$S_{\eta\eta}(\omega)$ = wave spectrum function   (m2.s/rad.)
η = fluctuation of water surface   (m)
ω = circular frequency   (rad/s)
H_s = significant wave height of the sea state (m)
T_z = zero up-crossing period   (s)

The probability of exceedence of individual wave heights in a sea state follows a 'Rayleigh' distribution. The only parameter in this formula is the surface under the spectrum shape m_0.

fig. 2   Deck structure of platform Dunlin A, partly with modules in-
          stalled

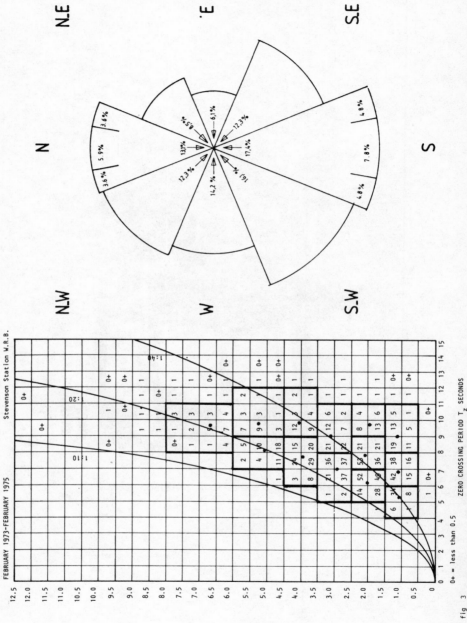

fig 3

SCATTER DIAGRAM : DESCRIBES ALL SEA STATES

FIG4 WAVE DIRECTION DISTRIBUTION NORTH SEA

$$P\ (H)\ =\ \exp\ \left(-\ \frac{H^2}{8m_o}\right)$$

$P\ (H)$ = probability of exceedence of a wave height H

$m_o\ =\ \int_o^\infty S_{\eta\eta}(\omega)d\omega$ = zero moment of the spectrum
= surface under the spectrum shape

A scatter diagram (fig.3) describes all sea states during a certain period. Each point in the diagram represents a sea state with its probability of occurance. For the analysis it is, for practical reasons, necessary to re-arrange the large number (121) of different sea states into a smaller number (e.g. 13). A wave roset describes (fig.4) the direction distribution of the waves. For the north- and the south direction a sub-division was necessary because of a resonance peak in this direction (see point 10).

5. DYNAMIC ANALYSIS

From the ANDOC platform a three dimensional 'lumped mass' member element computer model (fig.5) has been made with 136 dynamic degrees of freedom. Using STRUDL, 50 eigenvalues and eigenvectors were calculated which were necessary for the normal-mode-super-position method of dynamic response calculation. Fig. 6 to 8 show some typical mode shapes, namely the bending- twisting- and breathing mode.

6. HARMONIC DYNAMIC RESPONSE (fig.9)

The wave loading input data only consists of wave spectra to be presented as spectral densities for a number of frequencies. For each given frequency, using the highest wave amplitude to be expected per frequency, the wave force amplitude with relative phase relationship is calculated at each loaded point, using the Sinusoidal wave theory and the Morrison formule.

By means of normal mode superposition a harmonic structural dynamic response is calculated and divided by the input wave amplitude per frequency, resulting in a transfer function between wave and response spectrum for each required quantity (displacement, forces etc.).

This transfer function squared and multiplied with the wave spectrum gives the response spectrum for each quantity.

258

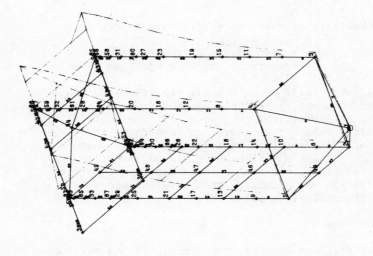

fig. 6   BENDING MODE   PERIOD 4.2 sec.

ANDOC SHELL DUNLIN A RENEWED SPECTRAL FATIGUE ANALYSIS
EIGVECT 2

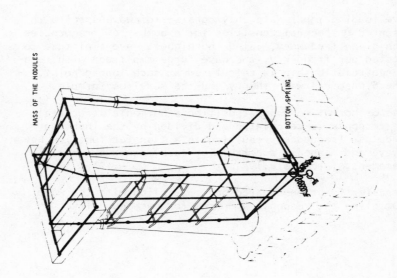

MASS OF THE MODULES

BOTTOM-SPRING

fig. 5   LUMPED MASS MEMBER ELEMENT COMPUTER MODEL

$$S_{rr}(\omega) = \left| H_{\eta,r}(\omega) \right|^2 * S_{\eta\eta}(\omega)$$

$S_{rr}(\omega)$ = response spectrum for displacement, forces etc.
$\quad\quad$ (...^{2}.s/rad)

$H_{\eta r}(\omega)$ = transfer function from wave amplitude to response
$\quad\quad$ (.../m)

$S_{\eta\eta}(\omega)$ = wave spectrum (m^2.s/rad)

Since all processes between the wave spectrum and the response of the structure are linear or made linear the response spectra also follows a Rayleigh distribution.

With this knowledge further calculations can be carried out.
A design value with an accepted probability of exceedence of
a fatigue life can be calculated.

7. LINEARISATIONS (fig.10)

The following linearisations have been applied:

1) First order sinusoidal wave theory instead of higher order
   Stokes theory. This affects the water particle velocities
   and acceleration, expecially near the water surface.

2) No vertical movement of the water surface. The increase of
   wave force due to innundation of an additional part of
   the structure as the wave top passes is neglected.

3) Linearised drag force in the Morrison equation. The drag
   force is made linear by calculating the drag force for the
   highest wave to be expected for each frequency and then
   assuming a linear relationship between drag force and wave
   height.

For high waves, the second linearisation expecially gives unsatisfactory results with the above described method. However
for these waves the transient response analysis can succes-
fully be used.

8. TRANSIENT DYNAMIC RESPONSE (fig.11)

The wave loading input data consists of time dependent wave
forcing functions for each loaded point. These forcing functions can be calculated with a higher order wave force theory
taking all the non-linearities in account. The transient structural response is calculated using normal mode super position
and time step integration of the equations of motion, resulting in a time dependent response in each point for each required quantity (forces, displacements etc.).

The structural movement must be initiated, therefore sufficient load cycles need to be passed before a steady state
solution will be achieved. This means that for each wave an
average of 7 cycles at 10 time steps/cycle = 70 time steps
must be computed. Because one time step costs approximately
the same as a harmonic response calculation of one frequency
this type of dynamic analysis is much more expensive than the
first one and therefore should only be carried out in cases
when the linearisations mentioned are unacceptable.

9. COMPARISON OF THE TWO METHODS (fig. 12-14)

Fig. 12 to 14 give a result of the two methods for a 1m/5.7 s
wave, a 9.5m/12.6 s wave and a 30m/17s wave. As can be seen
especially for the high waves the peak response of the time
step analysis is much larger because of the innundation of
the columns during the passing of the wave top.

10. RESONANCE COMPLICATION

With the ANDOC configuration, there was a complication that
for the basic resonance period of 4.2 sec. the wave length
was $1,56 \times (4,2)^2 = 27,5$ m which was exactly $\frac{1}{2}$ x the column
spacing of 55 m, so the wave forces on the four columns were
in phase, resulting in extra amplification. However, this
was only the case for wave directions perpendicular to the
platform.
For other wave directions this phenomenon quickly decreased,
as can be seen in fig. 15.

11. FATIGUE CALCULATION

The most probable lifetime of a construction detail is normally calculated with the Miner rule:

$$AD = \sum_{i=1}^{total} \frac{n(s_i)}{N(s_i)}$$

AD = Accumulated damage during the considered time
$n(s_i)$= number of cycles with stress level s_i

$N(s_i)$= number of cycles to failure at stress level s_i

Lifetime= considered time span
          accumulated damage

Since for each sea state the probability densities of the
stress amplitudes follow a Rayleigh distribution, the Miner
rule for one sea state can be expressed as:

$$AD = f.T. \int_0^\infty \frac{p(s)}{N(s)} \, ds$$

f   = dominant frequency of the sea state or resonance fre-
       quency of the structure (dependent on which one domi-
       nates the response spectrum) (s^{-1})

T   = duration of the considered sea state (s)

$$p(s) = \frac{s}{m_o} \exp\left(\frac{-s^2}{2m_o}\right)$$

p(s) = probability density of stress level s (Rayleigh distri-
                                                    bution)

$$m_o = \int_0^\infty S_{ss}(\omega) \, d\omega$$

$S_{ss}(\omega)$= stress response spectral density of the considered
       detail

N(s) = s -N fatigue curve of the considered detail

$$N(s) = N_1\left(\frac{s_1}{s}\right)^b$$

s_1, N_1= a point on the s -N curve

b   = slope of the s -N curve

By adding the accumulated damage of all sea states in the
scatter diagram, the total lifetime of the considered detail
is calculated.

12. REFERENCES

1. Krol H.R.
   Spectral analyses with ICES/STRUDL
   IBM Data Centre Services, Zoetermeer, January 1977

2. Zijp D, Pot B.J.G. v.d.
   Dynamic analysis of gravity type offshore platforms:
   experience, development and practical application.
   OTC paper 2433, May 1976

262

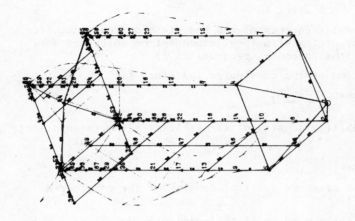

fig. 8   BREATHING MODE   PERIOD 0.93 sec.

ANOOC SHELL DUNLIN A RENEWED SPECTRAL FATIGUE ANALYSIS
EIGVECT 10

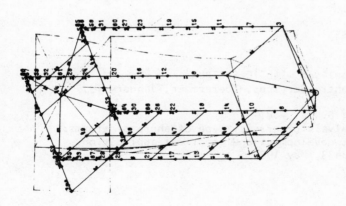

fig. 7   TWISTING MODE   PERIOD 2.1 sec.

ANOOC SHELL DUNLIN A RENEWED SPECTRAL FATIGUE ANALYSIS
EIGVECT 5

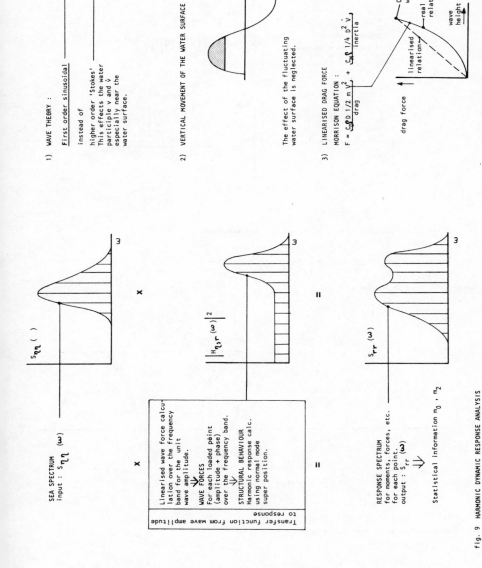

1) WAVE THEORY:

First order sinusoidal

instead of

higher order 'Stokes'
This effects the water particle v and $\dot{v}$ especially near the water surface.

2) VERTICAL MOVEMENT OF THE WATER SURFACE

This point feels only the wave tops

The effect of the fluctuating water surface is neglected.

3) LINEARISED DRAG FORCE

MORRISON EQUATION:

$$F = C_D \rho D \frac{1}{2} m v^2 + C_M \rho \frac{1}{4} D^2 \dot{v}$$

drag   inertia

Calculate drag force for the highest wave to be expected for the considered frequency, and assume a linear relation with the wave height.

linearised relation

real relation

drag force

wave height

Fig. 10 LINEARISED WAVE FORCE CALCULATIONS

SEA SPECTRUM
input : $S_{\zeta\zeta}(\omega)$

×

transfer function from wave amplitude to response

Linearised wave force calculation over the frequency band for the unit wave amplitude.

WAVE FORCES
For each loaded point (amplitude + phase) over the frequency band.

STRUCTURAL BEHAVIOUR
Harmonic response calc. using normal mode super position.

$|H_{\zeta r}(\omega)|^2$

=

RESPONSE SPECTRUM
for moments, forces, etc. for each point.
output: $S_{rr}(\omega)$

Statistical information m_0, m_2

$S_{rr}(\omega)$

fig. 9 HARMONIC DYNAMIC RESPONSE ANALYSIS

264

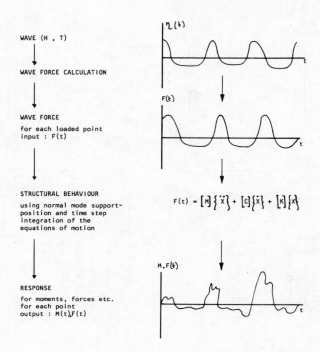

WAVE (H , T)

↓

WAVE FORCE CALCULATION

↓

WAVE FORCE

for each loaded point
input : F(t)

↓

STRUCTURAL BEHAVIOUR

using normal mode support-
position and time step
integration of the
equations of motion

↓

RESPONSE

for moments, forces etc.
for each point
output : M(t),F(t)

$$F(t) = [m]\{\ddot{x}\} + [c]\{\dot{x}\} + [k]\{x\}$$

fig. 11   TRANSIENT (TIME STEP) DYNAMIC RESPONSE ANALYSIS

Comparison harmonic - transient response

1.0 m ; 5.71 sec.
moment in   (MNm) at a column/deck joint

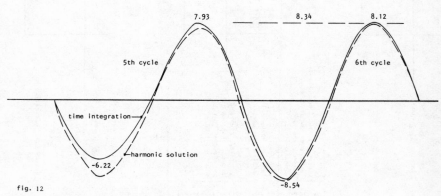

fig. 12

COMPARISON HARMONIC - TRANSIENT RESPONSE

Comparison harmonic _ transient response
9.5 m ; 12.57 sec
moment in(MNm) column/deck joint

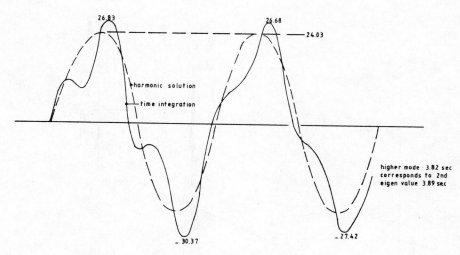

fig 13
COMPARISON HARMONIC _ TRANSIENT RESPONSE

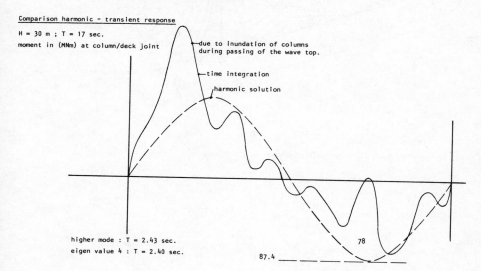

Comparison harmonic - transient response

H = 30 m ; T = 17 sec.
moment in (MNm) at column/deck joint

fig. 14

COMPARISON HARMONIC - TRANSIENT RESPONSE

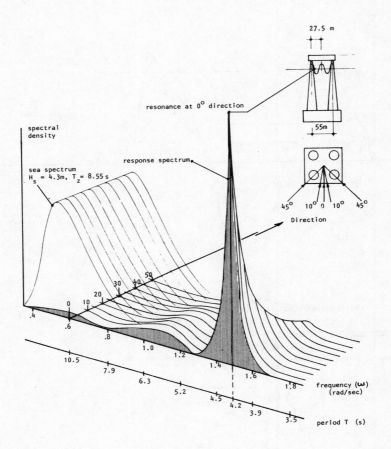

fig. 15

RESONANCE COMPLICATION AT THE PERPENDICULAR WAVE DIRECTION

A COMPUTATIONAL APPROACH FOR THE ANALYSIS OF FIXED OFFSHORE STRUCTURES INCLUDING PILING

A.J. Ferrante, E.C. Valenzuela and G.B. Ellwanger

COPPE-UFRJ and PETROBRAS S.A., Brazil

INTRODUCTION

The design and construction of piled steel offshore platforms poses several interesting problems for structural engineers. One of these is the analysis of such structures during transportation and installation, and when in operation.

Currently there are numerous computer programs, including some very sophisticated systems, for general structural analysis [1] . Many of those programs are capable to solve a wide variety of problems, and could be used for the analysis of offshore structures. In most cases, however, this would require extensive pre- and post-processing, unless some specific features of typical application for offshore structures are included. In practical terms, any organization heavily involved in the design of these structures must posses convenient computer programs, offering all the required specific options. In particular, it would be advantageous if those options are integrated within a general system, so that a wide range of other general capabilities are also available.

Consistent with previous ideas, several new procedures, specially tailored for the analyis of offshore structures, were added to the ICES STRUDL II language [2] . These procedures take the form of user-oriented commands, and are fully compatible with the original STRUDL commands.

The objective of the present paper is to describe the subset of the commands corresponding to the static analysis of foundation-structure interaction

problems, typical of small and medium size offshore
structures supported on pile foundations. The
commands were implemented as a part of a development
project carried out by Petrobrás S.A., in cooperation
with the Federal University of Rio de Janeiro. Efforts
to extend the procedure for dynamic analysis are
currently under way.

THE OVERALL ANALYSIS SCHEME

The analysis model adopted considers a system with
three components. The first component corresponds to
the structural assemblage located above the seabottom level. This includes the deck, the jacket and
a portion of the piles. The second component includes
the part of the piles which is below the seabottom
level. The third component is the soil.

The structural description of the first component is
done using mostly conventional STRUDL commands, and
can include any of the linear and finite element
types available. A new command was implemented to
represent more realistically the interaction between
the piles and the legs of the jacket. This command
constrains the lateral displacements of the piles to
be the same as those of the legs, at selected nodal
joints, but allows differential displacements in the
longitudinal directions.

The external actions applied to an offshore structure
are of various types, including loads due to dead
weight, buoyancy, waves, winds, operations,drilling,
earthquakes, etc. Excluding possibly earthquake loads,
all other type of loads act on the first component
only. Among those, the wave loads are of prime
importance, and have a dynamic nature. For small and
medium size platforms, however,the assumption that
the wave loads act statically may suffice to
determine the structural behaviour with reasonable
accuracy.

New commands were added to the STRUDL Language for
automatic generation of dead weight, buoyancy, and
wave loads. The command for wave loads,in particular,
includes several convenient features.It is possible
to use Stokes 5th order, Airy, cnoidal and solitary
wave theories. The command determines automatically
which members are partially or totally submerged,
and computes the forces exerted on those, using
Morison equation. For this the user can specify just
one particular wave, or can let the command vary the
offset to determine automatically the critical wave,
with regard to the total shear force. The existance

of marine growths can also be taken into account in those calculations.

Once the first component is completely defined, regarding both structural and load data, it must be condensed at seabottom level using substructure analysis techniques. Several substructure commands were implemented for this, allowing to use nested substructures, and to take advantage of similarities existing between different structural parts. As a result of condensation, this component is represented by a stiffness matrix and a load vector, relative to the nodes which connect it to the second component. Linear behaviour, both physically and geometrically, is assumed for the first component.

The second component includes the piles below seabottom level. The user defines each pile specifying the node which connects it to the first component, the coordinates of its tip, and the characteristics of its cross-section. Each pile is represented by a sequence of automatically generated space frame segments, which can have geometrically linear or non linear behaviour.

The soil is represented by a discrete model composed by lumped springs, connected to the piles at the intersection of the pile segment [3] . In the lateral directions the springs have non linear stiffnesses defined by P-Y curves [4,5] . These depend both on the soil characteristics and the depth. The soil characteristics are defined by layers. Skin friction and tip resistance can also be taken into account.

After the definition of the three-component model is completed, the non linear analysis procedure can be applied. This procedure is based on an incremental Newton-Raphson technique. The results of the analysis are generalized displacements and generalized forces at the pile nodal joints, and the soil reactions. Finally, knowing the displacements for the external nodes of the substructure representing the first component, results can be computed for its internal nodes and elements, thus completing the study.

THE SUBSTRUCTURING TECHNIQUE

Although the substructure concept is fairly well known, a brief summary will be included for consistency [6,7]. Any structure is composed by elements, interconnected by means of a given number of nodal points, or nodes. The elements of the structure can be divided into groups. Each group of elements can be represented in a compact form, using the substructure concept. In so doing the nodes joining together the elements of a substructure are considered to be of two types: internal nodes, which connects elements pertaining to the substructure only, and boundary nodes, which connect the substructure elements with other elements or substructures. The system of stiffness equations for the substructure can be written in matrix notation, as

$$\underset{\sim}{K}\ \underset{\sim}{U} = \underset{\sim}{P} \tag{1}$$

where $\underset{\sim}{K}$ is the stiffness matrix and $\underset{\sim}{U}$ and $\underset{\sim}{P}$ are the nodal displacement and load vectors, respectivelly. This equation can re-arranged, so that data relative to boundary nodes is separated from data relative internal nodes. Then equation (1) can be written in partitioned form, as

$$\begin{vmatrix} \underset{\sim}{K}_{bb} & \underset{\sim}{K}_{b1} \\ \underset{\sim}{K}_{ib} & \underset{\sim}{K}_{ii} \end{vmatrix} \quad \begin{Bmatrix} \underset{\sim}{U}_b \\ \underset{\sim}{U}_i \end{Bmatrix} = \begin{Bmatrix} \underset{\sim}{P}_b \\ \underset{\sim}{P}_i \end{Bmatrix} \tag{2}$$

where indices "b" and "i" indicate quantities related to the boundary and internal nodes respectively. Equation (2) is equivalent to

$$\underset{\sim}{K}_{bb}\underset{\sim}{U}_b + \underset{\sim}{K}_{bi}\underset{\sim}{U}_i = \underset{\sim}{P}_b \tag{3}$$

$$\underset{\sim}{K}_{ib}\underset{\sim}{U}_b + \underset{\sim}{K}_{ii}\underset{\sim}{U}_i = \underset{\sim}{P}_i \tag{4}$$

Pre-multiplying equation (4) by $\underset{\sim}{K}_{bi}\underset{\sim}{K}_{ii}^{-1}$ and substracting from equation (3) leads to

$$(\underset{\sim}{K}_{bb} - \underset{\sim}{K}_{bi}\underset{\sim}{K}_{ii}^{-1}\underset{\sim}{K}_{ib})\ \underset{\sim}{U}_b = \underset{\sim}{P}_b - \underset{\sim}{K}_{bi}\underset{\sim}{K}_{ii}^{-1}\underset{\sim}{P}_i \tag{5}$$

or

$$\underset{\sim}{K}_{eq}\ \underset{\sim}{U}_b = \underset{\sim}{P}_{eq} \tag{6}$$

where $\underset{\sim}{K}_{eq}$ and $\underset{\sim}{P}_{eq}$, the substructure equivalent stiffness matrix and load vector, respectively, are referred to the substructure boundary nodes only. Thus, the order of equation (6), number of boundary nodes, is in general much smaller than the order of equation (1), total number of nodes in the substructure.

Once $\underset{\sim}{K}_{eq}$ and $\underset{\sim}{P}_{eq}$ are available, the substructure can be treated as any common finite element. As such, it can be part of another substructure, or can be considered as a structural component in the analysis of a structure. However, before using K_{eq} and P_{eq} in the assembly of a system of equations, some transformations may be required. Equation (6) is normally set up with regard to the substructure own reference frame, which may differ from the structure global reference frame. Furthermore, the substructure degrees of freedom may differ from the structure degrees of freedom. In such cases, before using K_{eq} and P_{eq} in assembling systems of higher level, they will have to be transformed as follows

$$K_{eq}^{S} = B^{T} R^{T} K_{eq} R B \tag{7}$$

$$P_{eq}^{S} = B^{T} R^{T} P_{eq} \tag{8}$$

where R is a rotation matrix, and B is a boleean matrix which expands the substructure degrees of freedom into the full set of the structure degrees of freedom.

Several new commands were added to STRUDL, for substructure definition, saving, retrieval, and analysis. A susbtructure can be independent or dependent. To define an independent substructure the user must give the lists of its elements and boundary nodes, and the system will compute the corresponding K_{eq} and P_{eq} matrices. The list of elements can include substructures of lower level. Dependent substructures are those which present similarities with regard to other substructures, such that K_{eq}, and eventually also P_{eq}, need not to be computed again, but can be totally or partially copied from the same matrices corresponding to other substructures, directly or after a reference frame transformation.

After the K_{eq} and P_{eq} matrices were computed for a substructure, they can be saved for later use, or can be included as a component in a structural analysis of any type. After such analysis the displacements will be known for the substructure boundary nodes. The displacements for the substructure internal nodes can be computed using the expression

$$U_i = K_{ii}^{-1} (P_i - K_{ib} U_b) \tag{9}$$

which derives from equation (4). Knowing all the displacements allows to compute any other type of results desired.

As a very simple example, to illustrate some of the commands described above, let us consider the frame shown in Fig. 1. Members 4 to 10 may represent, very schematically, a jacket, while members 1 to 3 and 11 to 13 would be piles. The piles are solidly joined to the jacket at nodes 11 and 12 only. Assume that the following commands are issued, after the structural and loading definition,

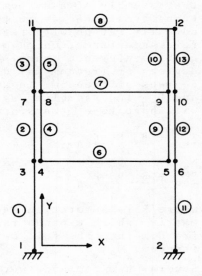

Figure 1

. . .

. . .

```
GENERALIZED CONSTRAINTS
8       7       U1      U1
9       10      U1      U1
4       3       U1      U1
5       6       U1      U1
SUBSTRUCTURE
20    ELEMENTS   2  TO  10  12  13  BOUNDARY  3  6
SAVE  STIFFNESS   20  FILE  'EXAMPLE'
STIFFNESS ANALYSIS
LIST  DISPLACEMENTS   JOINTS   3  6
SUBSTRUCTURE ANALYSIS   20
LIST  FORCES  ALL
FINISH
```

The GENERALIZED CONSTRAINTS command is used to link displacements of pairs of nodes. In the present case it is specified that the displacements U1, i.e. in the X direction, must be the same for nodes 7 and 8, 9 and 10, 4 and 3, and 5 and 6. Thus the lateral displacements will be the same for those pairs of nodes, but the vertical displacements will differ. This scheme can be used to simulate the iteraction of piles and the jacket legs.

The substructure defined, and called 20, includes elements 2 and 10, 12 and 13. Its boundary nodes are 3 and 6. After the SUBSTRUCTURE command is issued, the K_{eq} and P_{eq} matrices for substructure 20 become available. If desired, the substructure matrices can be saved for future utilization, as shown in the example. If an analysis is requested, the substructure will represent, by means of the $\underset{\sim}{K}_{eq}$ and $\underset{\sim}{P}_{eq}$ matrices,

all its elements and nodes. Thus for the STIFFNESS
ANALYSIS command indicated, the structure is
composed by elements 1 and 11, and substructure 20.
After that analysis the displacements for nodes 3
and 6 are known, and can be printed. The command
SUBSTRUCTURE ANALYSIS 20 causes the results for
the substructure internal nodes and elements to be
computed, based on the displacements of its
boundary nodes, 3 and 6. This phase can be omitted,
if the user is not interested in those internal
results.

GLOBAL ANALYSIS INCLUDING PILING

Once the stiffness and load characteristics of the
first component are known in condensed form, at
the seabottom level, the next steps correspond to
the definition of the soil and pile data, and the
application of the global analysis procedure.

A new command, called SOIL CHARACTERISTICS, was
added to the STRUDL Language for the definition
of the soil properties. The soil is divided
in as many layers as necessary. Each layer is
composed by a material of a given type, including
clay and sand, which can be loose, medium, or dense.
In addition to the material type, each layer is
characterized by its depth, and the material
specific weight, friction angle, and for clay,
cohesion factor. The data specified using this
command is simply added to the data structure.

Another new command, called PILE ANALYSIS, is used
both to define the pile data, and to apply the
global analysis procedure. The name of the node
connecting it to the substructure, the coordinates
of its tip, and the external diameters and
thicknesses, must be given for each pile. Several
options are available regarding the analysis
procedure. The pile behaviour can be geometrically
linear or non linear. The lateral soil resistance
is always taken into account, but consideration
of skin friction is optional. The length of the
segments in which the piles are subdivided is another
variable. The user can also define limits regarding
number of increments and interations, and the
tolerance, to be used in the non linear analysis
procedure. In order to simplify the input, default
values are adopted when a particular option is not
specified.

After all data are available, the piles are automatically subdivided into segments. The nodal points joining consecutive segments are arranged in a special internal order, selected to maximize the processing efficiency. Starting with the first pile, the nodes are ordered from the lower extreme to the top of the pile but without including the connection to the substructure. The substructure nodes are the last nodes in the list. Figure 2 shows this ordering, for a very simple case. The resulting structure for the stiffness matrix is shown in Figure 3.

The soil is represented by springs connected to the nodal points. The non linear behaviour of those springs, in the lateral directions, is defined in terms of the P-Y curves given by the API-RP-2 code [8]. If requested, longitudinal springs defined in terms of the P-Delta curves, representing skin friction, are also included. For each nodal point coefficients are computed in order to define the piece-wise representation for the soil curves. The stiffness coefficients for the springs are obtained evaluating the tangent to the soil curves for the corresponding pair of pressure-displacement values. In the case of the P-Y curves this is done for the total lateral displacement, obtained as the vector sum of the displacements along two normal lateral directions.

The global analysis scheme involves an incremental solution, such that the following operations are performed for each load increment:

a. - Solution of the linear system,

$$K_T \ U_i = P_i \tag{10}$$

where K_T is the total stiffness matrix, U_i is the displacement vector, and P_i is the load vector, for the load step i.

b. - Evaluation of the forces F_s in the load soil springs, and of the geometrical non linear forces F_g, due to displacements U_i.

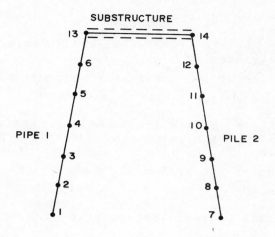

Figure 2 - Example of Nodal Ordering

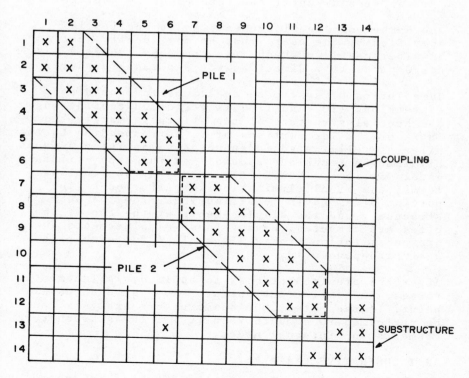

Figure 3 - Structure of Stiffness Matrix

c. - Evaluation of the unbalanced load vector

$$\Delta \underset{\sim}{P} = \underset{\sim}{P}_i - \underset{\sim}{K}_L \underset{\sim}{U}_i - \underset{\sim}{F}_s - \underset{\sim}{F}_g \tag{11}$$

where $\underset{\sim}{K}_L$ is the linear pile and substructure stiffness matrix.

d. - Evaluation of the total stiffness matrix

$$\underset{\sim}{K}_T = \underset{\sim}{K}_L + \underset{\sim}{K}_s + \underset{\sim}{K}_g \tag{12}$$

where $\underset{\sim}{K}_s$ is the non linear soil spring stiffness matrix and $\underset{\sim}{K}_G$ is the geometrical non linear stiffness matrix.

e. - Solution of the system

$$\underset{\sim}{K}_T \Delta \underset{\sim}{U} = \Delta \underset{\sim}{P} \tag{13}$$

f. - Correction of the displacements

$$\underset{\sim}{U}_i = \underset{\sim}{U}_i + \Delta \underset{\sim}{U} \tag{14}$$

g. - Convergence test. If the test is not satisfied steps b to g are repeated. Otherwise a new load increment is added and the process is re-started at step a, until the total load is applied.

The number of load increments to be applied is specified by the user. Thus he may specify just one load increment, in which case the above scheme corresponds to a direct Newton-Raphson procedure.

The solution of system of equations indicated by expressions (10) and (13) is performed considering the special structure of the stiffness matrix, shown in Figure 3. The portions of K_T corresponding to the piles are eliminated independently, applying a Gauss elimination technique specialized for hyper-tridiagonal matrices. Then, after reducing the coupling terms, the final elimination is performed to the portion of K_T having contributions from the substructure, and the displacements at the top of the piles are computed. Having those displacements the backsubstitution process is applied for each pile, again independently.

After the previous process is applied, displacements, forces, and soil reactions, are known at each nodal point. If requested, the substructure analysis can be performed to compute results for the substructure elements and internal nodes.

AN ILLUSTRATIVE EXAMPLE

The following listing shows the STRUDL data regarding

the foundation-structure interaction study for a
real platform.

```
STRUDL  'SIMPLE'  'FOUNDATION-STRUCTURE  INTERACTION'
   .        .        .
   .        .        .
   .        .        .
LOADING  1   'DEAD WEIGHT+BUOYANCY+WAVE'
DEAD WEIGHT
1  TO  170   PZ  -1.
BUOYANCY LOADS
1  TO  170  Z  -1.
WAVE LOAD PERIOD  10.5  HEIGHT  12.53  SWL  32.77 ANGW  315. -
SURFC  1.37  BOTTC  0.3  MEMBERS  1  TO  170
RETRIEVE STIFFNESS MATRIX  777  FILE  'CM-20x20'
SUBSTRUCTURE
888 ELEMENTS 1 TO 170  777 BOUNDARY NODES 91 98 97 94
SOIL CHARACTERISTICS LAYERS  2
1 CLAY DEPTH 11. WEIGHT 0.6 ANGLE 7. COHESION 0.4
2 MEDIUM SAND DEPTH 70 WEIGHT  0.9  ANGLE  32.
PILE  ANALYSIS WITH LATERAL FRICTION PILES 4 INCREMENTS 4 SUBST 888
1 NODE  91 X -15.6  Y-15.6  Z-65.  ED  0.762  THIC  0.044
2  NODE  98 X  15.6  Y -15.6  Z -65.  ED 0.762  THIC  0.044
3  NODE  97 X  15.6  Y  15.6  Z -65.  ED 0.762  THIC  0.044
4  NODE  94 X -15.6  Y  15.6  Z -65.  ED 0.762  THIC  0.044
SUBSTRUCTURE ANALYSIS  888
LIST DISPLACEMENTS FORCES ALL
FINISH
```

After the title the user would issue the commands
giving the structural definition for the first
component, including deck, jacket, and portion of the
piles above seabottom level. This definition is done
using mostly conventional STRUDL commands, and is not
shown due to space limitations.

Three types of loads are included in the loading
condition to be considered. The first type corresponds
to dead weight loads, for all elements of the first
component, applied in the Z axis negative sense. The
buoyancy loads are similarly defined. The wave loads
will be computed for a wave having a 10.5 seconds
period and a height of 12.53 meters. Data regarding
the still water level (SWL) attack angle (ANGW),
surface current (SURFC), and bottom current (BOTTC),
are also given. Since no mention is made the 5th
order Stokes theory will be used. The command will find
the critical wave varying the offset, and will compute
the member forces using Morison's equation.

The deck was condensed previously and saved on disk.
Its stiffness matrix is retrieved and is assigned to

superelement 777. The condensation of the first component, arbitrarily called 888, is done using the substructure command. Note that the deck substructure, 777, is included in the condensation performed. The soil is defined as a two-layers system, composed of a silty clay from 0 to 11. meters, and a medium sand from 11 to 55 meters.

The global analysis is requested for a system of 4 piles, connected to substructure 888. Consideration of lateral friction is specified and the load will be applied in 4 increments. Since no mention is made, the piles will be subdivided in segments of 1. meter length. Similarly, the pile behaviours will be geometrically linear.

Finally the analysis of substructure 888 is requested, and is performed using the displacements at its boundary nodes, computed in the previous commands. After result output the processing is ended.

CONCLUSIONS

The procedure described in the previous sections provides an integrated computational approach for the analysis of fixed steel platforms supported on pile foundations. It has been used for several months now, by engineers of the Engineering Division of the Exploration and Production of Petrobrás S.A., showing a good adaptability to the situations encountered in the analysis and design of real platforms. The fact that it is part of a problem oriented language makes the procedure very easy to learn and to use. Its compatibility with the conventional STRUDL commands allows the engineer to take advantage of other STRUDL capabilities for the analysis of offshore structures.

Current efforts are directed toward the implementation of a similar procedure for the dynamic analysis of foundation-structure interaction problems. This procedure is based on the modal superposition method, and includes an iterative algorithm for equilibrium checking at each time step, performed at the level of the modal coordinates.Preliminary tests performed with this technique have shown considerable efficiency advantages over direct numerical integration techniques, for the types of problems being considered.

REFERENCES

1. - A.J. Ferrante, E.C.P. de Lima, N.F. Ebecken,

'Problem Oriented Languages for Finite Element Analysis'. Proceedings of the International Conference on Applied Numerical Modelling, Madrid, 1978, Pentech Press (1979)

2. - R.D. Logcher et all, ICES STRUDL II - Engineering User's Manual, Vol. 1, Report R68-91,and Vol. 2, Report R70.71, Massachusetts Institute of Technology, Cambridge, Mass., USA.

3. - J. Penzien, 'Soil-Pile Foundation Interaction', Chapter 14, Earthquake Engineering (Edited by R.L. Wiegel), Prentice Hall, 1970,pp.349-381.

4. - H. Matlok, 'Correlations for Design of Laterally Loaded Piles in Soft Clay',Paper OTC 1204, Proceedings, Second Annual Offshore Technology Conference, Houston, Texas, 1970.

5. - R.C. Reese, W.R. Cox, F.D. Koop, 'Analysis of Laterally Loaded Piles in Sand', Paper OTC 2080, Proceedings, Sixth Annual Offshore Technology Conference, Houston, Texas, Vol. 2, pp. 473-483.

6. - J.S. Przemieniecki, 'Matrix Structural Analysis of Substructures, J.Am. Inst. Aeron. Astron.,pp. 138-147, January 1963.

7. - A.K. Noor, H.A. Kamel, R.E. Fulton, 'Substructuring Techniques-Status and Projection', Computers and Structures, Vol. 8, pp. 621-632, May, 1978.

8. - API Recommended Practice for Planning, Designing and Constructing Fixed Offshore Platforms, 7th Edition, New York, 1976.

SLAM LOAD HISTORIES ON CYLINDERS

I.M.C. Campbell, P.A. Weynberg

Engineers at the Wolfson Marine Craft Unit, University of
Southampton.

SUMMARY

Force histories and pressure distributions have been measured
from drop tests on a horizontal cylinder slammed into calm
water. From these measurements a mean hydrodynamic slam load
history has been derived. This history may be used together
with strip theory to predict loads for inclined impacts, and
the results have been confirmed experimentally. A structural
response to slam loading may be predicted from the load
histories using associated damping and added mass terms.

BACKGROUND TO THE PROBLEM

Experimental approach
Slamming is of a transient nature, and in the case of horizontal cylinders, the impact load rises suddenly. This makes
measurement of the load and pressure history difficult, since
they are masked by the response of the transducers. During
wave impact slamming, buoyancy, drag and inertia forces
successively load a body, and experiments provide a measure
of this resultant force history. Miller (1977) has considered
the loading regimes and concluded that slamming forces predom-
inate for Froude numbers greater than 0.6. Also in (1979) he
has reviewed the results of 11 slam experiments. These lack
complete agreement, and some show considerable scatter in the
data. Several experimenters have attributed the scatter to an
indeterminate rise in the slam load associated with unknown inclinations of the cylinder axis at impact. Tests using generated waves are particularly susceptible to this problem and may
also suffer because of an inability to reach high Froude numbers
which thereby makes it difficult to isolate the slam load from
the resultant force. High Froude numbers can be achieved with
drop tests, however the problem then becomes one of obtaining a
high enough response frequency from the force transducers in
order to follow the fast rise and decay of the slam load.

Furthermore, as Froude numbers and response frequencies increase smaller misalignments affect the response and even $\theta=5'$ was found to be significant in the present series of tests. Hence any slam experiment requires careful design if accurate measurements are to result.

Theoretical approach

Mathematical theories unfortunately have produced a variety of results for not only the well-discussed initial slam coefficient but also for the subsequent decay. The discrepancies are due to differences in assumptions, but their effect appears to have proved difficult to quantify in the absence of experimental evidence. Moran (1965) in his review of theories has considered most of the analytical potential flow solutions as well as numerical methods and the analysis of effects from variable entry speed, gravity, water, compressibility and air density.

He noted that all conventional solutions had sources of error. Briefly, Wagner's 'wetting correction' to the free surface shape, applied by Schnitzer and Hathaway (1953) to cylinders, doubled the initial slam load predicted by Von Karman (1929) using added mass considerations and linearised theory; however, flat plate fitting to body boundary conditions involved singularities at the spray root. These were avoided by Fabula (1957) using ellipse fitting, but there remained, as a result of the linearised theory, a singularity at the point of impact. Chou's method (1946), applied by Fabula and Ruggles (1955) to cylinders also yielded a non-singular pressure at the spray root and satisfied the dynamical free surface boundary conditions, but ignored the kinematic free surface boundary condition.

The analysis of some of the more recent slam investigations has in part been based on the foregoing theories. Miller (1977) considered the effect of rise times on structural response by extending Von Karman's analysis to inclined impact, and used this as an input to a dynamic analogue with which he compared his experimental results. Sarpkaya (1978) considered the effect on slam load of velocity changes during wave passage using the added mass of circular arc lens given by Taylor (1930). He also checked his experimental results against a dynamic analogue. Faltinsen et al (1977) extended Fabula and Ruggles' method using slender body theory to fit the body boundary conditions. However, the predicted slam load was less than that found in experiments by Sollied quoted in the same paper. The predicted slam load was then used in calculations on the dynamic structural bending of cylinders, and the results were compared with experiments, but again discrepancies were found.

Arhan et al (1978) sought to correlate the slam impulse, from experiments, with cylinder size and velocity, but without success.

So it appears that investigators have failed to confirm the

accuracy of any particular slam theory and in some cases have
resorted to assuming the fit of a theory in order to unravel
their experimental data. This indicates the difficulties involved in correlating hydrodynamic loads with the transient
responses measured in experiments.

EXPERIMENTAL DETAILS

Description of test rig

The test rig has been developed from that described by Campbell
et al (1977) and has been in regular use since 1975.

The test cylinders of 2, 3, 4 or 5in diameter x 32in length
were attached to a 4ft pivot arm via two compression force
transducers and a cross member, which for most tests was lead
filled to react to the transducer loads. The slam speed was
varied between 6 and 14ft/s by forcing the arm with a pneumatic
cylinder, and was controlled by hydraulics. The speed was monitored by a displacement and latterly an inductance transducer.
The water tank was 6ft long x 5ft wide x 4ft deep with overflow
troughs to enable rapid settling of the surface. The test
cylinder, force transducers and cross member could be inclined
to the water surface at angles to $\theta=15^{\circ}$ with an accuracy of $\pm\frac{1}{2}'$.
The rig was designed to be stiff and the minimum response
frequency was approximately 500Hz.

Pressures were measured using KULITE XTMS semiconductor transducers with 0.1in diameter diaphragms and a natural frequency
in water of approximately 40kHz.

The transducer signals were conditioned through instrumentation
with a 0-10kHz bandwidth to a 4 channel DATALAB DL905 digital
transient data recording system with 1 kbyte memories. The
recorders were triggered from a water probe, and most signals
had a noise level of less than the 0.4% F.S. digitising resolution.

Test procedure

The pressure data presented was obtained from a light alloy
cylinder machined to 4in diameter to eliminate a slight bow,
and tested without end plates. Records were typically of
5-20ms, enabling phase and rise times to be determined to 5µs.
The pressure transducers were statically calibrated following
each test series and were occasionally found to have changed
sensitivity by up to 15% as a result of possible gauge slip.
Prior to each test the cylinder was wiped dry as drips were
found seriously to affect the response. The water was also
checked to be free from surges of greater than 0.005in
amplitude.

The transient response of the rig was studied using accelerometer and force transducer measurements. The vibration was

found to be of a complex distributed mass nature. However, the force transducers were designed to provide an output associated with the resultant heaving motion of the cylinder with minimum interaction from other modes. By comparison of the force phasing with pressure measurements, accurate indications of the start of the slam were obtained which were particularly valuable in tests with end plates. Both force and pressure signals were at steady zero prior to impact, and this level was recorded. Daily static calibrations confirmed force transducer sensitivities within 2%.

Measurements during the slam period up to a value of Ut/D=1 showed changes in the mean velocity of 5% with oscillations superimposed. However, in horizontal impact tests a constant mean velocity was achieved during the initial slam period to Ut/D≈0.4.

Both the tests and method of analysis were continually refined, and from over 800 records only the most reliable data has been presented.

ANALYSIS AND DISCUSSION OF EXPERIMENTAL RESULTS

Slam description
A high speed film was taken at 40000 frames/sec of a cylinder slamming without end plates. This showed two steadily rising sheets of water projected out from the spray roots on either side of the cylinder, but with little disturbance from the ends. An underwater view showed that after approximately half-immersion streams of bubbles formed along the cylinder, apparently from discrete points, and indicating cavitation or ventilation. Measurements from pressure transducers indicated that the initial rise of the spray root was approximately twice that of static water level penetration. Faltinsen et al (1977) also reported photographing slamming and found the top of the cylinder to be dry at full static water level penetration, but described the sheet spray as waves.

Horizontal impact data
Pressure measurements   The responses from pressure transducers exhibited oscillations associated with the vibration of the test rig. The pressure histories were derived by fitting mean curves through the response as shown in Fig.1. Because the natural frequency of the transducer was high compared with the frequency of the oscillation there was little difficulty in fitting mean curves up to the initial peak response. The pressure histories were normalised with respect to the instantaneous mean velocity.

Careful analysis of a large number of pressure records has revealed the following points:-

(a) even though small pressure transducers were used, the

diaphragm size proved too large to enable resolution of the pressure distribution close to the spray root. Both the mean pressure distribution (Fig.3) and the predictions of Wagner (1931) and Fabula (1957) indicate rapid changes within a diaphragm width of the spray root.

(b) Accurate measurements of the rise of the spray root were possible from the phase of the rise in the pressure transducer response. The results in Fig.2 were recorded with an accuracy of ±0.0002 Ut/D, and the scatter at very early immersions may be attributable to small initial disturbances in the water surface or to a depression from air cushioning.

(c) The pressures from transducer measurements at the bottom of a cylinder, $\beta=0^{\circ}$, were characterised by a slight initial rise, $C_{P_A}<26$, followed by a rapid rise to the peak within 20μs. The initial rise was also noted by Arhan et al (1978) and was attributed to air cushioning. Measurements from transducers further around a cylinder at $\beta=6-30^{\circ}$ showed no initial rise, but exhibited increasing rise times, due to the passage of the spray root across the transducer diaphragm.

(d) Whilst analysed results from a particular series of transducer measurements exhibited little scatter as evidenced in Fig.4, considerable differences were found between the results from similar tests with different transducers. The differences in mean levels for the data presented are within ±15% and have been attributed to:

   (i) the low output (<10% rated) from the transducer following the rapid decay during the spray root passage;

  (ii) zero drift from thermal effects following impact;

 (iii) long term changes in transducer sensitivities.

Thus comparison of absolute pressure measurements was considered unreliable after immersions of Ut/D>0.036, although the relative pressures presented should be valid beyond this immersion.

(e) Spanwise pressure measurements indicated that end plates seriously affected the phase and character of the pressure distribution up to 10% of the length from each end. It is likely that the end plates disturbed the water surface prior to the cylinder impact, and force measurements exhibited an associated initial rise.

With end plates removed the phase from spanwise pressure measurements confirmed that alignment at impact was $\theta=\pm\frac{1}{2}'$, and within the limitations of (d) there were no significant pressure variations up to 4% of the length from each end.

(f) The pressure distributions given in Fig.3 were derived from

36 case histories taken at 3 Froude numbers, and were faired by cross plotting to average the errors discussed in (d). The output from a transducer was taken to represent the mean pressure on the diaphragm from the time it became fully covered. The slam load in Fig.7 was derived by graphical integration from Fig.3 and indicates the fairness of the distributions.

Force measurements   The responses from the force transducers exhibited oscillations associated with the vibration of the test rig. An example is shown in Fig.5. These varied in character depending on the condition of the test, although there was a dominant oscillation at approximately 500Hz, and generally higher frequency oscillations superimposed. In some instances long cycle beating was observed, and care was required when comparing with analogue responses to distinguish between beat decay and damping.

The hydrodynamic slam loading was derived from the measured responses by fitting mean curves, as for example in Fig.5. The comparatively slow response of the transducers compared with the sudden rise of the slam load caused the mean load to be least determinable over the first cycle of the vibration response, which corresponded to $Ut/D<0.05$, and is reflected in the greater scatter in slam coefficient data at the instant of impact than at subsequent immersions, as can be seen in Fig.7. The fit of the mean curves was checked by comparing measured responses with those predicted from a dynamic analogue presented in Appendix II.

This analogue idealised the test rig as a single mass connected to a rigid test cylinder by two springs representing the force transducers. Predictions of damping and added mass terms, associated with variable impact velocity, were included and the response in the force transducers was compared with an input slam load applied to a cylinder. A typical result from the analogue is given in Fig.6 and it can be seen to match well the corresponding measured response in Fig.5.

The derived slam load histories from tests on 2, 3, 4 and 5in diameter cylinders are given in Fig.7. The tests were at Froude numbers from 1.9-5.6 and Reynolds numbers from 0.8-4.4 x 10^5 with and without end plates, but no correlation was found between these parameters and the scatter in the load histories. The mass ratio factor of $m_1(m_1+m_2+m_o)$ discussed in Appendix II was applied in scaling the slam load to correct for the reduction in measured force, due to the finite rig mass. The factor, also considered by Watanabe (1933), varied from 1.065-1.582 and was found to improve the correlation of results from tests at different rig masses. The slam load integrated from the pressure distribution is within the scatter of slam loads from force measurements and close to the mean, as can be seen in Fig.7.

The maximum possible contribution of buoyancy to the slam coefficient was $\pi/2N_F^2$ i.e. $C_{S_B}=0.05-0.54$. However, the presence of spray root rise, sheet spray, and cavitation or ventilation, leaves doubt as to how buoyancy may be subtracted from the measured resultant load. Most previous experimenters have assumed the buoyancy load to rise with the penetration of static water level, although this can only be an approximation, and for example in ship motion calculations corrections are made to the hydrostatic pressures for attenuation due to the presence of waves. Also, most conventional theories assume zero gravity and neglect cavitation effects, as discussed by Moran (1965), and so offer no guidance. The slam load data given in Fig.7 for immersions $Ut/D>0.6$ were derived at a Froude number of $N_F=2.58$ with $C_{S_B}=0.24$, for which no correction has been made. The possible error in Fig.7 through neglecting buoyancy at early immersions of Ut/D 0.2, when the slam load is high, is likely to be <5%.

An empirical equation which describes the mean load history in Fig.7 is: $C_S = 5.15/(1+19Ut/D) + 0.55Ut/D$. \hfill (1)

Inclined impact data
Pressure measurements
The pressure histories from selected transducers were compared at various inclinations to $\theta=8^\circ$. The oscillations in the transducer responses reduced with inclination, rendering curve fitting almost unnecessary. The data from one representative transducer is shown in Fig.4, and it can be seen that there was no significant variation in the pressure history with inclination. The phase of the rise of spanwise pressures corresponded to that from the penetration of static water level to within 10%, and the phase of the rise of circumferential pressures was similar to that obtained from horizontal impacts, as can be seen in Fig.2.

This data was in agreement with the assumptions of the strip theory of inclined impact given in Appendix 1.

Force measurements
As the cylinder was inclined the rise of the slam load to a maximum was less sudden, until at approximately $\theta=15'$ it could be matched by the transducer response. The subsequent oscillations in the transducer response were far less severe than in the case of horizontal impacts, and the peak slam load was easily obtained by curve fitting. At inclinations from $\theta=1'$ to $15'$ the peak response steadily reduced.

Slam load histories derived from measurements at inclinations up to $\theta=8^\circ$ are given in Fig.8. The results were obtained principally at a Froude number of $N_F=2.58$, with some checks at $N_F=3.66$. It can be seen that the rise in slam load was not linear and that the peak load decreased with inclination. The strip theory in Appendix 1 was used to compute inclined impact load histories based on the empirical result from horizontal impacts equation (1). The results are given in Fig.7, and it can be seen that there is good correlation between the measured and predicted rate of rise and peak slam loads, with slight discrepancies for rise times and rates of decay.

Correlation with other results

Comparison with the 'conventional' mathematical theories discussed earlier with experimental results has indicated that only Wagner's flat plate fitting or Fabula's ellipse fitting solutions with wetting corrections, approximately match the measured initial spray root rise, and hence give comparable pressure distributions. The mean empirical slam load history has a lower initial value and decays faster than either theory predicts, but that of Fabula is closest. The differences could be due to the assumptions and approximations of the theory discussed earlier.

The responses measured in other experiments in which small inclinations or disturbances may have been present, could have resulted from the inclined load histories presented here. For example, from Fig.7 at an inclination of 0.58° the peak slam load is $C_S = \pi$, a value which several previous experimenters have attributed to the initial slam load for horizontal impacts, after due allowance for dynamic response and rise times, but neglecting strip theory, whereas, from this series of tests, the mean measured initial slam coefficient for horizontal impact was found to be just over 5.

CONCLUSIONS

1. The slam load history for the horizontal impact of a cylinder may be described by the empirical equation
$C_S = 5.15/(1 + 19Ut/D) + 0.55Ut/D$.

2. The initial value of the slamming coefficient was the most difficult part of the history to derive from force measurements, and the scatter in experimental data greatly reduced after immersions of $Ut/D=0.05$.

3. The initial spray root rise to $Ut/D=0.03$ was to approximately twice that of static water level penetration, and cavitation or ventilation was observed after $Ut/D \approx 0.50$; so there was uncertainty on how buoyancy forces increased with immersion.

4. Experimental results have confirmed that the empirical slam load equation may be extended to predict the load history for inclined impacts using strip theory.

5. Added mass and damping terms, associated with oscillating impact speeds, should be used when predicting a structural response to slam loading.

ACKNOWLEDGMENTS

The experiments were conducted with the assistance of Mr. J.
Robinson, and Dr. J. Wellicome provided advice on theoretical
aspects.

The project was funded by the Department of Energy through the
Offshore Structures Fluid Loading Advisory Group and the Offshore Energy Technology Board as part of an overall research
programme into fluid loading on offshore structures, and
resulted from initial research on behalf of British Petroleum
Trading Limited.

NOMENCLATURE

A,B,C	Constants
c_2	Damping coefficient
C_p	Slam pressure coefficient
C_{pA}	Slam air pressure coefficient
C_s	Slam coefficient for horizontal impact
C_{s_B}	Buoyancy contribution to slam coefficient
C_{s_θ}	Slam coefficient for inclined impact
D	Cylinder diameter
E	Kinetic energy
F_o	Slam force for constant impact velocity
F_S	Slam force = $\frac{1}{2}\rho U^2 L\ D\ C_s$
k_2	Stiffness coefficient
L	Cylinder length
m	Hydrodynamic added mass
m_o	Hydroelastic added mass
m_2	Effective mass of cylinder
m_1	Effective mass of rig
N_F	Froude number
t	Time from start of slam on an element
t_1	Time from start of slam for inclined impact
t_r	Rise time to peak slam load
u	Slam velocity oscillating component
U	Slam velocity
U_o	Slam velocity constant component
V	Slam speed component of U normal to cylinder
x,y,z	Rectangular Cartesian Coordinates
z'	Immersion of cylinder
$z_{1,2}$	Oscillatory displacements of cylinder and rig
β	Half angle subtended by water plane intersection at centre of cylinder
θ	Angle of entry of inclined cylinder
ρ	Mass density of water
σ	Standard deviation of normal distribution

REFERENCES

Arhan, M., Deleuil, G. and Doris, C.G. (1978) Experimental study of the impact of horizontal cylinders on a water surface. 10th Annual O.T.C., Houston, Texas.

Campbell, I.M.C., Wellicome, J.F. and Weynberg, P.A. (1977) An investigation into wave slamming loads on cylinders (OSFLAG 2A) Tech.Rep.Ctr.OT-R-7743.

Chou, P.Y. (1946) On impact of spheres upon water. Water Entry and Underwater Ballistics of Projectiles, Ch.8, Cal.Inst.Tech. OSRD Rpt.2251.

Chuang, S. (1967) Experiments on slamming of wedge shaped bodies. Journal Ship Res.

Fabula, A.G. (1957) Ellipse fitting approximation of two dimensional normal symmetric impact of rigid bodies on water. Proc. 5th Midwestern Conf. on Fluid Mechanics, Univ.Michigan.

Fabula, A.G. and Ruggles, I.D. (1955) Vertical broadside water impact of circular cylinder: growing circular arc approximation. Navord Rep.4947.

Faltinsen, O., Kjærland, O., Nottveit, A. and Vinje, T. (1977) Water impact loads and dynamic response of horizontal circular cylinders in offshore structures. 9th Annual O.T.C. Houston.

Greenberg, M.D. (1967) On the water impact of a circular cylinder: numerical results. Therm.Adv.Res.Rep. TAR-TR-6705.

Holmes, P., Chaplin, J.R., Flood, C. (1976) Wave slamming loads on horizontal members. Tech.Rep.Ctr. OT-R-7706.

Miller, B.L. (1977) Wave slamming loads on horizontal circular elements of offshore structures. Tech.Rep.Ctr. OT-R-7744

Miller, B.L. (1977) Wave slamming loads on horizontal circular elements of offshore structures. R.I.N.A.

Miller, B.L. (1979) Wave slamming loads on offshore structures. N.M.I. Report to Dept. of Energy (to be published).

Moran, J.P. (1965) On the hydrodynamic theory of water-exit and entry. Therm.Adv.Res.Rep. TAR-TR-6501.

Sarpkaya, T. (1978) Wave impact loads on cylinders. 10th Annual O.T.C., Houston.

Schnitzer, E. and Hathaway, M.E. (1953) Estimation of hydro-dynamic impact loads and pressure distributions on bodies approximating elliptical cylinders with special reference to water landings of helicopters. N.A.C.A. TN2889.

Taylor, J.L. (1930) Some hydrodynamical inertia coefficients. Phil.Mag.9 (series 7) 161-183.

Verley, R.L.P. and Moe, G. (1978) The effect of cylinder vibration on the drag force and resultant hydrodynamic damping. Symposium on Mechanics of Wave-induced Forces on Cylinders, Bristol, U.K.

Von Karman, T.L. and Wattendorf, F. (1929) The impact on sea-plane floats during landing. N.A.C.A. TN321.

Wagner, H. (1931) Landing of seaplanes. N.A.C.A. TN622.

Watanabe, S. (1933) Resistance of impact on water surface: Parts I-VII. Sc.Pap. I.P.C.R., Tokyo.

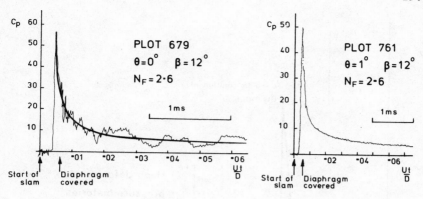

Figure 1: Typical Pressure Records

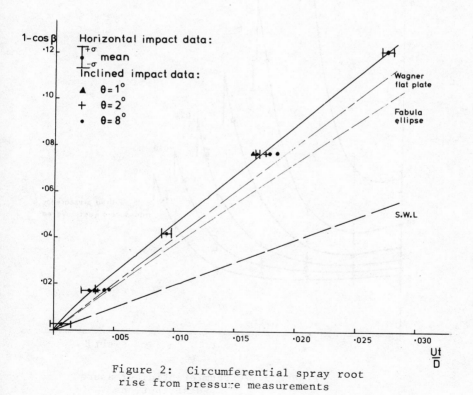

Figure 2: Circumferential spray root
rise from pressure measurements

292

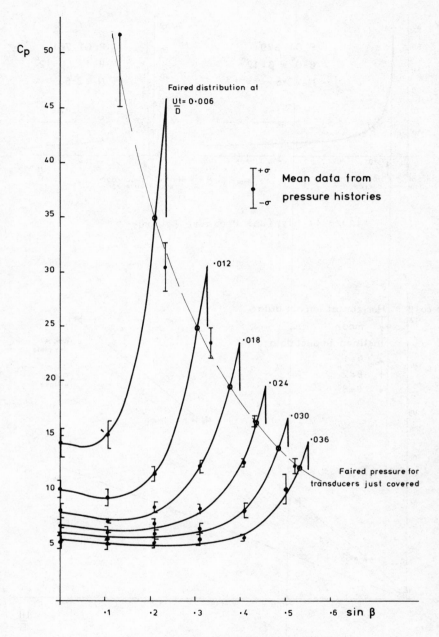

Figure 3: Circumferential pressure
distributions

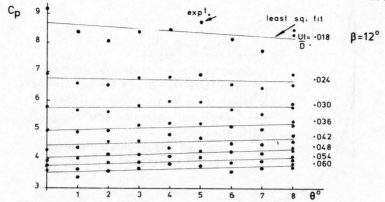

Figure 4: Comparative pressure histories at various inclinations

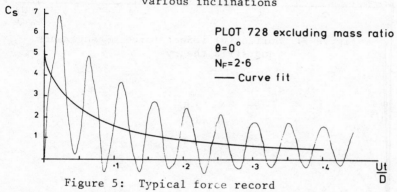

Figure 5: Typical force record

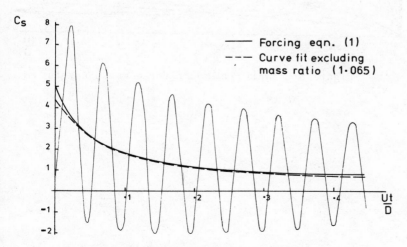

Figure 6: Analogue force record

294

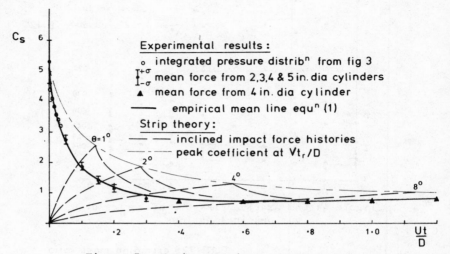

Figure 7: Horizontal impact force measurements
and strip theory

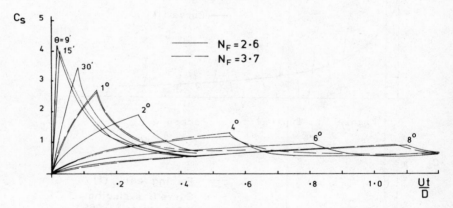

Figure 8: Inclined impact force measurements

APPENDIX I

Strip theory of inclined impact

Consider a cylinder of finite length L slamming into a water
surface at an inclination θ with a velocity U normal to the
water surface, as shown in Fig.9.

Assume that the normal force history on the cylinder element dy
may be determined from two-dimensional slam data which is unaffected by the axial component of the impact velocity U sin θ.
At time t_1 after impact ($t_1=0$ when $y_1=0$)

$$dF_s = \tfrac{1}{2}\rho V^2 D \, C_s(t) \, dy \quad \text{and} \quad L \tan \theta = Vt_r \qquad (2)$$

where t_r is the rise time to peak force ($y_1=L$).

Normalising with respect to V and L, the average inclined impact
slam coefficient

$$C_{s_\theta} = \frac{1}{L} \int_o^y C_s(t) \, dy$$

where t is the local time from the start of slam on element dy.

Putting

$$C_s = \frac{A}{1 + BVt/D} + \frac{CVt}{D}$$

gives for $\dfrac{Vt_1}{D} \leqslant \dfrac{Vt_r}{D}$ $\quad Vt = y \tan \theta$

$$C_{s_\theta} = \frac{A}{BVt_r/D} \ln(1+BVt_1/D) + \frac{C}{2Vt_r/D} (Vt_1/D)^2 \qquad (3)$$

and for $\dfrac{Vt_1}{D} \geqslant \dfrac{Vt_r}{D}$ $\quad y_1=L, \; Vt_1 = Vt + (L-y) \tan \theta$

$$C_{s_\theta} = \frac{A}{BVt_r/D} \ln \frac{(1 + BVt_1/D)}{(1+B(Vt_1/D - Vt_r/D))} + C(Vt_1/D - Vt_r/2D) \qquad (4)$$

The above results are plotted in Fig.7 based on the constants
from the mean empirical equation (1).

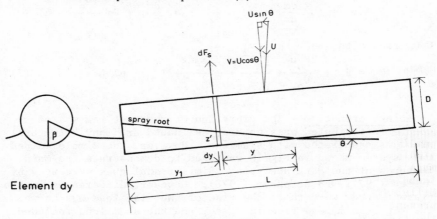

Figure 9: Geometry of inclined impact

APPENDIX II

Dynamic analogue of slam response

Effects of variable entry speed   The structural vibration
resulting from slamming is shown below to affect the slam loading. This effect is important, both in the analysis of data
from slamming experiments and in the design prediction of
structural response.

Using the general 'added mass' approach to the prediction of
slam loading:

$$\text{Kinetic Energy } E = \tfrac{1}{2}\rho m U^2 \tag{5}$$

where added mass $m = m(z')$ and immersion $z' = z'(t)$.

For the case of constant speed entry $U = U_o$

$$F_o \dot{z}' = \dot{E} = \tfrac{1}{2}\rho U_o^2 \frac{dm}{dz'} \dot{z}'$$

$$F_o = \tfrac{1}{2}\rho U_o^2 \frac{dm}{dz'} . \tag{6}$$

Conventional slam theories continue to find dm/dz' as a function
of the immersion z'.

For the case of a cylinder vibrating, as a result of the slam,
the speed can no longer be considered constant. Putting
$U = U_o + u(t)$,

$$F_s \dot{z}' = \tfrac{1}{2}\rho\frac{d}{dt}(mU_o^2 + 2mU_o u + mu^2)$$

from linearised theory

$$F_s \dot{z}' = \tfrac{1}{2}\rho\dot{z}'(U_o^2 \frac{dm}{dz'} + 2mU_o \frac{\dot{u}'}{\dot{z}'} + 2U_o u\frac{dm}{dz'} + u^2 \frac{dm}{dz'} + 2mu \frac{\dot{u}}{\dot{z}'})$$

$$F_s = \tfrac{1}{2}\rho(U_o^2 \frac{dm}{dz'} + 2U_o u \frac{dm}{dz'} + u^2 \frac{dm}{dz'} + 2m \frac{(U_o+u)}{\dot{z}'} \dot{u})$$

Comparing with (6)

$$F_s = F_o + \frac{2F_o u}{U_o} + \frac{F_o u^2}{U_o^2} + \frac{2\int F_o dz'}{U_o^2} \frac{1}{\dot{z}'}(U_o + u) \dot{u} \tag{7}$$

Equations of motion   The vibration of the test rig was of a
complex distributed mass nature, as indeed are many structural
problems associated with slamming. However, the simple lumped
mass system shown in Fig.10 was used to demonstrate the main
features of the data analysis. Similar analogues were used by
Miller (1977) and Sarpkaya (1978). However, differences in the
present study were that: the rise of the slam load was known,
the rig mass m_1 was free in space, and the slam load included
variable entry speed effects.

Using the notation in Fig.10

$$F_o - \frac{2F_o}{U_o} \dot{z}_2 - \frac{F_o \dot{z}_2^2}{U_o^2} - \frac{2\int F_o \, dz'}{U_o^2 \dot{z}'} (U_o + u)\ddot{z}_2 = m_2 \ddot{z}_2 + c_2(\dot{z}_2 - \dot{z}_1) +$$

$$+ k_2(z_2 - z_1) \tag{8}$$

$$0 = m_1 \ddot{z}_1 - c_2(\dot{z}_2 - \dot{z}_1) - k_2(z_2 - z_1) \tag{9}$$

From the above equations it can be seen that $2F_o/U_o \, \dot{z}_2$ and $F_o \dot{z}_2^2/U_o^2$ are in effect damping terms and were significant in reducing the response peaks. $2\int F_o dz'/U_o^2 \dot{z}'(U_o + u)\ddot{z}_2$ is the added mass in oscillation (m_o) and affected the response frequency.

The equations were solved by computer using the mean empirical equation for F_o and Wagner's estimate of $\int F_o dz'(U_o + u)/\dot{z}'$. No estimate was made of viscous damping since this could be accounted for in the measured value of F_o which included viscous drag. See Verley and Moe (1978) for further discussion of hydrodynamic damping.

A typical result is shown in Fig.6 and the discrepancy between the measured and predicted responses was attributed to physical differences from the simple lumped mass approach and the presence of long cycle beat decay.

The mean force level in the transducers $k_2(z_2 - z_1)$ was shown to differ from the slam load F_o by a mass ratio factor of $m_1/(m_1 + m_2 + m_o)$ due to the finite rig mass.

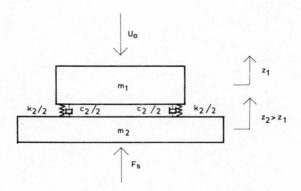

Figure 10: Dynamic analogue of experiment

THEORETICAL ANALYSIS ON THE TRANSVERSE MOTION OF A BUOY BY SURFACE WAVE

Kunihiro Ogihara

Prof. of Engineering, Civil Engineering Dept.
Toyo University, Japan

PREFACE

When a structure has a fairly high center of gravity, a transverse motion is observed in the direction of motion of the incident wave. In severe cases, this transverse motion has an amplitude of oscillation the same as the incident wave.
This phenomenon is observed in a column structure which has a center of gravity above the baesment level. This phenomenon can be regarded as a resonant oscillation between the wave action and the oscillation system of the structure.
Generally as a buoy system has a freedom of movement in all directions, the resonant oscillation in the direction of the incident wave is superior to the transverse one. But in the former described cases, the transverse motion becomes dominant. This phenomenon, therefore, may be understood as different to a forced oscillation. The author treats this phenomenon as an unstable problem of a two-dimensional and two-degree of freedom oscillation, and gets the results as Mathiu's unstable problem.

1. EQUATION OF MOTION

We now consider a case such as an anchored buoy where the anchor rope is always in tensile stress. This condition corresponds to the case that rope does not sag in oscillating motion.
For a column structure which has a circular section and the same rigidity of deformation in all directions, the result of this analysis can be applied after making a few changes.
Now the center of gravity of the buoy system moves around a point $(x,0,z)$, and the wave translates to an x-direction.

$$\left. \begin{array}{l} x_0 = l \sin \theta_0 \\ y_0 = 0 \\ z_0 = l \cos \theta_0 = h \end{array} \right\} \tag{1}$$

When an oscillation occurs, its center moves to a point (x,y,z)

that is determined by the change of angle θ and ϕ. In the case where the length of rope does not change, the following relations are derived.

$$
\left.\begin{array}{l}
x = l \sin(\theta_0 + \theta) \cos \phi - x_0 \\
y = l \sin(\theta_0 + \theta) \sin \phi - y_0 \\
z = l \cos(\theta_0 + \theta) - z_0
\end{array}\right\} \tag{2}
$$

Here the angle θ is shown as the angle between the rope and z-axis; this is not always the same as the direction of the incident wave. The angle ϕ is shown as the angle between the x-axis and a point on the x-y plane which is projected from the center of the buoy.
This angle is the same as the spherical coordinate.
The equations of motion in the θ and ϕ direction are derived as equation (3).

$$
\left.\begin{array}{l}
m l^2 \ddot{\theta} + R\dot{\theta} + k_\theta \theta = 0 \\
m(l \sin\theta_0)^2 \ddot{\phi} + R\dot{\phi} + k\phi = 0
\end{array}\right\} \tag{3}
$$

Here a mass m is added the virtual mass. The term k is the spring constant of deformation. It is considered to be smaller than the value k_θ and in the main comprises the variance of buoyancy.

$$
\begin{aligned}
k\theta &= wAz \cdot l\sin(\theta_0 + \theta) \\
&= wA \cdot l^2 [\cos(\theta_0 + \theta) - \cos \theta_0]\sin(\theta_0 + \theta)
\end{aligned} \tag{4}
$$

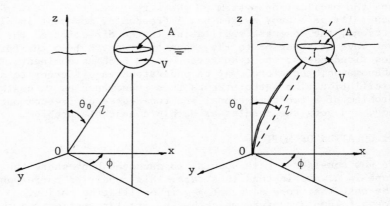

Fig.1 Coordinate system

In a case of small amplitude or small deformation, this equation is transformed to a linear type by the fact that $\theta \ll \theta_0$.

$$
k\theta = wAl^2\sin^2 \theta_0 \times \theta \tag{5}
$$

where A is the area of cross-section of buoy at the sea water surface. In the case that a spring constant of rope k_0 must be considered, the term k is written as follows.

$$
k = wAl^2 \sin^2\theta_0 + k_0 \tag{6}
$$

2. WAVE FORCE ACTS ON THE BUOY

Now the author assumes to analyze this problem as a linear equation. Wave equation and wave force are therefore also used as linear type. Airy's equation is used for wave equation and Morison's equation is used for wave force.

$$
\left.
\begin{aligned}
\eta &= \frac{H}{2} \sin \sigma t \\
u &= \frac{\pi H}{T} \frac{\cosh k'(h+z)}{\sinh k'h} \sin \sigma t \\
v &= \frac{\pi H}{T} \frac{\sinh k'(h+z)}{\sinh k'h} \cos \sigma t \\
\sigma &= \frac{2\pi}{T} , \quad k' = \frac{2\pi}{L}
\end{aligned}
\right\} \tag{7}
$$

$$
\left.
\begin{aligned}
Fx &= Cm \cdot \rho V \frac{\partial}{\partial t}(u-\dot{x}') + Cd \cdot \rho A \frac{(u-\dot{x}')}{2}|u-\dot{x}'| \\
Fz &= Cm \cdot \rho V \frac{\partial}{\partial t}(v-\dot{z}') + Cd \cdot \rho A \frac{(v-\dot{z}')}{2}|v-\dot{z}'|
\end{aligned}
\right\} \tag{8}
$$

As the equations are written for linear type, the term $\dot{x}', \dot{z}'$ is smaller than the term u and v and the second terms in equation (8) are written as form by fourier transform.

$$
\sin \sigma t \cdot |\sin \sigma t| = \frac{8}{\pi}\{\frac{1}{3} \sin \sigma t - \frac{1}{15} \sin 3\sigma t - \frac{1}{105} \sin 5\sigma t
$$
$$
\cdots\cdots - \frac{1}{(4m^2-1)(2m+3)} \sin(2m+1) \sigma t + \cdots \} \tag{9}
$$

$$
\cos \sigma t \cdot |\cos \sigma t| = \frac{8}{\pi}\{\frac{1}{3} \cos \sigma t + \frac{1}{15} \cos 3\sigma t - \frac{1}{105} \cos 5\sigma t
$$
$$
\cdots\cdots + \frac{(-1)^m}{(4m^2-1)(2m+3)} \cos(2m+1) \sigma t + \cdots \} \tag{10}
$$

Then the wave forces can be written by the equations from (7) to (10).
External force of oscillation equation is derived from the terms Fx and Fz as shown equation (11). (cf. Fig. 2)

$$
\left\{
\begin{aligned}
M_0 &= Fx \cdot \cos\phi \cdot l \cdot \cos(\theta_0 + \theta) \\
&\quad - Fz \cdot l \cdot \sin(\theta_0 + \theta) \\
&\qquad \cdots\cdots (11) \\
M_\phi &= -Fx \cdot \sin\phi \cdot l \sin(\theta_0 + \theta)
\end{aligned}
\right.
$$

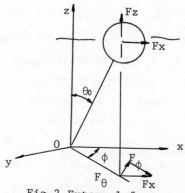

Fig.2 External force

And in the case the values ϕ and θ are small, next approximations can be done.

$$
\cos\phi \simeq 1, \quad \sin\phi \simeq \phi, \quad \cos(\theta_0 + \theta) \simeq \cos\theta_0 ,
$$
$$
\sin(\theta_0 + \theta) \simeq \sin\theta_0 \tag{12}
$$

And we get the following relations.

$$
\left.
\begin{aligned}
M_\theta &= Fx \cdot l \cos\theta_0 - Fz \cdot l \sin\theta_0 \\
M_\phi &= -Fx \cdot \phi \cdot l \sin\theta_0
\end{aligned}
\right\} \tag{13}
$$

3. EQUATION OF MOTION

In this report, in order to analyse the phenomenon as a linear relationship, all the equations hereafter are changed to a linear form. Therefore the terms of second order are neglected. The wave force and external forces of oscillation are written as a combination of the equations from (7) to (13). The results are shown as follows. Here the velocity of a water particle is taken at the mean water surface.

$$
\left.
\begin{aligned}
Fx &= \Sigma \; [Ax(n) \cos n\sigma t + Bx(n) \sin n\sigma t] \\
Fz &= \Sigma \; [Az(n) \cos n\sigma t + Bz(n) \sin n\sigma t]
\end{aligned}
\right\} \tag{14}
$$

$$
\left\{
\begin{aligned}
Ax(1) &= \frac{H}{2} \sigma^2 \coth k'h \cdot Cm \cdot \rho V \\[4pt]
Ax(n) &= 0, \quad n=2,3,4,\ldots \\[4pt]
Bx(2m+1) &= (\frac{\pi H}{T})^2 (\coth k'h)^2 \cdot Cd \cdot \rho A \; \frac{4}{\pi} \frac{1}{(1-4m^2)(2m+3)} \\[4pt]
Bz(1) &= -\frac{H}{2} \sigma^2 Cm\rho V + wA \frac{H}{2} \\[4pt]
Bz(n) &= 0, \qquad\qquad m = 0,1,2,\ldots \\[4pt]
Az(2m+1) &= (\frac{\pi H}{T})^2 Cd\rho A \cdot \frac{4}{\pi} \cdot \frac{(-1)^m}{(1-4m^2)(2m+3)}
\end{aligned}
\right. \tag{15}
$$

External moments of equation of oscillation are shown as follows.

$$
\left.
\begin{aligned}
M_\theta &= [\Sigma A_\theta(n)\cos n\sigma t + \Sigma B_\theta(n)\sin n\sigma t] \cdot l \\
M_\phi &= [\Sigma A_\phi(n)\cos n\sigma t + \Sigma B_\phi(n)\sin n\sigma t] \cdot (-l\phi)
\end{aligned}
\right\} \tag{16}
$$

$$
\left.
\begin{aligned}
A_\theta(n) &= Ax(n) \cdot \cos\theta_0 - Az(n) \cdot \sin\theta_0 \\
A_\phi(n) &= Ax(n) \cdot \sin\theta_0 \\
B_\theta(n) &= Bx(n) \cdot \cos\theta_0 - Bz(n) \cdot \sin\theta_0 \\
B_\phi(n) &= Bx(n) \cdot \sin\theta_0
\end{aligned}
\right\} \tag{17}
$$

To make the equations linear, they are written in series form. Therefore the equations of motion of this buoy system are derived as equation (18).

$$
\begin{aligned}
ml^2\ddot{\theta} + R\dot{\theta} + k_\theta\theta &= l \cdot \Sigma [A_\theta(n) \cdot \cos n\sigma t + B_\theta(n) \cdot \sin n\sigma t] \\
m(l \sin\theta_0)^2\ddot{\phi} + R\dot{\phi} + k_\phi\phi &= (-l\phi) \cdot \Sigma [A_\phi(n) \cdot \cos n\sigma t \\
&\qquad\qquad\qquad + B_\phi(n) \cdot \sin n\sigma t]
\end{aligned} \tag{18}
$$

4. SOLUTION OF EQUATION

An equation of motion in the θ-direction is the same as the equation of forced oscillation and its solution is given as follows.

$$\left.\begin{array}{l} \dfrac{R}{ml^2} = 2\gamma \;,\; \dfrac{k_\theta}{ml^2} = \omega_0^2 \;,\; h = \dfrac{\gamma}{\omega_0},\; u = \dfrac{n\sigma}{\omega_0} \\[2mm] F(u) = \dfrac{u^2}{\sqrt{(u^2-1)^2 + 4h^2 u^2}} \\[2mm] \tan^{-1}(n\sigma\tau) = \dfrac{2\gamma \cdot \sigma n}{[\,\omega_0^2 - (n\sigma)^2\,]} \end{array}\right\} \tag{19}$$

The solution is

$$\theta = \frac{1}{ml} \Sigma [A_\theta(n) \cdot \cos n\sigma(t-\tau) + B_\theta(n) \cdot \sin n\sigma(t-\tau)] \cdot F(u) \tag{20}.$$

The equation of motion for the ϕ-direction is determined by rearranging equation (18), and its result is shown in equation (21). This equation is the same as Hill's equation which has the term that the spring constant varies periodically.

$$m(l\sin\theta_0)^2\ddot{\phi} + R\dot{\phi} + \{k_\phi + l\Sigma[A_\phi(n) \cdot \cos n\sigma t$$
$$+ B_\phi(n) \cdot \sin n\sigma t]\} \cdot \phi = 0 \tag{21}$$

This equation has the series terms of trigonomerical function, but in regular wave, its first term may be considered to be larger than the other higher terms. A further approximation is introduced here, namely the higher order terms of the fourier series in equation (21) are neglected.
This approximation is considered to be satisfied sufficiently for a regular wave. To use the relation

$$\sqrt{A_\phi(1)^2 + B_\phi(1)^2} = C_\phi(1)$$

and to change the origin of time, equation (23) becomes Mathieu's equation of oscillation which has a dumping term.

$$\left.\begin{array}{l} 2\beta = \dfrac{R}{m(l\sin\theta_0)^2} \;,\; \tilde{\delta} = \dfrac{k_\phi}{m(l\sin\theta_0)^2} \\[2mm] \tilde{\varepsilon} = \dfrac{C_\phi(1)}{m(l\sin\theta_0)^2} \end{array}\right\} \tag{22}$$

$$m(l\sin\theta_0)^2\ddot{\phi} + R\dot{\phi} + [k_\phi + l(A_\phi(1) \cdot \cos \sigma t$$
$$+ B_\phi(1) \cdot \sin \sigma t)] \cdot \phi = 0 \tag{23}$$

The parameters in equation (22) apply to equation (23), next equation is derived.

$$\ddot{\phi} + 2\beta\dot{\phi} + (\tilde{\delta} + \tilde{\varepsilon}\cos \sigma t)\phi = 0 \tag{24}$$

304

Now to elminate the dumping term, the relations of equation (25) are introduced.

$$\phi = \text{Exp}(\frac{-\beta}{\sigma t}) \cdot \Phi(\tilde{t}), \quad \sigma t = \tilde{t} \tag{25}$$

Therefore equation (24) becomes a standard type of Mathiu's equation.

$$\ddot{\Phi} + (\delta + \varepsilon \cos \tilde{t})\Phi = 0,$$

$$\delta = \frac{(\hat{\delta} - \beta^2)}{\sigma^2} \quad , \quad \varepsilon = \frac{\tilde{\varepsilon}}{\sigma^2} \tag{26}$$

The solution of this equation can be seen in many books about vibration or oscillation, and its results are shown as figure 3. When the value ε is nearly equal to zero, its result is shown in figure 4.

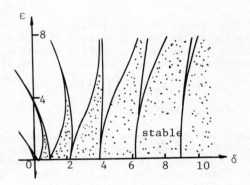

Figure 3 Mathiu's diagram

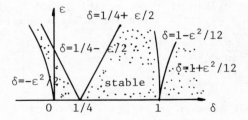

Figure 4

When the solution of equation (26) is assumed as Fourier series as equation (27), the coefficients of each term of this equation

A_j and B_j are able to be derived as equation (28).

$$\Phi = A_0 + \Sigma A_j \cos j\tilde{t} + \Sigma B_j \sin j\tilde{t} \tag{27}$$

$$A_1 = -\frac{A_0}{(\varepsilon/2)}, \quad A_2 = -\frac{[\varepsilon A_0 + (\delta-1)A_1]}{(\varepsilon/2)}$$

$$A_j = -A_{j-2} - A_{j-1} \cdot \frac{(\delta-j^2)}{(\varepsilon/2)}$$

$$B_2 = -\frac{B_1(\delta-1)}{(\varepsilon/2)}$$

$$B_j = -B_{j-2} - B_{j-1}\frac{(\delta-j^2)}{(\varepsilon/2)} \tag{28}$$

After the first terms A_0 and B_1 are determined for small amplitude, the other terms can be derived successively by these equations. But this solution is satisfied only in the case that the term ε is greater.

For the case where this value is small, another solution can be derived by a perturbation method as follows. Now we give a solution as equation (29) which consists of a series of functions for oscillatory equation (24).

$$\phi = \phi_0 + \varepsilon\phi_1 + \varepsilon^2\phi_2 + \dots \tag{29}$$

For this solution to be always satisfied in equation (24) all terms of any power of ε must be zero.

$$\varepsilon^0; \quad \phi_0 + 2\beta\phi_0 + \delta\phi_0 = 0$$

$$\varepsilon^1; \quad \phi_1 + 2\beta\phi_1 + \delta\phi_1 = -\varepsilon\phi_0 \cos \sigma t$$

$$\vdots$$

$$\varepsilon^n; \quad \phi_n + 2\beta\phi_n + \delta\phi_n = -\varepsilon\phi_{n-1}\cos \sigma t \tag{30}$$

The first equation corresponds to the natural damping oscillation. Therefore this is diminished for a long time. We then introduce a small oscillation term such as the disturbance instead of the first solution;

$$\phi_0 = C_0 \cos\sqrt{\delta-\beta^2}\, t \tag{31}$$

The solutions are given as follows.

$$\phi_1 = \frac{C_0\varepsilon}{\sigma(\sigma \pm 2\sqrt{\delta-\beta^2})} \text{Exp}(-\beta t) \cos[(\sqrt{\delta-\beta^2}\pm\sigma)t + \psi_1]$$

$$\vdots$$

$$\phi_n = \frac{C_0\varepsilon^n}{\sigma^n(\sigma \pm 2\sqrt{\delta-\beta^2}) \dots (n\sigma \pm 2\sqrt{\delta-\beta^2}) \cdot n!}$$

$$\text{Exp}(-\beta t) \cos[(\sqrt{\delta-\beta^2}\pm n\sigma)t + \psi_n] \tag{32}$$

Here ψ_1 and ψ_n are the phase angles.

5.APPLICATION TO MODEL TEST

Rearrangement of parameters

The stable or unstable zone is determined by two parameters ε and δ in Mathiu's diagram. It is clearer and more convenient in our case to derive these parameters from the following analysis.

$$\delta = (f_n/f_w)^2 = \frac{1}{\sigma^2} \cdot \frac{1}{m(l\sin\theta_0)^2} \cdot [k_\phi - \frac{R^2}{4m(l\sin\theta_0)^2}]$$

$$
\left.
\begin{aligned}
\varepsilon &= \frac{1}{\sigma^2} \frac{lC_\phi(1)}{m(l\sin\theta_0)^2} \\
C_\phi(1) &= \sin\theta_0 \sqrt{Ax(1)^2 + Bx(1)^2} \\
Ax(1) &= \frac{H}{2}\sigma^2 \coth k'h \cdot Cm \cdot \rho V \\
Bx(1) &= (\frac{H}{2})^2 \sigma^2 (\coth k'h)^2 \cdot Cd \cdot \rho A \cdot \frac{4}{3} \cdot \frac{1}{\pi}
\end{aligned}
\right\} \qquad (33)
$$

SOME INVESTIGATIONS ON MODEL TEST

The trace of the center of the buoy in the motion is observed by an 8mm cine camera in vertical and horizontal directions. The typical traces of these motions in plane view are shown in figure 5. The first one is the case that transverse motion does not occur and the second is the case that transverse motion occurs in the same direction as the motion of the incident wave. The frequencies of oscillation in the x and y direction in the second case are different and the former one is half of the latter. This means the parameter δ is 1/4 and this oscillation corresponds to the first unstable zone of Mathiu's diagram in figure 4.

The traces by calculation which are derived from a sine function such as Asin $2\pi t/T$ for x-direction and Asin$(4\pi t/T + \psi)$ for y-direction, are shown in figure 6. The values as −0.0312 and −0.0625 are the phase angle ψ in radians. These curves are very similar to the second trace of figure 5.

These phenomena about the circular cylinder are also observed by Dr. G.Sawaraghi of Osaka University in Japan and his data are shown in table 1.

Model	fn	fw	δ	
8	1.075	0.913	1.386	
		0.940	1.307	
6	1.351	1.112	1.476	
		0.885	2.33	Table 1
		0.666	4.11	
		0.436	9.60	
3	1.914	0.625	9.37	
		0.620	9.53	

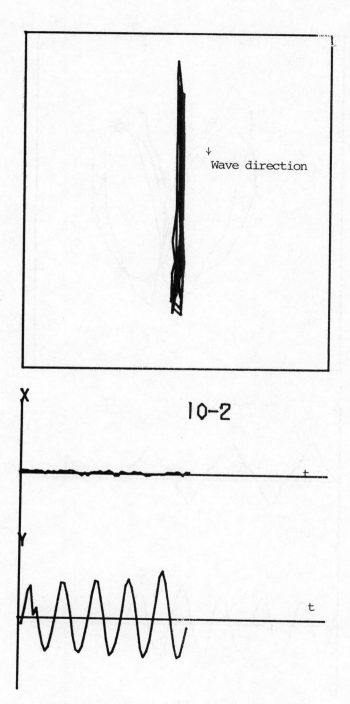

Fig.5.1 Trace of center of buoy

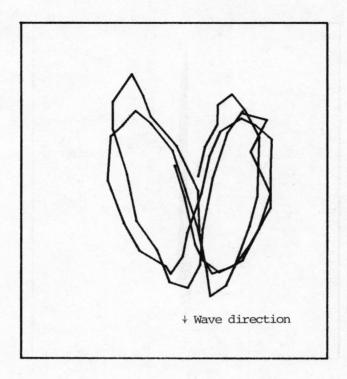

↓ Wave direction

9-1

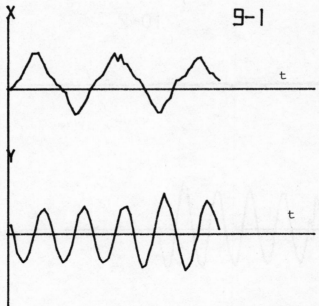

Fig.5.2 Trace of center of buoy

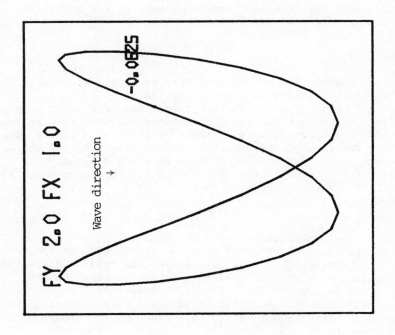

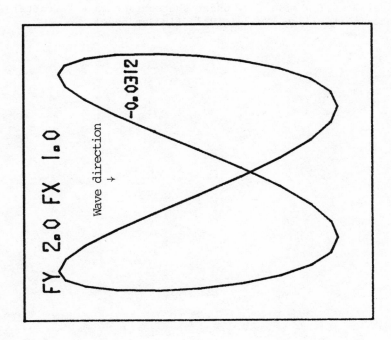

Fig.6

The values δ in this case are distributed around 1.0, 2.0, 4.0 and 9.0. This has also good correspondence to the unstable zone in Mathiu's diagram.

CONCLUSION

The transverse motion can be explained by Mathiu's unstable problem of oscillation in experimental data.
The author has some doubt whether these phenomena occur only in a model test because the field structures are more complex than the model structure such as a single cylinder or a single sphere buoy and the ratio δ between the natural frequency and the wave frequency may be very different fron Mathiu's diagram.
The author is now collecting field data for this phenomenon in order to study the problem further.

REFERENCES

T.Tuboi"Oscillation" Gendaikogakusha

G.Sawaraghi "Nonlinear Oscillation" Kyoritu shutphan

T. Sawaraghi, T. Nakamura and Hideki Miki "Oscillation of circular cylinder under the wave by vortex wake" Coastal engineering in japan no.23, 1976. (in Japanese)

Masashi Homma, Kunihiro Ogihara and Hiroya Emori " The motion of spherical buoy under the surface wave." Report of annual meeting of J.S.C.E. No 28, 1973

Kunihiro Ogihara " Theoretical analysis on the transverse motion of a buoy under the surface wave." Coastal Engineering in Japan. No.24, 1977. (in Japanese)

PROBABILISTIC DESIGN OF A FIBRE-REINFORCED CONCRETE FLOATING PONTOON

J.P. Rammant

E. Backx

Catholic University of Louvain

ABSTRACT

The use of fibre reinforced concrete has distinguished advantages for marine applications. This paper gives a design methodology which takes account of the uncertainty of material characteristics and the probabilistic loading. A floating pontoon is considered as an example. The analysis is simplified by the assumption that the sea behaves as a long-crested sea, so that the spatial distribution is excluded, leaving only the temporal wave height spectra as an input. The response of the floating pontoon for heave and pitch motions is then computed, following a stochastic sea loading. Subsequently, an attempt is made to predict the survival probability of the structure over a long time period.

INTRODUCTION

Fibre reinforced concrete is a young material which has distinguished properties for marine applications. The ductility, the impact resistance and the abrasion resistivity of the concrete increases significantly for a few added volume percentages of fibres. While experimental full scale tests on the use of steel fibre reinforced concrete in a marine environment have been performed only recently in different parts of the world, a need exists for design methods that take account of the specific conditions under which the material is behaving. In this paper the use of fibre-reinforced concrete for floating pontoons is investigated. Other marine applications may soon be expected: cover of off-shore pipelines, wave breaking units, constructional concrete (especially for thin sections).

312

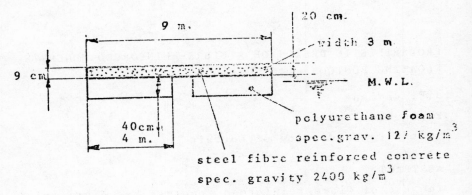

Fig.1 Pontoon

PONTOON MOTIONS IN A RANDOM SEA

1.Strip theory for a pontoon

In figure 1, a typical pontoon element is presented.
Of the six degrees of freedom, only two will be retained, nl. heave and pitch. Therefore, we model the
real sea through a long-crested sea. This also justi-
fies the use of the strip theory in which it is assumed that the time dependent water flow adjacent to the
wet surfaces of thin vertical slices of the floating
body is two-dimensional (Mc.Cormick,1973).
The differential system that describes the deterministic body motions is then:

$$[M]\{\ddot{x}\} + [B]\{\dot{x}\} + [C]\{x\} = \{F\} \tag{1}$$

wherein [M] ,[B] and [C] are 2 by 2 matrices.
The elements of the matrices are calculated from the
geometrical and hydrodynamical characteristics of
the pontoon in Appendix I. Essentially, it has been
assumed that the pontoon responds as a rigid body;
indeed, unlike fixed offshore structures, the own
dynamic motions are negligible compared with the rigid body motions.

2.Description of the sea state

The Pierson-Moskowitz spectrum was used:

$$S_{\eta\eta} = \frac{\alpha g^2}{\omega^5} \exp\left[-\beta(g/(W\omega))^4\right] \tag{2}$$

wherein: W, wind speed 20 m. above the sea surface
$\alpha = 8.1 \ 10^{-3}$ for the North Sea (Brebbia,
$\beta = 0.74$ 1975; Scott, 1965)
The significant wave height (H_s) and the mean wave
period (T_m) are often used to characterize the sea
state. The relation with the former sea parameters is

$$H_s = 2 \ W^2/g \ \sqrt{\alpha/\beta} \tag{3}$$

$$T_m = 2\pi \ W/g \ (1/\beta\pi)^{1/4} \tag{4}$$

Both H_s and T_m will vary. For a particular place however, there is a correlation between H_s and T_m. This
correlation can be obtained from experimental data
and is mostly given in the form of a bivariate histogram. For the present investigation, we will assume that α and β remain constant and that H_s is the
only variable quantity. This implies a unique relationship between H_s and T_m.
Data for the Belgian coast are available (Van Cauwen-
bergh, 1971). However, since it concerns here a harbour situation, these data have to be relaxed considerably. The statistical variation of H_s follows a
Weibull distribution given by:

$$p(H_s) = \frac{(H_s)^{\gamma-1}}{(H_c)^\gamma} \cdot \gamma \cdot \exp(-(H_s/H_c)^\gamma) \tag{5}$$

in which we used $H_c=1$ and $\gamma = 1.4$

The value of H_s can only be determined over a certain period. It will vary gradually. In the present analysis, we have chosen periods of 3 hours during which H_s remains constant. From equation 5, it follows that the average number of periods with a particular H_s during a year is given by:

$$\nu(H_s) = p(H_s) \cdot N_y \qquad (6)$$

wherin N_y is the number of periods in a year (365x8).

Since $\int_0^\infty p(H_s)\, dH_s = 1$, it follows $\int_0^\infty \nu(H_s)\, dH_s = N_y$

The sea behaviour differs from year to year. The expected number of periods having a significant wave height H_s in a span of t years is given by $\nu(H_s) \cdot t$. The probability distribution for the number of storms is given by a Poisson distribution:

$$p(n) = (\nu(H_s) \cdot t)^n/n! \cdot \exp(-\nu(H_s) \cdot t) \qquad (7)$$

3.Stochastic response analysis

Each equation of (1) can be written as:

$$m\, \ddot{\psi} + c\, \dot{\psi} + k\, \psi = f \qquad (8)$$

where ψ can be either z or θ and f either F or M. The transfer function is:

$$H_{\psi f}(i\omega) = 1/(k(1+2i\beta\omega/\bar{\omega} - (\omega/\bar{\omega})^2)) \qquad (9)$$

The response spectral density is then:

$$S_{\psi\psi}(\omega) = |H_{\psi f}|^2 \cdot S_{ff}(\omega) \qquad (10)$$

The right hand side of the equations (8) is deduced in Appendix 1 and is of the form:

$$f = (A+i.B) \int f(\xi)\eta\, d\xi$$

Taking the Fourier Transform and the conjugate, applying the power spectrum definition, one gets:

$$S_{ff} = (A+i.B)(A-i.B) \int f(\xi)\, \eta\, d\xi \cdot S_{\eta\eta}$$

$$S_{ff} = |H_{f\eta}|^2 S_{\eta\eta}$$

From equation (10) and (11), it follows:

$$S_{zz}(\omega) = |H_{zf}|^2 \cdot |H_{f\eta}|^2 S_{\eta\eta} = |H_{z\eta}|^2 S_{\eta\eta} \qquad (12)$$

The cross spectral densities can be found similarly:

$$\begin{vmatrix} S_{zz}(\omega) & S_{z\theta}(\omega) \\ S_{\theta z}(\omega) & S_{\theta\theta}(\omega) \end{vmatrix} = \begin{vmatrix} H_{z\eta}^{**} \\ H_{\theta\eta}^{**} \end{vmatrix} \cdot S_{\eta\eta}(\omega) \cdot \{H_{z\eta}\ H_{\theta\eta}\} \qquad (13)$$

wherein, S_{zz} the heave motion power spectrum, $S_{\theta\theta}$ the pitch motion power spectrum, $S_{z\theta}$ the heave-pitch cross spectral function, $S_{\theta z}$ the complex conjugate of $S_{z\theta}$, $H_{z\eta}$ the frequency transfer function for heave motion, $H_{\theta\eta}$ the frequency transfer function for pitch motion.

FIBRE REINFORCED CONCRETE

The mechanical characteristics of fibre reinforced concrete are reported in (Rilem,1975;ACI,1973;Swamy, 1975); one has mainly be concerned with the static ultimate strength behaviour of simple tests.

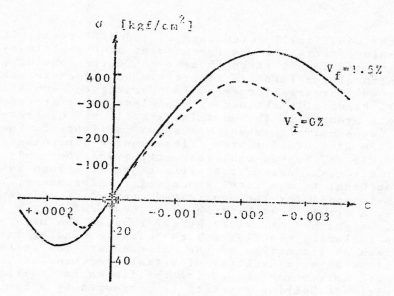

Fig.2 Uniaxial stress-strain curves

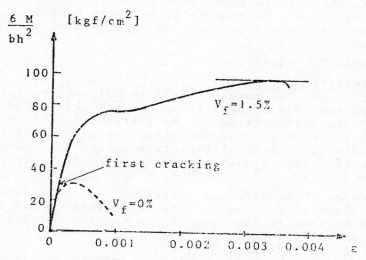

Fig.3 Experimental bending stress-strain curves

For the usual small percentage of added fibres (1-2%)
the main effects are the improvement of the ultimate
bending and impact strength over the corresponding
strength of unreinforced concrete. The effects of sea
water on fibre reinforcement is investigated in (Rider,
1978). Especially the use of stainless steel fibres
is very promising. The uniaxial tensile and compressive stress-strain curves are shown in figure 2. The
addition of 1.5Vol% fibres (fibre length 4cm, diameter 0.035 cm, hooked ends (Bekaert, 1975)) has only
minor effects. Indeed, the ultimate tensile load is
proportional to the total amount of reinforcement,
which is small for the considered steel fibre percentages. A simple beam bending test (beams of length
28cm, height h=7 cm, width h=7cm) reveals the major
characteristics: in figure 3 the elastic bending stress
is drawn as a function of the recorded strains at the
tensile side at midsection of a test-specimen.The
strength improvement by adding the fibres is drastic:
the ultimate bearing capacity is increased by 3 times.
The capability of absorbing energy is represented by
the area under the curves; as one observes the fibres
improve the impact resistance and the ductility by
major amounts. The cracking resistance which leads
to the improvements is linearly related to the large
specific surface of the fibre reinforcement.
The effect of fibre addition for other structures is
discussed in (Rammant, 1976,1977) where a general
theory for predicting the twodimensional static
strength behaviour of steel fibre reinforced concrete
is discussed. On the cyclic strength behaviour of
fibre reinforced concrete no experimental data are
known.

MECHANICAL BEHAVIOUR

1.Internal force

One is able to compute the internal forces, i.e. bending moment and shear force, at any section by expressing the equilibrium of the considered pontoon-part.
The forces acting on the structure were given in Appendix 1, only the integration limits are changed.
For the bending moment at midsection one gets:

$$M' = (M_o - M_1 - M_2 - M_3 - M_4 - M_w)_{x=0} \qquad (14)$$

Each of these force contributions can be expressed
in terms of the wave height at midsection $\eta(0,t)$ since
the forces are function of the pontoon motions. Equation (14) reduces to:

$$M'(t) = A(i\omega) . \eta(0,t) \qquad (15)$$

wherein $A(i\omega)$ expresses the dependance of all harmonic behaviour on the midsection wave height. The spectral density function for the bending moment is then:

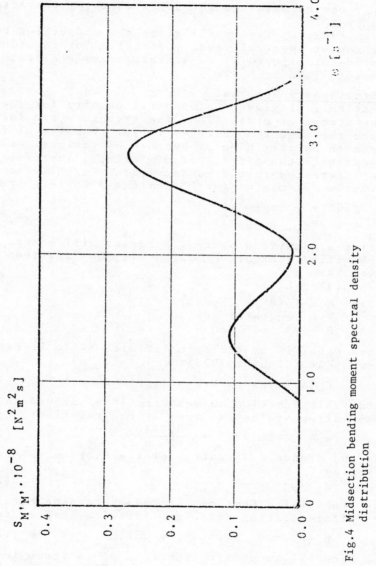

Fig.4 Midsection bending moment spectral density distribution

$$S_{M'M'}(\omega) = A^{\ddot{}}(i\omega) . A(i\omega) . S_{\eta\eta} \tag{16}$$

Similarly the internal force (or moment) at any section can be treated stochastically.

$S_{M'M'}$ can be obtained for any sea state, characterised by H_s.

As an example, figure (4) gives the midsection bending moment spectral density function for the floating pontoon of figure (1) in a stationnary sea state, characterized by $H_s = 1m$.

2.Stationnary sea state

Equation (16) gives the spectral density for the bending moment at midsection. The structure can fail because the moment M' becomes larger than the ultimate strength barrier M'=a or because of fatigue. We will investigate the first type of failure. Therefore, we are interested in the maximum values of M'. The distribution of maxima for M' follows a Rayleigh function:

$$p(Q) = \frac{Q}{\lambda_0} \exp(-\frac{Q^2}{2\lambda_0}) \tag{17}$$

$$\lambda_0 = \int_0^\infty S_{M'M'} \, d\omega$$

During a specified period, N maxima will occur. The probability that all N maxima will be less than a is given by:

$$P_N(a) = [P(a)]^N \tag{18}$$

where $N = 1/(2\pi) . (\lambda_2/\lambda_0)^{.5} . T$

T = length of period

$P = \int_0^a p(Q) \, dQ$

Since $P_N(a)$ is of the form 1-X with X<<1, it can be shown that,(Clough, 1975):

$$P_N(a) = \exp\left[-N.\exp(-\frac{a^2}{2\lambda_0})\right] \tag{19}$$

Using this function, Davenport (1964) showed that the mean extreme-value is given by the relation:

$$\bar{a} = (2 \ln N)^{.5} + \frac{0.5772}{(2 \ln N)^{.5}} \tag{20}$$

and the standard deviation of the extreme values is given by:

$$\sigma_{\bar{a}} = \frac{\pi}{\sqrt{6}} \frac{1}{(2 \ln N)^{.5}} \tag{21}$$

The probability density function p_N can be obtained by differentiating equation (19) with respect to a:

$$p_N(a) = P_N(a).N.\exp(-\frac{Q^2}{2\lambda_0}).\frac{Q}{\lambda_0} \tag{22}$$

The probability density function p_N is shown in figure (5) for different values of H_s. The reference period has been taken as 3 hours. Figure 6 gives p_N at $H_s=1$ for different reference periods. It is seen that as a function of the reference period, $\bar{a}$ becomes larger while $\sigma_{\bar{a}}$ decreases. If the material has a well specified allowable bending stress, it is

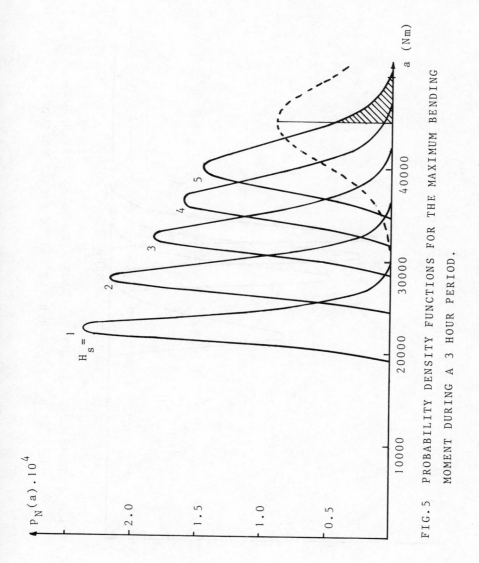

FIG.5 PROBABILITY DENSITY FUNCTIONS FOR THE MAXIMUM BENDING
       MOMENT DURING A 3 HOUR PERIOD.

320

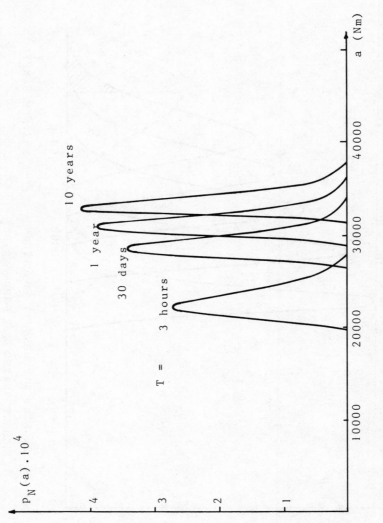

FIG.6 PROBABILITY DENSITY FUNCTION FOR THE MAXIMUM BENDING

MOMENT AT H_s =1m. FOR DIFFERENT REFERENCE PERIODS.

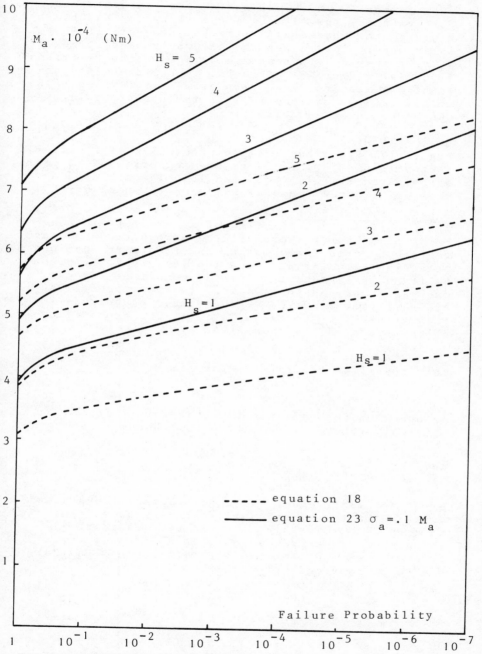

$M_a \cdot 10^{-4}$ (Nm)

$H_s = 5$

4

3

5

2

4

$H_s = 1$

3

2

$H_s = 1$

Failure Probability

----- equation 18

——— equation 23 $\sigma_a = .1 \, M_a$

FIG. 7 INFLUENCE OF THE MATERIAL PROPERTIES UNCERTAINTY ON
THE PROBABILITY OF FAILURE FOR A PERIOD OF 10 YEARS IN A
STATIONNARY SEA STATE

easy to verify which value of H_s can be withstood. For instance, the failure probability for $H_s = Ct.$ is given by the shaded area (fig.5). However, if there is an uncertainty about the allowable bending moment M_a, for instance represented by a Gaussian distribution $p(a)$ (figure 5), this has to be taken into account. It results, that also at relatively low values of H_s, failure can occur. This failure probability can be computed as follows:

$$P_{FAIL/H_s} = 1 - (1 - {_0\int^\infty} p(Q) [{_Q\int^\infty} p(a)\,da]\,dQ)^N \qquad (23)$$

Figure (7) gives the failure probability for different values of H_s, for a reference period of 10 years, as compared to $P_N(a)$ (equation 18). The standard deviation σ_a, used in $p(a)$ has been taken as $0.1\,M_a$.

3. Long term predictions

If one divides the lifetime of a structure into periods of 3 hours, there will be more periods characterized by a low value of H_s than there are periods at a high value of H_s. Equation (6) gives the probability density function for the number of periods during a year. Suppose we want to know the failure probability for 10 years. The number of periods for each value of H_s is taken as 10. . The probability of failure at H_s is then given by:

$$Pr_{H_s,1} = 1 - (P(a))^\nu \qquad (24)$$

The probability of failure for all H_s:

$$Pr_1 = {_{H_s=0}\overset{\infty}{\Sigma}} Pr_{H_s,1} \qquad (25)$$

In one also incorporates the uncertainty for the material properties, equation (24) becomes:

$$Pr_{H_s,2} = 1 - (1 - P_{FAIL/H_s})^\nu \qquad (26)$$

and thus:

$$Pr_2 = {_{H_s=0}\overset{\infty}{\Sigma}} Pr_{H_s,2} \qquad (27)$$

A further improvement is to incorporate the statistical character of the number of periods at a certain value of H_s. This is given by equation (7). The number of periods at a value H_s can now vary from 0 to n with a probability $p(n)$. Equation (26) has to be replaced by:

$$Pr_{H_s,3} = 1 - {_{n=0}\overset{\infty}{\Sigma}} (1 - P_{FAIL/H_s})^n \cdot p(n) \qquad (28)$$

so

$$Pr_3 = {_{H_s=0}\overset{\infty}{\Sigma}} Pr_{H_s,3} \qquad (29)$$

A comparison of equations (25), (27) and (29) is done in figure (8). The largest impact comes from the uncertainty in the material properties, while the difference between Pr_2 and Pr_3 is indistinguis-

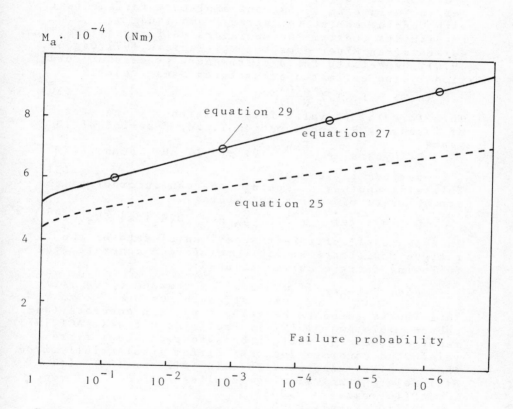

$M_a \cdot 10^{-4}$ (Nm)

FIG.8 PREDICTION OF LONG TERM FAILURE

hable. If one has designed a structure which is close
to the predefined limit of acceptance, from this figure it is possible to decide wether or not it is worth-
wile to make an extra effort (and therefore expense)
to improve the material characteristics or to make a
new design with a higher bending resistance.

4.Fatigue damage

The second type of failure is fatigue failure. The fatigue resistance of a material is known through the
$N(a)$ fatigue curve, wherein $N(a)$ is the number of cycles to failure in a constant-amplitude fatigue test
with bending moment (or stress) amplitude a.
Experiments confirm the engineers validity of the linear Palmgren-Miner damage law, (Tepfers,1977;Osgood,
1970); henceforth for a stationnary process of duration T, the estimated accumulated damage equals:

$$D(T) = f_e \cdot T \cdot \int_0^\infty \frac{p(Q)}{N(Q)} \, dQ \qquad (30)$$

This formula is valid for a stationnary sea state
of intensity H_s and duration T. For a period of 10
years:

$$D(10 \text{years}) = \int_0^{H \max} D. \, p(H_s). \, dH_s \text{ .number of}$$

periods in 10 years $\qquad (31)$

Following another reasoning, one can incorporate the
arrival rate of storms to obtain:

$$D'(10 \text{years}) = \int_0^{H \max} [\sum_{n=0}^\infty p(n).D(H_s).n] \, dH_s \qquad (32)$$

In absence of sufficient experimental data on the
fatigue resistance of fibre reinforced concrete, the
following fatigue curve was used:

$$N(a) = (a_u/a)^\alpha \text{ where } a_u \text{ is the one cycle ulti-}$$
mate strength; $\alpha=30$

This law is found to be valid for plain concrete (Ja-
cobsen, 1976;Award, 1974). Preliminary tests (ACI,1973)
reported that the fatigue behaviour of steel fibre
reinforced concrete does not differ substantially from
that of plain concrete. Formula (31) and (32) with
$N(a)$ relate the following results:

D(10 years) = 0.0031
D'(10 years)= 0.0030

CONCLUSIONS

The analysis technique presented in this paper allows
the design of a floating pontoon made of fibre reinforced concrete. The influence of uncertainty of the
material characteristics has been included in the
analysis. This allows to assess the increase in reliability when a better conditionned material (i.e. smaller σ_a) can be obtained.
At various points in the application of the proposed
analysis, simplifying assumptions had to be made.

More research leaves to be done to eliminate these uncertainties, i.e. concerning
-the determination of c_M and c_D
-the fatigue curve for fibre reinforced concrete
-the influence of T_m on the wave spectra
-the coefficients in the Weibull distribution for significant wave heights (equation (5)
-influence of crack growth on the fatigue damage.

REFERENCES

ACI Journal (Nov.1973) State-of-the-Art Report on Fibre reinforced concrete.

Award M.E., Hilsdorf H.K.(1974) Strength and Deformation
characteristics of Plain Concrete subjected to High Repeated and Sustained Loads, ACI publ. SP-41

Bazant Z.P., Bhat P.D. (1977) Prediction of Hysteresis
of Reinforced Concrete Members. ASCE, St.1

Bekaert N.V. (1975) Dramix, a new Concrete Reinforcement
(advertisement)

Brebbia C.A.(1975) Vibrations of Engineering Structures,
Comp.Mech.Ltd, Southampton.

Clough R.W., Penzien J.(1975) Dynamics of Structures,
Mc.Graw-Hill.

Mc.Cormick M.E.(1973) Ocean Engineering Wave Mechanics,
John Wiley&Sons.

Davenport A.G. (1964) Note on the distribution of the
Largest value of a random function with application to
Gust Loading, Proc.Inst.Civ.Eng. Vol.28,pp187-196.

Hogben N.(1976) Wave loads on Structures, BOSS'76,proc.
VOl.I

Jacobson A.B., Widmark (1976) Fatigue Properties of Reinforced Concrete Structures, BOSS'76, proc.VOLII.

Lewis E.V. (1976), The motion of ships in waves. Principles of Naval Architecture, Ed.Comstock J.P.,Soc.of
Naval Architects and Marine Engineers, N.Y.

Osgood C.C.(1970) Fatigue Design, Wiley Interscience,
John Wiley & Sons.

Rammant J.P. (1976) Rupture Calculations of Fibre Reinforced Concrete Continua with the F.E.M., Ph.D.thesis
Leuven.

Rammant J.P.,Van Laethem M.,Backx E. (1977) Steel fibre
Concrete, a safer material for reactor construction,
4th Int.Conf.on Struct.Mech.in Reactor Techn.,San Fr.

326

Rider R.G., Heidersbach R.H.(1978) The effects of
Seawater on the Structural Properties of metal fibre
reinforced concrete, OTC 3193, Offshore Technology
Conference, Houston.

Rilem Symposium (1975) Fibre Reinforced Cement and
Concrete, London, The Construction Press Ltd.

Schenck H.(1975) Introduction to Ocean Engineering.
Mc.Graw-Hill.

Scott J.R.(1965) A sea spectrum for model tests and
long-term ship prediction. Journal of Ship Research,
VOl.9,N°3.

Swamy R.N.(1975) Fibre Reinforcement of Cement and
Concrete. Evaluation of Fibre Reinforced cement based composites, Materials and Structures, Vol8,n°45.

Tepfers R.,Friden C.,Georgsson L.(1977) A study of
the applicability to the fatigue of concrete of Palmgren-Miner partial Damage Hypothesis, Magazine of
Concrete Research, Vol.29 N° 100.

Van Cauwenberghe C.(1975) Golfwaarnemingen vanaf 1918
tot 1971 aan boord van Belgische Lichtschepen, Tijd-
schrift der Openbare Werken van België.

APPENDIX I

Strip theory for a floating pontoon

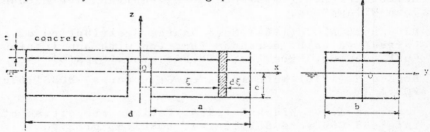

FIG. A-1

Table I	Dimensions in meter
d=9.	OG= .1088
a=4.	OB= .1469
b=3.	BM=2.553
c=0.2938	BM'=25.81
f=0.4	
t=0.09	

O = Still water position
G = Gravity center
B = Boyancy center
M = Metacenter for rolling
M'= Metacenter for pitching
w = displacement of G
θ = rotation around y
η = position of watersurface
$$\eta = f(\xi,t)$$

Equilibrium equations
A freely floating body is subjected to the following
forces (Brebbia,1975; Hogben,1976 ; Lewis,1967,

Schenck,1975): - the forces caused by the waves acting on a fixed body, - the forces caused by the movement of the body in still water.

For each strip we obtain the following forces:

1. the weight of the body: F_w
2. the hydrostatic restoring force: F_1

$$F_1 + F_w = - \gamma_w b \, d\xi \, (w + \xi\theta - \eta e^{-kc})$$

3. the Froude-Krylov force: F_2 4. Inertia forces: F_3

$$F_2 = - \rho b \, c \, d\xi \, (\ddot{\eta} \, e^{-kc}) \qquad F_3 = -c_M d\xi \, (\ddot{w} + \xi\ddot{\theta}e^{-kc})$$

5. drag forces: F_4

$$F_4 = - c_D \, d\xi \, (\dot{w} + \xi\dot{\theta} - \dot{\eta} \, e^{-kc})$$

When we integrate ξ from $-d/2$ to $d/2$, the following equation for heaving is obtained:

$$M_{\ddot{w}} = F_w + F_1 + F_2 + F_3 + F_4$$

$$(M + c_M . 2a)\ddot{w} + c_D . 2a \, \dot{w} + 2ab\gamma_w w = e^{-kc}(\gamma_w b - ic_D\omega - (c_M + \rho bc).$$

$$. \omega^2) {}_{-d/2}\!\!\int^{d/2} \eta \, d\xi$$

This is obtained with

$$\eta = a_\omega \, e^{i(k\xi - \omega t)} = a_\omega e^{ik\xi} e^{-i\omega t} = A(i\xi) \, \eta_0(t)$$

It should be noted that although the forces are function of θ, the equilibrium equation is not. The equilibrium equation for pitching is obtained similarly by writing the moments M_w to M_4. It results:

$$(I_y + I_{ya})\ddot{\theta} + c_\theta \, \dot{\theta} + \frac{\gamma_w}{12} b \, [d^3 - (d-2a)^3]\theta =$$

$$e^{-kc}(\gamma_w b - ic_D\omega - (c_M + \rho bc) \, \omega^2) \int_{-d/2}^{d/2} \xi\eta d\xi$$

where: $I_{ya} = c_M . [d^3 - (d-2a)^3] / 12$

$$c_\theta = c_D . [d^3 - (d-2a)^3] / 12$$

The determination of c_M and c_D

No attempt has been made to find the theoretical values of c_M and c_D. The values used have been taken from Brebbia (1975):

$$c_M = c_m \frac{\rho\pi D^2}{4} \quad \text{with } D=b \text{ (see Fig.A-1) and } c_m = 1$$

$$c_D = .5 \, c_d \, \rho D \sqrt{\frac{8}{\pi}} \, \sigma_{\dot{\eta}} \quad \text{with } D=b, \, c_d = 1 \text{ and } \sigma_{\dot{\eta}} = 0.2.$$

A STRUCTURAL PROBLEM OF PIERS RESTING ON PILES

Tuğrul Tankut

Middle East Technical University, Ankara, Turkey

ABSTRACT

The structural behaviour of a typical reinforced concrete pier resting on piles was studied, under horizontal loads caused by berthing ships. In one phase of the investigation, the normal, tangential and rotational components of the approach velocity of twenty four berthing ships were measured and the impact energy was computed for each berthing, using various expressions recommended by the earlier researchers. In another phase, the theory of a method for direct measurement of the impact force was developed. An excessive structural rigidity of the system attributed to the raking piles, was observed during the tests to verify the proposed theory. The probable consequences of this excessive rigidity were discussed in the last phase, and some recommendations were accordingly made.

NOTATION

a_x, a_y, b_y	Measurement components (Figure 5)
B	Beam of the ship
C_c	Construction coefficient (Taken 0.9) Lee proposes (0.8 ~ 1.0)
C_d	Deformation coefficient (Taken 0.6) Lee proposes (0.5 ~ 1.0)
C_g	Geometric coefficient (Taken 1.0) Lee proposes (0.85 ~ 1.25)
C_m	Hydro-dynamic mass coefficient (Taken 1.3)
C_s	Softness coefficient (Taken 0.9)
D	Draught of the ship
E	Impact Energy

E_L	E by Lee Method
E_P	E by Pages Method
E_S	E by Saurin Method
E_V	E by Vasco-Costa Method
F_{cr}	Cracking horizontal load
F_u	Ultimate horizontal load
K	Coefficient representing geometric and mechanical properties of energy absorbing elements
k	Radius of gyration of the ship (Taken 0.22L) Lee proposes (0.20L ~ 0.29L) Saurin proposes (0.20L ~ 0.22L)
L	Length of the ship
ℓ_x, ℓ_y	Dimensions of the platform
m	Impacting mass of the ship
P	Approximate impact force
p	Unit impact force
r	Distance of point I from point G
u, v, w	Displacement components
v_G	Translational velocity of the centre of gravity G
v_{Gx}, v_{Gy}	Tangential and normal components of v_G
v_I	Translational velocity of the point of impact I
v_{Ix}, v_{Iy}	Tangential and normal components of v_I
W	Impacting weight of the ship
$\bar{x}$, $\bar{y}$	Coordinates of the centre of rotation O
α	Angle of approach
γ	Angle between directions $\vec{GI}$ and $\vec{v}_G$
Δ	Displacement
Δ_{cr}	Displacement corresponding to cracking
Δ_o	Displacement corresponding to $0.4F_u$
ω	Rotational velocity of the ship

INTRODUCTION

The structural system consisting of a rigid platform supported on piles, is widely used in the design of shore structures. Some of such structures are not equipped with efficient fendering systems. This is the case in most of the shore structures in protected harbours, especially in developing

countries. In such a case, the impact energy is absorbed mainly
by the structure itself, and the magnitude of impact force
depends on the rigidity of the structure and ship's hull.

The berthing forces considered in the design of shore
structures and ship hulls are computed on the basis of
assumptions, regarding the ship displacement tonnage, approach
velocity, approach angle, etc. Some researchers have already
studied the problem from energy point of view(1,2,3,4,6,7).
However, the problem of determination of the actual impact
force which is essential for the structural design of shore
structures without sophisticated fenders, still deserves
attention.

Alsancak Pier in the protected Harbour of İzmir was
chosen for the present investigation. Figure 1 presents a plan
of the pier which rests on reinforced concrete piles of 600 mm
diameter and consists of numerous separate platforms.Reinforced
concrete slabs supported on very stiff grids of large beams
form the platforms which are equipped with simple wooden
fenders. The dredged depth around the pier is about 10 meters.
The ships are generally berthed by local expert captains with
the assistance of tug boats.

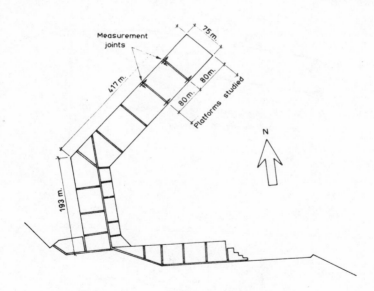

Figure 1 : General layout of the studied pier

The two major parts of the work carried out on the pier
are the evaluation of twenty four berthings studied from the
energy point of view, and the tests towards the verification
of the proposed method of direct measurement of the impact
force. These two approaches are briefly introduced in the

332

following two sections. The structural behaviour observed
during the tests are discussed in the next section, and in the
last section, some conclusions and recommendations are
presented.

ENERGY APPROACH

Experimental Work

The measurement of approach velocity components was essential
for the determination of impact energy. The simple mechanical
system specially developed[5] enabled the measurement of the
normal and tangential velocity components at two points S,
stern and B, bow (Figure 2) which were then used for the
determination of the angle of approach α, the rotational
velocity ω, the normal and tangential velocity components of
the centre of gravity v_{Gx}, v_{Gy} and of the point of impact v_{Ix},
v_{Iy} (Figure 3). The values obtained are presented in Table 1.

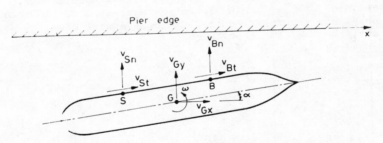

Figure 2 : Velocity components at the measurement points
and the centre of gravity

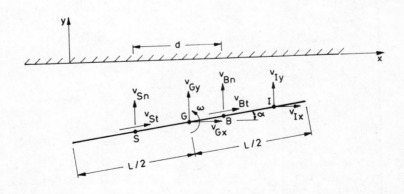

Figure 3 : Simplified model geometry

Berthing	W	v_{Gx}	v_{Gy}	v_I	α	γ	$\omega.10^5$	k	r	$1+2\frac{D}{B}$
	MN		mm/sec			degrees	rd/sc		m	-
1	44	9	88	89	-16	72	85	18.8	22.8	1.86
2	40	19	70	72	-14	42	-46	27.5	25.7	1.58
3	17	31	36	47	-11	20	-108	18.4	17.6	1.60
4	137	16	79	80	4	81	179	34.2	33.9	1.56
5	155	4	16	17	15	26	50	31.9	18.1	1.51
6	100	1	4	4	3	89	22	34.2	44.1	1.46
7	53	0	29	28	12	52	-48	33.0	23.1	1.80
8	11	16	119	120	2	72	-435	13.9	12.6	1.47
9	69	25	45	51	16	85	207	24.2	25.2	1.46
10	33	9	100	99	9	73	449	20.7	28.7	1.61
11	44	26	103	106	-10	73	213	20.8	18.4	1.68
12	47	2	179	180	23	46	-360	22.9	20.5	1.73
13	74	26	43	51	-22	58	-41	26.6	11.2	1.48
14	174	1	39	38	3	65	271	31.9	24.8	1.35
15	64	111	58	125	-20	11	-97	31.3	26.7	1.51
16	79	107	33	112	9	8	59	29.0	33.7	1.59
17	160	80	124	146	2	44	-245	33.0	42.3	1.65
18	38	7	84	84	2	64	137	28.8	15.8	1.97
19	72	100	87	133	22	89	-40	29.0	16.3	1.65
20	167	126	57	137	-19	26	-106	32.1	21.1	1.39
21	60	12	47	48	-4	87	-49	28.6	36.1	1.38
22	70	67	47	82	-6	109	-237	23.3	15.0	1.49
23	132	47	65	81	10	31	207	32.1	41.0	1.47
24	86	46	104	113	-12	35	-107	26.8	24.9	1.47

Table 1 : Recorded and computed information about the studied berthings

Impact Energy After a critical examination of the previous investigations, four methods   for determining the impact energy were found to be suitable for the evaluation of the test data. Impact energy expressions proposed in the selected methods are collected below. The expressions are presented in a form suitable for comparison.

Vasco-Costa[6] proposes,

$$E_V = \frac{mv_G^2}{2} (1 + 2 \frac{D}{B}) \frac{k^2 + r^2 Cos^2\gamma}{k^2 + r^2} \tag{1}$$

Lee[1] proposes,

$$E_L = \frac{mv_G^2}{2} (1 + 2 \frac{D}{B}) \frac{k^2}{k^2 + r^2} C_g C_d C_c$$

$$= \frac{mv_G^2}{2} (1 + 2 \frac{D}{B}) \frac{k^2}{k^2 + r^2} \times 1.0 \times 0.6 \times 0.9 \tag{2}$$

Saurin[3] proposes,

$$E_S = \frac{mv_I^2}{2} \cdot \frac{k^2}{k^2 + r^2} C_m C_s = \frac{mv_I^2}{2} \cdot \frac{k^2}{k^2 + r^2} \times 1.3 \times 0.9 \qquad (3)$$

Pages[2] proposes,

$$E_P = \frac{mv_{Gy}^2}{2} \cdot \frac{1}{1 + 16(\frac{r}{L})^2} \qquad (4)$$

The data were evaluated using the above expressions, and four different energy values   obtained for each berthing are presented in Table 2. The impacting mass of the ship was

Berthing	E_V	E_L	E_S	E_P	P_V	P_L	P_S	P_P	P_V	P_L	P_S	P_P
	kN-m				kN				$\times 10^{-5}$ sec/mm			
1	7	7	8	7	430	420	460	430	11	11	12	11
2	13	5	7	6	570	350	410	380	20	12	14	13
3	3	1	1	1	270	150	170	170	34	18	21	20
4	37	19	27	24	960	700	820	770	9	6	7	7
5	3	1	2	1	290	190	220	180	11	7	9	7
6	0	0	0	0	30	20	30	30	8	6	7	8
7	3	1	2	1	300	190	210	130	19	13	14	9
8	7	4	5	5	420	300	370	370	31	22	26	27
9	6	3	5	2	400	300	350	220	11	8	10	6
10	11	5	7	7	520	350	410	420	16	11	12	13
11	26	13	17	17	800	570	640	650	17	13	14	14
12	104	40	51	51	1610	1000	1120	1120	19	12	13	13
13	13	7	10	9	570	400	490	470	15	11	13	13
14	12	6	9	10	550	390	480	490	8	6	7	7
15	77	24	35	30	1380	780	930	860	17	10	12	11
16	80	19	26	27	1420	680	800	810	16	8	9	9
17	205	60	78	70	2270	1220	1390	1320	10	5	6	6
18	22	11	12	10	750	530	550	510	23	17	17	16
19	82	44	58	52	1430	1050	1200	1150	15	11	13	12
20	211	85	132	130	2300	1450	1810	1800	10	6	8	8
21	4	2	3	3	310	220	280	290	11	8	10	10
22	27	14	20	19	820	590	710	700	14	10	12	12
23	53	13	20	20	1150	570	700	700	11	5	7	7
24	71	24	35	37	1330	770	940	960	14	8	10	10

Table 2 : Energy, force and unit force values for the studied berthings

computed from the weight of the ship at berthing which was obtained as,

W = Displacement light + Present load or

W = Displacement loaded − Dead weight tonnage + Present load

<u>Approximate Impact Force</u> It was desired to find the order of the magnitude of the impact force. An approximate elastic analysis resulted in the following expression

$$P = K \sqrt{E}$$

(5)

The coefficient K was estimated on the basis of a few simple measurements and assumptions. Various impact force values obtained using various energy expressions are also included in Table 2.

<u>Unit Impact Force</u> Various investigators[1,3,5] observed that larger ships berth at lower approach velocities. This observations lead the author to the concept of "Unit Impact Force" which may be defined as the impact force per unit ship weight per unit approach velocity. Four different unit impact force values for each berthing were computed and presented in Table 2. A remarkable tendency was observed when these values were plotted against ship weight (Figure 4) : As ship weight

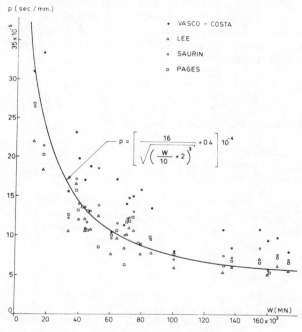

Figure 4 : Tonnage - unit impact force relationship

increases, unit impact force decreases and tends to make an asymptote to a limiting value. What is more important is the fact that this tendency is independant of the error in the approximate impact force since approximation is introduced through the coefficient K. A change in K affects the magnitude, but does not influence the trend. More reliable design guides can be developed on this basis if more reliable impact force values can be obtained. The curve expressed by

$$P = \left[\frac{16}{\sqrt{(\frac{W}{10} + 2)^3}} + 0.4 \right] \times 10^{-4} \qquad (6)$$

was considered a reasonable fit to the data available.

Observations

A careful study of Tables 1 and 2 revealed the following points which should obviously be restricted to the type of the berthing structure studied, the size of ships and the berthing conditions etc.

Approach Velocity The rotational velocity is generally quite small, and consequently the translational velocity components at the centre of gravity and the point of impact are not very different. The average value of the normal component is 69 mm/sec with a maximum of 179 mm/sec. These values agree with the measurements taken by Lee[1] and those obtained in Finnart, Scotland[4]. The tangential component was measured for the first time in this investigation. The average is 37 mm/sec with a maximum of 126 mm/sec. These values indicate the significance of the tangential component.

Approach Angle A considerable variation in the angle of approach was observed. The average is about 10^0, and there are a few values as high as 20^0. These values are somewhat higher than those measured in Finnart, Scotland[4].

Impact Energy Energy values obtained by Lee (E_L), Saurin (E_S) and Pages (E_P) methods are not very different, while those obtained by Vasco-Costa (E_V) method are somewhat higher.

Unit Impact Force The proposed curve representing Equation 6 fits very well to the points corresponding to Lee (p_L), Saurin (p_S) and Pages (p_P) methods. The points corresponding to Vasco-Costa (p_V) method have higher values.

FORCE APPROACH

In the structural design of shore structures and ship hulls, the berthing forces are considered in terms of equivalent static loads which are computed on the basis of a number of questionable assumptions. If the static equivalents of the actual berthing forces could be measured on existing shore structures, very useful and reliable information would be obtained for the design of similar shore structures. In the present investigation, a simple and practical method is proposed for measuring the static equivalents of the actual impact forces.

Theory

A pile supported pier is theoretically a flexible structural system which, if calibrated in some manner, can be used to assess the forces imposed on it. Various horizontal static loads of known direction and magnitude are applied to the rigid platform and the displacements caused are measured. The collected data are analysed to formulate the force-displacement relationship in a suitable form. When the displacements due to an unknown horizontal force are measured, the relationship already formulated can be used to determine the static equivalent of the unknown force. Impact forces thus determined during a number of berthings on a specific pier can be statistically analysed to state recommendations for the design of similar shore structures.

The following assumptions are implicitly made in the basic principle stated above :
a. The platform is assumed to be displaced as a rigid body in its horizontal plane.
b. The structural behaviour is assumed to be elastic (not necessarily linear). This assumption is justified since (i) the loads involved are temporary and much smaller than the ultimate capacity; (ii) the structure has been subjected to similar loads in the past, and (iii) the desired accuracy of measurement is not very high.

The basic principle explained above indicate that the rigid body displacement of the platform is fully defined by three displacement components (u, v, w). The platform obviously has a fixed centre of rotation the location of which is determined solely by the structural properties and independant of the loading. Referring to Figure 5, the displacement components can be expressed in terms of the measurement components (a_x, a_y, b_y) and coordinates of the centre of rotation $(\bar{x}, \bar{y})$ as follows :

$$w = (a_y - b_y)/ \ell_x \qquad (7)$$
$$u = a_x + w\bar{y} \qquad (8)$$
$$v = a_y - w\bar{x} \qquad (9)$$

These equations can be applied only if the coordinates of the centre of rotation are known. These are best determined experimentally by applying a horizontal couple to the platform which makes the translation components (u, v) zero. In this case, the above equations yield,

$$\bar{x} = a_y \ell_x/(a_y - b_y) \quad ; \quad \bar{y} = a_x \ell_x/(b_y - a_y) \qquad (10)$$

The proposed method can then be realised in two stages.

Calibration The coordinates of the centre of rotation is

338

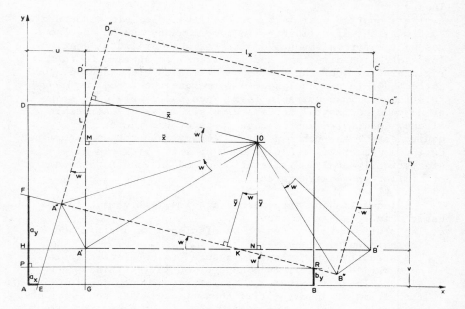

Fig.5 Displacement geometry

determined first, as explained above. Then, various
combinations of known loads are applied and the resulting
displacement components are measured and tabulated as follows :

Loading	Applied			Measured		
1	F_{x1}	F_{y1}	M_1	u_1	v_1	w_1
2	F_{x2}	F_{y2}	M_2	u_2	v_2	w_2
⋮						
n	F_{xn}	F_{yn}	M_n	u_n	v_n	w_n
	$[\bar{F}]$			$[\bar{u}]$		

$[\bar{F}]$ and $[\bar{u}]$ matrices are formed and a calibration matrix $[S]$
is obtained by the method of least squares in the following
form,

$$[S] = \left[([\bar{u}]^T [\bar{u}])^{-1} [\bar{u}]^T [\bar{F}] \right]^T \tag{11}$$

Evaluation   At the stage of evaluation, the displacement
components (u,v,w) measured during an actual berthing are used
in the determination of static force components (F_x, F_y, M)
equivalent to the impact force the ship has caused. This
procedure is based on the basic force-deformation relationship,

$$\{F\} = [S]\{u\} \tag{12}$$

Experimental Work

A small scale model, representing in a simplified manner one of the platforms of the studied pier was constructed and tested in the laboratory as an initial verification of the proposed theory, and encouraging results were obtained.

The two end platforms of the pier were then chosen for the actual in-service tests. The static horizontal loads needed for calibration were applied between these two adjacent platforms, either pushing them apart or pulling them towards each other by using loading units of approximately 1 MN capacity each. Displacement measurement components (a_x, a_y, b_y) were taken along two construction joints indicated in Figure 1.

Some of the piles (about 5%) supporting the pier were raking piles. The approximate calculations neglecting the effect of these had indicated that a reasonable amount of displacement (upto 10 mm) would be obtained by the loading system used. However, it was realized as soon as the loads were applied, that the increase in the structural stiffness due to the raking piles was enermous. Consequently, the measured displacements never exceeded a few hundredths of a millimeter even when both pull and push type loading systems were loaded upto their capacity.

EXCESSIVE STRUCTURAL RIGIDITY

Effect of Raking Piles

Approximate calculations[5] ignoring the raking piles indicated the following points :

a. A total horizontal load of approximately

$$F_{cr} = 4 \text{ MN} \tag{13}$$

would start cracking in the piles, and a displacement of approximately

$$\Delta_{cr} = 11 \text{ mm} \tag{14}$$

would be caused by this load.

b. The piles would fail in flexure at a load of approximately

$$F_u = 24 \text{ MN} \tag{15}$$

c. During transportation, the piles must have already been subjected to a bending moment (approximately equal to 40% of the ultimate capacity) which approximately corresponds to a displacement of

$$\Delta_o = 65 \text{ mm} \tag{16}$$

d. Therefore, a displacement of approximately

$$\Delta = 5 \text{ mm}$$

would be expected when both pull and push type loading systems are loaded to their capacity.

340

The impressive difference between the above values and
the measured displacements of a few hundredths of a millimeter,
is a very clear indication of excessive structural rigidity.
In other words, the pier is behaving as a rigid mass and
consequently absorbing practically no energy and unnecessarily
causing high impact forces since it is not equipped with an
efficient fender.

Thermal Stresses

The temperature recorder used during field work was modified
to record the relative displacements of the two adjacent
platforms. A number of berthings have taken place during the
recording period, but none of them has given any noticeable
indication on the displacement record. However, it was observed
that displacements caused by temperature changes between day
and night (about 20^0C above and almost zero below the pier)
were approximately 1.0 mm while a simple approximate
calculation of free thermal expansion indicated 8.0 mm.

The difference which corresponds to residual stresses
far greater than those caused by the most severe berthing, was
again attributed to excessive structural rigidity. The thermal
expansion was being restrained by the raking piles virtually
forming very rigid space trusses. The systems shown in Figure
6 can explain the phenomenon clearly. Expansion of the beam AB

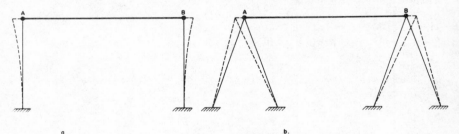

Fig.6 Thermal expansion modes of two different
systems: (a) vertical piles, (b) raking piles

is restrained by the flexural rigidity (EI/ℓ) of the very tall
and slender vertical piles in case a, and by the structural
rigidity of the trusses formed by the raking piles in case b.
The residual stresses developing in the two systems will
obviously be incomparable.

CONCLUSIONS

The studied pier which can be considered as a reasonably good
representative of its class (i.e, consisting of rigid
reinforced concrete platforms resting on piles ; accommodating
medium size general cargo and passanger ships; equipped with
inefficient simple fenders) was found excessively rigid. The
author is convinced that many structural problems can be

eliminated, stresses can be reduced and consequently the cost can be cut down if the structural system is designed somewhat more flexibly to allow a small amount of elastic deformation. A displacement of 20~25 mm would not cause any damage to the structure; on the contrary, it would reduce the danger of local damage to the structure or to the hull of the ship. In other words, it would allow a beneficial use of the latent structural strength.

Very high residual stresses were anticipated due to thermal expansion restrained by the raking piles. By making the structural system more flexible, thermal stresses would also be reduced.

None of the existing approaches can reliably express the impact energy. However, Lee's proposal is considered relatively reliable and practical. On the other hand, the statistical approach proposed by Svendsen appears to be powerful, in spite that no practical results has been presented yet.

The measured values of the normal approach velocity component agree with the values taken by Lee and those obtained in Finnart, Scotland. The tangential component measured for the first time appears to be rather significant. This observation supports Vasco-Costa[6] who emphasizes the importance of the tangential velocity component, and opposes Lee[1] who claims the contrary.

The concept of "Unit Impact Force" may be developed into a practical and reliable design guide. Its variation with ship tonnage displays a remarkable trend independant of the error in the approximate impact force.

REFERENCES

1. Lee, T.T., "Full Scale Investigation of Berthing Impacts and Evaluation of a Hydraulic-Pneumatic Floating Fender", presented at the Sept., 1966, Tenth Conference on Coastal Engineering, held at Tokyo, Japan.

2. Reeves, H.W., "Marine Oil Terminal For Rio de Janeiro, Brazil," Proceedings, American Society of Civil Engineers, Waterways and Harbors Division, Vol.87, No. WW1, Feb. 1961.

3. Saurin, B.F., "Berthing Forces of Large Tankers", presented at the June, 1963, Sixth World Petroleum Conference, held at Frankfurt/Main, Germany.

4. Svendsen, I.A., "Measurement of Impact Energies on Fenders", The Dock and Harbour Authority, Vol.51, Nos. 599, 600, Sept., Oct. 1970, London.

342

5. Tankut, A.T., "Berthing Forces Acting on Reinforced Concrete Shore Structures", habilitation thesis presented to the Middle East Technical University, in 1975, Ankara, Turkey.

6. Vasco-Costa, F., "The Mechanics of Impact and Evaluation of the Hydrodynamic Mass : Analytic Study of the Problem of Berthing", presented at the July, 1965, Nato Advanced Study Institute on Analytical Treatment of Problems of Berthing and Mooring Ships, held at Lisbon, Portugal.

7. Woodruff, G.B., "Berthing and Mooring Forces", Proceedings, American Society of Civil Engineers, Waterways and Harbors Division, Vol.88, No. WW1, Feb. 1962.

SECTION III

VIBRATIONS

FATIGUE LIFE OF STUD SHEAR CONNECTORS IN COMPOSITE BOX-GIRDER BRIDGES

K.S. Virdi, Lecturer

Department of Civil Engineering, The City University, London
Formerly, University of Melbourne, Australia

Y.S. Woo, Post Graduate Student
L.C. Schmidt, Reader and Chairman
L.K. Stevens, Professor

Department of Civil Engineering, University of Melbourne,
Australia

INTRODUCTION

In much of fatigue testing of stuctural components, because
of the high cost of conducting experiments on full-scale
structures, recourse is often taken to the collection of
design data through experiments on small scale models with a
simulated stress-loading.   One such example is the fatigue
behaviour of stud shear connectors as used in composite
box-girder bridge decks.   It has been a common practice
to conduct experiments on push-out specimens, incorporating
two or four stud connectors, in order to build up a
representative S-N diagram for use in design.   Investigations
of this type, mainly with applications to composite beams
rather than composite box-girders in mind, have been
reported by Viest (1956), Thürliman (1958), Slutter and
Fisher (1966), Mainstone and Menzies (1967), and Hallman
(1976).   The push-out specimens were typically as shown in
Figure 1.

The collapse of the West Gate bridge in Melbourne in 1970,
led to a change in the design of the bridge deck from a
composite slab to a stiffened plate.   The unused steelwork,
with stiffeners and stud connectors already intact, was
made available to research institutions in Melbourne.   At
the University of Melbourne, work was initiated to establish
a basic S-N diagram for the design of stud connectors.

346

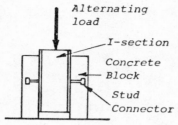

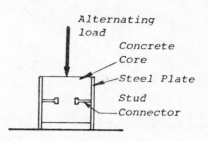

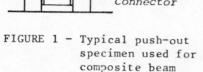

FIGURE 1 - Typical push-out   FIGURE 2 - Push-out
specimen used for   specimen used in
composite beam   Melbourne study
studies

Lo (1978) conducted numerous tests on small scale push-out
specimens, such as the one shown in Figure 2. It differs
from the push-out specimens used by previous investigators
in that instead of an I-beam placed between two concrete
blocks, the push-out specimens had a single concrete core
placed between two steel plates. Based on his experiments
Lo (1978) was able to obtain a representative S-N diagram
which can be used directly for the design of stud connectors
loaded predominantly in shear.

It has been recognised that the stress-state of shear
connectors in push-out specimens is different from that in a
composite beam or slab. Yet, in the past, design data
collected from push-out tests has been used in the design of
composite beams. The very few investigations aimed at
establishing the validity of this procedure, e.g. King,
Slutter and Driscoll (1963), Toprac (1965) and Mainstone and
Menzies (1967), have given inconclusive results. This lack
of general guidance led to the six experiments on beam
specimens described in this paper. For reasons of time, as
well as cost, it was decided to carry out this investigation
on beam specimens rather than on slab specimens.

TEST SPECIMENS

Figure 3 shows the dimensional detials of the beam specimens
tested. The web of the beam was formed by the bulb
stiffener of the steelwork. The overall dimensions of the
six specimens were identical. However, the number and
spacing of the stud connectors were different. The three
beams in Group A, namely Beams A-1, A-2 and A-3, had 7 rows
of stud connectors, 3 in a row. Beams B-1 and B-2 had 4
rows of stud connectors, the spacing between the rows being
twice that for Group A beams. This was achieved by first
taking a panel similar to that for Group A beams, and then

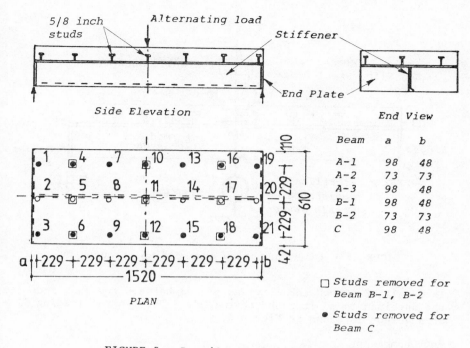

FIGURE 3 - Details of beam specimens tested

cutting and machining off alternate rows of stud connectors. Beam C had a single longitudinal line of stud connectors, located near the web (stiffener).

INSTRUMENTATION

The beams were instrumented for measurements of central deflection (dial gauge), flexural strains (strain gauges in steel and 'Demec' gauges in concrete), slip between the concrete and steel (dial gauges with brackets), local strain near the stud (strain gauges) and strain distribution across the width ('Demec' gauges).

TEST ARRANGEMENT

The specimens were tested as simply supported beams with a span of 1.520 m and subjected to an alternating concentrated load at the mid-span. The load was applied by means of a single-acting hydraulic jack of 500 kN capacity, connected to a servo-controlled Moog valve which regulated the hydraulic flow from the pump to the jack, making it possible to apply a constant load range during dynamic testing. A

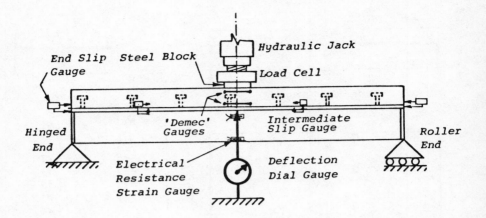

FIGURE 4 - Instrumentation and Loading Arrangement

load-cell of 500 kN capacity was placed between the jack and
the speciman to monitor the test load.   The general loading
arrangement is shown in Figure 4.

TEST PROCEDURE

Inititally, each beam was loaded statically to its maximum
load in small increments, and similarly unloaded, three times.
This was done to allow the beam to 'settle' in the rig.
Deflection, strain, and slip readings were recorded after
each load step.   Following this the load was applied
dynamically at a frequency of 1.0 Hz.   The load variation
was sinusoidal throughout.   At frequent intervals, the
dynamic loading sequence was interrupted to enable the
recording of deflections, strains, and slips.

EFFECT OF OPERATING FREQUENCY

Tests conducted by Lo (1978) on push-out specimens which form
the basis for the design S-N diagram, were carried out at an
operating frequency of 3.0 Hz.   It was found that the larger
beam tests could not be carried out at this frequency because
of the limitations in the capacity of the Moog valve available.
Woo (1979) describes some tests on push-out specimens,
identical in every respect to those tested by Lo (1978), but
carried out at a frequency of 1.0 Hz rather than 3.0 Hz.
Failure occurred at slightly shorter fatigue life, falling
within the 90% confidence interval based on Lo's results.
It was concluded that the operating frequency in the range 1.0
to 3.0 Hz has an insignificant effect on the fatigue life of
stud connectors.

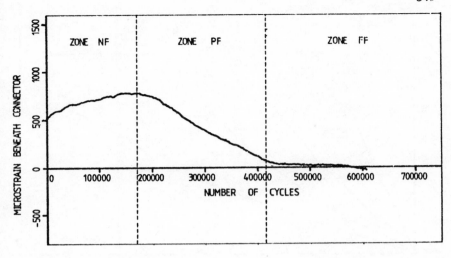

FIGURE 5 - Typical charatericstic of strain beneath a
connector

FATIGUE LIFE OF STUD CONNECTORS IN BEAM SPECIMENS

The strain history of the gauges pasted on the steel plate
at locations near the stud connectors gives a good indication
of the failure of a stud. This method of detecting
connector failure is based on the assumption that when the
stud connector is still intact, there exists a steady state
of strain near the connector. When the connector first
develops a crack, there occurs a change in the state of
strain around the stud connector. This change continues until
the connector has completely failed, after which once again
a steady state prevails. This is illustrated in Figure 5
which shows the strain on the opposite side of Connector 17,
Beam A-2, plotted against the number of cycles of loading.
Verification of this hypothesis was done by the following
procedure.

When sufficient number of connectors were deemed to have
failed according to the above criterion, in a given test, the
test was stopped and the concrete was removed carefully with
a hammer. Stud connectors were found in various stages of
failure. Invariably, studs still fully intact had strain
history as indicated in Zone NF, Figure 5. Studs in partial
failure had reached Zone PF, and fully failed studs were in
Zone FF. It is obvious that the transition from one zone to
another does not occur at a precise point in time. In many
cases, certain amount of judgement had to be exercised to
determine the fatigue life of the particular connector.
Table 1 summarises the results obtained, expressed as cycles
to failure of the most highly stressed stud connector.

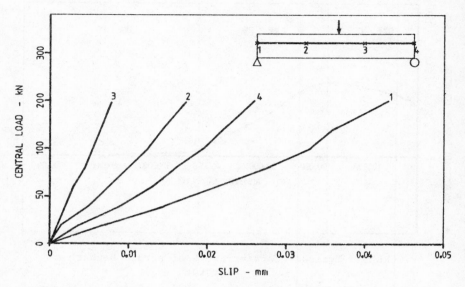

FIGURE 6 - Typical load-slip response in beam specimens

LOAD VERSUS SLIP CHARACTERISTIC OF THE STUD CONNECTOR

A typical load - slip response for the beams tested is shown
in Figure 6.   It is evident that the end connectors
transmitted more load than the intermediate connectors.
Also, the characteristics are, in the main, linear.   It is
interesting to compare these curves with the load-slip curves
obtained for the push-out specimens (Figure 7). Apart from
the initial low stiffness, explained by the early settlement
of the connector against the concrete, the response is again
mainly linear.   A direct comparison is made difficult
because the shear force in a connector in the beam specimens
is difficult to predict.   This is discussed in detail later.
It can, however, be concluded that the connector in a beam

TABLE 1 - FATIGUE LIFE OF CONNECTORS IN BEAM SPECIMENS

Beam Label	Central Load (Kn)	Fatigue life of first Connector to fail
A-1	140	575 000
A-2	170	240 000
A-3	150	180 000
B-1	150	21 750
B-2	190	10 000
C	150	11 250

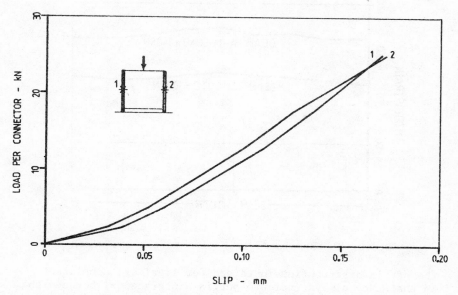

FIGURE 7 - Typical Load-Slip response in push-out specimen

specimen behaves in a manner broadly similar to that of a
connector in a push-out specimen.

DEGREE OF INTERACTION

The degree of composite action attained in a composite beam
depends on the effectiveness of the shear connectors in
preventing relative movement between steel and concrete.
Figure 8 shows the strain distribution at mid span for
three beams.   The experimental results are shown along
with the theoretical results predicted by the no-interaction
theory and the full-interaction theory.

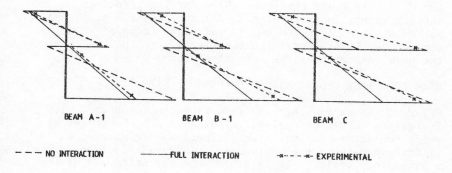

FIGURE 8 - Strain distribution in the beam specimens

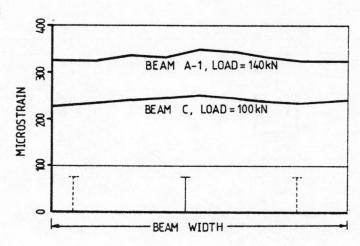

FIGURE 9 - Strain distribution across the width

The strain distributions obtained for steel are consistent
in that they always indicate partial interaction, between the
two extremes of full and no interaction, and further the
diminishing effectiveness of the shear connectors in beams of
Groups A, B and C as was to be expected. The strains in
concrete, on the other hand show a greater amount of scatter
explained by the fact that the gauge length for concrete
strains was 200 mm as compared with a gauge length of 5 mm
for steel. Another important observation is that in all the
beams tested, the lower end of the concrete slab had
significant tensile strains, supported by experimental
observation of very fine tensile cracks. In both the full
interaction and no interaction results given above, cracking
of concrete was ignored.

EFFECTIVE WIDTH

Figure 9 shows the strain distributions across the width
measured on the top surface of concrete for the beams A-1
and C. It is clear that that the shear lag effect is
insignificant, in spite of the fact that the two beams had a
marked difference in the number and disposition of shear
connectors. Thus, for further analysis, the full width of
the beam was taken as the effective width.

BEAM BEHAVIOUR UNDER FATIGUE LOADING

Figure 10 shows the increase in deflection for Beam C with
increasing number of cycles. Also shown are the stages at
which connectors failed. The deflections are plotted as the
difference between the maximum and minimum deflections during
a given load cycle. It is clear that the stiffness of the

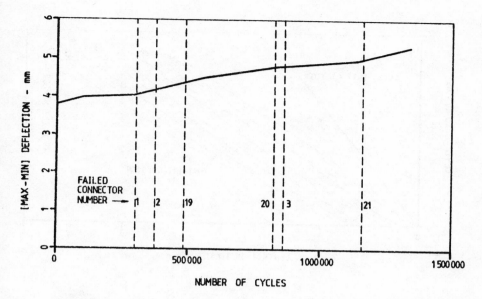

FIGURE 10 - Deflection characteristic for Beam B-1

beam did not reduce dramatically even after most of the connectors in the beam had failed completely, supporting the view that the degree of interaction in the beam at the start was low, so that the loss in stiffness with successive failure of connectors was marginal.

Figures 11 and 12 show examples of load-deflection response obtained for a beam with a large number of shear connectors (Beam A-1) and another with relatively fewer shear connectors (Beam C). The following conclusions emerge:

- All beams showed evidence of partial interaction.
- There was only a marginal loss in the beam stiffness even when some shear connectors had failed.
- The full-interaction theory grossly overestimates the stiffness of the beam.
- The no-interaction theory also, in general, over-estimates the stiffness of the beam mainly because the cracking of concrete is not accounted for.

CORRELATION BETWEEN BEAM AND PUSH-OUT TESTS

To predict the fatigue life of a stud connector in a beam specimen, using the statistical data obtained from push-out specimens, it is important to determine the shear stress existing in the shear connector under a given beam loading.

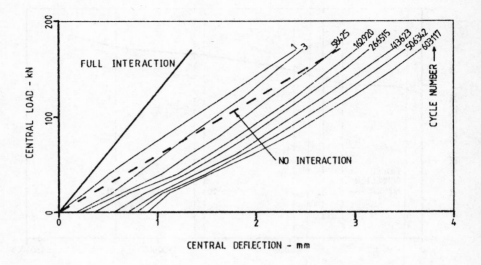

FIGURE 11 - Load-deflection response of Beam A-2

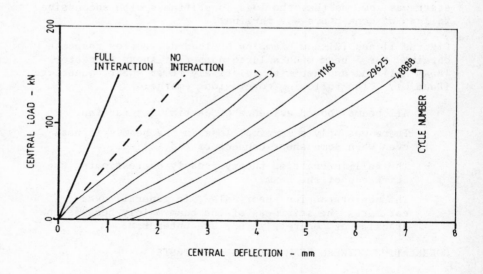

FIGURE 12 - Load-deflection response of Beam C

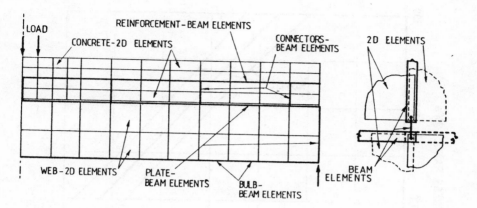

FIGURE 13 – Finite Element modelling of the composite beam

This can be done in many ways, a few of which are:

1. Transformed area method – full interaction assumed.

2. A method based on partial interaction, due to Newmark, Siess and Viest (1951).

3. Finite element method (Hirst and Schmidt, 1975).

The transformed area method is too well known to merit description here. The method due to Newmark, Siess and Viest, although published as early as 1951, has not been widely used. It makes use of the fact that the bond between steel and concrete is imperfect and assumes that the load-slip characterisation of shear connectors is linear. The finite element model used here is a two dimensional one. The concrete and the web of the bulb flat were represented by plane stress isoparametric elements, while beam elements were used for the steel reinforcement, stud connectors, the steel plate, and the bulb of the bulb flat (Figure 13). In the present analysis, the cracking of concrete has been ignored.

TABLE 2 – CONNECTOR STRESS GIVEN BY THREE ANALYTICAL METHODS

Beam	Max. Load (kN)	Extreme Connector Stress (MPa)		
		Transformed Area Method	Newmark's Method	Finite Element Method
A-1	140	214	113	112
A-2	170	260	137	136
A-3	150	230	121	120
B-1	150	460	156	178
B-2	190	583	197	225
C	150	690	196	206

356

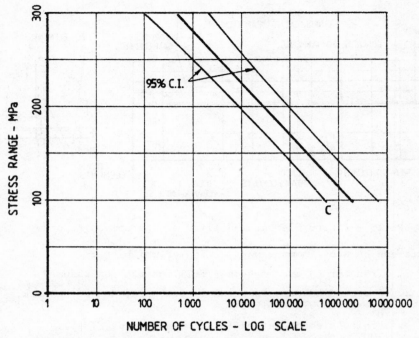

FIGURE 14 - S-N diagram based on push-out tests by Lo (1978)

For the six beams tested, the results from the three sets of
analysis are presented in Table 2. Using the shear stresses
so obtained, the fatigue life of the most highly stressed
connector in the beam, as predicted by the lower 95% confidence interval curve (Figure 14) for the data from push-out
specimens, has been listed in Table 3 for the three methods
of analysis. It is clear that the transformed area method
is grossly conservative. Both the Newmark's method and the
Finite element method give a far superior correlation with the
experimentally observed fatigue life. Further the results
from the Finite Element method and the Newmark analysis are

TABLE 3 - COMPARISON BETWEEN PREDICTED AND EXPERIMENTAL
FATIGUE LIFE

| | Theoretical Fatigue Life | | | Experimental |
Beam	T.A. Method	Newmark's Method	F.E. Method	Fatigue Life
A-1	4 570	302 000	288 000	575 000
A-2	620	112 200	114 800	240 000
A-3	2 240	229 000	234 000	180 000
B-1	1	53 700	20 900	21 750
B-2	1	9 100	2 800	10 000
C	1	10 000	6 600	11 250

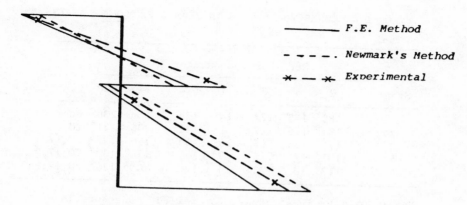

FIGURE 15 - Strain distribution in the beam

close to one another. Allowing for the uncertainty in
establishing the fatigue life of the stud connectors in beam
specimens, the correlation obtained in the order of magnitude
of fatigue life is excellent. It may,therefore, be concluded that a very good estimate of the fatigue life of the
stud connectors in beam specimens can be obtained by first
determining the shear stress in the most highly stressed
connector using either a finite element analysis or the
Newmark's analysis, and then obtaining the fatigue life from
the lower 95% confidence interval curve for the S-N data
obtained from push-out specimens. Newmark's method has the
additional merit of being a hand calculation method, whereas
the Finite Element method is computer dependent.

The above conclusions are further supported by the comparison
between the theoretical and observed strain distributions
in the beams. For example, Figure 15 shows the strain
distributions as given by the Finite Element method and the
Newmark's method together with the experimentally observed
strains by Beam A-3. Similarly good correlation was
obtained for the other beams.

PREDICTION OF STUD FAILURES IN BEAM SPECIMENS

In the Finite Element method the shear connectors were
treated individually, rather than by assuming them to be
continuous along the length of the beam as in Newmark's
analysis. Consequently, this method can be used to evaluate
the stress history of every connector in the beam at successive
stages of connector failure. Further, with the aid of Miner's
cumulative damage rule, (Miner, 1945) this load history can
be used to predict the sequence in which the connectors fail.
In applying this procedure, once a connector has 'failed',
it is assumed that it loses 50 percent of its stiffness.

TABLE 4 -   CONNECTOR STRESS IN BEAM C FROM F.E. ANALYSIS

| Stage | \multicolumn{7}{c}{Stress per Connector (MPa)} | Failed Connectors |
	2	5	8	11	14	17	20	
1	189	175	122	14	113	183	206	None
2	181	166	111	1	131	206	121*	20
3	171	155	97	16	153	119*	135*	20,17
4	101*	175	113	3	141	114*	130*	20,17,2
5	113*	102*	133	13	128	108*	124*	20,17,2,5

Note: 50% loss in stiffness for failed connectors (*)

TABLE 5   -   PREDICTION OF STUD FAILURE IN BEAM C

| | \multicolumn{7}{c}{Connector Number} | \multicolumn{2}{c}{Cycles} |
	2	5	8	11	14	17	20	n (cum.)	Observed
N_1	13200	25100	211000	∞	288000	17400	6600	6600	
n_1/N_1	0.50	0.26	0.03	0	0.02	0.38	1.00	(6600)	11250
N_2	18800	35100	316000	∞	141000	6300			
n_2/N_2	0.21	0.11	0.01	0	0.03	0.62		3900	
$\Sigma n/N$	0.71	0.37	0.04	0	0.05	1.00		(10500)	14250
N_3	27500	53700	∞	∞	58900				
n_3/N_3	0.29	0.15	0	0	0.13			7900	
$\Sigma n/N$	1.00	0.52	0.04	0	0.18			(18400)	25750
N_4		25100	288000	∞	100000				
n_4/N_4		0.48	0.04	0	0.12			12000	
$\Sigma n/N$		1.00	0.08	0	0.30			(30400)	12500
N_5			128800	∞	164000				
n_5/N_5			0.89	0	0.70			114800	
$\Sigma n/N$			0.97	0	1.00			(145200)	35700

This figure of 50% was arbitrarily chosen in view of the observation that the stud connectors in the beam specimens seldom, if ever, fail completely.   The fatigue crack propagates into the heat affected zone in the plate, but no total separation takes place.

As an illustration, Beam C was chosen for further analytical study.   Table 4 gives a summary of the stresses obtained

in various studs for each successive stage of stud failure.
Using the stress values so obtained, together with Curve C
in Figure 14, the sequence of stud failure predicted is
20, 17, 2, 5 and 14 (see Table 5) which is quite comparable
with the experimentally observed sequence of 20, 2, 5, 17 and
14. It is interesting to note that the fatigue life obtained
for each stud by the theoretical analysis is quite close to
the experimentally observed value, keeping in mind the
relatively large scatter in results always to be expected in
fatigue testing.

CONCLUSIONS

It has been shown that the fatigue life of stud connectors
in a beam specimen, built from the steelwork of a composite
box girder bridge, can be adequately predicted from basic
fatigue data compiled from fatigue tests on relatively less
expensive push-out tests. It was found that the crucial
step involved is the determination of the shear stress in the
stud connector. Ordinary beam theory, using the transformed
area of the composite section, was found to be far too
conservative. A simplified Finite Element anlysis was
found to give the best correlation not only in predicting
the fatigue life of the most highly stressed connector, but
also in predicting the sequence of failure of the studs.
Another method, proposed by Newmark, Siess and Viest (1951),
was found to have an accuracy comparable with the Finite Element
method. It also had the merit of being a hand calculation
method, whereas the Finite Element method is computer dependent.

ACKNOWLEDGEMENT

The work described in this paper was carried out in the
Department of Civil Engineering, University of Melbourne,
Australia, as part of a continuing programme of research on
the fatigue behaviour of structures.

REFERENCES

Hallam, M.W. (1976) The Behaviour of Stud Shear Connectors
under repeated Loading. Research Report No. R280, University
of Syndey School of Civil Engineering.

Hirst, M.J.S. and Schmidt, L.C. (1975) Static Behaviour of
Composite Box-Girder Bridge Decks. Fifth Australian
Conference on the Mechanics of Structures and Materials,
Melbourne.

King, D.C., Slutter, R.G. and Driscoll, G.C. (1965) Fatigue
Strength of $\frac{1}{2}$ in diameter Stud Shear Connectors. Highway
Research Record 103.

360

Lo, K.K. (1978) Fatigue Testing of Stud Shear Connectors. MEngSc Thesis. Department of Civil Engineering, University of Melbourne.

Mainstone, R.J. and Menzies, J.B. (1967) Shear Connectors in Steel-Concrete Composite Beams for Bridges. Part 1 - Static and Fatigue Tests on Push-out Specimens. Concrete, London. 1, 9. Part 2 - Fatigue Tests on Beams. Concrete, London. 1, 10.

Miner, M.A. (1945) Cumulative Damage in Fatigue. Journal of Applied Mechanics, 42, 9.

Slutter, R.G. and Fisher, J.W. (1966) Fatigue Strength of Shear Connectors. Highways Research Record. 147.

Thürlimann, B. (1958) Composite Beams with Stud Shear Connectors. Highway Research Record. 174.

Toprac, A.A. (1965) Fatigue Strength of 3/4 inch Stud Shear Connectors. Highway Research Record. 103.

Viest, I.M. (1956) Investigation of Stud Shear Connectors for Composite Concrete and Steel T-beams. Journal of the American Concrete Institute. 27, 2.

Woo, Y.S. (1979) Fatigue Behaviour of Composite Construction of Concrete and Stiffened Plate. MEngSc Thesis. Department of Civil Engineering, University of Melbourne.

VIBRATIONS OF FLOOD SPILLWAYS AND EFFECTS INDUCED BY THEM

L.A.Goncharov, L.A.Zolotov, L.D.Lentjaev, V.M.Semenkov

The Scientific Research Centre of the "Hydroproject" Institute, Moscow, USSR

Stringent requirements are imposed on the construction of large hydroelectric projects (especially in seismically active regions) as regards their security and safe operation. These requirements are placed not only upon dams and other hydraulic structures proper but also upon their foundations, and they should be provided at all stages of the project life, i.e. at the stages of construction and operation. But from experience it is already known that at an early stage hydraulic structure foundations (especially rock ones) begin to "degrade", i.e. their physical and mechanical properties inadvertently become impaired. This is because during the pit excavation, before the structures construction, a slacking layer of weathered rocks is stripped. The foundation is relieved and becomes jointed. Then as the concrete structure is completed, and the reservoir filled, the foundation subsides. For example, after filling the reservoir created by the 125m concrete dam, a vast territory adjacent to the dam subsided, the largest recorded subsidence amounted to 74mm (G.Soukhanov, 1978). Subsidence of the foundation causes the creation of tensile stresses and is accompanied by the development of vertical and horizontal joints,etc. The above mentioned and some other processes result in changes of elastic properties of dams and other hydraulic structure foundations and in decompaction of the contact zone (concrete-rock). This statement is confirmed by the results of direct sampling from the dam foundation. A case is known when, due to the variation with time of the physical and mechanical properties of the foundation, the parameters of forced vibration of 30m of concrete spillway under flood, measured under identical hydraulic conditions, have markedly varied for the two years of the structure's performance has been monitored:

- RMS displacements, or root mean square deviations from the mean have increased by 15-30%;
- "Leading frequencies" of the narrow-band vibrations, determined

from the first zero of the autocorrelation functions, increased
approximately by 20% (L.Goncharov, V.Semenkov, 1974).

For reliable forecast of long-term hydraulic structure security,
concurrent with the conventional evaluation methods, complementary information describing specific features of the structure
vibrations and effects induced by them, is obviously necessary.
In this connection the measurement data accumulated in the
"Hydroproject" institute during the prototype investigations of
two concrete dams is of particular interest.

The main retaining structures of the projects A and B are concrete gravity dams 125m and 120m high and 1430 and 1100m long,
respectively (Fig.1). Both dams are of triangular section with
the vertical upstream and the inclined downstream face (1:0.8
and 1:076). They are subdivided into separate blocks by construction joints spaced at 22m and 15m intervals, respectively.
Power houses 515m and 430m long, respectively, are located downstream beside the toe of the dam.They are equipped with 18 and
12 generating sets with unit capacities 225 and 500MW and speeds
125rpm and 93.8rpm, respectively. The foundation for project A
is composed of monolitic diabases 36-50m thick. The diabases are
underlain by siltstones and further by sandstones. Medium and
fine grained granites with different degrees of disintegration
occur in the project B foundation.

Instrumentation of two types was used for measuring the dam's
vibration: a conventional galvanometer registration with displacement recording (effective bandwidth 0.8-80Hz) and an analog
device developed and manufactured in the "Hydroproject" - vibration spectrum analyzer. It is intended for express analysis of
stationary random vibrations with resonance properties, operating
as an integral set with the seismoscope, the natural frequency
of which being 10Hz. It permits operation in one of three regimes; that of the vibrometer, velocity meter and accelerometer
and measurement of the double amplitude and RMS vibrations, mean
zero-crossing frequency, the frequency variancy in the bandwidth
from 0.5 to 120Hz and in eight octaves: 0.5-1-2-4-8 -30-60-120Hz.
The width of the resonance curve of vibration spectra, being unambiguously related to the logarithmic decrement, is evaluated
from the variance of the mean zero-crossing frequency, in octaves.

Vibration measurements at the hydroprojects A and B were planned
as an ingredient of comprehensive full-scale hydraulic investigations, control measurements and commissioning tests regularly
performed by the "Hydroproject" specialists at large hydraulic
structures. Further, one of the aims of the studies was experimental justification and check of usability, and an introduction
of a new non-destructive method of evaluation of dynamic characteristics of the dams and other hydraulic structure foundations
(The USSR Author's Certificate N.515053). The known evaluation

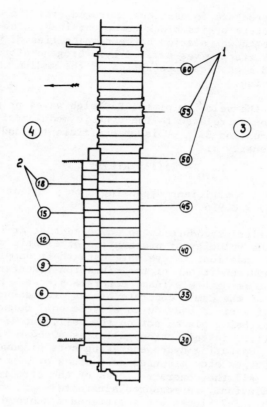

Fig.1. Scheme of observation at the project A, using the "vibro-
scopy" method: 1 - numbers of the concrete dam block;
2 - numbers of the generating sets; 3 - upstream; 4 -
downstream.

methods for existing structure foundations, i.e. methods of direct sampling, seismoacoustic, ultrasonic and other methods are
rather labour-consuming. Thus the information obtained is of a
local character. It describes properties of foundations within
a certain zone, the size of which is small as compared with the
entire structure base. To decrease the influence of local specificities of sampling points or soundings, to obtain some integral evaluations, a large number of the above mentioned sampling
points (in the broad sense) are required. However, this is un-
reasonable as the cost soars, the foundation is disturbed and its
security decreases.

The new method under development, conditionally termed "vibro-
scopy", is devoid of the above deficiencies and will enable
integral evaluations of dynamic characteristics of dams and other

hydraulic structure foundations averaged over the entire base of the structure or its block to be obtained. The method is based on the known relation between velocities of propagation of elastic surface waves excited by vibrogenerators and the physical and mechanical properties of the medium where they are propagating.

For example the velocity of the Rayleigh waves propagation V_R is unambiguously related to the dynamic modulus of elasticity for the foundation E_{dyn}, Poisson coefficient μ and the foundation soil density ρ:

$$E_{dyn} = 2\rho(1+\mu)K(\mu)V_R^2 ,$$

where $K(\mu)$ is coefficient depending upon μ and at $0 < \mu \leq 0.50$ $0.874 < K(\mu) < 0.956$.

The observation procedure is so composed that it is possible to change the velocity of propagation of elastic waves in the structure foundation, to evaluate both their damping with distance, and conditions of the vibration transfer from foundation to the structure's (dam) concrete blocks. Hence, the vibrations of the dam's concrete blocks are measured as the vibration of a rigid body on an elastic soil foundation; the vibropickups being placed not on the soil, but directly on the structure (i.e. in inspection galleries 40-60 m higher than the base). Operating hydrogenerating sets of power plants, water discharges etc. are used as sources or generators of vibration. All the concrete blocks of the structure (dam), without exclusion, are measured alternately and vibrations in not less than 5-7 blocks are registered synchronously.

The observation results indicated some structure portions where registered values of elastic constants of foundations and conditions of vibration transfer from foundation to the structure blocks etc., sharply differed from the average or from conventional values. Thus revealed portions of the structures are further subjected to careful control measurement for a number of years. Results of many-years observations reveal trends of variation of the controlled parameters, enabling the variation of the security of hydraulic structures with time to be evaluated.

The observed specificity of vibrations of large dams and other hydraulic structures as a system with a large number of vibration sources (operating generating sets of power plants, water discharges through hydraulic structures, etc.) turned out to be beneficial for the useful introduction of the non-distructive control method into practice. The observations show that with simultaneous action of several vibration sources the effective values of decrements of the system vibration decrease (the system dynamism increases). Attenuation of vibration i.e. due to interference, is not observed. On

the contrary, a system with several identical vibration sources,
i.e. operating generating sets of power plants, is capable of
tuning or self-synchronization "with delay" (the propagation of
elastic waves in foundations, for individual phase shifts in
the vibration transfer system of the type "generating set –
one power house block – soil foundation – another powerhouse
block – another generating set, etc."). Hence the whole complex
of power house blocks with operating generating sets is a
particular unique dispersed self-synchronizing vibration source
which generates elastic waves in the foundation. The waves
are interacting with the concrete structure i.e. with the concrete dam blocks and induce almost harmonic vibrations with
markedly traced inclined axes of cophasing (Fig. 2). The
analysis discloses that the waves generated in the foundation
are elastic Rayleigh waves and the vibrations of some concrete
dam blocks are induced in the vertical plane (being parallel
to flow). Vibrations mainly represent superposition of vertical
and rotational motions (as rigid bodies on elastic foundations)
around some horizontal axis which is parallel to the dam axis
and is arranged 10-20 m lower than the dam base. The observed
pattern of the dam concrete blocks vibrations promotes measurements of the phase velocity of elastic wave propagation (or of
its projection in the direction of the dam axis).

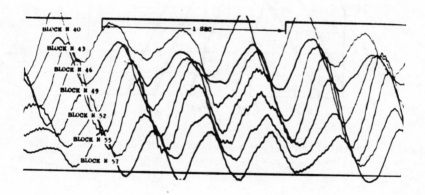

Fig. 2   Typical oscillogram of the dam block vibrations
         (project A).

The registered vibrations of the dam blocks are of a random
type despite the seeming orderliness, closely conforming to
the normal law of probability distribution (Gauss distribution).
Hence the known correlation and spectral methods of analysis
of stationary random processes are employed to obtain a higher
accuracy in evaluating the phase velocity of elastic waves
propagating in the dam foundation. Thus delay time for vibrations propagating from one dam block to another was determined

in two ways:

- by the time of the first maximum τ_{ij} of mutual-correlation functions of vibrations synchronously registered in two points (dam blocks) i and j (Fig. 3);

- by phase shift $\Delta\phi_{ij}(\omega_m)$ between the same processes, i.e. from the value of phase-frequency characteristics of the mutual spectrum at the frequency ω_m at which maximums of density are simultaneously observed in the autospectra of the processes i and j, i.e.

$$\tau_{ij} = \frac{1}{\omega_m} \cdot \Delta\phi_{ij}(\omega_m).$$

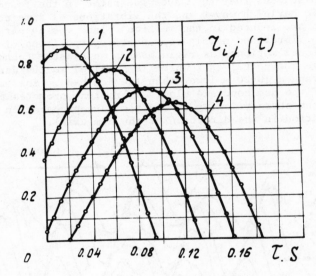

Fig. 3   Correlation between vibrations of different blocks (i,j) of the project A dam. (The dam block N.50 was adopted as the reference one, i.e. i = 50).
1. i = 50; j = 52.   2. i = 50; j = 55.
3. i 0 50; j = 57.   4. i = 50; j = 59.

Oscillograms with not less than 100-120 periods of the vibrations "fundamental tone" were used for data processing. Time delays calculated by the two above methods differ from one another by no more than ± 5%. The hodograph equation was calculated by the least squares method.

The following specific features were revealed in the analysis of experimental vibration studies of the concrete dams of the hydroprojects A and B:
(a) Values of phase velocities of elastic wave propaga; tion in the bases of both dams found by the above procedure, considerably differ from the universally known ones: at the

project A (1974) the measured value of velocity was 1200 m/sec (for massive diabase - within 2600-3900 m/sec), and at the project B (1973) it equalled 1270 m/sec (for fissured granite 1600-2900 m/sec).

(b) Phase velocities of elastic waves propagating in the dam bases markedly decrease with time: at the project A the measured value decreased from 1200 m/sec (1974) to 840 m/sec (1976) and at the project B from 1270 m/sec (1973) to 1085 m/sec (1976).

(c) The concrete dam blocks were found to feature abnormally low values of vibration decrements. For example the measurement of vibrations of some project A dam blocks, with angular displacements less than 10^{-7} rad, performed with the help of the vibration spectrum analyzer, yielded the following results: for octave bandwidths 0.5-1.0; 1-2; 2-4; 4-8; 8-16 the logarithmic decrements did not exceed respectively the values 0.040; 0.040; 0.030; 0.009 and 0.003. (It should be noted that the reported values of decrements of mass structures are as high as 0.8-2.5 and upwards).

(d) Being a sharp-resonance system, the concrete dam blocks extract energy from the wide-band spectrum mainly in the narrow bandwidth of frequencies (close to one of the natural frequencies). For example, the measurements in the project A dam reveal practical coincidence of the vibration frequency of the dam blocks with one of the frequencies of the determinate effect (with the double rotational frequency of the generating set rotation equalling 2 × 125 : 60 = 4.166 Hz). As a result of this, the effect of the "resonance" sway of the dam blocks occurs, and the "resonance contribution" is observed in the bandwidth from 4.15 to 4.19 Hz, comprising an appreciable share in the total vibration variance:

Dam Block	51	54	57	60	63	66
Per cent contribution of variance (RMS^2) in the bandwidth 4.17±0.02 Hz	38.7	33.4	29.4	52.0	15.4	65.0

(e) The repeated vibration measurements of the project A dam (1974 and 1976) did not reveal any decrease in the "resonance" sway effect, as might be expected on the assumption that the dissipative forces increase as the foundations deteriorate. On the contrary, effects of the "resonance" sway increase from 1.3 to 2.0 times were observed at 60% of the examined dam blocks.

The processes which develop in the concrete dam foundation and lead to its decompaction and deterioration, may be assumed in our opinion as one of the causes of small decrements in the mass structure vibrations, and of low velocities of elastic waves propagation in the foundations of concrete dams of the projects A and B (and of low effective values of dynamic moduli of elasticity, correspondingly), as well as the cause of their decrease with time. For example, seasonal temperature deformations of the dam's downstream face accompanied by the deformation of the contour of the entire dam block cause local breaks of contact in the upstream and downstream dam zones (i.e. from the upstream and downstream side, respectively). Results of piezometric measurements are a testimony to this. Fluctuations in the reservoir level, varying loads and wave loads, vibrational ones included, cause deterioration of the dam rock foundation, especially in the zones adjacent to ribs of the upstream and downstream toes. Due to deformation of these zones the foundation surface may assume a curvilinear ("barrel-type") shape. These effects can be responsible for high dynamism of concrete hydraulic structures on rock foundation. P. Koulmach (P. Koulmach, 1958) was first to pay attention to possible development of such effects. He analysed the results of research into the vibration of a breakwater in Algeria (J. Renaud, 1935) weighing 8060 t, its base area being 325 m. The breakwater rested on a stone bed 4 m thick underlain by 10 m layer of highly compacted fine-silted sand, below which conglomerates were occurring. In case of agitation both wave pressure (water level fluctuations) and forced breakwater block vibrations (intact with the wave) were recorded synchronously. Using the measurement results, P. Koulmach calculated the moments of the acting forces. Comparing the obtained values with the values of the breakwater blocks displacements (vibrations) he found the values of generalized elastic characteristics of foundations - C_z , C_y , C_x which turned out to be much smaller than universally adopted. Considering the fact that the breakwater started swaying at insignificant agitation at the wall (wave 0.3 m high or downward), P. Koulmach proposed, a later confirmed assumption, that the breakwater foundation had deformed and assumed a "barrel-type" shape.

The specific vibration features of concrete dams at large hydraulic projects have been insufficiently examined as yet, and further studies are under way. But even now the available experimental results enable us to assert that the above effects should be allowed for in stability analysis for hydraulic structures, especially for seismic effects acting on these structures.

REFERENCES

1. Goncharov, L.A., Semenkov, V.M. (1974) Flow-induced Structural Vibrations. Symposium IUTAM-IAHR, August 14-16, 1972, Karlsruhe. Springer-Verlag: 297-317.
2. Koulmach, P.P. (1958) "Ghidritechnicheskoe Stroitel'stvo", 4: 35-38.

3. Method of Determination of Dynamic Characteristics of Structure Foundations. USSR Author's Certificate N.515053 (Priority date 7 Jan., 1975). Bull. "Otkrytiya. Izobreteniya. Promyshlennye Obraztsy. Tovarnye Znaki", Moscow, 1976, 19 : 104.

4. Renaud, J.M. (1935) Annales des ponts et chausée, 5.

5. Soukhanov, G.K. et al. (1978) "Ghidrotechnicheskoe Stroitel'stvo", 4 : 12-18.

DETERMINATION OF NATURAL FREQUENCIES OF THE THIN ROTATIONAL
SHELLS BY FINITE ELEMENT METHOD.
R. Delpak.

Principal Lecturer in Civil Engineering, The Polytechnic of
Wales, Wales, U.K.

ABSTRACT

A curved parametric element, capable of giving the natural
unforced and undamped frequencies of thin rotational shells
has been developed. The new formulation possesses a number
of interesting features including true nodal conformity,
genuine curved generator, optional number of degrees of
freedom and variable thickness. The nature of the displacements and the generation of the stiffness and the mass
matrices are outlined in the text. Finally the accuracy of
the present formulation is examined by considering a selected
number of well known examples.

INTRODUCTION

The importance of allowing for time variable loads in
designing Civil Engineering structures has long been
appreciated. The incorporation of such allowances in the
design and construction of practical structures has generally
taken the form of increasing the safety margins by introducing the static equivalent of dynamic loads applied to the
structure. Recent well-known failures of structures at
various locations have intensified research work in understanding the nature of dynamic loading, the evaluation of
the ensuing response of the loaded structure, development
of economic means of predicting such response and finally
incorporating the findings into design processes.

Rotational shells owing to their appearance and inherent
strength have provided frequent solutions and challenges to
architects and designers and thus brought about the
application of new methods of analysis based on the thresholds
of research and development. The static analysis of such
shells, being a labour intesive area, has gathered
considerable attention and is a relatively cheap process.

A satisfactory dynamic analysis which may result in a full response - type solution could prove costly and is generally replaced by other more conventional methods. Some notion of dynamic behaviour can be gleaned by examining the unforced, undamped shell frequencies. The present work is confined to determining the above mentioned frequencies of thin elastic rotational shells by the finite element method.

ELEMENT FORMULATION

The element discussed here is of a parametric type, the full description of which is given elsewhere (1) (2). Certain features which are relevant to the text will be included.

Element Geometry

The parent element is linear with nodes 1 and 2 and a variable ξ so that $-1 \leq \xi \leq +1$ with extremities -1 and $+1$ which correspond to nodes 1 and 2 respectively, Figure 1. The Functional ϕ, defined over the parent element, is assumed to take the value of ϕ; at the nodes 1 and 2 so that,

$$\left. \begin{array}{l} \xi = -1 \\ i = 1 \\ \phi_i = \phi_1 \end{array} \right\} , \quad \left. \begin{array}{l} \xi = +1 \\ i = 2 \\ \phi_i = \phi_2 \end{array} \right\} ,$$

The variation of ϕ within the range $-1 \leq \xi \leq +1$ is calculated in terms of the nodal values ϕ_i above and a set of interpolating functions N which are normalised (i.e. assume a value of unity) at each node. Thus the value of the functional at any point ξ can be estimated from,

$$\phi_{i\xi} = \phi_i N_i (\xi).$$

A similar representation can be made with slope or derivative variable so that,

$$\phi'_{i\xi} = (\frac{d\phi}{d\xi})_i N'_i (\xi).$$

The total response within the parent element due to ordinary and slope nodal variable is evaluated by incorporating both functions, namely.

FIGURE 1

B1-B4 ARE BASIC FUNCTIONS

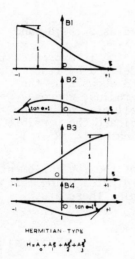

HERMITIAN TYPE

$$H = A_0 + A_1 \xi + A_2 \xi^2 + A_3 \xi^3$$

FIGURE 2

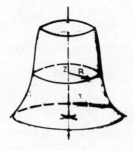

FIGURE 3

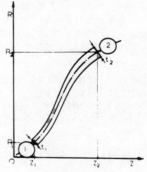

FIGURE 4

$$\phi\,(\xi) = \sum_{i=1}^{2} \left\{ \phi_i N_i\,(\xi) + (\tfrac{d\phi}{d\xi})_i N_i'\,(\xi) \right\}. \qquad\qquad \text{Eqn.1(a)}$$

Third order Hermitian Polynomials are used to represent the above variation which has the following explicit form,

$$\phi\,(\xi) = \left\{ \phi_1 \cdot \tfrac{1}{4} \cdot (\xi^3 - 2\xi + 2) + (\tfrac{d\phi}{d\xi})_1 \cdot \tfrac{1}{4} \cdot (1-\xi)^2(1+\xi) + \right.$$

$$\left. \phi_2 \cdot \tfrac{1}{4} \cdot (-\xi^3 + 3\xi + 2) + (\tfrac{d\phi}{d\xi})_2 \cdot \tfrac{1}{4} \cdot (1-\xi)(1+\xi)^2 \right\}. \qquad \text{Eqn.1(b)}$$

The interpolating functions associated with

ϕ_1, $(\tfrac{d\phi}{d\xi})_1$, ϕ_2 and $(\tfrac{d\phi}{d\xi})_2$

are called Basic Functions and are plotted as B_1, B_2, B_3 and B_4 in Figure 2.
The shell geometry in the present formulation is expressed in cylindrical coordinates and is uniquely defined by its principal parameters Z, R and t . Figure 3 shows Z as the axis of rotation and R the corresponding radius, whereas Figure 4 depicts a cut section of the shell element in the Z-R plane. Figure 4 also attempts to indicate the possible variation of thickness t and nodal coordinates (Z_1, R_1, t_1) and (Z_2, R_2, t_2).

The intermediate coordinates Z, R and t of the shell element are calculated in terms of the nodal coordinates using the Basic Functions of Equations 1 (a) and (b). These are represented symbolically as,

$$Z(\xi) = \sum_{i=1}^{2} \left\{ Z_i N_i(\xi) + (\tfrac{dZ}{d\xi})_i N_i'(\xi) \right\}, \qquad\qquad \text{Eqn.2(a)}$$

$$R(\xi) = \sum_{i=1}^{2} \left\{ R_i N_i(\xi) + (\tfrac{dR}{d\xi})_i N_i'(\xi) \right\}, \qquad\qquad \text{Eqn.2(b)}$$

$$t(\xi) = \sum_{i=1}^{2} \left\{ t_i N_i(\xi) + (\tfrac{dt}{d\xi})_i N_i'(\xi) \right\}. \qquad\qquad \text{Eqn.2(c)}$$

The variation with respect to ξ of coordinates Z, R and t is shown in Figure 5. Comparing Eqn. 1 (1) with Eqns. 2 (a) and (b) and taking note of the definition given in P. 137 of Reference (1) it is clear that the shell element falls within the isoparametric family of elements.

The following geometric facilities which are set out briefly, are characteristics of the present formulation.

(i) The element is truly conforming since the slope of the tangent $(\frac{dR}{dZ})$ at a common node of the neighbouring elements, shown in Figure 6, can be expressed by,

$$\left[(\frac{dR}{d\xi})_2 / (\frac{dZ}{d\xi})_2\right]_{elem.N} = \left[(\frac{dR}{d\xi})_1 / (\frac{dZ}{d\xi})_1\right]_{elem.N+1}$$

(ii) An abrupt change in the slope of the shell generator can be created in a like manner by inputting the slopes $\tan \alpha_N$ (node 2 of element N) and $\tan \alpha_{N+1}$ (node 1 of element N + 1) which are shown in Figure 7.

(iii) The coordinates of shells of revolution having branching members, shown in Figure 8, can be generated such that the coordinates Z and R are common, but the nodal slopes must be prescribed accordingly.

(iv) An abrupt change in shell thickness can be accommodated at the common node of elements N and N + 1 by an appropriate choice of the thickness values $(t_2)_N$ and $(t_1)_{N+1}$, Figure 9.

A maximum variation of up to a cubic (vide Eqn. 2 (c) can allow for any other change in thickness along the generator of a given element.

ELEMENT DISPLACEMENTS

The components of the displacement field for the present work are identical to the corresponding components adopted for use in classical functional methods. This ensures the validity of application of well established kinematic relationships to determine the strains in comparable situations.

The displacements at the nodes 1 and 2 are symbolised by δ_1 and δ_2 respectively, and the field elswhere within the element is determined by using the interpolation functions $B_1 \ldots B_4$ of Figure 2. The representation is therefore made identical to the functional variation of Eqn. 1.(a) and coordinate calculation of Eqn. 2 (a) and (b). Thus,

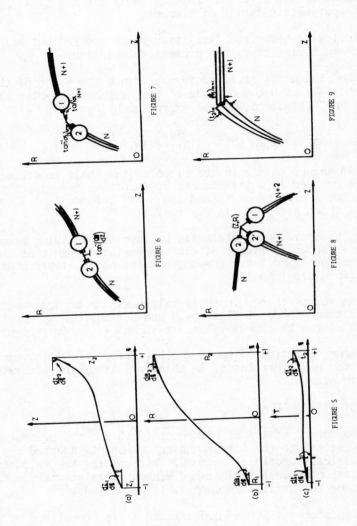

FIGURE 5

FIGURE 6

FIGURE 7

FIGURE 8

FIGURE 9

$$u(\xi) = \sum_{i=1}^{2} \left\{ u_i N_i(\xi) + (\tfrac{du}{d\xi})_i N_i^{'}(\xi) \right\} ,$$
Eqn. 3 (a)

$$w(\xi) = \sum_{i=1}^{2} \left\{ w_i N_i(\xi) + (\tfrac{dw}{d\xi})_i N_i^{'}(\xi) \right\} ,$$
Eqn. 3 (b)

$$v(\xi) = \sum_{i=1}^{2} \left\{ v_i N(\xi) + (\tfrac{dv}{d\xi})_i N_i^{'}(\xi) \right\} .$$
Eqn. 3 (c)

Owing to the very nature of the present formulation, the above displacements are globally directed, which is essential in maintaining continuity of displacements at the common node of two or more elements. However, the nodal rotation $(\tfrac{dw}{ds})$ must be expressed in local (meridional) coordinates for meaningful results.

The element also enjoys displacements which are not connected with nodes 1 nor 2 and are therefore completely independent of Basic Functions $B_1 \ldots B_4$. This new class of displacements is associated with internal or heirarchical nodes and their variation within the element is determined by generating Legendre' - type Functions (3), (4). The number and order of the above functions are optional depending on a number of considerations. For convenience of identification, these functions are labelled Surplus Functions - a specimen number of which namely S_o , $S_1 \ldots S_N$ are plotted in Figure 10.

In addition to their number and order, Surplus Functions can be generated in local or global coordinates to represent u, w and v displacements. Figures 11 to 14 are visual representation of B and S functions in their varied capacities i.e.,

B_1 as u (global) and w (global), Figures 11 (a) and (b),

S_o as u (global) and w (global), Figures 12 (a) and (b),

S_1 as u (local) and w (local), Figures 13 (a) and (b),

B_1 as v (global or local), Figure 14 (a),

S_1 as v (global or local), Figure 14 (b).

As can be seen, the element at its minimum capability is isoparametric and thus conforms to corollaries drawn in Reference (5) some of which are outlined below.

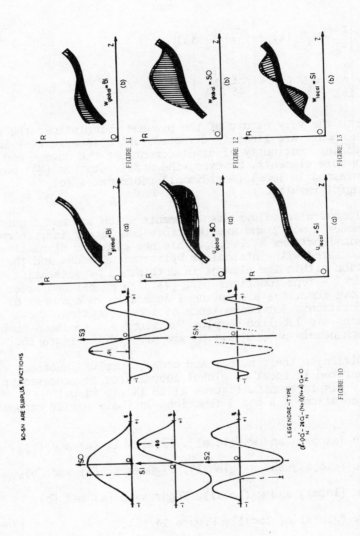

FIGURE 11

FIGURE 12

FIGURE 13

SO-SN ARE SURPLUS FUNCTIONS

LEGENDRE-TYPE

FIGURE 10

(i) Element characteristic could be altered as a matter of input and <u>not</u> of formulation.

(ii) Rigid body motions become readily available in an isoparametric element. This property does not depend on the choice of shape functions N_i.

(iii) It is stated that when the computer capacity is a limiting factor, the improved accuracy is achieved by relatively few sophisticated elements amounting to a given number of degrees of freedom than by a number of simpler elements totalling the same figure. The present formulation posesses all the desirable features discussed above.

GENERATION OF STIFFNESS AND MASS MATRICES

The stiffness and mass matrices which are generated from the will known relationships of Reference (1) are represented in their familiar form by,

$$[K] = \int_V [B]^T [D] [B] \ d(vol),$$
Eqn. 4 (a)

and

$$[M] = \int_V [N] [\rho] [N] \ d(vol).$$
Eqn. 4 (b)

The periodic displacements given below which portray the deformed state of the shell fall within two categories of θ - symmetric and anti - symmetric, namely,

$$\{\delta\}_{\theta\text{-sym.}} = \begin{Bmatrix} u.\cos(n\theta) \\ w.\cos(n\theta) \\ v.\sin(n\theta) \end{Bmatrix} e^{i\omega t} \ , \ \{\delta\}_{\text{anti-sym.}} = \begin{Bmatrix} u.\sin(n\theta) \\ w.\sin(n\theta) \\ v.\cos(n\theta) \end{Bmatrix} e^{i\omega t}$$
Eqn.5 (a)

The kinematic equations used are from Reference (6) which are particularly useful in their matrix and operator form. The generalised strain vector $\{\epsilon\} = [B] \{\delta\}$, obtained from the θ - symmetric set of displacements is given by ,

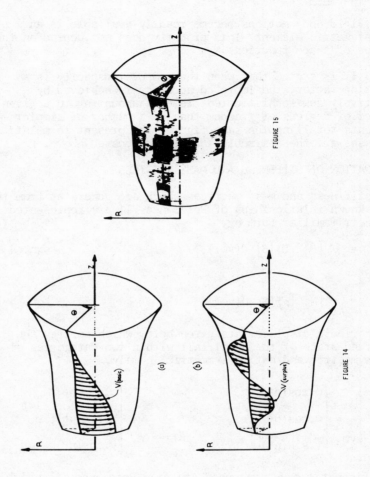

FIGURE 15

FIGURE 14

$$
\begin{Bmatrix} \varepsilon_s \\ \varepsilon_\theta \\ \varepsilon_{s\theta} \\ x_s \\ x_\theta \\ x_{s\theta} \end{Bmatrix} =
\begin{bmatrix}
(\cos\alpha\frac{\partial}{\partial s} & \sin\alpha\frac{\partial}{\partial s} & 0) & \cos n\theta \\
(0 & \frac{1}{R} & \frac{n}{R}) & \cos n\theta \\
((-\frac{n}{R})\cos\alpha & (-\frac{n}{R})\sin\alpha & -\frac{\sin\alpha}{R}+\frac{\partial}{\partial s}) & \sin n\theta \\
(\sin\alpha\frac{\partial^2}{\partial s^2} & -\cos\alpha\frac{\partial^2}{\partial s^2} & 0) & \cos n\theta \\
(-(\frac{n}{R})^2\sin\alpha+\frac{\sin\alpha}{R}\frac{\partial}{\partial s} & (\frac{n}{R})\cos\alpha-\frac{\sin\alpha\cos\alpha}{R}\frac{\partial}{\partial s} & (\frac{n}{R})\frac{\cos\alpha}{R}) & \cos n\theta \\
((\frac{n}{R})\sin\alpha\frac{\partial}{\partial s}+(\frac{n}{R})\frac{\sin^2\alpha}{R} & (\frac{n}{R})\cos\alpha\frac{\partial}{\partial s}+(-\frac{n}{R})\frac{\sin\alpha\cos\alpha}{R} & \frac{\cos\alpha}{R}\frac{\partial}{\partial s}-\frac{\sin\alpha\cos\alpha}{R^2}) & \sin n\theta
\end{bmatrix}
\begin{Bmatrix} u \\ w \\ v \end{Bmatrix}_{global} e^{i\omega t}
\qquad \text{Eqn.5(b.)}
$$

The expression for velocities $\{\dot{\delta}\} = \frac{d}{dt}\{\delta\}$ is considerably simpler since all three direct components are retained and the angular components are neglected, i.e.

$$
\begin{Bmatrix} \dot{u} \\ \dot{w} \\ \dot{v} \end{Bmatrix} = -i\omega
\begin{bmatrix} \cos n\theta & 0 & 0 \\ 0 & \cos n\theta & 0 \\ 0 & 0 & \sin n\theta \end{bmatrix}
\begin{Bmatrix} u \\ w \\ v \end{Bmatrix}_{global} \cdot e^{i\omega t}
\qquad \text{Eqn.5(c)}
$$

The elastic matrix $[D]$ used in eqn. 4(a) is given by

$$
[D] = \frac{E\,t}{1-\nu^2}
\begin{bmatrix}
1 & \nu & 0 & & & \\
\nu & 1 & 0 & & \text{zero} & \\
0 & 0 & \frac{1-\nu}{2} & & & \\
& & & \frac{t^2}{12} & \frac{\nu t^2}{12} & 0 \\
& \text{zero} & & \frac{\nu t^2}{12} & \frac{t^2}{12} & 0 \\
& & & 0 & 0 & \frac{(1-\nu)}{2}\frac{t^2}{12}
\end{bmatrix}
, \qquad \text{Eqn.6(a)}
$$

which is for the plane stress condition. Thus the product $\{T\} = [D] \cdot \{\varepsilon\}$ obtained from Equations 5 (b) and 6 (a) is the generalised stress resultant vector, the components and the directions of which is shown in Figure 15. The density matrix $[\rho]$ has the appearance of an identity matrix, namely

$$
\left[\rho\right] = \begin{bmatrix} \rho & 0 & 0 \\ 0 & \rho & 0 \\ 0 & 0 & \rho \end{bmatrix} \qquad\qquad \text{Eqn. 6 (b)}
$$

which together with velocities $\{\dot{\delta}\}$ of Eqn. 5 (c) results in the momentum vector $\{\mu\} = [\rho] \cdot \{\dot{\delta}\}$ for a unit volume of the shell element. It is now possible to carry out the triple multiplication of Eqns. 4 (a) and (b) which will ultimately result in stiffness and mass matrices.

SOME PRACTICAL DETAILS

The parametric formulation of the present element facilitates both codification and algorithm, two aspects of which is discussed below.

Change of variables

As discussed earlier, ξ is the only parameter of the functional ϕ for the parent element. It is natural to express all the coordinates and the displacements in terms of ξ as given in quations 1, 2 and 3. On the other hand the kinematic equations involve some trignometric variables consisting of the variable α, the generator angle, which in cylindrical coordinates can be obtained from $\tan \alpha = \frac{dR}{dZ}$.

In practice such expressions are easily re-written using chain rule i.e. $\tan \alpha = \frac{dR}{d\xi} / \frac{dZ}{d\xi}$, similar expressions to which can be written to embrace other terms. A fuller account of this is given in page 3.14 of Reference (2).

Numerical integration

The stiffness and mass matrices have been generated using Gaussian quadrature throughout this work. Displacements, velocities, stresses and momenta are enumerated for the parameter ξ corresponding to a Gauss-point. The integrals such as $\int (\ldots) \, dv$ are reduced to the following form,

$$I = \int_V (\dots) \, dv$$

$$= \int_S (\dots) t \, ds \qquad \text{(ds is an infinitesimal element along generator)}$$

$$= \int_{-1}^{+1} (\dots) t(\xi) \frac{ds}{d\xi} \, d\xi$$

$$I = \sum_{i=1}^{n} (\dots)_i t_i \left(\frac{ds}{d\xi}\right)_i H_i \cdot \qquad \text{(page 198, Ref (1))}$$

The use of quadrature in the present attempt has lead to an all-round improvement, the salient points of which are listed below.

(i) Shells with variable thickness along the generator present no added computational difficultes to the programmer.

(ii) Quadrature alleviates the precondition of generating elements with a fixed number of degrees of freedom. It is accepted that higher order terms would require more integrating points and therefore a marginal increase in computation time, but this is found to be a small price to pay for an improvement in flexibility and accuracy.

(iii) Unlike some previous formulations, the necessity for having additional end or cap elements does not arise here for the following reasons:(1) the shape functions used are in a polynomial form and are thus better conditioned and (2) possible singularities in functional representations are generally overcome or at worst avoided, since evaluations can be made in the vacinity of the singularity by sampling at appropriate Gauss-points.

EXAMPLES

The author has found it difficult to present examples which illustrate fully the different features of the present work without bordering on tedium. It was thought that only these examples which highlight the salient aspects of the formulations - such as the ability of a few elements to represent complex configurations, correct response to membrane and flexural modes, and relative accuracy in using fewer elements - need be included. Some such examples have been used for illustration by other researchers and thus do not appear in units of measurement currently accepted internationally.

THIN CIRCULAR PLATE
SUMMARY OF HARMONICS AND MODES

nodal circle C	n=0 Th.	n=0 F.E.	n=1 Th.	n=1 F.E.	n=2 Th.	n=2 F.E.	n=3 Th.	n=3 F.E.
0	17.30	17.13	35.60	35.65	58.50	58.49	—	85.59
1	66.67	66.71	102.10	102.06	148.50	141.95	—	188.41
2	149.00	149.57	202.00	204.30	265.00	277.83	—	345.31
3	—	289.49	—	270.00	—	465.85	—	746.25

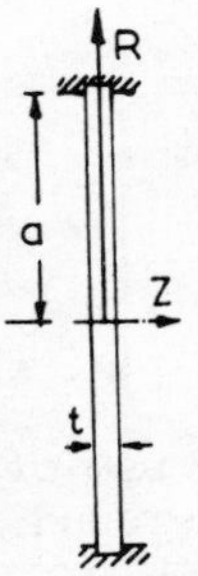

nodal circle C	n=0 Th.	n=0 F.E.	n=1 Th.	n=1 F.E.	n=2 Th.	n=2 F.E.	n=3 Th.	n=3 F.E.
0	6.295	6.333	—	3.402	8.80	8.81	20.50	20.496
1	35.09	35.32	34.40	37.90	50.10	59.10	88.60	88.73
2	112.00	102.29	100.22	107.24	—	141.61	—	190.55
3	200.00	202.88	—	210.66	—	279.88	—	350.72

nodal circle C	n=0 Th.	n=0 F.E.	n=1 Th.	n=1 F.E.	n=2 Th.	n=2 F.E.	n=3 Th.	n=3 F.E.
0	8.4 *	8.358	—	23.37	—	43.02	—	67.07
1	—	49.91	—	81.36	—	117.73	—	159.91
2	—	124.52	—	174.40	—	233.54	—	289.66
3	—	242.54	—	270.02	—	413.09	—	468.63

$*$ – See text $a = 6.0$ $E = 30 \times 10^6$

— Not available $t = 0.1$ $\nu = 1/3$

$\rho = 1.0$

FIGURE 16

(1) The circular plate

This geometry is ideal for examining the calculated frequencies against those predicted by functional means. All conventional boundary supports were attempted, illustrations of which, together with the geometrical and material properties of the plate, are given in Figure 16. The Finite Element and the theorectical frequencies (where available) are tabulated for easy comparison. The main sources of the theorectical results were Reference (7) and the Handbook of Plant Engineering.

The asterisk * in Figure 16 indicates that there is some difficulty in obtaining an "exact" analytical frequency for the case shown. An approximate set of calculations based on Rayleigh-Ritz, using an unrefined quadrature technique, was employed with apparently satisfactory outcome. One element represented the entire geometry with a maximum total of 14 degrees of freedom.

(2) The long thin cylinder

This example attempts to examine all three possible categories of response of a thin, long circular tube. These are extensional, torsional and flexural modes of vibration. The formulae from which the theorectical frequencies of examples were calculated are from Reference (7) which gives an adequate coverage for the determination of the above modes. Only one element with no more than six internal functions was used for the above solution. The dimensional and mechanical details are given in Figure 17 which also includes the tabulated natural frequencies. As can be seen the worst errors encountered were in the flexural mode.

(3) Torsion of a truncated cone

In this example the torsional vibrations of the truncated cone shown in Figure 18 were considered. The theoretical values (8) are again compared with the Finite Element results which consist of the improved frequencies owing to the presence of Surplus Functions $S_o - S_4$.

(4) Short Cylinder clamped at both extremities

The cylindrical shell depicted in Figure 19 typifies a certain class of vibration where the contributions to strain energy are not all due to flexure. The published results on the analysis (9) which appeared in 1949 were verified later experimentally (10). The shell was represented with one element having two Surplus Functions for each of the u, w and v displacements which totalled 10 degrees of freedom. It is interesting to note that the lowest frequency of the first

386

GENERAL VIBRATIONS OF A LONG CYLINDER

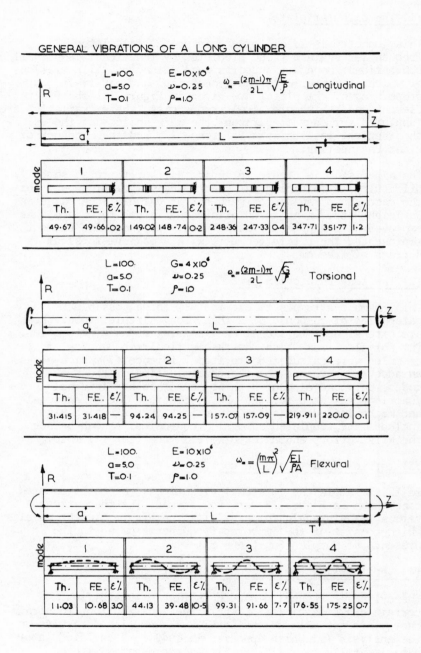

FIGURE 17

and the second modes corresponds to n = 6 and n = 9 respectively.

(5) Vibrations of a thin conical frustum

The conical shell of Figure 20 has been the subject of previous studies both experimentally (11) and numerically (12) The geometrical data of the shell analysed in the present work is identical to the data given in Reference (12) and the frequencies obtained are in good agreement. There is a slight disparity in the generator length of the cones discussed in the previous two references. The predicted frequencies given in Reference (11) which are reported to be based entirely on the flexural strain energy, are in some disagreement with the two Finite Element results. The total number of the degrees of freedom used here are 14 and 24 which correspond to 1 and 2 element representations respectively.

(6) Fixed hemispherical cap

The natural frequencies for the spherical caps have been worked out for clamped edge conditions generally. The results are normally tabulated in the form of $\frac{\omega}{\omega_0}$ where $\omega_0 = \frac{1}{a}\left[\frac{E}{\rho(1-\nu^2)}\right]^{\frac{1}{2}}$ and are taken from References (13) and (14). The details of the cap analysed and the plot of the frequencies obtained are given in Figure 21. Except for the fourth mode, the agreement between the functional and the numerical frequencies appear satisfactory.

(7) Natural frequencies of cooling towers

The last example in the present work is confined to the tower initially stressed by Albasiny and Martin. The tower has since become a classical test case for both static and dynamic analysis. The dynamic results consist of

(a) numerical integration method (15)

(b) the finite difference method (16)

(c) the finite element method (12)

For the tower shown in Figure 22, the first frequency corresponds to n = 5. A frequency/harmonic plot for the above shell displays a similar variation to the curves given in Figures 19 and 20. The tower is represented by 3 and 5 elements requiring 28 and 52 degrees of freedom respectively. Unlike the experience reflected in Reference (12), the need for increasing the number of elements with the increase in the number of nodal diameters was not felt necessary. The latter run (52 d.o.f) was the only occasion when the data was not processed on an IBM 1130 (16 K word) machine and had to

388

TRUNCATED CONE

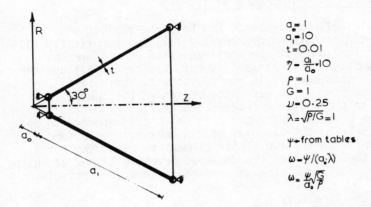

$a_o = 1$
$a_1 = 10$
$t = 0.01$
$\eta = \dfrac{a_1}{a_o} = 10$
$\rho = 1$
$G = 1$
$\nu = 0.25$
$\lambda = \sqrt{\rho/G} = 1$

$\psi \rightarrow$ from tables
$\omega = \psi/(a_o \lambda)$
$\omega = \dfrac{\psi}{a_o}\sqrt{\dfrac{G}{\rho}}$

1 ELEMENT

	Exact	SURPLUS DEGREES OF FREEDOM					
		—	1 v	2 v	3 v	4 v	5 v
Mode 1	0.39409	0.39462	0.39426	0.39419	0.39412	0.39410	0.39409
error %		0.1346	0.0432	0.0252	0.0077	0.0026	0.0000
Mode 2	0.73306	0.76176	0.74172	0.73342	0.73336	0.73308	0.73305
error %		3.9151	1.1814	0.0491	0.0410	0.0028	−0.0014
Mode 3	1.07483	*	1.15667	1.11281	1.07765	1.07625	1.07489
error %		—	7.6143	3.5336	0.2624	0.1322	0.0056
Mode 4	1.41886	*	*	1.60034	1.52036	1.42996	1.42449
error %		—	—	12.7906	7.1537	0.7842	0.3968

ω Htz.

* – choice of displacements **not compatible** with the mode

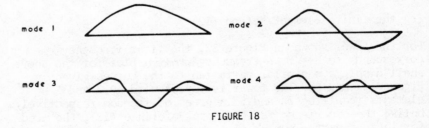

mode 1   mode 2

mode 3   mode 4

FIGURE 18

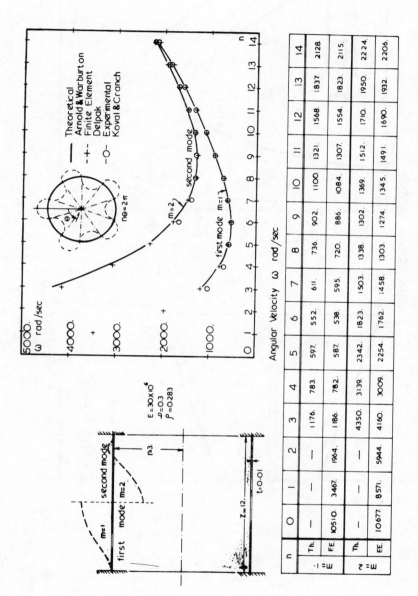

FIGURE 19

THIN CONICAL FRUSTUM

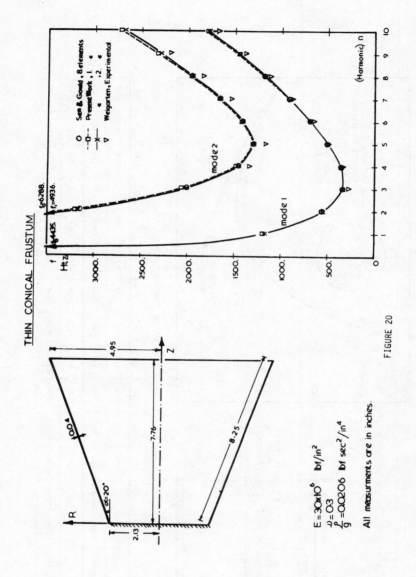

FIGURE 20

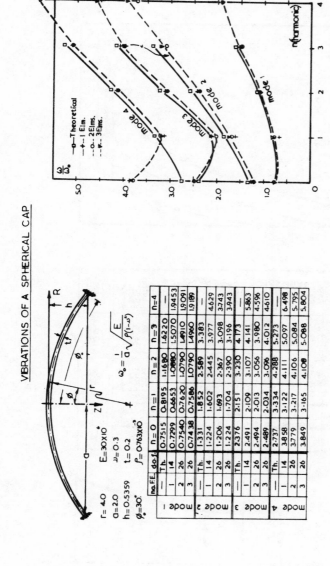

VIBRATIONS OF A SPHERICAL CAP

$r = 4.0$
$a = 2.0$
$h = 0.5359$
$\phi_0 = 30°$

$E = 30 \times 10^6$
$\nu = 0.3$
$t = 0.2$
$\rho = 0.763 \times 10^{-3}$

$$\omega_0 = \frac{1}{a}\sqrt{\frac{E}{\rho(1-\nu^2)}}$$

	no.F.E.	do.f.	n=0	n=1	n=2	n=3	n=4
mode 1	—	Th.	0.7515	0.8195	1.1680	1.6220	1.9453
	1	14	0.7299	0.6653	1.0680	1.5070	1.9453
	2	26	0.7540	0.7620	1.0790	1.4910	1.9091
	3	26	0.7438	0.7586	1.0790	1.4960	1.9189
mode 2	—	Th.	1.331	1.852	2.589	3.383	—
	1	14	1.224	1.602	2.445	3.977	4.629
	2	26	1.206	1.693	2.367	3.098	3.743
	3	26	1.224	1.704	2.390	3.196	3.943
mode 3	—	Th.	2.376	2.151	3.230	4.173	—
	1	14	2.491	2.109	3.107	4.141	5.863
	2	26	2.494	2.013	3.056	3.980	4.596
	3	26	2.489	2.034	3.096	4.012	4.610
mode 4	—	Th.	2.737	3.334	4.288	5.273	—
	1	14	3.858	3.122	4.111	5.097	6.498
	2	26	3.779	3.217	4.106	5.084	5.795
	3	26	3.849	3.165	4.108	5.088	5.804

FIGURE 21

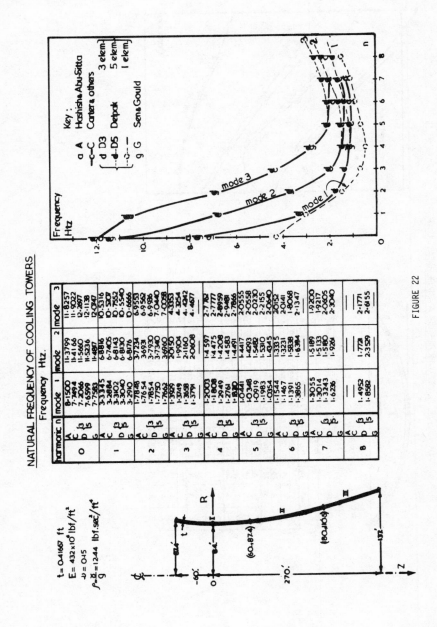

NATURAL FREQUENCY OF COOLING TOWERS

FIGURE 22

be referred elsewhere. The results obtained for the first mode using just 1 element (14. d.o.f) is included for general interest.

CONCLUDING REMARKS

A curved parametric element has been developed with a number of desirable features some of which are listed below.

(a) The element is doubly curved and possesses genuine nodal conformity.

(b) The element can support a change in thickness with a variation of up to a cubic.

(c) The number of elemental degrees of freedom is optional and can be chosen as a matter of input rather than formulation.

(d) Rigid Body Motions are readily available on this element.

(e) Numerical integration permits the flexibility of modification or change which is ideal for research and development work within the program.

(f) The present shape functions being polynomials have improved numerical conditioning.

(g) The need for incorporating special elements at the extremities, e.g. cap elements has been eliminated. The dfficulties of having singular values of stresses or stiffness matrices have also been avoided.

(h) The provision of being able to specify the required number of internal nodes have resulted in reducing the overall size of the problem. This has led to either a cheaper solution or the use of a very small computer (all the example with the exception of number (7) were processed on a 16 k word machine).

(i) The frequencies obtained from the above examples show satisfactory agreement with the comparable data published by other researchers.

REFERENCES

1 - O. C. Zeinkiewicz "The Finite Element Method", McGraw-Hill, 1977, London and New York.

2 - R. Delpak, "Rôle of the Curved Parametric Element in Linear Analysis of Thin Rotational Shells", Ph.D. Thesis, CNAA, 1975. London.

394

3 - B. M. Irons, " Shape Functions for Elements with Point Conformity", A. S. M. 1204, Rolls-Royce Library.

4 - R. Delpak, "Axisymmetric Vibration by the Curved Cone Element", p. 6-27. Proc. of Conf. on Recent Advances in Stress Analysis, the Joint Brit. Committee for Str. Anal., Roy. Aer. Soc., London, 1968.

5 - B. M. Irons and A. Razzaque, "The Mathematical Foundations of the Finite Element Method with Applications to Partial Differential Equations", p.557, 585, Academic Press, 1972, New Yrok and London (edited by A. K. Aziz).

6 - R. Delpak and B. W. Preece, "Linear Strain - Displacement interpretation of Thin Shells of Revolution ", p. 29-35, Strain (Jour. of the Brit.Soc. for Strain Measurement) January 1974, U.K.

7 - S. P. Timoshenko and D. H. Young,"Vibrations Problems in Engineering", Van Nostrand, 3rd Edn., 1964,

8 - H. Garnet, M. A. Golberg and V. L. Salerno, "Torsinal Vibrations of a Shell of Revolution", p. 571-573, Journal of Applied Mechanics, 28, 1961.

9 - R. N. Arnold and G. B. Warburton, "Flexural Vibrations of the Walls of Thin Cylindrical Shells Having Freely Supported Ends", p. 238-256, Proc. of Royal Soc. of London, 197, 1949, London.

10 - L. R. Koval and E. T. Cranch, "On the Free Vibrations of Thin Cylindrical Shells Subjected to an Initial Static Torque", p. 107-117, Proc. of the Fourth U.S. National Congress of App. Mech.,1962.

11 - V. I. Weingarten, "Free Vibrations of Conical Shells" , p. 69-87, Jour. of Eng. Mech. Div., ASCE, 1965.

12 - S. K. Sen and P. L. Gould, "Free Vibrations of Shells of Revolution Using FEM", p. 283-303, Jour. of Eng. Mech. Div., ASCE, 1974.

13 - H. Kraus,"Thin Elastic Shells", p. 237-331, J. Wiley and Son Inc., New York.

14 - M. S. Zarghamee and A. A. Robinson, "A Numerical Method for Analysis of Free Vibrations of Spherical Shells", p. 1256-1261, Jour. Am. Inst. Aero. and Astro, Vol.5, No. 7, 1967.

15 - R. L. Carter, A. R. Robinson and W. C. Schnobrich, "Free and Forced Vibrations of Hyperbolical Shells of

Revolution", Structural research Series, No. 334, Civil
Eng. Studies, Univ. Illinois, 1968.

16 - M. G. Hashish and S. H. Abu-Sitta, "Free Vibrations of
Hyperbolic Cooling Towers", p. 253 - 269, Jour. Eng. Mech.
Div., ASCE, 1971.

ACKNOWLEDGEMENTS

The author wishes to express his appreciation to the staff of
the Polytechnic of Wales Computer Centre for their help and
advice ,   to his Head of Department Mr. R. D. McMurray for
making Departmental facilities available for the present
work, and to Mrs. J. A. Morgan for typing a difficult
manuscript.

DYNAMICS AND STABILITY OF SHELLS OF REVOLUTION

D.J. Collington
Computational Mechanics Centre, Southampton

C.A. Brebbia
University of Southampton

INTRODUCTION

It is important in many practical applications to be able to predict the full dynamic and stability behaviour of shells of revolution. Buckling and vibrations are related phenomena (Leipholz, 1975) and should not be considered separately in the analysis of structures subject to dynamic loading. The combined behaviour may be studied by a full nonlinear, dynamic analysis of the structure, but this is an expensive procedure and cannot be used for a parametric study. In many cases the designer may obtain a better insight into the behaviour of shell type structures by considering their dynamic and buckling bifurcation behaviour.

The present study represents an attempt to quantify the relationship between free vibration frequencies and bifurcation loads for thin shells. The dynamic equilibrium equations for a system are derived from a generalised form of Hamilton's Principle. These equations are solved by a finite element discretisation which uses a non fully compatible, but rapidly convergent, set of displacement expansions. The shell strain-displacement relationships are based on the work of Sanders (1959, 1963) and are valid for linear and geometrically non-linear behaviour. Discretisation of the Lagrange equations produces a symmetric system in which dynamic, viscous and stiffness (linear and nonlinear) forces are taken into consideration. Specialising these equations leads to the normal form of the free vibration equation, or to the equation for static bifurcation from a linear fundamental path. In addition they contain the information required for static bifurcation from a nonlinear fundamental path, and the dynamic stability equations of Bolotin (1964). Examples are presented which illustrate the application of these equations to some practical problems.

THEORY

The equations of motion for a system can conveniently be derived from a generalised form of Hamilton's Principle (Meirovitch, 1970).

$$\int_{t_o}^{t_1} \delta L \, dt + \int_{t_o}^{t_1} \sum_{k=1}^{n} Q_{knp} \, \delta q_k \, dt = 0 \tag{1}$$

in which $L = T - \Pi_p$ is the Lagrangian, T is the total kinetic energy, Π_p is the total potential energy, Q_{knp} any forces not derivable from a potential function and the q_k are a set of generalised coordinates. Since L is dependent on q_k and $\dot{q}_k$ the variation in equation (1) may be expanded to give

$$\int_{t_o}^{t_1} \sum_{k=1}^{n} \left\{ \frac{\partial L}{\partial q_k} \delta q_k + \frac{\partial L}{\partial \dot{q}_k} \delta \dot{q}_k \right\} dt + \int_{t_o}^{t_1} \sum_{k=1}^{n} Q_{knp} \, \delta q_k \, dt = 0$$

$$\tag{2}$$

or integrating the second term of the first integral by parts.

$$\int_{t_o}^{t_1} \left\{ \frac{d}{dt} \left(\frac{\partial L}{\partial \dot{q}_k} \right) - \frac{\partial L}{\partial q_k} - Q_{knp} \right\} dt = 0 \tag{3}$$

Therefore for small displacements [i.e. $T \equiv T(\dot{q}_k)$ and $\Pi_p \equiv \Pi_p(q_k)$] the Euler-Lagrange equations of the system are

$$\frac{d}{dt} \left(\frac{\partial T}{\partial \dot{q}_k} \right) + \frac{\partial \Pi_p}{\partial q_k} = Q_{knp} \; ; \quad k = 1, 2, \ldots \tag{4}$$

The present work is concerned with small dynamic response about an equilibrium state $E \{q_k, \lambda\} \equiv E\{ 0, \lambda\}$, where λ is a loading parameter and the displacements q_k are measured relative to the equilibrium state. In this case the value of the potential function near to equilibrium may be written as a Taylor series in terms of the values at equilibrium (Croll and Walker, 1972).

$$\Pi_p(q_k, \lambda) = \Pi_p \Big|_E + \frac{1}{1!} \sum_k \frac{\partial \Pi_p}{\partial q_k} \Big|_E q_k + \frac{1}{2!} \sum_k \sum_j \frac{\partial^2 \Pi_p}{\partial q_k \partial q_j} \Big|_E q_k q_j + \cdots$$

$$\tag{5}$$

in which, from the definition of equilibrium, $\left.\dfrac{\partial \Pi_p}{\partial q_k}\right|_E = 0.$

Hence, neglecting high order terms

$$\frac{\partial \Pi_p}{\partial q_k} \simeq \frac{1}{2!} \sum_j \left.\frac{\partial^2 \Pi_p}{\partial q_k \partial q_j}\right|_E q_j \tag{6}$$

and the system equilibrium equations may be written as

$$\frac{d}{dt}\left(\frac{\partial T}{\partial \dot{q}_k}\right) + \frac{1}{2!} \sum_j \left.\frac{\partial^2 \Pi_p}{\partial q_k \partial q_j}\right|_E q_j = Q_{knp} \tag{7}$$

THE ELEMENT

The element used in this study is an isoparametric, doubly curved, quadrilateral, thin shell element based on the following displacement functions proposed by Bognor, Fox and Schmit (1966) and given in dimensionless form by Key and Beisinger (1968).

$$h_1(\chi) = \tfrac{1}{4}(\chi^3 - 3\chi + 2)$$

$$h_2(\chi) = -\tfrac{1}{4}(\chi^3 - 3\chi - 2)$$

$$h_3(\chi) = \tfrac{1}{4}(\chi^3 - \chi^2 - \chi + 1) \tag{8}$$

$$h_4(\chi) = \tfrac{1}{4}(\chi^3 + \chi^2 - \chi - 1)$$

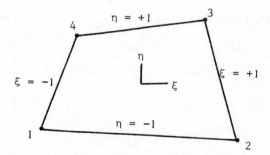

Fig. 1   Local coordinates and node numbering

For the element shown in figure 1 the expansion for w (u and v are similar) is given by Bognor et al (1966) as

$$w = h_1(\xi)h_1(\eta)w_1 + h_3(\xi)h_1(\eta)w_{1,\xi} + h_1(\xi)h_3(\eta)w_{1,\eta} + h_3(\xi)h_3(\eta)w_{1,\xi\eta}$$

$$+ h_2(\xi)h_1(\eta)w_2 + h_4(\xi)h_1(\eta)w_{2,\xi} + h_2(\xi)h_3(\eta)w_{2,\eta} + h_4(\xi)h_3(\eta)w_{2,\xi\eta}$$

$$+ h_2(\xi)h_2(\eta)w_3 + h_4(\xi)h_2(\eta)w_{3,\xi} + h_2(\xi)h_4(\eta)w_{3,\eta} + h_4(\xi)h_4(\eta)w_{3,\xi\eta}$$

$$+ h_1(\xi)h_2(\eta)w_4 + h_3(\xi)h_2(\eta)w_{4,\xi} + h_1(\xi)h_4(\eta)w_{4,\eta} + h_3(\xi)h_4(\eta)w_{4,\xi\eta}$$

$$(9)$$

This element has 12 degrees of freedom per node of which $u_{i,\xi\eta}$; $v_{i,\xi\eta}$ and $w_{i,\xi\eta}$ are not essential to the strain-displacement relations and may be eliminated by using the following finite difference expressions (Wilson and Brebbia, 1971)

$$w_{1,\xi\eta} = \tfrac{1}{2}\left[\tfrac{1}{2}(w_{2,\eta} - w_{1,\eta}) + \tfrac{1}{2}(w_{4,\xi} - w_{1,\xi})\right]$$

$$w_{2,\xi\eta} = \tfrac{1}{2}\left[\tfrac{1}{2}(w_{2,\eta} - w_{1,\eta}) + \tfrac{1}{2}(w_{3,\xi} - w_{2,\xi})\right]$$

$$w_{3,\xi\eta} = \tfrac{1}{2}\left[\tfrac{1}{2}(w_{3,\eta} - w_{4,\eta}) + \tfrac{1}{2}(w_{3,\xi} - w_{2,\xi})\right]$$

$$w_{4,\xi\eta} = \tfrac{1}{2}\left[\tfrac{1}{2}(w_{3,\eta} - w_{4,\eta}) + \tfrac{1}{2}(w_{4,\xi} - w_{1,\xi})\right]$$

$$(10)$$

which when substituted into equation (9) gives

$$w = h_1(\xi)h_1(\eta)w_1 + h_3(\xi)g_1(\eta)w_{1,\xi} + g_1(\xi)h_3(\eta)w_{1,\eta}$$

$$+ h_2(\xi)h_1(\eta)w_2 + h_4(\xi)g_1(\eta)w_{2,\xi} + g_2(\xi)h_3(\eta)w_{2,\eta}$$

$$+ h_2(\xi)h_2(\eta)w_3 + h_4(\xi)g_2(\eta)w_{3,\xi} + g_2(\xi)h_4(\eta)w_{3,\eta}$$

$$+ h_1(\xi)h_2(\eta)w_4 + h_3(\xi)g_2(\eta)w_{4,\xi} + g_1(\xi)h_4(\eta)w_{4,\eta}$$

$$(11)$$

where the two new displacement functions are defined as

$$g_1(\chi) = h_1(\chi) - \tfrac{1}{4}h_3(\chi) - \tfrac{1}{4}h_4(\chi)$$

$$= \frac{1}{8}(\chi^3 - 5\chi + 4)$$

$$g_2(\chi) = h_2(\chi) + \tfrac{1}{4}h_3(\chi) + \tfrac{1}{4}h_4(\chi)$$

$$= \frac{1}{8}(-\chi^3 + 5\chi + 4)$$

$$(12)$$

Hence from equation (11) the total midsurface displacements may be written as $u = \phi^T u^n$; $v = \phi^T v^n$ and $w = \phi^T w^n$ or

$$\underset{\sim}{u} = \begin{Bmatrix} u \\ v \\ w \end{Bmatrix} = \begin{Bmatrix} \phi^T & 0 & 0 \\ 0 & \phi^T & 0 \\ 0 & 0 & \phi^T \end{Bmatrix} \begin{Bmatrix} u^n \\ v^n \\ w^n \end{Bmatrix} = \Phi^T \underset{\sim}{U}^n \qquad (13)$$

Transformation of derivatives
For an arbitrary function, f,

$$\frac{\partial f}{\partial \xi} = \frac{\partial f}{\partial \theta} \frac{\partial \theta}{\partial \xi} + \frac{\partial f}{\partial z} \frac{\partial z}{\partial \xi}$$

$$\frac{\partial f}{\partial \eta} = \frac{\partial f}{\partial \theta} \frac{\partial \theta}{\partial \eta} + \frac{\partial f}{\partial z} \frac{\partial z}{\partial \eta}$$

or

$$\begin{Bmatrix} \dfrac{\partial f}{\partial \theta} \\[2mm] \dfrac{\partial f}{\partial z} \end{Bmatrix} = \begin{Bmatrix} \dfrac{\partial \theta}{\partial \xi} & \dfrac{\partial z}{\partial \xi} \\[2mm] \dfrac{\partial \theta}{\partial \eta} & \dfrac{\partial z}{\partial \eta} \end{Bmatrix}^{-1} \begin{Bmatrix} \dfrac{\partial f}{\partial \xi} \\[2mm] \dfrac{\partial f}{\partial \eta} \end{Bmatrix} = J^{-1} \begin{Bmatrix} \dfrac{\partial f}{\partial \xi} \\[2mm] \dfrac{\partial f}{\partial \eta} \end{Bmatrix} \qquad (14)$$

and

$$\frac{\partial^2 f}{\partial \xi^2} = \left(\frac{\partial^2 \theta}{\partial \xi^2}\right) \frac{\partial f}{\partial \theta} + \left(\frac{\partial^2 z}{\partial \xi^2}\right) \frac{\partial f}{\partial z} + \left(\frac{\partial \theta}{\partial \xi}\right)^2 \frac{\partial^2 f}{\partial \theta^2} + \left(\frac{\partial z}{\partial \xi}\right)^2 \frac{\partial^2 f}{\partial z^2}$$

$$+ 2 \left(\frac{\partial \theta}{\partial \xi}\right)\left(\frac{\partial z}{\partial \xi}\right) \frac{\partial^2 f}{\partial \theta \partial z}$$

$$\frac{\partial^2 f}{\partial^2 \eta} = \left(\frac{\partial^2 \theta}{\partial^2 \eta}\right) \frac{\partial f}{\partial \theta} + \left(\frac{\partial^2 z}{\partial^2 \eta}\right) \frac{\partial f}{\partial z} + \left(\frac{\partial \theta}{\partial \eta}\right)^2 \frac{\partial^2 f}{\partial \theta^2} + \left(\frac{\partial z}{\partial \eta}\right)^2 \frac{\partial^2 f}{\partial z^2}$$

$$+ 2 \left(\frac{\partial \theta}{\partial \eta}\right)\left(\frac{\partial z}{\partial \eta}\right) \frac{\partial^2 f}{\partial \theta \partial z} \qquad (15)$$

$$\frac{\partial^2 f}{\partial \xi \partial \eta} = \left(\frac{\partial^2 \theta}{\partial \xi \partial \eta}\right) \frac{\partial f}{\partial \theta} + \left(\frac{\partial^2 z}{\partial \xi \partial \eta}\right)\frac{\partial f}{\partial z} + \left(\frac{\partial \theta}{\partial \xi}\right)\left(\frac{\partial \theta}{\partial \eta}\right) \frac{\partial^2 f}{\partial \theta^2} + \left(\frac{\partial z}{\partial \xi}\right)\left(\frac{\partial z}{\partial \eta}\right) \frac{\partial^2 f}{\partial z^2}$$

$$+ \left\{ \left(\frac{\partial \theta}{\partial \xi}\right)\left(\frac{\partial z}{\partial \eta}\right) + \left(\frac{\partial z}{\partial \xi}\right)\left(\frac{\partial \theta}{\partial \eta}\right) \right\} \frac{\partial^2 f}{\partial \theta \partial z}$$

which since $\partial f/\partial \theta$ and $\partial f/\partial z$ are known from equation (14) gives the following expression for the global derivatives

$$
\left\{
\begin{array}{c}
\dfrac{\partial^2 f}{\partial \theta^2} \\[3mm]
\dfrac{\partial^2 f}{\partial z^2} \\[3mm]
\dfrac{\partial^2 f}{\partial \theta \partial z}
\end{array}
\right\}
=
\left\{
\begin{array}{ccc}
\left(\dfrac{\partial \theta}{\partial \xi}\right)^2, & \left(\dfrac{\partial z}{\partial \xi}\right)^2, & 2\left(\dfrac{\partial \theta}{\partial \xi}\right)\left(\dfrac{\partial z}{\partial \xi}\right) \\[4mm]
\left(\dfrac{\partial \theta}{\partial \eta}\right)^2, & \left(\dfrac{\partial z}{\partial \eta}\right)^2, & 2\left(\dfrac{\partial \theta}{\partial \eta}\right)\left(\dfrac{\partial z}{\partial \eta}\right) \\[4mm]
\left(\dfrac{\partial \theta}{\partial \xi}\right)\left(\dfrac{\partial \theta}{\partial \eta}\right), & \left(\dfrac{\partial z}{\partial \xi}\right)\left(\dfrac{\partial z}{\partial \eta}\right), & \left(\dfrac{\partial \theta}{\partial \xi}\right)\left(\dfrac{\partial z}{\partial \eta}\right) + \left(\dfrac{\partial \theta}{\partial \eta}\right)\left(\dfrac{\partial z}{\partial \xi}\right)
\end{array}
\right\}^{-1}
$$

$$
\times
\left\{
\begin{array}{c}
\dfrac{\partial^2 f}{\partial \xi^2} - \left(\dfrac{\partial^2 \theta}{\partial \xi^2}\right)\dfrac{\partial f}{\partial \theta} - \left(\dfrac{\partial^2 z}{\partial \xi^2}\right)\dfrac{\partial f}{\partial z} \\[5mm]
\dfrac{\partial^2 f}{\partial \eta^2} - \left(\dfrac{\partial^2 \theta}{\partial \eta^2}\right)\dfrac{\partial f}{\partial \theta} - \left(\dfrac{\partial^2 z}{\partial \eta^2}\right)\dfrac{\partial f}{\partial z} \\[5mm]
\dfrac{\partial^2 f}{\partial \xi \partial \eta} - \left(\dfrac{\partial^2 \theta}{\partial \xi \partial \eta}\right)\dfrac{\partial f}{\partial \theta} - \left(\dfrac{\partial^2 z}{\partial \xi \partial \eta}\right)\dfrac{\partial f}{\partial z}
\end{array}
\right\}
\qquad (16)
$$

SHELL THEORY

The strain-displacement relationships are based on the linear and nonlinear shell theories of Sanders (1959, 1963) as given by Brebbia and Connor (1973).

Linear part of membrane strain , κ

$$
e_{\theta\theta} = \frac{u_{,\theta}}{A_\theta} + \frac{A_{\theta,z}}{A_\theta A_z} v + \frac{w}{R_{\theta\theta}}
$$

$$
e_{zz} = \frac{A_{z,\theta}}{A_\theta A_z} u + \frac{v_{,z}}{A_z} + \frac{w}{R_{zz}} \qquad (17)
$$

$$
e_{\theta z} = \frac{A_{\theta,z}}{A_\theta A_z} u + \frac{u_{,z}}{A_z} + \frac{A_{z,\theta}}{A_\theta A_z} v + \frac{v_{,\theta}}{A_\theta} + 2\frac{w}{R_{\theta z}}
$$

(where A_θ and A_z are surface matrices and $R_{\theta\theta}$, R_{zz} and $R_{\theta z}$ are radii at curvature).

Nonlinear part of membrane strain, κ

$$\eta_{\theta\theta} = (\beta_\theta^2 + \beta^2)/2$$
$$\eta_{zz} = (\beta_z^2 - \beta^2)/2 \tag{18}$$
$$\eta_{\theta z} = \beta_\theta \beta_z$$

Bending strain

$$\chi_{\theta\theta} = \frac{\beta_{\theta,\theta}}{A_\theta} + \frac{A_{\theta,z}}{A_\theta A_z}\beta_z + \frac{\beta}{R_{\theta z}}$$

$$\chi_{zz} = \frac{A_{z,\theta}}{A_\theta A_z}\beta_\theta + \frac{\beta_{z,z}}{A_z} + \frac{\beta}{R_{\theta z}} \tag{19}$$

$$\chi_{\theta z} = \frac{\beta_{\theta,z}}{A_z} + \frac{A_{\theta,z}}{A_\theta A_z}\beta_\theta + \frac{\beta_{z,\theta}}{A_\theta} + \frac{A_{z,\theta}}{A_\theta A_z}\beta_z + \left(\frac{1}{R_{zz}} - \frac{1}{R_{\theta\theta}}\right)\beta$$

Rotations

The assumption of negligible transverse shear strain, $\gamma_\theta = \gamma_z = 0$, leads to definitions for rotations of the normal as

$$\beta_\theta = \frac{u}{R_{\theta\theta}} + \frac{v}{R_{\theta z}} - \frac{w_{,\theta}}{A_\theta}$$

$$\beta_z = \frac{u}{R_{\theta z}} + \frac{v}{R_{zz}} - \frac{w_{,z}}{A_z} \tag{20}$$

and the rotation about the normal is defined as (Sanders, 1959)

$$\beta = \frac{1}{2}\left\{ \frac{v_{,\theta}}{A_\theta} - \frac{u_{,z}}{A_z} - \frac{A_{\theta,z}}{A_\theta A_z}u + \frac{A_{z,\theta}}{A_\theta A_z}v \right\} \tag{21}$$

Constitutive relations

The stress-strain relationships given by Lekhnitskii (1963) for a three dimensional orthotropic body, when specialised to a two dimensional body with planes of symmetry coincident with the principal directions (θ,z), give

$$\underset{\sim}{\sigma} = \underset{\sim}{D}\,\underset{\sim}{\varepsilon} \tag{22}$$

where

$$
\underset{\sim}{D} = \left\{ \begin{array}{ccc}
\dfrac{E_\theta}{(1-n\nu^2)} & \dfrac{\nu E_\theta}{(1-n\nu^2)} & 0 \\[3ex]
\dfrac{\nu E_\theta}{(1-n\nu^2)} & \dfrac{E_\theta}{n(1-n\nu^2)} & 0 \\[3ex]
0 & 0 & G
\end{array} \right\}
$$

in which

$$
n = \nu_{\theta z}/\nu_{z\theta} = E_\theta/E_z \qquad : \qquad \nu = \nu_{z\theta}
$$

SYSTEM EQUATIONS

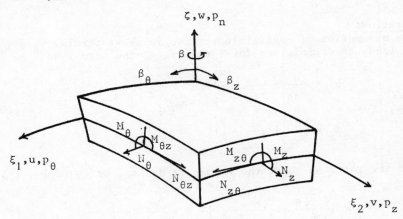

Fig. 2   Shell Notation

The total potential energy, Π_p, is the sum of the strain energy, V, and the potential energy of the loading, Ω. The strain energy of an elastic system is

$$
V = \tfrac{1}{2} \int_V \underset{\sim}{\sigma}^T \underset{\sim}{\varepsilon} \; dV \tag{23}
$$

in which the total strain, $\underset{\sim}{\varepsilon}$, is assumed to vary linearly over the shell thickness.

$$
\underset{\sim}{\varepsilon} = \underset{\sim}{\kappa} + \zeta \underset{\sim}{\chi} \qquad . \tag{24}
$$

Therefore, substituting equation (24) into equation (23) and introducing the stress resultants,

$$N = \int_{-h/2}^{h/2} \sigma \, d\zeta \text{ , and the stress couples, } M = \int_{-h/2}^{h/2} \sigma\zeta \, d\zeta \text{ ,}$$

gives the expression

$$V = \tfrac{1}{2} \int \{ N^T \kappa + M^T \chi \} \, dA \tag{25}$$

Introducing the constitutive equation (22) into equation (25)

$$V = \tfrac{1}{2} \int_A \{ (e + \eta)^T D_\ell (e + \eta) + \chi^T D_b \chi \} \, dA \tag{26}$$

where $D_\ell = h \, D$, $D_b = h^3 D/12$ and the total membrane strain, κ , has been expressed as the sum of linear and nonlinear components e and η .

The potential of the loading for a surface load vector $p^T = \{p_\theta, \ p_z, \ p_n\}$ is

$$\Omega = - \int_A p^T u \, dA \tag{27}$$

The contribution of the potential energy to the equilibrium equation (7) is in the form

$\dfrac{1}{2!} \sum\limits_j \dfrac{\partial^2 \Pi_p}{\partial q_k \partial q_j}\bigg|_E q_j$ which is equivalent to $\dfrac{1}{2!} \dfrac{\partial}{\partial q_k} (\delta^2 V)$. Hence

the potential of the loading makes no direct contribution to this term and equation (7) becomes

$$\frac{d}{dt} \left(\frac{\partial T}{\partial \dot{q}_k} \right) + \frac{1}{2!} \frac{\partial}{\partial q_k} (\delta^2 V) = Q_{knp} \text{ ; } \quad k = 1, 2, \ldots \tag{28}$$

where

$$\delta^2 V = \int_A \{ (\delta e + \delta\eta)^T D_\ell (\delta e + \delta\eta) + \delta\chi^T D_b \delta\chi \} dA + \int_A (N^{*,T} \delta^2\eta) \, dA \tag{29}$$

in which $N^* = D_\ell \, \kappa^* = D_\ell (e^* + \eta^*)$ represents the existing state of stress in the shell. If the total nodal displacement vector, U_T^n , is defined as $U_T^n = U^* + \Delta U$; where U^* represents the values at equilibrium and $\underset{\sim}{U}$ is an incremental vector. Then, substituting the displacement expansions of equation (13) into the strain displacement relationships (equations 17-19) gives

$$\delta\underset{\sim}{e} = \underset{\sim\ell}{B} \, \Delta\underset{\sim}{U}$$

$$\delta\underset{\sim}{\chi} = \underset{\sim b}{B} \, \Delta\underset{\sim}{U}^*$$

$$\delta\underset{\sim}{\eta} = \underset{\sim n}{B}(\underset{\sim}{U}^*) \, \Delta\underset{\sim}{U}$$

$$\delta^2\underset{\sim}{\eta} = \underset{\sim n}{B}(\Delta\underset{\sim}{U}) \Delta\underset{\sim}{U}$$

(30)

Hence, the first integral of equation (29) has the discretised form

$$\underset{\sim}{U}^T\{ \underset{\sim L}{K} + \underset{\sim 1}{K} + \underset{\sim 2}{K}\} \, \Delta\underset{\sim}{U}$$

in which

$$\underset{\sim L}{K} = \int_A \{\underset{\sim\ell}{B}^T \underset{\sim\ell}{D} \underset{\sim\ell}{B} + \underset{\sim b}{B}^T \underset{\sim b}{D} \underset{\sim b}{B} \}dA$$

(31)

is the linear stiffness matrix.

$$\underset{\sim 1}{K} = \int_A \{\underset{\sim\ell}{B}^T \underset{\sim\ell}{D} \underset{\sim}{B}(\underset{\sim}{U}^*) + \underset{\sim n}{B}^T(\underset{\sim}{U}^*)\underset{\sim\ell}{D} \underset{\sim\ell}{B}\}dA$$

(32)

is a nonlinear stiffness matrix dependent on $\underset{\sim}{U}^*$ and

$$\underset{\sim 2}{K} = \int_A \{\underset{\sim n}{B}^T(\underset{\sim}{U}^*)\underset{\sim\ell}{D} \underset{\sim n}{B}(\underset{\sim}{U}^*)\} \, dA$$

(33)

is a nonlinear stiffness matrix dependent on $(\underset{\sim}{U}^*)^2$. For the discretisation of the second integral it is necessary to define the rotations as $\beta_\theta = \alpha \, \underset{\sim}{U}^*$, $\beta_z = b \, \underset{\sim}{U}^*$ and $\beta = c \, \underset{\sim}{U}^*$ so that

$$\delta^2\underset{\sim}{\eta} = \left\{ \begin{array}{c} \Delta\underset{\sim}{U}^T(\underset{\sim}{a}^T \underset{\sim}{a} + \underset{\sim}{c}^T \underset{\sim}{c})\Delta\underset{\sim}{U} \\ \Delta\underset{\sim}{U}^T(\underset{\sim}{b}^T \underset{\sim}{b} - \underset{\sim}{c}^T \underset{\sim}{c})\Delta\underset{\sim}{U} \\ \Delta\underset{\sim}{U}^T(\underset{\sim}{a}^T \underset{\sim}{b} + \underset{\sim}{b}^T \underset{\sim}{a})\Delta\underset{\sim}{U} \end{array} \right\}$$

(34)

and since $\kappa^* = e^* + \eta^* = \underset{\sim\ell}{B} \, \underset{\sim}{U}^* + \underset{\sim n}{B}(\underset{\sim}{U}^*)\underset{\sim}{U}^*$ is known the geometric stiffness matrix may then be defined as

$$K_G = \int_A \{ N_\theta^*(a^T \underset{\sim}{a} + c^T \underset{\sim}{c}) + N_z^*(b^T \underset{\sim}{b} - c^T \underset{\sim}{c}) + N_{\theta z}^*(a^T \underset{\sim}{b} + b^T \underset{\sim}{a}) \} dA$$

Hence, assuming that K_G is directly proportional to the load level $\underset{\sim}{K}_G$ along the fundamental path

$$\delta^2 V = \Delta U^T \{ K_L + K_1 + K_2 + \lambda K_G^\lambda \} \Delta U \tag{36}$$

where $\underset{\sim}{K}_G^\lambda$ represents K_G evaluated at $\lambda = 1$.

The kinetic energy of the system is

$$T = \frac{1}{2} \int_V \rho \, \dot{\underset{\sim}{s}}^T \dot{\underset{\sim}{s}} \, dV \tag{37}$$

where $\underset{\sim}{s}^T = \{ s_\theta, \ s_z, \ s_n \}$ is the displacement vector of a generic point within the shell and may be represented in terms of the midsurface displacements and rotations by

$$s_1 = u + \zeta \beta_\theta$$

$$s_2 = v + \zeta \beta_z$$

$$s_n = w$$

which may be written in discretised form as $\dot{\underset{\sim}{s}} = \underset{\sim}{B}_m(\zeta) \Delta \dot{\underset{\sim}{U}}$, hence

$$T = \Delta \dot{\underset{\sim}{U}}^T \ \underset{\sim}{M} \ \Delta \dot{\underset{\sim}{U}}$$

where

$$M = \frac{1}{2} \int_V \rho \, \underset{\sim}{B}_m^T(\zeta) \ \underset{\sim}{B}_m(\zeta) \, dV$$

For the purposes of this study the only non-potential forces of interest are those which are due to viscous damping. These will be introduced by a generalisation of the Rayleigh dissipation function (Meirovitch, 1978)

$$F = -\frac{1}{2} \int_V c \, \dot{\underset{\sim}{s}}^T \dot{\underset{\sim}{s}} \, dV \tag{39}$$

in which c is the dissipation constant. Therefore by analogy with the kinetic energy

$$F = \Delta \dot{\underset{\sim}{U}}^T \left[-\frac{1}{2} \int_V c \, \underset{\sim}{B}_m^T(\zeta) \ \underset{\sim}{B}_m(\zeta) \, dV \right] \Delta \dot{\underset{\sim}{U}} \tag{40}$$

$$= \Delta \dot{\underset{\sim}{U}}^T \ \underset{\sim}{C} \ \Delta \dot{\underset{\sim}{U}}$$

If the incremental displacements, ΔU, are identified with the generalised coordinates of equation $\widetilde{}$ (7), then substitution of equations (36), (38) and (40) gives the discretised form of the system equations

$$\underset{\sim}{M}\, \Delta\ddot{U} + \underset{\sim}{C}\, \Delta\dot{U} + \{\underset{\sim L}{K} + \underset{\sim 1}{K} + \underset{\sim 2}{K} + (\lambda_o + \lambda_t\psi(t))\underset{\sim G}{K^{\lambda}}\,\}\Delta\underset{\sim}{U} = 0 \qquad (41)$$

in which the load level λ has been expressed in the more general form $\lambda = \lambda_o + \lambda_t\psi(t)$, where λ_o is the mean load level and λ_t is the amplitude of an arbitrary periodic function $\psi(t)$.

Equation (41) is the full dynamic equilibrium equation and reduces to several classical forms

    i)   <u>free vibration</u>

$$(\underset{\sim L}{K} - \omega^2\underset{\sim}{M})\Delta\underset{\sim}{U} = 0 \qquad (42)$$

    ii) <u>static bifurcation from a linear fundamental path</u>

$$(\underset{\sim L}{K} + \lambda_o\, \underset{\sim G}{K^{\lambda}})\Delta\underset{\sim}{U} \qquad (43)$$

    iii) <u>static bifurcation from a nonlinear fundamental path</u>

$$(\underset{\sim L}{K} + \underset{\sim 1}{K} + \underset{\sim 2}{K} + \lambda_o\, \underset{\sim G}{K^{\lambda}})\Delta\underset{\sim}{U} = 0 \qquad (44)$$

Whilst $\lambda_t = 0$ the solution of equation (41) is the classical eigenproblem and the frequencies and modes are best determined by inverse iteration or simultaneous iteration. When $\lambda_t \neq 0$ equation (41) represents a second order differential equation of the Mathieu-Hill type (Bolotin, 1964). Depending on the parameters of the equations the solutions are either stable (bounded in time) or unstable. The solution of these equations is generally reduced to locating the boundaries between stable and unstable regions in the parameter space (Bolotin, 1964 and Hutt et al, 1971). Boundaries to regions are marked by periodic solutions, two solutions of the same period bounding regions of stability and two solutions with different periods bounding regions of stability. For the specific case of $\psi(t) = \cos\theta t$ the solution of equation (41) is obtained by substituting the convergent series

$$\Delta\underset{\sim}{U} = \sum_{k=1,3,5}^{\infty} \{a_k\} \sin\frac{k\theta t}{2} + \{b_k\} \cos\frac{k\theta t}{2} \qquad (45)$$

so that the condition for the existence of solutions with period 2T reduces to the vanishing of an infinite determinant known as Hill's determinant. A first approximation to the principal instability region is obtained by retaining only the

first term of equation (45). Thus, neglecting the nonlinear
parts of equation (41) gives the two conditions for harmonic
solutions, combined with the ± sign as

$$| \underset{\sim}{K}_L + (\lambda_o \pm \lambda_t/2) \underset{\sim}{K}_G^\lambda - (\theta^2/4) \underset{\sim}{M} | = 0 \qquad (46)$$

APPLICATIONS

The thin shell element formulation presented in this paper has
undergone extensive testing in programs for linear and bifurcation analysis of shells.

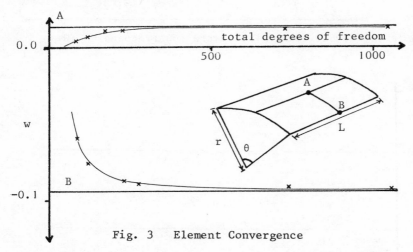

Fig. 3   Element Convergence

Convergence of the element is illustrated in Fig. 3. The
normal displacement, w, of points A and B on a cylindrical
panel with diaphragm end supports (r = 7.62 m, L = 15.24 m,
h = 0.0762 m, θ = 40º, E = 2.07 × 10^{10} N/m^2 , v = 0, p = 4310
N/m^2) is compared with the exact results (Tottenham and
Brebbia, 1970). Although representation of the diaphragm end
conditions, u = w = N_z = M_z = 0, is not possible using a displacement model, good convergence is still obtained using
u = w = 0 with the present formulation.

Free vibration
The two lowest symmetrical vibration frequencies of a similar
panel (r = 15.24 m, ρ = 2400 kg/m) were compared with their
exact values (Leissa, 1973) and a finite element analysis by
Clough and Wilson. From table 1 it is clear that the results
of Clough and Wilson are overstiff. This is probably a consequence of their using a flat plate type of element to model
the shell. Results from the current study show good agreement
with the exact values, although the second frequency converges
to a slightly low value due to the inadequate representation

of the boundary conditions and the lack of element compatibility.

Mesh	ω_1		ω_2	
	Clough	Present	Clough	Present
1 × 2	–	12.53	–	23.54
2 × 2	8.61	9.60	22.52	20.79
4 × 4	9.64	8.11	24.49	19.60
5 × 5	–	7.97	–	19.40
6 × 6	9.76	–	24.09	–
Exact	7.80		21.25	

Table 1   Comparison of Free Vibration Frequencies (Hz)

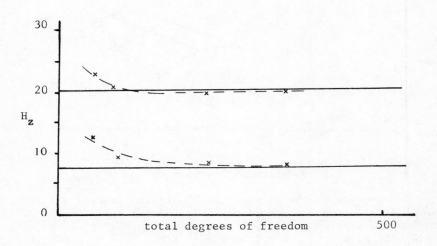

Fig. 4   Convergence of Free Vibration Results

Linear bifurcation

Tests were carried out on cylindrical shells of various lengths with radius, r, of 2m, thickness, h, of 0.02 m and Young's Modulus, E, of 10^5 Nm^{-2} with a Poisson's ratio, ν, of either 0.0 or 0.3.

The bifurcation results are given in table 2 in terms of a dimensionless parameter $\lambda = r^3 p^o/D$ where they are compared with the theoretical results of Wang and Billington (1975) and the finite element results of Cole (1973). The values of λ for both finite element schemes fall above those given in the theoretical work but this is to be expected since in both cases fairly coarse meshes were used, Cole used 25 three dimensional ring elements, and more accurate results are obtained with a

	Cole (1973)		Wang et al (1975)		Present	
ℓ/r	$\nu=0.0$	$\nu=0.3$	$\nu=0.0$	$\nu=0.3$	$\nu=0.0$	$\nu=0.3$
2	–	31.3(5)*	31.78(5)	30.78(5)	35.39(5)	36.52(5)
5	–	12.8(3)	16.95(3)	11.68(3)	15.04(3)	15.8(3)
10	–	7.69(2)	6.11(2)	6.05(2)	9.54(2)	10.0(2)

* Figures in parentheses represent the circumferential wave
number

Table 2   Dimensionless Buckling Parameter, λ, for Clamped-Free
          Cylindrical Shell under Uniform Load

finer grid. For all results the buckling patterns around the
rim of the shell were as predicted by theory, in agreement
with Cole's results, and in error at any point by less than
5%.

Dynamic Loading
Finally an example is given of the application of equation 46
to the case of the cylindrical panel with a point load at the
crown. The principal region of instability associated with
the lowest symmetric vibration frequency is shown on the
$(\lambda_o/\lambda_{cr},\ \theta/2\omega)$ plane (fig. 5). Regions are plotted for $\lambda_t =$
$0.2\lambda_{cr}$ (outer envelope) and $\lambda_t = 0.1\lambda_{cr}$ (inner envelope).
As $\lambda_t \to 0$ the region of instability degenerates into the back-
bone curve shown in fig. 5.

CONCLUSION

From the foregoing analysis it is seen that free vibrations and
bifurcation are related phenomena, and should be treated as
such. In fact fig. 5 shows that for $\lambda_o = 0.2\lambda_{cr}$ and $\lambda_t = 0.1\lambda_{cr}$
(i.e. loading $\le 0.3\lambda_{cr}$) the panel is unstable for load fre-
quencies of 2.4ω to 2.7ω . That is, unbounded response may
be obtained at loading frequencies far removed from the
natural frequency of the shell and at load levels significantly
below the critical load.
      The method presented here, being based on the finite
element technique, is completely general and may be applied to
any thin shell structure.

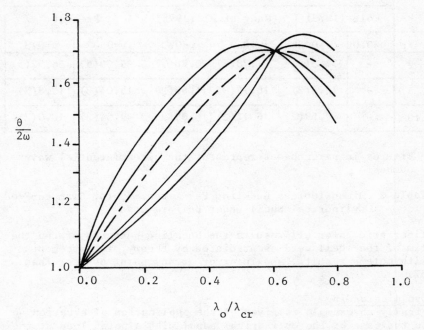

Fig. 5 Instability Region for the Cylindrical
Panel with a Point Load at the Crown

REFERENCES

Bognor, F.K., Fox, R.L. and Schmit, L.A. (1966) The generation of interelement compatible stiffness and mass matrices by the use of interpolation formulas. AFFDL-TR-66-80.

Bolotin, V.V. (1964) The dynamic stability of elastic systems. Holden-Day Inc.

Brebbia, C.A. and Connor, J.J. (1973) Fundamentals of the Finite Element Technique. Butterworths.

Cole, P.P. (1973) Buckling of hyperbolic cooling tower shells. Thesis, Princeton University, Princeton, N.J.

Clough, R.W. and Wilson, E.L. (1971) Dynamic Finite Element Analysis of Arbitrary Thin Shells. Proc. Conf. on Matrix Methods in Structural Mechanics, Wright Paterson AFB, Ohio.

Croll, J.G.A. and Walker, A.C. (1972) Elements of Structural Stability. Macmillan.

Hutt, J.M. and Ahmed S. Salam (1971) Dynamic Stability of Plates by Finite Elements. ASCE, EM3:879-899.

Key, S.W. and Beisinger, Z.E. (1968) The analysis of thin shells with transverse shear strain by the finite element method. AFFDL-TR-68-150.

Leipholz, H.H.E. (1975) Aspects of dynamic stability of structures, ASCE, EM2:

Leissa, A.W. (1973) Vibration of Shells. NASA SP-288.

Lekhnitskii, S.G. (1963) Theory of elasticity of an anisotropic elastic body. Holden-Day Inc.

Meirovitch, L. (1970) Methods of analytical dynamics. McGraw-Hill.

Sanders, J.L. (1959) An improved first approximation theory for thin shells. NASA TR-24.

Sanders, J.L. (1963) Nonlinear theories for thin shells. Quarterly Applied Math. 21,1:21-36.

Tottenham, H. and Brebbia, C.A. (1970) Finite Element Techniques in Structural Mechanics, Stress Analysis Publishers.

Wang, Yang-Shih and Billington, D.P. (1975) Buckling of cooling tower shells: State of the art. ASCE, J.Struct. Div. Vol.101.

VIBRATIONS OF SPATIAL BUILDING STRUCTURES

Riko Rosman,

Arhitektonski fakultet, Zagreb, Yugoslavia

INTRODUCTION

Vibration mode shapes and periods appreciably affect the response of spatial building structures when subjected to wind or earthquake forces.

The vertical structure is supposed to consist of shear walls and/or columns. Their cross-sectional properties are assumed to be constant along the height of the building; hence, the building has a vertical rigidity axis. The modulus of elasticity is supposed not to vary along the height of the structure. At their bottom ends, all the shear walls and columns are fixed into a rigid foundation, a grid of basement walls or the like. There can also be an arbitrary number of pin-ended columns, which do not contribute to the structure's lateral stiffness. The floor slabs are assumed to be pin-connected to the vertical supporting elements, so that there is no frame action. On all floors, the mass is distributed equally, so that the building has a vertical mass axis. The distribution of the mass along the height of the building can be arbitrary.

In general, the rigidity axis and the mass axis do not coincide.

For analysis purposes, Rosman (1968) has substituted the structure by an equivalent cantilever. The substitute cantilever is supposed to work primarily in bending and warping; because its shear and pure torsional stiffnesses are usually small and unreliable, they are not introduced into the analysis.

In the following, a hand method is developed, which enables a simple, rapid determination of the fundamental vibrations' mode shape and period. It represents a generalization of Rosman's (1968, 1973, 1974, 1975) previous investigations.

FLEXURAL AND TORSIONAL VIBRATIONS

The following notation is used in this chapter:

S shear center and rigidity axis of the substitute cantilever

x,y principal axes of the substitute cantilever's cross section

z vertical axis coinciding with the rigidity axis

I_x, I_y moments of inertia of the substitute cantilever for the x and y directions, respectively

I_w warping moment of inertia of the substitute cantilever with respect to S

H height of the structure

n number of stories

G mass center of the system's cross section and mass axis

i mass radius of gyration with respect to S

Q weight of the building

E modulus of elasticity

Dimensionless system parameters are defined according to

$$k_y = \frac{I_x}{I_y} \quad , \quad k_w = \frac{i^2 \, I_x}{I_w} \, . \tag{1}$$

In the case the mass axis coincides with the rigidity axis, the vibration modes are uncoupled. There exist n plane vibration modes in the x direction, n plane vibration modes in the y direction and n rotational vibration modes with respect to the z axis. The largest of the n periods corresponding to flexural vibrations in the x direction, the largest of the n periods corresponding to flexural vibrations in the y direction and the largest of the n periods corresponding to torsional vibrations about the z axis were found by Rosman (1973, 1974) to be

$$T_x = t' \cdot H^2 \sqrt{\frac{Q}{H \, E \, I_x}} \, , \tag{2}$$

$$T_y = \sqrt{k_y} \cdot T_x \quad , \quad T_z = \sqrt{k_w} \cdot T_x \quad . \tag{3}$$

The dynamic coefficient t' depends on the number n of stories and the distribution of the mass along the height of the system. Numerical values of the dynamic coefficient for the frequent case when all the story heights are equal and all the story masses are equal are listed in Table 1.

Table 1   Dynamic coefficients ($m^{-1/2}$ sec)

n	1	2	3	4	5
t'	1.158	0.859	0.762	0.714	0.685

n	6	7	8	9	10
t'	0.666	0.652	0.642	0.634	0.628

The mass of systems with a very large number of stories can be assumed to be uniformly distributed along the height; in this case $t' = 0.570$.

The system's largest, i.e. fundamental vibration period T is equal to the largest of the three uncoupled vibration periods T_x, T_y, T_z.

EQUATION OF PERIOD COEFFICIENT

If the mass axis G of the system does not coincide with its rigidity axis S, the flexural and torsional vibrations are coupled. The system vibrates in flexural-torsional modes.

To facilitate numerical analyses, the system's vibration period T is related to its uncoupled vibration period T_x according to

$$T = \sqrt{t} \cdot T_x \quad . \tag{4}$$

Herein t denotes the – dimensionless – period coefficient.

When the system vibrates, the floor slabs undergo plane motions. These may be regarded as a rotation about a zero-displacement point. Rosman (1979) has proved that, with the fundamental mode, the zero displacement points of all the floor slabs lie on a vertical line, the rotation axis R. Hence, the flexural-torsional vibration can be considered as a torsional vibration about a - hitherto unknown - rotation axis.

From Equations 1, 3 and 4 the period coefficient follows to be

$$t = \frac{\bar{i}^2}{\bar{I}_w} I_x \ , \tag{5}$$

where $\bar{I}_w$ and $\bar{i}$ denote the warping moment of inertia of the substitute cantilever and the mass radius of gyration, respectively, about R.

By Strength of materials, $\bar{I}_w$ and $\bar{i}$ can be expressed through the coordinates x_G, y_G of G and the coordinates x_R, y_R of R. By introducing these expressions into Equation 5, the period coefficient becomes a function of the location of the rotation axis,

$$t = \frac{i^2 + x_R^2 + y_R^2 - 2\,x_G\,x_R - 2\,y_G\,y_R}{\dfrac{i^2}{k_w} + \dfrac{x_R^2}{k_y} + y_R^2} \ . \tag{6}$$

The period T becomes maximal when t is maximal. The two extremum requirements

$$\frac{\partial t}{\partial x_R} = 0 \ , \qquad \frac{\partial t}{\partial y_R} = 0 \tag{7}$$

and Equation 6 uniquely determine the governing period coefficient and the corresponding mode shape.

The Differential calculus gives the coordinates

$$x_R = \frac{x_G}{1 - t/k_y} \ , \qquad y_R = \frac{y_G}{1 - t} \tag{8}$$

of the rotation axis.

On introducing these results into Equation 6, it

becomes the - cubic - period coefficient equation

$$(1 - t)(k_y - t)(k_w - t) -$$

$$\left[(1 - t) k_y \frac{x_G^2}{i^2} + (k_y - t) \frac{y_G^2}{i^2}\right] k_w = 0 . \tag{9}$$

Carrying out the multiplications it takes on the final form

$$t^3 - c_2 t^2 + c_1 t - c_0 = 0 . \tag{10}$$

The equation coefficients amount to

$$c_2 = 1 + k_y + k_w ,$$

$$c_1 = k_y + (1 - \frac{y_G^2}{i^2}) k_w + (1 - \frac{x_G^2}{i^2}) k_y k_w , \tag{11}$$

$$c_0 = k k_y k_w ,$$

where

$$k = 1 - \frac{x_G^2 + y_G^2}{i^2} . \tag{12}$$

All the coefficients are dimensionless and depend only on the system parameters and the relative excentricities of G with respect to S.

The governing period coefficient c is the largest of the three roots of the period coefficient equation 10. By analyzing the properties of this equation it is found, that c is larger than the largest of the three system parameters 1, k_y, k_w.

Hence, c may easily be determined solving Equation 10 by iteration, using a pocket calculator, proceeding from the largest of the three values 1, k_y, k_w to larger values.

PERIOD AND ROTATION AXIS. PLANE OF EQUIVALENT LATERAL LOAD

After the period coefficient has been determined, the vibration period is found by Equation 4; the uncoupled period T_x corresponding to plane vibrations in the x direction follows from Equation 2 and Table 1. The location of the rotation axis is given by Equations 8.

The - vertical - plane L of the equivalent lateral load which produces a deformation corresponding to the fundamental vibration mode shape intersects the x and y axes, respectively, at

$$x_L = -\frac{I_w}{I_y \, x_R} \quad , \quad y_L = -\frac{I_w}{I_x \, y_R} \, . \tag{13}$$

SYSTEM WITH A SYMMETRY PLANE

Let the system, i.e. both the substitute cantilever and the mass, be symmetric with respect to, say, the xz plane. Due to $y_G = 0$, the period coefficient equation 9 simplifies. One of its roots is

$$t = 1 \; ; \tag{14}$$

it corresponds to the vibration in the system's symmetry plane.

The larger of the other two roots is

$$t = \frac{k_y + k_w}{2} + \sqrt{(\frac{k_y + k_w}{2})^2 - k \, k_y \, k_w} \, . \tag{15}$$

The rotation axis lies in the system's symmetry plane and is determined by the first Equation 8. The equivalent lateral load is perpendicular to xz and acts at the coordinate given by the first Equation 13.

SYSTEM THE SUBSTITUTE CANTILEVER OF WHICH IS TWO-FOLD SYMMETRIC

Let the moments of inertia I_x and I_y of the substitute cantilever be equal. Then $k_y = 1$. The period coefficient equation 9 simplifies. One of its roots is again $t = 1$; it corresponds to the vibration in the SG plane, i.e. in the plane determined by the rigidity and the mass axes. The larger of the two other roots is found to be

$$t = \frac{1 + k_w}{2} + \sqrt{(\frac{1 + k_w}{2})^2 - k\, k_w} \quad . \qquad (16)$$

The rotation axis lies in the SG plane. The equivalent lateral load acts perpendicularly to SG.

SYSTEM THE RIGIDITY AND MASS AXES OF WHICH COINCIDE

With systems the rigidity and mass axes of which coincide, $x_G = y_G = 0$. The period coefficient equation 9 yields the three roots 1, k_y, k_w. The flexural vibrations in the x direction, the flexural vibrations in the y direction and the torsional vibrations about the z axis are uncoupled. The fundamental period is equal to the largest of the three uncoupled periods T_x, $\sqrt{k_y}\, T_x$ and $\sqrt{k_w}\, T_x$.

EXAMPLE

Determine the period and the rotation axis of the fundamental mode vibrations of the shear-wall structure shown in Figure 1. Assume that all floor masses are equal and that they are uniformly distributed over the plan area.

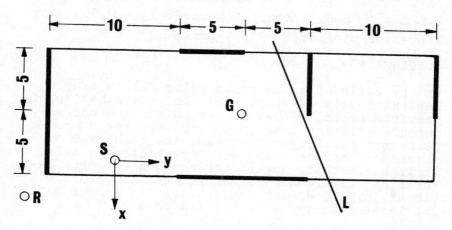

Figure 1   Shear-wall structure according to the example

The coordinates of the stiffness axis S with respect to the left bottom corner of the plan are 1.11 m and 5.00 m. Let I denote the moment of inertia of the smaller shear walls; the moment of inertia of the larger shear walls then amounts 8 I.

Moments of inertia of the substitute cantilever: I_x = 10 I, I_y = 9 I, I_w = 1139 I. Mass axis: x_G = -3.889 m, y_G = 10.000 m. Square of the mass radius of gyration: i^2 = 198.45 m^2. System parameters: k_y = 1.111, k_w = 1.742. Auxiliary value: k = 0.4199. Period coefficient equation and its largest root: t^3 - 3.8534 t^2 + 3.7638 t - 0.8129 = 0; t = 2.455. Rotation axis: x_R = 3.22 m, y_R = -6.87 m. Plane of the equivalent lateral load: x_L = -39.4 m, y_L = 16.6 m.

Suppose now that the number of stories is 10 and that the story height is 3.0 m. The thickness of the shear walls let be 0.20 m. The modulus of elasticity amounts to 2.1·10^4 MN/m^2.

Height of the structure: H = 30.0 m. Moment of inertia of the smaller shear walls: I = 2.083 m^4. Flexural stiffness of the substitute cantilever: E I_x = 4.374·10^5 MNm^2. Dynamic coefficient: t′ = 0.628 $m^{-1/2}$sec. Uncoupled period: T_x = 0.855 sec. Vibration period: T = 1.34 sec.

CONCLUSIONS

On the basis of an extremum principle, a simple hand method is developed for the determination of the fundamental vibration period and the corresponding rotation axis of a frequent type of contemporary building structures, the mass axis of which does not coincide with its stiffness axis. It is shown that, due to the coupling of the flexural and torsional vibrations, the structure′s vibration period might be appreciably larger than its largest uncoupled vibration period.

Moreover, the plane of the lateral load is determined, which produces a deformation corresponding to the fundamental mode vibration. The structure is the weakest when loaded in this plane. A corresponding wind load produces the largest dynamic effect and might, hence, be a governing factor when designing the building′s structure.

An example illustrates the practical application of the method.

REFERENCES

Rosman, R. (1968) Statik und Dynamik der Scheibensysteme des Hochbaues. Springer-Verlag Berlin Heidelberg New York.

Rosman, R. (1973) Eigenwertprobleme von Kragsystemen. Deutsche Bauzeitschrift, Gütersloh, 4: 703-714.

Rosman, R. (1974) Drehschwingungen und Drehknicklasten von Kragsystemen. Bautechnik, Berlin, 51, 4:120-129.

Rosman, R. (1975) Berechnung gekoppelter Stützensysteme im Hochbau. 2nd edition. W. Ernst & Sohn, Berlin München.

Rosman, R. (1979) Schwingungen im Grundriß unsymmetrischer Stützen- und Wandscheibensysteme. Beton- und Stahlbetonbau, Berlin, to be published.

RANDOM PROCESSES AND TRANSFER FUNCTIONS OF DYNAMIC SYSTEMS

L. Kus

CKD Praha, zav. Lokomotivka, Research Institute of Locomotives, Ceskomoravska 205, Praha 9, CSSR

INTRODUCTION

The finite element method has been in the past predominantly used for static calculations of beam structures. The most frequently applied calculations were for bridges, cranes, oil rigs, antenna towers etc. In recent years the application of this method also entered into the field of dynamic calculations of various mechanical assemblies, since the development of digital computers made the solution of larger systems of equations possible, which is vitally important for dynamic calculations.

In the Research Institute for Diesel Locomotives, CKD Praha, Lokomotivka Works, a program was written for the ICL 1905 computer. This program serves for the solution of dynamic problems in bogies and locomotive main frames. Basically, it is possible to solve general dynamic problems in a space arrangement, for systems having both displacement and torsional vibrations. The dynamic systems can be built up of masses, bogies, beam springs and dampers. Excitation can be introduced into all modal points of the system, either in force or kinematic forms. The program is designed so as to provide for a solution of the forced vibrations of dynamic systems with excitation of a stochastic nature.

DYNAMIC SYSTEM EXCITED BY RANDOM PROCESS

A locomotive is a very complicated dynamic system, excited both by the rail irregularities (kinematic excitation) and by the forces, caused by unbalanced masses in the motor, drive etc. (so called dynamic excitation). Both main groups of excitation are mutually non-correlated. Mutually non-correlated are also subgroups of kinematic excitation, namely those caused by the vertical or lateral irregularities and those caused by the rail superelevation - so called angular excitation. Previously mentioned types of excitation act upon all wheel-sets of the vehicle and when an ideally stiff track is assumed they are mutually fully correlated.

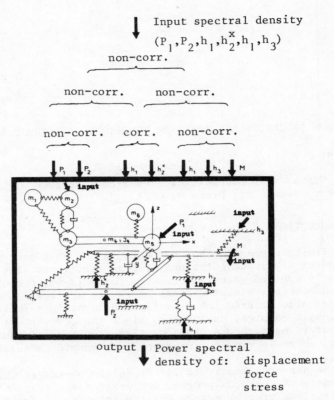

Fig.1.: Possibilities of the solved dynamic system, input exciting variables and the form of the computer output.

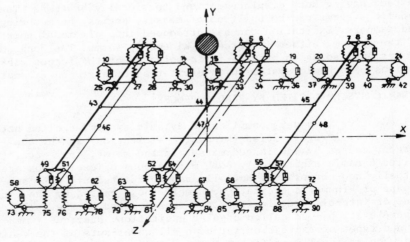

Fig.2.: Model of a vehicle – Finite Element Method

From the above mentioned facts it follows that a locomotive as a dynamic system is a system including excitation both mutually correlated and non-correlated. Both these types occur for kinematic and dynamic excitations. These conclusions served as a basis for the whole program structure. The objective was to provide for computation of the system transfer functions and power spectral densities at individual points. This is possible for systems with multiple inputs, both mutually correlated and non-correlated.

The complete program facilitates the solution of dynamic systems in a general space arrangement while using basic elements: mass, body, beam, spring, damper. The extent of the solved problem is in our case limited to 250 elements. This limitation is, however, caused by the computer core store size, not by the applied method.

The scope of utilisation of this program for the solution of general dynamic systems is shown in Fig. 1. The possibility for defining the input quantities, the dynamic system and form of the results obtained from the computer are shown there as well. The possibility for combining correlated and non-correlated inputs is unlimited and the excitation can be introduced in all modal points of the structure. The results obtained serve for the analysis of the dynamic system and for the computation of the running qualities of the vehicle as is shown in Fig. 2.

Power spectral densities of stresses in the individual modal points of the structure serve then for the computation of the vehicle component life expectation. One of the latest theories for life expectation computation is that worked out by Novarov. The formula for computation of the life expectation of a component, loaded by stress of certain spectrum width is defined:

$$
T = \frac{\left(\dfrac{\sigma_c}{\mathcal{H}(\varepsilon)}\right)^m \cdot N_0}{J_1 \left[\displaystyle\int_0^\infty f^{2/m} G_\sigma(f)\, df\right]^{m/2}} \quad ; \quad [s] \qquad (1)
$$

where

T — life expectation

$G_\sigma(f)$ — power spectral density of the process of component stress

f — frequency in Hz

σ_c — fatigue limit of given material

m — slope of the oblique arm of Woehler's line

N_0 — number of cycles in the point of fracture of Woehler's line

$\mathcal{H}(\varepsilon)$ — corrective coefficient for the fatigue limit, which is a function of the stress spectrum width

k — coefficient of the process irregularity = ration of the number of signal extremes to the number of zero level signal crossovers per unit of time

$$
J_1 = \int_{x_0}^\infty x^{m+1} \exp\left(-\frac{x^2}{2}\right) dx = 2^{m/2}\, \Gamma\left(\frac{m+2}{2}\right) P(x_0^2; m+2)
$$

$\Gamma(\frac{m+2}{2})$ x_0 – bottom value in the random process of the stress
$\Gamma(\frac{m+2}{2})$ – gamma function value, given by integral $\int_0^\infty e^{-t} \cdot t^{m/2} \, dt$
$P(x_0^2; m+2)$ – Pearson's distribution

Equation (1) has been defined for the power spectral density of the stress, having a certain frequency width. For loading the components by the wide-band process and for transformation of the life expectation computation to the kilometer runaway, equation (1) can be expressed in the following form.

$$L[\text{km}] = \frac{N_0 \cdot \varphi_6^m \cdot \delta_c^m}{3600 \, n_L \cdot \int_{\varphi_{6x_0}}^\infty x^{m+1} \exp\left(-\frac{x^2}{2}\right) dx} \left\{ \int_0^{v_{max}} \frac{p(v) \cdot S_6^m(v) \, \mathscr{R}^m(\varepsilon) \frac{dv}{v}}{\left[\int_0^\infty f^{2/m} g_6(f,v) \, df \right]^{-m/2}} \right\}^{-1} \quad (2)$$

where

φ_6 coefficient of the cycle asymmetry, the value of which, according to Gudmann, is $\varphi_6 = 1 - \frac{\delta_m}{\delta_{pt}}$

δ_m – static bias in examined point

δ_{pt} – strength limit for given material

$p(v)$ – function of probability density of occurence of operation speeds

v – track speed

n_L – corrective factor (linearity of hypotheses, inaccuracy of computational relations, inaccuracy in the value of fatigue limit....).

The advantage of the above mentioned procedure in computing the life expectation lies in the possibility of its introduction into the theoretical computations and thus the possibility of evaluating the life expectation of individual locomotive components in the design stage. For the computation of the output power spectrum the following relations, defining the dependence of the input and output power spectral density are available.

NON-CORRELATED INPUTS

$$G_y(f) = \sum_{i=1}^N |H_i(f)|^2 \cdot G_{ii}(f) \quad (3)$$

where

$|H_i(f)|^2$ – absolute value of the ith system transfer squared

$G_{ii}(f)$ – power spectral density of the ith input to the system

$G_y(f)$ – power spectral density at the output of the system

CORRELATED INPUTS

$$G_y(f) = \mathbf{H}(f) \cdot \mathbf{G}_{xx}(f) \cdot \mathbf{H}^{*T}(f) \quad (4)$$

where

$H(f)$ – matrix of individual system transfers
$G_{xx}(f)$ – matrix of the cross-spectral densities of the system inputs
$H^{*T}(f)$ – conjugate transposed matrix of the system

To obtain the input power spectral density requires experimental measurement of their behaviour. These characteristics are thus statistically defined; it does not matter whether the inputs are correlated or non-correlated. The unknown characteristics in equations (3) and (4) are the transfers $H_i(f)$ and $H(f)$ which describe the dynamic system and its quality. By means of a change or a suitable combination of the locomotive parameters (mass, stiffness of the spring suspension, damping, geometric characteristics etc.) an optimal form of the vehicle transfer function can be achieved and thus suitable output spectrum $G_y(f)$ can be obtained. By the term 'suitable output spectrum' is understood a spectrum which, from the point of view of the vehicle ride qualities, insures the minimum dispersion variance of the vehicle body acceleration and from the point of view of the vehicle life expectation insures the maximum life in connection with relations (1) and (2). The dispersion variance of the locomotive body acceleration is computed according to the relation

$$\overline{\sigma}^2(\ddot{y}) = \int_0^\infty G_y(\omega) \cdot \omega^4 \cdot d\omega \qquad (5)$$

The complete program facilitates the computation of the output spectra of the displacement, forces and stresses in all nodal points of the structure. The computation procedure is based upon finding out the transfers between the nodal points of the structure and hence computing the output spectrum according to equations (3) and (4).

MATHEMATICAL MODEL OF THE DYNAMIC SYSTEM

As was previously stated, the question of transforming the random input excitation to the statistical characteristic at some point of the vehicle is determined firstly by the processing of the random excitation into the form of power spectral density $G_{\ddot{u}}(f)$ and secondly by multiplying by $|H_i(f)|^2$, which is the square of absolute value of the transfer between the input and output (equation (3)). The power spectral density of the input irregularities is obtained from the acceleration measurement at the axle box, subsequent computation of the autocorrelation function and its Fourier transformation. The problem to be solved is the determination of the transfer function's absolute value squared, $|H_i(f)|^2$. For the computation of the vehicle transfer functions there is a possibility of working out the equations of motion for a certain model, while introducing the harmonic excitation with phase shift at individual wheel-sets having unity amplitude and determining the response in the individual output points. Since there is a wide range of vehicles, for each of which the mathematical model is different, having a different number of degrees of freedom, it is advisable to turn to the general computation method.

The finite element method using matrix notation proves to be very suitable. The matrix notation is very advantageous during the necessary computer processing. Further, by means of the finite element method it is possible to model general three dimensional structures and structures consisting of beams, springs and dampers, which for railway vehicle dynamic quality modelling is quite sufficient.

In the matrix notation, any structure can be described by an equation of the form:

$$M\ddot{u} + C\dot{u} + Ku = P + C\dot{h} + Kh \qquad (6)$$

The individual terms in the equation expressing:

M — structure mass matrix
C — structure damping matrix
K — structure stiffness matrix
P — matrix of the excitation forces acting upon the structure
h — matrix of the amplitudes of the excitation kinematic irregularities
$\dot{h}$ — matrix of the velocities of the excitation kinematic irregularities
u — matrix of displacements of individual nodal points of the structure
$\dot{u}$ — matrix of displacement velocities of individual nodal points of the structure
$\ddot{u}$ — matrix of displacement accelerations of individual nodal points of the structure

The excitation variables P and h are of a random nature. For railway vehicles they consist of excitation from the motor P and excitation by rail irregularities h and $\dot{h}$. After performing the Fourier transformation of equation (6) we obtain:

$$(K - \omega^2 M + i\omega C)\bar{U} = \bar{P}_0 + (K + i\omega C)\bar{H}_0 \qquad (7)$$

while $\bar{U}$, $\bar{P}_0$ and $\bar{H}_0$ are Fourier transforms of quantities u, P and h, expressed in complex form (the bar over the quantity). The complex form of the excitation quantities makes it possible during the solution of the vehicle transfer function to introduce the phase delay of the excitation function at the individual wheels of the vehicle. For further processing, equation (7) can be rewritten in the form:

$$\bar{D} \cdot \bar{U} = \bar{P}_0 + \bar{B}\,\bar{H}_0 \ , \qquad (8)$$

where

$$\bar{D} = K - \omega^2 M + i\omega C,$$
$$\bar{B} = K + i\omega C \tag{9}$$

Equation (8) constitutes an equation system, having complex co-efficients. To make the subsequent solution of the system easier, system (8) is transformed, using the comparison of real and imaginary parts, to the system with the number of equations doubled.

$$\begin{bmatrix} Re\,\bar{D} & -Jm\,\bar{D} \\ Jm\,\bar{D} & Re\,\bar{D} \end{bmatrix} \cdot \begin{bmatrix} Re\,\bar{U} \\ Jm\,\bar{U} \end{bmatrix} = \begin{bmatrix} Re\,\bar{P}_0 \\ Jm\,\bar{P}_0 \end{bmatrix} + \begin{bmatrix} Re\,\bar{B} & -Jm\,\bar{B} \\ Jm\,\bar{B} & Re\,\bar{B} \end{bmatrix} \cdot \begin{bmatrix} Re\,\bar{H}_0 \\ Jm\,\bar{H}_0 \end{bmatrix} \tag{10}$$

Using relation (9), equation (10) can be expressed in the form:

$$\begin{bmatrix} K-\omega^2 M & ; & -\omega C \\ \omega C & ; & K-\omega^2 M \end{bmatrix} \cdot \begin{bmatrix} Re\,\bar{U} \\ Jm\,\bar{U} \end{bmatrix} = \begin{bmatrix} Re\,\bar{P}_0 \\ Jm\,\bar{P}_0 \end{bmatrix} + \begin{bmatrix} K & ; & -\omega C \\ \omega C & ; & K \end{bmatrix} \cdot \begin{bmatrix} Re\,\bar{H}_0 \\ Jm\,\bar{H}_0 \end{bmatrix} \tag{11}$$

or, after rewriting:

$$D_1 \cdot \bar{U}_1 = \bar{P}_{01} + \bar{B}_1 \cdot \bar{H}_{01} \tag{12}$$

System (12) is the final system of equations, from which is possible, for given unity excitation $\bar{P}_{01}$ and $\bar{H}_{01}$ to compute the required transfer function of the structure for the correlated inputs, or the matrix of transfers for the non-correlated inputs.

$$\bar{U}_1 = \begin{bmatrix} Re\,\bar{U} \\ Jm\,\bar{U} \end{bmatrix} = D_1^{-1} (\bar{P}_{01} + B_1\bar{H}_{01}) \tag{13}$$

The transfer vector of the system, or the square of the transfer absolute value, is then determined by the following expression:

$$H(i\omega) = Re\,\bar{U} + i\,Jm\,\bar{U}$$
$$|H(i\omega)|^2 = |Re\,\bar{U}|^2 + |Jm\,\bar{U}|^2 \tag{14}$$

FINITE ELEMENT METHOD

From the equations for the system transfer function computation (equations (11), (12) and (13)), it is evident that for the required number of transfers it is necessary to compose the mass matrix M, stiffness matrix K and damping matrix C of the whole structure.

Equation (11) contains, apart from the previously mentioned matrices, the matrices of the loading forces $\bar{P}_0$, of the kinematic excitation $\overline{H}_0$ and matrices of the resultant displacements of individual nodal points of the structure.

The computation procedure consists, for certain structures, of dividing the structure into parts, defined in the main coordinate system (MCS) by the location of nodal points of the structure - (X,Y,Z). The quantities referenced in capital letters are related to the local coordinate system assigned to each structure element (the element being defined by the location of its two nodal points, by the element shape and by the element material constants).

In matrix form, the column matrices of forces and displacements are referred to as P, p, U, u. The whole structure will be described by coordinates within the main coordinate system (MCS). During the computation procedure it will be necessary to transform the quantities from the main coordinate system (MCS) to the local coordinate system (LCS). For this purpose the so-called transformation matrix T serves

$$p = T \cdot P \quad ; \quad u = T \cdot U \tag{15}$$

Since matrix T is a square matrix of orthogonal type $T^t = T^{-1}$

$$P = T^t \cdot p \quad ; \quad U = T^t \cdot u \tag{16}$$

If there is a linear dependence between the diaplacement matrix u and the force matrix P

$$p = k \cdot u \tag{17}$$

then, by means of relations (15) and (16), it is possible to rewrite equation (17) in the form:

$$TP = kTU \Rightarrow P = K \cdot U \quad ; \quad K = T^t \cdot k \cdot T \tag{18}$$

Characteristic quantities of individual structure elements are, for simplicity, input in the local coordinate system. It is evident from equation (18) that one of the basic input parameters is the matrix of element stiffnesses k_N, usually expressed in the following form:

$$k_N = \begin{bmatrix} k_{ii} & k_{ij} \\ k_{ji} & k_{jj} \end{bmatrix} \tag{19}$$

For space arrangement, the matrix k_N is of order $(12,12)$. Considering the notation in (19), the transformation matrix is input in the form:

$$T = \left[\begin{array}{cc|cc} \lambda & & & \\ & \lambda & 0 & \\ \hline & 0 & \lambda & \\ & & & \lambda \end{array} \right] = \left[\begin{array}{c|c} \lambda_N & 0 \\ \hline 0 & \lambda_N \end{array} \right] \tag{20}$$

where λ_N is the matrix of direction cosines, of order $(6,6)$. Provided that individual connections of basic elements within the whole structure are denoted by numbers $1, 2, 3, \ldots n$, the relation between the matrix of displacements U and that of forces P can be expressed using the stiffness matrix of the whole structure K in the following form:

$$
\begin{bmatrix} P_1 \\ P_2 \\ P_3 \\ P_4 \\ \vdots \\ P_n \end{bmatrix} = \begin{bmatrix} K_{11} & K_{12} & K_{13} & . & . & K_{1n} \\ K_{21} & K_{22} & K_{23} & . & . & K_{2n} \\ K_{31} & K_{32} & K_{33} & . & . & K_{3n} \\ K_{41} & K_{42} & K_{43} & . & . & K_{4n} \\ \vdots & \vdots & \vdots & & & \vdots \\ K_{n1} & K_{n2} & K_{n3} & . & . & K_{nn} \end{bmatrix} \cdot \begin{bmatrix} U_1 \\ U_2 \\ U_3 \\ U_4 \\ \vdots \\ U_n \end{bmatrix} \tag{21}
$$

$$
P = K \cdot U \tag{22}
$$

Equation (21) is derived from the fact that at a nodal point where a number of elements connect, a displacement of the nodal point occurs due to force P, equal to the sum of products of individual stiffness matrices multiplied by the displacement vectors of the corresponding elements. This results in the diagonal elements being determined as the sum:

$$
K_{11} = K_{12} + K_{13} + K_{14} + \ldots + K_{1n}
$$

$$
K_{21} = K_{21} + K_{23} + K_{24} + \ldots + K_{2n} \tag{23}
$$

$$
K_{nn} = K_{n1} + K_{n2} + K_{n3} + \ldots + K_{n,n-1}
$$

If the system (22) is divided into the matrices of known nodal load P with unknown displacements U and matrices of nodal load P_p with known displacements U, we obtain:

$$
\begin{bmatrix} P \\ P_p \end{bmatrix} = \begin{bmatrix} K_A & K_B \\ K_B^t & K_p \end{bmatrix} \cdot \begin{bmatrix} U \\ U_p \end{bmatrix} \tag{24}
$$

The relation for the computation of the unknown displacements is obtained from rearranged equation (24)

$$
U = K_A^{-1} \cdot P - K_A^{-1} K_B U_p \, , \tag{25}
$$

or the relation for the computation of forces P_p

$$
P_p = K_B^t \cdot K_A^{-1} P - K_B^t K_A^{-1} K_B U_p + K_p U_p \tag{26}
$$

From equations (25) and (26) it is possible to compute for a given structure load P the displacements of the nodal points of the structure U and further the load in individual nodal points of the structure, including the stresses in them (which is computed after substitution of the displacements into the elementary formula of elasticity for the calculation of stresses). Previously mentioned matrices of stiffnesses, damping and masses for the basic finite elements, namely for beams, springs and dampers, are

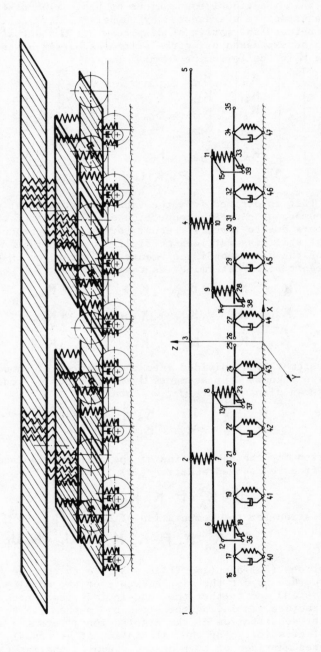

Fig.3.: Model of vehicle for horizontal excitation

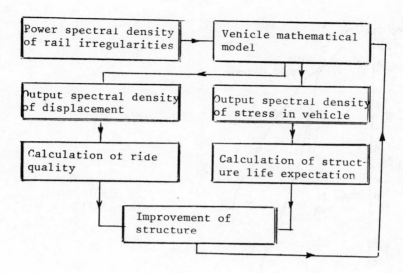

Fig.4.: Calculation of ride quality and life expectation

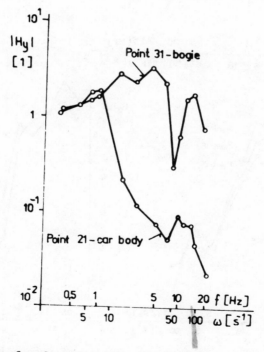

Fig.5.: Transfer function for displacement of points 21 and 31

436

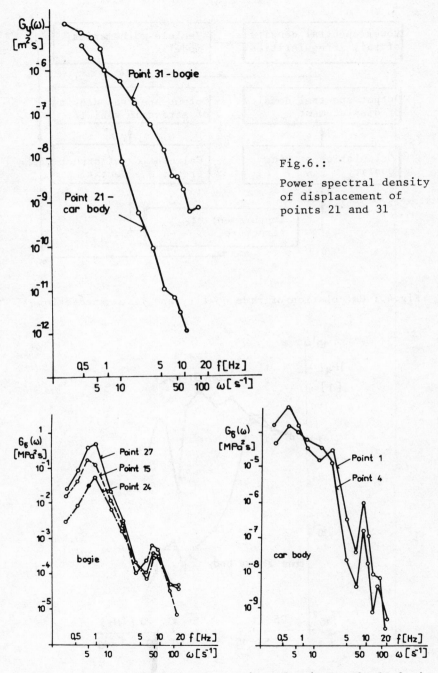

Fig.6.:

Power spectral density
of displacement of
points 21 and 31

Fig.7.: Power spectral stress density of points 1,27,15,24,4

expressed in equation (1).

RESULTS OF THE COMPUTATIONS

The above mentioned theory was applied to the solution of a locomotive, dynamic chart shown in Fig. 2. This model was used for vertical and horizontal excitation. Another case for calculation of horizontal excitation is shown in Fig. 3.

System excitation in the form of power spectral density was applied to all wheel-sets both in the vertical and horizontal directions. The resulting diagrams of transfer functions and power spectral densities of the displacements and stresses are shown in Figs. 5,6,7. These diagrams can be computed for all points of the structure. They serve for calculations of ride quality and characteristics of the dynamic load and stress in the vehicle (Fig.4). Further calculations of structure life expectation or the location of the dynamically most exposed point in the structure by formulae (1) and (2) are also possible.

REFERENCES

Kolar, J., 'Calculation of life expectation of structures', Informacni zpravodaj CKD, 1978, 3.

Novarov, V., 'Issledovanie ustalostnoj procnosti ram telezek elektrovozov s ucotom statisticeskich parametrov dinamiceskych nagruzok', MIIT, 1972.

Rus, L., 'Hunting motion of bogies of railway vehicles', Communications of VSD Zilina, Vol.12, 1971, p.155-168.

Rus, L., 'Finite Element Method - Used in Dynamics System Computation, Report VUML 645.37 2.

Rus, L., 'Running stability and railway vehicle transfer function, solved by method of statistical linearisation in The Dynamics of Vehicles', Vienna 1977.

Sixta, P., Programme for Calculation of Dynamical Responses, Finite Element Method, CKD Praha, VUML.

438

INTERPOLATION-BASED METHODS FOR THE EFFICIENT DETERMINATION OF THE DYNAMIC RESPONSES OF LINEAR STRUCTURAL SYSTEMS

Andrew G. Collings and Leith R. Saunders

Department of Civil Engineering, University of Auckland, New Zealand

INTRODUCTION

The dynamic response of a linear structural system, having m-degrees of freedom, when subjected to time-dependent loadings may be given by solutions to the following set of familiar coupled linear second-order ordinary differential equations:

$$M\ddot{D} + C\dot{D} + KD = H(t) \qquad (1)$$

In equation (1), $\ddot{D}$, $\dot{D}$, D and $H(t)$ are respectively m-vectors of generalised accelerations, velocities, displacements and forces while M, C and K are respectively m-order mass, damping and stiffness matrices. In practical cases, the system of equations (1) is usually *stiff*, in the sense that the ratio of the longest undamped period to the shortest is large.

Under suitable conditions equation (1) may be uncoupled by a similarity transformation that uses an n-column matrix, $(1 \le n \le m)$, of eigenvectors ϕ of the homogeneous system $K\phi - M\phi\Omega = 0$ where Ω is a diagonal matrix of real eigenvalues. Letting:

$$D = \phi W \qquad (2)$$

where W is measured with respect to a set of normalised coordinates, then a system of n-uncoupled scalar second-order ordinary differential equations results of which the rth may be written:

$$\ddot{w}_r + 2\beta_r \dot{w}_r + \bar{\omega}_r w_r = p_r \qquad (3)$$

Commonly the elements of the loading vector $H(t)$ are highly irregular functions of time and these are frequently specified in the form of discrete time-series at constant time intervals. This paper is concerned with that situation. The time-series are produced by sampling the continuous loading at small intervals of time.

There are a number of established methods for solving the equations of motion when the driving function is presented as a time series. One strategy is to solve (1) directly; for example, by the use of numerical integration or frequency domain techniques. Or, alternatively the modal equations (3) may be solved individually and then the original solution domain of equation (1) re-entered by use of equation (2). The solution of equation (3) may be accomplished in a variety of ways. One technique is to apply piece-wise linear or higher order interpolations to the input data and thence obtain closed form solutions by way of Laplace transforms. Alternatively, as with the solution of (1), numerical integration or frequency domain methods may be used. There is also the commonly used technique which requires that the response be expressed as a convolution or Duhamel integral, which is evaluated numerically. Implicit to any technique used however is the requirement that the original loading be matched exactly or approximated closely by a continuous time-function.

This paper presents two further techniques for solving the equations of motion, both of which depend upon interpolation of the input time-series. The first method requires that the input be expressed as a finite trigonometric series. This series is the global trigonometric interpolation due to Lanczos, (Lanczos, 1957): it reproduces the discrete input exactly and gives a very smooth function between the given data points. To find the coefficients of the trigonometric series a Goertzel or fast Fourier algorithm may be used. Since the loading has been transformed to a trigonometric function the solution to (3) becomes a straightforward matter. The particular integrals, which correspond to the trigonometric driving functions, give the steady-state responses, while the complementary functions give the evanescent transient responses. The solution is obtained in the form of a finite trigonometric series. If an input record is processed to appear as a trigonometric series it is usually found that significant frequencies will reside within some band, in the case of earthquake records this might typically be .1 → 50 Hertz. Moreover the structure whose dynamic response is required may have insignificant response to frequencies outside that band. In such circumstances a worthwhile saving in computation can be achieved by simply truncating the interpolation formula. The formula then loses its interpolation property in that it is no longer exact at the data points. Nevertheless it can remain an excellent approximation.

The second solution method subjects the input data to a local interpolation process and the continuous function so produced is solved exactly. The only source of error lies in the interpolation process itself and this can be made fairly accurate. The local interpolations presented are superior to those of the piece-wise linear and parabolic variety and there-fore give superior results whatever the mode of solution of the differential equation.

The local interpolation methods use the theory of sampled-data linear systems for which the Laplace transform, specially adapted for discrete systems, plays a significant role. The theory of sampled-data systems is concerned with the behaviour of systems which include continuous-time elements but have, in addition, at one or more stages, functions which are discrete in nature, i.e. pulsed sequences. This is indeed the case which confronts us. We have an input loading supplied as a time series, and also a continuous time-element which is the system transfer function $F(s)$ of equation (3), (expressed in the Laplace s domain as $1/(s^2 + 2\beta s + \overline{\omega^2})$). The output to be determined will also be specified as a time-series. It is inconceivable that the input pulse sequence should be applied directly to the system whose transfer function is $F(s)$. Instead one introduces an interpolator which converts the pulse sequence to a continuous function, intended to resemble the original loading. Considerations in the frequency domain indicate that an interpolator should approximate an ideal low-pass filter.

Now the complete system consisting of discrete input, inter-polator, transfer function and discrete output can be studied through the use of z transform theory; such transforms have a convenient relation to Laplace transforms. The application of z transforms provide the final response solutions in the form of simple recurrence relations, which are incidentally convenient for use with an electronic hand calculator.

THE GLOBAL TRIGONOMETRIC INTERPOLATION SOLUTION

Consider a sequence of input data $X(j)$, such as that which might be sampled from the time-dependent loadings on a structural system, having the following range:

$$j = -N \ -(N-1), \ \ldots, \ (N-1), \ N \ .$$

Then it may readily be shown that for these $2N + 1$ interger values of j, the following transform pair holds:

$$X(j) = \sum_{n=-N}^{N}{}' \ c(n)e^{inj\pi/N}$$

where

$$c(n) = \frac{1}{2N} \sum_{j=-N}^{N}{}' \ X(j)e^{-inj\pi/N}, \ (-N \leq n \leq N) \tag{4}$$

The dashes on the transform pairs indicate that the two extreme members of each sum are taken with half weight.

In general $C(n)$ is a complex number, but equation (4) reveals that $C(O)$ and $C(N)$ are real. It can also be seen that $C(-n) = C(n)$, the complex conjugate, which implies that $C(-n)e^{inj\pi/N}$ is the complex conjugate of $C(n)e^{inj\pi/N}$. Putting $C(n) = E(n) + iF(n)$, then:

$$X(j) = E(O) + \sum_{n=1}^{N-1} 2Re\left\{E(n) + iF(n)\right\}\left\{\cos(nj\pi/N) + i\,\sin(nj\pi/N)\right\}$$
$$+ E(N)\,\cos(N\pi j/N) \tag{5}$$

which may be rewritten:

$$X(j) = \sum_{n=0}^{N} A(n)\,\cos(nj\pi/N) + \sum_{n=1}^{N-1} B(n)\,\sin(nj\pi/N) \tag{6}$$

where $A(n) = 2E(n)$ and $B(n) = -2F(n)$. We now have $X(j)$ expressed as the sum of an even function and an odd function, defined for integer j. It is easy to express $X(j)$ as the sum of an even sequence and an odd sequence, in particular:

$$Y(j) = \left\{X(j) + X(-j)\right\}/2,\; Z(j) = \left\{X(j) - X(-j)\right\}/2 \tag{7}$$

so that $X(j) = Y(j) + Z(j)$. In addition from (4) $C(n)$ may be expanded as:

$$C(n) = \frac{1}{2N} \sum_{j=-N}^{N}{}' \; X(j)\left\{\cos(nj\pi/N) - i\,\sin(nj\pi/N)\right\}$$
$$= \frac{1}{2N}\left[X(O) + \sum_{j=1}^{N-1} \left\{X(j) + X(-j)\right\}\cos(nj\pi/N)\right.$$
$$+ \frac{1}{2}\left\{X(N) + X(-N)\right\}\cos(n\pi) - i\sum_{j=1}^{N-1}\left\{X(j) - X(-j)\right\}$$
$$\left. \sin(nj\pi/N)\right] \tag{8}$$

Since $C(n) = \frac{1}{2}\left\{A(n) - i\,B(n)\right\}$ we have:

$$\left. \begin{array}{c} A'(n) = \dfrac{2}{N}\sum_{j=0}^{N}{}' \; Y(j)\,\cos(nj\pi/N) \\[2mm] A(n)=A'(n),\,(n=1,2,\ldots,N-1);\; A(O)=\tfrac{1}{2}A'(O),A(N)=\tfrac{1}{2}A'(N) \end{array} \right\} \tag{9}$$

$$B(n) = \frac{2}{N} \sum_{j=1}^{N-1} Z(j) \sin(nj\pi/N) \quad n=1,2,\ldots,N-1 \quad (10)$$

So far we have relationships between the discrete sequences $X(j)$, $Y(j)$, $Z(j)$, $A(n)$ and $B(n)$. We now investigate the possibility of extending these ideas to the continous domain. We have:

$$Z(j) = \sum_{n=1}^{N-1} B(n) \sin(nj\pi/N) \quad (11)$$

where the $B(n)$ are given by (10). Consider the continuous function defined by:

$$z(t) = \sum_{n=1}^{N-1} B(n) \sin(n\pi t/NT) \quad (12)$$

where T is the time interval between adjacent members of the time series $X(j)$. At once we have:

$$z(jT) = Z(j) \quad j = -N, -(N-1),\ldots,(N-1), N$$

Provided that the boundary conditions $Z(O) = Z(N) = 0$ hold, $z(t)$ is the interpolation of a hypothetical function specified at the finite points jT. A general comment on trigonometical interpolation can be made. Consider the situation where a function $f(t)$ is given as a continuous function, but the integrals required for a standard Fourier analysis cannot be conveniently determined. Under such circumstances it is possible to sample the function to produce $f(jT)$ and then use trigonometric interpolation as an approximation to $f(t)$. In the case of a function which is suitably band-limited the approximation given by trigonometric interpolation may be exact.

Parallel to equation (12) we may write:

$$y(t) = \sum_{n=0}^{N} A(n) \cos(n\pi t/NT)$$

with

$$y(jT) = Y(j)$$

$$(13)$$

We may now write the interpolation formula:

$$x(t) = y(t) + z(t) \quad (14)$$

At this stage it is important to raise a matter so far neglected. Expression (12) necessarily implies that $Z(N) = 0$. However the application of (7) will in general yield a non-zero $Z(N)$. It is possible to overcome this difficulty by subtracting a straight line through the origin from $Z(j)$. Explicitly:

$$Z'(j) = Z(j) - (j/N)\, Z(N) \qquad (15)$$

hence the condition $Z'(N) = 0$ is observed and thus sine interpolation may be applied to $Z'(j)$. Equation (14) is now replaced by:

$$x(t) = y(t) + z'(t) + (t/NT)\, Z(N) \qquad (16)$$

From (13), (12) and (16) we obtain:

$$x(t) = \sum_{n=0}^{N} A(n)\cos(n\pi t/NT) + \sum_{n=1}^{N-1} B(n)\sin(n\pi t/NT)+(t/NT)Z(N) \qquad (17)$$

This formula applies for $- NT \le t \le NT$, so that the point $t = 0$ is at the centre of the range. It is more convenient in practice to have the point $t = 0$ at the beginning of an interval especially when initial conditions are involved. Accordingly the following transformation is used:

$$t = \tau - NT \qquad (18)$$

it is easy to show that (17) may be rewritten as:

$$\left.\begin{aligned}
x(\tau) &= \sum_{n=0}^{N} \hat{A}(n)\cos(n\pi\tau/NT)+ \sum_{n=1}^{N-1} \hat{B}(n)\sin(n\pi\tau/NT)+(\tau/NT-1)Z(N) \\
\text{where} \quad \hat{A}(n) &= (-1)^{n} A(n) \quad, \quad 0 \le \tau \le 2NT \\
\hat{B}(n) &= (-1)^{n} B(n)
\end{aligned}\right\} \quad (19)$$

Solution of the modal equations

The solution of the modal equation (3) proceeds as follows. The right-hand-side irregular loading p_r, which has been specified as a discrete time-series, is replaced by an interpolation similar to (19). The coefficients of the interpolation are generated by using the discrete data of the input loading. Since the interpolation formula is a linear combination of the functions $\cos(n\alpha\tau)$, $\sin(n\alpha\tau)$ and $a + b\tau$, where $\alpha = \pi/NT$ the solution of (3) becomes a straightforward matter. The particular integrals of (3) corresponding to the cosine, sine and linear terms are:

$$\left.\begin{aligned}
&\frac{(\omega^2-n^2\alpha^2)\cos(n\alpha\tau)+2\beta n\alpha\,\sin(n\alpha\tau)}{(\omega^2-n^2\alpha^2)^2 + (2\beta n\alpha)^2}, \frac{(\omega^2-n^2\alpha^2)\sin(n\alpha\tau)-2\beta\alpha n\,\cos(n\alpha\tau)}{(\omega^2-n^2\alpha^2)^2 + (2\beta n\alpha)^2} \\[2mm]
&\frac{a}{\omega^2} - \frac{2\beta b}{\omega^4} + \frac{b\tau}{\omega^2}
\end{aligned}\right\} \quad (20)$$

The complementary function of (3) has the form:

$$e^{-\beta\tau}(\kappa_1 \cos c\,\tau + \kappa_2 \sin c\,\tau) \qquad (21)$$

where $c^2 = \omega^2 - \beta^2$. The complete solution to (3) may thus be written:

$$w_{r}(\tau) = \sum_{n=1}^{N-1} \hat{\theta}_1(n)\cos(\alpha n\tau) + \sum_{n=1}^{N-1} \hat{\theta}_2(n)\sin(\alpha n\tau) + \hat{A}(0)$$

$$+ \hat{A}(N)\, \frac{\left\{\psi_1(N)\cos\left[\frac{\pi\tau}{T}\right] + \psi_2(N)\sin\left[\frac{\pi\tau}{T}\right]\right\}}{\psi_1(N)^2 + \psi_2(N)^2} + e + \frac{\beta\tau}{\omega^2}$$

$$+ e^{-\beta\tau}\left\{\kappa_1\cos(c\tau) + \kappa_2\sin(c\tau)\right\}$$

where

$$\hat{\theta}_1(n) = \frac{\hat{A}(n)\psi_1(n) - \hat{B}(n)\psi_2(n)}{\psi_1(n)^2 + \psi_2(n)^2}, \quad \hat{\theta}_2(n) = \frac{\hat{A}(n)\psi_2(n) + \hat{B}(n)\psi_1(n)}{\psi_1(n)^2 + \psi_2(n)^2}$$

$$\psi_1(n) = \omega^2 - n^2\alpha^2, \quad \psi_2(n) = 2\beta n\alpha, \quad e = \frac{a}{\omega^2} - \frac{2\beta b}{\omega^4} \quad (22)$$

The constants κ_1 and κ_2 may be determined by initial conditions on displacement and velocity. Equation (22) gives the displacements as continuous functions of τ and is therefore a general analytical representation of the response which incidentally may be differentiated to yield expressions for the velocities and accelerations.

To evaluate the sums appearing in equations (9), (10) and (22) it is natural to use one of the economical algorithms which have appeared in recent years. For long series it is possible to adapt the fast Fourier transform method used for discrete Fourier transforms (Collings and Saunders, 1978). For smaller series the splitting into even and odd series allows a simpler approach to be adopted. It can be shown (Lanczos, 1957), that the symmetry and periodic properties of trigonometric functions can be exploited to reduce the number of multiplications by a factor of four against direct evaluation of the sum. However, it is still necessary to generate a table of cosines and sines. It is considered that in many applications the method of evaluation of finite trigonometric series due to Goertzel (Goertzel, 1958) has much in its favour. This method uses recursion and does not require the prior setting up of tables. It is a straightforward algorithm and is very easily implemented. In a recent paper (Collings and Saunders, 1978) the Goertzel method, in a form due to Hamming (Hamming, 1973), is presented. In addition an ALGOL 60 procedure for summing the trigonometric series is given.

The input record is usually a long series and the possible

advantages of dividing the record into consecutive sections may
be considered.   To be specific suppose the earthquake record
to be supplied as 500 values at intervals of $T = 1/40$ second.
It would be possible to find a single interpolation for the
entire record.   However, such a procedure would ignore the
frequency spectrum of the record.   For example suppose that
the amplitudes are small for frequencies below 0.5 Hertz.
The lowest frequency representable by a section of M data with
interval T is $1/MT$ Hertz.   These bounds for minimum and
maximum representable frequencies provide a basis upon which
the section length may be chosen.

In the paper (Collings and Saunders, 1978), there is given a
detailed example of the application of trigonometric interpolation for the solution of a problem involving the seismic
excitation of a structural system.

THE LOCAL INTERPOLATOR METHODS

Local interpolators use a finite band of discrete data values
to produce a continuous function which is exact at the data
points and is a good approximation elsewhere.   In order to
apply standard mathematical techniques, such as integration, it
is expedient to replace each value $x(nT)$ of the input data by
the equivalent pulse $x(nT)\delta(t-nT)$, where $\delta(t)$ is the Dirac
delta function.   The pulse sequence is therefore written
$x^*(t) = x(t) \sum_{n=0}^{\infty} \delta(t - nT)$.   This pulse sequence enters a
system which is known as an *interpolator*, the output of which
is a continuous function, exact at $t = nT$ and a good approximation elsewhere.   Note that the interpolator has a transfer
function just as any system does: we use the symbol $Q(s)$,
which stands for Laplace transform of the response $q(t)$ of the
interpolator to a unit pulse $\delta(t)$.   The system we are dealing
with is characterised by the differential equation (3) with
transfer function $F(s)$.   Its input is the output of the interpolator.   *Figure 1* depicts this situation, and shows *samplers*

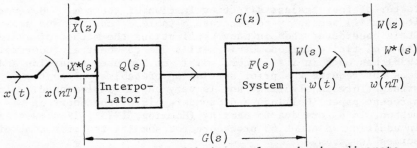

Figure 1 - Linear system with interpolator having discrete
input and output

at each end. The output sampler is present because we wish to represent the output as a discrete series. By the convolution theorem the transfer function between the samplers is:

$$G(s) = Q(s) \ F(s) \tag{23}$$

However our interest is the relation between the sampled input $x(nT)$ and the sampled output $w(nT)$. Accordingly we try to formulate a *pulse transfer function*, derivable from $G(s)$, to relate the input and output sequences. This distinct function shall be given the same symbol, but is a function of a special variable z: thus we write $G(z)$.

At this stage we introduce the z transform of a series. Thus:

$$X(z) = \sum_{n=0}^{\infty} x(nT)z^{-n} \tag{24}$$

(the two-sided transform has the lower limit $n = -\infty$). This transform is as fundamental to discrete systems theory as are the Laplace and Fourier transforms in the continuous domain. The z transform for the output $w(nT)$ is defined in the same manner. It can be shown (Ragazzini and Franklin, 1958, pp. 67–68) that:

$$W(z) = G(z) \ X(z) \tag{25}$$

The strategy which leads to the creation of computational algorithms requires the following steps:

1. Establish a high-quality interpolator $Q(s)$
2. Derive $G(z)$ from $Q(s) \ F(s)$
3. Convert equation (25) to a recursive algorithm on the input data $x(nT)$

Regarding step 1 it is proper to develop criteria for interpolators. This is best achieved in the frequency domain. Let the Fourier transform of the original continuous input be $X(i\omega)$. We regard $x(t)$ as defined (though zero) for $t < 0$. Thus the Fourier transform takes standard form:

$$X(i\omega) = \int_{-\infty}^{\infty} x(t)e^{-i\omega t} \ dt \tag{26}$$

The first step is to find a Fourier series for the unit pulse sequence $\sum_{n=-\infty}^{\infty} \delta(t-nT)$. This sequence is really a periodic function defined in the range $(-T/2, \ T/2)$, and having period T. The nth Fourier coefficient of this series is:

$$\frac{1}{T} \int_{-T/2}^{T/2} \delta(t) e^{-2\pi int/T} \, dt = \frac{1}{T} \tag{27}$$

Therefore, in a formal sense:

$$\sum_{n=-\infty}^{\infty} \delta(t-nT) = \frac{1}{T} \sum_{n=-\infty}^{\infty} e^{2\pi int/T} \tag{28}$$

The Fourier transform of $x^*(t)$ is thus:

$$X^*(i\omega) = \frac{1}{T} \int_{-\infty}^{\infty} x(t) \sum_{n=-\infty}^{\infty} e^{-i(\omega-2\pi n/T)t} \, dt$$

$$= \frac{1}{T} \sum_{n=-\infty}^{\infty} \int_{-\infty}^{\infty} x(t) e^{-i(\omega-2\pi n/T)t} \, dt$$

i.e. $\quad X^*(i\omega) = \frac{1}{T} \sum_{n=-\infty}^{\infty} X\{i(\omega-2\pi n)/T\} \tag{29}$

This expression shows that $X^*(i\omega)$ is an infinite collection of $X(i\omega)$ functions spaced at intervals for ω of $2\pi/T$.

When data is sampled to provide a discrete series it is usual to choose for T a value small enough to ensure that $1/2T$ is not less than the highest frequency present (Bath, 1974). The data has a finite time span, and we take the interval in which $x(t)$ is non-zero to be $(0, s)$. The general coefficient for a Fourier expansion of $x(t)$ is:

$$C_n = \frac{1}{s} \int_{0}^{s} x(t) e^{-2\pi int/s} \, dt \tag{30}$$

The Fourier transform of $x(t)$ is:

$$X(i\omega) = \int_{0}^{s} x(t) e^{-i\omega t} \, dt \tag{31}$$

Therefore $C_n = \frac{1}{s} X(i2\pi n/s)$. The fundamental frequency is $1/s$, and now let the highest frequency correspond to $n = N$ where $N/s = 1/2T$. It then follows from (30) that $C_N = \frac{1}{s} X(i\pi/T)$. It may therefore be inferred that $X(i\omega)$ vanishes for $|\omega| > \pi/T$. If it is the case that $X(i\omega)$ does not precisely have the property of band-limitation then discussion which follows is still apposite, the only errors being mild effects known as "aliasing" (Bath, 1974).

Now if it is true that $X(i\omega)$ has the above-mentioned feature of of being non-zero only in the band $-\pi/T \leq \omega \leq \pi/T$, then there is no overlap of the "images" in the right-hand-side of (29). Suppose now that the function $x^{\star}(t)$ is subjected to an ideal low-pass filter with gain T and bandwidth $(-1/2T, 1/2T)$ in frequency terms. Since $\omega = 2\pi$ x frequency, the ideal-low-pass filter we contemplate has a frequency response (Fourier transform) satisfying:

$$Q(i\omega) = \begin{cases} T, & -\pi/T \leq \omega \leq \pi/T \\ 0 & |\omega| > \pi/T \end{cases} \tag{32}$$

Referring now to (32) this low-pass filter after acting on $x^{\star}(t)$ eliminates every term on the right-hand side of (29) except that corresponding to $n=0$. The latter term is just $X(i\omega)$, from which we can infer that the ideal low-pass filter will convert its pulse-sequence input into the original function $x(t)$, provided T is small enough. If the data were supplied with a T value a little larger than desirable, the ideal low-pass filter will produce an acceptable approximation to $x(t)$.

We now have established one criterion for an interpolator: its frequency response must be a good approximation to that of a low-pass filter with pass-band ending at $\omega = \pi/T$.

The interpolator may be regarded as a system which has a response $q(t)$ to a delta function $\delta(t)$. Since for practical interpolators $q(t)$ is non-zero for t both negative and positive the word *response* could be misleading, (since the response is anticipatory). Therefore the term *weighting function* is sometimes preferred. The response to $\delta(t-nT)$ is clearly $q(t-nT)$. The general interpolator operates on a pulse sequence to produce a continuous output $y(t)$ given by superposition of pulse responses:

$$y(t) = \sum_{n=-\infty}^{\infty} x(nT)\, q(t-nT) \tag{33}$$

The weighting function is the inverse Fourier transform of $Q(i\omega)$, and for the ideal low-pass filter defined above is given by:

$$q(t) = \frac{1}{2\pi} \int_{-\pi/T}^{\pi/T} T\, e^{i\omega t}\, d\omega = \frac{\sin(\pi t/T)}{\pi t/T} \tag{34}$$

Given a general weighting function $q(t)$ the corresponding two-sided Laplace transform is $Q(s)$. We now investigate the nature of the corresponding z transform $Q(z)$. Since $y(t)$ of equation (33) is an interpolation we have $y(nT) = x(nT)$, which implies $Y(z) = X(z)$. The convolution theorem, equation (25), applied to (33) yields $Q(z) = 1$. *The problem of finding a practical interpolator is thus equivalent to finding a low-pass filter for which* $Q(z) = 1$.

Some interpolator designs

One approach is to take the transfer function (Cruickshank, 1961), as $Q(s) = F_{k+1}(z)/s^{k+1}$, k an integer. The condition $Q(z) = 1$ implies that $F_{k+1}(z)$ is the reciprocal of the z transform associated with $1/s^{k+1}$ (to be explained shortly).

For the case $k=1$, $Q(s) = z(1-z^{-1})/Ts^2$ the weighting fuction $q(t)$ has a triangular appearance and consists of two lines, one bounded to the points $(-T, 0)$ and $(0, 1)$ while the other is bounded by $(0, 1)$ and $(T, 0)$. Each impulse in the input produces intersecting lines to the effect that piece-wise linear interpolation is the result. Choosing higher values for k leads to unstable interpolators.

Although we introduce superior interpolators we shall consider the linear interpolator as an example of the general procedure. The overall system function (see *Figure 1*) is:

$$G(s) = Q(s) \, F(s) = \frac{z(1-z^{-1})^2}{T} \frac{1}{s^2(s^2 + 2\beta s + \overline{\omega}^2)} \qquad (35)$$

Incidentally $z = e^{sT}$, the shifting operator of Laplace transforms. The term on the extreme right of the above equation requires to be expressed as a z transform. Consider for a simple example $1/s^2$. The inverse Laplace transform is t. The sampling of t produces the sequence $0, T, 2T, \ldots$ and the z transform is $0 + T z^{-1} + 2T z^{-2} + \ldots$ which equals $T z^{-1}/(1-z^{-1})^2$. Pairs of such associated z and Laplace transforms have been tabulated, e.g. (Ragazzini and Franklin, 1958, Appendix I).

The partial fraction expansion can be written:

$$\frac{1}{s^2(s^2 + 2\beta s + \overline{\omega}^2)} = \frac{C}{s} + \frac{D}{s^2} + \frac{A_s + B}{(s + \beta)^2 + c^2} \qquad (36)$$

where $c^2 = \overline{\omega}^2 - \beta^2$ and the constants are given follows $D = 1/\overline{\omega}^2$, $C = -2\beta D/\overline{\omega}^2$, $A = -C$ and $B = -(2\beta C + D)$. We now transform the right-hand-side of (36) to the z domain; this is accomplished by the use of tables. Hence:

$$\left\{\frac{1}{s^2(s^2+2\beta s+\bar{\omega}^2)}\right\} = \frac{A + Q\,z^{-1}}{1-q_1 z^{-1} + q_2\, z^{-2}} + \frac{C}{1-z^{-1}} + \frac{D\,T\,z^{-1}}{(1-z^{-1})^2}$$

where

$$Q = \left\{(B - A\beta)/c\right\}e^{-\beta T}\sin(cT) - A\,e^{-\beta T}\cos(cT)$$

$$q_1 = 2e^{-\beta T}\cos(cT), \quad q_2 = e^{-2\beta T}$$

(37)

Equation (35) is thus transformed to:

$$G(z) = \frac{z(1-z^{-1})^2}{T}\left\{\frac{A + Q\,z^{-1}}{1-q_1\,z^{-1}+q_2\,z^{-2}} + \frac{C}{(1-z^{-1})} + \frac{D\,T\,z^{-1}}{(1-z^{-1})^2}\right\} \quad (38)$$

With appropriate manipulations equation (38) may be rearranged into a rational fraction of the form:

$$G(z) = \frac{p_0 + p_1\,z^{-1} + p_2\,z^{-2}}{T(1 - q_1\,z^{-1} + q_2\,z^{-2})}\quad,$$

where

$$p_0 = Q + C(1-q_1) + DT, \quad p_2 = Q - C\,q_2 + DT\,q_2\,,$$

$$p_1 = -2Q + C(q_1 + q_2 -1) - DT\,q_1$$

(39)

Since $G(z)$ is given explicitly by equation (39) the problem now becomes a matter of transforming the product $W(z) = G(z)\,X(z)$ expressed in the z domain to a recurrence relationship in the time domain. Clearing the rational fraction of (39) and transposing one obtains:

$$(1 - q_1\,z^{-1} + q_2\,z^{-2})\,W(z) = (p_0 + p_1\,z^{-1} + p_2\,z^{-2})/T\,X(z) \quad (40)$$

Because z^{-1} is an ordering variable we may write the following recurrence relationship:

$$w_n = q_1\,w_{n-1} - q_2\,w_{n-2} + \left\{p_0\,x_n + p_1\,x_{n-1} + p_2\,x_{n-2}\right\}/T \quad (41)$$

Note that for w_0 and w_1 some of the suffices of equation (41) are negative and these terms are ignored.

The shortcomings of the linear interpolator are revealed in *Figure 2*, where it is seen that its frequency response falls too quickly in comparison with the ideal low-pass filter.

In the search for better interpolators a novel set was discovered

(Brown, 1965). These were derived from the classical Lagrange polynomial interpolation and most contained two major terms. (The discussion which follows is an abbreviated form of a detailed development given in (Saunders and Collings, 1979).) We generalised one of Brown's interpolators as follows:

$$Q(s) = \frac{F_1(z)(1-z^{-1})^4}{T^3 s^4} + \frac{F_2(z)(1-z^{-1})^2}{T s^2} \tag{42}$$

One basis of this formulation is that $\underset{s \to 0}{\text{Lim}} \dfrac{(1-e^{-Ts})^k}{k}$ is finite, and if s is replaced by $i\omega$ it follows that the zero frequency response is finite. Referring to tables for associated z transforms of $1/s^4$ and $1/s^2$, we find that the condition $Q(z) = 1$ reduces to,

$$\frac{1}{6} F_1(z)\left\{z^{-1} + 4z^{-2} + z^{-3}\right\} + z^{-1} F_2(z) = 1 .$$

Putting $F_1(z) = az^2$; $F_2(z) = b_2 + c_2 z + d_2 z^2$ and solving we obtain that $b_2 = d_2 = -a/6$ and $c_2 = (6 - 4a)/6$. We thus have a family of interpolators with parameter a. For this family:

$$Q(s) = \frac{az^2(1-z^{-1})^4}{T^3 s^4} - \frac{\{a - (6-4a) z + az^2\}(1-z^{-1})^2}{6 T s^2} \tag{43}$$

The frequency response is:

$$Q(i\omega) = T\left[a\left(\frac{\sin\left(\frac{\omega T}{2}\right)}{\frac{\omega T}{2}}\right)^4 - \frac{(2a \cos(\omega T) - (6-4a))}{6}\left(\frac{\sin\left(\frac{\omega T}{2}\right)}{\frac{\omega T}{2}}\right)^2\right] \tag{44}$$

The case $a = 0$ gives the linear interpolator, and as a is increased the frequency response improves. When a is raised beyond 1 the response exceeds T before returning. A small amount of such ripple in the pass band has the advantage of a more rapid transition to the stop band. We selected $a = 1.5$, and as *Figure 2* shows this gives an excellent frequency response. The *overshoot* is acceptably small at $1.017\ T$.

To practically implement this interpolator the steps are the same as for the linear interpolator, though they are much more complicated, due largely to the presence of two major terms in (42). One finally develops a formula of the form:

$$G(z) = \frac{p_0 z + p_1 + p_2 z^{-1} + p_3 z^{-2} + p_4 z^{-3}}{1 - q_1 z^{-1} + q_2 z^2} \tag{45}$$

The resolution of this result into a usable algorithm follows the same course as for the linear interpolator. A special condition is that it is necessary to have $x_0 = 0$. The algorithm is:

$$w_n = q_1\, w_{n-1} - q_2\, w_{n-2} + p_0\, x_{n+1} + p_1\, x_n + p_2\, x_{n-1} + p_3\, x_{n-2} \quad (46)$$

For $n = 0,1$, the results are inaccurate; and also, terms of negative suffix are to be made zero. Numerical details of the $a = 1.5$ algorithm are given later.

A superior though more complex set of algorithms can be derived from the assumption:

$$Q(s) = \frac{f_1(z)(1-z^{-1})^4}{T^3\, s^4} + \frac{f_2(z)(1-z^{-1})^3}{T^2\, s^3} \quad (47)$$

with

$$f_1(z) = a_1 z + b_1 z^2 + c_1 z^3 \,, \quad f_2(z) = a_2 + b_2 z + c_2 z^2 + d_2 z^3$$

The $Q(z) = 1$ assumption leads to

$$\frac{1}{6}(z^{-1}+4z^{-2}+z^{-3})\, f_1(z) + \frac{1}{2}(z^{-1}+z^{-2})\, f_2(z) = 1 \,.$$

The constants b_1 and a_2 can be taken as basic, and the remaining constants expressed in terms of them. In order that $Q(i\omega)$ be real it is necessary that $a_1 = c_1$, $a_2 = d_2$, $b_2 = c_2$. All these conditions are simultaneously satisfied by the condition $b_1 + 6a_2 = 3$. With this condition all parameters can be expressed in terms of b_1, from which we shall now drop the suffix.

The transfer function of the general b interpolator is:

$$Q(s) = \frac{[-\{(3-b)/2\}(z+z^3)+bz^2](1-z^{-1})^4}{T^3\, s^4}$$

$$+ \frac{\{(3-b)(1+z^3)+(9-5b)(z+z^2)\}(1-z^{-1})^3}{6T^2\, s^3} \quad (48)$$

We have calculated the frequency response for b ranging from 0 to 4. As b is reduced from 4 the frequency response improves. The case $b = 3$ corresponds to a Lagrange case (Brown, 1965). When b is reduced below 1.7 the response initially rises as ω is increased from zero. If such a ripple is thought undesirable then $b = 1.8$ is the choice to make. A generally flatter passband and a steeper transition to the stop-band are obtainable if a ripple is permitted. We recommend $b = 0.6$, for which the "overshoot" is to $1.013T$.

For the b interpolators the pulse transfer function takes the form:

$$\sigma = \sqrt{\bar{\omega}^2 - \beta^2} \ , \quad q_1 = 2e^{-\beta T}\cos(\sigma T) \ , \quad q_2 = e^{-2\beta T}$$

$$A = \frac{4\beta}{\bar{\omega}^6 T^3}\left[\frac{2\beta^2}{\bar{\omega}^2} - 1\right] \ , \quad C = \frac{1}{\bar{\omega}^4 T^2}\left[\frac{4\beta^2}{\bar{\omega}^2} - 1\right] \ , \quad D = -\frac{2\beta}{\bar{\omega}^4 T} \ , \quad E = \frac{1}{\bar{\omega}^2} ;$$

$$B = \frac{e^{-\beta T}}{\bar{\omega}^4 T^3}\left\{\left[\frac{8\beta^4}{\bar{\omega}^4} - \frac{8\beta^2}{\bar{\omega}^2} + 1\right]\frac{\sin(\sigma T)}{\sigma} - \left[\frac{8\beta^3}{\bar{\omega}^4} - \frac{4\beta}{\bar{\omega}^2}\right]\cos(\sigma T)\right\} \ ,$$

$$F = \frac{e^{-\beta T}}{4\bar{\omega}^2 T}\left\{\left[\frac{2\beta^2}{\bar{\omega}^2} - 1\right]\frac{\sin(\sigma T)}{\sigma} - \frac{2\beta}{\bar{\omega}^2}\cos(\sigma T)\right\},$$

$$G = \frac{e^{-\beta T}}{\bar{\omega}^4 T^2}\left\{\left(3\beta - \frac{4\beta^3}{\bar{\omega}^2}\right)\frac{\sin(\sigma T)}{\sigma} + \left[\frac{4\beta^2}{\bar{\omega}^2} - 1\right]\cos(\sigma T)\right\} \ .$$

The coefficients for the $a = 1.5$ interpolator solution, equation (46), are as follows:

$$P_0 = 1.5A(q_1 - 1) + 1.5B + 1.5C + .25D(2 + q_1) - F \ .$$

$$P_1 = 1.5A(3 - 3q_1 - q_2) - 6B - 1.5C(2 + q_1) + .25D(1 - 4q_1 - q_2) + E + 2F \ ,$$

$$P_2 = 4.5A(q_1 + q_2 - 1) + 9B + 1.5C(1 + 2q_1 + q_2) + .25D(q_1 + 4q_2 - 4) - Eq_1 - 2F \ ,$$

$$P_3 = 1.5A(1 - q_1 - 3q_2) - 6B - 1.5C(q_1 + 2q_2) + .25D(1 + 2q_1 - q_2) + Eq_2 + 2F \ ,$$

$$P_4 = q_2(1.5A + 1.5C - .5D) + 1.5B - F \ .$$

The coefficients for the $b = .6$ interpolator solution, equation (50), are as follows:

$$P_0 = 1.2A(1 - q_1) - 1.2B - .4C(2 + q_1) - .2D + .4G \ .$$

$$P_1 = A(4.2q_1 + 1.2q_2 - 4.2) + 5.4B + C(3.2 + q_1 + .4q_2) + D(.9 + .2q_1) - .2G \ ,$$

$$P_2 = A(6.6 - 6.6q_1 - 4.2q_2) - 10.8B - C(4.2 + 2.4q_1 + q_2) - D(.9q_1 + .2q_2) + E - .8G \ ,$$

$$P_3 = 6.6A(q_1 + q_2 - 1) + 13.2B + C(2.4 + 4.2q_1 + 2.4q_2) - .9D(1 + q_2) - Eq_1 \ ,$$

$$P_4 = A(4.2 - 4.2q_1 - 6.6q_2) - 10.8B - C(1 + 3.2q_1 + 4.2q_2) + D(.2 - .9q_1) + Eq_2 + .8G \ ,$$

$$P_5 = A(1.2q_1 + 4.2q_2 - 1.2) + 5.4B + C(.4 + .8q_1 + 3.2q_2) - D(.2q_1 + .9q_2) + .2G \ ,$$

$$P_6 = -1.2Aq_2 - 1.2B - .8Cq_2 + .2Dq_2 - .4G \ .$$

Table 1 — Coefficients for the $a = 1.5$ and $b = .6$ interpolator solutions

$$G(z) = \frac{p_0 z^2 + p_1 z + p_2 + p_3 z^{-1} + p_4 z^{-2} + p_5 z^{-3} + p_6 z^{-4}}{1 - q_1 z^{-1} + q_2 z^{-2}} \tag{49}$$

The corresponding recursive algorithm is:

$$w_n = q_1 w_{n-1} - q_2 w_{n-2} + p_0 x_{n+2} + p_1 x_{n+1} + \ldots + p_6 x_{n-4} \tag{50}$$

For $n \leq 4$ terms with negative suffices are ignored. It is necessary to start with $x_0 = x_1 = 0$, and w_0, w_1 and w_2 are inaccurate.

The weighting function of the a and b interpolators amount to cubic splines. They resemble in shape the central portion of the curve $\sin(\pi t/T)/(\pi t/T)$. For the a interpolators $q(t)$ extends from $-2T$ to $2T$, while for the b interpolators it extends from $-3T$ to $3T$. Consult *Table 1* for the numerical details of the $a = 1.5$ and $b = .6$ interpolator solutions.

Results of an investigation
Three different local interpolator methods, the linear, $(a=0)$, the $a = 1.5$ and $b = .6$, were used to solve the modal equation (3). In attempting to anticipate the performance of the different interpolator solutions *Figure 2, the frequency response of various interpolators,* is useful. Since the only errors which arise in interpolator solutions, (excluding *roundoff*), are those associated with the quality of the interpolation itself we would expect the more sophisticated interpolators to produce superior solutions. Based on the indications of *Figure 2* we would not expect the linear interpolator, because of its relatively poor frequency response, to produce solutions as accurate as the $a = 1.5$ or $b = .6$ interpolators. *Figure 2* shows that the interpolators have a pass-band for ω in the interval $(0, \pi/T)$, the upper frequency corresponding to $1/2T$, see equation (32), which is the Nyquist frequency. The discretisation interval T of the input time-series might typically be $1/100$, $1/40$ seconds, etc. For example, suppose it were $1/40$ second, then the corresponding Nyquist frequency would be 20 Hertz. The accurate solution of a modal equation by a local interpolator method depends upon relations between frequency response of the interpolator, the frequency spectrum of the input data, and the characteristic period and damping of the mode. *Figure 2* shows that in the neighbourhood of the transition from pass to stop-band the frequency response of the interpolators departs markedly from the ideal low-pass band shape. It may therefore be expected that there will be a loss of solution accuracy for modal equations with periodic constants in this region.

Incidentally because of the superior frequency domain characteristics of the $b = .6$ interpolator it should be this interpolator which gives most accurate solutions - which indeed it does. *Table 2* shows solutions to the modal equation by the

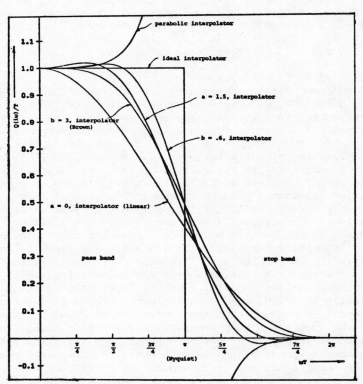

Figure 2 – The frequency response of various interpolators

Time $x \frac{1}{40}$ Seconds	Trigonometric Interpolation Solution $(M \times 10^{-6})$	Linear Interpolation Solution $(M \times 10^{-6})$	$a = 1.5$ Interpolation Solution $(M \times 10^{-6})$	$b = .6$ Interpolation Solution $(M \times 10^{-6})$
40	8.4132	6.5832	8.0605	8.2360
41	- 3.2733	- 3.2483	- 3.0752	- 3.0532
42	-15.9069	-14.2931	-15.7651	-15.9012
43	-12.9856	-12.6892	-13.2631	-13.1582
44	- 4.9568	- 6.0810	- 5.4260	- 4.9463
45	- 7.6861	- 6.8107	- 7.7321	- 7.5273
46	- 2.6734	- .5608	- 2.6325	- 2.9273
47	24.7501	24.1531	24.9021	24.5794
48	46.7381	43.4151	47.2850	47.2852
49	32.4656	30.7545	32.4881	32.4363
50	- .8119	2.1795	- .2884	- 1.0667
51	- 2.6714	- 2.3703	- 2.6193	- 2.4823
52	- .6456	- 3.7306	- 2.0394	- .7284
53	-30.3139	-26.3841	-30.1671	-30.5770
54	-18.6251	-15.0634	-17.6205	-18.6693
55	29.6142	22.8920	29.1545	30.2602
56	8.0095	4.5718	6.7661	7.6382
57	-48.6391	-38.6081	-46.7923	-49.0681
58	- 2.9232	- 3.3749	- 1.8663	- 2.0507
59	38.0581	24.6172	34.3162	37.6078
60	-57.4862	-48.2221	-56.2361	-57.9147

Table 2 — Results of various interpolator solutions

global trigonometric interpolation and the three local interpolation methods. The solutions range over .5 seconds and the
modal equation has natural frequency and critical damping constants of $\bar{\omega}$ = 71.8 rad/sec and ζ = 1.4% respectively. The
loading was specified in the form of a discrete time series with
interval T = 1/40 second. The trigonometric interpolation
method gives for input data sufficiently band-limited an exact
solution. Further if the condition of band-limitedness holds
then the trigonometric interpolation will have a frequency
response corresponding to the ideal low-pass band filter.
For these reasons the trigonometric interpolation solution may
be taken as a basis for comparison for the local interpolator
methods. The expectation that the linear interpolator would be
least accurate, the α = 1.5 interpolator of intermediate quality
with the b = .6 the best, was fulfilled. For the particular
modal equation solved it is clear that linear interpolation
gives quite large errors, even though the method involves an
exact solution of the differential equation. This is because
the linear interpolation attenuates frequencies in the
neighbourhood of the undamped characteristic frequency of the
modal equation. This frequency 11.4 Hertz corresponds to
ωT = .57π and reference to *Figure 2* with this abscissa clearly
indicates the inadequacy of linear interpolation.

REFERENCES

Bath, M. (1974) 'Spectral Analysis in Geophysics', Elsevier

Brown, B.M. (1965) 'The Mathematical Theory of Linear Systems',
Chapman and Hall, London

Collings, A.G. and Saunders, L.R. (1978) 'Use of Trigonometric
Interpolation for the Analytical Determination of the Dynamic
Response of Linear Systems to Arbitrary Inputs', Eng. Struct.
1, 1 : 41-52

Cruikshank, A.J.O. (1961) 'Time Series and z Transform Methods
of Analysis of Linear and Non-Linear Control Systems', Proc.
First Int. Congress of Int. Fed. of Automatic Control, 277-285

Goertzel, G. (1958) 'An Algorithm for the Evaluation of Finte
Trigonometric Series', Am. Math. Monthly 65, 34-35

Hamming, R.W. (1973) 'Numerical Methods for Scientists and
Engineers', McGraw-Hill, New York

Lanczos, C. (1957) 'Applied Analysis', Pitman, London

Ragazzini, J.R. and Franklin, G.F. (1958) 'Sampled Data Control
Systems', McGraw-Hill

Saunders, L.R. and Collings, A.G. (1979) 'Some Local Interpolation Based Computational Algorithms for the Determination of
the Dynamic Response of Linear Systems Subjected to Arbitrary
Loadings' (to appear)

ANALYTICAL COMPUTATIONS OF DYNAMIC BEHAVIOUR OF PIN JOINTED STRUCTURES

G. Ballio, A. Gobetti and P. Zanon

Istituto di Scienza e Tecnica delle Costruzioni - Università di Pavia - Italy

THE PROBLEM

The design of steel and concrete structures in seismic zones, according to the actual codes, is based on the following concepts. First of all, the structure has to be designed to resist in the elastic range a seismic event characterized by a return period ΔT_o of the same order of magnitude as the design life ΔT_s of the structure itself. The second step consists of checking that the structure may evolve into plastic range with enough ductility to avoid collapse during a seismic event with a return period ΔT_o much greater than ΔT_s.

The first check minimizes damage during an earthquake that has some probability of occurring during the life of the structure. It may be carried out with dynamic calculations based on modal analysis or with a static approach, loading the structure with conventional horizontal forces equivalent to the seismic action. The second check has the purpose of safeguarding human life during a destructive earthquake, permitting at the same time important damage in the structure. It cannot be made by current structural analysis because it requires a dynamic elastic-plastic sophisticated approach. For this reason, at the present time, actual codes limit such a check to ensuring that the structure is able to support the strains related to the real displacements v that can occur during a seismic event. Real displacements v may be evaluated by means of the relation:

$$v = \phi v_e \qquad (1)$$

where v_e are the elastic displacements calculated during the first check and ϕ is an amplification factor that may assume a value in the range from 3 to 7 depending on the type of structure and on the numerical approach adopted for evaluating v_e. In such an approximate way codes lead the check of the structural behaviour during an exceptional seismic event back to the one for a normal earthquake.

On the other hand the designer must be sure that each structural element and each connection is able to permit such displacements before brittle collapse occurs.

Buildings with steel structures are usually characterized by an isostatic scheme or by a redundant one with a limited number of hyperstatic constraints. For this reason each member is designed for a precise static function and in many cases bracings are considered to be the only structural elements able to support seismic actions. In multi-storey buildings bracings are also supposed to provide overall stability of the structure and their topologies are often related to occupancy problems.
On the contrary mill buildings and, more generally, many one- or two-floor buildings, are differently conceived in the two main directions of the building itself. In the transverse section the structure often may be studied as a plane frame. For evaluating its behaviour in the plastic range it is compulsory to evaluate destabilizing effects of vertical actions (P-Δ effects) due to horizontal displacements caused by the earthquake. This may be done only in the field of hypotheses that allowed the European Convention of Steel Construction (CECM-ECCS,1976) to develop buckling curves for steel profile (Setti-Zandonini, 1977). In the longitudinal sense often bracings are disposed in order to support horizontal forces and to prevent horizontal displacements of the columns, supposed hinged both at the basis and at the top. In such a case destabilizing effects of vertical loads are negligible but ductility control becomes essential. It means that the designer has to ensure that the diagonal members of bracing and their connections have sufficient ductility to prevent brittle fracture during an exceptional seismic event. For example let us design by means of static equivalent forces of 10% of the vertical ones the bracing of Fig.1a, placed between columns of 6-8 m high at a distance of 6-12 m. If we suppose that, due to the slenderness, only the tension member is active, the requirement (1) with $\phi = 6$ ensures that the diagonal bars may support before collapse an elongation of 4-5 times that at the elastic limit. For other types of bracings (Fig.1b,c) the ratio between the required elongations and the elastic ones may be even greater and may reach values very near to those of co-efficient ϕ.

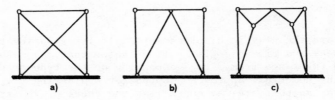

a)   b)   c)

Figure 1

Any kind of steel may reach such strains (of the order of 0.7-
1%) but, on the other hand, it is necessary to design connections in order to allow the tensioned members to yield before
the required elongation is reached.

More generally, it has to be observed that the behaviour of
structural members may be much more unfavourable than that of
the basis material. This is due, mainly, to the following reasons:
- the joint may be a weakening of the connected section and
  cause the break of the member before stresses reach yielding
  values;
- local stability of compressed parts of the section may reduce
  and sometimes eliminate the ductility of the member.

Fig.2a shows, in a schematic way, the relationship between the
applied load N and the total elongation ΔL of a tensioned steel
bar. Curve "1" deals with a bar connected with welded joints
designed for the ultimate value of the load acting in the bar.
This can reach the plastic range before collapse; so ductility
of the structural member (bar plus connections) is guaranteed.
Curve "2" refers to a bolted or welded connection that collapses
at the same value as the yielding load of the bar: the resistance
of the tensioned member is the same for the bar alone but no
important plastic deformations can occur. Finally Curve "3"
shows the behaviour of a bar connected with a bolted or welded
connection whose resistance is lower than that of the bar. For
both cases "2" and "3" only the end zones of the members reach
the plastic range while the bar maintains itself in the elastic
range: ductility is thus very poor.
Fig.2b shows the relationship between axial load and axial
elongation of a compressed bar. Curve "1" refers to a bar for

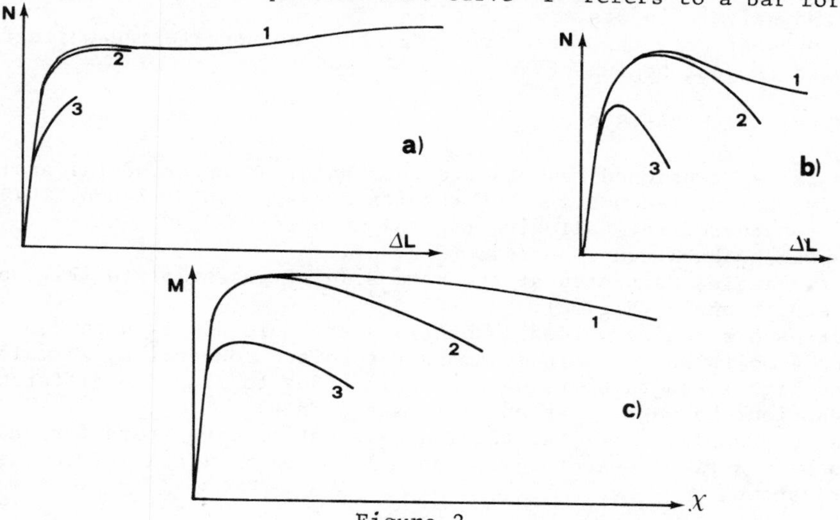

Figure 2

which local instability does not occur while curve "2" refers to
a compressed member in which parts of the section are subjected
to local instability but for values of deformation greater than
the ones that correspond to overall stability. Finally curve "3"
refers to a compressed bar for which overall and local stability
interact with each other, reducing the local carrying capacity.
In the first and second case ductility may be sufficient, in the
third case often a brittle collapse appears.
In Fig.2c different behaviours of bent bars are shown, referring
to the relationship between bending moment M and curvature χ .
For limited values of b/t ratios between the width b and the
thickness t of the outstanding flanges of the compressed part of
the section we have curve "1" that allows for a high degree of
ductility, necessary for the maintenance of plastic moment in
the plastic hinges till the collapse mechanism occurs. For b/t
ratios greater than the former, a behaviour as shown in curve
"2" may appear: the ultimate value of the bending moment may be
reached but it may not be kept in the struts for greater values
of the curvature χ. Finally for b/t values even greater, the
beam collapses for a value of applied load lower than that corresponding to the ultimate bending moment of the section.

In the sphere of these problems the authors have tried to solve
the most simple but not least interesting case: the seismic behaviour of bracings built up with angles profiles. For this
purpose the following are needed:
- an experimental and numerical analysis in order to measure the
  ductility of tensioned and compressed members;
- a numerical approach for simulating with a computer the behaviour of bracing structure during an earthquake;
- an extensive research in order to measure the ductility necessary for avoiding brittle collapse changing types of structure
  and seismic inputs.
At present only the first two items have been performed while the
third is just beginning.

EXPERIMENTAL BASIS

So far as tensioned members are concerned, an experimental analysis was carried out and the results are available (Zanon, 1979).
It considered the following typical members:
- back to back equal angles; (Fig.3a);
- two angles connected at the same side of gusset plate (Fig.3b);
- single angle (Fig.3c).
For each solution welded connections and bolted ones with 2,3
and 4 bolts were examined. Among the bolted connections normal
and high strength bolts were used in order to find the different
behaviour between shear and friction joints.
As an example, in Fig.4, the load elongation laws found for back
to back angles connected with normal bolts or with a welded joint
are shown.

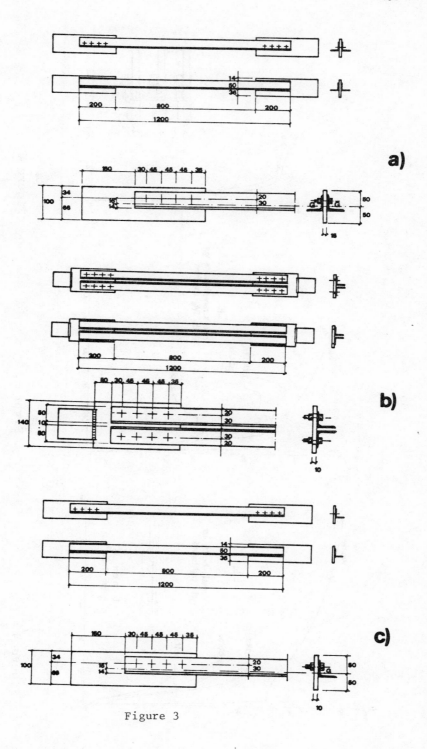

Figure 3

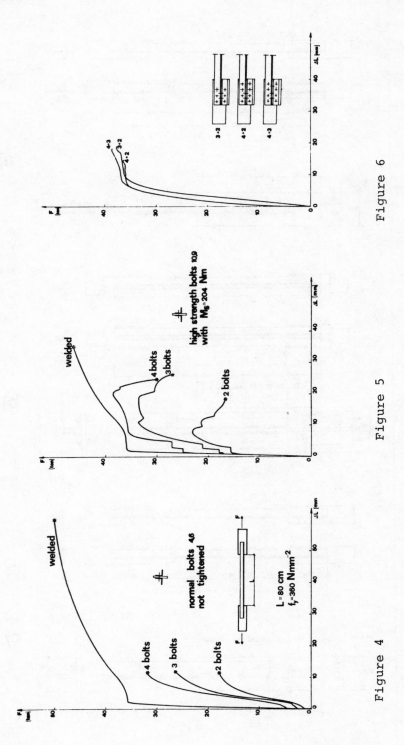

Figure 6

Figure 5

Figure 4

It can be observed that the bar with bolted joints is always in
the elastic range and that collapse occurs when the connection
is ruptured: the member has no ductility. In Fig. 5 the same bar
is connected with tightened high strength bolts (10.9). It can
be observed that tensioned members with 3 and 4 bolts present a
wide excursion in the plastic region before collapse.
The behaviour is much better for a member connected by a
joint not very common at the present time but typical of riveted
steel constructions (Fig. 6) of fifty years ago.

The behaviour of compressed bars was examined by a numerical simulation computer program (Ballio, Petrini, Urbano, 1973) based
on the hypotheses common to ECCS studies (CECM–ECCS, 1976). Fig.
7 shows in a nondimensional form the applied load versus shorten
ing of the bar diagrams for members built up with two back to
back equal angles, of different slenderness λ and with an initial
geometrical imperfection $v_o/L = 1/1000$. Yielding of steel is sup
posed 235 N mm^{-2}. It can be seen how the slopes of the unloading
path decrease at the increasing of strain. For this reason it does
not appear realistic to consider a compressed member as an ideal
elastic plastic body.

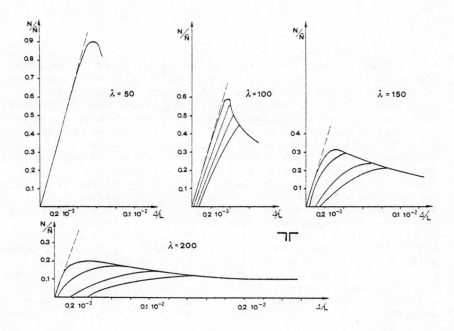

Figure 7

CONSTITUTIVE LAWS

In the dynamic analysis slipping of bolts may be neglected. In
fact, if, at the end of computations, the maximum value of elongation compatible with the behaviour of the connection, is verified not to be exceeded,the elastic-perfectly plastic constituti_
ve law may be assumed as adequate.
The modelling of a compressed bar is more complicated. Referring
to the typical behaviour shown in Fig. 7, for very high slenderness it may be reasonable to assume the bar elastic but with a
modulus E_c very small if compared with the one of   material
E_t ($E_c/E_t \simeq 1/100$). Such an assumption (Fig. 8a) is certainly
safe: it is equivalent to supposing that no dissipation of energy
can occur in the compressed member.

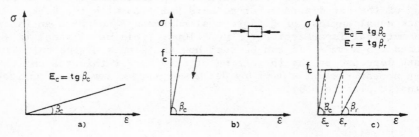

Figure 8

On the other hand,bars with very low slenderness may be model_
led by an elastic-perfectly plastic relationship with a maximum
value equal to the average stress f_c = N_c/A induced by the maximum load N_c = $N_c(\lambda)$ compatible with the slenderness of the bar
(Fig. 8b). For bars with normal values of slenderness a similar
criterion (Fig. 8c) may be adopted but the slope of the unloading
path may not be considered the same value as that of the
material (Fig. 7). Such a slope has to be assumed depending on
the value ε_t of the strain at the reversal point. If the ideal
elastic plastic law is superposed on the real diagram (Fig. 9a)
and the ideal unloading path is drawn,   compensating the
areas of the effective one, the graph of Fig. 9b may be obtain_
ed. It shows the ratio between the values of longitudinal modulus E_r of unloading paths and the E_c of the material versus the ratio between the value of the strain ε_t at the reversal
point and the ε_c at the elastic limit of the ideal elastic
plastic law. It can be seen that the same curve fits all the numerical values independently of the slenderness of the compres_
sed beam. Such a result allows a model of a compressed bar,
like the one shown in Fig. 8c, with a ratio E_t/E_c given by Fig.
9b to be used.

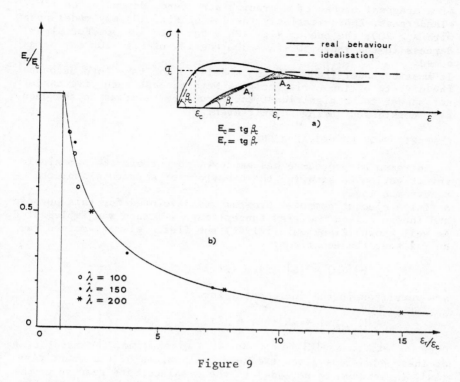

Figure 9

Combining the behaviour in tension and compression, the three constitutive laws of Fig. 10 may be judged as a realistic model

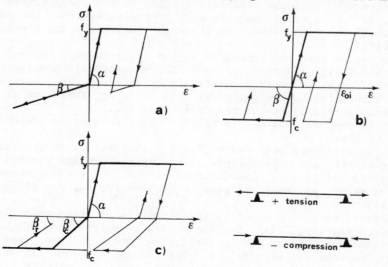

Figure 10

of a diagonal member of a bracing structure, depending on its
slenderness. Approximatively the law of Fig. 10a may model a bar
with $\lambda > 300$; the one of Fig. 10b a bar with $\lambda < 50$. For slenderness in the range $50 < \lambda < 300$ the law of Fig. 10c may be as
sumed.
It must be underlined that the models do not take into account
the effects of low cycle fatigue. Nevertheless such effects do
not appear to be significant for structures similar to the ones
here considered during seismic events.

COMPARISON OF NUMERICAL RESULTS

An incremental procedure was employed to perform the numerical
investigation in studying the behaviour of structure subject
to dynamic loads.
A finite element computer program was developed for this purpose
and the so called modified tangent matrix method was implemented.
As well known (Zienkiewicz, 1971) the finite element method requires that the equations:

$$|M| \ \{x\} + |K| \ \{x\} = \{F(t)\}$$

are rewritten in the incremental form:

$$|M| \ \{\Delta x\} + |K| \{\Delta x\} = \{\Delta I\}$$

Once the stiffness $|K|$ is known at a given time, the resolution
of these equations gives the numerical values of the quantities
$\{\Delta x\}$; these have to be added to the displacements reached at the
previous step of the load history. In fact suitable relations,
assigned in closed form at a finite number of points in the struc
ture, assure that the stiffness depends on the local stress ten-
sor. In such a way it becomes easy to evaluate either the excess
of the loads with respect to the constitutive law and the law
itself.
In order to test the computer program performance , the bracing
system of Fig. 11 was considered. The characterizing parameters
of the structure are also quoted in this figure. They were chosen with reference to a typical mill building. The value of the
natural period of the structure is 0.34 sec.
All the computations were performed for the Tolmezzo (Friuli)
earthquake.
For the different proposed models the variability in time of the
most significant   parameters of the phenomenon are plotted in
Figs. 12 and 13 (i.e. displacement at the top, total stresses
and strains, and residual plastic strains for the two diagonal
bars).
The first presented model is the symmetric elastic-perfectly
plastic law (Fig. 12b). Because of the peculiar shape of the
earthquake the first strong plasticization is the most important
one. The residual deformation never changes sign and its value
oscillates around that reached at the end of the initial

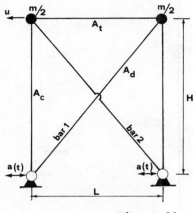

$$L = 600 \text{ cm}$$
$$H = 600 \text{ cm}$$
$$A_c = 64.3 \text{ cm}^2$$
$$A_d = 14.38 \text{ cm}^2$$
$$A_t = 200 \text{ cm}^2$$
$$m = 51000 \text{ kg}$$
$$E = 210000 \text{ N mm}^{-2}$$
$$f_y = 235 \text{ Nmm}^{-2}$$

Figure 11

shock. The constitutive model performs the role of keeping symmetric the response of the structure, i.e. the stresses and strains
for the two transversal bars only differ in sign.

The second model (Fig. 12a) introduces non-symmetry in the struc
tural behaviour. In this case the statement that one bar always
works in tension while the other one is compressed at every time
is not possible.Moreover, even if the first plasticization occurs in the compressed bars because of this lower yield level,
the absolute value of residual deformation can be smaller than
the one in the symmetric case (for the same earthquake). The
elastic in compression and elastic-perfectly plastic in tension
model (Fig. 13a) is probably the most severe one. The level of
plastic deformations increases as the earthquake released energy
increases. Both the bars of the bracing system are strongly
stretched in tension beyond the elastic range.

The last model (Fig. 13b),with variable Young modulus, gives an
intermediate response among the ones examined.

In order to have significant   representation of the differences among the proposed models, computations were performed assuming a variable scaling factor for the exciting earthquake.
Fig. 14 shows a plot of the total plastic deformation versus
the earthquake scale factor for the four models.

470

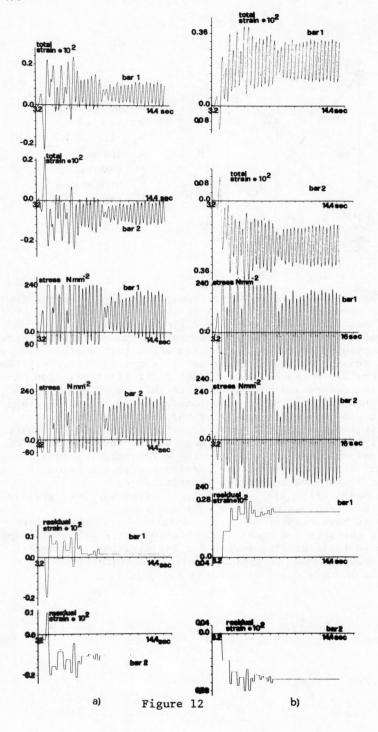

Figure 12

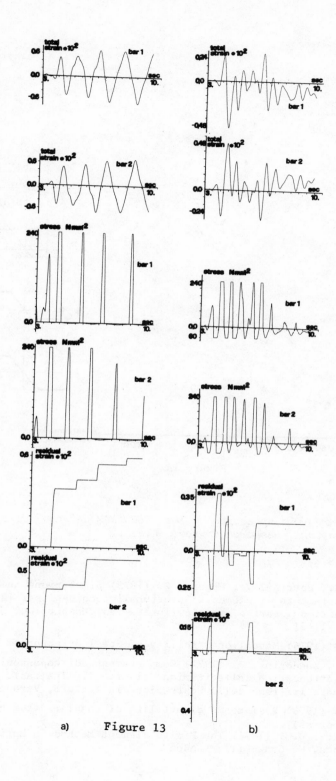

a)　　Figure 13　　b)

472

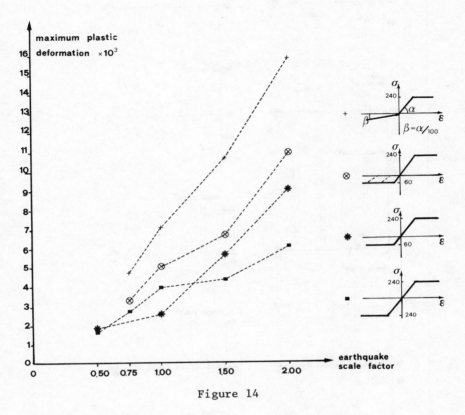

Figure 14

ACKNOWLEDGMENT

*This research was sponsored by the "GEODINAMICA" project of the
Italian National Research Council (C.N.R.).*

REFERENCES

Ballio, G., Petrini, V., Urbano, C. (1973) Simulazione numerica
del comportamento di elementi strutturali compressi per incre-
menti finiti del carico assiale (in Italian) Costruzioni Metalliche, 1: 11-21

CECM-ECCS (1976) Manual on Stability of Steel Structures, Paris

Setti, P., Zandonini, R. (1977) Comportamento di capannoni in
acciaio sottoposti a sollecitazione sismica (in Italian), Report
at "Giornate Italiane della Costruzione in Acciaio, Verona

Zanon, P. (1979) Resistenza e duttilità di angolari tesi bullo-
nati (In press)
Zienkiewicz, D.C. (1971) The Finite Element Method in Engineering Science, Mc Graw-Hill, London

SECTION IV

OTHER EFFECTS

SHAKEDOWN STRESSES DUE TO CREEP AND CYCLIC TEMPERATURES IN CONCRETE STRUCTURES

K.R.F. Andrews, A. Moharram and G.L. England

Civil Engineering Department, King's College, University of London.

INTRODUCTION

The significant redistribution of stress due to creep in concrete structures subjected to non-uniform temperatures is well known. An analogous elastic modulus for the determination of the long term stress distributions has been proposed by England (1966). The use of this elastic analogy enables bounds on the stress distributions to be found without the necessity of lengthy transient calculations.

Many types of concrete structures are subjected to time varying temperatures. Of particular interest are those likely to undergo cyclic heating and cooling such as bridge decks and oil storage tanks. England (1977) formulates a complementary average power functional for the analysis of such structures and solves restrained beam problems using a flexibility approach with a Ritz method.

This paper presents an elastic stiffness analogy for the cyclic thermal creep problem. It is developed from a minimum average power principle and is readily utilised in standard finite element programs.

CREEP WITH SUSTAINED TEMPERATURE DISTRIBUTIONS

In transformed or pseudo-time (England and Jordaan, 1975), the three dimensional constitutive relationship for concrete may be written (England, 1979a):

$$D[\varepsilon] = \left[\frac{D}{E} + \phi(T)\right] [V] [\sigma] \qquad (1)$$

where $D = d(\ )/dt'$ (t' is pseudo-time), $[\varepsilon]$ and $[\sigma]$ are the three dimensional strain and stress vectors, $\phi(T)$ is the normalising creep-temperature function (England and Jordaan, 1975), E is Young's Modulus and $[V]$ the appropriate Poisson's ratio matrix (England, 1979b).

In the work that follows, $\phi(T)$ is taken to be $\phi(T) = T$ and the uniaxial equation corresponding the Equation (1) reduces to:

$$\dot{\varepsilon} = \frac{\dot{\sigma}}{E} + \sigma T \tag{2}$$

The dot notation refers to differentiation with respect to pseudo-time and clearly at steady state when $\dot{\sigma} = 0$,

$$\dot{\varepsilon} = \sigma T \tag{3}$$

An equivalent elastic modulus $1/T$, relating stresses and strain rates is apparent in Equation (3). This analogy can be extended, without formality, to two and three dimensions.

With regard to the determination of the time-dependent stresses and strain rates, England (1979b) gives two variational equations:

$$\int_{Volume,V} [\dot{\varepsilon}]^T [\delta\sigma] \, dV - \int_{Surface,S} \dot{u}_s \delta R_s \, dS = 0 \tag{4}$$

$$\int_{Volume,V} [\sigma]^T [\delta\dot{\varepsilon}] \, dV - \int_{Surface,S} P_e \delta\dot{u}_e \, dS = 0 \tag{5}$$

In these equations, $\dot{u}_e$ are the displacement rates associated with the applied loads P_e and $\dot{u}_s$ are the displacement rates associated with support reactions R_s. Referring back to Equation (1), it can be seen that Equation (4) may be written as:

$$\delta \left(\int_{Volume,V} (\dot{U} + \tfrac{1}{2}\dot{D}) dV - \int_{Surface,S} \dot{u}_s R_s \, dS \right) = 0 \tag{6}$$

where $\dot{U}$ is the rate of change of complementary strain energy and $\dot{D}$ is the complementary energy dissipation rate in creep. At the steady state $\dot{U} = 0$, and:

$$\delta \left\{ \int_{Volume,V} \tfrac{1}{2}\dot{D} dV - \int_{Surface,S} \dot{u}_s R_s \, dS \right\} = 0 \tag{7}$$

which is a statement of the complementary energy dissipation variational principle at the steady state.

Using this principle and a Ritz method, England and McLeod (1978) investigated the steady state stresses in a prestressed concrete reactor containment vessel. In general, the stress distribution at any time is written as a weighted summation of self-equilibrating distributions and a distribution in equilibrium with the applied loading. These weighting functions, determined so as to satisfy Equation (7) are, in general, functions of time but are constants at the steady state. Good results are obtained with very few distributions.

One particular advantage of using Equation (7) is that for most practical problems $\dot{u}_s = 0$ and the second integral disappears.

In a similar way, Equation (5) may be written as:

$$\delta \left(\int\limits_{\text{Volume,V}} \tfrac{1}{2}\dot{D}dV - \int\limits_{\text{Surface,S}} \dot{u}_e P_e dS \right) \quad = \quad 0 \qquad (8)$$

which is a statement of the energy dissipation variational principle at the steady state. Since $\dot{u}_s$ is not zero, this principle is only suitable for a stiffness approach and can be used to determine the relevant finite element stiffness matrices in conjunction with the analogous elastic modulus $1/T$. It should be noted that any displacement rate field $(\dot{u})$ together with the consistent strain rates $(\dot{\varepsilon})$ may be used in determining stresses from Equation (8).

CREEP WITH CYCLICALLY VARYING TEMPERATURES - HOMOGENEOUS CYCLING

In this section problems are investigated in which the periods of application of the higher and lower temperatures $(T$ and $T_0)$ are the same throughout the structure. For simplicity of presentation, one dimensional stress and strain is considered. Generalisation to three dimensions can be readily made by use of Equation (1).

Flexibility Analysis

Considering some part of a structure subjected to temperature cycling of the form shown in Figure 1(b), England (1979a) proposed that at the steady state, the average internal power dissipated in one cycle is:

$$P = \frac{1}{1+k} \left(\int\limits_{\text{Volume V}} (\sigma^*T) \; \sigma^* dV + k \int\limits_{\text{Volume,V}} (\sigma^*-\sigma_\alpha)T_0(\sigma^*-\sigma_\alpha)dV \right) \quad (9)$$

Here σ^* are the steady state stresses and are the mean stresses over the part of the cycle for which the temperature is T, σ_α is the thermo-elastic stress change corresponding to the temperature change from T_o to T and T_o is applied for a proportion k of the time for which T is applied. Equation (3) has been used to express strain rates as stresses.

The stress formulation of Equation (9) gives rise to a variational equation based on the complementary average power:

$$\frac{1}{1+k} \int_{\text{Volume, V}} \left(\sigma^* T \delta \sigma^* + k \, (\sigma^* - \sigma_\alpha) \, T_o \delta (\sigma^* - \sigma_\alpha) \right) \, dV$$

$$- \int_{\text{Surface,S}} \dot{u}_s \delta R_s dS = 0 \tag{10}$$

England (1977) used this principle to solve restrained beam problems ($\dot{u}_s = 0$) using the Ritz method referred to earlier.

Stiffness Analysis

For the development of a stiffness analysis that can be readily incorporated into existing finite element programs, it is essential that $\dot{u}_e$ (Equation 8) and the consistent internal strain rates are continuous over a complete cycle. This suggests that an average displacement rate be defined thus:

$$\dot{u}_e \equiv \dot{u}_a = (\dot{u} + k\dot{u}_o)/(1 + k) \tag{11}$$

where $\dot{u}$ is the displacement rate in the portion of the cycle defined by T and $\dot{u}_o$ is the displacement rate in the rest of the cycle. It follows that the consistent strain rate is:

$$\dot{\varepsilon}_a = (\dot{\varepsilon} + k\dot{\varepsilon}_o)/(1 + k) \tag{12}$$

in which typically $\dot{\varepsilon} = d\dot{u}/dx$ and $\dot{\varepsilon}_o = d\dot{u}_o/dt$ in one dimension. Using relations of the form shown in Equation (3), this becomes:

$$\dot{\varepsilon}_a = (\sigma^* T + k \, (\sigma^* - \sigma_\alpha) T_o)/(1 + k)$$

and hence

$$\sigma^* = \frac{1}{T_a} \left(\dot{\varepsilon}_a + \frac{kT_o \sigma_\alpha}{1 + k} \right) = \frac{1}{T_a} (\dot{\varepsilon}_a - \dot{\varepsilon}_o) \tag{13}$$

where $T_a = (T + kT_o)/(1 + k)$ is the average cycle temperature.

The form of this stress-strain relation suggests an equivalent elastic modulus of $1/T_a$ and an analogous "initial strain rate" $\dot{\varepsilon}_o = -kT_o\sigma_\alpha/(1 + k)$ for the stiffness analysis of steady state stress.

For the appropriate variational principle, the power dissipation in one cycle given in Equation (9) is returned to its basic form using Equation (13):

$$P = \left\{ \int\limits_{\text{Volume},V} (\sigma^*T)\frac{1}{T_a}(\dot{\varepsilon}_a - \dot{\varepsilon}_o)dV + k\int\limits_{\text{Volume},V} (\sigma^* - \sigma_\alpha)T_o\left(\frac{1}{T_a}(\dot{\varepsilon}_a - \dot{\varepsilon}_o) - \sigma_\alpha\right)dV \right\} \times$$

$$\times 1 /(1 + k) \tag{14}$$

After manipulation this becomes:

$$P = \int\limits_{\text{Volume, }V}\left(\sigma^*(\dot{\varepsilon}_a + \dot{\varepsilon}_o) + \frac{\sigma_\alpha^2\, kT_o}{1 + k}\right)dV \tag{15}$$

The minimum average power dissipation principle can be determined from the first variation of Equation (15) with respect to $\dot{\varepsilon}_a$. With boundary terms this is:

$$\int\limits_{\text{Volume},V}\sigma^*\delta\dot{\varepsilon}_a\, dV + \int\limits_{\text{Surface},S} P_e\,\delta\dot{u}_a\, dS = 0 \tag{16}$$

Substitution of Equation (13) gives:

$$\int\limits_{\text{Volume},V}\frac{1}{T_a}\left(\dot{\varepsilon}_a + \frac{kT_o\sigma_\alpha}{1+k}\right)\delta\dot{\varepsilon}_a\, dV + \int\limits_{\text{Surface},S} P_e\,\delta\dot{u}_a\, dS = 0 \tag{17}$$

This confirms the analogous modulus and initial strain rate.

The first variation of Equation (15) with respect to σ^* will give a complementary average power dissipation principle:

$$\int\limits_{\text{Volume},V}\left(\dot{\varepsilon}_a - \frac{kT_o\sigma_\alpha}{1 + k}\right)\delta\sigma^*\, dV + \int\limits_{\text{Surface},S} \dot{u}_a\,\delta R_s\, dS = 0 \tag{18}$$

It is possible to extend the elastic analogy to cycling involving several temperature changes to temperatures T_i in a given cycle. Again an average temperature may be defined:

$$T_a = \frac{T + \sum_i k_i T_i}{1 + \sum_i k_i}$$

and the equivalent initial strain rate will be the summation of terms similar to those in Equation (19) for each temperature change.

The plate problem shown in Figure 1 has been analysed using the stiffness analogy with the finite element method. The resulting steady state stresses σ^* are tabulated for various k values in Table 1 and may be compared with analytical results based on work by England (1977). Also tabulated in Table 1 are the results of a flexibility analysis of the same plate with a different temperature gradient. These may also be compared with the analytical solution. Both stiffness and flexibility solutions are plotted in Figure 2.

NON-HOMOGENEOUS CYCLIC THERMAL CREEP

The analogy established above may, without formality, be extended to problems where k is a function of position, i.e. the temperature cycle varies spatially. The variable cycling may be due to an oscillating hot-cold interface or diurnal atmospheric temperature changes and so on. As before the analogy is one of a non-homogeneous elastic medium with spatially varying initial strains due not only to the variation of σ_α but also k. A flexibility formulation of such problems is difficult due to the required determination of self-equilibrating stress fields, though these may be determined by finite element analysis.

The determination of σ_α requires careful consideration in such non-homogeneous problems. Due to the time varying temperature changes within the body, the thermal stresses may not remain constant within the sustained temperature periods of a cycle. Typically, for thermal cycling due to a moving hot-cold interface, perturbations of the steady state stress cycle of the form shown in Figure 3b are encountered (England, Andrews, Moharram and Macleod, 1979). These perturbations can be shown to affect only the value of the initial strains used in the analogy, the equivalent modulus is not affected.

As an example, consider the analysis of a short section of a thin walled cylindrical oil storage tank within which there is a constant filling and emptying rate (Figure 3). The tank is prestressed axially and circumferentially. Each horizontal section is subjected to different periods of heating within a constant cycle and therefore k is a linear function of depth. Due to the thermal incompatibility in the wall at

k	0		1		2		5		∞	
X	A	B	A	B	A	B	A	B	A	B
0.933	5.22	5.22	8.94	8.94	10.67	10.65	12.76	12.73	15.32	15.32
0.867	5.41	5.41	8.70	8.70	10.17	10.17	11.93	11.92	14.04	14.04
0.733	5.83	5.83	8.19	8.19	9.17	9.16	10.26	10.25	11.48	11.48
0.667	6.06	6.06	7.91	7.92	8.62	8.63	9.39	9.40	10.20	10.20
0.533	6.59	6.59	7.32	7.32	7.50	7.51	7.62	7.63	7.64	7.64
0.467	6.89	6.89	6.99	7.00	6.90	6.91	6.70	6.72	6.36	6.36
0.333	7.57	7.57	6.30	6.30	5.64	5.66	4.82	4.84	3.80	3.80
0.267	7.97	7.97	5.92	5.92	4.97	4.99	3.85	3.87	2.52	2.52
0.133	8.91	8.91	5.09	5.09	3.55	3.56	1.85	1.87	-0.04	-0.04
0.067	9.47	9.47	4.63	4.64	2.79	2.81	0.82	0.83	-1.32	-1.32

Column 'A': Results from finite element using elastic analogy
Column 'B': Results from flexurally restrained beam analysis
$T = 80^\circ C$; $T_c = 40^\circ C$, $E = 40000$ N/mm^2

k	0		1/9		1		9		∞	
X	A	B	A	B	A	B	A	B	A	B
0.95	2.69	2.64	2.89	2.86	4.34	4.32	10.29	10.28	18.01	18.01
0.85	2.98	2.92	3.18	3.14	4.57	4.55	9.81	9.80	15.57	15.57
0.75	3.34	3.27	3.53	3.48	4.84	4.82	9.27	9.26	13.12	13.12
0.65	3.80	3.72	3.97	3.92	5.18	5.16	8.66	8.65	10.67	10.67
0.55	4.40	4.31	4.55	4.50	5.62	5.60	7.95	7.94	8.22	8.22
0.45	5.22	5.12	5.34	5.28	6.20	6.17	7.14	7.13	5.77	5.77
0.35	6.42	6.30	6.51	6.43	7.01	6.97	6.21	6.18	3.33	3.33
0.25	8.35	8.19	8.34	8.24	8.20	8.16	5.10	5.06	0.88	0.88
0.15	11.93	11.70	11.68	11.54	10.15	10.09	3.75	3.71	-1.57	-1.57
0.05	20.87	20.50	19.67	19.47	13.87	13.80	2.09	2.05	-4.02	-4.02

Column 'A': Results from flexibility analysis
Column 'B': Results from flexurally restrained beam analysis
$T = 50^\circ C$; $T_o = 10^\circ C$; $E = 34000$ N/mm^2.

Table 1. Comparison of steady-state-cyclic stresses, σ^*, for section Y-Y of plate in Figure 1 obtained from stiffness, flexibility and analytical methods. Nine self-equilibrating stress distributions used in flexibility analysis. Stresses in MN/m^2, compression positive.

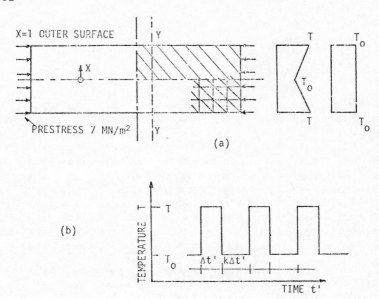

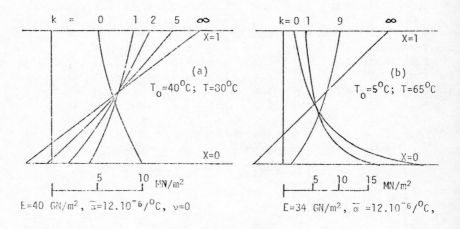

Figure 1: Prestressed plate problem. Quarter section (shaded) equivalent to restrained beam. (a) Plate with typical finite element mesh, prestress and temperature limits. (b) Temperature cycle in pseudo-time.

Figure 2: Steady-state-cyclic stresses (σ^*) for section Y-Y shown in Figure 1. Stresses are symmetric about centre line of plate. For steady-state stresses in temperature state T_o subtract thermo-elastic solution ($k = \infty$). $k = 0$ gives sustained temperature solution.

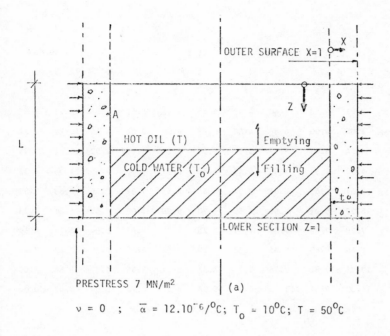

PRESTRESS 7 MN/m² (a)

$\nu = 0$; $\bar{\alpha} = 12.10^{-6}/{}^{\circ}C$; $T_o = 10^{\circ}C$; $T = 50^{\circ}C$

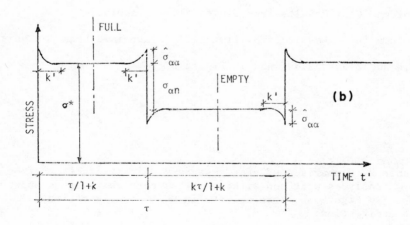

Figure 3: (a) Short section of cylindrical oil storage tank with data. (b) Change of stress with time at general location A during complete filling and empyting operation. $\phi = k'/\tau$ is a constant for uniform filling and empyting rate.

Z	1/60		1/4		1/2		3/4		59/60	
k	0.017		0.333		1.000		3.000		59.00	
X	C	B	C	B	C	B	C	B	C	B
0.05	3.81	3.68	4.60	4.43	5.85	5.91	8.47	8.36	13.73	13.67
0.15	4.05	4.00	4.86	4.80	5.98	6.05	8.24	8.15	12.28	12.24
0.25	4.44	4.40	5.17	5.12	6.13	6.21	7.97	7.92	10.81	10.79
0.35	4.92	4.87	5.55	5.49	6.30	6.39	7.70	7.66	9.33	9.32
0.45	5.51	5.47	6.01	5.97	6.51	6.61	7.39	7.36	7.83	7.83
0.55	6.27	6.24	6.59	6.55	6.90	6.88	7.02	7.03	6.31	6.33
0.65	7.28	7.26	7.34	7.31	7.25	7.21	6.61	6.64	4.77	4.80
0.75	8.69	8.68	8.35	8.33	7.70	7.62	6.13	6.20	3.23	3.25
0.85	10.81	10.81	9.78	9.78	8.31	8.17	5.56	5.67	1.66	1.68
0.95	14.30	14.34	11.95	11.97	9.08	8.91	4.90	5.04	0.07	0.09

Column 'C' : Results from finite element analysis

Column 'B' : Results from flexurally restrained beam analysis

k,X and Z are as defined in Figure 3

Table 2. Comparison of steady-state-cyclic stresses, σ^*, from analyses with and without stress perturbations as shown in Figure 3b. Stresses are in MN/m^2 ; circumferential direction.

the hot-cold interface, the thermal stresses due simply to temperature changes ($\sigma_{\alpha n}$) are perturbed, to a decreasing extent, above and below the interface. In consequence, a typical steady state stress cycle is of the form shown in Figure 3b.

By consideration of average power and assuming the perturbations to be parabolic within a cycle (maximum value $\hat{\sigma}_{\alpha\alpha}$) it is found that the variational equation becomes:

$$\delta P = \int\limits_{Volume, V} \frac{1}{T_a}\left[\dot{\varepsilon}_a + \frac{kT_o\sigma_{\alpha n}}{(1 + k)} + \frac{2}{3}\hat{\sigma}_{\alpha\alpha}(T - T_o)\phi \right]\delta(\dot{\varepsilon}_a)dV$$

$$- \int\limits_{Surface, S} P_e \delta(\dot{u}_a)dS = 0 \tag{19}$$

where ϕ is constant for constant filling and emptying rates (England, Andrews, Moharram and MacLeod, 1979). The perturbations have thus provided an additional initial strain rate term.

Table 2 shows the results of a finite element analysis of such a thin walled large radius (inside radius/wall thickness ratio of 50) tank subjected to a   prestress of 7 MN/m^2 with other quantities as shown in Figure 3. Steady state values of the   circumferential stresses are tabulated for various positions (X) across horizontal sections at various depths (Z). At the upper end of the cylindrical section (Z = 1/60), k is small and the stresses are in close agreement with the sustained temperature steady state stresses for the analagous restrained beam with the same prestress and k value. In this latter case the thermal stress perturbations are not present. At the lower end where k is large, close agreement with thermo-elastic solution for the equivalent restrained beam is obtained. For the actual stress levels at any point in the cycle, the appropriate thermal stresses must be added as discussed by England, Andrews, Moharram and MacLeod (1979).

It would appear that the perturbations have little effect on the steady state stress calculations for this problem. In general this may not be the case.

CONCLUSIONS

The elastic stiffness analogy developed in this paper has wide application in cyclic thermal creep problems. Restrictions on geometry and boundary conditions are removed as the analogy can be readily incorporated in standard

finite element programs. This gives certain advantages over
the flexibility method which has been shown to give good
results for some problems with only a few self-equilibrating
stress distributions.

REFERENCES

Andrews, K.R.F. and England, G.L. (1979). Elastic analogy
to steady-state stresses for creep and cyclic temperatures -
a stiffness approach. Seventh Canadian Conference on Applied
Mechanics, Sherbrooke, 1979.

England, G.L. (1966). Steady-state stresses in concrete
structures subjected to sustained loads and temperatures.
Nuclear Engineering and Design, 3, 54 - 65 & 246 - 255,
January and February/March, 1966.

England, G.L. (1977). Steady-state stresses in concrete
structures subjected to sustained and cyclically varying
temperatures. Nuclear Engineering and Design, 44, 97 -
107, October 1977.

England, G.L. (1979a). The calculation of stresses in
concrete ocean structures subjected to steady and time
varying temperatures as influenced by creep. Journal of
Applied Ocean Resarch, 1, 1 - 9, January 1979.

England, G.L. (1979b). Temperature-creep stresses in
concrete structures: minimum power formulation. Proc.Conf.
on Numerical Methods in Thermal Problems, Swansea 1979.

England, G.L., Andrews, K.R.F., Moharram, A. and MacLeod,
J.S. (1979). The influence of creep and temperature on
the working stresses in concrete oil storage structures.
Proc.Conf. on the Behaviour of Off-Shore Structures.
London 1979.

England, G.L. and Jordaan, I.J. (1975). Time-dependent
and steady-state stresses in concrete structures with
steel reinforcement at normal and raised temperatures.
Magazine of Concrete Research, 27. No. 92, 131-142, September
1975.

England, G.L. and MacLeod, J.S. (1978). Creep analysis
for prestressed concrete reactor containment vessels.
Proc.Conf. Structural Analysis, Design and Construction
in Nuclear Power Plants. Brazil, 1978, 2, 663 - 680.

TEMPORAL AND SPACIAL PATTERNS IN THE DEVELOPMENT OF BIOLOGICAL FOULING

B.Benham & E.Bellinger

Pollution Research Unit, Manchester University

INTRODUCTION

A large number of marine organisms have evolved the ability to attach to surfaces and thus withstand wave action. They have not, however, evolved the ability to differentiate between their natural substrates and a range of objects which are introduced into the marine environment by man. In the shipping industry the increase in frictional resistance of vessels due to fouling and the resultant increases in fuel consumption, machinery wear and docking costs are well known. Marine growths may also block sea-water mains and condenser tubes, appreciably increase the diameter, and hence water resistance, of under-water structures, hide faulty or failing welds and disguise or even encourage corrosive processes. Thus, marine fouling may also have considerable relevance to the stability of off-shore structures. Control of fouling relies upon the regular application of antifoulant paints and it is undoubtedly true that protection of ships from biological fouling remains an area where further developments are desirable. The work reported here was initiated in 1972 as an attempt to describe the contemporary ship fouling assemblage in the hope that it would provide both base-line data for the paint industry, so that service performance of paints could be more accurately assessed and also so that biologists could ascertain whether or not the organisms on which they were working were actually relevant to present day ship fouling.

METHODS

Samples of fouling and its paint substrate were collected from vessels at dry-docking. The samples were usually obtained at regular intervals along three transects down the hull at the bow, mid and stern areas. The collections were made using the head of a dutch hoe mounted on an extending rod which could take samples up to eight metres above the floor of the dock. Areas approximately 15 cms. x 25 cms. were scraped and the material collected in a polythene bag held onto the hoe-head.

In the laboratory the samples were preserved in a 5% formaldehyde solution and examined so that, so far as was practicable, the individual fouling species were identified and their cover of the surface assessed on a six point scale, Table 1.

Table 1

The Scale Used To Record Species Cover.

Scale Class	Percent Cover
1	0- 1
2	1- 5
3	5- 25
4	25- 50
5	50- 75
6	75-100

Not all of the organisms found could be identified to their biological species and the term 'taxon' is used when reference is made to these organisms.

RESULTS

The results of the examinations of 252 samples obtained from nine vessels which operated exclusively in British and Northern European waters for periods ranging between four and twenty five months are reported here. A total of 46 taxa were identified on the vessels and their presence and maximum abundance on each ship is recorded in Table 2. The overall pattern shown by the vessels in this survey is one of an increased number of fouling taxa with time since the last application of antifoulant, regardless of season of painting or sampling. Regression analysis of these data results in a best fit line with a slope of 1.36 which is highly significant $P < 0.001$; No : $b = \beta = 0$. (Figure 1).

When coefficients of similarity (Sorensen, 1948) are calculated for the presence of the taxa on each of the vessels (see Table 3) it is clear, where the coefficients are expressed as percentages of similarity, that the vessels with the most similar fouling complements were those which had been in service for similar periods. This is particularily important as it indicates that operation on

Table 2   The maximum cover values of the organisms found
on the vessels.

	VESSEL NUMBER								
	1	2	3	4	5	6	7	8	9
	MONTHS IN SERVICE								
	4	7	7	12	16	16	24	24	25
1) GREEN ALGAL SLIME	2								
2) ULOTHRIX IMPLEXA	5	2	4						
3) DIATOM SLIME	6	6	6	6	4	6	2	2	
4) BACTERIAL SLIME	6	6	6	6	6	6			
5) ENTEROMORPHA SPECIES	1		1	4	6	6	6	6	6
6) ULOTHRIX FLACCA		1	6	1			1	2	4
7) UROSPORA ISOGONA		4	4	1	2	1	1	2	3
8) ECTOCARPUS SILICULOSUS		1	4	6	2	3	1	2	2
9) ELMINIUS MODESTUS			1	1		2	4	6	4
10) CLADOPHORA RUPESTRIS			2		1	3		2	1
11) POLYSIPHONIA FIBRATA			1				3	1	1
12) BANGIA FUSCO-PURPUREA				1	2	1			
13) NEREIS PELAGICA				1		1	1	1	1
14) PHYLLITIS FASCIA				1					
15) COROPHIUM SPECIES				1		2	1	1	2
16) PYGOSPIO ELEGANS				1			1	1	2
17) BALANUS CRENATUS					6	4	5	6	6
18) BOWERBANKIA GRACILIS					5	3	1	1	1
19) BALANUS IMPROVISUS					4	6	3	4	3
20) CONOPEUM RETICULUM					1	4	3	4	4
21) OBELIA DICHOTOMA					1	1	1	1	1
22) ANTITHAMNION PLUMULA						1			
23) RHODOCHORTON ROTHII						1			
24) MYTILUS EDULIS						1	1	2	2
25) TURBULARIA LARYNX						1			
26) MOLGULA MANHATTENSIS						1	1	1	1
27) BLIDINGIA MARGINATA							6	6	6
28) POLYSIPHONIA NIGRA							2		1
29) CERAMIUM PENNATUM							1		
30) CERAMIUM RUBRUM							1	1	1
31) CERAMIUM DESLONGCHAMPSII							1	1	1
32) PILAYELLA LITTORALIS							1	1	2
33) PORPHYRA UMBILICALIS							1	2	1
34) BALANUS BALANOIDES							1	1	5
35) ACINETOSPORA PUSILLA							2	2	2
36) ELECTRA PILOSA							3	4	4
37) ALCYONIDIUM POLYOUM							3	4	3
38) PHYLLODOCE MACULATA							1	1	1
39) DYNAMENA PUMILA							1		
40) BUGULA STOLONIFERA							1		1
41) CRYPTOSULA PALLASIANA								1	2
42) CHAETOMORPHA AEREA								1	1
43) POLYSIPHONIA MACROCARPA								1	
44) PRASIOLA STIPULATA									1
45) CLADOSTEPHUS VERTICILLATUS									1
46) DUDRESNAYA VERTICILLATA									1

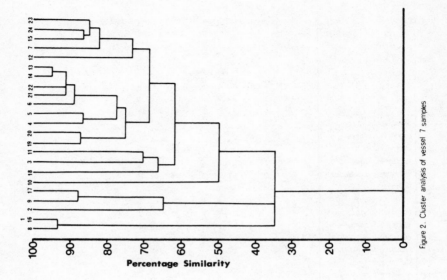

Figure 2. Cluster analysis of vessel 7 samples

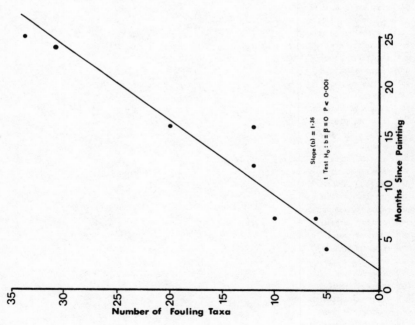

Figure 1. Regression analysis of fouling taxa against service period

routes within the Northern European area results in, more or less, constant fouling complements developing and that the trend towards increased fouling with increased period of service is due to a steady change in the individual taxa present rather than to the vessels acquiring equal numbers of widely different species.

Table 3

The Percentages of Similarity of the Fouling Complements.

Vessel number

Vessel number	2	3	4	5	6	7	8	9
1	55	53	35	39	24	11	11	5
2		75	56	56	38	22	22	15
3			64	55	47	34	39	32
4				50	56	42	42	35
5					75	42	47	39
6						55	59	52
7							87	86
8								89

Percentage similarity

The earliest colonists of a freshly applied paint surface appeared to be bacteria, which may accumulate to form slimes and this phenomenon has been well documented in the work of Marine Corrosion Sub-Committee (1943) where it was shown that film development can occur within ten days of submersion. The bacteria were followed by a number of algal taxa such as Ulothrix implexa, Diatom Slime, Green Algal Slime, Enteromorpha spp., Ulothrix flacca, Urospora isogona, Ectocarpus siliculosus, Cladophora rupestris and Polysiphonia fibrata, which all occurred on vessels with less than twelve months in service. The only animal species present during this period was the barnacle Elminius modestus which occurred in a single sample from Vessel 3.

On the vessels with between 12 and 16 months service only 4 more algal species occurred while some, Green Algal Slime, Ulothrix implexa and Polysiphonia fibrata, were lost. On these ships 11 animal species were found for the first time. Notable amongst these were the barnacles Balanus crenatus and Balanus improvisus, the tube-living Corophium spp. and

<u>Pygospio elegans</u> and the polyzoans <u>Bowerbankia gracilis</u> and
<u>Conopeum reticulum</u>.

On vessels 7, 8 and 9 with 24 or 25 months of service an
additional 20 species of fouling organisms were recorded and
5 of the taxa present on the vessels with between 12 and 16
months service were lost. Of these additional species 13
were algal. Amongst these species particular attention is
drawn to the green alga <u>Blidingia marginata</u> which dominated
the areas of the hulls above the low-water line and a large
number of members of the Phaeophycae and Rhodophycae which
occurred at low abundance on the non-toxic surfaces formed
by the shell plates of the barnacles. Of the 7 new animal
species, <u>Electra pilosa</u>, <u>Alcyonidium polyoum</u>, <u>Bugula</u>
<u>stolonifers</u> and <u>Cryptosula pallasiana</u> are polyzoans and the
above vessels provided the only specimens of the common
intertidal barnacle, <u>Balanus balanoides</u>, which were found
in the entire study. This pattern of colonization forms
a clear trend from algal to animal fouling and this is
illustrated in Table 4 in which the percentages of animal
fouling on each of the vessels are shown.

Table 4

Percentages of Animal Fouling

Vessel number	Presence or absence	Cover values
1	0	0
2	0	0
3	10	1
4	33	8
5	42	45
6	55	54
7	55	76
8	52	73
9	50	75

The coeffiecients of similarity previously used to compare
the fouling complements of entire vessels can also be used to
compare the individual samples from each of the vessels on
a quantitative basis. The formula (Equation 1) in which;

a = the sum of all of the quantitative measures of the

species in one sample

b = the sum of all of the quantitative measures of the species
in a second sample

J = the sum of the lesser values of all of the species common
to both samples,

"Equation 1"

$$\text{percent similarity} = \frac{2J}{a + b} \ 100$$

is applied to each pair of samples in turn and the results
expressed in the form of a matrix, as in Table 3.

Additional information can be obtained from these data by
the application of cluster (see Everitt, 1974) and ordination
(see Bray and Curtis, 1957) analyses to the data matrix. The
first of these techniques groups together the most similar
samples and displays them in the form of tree diagrams
(dendrograms), whilst the second depicts the samples
graphically and the major axes of the graph are generally
related to gradients in environmental factors. Typical
results of cluster and ordination analyses are shown by the
results from Vessel 7 which had been on cross channel service
for 24 months since its last painting.

The results of the cluster analysis, Figure 2, indicate the
division of the samples into six groups (samples 8, 1 and 16,
samples 2, 9 and 17, sample 10, samples 18, 3 and 11, samples
19, 20, 4, 5, 6, 21, 22, 14 and 13 and samples 12, 7, 15,
24 and 23) which occur on the vessel in bands extending along
the length of the ship so that clear vertical gradients are
displayed, Figure 3.

However, Figure 4, in which the ordination analysis is
displayed, shows that the samples 8, 1, 16, 2, 9 and 17 occur
along the x axis whilst the remaining samples are arranged
along the y.

This division suggests that two main lines of environmental
variation occur. It is suggested that the uppermost samples
(which fall along the x axis) are controlled mainly by the
availability of water. These samples occur in areas above
the light-load line and have as their dominant fouling types
the algae Blidingia marginata and Enteromorpha spp. The
remaining samples, which occur along the y axis of the
ordination, were probably continuously submerged and their
species composition appeared to show an increased importance
of animal species at the lower levels of the hull. It is,

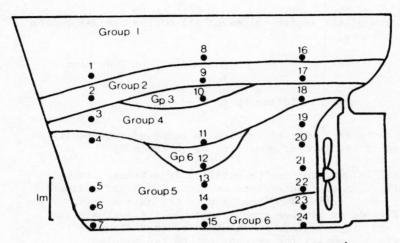

Fig.3 The positions of the groups of fouling
on Vessel 7

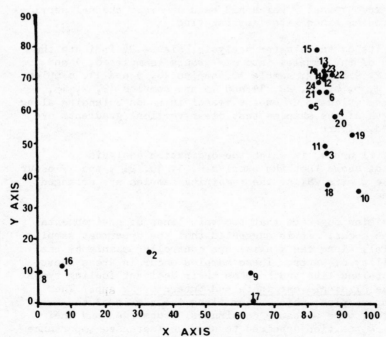

Fig.4 Ordination analysis of samples from
Vessel 7

therefore, suggested that this change and the gradient itself are determined by illumination levels which are rapidly reduced as light passes through the water so that plant fouling is precluded.

DISCUSSION OF THE RESULTS

This study appears to have succeeded in defining, firstly, the sequence of colonising species, secondly the patterns of colonization of the individual species (see Benham 1978), and thirdly, the division of the fouling complements into distinct communities. Thus, a basis for the comparison of future changes in the composition of marine fouling communities has been obtained. It is, therefore, unfortunate that relevant studies of the biology of many of the early and most important colonizing species do not exist and that data on the capacity for resistance to antifoulant toxins are also wanting for the majority of the species. Much of the blame for this situation may be ascribed to the problems involved in the sampling from the hull of a vessel and the inadequacy of previous descriptions of the ship fouling assemblage. Comparisons with earlier studies are complicated by the dissimilarities between the results of this and the earlier works due to the changes in service speeds, areas of operation and composition of antifoulant paints which have occurred. In addition, there has been little overall consideration of the many environmental factors likely to determine the fouling developed. Hentschel (1923) paid particular attention to the waters cruised, but little to the seasons of sampling or the service periods of the individual ships. Visscher (1927) devoted much attention to the service period but somewhat neglected the other two factors. Skerman (1960) concentrated attention on the area cruised and (like Millard, 1951) considered the effects of seasonal settlement. Such seasonal effects were, however, based largely on the study of test plates and not the vessels themselves. Woods Hole (1952) considered each of the factors in turn and appear to bring together what (little) information was available, but made little progress towards the integration of these factors.

Figure 1, in which the number of fouling taxa are plotted against the period of service since the last application of antifoulant protection, shows no slowing down of the rate of colonization after 25 months. This does not bode well for the prevention of fouling on off-shore structures which may remain on station for many years. On such structures the development of a complex and dense fouling community at certain depths appears inevitable, with little prospect of data obtained from studies of ship fouling being of direct relevance due to the restricted period between the replacement of the antifoulant protection of the vessels.

Frictional resistance is the greatest problem caused by the fouling of vessels, but the power increase to overcome this eventually stabilizes due to the irregularities of the uppermost layers only impeding the water flow. In contrast, the resistance due to increased diameter of the substrate, which is much less important on vessels than off-shore structures, will continue to increase as the fouling succession progresses. Moreover, relatively thin layers of fouling may cover faulty welds and the links between fouling and enhanced rates of corrosion remain poorly understood. Because of this, it must be considered unlikely that conventional techniques will provide a solution to the problem and novel approaches such as particulate toxins, exfoliating surfaces or renewable cladding are worthy of investigation, whilst allowances for fouling at the design stage and increased attention to the removal of marine growths as they occur offer the best interim protection.

REFERENCES

Benham, B.R. (1978) A Sudy of 'Fouling' on Marine Shipping.
Ph.D. thesis, University of Salford.

Bray, J.R. and Curtis, J.T. (1957) An Ordination of the
Upland Forest Communities of Southern Wisconsin. Ecol.
Monogr., 27 : 325-349.

Everitt, B. (1974) Cluster Analysis. Heinemann Educational
Books, London.

Hentschel, E. (1923) Der Bewuchs an Seeschiffen. Inter. Rev.
ges. Hydrobiol. u. Hydrogr., 11 : 238-264.

Marine Corrosion Sub-Committee (1943). First Report of the
Marine Corrosion Sub-Committee. J. Iron Steel Inst., 147:
339-451.

Millard, N. (1951). Observations and Experiments on Fouling
Organisms in Table Bay Harbour, South Africa. Trans. R. Soc.
S. Afr., 33 : 415-445.

Skerman, T.M. (1960) Ship-Fouling in New Zealand Waters: A
Survey of Marine Fouling Organisms From Vessels of the
Coastal and Overseas Trades. N.Z. J. Sci., 3 : 620-648

Sorensen, T. (1948) A Method of Establishing Groups of Equal
Amplitude in Plant Sociology Based on Similarity of Species
Content and its Application to Analyses of the Vegetation on
Danish Commons. Biol. Skr., 5 : 1-34.

Visscher, J.P. (1927) Nature and Extent of Fouling of Ship's
Bottoms. Bull. U.S. Bur. Fish., 43 : 193-252.

Woods Hole Oceanographic Institution (1952) Marine Fouling
and its Prevention. Contribution No. 580 from the Woods Hole
Oceanogr. Instn, prepared by the United States Bureau of
Ships, Navy Department. United States Naval Institute,
Annapolis.

SAFETY ANALYSIS FOR RANDOM ELASTIC-PLASTIC FRAMES IN THE
PRESENCE OF SECOND-ORDER GEOMETRICAL EFFECTS

F. Casciati and L. Faravelli

Istituto di Scienza e Tecnica delle Costruzioni – Università di
Pavia (Italy)

INTRODUCTION

Reliability analysis approaches were generally developed with
reference to structures whose behaviour may be studied in the
spirit of the first order theory (*i.e.* stresses calculated over
the initial geometry of the structure). Under this assumption,
the reliability $\mathbb{R}$ of a structure can be evaluated once the conditional probability of failure $F_R(\underline{w})$ (C.P.F.), given that the
external loads $\underline{W}$ assume the values $\underline{w}$, is known (Augusti-Baratta,
1973)(Casciati-Sacchi, 1975) (Augusti-Baratta-Casciati, 1979).
The C.P.F. represents a generalization of the safety domain
usually introduced in the deterministic structural analysis.
This function depends on the randomness of the geometrical and
mechanical structure properties. It is generally evaluated via
suitable Monte Carlo techniques (Casciati-Faravelli-Sacchi,1977)
(Casciati-Faravelli, 1977).
However, if environmental forces on steel frames are considered
(*i.e.* wind load, snow load and, in some regions, earthquakes actions), geometrical non-linearities may not be neglected in order to obtain an estimation of the actual structural carrying
capacity. The steel frames are generally designed so that the
serviceability conditions are satisfied in the elastic field.
Nevertheless these serviceability limit-states are actually attained, even if seldom, during the structure lifetime. Hence, a
steel frame subject to environmental forces may reach the inelastic field and large displacements may occur.

The aim of this paper is to perform a reliability analysis of
elastic-plastic frames taking into account the geometrical non-
linearities of the structural behaviour. Particularly the required modifications in the usual approach to safety analysis
of structures fully described by a first-order model, are emphasized.
At first the problem is investigated with reference to structures having random mechanical properties and subjected to random

live loads and environmental forces varying in time in proportion to one stochastic function. In this case structural reliability may still be evaluated by the classical convolution integral. The C.P.F. can be approximated by using Monte Carlo techniques. For this purpose the whole elastic-plastic analysis, with iterations on the stiffness matrix (because of the second-order effects), must be repeated over again for each of the simulated experiments.

In order to reduce the required computational effort, the parametric Monte Carlo technique able to analyse the randomness of the first-order elastic-plastic behaviour of frames (Casciati-Faravelli, 1979) is generalized taking into account the second-order effects by a suitable set of *"fictitious shears"*.

Safety analysis is then performed with reference to structural problems for which more stochastic processes in time must be introduced to describe the acting environmental forces. The geometrical non-linearities involve that the structure failure depends on the whole loading path. In fact cycles of loading and unloading may produce,along the structure,permanent deformations that vary the initial geometry at each loading application and decrease (or increase) the actual structural strength. Hence, in the load space, only a safety domain referred to a given loading path is meaningful. Moreover, if the external actions are described by stochastic processes, also the C.P.F. becomes a stochastic field in time. Therefore a correct estimation of structural reliability would require the knowledge of the joint probability density function (J.P.D.F.) of the load parameters and the C.P.F.. In order to avoid cumbersome calculations, some approaches, that take into account the *"memory"* of the structure but allow to consider the C.P.F. constant during the structural lifetime, are presented.

Finally a numerical example is developed to illustrate the computational aspects of the method.

GOVERNING RELATIONS

Reliability analysis

Consider a structure subject to a set of simultaneous loading conditions that do not vary in time and are fully defined by n random parameters W_i (i = 1,..., n) with given J.P.D.F. $f_{\underline{W}}(\underline{w})$. Moreover let the structural behaviour be described by a — first-order model. The structural reliability R can be evaluated by the classical convolution integral (Augusti-Baratta-Casciati, 1979):

$$R = 1 - p_{fail} = 1 - \int_{\Omega} f_{\underline{W}}(\underline{w})\, F_R(\underline{w})\, d\underline{w} \qquad (1)$$

where p_{fail} denotes the probability of failure of the structure and Ω is the definition field of the load parameters W_i. The function $F_R(\underline{w})$ is the conditional probability of failure (C.P.F.) given that $\underline{W} = \underline{w}$ and depends on the randomness of the geometrical and mechanical structural properties. The C.P.F. is generally evaluated making use of suitable Monte Carlo techniques (Casciati-Faravelli-Sacchi, 1977)(Casciati-Faravelli, 1977).

A direct estimation of structural reliability by simulation procedures is impossible in practice because of the very large number of experiments required to evaluate low frequencies. Therefore only the structural response randomness is generally investigated by Monte Carlo techniques, while the reliability is estimated by introducing a suitable probabilistic analytical model (Casciati-Faravelli-Gobetti, 1979). For the studied problem R is estimated by Eq. (1).

Assume now the structure subject to (n-1) permanent actions W_i (i = 1,...,n-1) and to one load S(t) varying in time. Eq. (1) still holds if the n-th component of the vector W is defined as the random maximum value S^+ of S(t) during the structure lifetime. Furthermore, if S^+ is independent of W and its coefficient of variation (c.o.v.) is much greater than the ones of the permanent loads W (characteristics that snow and wind loads generally present), Eq. (1) may be written in the form (Casciati-Faravelli, 1977):

$$R = 1 - P_{fail} = 1 - \int_{-\infty}^{+\infty} f_{S^+}(s)\, F_{R,W}(s)\, ds \qquad (2)$$

where $F_{R,W}(s)$ is the C.P.F. given that $S^+ = s$. In Eq. (2) the randomness of the actions W is taken into account by the C.P.F. because these loads are regarded as elements that decrease (or increase) the actual resistance of the structure to the load S(t).

However this approach may not be generalized to the case in which two or more loading conditions $S_j(t)$ (j = 1,...,h) varying in time act upon the structure. In fact the maximum values S_j^+ generally do not occur simultaneously.

If $S(t)$ means the environmental force vector, the variability in time of every $S_j(t)$ may be idealized by a filtered Poisson process (F.P.P)(Parzen, 1972)(Krée, 1976) with interarrival times much larger than the average duration d_j of a single action pulse (see Fig. 1). For such problems a lower bound of the structural reliability in a suitable time interval τ can be calculated by the expression (Casciati-Faravelli-Zanon, 1978):

$$R(\tau) = 1 - P_{fail}(\tau) \geq 1 - \left\{ \sum_{j=1}^{h} P_{fail}(\underline{W}, S_j|\tau) + \sum_{j=1}^{h-1} \sum_{k=j+1}^{h} \Pi(S_j, S_k|\tau) P_{fail}(\underline{W}, S_j, S_k) \right\} \qquad (3)$$

where $P_{fail}(\underline{W}, S_j|\tau)$ denotes the probability of failure in τ of the structure acted upon by the permanent loads W and the j-th varying in time load $S_j(t)$; $P_{fail}(\underline{W}, S_j, S_k)$ is the probability of failure of the structure acted upon by W, S_j and S_k. Both these probabilities can be evaluated making use of Eq. (1) (or (2)). Moreover $\Pi(S_j, S_k|\tau)$ denotes the probability that the loads S_j and S_k occur simultaneously in the time interval $(0,\tau)$

502

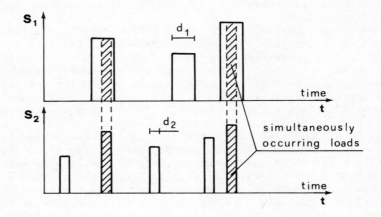

Figure 1 - Idealization of the stochastic load processes $S_1(t)$
and $S_2(t)$

and it can be calculated under suitable hypotheses on the stochastic processes $S_j(t)$ and $S_k(t)$. In Eq. (3) the probabilities
that the failure occurs when more than two loads act simultaneously on the structure are neglected because the probabilities
$\Pi(S_j, S_k, S_\ell|\tau)$ $(j \neq k \neq \ell)$ are generally very small.
A lower bound of the structural reliability during the lifetime T
is then obtained by the expression:

$$\mathbb{R}(T) = (\mathbb{R}(\tau))^{T/\tau} \qquad (4)$$

Structural response

With reference to a plane frame, the following assumptions are
introduced:

a) structural behaviour is geometrically non-linear (*i.e.* dis-
splacements from the initial geometry are not negligible);
b) flexural action predominates;
c) shear, axial force, axial shortening are neglected;
d) inelastic deformations concentrate in single sections (critical sections) and give rise to rotations θ(plastic hinge hypothesis);
e) the flexural characteristics are idealized to be elastic-perfectly plastic and non-holonomic.

Consider the plane frame discretized in perfectly elastic elements connected by N critical sections. Moreover let S_o and
θ_o denote the load level and the inelastic rotations that define a known initial solution.

Under the above assumptions, the hinge-by-hinge elastic-perfectly plastic analysis, as the load S increases (or decreases),
may be performed by solving the following parametric linear programming (P.L.P.) (Casciati-Faravelli, 1978):

$$\max - (\; \tilde{\underline{\xi}}\; \underline{\Delta\theta}^+ + \tilde{\underline{\zeta}}\; \underline{\Delta\theta}^- + \tilde{\underline{\mu}}\; \underline{\omega}^+ + \tilde{\underline{\nu}}\; \underline{\omega}^-) \qquad \text{a)}$$

$$\underline{Z}_G\; \underline{\Delta\theta}^+ - \underline{Z}_G\; \underline{\Delta\theta}^- + \quad \underline{\omega}^+ \quad =$$

$$= \underline{m}_L^+ - \underline{m}_{WG} - S_o\; \underline{m}_{EG} - \underline{Z}_G\; \underline{\theta}_o - \Delta S\; \underline{m}_{EG} \qquad \text{b)}$$

$$-\underline{Z}_G\; \underline{\Delta\theta}^+ + \underline{Z}_G\; \underline{\Delta\theta}^- + \quad \underline{\omega}^- = \qquad\qquad (5)$$

$$= \underline{m}_L^- + \underline{m}_{WG} + S_o\; \underline{m}_{EG} + \underline{Z}_G\; \underline{\theta}_o + \Delta S\; \underline{m}_{EG} \qquad \text{c)}$$

$$\underline{\Delta\theta}^+ \geq 0\; ;\; \underline{\Delta\theta}^- \geq 0\; ;\; \underline{\omega}^+ \geq 0\; ;\; \underline{\omega}^- \geq 0 \qquad \text{d)}$$

In Eq. (5) $\underline{m}_{EG}$ and $\underline{m}_{WG}$ denote the vectors of the elastic bending moments at the N-critical sections due to the load S, made equal to the unit, and to the permanent actions $\underline{W}$ respectively; $\underline{m}_L$ and $-\underline{m}_L$ are the vectors of the positive and negative yielding moments; $\underline{\omega}^+$ and $\underline{\omega}^-$ are vectors of "slack" variables. The NxN matrix $\underline{Z}_G$ is the matrix of the influence coefficients. The vectors $\underline{\Delta\theta}^+$ and $\underline{\Delta\theta}^-$ denote the positive and negative increments of the inelastic rotations at the N-critical sections: the actual inelastic rotations $\underline{\theta}$ are given by $\underline{\theta} = \underline{\theta}_o + (\underline{\Delta\theta}^+ - \underline{\Delta\theta}^-)$. The orthogonality relations due to the constitutive law are satisfied by defining the components of the vectors $\underline{\xi}$, $\underline{\zeta}$, $\underline{\mu}$ and $\underline{\nu}$ in the following way: $\xi_i = 0$, $\zeta_i = 1$, $\mu_i = 1$, $\nu_i = 0$ at the positive plastic hinges; $\xi_i^i = 1$, $\zeta_i^i = 0$, $\mu_i^i = 0$, $\nu_i = 1$ at the negative plastic hinges; $\xi_i^i = 1$, $\zeta_i^i = 1$, $\mu_i^i = 0$ and $\nu_i = 0$ at the remaining sections.

In Eq. (5) $\underline{m}_{EG}$, $\underline{m}_{WG}$ and $\underline{Z}_G$ are calculated taking into account both the elastic and the geometrical stiffness matrices. Therefore each step of this incremental structural analysis may be summarized in the following three phases:

I) to determine the critical value ΔS_c to which load S can be increased (or decreased) before the initial "basic" solution will non longer be "optimal";

II) to iterate phase I) until the axial forces along the structural elements calculated for $S = S_o + \Delta S$ coincide with the ones considered in the evaluation of the geometrical stiffness matrix;

III) to find the new stress-strain solution of Eq. (5) for $S = S_o + \Delta S_c$, taking into account that a new plastic hinge forms or a local unloading occurs at this load level, as determined by phase I).

In order to avoid the computational effort required by this iterative procedure, the authors proposed (Casciati-Faravelli, 1978) an approximate method capable of performing the second-order elastic-plastic analysis of a given frame without iterations. This approach consists in calculating the stresses over the initial geometry and in taking into account the geometric non-linearities by a suitable set of "fictitious shears". The solution of problem (5) is approximated by the one of the following P.L.P. problem:

504

$$\max - (\ \tilde{\underline{\xi}} \ \underline{\Delta\theta}^+ + \tilde{\underline{\zeta}} \ \underline{\Delta\theta}^- + \tilde{\underline{\mu}} \ \underline{\omega}^+ + \underline{\nu} \ \underline{\omega}^-) \qquad \text{a)}$$

$$\underline{Z} \ \underline{\Delta\theta}^+ - \underline{Z} \ \underline{\Delta\theta}^- + \quad \underline{\omega}^+ \qquad = \underline{m}_L^+ - \underline{m}_W^+ +$$

$$- S_o \ \underline{m}_E - \underline{Z} \ \underline{\theta}_o - \underline{M}_T \ \underline{t}_o - \Delta S \ \underline{m}_E - \underline{M}_T \ \underline{\Delta t} \qquad \text{b)}$$

$$\qquad\qquad\qquad\qquad\qquad\qquad\qquad\qquad (6)$$

$$-\underline{Z} \ \underline{\Delta\theta}^+ + \underline{Z} \ \underline{\Delta\theta}^- \qquad + \quad \underline{\omega}^- = \underline{m}_L^- + \underline{m}_W^- +$$

$$+ S_o \ \underline{m}_E + \underline{Z} \ \underline{\theta}_o + \underline{M}_T \ \underline{t}_o + \Delta S \ \underline{m}_E + \underline{M}_T \ \underline{\Delta t} \qquad \text{c)}$$

$$\underline{\Delta\theta}^+ \geq 0 \ ; \ \underline{\Delta\theta}^- \geq 0 \ ; \ \underline{\omega}^+ \geq 0 \ ; \ \underline{\omega}^- \geq 0 \qquad \text{d)}$$

where $\underline{t}_o$ is the fictitious shear vector of length K (K denotes
the storey number of the frame) and $\underline{M}_T$ is the NxK matrix whose
j-th column means the bending moments at the N critical sections
due to the j-th fictitious shear made equal to the unit. The matrix
$\underline{M}_T$ is calculated over the initial geometry of the structure.
Moreover Δt_j is the increment of the j-th fictitious shear t_j
corresponding to the increment ΔS of the external load S; every
Δt_j can be expressed by a rational function of ΔS. Therefore
the load increment ΔS is the only free parameter of the R.H.S.'s
of Eqs. (6b) and (6c). If h is the number of stochastic functions that define the environvental forces acting upon the structure, h becomes the number of these free parameters.
In Eq. (6) $\underline{m}_E$, $\underline{m}_W$ and $\underline{Z}$ (as well as $\underline{M}_T$) are evaluated over the
initial geometry of the structure. Hence the technological ma-
trix of problem (6) (i.e. the L.H.S.'s of Eqs. (6b) and (6c))
does not vary as the load S is increased (or decreased) and phase
II) of the incremental analysis (5) is not required.
For both the approaches (5) and (6), the failure of the frame
occurs when suitable stability conditions are not satisfied (Ca
sciati-Faravelli, 1978).

ENVIRONMENTAL FORCES DEFINED BY ONE STOCHASTIC PROCESS

Consider a structure having random material strength and assume
that random permanent loads and environmental forces varying in
time in proportion to one stochastic function S(t) act upon this
structure. Whether the structural behaviour is geometrically
linear or non-linear, its resistance to the environvental forces
is not influenced by the inelastic deformations that may develop
along the structural members during the lifetime T. Hence this
carrying capacity does not depend on the actual loading path,
but only on the random material strength ($\underline{m}^+$ and $\underline{m}^-$ in Eq. (5))
and on the random values of the live loads ($\underline{m}_{WG}$ in Eq. (5)).
Therefore, for both the analysed structural behaviours, the reliability of the structure can be evaluated by Eq. (2).
However, if the second-order effects must be taken into account,
the estimation of the actual structural strength involves the
performance of the whole stress-strain analysis as the environmental forces are monotonically increased up. The numerical na
ture (summarized in Eq. (5) for elastic-perfectly plastic frames)

of this analysis leads to approximate the C.P.F. $F_{R,W}(s)$ by using
Monte Carlo techniques. For this purpose the hinge-by-hinge
second-order elastic-plastic analysis must be repeated for each
of the yielded experiments.

In order to obtain a computational advantage (and hence to make
possible the analysis of large samples), it is possible to modify for problem (6) the parametric Monte Carlo technique proposed by the authors (Casciati-Faravelli, 1979) with reference
to geometrically linear elastic-plastic frames. This procedure
avoids that the structural analysis is performed over again for
each artificially yielded experiment and it is available only
if the technological matrix of the solving P.L.P. is the same
for all the experiments. Therefore the method can be extended
to problem (6), but not to problem (5) whose technological matrix depends on the geometrical stiffness matrix of the structure.

If the geometrical non-linearities are taken into account by
introducing a suitable set of fictitious shears, the "ad hoc"
simulation technique is formulated as follows. At first the
deterministic analysis of the average structure ($i.e.$ the structure defined by the average values of the random design parameters) is performed and the piecewise-linear relationships $\bar{\theta}(S)$
between the inelastic rotations θ and the load level S are
stored. Then a first structure (whose elastic solution coincides, under the made hypotheses, with the one of the average
structure) is simulated. For this structure phase I) of the
deterministic structural analysis is developed, so that the position of the first formed plastic hinge is determined. If this
plastic hinge coincides with the first one of the average structure, phase III) of the deterministic analysis is avoided because the new searched stiffness matrix coincides with the
stored one of the average structure. This statement is true
for all the successive branches of the piecewise-linear relationship $\bar{\theta}(S)$ as long as the found plastic hinge is the same
that forms in the average structure. Otherwise phase III) of
the deterministic structural analysis must be developed. For
each of the artificially yielded experiments, the whole procedure must be repeated and phase III (that requires the more
cumbersome calculations) is avoided as long as it is possible.
It is worth noting that phases I) of the geometrically linear
and non-linear analyses are different. In fact, in the first
case, ΔS_c is the lower value of the roots of a set of linear
equations, while, if the structural behaviour is geometrically
non-linear, ΔS_c is the lower value of the roots of a set of
equations of degree $(K + 1)$ (being K the storey number of the
frame).

The frame of Fig. 2, whose structural response randomness was
already analysed by the authors (Casciati-Faravelli, 1979) in the
spirit of the first-order theory, is considered in order to
illustrate the computational advantage attainable making use of
the above simulation procedure. Beams and columns of the frame

506

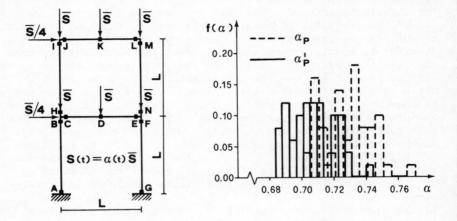

Figure 2 - Analysed structural
        problem and its dis‍
        cretization

Figure 3 - Histograms of the
        failure load factors
        α_p and α_p' for the
        frame of Fig. 2

are of the same uniform ideal I-section. They were designed
such that the rigid-plastic collapse of the average frame occurs when the load factor α is equal to the unit. The slenderness λ of the columns was taken equal to 120 and the following
values were assumed for the mechanical parameters: mean yield
stress $E\{\sigma\}$= 2.528 t/cm^2; elastic modulus E = 2107 t/cm^2. The
frame was discretized in eight perfectly elastic elements and
fourteen critical sections were considered (*see* Fig. 2). The
yielding stresses of the material at the critical sections were
assumed to be uncorrelated normally distributed random vari-
ables with the same mean value $E\{\sigma\}$ and the same c.o.v. c_σ = 5%.
Moreover the interaction between axial force and bending moment
was taken into account. The randomness of the actual failure
load factor α_p was studied by an usual Monte Carlo technique.
For this purpose the whole stress-strain analysis summarized in
Eq. (5) was performed for each of the simulated structures. The
histogram of α_p is drawn by dashed line in Fig. 3. This histogram, that approximates the C.P.F. of the frame, was obtained
from the analysis of 50 structures. Then the failure load factor
α_p' was determined taking into account the second-order effects by
a suitable set of fictitious shears. Its histogram, that is
drawn in solid line in Fig. 3, was obtained by using the above
improved simulation method. From a computational point of view,
if one assumes equal to the unit the computer time required by
the evaluation of the histogram of α_p, the histogram of α_p' required 0.51 making use of an usual Monte Carlo technique and
0.22 via the "ad hoc" simulation procedure.
Table 1 compares the average values ($E\{\cdot\}$) and the coefficients
of variations (c) of α_p, α_p' and α_p° (α_p° is the plastic collapse
load factor obtained in the spirit of the first-order theory

(Casciati-Faravelli, 1979)).
If steel structures whose behaviour is well described by a first
order approach are considered, the structural randomness is generally negligible in the estimation of their reliability. That
is due to the low values of the coefficient of variation of α_P° in
comparison with the ones of the environmental forces. The previous statement is also true for the structure of Fig. 2 whose
behaviour is geometrically non-linear (see the coefficients of
variation of α_P and α_P' in Table 1).

Table 1 - Results obtained for the structure of Fig. 2

Load factor	α_P	α_P'	α_P°
$E\{\cdot\}$	0.728	0.707	0.956
$c(\%)$	2.070	1.960	2.110

ENVIRONMENTAL FORCES DEFINED BY MORE STOCHASTIC PROCESSES

Suppose now that the environmental forces acting on the structure are described by more than one stochastic process. Moreover
assume that these forces may be idealized as F.P.P.'s with interarrival times much larger than the average duration of a
single action pulse.
If the structural behaviour is geometrically non-linear, the inelastic deformations that may develop along structural members
before any time t, can influence the structural resistance to
the environmental forces occurring at time t. It follows that:
  i) the structural lifetime T cannot be discretized in a finite
     number of intervals such that the structural behaviours in
     two successive intervals are independent between them. There
     fore Eq. (4) is no longer valid and the reliability of the
     structure can be evaluated by Eq. (3) only for $\tau = T$.
 ii) the C.P.F. becomes a stochastic process in time $F_{R,W}(s,t)$
     that depends on the actual loading path. Hence the probabil-
     ities of failure of the R.H.S. of Eq. (3) cannot be calcu-
     lated by Eq. (1) or (2).
Therefore a correct safety analysis of the structure requires
the evaluation of the probability that the point representing,
in the load space, the multivariate stochastic process of the
environmental forces at any time $\bar{t}$ $(0 < \bar{t} < T)$, is not outside
the safety domain defined by the realization of $F_{R,W}(s,\bar{t})$ at
time $\bar{t}$. However the solution of this problem is quite difficult
because $F_{R,W}(s,t)$ depends on the loading stochastic process and
on the randomness of the structural response. From a computational point of view a rough estimation of the structural reli
ability can be obtained by performing a sufficiently large number of simulated time-history analyses in (0,T). In fact, for such
a problem, as well as for structures subject to dynamical loads,
it is not possible to estimate the reliability independently of
the structural response analysis.

A different approach to safety analysis consists in determining
lower and upper bounds to the probability of failure p_{fail}.
This approach is possible if the path along which the loads
reach their final values is known and is always the same. Under
these hypotheses, in fact, the safety domain in the load space
is meaningful. If some simulated time-history analyses in (O,T)
are performed, for each of them several successive safety domains are determined as the analysis proceeds. The inward and
outward envelopes of the boundaries of these safety regions define the more dangerous domain Γ^U and the less dangerous one
Γ^L respectively. The set of the simulated Γ^U allows to approximate the C.P.F. $F_{R,W}^U(\underline{s})$ given that the failure occurs when the
structure carrying capacity assumes its lowest value in (O,T),
whereas the set of the safety domains Γ^L approaches the C.P.F.
$F_{R,W}^L(\underline{s})$ given that the failure occurs when the structural strength
assumes its highest value in (O,T). The functions $F_{R,W}^U(\underline{s})$ and
$F_{R,W}^L(\underline{s})$ do not vary in time and do not depend on the actual loading path. Therefore, by using Eq. (2) or (1) and then Eq. (3),
it is possible to evaluate the probabilities of failure p_{fail}^U
and p_{fail}^L respectively. Then the actual probability of failure
of the structure is bounded by:

$$p_{fail}^L \leq p_{fail} \leq p_{fail}^U \tag{7}$$

The calculation of p_{fail}^L and p_{fail}^U also requires the performance
of simulated time-history analyses in (O,T), but a small sample
is sufficient to obtain close results. In fact, unlike the direct
approach, the simulation technique is only employed to analyse
the randomness of the structural response, while the probabilities of failure are estimated by analytical models. Moreover as
the carrying capacity of the usual structures does not vary much
during the lifetime T, the difference $(p_{fail}^U - p_{fail}^L)$ is generally small and therefore a good approximation of the structural
reliability is obtained by Eq. (7).
In many cases steel structures are designed taking into account
that, if it will be necessary, they can be repaired during their
structural lifetime. The repair re-establishes the initial conditions and it is performed at intervals that may be regular,
f.i. once a year, or irregular, f.i. whenever the excessive permanent displacement limit state is attained. For both these cases
the evaluation of the C.P.F.'s $F_{R,W}^U(\underline{s})$ and $F_{R,W}^L(\underline{s})$ requires to
simulate a sample of time-history analyses only between two suc-
cessive repairs.
Furthermore, if the environmental forces may not change sign and
the possible inelastic deformations decrease the actual resistance of the structure, $F_{R,W}^L(s)$ coincides with the C.P.F. calculated for the structure at the initial conditions and $F_{R,W}^U$ is
the C.P.F. of the structure just before the repair. In order to
evaluate the last one, it is possible to simulate a sample of
time-history analyses between two successive repairs so that the
mean value vector $E\{\underline{\theta}\}$ and the covariance matrix Σ_θ of the inelastic rotation $\underline{\theta}$ just before the repair, are determined. Then

$F_{R,W}^U$ is calculated by the improved simulation technique discussed in the previous section, under the hypothesis that the initial conditions are defined by random inelastic rotations θ_o with mean values $E\{\underline{\theta}\}$ and covariance matrix $\underline{\Sigma}_\theta$.

NUMERICAL EXAMPLE

A numerical example was developed with reference to a two-storey single-span steel frame (*see* Fig. 4) subject to the live loads W_1 and W_2 and to the environmental forces S_1 (wind load) and S_2 (snow load). Also this frame was discretized in eight perfectly elastic elements with fourteen critical sections. The yielding stresses σ of the material at the critical sections were assumed to be uncorrelated normally distributed random variables with the same mean value $E\{\sigma\} = 2.4$ t/cm^2 and the same c.o.v. $c_\sigma = 5\%$. Beams and columns were chosen among the standard European I-sections (HEA and HEB series) with elastic modulus $E = 2100$ t/cm^2. The assumed axial force-bending moment interaction profile is drawn in Fig. 4b).

The permanent loads W_1 and W_2 acting upon the girders were assumed to be strictly correlated normally distributed random variables with mean values $E\{W_1\} = 1.5$ t/m and $E\{W_2\} = 3.6$ t/m and c.o.v. $c_W = 0.05$.

The maximum yearly wind and snow loads have an extreme type I probability distribution. The mean values and the coefficients of variations are: $E_1\{S_1\} = 1.275$ t and $c_{S1} = 0.20$ for the wind load; $E_1\{S_2\} = 0.72$ t/m and $c_{S2} = 0.30$ for the snow load. The intensities of the Poisson processes that describe the inter-arrival time of these actions are $\Lambda_{S1} = 12$ per year and $\Lambda_{S2} = 6$ per year. The snow load is assumed to occur only during the six winter months. Furthermore the average durations of the single action pulses are $d_{S2} = 10$ days and $d_{S1} \ll d_{S2}$. Therefore the

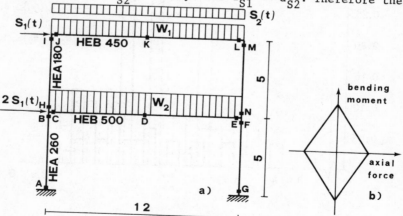

Figure 4 - Analysed structural problem and its discretization (lengths in meters); b) axial force-bending moment interaction profile

probability that the two environmental forces occur simultaneous ly during the time interval τ = 0.5 year is (Casciati-Faravelli-Zanon, 1978):

$$\Pi(S_1, S_2|\tau) = 1 - \exp(-\Lambda_{S1}\ \Lambda_{S2}\ d_{S2}\ \tau) = 0.86 \qquad (8)$$

For the considered problem $\Pi(S_1,S_2|0.5)$coincides with the probability of simultaneousness in one year. Moreover, if the loads occur simultaneously, the wind load is assumed to act on the structure after the snow load has reached its final value. Finally the hypotheses that the structure is repaired once a year and the actions do not change sign are made.

Fifty time-history analyses in the interval between two success ive repairs were performed. The obtained results showed that, for the considered problem, the developed inelastic deformations are very small so that the upper bound p_{fail}^U coincides with the lower bound p_{fail}^L and hence with the actual probability of failure.
In Fig. 5 the histograms, that approximate the C.P.F.'s $F_{R,W}^L(s_1|s_2)$ for given values of the snow action, are drawn.These histograms were obtained by using the above improved Monte Carlo technique for the given loading path. Table 2 shows the yearly probabilities of failure whose values are required by the R.H.S. of Eq. 3. The reliability of the considered frame during the lifetime T = 50 years is then bounded by Eq.(4):

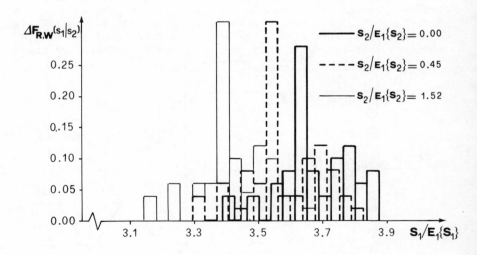

Figure 5 - Histograms whose cumulative frequency distributions approximate the C.P.F's $F_{R,W}^L(s_1|s_2)$ for given values of the snow action

Table 2 - Probabilities of failure for the structure of Fig. 4
          in one year

$P_{fail}(S_1\|1)$	$P_{fail}(S_1,S_2)$	$P_{fail}(S_2\|1)$	$P_{fail}(1)$
3.055 E-06	7.073 E-07	8.575 E-07	4.521 E-06

$$\mathbb{R}(T) = (\mathbb{R}(1))^{50} \geq$$

$$\geq \left[1 - P_{fail}(S_1|1) - P_{fail}(S_2|1) - \Pi(S_1,S_2|1) \, P_{fail}(S_1,S_2) \right]^{50}$$

CONCLUSIONS

This paper deals with the reliability analysis of structures
whose behaviour is geometrically non-linear. More independent
loads varying in time are considered to act on the structure.
As the inelastic deformations developing along structural members
can influence the actual carrying capacity of the structure, the
structural strength is a stochastic process in time depending
on the actual loading path. Therefore the usual reliability
approaches are not available. However, under suitable hypotheses,
these procedures can be employed to determine upper and lower
bounds to the actual failure probability of the structure.

ACKNOWLEDGMENT

*This research was supported by grants from the Italian Research
Council (C.N.R.).*

REFERENCES

Augusti, G. and Baratta, A. (1973)Theory of Probability and Limit Analysis of Structures under Multi-Parameters Loading; in Sawczuk, A. (ed.), Foundation of Plasticity, Noordhoff, Leyden: 347-364

Augusti, G., Baratta, A. and Casciati, F. (1979) Structural Response under Random Uncertainties; Guest Lecture for Session 5A, ICASP 3, Sydney

Casciati, F. and Faravelli, L. (1977) La sicurezza strutturale nei confronti degli stati-limite descrivibili come problemi di ottimizzazione vincolata (in Italian); Costruzioni Metalliche, 29, 3: 153-158

Casciati, F. and Faravelli, L. (1978) Analisi elasto-plastica di telai piani geometricamente non lineari (in Italian); Proc. of 4° Congresso AIMETA, Firenze: 141-151

Casciati, F. and Faravelli, L. (1979) Elasto-Plastic Analysis of Random Structures by Simulation Methods; Proc. of IMACS Congress, Sorrento (In press)

Casciati, F., Faravelli, L. and Gobetti, A. (1979) Reliability Analysis of Frames with Geometric Non-Linearity Subject to Stochastic Ground Accelerations; Proc. of ICASP 3, Sydney: 756-764

Casciati, F., Faravelli, L. and Sacchi, G. (1977) Etude probabiliste des structures par programmation paramétrique (in French) Bulletin Technique de la Suisse Romande, 14: 181-186

Casciati, F., Faravelli, L. and Zanon, P. (1978) Criteri di combinazione dei carichi accidentali di strutture metalliche in zona sismica (in Italian); Costruzioni Metalliche, 30, 3: 128-135

Casciati, F. and Sacchi, G. (1975) On the Longevity of Structures having Random Resistance; Proc. of ICASP 2, Aachen:101-114

Krée, P. (1976) Utilisation des probabilités et des statistiques pour l'évaluation de la securité des ouvrages (in French); in La Securité des Constructions, Eyrolles, Paris

Parzen, E. (1972) Stochastic Processes; Holden Day Inc.

PRACTICAL APPLICATION OF SEMI-RIGID BEAM TO COLUMN END-PLATE CONNECTIONS

Socrates A. Ioannides
Structural Engineer
S. D. Lindsey & Associates, Ltd.
Nashville, Tennessee USA

Thomas S. Tarpy, Jr.
Vanderbilt University
Nashville, Tennessee USA
Structural Engineer
S. D. Lindsey & Associates, Ltd.
Nashville, Tennessee USA

INTRODUCTION

Considerable expense is associated with the fabrication, erection, and inspection of steel rigid frame connections to withstand lateral loads caused by wind, seismic action, etc. This is especially true with structures of less than ten stories located in zones of low to moderate wind velocity. For this case, the structural engineer is faced with the decision of how to economically resist these wind forces without the expense normally associated with rigid connections.

One approach the writers have found to be economical from both a fabrication and erection viewpoint is the use of semi-rigid bolted end-plate connections with unstiffened column flanges (Figure 1). The analysis of frames with semi-rigid connections is achieved by the use of a computer code, utilizing the direct stiffness method of analysis and modeling each of the semi-rigid connections as a rotational spring. The stiffness of this spring is obtained from moment-rotation curves.

Such moment rotation curves are, unfortunately, not widely available. The authors have used the experimental and analytical curves reported in Reference 4 for beam-to-column connection. These curves are part of an on-going research project on end-plate beam-to-column connections with unstiffened column flanges at Vanderbilt University.

The basic theory behind the method of analysis and the development of the computer code will be briefly presented in this paper.

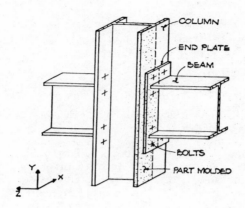

Figure I.   Beam-To-Column End
             Plate Connection.

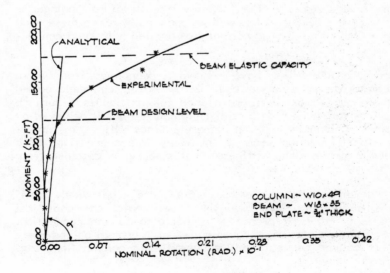

Figure 2.   Moment-Rotation Curve
             For Test No. 2.

Examples of frames analyzed utilizing the method will be discussed and comparisons will be made with results from analysis of rigidly connected frames. The background of the development of the moment-rotation curves used will be presented first, in order that a better understanding of the method is achieved.

MOMENT-ROTATION CURVES

Table I gives a partial summary of the experimental program (4) conducted on beam-to-column end plate connections. The steel used was mild steel of 36,000 psi yield strength (A36) and the bolts were friction type (A325). The bolt pattern used can be seen in Figure I. Each test specimen was instrumented with strain gauges and dial gauges and the stresses and deflections were monitored at various beam moment load increments until failure of the connection occurred.

TABLE I

EXPERIMENTAL PROGRAM

		TEST NUMBER	
BEAM	COLUMN	END PLATE THICKNESS	
W14x22	W 8x35	TEST 1 EP = 5/8"	TEST 4 EP = 7/8"
W18x35	W10x49	TEST 2 EP = 3/4"	TEST 5 EP = 1 1/4"
W24x55	W14x48	TEST 3 EP = 7/8"	TEST 6 EP = 1/14"

Figure 2 shows the moment-rotation curve for Test 2 (W18 x 35 beam, W10 x 49 column and 3/4" end plate). The center of rotation of the connection is not readily definable and will even shift thru the elastic plastic transition. In establishing these curves, the "nominal rotation" was used. That is, the displacement of the end plate at the location of the beam tension flange divided by the depth of the beam. Similar curves have been developed for the other tests.

Experimental moment-rotation curves for various connections are costly to obtain and the process is lengthy. A numerical finite

516

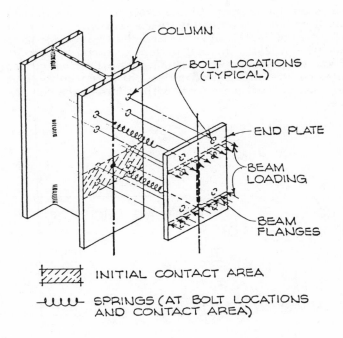

Figure 3. Initial Spring Locations.

517

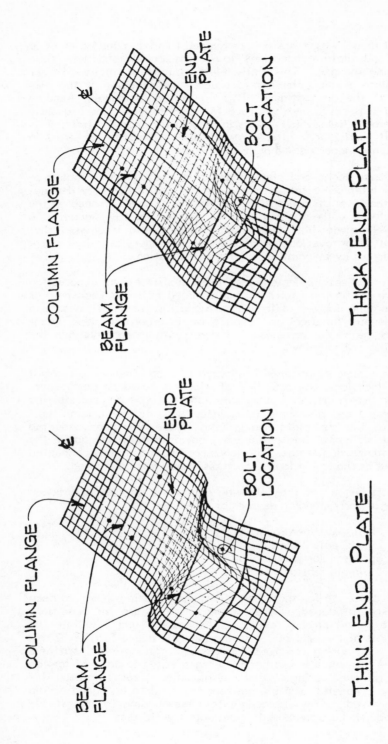

THIN-END PLATE

THICK-END PLATE

Figure 4.   End Plate-Column Flange Deformation Patterns.

element model was, therefore, developed to predict the behavior of bolted end plate connections in a less costly and less time consuming manner. This finite element model utilizes a higher order plate bending element to predict the behavior of the elements of the connections. This conforming triangular bending element was developed by Bruce M. Irons (6), and has a quartic displacement field which results in a quadratic moment field. This makes it particularly suitable for predicting the state of stress in the column flange and end plate.

The column flange-end plate interaction is a complex problem in that the interface conditions between the two bodies vary depending on the level of interaction and loading. For every node on the end plate, a corresponding node is defined on the flange. To effectively model this connection, a mechanism is devised that does not allow overlapping of the two members but does allow separation at every point except the bolt locations.

A theoretical spring element with infinite stiffness in compression and zero stiffness in tension is introduced between each pair or corresponding degrees of freedom. Since the spring stiffness is a discontinuous function and cannot be incorporated in a linear stiffness method of analysis, it is dealt with interactively in the manner described below.

The two bodies are assumed connected to each other via springs at certain locations, the selection of which is based on past experience or experimental observations. In this case a good starting point would be to assume that, in addition to the original bolt locations, the end plate and column flange are also connected below a horizontal line defined by the inside bolt line above the beam compression flange. The loads are applied to this connected structure as shown in Figure 3, and displacements calculated.

At this stage one of the following conditions may exist between any pair of corresponding degrees of freedom:

1) No displacement overlap,
2) Displacement overlap,
3) Spring in compression, or
4) Spring in tension.

Cases 1 and 2 are for those conditions where springs have not been previously introduced, whereas cases 3 and 4 are for conditions where a spring already exists. As cases 2 and 4 violate the constraints of the problem, a spring must be added for case 2 and the existing spring removed for case 4. The original stiffness matrix is then modified rather than regenerated to account for the new or removed springs and the displacements reevaluated. This process is repeated until no new springs need to be added or old ones removed. The displacements corresponding to this steady state condition are then used to evaluate the stresses.

The beam flange is not part of the model, but its contribution to the bending behavior of the end plate is included. The in-plane deformation of the beam flanges are rather small compared to the bending deformation of the end plate; thus, the nodes that lie on the lines that define the position of the beam flanges displace longitudinally essentially the same amount. This effect is modeled by introducing high internodal stiffnesses for these lines of nodes.

The bolts greatly affect the bending behavior of the connection in addition to connecting the column flange and the end plate. As such, the effect of the bolt head on the bending behavior must be taken into account as far as possible to accurately model the connection. In the model, a node is located at the center of each bolt, on both the flange and the end plate. This pair of nodes is restrained relative to each other in all three degrees of freedom (displacement and rotations). Although the actual clamping effect could not be simulated with the plate bending model, the resistance offered by the bolt and bolt head to the separation and bending of the end plate and column flange was approximately represented by forcing the lateral displacements of the flange and the end plate nodes lying within the confines of the bolt head to be the same for each pair of nodes.

Figure 4 shows the displacement results of this model plotted in three dimensional view for both a thin and a thick end plate. The analytical-moment rotation curve developed from the use of this model for the data of Test 2 is shown in Figure 2. To further demonstrate the end plate thickness effect on the column flange behavior and rotational capacity of the connection, Figure 5 shows the variation of maximum flange displacement and stress for different end plate thickness.

It should be noted from Figure 2 that the model is linear. The analytical results are in good agreement with the experimental results in the design range. Use of the rotational stiffness of the connection as predicted by the analytical model could, therefore, be ascertained for this beam-column-end plate combination as long as the beam end moments are kept below the design range.

Current research centers on parameter studies to develop simplified equations for the prediction of the linear moment-rotation curve and the moment level up to which such linear approximations could be safely used. Some of the basic parameters in this study are the thickness of the end plate, thickness of the column flange, pitch and gauge of the bolts and depth of the beam.

For the present, the authors have been using the results reported in Reference 4 and some additional runs of the finite element model for other beam-to-column-end plate combinations. Use of beam-to-column-end plate combinations for which analytical results are not available has been achieved by "careful interpolation"

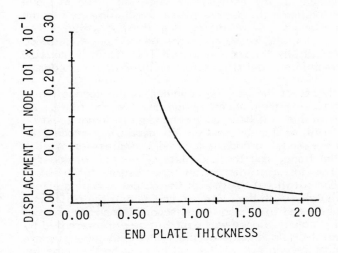

a)   Displacement Versus End Plate
Thickness.

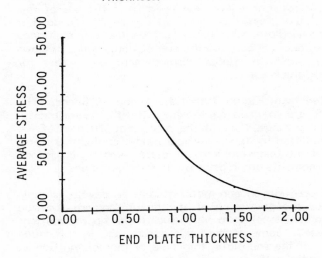

b)   Maximum Bending Stress Versus End
Plate Thickness.

Figure 5.   Influence of End Plate Thickness
on Displacements and Stresses.

between the known cases. When the results of the parameter studies become available such "educated guesses" will no longer be necessary.

MODELING OF SEMI-RIGID CONNECTIONS

The stiffness method is used to model frames with semi-rigid connections. The beams and columns are modeled using the familiar beam elements shown in Figure 6. The semi-rigid connections are modeled by the rotational springs shown on each end of the beam. In both of the models above, axial deformations have been neglected to minimize the core requirements.

The stiffness matrix of the beam element with respect to the local coordinate axis of the element can be written as:

$$
(K_b) = \begin{bmatrix}
K_L^s & & & & & \\
0 & K_R^s & & & \text{(SYM.)} & \\
-K_L^s & 0 & K_{33}^b + K_L^s & & & \\
0 & -K_R^s & K_{43}^b & K_{44}^b + K_R^s & & \\
0 & 0 & K_{53}^b & K_{54}^b & K_{55}^b & \\
0 & 0 & K_{63}^b & K_{64}^b & K_{65}^b & K_{66}^b
\end{bmatrix}
\begin{matrix}
1 \\
2 \\
3 \\
4 \\
5 \\
6
\end{matrix}
$$

where the subscripts L and R denote the left and right springs, respectively, the superscript b refers to the conventional beam element (2), and the superscript s refers to the rotational spring, K denotes a stiffness coefficient. The subscripts ij specify the stiffness coefficient with respect to the basic displacements identified in Figure 6(a).

When establishing the stiffness matrix for a beam element, if the springs are assumed to be infinitely stiff $\theta_1 = \theta_3$ and $\theta_2 = \theta_4$,

522

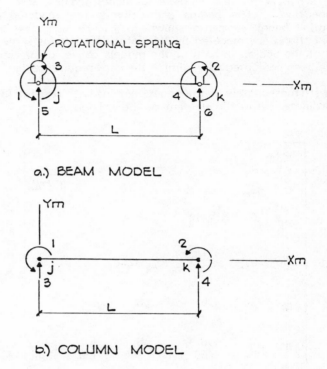

a.) BEAM MODEL

b.) COLUMN MODEL

Figure 6.   Identification of Components of End Displacements and End Actions for Basic
Structural Elements.

resulting in the conventional beam-column element with both ends restrained; this amounts to assuming a rigid connection. if the springs are assumed to have no stiffness, i.e. $K_L^s = K_R^s = 0$, then the rotations θ_1 and θ_3, as well as θ_2 and θ_4, are uncoupled; the stiffness matrix reduces to that for a member with pin connections at both ends.

The stiffness of the rotational springs is obtained from the moment-rotation curves, for the particular connection in use. If the linear analytical M-θ curve is used, the spring stiffness is simply the slope of the curve.

$$K^s = \tan^{-1}(\alpha) \quad = \frac{M}{\theta}$$

However, if non-linear moment rotation curves are available (experimental or from more sophisticated models), then the iterative method described in Reference 3 and 5 may be used. The linear approximation was chosen here because of its simplicity and relative availability.

The problem of analyzing frames with end plate connections and unstiffened column flanges then reduces to one using the above mentioned computer code. In designing the elements of such frames, in addition to checking the beams and columns, the connections must now be checked to insure that the moments in these end plate connections do not exceed the allowable ones.

DESIGN EXAMPLE

The frame shown in Figure 7a can be used for demonstrating the practicality of the method. The sizes shown are those obtained from a preliminary design. Lateral and gravity loads are shown in Figure 7b. The frame was analyzed both with rigid and semi-rigid connections for comparison purposes.

The stiffness of the semi-rigid connections was computed from Fig. 2 as:

$$K = \tan^{-1}(\alpha) \quad = \frac{150 \times 12}{0.00245} = 734,690 \text{ k-in/rad.}$$

This value was rounded to 750,000 k-in/rad.

Wind and gravity loading conditions were analyzed separately so that the effect of the semi-rigid connections on stresses and displacements could be seen separately. The results of these two loadings were also combined to demonstrate the effect of the semi-rigid connections on the combined loading case. The following behavior was observed.

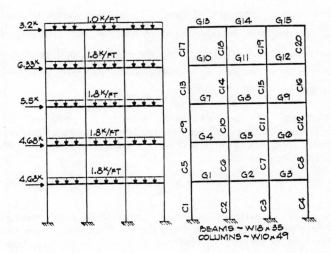

a) Frame Geometry

b) Member Designations

Figure 7. Design Example.

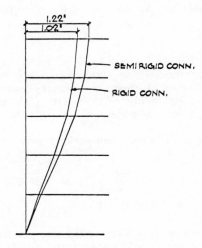

Figure 8. Lateral Drift for Design Example.

Table 2: BEAM END MOMENTS

		G1		G2		G3	
		ML	MR	ML	MR	ML	MR
Gravity	R	70.9	-103.1	95.2	-95.2	103.1	-70.9
Gravity	SR	66.2	-88.7	83.7	-83.7	88.7	-66.2
Wind	R	39.4	-34.9	-30.4	-30.4	-34.9	-39.4
Wind	SR	-36.9	-32.9	-30.1	-30.1	-32.9	-36.9
0.75 (W + GR)	R	23.7	-103.4	48.6	-94.1	51.2	-82.7
0.75 (W + GR)	SR	22.0	-91.2	40.2	-85.3	41.8	-77.3

Table 3: COLUMN END MOMENTS

	Gravity		Wind		0.75 (GR + W)	
	R	SR	R	SR	R	SR
C1 top	-29.4	-27.5	20.6	18.2	-6.6	-7.0
C1 bott	-14.7	-13.7	38.3	40.6	17.7	20.2
C5 top	-38.8	-36.6	21.2	21.0	-13.2	-11.3
C5 bott	-41.5	-38.7	18.8	18.7	-17.0	-15.0
C2 top	3.4	2.2	31.5	29.2	26.2	23.5
C2 bott	1.7	1.1	43.8	46.1	34.1	35.4
C6 top	3.9	2.5	34.7	34.8	29.0	28.0
C6 bott	4.5	2.9	33.7	33.8	28.7	27.5

Girder Moments

The gravity girder end moments for the semi-rigidly connected frame were lower than those for the rigidly connected frame by 7% to 14%. For wind loading, the moments in the exterior girders decreased from 2% to 6%; the change in moments in the interior girders ranged from a 1% decrease at the lower floor to a 4% increase at the top floor. For the combined gravity and wind loading case (which is the actual design case rather the wind or gravity load alone), the maximum girder end moments decreased from 6% to 12%. Table 2 shows the comparisons for the beam line at the second floor.

Column Moments

The gravity column moments were lower for the semi-rigid connected frame than for the one rigidly connected by 7% to 8%. The wind moments ranged from 6% higher at the first floor to 7% lower at the higher floor. The column end moments for the combined gravity and wind loading case ranged from 12% lower to 14% higher for the semi-rigidly frame than for the rigid frame. Table 3 shows the comparison for one exterior and one interior column for the first two floors.

Deflections.

The comparison of wind drift between the two frames is shown in Figure 8. The semi-rigid frame deflects laterally about 20% more than the rigid frame. Beam deflections for gravity loads are about 30% higher for exterior beams and about 50% higher for the interior beams. However, they are still within acceptable design limits.

The results of the example, show that the use of semi-rigid end plate connections tend to lower the girder end moments and gravity column moments, whereas they tend to increase wind drift and beam deflections. Wind moments in columns tend to become larger for the lower stories and smaller for the upper stories because of the increase in lateral deformation.

CONCLUSIONS

A method of analyzing frames with semi-rigid connections, such as the end plate type, has been presented and the computer code developed for such frame has been demonstrated. The usability of this code relies on the availability of moment-rotation curves for the semi-rigid connections.

The linear elastic finite element model used to develop the moment-rotation curves used by the authors has been presented. This model includes the interaction between the end plate and the column flange using a contact problem solution technique. The capabilities of the analytical model have been demonstrated thru comparisons with experimental tests.

The practical application of beam-to-column end plate connections has been demonstrated with the use of a design example. It has been shown how end plate connections can be used in lieu of welded connections and how the utilization of end plate connections can result in reduction of member sizes without detrimental increase of deflections.

ACKNOWLEDGEMENTS

The project described herein is being conducted at Vanderbilt University under the sponsorship of The American Institute of Steel Construction and the Research Council on Riveted and Bolted Structural Joints. The writers are extremely grateful to the AISC Task Advisory Group and the members thereof for their advice and assistance.

The contents of this paper reflect the view of the writers who are responsible for the facts and accuracy of the data presented herein. The contents do not necessarily reflect the official views of the American Institute of Steel Construction and do not constitute a standard, specification, or regulation.

REFERENCES

1. _____, Manual of Steel Construction, American Institute of Steel Construction, Seventh Edition, New York, 1970.

2. Beaufait, F. W., et.al., Computer Methods of Structural Analysis, Prentice Hall, Englewood Cliffs, N.J.

3. Gerlein, M. A., "An Interactive Approach to the Nonlinear Analysis and Design of Reinforced Concrete Structures", Ph.D. dissertation, Vanderbilt University, 1978.

4. Ioannides, S. A., "Column Flange Behavior in Bolted End Plate Moment Connections", Ph.D. Dissertation, Vanderbilt University, 1978.

5. Ioannides, S.A., Gerlein, M.A., Beaufait, F.W., Lindsey, S. D., "Nonlinear Semi-rigid Connections and PΔ Effects in Steel Design", Unpublished work.

6. Irons, B. M., "A Conforming Quartic Triangular Element for Plate Bending," International Journal for Numerical Methods in Engineering, Vol. I, pp. 29-45, 1969.

ON STOCHASTIC DYNAMICS OF FLOATING BRIDGES

Ivar Langen
The Norwegian Institute of Technology, Trondheim, Norway.

Ragnar Sigbjörnsson
SINTEF, the Foundation of Scientific and Industrial Research
at the Norwegian Institute of Technology, Trondheim, Norway.

INTRODUCTION

During the last years there has been a growing interest in
floating bridges in Norway as alternatives to long-span suspension bridges for spanning the many deep fiords on the
western coast of the country.

Different design concepts have been proposed: The most common
design is the surface bridge consisting of a continous straight
or curved box pontoon supporting the bridge deck either directly or through columns. In shallow and medium deep water the
pontoon is usually anchored to the sea bed at several points
along the bridge. In case of deep water where anchoring is
not feasible it is advantageous to use curved pontoon. Examples
of this type of bridge are the well-known Hood Canal Bridge
located near the Olympic Peninsula in the State of Washington
and the proposed Salhus Bridge outside Bergen which will be
described below. Common for bridges of this type is that the
pontoon emerges the water surface and therefore is exposed to
large wave forces. The cutting of the free water surface may
also have negative ecological consequences for the marine
biology. Further, special arrangements have to be made for
possible sea way traffic for instance by incorporation of a
moveable draw span or connecting the floating structure to a
fixed elevated bridge preferably at one end.

These drawbacks are not present in the semi-submersible design
concept where the bridge deck is supported by columns resting
on submerged pontoons. Only the columns emerge the water surface. Necessary stability and buoyancy margin are obtained
by using tensioned anchor cables.

As a third concept the submerged buoyant tubular bridge may be
mentioned. Here the traffic runs within a submerged tube positioned by a system of buoyancy tanks and tensioned anchor
cables.

The structures described above are exposed to random natural
forces such as waves, current and in some cases wind. The severity of these forces is underlined by the Hood Canal Bridge
disaster on the 15th of February, 1979. Therefore, a realistic
stochastic dynamic analysis is required to predict the behaviour
of the bridges. Dynamic analysis of floating bridges and related structures has been treated in the literature by Mukherji
1972, Hartz and Mukherji 1977, Wen and Shinozuka 1972, Wen 1974,
Adee 1975, Holand, Langen and Sigbjörnsson 1977, Clough,
Sigbjörnsson and Remseth 1977 and Sigbjörnsson 1978.

This paper deals with theoretical analysis of the dynamic behaviour of floating bridges with the main emphasis on the wave-
induced response. The proposed Salhus floating bridge is used
as an example.

THE SALHUS BRIDGE PROJECT

The Norwegian Public Roads Administration considers a floating
bridge to replace the ferry connection across the Salhus Fiord
north of Bergen. The bridge is designed as a horizontal arch,
see Figure 1. The traffic runs on a concrete slab supported
by columns resting on a continuous pontoon. The cross-section
is shown in Figure 2. The given minimum dimensions are used
except at the southern end. Here the slab has been raised to
allow sufficient free sailing space, see Figure 1. The number
of cells (three in Figure 2) is supposed to be increased to
maximum 7 at the southern end. To allow the bridge to follow
the tide variation, hinge sections are implemented at the ends.

The Division of Structural Mechanics and the Foundation of
Scientific and Industrial Research at the Norwegian Institute
of Technology have been engaged to carry out dynamic analysis
of the bridge. The results have been reported by Holand and
Langen 1972, Sigbjörnsson and Langen 1975 and 1979. The project is still under evaluation and possible redesigns to improve the dynamic behaviour are being considered.

STRUCTURAL MODELLING

The structural idealization is based on the finite element
method. The curved bridge is modelled by 20 straight beam
elements (see Figure 1). The number of element is determined
so as to represent the continuous structure with sufficient
accuracy.

The equation of motion may be written

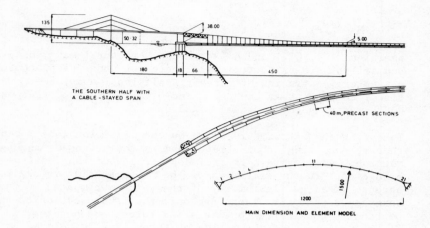

Figure 1   The Salhus floating bridge. Dimensions in m.

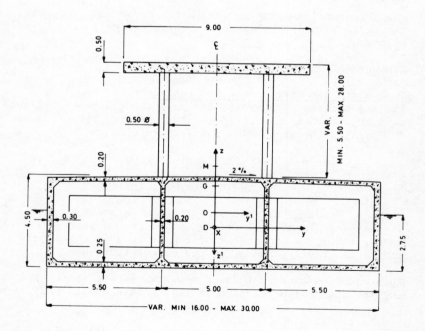

Figure 2   Cross-section. Dimensions in m.

$$M^{(s)}\ddot{r}(t) + C^{(s)}\dot{r}(t) + K^{(s)}r(t) = Q^{(r)}(t) \qquad (1)$$

in which $M^{(s)}$, $C^{(s)}$ and $K^{(s)}$ are the structural mass, damping and stiffness matrix, $r(t)$ is a vector of nodal displacement parameters consisting of three translational and three rotational components at each node and $Q^{(r)}(t)$ is the corresponding hydrostatic and hydrodynamic load vector.

The stiffness matrix K is based on the traditional beam bending theory, excluding shear strain, and on the St.Venant theory of torsion. Axial strain is included. The contribution from the bridge deck to the stiffness of the cross-section is neglected because of large flexibility of the columns compared to the box-shaped pontoon. Therefore only the cross-section of the pontoon is considered to take axial forces and bending and torsional moments, and the motion of the bridge is related to the centre of shear D of the pontoon alone (see Figure 2).

A consistent model is used for the mass matrix M. The whole cross-section is considered for the mass calculation, and thus the centre of gravity G is located above the centre of shear D.

The structural damping is considered to be small compared with the hydrodynamic damping and is therefore neglected in the calculations.

HYDRODYNAMIC MODELLING

Stochastic geometry of the sea surface

The sea surface is approximated as a three-dimensional zero mean ergodic Gaussian field which is assumed specified by the following cross-spectral density (Sigbjörnsson 1979)

$$S_{\eta_m \eta_n}(\omega) = S_{\eta\eta}(\omega) \int_\theta \Psi(\theta,\omega) \exp\{-i\,\mathrm{sign}(\omega)\frac{\omega^2}{g}((x_m - x_n)\cos\theta + (y_m - y_n)\sin\theta)\}d\theta \qquad (2)$$

Here, η_m denotes the wave amplitude in point (x_m, y_m), $S_{\eta\eta}(\omega)$ is the one-dimensional frequency spectral density of the waves, $\Psi(\theta,\omega)$ is the spreading function, g is the acceleration of gravity, and ω is the circular frequency.

This equation applies to deep water waves and is derived applying the dispersion relation to relate wave number and frequency.

Stochastic loading

The hydrodynamic loading is assumed obtained applying the linearized probabilistic potential theory. Then the total loading may be expressed as

$$Q^{(r)}(t) = Q(t) - \int_{-\infty}^{\infty} M^{(h)}(t-\tau)\ddot{r}(\tau)d\tau - \int_{-\infty}^{\infty} C^{(h)}(t-\tau)\dot{r}(\tau)d\tau - K^{(h)}r(t) \tag{3}$$

Here, $M^{(h)}$ is the hydrodynamic mass, $C^{(h)}$ is the hydrodynamic radiational damping and $K^{(h)}$ represents the hydrostatic restoring (buoyancy) where influence from the wavy sea surface has been neglected. $Q(t)$ is a vector of nodal point wave excitation forces given as

$$Q(t) = \sum_n a_n^T \int_{\ell_n} N^T \int_{-\infty}^{\infty} q(\omega)e^{i\omega t}dZ_\eta(\omega)ds \tag{4}$$

where $Z_\eta(\omega)$ is the spectral process associated with η and $q(\omega)$ is a complex valued frequency response function relating wave amplitude and distributed wave forces on the element, N is a vector of shape functions describing the displacement fields in the finite element idealization by nodal point values

$$u = Nv_n \tag{5}$$

a_n is the connectivity matrix relating the element displacement vector v_n to r

$$r = a_n v_n \tag{6}$$

Furthermore,

$$M^{(h)}(t) = \frac{1}{2\pi} \int_{-\infty}^{\infty} m^{(h)}(\omega)e^{i\omega t}d\omega \tag{7}$$

$$C^{(h)}(t) = \frac{1}{2\pi} \int_{-\infty}^{\infty} c^{(h)}(\omega)e^{i\omega t}d\omega \tag{8}$$

Hence it is seen that frequency independent hydrodynamic mass or damping is described in the time space in terms of the Dirac delta function.

It is seen from Equation 3 and 4 that the hydrodynamic loading is an ergodic Gaussian process provided that the structural response r and the sea surface are ergodic Gaussian processes.

Of special interest in the present study is the rectangular continuous pontoon. Frank 1967 and Vugts 1968 and 1970 have calculated the wave force amplitudes and the hydrodynamic quantities for sway, heave and roll for different types of ship-like sections using two-dimensional potential theory and different numerical schemes. Faltinsen 1969 has compared the methods for rectangular sections. The results show good agreement. Vugts 1968 and 1970 has also compared his theoretical results with experimental results over a large range of frequencies. The comparison shows that all the hydrodynamic quantities can be computed with sufficient accuracy by potential theory. One important exception is the roll damping, where the viscous

effects are much more important than in heave and sway.

Model tests designed for the Salhus floating bridge support these findings (Holand and Langen 1972).

The hydrodynamic mass and damping as well as wave excitation transfer function applied in the present study are derived from the results given by Vugts 1968 and 1970. Note that all these quantities are frequency dependent.

RESPONSE ANALYSIS

Equation of motion

The response of the structure-fluid system is governed by the following equation obtained from Equation 1 and 3

$$\int_{-\infty}^{\infty} M(t-\tau)\ddot{r}(\tau)d\tau + \int_{-\infty}^{\infty} C(t-\tau)\dot{r}(\tau)d\tau + Kr(t) = Q(t) \qquad (9)$$

where the mass, damping and stiffness matrices M, C and K include both structural, hydrostatic and hydrodynamic contributions.

It is seen that this equation represents a linear system. Hence it follows by applying Equation 4 that the response r is a zero mean ergodic Gaussian process as the sea surface is represented as a zero mean ergodic Gaussian field. Then the response statistics is completely specified in terms of the second order statistics.

Solution procedures

Frequency response method. A direct approach to solve Equation 9 is furnished by the frequency response method yielding the following expression for the spectral density of the response

$$S_{rr}(\omega) = H(\omega)S_{QQ}(\omega)H^{T*}(\omega) \qquad (10)$$

Here, the virtual frequency response function is given as

$$H(\omega) = [K - \omega^2(M^{(s)} + m^{(h)}(\omega)) + i\omega(C^{(s)} + c^{(h)}(\omega))]^{-1} \qquad (11)$$

while the spectral density of the wave excitation force is given as

$$S_{QQ}(\omega) = \sum_{mn} a_m^T [\int_{\ell_m} \int_{\ell_n} N^T q_i(\omega) q_j^{T*}(\omega) N S_{\eta_i \eta_j}(\omega) ds_i ds_j] a_n \qquad (12)$$

It is seen that in this method it is easy to treat the frequency dependent system properties. However, nonlinear properties may cause difficulties even if an equivalent linearization procedure is used.

Step-by-step integration. An alternative approach is furnished

by the step-by-step integration procedure which combined with a
properly simulated load history represents a sample treatment of
the response. The sample treatments are generally preferable if
the time series of the response are desirable in design. Furthermore, in the analysis of nonlinear systems it is often the only
tractable procedure for practical problems.

The step-by-step integration is especially feasible in connection with Equation 9 in either of the following cases: Firstly,
the retardation time, defined as

$$\tau_r = \max_{i,j} \{a_{ij}^{-1}(0) \int_0^t |a_{ij}(t)| \, dt\} \quad \forall a_{ij}(t) \in \{M(t), C(t)\} \tag{13}$$

is so short that the time space system functions may be approximated by the Dirac delta function. As a rule of thumb it is
sufficient that τ_r^{-1} is less than the sample frequency required
for a proper description of a sample function. Secondly, the
bandwidth of the response spectral density is so small that
the hydrodynamic mass and damping may be considered as constant
within the corresponding frequency range. This may be the case
when the wave excitation is governed by a narrow banded one-
dimensional frequency spectral density in resonance with a
single natural mode.

If either of these cases are fulfilled Equation 9 is reduced to
a second order stochastic differential equation with constant
coefficients. Hence, conventional step-by-step integration
methods are directly applicable.

It is worth pointing out that in the case of the continuous
pontoon of the Salhus bridge the equations of motion can strictly only be decoupled applying the damped normal mode due to the
hydrodynamic coupling between sway and roll. Therefore, the
direct step-by-step integration is used in analysis of the
Salhus bridge.

In the solution of the decoupled equations of motion obtained
by using normal modes, a conditional stable explicit integration scheme like the second central difference may be feasible
since the higher insignificant modes which cause instability
can be excluded from the numerical solution. The size of the
time step is then usually determined by the stability requirement for the highest included mode. Note that for a continuous
floating pontoon the natural periods are very close to each
other and therefore a large number of modes have to be included
in the superposition. This is particularly the case for the
moment and stress calculation which may be sensitive to the
exclusion of higher modes.

In the direct step-by-step integration an unconditional stable
scheme like the trapezoidal rule should be used. The time step
is then determined from accuracy requirements only. To ensure

stability with a conditional stable method the time step has to be a fraction of the smallest natural period of the model which leads to very short time steps and unnecessary computation compared to the use of an unconditional stable method. In a stochastic analysis where it is necessary to obtain relatively long time series in order to get proper statistics, the step size and the amount of computation per step are important.

Applied to free vibration the integration methods shows more or less two types of errors: Amplitude decay and period elongation. In a stochastic analysis, however, we are not primarily concerned with a response history as such but with an accurate estimation of the statistical properties of the response. For the first order statistics and the estimation of extreme values it is important that the method shows small amplitude error, while the period elongation may be less important. The period error may, however, be serious for the second order statistics of the response, e.g. the spectral densities.

Monte Carlo solution

Generation of load histories. To facilitate the sample treatment of the response based on the step-by-step integration it is necessary to generate representative sample functions of the excitation process. In the case of an ergodic multivariate Gaussian excitation this may be done as follows (see for instance Shinozuka 1972)

$$Q_j(t) = \text{Re}[\sqrt{2\Delta\omega} \sum_{\ell=1}^{j} \sum_{k=1}^{N} B_{j\ell}(\omega_k)\exp\{i(\omega t + \phi_{\ell k})\}]] \qquad (14)$$

Here, $\Delta\omega$ is the frequency resolution, $N=\max(\omega)/\Delta\omega$, ϕ is a random phase angle uniformly distributed between 0 and 2π, and $B_{j\ell}(\omega)$ is an element of the triangular matrix $B(\omega)$ obtained by Cholesky factorization of the spectral density matrix $S_{QQ}(\omega)$

$$B(\omega)B^{T}*(\omega) = S_{QQ}(\omega) \qquad (15)$$

Further discussion may be found in Sigbjörnsson 1979 b.

Statistical analysis of response histories. Once the response histories have been obtained the desired response quantities may be derived using standard statistical methods.

Generally, a fair estimate of the rms response may be obtained from rather short time series. This is also the case for the first order parent distribution and the maxima distribution of the response as well as the rate of zero crossings and maxima. This is, however, not the case for the extreme response commonly applied in design. The simulation of long time series to yield direct estimates of the extremes may be avoided by applying extrapolation. The expected largest maximum in a record

of independent maxima may be obtained by the following expression (Cartwright and Longuet-Higgings 1956)

$$E[r_{max}] = \sigma_r [\sqrt{2\ln(m\sqrt{1-\epsilon^2})} + \frac{0.5772}{\sqrt{2\ln(m\sqrt{1-\epsilon^2})}}] \qquad (16)$$

while the standard deviation of the largest maximum is given by

$$\sigma[r_{max}] = \sigma_r \frac{\pi}{\sqrt{6}} \frac{1}{\sqrt{2\ln(m\sqrt{1-\epsilon^2})}} \qquad (17)$$

Here, σ_r is the rms response, m denotes the expected number of maxima during the design storm, and ϵ is the rms bandwidth parameter given as

$$\epsilon = \sqrt{1 - (\frac{n}{m})^2} \qquad (18)$$

where n is the expected number of zero-crossings.

The design response may be calculated by scaling up m and n obtained from a short time series, in accordance with the length of the design storm. This procedure applies also when estimating the extreme distribution, which in the Gaussian case is completely defined by the rms response and the rate of zero-crossings.

NUMERICAL RESULTS

The theory presented has been applied to evaluate the response of the proposed Salhus floating bridge (see Figure 1 and 2) in sinusoidal waves (Sigbjörnsson and Langen 1975a), long crested stochastic waves (Sigbjörnsson and Langen 1975b) and short crested stochastic waves (Sigbjörnsson and Langen 1979).

The analysis of wave measurements indicates that the one-dimensional wave spectral density in the Salhus Fjord is more narrow and more peaked than wave spectral densities in open sea. The JONSWAP wave spectral density (Hasselmann et al 1973) furnishes these desirable feature

$$S_{\eta\eta}(\omega) = \alpha g^2 \omega^{-5} \exp\{-\frac{5}{4}(\frac{\omega}{\omega_p})^{-4}\}\gamma^{\exp\{-\frac{(1-\omega/\omega_p)^2}{2\sigma^2}\}} \qquad 0<\omega<\infty \qquad (19)$$

Here, α is the Phillips parameter, g is the acceleration of gravity, ω_p is the spectral peak frequency, γ is the peakedness parameter, and σ is a parameter defining the width of the spectral peak.

Two different values are used for the spectral peak frequency: (a) ω_p = 1.083 rad/s, which is close to several natural frequencies and corresponds to the period 5.8 s used for the

sinusoidal waves, and (b) $\omega_p = 1.419$ rad/s, which also corresponds to a natural frequency and is close to the estimated average frequency of the waves in the fjord. σ is taken as

$$\sigma = 0.07 \text{ if } \omega \leq \omega_p \qquad \text{and} \qquad \sigma = 0.09 \text{ if } \omega > \omega_p$$

The peakedness parameter γ is chosen equal to 7.0, which corresponds to a very sharply peaked spectrum. The Phillips parameter α is adjusted so that the significant wave height is equal to 1 m.

The spreading function $\Psi(\theta,\omega)$ used in Equation 2 to describe the cross-spectral density of the sea surface, is assumed as

$$\Psi(\theta) = C\cos^n(\theta - \bar{\theta}) \qquad -\frac{\pi}{2} \leq \theta - \bar{\theta} \leq \frac{\pi}{2} \tag{20}$$

where n is a parameter defining the width of the distribution and C is a normalization factor defined by

$$C = (\int_{-\pi/2}^{\pi/2} \cos^n\theta d\theta)^{-1} \tag{21}$$

For actual sea states the value of the parameter n is supposed to be between 2 and 10.

The equation of motion (Equation 9) is solved in the time domain using direct step-by-step integration. Due to the peaked form of the wave spectral density and the correspondance between ω_p and natural frequencies of the bridge, the hydrodynamic mass and damping as well as the wave exitation transfer functions are approximately constant with values corresponding to ω_p. The unconditional stable trapezoidal rule (Newmark integration method with $\beta=1/4$) is used. The time step is 0.2 second which is supposed to give sufficient accuracy in the integration of modes with period larger than 3 seconds.

A time series for the wave excitation force is simulated for each nodal point of the bridge using Equation 14. The length of a series is approximately 400 seconds. A typical nodal point time series for $T_p=4.427$ s is shown in Figure 3. The series is analysed and the results in form of probability density, autospectral density and autocorrelation function are shown in the same figure.

The curvature of the bridge and the phase angle between sway force, heave force and roll moment yield time lags in the time series.

Figure 4 gives an example of a calculated response series and the corresponding probability densities, autospectral density and autocorrelation function.

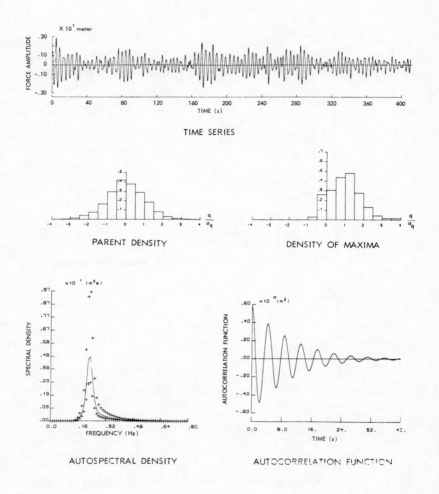

Figure 3 Typical nodal point wave force in a three-dimensional sea state.

The observed peak response at the midpoint of the bridge for two different sea states (n=2 and n=10) is shown in Table 1. The observed values are compared with the expected larges value. The agreement is satisfactory. The more long crested wave situation (n=10) shows between 4 and 12 per cent higher response values than the very short crested situation n=2. For T_p=5.8 s the difference is between 12 and 20 per cent.

A deterministic analysis of the bridge has also been performed for sinusoidal waves with the wave height equal to 2 m,

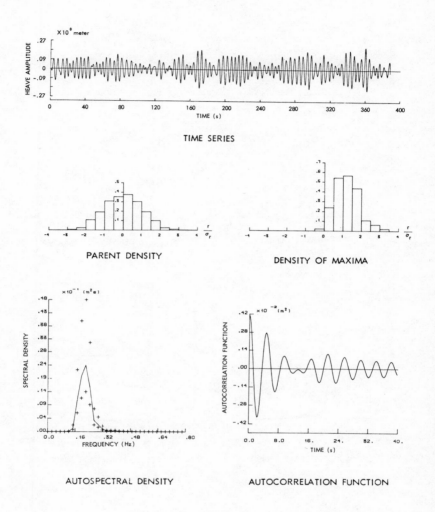

Figure 4 Heave response at the midpoint of the bridge induced by a three-dimensional sea state.

period T = 5.8 s and crest-length equal to the bridge span. Comparison of those results to the present results can be obtained by extrapolation based on Equation 16, assuming the time series to be long enough to yield the expected largest wave amplitude equal to 1 m. The extrapolated response values for T_p = 5.8 s is shown in Table 2 together with the response induced by sinusoidal waves. Table 2 shows also results for two-dimensional irregular waves described by the same spectral density as above and with crest-length equal to the bridge span. In this case only one time series is generated to simulate the

Table 1 Largest observed response at the midpoint of the bridge compared with the expected largest response

Response	Sea state $T_p = 4.427$ $n = 10$				Sea state $T_p = 4.427$ $n = 2$			
	Standard deviation	Largest observed	Theoretical $E[r_{max}]$	$\sigma[r_{max}]$	Standard deviation	Largest observed	Theoretical $E[r_{max}]$	$\sigma[r_{max}]$
Axial def(m)	0.0079	- 0.027	0.025	0.003	0.0073	- 0.025	0.023	0.003
Sway (m)	0.077	0.256	0.244	0.033	0.068	0.233	0.216	0.029
Heave (m)	0.072	0.220	0.229	0.031	0.064	0.196	0.204	0.027
Roll (rad)	0.0028	0.0081	0.0091	0.0012	0.0027	0.0077	0.0088	0.0011

Table 2 Comparison of the response induced by sinusoidal and stochastic waves. Significant wave height $H_{1/3} = 1.0$ m, $E[H_{max}] = 2.0$ m,
$T_p = 5.8$ s. The values refer to the midpoint of the bridge.

Response	Sinusoidal waves	Two-dimensional irregular waves $n = \infty$		Three-dimensional irregular waves $n = 10$	
		St.dev.	$E[r_{max}]$	St.dev.	$E[r_{max}]$
Axial def(m)	0.12	0.006	0.022	0.018	0.074
Sway (m)	1.52	0.33	1.30	0.16	0.64
Heave (m)	1.50	0.33	1.30	0.14	0.56
Roll (rad)	0.0050	0.0028	0.0113	0.0019	0.0076

wave situation.

The two-dimensional irregular waves show only minor reduction
in the sway and heave response compared to sinusoidal waves,
while the three-dimensional irregular wave model induces sway
and heave displacements at the midpoint which is approximately
1/3 of the response for the sinusoidal waves. For the moments
the reduction is approximately 50 per cent. In axial deformation and roll motion, however, the response is larger in irregular waves. This is mainly due to the fact that the irregular wave situation includes some higher frequency components
which correspond to natural modes with larger content of axial
deformation or roll motion.

ACKNOWLEDGEMENT

The bulk of the results reported in this paper have been obtained under several projects sponsored by the Norwegian
Public Roads Administration. The authors express their gratitude to Director P. Tambs-Lyche for the permission to publish
the results. Furthermore, the authors are indepted to Chief
Engineer H. Öderud, head of the Salhus project, and Director
emeritus A. Arild who has worked out the floating bridge design proposals presented herein.

REFERENCES

Adee, B.H. (1975) Analysis of Floating Breakwater Mooring
Forces, Ocean Engineering Mechanics, ed. N.T. Monney, OED
Vol. 1, ASME.

Cartwright, D.E. and Longuet-Higgins, M.S. (1956) The Statistical Distribution of the Maxima of a Random Function. Proceeding of the Royal Society of London, Series A, 237, Oct.

Clough, D., Sigbjörnsson, R. and Remseth, S.N. (1977) Response
of a Submerged, Buoyant Tubular Bridge Subjected to Irregular
Sea Waves, SINTEF Report STF 71 A77028, Trondheim.

Faltinsen, O. (1969) A Comparison of Frank Close-Fit Method
with some other Methods used to find Two-Dimensional Hydrodynamic Forces and Moments for Bodies which are Oscillating
Haromincally in an Ideal Fluid, Det norske Veritas, Report
No. 69-43-3, Oslo.

Frank, W. (1967) Oscillation of Cylinders in or below the Free
Surface of Deep Fluids, Naval Ship Research and Development
Center, Washington DC, Report No. 2375.

Hartz,B.J. and Mukherji, B. (1977) Dynamic Reponse of a
Floating Bridge to Wave Forces, International Conference on
Bridging Rion-Antirrion, Patras, Greece.

Hasselmann, K. et al. (1973) Measurements of Wind-Wave Growth and Swell Decay during the Joint North Sea Wave Project (JONSWAP). Ergänzungsheft zur Deutschen Hydrographischen Zeitschrift, Reihe A (8°), Nr. 12.

Holand, I. and Langen, I. (1972) Salhus Floating Bridge : Theory and Hydrodynamic Coefficients, SINTEF Report, Trondheim.

Holand, I., Langen, I. and Sigbjörnsson, R. (1977) Dynamic Analysis of a Curved Floating Bridge, IABSE Proceedings P-5/77.

Langen, I. and Sigbjörnsson, R. (1979) FLOATB - Program for dyanmisk analyse av kontinuerlige flytende konstruksjoner, Brukermanual, SINTEF Report in preparation.

Mukherji, B. (1972) Dynamic Behaviour of a Continuous Floating Bridge, Ph.D. Dissertation, University of Washington, Seattle.

Shinozuka, M. (1972) Monte Carlo Solution of Structural Dynamics. Computers and Structures, 2: 855-874.

Sigbjörnsson, R. and Langen, I. (1975) Wave-induced Vibrations of a Floating bridge: The Salhus Bridge, SINTEF Report STF71 A75018, Trondheim.

Sigbjornsson, R. and Langen, I.(1975) Wave-induced Vibrations of a Floating Bridge: A Monte Carlo Approach, SINTEF Report STF71 A75039, Trondheim.

Sigbjörnsson, R. (1978) Om dynamiske beregninger av flytebruer, SINTEF Report STF71 A78014, Trondheim.

Sigbjörnsson, R. (1979) Stochastic Theory of Wave Loading Processes, J. Eng. Structures, 1,2: 58-64.

Sigbjörnsson, R. (1979) Stochastic Analysis of Wave Load Effects for Probabilistic Design of Offshore Structures, Proc. International Conference on Environmental Forces on Engineering Structures, Imperial College, London.

Sigbjörnsson, R. and Langen, I. (1979) Dynamisk analyse av flytebro i irregulære kortkammede bølger, SINTEF Report STF71 A79003.

Vugts, J.H. (1968) The Hydrodynamic Coefficients for Swaying, Heaving and Rolling Cylinders in a Free Surface, Int. Shipbuilding Progr. 15, 251-276.

Vugts, J.H. (1970) The Hydrodynamic Forces and Ship Motions in Waves, Dr. thesis, Delft.

Wen, Y-K. and Shinozuka, M. (1972) Analysis of Floating Plate under Ocean Waves, J. Waterways and Harbors Div., ASCE, 98, WW2 177-190.

Wen, Y-K. (1974) Interaction of Ocean Waves with Floating Plate, J.Eng. Mech. Div., ASCE, 100, EM2, 375-395.

STOCHASTIC ANALYSIS OF WAVE LOAD EFFECTS FOR PROBABILISTIC
DESIGN OF OFFSHORE STRUCTURES

Ragnar Sigbjörnsson

SINTEF, the Foundation of Scientific and Industrial Research
at the Norwegian Institute of Technology, Trondheim, Norway.

INTRODUCTION

The rapidly expanding oil industry offshore has enforced development of structures for operations in deep waters and hostile environments. Under these conditions the structures are
exposed to action of random natural forces including wind,
waves, current and even strong motion earthquakes. To achieve
optimal structural safety and serviceability it is of vital
importance to predict these loads and their load effects realistically. Such predictions call for rational and reliable
methods of analysis which have been treated in the literature
by Borgman 1972, Nath and Harleman 1969, Forster 1970,
Malhotra and Penzien 1970, Berge and Penzien 1974, Eatock
Taylor 1975, Moan, Syvertsen and Haver 1977, among others.

The purpose of this paper is to present a consistent approach
for estimating wave induced vibrations of fixed offshore
structures. The basic principles are, however, general and
can be applied to other types of structures. Since the eventual objective of this approach is the prediction of safety
and serviceability, the main emphasis is put on the stochastic
aspects of the problem.

OCEAN WAVES AS A STOCHASTIC PROCESS

Long term model
The most characteristic feature of ocean waves is randomness
in time and space. Hence, the most suitable mathematical description of their behaviour and effects is furnished by the
theory of stochastic processes. Heuristically it seems necessary to apply non-stationary processes to be able to account
for their intrinsic long term variability.

The development of a non-stationary model requires, strictly

speaking, a finite ensemble of sample functions describing the same time sequence, which unfortunately is generally not obtainable in the case of geophysical processes. This difficulty may, however, be overcome by recognizing that the dominating long term fluctuations in many cases, for instance in the North Atlantic Ocean and the North Sea, are primarily governed by movements of whole weather systems with a characteristic period which is much longer than the average wave periods in question.

A suitable engineering model may then be obtained from a special class of non-stationary processes characterized by evolutionary spectral densities (Priestley 1965). The evolutionary spectral density is a time dependent spectral density changing continuously with time. The advantage of this model is, that it is possible to estimate the spectral properties of the wave process from a single sample function by time-averaging over short segments provided that the spectral density evolves sufficiently slowly with time.

This approach yields, of course, time dependent spectral moments and hence time dependent significant wave height, expected zero-crossing wave period (or an alternative statistical measure of wave periods) and average wave direction. The most common standard wave spectral densities can be specified completely in terms of these parameters (see for instance Price and Bishop 1974). The essentials of the long term model of ocean waves is therefore the joint statistical description of these parameters.

The structural response is sensitive to the spectral composition of ocean waves. This is especially the case for fixed offshore structures subjected to high frequency wave action. It is therefore of importance to establish an upper boundary of the evolutionary wave spectral density. Intuitively, the existence of such a boundary seems plausible due to limited fetch length and breaking of individual waves. This boundary is given by the envelope of wave spectra put forward by Bretschneider (see for instance Bretschneider 1975). Together with proper standard wave spectral densities the envelope of wave spectra yields also a lower boundary of the expected zero-crossing wave period T_z and an upper boundary of the significant wave height H_s. This is of great importance for fixed offshore platforms as sea states described by small T_z and large H_s normally induce relatively large structural response. These boundaries should be accounted for properly in the estimation of the joint probability density of H_s and T_z.

Short term model
The aforementioned evolutionary wave model may be approximated by a stationary (ergodic) process within a sufficiently short time interval (Priestley 1965,p.212). For these short term sea states the notion of generalized harmonic analysis of zero

mean ergodic processes applies. This leaves the following expression for the wavy sea surface (see for instance Phillips 1977):

$$\eta(\mathbf{x},t) = \int_{-\infty}^{\infty} e^{i(\omega t - \kappa \cdot \mathbf{x})} dB(\kappa,\omega) \tag{1}$$

Here, $\mathbf{x} = \{x,y\}^T$ denotes the spatial coordinates; κ is the two-dimensional wave number vector, t is the time; ω is the circular frequency; and $B(\kappa,\omega)$ is the spectral process associated with η.

This equation will represent a stationary and homogeneous stochastic field (at least in the wide sense) if the spectral process has zero mean and orthogonal increments. Furthermore emphasizing the second order statistics it follows that:

$$E\left[\left|dB(\kappa,\omega)\right|^2\right] = dG_{\eta\eta}^{(3)}(\kappa,\omega) \tag{2}$$

where $G_{\eta\eta}^{(3)}$ denotes the three-dimensional wave spectral distribution. Unfortunately, little is known concerning the general properties of the three-dimensional wave spectral distribution, which is mainly due to difficulties in measuring the spatial structure of waves. It is therefore common to simplify the problem by assuming a deterministic one-to-one relationship between the modulus of the wave number vector and the frequency, i.e. $\kappa=f(\omega)$, where $f(\cdot)$ must be an antimetric function. A suitable expression in this context is the dispersion relation known from the linear potential theory. This implies that the three-dimensional wave spectral distribution under rather general conditions can be expressed as follows applying cylindrical coordinates:

$$dG_{\eta\eta}^{(3)}(\kappa,\theta,\omega) = S_{\eta\eta}^{(2)}(\omega,\theta)\delta(\kappa-f(\omega))d\kappa d\theta d\omega \tag{3}$$

Here, $\delta(\cdot)$ is the Dirac delta-function and $S_{\eta\eta}^{(2)}$ is the socalled directional frequency spectral density.

In the structural analysis it is generally most convenient to express the second order statistics of the wave field in terms of the cross-spectral density, which is related to the three-dimensional wave spectral distribution as follows

$$S_{\eta_m \eta_n}(\omega)d\omega = \int_\kappa e^{-i\kappa \cdot \rho} dG^{(3)}(\kappa,\omega) \tag{4}$$

Here, m and n refer to two points on the sea surface; and $\rho = \{\Delta x, \Delta y\}^T = \{x_m - x_n, y_m - y_n\}^T$. This expression can be rewritten as follows applying Equation 3:

$$S_{\eta_m \eta_n}(\omega) = S_{\eta\eta}(\omega)\int_\theta \psi(\theta,\omega)\exp\{-if(\omega)(\Delta x\cos\theta + \Delta y\sin\theta)\}d\theta \tag{5}$$

548

which is a similar result as given in Berge and Penzien 1975.
Here, $S_{\eta\eta}(\omega)$ denotes the one-dimensional frequency spectral
density of the wave field; and $\psi(\theta,\omega)$ is the socalled spreading function, which is commonly assumed frequency independent.
A variety of models for $S_{\eta\eta}$ and ψ can be found in the literature (see for instance Price and Bishop 1974, and Berge and
Penzien 1975), which makes Equation 5 immediately applicable in
engineering analysis

In order to study the correlation structure of the wave field
expressed in the frequency space by Equation 5, the coherence
spectrum is applied. That is:

$$\text{Coh}_{\eta_m\eta_n}(\omega) = |S_{\eta_m\eta_n}(\omega)|^2/S_{\eta\eta}^2(\omega) \qquad (6)$$

Figure 1 displays some results obtained applying spreading functions based on the $\cos^n\theta$ law (Sigbjörnsson 1979). It is seen
that the loss in coherence is substantial compared to the theo-

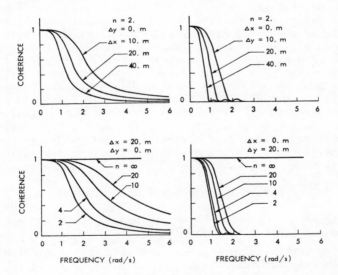

Figure 1 Coherence spectra of wave elevation. Spreading function is given by $\cos^n\theta$; and average directions of wave propagation is in direction of x-axis

retical values for idealized long crested waves (n=∞), except
for low frequencies and for small distances. The loss in coherence increases with increasing frequency and increasing distance between the points as expected. It is worth pointing out
that the loss in coherence is greatest for points on a line
perpendicular to the mean direction of wave propagation. This
is due to the fact that the quadrature spectral density vanishes for these points.

WAVE LOADING PROCESSES

Following the notion of the preceding section the wave loading may be approximated by a stationary (ergodic) stochastic process within a sufficiently short interval in time and space.

In the case of large volume structures this stochastic process may be derived applying the linearized probabilistic potential theory. This approach assumes the existence of a stochastic velocity potential which satisfies the Laplace equation with appropriate linearized boundary conditions. By applying the principle of superposition this potential is assumed given as the sum of the velocity potentials of, respectively, the incident waves and the waves diffracted by a restrained structure and the waves generated by the various modes of structural vibrations (see for instance Faltinsen and Michelsen 1974). It follows that each of these potentials must satisfy the Laplace equation.

The velocity potential of incident waves can be expressed as:

$$\phi^{(w)}(x,t) = \int_{-\infty}^{\infty} \phi^{(w)}(\kappa,\omega)e^{i(\omega t - \kappa \cdot x)}dZ_\eta(\kappa,\omega) \tag{7}$$

where, $dZ_\eta = idB$; and the transfer function is given as follows applying a Cartesian coordinate system placed on the sea floor with the z-axis pointing upwards:

$$\phi^{(w)}(\kappa,\omega) = \frac{g}{\omega}\frac{\cosh\{\kappa z\}}{\cosh\{\kappa d\}} \tag{8}$$

Here, g is the acceleration of gravity; and d is the water depth. The water particle velocity and acceleration in the undisturbed stochastic wave field are readily obtained from Equations 7 and 8 by differentiation (see for instance Sigbjörnsson 1979).

The remaining stochastic velocity potentials can be expressed formally applying equations similar to Equation 7. In this case it is generally required to use numerical methods to calculate the transfer functions (see for instance Garrison 1978).

When all the velocity potentials have been determined, the stochastic wave loading process may be derived applying the Bernoulli equation. This yields, disregarding second order terms:

$$q^{(p)}(t) = q^{(I)}(t) - \int_{-\infty}^{\infty} m^{(h)}(\omega)e^{i\omega t}dZ_{\ddot{r}}(\omega) - \int_{-\infty}^{\infty} c^{(h)}(\omega)e^{i\omega t}dZ_{\dot{r}}(\omega) \tag{9}$$

where $q^{(I)}$ is the inertia force due to incident and diffracted waves; $m^{(h)}$ and $c^{(h)}$ are the hydrodynamic mass and the radiational damping due to wave generation by the vibrating structures (see for instance Faltinsen and Michelsen 1974); and $Z_{\ddot{r}}$ and $Z_{\dot{r}}$ are the spectral processes associated with the accele-

ration and the velocity of the structure.

In the important case of a single vertical circular cylinder the wave excitation force can be expressed by a closed form solution (MacCamy and Fuchs 1954). In the stochastic case this leads to the following expression for the force per unit length:

$$q^{(I)}(x_0,z,t)$$

$$= \int_{-\infty}^{\infty} \frac{\pi}{4}\rho D^2 C_M(\kappa D) \begin{bmatrix} \cos\theta \\ \sin\theta \end{bmatrix} \frac{\cosh\{\kappa z\}}{\sinh\{\kappa d\}} e^{i(\omega t - \kappa \cdot x_0 + \alpha(\kappa D))} dZ_\eta(\kappa,\omega)$$

(10)

where, D is the diameter of the cylinder; ρ is the density of sea water; x_0 is the position vector of the cylinder; $C_M(\cdot)$ and $\alpha(\cdot)$ are respectively the coefficient of inertia and the angle of phase lag (see for instance Sigbjörnsson 1979). In the one-dimensional case this expression equals the inertia term in the Morison equation (Morison et al. 1950) as κD approaches zero.

The potential theory is not adequate for slender bodies due to viscous effects like drag forces and lift forces. As these forces are nonlinear, a reliable prediction is difficult. It seems unlikely that lift forces should yield a significant contribution to estimates of the total loading on fixed off-shore platforms. This is, however, not necessarily the case for local load effects. They are therefore neglected in the present study.

The drag forces per unit length are commonly assumed given by the following expression:

$$q^{(d)}(t) = \frac{1}{2}\rho C_D D\{(\dot{u}-\dot{r})|\dot{u}-\dot{r}|\}$$

(11)

Here, C_D is the drag force coefficient; $\dot{u}$ is the water particle velocity; and $\dot{r}$ is the structural velocity. This equation relates the drag forces to the relative velocity, which seems reasonable as long as the structural member in question is sufficiently slender in the hydrodynamic sense. On the other hand, if the structural member is not slender, the drag forces are of minor importance and Equation 11 may therefore still apply with reasonable accuracy.

Due to the fact that Equation 11 is an empirical approximation rather than a physical law, it is of importance in the analysis to apply a C_D value derived from measurements using an estimator consistent with the method of analysis. Thus, if the stochastic linearization method is used in the analysis, it should be preferred in the estimation of C_D (see Borgman 1972 for detailed discussion).

RESPONSE PROCESSES

Analysis

It is assumed that the structure can be modelled as an assembly of finite (beam) elements applying boundary spring-dashpot elements to idealize the soil-foundation system. These spring-dashpot elements will generally possess time (frequency) dependence. This implies that the equation of motion may be expressed as follows by introducing the Dirac delta function in the time space for frequency independent system parameters:

$$\int_{-\infty}^{\infty} M(t-\tau)\ddot{r}(\tau)d\tau + \int_{-\infty}^{\infty} C(t-\tau)\dot{r}(\tau)d\tau + \int_{-\infty}^{\infty} K(t-\tau)r(\tau)d\tau = Q(t) \qquad (12)$$

Here, r is the nodal response process; M is the virtual mass of the system including the hydrodynamic mass; C is the virtual damping including hydrodynamic effects; K is the stiffness matrix; and $Q(t)$ is the wave loading process, which strictly speaking is a function of $\dot{r}$ (see Equation 11). This coupling is assumed removed in the following by applying the stochastic linearization method including Gaussianization of the excitation as well as the response. Other nonlinearities are assumed treated in the same way. Hence, a drag damping proportional to the rms relative velocity is introduced into the C matrix, i.e. $\frac{1}{2}\rho C_D(E[(\dot{u}-\dot{r})^2]8/\pi)^{1/2}$.

In the solution of this equation the notion of the preceding sections is carried a step further by assuming the evolution of the sea state slow enough to neglect the transient structural response. Then, the long term probability density of the response can be expressed as:

$$p_r(r) = \int_{H_S} \int_{T_Z} \int_{\bar{\theta}} p_{r|H_S T_Z \bar{\theta}}(r|h,t,v) p_{H_S T_Z \bar{\theta}}(h,t,v) dv dt dh \qquad (13)$$

Here, $p_{H_S T_Z \bar{\theta}}$ denotes the long term joint probability density of $H_S T_Z \bar{\theta}$ significant wave height and expected zero-crossing wave period and average wave propagation; $p_{r|H_S T_Z \bar{\theta}}$ is the conditional probability density of the response derived applying the statistics of the response process as approximated by:

$$r(t) = \int_{-\infty}^{\infty} e^{i\omega t} dZ_r(\omega) \qquad (14)$$

where Z_r is the spectral process of r possessing orthogonal increments.

It follows that the response process has zero mean and

spectral density given as:

$$S_{rr}(\omega) = H(\omega)S_{QQ}(\omega)H^{T*}(\omega) \tag{15}$$

where

$$H(\omega) = [K(\omega) - \omega^2 M(\omega) + i\omega C(\omega)]^{-1} \tag{16}$$

is the virtual frequency response function of the system; and S_{QQ} is the spectral density of the wave excitation.

By this approach the system given in Equation 12 is in the general case approximated by a hierarchy of linear systems due to the fact that $H(\omega)$ is a function of sea state severity. For each of these linear systems the response is approximated by an equivalent Gaussian process. Hence, the response statistics are completely defined in terms of Equations 13, 15 and 16.

It is worth pointing out that this direct frequency response method does not put any restrictions on the damping matrix, and the frequency dependent system matrices cause no extra problems. Further, it is generally most convenient to use complex arithmetic in the evaluation of $H(\omega)$ (further discussion may be found in Sigbjörnsson et al. 1978).

In the important case of multilegged platforms the spectral density of wave excitation can be derived applying Equations 10 and 11 assuming the interaction effects negligible. This yields applying the stochastic linearization method:

$$S_{QQ}(\omega) = F(\omega)S_{\eta\eta}(\omega) \tag{17}$$

Here

$$F(\omega) = \omega^2 \sum_m \sum_n a_m^T D(m,n) \left[A_D(m)A_D^T(n) \right.$$

$$+ i\omega[A_D(m)A_M^T(n)e^{-i\alpha_n} - A_M(m)A_D^T(n)e^{i\alpha_m}]$$

$$\left. + \omega^2 A_M(m)A_M^T(n)e^{i(\alpha_m - \alpha_n)} \right] a_n \tag{18}$$

where

$$D(m,n) = \int_\theta \begin{bmatrix} \cos^2\theta & \cos\theta\sin\theta \\ \cos\theta\sin\theta & \sin^2\theta \end{bmatrix} \psi(\theta,\omega)$$

$$\times \exp\{-if(\omega)((x_m - x_n)\cos\theta + (y_m - y_n)\sin\theta)\}d\theta \tag{19}$$

and

$$
\begin{bmatrix} A_D(n) \\ ----- \\ A_M(n) \end{bmatrix} = \int_{\ell_n} N_n^T \begin{bmatrix} k_D(n)\sigma_{v_x}(n) \\ k_D(n)\sigma_{v_y}(n) \\ k_M(n) \\ k_M(n) \end{bmatrix} \frac{\cosh\{\kappa z_n\}}{\sinh\{\kappa d\}} \, dz_n \tag{20}
$$

The following notation is applied:

$$
\left. \begin{aligned} k_D(n) &= \tfrac{1}{2}\rho C_D(n) D_n \sqrt{8/\pi} \\ k_M(n) &= \tfrac{\pi}{4}\rho C_M(\omega,n) D_n^2 \end{aligned} \right\} \tag{21}
$$

Further, $\sigma_v^{\bullet} = (E[(\dot{u}-\dot{r})^2])^{1/2}$; $N_n = \{N_1, N_2, \ldots\}$ denotes the shape functions assumed identical for the x- and y-direction; a_n denotes the connectivety matrix; and ℓ_n is the element length. The summation in Equation 17 is carried out over all wetted elements.

It is seen that $F(\omega)$ is a Hermitian matrix, which in the following will be termed the hydrodynamic transfer function of the system. It is worth noting that terms due to correlation between inertia and drag forces will only vanish for the diagonal term, i.e. the autospectral densities, in the case of slender structural members ($\alpha=0$) or in the trivial case when the drag forces are negligible. Further, it is seen that terms purely governed by the inertia forces only will be independent of the phase spectrum α for autospectral densities and cross-spectral densities of forces on cylinders with identical diameters.

It is assumed that Equation 17 holds formally in more general cases if the hydrodynamic transfer function is redefined.

The hydrodynamic transfer function as well as the virtual frequency response function depend on the rms relative velocity, which can be obtained by:

$$
\sigma_v^{\bullet} = \sqrt{\sigma_{\dot{u}}^2 + \sigma_{\dot{r}}^2 - 2\sigma_{\dot{r}\dot{u}}} \tag{22}
$$

Here, the covariance $\sigma_{\dot{r}\dot{u}}$ is derived applying the following cross-spectral density:

$$
S_{\dot{r}\dot{u}}(\omega) = i\omega S_{\eta\eta}(\omega)H(\omega)T(\omega) \tag{23}
$$

where

$$
T(\omega) = \omega^2 \frac{\cosh\{\kappa z_n\}}{\sinh\{\kappa d\}} \sum_m a_m^T D(m,n)[A_D(m) + i\omega A_M(m)] \tag{24}
$$

The iteration procedure required to evaluate Equation 22 is started by a proper selection of $\sigma_{\overset{..}{v}}$, which yields estimates of $\sigma_{\overset{.}{r}}$ by applying Equations 15 and 17 and $\sigma_{\overset{..}{ru}}$ by applying Equation 23 . This sequence is then repeated until satisfactory accuracy is obtained.

In many practical cases it is observed that the water particle velocity is much greater than the structural velocity, i.e. $\overset{.}{u} >> \overset{.}{r}$. Then it holds that $\sigma_{\overset{.}{v}} \approx \sigma_{\overset{.}{u}}$ which simplifies the calculation considerably.

<u>Synthesis</u>
In some cases it is desirable to carry out synthesis of the response process, i.e. to generate a time space sample function from a given response statistics. In the case of stationary multivariate Gaussian response processes this may be done effectively by Monte Carlo simulation, for instance combined with (a) direct time space filtering (see for instance Borgman 1969), (b) spectral factorization combined with Equation 14 (see for instance Shinozuka 1972), and (c) autoregressive moving-average time series model (see for instance Gersch and Luo 1972).

The factorization of the response spectral densities combined with Equation 14 yields:

$$r_m(t) = \text{Re}[\sqrt{2\Delta\omega} \sum_{\ell=1}^{m} \sum_{k=1}^{N} B_{m\ell}(\omega_k) \exp\{i(\omega t + \phi_{\ell k})\}] \qquad (25)$$

Here, $\Delta\omega$ is the frequency resolution; $N = \max(\omega)/\Delta\omega$; ϕ is a random phase angle uniformly distributed between 0 and 2π; and

$$B(\omega)B^{T*}(\omega) = S_{rr}(\omega) \qquad (26)$$

where B is a triangular matrix which may be determined by a Cholesky factorization.

This procedure can be made very effective by applying the fast Fourier transformation algorithm. However, it is a drawback that the series will be periodic with the period $2\pi/\Delta\omega$. Therefore, a limitation on the length of the time series is set by a finite computer memory. It is also worth pointing out that this method yields no random variation in the sample spectral densities. These problems may be avoided at least partly by applying a related method (Shinozuka 1971), which however will be more time consuming.

An attractive alternative to this method as well as the direct time space filtering is in some cases furnished by the autogressive moving-average time series. In this case a limitation in computer memory is generally no problem and the periodicity of the sample function follows the periodicity of the pseudo-random number generator. The estimation of the model

parameters may, however, be rather tedious. The method is
therefore most tractable to generate long sample functions.

On the effects of short-crestedness of waves

To exemplify the effects of the short-crestedness of the wave
field on the structural response, a gravity platform of
typical design has been selected. A sketch of the platform is
shown in Figure 2 indicating the finite element model applied.

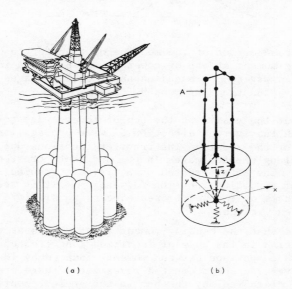

(a)   (b)

Figure 2   An example platform: (a) A sketch of the platform;
(b) a finite element idealization.

The foundation-soil system is modelled by frequency dependent
spring-dashpot elements using the half space analogy (see for
instance Veletsos and Wei 1971). It can be argued that this
model yields somewhat too high radiational damping in the case
of layered soil. Therefore the material (hysteretic) damping
in the soil is neglected. The wave loading on the towers is
obtained applying the stripe approach combined with Equations
17 to 21 and neglecting interaction effects. The importance
of mutual interaction can be evaluated applying results given
in Chakrabarti 1978. The drag forces are found to be almost insignificant in the whole frequency range in question. The
wave forces on the caisson are based on potential theory and
lumped as moment and shear force at mudline. More detailed
discussion on the modelling including numerical results may
be found in Sigbjörnsson et al. 1978.

The transfer function of the structure-soil-sea system is
defined as $H(\omega)F(\omega)H^T*(\omega)$ in accordance with Equations 15 and
17. Figure 3 displays transfer functions for the bending

556

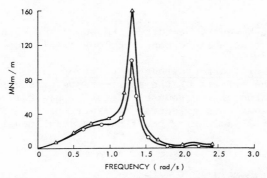

Figure 3   Square root of the modulus of transfer functions
for bending moment at top of tower A: -△- Based on idealized
long crested waves; -o- based on short crested waves described
by spreading function $(2/\pi)\cos^2\theta$.

moment about the y-axis at the top of tower A due to wave
fields with the average direction of wave propagation in the
direction of the x-axis. The transfer functions due to
idealized long crested waves ($n = \infty$) and short crested waves
are shown. The short crested wave field is assumed specified
in terms of spreading function $(2/\pi)\cos^2\theta$. It is seen that
the influences of the short crestedness are significant in the
high frequency range.

It is found that the bending moment about x-axis at the top of
tower A vanish in the case of idealized long crested waves. In
this case a comparison between moments induced by idealized long
crested waves and short crested waves seems therefore irrelevant
and is not discussed any further in the present context.

Figure 4 displays the response of the example platform induced
by wave fields specified in terms of the parameterized JONSWAP
wave spectrum as given by Houmb and Overvik 1976. The statistical response quantities (definition may be found in
Sigbjörnsson et al. 1978 p. 270) are expressed as functions
of the expected zero-crossing wave period for a given significant wave height (defined as $4(E[\eta^2]^{1/2})$. It is seen that the
reduction in the rms moment due to short-crestedness is
greatest close to resonance but decreases with increasing low
frequency quasi-static response, while the expected zero-
(up)crossing period of the moment shows the opposite effect.
The rms bandwidth of the moment shows only small increase
due to short-crestedness. The increase is, however, barely
significant in an intermediate range with approximately equal
contributions of resonance and quasi-static response.

Figure 5 shows the long term distribution of maxima of moment
at the top of tower A induced by unidirectional evolutionary
sea state. This distribution is based on Gaussianization of
the response process applying the Rice distribution for maxima.

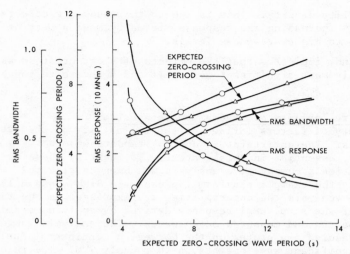

Figure 4 Bending moment at top of tower A as a function of expected zero-crossing wave period: Significant wave height equal to 3.25 m; -Δ- based on idealized long crested waves; -o- based on short crested waves described by spreading function $(2/\pi)\cos^2\theta$.

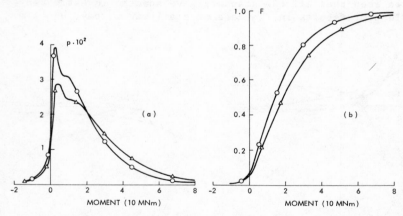

Figure 5 Long term distribution of maxima of moment at top of tower A: (a) Probability densities; (b) probability distribution; -Δ- based on idealized long crested waves; -o- based on short crested waves described by spreading function $(2/\pi)\cos^2\theta$.

The joint distribution of significant wave height and expected zero-crossing period is derived applying wave data from the weather ship Famita. The effects of short-crestedness are especially pronounced for the probability density. It is also worth noting the strange shape of the peak of the

probability density. This is due to the resonance effects, which by increasing the response virtually move a part of the density to higher response levels.

This study indicates that the effects of short-crestedness is generally most significant for small and moderate response levels.

On the reliability of standard wave spectra

A variety of factors influences the reliability of the response analysis. The uncertainties due to standard wave spectra are, however, essential in this context. To quantify these uncertainties a comparative analysis has been carried out applying the example platform (Spidsöe and Sigbjörnsson 1979). In this analysis the uncertainties are expressed as the relative error of the structural response derived by applying standard wave spectra and measured ocean waves. The analysis includes 1008 measured sea states on the Norwegian Continental Shelf with significant wave heights up to roughly 9 m. The relative error of the moment at the top of tower A is evaluated and quantified by the mean value μ and the standard deviation σ. These values are plotted in Figure 6 as functions of the significant wave height H_S.

It is seen that all the standard wave spectra investigated yield poor results in the case of small waves, say, $H_S < 4$ m.

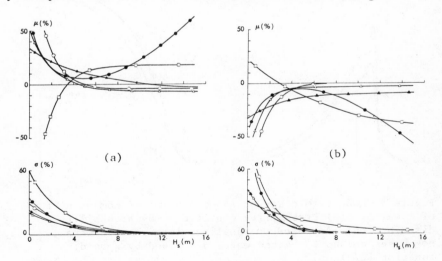

Figure 6 Moment at top of tower A: Mean value and standard deviation of (a) rms moment; (b) rms bandwidth of moment; -▲- based on ISSC wave spectrum; -Δ- based on parameterized JONSWAP wave spectrum; -●- based on Darbyshire-Scott wave spectrum (Scott 1965); -o- based on Darbyshire-Scott wave spectrum (Wiegel 1975); -□- based on six parameter wave spectrum (Ochi 1976).

For moderate and severe sea states (the last based on extra-polations) the results indicate that the most favourable standard wave spectra is the parameterized JONSWAP wave spectrum and the Darbyshire-Scott wave spectrum applying the parameters proposed by Wiegel. It is worth stressing that these findings are not necessarily valid for other regions.

APPLICATIONS IN PROBABILISTIC DESIGN

Extremes

The basic problem in structural design concerns first excursion failures and requires traditionally information concerning the characteristic extreme response and in modern approaches even the extreme value distribution (Flint and Baker 1978).

An estimate of the "most probable" largest maximum during the time interval T may be obtained from the long term distri-bution of maxima as:

$$Pr[r>\xi] = (\nu T)^{-1} \tag{27}$$

Here, ν denotes the expected zero-crossing frequency given as follows:

$$\nu = \frac{1}{2\pi} \int_{H_S} \int_{T_Z} \int_{\theta} \sqrt{\frac{E[\ddot{r}^2]}{E[r^2]}} \; p_{H_S T_Z \bar{\theta}}(h,t,v) \, dv dt dh \tag{28}$$

applying the notion of the slowly evolving response. The reliability function given in Figure 7 is based on this approach and derived from the results given in Figure 5.

An extreme value distribution for stationary processes put forward by Ditlevsen is:

$$F(\xi,T) \simeq F(\xi,0) e^{-\nu_\xi T/F(\xi,0)} \tag{29}$$

where ν_ξ denotes the expected number of upcrossings per unit time at level ξ. The validity of this distribution has been tested in an extensive simulation study applying a variaty of rather broad banded processes including non-Gaussian processes (Ditlevsen 1971). Heuristically, it seems possible to generalize Equation 27 to the semi-stationary processes discussed in Priestley 1965. Then, ν_ξ is given by an expression similar to Equation 28.

Fatigue

The continuous wave action subjects the structures to an extremely large number of oscillations. It is therefore necessary to consider fatigue in the structural design.

A variety of theories have been proposed to account for fatigue

560

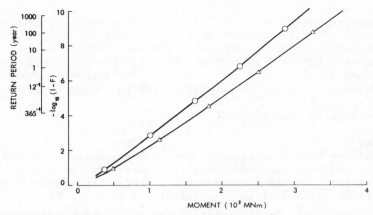

Figure 7 Long term probability distribution of maxima of moment at top of tower A: -Δ- Based on idealized long crested waves; -o- based on short crested waves described by spreading function $(2/\pi)\cos^2\theta$.

under complex stressing, but none has apparently received universal acceptance. In the present study the concept of the maximum principal stress as a fatigue governing parameter is applied. At least for in-phase stresses this concept seems to give reasonable results (Toor 1975). Unfortunately, most structural elements in an offshore structure virtually experience multiaxial out-of-phase stochastic stressing.

To facilitate a proper counting of stress cycles a generation of sample functions of stresses is desirable. The first step is the computation of the spectral densities of hot spot stresses, which generally requires refined element analysis.

Figure 8 shows the long term distribution of stress cycles and maxima in a given "hot spot" at the deck/tower connection of the example platform subjected to unidirectional long crested waves described by the parameterized JONSWAP wave spectrum (Sigbjörnsson and Syvertsen 1978). The stress cycles are counted applying the rain-flow method. Fortunately, in this case it is found that the principal axis corresponding to maxima of principal stresses are almost fixed in space. The distribution of partial damage is obtained applying the standards of the Norwegian Petroleum Directorate (Curve No.D). This yields, applying the Palmgren-Miner rule, an estimate of the fatigue life equal to 4 years. In the case of short crested waves this estimate is roughly twice as high. The distribution of stress cycles derived in this way may directly be applied in a more refined probabilistic analysis (Wirsching and Haugen 1973) or even in fracture mechanics approaches.

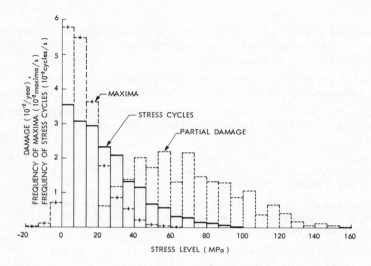

Figure 8 Long term distribution of stress cycles and maxima and partial damage.

FINAL REMARKS

A consistent stochastic analysis of waves and wave load effects is presented. The potential of the analysis is demonstrated by numerical examples. Uncertainties in the model of the wave field are discussed emphasizing the effects of short-crestedness on the structural response. The reliability of standard one-dimensional wave spectra is also touched upon.

It is found that the response induced by light and moderate sea states may be quite inaccurate, implying a serious draw-back for the fatigue analysis. The accuracy of the extreme response appears on the other hand to be satisfactory. This may, however, not be so surprizing bearing in mind that the wave climate on the Norwegian Continental Shelf is governed by very instable wind fields concerning homogeneity, direction and fetch. Therefore, more research is needed before wave in-duced response can be predicted with a desirable accuracy.

ACKNOWLEDGEMENTS

The cooperation with Senior Structural Engineers S.-E. Jensen and E.T. Moe at A.S. Bergens Mekaniske Verksteder on fatigue problems is greatly acknowledged. Further, the author is indebted to Assist. Prof. K. Syvertsen and Research Assist. N. Spidsöe at the Norwegian Institute of Technology for assistance in preparing some of the numerical results reported herein.

REFERENCES

Berge,B. and Penzien,P. (1974) Three-dimensional Stochastic
Response of Offshore Towers to Wave Forces. Offshore Technology Conference, Houston.

Borgman,L.E. (1969) Ocean Wave Simulation for Engineering Design. J.Waterways Harbors Division, ASCE, 95, WW4: 556-583.

Borgman,L.E. (1972) Statistical Models for Ocean Waves and
Wave Forces. Advances in Hydroscience, Academic Press, 8:
139-181.

Bretschneider,C.L. (1975) The Envelope Wave Spectrum. The
Third Int.Conf. on Port and Ocean Engineering under Arctic
Conditions, POAC-75, Fairbanks.

Chakrabarti,S.K. (1978) Wave Forces on Multiple Vertical
Cylinders. J.Waterway, Port, Coastal and Ocean. Division,
ASCE, 104: WW2, 147-161.

Ditlevsen,O. (1971) Extremes and First Passage Times (doctoral dissertation), Copenhagen.

Eatock Taylor,R. (1975) Structural Dynamics of Offshore Platforms. Offshore Struct. Conf., London, 125-132.

Faltinsen,O. and Michelsen,F.C. (1974) Motion of Large Structures in Waves at Zero Froude Number. Int.Symp. on the Dynamics of Marine Vehicles and Structures in Waves, London.

Flint,A.R. and Baker,M.J. (1978) Safety Approach for Structures subjected to Stochastic Loads. Int. Symp. on the Integrity of Offshore Structures, Glasgow.

Forster,E.T. (1970) Model for Nonlinear Dynamics of Offshore
Towers. J. Engineering Mechanics Division, ASCE, 86, EM1:
41-53.

Garrison,C.J. (1978) Hydrodynamic Loading of Large Offshore
Structures: Three-Dimensional Source Distribution Methods,
in Numerical Methods in Offshore Engineering, ed. Zienkiewicz
et al., John Wiley & Sons, 87-140.

Gersch,W. and Luo,S. (1972) Discrete Time Series Synthesis
of Randomly Excited Structural Response. J. Acoustical
Society of America, 51: 402-408.

Houmb,O.G. and Overvik,T. (1976) Parameterization of Wave
Spectra and Long Term Joint Distribution of Wave Height and
Period. The First Int. Conf. on the Behaviour of Offshore
Structures, BOSS-76, Trondheim.

MacCamy,R.C. and Fuchs,R.A. (1954) Wave Forces on Piles: A
Diffraction Theory. U.S.Army Corps of Engineers Beach Erosion
Board, 69.

Malhotra,A.K. and Penzien,J. (1970) Nondeterministic Analysis of Offshore Structures. J. Engineering Mechanics Divi-

Acknowledgements

Editing a volume of this scale is an intensely collaborative process. We are indebted to all the contributors for their support, collegiality, and cooperation throughout this process. We appreciate the quality of their work and the time dedicated to producing this volume in already crowded schedules. David Musson at Oxford University Press was a pleasure to work with and a source of great wisdom and sound advice. We also acknowledge the contributions of Emma Booth, Rachel Platt, and other colleagues at the Press who all played their parts with professionalism and unstinting conscientiousness. Savita Kumra would like to dedicate this book to her parents: her mother, who though sadly no longer with us will always be a great source of inspiration and strength, and her father who challenged gender stereotypes on a daily basis with calm determination. Thanks also to her sons Kieran and Aron and husband Julian for their constant love and support. She would like to acknowledge the support of colleagues at Brunel Business School and also those in the BAM GiM SIG who have been a constant source of friendship, support, and professional inspiration. Ruth Simpson would like to thank colleagues at Brunel Business School for their support and dedicates this book to her children Rachel, Matthew, and Alex and grandchildren Josh and Molly. Ron Burke would like to acknowledge the support provided by dear colleagues, including Tamara Weir, Carol McKeen, Mary Mattis, Debra Nelson, Susan Vinnicombe, and Lyn Davidson. He would like to thank York University for supporting his contributions and he would also like to thank Savita and Ruth for their leadership on the project. It is also his hope that his daughters, Sharon, Rachel, and Liane benefit from our work.

Contents

PART IV MASCULINITIES IN ORGANIZATIONS

PART IV: MASCULINITIES IN ORGANIZATIONS

List of Figures

2000), 'Beyond Body-Counting' (in *Gender, Identity and the Culture of Organization*, Routledge, 2002), 'Gender and Organization: Toward a Differentiated Understanding' (*Organization Studies*, 1992), all with co-author Mats Alvesson. Other publications include 'Are Women in Management Victims of the Phantom of the Male Norm?' (*Gender, Work & Organization*, 2011) and *The Will to Power* (Studentlitteratur, 2006).

Alice H. Eagly is Professor of Psychology, James Padilla Chair of Arts and Sciences, Professor of Management & Organizations, and Faculty Fellow in the Institute for Policy Research, all at Northwestern University. Her research interests include the study of leadership, gender, attitudes, prejudice, and stereotyping. She has won numerous awards for her research and writing, including the Distinguished Scientific Contribution Award from the American Psychological Association.

Sandra L. Fielden is Senior Lecturer in Organizational Psychology at Manchester Business School at the University of Manchester, and a Chartered Occupational Psychologist. This year she was involved in the Cabinet Office's campaign for the promotion of diversity on public boards. She is well published with numerous journal papers and book chapters and co-editor of three books and one recently published authored book *Minorities in Entrepreneurship* (with M. J. Davidson and G. Wood).

Leire Gartzia is Professor of Leadership at Deusto Business School, Spain. She earned a European PhD in organizational psychology and was a postdoctoral fellow at Northwestern University. Her work has been recognized by awards including the Academy of Management Best Paper and the Dorothy Harlow Distinction in Gender Studies. She recently co-edited for *Sex Roles* a special issue about gender research in Spain. Her research interests include gender roles, leadership, crisis management, emotions, and stereotypes.

Silvia Gherardi is Full Professor of Sociology of Work and Organization at the University of Trento, Italy, where she is responsible for the Research Unit on Communication, Organizational Learning, and Aesthetics (RUCOLA). Her research activities focus on practice-based studies and her theoretical background is in qualitative sociology, organizational symbolism, and feminist studies.

Gina Grandy is Associate Professor with the Commerce Department, Ron Joyce Centre for Business Studies at Mount Allison University located in New Brunswick, Canada. Her research focuses upon competitive advantage, leadership, identity, organizational change, gender, and stigmatized occupations. Gina has published in such journals as *Organization, Gender, Work and Organization, Journal of Management Studies, Gender in Management: An International Journal, Qualitative Research in Organizations and Management: An International Journal, The Learning Organization*, and *Journal of Strategy and Management*.

S. Alexander Haslam is Professor of Psychology and Australian Laureate Fellow at the University of Queensland. His work with colleagues around the world focuses on the study of social identity in social and organizational contexts. This is represented in his most recent books: *The New Psychology of Leadership: Identity, Influence and Power* (with Stephen Reicher and Michael Platow, Psychology Press, 2011) and *The Social Cure: Identity, Influence and Power* (with Jolanda Jetten and Catherine Haslam, Sage, 2012). He is Fellow of the Canadian Institute of Advanced Research and a former recipient of EASP's Lewin Medal for research excellence.

Jeff Hearn is Professor in Management and Organization at Hanken School of Economics, Finland, Professor in Gender Studies at Linkoping University, Sweden, and Professor of Sociology at the University of Huddersfield, UK. His books include *'Sex' at 'Work'* (with Wendy Parkin, 1987/1995), *Men in the Public Eye* (1992), *Men as Managers, Managers as Men* (edited with David Collinson, 1996), *Gender, Sexuality and Violence in Organizations* (with Wendy Parkin, 2001), *Information, Society and the Workplace* (edited with Tuula Heiskanen, 2004), and *Rethinking Transnational Men* (edited with Marina Blagojavic and Katherine Harrison).

Jean Helms-Mills is Professor of Management at the Sobey School of Business, Saint Mary's University, Canada, and Professor (part time) at Jyväskylä School of Business and Economics, Jyväskylä University, Finland. She is Associate Editor of *Gender, Work and Organization* and serves on the editorial boards of several other journals. Her books include *Making Sense of Organizational Change* (Routledge, 2003), *Understanding Organizational Change* (Routledge, 2009), and the forthcoming *Routledge Companion to Critical Management Studies*.

Evangelina Holvino is President of Chaos Management, Ltd and an Affiliate Research Faculty at the Center for Gender in Organizations at the Simmons School of Management, Boston, MA, USA. She has over twenty-five years of experience as an organizational consultant, educator, and action researcher in the United States and internationally. Her conceptual and empirical work uses intersectionality and transnational feminist theories to study the simultaneity of race, gender, class, ethnicity, sexuality, and nation in organizations. She currently leads a research grant on the applications of the simultaneity model of social differences in teaching and management.

Heather Höpfl is Professor of Management Psychology and Director of Executive Education at Essex Business School, University of Essex. She is Adjunct Professor of the University of South Australia and has had a range of visiting appointments inter alia in Trento, Utrecht, and Warsaw. Recent publications include *Belief and Organization* (Palgrave, 2012) with Peter Case and Hugo Letiche, 'Objects in Exile' with Ricky Ng (*Journal of Organizational Change Management*, 2011), and five special issues on architecture, the visual, and theology. She is currently working on Irigaray and the links between her work and organizational theory.

Carianne Hunt is Knowledge Transfer Research Fellow for the Collaboration for Leadership in Applied Health Research and Care (CLAHRC) for Greater Manchester. She has worked on research studies examining equality and diversity in the workplace and female entrepreneurship.

David Knights is Professor at Bristol Business School and Swansea University's College of Business, Economics and Law, and Visiting Professor at Stockholm University and Lancaster University.

Alexei Koveshnikov is Doctoral Candidate at the Department of Management and Organization at Hanken School of Economics, Finland. His research and teaching interests include diverse issues related to power and legitimacy in organizations and, more specifically, critical perspectives on managing a multinational corporation, in terms of cultural stereotyping, nationalism, and gender.

Carol T. Kulik (PhD, University of Illinois at Urbana-Champaign) is Research Professor in the School of Management, University of South Australia. Her interests encompass cognitive processes, demographic diversity, and organizational fairness, and her research focuses on explaining how human resource management interventions affect the fair treatment of people in organizations.

Savita Kumra is Senior Lecturer at Brunel Business School. She completed her doctorate at Cranfield School of Management where she is Visiting Fellow in the Centre for Developing Women Business Leaders. Savita is also International Research Fellow in the Novak Druce Centre for Professional Services at the Said Business School, University of Oxford, and has held academic posts at Oxford Brookes and Keele Universities. She is Co-track Chair of the Gender in Management Track at the British Academy of Management, *Associate Editor of Gender, Work and Organization*, and Editorial Board Member of *Gender in Management: An International Journal*. She is a member of the Chartered Institute of Personnel and Development. Savita's research interests focus on diversity, the gendered nature of the career development process in the professional services, and the importance of developing and deploying key career enhancement strategies, e.g. impression management, and building and leveraging social capital. She has published in the *British Journal of Management, Gender, Work and Organization, Journal of Business Ethics*, and *Gender in Management: An International Journal*. She has recently published her first book, co-authored with Dr Simonetta Manfredi, *Equality and Diversity Management: Theory and Practice* (Oxford University Press).

Patricia Lewis is Reader in Management in the Kent Business School, University of Kent, UK. Her current research interests include post-feminism, femininity and entrepreneurship, entrepreneurial identity, and gender and entrepreneurship in general. She has published in a range of journals including *Gender, Work and Organization, British Journal of Management*, and *Human Relations*.

Kate Lockwood Harris is a doctoral candidate in the Department of Communication at the University of Colorado Boulder. She studies the communicative aspects of gender, violence, and sexuality. Her publications have appeared in *Women's Studies in Communication, Women and Language*, and *thirdspace: a journal of feminist theory and culture*. Her current research develops an intersectional approach to the relationship between sexual violence and organization.

Debra A. Major is Professor of Psychology and Associate Chair for Research at Old Dominion University. Her research focuses on career development issues, including barriers faced by women and minorities, developmental relationships at work, and the work–family interface. She is a fellow of APA, APS, and the Society for Industrial and Organizational Psychology.

Sharon Mavin is Professor of Organization and Human Resource Management at Newcastle Business School, Northumbria University, UK. Her research interests focus upon doing gender well and differently; women's intra-gender relations and management careers; leadership, identity, and emotion work in organization management. Sharon is Co-editor of *Gender in Management: An International Journal* and has recent publications in *Gender Work and Organization, Organization, International Journal of Management Reviews, Gender in Management: An International Journal*, and the *British Journal of Management*.

Isabel Metz (PhD, Monash University, Australia) is Associate Professor of Organizational Behaviour at the Melbourne Business School, University of Melbourne. Her interests encompass gender and careers, diversity management, work and family, and employment relationships. Current research projects focus on diversity practices and organizational outcomes, women in leadership, work–family conflict, and psychological contracts.

Albert J. Mills is Professor of Management and Director of the Sobey PhD (Management) at Saint Mary's University, Canada. He has served as Senior Research Fellow at Hanken University, Finland, and more recently as Distinguished Visiting Fellow at Queen Mary University, UK. He has published over 20 books, including *Sex, Strategy and the Stratosphere* (Palgrave/MacMillan, 2006) and in the forthcoming *Oxford Handbook of Diversity in Organizations* (with R. Bendl, I. Bleijenbergh, and E. Henttonen).

Mariana Ines Paludi is a PhD student at the Sobey School of Business, Saint Mary's University, Canada, and Teaching and Research Assistant at Universidad Nacional de General Sarmiento, Argentina. Her areas of research include critical management, gender, culture, Latin America, and postcolonialism. In 2011, she was awarded with Vanier Canada Graduate Scholarship to pursue her PhD studies.

Gary N. Powell, PhD, is Professor of Management at the University of Connecticut. He is author of *Women and Men in Management* (fourth edition), editor of the *Handbook of Gender and Work*, and author of *Managing a Diverse Workforce: Learning Activities* (third edition).

Floor Rink is Associate Professor at the University of Groningen in the Netherlands. She examines the social psychological mechanisms underlying individual and group behaviour. Her research topics include gender inequality and the upward mobility of minorities in the workplace, group receptivity to deviance (i.e. moral rebels, newcomers), and status differences within groups.

Nick Rumens is Reader in Management and Organization at University of Bristol, UK. His current research interests include queer theories and the disruptions they might generate within the field of organization studies. Nick's research mobilizes queer theories to examine workplace friendships and intimacies, genders and sexualities in organization, and critical management research. He has published articles on these topics in journals including *Human Relations, The Sociological Review*, and *Human Resource Management Journal*, and in books such as *Queer Company: Friendship in the Work Lives of Gay Men* (Ashgate, 2011) and, co-authored with Mihaela Kelemen, *An Introduction to Critical Management Research* (Sage, 2008).

Michelle K. Ryan is Professor of Social and Organizational Psychology at the University of Exeter, UK, and Professor of Diversity at the University of Groningen, the Netherlands. Her major area of research is the study of gender and gender differences in context. She is particularly interested in gender discriminatory practices in the workplace, such as the glass cliff and the gender pay gap. Other research interests include the study of complex and stigmatized social identities, such as those based on race, sexuality, and disability. She works closely with industry and policymakers to translate her research into practical interventions.

Ruth Simpson is Professor of Management at Brunel Business School, UK. She has published widely in the area of gender and management, gender and emotions, and gender and careers. Recent books include *Men in Caring Occupations: Doing Gender Differently, Gendering Emotions in Organizations, Revealing and Concealing Gender in Organizations, Dirty Work: Concepts and Identities*, and *Emotions in Transmigration*.

Linda Smircich is Professor of Organization Studies in the Department of Management at the Isenberg School of Management, University of Massachusetts, Amherst. Since the late 1980s she and Professor Marta Calás have collaborated to explore the epistemological roots and gendered features of contemporary issues in management and organizations. For their academic leadership, and the impact of their body of scholarship in the area of gender and diversity, they received the SAGE Award for distinguished scholarly contribution from the Gender, Diversity, and Organization division of the Academy of Management. In 1994, Professors Smircich and Calás, with colleagues from the UK, were part of the founding editorial team of the interdisciplinary journal *Organization*, serving in editor capacity for more than fifteen years.

Valerie N. Streets is a doctoral student in the Industrial/Organizational Psychology programme at Old Dominion University. Her research is centred upon gender issues within organization, with special emphasis on the role of stereotyping and stereotype threat.

Janne Tienari is Professor of Organizations and Management at Aalto University, School of Business, Finland. He also works as Guest Professor at Stockholm University, School of Business. Tienari's research and teaching interests include managing multinational corporations, strategy work in the global context, cross-cultural communication, gender and diversity, media and the language of global capitalism. His latest passion is to understand management, new generations, and the future.

Maria Tullberg is Associate Professor and Senior Lecturer at the School of Business, Economics, and Law, University of Gothenburg, Sweden. Her research focus is on gender in business, with special attention to management.

Stephen M. Whitehead is Visiting Professor of Gender Studies at Shih Hsin University, Taipei, and Asia Programme Coordinator for Keele University, UK. He has undertaken worldwide research into gender, men, and masculinities. His tenth book *Gender Identity* (co-authored) was published by Oxford University Press in 2013.

Jannine Williams is Lecturer in Organizational Behaviour at Newcastle Business School, Northumbria University, UK. Her research interests encompass processes of organizing; categories of social relations and constructions of difference, particularly disability and gender; women's intra-gender relations and friendship at work; career studies with a focus upon boundaryless careers. She has co-edited a book *Deaf Students in Higher Education: Current Research and Practice* and published in the *International Journal of Management Reviews*.

INTRODUCTION

SAVITA KUMRA, RUTH SIMPSON, AND
RONALD J. BURKE

THE issue of gender in organizations has attracted much attention and debate over a number of years. The focus of examination is inequality in opportunity between the genders and the impact this has on organizations, individual men and women, and society as a whole. It is undoubtedly the case that progress has been made with women participating in organizational life in greater numbers and at more senior levels than has been historically the case—challenging notions that senior and/or influential organizational and political roles remain a masculine domain. In 2013, 12.7 million women were in the workforce, compared with 15.3 million men. According to the 2011 *Sex and Power Report* published by the Equality and Human Rights Commission (EHRC), women comprise 22.2 per cent of members of parliament in 2011, compared with 18.1 per cent in 2003. They constitute 17.4 per cent of members of cabinet, compared with 23.8 per cent in 2003. In business, 12.5 per cent of directors (executive and non-executive) in FTSE 100 companies were women, up from 8.6 per cent in 2003, and 9.5 per cent of editors of national newspapers were women in 2011, compared with 9.1 per cent in 2003. Small progress has been made in public appointments, with women comprising 29.2 per cent of civil service top management in 2011, compared with 22.9 per cent in 2003, and 12.9 per cent of senior judiciary (high court judges and above) in 2011, rising from 6.8 per cent in 2003. There has been some improvement for women represented in senior positions in education, with women constituting 14.3 per cent of university vice-chancellors in 2011, up from 12.4 per cent in 2003. Women also represent 31.4 per cent of health service chief executives in 2011, compared with 28.6 per cent in 2003. The report describes the progress of women to positions of authority in Britain as 'tortuously slow'. As we can see from these statistics, progress is not even—it waxes and wanes or even reverses in some areas. The report explains that women are on a level pegging with men in their twenties but several years later a very different picture emerges. Many have disappeared from the paid workforce 'or remain trapped in the "marzipan layer" below senior management, leaving the higher ranks to be dominated by men'. It estimates that, if women were to achieve equal representation among Britain's 26,000 top jobs, approximately 5,400 missing women could rise through the ranks to positions of influence.

Women now comprise over 40 per cent of practising solicitors and 42 per cent of members of chartered accountants in England and Wales. As Bolton and Muzio (2008: 282) note: 'There seems little doubt that women have made huge progress—numerically dominating areas of the labour market and entering and succeeding in previously male dominated occupations and professional groups.' They also comprise 15 per cent of directors of FTSE 100 companies and 6.6 per cent of FTSE 100 executive directors (Sealy and Vinnicombe 2012).

Therefore, while women have made inroads into key areas of public life, advancement has been slow, leading some commentators to argue that, in the context of management and organizations, progress has 'stalled' (Cohen *et al.* 2009).

Given these observable differences in organizational outcomes between men and women, this *Handbook* focuses on organizations and the way in which their processes and practices systematically work to produce gender inequities. Drawing on Sturm's (2001) conceptualization of 'second generation' gender discrimination, it seeks to problematize organizational work cultures and practices which appear superficially neutral, but result in differences in experience and treatment between men and women (Kolb and McGinn 2009). In contrast to first generation gender discrimination which typically involved intentional acts of discrimination, second generation gender issues appear objectively based and to contain no intentional bias. However, through their reflection of masculine values and life situations they inevitably contain a disadvantage for those with different values and life experiences (Kolb and Blake-Beard 2009). Here, Ely and Meyerson (2001) have identified a number of practices that serve to perpetuate gender inequity in organizations, for example, gendered jobs, gendered work, gendered definitions of leadership, and the gendered structure of social capital. These gendered processes can be concealed within taken-for-granted norms, practices, and values. Thus, it has been shown that equal opportunity policies, through discourses of meritocracy and gender justice, can serve to conceal continuing gender disadvantage beneath a persuasive rhetoric of 'best person for the job'—institutionalizing invisible forms of gendered normativity (Lewis and Simpson 2010).

The *Handbook* develops and discusses a number of these issues through critical examination by leading authors in the area of gender and organizations. In so doing, it provides a basis for elaboration and elucidation on questions and challenges confronting those interested in better understanding and tackling gender issues. We are of course aware that the contributions to the text cannot address conclusively all the questions that concern the field. However, we can capture some of the disparate voices and provide a focus for ongoing examination, dialogue, and debate of critical issues related to gender in organizations.

We have grouped chapters into four comprehensive sections that highlight some of the key developments in the field: (1) Theorizing Gender and Organizations, (2) Gender in Leadership and Management, (3) Gender and Careers, and (4) Masculinities in Organizations. We recognize that many of the dynamics discussed within each section go beyond one specific dimension and address, simultaneously, different themes. However, they provide a good starting point in terms of highlighting key contemporary issues relating to men's and women's experiences at work and serve to map out what has now become a broad terrain.

The first section, Theorizing Gender and Organizations, examines some of the major theoretical developments and provides an overview of how work in gender and organizations has evolved. The increasingly diverse literature on gender and work can be charted from the 1970s, which sought, at a fundamental level, to bring women into theorizing on management and organization. Such literature often explored differences (e.g. in career progress, career experiences) between women and men—where those differences were assumed to be both observable and stable. Recent work has focused on gender as a process—reproduced through performance and practice as well as on how gender and other categories of difference may be mobilized and resisted, and how they intersect. This broad movement is reflected to a large extent in the six chapters in this section which draw out the different ways of seeing and understanding gender and how these understandings both inform and are informed by contemporary organizational contexts.

Our first chapter: 'Theorizing Gender and Organizations: Changing Times... Changing Theories?', by Marta Calás, Linda Smircich, and Evangelina Holvino, provides a comprehensive and thought-provoking account of the theoretical perspectives that form the conceptual contours of the gender and organization literature as it has taken shape over the last forty years. Based on an evaluation of these conceptual framings, they argue that we need to incorporate more fully the changing conditions within the contemporary landscape of gender disadvantage into these understandings. In particular, they address the changing contours of inequality under current global circumstances, which they see as requiring new theorizing. They raise thought-provoking issues regarding the role that some theorizing has had in the perpetuation of inequality and how contemporary literatures outside the realm of gender-and-organization may become the framing for new organizational theorizing on sex/gender inequality.

In 'Disturbing Thoughts and Gendered Practices: A Discursive Review of Feminist Organizational Analysis', Mariana Ines Paludi, Jean Helms-Mills, and Albert Mills take a historical and discursive approach to explore how feminist organizational analysis has developed over time. Here, they move away from seeing this development as a linear and/or progressive history and instead focus on how and why particular accounts were/are received and how dominant (e.g. positivist, managerialist) accounts have influenced the development of the field. They highlight some of the historically grounded politics of exclusion as women's voices were ignored, whilst showing how different discursive contexts and positions have created a rich array of (plural, contextual) feminist contributions. Here, they argue that, in order to deal with discrimination *within* organizations, we need to question and problematize the underlying histories that serve to support and legitimize management and organization studies as a discipline, and treat both MOS and feminist theory as a reaction to certain discursive realities. In this way, attention can be focused on the discursive nature of organizational analysis and the implications for feminist change strategies.

In 'Organizations as Symbolic Gendered Orders', Silvia Gherardi adopts a symbolist approach to grasp the ambiguity and diversity of gendered meanings in organizations, e.g. around emotionality, subjectivity, sexuality. Drawing on a cultural approach, she shows how the symbolic realm is a dimension of reality and helps us understand how gender is 'done' at work, and how organizations as cultural products 'do' gender. Both play an active role in gender performativity through the interplay of ceremonial and remedial work that underpin the symbolic ordering of an organizational culture. Through the rich symbology of the dragon, she provides fascinating insight into organizations as symbolic gendered orders. In so doing, she highlights how gender symbolism is maintained, reproduced and culturally transmitted through the ceremonial work that takes places within organizations—while remedial work restores the symbolic order of gender following instances when it has been challenged.

Heather Höpfl, in her aptly titled 'Was will der Mann?' seeks to reverse Freud's famous question about what women want. She draws on Freud's veneration of the phallus and the orientation to women as 'deficiency' and 'lack' to speculate on the possibility of a discourse of maternity which might bring the possibility of social change. The chapter addresses how the relationship between the valorized and the deficient can be seen in organizations and their practices. As she points out, conventional patriarchal representations of the organization reduce *organization* to mere abstract relationships, rational actions, and purposive behaviour (forecasting, monitoring) that are synonymous with the phallus and with masculinized regulation and control. This renders all organizational members deficient in relation to targets which have to be achieved—with the double consequence for women, since there is already the acknowledgement of a lack which precedes entry. Through evocative stories and examples, Hopfl shows how, through the privileging of metrics, potentially transformative maternal values such as care and nurture are set aside in favour of (disembodied) volume and frequency—both of which are primarily patriarchal, phallogocentric measures.

In a her stimulating chapter, 'Feminism, Post-feminism, and Emerging Femininities in Entrepreneurship', Patricia Lewis draws on cultural phenomenon of post-feminism to explore the neglected issue of femininity and entrepreneurship. As she argues, the potential alignment between the agentic, self-reinventing post-feminist subject and the autonomous, self-regulating entrepreneur means that post-feminism is a useful (but often neglected) cultural frame for exploring the issue of femininity and entrepreneurship. The fluidity regarding what femininity means in contemporary times allows us to examine contemporary changes in the relationship between home and work and how this contributes to the emergence of different modes of entrepreneurial femininity. Lewis draws on three post-feminist factors: individualism, choice and empowerment; notions of 'natural' sexual difference; and retreat to the home. In a highly persuasive analysis, she connects these factors to four emerging entrepreneurial femininities including individualized entrepreneurial femininity, maternal entrepreneurial femininity, relational entrepreneurial femininity, and excessive entrepreneurial femininity. As she argues, these are embedded within a historically specific post-feminist context and incorporate transformations in popularly available understandings of femaleness and women's positioning in the contemporary world of work. In so doing, Lewis articulates how post-feminism can help us develop a more nuanced understanding of gender dynamics in organizations.

In our final chapter in this section, 'Meaning that Matters: An Organizational Communication Perspective on Gender, Discourse, and Materiality', Karen Lee Ashcraft and Kate Lockwood Harris develop new understandings regarding a 'constitutive' view of discourse. This departs from a more conventional view of discourse as merely expressing, transmitting, or maintaining already existing realities and better accounts for contemporary concerns regarding materiality. Referring to the split in current work between discourse and matter, they advance the development of a new (fifth) frame, *communication as discursive-material evolution*, which highlights the ongoing inter-penetration of discourse and matter in everyday life. This 'post-humanist' conception of communication fuses discourse and materiality into 'meaning that matters'. To facili-tate this development, they draw on organizational communication theory and theorize communication as the process through which symbol and material meet and meld into the tangible realities of work and organizational life. To demonstrate the potential of this lens, they present a powerful analysis of the material-discursive interpenetrations of gendered violence, drawing specifically on the recent Penn State scandal in the US. They show how the frame 'reorganizes' the discursive and material complexities of gendered violence and offers conceptual resources, not yet shared among gender and organiza-tion scholars, that can help us reimagine how discourse 'matters'.

In our second section, examining Gender in Leadership and Management, we argue experiences of advantage and disadvantage manifest more clearly at higher levels of the organization. Here, dominant notions of 'good leadership' often support and are underpinned by masculine values—further entrenching traditional gender based dif-ference. This has been both supported and refuted by research that has explored gen-der differences in management and leadership; that has considered leadership as part of masculine discourse; that has promoted a feminization of management thesis; and that has suggested a colonization of the feminine within the masculine leadership domain. This section explores these key issues and others in the context of contempo-rary organizations.

In the first chapter in this section 'Female Advantage: Revisited', Alice Eagly, Leire Gartzia, and Linda Carli provide a critical review of evidence to date that women have particular advantages and disadvantages as leaders. They focus on a key contention evident in the literature that women more than men evince a transformational leader-ship style, which has been linked to enhanced leader and organizational performance. Women's higher emotional intelligence, ethical standards, and endorsement of benevo-lent and universalistic values may also confer benefits in some contexts. However, they reflect against this backdrop that women leaders continue to experience organizational prejudice, discrimination in pay and advancement, and difficulty in obtaining desir-able developmental job opportunities. Given this blend of advantage and disadvantage, prevailing evidence in respect of women's leadership effectiveness provides a mixed picture. Women leaders are more effective than men; particularly in less masculine set-tings. Gender diversity enhances team performance, but only when teams are managed to overcome group conflict, and enhances corporate financial outcomes only in firms that are poorly governed or that emphasize innovation. They conclude that increasingly favourable attitudes towards women leaders and the emergence of a more androgynous cultural model of leadership bode well for women leaders in the future.

In the second chapter in this section, 'The Rocky Climb: Women's Advancement in Management', Isabel Metz and Carol Kulik argue that women's advancement is impeded by both traditional and modern barriers. They thus liken women's advancement to a 'rocky climb', involving a great deal of effort relative to the amount of upward progress, and with significant opportunities for backsliding. They identify 'traditional' barriers as those which have persisted over time, including decision-makers' denial of gender discrimination, social gender roles, stereotypes and perceptions, and organizational culture. Moving on to consider contemporary issues, they observe that overt gender discrimination has been replaced by a more covert form, modern sexism, and 'gender fatigue' have emerged as new barriers to women's advancement. To address the complexity of the issues identified, the authors advocate a customized, step-by-step approach to assist in overcoming 'gender fatigue', thereby boosting 'gender equity' in organizations. The anticipated outcome of such an approach is to change the 'rocky climb' to just a plain old 'climb' up the hierarchical ladder for those women who wish to advance in management.

In their thought-provoking chapter, 'Leadership: A Matter of Gender?', Yvonne Due Billing and Mats Alvesson tackle the long-lasting and, they argue, possibly not very useful question of whether women's and men's leadership are similar or different. They further explore possible explanations as to why we do not yet have conclusive answers to this fundamental question. Their chapter discusses different stances of the no-difference and the gender-stereotypical views and the problems in investigating the subject— including complications in how to determine and motivate whether there is a difference in women's and men's ways of leading. Arguing that it is not necessarily very fruitful to seek general patterns in terms of gender and leadership, the authors posit that understanding gender and leadership calls less for a strict adherence to a specific theory and law-like patterns than an ability to consider a variety of key aspects and dynamics and a variation of tendencies and empirical outcomes.

Considering gender and leadership from an alternative perspective, Sharon Mavin, Jannine Williams, and Gina Grandy in their chapter 'Negative Intra-Gender Relations between Women: Friendship, Competition, and Female Misogyny', tackle the thorny issue of the nature of women's negative intra-gender social relations, with a view to offering new insights into gendered organizations and how gendered organizing processes impact upon social interactions and relationships between women. They theorize women's negative intra-gender relations by fusing theory in the areas of doing gender well and differently; gendered contexts; homophily and homosociality; women's intra-gender competition and processes of female misogyny, as complex interlocking gendered practices and processes. The chapter contributes a conceptual framework of women's intra-gender relations, which reveals under-researched, often hidden forms of gender in action. Through their analysis, the authors seek to extend the theoretical development of women's negative relations by recognizing that they have the power to limit the potential for homosocial and homophilous relations between women and conclude by offering questions to guide future research agendas.

'Sex, Gender and Leadership: What do Four Decades of Research Tell us?' is a comprehensive and insightful contribution from Gary N. Powell. In the chapter he reviews four decades of research on the linkages among sex, gender, and leadership by examining status, preferences, stereotypes, attitudes, behaviours, and effectiveness associated with the leader role in relation to gender stereotypes and roles. The chapter begins by reviewing women's status over recent decades. It then considers preferences for male versus female leaders in general, moving on to compare leader stereotypes with gender stereotypes, examining whether leader stereotypes have changed over time. Attention then turns to a review of attitudes toward female leaders, with a consideration of whether (and if so, how) female and male managers differ in their behaviour and overall effectiveness as leaders. The chapter concludes with a discussion of implications for future theory, research, and practice.

In the final chapter in this section 'Gendered Constructions of Merit and Impression Management within Professional Service Firms', Savita Kumra argues women may be disadvantaging their career advancement opportunities within professional services firms by ignoring the benefits impression management activities can bring to them, preferring to rely instead on principles of meritocracy. The chapter considers the concept of meritocracy and assesses its key characteristics. Also considered are some of the inherent contradictions within the concept and the gender implications of these. Attention then turns to an analysis of impression management, contrasting the key characteristics of the two concepts. The chapter concludes with consideration of the nature of career advancement processes within the professional services, with particular emphasis on the performative nature of this process; gender implications are then assessed and discussed.

Our third section on Gender and Careers highlights the different obstacles and challenges that managerial and professional women face largely because of the gendered or masculine nature of organization (Catalyst 1998). These include long work hours and 'extreme jobs' that are highly demanding in terms of time and effort and which women may be unable or unwilling to undertake (Hewlett and Luce 2006); job responsibilities that 'support' (e.g. human resources, public affairs) rather than those which are central to the business or organization and hence may offer limited opportunities for career advancement; gendered notions of mobility that prioritize availability and international movement as prerequisites for some global careers and the paradoxical nature of career gaps or career breaks which, though provided and promoted by organizations, have been found to be detrimental in terms of promotion and pay (Schneer and Reitman 2006).

Women may also face cultural barriers to building human and social capital that have detrimental effects. Here, two kinds of social relationships have been shown to facilitate the career development and advancement of managerial and professional women and men: mentoring relationships and networking. Mentoring improves both protégé and mentor job performance advantages in the form of protégé socialization, support for long-term human resource planning, informing succession planning, and leadership development (Allen *et al.* 2004). However, most mentors are male and women find it much more difficult to develop mentor relationships than men (O'Neill and Blake-Beard 2002). Further, cross-gender mentoring relationships raise unique issues, including questions surrounding the appropriate levels of intimacy or distance, perceptions of bias towards the protégé held by outsiders, perceptions of sexuality, and actual sexual relationships (Clawson and Kram 1984).

Networks have many benefits to both individuals and organizations (Cross and Parker 2004). Individuals get access to information, greater visibility, more support, and increased chances for advancement. Effective networking is associated with career and business success. But organizational culture will clearly impact the effectiveness of women and men's networking activity. Much of networking occurs after work hours and involves socializing after work; often including drinks, sports, and sometimes strip clubs. Women and men have different networks that are not well integrated with each other (Ibarra 1992). Women may lack access to men's networks (even though they prefer and place greater trust than men in their networks for high-risk issues) and their networks are smaller and can contain members with less power and centrality. Women and men also see different benefits and values in networking—with women perhaps less likely to appreciate the career-enhancing benefits of the activity and hence needing to reconsider their relationship with this key organizational activity.

Not surprisingly, our contributions addressing gender in organizations and careers are consistent with and support previous writing. Valerie Streets and Debra Major in their chapter 'Gender and Careers: Obstacles and Opportunities', directly address such career issues—what they term the 'leaky pipeline' for women—at two stages: as girls in school and women entering the workforce. Girls in school have limited access to successful role models, are low in self-efficacy, and low in self-image. As a result they tend to perceive low opportunities and have low expectations. Women entering the workforce are less likely to enter male-dominated occupations, and because of 'stereotype threat' perform below their talents. They also encounter sexism (think manager, think male). Gender-based discrimination manifest through mechanisms such as wage, performance evaluations, and promotion gap (the glass ceiling). Women may also be 'punished' for displaying agentic characteristics (i.e. behaving in non-stereotypical ways). The authors offer suggestions for addressing these obstacles for both girls and women entering the workforce.

Susanne Bruckmüller, Michelle Ryan, Floor Rink, and Alexander Haslam in 'The Glass Cliff: Examining Why Women Occupy Leadership in Precarious Circumstances', document still another challenge managerial and professional women face—the glass cliff. Instead of 'think manager, think male' the glass cliff represents 'think crisis, think female'. When women get 'promoted', more often than not it is to a high-risk job. In these jobs women receive less reward, face greater scrutiny, and are at greater risk of being unsuccessful. The glass cliff is another form of gender discrimination, reflecting both sexism and the expendability of women. It also reflects the stereotype that women are better able to handle crises than are men. In addition, in-group favouritism by men supports the preference to appoint men to low-risk jobs, with a greater likelihood of success.

In 'Power and Resistance in Gender Equality Strategies: Comparing Quotas and Small Wins', Yvonne Benschop and Marieke van den Brink consider two strategies for improving both the work experiences of women in the workplace and their career advancement possibilities. As they point out, progress in these areas has been very slow. They consider the use of quotas and 'small wins' as possible change strategies and focus on power and resistance as central to change efforts. Quotas address issues of structural inclusion of women (their numbers) while small wins attempts to bring about structural change. Resistance today tends to encompass both overt and subtle forms, depending on the situation. The use of quotas tends to engender resistance. They usefully examine the pros and cons of each strategy, advocating in the end a combination of both.

Sandra Fielden and Carianne Hunt in 'Sexual Harassment in the Workplace' consider the highly emotive issues of sexual harassment in the workplace. Sexual harassment appears to be a fairly common occurrence, though often under-reported. Sexual harassment reflects male dominance and power over women. In a unique account, they review and analyse this behaviour from the perspective of perpetrators as well as victims. They offer primary, secondary, and tertiary interventions to reduce levels of sexual harassment in organizations.

In 'Organizational Culture, Work Investments, and the Careers of Men: Disadvantages to Women?', Ronald Burke, using gendered or masculinized organizations as the setting, considers ways in which men are advantaged (women disadvantaged) in terms of their work experiences and career progress. Organizations were designed by men for men; women are less comfortable in them and less likely to fit. The literature has documented the following barriers facing women managers and professionals: women have difficulty travelling, women are not interested in international assignments, women are adversely affected by career breaks or career gaps to raise children, women face sexual harassment, women are less likely to take part in after-hours socializing, women are given more risky assignments, women are scrutinized more closely, they have greater levels of home and family responsibilities, less access to mentoring and networking relationships, they are excluded from the 'old boys' club', unable or unwilling to work the long hours required in some jobs, and there is little agreement between women and men about the obstacles facing women. Burke reviews some of the solutions that address these concerns

Barbara Bagilhole, in her chapter 'Challenging Gender Boundaries: Pressures and Constraints on Women in Non-Traditional Occupations', considers the experiences of women in non-traditional occupations. These occupations exhibit both horizontal and vertical segregation—women are typically at lower organizational levels and clustered in a small number of occupations. She documents pervasive barriers for these women. These include obstacles in recruitment, career development, selection and promotion, working conditions, lack of geographic mobility, hostile reactions from men, men preferring to work with other men, limited access to men's networks, limited socialization to the occupation, being given harder jobs, few other female role models, a chilly climate, sexual harassment, and potential threats to masculinity of male colleagues. She paints a bleak picture of women's experiences in non-traditional occupations, concluding that dramatic change is needed if improvement is to be achieved.

Our final section, Masculinities in Organizations, explores some of the ways in which masculinity and masculine values have been concealed in the context of managing and organizing and some of the implications this can have for the experiences of women. As we have seen in different chapters in this book, organizations are sites of gendered power and reflect gendered values and practices including rationality, instrumentality, measurement and, through homosociability, a preference by men for men. Further, organizational practises such as target setting, skills audits, and the production of performance metrics can be seen to reflect and reinforce masculine discourses based on a desire for control. Recent accounts of gender in organizations have shown how such issues may be concealed under a rhetoric of gender neutrality or implicit assumptions concerning the validity of some forms of knowledge and of the 'appropriate' positioning of women and men. In this respect, men are often blind to issues of gender in their own behaviours and practices as well as how the advantages and privileges of masculinity are normalized and hence obscured. In Whitehead's (this volume) terms, men are invisible as gendered and privileged subjects—and gender is accordingly only seen in the context of harm that attaches to women. Against this background, the chapters in this section variously undertake a 'gendering' of men and masculinity in order to develop a greater understanding of men's practices and men's power.

The first chapter, 'Contextualizing Men, Masculinities, Leadership, and Management: Gender/Intersectionalities, Local/Transnational, Embodied/Virtual, Theory/Practice', by Jeff Hearn, provides a welcome overview of some of the key developments in the area of masculinity and leadership/management research—pointing to the diversity of 'leadership masculinities' and to the gendering as well as the often non-gendered status of understandings of the field. Hearn offers personal reflection on the area as a prelude to examining recent developments in Critical Studies on Men and Masculinities (CSMM)—an area that has been very much shaped by his work. Hearn identifies three 'absences' within critical accounts of gender and leadership: gender and intersectionalities which moves away from a singular focus on gender and which can incorporate multiple divisions such as age, class, ethnicity, and sexuality; localization and transnationalization that rejects 'methodological nationalism' for a focus on globalization, transnationalization, and postcolonialism and how these are implicated in gendered leadership behaviours, and practice; and embodiment and virtualization and how developments such as the spread of ICT in the doing, experiencing, and being affected by leadership. Hearn concludes the chapter with some illuminating insight into the importance of these developments for the relations of theory and practice.

In 'Masculinities in Management: Hidden, Invisible, and Persistent', Stephen Whitehead draws on some of the issues raised by Hearn and critically explores the reasons why masculinities persist in management, highlighting the ways they remain hidden and invisible even while heavily influencing the wider organizational culture. As he acknowledges, despite the increasing presence of women in management, most organizations continue to be heavily masculinist in culture. Here he highlights how a prevailing 'masculinist' organizational culture is largely sustained through managers taking up those gender discourses which reflect masculine ways of being. Drawing on feminist poststructuralist theories of gender identity and particularly the masculine subject, the chapter examines managerial practices such as performativity and how such practices serve to sustain both a masculinist organizational paradigm and many managers' sense of self—both influencing and being replicated within organizational cultures and management practices and reinforcing masculine-management subjectivities. Whitehead concludes by recognizing that while women managers do, for the most part, occupy a different discursive place to most men managers, their experiences of becoming a leader/manager invariably require them to convincingly replicate masculine discourses as gendered subjects.

Nick Rumens opens his chapter 'Masculinity and Sexuality at Work: Incorporating Gay and Bisexual Men's Perspectives' with the acknowledgment that gender has often been afforded priority over other aspects of difference such as sexuality. Here, he points to the heteronormative nature of critical organizational scholarship on men and masculinities and the tendency to normalize white, middle-class, heterosexual, able-bodied gender norms. These, he argues, unhelpfully maintain dualistic modes of understanding sexuality and gender and routinely overlook, despite the poststructuralist emphasis on plural masculinities, 'gay' and 'bi' (bisexual) men. As he observes, while the organizational literature on men and masculinities directs attention to how men are expected to work as 'men', there is gap in terms of the examination of how and why men are expected to be *particular kinds of men*. Gay and bi-identified men do not regularly figure in the critical masculinities literature and Rumens identifies some of those missing men and masculinities by drawing on the sexuality of organization scholarship, focusing specifically on gay and bisexual men. Here he argues persuasively for a greater dialogue between organizational scholars interested in men and masculinities and those scholars involved in examining gay and bi sexualities at work. Queer theory, that underscores the oppressive effects of heteronormativity and allows us to see how lives and identities can be constructed differently, is presented as one way forward here and, through this, Rumens calls for a richer organizational literature on gender that can account fully for how gender is shaped by a range of sexualities.

In 'Doing Gender Differently: Men in Caring Occupations' and drawing on her research in Australian and the UK, Ruth Simpson explores some of the ways in which men 'do' gender in a non-traditional occupational context. The chapter looks at the challenges men face in a non-traditional (e.g. service and/or caring) role and the strategies adopted to manage gender and occupational identity. As she argues, the context of 'feminized' work, where masculinity is highly visible and vulnerable to challenge, is a powerful one for exploring masculinities and how they are constructed, resisted, and maintained. The chapter highlights some of the challenges men face in a non-traditional role. Further, through accounts of the significance of bodies and embodiment, of the gendering of service and care and of gendered spaces, Simpson illustrates the diverse ways in which men manage gender in these contexts. As she argues, a focus on distancing strategies, where men withdraw from meanings attached to femininity, may hide more complex and often contradictory processes as, for example, men negotiate the marking of their bodies as 'out of place' in ways that then support essentialized notions of men's non-suitability for the job and as men invoke masculinity through strategies of 'new manhood' that seek to reject its status while, paradoxically, reifying its normativity. The chapter highlights how, by entering a non-traditional occupation, men simultaneously 'do' and 'undo' gender acting to reinforce as well as to destabilize gender in its stereotypical forms.

David Knights and Maria Tullberg, in 'Masculinity in the Financial Sector', problematize 'malestream' writing that generally denies or denigrates the significance of gender in the study of management and organizational practice. As they point out, while men and masculine discourses dominate management and organization, their conceptualization and interrogation are neglected even in critical studies of work and organization. Through an examination of the highly topical context of the financial sector (given the numerous scandals that followed the global financial crisis of 2007–8), they show how homosocial constructions and masculine performances reproduce power and privilege for men. The feminine is kept at a safe distance and women are prevented from invading a protected space that is characterized by a particularly aggressive macho competitiveness and preoccupation with financial reward. As they argue, such distancing minimizes the threat of the Other to masculine discourses and their performative outcomes in a context where the greatest fear is to fail to be a 'real man'. The chapter aptly illustrates this dynamic through the aggressively instrumental and competitive pursuit of the highest level of economic reward within the sector—drawing links between the gendered dynamics of the sector and discourses of masculinity in contemporary management and organizational contexts.

In our final chapter, 'Masculinities in Multinationals', Janne Tienari and Alexei Koveshnikov draw on insights from critical studies of men as well as feminist theory to revisit three influential texts on the management of multinational corporations. As they point out, management in multinationals is infused with specific masculinities as the top echelons of these organizations are taken up by a particular type of man and the ways in which they operate routinely exclude others from positions of influence. A masculinist image of the 'ideal' corporate executive is accordingly reproduced based on a form of 'geocentric' man who is free of national sentiment, constantly available, and willing to move at short notice. Through their fascinating and insightful analysis, that draws in part on telling stories of heroism and masculine success, they show how the texts (re) construct an idealized image of management based on problematic assumptions, e.g. around mobility and about family status and arrangements, and that they carry meanings far beyond the openness and cultural sensitivity which is claimed. As they show, assumptions about management in multinationals have repercussions far beyond the boundaries of these organizations. Privileged key people within multinationals draw from and reproduce a global system of inclusion and exclusion, which serves to perpetuate inequality within and across societies.

In conclusion, through the chapters summarized briefly above, this *Handbook* captures the diverse and complex ways in which organizational processes and practices systematically work to produce gender (dis)advantage. Leading authors in the field of gender and organizations have come together in this unique edited volume to highlight some of the key contemporary issues facing men and women in organizations and how gender dynamics can be understood. Through their insightful and thoughtful commentaries we provide a basis for discussion and elaboration on some of the key questions and issues that underpin a better understanding of gender in organizations.

REFERENCES

Allen, T. D., Eby, L. T., Poteet, M. L., Lentz, E., and Lima, L. (2004). 'Career Benefit Associated with Mentoring for Protégés: A Meta-Analysis', *Journal of Applied Psychology*, 89: 127–36.

Bolton, S., and Muzio, D. (2008). 'The Paradoxical Process of Feminization in the Professions: The Case of Established, Aspiring and Semi-Professions', *Work, Employment and Society*, 22(2): 281–299.

Catalyst (1998). *Advancing Women in Business: The Catalyst Guide*. San Francisco: Jossey-Bass.

Clawson, J. G., and Kram, K. E. (1984). 'Managing Cross-Gender Mentoring', *Business Horizons*, 1176: 22–32.

Cohen, P. N., Huffman, M. L., and Knauer, S. (2009). 'Stalled Progress? Gender Segregation and Wage Inequality among Managers: 1980–2000', *Work and Occupations*, 36: 318–42.

Cross, R., and Parker, A. (2004). *The Hidden Power of Social Networks: Understanding How Work Really Gets Done in Organizations*. Boston: Harvard Business School Press.

Ely, R., and Meyerson, D. (2001). 'Theories of Gender in Organizations: A New Approach to Organizational Analysis and Change', *Research on Organizational Behaviour*, 22: 103–51.

Hewlett, S. A., and Luce, C. B. (2006). 'Extreme Jobs: The Dangerous Allure of the 70-Hour Work Week', *Harvard Business Review* (Dec.): 49–59.

Equality and Human Rights Commission (2011). *Sex and Power Report*. London: EHRC. http://www.equalityhumanrights.com/uploaded_files/sex+power/sex_and_power_2011_gb__2_.pdf, 4–6 (accessed Jan. 2013).

Ibarra, H. (1992). 'Homophily and Differential Returns: Sex Differences in Network Structure and Access in an Advertising Firm', *Administrative Science Quarterly*, 17: 422–70.

Kolb, D. M. and Blake-Beard, S. (2009). 'Navigating the System and Coalescing for Change: Second Generation Issues and Leadership Programmes for Women', Paper presented at the Academy of Management, Chicago.

Kolb, D. M. and McGinn, K. (2009). 'From Gender and Negotiation to Gendered Negotiation', *Negotiation and Conflict Management Research*, 2(1): 1–16.

Lewis, P., and Simpson, R. (2010). *Revealing and Concealing Gender: Issues of Visibility in Organizations*. Basingstoke: Palgrave.

O'Neill, R. M., and Blake-Beard, S. (2002). 'Gender Barriers to the Female Mentor–Male Protégé Relationship', *Journal of Business Ethics*, 37: 51–63.

Sealy, R., and Vinnicombe, S. (2012). 'The Female FTSE Board Report 2012: Milestone or Millstone', Cranfield School of Management, Cranfield University.

Schneer, J., and Reitman, F. (2006). 'Time Out of Work: Career Costs for Men and Women US Managers', *Equal Opportunities International*, 25: 285–98.

Sturm, S. (2001). 'Second Generation Employment Discrimination: A Structural Approach', *Columbia Law Review*, 101: 458–568.

PART I

THEORIZING GENDER AND ORGANIZATIONS

CHAPTER 1

....

THEORIZING GENDER-AND-ORGANIZATION

Changing Times... Changing Theories?

....

MARTA B. CALÁS, LINDA SMIRCICH, AND
EVANGELINA HOLVINO

THIS chapter maps theoretical perspectives forming the conceptual contours of the gender and organization literature as it has taken shape over the last forty years. Our concern is with the trajectory of this literature and how it has changed—or not—with the purpose of reclaiming the social change potential that marked its beginning.

As we see it, *the necessary motivating condition* for the continued existence of the gender-and-organization literature *is the persistence of sex/gender inequality in organizations and society*. We cannot imagine a more general explanation for the staying power and proliferation of this literature as well as for the variety of theories that have appeared. With this as our point of departure, we focus critically on how this theorizing has been and continues to be done in organization and management studies, and what it may have accomplished so far.

The initial turn to gender within organization and management studies occurred during the 1960s and 1970s when feminist and other social movements brought about heightened attention to sex/gender inequality in public life. It was not the first time. Earlier feminist and abolitionist movements had already done so. However, 'second wave' feminism in the 1960s articulated the limits to women's rights in relation to men's rights as a significant social concern with economic consequences. In most affluent nations organizations were salient settings where these limits were evident and remedies were to be applied, in particular through legal mandates such as EEO/Affirmative Action policies. Against this background of cultural and political change, researchers began to explore obstacles facing women as they entered various areas of the economy. Yet, despite enacting mandates, and establishing several forms of remedies, changes in policies, organizational programmes, assorted managerial strategies, and regardless of changes in social and economic contexts, sex/gender inequality in labour markets and organizations remains. How have scholars of organization studies accounted for this persistence?

That the organizational landscape continues to be marked by persistent patterns of gender inequality is not in dispute. Documentation on the enduring sex segregation of organizations and occupations, on the stubborn fact of pay disparities between women and men, on the continued devaluation of women's work, and on women's absence at higher levels of organizations, is voluminous. Yet, we question, can the scholarly literature do more than document such facts?

Further, the backdrop against which sex/gender inequality exists today has fundamentally changed since the 1960s, including altered geopolitics, economic and cultural changes articulated in globalization processes, and, in particular, the expansion of neo-liberal market ideology worldwide. Has gender and organization theorizing changed sufficiently to address these changing circumstances?

These questions guide us as we examine gender and organization theorizing, considering its location in time and place. The conceptual framings developed and analyses conducted in the 1960s or 1970s are of interest both in terms of the history of the theoretical premises they employ, as well as the context in which they were deployed. But it is possible that some theoretical perspectives have become of little value for addressing sex/gender inequality under contemporary conditions—unless we assume that since their appearance nothing has changed in culture and society and in conceptualizing 'gender' and 'organization'.

Our approach is not a neutral evaluation of which theoretical explanations may be better—in general—to account for the persistence of sex/gender inequality in organizations. Rather, it asserts that this persistence is an *outcome—a manifestation—of various social dynamics and social processes changing over time*, and thus *requires understanding the changing conditions of its reproduction*. It requires asking, as well, what theoretical premises about gender-and-organization may be needed now, and which may require rethinking or abandoning.

While these may seem to be bold claims, some facts are worth considering. For instance, the 'gender gap', a catch-all phrase now circulating widely to signify sex/gender inequality, refers to both systematic sex differences in numbers of people participating in economic activity and systematic differences in economic rewards (Goldin 1990). As a general trend, the percentage rate of women's labour force participation worldwide has steadily increased, but occupational and income disparities associated with sex/gender inequality continue (e.g. Blau and Khan 2000; World Bank 2013). In management studies some see the situation today as 'stalled progress' (Cohen, Huffman, and Knauer 2009).

Meanwhile important but by no means radical transnational institutions as diverse as the UN, the World Economic Forum (WEF), and the International Labor Organization (ILO) have acknowledged that the problem is fundamental and are increasingly vocal in their concerns about the economic and social consequences of the gender gap and its persistence worldwide. As an example, a recent WEF report admonishes countries and companies as follows:

> Women...constitute a burgeoning portion of the talent pool available to companies today. Over time, therefore, a nation's competitiveness depends significantly on

whether and how it utilizes its female talent. . . .. Governments play an important role in helping create the right environment for improving women's economic participation . . . However, it is then primarily the imperative of companies to create ecosystems where the best talent—both male and female—can flourish.

(Zahidi and Ibarra 2010: p. v)

These ongoing conversations—inside and outside the academic realm—offer an occasion for assessing what may have been accomplished in management and organization studies to address these issues, and to reflect on what else might be needed now. Ours is not a comprehensive review of the literature (for a recent review, see Broadbridge and Simpson 2011). Rather, we illustrate the literature's contours seen from our vantage point, as critical gender-and-organization scholars and practitioners located in the contemporary context of the USA.

The chapter is organized as follows: The next section, Contours, distinguishes the literature meta-theoretically. It contains two subsections, Theorizing 'Gender *in* Organizations' and Theorizing 'Gendering Organizations', contrasting theoretical approaches which developed sequentially. Nonetheless, both approaches continue to be followed at present and we note that this is in itself remarkable, for one has tended to obscure conditions promoting inequality in organizations, while the other has made these even more visible as well as more complicated. In light of this we ask, up to what point does certain gender and organization literature still produced today contribute to perpetuating the situation it purports to be studying? A second section, 'The Changing Contours of Inequality', addresses how contemporary literatures mostly outside the realm of gender-and-organization may become, of necessity, the framing for new organizational theorizing on sex/gender inequality. Their explicit focus on social change under current global circumstances may provide a new and much needed impetus for re-examining relationships between scholarly production on this topic and its effects in 'the real world'. In a brief concluding section we offer further reflections on institutional conditions bearing on the production of knowledge on gender and organizations and more generally.

CONTOURS: THEORIZING 'GENDER *IN* ORGANIZATIONS' OR THEORIZING 'GENDERING ORGANIZATIONS'?

At the most general level we can identify two main meta-theoretical approaches in the gender and organization literature. The first and older approach—theorizing gender *in* organizations—follows a more 'naturalistic' or 'common-sense' orientation towards gender; understands *sex* as biological characteristics—male and female—and *gender* as social or cultural categorization usually associated with a person's sex, i.e. masculinity and femininity, often conceived as stable traits or roles (Alsop, Fitzsimmons, Lennon, and Minsky 2002). Issues of gender and organizations are typically framed as having to do with conditions for women, often in comparison to men. Thus, from within this perspective the title originally assigned for this chapter by the editors, 'Theorizing Gender *in* Organizations', makes sense, as the literature tends to assume people as sexed/gendered beings acting within the confines of a neutral 'container'—'the organization'. The organization functions as a stage on which individuals act but—with few exceptions—the stage is rarely examined.

The second approach—gendering organizations—'de-naturalizes' the common sense of gender using processual, social constructionist theoretical approaches. Rather than positioning individuals at the centre of inquiry and assuming binary notions of 'women' and 'men', or 'feminine' and 'masculine' roles, the focus instead is on gender as a social institution which is socially accomplished through gender relations. This brings the analyst's attention not simply to the sex of participants as embodied actors, but to the cultural production of their subjectivities and the material production of their lives, including the ongoing production/reproduction along gender lines of social structure/ing. In contrast to thinking of gender as a possession or attribute of people working in organizations, social constructionist accounts consider the ways gender(ing) is an outcome or a co-production of organizing processes. If this chapter were exclusively concerned with this second approach the suggested chapter title would change to 'Theorizing "*Gendering Organizations*"'.

Thus the title we ultimately chose, 'Theorizing Gender-and-Organization', is a way for us to highlight distinctions between the two approaches as we recount their basic tenets and expand on their differences.

Theorizing '*Gender in Organizations*': Gender = Sex = Women (and Men)

In the 1970s, as a consequence of the changing aspirations and opportunities for women in industrialized societies, and their greater presence and influence in the economy and the academy, a field of academic inquiry began to take shape under the umbrella term 'women-in-management'. With business organizations no longer legitimately the exclusive provenance of men, this research, much of it conducted by those affiliated with the Women in Management division of the US Academy of Management, explored issues related to the status of women in organizations, including their under-representation at higher levels. Most attempted to establish whether or not there were sex differences across an array of topics of concern to the field and what accounted for women's secondary status in organizations. The underlying question was 'why don't women achieve?'

There were echoes of liberal feminism in this question, for underneath it all women were assumed to be as good as men (i.e. the assumption of abstract individualism) and therefore deserving of the same rewards in a meritocratic society. As empirical research on these themes continued into the 1980s, the term 'gender' replaced the term 'sex'. Yet there was no shift in research logic; research continued to be about women in management and the problems they face. Much contemporary gender research in organization studies follows the path initiated at its beginning: the study of 'gender in organizations' is still primarily the study of the conditions women face in organizations centred on a fulcrum of 'difference'.

One of the longest lines of research in this tradition closely associated with the gender gap focuses on women and positions of authority. A pattern of inquiry was established when psychologically oriented research showed that respondents' general conceptions of 'a successful manager' included characteristics identified with men—but not women (Schein 1973, 1975). Numerous replications, across multiple countries, supported what came to be called the 'think manager—think male' association (e.g.Heilman, Block, Martell, and Simon 1989; Dodge, Gilroy, and Fenzel 1995; Schein, Mueller, Lituchy, and Liu 1996; Schein 2001, 2007; Ryan, Haslam, Hersby, and Bongiorno 2011) .

Other research at this time showed that women leaders were held to different standards of evaluation, even when under experimental conditions their behaviour was identical to men's (Bartol and Butterfield 1976). Subsequently this 'double standards' finding has also been reproduced numerous times, for example in assessment of job candidates, evaluations of performance, or awards of pay or promotion (e.g. Eagly, Makhijani, and Klonsky 1992; Foschi, Lai, and Sigerson 1994; Kulich, Ryan, and Haslam 2007; Roth, Purvis, and Bobko 2012).

Using somewhat different terminology, researchers echoed and extended similar observations over the decades, painting an ever sharper picture of women being disadvantaged because they did not 'fit the mould' within organizations (e.g. Heilman 1983, 2001; Morrison, White, and Van Velsor 1987; Biernat and Fuegen 2001; Eagly and Carli 2007), a situation presumed to be responsible for their being blocked from advancement by a 'glass ceiling'. In contrast, men who entered female-dominated professions often benefited from an apparent 'glass escalator' on which they could ascend to the higher ranks (Williams 1992; Maume 1999). More recently, some researchers observe that women who do 'break through' are over-represented in situations of precarious leadership, a situation dubbed the 'glass cliff' (Ryan and Haslam 2007) where female leaders are 'more likely to be appointed in a time of poor performance or when there is an increased risk of failure' (Ryan et al. 2011: 472). Under such contextual conditions the 'think manager—think male' association becomes a 'think crisis—think female' association.

How do researchers in this tradition explain continuing differential conditions? Where do they direct their attention for explanations?

Theorizing Gender in Organizations through Cognitive Processes: An Individual Lens

Research in the 'gender in organizations' tradition tends to theorize these conditions as outcomes of cognitive mechanisms—social judgement processes, involving implicit theories of leadership, in which stereotypes about traits and abilities interfere with, or shortcut, accurate perceptions of women. Typically they theorize the underlying dynamics via *social role theory* (Carli and Eagly 1999), where gender is associated with roles appropriate for each sex, and via *status characteristics theory* (Ridgeway 2001; Ragins and Winkel 2011) where gender is associated with desirable characteristics, such as competence, attributed differently by virtue of sex category membership.

Role Theory: Gender as a Social Role

Early research by organizational psychologists invoked the notion of sex roles (Parsons 1942) to account for the ways women were assessed as leaders. Subsequently the preferred terminology shifted to gender roles and then more recently to the more general 'social roles' (Eagly 1987; Powell and Butterfield 2003), but in any version these accounts trace differences between women's and men's social influence and achievement in organizations to the historical division of labour from which specific norms and expectations for the behaviour of each developed. As noted by Carli and Eagly, '[T]he tendency of men and women to occupy different roles, which require somewhat different behaviors, fosters gender roles by which people expect each sex to have characteristics that equip it for its sex-typical roles' (1999: 207). Gender roles are thus understood to be descriptive of the ways people act and prescriptive of the ways they should act. Hence, in practice, researchers note that women are disadvantaged because stereotypes suggest they 'don't fit'.

Status Processes: Gender as a Diffuse Status Characteristic

In contrast, sociological research identified 'status generalization' as an alternate mechanism behind the creation and recreation of gender inequality (Rashotte and Webster 2005). In the 1970s, sociologists studying small task-focused groups observed that performance expectations, and expectations for the legitimacy of competitive or dominating behaviour, were affected by status processes (Berger, Cohen, and Zelditch 1972; Meeker and Weitzel-O'Neill 1977). Their 'expectation states' research indicated that when people interact in goal centred situations, such as in the workplace, their beliefs about status shape their interaction and the enactment of social hierarchies (Ridgeway 2001). Within this formulation gender, as well as age and race, is understood to function as a diffuse status characteristic, carrying cultural beliefs about the relative competence of group members in terms of knowledge, ability, or influence, with more status being attributed to men (Carli and Eagly 1999; Ridgeway 2001; Roth *et al.* 2012). Status beliefs thus explicitly imply both difference and inequality, with inequality being grounded in group membership itself (Ridgeway 2001).

Whereas role theorists and status theorists make similar observations about the difficulties women face in exerting influence, for status theorists this is not a problem of role incongruence or lack of fit, but a problem of legitimacy. Further, while role theorists are more inclined to talk about gender differences, status theorists are more likely to talk about hierarchical inequality.

Gender as a Primary Cultural Frame

Ridgeway brought together these two traditions by arguing that gender is a 'primary cultural frame for coordinating social relations' (2011: 88). As people automatically and routinely sex-categorize one another they implicitly draw on gender stereotypes in which status beliefs and beliefs about traits and roles are embedded. She notes that in the ongoing need to coordinate social relations in the workplace, the gender frame 'offers a too convenient cultural device' (2011: 122). Its activation provides 'an ever-available framework for filling in the details of an uncertain work task, setting, or person and for providing an overarching, simplifying interpretation of complex circumstances' (2011: 93).

Where does this Leave Women?

Though offering somewhat differing views on the cognitive mechanisms through which women are judged, whether their actions are filtered through conceptions of roles, status, or cultural beliefs, this scholarship continually points to a dilemma facing women in situations of authority. As Rudman and Phelan (2008: 70) note: 'the research evidence clearly points to negative consequences for female agency'. The implicit stance in the literature, as Kolb (2009) aptly puts it, is: 'too bad for the women'. They confront a 'catch 22' or 'double bind'. In order to be seen as a 'proper leader' they must act in ways to disconfirm female gender stereotypes, but in so doing they risk coming across as socially deficient, and not as a 'proper woman' (e.g. Marshall 1984; Ryan and Haslam 2007: 551). As many have observed, women especially face a 'backlash'—social and economic reprisal for acting counter-stereotypically (Rudman 1998; Eagly and Karau 2002; Brescoll 2012). Accordingly as individuals they encounter 'an impression management dilemma' (Rudman and Phelan 2008: 62) whereby they must remain vigilant about appearing 'too emotional', 'too assertive', 'too angry' (Brescoll and Ullman 2008), and so on.

Unfortunately, such reckoning places the problem of backlash on the laps of women. They are the ones who must 'work around double binds' (Kolb 2009: 12) and engage in the extra effort of 'self-monitoring' (Rudman and Phelan 2008) by, for instance, adopting stereotypically low-status, more indirect approaches in negotiation (Bowles and Flynn 2010) or, in the case of high-power women, adjusting how much they speak so their level of volubility matches that of 'low-power women and low-power men' (Brescoll 2012: 636).

This last example is said to demonstrate that powerful women are correct in assuming they will suffer backlash if they talk more than others. The article concludes by saying 'from the vantage of women's ability to achieve success in an organization [the study] suggests that existing power hierarchies may be quite difficult for women to navigate and may require some creative strategies that may work better for them than for men' (Brescoll 2012: 637–8). Another instance of 'too bad for the women'!

To really address gender inequality, however, a research study would not end there but rather this would be the starting point for raising more questions. For instance, a logical question would be: if the amount of talking powerful men do is different from women's *then how is a greater amount of talking associated with actual capabilities for performing the job?*—i.e. the focus of the study would turn to demonstrating the absence or presence of such an association.

Unfortunately, under these approaches such questions are rarely raised. Rather the system is assumed to be gender neutral, and so are the norms by which members are judged and the relationship of these norms to desirable outcomes. Where does this leave women? They remain the ones who have to engage in 'creative strategies', carrying the burden of being 'different'. Instead of highlighting *how* women are disadvantaged under assumed systemic gender neutrality, in this research tradition commonly called for remedies include exhortations for 'more education', or vague calls for 'changing the culture'—as if were possible to stop the impossible: sex/gender categorizing.

Epistemologically, these approaches stem from sociological theories of the 1970s, including role theory, when functionalism and positivism dominated scholarship in the USA. As well, the strong presence of psychological perspectives in US business schools, supporting the emergence of organizational behaviour as a subdiscipline during the 1950s and 1960s, continues to influence explanations for 'sex/gender differences' in organizational outcomes while seldom mentioning 'inequality'. Theorizing and research from these perspectives emphasize neutrality and generalizability, with findings to be translated into actionable practices based on cognitions, as if remedies could be located inside the heads of people. That these theoretical approaches have been followed for so many years while there has been so little progress in remedying the situation of gender inequality in organizations is sufficient to make us wonder whether these lines of research should continue. In particular, some of this work has been critiqued (see e.g. Ely and Padavic 2007) as falling into tautological explanations, where sex differences in outcomes attach to a priori assumptions of women's differences. This almost comes full circle to the starting point of this literature, which under assumptions of abstract individualism is fundamentally asking if women are as good as men.

Organizational Mechanisms of Stratification: An Organizational Lens

The literature discussed so far concentrates on what are understood as mostly cognitive processes of categorization and stereotyping engaged in by individuals in judgment situations. Researchers explain conditions disadvantageous for women by referring to gender beliefs resulting in 'glass ceiling' effects. As indicated earlier -and with some exceptions (e.g. Ridgeway 2011), in these approaches 'the organization' remains in the background. As noted by some, this literature fails to fully attend to the organizational mechanisms linking beliefs held by individuals to unequal workplace outcomes (Reskin 2005; Stainback, Tomaskovic-Devey, and Skaggs 2010).

Recently, an alternative, organization-centred approach, mostly pursued by organizational sociologists in the US—attends to the organizational processes through which gender (and racial) stratification occurs (e.g. Bielby 2012; Castilla 2012). Research in this tradition investigates the everyday organizational practices through which organizations recruit, evaluate, compensate, and promote employees to assess whether—and how—gender (and racial) disparities are produced, for example, how organizations become stratified by sex. Unfortunately, this type of theoretical approach is seldom used in US business schools.

An exception is Castilla (2012) who, using longitudinal personnel data on over 8,000 employees of one large service organization, investigated the effects of its merit-based performance management system on race and gender inequality at three key stages. His analyses identified significant gender and racial disparities across multiple stages of assessment, such as the performance evaluation, salary, and career setting, despite the firm's supposed merit-based practices. In the long run, employees who had received *the same performance* ratings received different salary increases, as well as different outcomes in key career decisions such as terminations and transfers, depending on their gender, race, or nationality.

In another study, notable because it involved objective measurements of performance, Madden (2012) investigated the gender wage gap among stockbrokers. With access to data from two of the largest US stockbrokerage firms in the 1990s, she conducted a 'natural experiment' to test for the sources of differences in compensation by sex. Were women's lower wages within formalized merit-based pay plans due to their lesser sales capacity, i.e. attributable to 'true performance' differences, or were they due to some other organizational factor? Madden tracked internal transfers of customers' accounts among stockbrokers, demonstrating that women in these firms received 'inferior account assignments' (i.e. accounts with lower historic commissions and/or asset values). At the same time, her analyses showed that when accounts with equivalent prior sales histories were transferred, there were no gender differences. Madden concluded that the gender pay gap among stockbrokers in these firms was attributable to management's discretionary assignments of sales opportunities.

When it comes to theorizing gender in organizations with a concern for reducing sex/gender inequality we see more promise in these approaches. Instead of offering social or cognitive psychological explanations articulating gender differences, this literature focuses on the non-neutrality of organizational decision making and organizational practices, and goes to the heart of organizational conditions creating inequality. Therefore, potential remedies are directed towards organizational change.

Nonetheless, scholars in all 'gender in organizations' theoretical traditions are seeking to chart generalizable patterns resulting in observed disparities between women and men. Though the term gender is used, all these approaches rely on 'body-counting' as evidence that something may not be right when sex inequality persists (Martin 1994). In fact, none of these approaches theorize gender. They are primarily applications of conventional psychological and sociological concepts to explain the 'problem' of low numbers of women in positions of authority in organizations, and not about understanding the reproduction and persistence of sex/gender inequality. What else to do?

From Theorizing 'Gender in Organizations' to Theorizing 'Gendering Organizations': Produced, Producing, and Reproducing

The theoretical approaches in this section originally appeared during the late 1970s and 1980s in feminist sociology, mostly in the US and the UK, and by the 1990s in organization and management studies primarily in European locations, and also Australia, Canada, and New Zealand. Still, they remain under-represented in organization and management studies in US business schools.

To fully appreciate the distinguishing characteristics of these approaches it is important to introduce them in contrast to those in the prior section. Their differences go beyond the matter of the time in which they first appeared and the geographical and disciplinary locations where they are mostly adopted. It is in a discussion of their meta-theoretical premises where their fundamental differences would be best understood. These differences have important implications when examining sex/gender inequalities in organizations.

First, different from the functionalist and positivist orientation of the theories framing the literature in the prior section, with gender and sex usually understood as discrete variables, the theoretical approaches discussed in this section share a social constructionist understanding of gender/sex, conceiving of these as social processes. That is, ontologically, gender is an emergent feature of social situations, not something one 'has' as an individual, but something humans 'do' in relation to each other as an ongoing accomplishment in social life (West and Zimmerman 1987). Similarly, sex is not 'naturally' meaningful as a relevant social category; it is *produced* as socially relevant by agreeing, in West and Zimmerman's words, 'upon biological criteria for classifying persons as females or males' (1987: 127).

Understandings of sex and gender as two intertwined, stable, and usually binary properties of people *in* organizations in the prior section are instead conceived as simultaneously enacted social processes contributing to the production of institutional contexts (e.g. organizations) as gendered spaces. As well, the notion of levels of analysis (e.g. micro or macro) pervading the prior theoretical approaches and allowing for the separation of persons and organizations is recast. Processual understandings assume the mutually constitutive outcomes of social practices: humans as social beings who produce and reproduce what then is reified as social structure and experienced as resources or constraints for human actions. As such 'structures' are precarious, contingent, and potentially ephemeral; maintaining their appearance of durability relies on much social effort, but it also means they can be changed.

Second, rather than assuming the existence of gender mechanisms (e.g. Ridgeway 1997), be they psychological and/or sociological, with an aim towards obtaining generalizable and predictable understandings of gender inequality in organizations, the question of inequality is turned around towards understanding its production and reproduction in situated circumstances where gender is done as power relations. In the traditions discussed in this section gender is understood as a historical and culturally institutionalized system (Lorber 1994). It is produced and reproduced through relations of subordination and domination: hierarchical relations which precede any particular person in any particular organization. Organizations are thus seen as 'gender factories' (Williams 2010) or 'inequality regimes' interconnecting organizational processes that produce and maintain racialized and gendered class relations (Acker 2006a).

In these theoretical approaches gender differences—and attributions made about them—are not a matter of perceptions when people face each other in organizations; rather organizations are already gendered (and racialized and classed) as people enter them—for they (people and organizations) are part and parcel of a social system based on historical hierarchical differentiation by sex, class, and race (Anderson 1996; Acker 2006b). This argument is currently further articulated in notions of intersectionality, which we discuss in the next section.

Third, most *gender in organization* theorizing is framed through liberal humanist assumptions, where abstract individualism sustains the possibility of a meritocratic society, and where gender neutrality and just outcomes are assumed to be the norm. In contrast, *gendering organization* theorizing is framed through critical, mostly feminist philosophies, including socialist and poststructuralist theorizations often addressing the historical and cultural reproduction of a patriarchal system where domination and inequality are the norm. They are gendering process theories. In light of these different framings, basic conceptualizations of 'sex/gender inequality' in organizations will necessarily change.

Assuming abstract individualism leads to inquiries about numerical differences between men and women in hierarchical positions. If unbalanced numbers are found, this outcome is attributed to potential deficiencies in particular individuals, for instance cognitive barriers of those judging women as inferior and/or explicit discriminatory practices by some in organizations. Under these approaches sex/gender inequality is mostly a disturbance of a neutral system. Explanations must reside in antecedents of the disturbance (stereotypes; roles; status characteristics) which may lead to apparent discrimination, but which eventually can be used as justifications to dispel such beliefs— e.g. people do not discriminate intentionally on the basis of sex; it is that people who seem 'different' are imputed stereotypical characteristics affecting how they are perceived as capable or not. However, the way 'difference' is judged within the logic of this argument is based on comparisons with an accepted value norm—the 'not-different'— which is seldom explicitly stated. Leaving this norm unstated often makes for a tautological argument, sex differences explaining sex differences, which comes very close to blaming the victim through abstractions. This type of theorizing and research rarely leads to dispelling stereotypes, and even less to reducing inequalities.

Assuming gendering processes, in contrast, leads to addressing how humans' relational practices may produce different outcomes for men and women, opening a way for observing the production and reproduction of socially systemic inequalities as they thicken into hierarchical institutional forms. These approaches pay immediate attention to justifications of any taken-for-granted norm. How is it possible that 'gender differences' have become an explanation for sex/gender-based inequality in organizations? From the 'gendering organizations' perspective '(t)he point is not that women are different, but that gender difference is the basis for the unequal distribution of power and resources' (Wajcman 1998: 159–60). A central question to be asked then is: how is 'difference' done and maintained, and to what effect? (e.g. West and Fenstermaker 1995).

Gendering, thus, is constitutive of social and organizational processes where practices, images, and ideologies, as well as distribution of power, contribute to the production and maintenance of gender inequality. This inequality is based on a historically located but persistent gendered substructure of society—the division of production and reproduction into to different spheres of social life (public or private) and belonging to different people (men or women). Organizational practices and activities are concrete relational contexts where the creation and recreation of the gendered substructure is negotiated and often contested in everyday life and where gendering processes may become visible (Acker 1990, 1992b).

Fourth, as these two different approaches to theorizing gender-and-organization start from fundamentally different ontological and epistemological assumptions about sex and gender, organizational research on sex/gender inequalities would also be framed through substantially different methodologies (Calás and Smircich 2009). Both approaches may start from similar general questions. How can we know what is happening? How can we know how it happens? What to do next? But, as discussed, the specific questions asked would arise from different understandings of the notion of inequality, both theoretically and empirically.

These different understandings would lead to different research design and methods. Research from the gendering organizations perspective would proceed by scrutinizing how 'gender' and 'difference' are done, how organizations become gendered under these processes, and what are their power effects. This requires *closeness to the actual happenings* for it is there that gendering is done. In this case, methodologies and methods such as those of ethnomethodology, and institutional ethnography would be appropriate. This stands in contrast to the *distancing conceptual abstractions* (e.g. stereotypes; roles) and aims of generalization typifying most approaches discussed in the prior section where appropriate methods include, for instance, laboratory studies and statistical analysis of archival data.

As well, research outcomes would lead to different implications and contributions. Research in the prior section may treat its findings as an objectively observed *problem* addressed by noting how meritorious individuals may not be justly treated in a hierarchical but presumably gender-neutral institutional domain. Implications are often subtly stated as generalized cognitive or structural limitations but little is said about how to remedy the injustices which these may create. By contrast, research in this section observes reflexively how 'gender/sex inequality' is *ingrained in the reproduction* of a hierarchical power relations system (organizations) which often becomes normalized under meritocratic ideologies. Understanding gendering processes might then lead to making visible power relations maintaining inequality as a 'natural' or 'inevitable fact'. Such knowledge would open space for potential, even if contested, systemic change.

Finally, observing who the subjects of the research are is also revealing of the different orientations in these two approaches for theorizing gender-and-organization. The literature on *gender in organizations* focuses primarily on subjects deemed likely to occupy management positions. Its managerialist orientation ignores the existence of other members of the organization as relevant for understanding the production and consequences of sex/gender inequality. No surprise that leadership is an important topic, for this is mostly an elitist organizational literature.

In contrast, *gendering organization* approaches address a range of subjects as they 'do' gender. The focus is on how gender differences and inequalities may be produced and reproduced in many different contexts by various people engaged in everyday relational processes occurring through organizing practices. Mostly the interest is to observe relationships, but not necessarily those predefined by formal hierarchical structural positions. Rather, the focus on gendering processes provides a lens for understanding how formal structures become hierarchically organized along gender lines as if it were a natural objective fact.

Below we illustrate *gendering organization* theoretical approaches with research on two of its key themes: *doing/(un)doing gender* and *embodiment*. These themes are central for understanding the production and reproduction of gender inequality in organizations.

Doing/(un)doing Gender

Organization and management scholars have produced a considerable amount of theoretical and empirical work under this theme since the 1990s by following the original insights of West and Zimmerman (1987) on 'doing gender' as well as Acker's theories (1990) of gendered organizations. More recently Butler's work (1990, 2004*a*), in particular her notions of performativity and (un)doing gender, has joined this scholarship. We offer general illustrations of research under this theme, and follow them with two subthemes addressing how gender is done in organizational practices and organizational knowledge. All examples focus on the production of gender inequality.

In early examples, Gherardi (1994), addressed how 'doing gender' at work produced an organizational culture governing what is fair in the relationship between the sexes; while Benschop and Doorewaard (1998) argued that both the persistence of gender inequality and the perception of equality emerge from a set of often concealed power-based gendering processes, systematically reproducing gender distinctions, a gender subtext. More recently, Kantola (2008) used Acker's theories to examine the ways in which hidden discrimination and the gendered organization work together, and Benschop, Mills, Mills, and Tienari (2012) called for a revaluation of the hierarchical order in organizations. Their call aimed at *changing gender as a structure* by changing everyday organizational routines and interactions (re)producing gender inequalities.

The notion of (un)doing gender (Butler 2004*a*) has also appeared in this research. While some have confused this notion with the possibility of transcending gender, the bulk of this work has followed Butler's insights, observing what happens when trying to transgress the norms of gender. Confusion has also occurred when Butler's performativity is interpreted as another ethnomethodological approach to 'doing gender'. Yet, Butler (2004*a*, 2004*b*) has provided clarifications that the basic insight of performativity is not 'doing gender'; rather, performativity aims to describe what makes gender intelligible, its conditions of possibility, within a context as well as addressing the norms of gender in such a context.

Following these insights, Pullen and Knights (2007) addressed how gender gets done and undone in organizing and organizations and with what consequences. Other examples include Hancock and Tyler (2007), Jeanes (2007), and Powell, Bagilhole, and dainty (2009). Further, Kelan (2010) contrasted ethnomethodological (e.g. West and Zimmerman) and performative (Butler) on doing/(un)doing gender, recognizing that (un)doing gender as such is a way of doing gender.

Gendering Practices

Sex/gender inequality as the outcome of 'practising gender' in everyday life, i.e. the micro-practices of 'doing gender', has also received much attention in these literatures (e.g. Poggio 2006). In several articles Martin (2001, 2003, 2006) has theorized and further demonstrated the productive work of gender practices by differentiating their content (what one says or does) from the literal practising (the processes involved in actual sayings and doings). She argues that attending to their interface contributes to understanding how gender inequities are accepted and perpetuated in organizations. Similarly, Tienari, Søderberg, Holgersson, and Vaara (2005) focused on language as carrier of social practices of doing gender through specific worldviews when studying how male executives accounted for gender inequality in the higher echelons of merged organizations. Yet, Benschop (2009) sees a more optimistic outcome of these dynamics in a study of networking as gendering practices in organizations. Her micro-political analysis suggests that networking may open possibilities for changing the gender order and does not necessarily reinforce gender inequality.

Gendering Knowledge

Some literature addresses how linguistic categories, used in scholarship and actual organizational practices, reflect and produce gendering effects. Most of these works attest to the reproduction in workplaces of the gender substructure of society as noted by Acker, and to its persistence over time. For instance, Smithson and Stokoe (2005) studied how trying to use 'genderblind' terms for promoting equality in organizations, in particular neutralizing terms already associated with women's assumed cultural roles as mothers, may not work in masking or minimizing their gendered associations. Rather, terms such as 'flexible working' became recoded as female gendered and had little impact as a means for advancing gender equality. Similarly, Kugelberg (2006) examined the construction of motherhood and fatherhood in workplaces as important to the production of gender stereotyping. She described the production of stereotypes as an inter-discursive contest constituting focal points for separating women from men in the workplace, with effects that reduced work opportunities for women. See also Nentwich (2006) on the construction of meaning for 'equal opportunities'; and Bendl and Schmidt (2010) who analysed the value of two different metaphors, 'glass ceiling' and 'firewalls', for describing potential gender discrimination in organizations.

Embodiment

Not only is gender done through practices and linguistic categories; this second theoretical theme brings in aspects of gendering processes where the materiality *and* materialization of the human body as gendered play an important part in the production of sex/gender inequalities in society and organizations. The body is both a visible cultural signifier for 'gender differences' and implicated in practices and processes that produce, reproduce, and naturalize those 'differences'. This theme is amply theorized in what is known as corporeal feminist scholarship, which has also influenced the gender-and-organization literatures (see Calás and Smircich 2006: 311–12). Here we focus first on embodiment as a general gendering process in the production of inequality, and later we discuss research examples in two subthemes where sexuality and masculinity also play a part.

A well-known example under this theme is Sinclair (2005), where she contrasts the general 'disembodiment' of management theory with conventional associations of women as bodies, and argues that this renders invisible the male bodies in organizations while casting women's bodies as problems. Her reflections on the body and management pedagogy speak to the multiple meanings of bodies, seeking to emphasize liberating possibilities of attending to bodies in management practices and in classrooms. Yet, Davies, Browne, Gannon, Honan, and Somerville (2005: 351) highlight the difficulties of doing so for a group of women academics under the effects of neo-liberal discourses, where primacy is given to flexible, self-controlled individuals who propel themselves into the 'ever-re-invented demands of the institution'.

Other research addressing the production of sex/gender inequality examines embodiment in non-academic workplaces. In one example, Pettinger (2005) studied how gendered attributes of workers' bodies function as important dimensions for understanding stratification, such as when they are used to promote and differentiate store brands.

Others have studied women's embodiment as professionals. For instance, Haynes (2012) notes that women's embodiment in the context of accounting and law firms produces a particular form of commodification and physical capital associated with the traditional masculine norm of these professions.

The pregnant body at work has also received attention. A recent example is Gatrell (2011), who focused on the experiences of pregnant workers as their bodies continue to reaffirm the association of women with reproduction and the home. While they are seen as transgressing the boundaries of the workplace by being pregnant, they try to survive as workers; and they do so by having recourse to strategies of secrecy, silence, and overperformance. Studies such as this demonstrate that the pregnant body in management and organizations, as noted by Martin (1990), continues to be a very visible sign of women as 'matter-out-of-place' (e.g. Game 1994). It reiterates the persistent reproduction of a taken-for-granted private/public 'divide' as a gendered substructure of society and organizations.

Sexuality

As noted by Pullen and Thanem (2010), sexuality in the workplace has been part of the management and organization literature at least since the mid-1980s. However, in recent years its appearance in organizational literatures has increased, gaining traction through Butler's theorizations on questions of sexuality and the body, in particular the association of her work with the heterosexual matrix and with 'queer theory'.

Sexuality and embodiment come together in Warhurst and Nickson (2009), who examine interactive services, such as bartending. They observed that employees' interactions with customers may include some sexual connotations personally initiated by the employees but that these modes of interacting may then become appropriated by their organizations. At the end, to gain a desirable corporate look, the worker's corporeality may become commodified through aesthetic labour strategically prescribed by the organization. This point was also made by Bruni (2006), who further observed that when sexuality is commodified in organizations, for instance in relationships with customers, it is done so according to a model of heterosexual relationships which often underlie hegemonic masculinity.

Others have noted the presence of heterosexual norms in organizations as well as their reiteration by bringing them to visibility in reference to non-heterosexual organizational members. For example, Pringle (2008) reframes gender as 'heterogender' to highlight how gender as done in organizations is intertwined with heterosexuality. Specifically, she argues that gender has been constructed as stereotypical heterosexual femininity and masculinity binary, and wonders if visible lesbian managers would expose the heterosexuality of organizations in the same way the presence of women exposed the 'man' in management. However, Schilt and Connell (2007) show how the context of the workplace limits the potential impact of transsexual/transgender people for challenging gender inequality and binary views on gender in organizations. See also Rumens (2008) and Tyler and Cohen (2008).

Masculinity, Hierarchy, and the Gender Order

While masculinity has been amply theorized in organization studies since the 1990s, in particular in the formulation of hegemonic masculinity and beyond (e.g. Connell 1987; Collinson and Hearn 1994), the few examples we discuss are explicit in addressing masculinity as embodied, e.g. the hegemony of men (Hearn 2004), and related to the production of sex/gender inequality in organizations.

Hall, Hockey, and Robinson (2007), for instance, consider how men's masculine identity is done in different occupational cultures usually stereotyped as masculine (e.g. fire-fighting) or feminine (e.g. hairdressing). In particular, the study explores the embodied experience of members of these occupations showing how they conform to, draw upon, and resist these gendered stereotypes. However, the authors also note the possibility that men involved in stereotypical feminine practices of care and empathy within occupations traditionally done by women represent an appropriation of women's resources. See also Simpson (2011). In contrast Blomberg's (2009) study focused on the gender regime in the world of finance where the processes he observed among brokers and analysts were less about subordinating women than about changing status relationships among categories of men.

At the end, it is important to highlight how masculinity performs by symbolic excess, whether or not the bodies of men are the majority of organizational members. In a relevant example, Gregory (2009) studied homosociability processes in advertising agencies over time to address the maintenance of a gendered labour market. Using the metaphor of the 'locker room'—as men's spaces constructing workplace opportunities—she observed relationships between gender, discrimination, and corporate work. Her findings noted that hegemonic masculinity is still prevalent in the discrimination encountered by female advertisers despite the fact that women are half of the industry's labour force, which also caters to a 50 per cent female clientele. See also Pacholok (2009).

In Summary: Is 'Gendering Organizations' Enough?

We have contrasted two approaches to theorizing gender and organization, focusing on their potential contributions for addressing the persistence of sex/gender inequality: *'gender in organizations'* and *'gendering organizations'*. Our starting argument was that these theoretical approaches are not necessarily of equal value for addressing such persistence today. Some have lost relevance, partially on account of changes in the context where they first developed, and partially on account of limits in their original theoretical inspirations, such as role theory, which has been amply criticized (Connell 1987; Acker 1992*a*; Ferree, Kahn, and Morimoto 2007).

Specifically, we see much of the sex/gender differences literature based on cognitions within *gender in organizations* as fairly exhausted. It is as if the original focus of the research—a contribution to noting organizational conditions predominantly affecting women—has been forgotten and researchers have become content with repeating what we already know: women face difficulties in organizations and they are judged as inferior to men. Meanwhile, the assumed scientific neutrality of the research has increased with ever more methodological prowess and theoretical abstractions while obscuring the actual conditions promoting inequality in organizations. This is, at its most fundamental, a conservative literature with a managerial orientation which, under a mantle of 'objectivity', may in fact be hurting women while becoming an obstacle for organizational change.

Nonetheless, the literature on gender stratification, still within *gender in organizations*, offers a glimmer of hope by focusing on actual organizational activities, some standard practices but often clearly discriminatory practices, which affect women (and others such as non-white people) in disproportional numbers. Documenting these conditions is a minimal requirement for any process of organizational change. Note, however, that this literature seems to be a recent response to the limits observed in the cognitive approaches after many years of repeating the same explanations without much success for changing conditions of inequality (e.g. Reskin 2005).

In our view, therefore, it is the *gendering organizations* literature that holds the most potential for intervening on gender inequality in organizations and society at present, even with several caveats discussed in the next section. First, this is a critical literature which places itself on the side of those women and others wrestling for recognition and justice in organizations. It is not a 'neutral' literature but one centrally located in the power and politics of organizational life. In this work gender stratification in organizations, as noted above, would serve as the evidence for then observing everyday struggles holding together apparently neutral organizations, benefiting some members at the expense of others.

Documenting how this happens, from micro-practices to macro-structures producing and reproducing these conditions, is however a laborious process. Researchers must be embedded in these happenings, documenting their ongoing outcomes as they occur. With these approaches organizational changing towards reducing inequality would always proceed without guarantees, for there is no end to the processes producing and reproducing it. Paradoxically then, 'gendering' theoretical approaches and research have made the reproduction of gender inequality in organizations more visible but they have also highlighted the immense complications entailed in its eradication.

Table 1.1 summarizes and illustrates the contrast between the two theoretical approaches discussed in this section. But this does not complete the story. As there is no end to the processes reproducing inequality in organizations, there is no end to theorizing its contours. Thus, contemporary conditions, including globalizing processes furthering human inequalities, and the increasing non-responsibility in organizations for human reproduction and survival (Acker 2004), require additional theoretical considerations which we discuss next.

Table 1.1 Theorizing gender and organization

	Gender in Organizations	Gendering Organizations
Origins and Disciplinary Orientations	Late 1960s: the US civil rights and women's movement; civil rights legislation, like Title VII and EEO/AA policies Social psychology theoretical frameworks Some sociological approaches to organizational stratification Favoured gender research in US business schools.	Late 1970s, 1980s: common roots, insights by West and Zimmerman (1987), Acker (1990; 1992a,b) Critical sociological frameworks Favoured organization and management research on gender in Europe, Australia, New Zealand
Sex and Gender	Two intertwined and stable 'human properties' used interchangeably and synonymous with women (and men); Understood as something one is/has as an individual, or identities one acquires.	Something humans *do* in social relations; an ongoing accomplishment in social life Sex: produced as a social category by agreeing upon biological criteria for classifying persons as females or males Gender: a culturally institutionalized system produced and reproduced through relations of subordination and domination based on historical hierarchical differentiations by sex (and also by class and race)
Organizations	Neutral containers for the activities of men and women; remains in the background. Some note the appearance of gender stratification as an organizational feature	'Inequality regimes' interconnecting organizational processes and practices producing and maintaining racialized, gendered, and classed relations
Feminist Theoretical Underpinnings?	Echoes of liberal feminist theorizing, including normative dualism and abstract individualism; tend to neutralize or rationalize women's disadvantages in organizations.	Materialist and poststructuralist feminist theorizing; focus on material relations; and/or language, discourse and the cultural meanings of gendering. Focus on implications for inequality and subordination.
Ideological Underpinnings	Mostly conservative and instrumental research. Underlying belief that inequality in organizations is to be expected in meritocratic hierarchical societies	Mostly critical research. Gender relations are power relations negotiated in everyday organization processes and practices Societies are basically unjust; conflictual sites of power and domination

(Continued)

Table 1.1 (Continued)

	Gender in Organizations	Gendering Organizations
Theoretical Frameworks	Psychological and socio-psychological theories; cognitive processes seen as explanations for observed sex differences in organizations. Generic psychological or sociological theories adapted to explain gendered phenomena	Social constructionist gender theorizing; gendering process theories. Ethno-methodological and poststructuralist approaches contribute to understanding how taken-for-granted social categories are naturalized and maintained with gendering effects
Justification for the Research	Important as organizations may benefit from promoting a variety of qualified individuals to positions of authority (e.g. reduce wasted talent; diversity as source of innovation). It is managerialist research	Important for denaturalizing and understanding how formal structures become hierarchically organized along gender lines. Engages with a wide range of organizational actors whether or not in positions of authority. It is critical research
Research Contributions	Documents the existence of disparity between women and men in organizations, and its persistence over the years. Emphasis on 'gender differences' has sometimes led to tautological explanations	Offers situated understandings of processes and practices leading to gender and other inequalities. Addresses gender inequality directly; offers analyses of its production and reproduction as these occur
Value for Addressing Sex/Gender Inequality	Has become mostly ineffective. Offers explanations for the status quo with little critical analysis. Primary focus is on the 'numbers problem' Psychological inspired research suggest that individuals—rather than organizations—are expected to change Some gender stratification research is more promising in addressing gender disparity as an organizational issue which could be changed	Somewhat effective. Repositions gender/sex inequality as an outcome of socially constituted gendering processes and practices of societies and organizations; problematic precisely for its persistence Gender inequality as social construction which can be changed; yet acknowledged as a contested situation embedded in hierarchical power relations Shows the complexity of these processes

THE CHANGING CONTOURS OF INEQUALITY: GENDERED ORGANIZATIONS, GENDERED GLOBALIZATION, GENDERED WORK

As we argued at the start, the persistence of sex/gender inequality in organizations and society is a manifestation of changing social dynamics requiring scholars to attend to the changing conditions of its reproduction. Globalization and a transnational work system where 'capital, production, finance, trade, ideas, images, people and organizations... flow transnationally across the boundaries of regions, nation-states and cultures' (Chow 2003: 444) are changing the contours of inequality and imply a need for different approaches to analysis. Feminist analysts, in particular, have 'gendered globalization'. Their work demonstrates the ways gender is embedded in the logic, structure, and processes of globalization, constructing unequal power relations between men and women, and documents the differential consequences, the gendered effects, of globalization.

Important questions to be asked when gendering globalization include, for instance, where do women do paid work around the world? Which women do what paid work around the world? What are the interconnections among organizations and other social spaces where these women do their work? And to whom and to where do the benefits from these interconnections accrue? (Dallafar and Movahedi 1996; Acker 1998, 2004; Chow 2003; Pyle and Ward 2003; Metcalfe and Rees 2010; Anthias 2012). These analyses thus imply that understanding gender-and-organizations under conditions of globalization requires understanding the structural positions of both gender and organizations in the context of Global North and South political economies and the development continuum gap, especially its intersections with sex, race, class, and other social relations of power, and including the location from where the theoretical perspectives for analyses emerge.

Acker (2004) has argued that theoretical discourses of globalization are represented as gender- and often race-neutral, making invisible the dominant masculine standpoint of much social theory. To produce a better understanding of global issues may require gendering, racializing, and ethnicizing the discourses of globalization. In doing so discontinuities between the ways mainstream scholarship articulates global processes and the realities of actual people's lives under these processes would be revealed. In her relational analyses, Acker shows how gender is embedded in globalizing capitalism and neo-liberal processes, which construct a separation between production and reproduction and sustain claims of non-responsibility for the latter.

This means, for instance, transnational corporations recognizing only the value of profit-making activities of workers while disregarding their needs, as well as ignoring the value of other activities non-paid members of society may perform (e.g. homemakers, children) and from which they also derive benefits. Stated in these terms, non-responsibility is an incentive for moving production from high wage countries to poorer, low wage, ones and contributes to the further realignment of poverty and inequality along gender and race lines worldwide. In these analyses Acker notes the emergence of a competitive global form of hegemonic masculinity in which a few powerful women are also implicated and, by contrast, she notes as well how gender is a resource for globalizing capital in the form of existing local gender relations. For instance, in some countries contract workers may work from home as a whole-family enterprise but still organized through patriarchal control.

Examples of these processes and new forms of inequality abound. While often the supposed benefits of globalization rely on a discourse of women's empowerment and progress, these assumptions about women's progress under globalization obscure the devastating effects of a neo-liberal agenda sustaining the 'inevitability of unrestrained global capitalism' (Fairclough 2000: 147). 'Globalization is a double-edged process as far as women are concerned', argue Fernández-Kelly and Wolf (2001: 1246). For instance, free trade zones in developing countries offer intensive manufacturing opportunities that 'liberate' women from subsistence conditions and provide 'choices' for working outside the home and fields, often redefining women's roles under traditional patriar-chal control; but these same 'opportunities' also generate new patriarchal and capitalist controls (Williams 1988; Ward 1990; Wichterich 2000).

From this gendered perspective, one also observes the unequal and contradictory results of globalization on diverse women, especially as forms of 'feminisms' are coopted under neo-liberal corporate agendas to mean 'individualism and the right to participate in the market economy as a worker or entrepreneur in one's own name, separated from one's roles as a wife and/or mother' (Eisenstein 2005: 498). These ideologies conceal, for instance, the barely subsistence wages received by many women in manufacturing activi-ties in the Global South often supplying goods for affluent consumers in the Global North. They also conceal the failures of microenterprises, another favoured Global North recipe for 'lifting' poor women out of poverty. These schemes have not lived up to their promises, except for the benefits accrued by the many financial institutions around the world now engaged in facilitating micro-loans to the poor (e.g. Keating, Rassmussen, and Rishi 2010)

This brief discussion merely illustrates the changing contours of inequality, which require resituating gender differences and gendered organization theorizing within an evolving globalization which is also gendered. Below we suggest two major tasks mov-ing forward: evolving theories of differences and power with intersectional and transna-tional feminism at the level of organization, and from there, retheorizing transnational organizational change and praxis.

CHANGING THEORIES OF GENDER INEQUALITY IN ORGANIZATIONS: TOWARDS INTERSECTIONAL AND TRANSNATIONAL FEMINIST ANALYSES

While there are important differences in the histories and focus of transnational and intersectional feminisms, in this chapter we want to bring them together. Considered together it is possible to appreciate how these approaches offer important relational understandings for analysing changing forms of inequality under globalization processes, as well as for intervening in them (e.g. Anthias 2012). Thus, we treat these intellectual influences as mutually supportive for advancing 'gendering organizations' theorizing and for reclaiming the focus on gender equality in a globalized world.

Intersectionality studies and transnational feminisms have emerged as important conceptual spaces for theorizing subordination and inequalities of differently situated women and men. They share many ontological and epistemological premises such as locating at the centre of analysis the experiences of 'non-white' women and the interactions of gender, race, ethnicity, class, sexuality, and other social differences implicated in the production of inequality. These approaches treat gender as an unstable, but useful historical and socially constructed category, which serves to illuminate differences among women and to problematize theory and praxis forged on assumptions about women's solidarity and universality.

These approaches highlight as well that gender is not only about women but rather how different forms of gender relations are implicated in the production of inequality and subordination. Both of these approaches are committed to analyses of inequality oriented towards social justice and change; both understand differences as power relations with subjective and material effects, but that are also amenable to disruption and change; and both articulate the complex interrelations among sites of differences such as sex, race, ethnicity, and sexuality and the micro and macro practices that form and transform them (Ifekwunigwe 1998; Mendoza 2002; Choo and Ferree 2010; Holvino 2010; Weber 2010).

Transnational feminist analyses centre their study around social differences and the ways in which differences are constituted through various gender, power, and work relations among and between different people in the world. Under globalization, these relations change work and organizations (Kim-Puri 2005; Tsing 2005; Calás and Smircich 2006, 2011; Kim 2007). These analyses also highlight the interconnections among gendered textual and cultural representations and the material consequences of these discursive positions (Desai 2007). Mapping these relations and connecting them to a neo-liberal corporate agenda gone global reveals (pre)dominant constructions of global subjects in organization studies and the material consequences of these constructions (Holvino, Özkazanç-Pan, and Scully 2012).

Under these premises, theorizing gender shifts from the never-ending search and explanation of differences between men and women ('*gender in organizations*') to tracing how privileged gendered subjects in organizational studies, i.e. men and women managers, relate to other actors in interconnected systems of labour, which are also raced, classed, sexualized, and so on, within and between heterogeneous social fields and networks stretching around the world. Analyses of gender practices ('*gendering organization*') become analyses of maps of relations of ruling (Smith 1987; Mahler and Pessar 2001; Nagar, Lawson, Mcdowell, and Hanson 2002; Mohanty 2003; Acker 2004) and of global inequality work regimes (Acker 2004, 2006*b*).

In all, transnational feminist analyses decentre privileged subjects in organizations such as managers and expatriates, and focus instead on relations among the multitude of subjects and processes which labour in the global economy in specific times and spaces. A constellation of unequal societal relations are charted, such as those between transnational workers and managers and other unlikely management subjects like pimps, caretakers, soldiers, and 'higglers and hacklers' (Freeman 1997; Banerjee and Linstead 2001; Holvino 2003; McDowell 2008; Berry and Bell 2012).

Meanwhile, intersectionality, an analytical tool, a metaphor (Cuadraz and Uttal 1999; Acker 2011), an ideograph (Alexander-Floyd 2012), a research paradigm (Hancock 2007; Dhamoon 2011), refers to how social and cultural categorizations such as gender, race, class, sexuality, interact on multiple and often simultaneous levels, contributing to systematic social inequality. Intersectionality theorizations indicate that traditional conceptualizations of oppression based, for instance, on gender or on race, are not independent of one another; rather they interrelate, creating a system of oppression reflecting the simultaneous effects of multiple forms of discrimination. Intersectionality also refers to the interlocking and mutually constitutive relations of gender, race, class, and other categories of difference in individual lives, social practices, institutional arrangements, and cultural ideologies and their differential outcomes in the lives of women and men (Crenshaw 1989; Collins 2000; Knudsen 2006; Dill, McLaughlin, and Nieves 2007; Valentine 2007; Davis 2008; Weber 2010).

Intersectionality studies identify social processes where race, class, gender, ethnicity, sexuality, caste, and nation, for example, are salient in social relations and in representations of subjects and identities at work, attending, as well, to those identities that are invisible, subjugated, or marginalized at particular points of intersection. Similar to transnational feminist arguments, these analyses surface invisible actors in a global economy seldom mentioned in the organizational literature, such as immigrant, sex, domestic, and microenterprise workers, and help to explore how their 'disadvantaged' subject positions are mutually constituted by other social processes and categories of difference, maintaining dominant actors in their positions of privilege (Collins 2000; Chow 2003; Berger and Guidroz 2009; Dill and Zambrana 2009; Bell, Kweisga, and Berry 2010).

Thus, analyses of intersectionalities combined with a transnational feminist framing would highlight the co-constitution of identities, subjectivities, and organizational practices through societal processes of globalization (Winker and Degele 2011; Bose 2012; Purkayastha 2012; Tatli and Özbilgin 2012). Such an approach would help identify the myriad ways in which identities, subjectivities, and modes of organizing are challenged, reformed, and reshaped through ongoing interconnections occurring in transnational contexts (Calás and Smircich 2011; Yuval-Davis 2011; Choo 2012).

The importance of bringing together transnational feminist and intersectionality analyses is highlighted in Poster's (2008) ethnography of a multinational corporation. Here she noted complex hierarchical relationships between the US headquarters and an India subsidiary during the transfer of diversity programmes from headquarters to subsidiary. These relationships were mediated by the North–South locations of these countries in the global political economy and also by a decentralized organization structure of the multinational giving some voice to the local Indian managers. Poster notes the deployment of 'global/local' rhetoric by both management sides involved in the transfer of these programmes, where a gendered rhetoric dominated the US side even when employees reported ethnic/racial discrimination, and a racialized rhetoric dominated in India even when employees reported gender discrimination. Both sides tended to ignore the intersections of gender, race, and class inequalities; instead, the diversity policy became 'the symbolic focal point for expressing tensions managers have with these various global and local actors' (Poster 2008: 336), while transnational power relations of managerial domination/subordination were negotiated.

Bringing together analyses of intersectionality and transnational approaches has further contributed to retheorizing both of them. In particular, contemporary transnational literatures addressing the mobility of populations around the world have also brought to visibility that transnational spaces exist beyond nation-states—e.g. transnational social fields (Glick-Schiller, Basch, and Blanc-Szanton 1992; Levitt and Glick Schiller 2004; Levitt and Jaworsky 2007). Observing the mobility of populations across transnational spaces has facilitated rearticulating conceptualizations of gender, race, class, and so on outside the dominant Euro-American conceptualizations of intersectionality. For example, Purkayastha (2012) describes how both a Ugandan Black immigrant woman and a Ugandan Indian immigrant woman would be racially marginalized in the US but in different ways consistent with racist ideologies, interactions, and institutional arrangements of this country. Yet, if they return home to Uganda, the Black Ugandan, as part of the majority population in her country, may experience a more privileged situation than her Indian counterpart. In Purkayastha's words, '[t]here are variations of who is part of the privileged majority versus the marginalized minority *within* a country, and these hierarchies do not always fit the white-yellow/brown-Black hierarchy extant in Western Europe and North America' (2012: 59).

Anthias (2012) extends this argument. Still with a focus on transnational mobility, she notes that all people inhabit transnational spaces, for all are interconnected in a network of global processes. Thus a transnational lens must pay attention to the hierarchical positions of different nations as well as how all actors are hierarchically positioned through global dimensions of power. Such an understanding also calls for repositioning 'the concept of intersectionality away from the idea of an interplay of people's group identities of class, gender, ethnicity,...to intersectionality being seen as a process' (2012: 107). Intersectionality as social process allows for observing actors' contradictory locations of domination and subordination at different times in different places. In Anthias' view, the notion of 'translocational positionality', which she developed (2002, 2008) to make sense of intersections between different structures and processes, can be a useful accompaniment to notions of intersectionality for it brings to the fore social context and temporality. Analyses from these perspectives would help make visible 'the existence of contradictory and shifting social locations where one might be in a position of dominance and subordination simultaneously on the one hand or at different times and spaces on the other' (2012: 108). In her argument, seeing these contradictions opens possibilities for transforming social locations of subordination.

Unfortunately, these advances in intersectionality theory and transnational feminism are seldom noticed in management and organization studies despite their importance for analyzing complex networks of social power relations at present.

What Kind of Theorizing on 'Gender and Organizations' is Necessary Next?

A major contribution of both of these analyses is confirming that, under conditions of globalization, continuing to describe differences between universal men and women without tying them to relations of privilege and disadvantage across a whole set of social relations will no longer do. Some progress has been made bringing intersectionality to organization studies (e.g., Adib and Guerrier 2003; Styhre and Ericksson-Zetterquist 2008; Baines 2010; Holvino 2010; Zander, Zander, Gaffney, and Olsson 2010; Healy, Bradley, and Forson 2011; Tatli and Özbilgin 2012); less so with transnational feminism. Still additional efforts must be made to bring them more fully together into the field.

In order to advance this agenda Holvino (2010) proposes three specific methodological approaches which she calls the 'simultaneity of differences': exploring identities, subjectivities, and identity practices at the intersections of social differences of dominance and subordination/marginalization; articulating the differential material impact of social differences and their representations in organizations; and understanding/interpreting organizational dynamics, such as the discourses of organizational theory, within the broader context of social, economic and geopolitical processes and interconnections in globalization.

Nonetheless, such arguments beg the question: what interventions for organizational and societal change flow from these analyses? How to theorize organizational change in ways that focus on change agendas for justice and equality in organizations in the context of globalization?

While the call for 'gendering change' has already been made (Benschop *et al.* 2012), more complex moves that complicate gender with intersectional and transnational analyses suggest that 'gendering change' is not enough. Perhaps more needed now are ways to rethink organization change strategies that can move us closer to social justice with an expanded understanding of social inequality beyond those allowed by theoretical framings from the Global North.

How to retheorize organizational change in ways which develop positive-aspirational visionsofgender-race-sexual-religious-class-ethnicjusticeinatransnational-intersectional world? For instance, how can the work on gender mainstreaming, originating in the field of international development with a gender equality and institutional and organization change focus, be rethought by incorporating its critique (Hawthorne 2004; Squires 2005; Verloo 2006; Eveline, Bacchi, and Binns 2009) to further the kind of complex organizational transnational-intersectional change we hope for?

Truthfully, we could not come up with a term to accurately express the complex meanings of doing intersectional-transnational feminist analysis and practice for organizational change and social justice. Evidently, we still need more interdisciplinary work, more creative language, and more theory–praxis connections amongst us and many others in the world.

REFLECTIONS

This chapter has mapped theoretical perspectives forming the conceptual contours of the gender and organization literature with a focus on gender inequality as it has taken shape over the last forty years. The landscape is made up of scholarly communities existing in almost separate worlds, sharing an interest in 'gender' and 'organization' whatever that means for them, and therefore focused on different questions and vastly different approaches to theorizing and research.

The theoretical and epistemological position of researchers producing the literature we have discussed cannot be easily separated from the locations from where their work is done. That 'sisterhood' is not global is quite clear. Differences in who and what is the subject of research—e.g. which women are included, what questions are asked in their names—are evident across different institutional domains, with those in more privileged locations in the US also addressing more privileged populations. They are as well more willing to adopt non-critical implications for their findings, and more willing to state scientific claims for them. If anything, they are more willing to make 'the business case'.

For us, making these remarks serves a purpose broader than issues about gender and organization research and theorizing. As we see it, the fact that the US literature, from business schools in particular, is marked by a state of repetition, with little awareness of other possibilities, speaks both to the insularity and ethnocentric orientation of much of this research. It also speaks to the consequences of such ethnocentrism in other ways beyond limitations in modes of theorizing and research. Provincializing this literature—i.e. showing how it passes as 'universal' when it is in fact a very 'local' set of ideas—is not difficult in a day and age after 'the postcolonial'. More difficult, however, is noting how other places in the world which either had abandoned or were never part of this mode of scholarship now find themselves involved in situations which may require them to adopt it.

Increasingly, it seems scholars outside the US are being asked to publish in dominant US management academic publications as part of worldwide institutional isomorphism, copying the North American model of the university (or at least the business school) as 'global' with the intent of becoming 'globally ranked'. What this means for gender research and theorizing (and for many other areas of theory and research, critical research in particular) is that the limited US modality will continue to be, and is likely to become, even more dominant.

That this theorizing and research is limited and exhausted—and even anachronistic—for addressing sex/gender inequality at present is no longer the only relevant issue. Rather, more relevant is that scholars working in the more processual gendering organizations tradition, and who would be interested in newer theorizing and research addressing the changing contours of inequality, may be unable to do so. Thus, keeping an eye on changing academic institutional contours worldwide and resisting their limitations may become an added task for those of us wanting to further the possibilities of social justice and contribute to reducing gender and other inequalities. A never-ending task if there ever was one!

References

Acker, J. (1990). 'Hierarchies, Bodies, and Jobs: A Theory of Gendered Organizations', *Gender and Society*, 4(1): 139–58.

Acker, J. (1992*a*). 'From Sex Roles to Gendered Institutions', *Contemporary Sociology*, 21(5): 565–9.

Acker, J. (1992b). 'Gendering Organization Theory', in A. J. Mills and P. Tancred (eds), *Gendering Organization Theory*, 248–260. Newbury Park: Sage.

Acker, J. (1998). 'The Future of Gender and Organization: Connections and Boundaries', *Gender, Work and Organization*, 5(4): 195–206.

Acker, J. (2004). 'Gender, Capitalism and Globalization', *Critical Sociology*, 30(1): 17–41.

Acker, J. (2006*a*). *Class Questions: Feminist Answers*. Lanham, MD: Rowman & Littlefield.

Acker, J. (2006b). 'Inequality Regimes: Gender, Class and Race in Organizations', *Gender and Society*, 20(4): 441–64.

Acker, J. (2011). 'Theorizing Gender, Race, and Class in Organizations', in E. L. Jeanes, D. Knights, and P. Y. Martin (eds), *Handbook of Gender, Work, and Organization*, 65–80. Chichester: Wiley.

Adib, A. and Guerrier, Y. (2003). 'The Interlocking of Gender with Nationality, Race, Ethnicity and Class: The Narratives of Women in Hotel Work. *Gender, Work and Organization*, 10: 413–32.

Alexander-Floyd, N. G. (2012). 'Disappearing Acts: Reclaiming Intersectionality in the Social Sciences in a Post-Black Feminist Era', *Feminist Formations*, 24(1): 1–25.

Alsop, R., Fitzsimmons, A., Lennon, K., and Minsky, R. (2002). *Theorizing Gender*. Cambridge: Polity Press.

Anderson, C. D. (1996). 'Understanding the Inequality Problematic: From Scholarly Rhetoric to Theoretical Reconstruction', *Gender and Society*, 10(6): 729–46.

Anthias, F. (2002). 'Where Do I Belong? Narrating Collective Identity and Translocational Positionality', *Ethnicities*, 2(4): 491–514.

Anthias, F. (2008). 'Thinking through the Lens of Translocational Positionality: An Intersectionality Frame for Understanding Identity and Belonging', *Translocations, Migration and Change*, 4(1): 5–20.

Anthias, F. (2012). 'Transnational Mobilities, Migration Research and Intersectionality', *Nordic Journal of Migration Research*, 2(2): 102–10.

Baines, D. (2010), 'Gender Mainstreaming in a Development Project: Intersectionality in a Post-colonial Un-doing?' *Gender, Work and Organization*, 17(2): 119–49.

Banerjee, S. B., and Linstead, S. (2001). 'Globalization, Multiculturalism and Other Fictions: Colonialism for the New Millennium?', *Organization*, 8(4): 711–50.

Bartol, K. M., and Butterfield, D. A. (1976). 'Sex Effects in Evaluating Leaders', *Journal of Applied Psychology*, 61: 446–54.

Bell, M. P., Kwesiga, E. N., and Berry, D. P. (2010). 'Immigrants: The New "Invisible Men and Women" in Diversity Research', *Journal of Managerial Psychology*, 25(2): 177–88.

Bendl, R., and Schmidt, A. (2010). 'From "Glass Ceilings" to "Firewalls": Different Metaphors for Describing Discrimination', *Gender, Work and Organization*, 17(5): 612–34.

Benschop, Y. (2009). 'The Micro-Politics of Gendering in Networking', *Gender, Work and Organization*, 16(2): 217–37.

Benschop, Y., and Doorewaard, H. (1998). 'Covered by Equality: The Gender Subtext of Organizations', *Organization Studies*, 19(5): 787–805.

Benschop, Y., Mills, J. H., Mills, A., and Tienari, J. (2012). 'Editorial: Gendering Change: The Next Step', *Gender, Work and Organization*, 19(1): 1–9.

Berger, J., Cohen, B. P., and Zelditch, M. Jr. (1972). 'Status Characteristics and Social Inter-action', *American Sociological Review*, 37: 241–55.

Berger, M. T., and Guidroz, K. (2009). *The Intersectional Approach: Transforming the Academy through Race, Class, and Gender*. Chapel Hill, NC: University of North Carolina Press.

Berry, D. P., and Bell, M. P. (2012). '"Expatriates": Gender, Race and Class Distinctions in International Management', *Gender, Work and Organization*, 19(1), 10–28.

Bielby, W. (2012). 'Minority Vulnerability in Privileged Occupations: Why Do African American Financial Advisers Earn Less than Whites in Large Financial Services Firms?', *Annals of the American Association of Political and Social Science*, 639: 13–32.

Biernat, M., and Fuegen, K. (2001). 'Shifting Standards and the Evaluation of Competence: Complexity in Gender-Based Judgment and Decision Making', *Journal of Social Issues*, 57: 707–24.

Blau, Francine D., and Kahn, Lawrence M. (2000). 'Gender Differences in Pay', *Journal of Economic Perspectives*, 14 (Autumn): 75–99.

Blomberg, J. (2009). 'Gendering Finance: Masculinities and Hierarchies at the Stockholm Stock Exchange', *Organization*, 16(2): 203–25.

Bose, C. E. (2012). 'Intersectionality and Global Gender Inequality', *Gender and Society*, 26(1): 67–72.

Bowles, H. R., and Flynn, F. (2010). 'Gender and Persistence in Negotiation: A Dyadic Perspective', *Academy of Management Journal*, 53(4): 769–87.

Brescoll, V. L. (2012). 'Who Takes the Floor and Why: Gender, Power, and Volubility in Organizations', *Administrative Science Quarterly*, 56: 622–41.

Brescoll, V. L., and Ullman, E. L. (2008). 'Can an Angry Woman Get Ahead? Status Conferral, Gender, and Expression of Emotion in the Workplace', *Psychological Science*, 19(3): 268–75.

Broadbridge, A., and Simpson, R. (2011). '25 Years On: Reflecting on the Past and Looking to the Future in Gender and Management Research', *British Journal of Management*, 22: 470–83.

Bruni, A. (2006). '"Have you Got a Boyfriend or are you Single?": On the Importance of Being "Straight" in Organizational Research', *Gender, Work and Organization*, 13(3): 299–316.

Butler, J. (1990). *Gender Trouble: Feminism and the Subversion of Identity*. New York: Routledge.

Butler, J. (2004a). *Undoing Gender*. New York: Routledge.

Butler, J. (2004b). 'Preface from Gender Trouble Anniversary Edition'. Reproduced in S. Salih (ed.), *The Judith Butler Reader*, 94–103. Malden, MA: Blackwell.

Calás, M. B., and Smircich, L. (2006). 'From the "Woman's Point of View" Ten Years Later: Towards a Feminist Organization Studies', in S. R. Clegg, C. Hardy, T. B. Lawrence, and W. L. Nord (eds), *The Sage Handbook of Organization Studies* (2nd edn), 284–346. Thousand Oaks, CA: Sage.

Calás, M. B., and Smircich, L. (2009). 'Feminist Perspectives on Gender in Organizational Research: What is and is Yet to Be', in D. Buchanan and A. Bryman (eds), *Handbook of Organizational Research Methods*, 246–69. London: Sage.

Calás, M. B., and Smircich, L. (2011). 'In the Back and Forth of Transmigration: Re-Thinking Organization Studies in a Transnational Key', in D. Knights, P. Y. Martin, and E. Jeanes (eds), *Handbook of Gender, Work and Organization*, 411–28. London: Wiley-Blackwell.

Carli, L. L., and Eagly, A. H. (1999). 'Gender Effects on Social Influence and Emergent Leadership', in G. N. Powell (ed.). *Handbook of Gender and Work*, 203–22. Thousand Oaks, CA: Sage.

Castilla, E. (2012). 'Gender, Race, and the New (Merit-Based) Employment Relationship', *Industrial Relations*, 51(s1): 528–62.

Choo, H. Y., and Ferree, M. M. (2010). 'Practicing Intersectionality in Sociological Research: A Critical Analysis of Inclusions, Interactions, and Institutions in the Study of Inequalities', *Sociological Theory*, 28(2): 129–49.

Choo, H. Y. (2012). 'The Transnational Journey of Intersectionality', *Gender and Society*, 26(1): 0–45.

Chow, E. N. L. (2003). 'Gender Matters: Studying Globalization and Social Change in the 21st Century', *International Sociology*, 18(3): 443–60.

Cohen, P. N., Huffman, M. L., and Knauer, S. (2009). 'Stalled Progress? Gender Segregation and Wage Inequality among Managers: 1980–2000', *Work and Occupations*, 36: 318–42.

Collins, P. H. (2000). *Black Feminist Thought* (2nd edn). New York: Routledge.

Collinson, D., and Hearn, J. (1994). 'Naming Men as Men: Implications for Work, Organization and Management', *Gender, Work and Organization*, 1(1): 2–22.

Connell, R. (1987). *Gender and Power*. Oxford: Polity Press.

Crenshaw, K. (1989). 'Demarginalizing the Intersection of Race and Sex: A Black Feminist Critique of Discrimination Doctrine, Feminist Theory and Antiracist Practice', *University of Chicago Legal Forum*, 89: 139–67.

Cuadraz, G. H., and Uttal, L. (1999). 'Intersectionality and In-Depth Interviews: Methodological Strategies for Analyzing Race, Class, and Gender', *Race, Class, and Gender*, 6(3): 156–86.

Dallafar, A., and Movahedi, S. (1996). 'Women in Multinational Corporations: Old Myths, New Constructions, and Some Deconstructions', *Organization*, 3(4): 546–59.

Davies, B., Browne, J., Gannon, S., Honan, E., and Somerville, M. (2005). 'Embodied Women at Work in Neoliberal Times and Places', *Gender, Work and Organization*, 12(4): 343–62.

Davis, D.R. (2010) Unmirroring Pedagogies: Teaching with Intersectional and Transnational Methods in the Women and Gender Studies Classroom. *Feminist Formations*, 22(1), 136–62.

Davis, K. (2008). 'Intersectionality as Buzzword: A Sociology of Science Perspective on What Makes a Feminist Theory Successful', *Feminist Theory*, 9(1): 67–85.

Desai, M. (2007). 'The Messy Relationship between Feminisms and Globalization', *Gender and Society*, 27: 797–803.

Dhamoon, R. K. (2011). 'Considerations on Mainstreaming Intersectionality', *Political Research Quarterly*, 64(1): 230–43.

Dill, B. T., and Zambrana, R. E. (eds) (2009). *Emerging Intersections: Race, Class, and Gender in Theory, Policy, and Practice*. New Brunswick, NJ: Rutgers University Press.

Dill, B. T., McLaughlin, A. E., and Nieves, A. D. (2007). 'Future Directions of Feminist Research: Intersectionality', in S. N. Hesse-Biber (ed.), *Handbook of Feminist Research*, 629–37. Thousand Oaks, CA: Sage.

Dodge, K. A., Gilroy, F. D., and Fenzel, L. M. (1995). 'Requisite Management Characteristics Revisited: Two Decades Later', *Journal of Social Behavior and Personality*, 10(6): 253–64.

Eagly, A. H. (1987). *Sex Differences in Social Behavior: A Social-Role Interpretation*. Hillsdale, NJ: Erlbaum.

Eagly, A. H., and Carli, L. L. (2007). *Through the Labyrinth: The Truth about How Women Become Leaders*. Boston: Harvard Business School Press.

Eagly, A. H., and Karau, S. (2002). 'Role Congruity Theory of Prejudice towards Female Leaders', *Psychological Review*, 109(3): 573–98.

Eagly, A. H., Makhijani, M. G., and Klonsky, B. G. (1992). 'Gender and the Evaluation of Leaders: A Meta-Analysis', *Psychological Bulletin*, 111: 3022.

Eisenstein, H. (2005). 'A Dangerous Liaison? Feminism and Corporate Globalization', *Science and Society*, 69(3): 487–518.

Ely, R., and Padavic, I. (2007). 'A Feminist Analysis of Organizational Research on Sex Differences', *Academy of Management Review*, 32(4): 1121–43.

Eveline, J, Bacchi, C., and Binns, J. (2009). 'Gender Mainstreaming versus Diversity Mainstreaming: Methodology as Emancipatory Politics', *Gender, Work and Organization*, 16(2): 198–216.

Fairclough, N. (2000). 'Language and Neo-Liberalism', *Discourse and Society*, 11(2): 147–8.

Fernández-Kelly, M. P., and Wolf, D. L. (2001). 'A Dialogue on Globalization', *Signs*, 26(4): 1243–9.

Ferree, M. M., Khan, S. R., and Morimoto, S. A. (2007). 'Assessing the Feminist Revolution: The Presence and Absence of Gender in Theory and Practice', in C. Calhoun (ed.), *Sociology in America: A History*, 438–79. Chicago: University of Chicago Press.

Foschi, M., Lai, L., and Sigerson, K. (1994). 'Gender and Double Standards in the Assessments of Job Applicants', *Social Psychology Quarterly*, 57(4): 326–39.

Freeman, C. (1997). 'Reinventing Higglering in Transnational Zones: Barbadian Women Juggle the Triple Shift', in C. L. Springfield (ed.), *Daughters of Caliban: Caribbean Women in the 20th Century*, 68–95. Bloomington, IN: Indiana University Press.

Game, A. (1994). ' "Matter Out of Place": The Management of Academic Work', *Organization*, 1(1): 47–50

Gatrell, C. (2011). 'Policy and the Pregnant Body at Work: Strategies of Secrecy, Silence and Supra-Performance', *Gender, Work and Organization*, 18(2): 158–81.

Gherardi, S. (1994). 'The Gender we Think, the Gender we Do in our Everyday Organizational Lives', *Human Relations*, 47(6): 591–610.

Glick-Schiller, N., Basch, L., and Blanc-Szanton, C. (1992). *Towards a Transnational Perspective on Migration: Race, Class, Ethnicity and Nationalism Reconsidered*. New York: New York Academy of Sciences.

Goldin, C. (1990). *Understanding the Gender Gap: An Economic History of American Women*. New York: Oxford University Press.

Gregory, M. R. (2009). 'Inside the Locker Room: Male Homosociability in the Advertising Industry', *Gender, Work and Organization*, 16(3): 323–47.

Hall, A., Hockey, J., and Robinson, V. (2007). 'Occupational Cultures and the Embodiment of Masculinity: Hairdressing, Estate Agency and Firefighting', *Gender, Work and Organization*, 14(6): 534–51.

Hancock, A. M. (2007). 'When Multiplication Doesn't Equal Quick Addition: Examining Intersectionality as a Research Paradigm', *Perspectives on Politics*, 5(1): 63–79.

Hancock, P., and Tyler, M. (2007). 'Un/doing Gender and the Aesthetics of Organizational Performance', *Gender, Work and Organization*, 14(6): 512–33.

Hawthorne, S. (2004). 'The Political Uses of Obscurantism: Gender Mainstreaming and Intersectionality', *Development Bulletin*, 64: 87–91.

Haynes, K. (2012). 'Body Beautiful? Gender, Identity and the Body in Professional Services Firms', *Gender, Work and Organization*, 19(5): 489–507.

Healy, G., Bradley, H. and Forson, C. (2011). 'Intersectional Sensibilities in Analysing Inequality Regimes in Public Sector Organizations', *Gender, Work and Organization*, 18: 467–87.

Hearn, J. (2004). 'From Hegemonic Masculinity to the Hegemony of Men', *Feminist Theory*, 5(1): 49–72.

Heilman, M. E. (1983). 'Sex Bias in Work Settings: The Lack of Fit Model', in B. Staw and L. Cummings (eds), *Research in Organizational Behavior*, v. 269–298. Greenwich, CT: JAI.

Heilman, M. E., Block, C., Martell, R., and Simon, M. (1989). 'Has Anything Changed? Current Characterizations of Males, Females and Managers', *Journal of Applied Psychology*, 74(6): 935–42.

Heilman, M. E. (2001). 'Description and Prescription: How Gender Stereotypes Prevent Women's Ascent up the Organizational Ladder', *Journal of Social Issues*, 57(4): 657–74.

Holvino, E. (2003). 'Globalization: Overview', in R. Ely, E. Foldy, and M. Scully and the Center for Gender in Organizations (eds), *Reader in Gender, Work, and Organization*, 381–6. Malden, MA: Blackwell Publishing.

Holvino, E. (2010). 'Intersections: The Simultaneity of Race, Gender, and Class in Organization Studies', *Gender, Work and Organization*, 17(3): 248–77.

Holvino, E., Özkazanç-Pan, B., and Scully, M. (2012). 'The Global Woman Manager: The Story Behind the Story', CGO Distinguished Speaker Series, 6 Dec. Center for Gender in Organizations, Simmons School of Management, Boston, MA.

Ifekwunigwe, J. O. (1998). 'Borderland Feminisms: Toward the Transgression of Unitary Transnational Feminisms', *Gender and History*, 10(3): 553–7.

Jeanes, E. L. (2007). 'The Doing and Undoing of Gender: The Importance of Being a Credible Female Victim', *Gender, Work and Organization*, 14(6): 552–71.

Kantola, J. (2008). ' "Why Do All the Women Disappear?" Gendering Processes in a Political Science Department', *Gender, Work and Organization*, 15(2): 202–25.

Keating, C., Rassmussen, C., and Rishi, P. (2010). 'The Rationality of Empowerment: Microcredit, Accumulation by Dispossession, and the Gendered Economy', *Signs*, 36(1): 153–76.

Kelan, E. K. (2010). 'Gender Logic and (Un)doing Gender at Work', *Gender, Work and Organization*, 17(2): 174–94.

Kim, H. J. (2007). 'The Politics of Border Crossings: Black, Postcolonial, and Transnational Feminist Perspectives', in S. N. Hesse-Biber and D. Piatelli (eds), *Handbook of Feminist Research: Theory and Praxis*, 107–22. Thousand Oaks, CA: Sage.

Kim-Puri, H. J. (2005). 'Conceptualizing Gender-Sexuality-State-Nation: An Introduction', *Gender and Society*, 19(2): 137–59.

Kolb, D. M. (2009). 'Too Bad for the Women or does it have to Be? Gender and Negotiation Research over the Past Twenty-Five Years', *Negotiation Journal* (Oct.): 515–31.

Knudsen, S. (2006). 'Intersectionality? A Theoretical Inspiration in the Analysis of Minority Cultures and Identities in Textbooks', in *Caught in the Web or Lost in the Textbook*, 61–76. Proceedings of 8th International Conference on Learning and Educational Media. Caen: IUFM de Caen, France.

Kugelberg, C. (2006). 'Constructing the Deviant Other: Mothering and Fathering at the Workplace', *Gender, Work and Organization*, 13(2): 152–73.

Kulich, C., Ryan, M. K., and Haslam, S. A. (2007). 'Where is the Romance for Women Leaders? The Effects of Gender on Leadership Attributions and Performance-Based Pay', *Applied Psychology*, 56: 582–601.

Levitt, P., and Glick Schiller, N. (2004). 'Conceptualizing Simultaneity: A Transnational Social Field Perspective on Society', *International Migration Review*, 38(3): 1002–39.

Levitt, P., and Jaworsky, B. N. (2007). 'Transnational Migration Studies: Past Developments and Future Trends', *Annual Review of Sociology*, 33: 129–56.

Lorber, J. (1994). *Paradoxes of Gender*. New Haven: Yale University Press.

McDowell, L. (2008). 'Thinking through Work: Complex Inequalities, Constructions of Differences and Transnational Migrants', *Progress in Human Geography*, 32(4): 491–507.

Madden, J. F. (2012). 'Performance-Support Bias and the Gender Gap among Stockbrokers', *Gender and Society*, 26(3): 488–518.

Mahler, S. J., and Pessar, P. R. (2001). 'Gendered Geographies of Power: Analyzing Gender across Transnational Spaces', *Identities: Global Studies of Culture and Power*, 7(4): 441–59.

Marshall, J. (1984). *Women Managers: Travellers in a Male World*. Chichester: Wiley.

Martin, J. (1990). 'Deconstructing Organizational Taboos: The Suppression of Gender Conflict in Organizations', *Organization Science*, 1(4): 339–59.

Martin, J. (1994). 'The Organization of Exclusion: Institutionalization of Sex Inequality, Gendered Faculty Jobs and Gendered Knowledge in Organizational Theory and Research', *Organization*, 1(2): 401–31.

Martin, P. Y. (2001). '"Mobilizing Masculinities": Women's Experience of Men at Work', *Organization*, 8(4): 587–618.

Martin, P. Y. (2003). '"Said and Done" vs. "Saying and Doing": Gendered Practices/Practicing Gender and Work', *Gender and Society*, 17(3): 342–66.

Martin, P. Y. (2006). 'Practising Gender at Work: Further Thoughts on Reflexivity', *Gender, Work and Organization*, 13(3): 254–76.

Maume, D. J. (1999). 'Glass Ceilings and Glass Escalators: Occupational Segregation and Race and Sex Differences in Managerial Promotions', *Work and Occupations*, 26: 483–509

Meeker, B. F., and Weitzel-O'Neill, P. A. (1977). 'Sex Roles and Interpersonal Behavior in Task-Oriented Groups', *American Sociological Review*, 42(1): 91–105.

Mendoza, B. (2002). 'Transnational Feminism in Question', *Feminist Theory*, 3(3): 295–314.

Metcalfe, B. D., and Rees, C.J. (2010). 'Gender, Globalization and Organization: Exploring Power Relations and Intersections', *Equality, Diversity and Inclusion*, 29(1): 5–22.

Mohanty, C. T. (2003). *Feminism without Borders: Decolonizing Theory Practicing Solidarity*. Durham, NC: Duke University Press.

Morrison, A. M., White, R. P., and Van Velsor, E. (1987). *Breaking the Glass Ceiling*. Reading, MA: Addison Wesley.

Nagar, R., Lawson, V., McDowell, L., and Hanson, S. (2002). 'Locating Globalization: Feminist (Re)Readings of the Subjects and Spaces of Globalization', *Economic Geography*, 78(3): 257–84.

Nentwich, J. C. (2006). 'Changing Gender: The Discursive Construction of Equal Opportunities', *Gender, Work and Organization*, 13(6): 499–521.

Pacholok, S. (2009). 'Gendered Strategies of Self: Navigating Hierarchy and Contesting Masculinities', *Gender, Work and Organization*, 16(4): 471–500.

Parsons, T. (1942). 'Age and Sex in the Social Structure of the United States', *American Sociological Review*, 7: 604–16.

Pettinger, L. (2005). 'Gendered Work Meets Gendered Goods: Selling and Service in Clothing Retail', *Gender, Work and Organization*, 12(5): 460–78.

Phillips, M., and Knowles, D. (2012). 'Performance and Performativity: Undoing Fictions of Women Business Owners', *Gender, Work and Organization*, 19(4): 416–37.

Poggio, B. (2006). 'Editorial: Outline of a Theory of Gender Practices', *Gender, Work and Organization*, 13(3): 225–33.

Poster, W. R. (2008). 'Filtering Diversity: A Global Corporation Struggles with Race, Class, and Gender in Employment Policy', *American Behavioral Scientist*, 52(3): 307–41.

Powell, G. N., and Butterfield, D. A. (2003). 'Gender, Gender Identity, and Aspirations to Top Management', *Women in Management Review*, 18(1–2): 88–96.

Powell, A., Bagilhole, B., and Dainty, A. (2009). 'How Women Engineers Do and Undo Gender: Consequences for Gender Equality', *Gender, Work and Organization*, 16(4): 411–28.

Pringle, J. K. (2008). 'Gender in Management: Theorizing Gender as Heterogender'. *British Journal of Management*, 19(s1): S110–S119.

Pullen, A., and Knights, D. (2007). 'Editorial: Undoing Gender: Organizing and Disorganizing Performance', *Gender, Work and Organization*, 14(6): 505–11.

Pullen, A., and Thanem, T. (2010). 'Editorial: Sexual Spaces', *Gender, Work and Organization*, 17(1): 1–6.

Purkayastha, B. (2012). 'Intersectionality in a Transnational World', *Gender and Society*, 26(1): 55–66.

Pyle, J. L., and Ward, K. B. (2003). 'Recasting our Understanding of Gender and Work during Global Restructuring', *International Sociology*, 18(3): 461–89.

Ragins, B. R., and Winkel, D. E. (2011). 'Gender, Emotion and Power in Work Relationships', *Human Resource Management Review*, 21: 377–93.

Rashotte, L. S., and Webster, M. Jr. (2005). 'Gender Status Beliefs', *Social Science Research*, 34: 618–33.

Reskin, B. F. (2005). 'Including Mechanisms in Our Models of Ascriptive Inequality', in L. B. Nielsen and R. L. Nelson (eds), *Handbook of Employment Discrimination Research*, 75–99. Amsterdam: Springer.

Ridgeway, C. L. (1997). 'Interaction and the Conservation of Gender Inequality: Considering Employment', *American Sociological Review*, 62: 218–35.

Ridgeway, C. L. (2001). 'Gender, Status and Leadership', *Journal of Social Issues*, 57: 637–55.

Ridgeway, C. L. (2011). *Framed by Gender*. New York: Oxford University Press.

Roth, P. L., Purvis, K. L., and Bobko, P. (2012). 'A Meta-Analysis of Gender Group Differences for Measures of Job Performance', *Journal of Management*, 38: 719–39.

Rudman, L. A. (1998). 'Self-Promotion as a Risk Factor for Women: The Costs and Benefits of Counterstereotypical Impression Management', *Journal of Personality and Social Psychology*, 74: 629–45.

Rudman, L. A., and Phelan, J. E. (2008). 'Backlash Effects for Disconfirming Gender Stereotypes in Organizations', *Research in Organizational Behavior*, 28: 61–79.

Rumens, N. (2008). 'Working at Intimacy: Gay Men's Workplace Friendships', *Gender, Work and Organization*, 15(1): 9–30.

Ryan, M. K., and Haslam, S. A. (2007). 'The Glass Cliff: Exploring the Dynamics Surrounding the Appointment of Women to Precarious Leadership Positions', *Academy of Management Review*, 32(2): 549–72.

Ryan, M. K., Haslam, S. A., Hersby, M. D., and Bongiorno, R. (2011). 'Think Crisis—Think Female: The Glass Cliff and Contextual Variation in the Think Manager—Think Male Stereotype', *Journal of Applied Psychology*, 96(3): 470–84.

Schein, V. E. (1973). 'The Relationship between Sex Role Stereotypes and Requisite Management Characteristics', *Journal of Applied Psychology*, 57: 95–105.

Schein, V. E. (1975). 'The Relationship between Sex Role Stereotypes and Requisite Management Characteristics among Female Managers', *Journal of Applied Psychology*, 60: 340–4.

Schein, V. E. (2001). 'A Global Look at Psychological Barriers to Women's Progress in Management', *Journal of Social Issues*, 57: 675–88.

Schein, V. E. (2007). 'Women in Management: Reflections and Projections', *Women in Management Review*, 22(1): 6–18.

Schein, V. E., Mueller, R., Lituchy, T., and Liu, J. (1996). 'Think Manager—Think Male: A Global Phenomenon?', *Journal of Organizational Behavior*, 17: 33–41.

Schilt, K., and Connell, C. (2007). 'Do Workplace Gender Transitions Make Gender Trouble?', *Gender, Work and Organization*, 14(6): 596–618.

Simpson, R. (2011). 'Men Discussing Women and Women Discussing Men: Reflexivity, Transformation and Gendered Practice in the Context of Nursing Care', *Gender, Work and Organization*, 18(4): 377–98.

Sinclair, A. (2005). 'Body and Management Pedagogy', *Gender, Work and Organization*, 12(1): 89–104.

Smith, D. (1987). *The Everyday World as Problematic: A Feminist Perspective*. Boston: Northeastern University Press.

Smithson, J., and Stokoe, E. H. (2005). 'Discourses of Work–Life Balance: Negotiating "Genderblind" Terms in Organizations', *Gender, Work and Organization*, 12(2): 147–68.

Squires, J. (2005). 'Is Mainstreaming Transformative? Theorising Mainstreaming in the Context of Diversity and Deliberation', *Social Politics: International Studies in Gender, State and Society*, 12(3): 366–88.

Stainback, K., Tomaskovic-Devey, D., and Skaggs, S. (2010). 'Organizational Approaches to Inequality: Inertia, Relative Power, and Environments', *Annual Review of Sociology*, 36: 225–47.

Styhre, A., and Eriksson-Zetterquist, U. (2008). 'Thinking the Multiple in Gender and Diversity Studies: Examining the Concept of Intersectionality', *Gender in Management*, 23(8): 567–82

Tatli, A., and Özbilgin, M. F. (2012). 'An Emic Approach to Intersectional Study of Diversity at Work: A Bourdieuan Framing', *International Journal of Management Review*, 14(2): 180–200.

Tienari, J., Søderberg, A.-M., Holgersson, C., and Vaara, E. (2005). 'Gender and National Identity Construction in the Cross-Border Merger Context', *Gender, Work and Organization*, 12(3): 217–41.

Tsing, A. L. (2005). *Friction: An Ethnography of Global Connection*. Princeton: Princeton University Press.

Tyler, M., and Cohen, L. (2008). 'Management in/as Comic Relief: Queer Theory and Gender Performativity in *The Office*', *Gender, Work and Organization*, 15(2): 113–32.

Valentine, G. (2007). 'Theorizing and Researching Intersectionality: A Challenge for Feminist Geography', *Professional Geographer*, 59(1): 10–21.

Verloo, M. (2006). 'Multiple Inequalities, Intersectionality and the European Union', *European Journal of Women's Studies*, 13(3): 211–28.

Wajcman, J. (1998). *Managing like a Man*. University Park, PA: Penn State Press.

Ward, K. (ed.). (1990). *Women Workers and Global Restructuring*. Ithaca, NY: ILR Press.

Warhurst, C., and Nickson, D. (2009). '"Who's Got the Look?" Emotional, Aesthetic and Sexualized Labour in Interactive Services', *Gender, Work and Organization*, 16(3): 385–404.

Weber, L. (2010). *Understanding Race, Class, Gender and Sexuality: A Conceptual Framework* (2nd edn). New York: Oxford University Press.

West, C., and Fenstermaker, S. (1995). 'Doing Difference', *Gender and Society*, 9(1): 8–37.

West, C., and Zimmerman, D. (1987). 'Doing Gender', *Gender and Society*, 1(2): 125–51.

Wichterich, C. (2000). *The Globalized Woman: Reports from a Future of Inequality*, tr. Patrick Camiller. London: Zed Books.

Williams, N. (1988). 'Role Making among Married Mexican American Women: Issues of Class and Ethnicity', *Journal of Applied Behavioral Science*, 24: 203–17.

Williams, C. L. (1992). 'The Glass Escalator: Hidden Advantages for Men in the "Female" Professions', *Social Problems*, 39: 253–68.

Williams, J. C. (2010). *Reshaping the Work–Family Debate: Why Men and Class Matter*. Cambridge, MA: Harvard University Press.

Winker, G., and Degele, N. (2011). 'Intersectionality as Multi-Level Analysis: Dealing with Social Inequality', *European Journal of Women's Studies*, 18(1): 51–66.

World Bank (2013). 'Labor Participation Rate, Female (% of Female Population Ages 15+) 2008-2012.' http://data.worldbank.org/indicator/SL.TLF.CACT.FE.ZS

Yuval-Davis, N. (2011). *The Politics of Belonging: Intersectional Contestations*. London: Sage.

Zahidi, S., and Ibarra, H. 2010. *The Corporate Gender Gap Report*. Geneva: World Economic Forum.

Zander, U., Zander, L., Gaffney, S., and Olsson, J. (2010). 'Intersectionality as a New Perspective in International Business Research', *Scandinavian Journal of Management*, 26(4): 457–66.

CHAPTER 2

DISTURBING THOUGHTS AND GENDERED PRACTICES

A Discursive Review of Feminist Organizational Analysis

MARIANA INES PALUDI, JEAN HELMS-MILLS, AND ALBERT J. MILLS

INTRODUCTION

IN this chapter we undertake a discursive review of selective feminist contributions to organizational studies, to assess the issues involved in developing feminist organizational analysis and the implications for addressing workplace discrimination.

Our lens is feminist and discursive, which indicates something of our initial starting point and biases. Although this reveals our sympathies with feminist poststructuralist accounts (Calás and Smircich 1992), we are also open to the manifold understandings of gender and organizations that other feminist approaches have to offer (Calás and Smircich 2005), as well as the *strategic* uses that feminists have made of Foucauldian discourse analysis (Ferguson 1984).

Nonetheless, in seeking to provide something of a historical account of feminist organizational analysis, we are aware of the need to take a reflective and transparent approach to history and the past (Munslow 2010) in order to avoid the pitfalls of the ahistorical and unreflective accounts of the past that dominate much of the extant literature of management and organization studies (Booth and Rowlinson 2006; Durepos *et al.* 2012). To that end, we draw on new historicism and Foucault's notion of discourse (Foucault 1979, 1982) as a way of framing our review of feminist organizational analysis over time.

By adopting a discursive lens, we avoid the problem of trying to *represent* a series of events (i.e. the development of feminist organizational analysis) as a linear and/or progressive history. This allows us to focus not only on what is said but the context in which it is said, thus facilitating analysis of how and why particular accounts were/are received. We also avoid the problem of presenting feminist organizational analysis as more or less true. Like all accounts, our approach is selective in its focus and should be read at best as one account of the development of the field. Part of the selective process includes our own starting point (in the construction of histories of management and organization studies) and the various researches we choose to discuss. Our selections were chosen not only in terms of time and space but also in terms of our own discursively shaped interests.

The end result is the development of a greater understanding of the contribution of feminist organization analysis to employment equity. In particular, we set out to show how different discursive contexts and positions have created a rich array of feminist contributions, whose paradigmatic differences may have served to obfuscate rather than enlighten a way forward for feminist organizational analysis. To that end, our conclusions centre on discussions on paradigmatic interplay as a way forward for feminist research.

DEFINING THE FIELD: HISTORY AND MANAGEMENT AND ORGANIZATION STUDIES

To understand something about the contribution of feminist theory to Management and Organization Studies (MOS) we need to understand something about the field itself, specifically how it came to be defined; and the role of 'history' in the process of definition.

Management History and the Development of MOS

Historical accounts of MOS exist in several forms. One (formal) form exists as conscious attempts to write a history of the field. Examples include various works by George (1968, 1972), Urwick (1938, 1963; Urwick and Brech 1957b), and Wren (1979, 1994, 2005; Wren and Bedeian 2009). Another (informal) form can be found in textbook accounts that discuss specific aspects of management (e.g. motivation and leadership theory) by tracing their development or evolution (e.g. in the respective work of Maslow and Lewin, Lippitt and White).

To state the obvious, history is written backwards, i.e. the historian writes about the past while located in the present. As such, he or she is subject to the discursive processes in which they are located (Foucault 1973). Such processes, arguably, serve to *create* rather than uncover the phenomenon under research (White 1973; Jenkins 1991). Thus, it can be argued that the historian of Management and Organization Studies has not so much written a history of the field as constructed a powerful account of what constitutes MOS.

This is not to argue that historical accounts are wholly or centrally responsible for defining a given disciplinary field. Clearly they draw on existing definitions of the field (see e.g. Drucker 1954; Koontz 1962; Astley and Van de Ven 1983) and associated practices (Khurana 2007). What we are arguing is that the definition of a field is an iterative process that relies on historical accounts, theoretical debates, and existing practices (Khurana 2007). Nor are we contending that notions of what constitute MOS as a field of study are uncontested (Burrell and Morgan 1979; Clegg and Dunkerley 1980) but rather that there are dominant (positivist, managerialist) accounts that have strongly influenced the development of the field and understandings of its history.

Although most histories of MOS (either formal or informal accounts) locate its 'beginnings' at the start of the twentieth century there is some indication that the historical accounts themselves were largely developed in the 1970s (George 1968, 1972; Burrell and Morgan 1979; Clegg and Dunkerley 1980; Khurana 2007) and were based largely on practices developed during the Cold War era (Cooke 1999, 2006; Cooke, Mills, and Kelley 2005; Kelley, Mills, and Cooke 2006; Runté and Mills 2006). This may help to explain a dominant view of the field that was developing throughout the 1970s and has influenced accounts since: namely, that Management and Organization Studies is a disciplinary field that is characterized as managerialist (focused largely on business organizations and outcomes of efficiency and effectiveness), scientistic (drawing primarily on the methods of the natural sciences), objectivist (taking a neutral, value-free stance to the research process), ahistorical (adopting a decontextualized, and universalistic approach to the research process), profoundly gendered, and Westernized (Burrell and Morgan 1979; Hearn and Parkin 1983; Donaldson 1985; A. Prasad 2003; Booth and Rowlinson 2006).

Drawing on these various cues, historical accounts have helped to solidify a particular view of MOS as a discipline that largely excluded research focused on government and not-for-profit agencies, humanist concerns that do not rely on bottom-line outcomes, and methods of research centred around subjectivity and meaning at work (Donaldson 1985, 1988). In the process, 'the founders' of MOS are, with few exceptions, characterized as men whose research contributed to the making of the field; in particular those men whose research in the major corporations of the United States (e.g. Bethlehem Steel; the Hawthorne Works of Western Electric) were seen as developing important 'schools of thought' (Koontz 1962; Astley and Van de Ven 1983), including 'scientific management' and the 'human relations' school in the pre-Second World War era (Urwick and Brech 1957a, 1957b; Rose 1978), and human relations, socio-technical systems, contingency theory, population ecology, New Institutional Theory, organizational culture, and several other theoretical developments in the post-Second World War era (Burrell and Morgan 1979; Clegg and Dunkerley 1980; Buchanan and Huczynski 1985).

These various accounts and related practices (Khurana 2007) have served to present a particular image of a discipline that rarely included female theorists (Genoe McLaren and Mills 2008; Tancred-Sheriff and Campbell 1992), ignored gender (Hearn and Parkin 1983; Wilson 1996) and race (Cox 1990; Nkomo 1992), and excluded as 'schools of thought' those approaches, such as feminism, that were not seen as contributing to explanations of organizational and managerial efficiency and effectiveness (Mills 2004).

The Problem with History

Like much of MOS itself, histories of the field have been dominated by positivist accounts that view history as a discovery and recounting of facts, helped no doubt by narrative, but narratives that are viewed as ultimately subservient to the facts themselves (Iggers 1997). This view of history has consequences for how people come to understand not only a field of study but also constructions of the people who inhabit that field. In the process, this can reinforce supposedly existing workplace philosophies and the role of women (Clegg and Dunkerley 1977; Alvesson and Billing 1997).

Drawing on poststructuralist historiography, we argue that the 'past' is ontologically unavailable and is thus only mediated by 'historiography, language, employment, voice, ideology, perspective or physical and/or mental states of tiredness, ennui, and so on' (Munslow 2010: 37). Facts are invented not discovered (White 1985) and rely as much on the narrative style of the teller (historian) and the context in which he or she is writing (White 1973; Jenkins 1994; Green 2007).

In terms of context, as suggested above, it is interesting to note that much of our sense of MOS as a field of study comes from the post-Second World War era during the height of the Cold War, when it was considered politically unsafe to associate with or write about political projects such as women's rights, racial equality, and militant labour union activities (Filippelli and McColloch 1995; Schrecker 1998; Cherny *et al.* 2004). These concerns not only affected labour activists but also reached into the heart of the universities, influencing what didn't get written as much as what did (Schrecker 1986). Some historians have argued that the internal Cold War, or McCarthyism, was in large part a delayed reaction to the New Deal era and the growth of militant labour unionism (Schlesinger 1959; Morgan 2003). That may go some way to explaining the absence of references to the New Deal in management textbooks (Foster *et al.* 2012). In any event, the work of a number of feminist activists, including Mary Van Kleeck and Betty Friedan, was disrupted by McCarthyism (Horowitz 1998; Nyland and Heenan 2005) and MOS took on a largely managerialist focus, devoid of political commentary on the nature of organizational work and its impact on people (Mills and Helms Hatfield 1998; Cooke 1999, 2006). To take but one example, the North American management textbook has had little to say about women at work, race/ethnicity, class or trade unions for much of the post-Second World War era. Nor, as we shall discuss later, has the management textbook acknowledged such major socio-political events as the Cold War, the Korean and Vietnam Wars, the March on Washington, the advent of women's liberation or various anti-colonialist struggles (Mills and Helms Hatfield 1998; Mills 2004).

The outcome was a field of study that Burrell and Morgan (1979) described as dominated by apolitical, ahistorical and managerialist accounts that largely neglected the impact of business organizations on people's social, psychological and material well-being (except where it affected the bottom line), and which rarely included as management theory those studies that utilized other than positivist methods. Burrell and Morgan failed to add, however, that the field was also highly gendered. Such was the condition of MOS that confronted feminist theorists in the 1970s.

ALTERNATIVE READINGS OF THE DEVELOPMENT OF MOS

In assessing the role of history in the construction of MOS, the fate of Lillian and Frank Gilbreth; Max Weber and Mary van Kleeck; Frances Perkins, Harry Hopkins, and Harold Ickes; and Mary Parker Follett are instructive.

Lillian and Frank

The role of Lillian Gilbreth is often played down in comparison with that of her husband Frank (Tancred-Sheriff and Campbell 1992). As Graham (2000: 285) points out, Lillian Gilbreth generally 'remembered as a mother of 12 children . . . was in fact not only a wife and mother, but also a pioneer of scientific management who earned a PhD in psychology and then worked as a partner with her husband, Frank Bunker Gilbreth . . . in their industrial engineering firm'. Graham (2000: 285) goes on to argue that, along with husband Frank and Henry Gantt, Lillian Gilbreth was the leader of 'a revisionist movement in scientific management which aimed to make it more responsive to human needs and concerns'. As we shall argue below, this focus on the human factor was creating a bridge between scientific management and the so-called Human Relations School. Finally, Graham (2000: 285) goes on to suggest that the marginalization of Lillian Gilbreth's contribution to MOS was in part due to 'sex discrimination from the business and engineering communities' following the death of her husband in 1924.

Max and Mary

Innumerable accounts of Max Weber contend that he was 'the father' of organization theory (in North America), playing a pivotal role in the development of MOS as a contemporary of Taylor and the early theorists of scientific management (Durepos *et al.* 2012). In particular, Weber's theory of bureaucracy is singled out as an important contribution to our understanding of the relationship between efficiency and organizational structure. Regardless, a number of recent accounts have argued that, far from being a pioneer in the development of MOS, Weber's work did not really have an influence until at least the late 1940s, with prominent translations by Talcott Parsons (Clegg and Lounsbury 2009). Furthermore, it is argued that many accounts of Weber's notion of bureaucracy are misplaced and just plain wrong in reducing Weber's social action philosophy to a simple reductionist framework (Burrell and Morgan 1979; Clegg and Dunkerley 1980; Cummings and Bridgman 2011).

In 1904 Weber, the future 'founder of organization theory', made his only trip to the United States—to present a paper at the Congress of Arts and Sciences in St Louis. His visit had little or no impact on US sociology, let alone Management and Organization Studies for another three decades or so (Swatos *et al.* 1998; Scaff 2011). That same year Mary Van Kleeck graduated from Smith College. The following year she engaged in several social work activities, involving work with feminist reformers. Over the years she combined her feminist concerns with an interest in scientific management and its potential for assisting social change. Her pioneering efforts on behalf of women at work led to important legislative changes and appointments to high levels of government, including director of the US Department of Labor's Women in Industry Service, President Harding's Conference on Unemployment, and the Labor Department's Federal Advisory Committee of the United States Employment Service of the Franklin Roosevelt administration. Van Kleeck also went on to serve as President of the Taylor Society, advocating a number of sweeping socio-economic changes, including the need for socio-economic planning and social insurance.

Yet, despite her prominence, van Kleeck's contribution to MOS has gone unheeded until very recently (Nyland and Rix 2000; Nyland and Heenan 2005) and has still to be recognized. The reason for the neglect of van Kleeck's work is likely two-fold. First, much of her work did not fit the narrow confines of what became known as MOS. Van Kleeck's central concerns were the impact of technology on the workers, and women workers in particular, and the potential misuse of scientific management efficiencies to create unemployment rather than improve working conditions. Her ideas of socio-economic planning and social insurance, despite or perhaps because of the New Deal, ran counter to pro-business sentiments of the day. Secondly, from the onset of the Depression van Kleeck's political views became increasingly leftist and from this point onwards she became an admirer of the Soviet Union and her views were steadily associated with those of the United States Communist Party. In 1953 her political views brought her to the attention of Senator Joseph McCarthy's Senate Permanent Committee on Investigations. In the process van Kleeck's work may have not so much have been written out of history as not included in what were to become histories of MOS (Cooke 1999; Nyland and Heenan 2005).

Frances, Harry, and Harold

Frances Perkins, Harry Hopkins, and Harold Ickes all played pivotal roles in the 'New Deal' polices and programs of the Franklin Roosevelt administration. Ickes ran the Public Works Administration (PWA), which oversaw massive expenditures in jobs and infrastructure; financing over 35,000 projects, including some of the largest infrastructure projects in the history of the US (Graham and Wander 1985; Taylor 2008). Hopkins ran the Works Progress Administration (WPA), which in the span of eight years—1935–43—spent $10.5 billion and employed 8.5 million men and women (Taylor 2008). Perkins, the first woman in US history to be appointed to a cabinet position, as Secretary of Labor developed and oversaw the introduction of far-reaching social security measures, including the Social Security Act (Schlesinger 1959; Hiltzik 2011). All three were part of Roosevelt's 'inner cabinet' and had a profound influence on the concept of work, employment and society through to the present day (Taylor 2008).

Nonetheless, the work of Ickes, Hopkins, and Perkins has gone virtually unnoticed in MOS (Foster *et al.* 2012). This is quite remarkable given the sweep of their accomplishments, their writings (Ickes 1934, 1953, 1954; Hopkins 1936; Perkins 1947) and the sheer scale of the capital expenditures and employment involved. Undoubtedly Bethlehem Steel (the second largest steel producer in the United States in Taylor's day) and Western Electric (supplier of 90 per cent of all telephone equipment in the United States at the onset of the Hawthorne Studies) occupied an important role in the US economy of the day, and their sheer size and importance were factors of legitimation of the management theorizing by Taylor, Mayo, Roethlisberger and Dickson. Yet the WPA, PWA, and the various programmess of the Social Security Act dwarf them in significance as, arguably, do the contested achievements of the New Deal, when compared to the speculative theories of Mayo, Roethlisberger and Dickson (Foster *et al.* 2012).

There are various theories to explain the absence of Perkins, Ickes and Hopkins in the emergent post-Second World War Management and Organization Studies. For one thing, the field was being defined in and around the needs of, and developments within, the business schools and so was less likely to include activities deemed outside of the business realm (e.g. government, social work). Government agencies were seen as either the province of public administration and/or political science and the social work roots of Perkins, Ickes, and Hopkins likely put them outside the realm of business research. Another equally powerful reason for ignoring Perkins *et al.* was their association with the New Deal, which was increasingly unpopular with the business community that became associated with the post-war backlash against the left and formed part of the McCarthyite atmosphere (Morgan 2003): even during the pre-war era Perkins and Hopkins came under fire from the Dies 'House Un-American Activities Committee', accused of supporting communistic activities through their New Deal programmes (Taylor 2008).

Mary Parker Follett

In terms of historical recognition, the fate of Mary Parker Follett has varied widely over the years. In recent years there has been considerable debate—largely among feminists—as to the contribution of Follett to MOS (and associated disciplines) and explanations as to her fluctuating reputation in the field. At the time of writing, it is fair to say that there is considerable recognition of Follett's contribution to MOS, but we note two things: first, this was arguably due in large part to the efforts of numerous feminist analyses, and second, there is still some sense, at least among feminists, that Follett's reputation remains unstable (Ansell 2009; Phipps 2011). Over the past two decades or so feminist organizational scholars have pointed out the relative neglect of Follett's contribution to MOS. Tancred-Sheriff and Campbell (1992: 32), for example, complain that despite Follett's early emphasis on 'the motivating desires of the individual and the group', and the fact that her 'writings bridge the transition from the Scientific Management School to the Human Relations School... [she is] hardly a household name among organizational researchers'. Similarly, Schilling (2000: 224) contends that, although Follett's writing 'predates the field of stakeholder theory by almost 60 years... no reference to her work is found in the stakeholder literature'. Eylon (1998: 16) argues that Follett was in fact a pioneer in the identification of workplace empowerment and decries the fact that, despite 'her innovative ideas, Follett has been a relatively unrecognized contributor in the social sciences'. Phipps (2011: 273) discovers in Follett's work the roots of the current interest in spirituality at work but argues that Follett and other 'female "laborers" that have helped to cultivate the field have received much less attention' than their male colleagues'. Godwin and Gitten (2012: p. xvi) view Follett as 'arguably the founder of the interactionist approach' to organizational studies and set out to 'restore [her] to her rightful place as a foundational thinker in the sociology of organizations' (p. xxiii). Others, however, speak of a 'revival' of Follett's work (Kolb *et al.* 1996: 16) and suggest that the 'significant attention' currently being paid to Follett's work is of recent vintage (Ryan and Rutherford 2000: 207).

What is of interest here is how and why feminist organizational theorists felt the need to confront the historical accounts of MOS. Simply put, it can be argued that depictions of what constitutes the field of MOS and legitimate areas of study within that field are imbued with embedded historical accounts (Booth and Rowlinson 2006; Burrell and Morgan 1979). In various ways feminists have revealed the limits of those accounts. In the process they have opened the field to gendered accounts and concerns but in some cases may have strengthened some of the other problematic areas of MOS history.

Following in the footsteps of Rowbotham (1974), a number of feminist have set out to 'rediscover' women in MOS and its history, largely through accounts written by and about women but also by those written from a feminist point of view (Humphreys 1994). The first approach owes much to the consciousness-raising philosophy of the second wave women's movement of the 1960s, and focuses on bringing a 'women's point of view' to organizational analysis. This is seen in several accounts where it is argued that the gendered character of MOS is linked to the male dominance of the field, which has led to the neglect not only of gender at work but also of key female writers. Phipps (2011: 279), for example, concludes her analysis of Mary Parker Follett and Mary Gilson by stating, from 'an historical viewpoint further research should be conducted to uncover other worthy female contributors'; arguing that in 'terms of practice, organizations would benefit from incorporating the values endorsed by these women in daily life to achieve the best results from their employees'. The second approach centres on 'including women in the historical record' (Davis 1994) and permeates various feminist accounts that discuss the neglect of female theorists and the loss of a women's perspective from MOS. Stivers, for example, argues that what is interesting about Follett is not only what we can learn about theories of work and organization but how those theories were shaped by her experiences as a woman: 'Considering Follett's own life and work for a moment, certain aspects do seem to have been shaped by her sex; indeed, given the sharp gender differentiation of the time, it would have been quite extraordinary for any person of her era not to have been so affected' (Stivers, 1996: 161–2). The third approach focuses on deconstructing historical (and contemporary) accounts of MOS in terms of the 'history of the conceptions of gender (i.e., of "men" and "women" as social not natural beings) and of the social relationships and experiences to which gender ideologies are tied, rather than as the history of "women" in isolation' (Humphreys 1994: 87). Calás and Smircich (1996b: 147) provide an illustrative example by emphasizing the hope that 'an ongoing reading of Follett... will lead us to thinking about management and organizations in very different terms from that of the contemporary organizational "mainstream"'. In particular they contend that 'a reading of Follett from a particular feminist *standpoint*... might result in a heightened appreciation for her work' and that by reading 'from the *intersections of class and gender* at a particular point in time we would notice more of Follett's view of the world, and the potential of her work for today's organization studies' (p. 150, emphases in the original).

In addition to issues of gender, explanations for the neglect of Follett's work include the observation that her writings fell outside the norm of business concerns, focusing on issues of the role of the state and the democratic process (Calás and Smircich 1996b; Stewart 1996; O'Connor 2000); that her views were 'politically incorrect', becoming 'increasingly associated with socialism, which itself fell increasingly out of favour' (O'Connor 2000: 187); and that her work fell out of favour when the American Business School was overtaken by the 'winds of positivism' (Calás and Smircich 1996b: 149).

History, Feminism and MOS

These various feminist accounts reveal a number of problems with existing histories of MOS. As we have seen above, a number of feminists have raised the issue of gender and how that has served to neglect or exclude the work of female theorists from mainstream histories of MOS (Tancred-Sheriff and Campbell 1992; Stivers 1996; Phipps 2011). These accounts question the problem of designating founders (Eylon 1998; Godwin and Gitten 2011; Phipps 2011); the periodization and construction of schools of thought; and defining the boundaries of a discipline built around paradigmatic difference (Kuhn 1962; Burrell and Morgan 1979).

Arguably the first two problems are linked to the latter. For example, the characterization of scientific management as the first major approach to MOS (Rose 1978; Burrell and Morgan 1979; Buchanan and Huczynski 1985) privileges people like Frederick Taylor with foundational qualities but excludes people like Mary Parker Follett, who, using a similar kind of logic, could conceivably have been seen as presaging the Human Relations school (Tancred-Sheriff and Campbell 1992). Indeed, it may be argued that the social work background of Follett, van Kleeck, Ickes, Hopkins, and Perkins could just as easily be seen as constituting a Social Security school of thought that overshadowed the much more limited Human Relations studies at the Hawthorne Works.

Several feminist theorists have argued—both directly and indirectly—that the definition of a field of interest depends in large part on who is involved and the context in which those definitions are being developed. This point is made indirectly by Ryan and Rutherford (2000), who argue that in her time Follett's work was extremely influential because it was in line with the dominant thinking of the time but fell out of favour as that thinking changed. Similarly, Frye and Thomas (1996: 18) contend that, as seen from today's perspective, 'Follett's writings are driven by a questionable philosophical premises and her perspective is, at best, partial'. Both sets of authors fail, however, to explain how the work of other (male) writers, who fit a similar description, continued to be popular through to today. In a similar vein, Kolb *et al.* (1996: 153) trace the 'resurgence of interest in Follett, in part, to societal changes regarding the value of negotiation as a means of dealing with differences'. More directly, as we have seen above, Calás and Smircich (1996*b*: 149) trace the demise of interest in Follett's work to the dominance of positivist thinking and the 'difficulties in twisting her ideas to fit smoothly into the positivist camp. It was perhaps easier to consign her to the category of eccentric woman.' They go on to suggest that Follett's demise may also have had something to do with her 'dissenting theoretical and philosophical concerns' (p. 150). The latter suggestion is shared by O'Connor (2000) who concludes that perhaps 'with the end of the Red Scare and the Cold War, Follett has finally found a congenial time, for her work in increasingly seen as relevant' (p. 187). In other words, we are in a different discursive space where evaluations of the gendered history of MOS can be undertaken.

However, these various feminist attempts to revise neglected women theorists are not without problems and in some aspects may reinforce the mainstream philosophical underpinnings of MOS histories. Calás and Smircich (1996*b*) caution against attempts to fit female theorists into existing ways of thinking. Calling Follett a prophet of management, or someone 'ahead of her time', for example, may actually fulfil 'various purposes *in support of the ideas of contemporary theorists while often submerging the ideas of "original authors"*' (p. 148, emphasis in the original). They also warn against the associated rhetorical problem of fostering 'a sense of "progress of knowledge" [where] . . . a line of thought is shown to have *evolved* to higher levels of sophistication, i.e., *as exemplified by contemporary theorists*' (p. 148, emphasis in the original). They also warn against the tendency to treat Follett and other female researchers of her generation like historic figures; noted and then 'put back on the (library) shelf once the celebration has passed' (p. 147).

These debates around the role of women and feminist theory in MOS, however, did not take place until the 1990s, long after the field had experienced the emergence of a series of challenges to the dominant positivist paradigm (Burrell and Morgan 1979;

Clegg and Dunkerley 1980), including various feminist challenges (Hearn and Parkin 1983; Mills and Murgatroyd 1990; Mills and Tancred 1992).

Feminist Theory, Organizational Analysis, and Changing Discursive Grounds

With the approach of the 1970s a number of socio-political struggles rocked the Western world and revealed a profound questioning of many of the established tenets of a modernist world in turmoil. In the United States, race riots, desegregation marches, and even armed conflict questioned the basis of racist thinking (King 1968; Seale 1970; X and Haley 1965). Anti-colonialist struggles were culminating in the ending of the Vietnam War and the widespread protests against US involvement (Fanon 1970; Zinn 1999). Across Europe and North America new traditions of postmodernist thought began to appear, including the work of Foucault (1965), Lyotard (1979), and Said (1979), and new forces for radical change were emerging to challenge the status quo, East and West (Marcuse 1969; Klimke *et al.* 2011), including 'second-wave feminism', or the women's liberation movement, which burst onto the scene questioning all aspects of patriarchal thinking (Friedan 1963; Greer 1970).

Critique and Application: 'A la recherche du genre perdu

In this context, the work of sociologists Joan Acker and Donald van Houten (1974) appeared (in the *Administrative Sciences Quarterly*) as a critique, in line with radical feminist in politics, against male bias in the analysis of organizations.

Although feminists in other fields of study (e.g. sociology) had, for some time explored the relationship of women to paid work (Hearn and Parkin 1983) such contributions were absent from established MOS journals and textbooks until this point. The work of Acker and van Houten (1974) was important for its 'revisiting' of 'classic' (Rose 1978; Wren 1979) organization studies, including the Hawthorne Studies, to suggest that gender had been neglected in the collection and analysis of data in the original studies. They contended that from the evidence available there was a 'sex power differential' at work in the relationships, not only between the male managers and the female employees but also between those same employees and the male researchers. In making their argument, Acker and van Houten also made reference to the socio-economic climate of the time and place of the Hawthorne Studies. This article was important not only in raising questions about gender and organizations but also about the gendered character of organizational studies (i.e. the way they were theorized). It also pointed to the need to take into account the sociological context in which organizations operate. Another important factor was that the work suggested a shift away from essentialist notions of men and women to more of a focus on the social construction of gender at work. Nonetheless, Acker and van Houten (1974) did not question the validity of the Hawthorne Studies as classic management studies. It was the first of a number of similar critiques that raised the gendered nature of particular aspects of so-called classic studies of management and organization (Feldberg and Glenn 1979; Cullen 1992; Benschop and Doorewaard 1998) but those critiques were few and far between.

The article by Acker and Van Houten represents a milestone in the field for several reasons. First, and not least, because they managed to get it published in a mainstream MOS journal. Second, they used a historical approach to draw attention to contextual factors in the making of gendered relationships. Third, they took issue with the gendered character of MOS theorizing. Fourth, they drew attention to sexuality at work, an issue that would remain relatively neglected even in feminist analyses of organization for some time (Hearn *et al.* 1989; Hearn and Parkin 2001). Fourth, they contended that organizational context needed to include gendered relationships that are reproduced across the domestic–public divide. Fifth, they also suggested that organizations are sites of the reproduction of gender inequities.

Accommodation: Women and Management Theory and the Modern Era

In 1977, still at a time when scholarly books gained as much, if not more, attention than scholarly articles, *Men and Women of the Corporation*, written by another feminist sociologist Rosabeth Moss Kanter, made an important impact on the emergent field of feminist organizational analysis.

Although coming from similar sociological roots, Kanter's work differs markedly from Acker and van Houten in its approach. For one thing, Kanter's approach eschews history, preferring a more decontextualized view of the internal working on a corporation. For another, Kanter moves from critique to application in attempting to spell out an agenda for change *within* organizations. This stands in opposition to Acker and van Houten's critique of organizational analysis, designed to effect broader social change. The one area of overlap between the two studies was a focus on structure and its relationship to gendered outcomes but even here Kanter's more managerialist position fit more neatly into Burrell and Morgan's (1979) mainstream functionalist paradigm, while the position taken by Acker and van Houten (1974) is more appropriately described as radical structuralist.

Despite its accommodation with existing mainstream approaches to organizational analysis and Kanter's own career development, her 1977 book has rarely been cited in business textbooks (Mills 2004). It did, however, but give birth to the development of a body of research that became known as the 'Women in Management' (WIM) perspective.

Largely essentialist in nature, WIM or 'liberal feminism' (Calás and Smircich 1996a) tends to focus on ways to improve women's position in the corporate sectors of society. In other words, WIM research has been oriented to develop comparative analyses of men and women in managerial positions, albeit ignoring how those managerial positions have been designed. It pursues equity (or gender justice) rather than the elimination of inequality (i.e. acknowledging issues in the premises that support the institutions which disadvantage particular groups). Issues that have drawn attention include different leadership styles (e.g. Eagly *et al.* 1992); job stress; job satisfaction; organizational commitment; and gender stereotypes (Brenner *et al.* 1989; Schein and Mueller 1992; Schein *et al.* 1996; Schein 2007); the glass ceiling phenomenon (Bullard and Wright 1993; Catalyst 1993; Powell and Butterfield 1994; Ragins *et al.* 1998; Daily *et al.* 1999; Meyerson and Fletcher 2000; Wright and Baxter 2000; Dreher 2003); organization demography; careers (Heller 2004, 2011) and social networks (Ibarra 1993, 1997), equal opportunity, affirmative action, and discrimination; sexual harassment and work/family issues (Baylin 2007—see Calas and Smircich, 2005 for a review).

Even multinational organizations like BID, OECD, the World Bank adopt the WIM perspective to develop international programmes encouraging gender equality (see OECD 2008; BID 2009; Rodriguez Gusta 2009). Women in Management scholars want to explain the causes of sex segregation in organizations using mainly quantitative methodologies. Here, the sex and gender concepts are variables, not an analytical framework (Smircich 1985; Alvesson and Due Billing 1997). Among other assumptions, gender neutrality and universality of organizations and management theory rest in this tradition. In this regard, it has been also criticized for being a hegemonic theory based on the experience of white women (Holvino 2010: 250).

Postmodern and Poststructuralism: Work in Progress

During the 1980s some researchers were noting a number of rapid changes in the socio-political and economic landscape, including the end of the Cold War, the decline and fragmentation of the left and the eventual collapse of the Eastern Bloc, the development of globalized technologies and economies, and the fragmentation of numerous aspects of social life and attendant thinking (Toffler 1981; Burrell *et al.* 1994; Adler *et al.* 2006). This may, in large part, explain the growing interest of organizational scholars during this period in postmodernism and poststructuralist theory (Burrell 1988).

In this context two works appeared in 1984 that drew on poststructuralist thought to address feminist concerns: one was by Gibson Burrell (1984) and the other by Kathy Ferguson (1984). Both drew on the work of Foucault, Ferguson more 'strategically' than Burrell's committed poststructuralist position.

Burrell's (1984) work was important for its focus on sexuality and attempt not only to 'elevate "sexuality" to a position of theoretical relevance in the analysis of organizations' but also to view it as a construct, which needs to be understood historically (p. 98). In other words, 'sexuality' should be viewed as discursive, i.e. the outcome of interrelated practices and ideas at a given point in time.

In a similar vein, Ferguson (1984) argued that bureaucracy was not so much about structure as about discourse; a discourse that equates bureaucracy with masculinity and, in the process, marginalizes women. Ferguson's (1984) work was also important for both its embrace of politics within organizational analysis and the 1960s feminist notion that 'the personal is the political'. In contrast to Kanter's (1977) appeal to give managerial tools to women to gain equity in the workplace (by changing the structure), Ferguson forces us to think about the impact of organizational discourse on identity and a sense of self. One of her main suggestions was the development of alternatives to the bureaucratic organization, embedded in feminist discourse.

Beyond Modernity and Structure: A Postcolonial View

Since the onset of the 1990s we have seen the development of postcolonialist theorizing in MOS (A. Prasad 1997; P. Prasad 1997, 2005), including feminist postcolonialist work (Calás 1992; Mirchandani 2004). This work was a long time coming in Western MOS journals and followed on from a series of events that included the process of decolonization since the end of the Second World War; the development of globalized technologies; the collapse of the Cold War notion of the Third World, with the inclusion of associated countries in one or other of the two world powers (the USA and the USSR); the continued growth of the multinational corporation and the advent of powerful economic unions (e.g. the European Economic Community); post-war immigration trends; the advent of a series of global wars (e.g. Iraq, Afghanistan); civil rights struggles in the United States; the collapse of Apartheid in South Africa; the events and aftermath of September 9/11 in the United States; and a multitude of other events that provided the space to question the Western project and the Othering of so-called non-Western people (Said 1979, 1993; Mignolo 1991; Bhabha 1994).

To have a postcolonial framework in research means to acknowledge global capitalism and its implications within organizational analysis (Calas and Smircich, 2006). In contrast with feminist poststructuralist research that acknowledges the importance of knowledge production and the representation of the 'others' within Western hegemonic discourses, postcolonialism goes further and emphasizes the Western imperialism in global institutions (P. Prasad 2005).

This tradition questioned the male dominance of national liberation movements; feminist movements dominated by white Western women; Eurocentric academic discourses; movements that privilege heterosexuality; and Marxist analyses privileging class struggle (Holvino 2010).

At the same time as Ferguson's (1984) book was released, a ground-breaking article from Mohanty (1984) criticized the way that Western scholars develop their discourse on Third World women as a single monolithic subject. Nonetheless, it would take a further six years before issues of race (Cox 1990; Nkomo 1992), let alone coloniality, would be raised in MOS in any substantial and coherent way. These studies include Calás's (1992) work on the absence (and possible representation) of 'Hispanic Women' in organizational texts, arguing that such texts need to be recognized as 'produced by privileged "voices of the West" ... [and] published in the privileged spaces of Western scholarship, inspired by a *Western* critique of *Western epistemology*' (p. 219, emphasis in the original).

Intersectionality: The Complexity Forgotten

With the opening of various spaces in feminist debates attention turned to intersectionality, or the study of the impact of overlaying discriminatory experiences due to any combination of race, ethnicity, sexual orientation, ethnicity, national origin, and gender (hooks 1994; Acker 2006). In an early iteration, Crenshaw (1991), exploring race and gender dimensions of violence against women of colour, argued that to understand racism and sexism it is necessary to understand how they intersect in the identity of 'women of color'.

As a theoretical framework intersectionality allows researchers to emphasize those identities neglected by others. Also, the notion of organization becomes complex and emergent to the local context. For example, the work by Pringle (2008) on lesbian managers in New Zealand makes evident that to manage gender means to manage heterosexuality. She argues that a reframing of gender as 'heterogender' foregrounds heterosexuality and gender as intertwined, thus providing another understanding as to how gender is 'done' in management. This means that organization becomes a process of heterosexualization. Moreover, globalization exacerbates heterogeneity and brings difference together. The continued growth of multinational corporations, with headquarters in developed countries and branches overseas, or outsourced organizations like call centres in underdeveloped countries—e.g. India, Argentina, Mexico, Chile—allows different ways for people to relate to each other through organizational forms. One potential outcome is that tensions will emerge when those different identities converge in space and time (see Kosmala 2008).

Recent work by Holvino (2010) argues that the intersections of race, gender, and class should be reconceptualized as simultaneous processes of identity, institutional and social practice. Her work makes two major contributions to the study of gender, organizations and intersectionality. On the one hand, she proposes a theoretical change where gender, class, and race are reconceptualized as simultaneous processes of identity, institutional and social practice. On the other hand, she suggests a methodological intervention, expanding the simultaneity of race, class, and gender to include new categories of ethnicity, sexuality, and nation in organizational analyses.

FINAL THOUGHTS

In this chapter, we set out to locate the advent and development of feminist organizational analysis in a series of discursive spaces over time. In so doing, we have attempted to make a number of observations.

First, and foremost, we have raised questions about the role of history in the development of MOS and the implications for feminist organizational analyses. Here our central point is that, in order to deal with discrimination *within* organizations, we need to retheorize the disciplinary fields in which the notion of organization is conceptualized. Acker (1990) contends that feminists need to deal simultaneously with the gender substructure of organizations and the organizational logic out of which understanding of organizations is generated. In so doing, however, we further argue that process should include a coherent questioning of the underlying histories that serve to support and legitimize MOS as a discipline.

Second, we suggest that feminist organizational analysis needs to be viewed as plural, contextual, and, hence, always in a state of transformation (Calás and Smircich 2005).

Third, we contend that feminist theories of MOS arise within and as a reaction to certain discursive realities. As such, they need to treat both MOS and feminist research itself as reflections of and contributions to particular discourses of managing and organizing. In the process this may encourage focus more on the discursive than the supposed truths of the situation. For example, Dye and Mills (2012) suggest that changes in the gender regime of a major airline company may in large part be due to the advent of competing organizational discourses that were the outcome of both the widespread activities of the women's liberation movement and the changing economic realities in which the company was operating. This may entail strategies for fusing particular feminist positions with understanding of the role of discourse in the construction of particular gendered subjectivities (Calás and Smircich 1992).

Fourth, and finally, we have attempted to make sure that our own account should not be seen as a definitive history of the advent of feminist organizational analyses. Our concern has been to problematize history, in the role of gendered relations at work, by showing how analyses of the past, including our own selected account, are contested narratives (White 1973). This, nonetheless, is set to focus attention on the discursive nature of organizational analysis and the implications for feminist change strategies.

REFERENCES

Acker, Joan (1990). 'Hierarchies, Jobs, Bodies: A Theory 'of Gendered Organizations', *Gender and Society*, 4(2): 139–58.

Acker, Joan (2006) 'Inequality Regimes: Gender, Class and Race in Organizations', *Gender and Society*, 20(4): 441–64.

Acker, Joan, and Van Houten, Donald R. (1974). 'Differential Recruitment and Control: The Sex Structuring of Organizations', *Administrative Science Quarterly*, 19(2), 152–163.

Adler, Paul S., Forbes, Linda C., and Willmott, H. (2006). 'Critical Management Studies: Premises, Practices, Problems, and Prospects', in J. P. Walsh and A. Brief (eds), Annals of the Academy of Management, Volume 1, pp. 119–179, New York: Lawrence Erlbaum Associates.

Alvesson, Mats, and Billing, Yvonne Due (1997). *Understanding Gender and Organizations*. London, and Thousand Oaks, CA: Sage.

Alvesson, Mats, and Due Billing, Yvonne (1997). *Understanding Gender and Organizations*. London: Sage.

Ansell, Christopher (2009). 'Mary Parker Follett and Pragmatist Organization', in P. S. Adler (ed.), *The Oxford Handbook of Sociology and Organization Studies: Classical Foundations*, 464–85. Oxford: Oxford University Press.

Astley, W. G., and Van de Ven, A. H. (1983). 'Central Perspectives and Debates in Organization Theory', *Administrative Science Quarterly*, 28: 245–73.

Baylin, Lotte (2007). 'Breaking the Mold: Family as a Complicating Issue', *MIT Sloan* (Spring/Summer): 18–21.

Benschop, Yvonne, and Doorewaard, Hans (1998). 'Six of One and Half a Dozen of the Other: The Gender Subtext of Taylorism and Team-Based Work', *Gender, Work and Organization*, 5(1): 5–18.

Bhabha, H. (1994). *The Location of Culture*. New York: Routledge.

BID. (2009). 'Género y negocios. Casos exitosos en cuatro continentes', in A. I. Piazze (ed.), Banco Interamericano de Desarrollo. Retrieved from http://idbdocs.iadb.org/wsdocs/get-document.aspx?docnum=1974116

Booth, Charles, and Rowlinson, Michael (2006). 'Management and Organizational History: Prospects', *Management and Organizational History*, 1(1): 5–30.

Brenner, O. C., Tomkiewicz, Joseph, and Schein, Virginia E. (1989). 'The Relationship between Sex Role Stereotypes and Requisite Management Characteristics Revisited', *Academy of Management Journal*, 32(3): 662–9.

Buchanan, D. A., and Huczynski, A. (1985). *Organizational Behaviour*. London: Prentice Hall.

Bullard, Angela M., and Wright, Deil S. (1993). 'Circumventing the Glass Ceiling: Women Executives in American State Governments', *Public Administration Review*, 53(3): 189–202.

Burrell, Gibson (1984). 'Sex and Organizational Analysis', *Organization Studies*, 5(2): 97–118.

Burrell, Gibson (1988). 'Modernism, Postmodernism and Organizational Analysis 2: The Contribution of Michel Foucault', *Organisation Studies*, 9: 221–35.

Burrell, Gibson, and Morgan, Gareth (1979). *Sociological Paradigms and Organizational Analysis*. London: Heinemann.

Burrell, Gibson, Reed, Michael I., Calás, Marta B., Smircich, Linda, and Alvesson, Mats (1994). 'Why Organization? Why Now?', *Organization*, 1(1): 5–17. doi: 10.1177/135050849400100101

Calás, Marta B. (1992). 'An/Other Silent Voice? Representing "Hispanic Woman" in Organizational Texts', in A. J. Mills and P. Tancred (eds), *Gendering Organizational Analysis*, 201–21. London: Sage.

Calás, Marta B., and Smircich, Linda (1992). 'Using the "F" Word: Feminist Theories and the Social Consequences of Organizational Research', in A. J. Mills and P. Tancred (eds), *Gendering Organizational Analysis*, 222–34. Newbury Park, CA: Sage.

Calás, Marta B., and Smircich, Linda (1996a). 'From "The Woman's" Point of View: Feminist Approaches to Organization Studies', in S. R. Clegg, C. Hardy, and W. R. Nord (eds), *Handbook of Organization Studies*, 218–57. London: Sage.

Calás, Marta B., and Smircich, Linda (1996b). 'Not Ahead of her Time: Reflections on Mary Parker Follett as Prophet of Management', *Organization*, 3(1): 147–52. doi: 10.1177/135050849631008

Calás, Marta B., and Smircich, Linda (2005). 'From the "Woman's Point of View" Ten Years Later: Towards a Feminist Organization Studies', in S. Clegg, C. Hardy, T. Lawrence, and W. Nord (eds), *The Sage Handbook of Organization Studies*, 284–346. London: Sage.

Catalyst (1993) Successful Initiatives for Breaking the Glass Ceiling to Upward Mobility for Minorities and Women (A Report on the Glass Ceiling Initiative). Washington, DC: US Department of Labor.

Cherny, Robert W., Issel, William, and Taylor, Kieran Walsh (eds) (2004). *American Labor and the Cold War*. New Brunswick, NJ: Rutgers University Press.

Clegg, Stewart, and Dunkerley, David (1977). *Critical Issues in Organizations*. London and Boston: Routledge & Kegan Paul.

Clegg, Stewart, and Dunkerley, David (1980). *Organization, Class and Control*. London: Routledge & Kegan Paul.

Clegg, Stewart, and Lounsbury, Michael (2009). 'Sintering the Iron Cage: Translation, Domination, and Rationality', in P. S. Adler (ed.), *The Oxford Handbook of Sociology and Organization Studies: Classical Foundations* (118–45). Oxford: Oxford University Press.

Cooke, Bill (1999). 'Writing the Left out of Management Theory: The Historiography of the Management of Change', *Organization*, 6(1): 81–105.

Cooke, Bill (2006). 'The Cold War Origin of Action Research as Managerialist Cooptation', *Human Relations*, 59(5), 665–93.

Cooke, Bill, Mills, Albert J., and Kelley, Elizabeth S. (2005). 'Situating Maslow in Cold War America: A Recontextualization of Management Theory', *Group and Organization Management*, 30(2): 129–52.

Cox, Taylor H., J. (1990). 'Problems with Organizational Research on Race and Ethnicity Issues', *Journal of Applied Behavioral Sciences*, 26: 5–23.

Crenshaw, Kimberle. (1991). Mapping the Margins: Intersectionality, Identity Politics, and Violence Against Women of Color', Standard Law Review, 43: 1241–91.

Cullen, D. (1992). 'Sex and Gender on the Path to Feminism and Self-Actualization', paper presented at the Administrative Sciences Association of Canada, Quebec.

Cummings, Stephen, and Bridgman, Todd (2011). 'The Relevant Past: Why the History of Management Should Be Critical for our Future', *Academy of Management Learning and Education*, 10(1): 77–93.

Daily, Catherine M, Certo, S Trevis, and Dalton, Dan R. (1999). 'A Decade of Corporate Women: Some Progress in the Boardroom, None in the Executive Suite', *Strategic Management Journal*, 20(1): 93–9.

Davis, Natalie Zemon (1994). 'What is Women's History?', in J. Gardiner (ed.), *What is History Today?*, 85–7. London: Macmillan.

Donaldson, Lex (1985). *In Defence of Organization Theory*. Cambridge: Cambridge University Press.

Donaldson, Lex (1988). 'In Successful Defence of Organization Theory: A Routing of the Critics', *Organization Studies*, 9 (1): 28–32.

Dreher, George F. (2003). 'Breaking the Glass Ceiling: The Effects of Sex Ratios and Work–Life Programs on Female Leadership at the Top', *Human Relations*, 56(5): 541–62.

Drucker, Peter (1954). *The Practice of Management*. New York: Harper & Brothers.

Durepos, Gabrielle, Mills, Albert J., and Weatherbee, Terrance G. (2012). 'Theorizing the Past: Realism, Relativism, Relationalism and the Reassembly of Weber', *Management and Organizational History*, 7(3): 267–81.

Dye, Kelly, and Mills, Albert J. (2012). 'Pleading the Fifth: Re-focusing Acker's Gendered Substructure through the Lens of Organizational Logic', *Equality, Diversity and Inclusion: An International Journal*, 31(3): 278–97. doi: 10.1108/02610151211209126

Eagly, Alice H., Makhijani, Mona G., and Klonsky, Bruce G. (1992). 'Gender and the Evaluation of Leaders: A Meta-Analysis',. *Psychological Bulletin*, 111(1): 3–22. doi: 10.1037/0033–2909.111.1.3

Eylon, Dafna (1998). 'Understanding Empowerment and Resolving its Paradox: Lessons from Mary Parker Follett', *Journal of Management History*, 4(1): 16–28. doi: 10.1108/13552529810203905

Fanon, Frantz (1970). *Toward the African Revolution*. Harmondsworth: Penguin.

Feldberg, R. L., and Glenn, E. N. (1979). 'Male and Female: Job versus Gender Models in the Sociology of Work', *Social Problems*, 26(5): 524–38.

Ferguson, Kathy E. (1984). *The Feminist Case Against Bureaucracy*. Philadelphia: Temple University Press.

Filippelli, Ronald L., and McColloch, Mark D. (1995). *Cold War in the Working Class: The Rise and Decline of the United Electrical Workers*. Albany, NY: State University of New York.

Foster, Jason, Mills, Albert J., and Weatherbee, Terrance G. (2012). 'The New Deal, History, and Management and Organization Studies II: Constructing Disciplinary Actors and Theories', paper presented at the Academy of Management annual conference, Boston.

Foucault, Michel (1965). *Madness and Civilization: A History of Insanity in the Age of Reason*. New York: Pantheon Books.

Foucault, Michel (1973). *The Order of Things: An Archaeology of the Human Sciences*. New York: Vintage Books.

Foucault, Michel (1979). *Discipline and Punish: The Birth of the Prison*. New York: Vintage Books.

Foucault, Michel (1982). *The Archeology of Knowledge and the Discourse on Language*. New York: Pantheon Books.

Friedan, Betty (1963). *The Feminist Mystique*. New York: Dell.

Fry, Brian R., and Thomas, Lotte L. (1996). 'Mary Parker Follett: Assessing the Contribution and Impact of her Writings', *Journal of Management History*, 2(2): 11–19. doi: 10.1108/13552529610106824

Genoe McLaren, Patricia, and Mills, Albert J. (2008). 'A Product of "his" Time? Exploring the Construct of Managers in the Cold War Era', *Journal of Management History*, 14(4): 386–403.

George, Claude S. (1968). *The History of Management Thought* (1st edn). Englewood Cliffs, NJ: Prentice-Hall.

George, Claude S. (1972). *The History of Management Thought* (2nd edn). Englewood Cliffs, NJ: Prentice-Hall.

Godwin, Mary, and Gitten, Jody Hoffer (2011). *Sociology of Organizations: Structures and Relationships*. Thousand Oaks, CA: Sage.

Godwyn, Mary, and Gitten, Jody Hoffer (eds) (2012). *Sociology of Organizations: Structures and Relationships*. Thousand Oaks, CA: Sage.

Graham, Laurel. (2000). 'Lillian Gilbreth and the Mental Revolution at Macy's, 1925–1928', *Journal of Management History*, 6(7): 285–305. doi: 10.1108/13552520010359306

Graham, Otis L., and Wander, Meghan Robinson (1985). *Franklin D. Roosevelt: His Life and Times. An Encyclopedic View*. Boston: G. K. Hall.

Green, Anna (2007). *Cultural History*. London: Palgrave Macmillan.

Greer, Germaine (1970). *The Female Eunuch*. London: MacGibbon & Kee.

Hearn, Jeff, and Parkin, P. Wendy (1983). 'Gender and Organizations: A Selective Review and a Critique of a Neglected Area', *Organization Studies*, 4(3): 219–42.

Hearn, Jeff, and Parkin, Wendy (2001). *Gender, Sexuality and Violence in Organizations: The Unspoken Forces of Organization Violations*. London and Thousand Oaks, CA: Sage.

Hearn, Jeff, Sheppard, Deborah, Tancred-Sheriff, Peta, and Burrell, Gibson (eds) (1989). *The Sexuality of Organization*. London: Sage.

Heller, Lidia (2004). *Nuevas voces del liderazgo*. Buenos Aires: Nuevo Hacer.

Heller, Lidia (2011). 'Mujeres en la cumbre corporativa, el caso de la Argentina', *Revista del Centro de Estudios de Sociología del Trabajo*, 3: 68–96.

Hiltzik, Michael (2011). *The New Deal: A Modern History*. New York: Free Press.

Holvino, Evangelina (2010). 'Intersections: The Simultaneity of Race, Gender and Class in Organization Studies', *Gender, Work and Organization*, 17(3): 248–77. doi: 10.1111/j.1468–0432.2008.00400.x

hooks, bell (1994). *Outlaw Culture: Resisting Representations*. New York: Routledge.

Hopkins, Harry L. (1936). *Spending to Save: The Complete Story of Relief*. New York: W. W. Norton & Co.

Horowitz, Daniel (1998). *Betty Friedan and the Making of the Feminine Mystique*. Amherst, MA: University of Massachusetts Press.

Humphreys, Sally (1994). 'What is Women's History?', in J. Gardiner (ed.), *What is History Today?*, 87–9. London: Macmillan.

Ibarra, Herminia (1993). 'Personal Networks of Women and Minorities in Management: A Conceptual Framework', *Academy of Management Review*, 18(1): 56–87.

Ibarra, Herminia (1997). 'Paving an Alternative Route: Gender Differences in Managerial Networks', *Social Psychology Quarterly*, 60(1): 91–102.

Ickes, Harold L. (1934). *The New Democracy*. New York: W. W. Norton & Co.

Ickes, Harold L. (1953). *The Secret Diary of Harold L. Ickes: The First Thousand Days, 1933–1936*. New York: Simon & Schuster.

Ickes, Harold L. (1954). *The Secret Diary of Harold L. Ickes: The Inside Struggle 1936–1939*. New York: Simon & Schuster.

Iggers, Georg G. (1997). *Historiography in the Twentieth Century*. Middletown, CT: Wesleyan University Press.

Jenkins, Keith (1991). *Re-Thinking History*. London and New York: Routledge.

Jenkins, Keith (1994). *Re-Thinking History*. London: Routledge.

Kanter, Rosabeth Moss (1977). *Men and Women of the Corporation*. New York: Basic Books.

Kelley, Elizabeth S., Mills, Albert J., and Cooke, Bill (2006). 'Management as a Cold War Phenomenon?', *Human Relations*, 59(5): 603–10.

Khurana, Rakesh (2007). *From Higher Aims to Hired Hands: The Social Transformation of American Business Schools and the Unfulfilled Promise of Management as a Profession*. Princeton: Princeton University Press.

King, Martin Luther (1968). *Chaos or Community?* London: Hodder & Stoughton.

Klimke, Martin, Pekelder, Jacco, and Scharloth, Joachim (2011). *Between Prague Spring and French May: Opposition and Revolt in Europe, 1960–1980*. New York: Berghahn Books.

Kolb, Deborah M., Jensen, Lisa, and Shannon, Vonda L. (1996). 'She Said it All Before, or What did we Miss about Ms Follett in the Library?', *Organization*, 3: 153–60.

Koontz, Harold (ed.) (1962). *Toward a Unified Theory of Management*. New York: McGraw-Hill.

Kosmala, Katarzyna (2008). 'Women on Work, Women at Work: Visual Artists on Labour Exploitation', *British Journal of Management*, 19: S85–S98. doi: 10.1111/j.1467-8551.2008.00 574.x

Kuhn, T. S. (1962). *The Structure of Scientific Revolutions*. Chicago University of Chicago Press.

Lyotard, Jean-François (1979). *La Condition postmoderne. Rapport sur le savoir*. Paris: Éditions de Minuit.

Marcuse, Herbert (1969). *Reason and Revolution*. London: Routledge & Kegan Paul.

Meyerson, Debra E., and Fletcher, Joyce K. (2000). 'A Modest Manifesto for Shattering the Glass Ceiling', *Harvard Business Review*, 78(1): 127–36.

Mignolo, Walter D. (1991). *The Idea of Latin America*. Oxford: Blackwell.

Mills, Albert J. (2004). 'Feminist Organizational Analysis and the Business Textbook', in D. E. Hodgson and C. Carter (eds), *Management Knowledge and the New Employee*, 30–48. London: Ashgate.

Mills, Albert J., and Helms Hatfield, Jean C. (1998). 'From Imperialism to Globalization: Internationalization and the Management Text', in S. R. Clegg, E. Ibarra, and L. Bueno (eds), *Global Management: Universal Theories and Local Realities*, 37–67. Thousand Oaks, CA: Sage.

Mills, Albert J., and Murgatroyd, Stephen (1990). *Organizational Rules: A Framework for Understanding Organizational Action*. Milton Keynes: Open University Press.

Mills, Albert J., and Tancred, Peta (eds) (1992). *Gendering Organizational Analysis*. Newbury Park, CA: Sage.

Mirchandani, Kiran (2004). 'Webs of Resistance in Transnational Call Centres: Strategic Agents, Service Providers and Customers', in R. Thomas, A. J. Mills, and J. Helms Mills (eds), *Identity Politics at Work: Resisting Gender, Gendering Resistance*, 179–95. London: Routledge.

Mohanty, C. T. (1984). 'Under Western Eyes: Feminist Scholarship and Colonial Discourses', *Boundary*, 2, 12/13(3/1), 333–358.

Morgan, Ted (2003). *Reds: McCarthyism in Twentieth-Century America*. New York: Ramdom House.

Munslow, Alun (2010). *The Future of History*. London: Palgrave Macmillan.

Nkomo, Stella (1992). 'The Emperor has No Clothes: Rewriting "Race in Organizations"', *Academy of Management Review*, 17(3): 487–513.

Nyland, Chris, and Heenan, Tom (2005). 'Mary van Kleeck, Taylorism and the Control of Management Knowledge', *Management Decision*, 43(10): 1358–74.

Nyland, Chris, and Rix, Mark (2000). 'Mary van Kleeck, Lillian Gilbreth and the Women's Bureau Study of Gendered Labor Law', *Journal of Management History*, 6(7): 306–322.

O'Connor, Ellen S. (2000). 'Integrating Follett: History, Philosophy and Management', *Journal of Management History*, 6(4): 167–90.

OECD (2008). *Gender and Sustainable Development: Maximising the Economic, Social and Environmental Role of Women*. Paris: OECD.

Perkins, Frances (1947). *The Roosevelt I Knew*. New York: Viking Press.

Phipps, Simone T. A. (2011). 'Mary, Mary, Quite Contrary', *Journal of Management History*, 17(3): 270–81.

Powell, Gary N., and Butterfield, D. Anthony (1994). 'Investigating the "Glass Ceiling" Phenomenon: An Empirical Study of Actual Promotions to Top Management', *Academy of Management Journal*, 37(1): 68–86.

Prasad, Anshuman (1997). 'The Colonizing Consciousness and Representations of the Other: A Postcolonial Critique of the Discourse of Oil', in P. Prasad, A. J. Mills, M. Elmes, and A. Prasad (eds), *Managing the Organizational Melting Pot: Dilemmas of Workplace Diversity*, 285–311. Thousand Oaks, CA: Sage.

Prasad, Anshuman (ed.) (2003). *Postcolonial Theory and Organizational Analysis: A Critical Engagement*. London: Palgrave.

Prasad, Pushkala (1997). 'The Protestant Ethic and the Myths of the Frontier: Cultural Imprints, Organizational Structuring, and Workplace Diversity', in P. Prasad, A. J. Mills, M. Elmes, and A. Prasad (eds), *Managing the Organizational Melting Pot*, 129–47. Thousand Oaks, CA: Sage.

Prasad, Pushkala (2005). *Crafting Qualitative Research: Working in the Postpositivist Traditions*. Armonk, NY: M. E. Sharpe.

Pringle, Judith K. (2008). 'Gender in Management: Theorizing Gender as Heterogender', *British Journal of Management*, 19: S110–S119. doi: 10.1111/j.1467-8551.2008.00576.x

Ragins, Belle Rose, Townsend, Bickley, and Mattis, Mary (1998). 'Gender Gap in the Executive Suite: CEOs and Female Executives Report on Breaking the Glass Ceiling', *Academy of Management Executive (1993–2005)*, 12(1): 28–42.

Rodriguez Gusta, Ana Laura (2009). 'Negocios que promueven la igualdad. Cómo poner en práctica programas de certificación de sistemas de gestión de calidad con equidad de género', in S. Izquierdo (ed.), *Compartir Conocimiento*, vi. PLACE: Programa de las Naciones Unidas para el Desarrollo Centro Regional para América Latina y el Caribe. Retrieved from http://www.undp.org.uy/getFile.asp?File=Negociosquepromuevenlaigualdad.pdf&Alias=06-15-2010-08-30-55a.m.-754.pdf

Rose, Michael (1978). *Industrial Behaviour*. Harmondsworth: Penguin.

Rowbotham, Sheila (1974). *Hidden from History: Rediscovering Women in History from the Seventeenth Century to the Present*. New York: Pantheon.

Runté, Mary, and Mills, Albert J. (2006). 'Cold War, Chilly Climate: Exploring the Roots of Gendered Discourse in Organization and Management Theory', *Human Relations*, 59(5): 695–720.

Ryan, Lori Verstegen, and Rutherford, Matthew A. (2000). 'Mary Parker Follett: Individualist or Collectivist? Or Both?', *Journal of Management History*, 6(5): 207–23. doi: 10.1108/13552520010348362

Said, Edward W. (1979). *Orientalism*. New York: Vintage.

Said, Edward W. (1993). *Culture and Imperialism*. New York: Vintage.

Scaff, Lawrence A. (2011). *Max Weber in America*. Princeton: Princeton University Press.

Schein, V. E. (2007). 'Women in Management: Reflections and Projections', *Women in Management Review*, 22(1): 6–18.

Schein, V. E., and Mueller, Ruediger (1992). 'Sex Role Stereotyping and Requisite Management Characteristics: A Cross Cultural Look', *Journal of Organizational Behavior*, 13(5): 439–447.

Schein, V. E., Mueller, Ruediger, Lituchy, Terri, and Liu, Jiang (1996). 'Think Manager—Think Male: A Global Phenomenon?', *Journal of Organizational Behavior*, 17(1): 33–41.

Schilling, Melissa A. (2000). 'Decades Ahead of her Time: Advancing Stakeholder Theory through the Ideas of Mary Parker Follett', *Journal of Management History*, 6(5): 224–42.

Schlesinger, Arthur M. (ed.) (1959). *The Age of Roosevelt: The Coming of the New Deal*. Boston: Houghton & Mifflin Co.

Schrecker, Ellen (1986). *No Ivory Tower: McCarthyism and the Universities*. New York: Oxford University Press.

Schrecker, Ellen (1998). *Many are the Crimes: McCarthyism in America*. Boston: Little, Brown & Co.

Seale, Bobby. (1970). *Seize the Time: The Story of the Black Panther Party and Huey P. Newton* (1st edn). New York: Random House.

Smircich, Linda (1985). 'Toward a Woman-Centered Organization Theory', paper presented at the Annual Meetings of the Academy of Management, San Diego, CA.

Stewart, R. (1996). 'Why the Neglect?', *Organization*, 3(1): 175–9. doi: 10.1177/135050849631012

Stivers, C. (1996). 'Mary Parker Follett and the Question of Gender', *Organization*, 3(1): 161–6. doi: 10.1177/135050849631010

Swatos, William H., Kivisto, Peter, and Gustafson, Paul M. (1998). 'Weber, Max', in W. H. Swatos (ed.), *Encyclopedia of Religion and Society*. Walnut Creek, CA: AltaMira Press. http://hirr.hartsem.edu/ency/Weber.htm).

Tancred-Sheriff, P., and Campbell, E. Jane (1992). 'Room for Women: A Case Study in the Sociology of Organizations', in A. J. Mills and P. Tancred (eds), *Gendering Organizational Analysis*, 31–45. Newbury Park, CA: Sage.

Taylor, Nick (2008). *American-Made: The Enduring Legacy of the WPA*. New York: Bantam Dell.

Toffler, A. (1981). *The Third Wave*. Glasgow: Pan.

Urwick, Lyman (1938). The Development of Scientific Management in Great Britain. A report distributed to members of the 7th International Management Congress, 1938, London. *British Management Review*, 3(4). Reprinted as a separate booklet.

Urwick, Lyman (1963). *The Golden Book of Management*. London: Newman Neame.

Urwick, Lyman, and Brech, E. F. L. (1957a). *The Making of Scientific Management: The Hawthorne Investigations*, iii. London: Sir Isaac Pitman & Sons.

Urwick, Lyman, and Brech, E. F. L. (1957b). *The Making of Scientific Management: Thirteen Pioneers*, i. London: Sir Isaac Pitman & Sons.

White, Hayden (1973). *Metahistory: The Historical Imagination in Nineteenth-Century Europe*. Balitimore, MD: Johns Hopkins University Press.

White, Hayden (1985). *Tropics of Discourse: Essays in Cultural Criticism*. Baltimore, MD: Johns Hopkins University Press.

Wilson, Fiona M. (1996). 'Research Note. Organization Theory: Blind and Deaf to Gender?', *Organization Studies*, 17(5): 825–42.

Wren, Daniel A. (1979). *The Evolution of Management Thought*. New York: Ronald Press.

Wren, Daniel A. (1994). *The Evolution of Management Thought* (4th rev. edn). New York: Wiley.

Wren, Daniel A. (2005). *The History of Management Thought*. Hoboken, NJ: Wiley.

Wren, Daniel A., and Bedeian, A. G. (2009). *The Evolution of Management Thought* (6th edn). Hoboken, NJ: John Wiley and Sons.

Wright, Erik Olin, and Baxter, Janeen (2000). 'The Glass Ceiling Hypothesis: A Reply to Critics', *Gender and Society*, 14(6): 814–21.

X, Malcolm, and Haley, Alex (1965). *The Autobiography of Malcolm X*. New York: Grove Press.

Zinn, Howard (1999). *A People's History of the United States: 1492–Present*. New York: HarperCollins.

CHAPTER 3

··

ORGANIZATIONS AS SYMBOLIC GENDERED ORDERS

··

SILVIA GHERARDI

GENDER AS THE SYMBOL OF DIFFERENCE

··

GENDER is one of the most powerful of symbols; indeed, the very word 'gender' encapsulates all the symbols that a culture elaborates to account for biological difference (Gherardi 1994, 1995). Therefore a symbolist approach is particularly able to grasp the ambiguity of gender relations, since the function of a symbol is to express a polysemy, to contain and to convey ambiguity. Symbols signify what something is and what it is not.

For example, in Chinese culture the Yang-Yin symbol represents the dualistic distribution of forces between the active, masculine principle ('yang') and the passive, feminine one ('yin'). This distribution is symbolized by a circle divided by a sigmoid line indicating the dynamic interpenetration of the two principles. The light half of the figure is the 'yang' force and the dark one the 'yin' force, but each half contains a small circle of the opposite shade which symbolizes that each principle contains the germ of the other. This symbol is a cross-section of a helicoid structure which links opposites and generates constant movement; a metamorphosis through contrary positions and situations. The vertical axis at its centre constitutes the 'mystical centre' where there is neither turbulence, nor impulse, nor suffering. The three levels of signification are therefore present in this symbol, just as they are in other Hindu or Hebrew symbols which elaborate sexual difference as separation and inseparability.

The symbolic realm is a dimension of reality and a dimension of signification, but it does not replace other levels of reality, nor does it deny them. Indeed, as Mircea Eliade writes:

> There is no need to believe that symbolic implication annuls the concrete and specific value of an object or of an operation. Symbolism adds a new value to an object or an action without affecting its own, immediate or historical, values.

(Eliade 1952: 57)

Taking an example from Cirlot (1971: 7), and positing an analogy between an organization and the facade of a monastery, we can see in the latter: (*a*) the beauty of the whole; (*b*) the architectural technique used in its creation; (*c*) its architectural style and its geographical and historical implications; (*d*) its implicit or explicit cultural and religious values; but also (*e*) the symbolic significance of its shapes. The understanding of what is symbolized by an ogival arch beneath a rose window is a form of knowledge very different from the others listed above.

Likewise, we may possess aesthetic, technological, historical, cultural, and symbolic knowledge of an organization: paradigm plurality is implicitly assumed by the symbolic approach. Likewise, we may explore gender in relation to power, dominion, inequality, or difference.

Unlike science based on the dichotomous code of true/false, mythical expression is grounded on that of sacred/profane, which constitutes one of the ordering principles of social reality (Durkheim 1912). This, according to Cassirer (1923), is the 'fundamental antithesis' which patterns reality according to a qualitative characterization. Cassirer (1923) described humans as symbolic animals in that symbols mediate between objective reality, the world or perception, and the knowing subject. Science is a specific form of expression of culture which employs signs to refer to things and universes of objects, and which performs a logical denotative function. Yet cultures are symbolic textures in which other expressive forms are at work, myth especially. Myth is a symbolic form in which the signifying function entails a relationship of reciprocal reference between signifier and signified. The symbolic medium

> thus assumes a density, a concreteness which the symbols of other expressive forms lack. Hence its structuring function of social experience: the values which regulate our behaviour, which indicate communalities and divergences in our systems of belonging, have this mythical cultural configuration.
>
> (Bolognini 1986: 88)

Symbolist theory is based on the following principles (Cirlot 1971: 32):

(*a*) There is nothing that does not matter. Everything expresses something and everything is meaningful.

(*b*) No form of reality is independent: everything stands in relation to something else.

(*c*) The quantitative is transformed into the qualitative at certain essential points which constitute the signification of quantity.

(*d*) Everything is serial. Seriality includes both the physical world (the spectrum of colours, sounds, shapes, landscapes) and the spiritual world (virtues, vices, sentiments).

(*e*) There are correlations of situation among different series, and of meaning among series and their constitutive elements.

If we take any particular symbol—the sword, for example—and analyse it, we first find the object in itself stripped of every relation, and then the object in its instrumental function. Finally we come to its symbolic function, which is the dynamic tendency of its quality equivalent to those located at the corresponding points in every analogous series. The symbolic function is to denote a general meaning, one which is often ambivalent and allusive. The multiplicity of which is never chaotic, however, because it moves to a shared rhythm. Thus the sword, iron, fire, the colour red, the god Mars, the Rocky Mountains, are interconnected and meet in a symbolic direction of equal significance: the desire for psychic determination and for physical destruction (Cirlot 1971: 33). These symbols unite with each other, they call upon each other because of the inner affinity that unites them, the shared rhythm that enables connections to be established among the diverse levels of reality. For example, the ovoid shape, undulating rhythm, high sounds conjure up a female rhythm which projects itself onto nature by taking things and shapes and translating them into personages in the drama.

Returning to organizations, let us take one of these rhythms and examine what it is that ties gender as a symbol to organizational cultures.

GENDER AND ORGANIZATIONAL CULTURES

In this section I shall set out a conception of the cultural approach to organizations in the tradition of symbolist thought, with principal reference to the cultural and philosophical legacy of symbolic interactionism (Mead 1934; Goffman 1977; West and Zimmerman 1987; Denzin 1992) at first and the linguistic turn after (Martin 1990, 2002; Tyler 2011).

It is extremely difficult to define the cultural approach, for it has become a field in which it is easier to draw distinctions than to unify. Corporate culture, organizational cultures or subcultures, cultural organization, postmodern approach to organizational culture: these are some of the labels adroitly deployed by Linstead and Grafton-Small (1992). In the postmodern perspective organizational cultures are seen as a textual space and the application of intertextuality to organizational analysis means treating culture, identity, symbols, and actions as interwoven texts that create one another via mutual referencing.

For the moment, I am interested in the features shared by the many approaches to cultural production—and organizations are a cultural product—and which differentiate them from others which reify culture and search for its properties. Appropriate here is a definition as broad in its scope as the title of an article by Czarniawska-Joerges (1991): culture is the medium of life. Drawing on Latour's (1986) distinction between an ostensive and a performative definition of society, Czarniawska-Joerges draws a parallel distinction between an ostensive definition of culture which assumes that, in principle, it is possible to discover properties that are typical of a given culture and which can explain its evolution, although in practice they might be difficult to detect, and a performative definition which assumes that, in principle, it is impossible to describe properties characterizing any given culture, but in practice it is possible to do so. Under an ostensive conception of culture, actors are useful informants and social researchers, using appropriate methodology (what Denzin (1992) calls 'ethnomethodological voyeurism'), uncover opinions, beliefs, myths, and rites and arrange them into a picture. Under a performative conception, there are no actors who know any more or any less, and researchers ask the same questions as any other actor, although they might use a different rhetoric in formulating their answers. Thus 'ostensive definitions are attempts to explain principles, whereas performative definitions explore practices' (Czarniawska-Joerges 1991: 287).

I therefore use the term 'cultural approach' to refer to a performative definition of organizational culture as the system of meanings produced and reproduced when people interact. An organizational culture is therefore the end-product of a process which involves producers, consumers, and researchers. Thus the construction of meaning is purposive, reflexive, and indexical.

Within the broader cultural approach, organizational symbolism is an area of research more sketched than thoroughly explored (Gagliardi 1990; Turner 1990; Frost *et al.* 1991; Alvesson and Berg 1992). It is a set of intuitions more than a methodology, and as such is graphically depicted by a dragon tearing up an organization chart, the symbol of organizational rationality. The dragon is a root metaphor for the symbolic approach to organizational cultures. The dragon is a potent symbol, one common in both Western and Eastern cultures and which represents the beast *par excellence*, the adversary, the devil. Combat with the dragon is the supreme test. Yet, on the other hand, the tamed dragon with five legs is the Chinese emblem of imperial power, of wisdom, and of rhythmic life.

Ambiguity and duality are the distinctive features of every symbol, since the symbolic function resides simultaneously in the force of coagulation (i.e. in the synthesis, by images and correspondences among symbols, of a multiplicity of meanings into one) and in the force of dissolution (i.e. in a return to chaos, to the mixing of meanings, to dissolution).

Dragons were conventionally portrayed with the bust and legs of an eagle, the body of an enormous serpent, the wings of a bat, and a coiled tail with an arrow-shaped tip. These images represented the fusion and confusion of all the elements and all the faculties: the eagle stood for celestial power, the serpent for occult and subterranean power, the wings for the flight of the intellect, and the tail for submission to reason.

Since 1984 the SCOS dragon is the logo of the Standing Conference on Organizational Symbolism and it 'was meant to symbolize the ambiguity of corporate or organizational cultures. On the one hand there was the terrifying, collective "beast" lurking beneath the smooth corporate surface; on the other hand, the dragon was to symbolize the ancient and inherited wisdom built into social structure and artifacts' (Alvesson and Berg 1992: 3). SCOS folklore has developed a real and proper 'draconological discourse' (Sievers 1990). And from this organizational symbolism we may deduce that the dragon is present to the consciousness of those who study organizations using a cultural approach as the intellectual unease provoked by the fact that, although rational explanation and refined theory have their logical and empirical foundations, there still remains the unexplored continent of shadowland, where the most interesting phenomena of organizational life occur, and to which the concepts and languages of normal science do not apply. Science and scientific discourse are based on distinction, on separation, on analyticity, and on logico-temporal sequence; their subject matter, by contrast, is untamed, its causations are multiple and reciprocal, its boundaries are uncertain and constantly shifting, and the very action of studying such matters transforms them before our eyes. On the contrary, the organization-as-dragon may provide a metaphor for what is hidden, suppressed, slumbering beneath the surface, the irrational, the feminine, the devouring mother (Höpfl 2003).

The symbology of the organizational dragon as the beast of dread condenses everything that is unconscious, everything that lies in the deeps, within the bowels of the structure, everything that may rise up to assault the Conscious Ego, the seat of rationality. Organizational scholars have always been aware of the dark side of organizational life, as expressed in the dichotomies of formal/informal, on the stage/behind the scenes, upper world/underworld; or in the spatial symbolism where above = managerial world = planning rational, below = workers' world = resistance = irrationality; or in the cognitive patterns where top–down = rationality moving downwards towards its implementation, bottom–up = institutionalization of social practices. In its battle to repel chaos and the irrational, management reincarnates St Michael or St George, although it is less aware of the gender symbolism implicit in the dragon.

In its positive symbology, the dragon blends the Ego with the richness and the creativity of the unconscious to produce a richer 'subjectivity'. The dragon ('culture' for Smircich 1983: 347–8):

> promotes a view of organizations as expressive forms, manifestations of human consciousness. Organizations are understood and analyzed not mainly in economic or material terms, but in terms of their expressive, ideational, and symbolic aspects. Characterized very broadly the research agenda stemming from this perspective is to explore the phenomenon of organization as subjective experience.

This is the romantic dragon (Ebers 1985) that we have inherited from the cultural tradition of the nineteenth century; the healer of profound conflicts because it shows 'the organization's expressive and affective dimensions in a system of shared and meaningful symbols' (Allaire and Firsirotu 1984: 213) and because it has transcendental functions for a humankind 'emotional, symbol-loving and needing to belong to a superior entity or collectivity' (Ray 1986: 295).

Culture conveys into organizational analysis subjectivity, emotionality, ambiguity, and sexuality, all themes associated with the symbolism of the female in its fundamental psychological ambivalence: the good mother and the devouring mother.

We now know a great deal about the organization/dragon, but one intriguing question is still unanswered: what sex is the animal? Very little is known about the sex of the dragon; draconology is somewhat reticent on the matter. With so little known directly about the sexual and reproductive life of dragons, we may indirectly deduce their gender by considering the relationships that humans have established with these strange beasts.

By far the best known relationship with the dragon is heroic combat and the dragon's slaughter (Dégot 1985), with the victor then absorbing its strength or, through a drop of its blood, achieving supreme knowledge. A broad array of Christian male saints, apart from St George and St Michael, have fought with dragons; but only two female ones: St Martha, who vanquished the dragon with holy water, and St Margaret, whose burning cross slew the monster. Male saints instead confront the dragon with a variety of weapons and in open combat. Combat is generally a type of social relation which arises among men, and it is valued more highly, the more it takes place between equal adversaries and according to the chivalric code. A man and a beast cannot share the same code of honour (cultural product) in combat, and there is nothing to prevent the beast from being female but ferocious and wicked. Yet combat is an activity which is assumed to be male and generally conceptualized within a male symbolic universe. Even the magic solution presupposes that it is a male dragon which is tamed—either by the Russian sorceress Marina or by the French ghost Lady Succube. The dragons of science fiction, too, are tamed, albeit by other means.

There are also good grounds for arguing that the dragon symbolizes the female gender: 'since the Middle Ages the dragon became a container for the often conscious anxieties related to sexuality..., a symbol of the pleasure of the flesh and lasciviousness which then had to be projected by men into women' (Sievers 1990: 217), In the Jungian psychoanalytic tradition, the dragon is the archetype of the 'great mother', of the most inaccessible level of the collective unconscious. The image of the Madonna with the dragon subdued beneath her feet is a symbol of the wholeness of the female self, and so too in the Christian tradition is the image of Mary crushing the head of the serpent (synonymous with the dragon).

The dragon ripping up the organization chart in the SCOS logo belongs to the subterranean world of shadows, of the intuitive, of the female, and of what has been erased. Corporate identity belongs to the domain of the conscious, of the public, and of the rational, whereas the dragon is the Jungian shadow, the unaccepted split-off part of it, irrational and emotional reality. The organization as dragon metaphor leads to several meanings. A possible interpretation is that the female hides behind the organization chart, but the female is both seductive and terrifying. An alternative interpretation is that the cosmic dragon represents chaos; it has no sex, it is Uroborus, the eternal flux, indeterminacy, and symbolizes process, becoming, the passage from organization to organizing.

Bearing in mind the three ways to handle the dragon (Sievers 1990)—slay it, tame it, or ingratiate oneself with it—let us look very briefly at their treatment in the literature on gender and the organization.

First of all, it is extremely difficult to take seriously the contention that 'gender and organization' is a neglected topic, given that so many articles have been written to make precisely this point. It may be that this view is only the romantic expression of nostalgia or, even worse, the grumbling of those who have been excluded. Broadly speaking, the literature adopts one of two equally good strategies to cope with the problem of gender: the functionalist strategy of treating gender as just one variable amongst others (Harding 1987), and therefore to be considered only when the need arises (Hearn and Parkin 1983), and the emancipationist strategy which emphasizes the fundamental 'sameness' of men and women and which stresses gender as a social structure that accounts for sex differences in work and positions within organizations (Kanter 1977; Risman 2004).

Equal opportunities and equal rights are consequently the preconditions for women to become as good as men. The literature contains a broad strand of prescriptive recipes on how to tame the dragon. I refer to the 'fit-in' school of thought, which instructs women on how to enter organizations and management. Evidently it is taken for granted that women and organizations do not 'fit' together naturally, especially at managerial levels, and that women must therefore be socialized to roles, jobs, and organizations that are by definition neuter. Another way of taming the dragon is to exploit, to the organization's advantage, the sexual division of labour in a society which differentially socializes men and women to diverse roles in family life, in order to obtain cheap labour from women (Calás and Smircich 1993, 2011) and a stable male labour force to be assigned the best jobs.

There is, finally, the strategy of ingratiating oneself with the dragon by recognizing the increasing feminization of all work, especially white-collar occupations. This strategy acknowledges the strategic importance of service, understood both as the tertiary sector and as the factor 'service' within the industrial sector, and therefore positively evaluates the different skills deployed by women because they have been socialized differently and because their skills are valuable to organizations. Following Chodorow (1978) and Gilligan (1982), the difference between the sexes which appoints women as care-givers and assigns to men a greater 'denial of relation' is an incentive to organizations to appropriate what is good (for them) in women and to preserve it.

I have employed the symbology of the dragon to convey multiple messages, but mainly to provide the reader with a first insight, more empathic than analytical, into organizations as symbolic gendered orders. I have sought to give an idea of the plurality and fragmentation of the subject, to show that various textual strategies can be used to address gender, depending on how the relationship between gender and organization is conceived. I have moved on various levels because symbolic understanding allows exploration of the area that lies between being and non-being. In the next section I shall give analytical treatment of what is meant by symbolic approach.

THE TOOL KIT OF THE SYMBOLIC RESEARCHER

It is not always easy to understand symbolic representations, nor to define the way in which they structure social experience. The difficulty stems from the nature itself of the symbol, which is so much the *significans* as to be indeterminate and constantly to defer its *significandum*, and which requires an indirect language, one which establishes relations and conserves transformative power. The invention of the symbol is a creative act which rests upon the ability to see a thing as what it is not (Castoriadis 1987: 137). Symbolic understanding is therefore generated on the borders of ambiguity, where being and non-being merge, where the indeterminate is about to transform itself into the determinate, and where possibilities are *in nuce*.

Is it possible to speak unambiguously about ambiguity, non-symbolically about symbols? Is it possible to identify a symbolist methodology and to specify symbolist research techniques? A radical answer would point to hermetic language and aesthetic experience as the only possibilities (indeed, for that matter, symbolism was born of the arts). However, linguistic mediation is possible if one wishes to remain in the field of organizational studies, even more so now that new metaphors taken from literary criticism are being used to rethink and rewrite the organization as a text.

Let us look more closely at the distinction between these two forms of analysis in organizational studies; a distinction drawn with extreme clarity by Alvesson and Berg (1992: 118–26) and which can be summarized as follows:

(*a*) Symbolists do not start from the assumption that there is a culture in a company; they instead study organizations from a symbolic perspective;

(*b*) Symbolists also differ from many culture researchers in that they emphasize the aesthetic, ethical and emotional dimensions of human life, rather than simply examining its cognitive and axiological dimensions;

(*c*) Symbolists dissolve the difference between the subjective and the objective; they link symbols, images, metaphors, etc. together: the mythic mode of symbolling;

(*d*) Symbolists assume that there are clear and impassable parameters within which reality can be moulded. Despite considerable lack of agreement as to what these parameters actually are, they have been communicated across generations and across cultural frontiers, and they express relationships with other human beings, the purpose of existence, of nature, etc.

In other words the symbolist has the following distinctive features:

- S/he is a qualitative researcher who prefers to see things through the eyes of the subject. S/he is interested in meanings, in the process of their attribution, in how they are sustained, in the way that some meanings prevail while others disappear.
- S/he is a participative researcher, who knows that s/he is part of the production of meaning and of the narration of stories, as both the narrating and the narrated subject.
- S/he is the product of contextual understanding of actions and symbols, not only because they are inseparable but because all symbols are value-laden and meaningful only in terms of their relationship to other symbols.
- S/he is a wanderer among the realms of knowledge seeking to reconstruct the links among the various levels of reality created by a symbol through individual symbolic production, the collective unconscious and artistic production: the

immanent with the transcendent, the mental with the physical, with action, with transformation.

The symbolist is interested in three principal areas of enquiry (Turner 1992):

- matters of style, the aesthetic qualities of a way of life, the generation and maintenance of identities, the symbolic construction of community;
- archetypes, allegories, myths, as the ways in which the collective unconscious—or the transpersonal production of culture—enter the phenomenal world of the organization and its processes;
- the imaginary capacity, as the mental and social ability to create symbols, evoke images, create a social imaginary as the context in which meanings, values, and the prefigurations of will and action take shape. This is the locus of prospective symbols, those which create and develop a vision.

In order to see how symbolic orders of gender are created and recreated within organizational cultures we need to focus on the centrality of language.

THE SYMBOLIC APPROACH AND THE LINGUISTIC TURN: FROM DIFFERENCE TO DIFFÉRE(A)NCE

The use of language by definition involves separation and differentiation, but also power. Male and female stand in a dichotomous and hierarchical relation: the first term is defined in positive as the One, the second is defined by difference, by default, as the non-One, that is, the Other. This was the lesson taught by Simone de Beauvoir (1949), from whom we have inherited the concept of second sex; an extremely useful analytical category both to describe female experiences of subordination and, by extension as in Ferguson (1984), to describe the clients of the bureaucracy, who are second-sexed whether they are men or women. When, immediately after the Second World War, de Beauvoir described the woman as the Other, the problem of language was not yet paramount, although the ontological problem was. Subsequently both issues were to be radically problematized by feminism and by other currents of thought which came under the label of 'postmodernism' and 'poststructuralism'.

It is not my intention to review the long debate on language conducted by feminism, and how it was revitalized by the impact of French poststructuralism. What is certain is that the feminist critique of language, and through this of the concept of person, of selfhood, and of subjectivity, rapidly and importantly developed a wide range of themes over a period of around twenty years.

Language was denounced for a masculine bias which underpinned a form of 'power over' others and expressed the experiences of the oppressors and their construction of reality. In parallel with this assault on language as a form of domination, the question arose as to whether a language of liberation could be created. French feminism was closely involved in this linguistic enterprise, for example, in the wide-ranging work of Irigaray (1974, 1977), who examined the concept of woman as Other from a psychoanalytic perspective and proposed the body as a writing instrument. Research at the end of the 1970s sought to create space for a *relecture interpretante* (interpretative rereading) of theory on the female subject, in which to 'speak female' (*parler femme*) and speak to each other without the interference of men (Irigaray 1977; Kristeva 1988; Höpfl 2011).

Another point of view, which argued against the project of a feminist discourse, described language users as hiding behind a variety of irremovable masks. Some of these masks are (Elshtain 1982): the mask of purity (presuming the victim's language to be untainted by her world), the mask of orthodox Marxism (presuming that discourse is nothing but rationalization for exploitative relations), the mask of militancy (the language of grim personal renunciation), the mask of systematic know-it-allism, or of unquestioned inner authenticity based on claims to the ontological superiority of the female being-in-itself. This latter point is particularly important because it leads directly to the trap of language as the grammar constitutive of human experience.

If male and female stand in an oppositional and hierarchical relationship, and if they constitute the One and the Other, can the Other be defined in positive without running the risk of reversing the relation while failing to resolve the contradiction? The question is anything but grammatical, for as soon as the political project predicated on difference—as opposed to the project in pursuit of equality—seeks to define difference, to valorize the female, to empower women with assertiveness and pride in female principles, it is in danger of reproposing the self-same relation by defining man as non-woman. It may be politically useful to reverse the relation in order to redress the balance of power, but theoretically it does not provide a way out of the gender trap. One way to reframe the question is to accept the position aptly expressed by Kristeva (1981) as 'women can never be defined'.

In the 1980s, French feminism continued its critique of language in semiological terms, adopting a position coherent with the school of thought that launched the attack on structuralism—Lacan, Derrida, Foucault—and called itself poststructuralist. Its essential thesis was that all social practices, including the meaning of subject and subjectivity, are not simply mediated by language but are constituted in and through language. Hence it follows that one must examine the tradition by which language has been understood, and to deconstruct that tradition in order to understand how persons are constituted in social and linguistic practices. The self as the centre of consciousness, the person as a distinctive whole and as a bounded and integrated unity, are linguistic inventions, artifices with which to give spatial and temporal location to the Self which speaks and has one body.

Social psychology joined this current of critical thought by dissolving the boundaries between individual and society and analysing the role of language in sustaining self-construction and the social construction of personhood (Harré 1984; Shotter and Gergen 1989; Butler 1990). Concepts such as the multiple self (Elster 1986), the saturated and populated self (Gergen 1991), the masquerade or pastiche personality, completed the decentralization of subjectivity, and the postmodern revolution was therefore accomplished. The autonomous self of the romantic and modernist tradition, the centre of consciousness, the agent *par excellence*, was relativized and dismissed as conviction, a way of talking and a product of conversation. Language was a form of relatedness, sense derived only from coordinated effort among people, and meaning was born of interdependence. In Baudrillard's (1981) words, 'we are terminals of multiple networks'. Our potential is realized because there are others who sustain it, who possess an identity deriving from the social processes in which we participate, and who are the type of person that the linguistic games we play enable them to be.

Several currents of thought were involved in the project to deconstruct the self and to create a relational self (Sampson 1989). Alongside feminism, social constructionism contributed analysis of the individual as a social and historical construction; systems theory with the ontological primacy granted to relations rather than to individual entities; critical theory—the Frankfurt School—with its unmasking of the ideology of advanced capitalism; and deconstructionism as a perspective internal to post-structuralism. Although these approaches belonged to very different disciplinary traditions, they converge on a conception of subjectivity in which 'the subjects are constituted in and through a symbolic system that fixes the subject in place while remaining beyond the subject's full mastery. In other words, persons are not at the centre . . . but have been decentred by these relations to the symbolic order' (Sampson 1989: 14).

The symbolic order of gender that separates the symbolic universes of the female and the male sanctions a difference whereby what is affirmed by the One is denied by the Other. The One and the Other draw meaning from this binary opposition, which forms a contrast created ad hoc which maintains a hierarchical interdependence (Derrida 1967, 1971). The interdependence-based symbolic order is a relational order which rests upon difference and the impossibility of its definition. Male and female are undecidable, their meaning is indeterminate and constantly deferred.

The origins of the widely used concept of 'difference' (Derrida 1971) warrant examination. By 'difference' is meant a form of self-reference 'in which terms contain their own opposites and thus refuse any singular grasp of their meanings' (Cooper and Burrell 1988). In order to stress the processual nature of difference, Derrida invented the term *différance*, which in French is pronounced the same as *différence* and incorporates the two meanings of the verb *différer*: defer in time, and differ in space. Male and female are not only different from each other (static difference) but they constantly defer each other (processual difference), in the sense that the latter, the momentarily deferred term, is waiting to return because, at a profound level, it is united with the former. The difference separates, but it also unites because it represents the unity of the process of division. There are therefore two ways of conceiving gender difference: as two separate terms—male and female—and as a process of reciprocal deferral where the presence of one term depends on the absence of the other. Derrida calls these two modes of thought 'logic of identity' and 'logic of the supplement' respectively.

Because of their multi-individual dimension and supra-individual duration, male and female as symbolic systems possess a static aspect, which creates a social perception of immutability, of social structure and institution. But male and female is also a social relation dynamic whereby meaning is processually enucleated within society and individual and collective phenomena. The symbolic order of gender is static difference and processual difference. Put better, it is the product of their interdependence: the impossibility of fixing meaning once and for all sanctions the transitoriness of every interpretation and exposes the political nature of every discourse on gender.

Feminism is a discourse on gender, but internally to it there exist many different voices (Calás and Smircich 1996, 2006). Whence derive various proposals for the maintenance of ambiguity and instability in the analytical categories of feminism (Ferguson 1991). Ferguson compares the hermeneutic account, to which she attributes an ontology of discovery in order to interpret the patriarchal domination of women, with the deconstructionist account based on genealogy, i.e. a posture of subversion towards fixed meaning claims. Interpretation, as a project which articulates the voice of women, and genealogy must be taken with a pinch of pragmatic irony, because 'affirmations are always tied to ambiguity and resolutions to endless deferral' (Ferguson 1991: 339). Hence it follows that differences must be articulated and contextualized.

Positioning gender is an approach which does not seek to posit a subjectivity of women or men in oppositional terms. It is instead an approach which reflects the essential indeterminacy of the symbolic order of gender, governed as it is by the endless process of the difference and deferral of the meaning of male and female. Positioning gender introduces a concept of subjectivity in which the subject is open-ended and indeterminate except when it is fixed in place by the culturally constituted symbolic order of gender. For Davies and Harré (1990), the concept of positioning belongs to social psychology, and their use of the term 'positioning' contrasts with the concept of human agency as role player. It is therefore useful for analysis of the production of self as a linguistic practice within the dynamic occasions of encounters. A subject position incorporates both a conceptual repertoire and a location for persons within the structure of the rights pertaining to those who use the repertoire. A position is what is created in and through conversations as speakers and hearers construct themselves as persons: it creates a location in which social relations and actions are mediated by symbolic forms and modes of being. It is within a particular discourse that a subject (the position of a subject) is constructed as a compound of knowledge and power into a more or less coercive structure which ties it to an identity. Positionality presupposes a discursive order where gender relations are the outcome of discourse practices; that is, they derive from the way in which people actively produce social and psychological realities.

In the next section I use this concept because I want to account for the dynamic that unites the production of gender relations at the level of interaction with the cultural structures that transcend concrete behaviour. This dynamic both produces and is produced by a symbolic order of gender. The concept of positionality recognizes the constitutive force of a symbolic order of gender which shapes discourse practices and also people's ability to exercise choice in relation to those practices.

POSITIONING THE SYMBOLIC ORDER OF GENDER WITHIN AN ORGANIZATIONAL CULTURE

Doing gender involves symbols, using them, playing with them, and transforming them: shuttling between a symbolic universe coherent with a static difference of gender identity and the symbolic realm of a processual gender difference. We do gender through ceremonial work and through remedial work. In the former kind of behaviour we stress the difference between the symbolic universes of gender; in the latter we defer the meanings of gender to situated interactions. It is in the way that we weave these two forms of behaviour together that resides the possibility of doing gender as a situated performance.

Men and women are engaged in the ceremonial work of giving proper representation to the attributes and behaviour of their own gender, of acknowledging and anticipating that others will do the same, and of legitimating this ceremonial in appropriate discourse strategies. It is impossible to avoid ceremonial work, because our first act of social categorization when encountering the other is to cast gender on the person. Consider the relational difficulty that arises when we must telephone or write to somebody whose gender we do not know because she or he has an uncommon first name. Or, even more embarrassingly, when we have engage in face-to-face interaction with an interlocutor who does not provide clear codes for the attribution of gender and does not emit appropriate gender display (Goffman 1976).

In the rituals demanded by etiquette, the sexes pay homage to the prerogatives of their gender (e.g. gentleness or strength), and in so doing they sanctify it and demonstrate competent gender behaviour. Recognition of the gender position of another comes about both through precedence-granting and other courtesy gestures, and through the verbal appreciation of the 'gifts' of the other gender. A well-known example is the way in which thanks are expressed for help given even though it was not compulsory: a male colleague who fixes the lock on a desk is thanked by having his skills as a handyman praised—'a real man about the house' (and office!)—while the female colleague who suggests an appropriate present for a child's birthday is thanked for her female intuition.

The courtesy system, in fact, provides an interesting example of a 'double-bind situation' (Watzlawick *et al.* 1967). In effect, there is a paradox in relations which is not to be found in analogous situations of social inequality: unlike other groups of disadvantaged adults, women are held in high consideration (Goffman 1977), and this consideration is manifest in social situations as due acknowledgement of their gender identity and therefore as an expression of their inequality. Failure to respect the courtesy system marks an adult male as socially incompetent, but it establishes a asymmetric relation in which the man is in the one-up position.

The courtesy system, in fact, is not as innocent as it may appear when considered solely as an occasion for the reciprocal recognition of gender identity or for the display of considerateness towards idealized femaleness. It is the prime arena for conveying relative rank: deciding who decides, who leads and who follows, who speaks and who listens, who has the power to position the other, and so forth. Most organizational structures already contain a definition of the relative ranks of those who engage in interaction, and they often provide constructs organized along gender lines where men occupy a 'naturally' higher position, one which is the 'natural' expression of their natural abilities. We may take this as an extreme form of a social situation legitimated by social beliefs about gender: the position of the woman is clearly a female position which requires considerateness and care from a man occupying the one-up position in every respect and who positions himself as the provider and the honourer. The gender performance that thus takes place reconfirms beliefs about the differing human natures of the two sexes. We may deduce that it is relatively simpler for women to assume a one-down position; to be those who receive orders, listen, execute, follow, and so on, since this position is inscribed in social situations as institutional reflexivity. In other words, the one-down situation does not contradict gender identity; indeed it may bring advantages or specific resources in interaction, and it does not necessary require all the face-work necessary to overcome the asymmetry of the relationship. The opposite is the case for men, who interpret subordination and the one-down position as a slight on their virility.

The political nature of gender rituals is highlighted by interactions among colleagues belonging to a peer group, or when the authority structure is at odds with gender lines. Patricia Martin (2001, 2003) interprets these kind of interactions in terms of 'mobilizing masculinities' (practices wherein two or more men jointly bring into play masculinity/ies) and 'practicing gender'. She defines practising gender as a moving phenomenon that is done quickly, directionally (in time), (often) non-reflexively, informed (often) by liminal awareness, and in concert with others. The aim of the approach to gender as a social practice (Bruni and Gherardi 2001; Poggio 2006; Martin 2006; Simpson and Lewis 2007) is to show gender dynamics as they happen, through what is said and done in situated interaction and within a wider cultural frame that see gender as a social institution. The situated dynamic of practising gender takes on Connell's (1995) claim that behaviour is gendered only because and when enacted within a gender order (or institution) that gives it meaning as gendered.

Genderism is competence in organizing social situations, and it may constitute a resource for a deliberately manipulative strategy, since celebration of the attitudes and activities of each sex is normative-complementary in character. The strength of the man is complementary to the frailty of the woman, the mechanical ability of the man to the ineptness of the woman, his determination and assertiveness to her care and compliance, his belligerence to her diffidence, and so on. If women assume the femaleness position (and the same applies to those men who do not adhere to maleness values), they automatically withdraw from the competition, and leave room for the 'tough who get going when the going gets tough'. Gender prerogatives can be used to mark out the boundaries of competition. The threat used to exercise social control is directed towards sexual identity. Or else it operates both as a threat and as a system of self-exclusion. It is in this situation that the organizational culture may make a major difference in the structuring of these social situations: the more the description of the work, of the talents required to perform it well, of the characteristics necessary to get ahead, of the dedication required to the work and/or to the organization are defined within a symbolic universe of the masculine, the more female competition is discouraged, is self-selected, is defined as unfair competition, and the less the men who do not support this value system are 'real men'. The corollary to this organizational culture is that when a woman competes or when she achieves high status, it becomes legitimate to suspect or to insinuate that she has lost out on femininity. She's married, she has children, she has a happy and female domestic life; what has she had to give up in order to become an organizational woman?

There is a certain class of celebration work which comprises ratification rituals (Goffman 1971). These are performed for or towards someone who has changed his or her status in some way, and they function as reassurance displays. The examples range from congratulations, through expressions of welcome to a new situation, to friendly (or apparently friendly) teasing.

These ratification rituals may disguise an ambiguous position if they are repetitive and if they involve people who meet frequently. A recurrent case is one in which a work group is predominantly single sex: here the lone woman or the lone man usually receives expressions of welcome, insistence on the exceptional nature of the situation, reassurances of non-aggression, and protective displays—as long as s/he is not in a position of authority. If the person is a woman and if the ratification ritual is repeated, elaborate and lengthy, the exceptional nature of her presence, and the benevolence of the group which accepts her as an equal and pays homage to her femininity, is insinuated even more strongly. In this case, what is ratified is a condition of symbolic subordination and the object of the reassurance is the group, which feels threatened by diversity (Gherardi and Poggio 2007).

In our working lives we create both material products and the symbolic product of a role assumed by a sexed body and performed by a gendered actor for an audience which not only judges the appropriateness and coherence of the performance with the symbolic universes of gender, but actively participates in the production of competence rules.

We might say that the first competence rule imposes the celebration of the symbolic order of gender as the archetype of separateness and univocality between what is male and what is female and the symbolic subordination of the latter to the former. The second competence rule prescribes the behaviour required to overcome the ambiguity of the experience that reveals the dual presence and which historicizes the forms of subordination according to the place, according to the circumstances, and according to the local culture which expresses what is 'fair' in the relationship between the sexes and in the discourses that constitute the social representation of gender.

The ceremonial work which sustains the symbolic order of gender (where male is male, and the female is second-sexed) is one of Durkheim's (1912) positive rituals; namely, those rituals which involve doing, paying homage, recognizing, or celebrating. Negative rituals, by contrast, involve avoiding, maintaining distance, forbidding; and when in interaction the symbolic order is broken, then remedial work is required (Goffman 1971; Owen 1983). The dual presence, that is, the transverse experience of

Table 3.1 The deferral of the symbolic order of gender (Gherardi 1995)

1st STAGE	Symbolic order of gender based on the separation of male and female
ACTION	Ceremonial work
COMPETENCE RULE	Acknowledgement of gender as separatedness and univocality
2nd STAGE	Symbolic order of gender based on the dual presence
ACTION	Remedial work
COMPETENCE RULE	Acknowledgement of gender as ambiguity and the discursive construction of a situated meaning of gender

gender, is a breach of the order, and as such requires a ceremonial which is simultaneously 'supportive of the symbolic order of gender' and 'remedial' of the offence.

This collusive manœuvre is a celebration of the symbolic order of gender, but the kinds of interaction in which this manœuvring takes place differ, because 'doing gender' stands at the intersection between the difference/deferral of the meanings of male and female and the material and symbolic structures which create inequalities on the basis of difference (second-sexing).

When we speak of 'doing gender' as a social practice, situationally and historically constructed, we are defining the rules and norms which regulate equal gender citizenship in a particular culture and therefore determine the amount of remedial work required—along a continuum ranging from gender 'play' (and playfulness) to the open conflict of the war between the sexes. An organizational culture expresses the ceremonial work and the remedial work implicit in doing gender.

In Table 3.1 I have tried to sum up the dynamics that sustain the practice of doing gender. I have started from the assumption that the presence of women in the workplace breaks with the symbolic order of gender based on the separation between male and female, public and private, production and reproduction. The co-presence of the sexes in the spheres of the social gives greater ambiguity to gender-based social differentiation, and the way in which people handle the dual presence—and the dilemma between diversity and inequality—creates the space for a historic transformation of gender. In interaction, gender is celebrated as a ritualization of separateness; in the second, the ambiguity of the dual presence is remedied-each ritual requires a specific competence rule.

CONCLUSIONS

A symbolic approach to organizational cultures as symbolic orders of gender aims to articulate a performative espistemology based on a poststructural understanding of gender as a discursive, situated, and organizational practice.

In the previous sections I illustrated the contribution of organizational symbolism to the analysis of gender cultures and highlighted both how gender is 'done' at work, and how organizations 'do' gender. Both play an active role in gender performativity through the interplay of ceremonial and remedial work underpinning the symbolic ordering of an organizational culture. Gender symbolism is maintained, reproduced, and culturally transmitted through the ceremonial work that takes places within organizations, while remedial work restores the symbolic order of gender following instances when it has been challenged.

Within a symbolic approach to gender as a social practice the issue of gender as difference and gender as power become entangled in a dynamic of practising gender as the site of continual contestation, struggle, and deferral. In fact power is seen as both a repressive and productive multiplicity sustaining static and processual gender difference.

References

Allaire, Y., and Firsirotu, M. (1984). 'Theories of Organizational Culture', *Organization Studies* 5(3): 193–226.

Alvesson, M., and Berg, P. O. (1992). *Corporate Culture and Organizational Symbolism*. Berlin: De Gruyter.

Baudrillard, J. (1981). *For a Critique of the Political Economy of the Sign*. St Louis: Telos Press.

Bolognini, B. (1986). 'Il mito come espressione dei valori organizzativi e come fattore strutturale', in P. Gagliardi (ed.), *Le Imprese come cultura*. 89–101, Milan: Isedi.

Bruni, A., and Gherardi, S. (2001). 'Omega's Story: The Heterogeneous Engineering of a Gendered Professional Self', in M. Dent and S. Whitehead (eds), *Knowledge, Identity and the New Professional*, 174–98. London: Routledge.

Butler, J. (1990). *Gender Trouble: Feminism and the Subversion of Identity*. London: Routledge.

Calás, M., and Smircich, L. (1993). 'Dangerous Liaisons: The "Feminine-in-Management" Meets Globalization', *Business Horizons* (Mar.–Apr.): 73–83.

Calás, M., and Smircich, L. (1996). 'From "the Woman's Point of View": Feminist Approaches to Organization Studies', in S. R. Clegg *et al.* (eds), *Handbook of Organization Studies*, 218–57. London: Sage.

Calás, M. and Smircich, L. (2006). 'From "the Women Point of View" Ten Years Later: Towards a Feminist Organization Studies', in S. R. Clegg *et al.* (eds) *Handbook of Organization Studies*, 284–346. London: Sage.

Calás, M., and Smircich, L. (2011). 'In the Back and Forth of Transmigration: Rethinking Organization Studies is a Transnational Key', in E. L. Jeanes, D. Knights, and P. Y. Martin (eds), *Handbook of Gender, Work and Organization*, 411–28. Chichester: Wiley.

Cassirer, E. (1955). *The Philosophy of Symbolic Forms*. New Haven: Yale University Press (orig. 1923).

Castoriadis, C. (1987). *The Imaginary Institution of Society*. Oxford: Polity Press.

Chodorow, N. (1978). *The Reproduction of Mothering*. Berkeley, CA: University of California Press.

Cirlot, J. E. (1971). *A Dictionary of Symbols*. New York: Philosophical Library.

Connell, R. (1995). *Masculinities*. Berkeley, CA: University of California Press.

Cooper, R., and Burrell, G. (1988). 'Modernism, Postmodernism and Organizational Analysis: An Introduction', *Organization Studies*, 9(1): 91–112.

Czarniawska-Jorges, B. (1991). 'Culture is the Medium of Life', in P. J. Frost, L. F. Moore, M. R. Louis, C. C. Lundberg, and J. Martin (eds), *Reframing Organizational Culture*, 285–97. Newbury Park, CA: Sage.

Davies, B., and Harré, R. (1990). 'Positioning: The Discursive Production of Selves', *Journal of the Theory of Social Behaviour*, 1: 43–63.

De Beauvoir, S. (1949). *Le Deuxième sexe*. Paris: Les Éditions de Minuit.

Dégot, V. (1985). 'Editorial', *Dragon*, 1: 3–6.

Denzin, N. (1992). *Symbolic Interactionism and Cultural Studies*. Cambridge: Blackwell.

Derrida, J. (1967). *De La Grammatologie*. Paris: de Seuil.

Derrida, J. (1971). *L'Écriture et la differance*. Paris: de Seuil.

Durkheim, E. (1912). *Les Formes élémentaires de la vie religieuse. Le Système totemique en Australie*. Paris: Alcun.

Ebers, M. (1985). 'Understanding Organizations: The Poetic Mode', *Journal of Management Studies*, 11(2): 51–62.

Eliade, M. (1952). *Images et symboles. Essais sur le symbolisme magico-religieux*. Paris: Gallimard.

Elshtain, J. (1982). 'Feminist Discourse and Its Discontents: Language, Power, and Meaning', *Signs*, 7(3): 603–21.

Elster, J. (ed.) (1986). *The Multiple Self*. Cambridge: Cambridge University Press.

Ferguson, K. (1984). *The Feminist Case Against Bureaucracy*. Philadelphia: Temple.

Ferguson, K. (1991). 'Interpretation and Genealogy in Feminism', *Signs*, 16(2): 322–39.

Frost, P. J., Moore, L. F., Louis, M. R., Lundberg, C. C., and Martin, J. (eds) (1991). *Reframing Organizational Culture*. Newbury Park, CA: Sage.

Gagliardi, P. (ed.) (1990). *Symbols and Artifacts*. Berlin: de Gruyter.

Gergen, K. J. (1991). *The Saturated Self: Dilemmas of Identity in Contemporary Life*. New York: Basic Books.

Gherardi, S. (1994). 'The Gender we Think, the Gender we Do in Everyday Organizational Life'. *Human Relations*, 47(6): 591–609.

Gherardi, S. (1995). *Gender, Symbolism and Organizational Cultures*. London: Sage.

Gherardi, S., and Poggio, B. (2007). *Gendertelling in Organizations: Narratives from Male Dominated Environments*. Copenhagen: Liber.

Gilligan, C. (1982). *In a Different Voice*. Cambridge, MA: Harvard University Press.

Goffman, E. (1976). 'Gender Display', *Studies in the Anthropology of Visual Communication*, 3: 69–77.

Goffman, E. (1971). *Relations in Public*. New York: Harper & Row.

Goffman, E. (1977). 'The Arrangement between the Sexes', *Theory and Society*, 4: 301–31.

Harding, S. (1987). 'Introduction: Is there a Feminist Method?', in S. Harding (ed.), *Feminism and Methodology*, 1–14. Bloomington, IN: Indiana University Press.

Harré, R. (1984). *Personal Being*. Cambridge, MA: Harvard University Press.

Hearn, J., and Parkin, W. (1983). 'Gender and Organizations: A Selective Review and a Critique of a Neglected Area', *Organization Studies*, 4(3): 219–42.

Höpfl, H. (2003). 'Maternal Organization: Deprivation and Denial', in H. Höpfl and M. Kostera (eds), *Interpreting the Maternal Organization*, 23–39. London: Routledge.

Höpfl, H. (2011). 'Women's Writing', in E. L. Jeanes, D. Knights, and P. Y. Martin (eds), *Handbook of Gender, Work and Organization*, 25–35. Chichester: Wiley.

Irigaray, L. (1974). *Speculum. De l'autre femme*. Paris: Les Éditions de Minuit.

Irigaray, L. (1977). *Ce sexe qui n'en est pas un*. Paris: Les Éditions de Minuit.

Kanter, R. M. (1977). *Men and Women of the Corporation*. New York: Basic Books.

Kristeva, J. (1981). 'Women Can Never Be Defined', in E. Harks (ed.), *French Feminism*. New York: Slocken.

Kristeva, J. (1988). *Etrangers à nous mêmes*. Paris: Fayard.

Latour, B. (1986). 'The Powers of Association', in J. Law (ed.), *Power, Action, and Belief: A New Sociology of Knowledge?*, 264–80. London: Routledge & Kegan Paul.

Linstead, S., and Grafton-Small, R. (1992). 'On Reading Organizational Culture', *Organization Studies*, 13(3): 331–55.

Martin, J. (1990). 'Deconstructing Organizational Taboos: The Suppression of Gender Conflict within Organizations', *Organization Science*, 1(4): 339–59.

Martin, J. (2002). *Organizational Culture: Mapping the Terrain*. Thousand Oaks, CA: Sage.

Martin, P. Y. (2001). '"Mobilizing Masculinities": Women's Experiences of Men at Work', *Organization*, 8: 587–618.

Martin, P. Y. (2003). '"Said and Done" vs. "Saying and Doing": Gendering Practices, Practicing Gender at Work', *Gender and Society*, 17: 342–66.

Martin, P. Y. (2006). 'Practising Gender at Work: Further Thoughts on Reflexivity', *Gender, Work and Organization*, 13(3): 254–76.

Mead, G. H. (1934). *Mind, Self and Society*. Chicago: University of Chicago Press.

Owen, M. (1983). *Apologies and Remedial Interchanges*. Berlin: Mouton.

Poggio, B. (2006). 'Outline of a Theory of Gender Practice', *Gender, Work and Organization*, 13(3): 232–3.

Ray, C. A. (1986). 'Corporate Culture: the Last Frontier of Control?'. *Journal of Management Studies*, 23(3): 287–96.

Risman, B. (2004). 'Gender as a Social Structure: Wrestling with Activism', *Gender and Society*, 18(4): 429–50.

Sampson, E. (1989). 'Foundations for a Textual Analysis of Selfhood', in J. Shotter and K. Gegen (eds), *Texts of Identity*, 54–65. London: Sage.

Shotter, J., and Gergen, K. (eds) (1989). *Texts of Identity*. London: Sage.

Sievers, B. (1990). 'Curing the Monster: Some Images of Considerations about the Dragon', in P. Gagliardi *et al.* (eds), *Symbols and Artifacts: Views of the Corporate Landscape*, 135–153. Berlin: de Gruyter.

Simpson, R., and Lewis, P. (2007). *Voice, Visibility and the Gendering of Organizations*. Basingstoke: Palgrave Macmillan.

Smircich, L. (1983). 'Concepts of Culture and Organizational Analysis', *Administrative Science Quarterly*, 28(3): 339–58.

Turner, B. A. (1990). *Organizational Symbolism*, Berlin: de Gruyter.

Turner, B. A. (1992). 'The Symbolic Understanding of Organizations', in M. Reed and M. Hughes (eds), *Rethinking Organization*, 218–57. London: Sage.

Tyler, M. (2011). 'Postmodern Feminism and Organization Studies: A Marriage of Inconvenience?', in E. L. Jeanes, D. Knights, and P. Y. Martin (eds), *Handbook of Gender, Work and Organization*, 9–24. Chichester: Wiley.

Watzlawick, P., Beavin, P., and Jackson, D. (1967). *Pragmatic of Human Communication: A Study of Interactional Patterns, Pathologies and Paradoxes*. New York: Norton.

West, C., and Zimmerman, D. (1987). 'Doing Gender', *Gender and Society*, 1(2): 125–51.

CHAPTER 4

....

WAS WILL DER MANN?

....

HEATHER HÖPFL

If Nancy Meyer's 2000 film, *What Women Want*, has any contribution to make, the answer to Freud's famous question 'Was will das Weib?'[1] is 'not very much'. According to the Hollywood view of gender, it seems that women want little more than attention, romance, someone to listen to them. To Freud, women were something of an enigma: dominated by their reproductive function, by penis envy, and hysteria rooted in sexual fantasy. Well, of course, this is a gross simplification and a much contested terrain. Irigaray has said of Freud that he is not speaking of two sexes but of one: that when speaking of women, 'The "feminine" is always described in terms of deficiency or atrophy' (Irigaray 1991: 119). Thus, for Freud, the feminine is always defined by lack, by the desire for the male organ. As Irigaray contends, in Freudian theory, the feminine is defined as the necessary complement to the operation of male sexuality. The masculine defines, the feminine complies. The masculine creates and stages a text. It enacts via a verbal fiat. The feminine is placed in a position of deference to it: always the lesser, always in a position of supplication. Made to submit, women are repeatedly punished for their irremediable lack.

Cixous (1981) relates in her well-known paper *Castration or Decapitation* the story of the Chinese king who calls on one of his military generals and asks him to train his 180 wives how to be 'male soldiers'. The story does not make clear why he wants to do this but the general, Sun Tse, lines up the women and puts in charge of each column two of the king's favourite wives. Sun Tse explains that when beats a drum two beats they are to move in one direction and when three beats another. The women listen to the instructions but when the drum is beaten, they all laugh and do not do as they are instructed. This happens several times. Then the general, with determination, calls for the two favourite wives to come forward and, in front of all the other wives, they are decapitated. The training is continued. He beats the drum. The women responded in silence and did not make a single mistake. However many times I read this story, I still find it chilling.

So, what do men want: order, control, power? Well, to a great extent these are just the things that, according to Freud, men desire:

> men are not gentle creatures, who want to be loved, who at the most can defend themselves if they are attacked; they are, on the contrary, creatures among whose instinctual endowments is to be reckoned a powerful share of aggressiveness. As a result, their neighbour is for them not only a potential helper or sexual object, but also someone who tempts them to satisfy their aggressiveness on him, to exploit his capacity for work without compensation, to use him sexually without his consent, to seize his possessions, to humiliate him, to cause him pain, to torture and to kill him. *Homo homini lupus* [man is wolf to man].
>
> (Freud 1961:58–9)

The phallocratic order establishes a powerful presence which demands deference. Such stories as that of Sun Tse function to regulate and reinforce a set of values, to remind us of the disciplined regime of the father, to show hierarchy and its rewards, to show who is to be rewarded and for what. The point is that such stories hold a terrible and implicit notion of male power. This is entirely taken for granted.

We are surrounded by images of powerful men: politicians, sports stars, business leaders. They are frequently defined by their commitment to ruthless behaviour, determination, power, ambition, relentlessness. There is no malice or conscious intent in these constructions yet they operate nevertheless in a way which reminds all of us of our place and of who has the power to define what is valued and worthy and what is not, what is lacking and deficient. However, by venerating 'the phallus' in this way, the feminine is rendered not only lacking in terms of a physical organ, but also in terms of the qualities which the valorization of the phallus represents. In effect, the emphasis on the father leaves us motherless and bereft of the values that remind us of a common humanity, of community, and of nurture.

One of the questions which this chapter seeks to address is how this relationship between the valorized and the deficient can be seen in organizations and their practices. Conventional patriarchal representations of the organization reduce *organization* to mere abstract relationships, rational actions, and purposive behaviour. Under these constraints, organization becomes synonymous with regulation and control. This is achieved primarily by the imposition of definition and location: in other words, by the defining characteristic of the masculine. Under such circumstances, organization comes to function in a very specific sense to establish a notion of what it considers to be *good* order, and to establish what can be taken for granted in administrative and managerial practice.

The presentation of the organization as an abstract entity requires the construction of the organization as a purposive entity with a trajectory towards a desired future state (Höpfl 2007). Strategy, as *fore*-casting, involves the representation of strategic intentions as future realities, as imaginings of future states and achievements. For this reason, many of the organizational metrics which regulate organizational life and with which

we are so familiar are concerned with the achievement of this future state, the measurement of progress towards this state, and the use of corrective measures to modify
and improve deficient performance. Such aspirations are always premised on the notion
that the future will be a progression from the present—once targets are achieved, goals
met, rewards assured, promises fulfilled. There is a privileging of the future over the present: a vicarious attachment to immediacy. Moreover, this means that not only is the
present itself lacking what achievement will bring but also that, by implication, all members of an organization, a society, are deficient relative to the targets which have to be
achieved (Höpfl 2007). For women this has a double consequence since there is already
the acknowledgement of a lack which precedes entry. Without a phallus, membership is
at best only partial and conferred only at the discretion of those who have the power to
define the boundaries of the organization and the conditions of entry.

This might be compared to the early pagan celebrations of May Day when the maypole was erected on the village green as a symbol of the Great God and crowned with
a wreath of hawthorn as a symbol of the Great Goddess. This was followed by dancing which weaved together the numerous ribbons which were attached to the top of the
maypole as a symbol of the fruitful union of male and female. Nowadays the symbolism is rather more one-sided. Metaphorically speaking, the maypole still presides over
the celebrations but it is no longer a symbol of union. In the privileging of rationality
and logic, and more particularly in the obsessive commitment to abstraction, there is a
loss of the flowering hawthorn, a loss of the physical and consequently a very distorted
and partial notion of reproduction. This is reproduction as text and idea, concept as
against conception. As Irigaray says, 'the whole of our society and culture, . . . at a primal
level, . . . function[s] on the basis of matricide' and she continues that the maternal function 'is always kept in a dimension of need' (Irirgaray 1991: 76), is censored, repressed.

But what of the power of the erection? The erection is the focus of all attention, around
which all must dance and to which deference must be shown. According to legend, the
Greek god Priapus' appearance was so unappealing that he was cast out from Mount
Olympus and cursed by the Goddess Hera with impotence. He found consolation with
Pan and the satyrs who are usually associated with dancing and drinking and lewd
behaviour. However, as the story goes, his persistent and unsatisfied lust gave him a permanent and large erection. Therefore, Pan gave Priapus the task of guarding the entry
to forest, and when intruders came along, Priapus would raise his tunic and scare them
away with the sight of his enormous appendage. Consequently, Priapus became a god
who guarded the entry to places—doorways, gardens, woods, vineyards, cross-roads.
Statues were erected to serve as a warning and those who encountered these statues
would touch the statue's phallus for protection, to appease the angry phallus as they
passed.

The main point of this short account of Priapus is not so much to draw attention to the
enormity of the phallus but rather to consider the implications of the status of the enormous phallus as a warning. It is interesting in this context to consider the etymology, of
the word *monition*. In organizations obsessed with monitoring it is salutary to consider
that *to monitor* comes from the Latin word, *monere*, meaning, to warn. Monitoring,

which as an activity has become increasingly prevalent in organizations in the last twenty or so years, is fundamentally about deference to the phallus. It is undertaken in order to warn whether or not performance is adequate to the achievement of goals but it is used widely and generally. It is an aspect of a culture based on the reciprocal relationship between definition, that is the power to define, and deference, the requirement to submit. Therefore, in order to be admitted—to the organization, to the role, to the occupation, to the club—it is necessary to accept the warning: to defer to the power of the enormous erection: to acknowledge the phallus which marks the point of entry.

Elsewhere (Höpfl 2007), I have examined what a commitment to metrics and other abstract representations of the organization have excluded from organizational life. It is apparent to anyone who works in an organization, picks up a textbook, watches television or reads a newspaper that there are numerous ways in which the organization constructs itself in textual and representational terms: the explicit use of rhetoric in marketing the products and images of organizations is one such construction. It is also present in the construction of statements, strategies, and structures, in its use of representation for regulation of staff: the endess self-evaluation, self monitoring, peer-reviews, line reviews, external reporting, responding, explaining, evaluating, accounting—endless accounting for one's position and evaluating one's performance. As much as a quarter of activities in organizations *can be accounted for* by monitoring activities and reporting: perhaps more. The fundamental characteristic of the organization as a purposive entity is its *directedness* and, clearly, there is a relationship between the direction (as in course of action) and direction (as command) of the organization and the rhetorical trajectory.

Of course, this manic obsession with monitoring is all about regulation by male discourse and male reflections, male theorization and male constructions. In other words, it is about the phallus. The motif of the erection stands as a powerful emblems of corporate constructions. However, this is not merely a reference to phallic symbolism. It is about the role of *standing* and authority, about the power to construct, about location and definition, and it is about the power to demand that women account for themselves and confront the erection. At the same time, it is true to say that both men and women are required to defer to the erection. The erection alone stands as a symbol of reproduction because it has usurped the site of physical fertility and replaced it with text, hopes, aspirations, and rhetoric. The phallic tower (Figure 4.1) is Babel itself and likewise doomed to collapse.

In conventional terms, the strategic direction of the organization involves the construction of the organization as a purposive entity with a trajectory towards a desired future state. Consequently, many of the organizational metrics with which we are all so familiar are concerned with the achievement of this future state, the measurement of progress towards this state and the use of corrective measures to modify and improve deficient performance. There is a privileging of the future over the present. Moreover, the aspirational goals of the trajectory mean that, in the present, we are all deficient relative to the targets which we have to achieve. In such movement into the future, the organization in its actions takes precedence over the individual and, therefore, any

FIGURE 4.1 Swiss Re Tower, Financial District, London

Source: 2. http://www.copyright-free-pictures.org.uk/london-england/10-gerkin.htm

ambivalence experienced by the individual about the purpose of the action must be concealed both as the price of membership and as a demonstration of commitment.

In organizational terms, this can be seen in the frenetic formulation and definition of future and idealized states: in strategic planning. In seeking to construct themselves both as sublime manifestations of male desire and as unattainable ideals, organizations lay themselves open to a range of problems. The therapeutic project of *saving* the organization, via the rule of logic, via insistent authority, and via psychology, is a process of mortification and the first victim is 'the mother'. Moreover, it is founded on a masculine sublime fabricated to reflect the male ego, narcissistic and inevitably melancholic. Women have no place, no reflection, no role in this construction other than to the extent that, in an entirely selective way, they serve as objects within the construction. In this construction, women are hysterical and have to be kept out because, by posing a threat to such representational forms (to mimesis), they threaten good order. Only if women are prepared to submit themselves to the symbol of the erection, primarily as objects of desire but also as homologues, can they enter into reflection. However, they must show proper deference. If they lack propriety they are nothing. If they reject the phallocratic order they have no part to play. They must enter what Irigaray has termed 'the scenography that makes representation feasible...that is, the architectonics of its theatre, its framing in space-time, its geometric organizations, its props, its actors…. That allows the logos, the subject to reduplicate itself, to reflect itself by itself' (1991: 123).

These are the defences which protect philosophy, protect apparent coherence, from failure and subversion by women; a phallocentric psychology which credits itself with the initiative and defends its position by either relegating women to its borders or making 'them homologues of men when it educates them' (Lyotard 1989: 114): an education in the style of Sun Tse. Clearly, part of such a strategy of defence rests on power over the control of reflection, theorization and discourse, and on the control of categories and their meanings. Lyotard argues that this desire to control women and to neutralize difference is exercised by making women into men (making women into objects), 'let her confront death, or castration, the law of the signifier. Otherwise, she will always lack the sense of lack' (Lyotard 1989: 113).

Organizations want to create the heroic sublime and this is inevitably male. Perhaps this account goes some way towards providing *insights* into the process. However, the author too is rendered melancholic by the process of theorization and caught in the 'trap of insight'. The very theorization that seeks to address the issue is the demonstration of the fall into homologation. Perhaps the appropriate re*version* (L. *vertere*, to turn) is to the *hystera* but Irigaray cautions that if we turn that way 'a dizzy delusion answers to the optical illusion of the fakes' (Irigaray: 1985: 278). Rather, Irigaray says that women privilege touch over vision and that to get away from the (male) speculations requires a language of the body. Woman is required to remain silent or to present herself according to the representation of herself as viewed through the male gaze: to produce a version of herself in a way which effaces her. There is no place to turn that is, in itself, not subject to capture.

In the organizational world, women's deficiencies can, it seems, be corrected by reason. In this sense, conformity requires submission to *psychology* (regulation of the psyche by the logos). If that which is defined as deficient will only submit to superior logic she will *realize* the extent of her disorder. She can be saved by surrendering to the logos, by abandoning her threatening femininity. When she is truly converted she might be permitted to play a role as long as she plays it in compliance with male expectations. When she does this, she will be rendered impotent as the price of membership. She will then be conformed to psychology: 'the wisdom of the master. And of mastery' (Irigaray 1985: 274). If the logic of organization can convert women to the power of the logos, it is able to demonstrate control over hysteria and disorder. In other words, women are permitted entry precisely because they are no longer women.

To return to the question *Was will der Mann?* De Certeau (1986) has said that a particular characteristic of rhetoric is that its trajectory is completed by *the other*. That is to say that rhetoric requires something from the audience to which it is directed. It is completed by a response. In a specific sense, the organization then, first as an expression of male desire and second, as a rhetorical entity *wants something* of the other—the employee, the customer, the competitor, the supplier, the general public. Ironically, the organization as an abstract entity transfers its own lack to its members who are thereby rendered deficient in relation to the abstract desires of the organization. This is the subtle fear of falling below some expectation: of failing to have a big enough member. It is a male fear. In response to the endless rhetoric which outlines the implications of failure and the endless commitment to monitoring, it is not surprising then that the consequence is a paralysing state that sets the individual at odds with the world, alienated from it, terrified of it, unable to act, react, or to adopt a moral disposition: unable to sustain the erection. So everything falls, institutions collapse, membership is lost, the erection falls. It is what happens in the face of the fear of the disordered other. It is the fear of disorder brought by women as the already deficient other (Lacoue-Labarthe 1989: 129). Irigaray maintains that 'All desire is connected to madness. But apparently one desire has chosen to see itself as wisdom, moderation, truth and has left the other to bear the burden of madness it did not want to attribute to itself, recognize in itself' (Irigaray 1991: 35). To refuse to conform to logic is to be declared 'mad' and the remedy for madness it seems is death. As Cixous (1981) has argued, the remedy is decapitation. So whereas men might fear madness and castration, physical and symbolic, so women are confronted with the power of logic and the threat of the severed head.

The fear of castration is the most primitive of fears. In his discussion of the role of the phallus, Žižek says of those symbols which confer power such as the insignia of office, the sceptre or the crown, are external to the wearer and worn only in order to exert power. As a consequence, he argues, they 'castrate', and deprive him of his virility. So that symbolic castration

> introduces a gap between what I immediately am and the function that I exercise (i.e., I am never fully at the level of my function).... [this is] the castration that occurs by the very fact of me being caught in the symbolic order, assuming a symbolic mandate. Castration is the very gap between what I immediately am and the symbolic mandate that confers on me this 'authority'.
>
> Žižek (2003):87

Leaving aside for a moment the gender implications of this statement, of course, *my* symbolic mandate is more equivocal, authority in such terms is like a mantle that signifies office. Authority then resides in the symbolic order and likes to make its presence felt in the behaviour of a community, displays of potency, the destruction of difference. After all, as Žižek argues, this constructed phallus is an 'organ without a body.... an excessive supplement' (Žižek 2003: 87) and it demands deference. We cower under the power of the erection. To act with authority requires that contradictions, discontinuities, collisions of meaning, and so forth must be regulated to preserve the appearance of order: the need to maintain the coherence of the *scenography*. This makes monitoring and measuring important activities which preserve the erection of order. Hence, in the service of the organizational visions and plans, in other words, in the service of the organization as a rhetorical entity, a man can be potent and effective (that is, have the power to bring about 'results') or he else he is rendered effete (worn out, exhausted, no longer fertile, from the Latin, *effetus*, meaning worn by breeding, *ef-fetus*). Likewise, a woman can commandeer phallic power and become a man (albeit a man manqué) or be herself, and demonstrate her own power, in which case she bears the risk of expulsion from the site of performance. It is this issue, which concerns the control of reproduction, which is the basis of the theoretical notions which underpin this chapter.

The organization comes to be regulated by the phallus and the phallus is the single most important determinant of membership. At the same time, the emphasis on rigidity and tumescence is a symbolic one. As such, ironically, it signifies that the organization has lost contact with the physical bodies of which it is made up. To repeat, it is an 'organ without a body... an excessive supplement' (Žižek 2003: 87). Not surprisingly, women can only enter into membership by the acquisition of a metaphorical phallus: by becoming ciphers of men and *real member*ship has status over symbolic membership on every count. 'A woman will only have the choice to live her life either *hyper-abstractly* (original italics)....in order thus to earn divine grace and homologation with the symbolic order; or merely *different* (original italics), other, fallen....But she will not be able to accede to the complexity of being divided, of heterogeneity, of the catastrophic-fold-of-'being' (Kristeva 1983, in Moi 1986: 173). What Kristeva is saying here is that women must either live as male constructions or *be found wanting*. In this sense, to lack a phallus is a very serious deficiency from the point of view of the male subject. It induces anxieties that the same fate might befall him. Therefore, it is a necessary condition of the male delusion of wholeness for women to be construed as castrated. She is always the deficient other.

Consequently, the need for performance measures suggests a fundamental need for reassurance that the organization is, after all, a sustainable erection. Of course, the trajectory of strategic development is not only about an organization establishing parameters of *normal* expectations but also about improvement: about bigger, better, and more. It is abundantly clear from the study of any organization that there is a contemporary obsession with measurement. Organizations are preoccupied with the idea that planning is the means of achieving targets; that metrics are the way of ensuring that progress into the future is being achieved. The league-table is an established way of understanding the pecking order of any number of institutions: universities, schools, hospitals, and, of course, football teams. There is a belief that destinies can be determined by analysis, prediction, and monitoring and an emphasis on progress and improvement leads to a concern with measurement.

Organizational life is replete with the exhortation to increase and extend, to do more, achieve more, to 'be the best', to seek improvement without end. So, all aspects of organizational life are subjected to the discourse of strategic management, human resource management, culture change, or whatever is the current vogue in change terminology. Management *tools*, the surrogates of the phallus, are one means by which this discourse reinforces itself. Its logic is not only phallogocentric, it is self-reinforcing phallogocentric reassurance. Consequently, if women do not defer to the logic of measurement and comparison it is sufficient reason for their exclusion. The construction of tools which facilitate measurement and comparison demands admiration and approval and this should preferably be both visible and sustained. However, the situation is worse than this. In order for the organization to sustain itself it must erect for itself symbols of what is lost to logic and rationality. It must erect the feminine emblematically. If real women cannot be admitted, what is lost to the organization must be elevated to fill the gap. Instead of flesh and blood, the organization creates for itself a representational version of what is no longer there. So, the organization erects itself in diagrams and charts, texts, and metrics which seek to uphold the representation of the feminine, the body, but which inevitably achieve a cancellation. It is little wonder, therefore, that notions of quality and care, the ubiquitous valorization of staff, the commitment to service improvement, and so forth, have more in them of absence rather than presence. To repeat Irigaray's point, such constructions reassure themselves that they are not mad by appealing to logic and the therapeutic quest, madness is all that is other, madness is alterity: the feminine. Hence, the phallocratic organization is concerned with logic and order and rationality, with location and hierarchy, with allocation and definition. It is organization which has lost contact with the body.

So what can the feminine do in response to definition and mortification, perhaps deal with the mother, seize phallic power and become a man as the price of entry but in so doing, she renders all relationships with men homomorphic. However, to do otherwise or to seek to be other or different requires the *patronage* of men. As Eagleton argues, 'The law is male, but hegemony is a woman; this transvestite law, which decks itself out in female drapery is in danger of having its phallus exposed' (Eagleton 1990: 58). This is apparent in the familiar ways in which organizations seek to create the lost *feminine*. This is rather similar to the idea put forward by Baudrillard in his critique of rationality in which he argues that the reduction of male and female to categories has produced an artificial distinction which *objectifies* the feminine. By this line of argument, the feminine is now constructed as a category of the masculine and, by implication, the power of the feminine to manifest itself in ambivalence is lost. Baudrillard sees *feminism, per se*, as ensnared within the construction of a phallic order (Baudrillard 1990). In organizational terms, these constructions of the feminine are intended to console. The vicarious and representational has more seductive power than the physical and disordered other. These emblems function as an anamnesis to register what is no longer present as representation. For this reason alone, the emblem of loss is melancholic and pervades the organization with melancholy. It cannot offer consolation because it can only recall that there is a loss. It cannot reassure because it arises from a mere *erection*. This is only a shadow of the feminine and it is a travesty. It is the feminine constructed in the image of masculine desire to meet the needs of sterile perfectionism and rationality. It is a feminine which in this form is tidy, logical, entirely representation and without power, ambivalence, and sexuality.

So, the organization constructs itself in diagrams and charts, texts and metrics which seek to uphold the representation of the feminine, the anima, the body, but which inevitably achieve a cancellation. It is little wonder, therefore, that notions of quality and care, the ubiquitous valorization of staff, have more in them of melancholy abstractions than of physicality. Embodied reproduction is then replaced by the reproduction of concepts and the fertility of the site is given up to the fertility of concepts and theory. The physical matrix reproduces from itself and matter is made incarnate. The patriarchal matrix, however, deals on the level of the abstract alone. Here, perfection comes from striving. In contrast to the maternal process, the 'patriarchal consciousness' has a preference for rationality and the direction of the will, the 'patriarchal order of society' (Dourley 1990: 50) and with this comes a commitment to 'sterile perfection in the divine as a hallmark of patriarchal consciousness' (Dourley 1990: 51) 'which could easily have been avoided by paying attention to the feminine idea of completeness' (Dourley 1990: 50). Consequently, the matrix gives birth into a world of obsessive reproduction and insatiable desire. Paternal reproduction arises from the sense of lack that only the acknowledgement of the unconscious, of the maternal matrix, could satisfy and give a sense of completion. Hence, the patriarchal matrix is concerned with logic and order and rationality, with location and hierarchy, with allocation and definition. The maternal matrix, however, *knows* in embodied experience and this knowledge is sufficient to itself when it finds expression in embodied action.

What then might be possible in relation to the maternal function? Is there a place for the maternal matrix? I have been intrigued in recent years and in the time of the TV programme *Sex in the City* to read in glossy magazines of the new sexuality and the behaviour. Advocates of this have been called 'metrosexuals'. Perhaps it doesn't matter but it does seem strange to speak in terms of new sexual values in terms of the sexuality of the *mother*, for this is precisely what metrosexual means. In the late nineties, I published a paper, *The Mystery of the Assumption: Mothers and Measures* (Höpfl 2001) in which I tried to examine the two terms which derive from the Greek word meter, that is, *metros*, meaning Mother and *metron*, meaning Measure. So, for example, one might compare, metronome, metronomos as measure/law with metropolis, mother/city. The metropolis is the mother city, the place of the mother, the womb, the *locus amoenus*. It is the place of polity of process and order. *Polis* is normally rendered city or city state yet, as Heidegger (1976) points out, the term *polis* means more than this, it is 'the place, the there, wherein and as which the historical being-there *is*' (Heidegger 1976: 152). It is a coincident word that brings together the body and the law, mother and order: metropolis. The trouble is that organizations have become more metron than metros, more measure than mother and consequently they have not valued the people who work in them.

By privileging metrics, maternal values such as care and nurture have been set aside in favour of volume and frequency both of which are primarily patriarchal measures. This can be seen in the obsessive commitment to quantification and taxonomic structures for collecting data on virtually everything. In universities these days the staff are more assessed than the students. Metrics function to regulate behaviour in the present and to guide actions towards specific targets in the future. Yet, it is worse than this. The representations of metros produce counterfeit structures, simulated qualities which derive from absence, from unacknowledged male deficiencies, and which have little in common with embodied experience. As Žižek argues, they are disembodied, empty symbols of power: castrated but without the insight to acknowledge the extent of their emasculation.

Following Kristeva's (1987) concern to establish a discourse of maternity and Irigaray's (1991) concern for the primacy of the maternal function as the basis of social order, it seems appropriate to attempt to attempt to say something about what men might actually want: matricide. However, the act of matricide must not be perceived as a mad act in order to preserve the patriarchal order. It must be subject to the rule of logic and rationality. The killing of the mother and her subjection to the rule of patriarchy, which incidentally involves the usurpation of her arcane powers, requires the substitution of the mother with a representation and deferential emblem which retains a captive version of the feminine albeit in a now entirely effete way.

A maternal discourse is concerned with the very ambivalence which is concealed and regulated by such actions. The notion of a maternal discourse thus releases itself from the deferential position it holds in relation to the dominant social discourse to challenge order, rationality, and patriarchal regulation. What this contributes to organizational theory is the capacity to make transparent the effects of the production of meaning, to render explicit the paternalistic quest of the organization, and to make problematic the notion of trajectory, strategy, and purpose.

By dealing with the *conception* of the organization as maternal, the notion of maternal organization seeks to break the body of the text in order to allow reflections on the mother/motherhood/maternal imagery to enter the text. Thus, the embodied subject speaks of division, separation, rupture, tearing, and blood whereas the text of the organization speaks of regulation and representation, of rational argument and rhetorical trajectory. By breaking the text, the implications of 'the sterile perfectionism' of the patriarchal consciousness (Dourley 1990: 51) is made transparent. So, maternal organization stands against the way in which conventional accounts of management are presented, poses alternative ways of understanding organization, and offers insights into the organization as embodied experience. Moreover, to give voice to the view that, despite the feminine being always defined in terms of its relationship to the phallocratic order, it might have what Irigaray has termed 'its own specificity' (Irigaray 1991: 119).

The notion of a maternal discourse of organization implicitly threatens the process of definition, that is to say, *finalization*, the taxonomic thinking which attaches to phallocratic order, and so inherently subverts power relations in the organization. This is the greatest threat. It is what Docherty calls 'indefinition' (Docherty 1996: 67). Lacoue-Labarthe (1989: 129) indicates what Plato has identified as the major threats to representation as being women and madness and, indeed, women and madness as themes spiral together as surely as *hystera* (Gk. *womb*) and the psycho*logical* condition of hysteria (as a disturbance of the nervous system thought to be brought about by uterine dysfunction) find a common origin in the function of reproduction. The fear is that in Kristeva's terms (1987) the Law will be undermined by the intrusion of the Body.

Well then, what do men want? Certainly, not what is under discussion here. A discourse of maternity establishes a *metro*politan order, a place of the mother: a place from which to consider 'the possibility of politics', alterity, and ethics. A maternal discourse runs counter to the frenetic quest for sterile perfectionism which is 'the hallmark of the patriarchal consciousness' (Dourley 1990: 51). It resists *de-finition* which is the root of so many social problems, political and economic objectives and organizational drives. It presents political possibilities for fundamental change in social life. However, the drum beats and survival depends on compliance.

NOTE

1. What does a woman want? 'The great question that has never been answered, and which I have not yet been able to answer, despite my thirty years of research into the feminine soul, is "What does a woman want?"' *Die grosse Frage, die nie beantwortet worden ist und die ich trotz dreißig Jahre langem Forschen in der weiblichen Seele nie habe beantworten können, ist die: Was will das Weib?* Letter to Marie Bonaparte, in Ernest Jones, *Sigmund Freud: Life and Work*, ii/3, ch. 16. London: Hogarth Press.

REFERENCES

Baudrillard, J. (1990). *Seduction*. London: Macmillan.

Cixous, H. (1981). 'Castration or Decapitation', *Signs*, 7(1): 41–58.

de Certeau, M. (1986). *Heterologies, Discourse on the Other*. Manchester: Manchester University Press.

Docherty, T. (1996). *Alterities: Criticsm, History, Representation*. Oxford: Clarendon Press.

Dourley, J. P. (1990). *The Goddess, Mother of the Trinity*. Lewiston, NY: Edwin Mellen Press.

Eagleton, T. (1990). 'The Ideology of the Aesthetic' in P. Hernadi (ed.), *The Rhetoric of Interpretation and the Interpretation of Rhetoric*. Durham NC: Duke University Press, pp 75–87.

Freud, S. (1961). *Civilisation and its Discontents*. NY: Norton Inc.

Heidegger, M. (1976). *An Introduction to Metaphysics*. New Haven: Yale University Press.

Höpfl, H. J. (2001). 'The Mystery of the Assumption: Of Mothers and Measures', in N. Lee and R. Monro (eds), *The Consumption of Mass*. Sociological Review Monograph Series. Oxford: Blackwell, pp. 46–60.

Höpfl, H. (2007). 'The Codex, the Codicil and the Codpiece: Some Thoughts on Diminution and Elaboration in Identity Formation', *Gender Work and Organization*, 14(6): 619–32.

Irigaray, L. (1985). *Speculum of the Other Woman*, tr. G. Gill. Ithaca, NY: Cornell University Press.

Irigaray, L. (1991). *Feminist Theory: Psychoanalysis and Feminism*. London: Routledge.

Kristeva, J. (1987). *Tales of Love*, tr. L. Roudiez. New York: Columbia University Press.

Lacoue-Labarthe, P. (1989). *Typography*. Stanford: Stanford University Press.

Lyotard, J.-F. (1989) *The Differend Phrases in Dispute*. Minneapolis, MN: University of Minnesota Press.

Moi, T. (1986). *The Kristeva Reader*. Oxford: Blackwell.

Žižek, S. (2003). *Organs without Bodies: Deleuze and Consequences*. London: Routledge.

CHAPTER 5

··

FEMINISM, POST-FEMINISM, AND EMERGING FEMININITIES IN ENTREPRENEURSHIP

··

PATRICIA LEWIS

INTRODUCTION

WOMEN-OWNED businesses in America are now the fastest growing segment of the small business sector with the number of female run businesses growing faster than businesses owned by men throughout the 1990s and into the 2000s (Loscocco and Bird 2012). In contrast the increase in female businesses in the UK has been more restrained, with the proportion of women-owned firms 'resting' at 29 per cent of the self-employed total, with limited fluctuations since 1984 (Carter *et al.* 2012). A 2009 report (*Greater Return on Women's Enterprise*) produced by the Women's Enterprise Task Force set up in 2006 by Gordon Brown, the then British Prime Minister, stated that 900,000 more businesses could be created in the UK resulting in £23 billion to the economy, if British women set up businesses at the same rate as their American counterparts. Further, if British women's start-up rate was at the same level as their male colleagues, this would result in 150,000 extra businesses per year in the UK.

The growth in the number of women setting up their own business, alongside concern at the reserved nature of this growth in countries such as the UK, has contributed to a significant increase in the amount of research attention directed at the relationship between gender and entrepreneurship. As this work is largely concerned with the disadvantages faced by women when embarking on business ownership, with attention directed at how the level of disadvantage experienced can be reduced, much of this research can be characterized as feminist in nature. From this research it would appear that the most consistent finding is that women-owned businesses are smaller in size and less lucrative in terms of earnings and profitability than those owned by men but the research is less uniform in terms of explaining why this is the case, possibly because the degree and nature of critique varies across different feminist perspectives (Marlow 2002; Calas *et al.* 2007, 2009; Loscocco and Bird 2012). A dominant strand of this research adheres to the tenets of liberal feminism, a second-wave feminist perspective which focuses on the 'woman entrepreneur' and understands women business owners as 'carriers' of gender into the entrepreneurial arena (Green and Cohen 1995; Berg 1997; Calas *et al.* 2007). From this perspective the entrepreneurial realm is perceived to be non-gendered or gender neutral, such that differences in the entrepreneurial experiences of male and female business owners are attributed to factors such as women's lack of business experience, as opposed to ascribing it to entrepreneurship *per se* (Marlow 1997; Greer and Greene 2003; Lewis 2006). Calas *et al.* (2007: 80) also identify the radical and psychoanalytic feminist perspectives as focusing on the 'woman entrepreneur' because like liberal feminism these perspectives also work from the assumption that women's disadvantage is a consequence of their condition as women.

In contrast, an alternative strand of feminist research asserts that gender in the entrepreneurial sphere does not simply originate from women's presence, rather gender is a principle of social organization and should be understood in relation to structures, institutional, and cultural practices, and discourses connected to entrepreneurship and not simply as something connected to individuals. From this perspective what is highlighted is the gendered nature of entrepreneurship and that this gendered nature can be characterized as masculine (Allen and Truman 1993; Green and Cohen 1995; Mulholland 1996; Reed 1996; Berg 1997; Gamber 1998; Mirchandani 1999; Ogbor 2000; Bruni *et al.* 2004a; Ahl 2006; Lewis 2006; Marlow *et al.* 2009; Bourne and Calas 2012. Calas *et al.* (2007: 90–1) locate the second-wave socialist feminist perspective and the third-wave poststructuralist/postmodern and transnational/(post)colonial feminist perspectives within this strand as all three understand gender in terms of process and practice 'produced and reproduced through relations of power among differently positioned members of society, including relations emerging from historical processes, dominant discourses and institutions, and dominant epistemological arguments, all of which become naturalized as *the way it is*'. Specified as a gendered activity, entrepreneurship has been identified as an economic arena within which gender is 'done', with gender and entrepreneurial activity being mutually constitutive, that is, gender is part of entrepreneurship and entrepreneurship is involved in the enactment and shaping of gender relations (Bruni *et al.* 2004*a*). This 'doing' of gender is evident in the symbols, images, rules, and values that overtly and covertly guide, validate, and rationalize gender distinctions in the entrepreneurial arena, with feminist studies of entrepreneurship exposing how masculinity is central to the enactment of successful entrepreneurship.

While there has been a strong focus on the masculinity of entrepreneurship and the implications of this for female business owners, less attention has been directed at the issue of femininity and its relationship to entrepreneurship. This is not unusual as a similar situation pervades the general area of gender and cultural studies where masculinity is a central area of study but femininity is not (Reay 2001; Gill and Scharff 2011).

An exception here is the work of Caroline Essers and Yvonne Benschop (2007, 2010) which looks at the identity work of female entrepreneurs of Moroccan or Turkish origin in the Netherlands. Through the development of the notion of female ethnicity, Essers and Benschop (2010) explore the meanings and positioning of femininity in entrepreneurial identity constructions for Muslim immigrant business women. In doing this they place an emphasis on the intersectionality manifest in a concept such as female ethnicity, arguing that if we are to understand the multifaceted impact of gender and ethnic processes in the construction of an entrepreneurial identity and if we want to achieve a better specification and understanding of entrepreneurship, an intersectional approach is crucial. Nevertheless while this chapter also explores the issue of femininity and its relationship to entrepreneurship and entrepreneurial identity it does not include a notion of intersectionality. Rather it explores the issue of femininity and its relationship with entrepreneurship through the lens of post-feminism and its connection to second- and third-wave feminism. In focusing only on gender I am not seeking to postulate that within the context of entrepreneurship gender is the privileged mark of difference. Instead the theoretical privileging of gender is adopted to facilitate a focus on femininity and its changing (post-feminist) form given its neglect in the gender and entrepreneurship literature.

Post-feminism as a cultural phenomenon is a useful frame for exploring the issue of femininity and entrepreneurship given its focus on the factors which contribute to a greater degree of fluidity regarding what femininity means in contemporary times. Femininity can be understood in terms of Connell's (1987) notion of emphasized femininity or femininity as 'super girly'. Emphasized femininity manifests as a set of embodied traits and practices—passive, supportive of men, attractive to men, infantile—that women, from the viewpoint of second-wave feminism in particular, are compelled to take up and which denote their subordinate status to men. Thus for second-wave feminist perspectives such as liberal feminism, emphasized femininity as traditionally articulated is something which must be rejected and overcome as it does not benefit women (Schippers and Sapp 2012). However, over recent decades 'feminist ideas (e.g. equality of opportunity in education and the workplace) have slowly worked their way into the material and ideological structures of society, and have become part of the general culture of femininity' (McRobbie 1993: 409). The cultural presence of feminism contributes to the celebration of qualities such as individuality, autonomy, dynamism, and agency in women—characteristics which are associated with entrepreneurship. Indeed the 'autonomous, calculating, self-regulating subject of neoliberalism' in the person of the entrepreneur has much in common with the 'active, freely choosing, self-reinventing subject of post-feminism' (Gill and Scharff 2011: 7). The insertion of feminist ideas into the culture of femininity therefore means that emphasized femininity is no longer 'the constrained embodiment of oppression imposed by one group (men) upon another (women) or a fixed set of bodily practices to signify subordination' (Schippers and Sapp 2012: 30). Instead, this evolution of the general culture of femininity means that women who embark on entrepreneurship are celebrated as living out the ideal of economic liberation fought for by second-wave liberal feminists.

Alongside this feminist egalitarianism there remains a strong current of traditional familialism which emerges through the 'well-disguised rearticulation of traditional gender stereotypes', particularly in connection to the domestic realm (Thornham and McFarland 2011: 66). The conflicting currents of feminist egalitarianism and traditional familialism which are at the centre of contemporary manifestations of femininity in general (McRobbie 2004, 2009; Cotter *et al.* 2011) are also present in the context of entrepreneurship. Issues of family responsibility and domesticity and the tension between this and entrepreneurial achievement have always been a key focus of considerations of women's entrepreneurship and used to explain the perceived weakness of women's business performance (Ahl 2006). Exploring this tension and other aspects of women's entrepreneurship through the lens of post-feminism allows us to examine contemporary changes in the relationship between home and work and how this contributes to the emergence of different modes of entrepreneurial femininity. Such an examination is timely for two reasons: first, focusing on femininity and entrepreneurship allows us to examine how gender in the context of entrepreneurship is lived, experienced and represented by women business owners. We can also consider how women's entrance into the realm of entrepreneurship is characterized not only as a manifestation of change but also as one of continuity. Women's lives are increasingly characterized by transformation *and* stability such that, while they are now free to enter the world of work, they have not yet managed to exit the world of home. The second reason why our attention should be directed at the issue of femininity is the increasing level of diversification which characterizes women's entrepreneurship. Heightened diversification among women business owners and the emphasis placed on multiplicity, plurality, and difference signals that we should be considering modes of entrepreneurial femininity. Thus this chapter will explore the variety of modes of entrepreneurial femininity constructed and taken up in entrepreneurship in post-feminist times and the variations in 'doing' business associated with different entrepreneurial femininities. In the next section the notion of post-feminism is explored further and used as a means in the following section to identify four modes of entrepreneurial femininity based on existing studies of women's entrepreneurship.

POST-FEMINISM

Why does being a feminist mean you can't be somebody's wife? I'm very happy to be 'the wife of...'. (Livia Firth, the wife of the actor Colin Firth, *The Sunday Times*, 4 Dec. 2011)

Despite its ubiquitous usage, post-feminism is a contested term with a range of meanings attached to it. As an expression it was first used in a *New York Times* magazine article titled 'Voices from the Post-feminist Generation', with debates about post-feminism largely being played out in America and Britain (Holmlund 2005). At this point in time four different interpretations of this cultural phenomenon can be identified. First, it has been understood as an epistemological break within feminism in the wake of its encounter with difference. In this sense it refers to the intersection of feminism with other 'posts' such as postmodernism, postcolonialism, and poststructuralism, with the 'post' prefix signalling change and ongoing transformation within feminism itself (Budgeon 2001; Gill 2007a). Second, post-feminism has been used in a historical sense to refer to the passing of a particular period of feminist activism (the 1970s) but not the passing of feminism itself. Third, post-feminism has been understood in terms of a backlash against feminism with backlash discourses taking a number of contradictory forms (Gill and Scharff 2011). The work of Susan Faludi (1991) is well known for its claim that, after a period of gains, women are now experiencing a backlash led by the new right and fundamentalist conservative groups. Within this context feminism is blamed for causing women's unhappiness by trying to 'have it all' (Budgeon 2001; Braithwaite 2002; Gill and Scharff 2011).

For the purposes of exploring entrepreneurial femininities this chapter will adopt the fourth interpretation of post-feminism developed by Rosalind Gill (2007a). Gill (2007a) characterizes post-feminism as a distinctive sensibility which incorporates McRobbie's (2004) suggestion of a double entanglement manifest in the coexistence of neoconservative values and processes of liberalization in relation to gender, sexuality, and family life. According to Gill (2007a: 163):

> What makes a post-feminist sensibility quite different from both prefeminist constructions of gender and feminist ones is that it is clearly a response to feminism (but not just that). In this sense, post-feminism articulates a distinctively new sensibility...because of its tendency to entangle feminist and anti-feminist discourses...(making its) construction of contemporary gender relations...profoundly contradictory...Yet these contradictions are not random, but contain the sediments of other discourses in a way that is patterned...The patterned nature of the contradictions is what constitutes the sensibility, one in which notions of autonomy, choice and self-improvement sit side-by-side with surveillance, discipline and the vilification of those who make the 'wrong' choices...

The notion of a sensibility facilitates the identification of a range of stable features—femininity as a bodily property; the shift from objectification to subjectification; the emphasis upon self-surveillance; the prominence given to individualism, choice, and empowerment; the ascendancy of a make-over paradigm; the revival and reappearance of notions of 'natural' sexual difference; the resexualisation of women's bodies and finally the retreat to the home as a matter of choice not obligation—that constitute a post-feminist discourse (Gill 2007a; Negra 2009; Gill and Scharff 2011). In addition, and as pointed out in the introduction, feminism now has a general cultural presence and this permeates post-feminism. However, this takes a specific form, that of second-wave liberal feminism. As a perspective liberal feminism argues that as human beings women and men are equal to one another and so therefore should be equal in the context of work, electoral institutions, and the private sphere of the home (Changfoot 2009). A 'good' society from the viewpoint of liberal feminism is one based on the notion of 'abstract individualism independent of social context' where all individuals can exercise autonomy to secure access to the resources that a society has to offer (Calas *et al.* 2007: 81). The aim is to promote equality within what are generally believed to be neutral societal and business structures, with an emphasis placed on the notion of equal rights to scarce resources and the securing of 'gender justice' (Calas and Smircich 1996, 2006; Simpson and Lewis 2005, 2007). As a perspective, liberal feminism is reflective of a 'politics of optimism' whereby gender differences can be eradicated allowing women to advance on a non-conflictual basis, inciting little response from men (Blum and Smith 1988; Childs and Krook 2008). However liberal feminism has been criticized, first, because it establishes a 'gender dichotomous logic' which sets up a zero sum tension between male and female, i.e. if one group is doing well, the other must be doing badly (Ringrose 2007). Second, liberal feminism overlooks a gender bias that favours masculinity as a source of cultural priority and relative advantage (Lewis and Simpson 2012), an issue we will come back to below.

As liberal feminism is the feminist perspective which permeates the contemporary cultural milieu, three of the post-feminist features highlighted by Gill (2007a) have particular relevance for female entrepreneurship. These are individualism, choice, and empowerment; notions of 'natural' sexual difference and retreat to the home as a matter of choice not obligation. These three elements capture the tension between feminism (understood in terms of entrepreneurial achievement in the public world of work) and femininity (understood in terms of domestic responsibilities in the private world of the home) between which, I suggest, entrepreneurial femininities emerge.

Individualism, Choice, and Empowerment

Commentators such as Giddens (1991) and Beck and Beck-Gernsheim (2001) argue that changes in economic and social life have contributed to the disintegration of previously existing institutions and social categories such as family and class, which no longer shape and determine the form that people's lives will take. This has contributed to an increased importance being attached to the individual, with a particular emphasis placed on the role individual choice plays in the outcomes of people's lives. There is a contemporary assumption that individuals shape their own destinies through the choices they make and as such are required to construct, perform, and pull together their biographies themselves (Beck 1992). Thus, in contemporary times the 'self' is understood to be something which is produced and constantly reflected upon so that 'we are, not what we are, but what we make of ourselves' (Giddens 1991: 75). From this perspective inequalities connected to position in the social structure, such as whether an individual is male or female, are understood as individual problems and attached to this is the belief that individual energy, effort, and achievement is enough to surmount social constraints (Rich 2005; Scharff 2011). This obscuring of underlying social structures by individualism, translates into a pervasive belief, particularly among young women, that while gender inequalities may persist they believe that they do not impact them. Alternatively if they do ever face gender constraints they assert that they can overcome such restrictions through their own personal effort and determination (Gonick 2004; Rich 2005; Scharff 2011). The demonstration of entrepreneurship through the setting up and running of a business is a particular manifestation of individualization which is perceived to emphasize independence, self-reliance, energy, and control, which fits neatly with contemporary assertions that through individual effort limitations connected to gender can be circumvented (Skeggs 1997).

Post-feminism and neo-liberalism both place a strong emphasis on individualism and completely disregard the possibility that individuals may be subject to constraints and restrictions which are socially located, that is, emerge from outside of them and which may have the potential to hinder individual activities. As such neither post-feminism nor neoliberalism recognize that individualization can be characterized as masculine, in that the notion of 'the individual' associated with it derives from discourses of Enlightenment rationality which discount and ignore women. According to Cronin (2000: 274), gender, race and other forms of difference have been 'structured out' as 'the individual' is 'an exclusive and politically privileged category'. Access to this category is restricted to men who have discursive admittance 'to the ideal of a unitary, temporally and spatially bounded selfhood'. What this means is that post-feminism, and the notion of choice associated with it, relies upon an unacknowledged masculine subjectivity which women must adopt if they are to succeed 'in carving out (a) definitive path towards freedom' (Changfoot 2009: 18). In addition, through the emergence of a post-feminist discourse women are required to self-manage and self-discipline to a much greater extent than their male colleagues. Women are required to transform the self, pay close attention to their every move, and to convey the notion that everything they do is freely chosen, leading Gill (2007a) to ask if women are constructed as the ideal subjects of neo-liberalism and the entrepreneurial activity associated with it. In exploring how processes of individualization are experienced by women, McRobbie (2009) coined the phrase 'female individualization' to highlight that it is a gendered concept. Specifically she argues that, while women are now able (or required) to 'write' their own biographies and to create their own life in similar ways to their male colleagues, they must do so while still maintaining a foothold in the domestic realm, a positioning which they cannot easily 'shake off' and which, through the theme of retreatism (discussed below), they are encouraged to return to.

'Natural' Sexual Difference

The pursuit of equality in the 1960s and 1970s was justified on the grounds that men and women were the same, i.e. they were both 'human' and as such both were entitled to have access to the resources that society had to offer. In the 1980s and 1990s this gave way to the notion of a fundamental distinction between male and female, connected to the emergence of a discourse of difference based on the notion of absolute 'natural' sexual variation, leading to an emphasis on particular features which distinguish masculinity from femininity. This notion of variation has given rise to a number of feminisms of 'difference', including a liberal feminism of difference which while continuing to place an emphasis on equality of rights also stresses the complementarities of 'natural' difference (Calas *et al.* 2007). 'Natural' difference is also central to gender-cultural feminism which accentuates a feminine ethic of care and nurture and has been taken up within organization studies through the women in management literature, a body of work which is consistent with liberal feminist thinking. This corpus of research argues that women's difference should not be seen as a defect or fault but rather as an advantage for corporate effectiveness in the twenty-first century (Calas *et al.* 2007). While some writers (e.g. Rosener 1990) have suggested that 'women's ways of knowing' should inform contemporary management practice, the reality is that organizational and management rituals and procedures informed by a female ethic exist (if they manifest a presence at all) alongside those connected to men's traditional dominance of organizations, with interpretations of this coexistence varying between feminist perspectives.

For feminist perspectives such as radical feminism or poststructuralist feminism, this coexistence is not characterized by equality or complementarity but an ongoing privileging of masculinity. These feminist perspectives connect women's subordination to their difference from the masculine norm and the hierarchical relationship between masculinity and femininity, with the former being prioritized over the latter which *is* '*constituted within binary logic as nothing*' (Paechter 2006a: 123). Within this dualism masculinity is associated with the mind, rationality, science, the economy, politics, and public life in general, while femininity is linked to the body, irrationality, emotion, the domestic, and the private realm. One consequence of the continued privileging of masculinity is the asymmetrical change in men and women's performance and experience of activities associated with one or other gender. While women have made significant inroads into educational arenas traditionally dominated by men, integrating into occupations which were in the past male-dominated, such as management, law, or medicine, men have not entered into 'female' fields in the same numbers largely due to the ongoing devaluation of activities and arenas associated with the feminine (England 2010). In the personal sphere, conventions symbolizing male dominance are even more entrenched—girls play with boys' toys but not vice versa; women routinely wear trousers but men rarely wear skirts; men rarely take women's last name upon marriage while women often take the name of their male partner. This asymmetry is connected to the ongoing disparagement of those aspects of social life which are culturally defined as feminine (England 2010). As Changfoot (2009: 22) argues, equality 'won for women, while substantive in the moment, could be short-lived especially in a context where masculine subjectivity predominates and its inherent constitutive dynamic predicts a relentless dynamic that "others" women'. Nevertheless, despite the extensive research which highlights the hierarchical relationship between masculinity and femininity, this hierarchy is not recognized within post-feminism. Rather what is emphasized is the notion of complementarity which connects to the issue of choice, with 'difference' understood as a key reason for variations in women's 'choices' when compared to those of their male colleagues.

Female Retreatism—Return to the Domestic

In her article 'Come Back Superwoman, the World Needs you!', Zoe Williams (2011) asks 'Didn't our mothers just toss us into the garden and go back to chatting to their friends? When did parenting become so hands-on?' The answer to this question is connected to the phenomenon of retreatism (also referred to as 'opt-out') which is one of the key social practices of post-feminist culture and is connected to the repositioning of women in the home. Retreatism is characterized by less continuous condemnation of the 'working mother' and more emphasis placed on the notion of 'choice'. From this perspective women's relocation back within the terms of traditional gender hierarchies by staying at home with children is interpreted as a matter of *choice* rather than obligation. The retreatist 'plot' entails the presentation of professional work as unrewarding and location in the domestic sphere as a form of salvation—summed up by Probyn (1990: 151) as follows: 'the world's a crazy place and you have to fight for yourself but at the end of the day you can always go home'. Embedded in this 'plot' are themes of 'miswanted' professional aspirations and 'adjusted ambition' which seek to discredit the meaning and value of work for women. This important element of post-feminism is a significant break with second-wave feminist perspectives such as liberal feminism which has at its centre the polemically and historically formative opposition between feminist and housewife (Brunsdon 2000; Johnson and Lloyd 2004). This opposition is succinctly summed up in the film *The Iron Lady* when Margaret Thatcher (though she did not align herself with feminism), says in the context of being proposed to by her future husband:

> I will never be one of those women Dennis who stays silent and pretty on the arm of her husband or remote and alone in the kitchen doing the washing up for that matter... One's life must matter Dennis beyond all the cooking and the cleaning and the children, one's life must mean more than that. I cannot die washing up a teacup....

The tension between highly valued public achievement and less valued domesticity is clear here, with the solution from the perspective of second-wave liberal feminism being to 'leave home and the domestic behind'. In this sense liberal feminism didn't allow for 'choice' as, from its perspective, the 'choice' to work was obvious, leading it to advocate that opting for independence through work, at whatever price, was worth it (Brunsdon 2000). In contrast, post-feminism permits what Probyn (1990) refers to as 'choiceoisie', that is, the possibility of choosing between home *or* career, family *or* success in the world of work. However 'choice' here has become more than an 'either/or'. Unlike feminism which prioritizes work, the cultural expectation surrounding work in a post-feminist era is that women don't choose home *or* work but home *and* work. In situations where women remain in the workforce, a psychological distance from work is established, contributing to the downgrading of the significance of employment such that engagement with the workplace is characterized by social compromise (McRobbie 2009). Thus as Zoe Williams (2011) argues it is no longer thought sophisticated to be the type of woman who does an important job well and does motherhood less well. Rather women are now expected to *adjust* their ambition and work at a job which interests them and they do well but does not interfere with or negatively impact on any aspect of home life, such that motherhood and home-making are also done well. This is not 'having it all' but 'having just enough', an orientation which has profound implications for patterns of female labour-force participation and gender equality, requiring that limits be placed on women's engagement with the world of work. In addition it is important to highlight that this 'choice rhetoric' connects women's position (or lack of) in the workforce to their personal tastes and preferences, disconnecting individual women from social structures and presenting their decisions as operating outside any system of constraints (Stone and Lovejoy 2004; Negra 2009). Thus in answer to the question 'what do women want', post-feminism suggests that they want choice but, according to Probyn (1990: 156), 'it is choice freed of the necessity of thinking about the political and social ramifications of the act of choosing'.

Post-feminism and Emerging Entrepreneurial Femininities

All three of the post-feminist themes outlined above connect into the 'doing' of entrepreneurship between feminism and traditional femininity (Hollows 2003) and contribute to the emergence of entrepreneurial femininities within a post-feminist gender regime. While the feminist element is associated with individualization and the unacknowledged privileging of masculinity within the theme of 'natural' sexual difference, the femininity element is associated with retreatism and the devaluing of the feminine. This coexistence of feminism and femininity is striking given that second-wave feminism has always been critical of femininity and has always called for women to unreservedly reject its imposition as it signifies women's subordinate status to men (Schippers and Sapp 2012). However the emergence of a post-feminist gender regime has not only led to the move of liberal feminist principles into mainstream culture but has also facilitated the embodiment and enactment of femininity by women not as something which is imposed on them, but rather as something connected to individual choice. Here choosing to embody femininity can be a source of confidence which within a post-feminist gender regime can happily coexist, to various degrees, with the pursuit of ambition in the public world of work. The relationship between these three themes and how they link feminism and femininity are summarized in Figure 5.1.

Following this, in Figure 5.2, the entrepreneurial femininities are summarized. These, I suggest, are embedded in a historically specific post-feminist context and incorporate

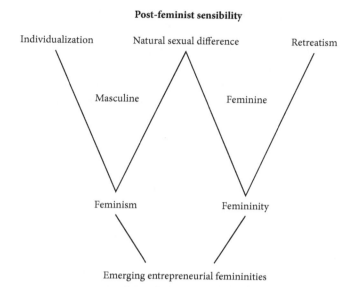

FIGURE 5.1 Positioning of emerging entrepreneurial femininities in a post-feminist representational regime

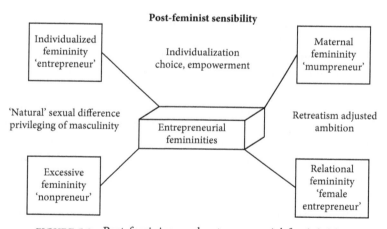

FIGURE 5.2   Post-feminism and entrepreneurial femininities

transformations in popularly available understandings of femaleness and women's positioning in the contemporary world of work. These femininities include individualized entrepreneurial femininity connected to the character of the 'entrepreneur', maternal entrepreneurial femininity encapsulated in the figure of the 'mumpreneur', relational entrepreneurial femininity represented by the 'female entrepreneur', and finally excessive femininity understood in terms of the 'nonpreneur'.

Individualized Entrepreneurial Femininity—Entrepreneur

In considering women's current levels of entrepreneurship in the UK, the Women's Enterprise Task Force identified a range of barriers—age, cultural issues, the delivery of public support services, taking time out to care for children—which it argued impede women's entrepreneurial progress. Glenda Stone, one of the co-chairs of the task force stated that her aim was 'to take women's enterprise in Britain from primarily "cottage industries" to ones that actually generate wealth, jobs and make a significant contribution to the economy' (Tyler 2007). She argued that this can be achieved by improving women's skills in business finance and technology and encouraging them to put making money and profit first. Stone's perception is that women must be empowered to make the right 'choice' of business through the removal of barriers and the provision of support and education. This will enable them to develop the 'right' skills to facilitate the 'choice' to establish a growth business which is focused on profit, as opposed to a 'hobby' business which makes no money.

In articulating such a position Glenda Stone can be characterized as manifesting an individualized entrepreneurial femininity which emphasizes the gender neutrality and meritocracy of the world of entrepreneurship, and asserts that the 'doing' of business basically requires 'the ability to abide by "universal" standards of good business' practice (Lewis 2006: 458). From this perspective women aren't as successful as they could be because they set up the wrong type of business or don't have the appropriate skills. Success in entrepreneurship is therefore understood as the effective performance of fundamentally gender-neutral methods, routines, and rituals, allied to a strong belief that individuals have an equal chance to succeed if they are ambitious and hard-working. The post-feminist subtext embedded in this mode of entrepreneurial femininity is that women are no longer 'oppressed' but instead are active, dynamic individuals who can 'choose' the nature of their entrepreneurial activities. Women who 'do' individualized entrepreneurial femininity establish a strict separation between home and work by living a 'dual sphere life'. According to Bourne and Calas (2012) this requires that an active ongoing separation is performed between the domestic sphere which contains 'not-real work' and the public sphere which contains 'real work'. Thus, instead of bringing home motherhood, and entrepreneurial activity together as mumpreneurs do (see below), these women are involved 'in symbolic forms of behavior to differentiate between their domestic and market-based activities, no matter how intertwined these may [seem] to a casual observer' (Bourne and Calas 2012: 5).

The emphasis placed on 'choice' and 'neutrality' in terms of 'doing' business in the form connected to this particular entrepreneurial femininity offers a rational, unified, and deliberate image of the women who perform it. Indeed, the autonomous, freely choosing subject who 'does' this individualized entrepreneurial femininity and 'does' business as a neutral activity 'appears peculiarly affectless, apparently not governed by any force' except herself (Gill 2007b: 76). This belief in neutrality alongside the post-feminist claim that prospects have never been better for women since gender 'issues' have been 'solved' means that these women are silent about the masculinist paradigm in which entrepreneurship is embedded (Bruni *et al.* 2004a; Lewis 2006). In this sense a woman who emphasizes neutrality and 'does' this mode of entrepreneurial femininity acts as if she is 'contextless' and cannot openly acknowledge the inherent masculinity of the entrepreneurship she does nor how the power of this works through her as a subject 'not in terms of crude manipulation, but by structuring (her) sense of self, by construct- ing (a) particular kind of (masculine) subjectivity' (Gill 2007b: 76). Nevertheless while the women who do entrepreneurship through the doing of an individualized entre- preneurial femininity do not acknowledge their doing of masculinity, they do so indi- rectly by distancing themselves from the practices and traits of traditional femininity and from women (e.g. mumpreneurs) who engage in conventional feminine behaviour. This highlights their understanding of the gender rules they are subject to. Distancing behaviour is engaged in so that women who 'do' individualized entrepreneurial femi- ninity will not be perceived as being different from the (unacknowledged) masculine norm of entrepreneurship (Lewis 2006). Thus in emerging between feminism and femininity, this individualized entrepreneurial femininity is located more on the side of (liberal) feminism. Women who 'do' individualized entrepreneurial femininity are trying to forge a way of being an entrepreneurial woman who is not associated with the domestic, private world of home and conventional femininity. Thus femininity is present in the mode of individualized entrepreneurial femininity largely through its repudiation secured through the stress placed on the neutrality and meritocracy of the entrepreneurial world and the establishment of clear boundaries between home and entrepreneurial work.

Maternal Entrepreneurial Femininity—Mumpreneur

The mumpreneur represents a new mode of entrepreneurial femininity based on the maternal and epitomizes the contemporary post-feminist entrepreneur being a 'willing subject of economic capacity while also undertaking to retain (her) traditionally marked out role in the household' (McRobbie 2009: 81). The mumpreneur can be connected to the long-standing view that a key reason why women set up a business is to secure flexibility around the performance of domestic and childcare responsibilities, a neces- sity referred to as the 'double shift' (Marlow 1997). Nevertheless, it is important to note that, with the emergence of a post-feminist sensibility and its elements of individualism, retreatism, and 'natural' sexual difference, the relationship between home and work for women in general and mumpreneurs in particular has different cultural connotations attached to it. Conventionally much of the research which explores the tension between domesticity and entrepreneurial achievement highlights the negative impact that the former has on the latter (Hamilton 2006; Rouse and Kitching 2006). For example, Hundley (2000, 2001) has outlined how marriage has a negative effect on the earnings of women business owners, with income decreasing when family size and time spent on domestic duties increase. Similarly research by Jennings and McDougald (2007), based on a comparison of male and female business owners and the economic performance of their businesses, concluded that differences between men and women around the work–family interface can account for the smaller number of employees, the lower levels of revenue, and the depressed income levels characteristic of many female-owned firms. Women in this position are normally advised to establish a clear division between home and the business, that is, to leave their 'home selves' behind so that they can be more ambitious, take risks, and expand their entrepreneurial activities. However, what appears to be a choice between independence and femininity in general or entrepreneurial achievement and domesticity in particular is one which assumes a unified and coherent feminist subject (see individualized entrepreneurial femininity above) connected to second-wave liberal feminism. This unified subject is held out to women as something which they are required to pursue but is a problem for women today, given the emergence of changing modes of femininity (Johnson and Lloyd 2004). For example, the maternal femininity manifest in the figure of the mumpreneur in the entrepreneurial realm is a clear rejection of such a choice, as central to the activities of the mumpreneur is the establishment of a specific link between her motherhood and entrepreneurial activities, not the rejection of home-making and mothering endeavours.

The emergence of the 'mumpreneur' as an entrepreneurial character is a fairly recent phenomenon. However, despite positive accounts of the mumpreneur experience, the contradictions associated with being a business person and mother manifest in this particular entrepreneurial femininity means that women who seek independence and self-reliance through entrepreneurship do so by setting limits on their participation in the business world. Thus domestic and childcare responsibilities are not something which are overcome on the way to entrepreneurial success but rather create tensions and determine the limits of entrepreneurial endeavours (Lewis 2010; Duberley and Carrigan 2012). Nevertheless the living of this femininity means that mumpreneurs are engaged in what Carlson (2011: 88) refers to as 'the strategic crossing of gender boundaries', being engaged in both masculine and feminine-marked realms. Thus while the symbolic space for entrepreneurship and the self-actualization associated with it is normally the public world of work, for mumpreneurs the symbolic space for 'doing' maternal entrepreneurial femininity is the spaces and places of motherhood such as the home and school gate. A key trait of mumpreneurial activity is the valorization of mothering and the home as a source of entrepreneurial skills, creativity, and innovation. This contributes to the emergence of an alternative mode of entrepreneurship connected to an ethic of care, with mumpreneur businesses tending to offer a product or service which is associated with family and motherhood. Thus the mumpreneur's focus is on what will not only fill a market gap but also connect to women's traditional caring responsibilities of looking after home and children (Lewis 2010; Ekinsmyth 2011).

Relational Entrepreneurial Femininity—Female Entrepreneur

Commentators such as Fletcher (1998) have argued that the dominance of masculinity, built into the structure of work and organizations, has silenced ways of working connected to a relational or stereotypically feminine belief system which emphasizes that development is more likely to flourish in a context of connection. According to Fletcher (1998: 167), 'it is the preeminence of connection over individuation in the developmental process that gives relational theory its potential as a destabilizing alternative voice in organizational discourse and marks it as feminine'. The post-feminist emphasis on 'natural' sexual variation has, within the context of organizations, contributed to increased recognition of and value placed on the skills, attributes, and leadership styles of women. The emergence of a discourse of difference highlights the special contribution that women's so-called uniquely feminine viewpoint can bring to management and organizations, emphasizing its dissimilarity from the conventional masculine character of managing and organizing (Calas and Smircich 1993; Fletcher 1994; Metcalfe and Linstead 2003). By drawing on this discourse women's difference is produced through suggestions that women lead through participation, power-sharing, and information exchange; demonstrated by a preference for putting relationships with people ahead of money and possessions; rejecting hierarchical relationships in favour of a team-based approach characterized by understanding and sympathy for others; valuing intuition over reason and decisive argument; favouring interdependence over an ideal of independence and putting an emphasis on cooperation and networking over competition and growth (Simpson and Lewis 2007).

This position is present in the entrepreneurship literature where, according to Bird and Brush (2002: 43), 'it is reasonable to assume that masculine and feminine aspects will be incorporated into...new venture(s)...(but) the feminine aspects of organizational creation and feminine dimensions of new ventures are not well articulated, and if articulated, not identified as feminine'. Nevertheless Bruni *et al.* (2004*b*) point out that there are studies which claim that an enterprise culture for women entrepreneurs has emerged which advocates that women should adopt a transformational leadership style that aims to share power, promote relations of trust with employees, and pursue collective as opposed to individual goals. Women who enact such an approach can be categorized as 'doing' relational entrepreneurial femininity and can be differentiated from those who do individualized entrepreneurial femininity by their rejection of a growth-dominated orientation to business. Research which examines such women suggests that this 'feminine' way of doing business challenges conventional definitions of what it is to be a successful entrepreneur by demonstrating commitment to an entrepreneurial model which advocates 'small and stable' business. As part of this such women also reject the implementation of management strategies which establish a divide between home and business, arguing instead for a more equal balance between the two realms (Chaganti 1986; Lee-Gosselin and Grise 1990; Buttner 2001; Fenwick 2002). While this latter element of relational entrepreneurial femininity connects those women who 'do' this femininity to the maternal entrepreneurial femininity of the mumpreneur, they are not primarily concerned with bringing motherhood and entrepreneurship together as the mumpreneur is. Rather there is greater emphasis placed on relational interaction, mutual empathy, and mutual empowerment in the way women do business. While Bruni *et al.* (2004*b*: 264) are critical of such research, arguing that it is based on 'essentialist or culturalist assumptions', it has contributed to the notion of a masculine versus a feminine way of 'doing' business. The suggestion is that women can draw on masculine or feminine discourses to either 'masculinize' (individualized entrepreneurial femininity) their identity or go through a process of 'femalization' by doing relational entrepreneurial femininity.

Yet despite the emphasis on a feminine way of doing business, Lewis (2011) highlights how women who can be categorized as doing relational entrepreneurial femininity also activate a project of masculinity. This is done by additionally drawing on a discourse of professionalism as a means of countering the danger that they will not be perceived as serious business people, due to emphasizing their difference from the masculine norm of entrepreneurship. Combining the notions of professionalism and difference allows these women to distinguish themselves from the traditional masculine approach to business which they shun, while also demonstrating that they embrace the values of a competitive market world.

Excessive Entrepreneurial Femininity—'Nonpreneur'

According to McRobbie (1993) what is notable about contemporary processes of individualization is that, while they have contributed to the extraordinary 'unfixing' of women's position in society, leading to a higher level of variability in terms of what femininity means and its location in social reality, the continued dominance of masculinity means that the rearticulation of femininity incorporates and embodies both masculine and feminine elements such that women now have to inhabit both realms and juggle both sets of attributes. This means that the work of 'doing' femininity requires that women perform and embody the feminine characteristics of nurture, emotion, passivity, attractiveness alongside the masculine traits of assertiveness, rationality, autonomy, economic and emotional independence (Gonick 2004; Walkerdine and Ringrose 2006; Ringrose 2007; Ringrose and Walkerdine 2008; McRobbie 2009). This new negotiation between the masculine and the feminine is now argued by some commentators to be *constitutive* of contemporary femininity. Thus, 'while there are sanctions for women who exclusively perform masculinity, doing femininity without doing (or having done) masculinity seems increasingly less tenable, less desirable and less (economically) liveable; rather an "independent" woman who can balance masculinity and femininity seems increasingly culturally validated' (Carlson 2011: 80).

This hybridized notion of femininity is important when considering excessive entrepreneurial femininity, which refers to those women who are deemed to display the wrong amount and wrong kind of femininity (Lawler 2001) within a business context. In recent years femininity has increasingly being constructed as valuable in both existing and new organizational contexts. While our exploration of maternal and relational entrepreneurial femininities demonstrates how femininity has been inserted very firmly into the realm of entrepreneurship (Lewis 2011), it would appear that acceptance of 'doing' femininity in the entrepreneurial arena is dependent on two things. First, the enactment of femininity within the world of business must be 'measured' and not perceived as disruptive. In other words women must be 'properly' feminine, that is, be feminine enough to benefit business but not engage in unnecessary or unwarranted feminine displays (Lewis 2012). Second, when women do enact stereotypically feminine behaviours within the domain of entrepreneurship, this encroachment must be compensated for by other behaviours which conform to the masculine norms of this arena, for example, displays of masculine professionalism. Thus while femininity has certainly established a presence within the world of business, conventional masculinity—even manifest in extreme forms—is still firmly anchored and constructed as valuable within the entrepreneurial context. Those who are deemed as being excessively feminine without the counterweight of masculinity will be interpreted as violating the hegemonic norms that are valued in an entrepreneurial context and risk rejection, which can take extreme forms. What is notable here is that, within a post-feminist representational regime, it is likely that such rejection will be seen as connected to the shortcomings of individual women and not associated with their minority presence in the ranks of entrepreneurs (Lewis 2012). This suppression 'of tracking down the links connecting individual fate to the ways and means by which society as a whole operates' (Bauman 2001: 9) means women who display what are perceived to be extreme feminine behaviours will be designated as 'nonpreneurs', i.e. not legitimate business people, and blamed for their own exclusion, with little attention directed at the structural and cultural constraints which act on them (Lewis 2012).

CONCLUSION

This chapter began by identifying two different approaches to the study of female entrepreneurship—a strand which concentrates on the individual woman entrepreneur and an alternative perspective which explores how gender, specifically masculinity, is a principle of the social organization of entrepreneurship. While women's entanglement with masculinity has been extensively examined in the female entrepreneurship literature, a neglected aspect of women's experience of business ownership is the 'doing' of femininity within the entrepreneurial arena. This chapter has sought to redress the inattention to femininity and entrepreneurship by mapping out the emergence of four entrepreneurial femininities, that is, individualized entrepreneurial femininity, maternal entrepreneurial femininity, relational entrepreneurial femininity, and excessive entrepreneurial femininity. Identification of these femininities is novel within research on female entrepreneurship, particularly within the context of a post-feminist sensibility (Gill 2007a). Extensive consideration of both femininity and post-feminism is currently absent from the literature on gender and entrepreneurship in particular and gender and organization studies in general and the contention here is that attention to both is valuable.

The value of inserting the notion of post-feminism into considerations of female entrepreneurship is that it has facilitated the depiction of multiple entrepreneurial femininities by making connections between feminism and femininity, highlighting the diversity among women business owners in terms of how they 'do' entrepreneurship. Balancing home (femininity) and work (feminism) and the diverse strategies adopted to do this by different women may at first glance appear to be nothing more than a mundane necessity which gives rise to different entrepreneurial femininities which are freely 'chosen' by women and simply coexist alongside each other. Post-feminism here is useful in highlighting the multiple ways by which women deal with the transformative impact of work alongside the continuity of domestic responsibilities through the enactment of these different entrepreneurial femininities. However, while advocating the insertion of the concept of post-feminism into an area such as female entrepreneurship to facilitate exploration of the femininity associated with women's entrepreneurial activity, it is also important to adopt a critical stance to post-feminism itself. Adopting a critical orientation to post-feminism ensures that femininity and its relationship to entrepreneurship is not simply understood as individual 'choice', with the embodiment and enactment of feminine behaviours chosen for reasons of individual confidence and self-empowerment. Rather a critical orientation involves understanding that the performance of any particular entrepreneurial femininity does not occur in a vacuum but is connected to broader relations of inequality within a particular social context. What this means is that 'access' to any particular entrepreneurial femininity is influenced by the structural position of individual women and not simply dependent on women's individual choice. So for example access to maternal entrepreneurial femininity and the take-up of the identity of 'mumpreneur' can be dependent on the class position of the woman involved. Specifically possession of a partner/spouse who can provide financially for the family without requiring input from the 'mumpreneur' is more likely to be associated with middle-class women.

As post-feminism does not take into account ongoing systemic and structural inequalities within a field such as entrepreneurship, both second-wave (e.g. socialist feminist theorizing) and third-wave feminist accounts (such as post-structuralism and transnational/(post-)colonial feminism) of women's entrepreneurial experiences continue to be important as they recognize that the take-up of any mode of femininity occurs within the structural conditions of masculine dominance and other forms of social power (Schippers and Sapp 2012). Thus the suggestion is that there is a need for post-feminism as a concept within the female entrepreneurship research arena, as it is useful in exploring the notion of femininity, but that to secure full benefit from its introduction into the female entrepreneurial research arena it should coexist with the range of feminist perspectives (Calas *et al.* 2007) already in use in the entrepreneurship field.

References

Ahl, H. (2006). 'Why Research on Women Entrepreneurs Needs New Directions', *Entrepreneurship, Theory and Practice*, 30(5): 595–621.

Allen, S., and Truman, C. (1993). 'Women and Men Entrepreneurs: Life Strategies, Business Strategies', in S. Allen, and C. Truman (eds), *Women in Business: Perspectives on Woman Entrepreneurs*, 1–19. London: Routledge.

Bauman, Z. (2001). *The Individualized Society*. Cambridge: Polity Press.

Beck, U. (1992). *Risk Society: Towards a New Modernity*. London: Sage.

Beck, U., and Beck-Gernsheim, E. (2001). *Individualisation: Institutional Individualism and its Social and Political Consequences*. London: Sage.

Berg, N. (1997). 'Gender, Place and Entrepreneurship', *Entrepreneurship and Regional Development*, 9(4): 259–68.

Bird, B., and Brush, C. (2002). 'A Gendered Perspective on Organizational Creation'. *Entrepreneurship: Theory and Practice*, 26(3): 41–65.

Blum, L., and Smith, V. (1988). 'Women's Mobility in the Corporation: A Critique of the Politics of Optimism', *Signs*, 13(3): 528–45.

Bourne, K. A., and Calas, M. B. (2012). 'Becoming "Real" Entrepreneurs: Women and the Gendered Normalization of "Work"', *Gender, Work and Organization*, doi: 10.1111/J-1468-0 432.00591.X

Braithwaite, A. (2002). 'The Personal, the Political, Third-Wave and Post-feminisms', *Feminist Theory*, 3(3): 335–44.

Bruni, A., Gherardi, G., and Poggio, B. (2004a). 'Doing Gender, Doing Entrepreneurship: An Ethnographic Account of Intertwined Practices', *Gender, Work and Organization*, 11(4): 406–29.

Bruni, A., Gherardi, G., and Poggio, B. (2004b). 'Entrepreneurship-Mentality, Gender and the Study of Women Entrepreneurs', *Journal of Organizational Change Management*, 17(3): 256–69.

Brunsdon, C. (2000). *The Feminist, the Housewife, and the Soap Opera*. Oxford: Oxford University Press.

Budgeon, S. (2001). 'Emergent Feminist (?) Identities: Young Women and the Practice of Micropolitics', *European Journal of Women's Studies*, 8(1): 7–28.

Buttner, E. H. (2001). 'Examining Female Entrepreneurs Management Style: An Application of a Relational Frame', *Journal of Business Ethics*, 29(3): 253–69.

Calas, M. B., and Smircich, L. (1993). 'Dangerous Liaisons: The "Feminine-in-Management" Meets "Globalization"', *Business Horizons*, 36(2): 71–81.

Calas, M. B., and Smircich, L. (1996). 'From the "Woman's Point of View": Feminist Approaches to Organization Studies', in S. R. Clegg, C. Hardy, and W. Nord (eds), *Handbook of Organization Studies*, 218–57. London: Sage.

Calas, M. B. and Smircich, L. (2006). 'From the "Women's Point of View" Ten Years Later: Towards a Feminist Organization Studies', in S. R. Clegg, C. Hardy, T. B. Lawrence, and W. R. Nord (eds), *The Sage Handbook of Organization Studies*, 284–346. London: Sage.

Calas, M.B., Smircich, L., and Bourne, K. A. (2007). 'Knowing Lisa? Feminist Analyses of Gender and Entrepreneurship', in D. Bilimoria and S. K. Piderit (eds), *Handbook on Women in Business and Management*, 78–105. Cheltenham: Edward Elgar.

Calas, M. B., Smircich, L., and Bourne, K. A. (2009). 'Extending the Boundaries: Reframing Entrepreneurship as Social Change through Feminist Perspectives', *Academy of Management Review*, 34(3): 552–69.

Carlson, J. (2011). 'Subjects of Stalled Revolution: A Theoretical Consideration of Contemporary American Femininity', *Feminist Theory*, 12(1): 75–91.

Carter, S., Marlow, S., and Bennett, D. (2012). 'Gender and Entrepreneurship', in S. Carter and D. Jones-Evans (eds), *Enterprise and Small Business: Principles, Practice and Policy*, 218–31. Harlow: Pearson Education.

Chaganti, R. (1986). 'Management in Women-Owned Enterprises', *Journal of Small Business Management*, 24(4): 18–29.

Changfoot, N. (2009). 'The Second Sex's Continued Relevance for Equality and Difference Feminisms'. *European Journal of Women's Studies*, 16(1): 11–31.

Childs, S., and Krook, M.L. (2008). 'Critical Mass Theory and Women's Political Representation', *Political Studies*, 56(3): 725–36.

Connell, R. W. (1987). *Gender and Power*. Cambridge: Polity Press.

Cotter, D., Hermsen, J. M., and Vanneman, R. (2011). 'The End of the Gender Revolution? Gender Role Attitudes from 1977 to 2008', *American Journal of Sociology*, 117(1): 259–89.

Cronin, A. M. (2000). 'Consumerism and Compulsory Individuality', in S. Ahmed, J. Kilby, C. Lury, M. McNeil, and B. Skeggs (eds), *Transformations: Thinking Through Feminism*, 273–87. London: Routledge.

Duberley, J., and Carrigan, M. (2012). 'The Career Narratives of Mumpreneurs: Women's Experiences of Combining Enterprise and Motherhood', *International Small Business Journal*, doi: 10.1177/0266242611435182.

Ekinsymth, C. (2011). 'Challenging the Boundaries of Entrepreneurship: The Spatialities and Practices of UK "Mumpreneurs"', *Geoforum*, 42(1): 104–14.

England, P. (2010). 'The Gender Revolution: Uneven and Stalled', *Gender and Society*, 24(2): 149–66.

Essers, C., and Benschop, Y. (2007). 'Enterprising Identities: Female Entrepreneurs of Moroccan and Turkish Origin in the Netherlands', *Organization Studies*, 28(1): 46–69.

Essers, C., and Benschop, Y. (2010). 'Female Ethnicity: Understanding Muslim Migrant Business Women in the Netherlands', *Gender, Work and Organization*, 17(3): 320–40.

Faludi, S. (1991). *Backlash: The Undeclared War Against American Women*. New York: Crown.

Fenwick. T. (2002). 'Transgressive Desires: New Enterprising Selves in the New Capitalism', *Work, Employment and Society*, 16(4): 703–23.

Fletcher, J. K. (1994). 'Castrating the Female Advantage: Feminist Standpoint Research and Management Science', *Journal of Management Inquiry*, 3(1): 74–82.

Fletcher, J. K. (1998). 'Relational Practice: A Feminist Reconstruction of Work', *Journal of Management Inquiry*, 7/2: 163–86.

Gamber, W. (1998). 'A Gendered Enterprise: Placing Nineteenth-Century Business Women in History', *Business History Review*, 72 (Summer): 188–218.

Giddens A. (1991). *Modernity and Self-Identity: Self and Society in the Late Modern Age*. Cambridge: Polity Press.

Gill, R. (2007a). 'Postfeminist Media Culture: Elements of a Sensibility', *European Journal of Cultural Studies*, 10(2): 147–66.

Gill, R. (2007b). 'Critical Respect: The Difficulties and Dilemmas of Agency and "Choice" for Feminism', *European Journal of Women's Studies*, 14(1): 69–80.

Gill, R., and Scharff, C. (2011). 'Introduction', in R. Gill and C. Scharff (eds), *New Femininities: Postfeminism, Neoliberalism and Subjectivity*, 1–17. Basingstoke: Palgrave Macmillan.

Gonick, M. (2004). 'Old Plots and New Identities: Ambivalent Femininities in Late Modernity', *Discourse: Studies in the Cultural Politics of Education*, 25(2): 189–209.

Green, E., and Cohen, L. (1995). 'Women's Business: Are Women Entrepreneurs Breaking New Ground or Simply Balancing the Demands of "Women's Work" in a New Way?', *Journal of Gender Studies*, 4(3): 297–314.

Greer, M. J., and Greene, P. G. (2003). 'Feminist Theory and the Study of Entrepreneurship', in J. E. Butler (ed.), *New Perspectives on Women Entrepreneurs*, 1–24 Greenwich CT: Information Age Publishing.

Hamilton, E. (2006). 'Whose Story is it Anyway? Narrative Accounts of the Role of Women in Founding and Establishing Family Business', *International Small Business Journal*, 24(3): 253–71.

Holmlund, C. (2005). 'Post-feminism from A to G', *Cinema Journal*, 44(2): 116–22.

Hollows, J. (2003). 'Feeling like a Domestic Goddess: Post-feminism and Cooking', *European Journal of Cultural Studies*, 6(2): 179–202.

Hundley, G. (2000). 'Male/Female Earning Differences in Self-Employment: The Effects of Marriage, Children and the Household Division of Labour', *Industrial and Labour Relations Review*, 54(1): 95–114.

Hundley, G. (2001). 'Why Women Earn Less than Men in Self-Employment', *Journal of Labour Research*, 22(4): 817–29.

Jennings, J. E., and McDougald, M. S. (2007). 'Work–Family Interface Experiences and Coping Strategies: Implications for Entrepreneurship Research and Practice', *Academy of Management Review*, 32(3): 747–60.

Johnson, L., and Lloyd, J. (2004). *Sentenced to Everyday Life: Feminism and the Housewife*. Oxford: Berg Publishers.

Lawler, S. (2001). 'Mobs and Monsters: Independent Man Meets Paulsgrove Woman', *Feminist Theory*, 3(1): 103–13.

Lee-Gosselin, H., and Grise, J. (1990). 'Are Women Owner-Managers Challenging our Definitions of Entrepreneurship? An In-Depth Survey', *Journal of Business Ethics*, 9(4–5): 423–33.

Lewis, P. (2006). 'The Quest for Invisibility: Female Entrepreneurs and the Masculine Norm of Entrepreneurship', *Gender, Work and Organization*, 13(5): 453–69.

Lewis, P. (2010). 'Mumpreneurs: Revealing the Post-feminist Entrepreneur', in P. Lewis and R. Simpson (eds), *Revealing and Concealing Gender: Issues of Visibility in Organizations*, 124–38. Basingstoke: Palgrave Macmillan.

Lewis, P. (2011). 'The Search for an Authentic Entrepreneurial Identity: Difference and Professionalism among Women Business Owners', *Gender, Work and Organization*, Early View, doi:10.1111/j.1468-0432.2011.00568.x

Lewis, P. (2012). 'Post-feminism and Entrepreneurship: Interpreting Disgust in a Female Entrepreneurial Narrative', in R. Simpson, N. Slutskaya, P. Lewis, and H. Höpfl (eds), *Dirty Work: Concepts and Identities*, 223–38. Basingstoke: Palgrave Macmillan.

Lewis, P., and Simpson, R. (2012). 'Kanter Revisited: Gender, Power and (In)visibility', *International Journal of Management Reviews*, 14(2): 141–58.

Loscocco, K., and Bird, S. R. (2012). 'Gendered Paths: Why Women Lag Behind Men in Small Business Success', *Work and Occupations*, 39(2): 183–219.

McRobbie, A. (1993). 'Shut up and Dance: Youth Culture and Changing Modes of Femininity', *Cultural Studies*, 7(3): 406–26.

McRobbie, A. (2004). 'Post-feminism and Popular Culture: Bridget Jones and the New Gender Regime', *Feminist Media Studies*, 4(3): 255–64.

McRobbie, A. (2009). *The Aftermath of Feminism*. London: Sage.

Marlow, S. (1997). 'Self-Employed Women—New Opportunities, Old Challenges?', *Entrepreneurship and Regional Development*, 19(3): 199–210.

Marlow, S. (2002). 'Self-Employed Women: A Part of or Apart from Feminist Theory?', *International Journal of Entrepreneurship and Innovation*, 2(1): 83–91.

Marlow, S., Henry, C., and Carter, S. (2009). 'Exploring the Impact of Gender upon Women's Business Ownership', *International Small Business Journal*, 27(2): 139–48.

Metcalfe, B., and Linstead, A. (2003). 'Gendering Teamwork: Rewriting the Feminine', *Gender, Work and Organization*, 10(1): 95–119.

Mirchandani, K. (1999). 'Feminist Insight on Gendered Work: New Directions in Research on Women and Entrepreneurship', *Gender, Work and Organization*, 6(4): 224–36.

Mulholland, K. (1996). 'Entrepreneurial Masculinities and the Self-Made Man', in D. L. Collinson and J. Hearn (eds), *Men as Managers, Managers as Men*, 123–49. London: Sage.

Negra, D. (2009). *What a Girl Wants? Fantasizing the Reclamation of Self in Post-feminism*. London: Routledge.

Ogbor, J. O. (2000). 'Mythicizing and Reification in Entrepreneurial Discourse: Ideology-Critique of Entrepreneurial Studies', *Journal of Management Studies*, 37(5): 605–35.

Paechter, C. (2006a). 'Masculine Femininities/Feminine Masculinities: Power, Identities and Gender', *Gender and Education*, 18(3): 253–63.

Probyn, E. (1990). 'New Traditionalism and Post-feminism: TV Does the Home', *Screen*, 31(2): 147–59.

Reay, D. (2001). '"Spice Girls", "Nice Girls", "Girlies" and "Tomboys": Gender Discourses, Girls' Culture and Femininities in the Primary Classroom', *Gender and Education*, 13(2): 153–66.

Reed, R. (1996). 'Entrepreneurialism and Paternalism in Australian Management: A Gender Critique of the Self-Made Man', in D. L. Collinson and J. Hearn (eds), *Men as Managers, Managers as Men*, 99–122. London: Sage.

Rich, E. (2005). 'Young Women, Feminist Identities and Neo-Liberalism', *Women's Studies International Forum*, 28(6): 495–508.

Ringrose, J. (2007). 'Successful Girls? Complicating Post-Feminist, Neoliberal Discourses of Educational Achievement and Gender Equality', *Gender and Education*, 19(4): 471–89.

Ringrose, J., and Walkerdine, V. (2008). 'Regulating the Abject: The TV Make-Over as Site of Neo-Liberal Reinvention toward Bourgeois Femininity', *Feminist Media Studies*, 8(3): 228–46.

Rosener, J. B. (1990). 'Ways Women Lead', *Harvard Business Review* (Nov.–Dec.). Reprint number 90608. 3–10

Rouse, J., and Kitchings, J. (2006). 'Do Enterprise Support Programmes Leave Women Holding the Baby?', *Environment and Planning C: Government and Policy*, 24(1): 5–19.

Scharff, C. (2011). 'Disarticulating Feminism: Individualization, Neoliberalism and the Othering of "Muslim Women"', *European Journal of Women's Studies*, 18/2: 119–34.

Schippers, M., and Sapp, E. G. (2012). 'Reading Pulp Fiction: Femininity and Power in Second and Third Wave Theory', *Feminist Theory*, 13(1): 27–42.

Simpson, R., and Lewis, P. (2005). 'An Investigation of Silence and a Scrutiny of Transparency: Re-examining Gender in Organization Literature through the Concepts of Voice and Visibility', *Human Relations*, 58(10): 1253–75.

Simpson, R., and Lewis, P. (2007). *Voice, Visibility and the Gendering of Organizations*. Basingstoke: Palgrave Macmillan.

Skeggs, B. (1997). *Class, Self, Culture*. London: Routledge.

Stone, P., and Lovejoy, M. (2004). 'Fast-Track Women and the "Choice" to Stay Home', *Annals of the American Academy of Political and Social Science*, 596(1): 62–83.

Thornham, H., and McFarland, A. (2011). 'Cross-Generational Gender Constructions: Women, Teenagers and Technology', *Sociological Review*, 59(1): 64–85.

Tyler, R. (2007). 'Enterprise Unleashing Britain's "Mumpreneurs": The Women in Charge of Getting Others to Run their own Firms Talks to Richard Tyler about Encouraging Risk-Taking', *Daily Telegraph*, 23 Jan.

Walkerdine, V., and Ringrose, J. (2006). 'Femininities: Reclassifying Upward Mobility and the Neoliberal Subject', in B. Francis and C. Skelton (eds), *The Sage Handbook of Gender and Education*, 62–80. Thousand Oaks, CA: Sage.

Williams, Z. (2011). 'Come Back Superwoman: The Lost Ideal of Combining Motherhood and Work', *Guardian*, 17 Sept.

Women's Enterprise Task Force (2009). *Greater Return on Women's Enterprise: The UK Women's Enterprise Task Force's Final Report and Recommendations*. London: Women's Enterprise Task Force.

CHAPTER 6

....

'MEANING THAT MATTERS'

An Organizational Communication Perspective on Gender, Discourse, and Materiality

....

KAREN LEE ASHCRAFT AND
KATE LOCKWOOD HARRIS

FOR the last twenty years, gender and organization scholars have treated discourse as a consequential matter. In its broadest sense, 'discourse' is used to signify a range of symbolic activities, such as language, interaction, imagery, story-telling, and loose amalgamations of texts and practices. Typically, discourse has been ascribed to the ideational and social rather than material realms. Increasingly, however, discursive activity is said to manifest materially—to constitute not only social and psychological realities, but also corporeal, temporal, spatial, economic, and institutional realities. Gender and organization scholarship that explicates the material force of discourse is aligned with a 'constitutive' view, in that it departs from a more conventional view of discourse as merely expressing, transmitting, or maintaining already existing realities (Ashcraft et al. 2009).

A constitutive view is evident, for example, in the common definition of discourse among poststructuralist feminists as 'practices that form the object of which they speak' (Foucault 1972: 49). This definition asserts the 'performativity' of organizing— a concept meant to capture not only the dramaturgical texture of everyday life, but also how mundane performances serve a generative function (Butler 1999). In brief, discourse brings into being the material of organizing. Gendered organization occurs as we enact narrative formations—like entrepreneur, stewardship, professionalism, the boss–secretary pairing, feminine leadership, or sexual harassment—which make it possible for us to know, inhabit, and configure 'work' and 'organizations' in circumscribed ways. In gender and organization studies, poststructuralist feminism has become the primary way to explicate the material force of discourse, though it is not the only means.

This chapter aims to further develop a constitutive view for the context of gender and organization studies. It proceeds along two tracks: introducing those less familiar with the discourse literature to the contours of a constitutive view while advancing constitutive theorizing such that it contends more fully with materiality. We begin by reviewing available claims that discourse 'en/genders' work and organization; and we argue for a more nuanced account that renders the discourse–materiality relation such that neither is privileged and their persistent division dissolves. To facilitate this development, we draw on our 'home' discipline of organizational communication theory, which offers conceptual resources that are not yet shared among gender and organization scholars but can help us reimagine how discourse 'matters'. Specifically, we theorize communication as the process through which symbol and material meet and meld into the tangible realities of work and organizational life. To demonstrate the potential of our proposal, we analyse how it 'reorganizes' the discursive and material complexities of gendered violence. Finally, we sift this analysis for the future tasks it signals.

Discourse Makes a Difference: Constitutive Claims in Gender and Organization Studies

By way of context, we briefly consider how scholars have conceptualized discourse in gender and organization studies. Addressing this question nearly ten years ago, Ashcraft (2004) identified four major frames—or ways of seeing the relationship among gender, discourse, and organization—that characterize the literature. The first frame, 'discourse as outcome', holds that different styles of interacting stem from gender identity. Those identified with masculinity tend to share a transactional, outcome-centred view of communication, whereas those identified with femininity lean towards a relational, process-centred view. This binary split in perception yields predictable habits of interacting in organizations: contrasting modes of leading, engaging in conflict, working in teams, and managing emotion, for example.

The causal direction declared by frame one is reversed in frame two, 'discourse as performance'. Instead of casting one's interaction style as a stable byproduct of the gender binary, proponents of this frame maintain that apparently stable, binary difference is an effect of interaction. Put another way, we constantly manage our organizational behaviour in light of normative expectations for gender difference, and we do so in concert and negotiation with others—in short, through interaction (West and Zimmerman 1987). Gender here is an ongoing accomplishment, achieved through audience participation. In this light, the so-called feminine style of leading is an organizational expectation that women managers face. Whatever aptitude some may display stems from years of situated pressure and responsive rehearsal, not from a psychological trait they possess and express. Two strands of this frame are apparent in gender and organization studies: (*a*) the 'doing gender, doing difference' model associated with West and colleagues' socio-logical, ethnomethodological approach (see Fenstermaker and West 2002); and (*b*) the performativity model associated with Butler's (1993, 1999) poststructuralist approach.

Whereas frames one and two stress micro-interaction, the third frame, 'text–conver-sation dialectic', refigures organization structure or form of governance (e.g. bureau-cracy) as a regulatory 'text' (i.e. durable formation spanning space and time) that scripts 'conversation' (i.e. lively, transitory local interaction).[1] According to frame three, gender difference is an organizing principle of such scripts, which prescribe gendered divisions, hierarchies, enactments, and controls of labour before members enter the door. Like frame two, this premise upends frame one's suggestion that people bring fully formed gender skills with them into gender-neutral organizations. Frames two and three agree that mundane interaction may tweak the very scripts it follows, yielding innovations in organizational form. But while frame two highlights such performances, frame three brings into relief how organizations are culpable for *producing*, not only profiting from, so-called gender skills. A range of conflicting theoretical influences—radical and post-structuralist feminisms, for example—co-mingle in the literature associated with this frame (for more, see Ashcraft in press).

The final frame, 'discourse as social text', can be called extra-organizational, in that it leaves the physical sites of work to examine where else scripts for gendering work and organization develop. Scholars adopting this frame have explored representations of gendered organizing in science (e.g. management and organization theory) as well as trade and 'popular' culture (e.g., industry and occupational imagery, television and film), motivated by the performativity premise that these cultural formations condition our participation in work and organizing. Social texts are gendering agents that actu-ate organizational possibilities; they do not simply and passively reflect extant practice. Most studies emanating from this frame are avowedly poststructuralist.

Research rooted in these frames has yielded rich theoretical, empirical, and practi-cal insight. Frame one, for example, exposed how preferred modes of organizational interaction (e.g. 'competent' professional, 'authoritative' manager) privilege norms asso-ciated with masculinity. By challenging the ontological security of the gender binary, frame two surfaced the remarkable variety of gender performances (i.e. masculini*ties* and feminini*ties*) possible as gender entangles with other social identity norms in situ-ated organizational interaction. Frame three revealed the vital role of organizational form-as-text in the pursuit of gender equity at work and advanced creative hybrid scripts (e.g. feminist bureaucracy) to help members navigate associated dilemmas of power and participation. Frame four instigated awareness that gendered organizing has many cultural stages, not only within but well beyond 'the workplace'.

Pertinent to our effort here, this review prompts several interlinked observations about how the four frames conceive of discourse *per se*. First, (*a*) as a phenomenon or object of study, discourse shifts rather wildly in meaning across the frames—from a personal communication style determined by one's gender disposition (outcome), to ongoing interaction that creates the seeming 'reality' of gender difference (perfor-mance), to enduring local scripts that members follow and occasionally improvise (text-conversation), to societal narratives that activate working subjectivities (social text). As this hints, (*b*) the discursive phenomena of interest are increasingly abstracted across the frames, moving from 'micro' to 'macro', or from lower to upper case d/ Discourse (Alvesson and Karreman 2000). Specifically, the spotlight turns from the individual expression of an internal state (outcome), to the daily negotiation of highly situational scripts (performance), to the heavy influence of isomorphic 'meso' scripts (text-conversation), to the potent formation of broadly shared cultural constructions (social text). Of course, the progression implied here is complicated in practice. Several studies, for instance, reach across levels of discursive activity (e.g. Nadesan and Trethewey 2000). Our point is that what counts as 'discourse' in gender and organization studies has grown in scope over time.

A third observation follows from this expansion of level and scope: (*c*) the 'containment' of discourse also varies across the frames, as does curiosity about the 'container' itself. Frames one (outcome) and two (performance) highlight gender *in* organizational settings. Even as they differ, both accept 'the organization' as a kind of housing for interaction. Frame three (text-conversation) emphasizes the gender *of* organization, bringing the container under scrutiny such that it is exposed as an active, evolving textual agent, not a neutral, static receptacle. To date, this is the frame most engrossed with the ontology of organizing, at pains to show how gender and organization are mutually constitutive, though the demonstrations often lack precision (Britton 2000). Frame four (social text) shatters the container, or severely stretches its boundaries, by insisting on the relevance of discursive formations that seem detached from actual workplaces. This frame maintains that discourse permeates and transcends site—a premise captured by the concept of 'intertextuality' (see Taylor 1999). In this sense, frame four challenges the previous three by problematizing organizational boundaries. Yet its explanation of how sites and levels of discursive activity become coupled (i.e. the ways in which texts 'travel' over place and time) often remains vague as well.[2]

Fourth, we can observe that (*d*) the dimensionality of discourse—to what extent it is 'flat' or multifaceted, fixed or moving—differs across the frames. Discourse in the first frame (outcome) is a practice born of reflex, steady and unidimensional, caused by a predisposition resistant to change and context. In the second frame (performance), discourse appears most fluid and animated. Indeed, research in the vein of frame two has drawn fire for risking the impression that we make ourselves anew in each idiosyncratic interaction—a criticism refuted, for example, by West and Fenstermaker (1995) with the notion of accountability, and by Butler (1993, 1999) with a conception of historicity. In frames three (text-conversation) and four (social text), discourse seems to ossify somewhat. It congeals into familiar, imposing formations that transcend time and space, even as they are vulnerable to the local performances that interest frame two. This sort of ossification (i.e. solidity or steadiness) is of an entirely different quality, not merely degree, than that depicted by frame one. Whereas frame one employs the metaphor of variable relations (i.e. cause–effect, stability as predictability), frames three and four evoke the metaphor of crystallization (i.e. multifaceted, not wholly stable, morphing on the liquid–solid border) or, especially in frame four, the metaphor of rhizome (i.e.

multiple, roving connections with no traceable origin, organization, or end; see Deleuze and Guattari 1987; Chia 1999).

Finally, (*e*) the potency,[3] or adherence to a constitutive view, of discourse varies across the frames. As evident in its title (i.e. discourse as outcome), frame one rejects a constitutive view; but the remaining three (performance, text-conversation, social text) assert strong claims to the generative force of discourse. For these frames, discourse continually delivers the gender binary as a patent 'fact', a 'natural' means to organize work. Discourse is so powerful that it (re)fashions our identities, relations, and behaviours, our anxieties and pleasures, currencies and institutions, even the 'undeniable imperatives' of bodies, time, and space. Discourse, in short, has material effects.

The upshot of a constitutive view for gender and organization studies is this: it is no longer tenable to speak of gender as a physical, psychological, or cultural trait, skill set, or vector of bias that individuals import into the workplace. Rather, organizing and working 'en/gender' systems of self and power; and discourse—not human will, itself a discursive product—is the engine propelling this condition. Phrased in a stronger poststructuralist voice of performativity, it *is* the condition. Discourse does not merely *produce* the materiality of work and organization, including the tangible limits of our agency as 'employees' and 'managers'. It *is* that material, the 'practices that form the object of which they speak' (Foucault 1972: 49). Literally, then, discourse 'matters'.

As it places discourse in a primary or privileged position, this kind of constitutive view, increasingly popular in gender and organization studies, is vulnerable to charges of naïve constructivism, text positivism, and nominalism, well abridged in Barad's (2003: 801) pithy declaration, 'Language has been granted too much power'. Put simply, a constitutive view risks the reductionistic claim that work and organizations are whatever we say they are. It turns the tables to manœuvre a better seat for discourse yet minimizes 'how matter makes itself felt' (Barad 2003: 810) and perpetuates the division and conflict of materialism versus idealism. It can be read as a reaction to the enduring Enlightenment hierarchy that splits real/material from thought/ideational and prioritizes both over symbol/discursive, such that 'mere' talk seems separate from and subordinate to action and philosophy. Such a reaction is understandable, indispensable, productive, *and* ultimately unsatisfactory. Is it not the case, for instance, that physical environments and bodily facts profoundly narrow our discursive options for gendering work and organization?

Comparable questions are finding traction across organization studies, prompting some to declare that the constitutive view of discourse goes too far and must be reined in by robust theorizing of materiality (for a recent example, see Alvesson and Karreman 2011). A rousing conversation about the discourse–materiality relation is gaining steam, particularly among those affiliated with critical management studies (e.g. Chia 2000, 2003; Reed 2000, 2004, 2005; Ford and Harding 2004; Fleetwood 2005; Mumby 2005; Willmott 2005; Cheney and Cloud 2006). Little has been said about the implications of this conversation for gender and organization studies, and pertinent developments in feminist theory are often absent from the debate. We seek to address both silences next, as we outline and illustrate the prospect for a new, fifth frame.

Toward a Fifth Frame on Gender, Discourse, and Organization: 'Communication as Discursive–Material Evolution'

For a starting point, we turn to the field of organizational communication studies. We understand that most gender and organization scholars are likely to be unfamiliar with this field of inquiry. Hence, we hope to share emerging resources from a field that has undertaken the most explicit efforts to theorize a constitutive view (e.g. Taylor 1993; Taylor *et al.* 1996; Cooren 2000; Kuhn 2008; Putnam and Nicotera 2008), and where recent work attempts the precise task of interest here: developing constitutive claims that are responsive to criticisms regarding materiality (e.g. Ashcraft *et al.* 2009). For those less familiar with the terrain of discourse studies, 'organizational communication' and 'organizational/business discourse' operate as subfields within larger disciplines—communication and management studies, respectively. These subfields share many common interests, theoretical influences, and analytical approaches. Not surprisingly, there is much productive overlap and exchange among them. As in most interdisciplinary relations, useful differences can also be drawn (for more, see Ashcraft *et al.* 2009).

In particular, many communication scholars, ourselves included, are inclined to theorize 'communication' as a specific kind of discursive process, and to distinguish it from 'discourse' for several reasons. As Cooren (2006*a*) suggests, this is due in part to our abiding concern for symbols in use *and* in abstraction. In subtle yet significant contrast, many 'organizational discourse' scholars are prone to emphasize discourse in the social text sense outlined above (see frame four), wherein 'a *Discourse*' is a loosely coalescing formation of texts, a crystallized narrative construction in wide circulation (see Grant *et al.* 2004). Meanwhile, scholars affiliating with 'business discourse' often prioritize detailed analyses of highly localized *d*iscourse (e.g. a meeting transcript), excluding or downplaying larger Discourses that may be at play in the interactions studied (see Bargiela-Chiappini 2009). Communication scholars are more apt to hold these together and interrogate their meeting through a conception of communication as the dynamic site wherein local, fleeting discussions encounter enduring, translocal constructions (Barge and Fairhurst 2008; Jian *et al.* 2008). Taylor and Van Every (2000), for instance, theorize that organization comes to be as communication continually 'tiles' textual and conversational modalities.[4]

Specific to gender and organizational communication studies, this distinctive way of treating the discourse–communication relation is evident in Ashcraft and Mumby's (2004: 116) definition of communication 'as the dynamic, situated, embodied, and contested process of creating systems of gendered meanings and identities by invoking, articulating, and/or transforming available discourses'. This definition, however, exemplifies the problems with contemporary constitutive views identified above. Namely, it gives a starring role to symbolism and nods to materiality (i.e. with 'situated, embodied', and possibly 'contested'), the muscle of which remains underdeveloped. Human communication 'creates', while environmental and bodily exigencies cling to the creative process like inert qualifiers rather than powerful actors. Though the authors go on to explicitly theorize communication as a dialectic of discourse and materiality (and thus move closer to integrating discursive and material forces), the spirit of reciprocity fades out in definition. In any case, the divide of notion and symbol from real and material remains largely intact.

In her appreciative critique of poststructuralist performativity, Barad (2003) applauds such theorizing for taking major strides towards challenging the persistent dualisms that form a representationalist trap: words divided from things, culture from nature, social construction from biology, knower from known, and subject from object—to name a few. And yet, she contends, poststructuralist performativity simultaneously preserves these binaries, albeit through more subtle means. In particular, it continues to split the discursive and material, ironically, by conceptualizing their union in a way that grants the former constitutive force at the expense of the latter: 'materiality itself is always already figured within a linguistic domain as its condition of possibility' (801), inviting an 'implicit reinscription of matter's passivity' (809). Hence, 'it is vitally important that we understand how matter matters' (803).

This call to more fully rupture the dualism of ideation and material in a way that rehabilitates the agency of matter is echoed in a number of diverse developments across feminist/gender and queer/sexuality studies. For example, drawing on de Beauvoir, Merleau-Ponty, and Moi, among other influences, Young (2005) theorizes the 'lived body' as a way to upend the nature–culture binary embedded in the distinction between sex and gender. For Young, the lived body is 'a unified idea of a physical body acting and experiencing in a specific sociohistorical context' (16). Bodies always live in situation, which she posits as the meeting of 'facticity' (i.e. the concrete relations of bodily existence in physical-social environments) and 'freedom' (i.e. the ontological capacity to construct oneself in relation to facticity—to carry out bodily projects like purposive movement). Young's classic analyses of 'breasted experience' and 'throwing like a girl' demonstrate keen sensitivity to how matter not only matters but also moves.

Likewise, Grosz (1994) develops a model of the body at the confluence of material, discourse, and idea. She synthesizes corporeal depictions from poststructuralism, psychoanalysis, neuropsychology, phenomenology, and other historical philosophies. Grosz too is concerned with movement and dimension, though of a different sort: from 'inside out' and 'outside in'. Her model hinges around the metaphor of a Möbius strip—not unlike the aforementioned rhizome, but with particular regard for depth. As Grosz (1994: 210) explains, this metaphor

> enables subjectivity to be understood not as the combination of a psychical depth and a corporeal superficiality but as a surface whose inscriptions and rotations in three-dimensional space produce all the effects of depth. It enables subjectivity to be understood as fully material and for materiality to be extended and to include and explain the operations of language, desire, and significance.

Although such developments further erode the division of discursive and material, offering compelling demonstrations of their entangled incarnation, Barad's (2003) analysis suggests an additional problem. Namely, these projects maintain 'human' as the custodian of agency, thereby continuing to prioritize discursive over material agency. This is a charge Barad levels at poststructuralist performativity as well; and the critique is especially ironic in light of poststructuralist aims to decentre the pre-existing, autonomous subject of classic humanism. Like the accounts of Young and Grosz, however, Butler's (1993) treatment of the 'materialization' of bodies is premised on an a priori, rather than enacted, distinction of human from other kinds of matter. What we need in order to refurbish the agency of matter, says Barad, are *post*-humanist accounts of performativity that recognize the inevitable inseparability of material and discursive without reducing the agency of the former to that of the latter or, conversely, returning to nostalgia for the omnipotent material. As Barad (2003: 824, original emphasis) reminds us, the 'material is always already material-discursive—*that is what it means to matter*'.

Ashcraft *et al.*'s (2009: 34) effort to redress such problems in the context of organization studies develops a post-humanist conception of communication as an evolutionary process that belongs neither to the ideational nor material realm but, rather, has a stake in both:

> Instead, communication is the mechanism whereby the material and ideational co-mingle and transform accordingly. In communication, symbol becomes material; material becomes symbol; and neither stay the same as a result. Applied to the case of bodies, it is insufficient to say that there are actual bodies and notions of bodies, and that communication employs the former to express the latter. Rather, in communication, ideas materialize in bodies in un/expected ways; ideas take root or shift in response to bodily resistance; and bodies are experientially and literally altered.[5]

In this view, 'facts' and 'constructions' can never be entirely disentangled. Each is already embedded in the other; they evolve and can only be apprehended in relation. As the excerpt suggests, the split of material and ideational begins to dissolve in this new understanding of communication, into shape-shifting permutations that rest on their ever-unfolding fusion rather than their dialectical relation. The motif here is not interaction, but interpenetration.[6]

It is in this commitment to interpenetration that material agency arrives on the scene intact. Meaning is not the sole purview of human talk or thought; it is diffused across a variety of matter dubbed human and non-human. Because the discourse–material relation unfolds through this multiplicity of agencies, we need a fuller accounting of participants in the co-construction of organizational realities (Cooren 2006*b*); but Barad (2003) cautions that such 'agents' are indeterminate until enacted. In this sense, the discursive–material relation plays out in *intra*-action, not *inter*action. Whereas the latter 'presumes the prior existence of independent entities' acting on one another (Barad 2003: 815), the former suspends such demarcations for given performances:

> Agency is a matter of intra-acting; it is an enactment, not something that someone or something has. Agency cannot be designated as an attribute of 'subjects' or

'objects' (as they do not preexist as such). Agency is not an attribute whatsoever—it is 'doing'/'being' in its intra-activity. Agency is the enactment of iterative changes to particular practices through the dynamics of intra-activity. (826–7)

We suggest that, for discourse scholars in organization studies, and gender and organization scholars in particular, 'communication' is a productive way to conceptualize and explore the dynamic process of agency evolving in intra-activity. Here, communication is redefined as *the ongoing, situated, and embodied process whereby agencies are enacted that interpenetrate discourse and materiality toward 'meaning that matters', constituting the very existence and relation of work, worker, organization, member, and affiliated phenomena like gender* (for more, see Ashcraft *et al.* 2009). Phrased in more straightforward terms, communication is not about humans wielding symbols towards material effects. It is, rather, the agentic process through which matter and discourse intertwine and become felt and known. Communication is how discursive–material permutations become 'real-ized'.

This conception of communication, we argue, has the potential to avoid the twin pitfalls of prioritizing discursive agency and sustaining the discourse–materiality split. As it foregrounds the unfolding formation, transformation, and dissolution of material-discursive permutations, it brings a multitude of agencies—'human', 'non-human', and their many mergers—into the picture. Equally vital, it stays grounded in the concrete, vibrant, and erratic—in a word, real—environments of everyday life, resisting the temptation to drift into abstractions of infinite possibility. Thus far, however, it may not be evident how this model of communication informs gendered organizing per se. To render our proposal directly applicable and accessible to gender and organization studies, we demonstrate next its potential to 'reorganize' gendered violence.

APPLYING THE FIFTH FRAME: THE CASE OF VIOLENCE, GENDER, AND ORGANIZATION

The interpenetrating real-izations of the material and the ideational are especially consequential in violence. Because of the presumed physicality of violence and its associated symbolic weight, those who make efforts to locate violence and its gendering grapple with definitional tensions that the fifth frame brings to the fore. In this section, we highlight these material–discursive interpenetrations of gendered violence and, in so doing, complicate the organization of violent agency.

By way of drawing out these definitional tensions, imagine two scenarios along with us. First, conjure an image of a fistfight in a bar. One of the brawlers throws a punch, and in the moment the fist meets the cheek, the violence is easily visible. The bar-room brawler who leaves with a black eye bears the physical evidence of the violence on his body. This episode may be easily categorized as violence, perhaps because the most readily available ideas of what counts as violence tend to be tethered to materially observable phenomena and outcomes. We invite you now to consider an example that, for some, may be more ambiguous. A male supervisor tells a female subordinate that the only way she can earn a pay bonus is by offering him sexual favours. In the US, this type of interaction could be considered quid pro quo sexual harassment. Unlike the bar-room brawl, this example involves no immediately visible, *physical* impact. Other material effects, such as the possible loss of an employment benefit, are not in themselves enough to ground a claim to violence (e.g. an employee who is told that her annual bonus is contingent upon satisfactory job performance reviews is not experiencing violence). In this case, a claim to violence seems rooted in coercive interaction: the supervisor uses the threat of money as a type of force that renders free consent to sex impossible. Thus the symbolic reference to material outcomes (in the form of loss of pay or non-consensual sex) is part of the violence of this episode.

Tensions around claims of violence, as illustrated in these two examples, resonate with the dilemmas that motivate a fifth frame on gender, discourse, and organization. Although the recognition of violence is often attached to its physicality in person-to-person interactions, the fifth frame loosens this hold. Instead, it leads us to notice that (*a*) questions about what 'really happened', or what counts as violence, are symbolic-material contests; (*b*) the coercive elements of violence are embedded in its discursive physicality; and (*c*) the organization of these contests and elements exceeds any individual.

Making these symbolic–material contests legible is especially important in the context of *gendered* violence where understandings of rape, other forms of sexual assault, and intimate partner violence have both pragmatic and political consequences. Those who study gendered violence, especially through critical and/or feminist lenses, are at pains to establish the realities of the gendered perpetration of these episodes of violence. Most scholars of violence agree that males are the primary perpetrators of intimate partner violence and other forms of sexual assault (e.g. Kimmel 2002; Katz 2006). The same community of scholars who ground their analysis of violence in a sexed material world is vexed around how to generate the *symbolic* space for non-violence (e.g. Marcus 1992; Cahill 2001; Mardorossian 2002). By repeatedly asserting the maleness of violent perpetration, scholars risk reiterating an association between masculinity and the ability to enact violence. Under this rubric, both recognizing the masculinity of non-violent men and rendering a politically adept analysis of women who perpetrate violence against men become difficult. Further, theoretical approaches to gendered violence may reinforce a problematic discursive–material binary that casts femininity as anti- or non-violent (and, consequently, easily associated with victimization) and masculinity as always violent (and, thus, readily aligned with agency).

The tangle between discourse and materiality around gender-based violence is clear in critiques of research findings that men and women commit intimate partner violence (IPV) at equal rates. Johnson (1995, 2005), for example, points out that these studies have confused common couple violence (which is not considered IPV) with intimate terrorism (which is considered to be IPV). The difference between the two is the extent to which the violence is part of a more comprehensive system of control and abuse within a relationship, and that system of control and abuse includes elements that are not purely material or exceed the material: verbal and emotional abuse, intimidation, social isolation, and threats. Here, Johnson's comments reflect an assumption embedded in the most commonly used model for this type of gender-based violence (the one used by the US Centers for Disease Control and Prevention as well as numerous women's shelters and violence prevention programmes). For an episode of violence to be considered gender-based, it is not enough merely for physical violence to be present. Rather, the ways in which the material and discursive elements interact determines the gendered nature of the violence.

What exactly, though, is the relationship between the material and the discursive in gender-based violence? Some scholars figure the discursive and the material as a continuum where sexual harassment (like the kind described in our second scenario) lies at one end and gang rape on the other (e.g. Kelly 1988). Others suggest that physical and discursive actions have similar effects, such as Projansky (2001: 2–3, original emphasis) who argues that 'rape discourse is *part of* the fabric of what rape is...Like physical actions, rape discourses have the capacity to inform, indeed embody and make way for, future actions, even physical ones'. Others still suggest that utterances and configurations of the discursive cause physical manifestations of gender-based violence or make physical violence more likely (e.g. Burt 1980; Edwards *et al.* 2011). These varied theorizations of the discourse–material relationship have impacts on the recognition of gender-based violence. That is, explanations that slide too completely towards the material *or* the discursive often exist at odds with projects to eradicate these forms of violence.

Feminist activist groups, such as the US-based Feminist Majority Foundation, have advocated for rape kits that gather evidence following an assault to be available at community centres, including hospitals, for free. They have also agitated for the processing of the massive backlog of evidence collected with these kits. This line of action focuses attention on the material effects of rape. These effects, of course, are crucially important, and the documentation of these effects often is a prerequisite for entry into the justice system. An emphasis on the physicality of rape, however, may reinforce the myth that rape occurs in the presence of overt force and may make sexual violence involving the complexities of coercion and manipulation—the far more common type of sexual violence—less intelligible as such. On the other hand, attention to the interpretive, discursive elements of sexual violence can lead to problematic arguments in which lack of clarity about how to name and define violence is taken to be evidence that violence did not occur (e.g. Roiphe 1993).

This kind of argument played out in public conversations around the allegations of sexual assault brought against former director of the International Monetary Fund, Dominique Strauss-Kahn. The hotel employee who alleged that Strauss-Kahn assaulted her was discredited when her description of the events changed. That is, if the material facts of the case were not clear, then serious doubts about whether violence occurred came to dominate public impressions. Here we see part of why activists work so hard, then, to ground claims of violence in the real. Those committed to the eradication of gender-based violence engage the relationship between the material and the discursive in concert with pragmatic considerations about how theories of the verbal-physical aspects of gender-based violence play out in a collective ability to talk about, respond to, and prevent these assaults.

Thus far, our discussion of gendered violence has focused on its discursive–material interpenetrations. To inflect violence through the fifth frame in relationship to *organization*, however, not only makes these interpenetrations legible but, in so doing, problematizes violent agency. Common conceptualizations of violence—in part because of their focus on material, bodily injury and a bounded moment in which that injury occurs—tend to be tethered to an individual, human actor. To think of a person who commits violence, a type of violence Žižek (2008) calls 'subjective' violence, is relatively straightforward. Yet understandings of violence untethered from an individual person, or violence rooted in complex, diffuse systems and processes are more difficult to conceptualize (Catley and Jones 2002; Presser 2005; Smith 2005). In an assertion that those individual moments of violence committed by a person constitute what violence *is*, we assume that humans are the sole violent agents.

Yet this emphasis on individuals who commit violence is difficult to reconcile with a communicative approach to materiality which claims that 'humans and other living creatures do not own action' (Ashcraft *et al.* 2009: 36). In what follows, we draw two lines in a sketch for theorizing the organizational aspects of gendered violence in relationship to a dominant tendency to locate violent agency in individuals. This sketch requires that we render violence such that it exceeds an intentional subject yet also accumulates in organizational embodiment.

If, following Barad (1998: 112), we trouble the bifurcation between discourse and materiality and conceptualize agency as 'an enactment and not something one has', understandings of violence must move away from a total focus on intentionality.[7] That is, to think through organization and violence, we cannot think only of a discrete subject *in* an organization. Systems, policies, and practices become central concerns, and individual episodes of injury look less like the beginning and end of violence. Ashcraft *et al.'s* (2009) discussion of knowledge offers a model for how violence would thus be reconceptualized. In this model, knowledge is seen as

> an attribution made about practice, not as a discrete entity. Analytical interest thus shifts to *processes of knowing*—to the activity of problem-solving, which is always embodied, embedded in sites, and connected to the material circumstances through which it emerges...This approach holds that work is not so much interdependent lines of action among autonomous agents...as it is ongoing problem-solving across intra-organizational sites. (38, original emphasis)

Similarly, we might think of violence as 'an attribution made about a practice' (i.e. a claim made about gendered interactions and processes), 'not as a discrete entity'. Some of these attributions are rooted in a moment of intense physicality (such as rape) where others are temporally and physically diffuse (such as hostile environment sexual harassment). This understanding of violence would shift our analytic attention to processes of 'violence-ing'. We could argue that the activity of violence-ing is always embodied: perhaps an individual has a bodily experience involving fear of violation or has physical reactions to using or hearing violent phrases or symbols.[8] To claim, as in the above excerpt, that the material–discursive aspects of violence are 'embedded in sites and connected to material circumstances through which it emerges' would call for an analytic attentive to the social, sexual, intellectual, and other forms of capital that circulate in particular organizations and around organizing.

Bourdieu's work offers a second line to this sketch. Discussing what he calls symbolic violence, or 'violence which is exercised upon a social agent with his or her complicity' (Bourdieu and Wacquant 1992: 167), Bourdieu is careful to note that the 'symbolic' does not refer only to the realm of language. That is, he does not assume a bifurcation of the material and the discursive. Bourdieu and Wacquant (1992: 172) say that 'we cannot understand symbolic violence and practice without forsaking entirely the scholastic opposition between coercion and consent, external imposition and internal impulse'. Thinking about violence, we have to conceptualize those individuals who commit violent acts as operating from *both* internal impulses *and* also conditions of external imposition. If responses to violence then focus only on the individual, part of the mechanism of violence has been ignored: embodied organizational elements. For Bourdieu and Wacquant (1992: 16), 'habitus' describes those 'historical relations "deposited" within individual bodies in the form of mental and corporeal schemata of perception, appreciation, and action'. It is a 'structuring mechanism' that is 'transcendent to the individual' (18–19). Integrating Bourdieu with our communicative model, organizational violence is an attribution made about a practice or, in this case, an action that is itself embedded in embodied perceptions and appreciations. Enactments of violence can no longer be considered to emerge from an individual, but from an individual *and* an embodied accumulation of material–discursive relations (or, to borrow from Barad, intra-actions).

ILLUSTRATING THE FIFTH FRAME: COMMUNICATION AS DISCURSIVE–MATERIAL EVOLUTION IN THE PENN STATE SCANDAL

A close reading of the recent US controversy surrounding the Pennsylvania State University football programme—and, specifically, Assistant Coach Jerry Sandusky's sexual misconduct and assault of young boys—can demonstrate more vividly how the fifth frame recasts discourse, gender, and organization. Public discussion of the scandal features a persistent concern that knowledgeable individuals did not report information about Sandusky to the police and university administration. Here, communication is figured as solely instrumental, and the problem is cast as the breakdown of channels for relaying messages about a straightforward reality. The Freeh report, for instance, suggests that the incidents at Penn State would have played out differently if only police officials had included a 1998 incident involving Sandusky in the police log. From the constitutive view of discourse advanced by frames two, three, and four (performance, text-conversation, and social text), we would rightly critique the representationalist view of discourse—as mere expression and transmission, or relay of already existing reality—embedded in these discussions. We would underscore, for example, that knowing certain physical acts as 'sexual abuse' is a contested process, settled only through discursive negotiation. What we emphasize here, however, is how the fifth frame (communication as discursive–material evolution) retains this insight yet resists a total descent into discursivity.

Against a popular conception that locates violence in its physicality, the fifth frame asks us to consider the interpenetration of the discursive and the material in ways that move beyond a human inflicting bodily injury. A janitor who witnessed Sandusky abusing a child feared he would lose his job and did not want to challenge the powerful football programme. A child worried that he would no longer be invited to sit on the sidelines of football games if someone confronted Sandusky about the inappropriate interactions between Sandusky and the boy. Both of these individuals feared material loss (a job, an ability to attend games), and the material loss was tangled in the symbolic status of the team. The programme's impressive record includes being one of the winningest, most competitive Division I teams in the US. Using Department of Education statistics, one source estimates that Penn State's football programme made $31,619,687 in 2010–11, the second highest profit for a US collegiate football team (Jessop 2012). These successes are an element of what the Freeh report calls a 'reverence for the football program that is ingrained at all levels of the campus community' (Freeh *et al.* 2012: 17). This reverence is secured not only through material success, but also through the mobilization of dominant discourses. Part of the prestige of such a profitable team rests in its physical-symbolic *invulnerability* and *impenetrability*, key elements of a conquest-oriented, hegemonic heterosexuality that many scholars argue is embedded in athletic competition (e.g. Messner 2002; Enck-Wazner 2009). To acknowledge that a powerful male member of this community repeatedly desired and had sexual, physical contact with young male bodies disrupts the meaning, mattering, and enactment of violence that is woven through the fabric of reverence around Penn State.[9]

Finally, the fifth frame grounds us in the complexities of meaning and mattering that always exceed individuals. Unmooring violence from its nearly total association with physicality, while reasserting constitutive views of communication, requires that we think beyond centred, intentional, human actors. Public conversations and some aspects of the Freeh report have persistently sought to attach blame to individuals involved in a larger enactment of violence. Mike McQueary, Joe Paterno, Tim Curley, and Gary Schultz have all been criticized during a continual search for the *people* who let this happen. Although we find the actions of many of these individuals reprehensible and in some instances criminal, the search for discrete *origins* of violence allows important elements of its *enactment* to disappear. That is, if we expand the location of agency to encompass not only individual humans but also organizing, we cannot help but shift away from efforts to eradicate violence by removing single people (or the symbolic renderings of those people) from organizations.

A poignant example can be found in the debate over whether to remove a statue of Coach Paterno from the Penn State campus, a conflict that exemplifies our abiding cultural faith in individual freedom of action, especially the feats and failures of 'great men'. To remove the statue reinforces the rooting of both reverence and blame in a human whose very presence (at once matter *and* meaning) arguably evolves through an amalgam of forces that enacted 'his' agency. Rather than exorcize Paterno's physical-symbolic legacy, as embodied in this statue, the fifth frame leads us to argue for a refiguring of a meaning of the statue, perhaps through the addition of a reflective plaque that reads: 'In the wake of our knowledge of rampant abuse perpetrated on these grounds, this statue serves to remind us of the institutions we make; which exceed the sum of individual minds, hands, and hearts; which enable great feats of collectivity; and which in turn make *us*, inhuman and otherwise'.

Because our rendering of the fifth frame is intended to be both responsive to current academic dialogues and a heuristic for continuing conversations, our reading of violence through the fifth frame both resonates with and departs from some of the existing literature on organization and violence. First, unlike some of the work on organization and violence (especially that on bullying), this approach foregrounds a gender analytic. Consequently, our reading of violence in relationship to discourse and organization is critical of the circulation of power. We do not argue, as some literature on organization and violence does, that organizing is inherently violent (e.g. Bergin and Westwood 2003; Pelzer 2003). Although our approach questions and complicates what counts as violence, we do not want the category 'violence' to grow overly capacious, and its limit conditions are established here through feminism. Second, a reading of violence through the fifth frame departs from an extensive literature that considers violence *in* organizations (e.g. Chaberlain *et al.* 2008). This collection of scholarship takes violence to be an interpersonal phenomenon that accrues an organizational component because of the location or context in which it occurs. Violence, in this literature, is a fully human enactment. Studies that identify organizational features that make the human enactment of violence more or less likely (e.g. Salin 2003) move towards but do not fully articulate organizational agency, however that extra-human agency tends to take root through discourse (e.g. Conrad and Taylor 1994; Dougherty and Smythe 2004). Third, like Hearn and Parkin's (2001) work on organizational violation, this approach considers the ways in which violence exists not only in a moment of physical injury, but in a complex interaction between both the material and symbolic aspects of that violence, both of which have material effects. As such, this approach begins to establish agency in violence that includes and extends beyond an individual human perpetrator.

CONCLUSION

This chapter has worked along two simultaneous trajectories: first, (*a*) familiarizing gender and organization scholars who are less attuned to discourse studies with the nature and contribution of a 'constitutive' view; and second, (*b*) developing, for the specific context of gender and organization studies, a constitutive view of discourse that better accounts for contemporary conversations about materiality. We began with four available frames on the relationship among discourse, gender, and organization (i.e. discourse as outcome, performance, text-conversation, and social text). We then explained how the constitutive view of discourse advanced by the latter three has come under fire in recent years—chiefly, for reinscribing the split of discursive and material as it empowers discourse *at the expense* of matter, such that material agency is weakened, already prefigured in discursive terms.

In response, we developed a fifth frame—communication as discursive–material evolution—which emphasizes the ongoing interpenetration of discourse and matter in the intra-actions of everyday life. Based on key developments in organizational communication and feminist theory, the fifth frame advances a post-humanist conception of communication as the central process whereby discourse and material collide and fuse in agentic enactments that yield 'meaning that matters'.[10] These evolving fusions constitute the very existence and relation of work, organization, and gender.

To demonstrate how this lens can be particularly useful for gender and organization studies, we examined its potential to help us 'reorganize' (i.e. reconceptualize *and* develop a stronger account of how organization *per se* 'matters' in) gendered violence. Specifically, we showed how the fifth frame loosens gendered violence from the either-or trap that perpetuates the divide and conflict of material and discursive: on the one hand, (*a*) a representationalist view that casts sexual abuse as a primarily physical reality transmitted through discourse (i.e. omnipotent matter) and, on the other, (*b*) a constitutive view that casts sexual abuse as a reality made entirely through discursive negotiation (i.e. omnipotent discourse). Despite their apparent opposition, both place the a priori 'human' at the centre of violent agency, directing us to hunt for culpable individuals, the perpetrators and co-conspirators. In so doing, both neglect other evolving discursive–material agencies (e.g. the imperative interests of a flourishing football programme)—specifically, *organizational* agencies—which create and authorize capacities for violence-ing, including those heroic figures of 'great men' we rush to praise and blame. In this way, the fifth frame shifts attention from definite origins to distributed enactments of gendered violence—that is, towards performances of violence in *ongoing* communication, particularly as these come to appear to us like discrete interpersonal incidents between perpetrator and victim. The fifth frame asks not 'who is to blame' but, rather, 'what agencies are enacting this formation of violence', 'how are incidents and participants, and the evident boundaries around them, produced', and 'what are our investments in the quest for a guilty origin?'

Ultimately, it is our hope that reading gender and organizational phenomena, like violence, through the fifth frame can advance our understanding of discourse and materiality, allowing us to grasp their complex relation and more fully decentre the human. Rather than silence questions of agency, accountability, and transformation, such an approach beckons us to reimagine them in a thoroughly networked or systemic—in a word, organizational—way.

Notes

1. For an elaboration of 'text' and 'conversation', see Taylor and Van Every (2000). Ashcraft (2004) employs this framework as a sense-making device: Gender and organization scholars have not often utilized this framework directly, but it provides a useful way to capture what distinguishes frame three.
2. Read in light of this paragraph, the present volume's title summons only the first two frames.
3. Gender-sexual wordplay is intentional.
4. The depiction of subfield relations provided in this paragraph is intended as a 'broad brush' portrait for readers less familiar with discourse and communication studies. We readily acknowledge exceptions to every general trend observed here, but a 'close-up' rendering requires far greater nuance than space affords.
5. To avoid confusion, it is worth noting that 'symbol' and 'idea' are used interchangeably in this quote, and that 'discourse' may just as well suffice. The point is that all three constructs, often lumped together as idealist explanations of organizational life, are already imbricated in what we know as material, and vice versa. Moreover, while Ashcraft et al. (2009) refer to human bodies here, Barad's (2003) point in the previous paragraph guides us to reread this excerpt as including 'bodies' of all kinds.
6. Again, the gender-sexual wordplay is intentional. We find the notion of 'interpenetration' especially provocative for the mutuality and reciprocity of agency it evokes. Consider the contrastingly passive equality implied by 'interdependence', an alternate term we might have used. Also, much like the theoretical shift we trace here, the notion of interpenetration refuses persistent imagery of the symbolic as 'soft' against the 'hardness' of materiality.
7. A move away from intentionality is not a move away from accountability. This theorization should not be read as a move to excuse those individuals who perpetrate violence. To argue that violent enactment exceeds individuals is not to argue that violence does not involve individuals. Rather, it is simultaneously to condemn the individual perpetration of violence and to complicate our understanding of that perpetration by exploring how agency, for both individual and organization, is figured through discursive–material intra-actions.
8. The work of communobiologists, who document the bodily effects of communicative acts, could offer a useful post-positivist articulation of the material impact on individual bodies of particular utterances or symbolic actions.
9. We are not arguing that success in competitive sporting leads to violence, nor are we arguing that the multiple relationships between athletics of the bedroom and the football field are only detrimental or necessarily dangerous. This is a misreading that relies upon the perpetual search for a discrete origin of violence that we want to write beyond. We are seeking, here, to think broadly about how the physical moments of violence exist in relationship to the symbolic-material successes and status of the programme.

10. We could just as easily use the double entendre 'matters of meaning'. In other words, our intent here is to prioritize neither matter nor meaning, in keeping with our proposal.

References

Alvesson, M., and Karreman, D. (2000). 'Varieties of Discourse: On the Study of Organizations through Discourse Analysis', *Human Relations*, 53(9): 1125–49.

Alvesson, M., and Karreman, D. (2011). 'Decolonializing Discourse: Critical Reflections on Organizational Discourse Analysis', *Human Relations*, 64(9): 1121–46.

Ashcraft, K. L. (2004). 'Gender, Discourse, and Organization: Framing a Shifting Relationship', in D. Grant, C. Hardy, C. Oswick, N. Phillips, and L. L. Putnam (eds), *The SAGE Handbook of Organizational Discourse*, 275–98. Thousand Oaks, CA: Sage.

Ashcraft, K. L. (in press). 'Feminist Theory', in L. L. Putnam and D. K. Mumby (eds), *The SAGE Handbook of Organizational Communication* (3rd edn). Thousand Oaks, CA: Sage.

Ashcraft, K. L., Kuhn, T., and Cooren, F. (2009). 'Constitutional Amendments: "Materializing" Organizational Communication', *Academy of Management Annals*, 3(1): 1–64.

Ashcraft, K. L., and Mumby, D. K. (2004). *Reworking Gender: A Feminist Communicology of Organization*. Thousand Oaks, CA: Sage.

Barad, K. (1998). 'Getting Real: Technoscientific Practices and the Materialization of Reality', *Differences: A Journal of Feminist Cultural Studies*, 10(2): 87–128.

Barad, K. (2003). 'Posthumanist Performativity: Toward an Understanding of How Matter Comes to Matter', *Signs*, 28(3): 801–31.

Barge, J. K., and Fairhurst, G. T. (2008). 'Living Leadership: A Systemic Constructionist Approach', *Leadership*, 4(3): 227–51.

Bargiela-Chiappini, F. (ed.) (2009). *The Handbook of Business Discourse*. Edinburgh: Edinburgh University Press.

Bergin, J., and Westwood, R. I. (2003). 'The Necessities of Violence', *Culture and Organization*, 9(4): 211–23.

Bourdieu, P., and Wacquant, L. (1992). *An Invitation to Reflexive Sociology*. Chicago: University of Chicago Press.

Britton, D. M. (2000). 'The Epistemology of the Gendered Organization', *Gender & Society*, 14(3): 418–34.

Burt, M. R. (1980). 'Cultural Myths and Supports for Rape', *Journal of Personality and Social Psychology*, 38(2): 217–30.

Butler, J. (1993). *Bodies That Matter: On the Discursive Limits of 'Sex'*. New York: Routledge.

Butler, J. (1999). *Gender Trouble: Feminism and the Subversion of Identity* (10th anniversary edn). New York: Routledge.

Cahill, A. J. (2001). *Rethinking Rape*. Ithaca, NY: Cornell University Press.

Catley, B., and Jones, C. (2002). 'Deciding on Violence', *Philosophy of Management*, 2(1): 25–34.

Chamberlain, L. J., Crowley, M., Tope, D., and Hodson, R. (2008). 'Sexual Harassment in Organizational Context', *Work and Occupation*, 35(3): 262–95.

Cheney, G., and Cloud, D. (2006). 'Doing Democracy, Engaging the Material: Employee Participation and Labor Activity in an Age of Market Globalization', *Management Communication Quarterly*, 19(4): 501–40.

Chia, R. (1999). 'A "Rhizomic" Model of Organizational Change and Transformation: Perspective from a Metaphysics of Change', *British Journal of Management*, 10(3): 209–27.

Chia, R. (2000). 'Discourse Analysis as Organizational Analysis', *Organization*, 7(3): 513–18.

Chia, R. (2003). 'Ontology: Organization as "World-Making"', in R. Westwood and S. Clegg (eds), *Debating Organization: Point-Counterpoint in Organization Studies*, 98–113. Malden, MA: Blackwell.

Conrad, C., and Taylor, B. (1994). 'The Context(s) of Sexual Harassment: Power, Silences, and Academe', in S. G. Bingham (ed.), *Conceptualizing Sexual Harassment as Discursive Practice*, 45–58. Westport, CT: Praeger.

Cooren, F. (2000). *The Organizing Property of Communication*. Amsterdam: John Benjamins.

Cooren, F. (2006a). 'The Organizational Communication-Discourse Tilt: A Refugee's Perspective', *Management Communication Quarterly*, 19(4): 653–60.

Cooren, F. (2006b). 'The Organizational World as a Plenum of Agencies', in F. Cooren, J. R. Taylor, and E. J. Van Every (eds), *Communication as Organizing: Empirical and Theoretical Explorations in the Dynamic of Text and Conversation*, 81–100. Mahwah, NJ: Lawrence Erlbaum.

Deleuze, G., and Guattari, F. (1987). *A Thousand Plateaus: Capitalism and Schizophrenia*. Minneapolis: University of Minnesota Press.

Dougherty, D. S., and Smythe, M. J. (2004). 'Sensemaking, Organizational Culture, and Sexual Harassment', *Journal of Applied Communication Research*, 32(4): 293–317.

Edwards, K. M., Turchik, J. A., Dardis, C. M., Reynolds, N., and Gidycz, C. A. (2011). 'Rape Myths: History, Individual and Institutional-Level Presence, and Implications for Change', *Sex Roles*, 65(11–12): 761–73.

Enck-Wazner, S. M. (2009). 'All's Fair in Love and Sport: Black Masculinity and Domestic Violence in the News', *Communication and Critical/Cultural Studies*, 6(1): 1–18.

Fenstermaker, S., and West, C. (eds) (2002). *Doing Gender, Doing Difference: Inequality, Power and Institutional Change*. New York: Routledge.

Fleetwood, S. (2005). 'Ontology in Organization and Management Studies: A Critical Realist Perspective', *Organization*, 12(2): 197–222.

Ford, J., and Harding, N. (2004). 'We went Looking for an Organization But Could Find Only the Metaphysics of its Presence', *Sociology*, 38(4): 815–30.

Freeh, Sporkin, & Sullivan, LLP (2012). *Report of the Special Investigative Counsel Regarding the Actions of the Pennsylvania State University Related to the Child Sexual Abuse Committed by Gerald A. Sandusky*. Retrieved from http://i.usatoday.net/news/nation/2012-07-12-penn-state-freeh-report.pdf

Foucault, M. (1972). *The Archaeology of Knowledge and the Discourse on Language*, tr. A. M. S. Smith. New York: Pantheon.

Grant, D., Hardy, C., Oswick, C., Phillips, N., and Putnam, L. L. (eds) (2004). *The SAGE Handbook of Organizational Discourse*. London: Sage.

Grosz, E. (1994). *Volatile Bodies: Toward a Corporeal Feminism*. Bloomington, IN: Indiana University Press.

Hearn, J., and Parkin, W. (2001). *Gender, Sexuality and Violence in Organizations: The Unspoken Forces of Organization Violations*. Thousand Oaks, CA: Sage.

Jessop, A. (21 March 2012). 'Highest Net Income amongst Athletic Departments', *Business of College Sports*. Retrieved from http://businessofcollegesports.com/2012/03/21/highest-net-income-amongst-athletics-departments/

Jian, G., Schmisseur, A. M., and Fairhurst, G. T. (2008). 'Organizational Discourse and Communication: The Progeny of Proteus', *Discourse and Communication*, 2(3): 299–320.

Johnson, M. P. (1995). 'Patriarchal Terrorism and Common Couple Violence: Two Forms of Violence Against Women', *Journal of Marriage and the Family*, 57(2): 283–94.

Johnson, M. P. (2005). 'Domestic Violence: It's Not about Gender—Or is it?', *Journal of Marriage and Family*, 67(5): 1126–30.

Katz, J. (2006). *The Macho Paradox: Why Some Men Hurt Women and How All Men Can Help*. Naperville, IL: Sourcebooks.

Kelly, L. (1988). *Surviving Sexual Violence*. Cambridge: Polity Press.

Kimmel, M. S. (2002). '"Gender Symmetry" in Domestic Violence: A Substantive and Methodological Research Review', *Violence Against Women*, 8(11): 1332–63.

Kuhn, T. (2008). 'A Communicative Theory of the Firm: Developing an Alternative Perspective on Intra-Organizational Power and Stakeholder Relationships', *Organization Studies*, 29(8–9): 1227–54.

Marcus, S. (1992). 'Fighting Bodies, Fighting Words: A Theory and Politics of Rape Prevention', in J. Butler and J. W. Scott (eds), *Feminists Theorize the Political*, 385–403. New York: Routledge.

Mardorossian, C. M. (2002). 'Toward a New Feminist Theory of Rape', *Signs*, 27(3): 743–75.

Messner, M. A. (2002). *Taking the Field: Women, Men, and Sports*. Minneapolis: University of Minnesota Press.

Mumby, D. K. (2005). 'Theorizing Resistance in Organization Studies: A Dialectical Approach', *Management Communication Quarterly*, 19(1): 19–44.

Nadesan, M. H., and Trethewey, A. (2000). 'Performing the Enterprising Subject: Gendered Strategies for Success(?)', *Text and Performance Quarterly*, 20(3): 223–50.

Pelzer, P. (2003). 'The Dinner Party of Silent Gentlemen: The Intrinsic Violence of Organisations', *Culture and Organization*, 9(4): 225–37.

Presser, L. (2005). 'Negotiating Power and Narrative in Research: Implications for Feminist Methodology', *Signs: Journal of Women in Culture and Society*, 30(4): 2067–90.

Projansky, S. (2001). *Watching Rape: Film and Television in Postfeminist Culture*. New York: New York University Press.

Putnam, L. L., and Nicotera, A. (eds) (2008). *Building Theories of Organization: The Constitutive Role of Communication*. Oxford: Routledge.

Reed, M. (2004). 'Getting Real about Organizational Discourse', in D. Grant, C. Hardy, C. Oswick, N. Phillips, and L. L. Putnam (eds), *The SAGE Handbook of Organizational Discourse*, 413–20. London: Sage.

Reed, M. (2005). 'Reflections on the "Realist Turn" in Organization and Management Studies', *Journal of Management Studies*, 42(8): 1621–44.

Reed, M. I. (2000). 'The Limits of Discourse Analysis in Organizational Analysis', *Organization*, 7(3): 524–30.

Roiphe, K. (1993). *The Morning After: Sex, Fear, and Feminism on Campus*. Boston: Little, Brown.

Salin, D. (2003). 'Ways of Explaining Workplace Bullying: A Review of Enabling, Motivating, and Precipitating Structures and Processes in the Work Environment', *Human Relations*, 56(10): 1213–32.

Smith, A. (2005). *Conquest: Sexual Violence and American Indian Genocide*. Cambridge, MA: South End Press.

Taylor, B. C. (1999). 'Browsing the Culture: Membership and Intertextuality at a Mormon Bookstore', *Studies in Cultures, Organizations, and Societies*, 5(1): 61–95.

Taylor, J. R. (1993). *Rethinking the Theory of Organizational Communication: How to Read an Organization*. Norwood, NJ: Ablex.

Taylor, J. R., Cooren, F., Giroux, N., and Robichaud, D. (1996). 'The Communicational Basis of Organization: Between the Conversation and the Text', *Communication Theory*, 6(1): 1–39.

Taylor, J. R., and Van Every, E. J. (2000). *The Emergent Organization: Communication as its Site and Surface*. Mahwah, NJ: Lawrence Erlbaum.

West, C., and Fenstermaker, S. (1995). 'Doing Difference', *Gender & Society*, 9(1): 506–13.

West, C., and Zimmerman, D. (1987). 'Doing Gender', *Gender & Society*, 1(2): 125–51.

Willmott, H. (2005). 'Theorizing Contemporary Control: Some Post-Structuralist Responses to Some Critical Realist Questions', *Organization*, 12(5): 747–80.

Young, I. M. (2005). *On Female Body Experience: 'Throwing Like a Girl' and Other Essays*. New York: Oxford University Press.

Žižek, S. (2008). *Violence: Six Sideways Reflections*. New York: Picador.

GENDER IN LEADERSHIP AND MANAGEMENT

PART II

GENDER IN
LEADERSHIP AND
MANAGEMENT

CHAPTER 7

....

FEMALE ADVANTAGE

Revisited

....

ALICE H. EAGLY, LEIRE GARTZIA, AND
LINDA L. CARLI

HEADLINES in the popular press trumpet a female advantage theme: 'No Doubts: Women Are Better Managers' (Smith 2009); 'Women in Top Jobs Are Viewed as "Better Leaders Than Men"' (*Daily Mail* 2010); and 'The Spectacular Triumph of Working Women Around the World' (Thompson 2012). These and other newspaper and magazine articles praising women leaders often claim that women have superior leadership styles and that the inclusion of women executives improves companies' profitability.

The early proponents of such views included writers such as Loden (1985), Helgesen (1990), and Rosener (1990), whose opinions attracted attention, but who evidently failed to persuade the business community, given women's slow progress, especially at the higher levels of corporate leadership. But perhaps the situation is now changing. Recently, very positive assessments of women as leaders have come from influential voices within the business community. The most prominent example is McKinsey Consulting's series of four reports labelled *Women Matter* (described at http://www.mckinsey.com/Features/Women_Matter), whose authors argued that the inclusion of women in leadership groups enhances corporate competitiveness. These reports claimed positive effects of managerial women and ascribed these effects mainly to women's presumably superior leadership styles: 'If today's corporations lack the most effective leadership behaviors to meet future challenges, it is a matter of some urgency to reinforce and develop these behaviors. Developing gender diversity is a key lever to achieve this, since women adopt most of these behaviors more frequently than men. (Desvaux and Devillard 2008: 15).

Accompanying this increasing interest in women managers is academic research exploring women's contributions to management. High-impact journals relevant to business and management have featured numerous articles on women and gender, including *Academy of Management Journal* (e.g. Powell and Greenhaus 2010), *Journal of Management* (e.g. Roth *et al.* 2012), and *Journal of Applied Psychology* (e.g. Rosette and Tost 2010). Specialty journals on women and management have also appeared, such as *Gender in Management* in 2008 and *Gender, Work and Organization* in 1994. Furthermore, scanning book titles on women and management at the Amazon website reveals a plethora of recent trade books featuring female advantage themes, such as *How Remarkable Women Lead: The Breakthrough Model for Work and Life* by Barsh *et al.* (2009); *Women Lead the Way* by Tarr-Whelan (2009); and *The Female Vision: Women's Real Power at Work* by Helgesen and Johnson (2010).

Despite this expansion of popular and academic interest in understanding how gender may affect leadership and management, many questions remain. Previously, Eagly and Carli (2003*a*, 2003*b*) and Vecchio (2002, 2003) reviewed existing academic research in a scholarly debate about the possible advantages of women as leaders. Because considerable new research has accumulated in the decade since those articles appeared, the time seems right to revisit these questions.

Before launching into our analysis, we first note that discussions of a female leadership advantage are too often oversimplified. In reality, the phenomenon must be deconstructed into its component parts. One issue is whether women and men behave differently as leaders, either on average or in some particular situations, and if so, whether these differences occur in the types of behaviours that are known to be effective. Also, women may differ from men in consequential ways other than leadership style—in particular, in their values and attitudes. Addressing such issues about sex differences and similarities should not end an analysis.[1] As long as leadership remains culturally masculine, the context of leadership is different for women than men. In particular, prejudicial reactions may penalize women but not men even when their behaviours are equivalent. The cultural and organizational context of leadership may thus challenge women in ways that undercut their effectiveness. Nonetheless, given contemporary flux in gender roles and in norms about good leadership, the opportunities for effective female leadership may be greater than at any time in the past. This chapter evaluates the plausibility of this claim.

Sex-Related Leadership Styles

Given that leaders' style is an important influence on their effectiveness (Yukl 2010), much of the interest in women as leaders has centred on whether they lead differently than men, and, if so, whether these style differences contribute to their effectiveness. As Eagly and Carli (2003*a*) reported, research reveals some sex-related differences in leadership style. A meta-analysis of 162 studies of leadership style found that female managers overall adopted a more democratic and less autocratic leadership style than male managers did (Eagly and Johnson 1990). In contrast, no sex-related differences were found among managers in the gender-stereotypical tendencies for men to be more task-oriented and women more interpersonally oriented; these differences were stronger in non-managerial samples, especially among university students.

A subsequent meta-analysis added credence to Eagly and Johnson's findings by revealing stereotypical sex differences in democratic versus autocratic leadership style (van Engen and Willemsen 2004).

Eagly and Johnson (1990) argued that organizational norms about appropriate managerial behavior tend to suppress any sex differences in the aspects of style captured by measures of task and interpersonal orientation. Yet, these norms surely differ across organizations, given that women did tend to have a more interpersonally oriented style than men in leader roles that were less male-dominated. Eagly and Johnson also argued that women's preference for democratic and participative leadership styles made sense given strong gender norms prohibiting women from leading in an autocratic, directive manner. Yet, this sex-related difference in leadership style was also stronger in less male-dominated leader roles. It thus appears that there is more leeway for culturally feminine leadership styles in contexts where greater numbers of women inhabit leader roles.

Do these findings shed any light on the issue of female advantage? Judge *et al.*'s (2004) meta-analysis demonstrated that both task-oriented and interpersonally oriented styles were positively correlated with leaders' effectiveness, but the absence of consistent sex differences in these behaviours precludes them from creating an overall advantage for either sex. More pertinent is the sex difference in autocratic versus democratic leadership style, which pertains to leaders' sharing of their decision-making power with subordinates. Although earlier meta-analyses found that the effectiveness of autocratic and democratic leadership depended on a variety of situational factors (e.g. Gastil 1994; Foels *et al.* 2000), subsequent research has suggested that non-linear relationships between these styles and effectiveness may be common. In particular, research by Ames and Flynn (2007) focused on leaders' assertiveness, understood as actively pursuing and defending one's own interests versus cooperatively taking followers' interests into account. This research demonstrated that high levels of assertiveness tend to damage social relationships, whereas low levels limit goal achievement. Because of these trade-offs, the middle ground of moderate assertiveness is, in general, more effective than the high or low extremes. Additional research would be required to determine whether, consistent with the female advantage hypothesis, the tendency for women to lead somewhat more democratically than men typically places women in this advantageous middle ground.

In the last twenty years many researchers have turned their attention to *transformational leadership*, a style that is attuned to the complexities of leading in modern organizations (Bass 1998; Avolio 2010). Transformational leaders are characterized by their ability to serve as inspirational role models, generate good human relationships, develop followers, and motivate them to go beyond the confines of their job descriptions. Researchers have contrasted this type of leadership with *transactional leadership*, which involves using rewards and punishments to motivate subordinates, thus appealing to their self-interest. Researchers have also examined a *laissez-faire* style of leadership that is characterized by a general failure to take responsibility for managing.

A meta-analysis of forty-five studies examined research comparing men's and women's transformational, transactional, and laissez-faire leadership (Eagly *et al.* 2003). Results revealed small sex differences (reported in the *d* metric of standardized differences, with differences in the male direction given a positive sign). Women were generally more transformational than men (mean *d* = −0.10) and also more transactional in terms of providing rewards for satisfactory performance (mean *d* = −0.13). Women's transformational leadership differed most from men's in *individualized consideration*, that is, a focus on developing and mentoring followers and attending to their individual needs (mean *d* = −0.19). In contrast, compared with women, men showed more transactional leadership by emphasizing followers' mistakes and failures (mean *d* = 0.12) and waiting until problems become severe before intervening (mean *d* = 0.12). Men were also more likely than women to manifest laissez-faire leadership (mean *d* = 0.16). These results were replicated in a large-scale study of leadership by Antonakis *et al.* (2003) and also in research reported by Desvaux and Devillard (2008) in the second of the McKinsey *Woman Matter* reports. Other research has suggested that workplaces with more women in management may have practices that reflect women's greater transformational leadership. Specifically, Melero (2011) found more monitoring of employee feedback and concern with mentoring and developing employees with more women in management teams.

These findings have implications for female advantage. As confirmed by three meta-analyses (Lowe *et al.* 1996; Judge and Piccolo 2004; Wang *et al.* 2011), transformational style and the component of transactional style that involves providing rewards were correlated with effectiveness in contemporary organizational contexts. In contrast, managers' reliance on the more negative, punishing aspect of transactional leadership was only weakly associated with effectiveness. The transactional component involving delaying action until problems become severe and the laissez-faire style were negatively associated with effectiveness.

Given these effectiveness findings, an obvious implication of sex-related differences in transformational, transactional, and laissez-faire leadership is that the styles favoured by women relate positively to effectiveness, whereas the styles favoured by men either do not or do so only weakly. In support of this interpretation, female managers received higher effectiveness ratings than male managers in the subset of the studies in Eagly *et al.*'s (2003) meta-analysis that provided effectiveness data. Although the clearest evidence for female advantage in leadership style thus derives from research on transformational, transactional, and laissez-faire styles, the effect sizes for these differences are quite small, raising questions about their importance. Although small differences, when repeatedly enacted across time and settings, can produce relatively large effects (e.g. Abelson 1985), researchers should resist the temptation to label transformational leadership as *feminine* (see also Billing and Alvesson 2000). Instead, this culturally androgynous style can be comfortably and appropriately enacted by men as well as women.

Any advantage associated with women's styles could be magnified if people recognized that these sex differences in style exist and that the particular styles favoured by women enhance effectiveness. On the other hand, if people are unaware of the differ-ences in female and male styles or the effectiveness associated with the different styles, or if they evaluate the same styles less favourably when enacted by women than men, any female advantage would be undermined. Exploring such possibilities, Vinkenburg *et al.* (2011) found that participants asked to identify the leadership style of typical men and women (or typical male and female managers) correctly estimated the sex differ-ences in transformational and transactional style. This accuracy favours the idea of female advantage.

To further explore potential advantages of transformational and transactional leadership styles, a second study by these researchers examined the perceived impor-tance of these styles for women's and men's promotion to different levels in organiza-tions (Vinkenburg *et al.* 2011). One relatively agentic component of transformational style, *inspirational motivation*, was perceived as more important for men than women and especially important for promotion to CEO. In contrast, the more com-munal component of *individualized consideration* was perceived as more important for women than men and especially important for promotion from middle to senior management. These findings revealed conventional gender norms whereby partic-ipants prescribed behaviours for female leaders that are more communal, but less helpful for achieving the highest level of corporate success. Moreover, as suggested by Ayman *et al.*'s (2009) research, male subordinates of female transformational lead-ers may be less willing than female subordinates to perceive such women as effective. These subtleties of transformational leadership thus can complicate women's path to high positions.

OTHER SEX-RELATED DIFFERENCES

To complement findings on gender and leadership style, researchers have explored other potential bases for female advantage in leadership. One example is the disposi-tional trait of *emotional intelligence*, generally defined as the ability to perceive, use, understand, and manage emotions (Mayer *et al.* 2004). Business writers have argued that emotional intelligence is crucial to effective leadership (e.g. Goleman *et al.* 2002). According to a meta-analysis by Joseph and Newman (2010), women are superior to men on performance-based measures of emotional intelligence although not on self-reported ability measures. Consequently, women's apparently better emotional intelligence might yield female advantage if emotional intelligence enhances leaders' effectiveness. However, Joseph and Newman's data also showed that emotional intel-ligence was associated with better performance mainly in jobs that require considerable emotional labour (e.g. interacting with customers in a consistently cheerful, friendly manner), which may not include all or even most leader roles. Moreover, the relations between emotional intelligence and transformational leadership appear to be quite weak (see meta-analysis by Harms and Credé 2010). Therefore, the conditions under which women's emotional intelligence may enhance their performance as leaders remain to be discovered.

Also potentially relevant to a female leadership advantage are differences between women and men in values and attitudes. Consistent with Gilligan's (1982) claims about women's caring, relational qualities, cross-national surveys have shown that, in general, women accord more importance than men to the social values of benevolence and universalism (Schwartz and Rubel 2005). Benevolence refers to 'preservation and enhancement of the welfare of people with whom one is in frequent personal contact' and universalism to the 'understanding, appreciation, tolerance, and protection for the welfare of all people and for nature' (Schwartz and Rubel 2005: 1010–11). Similarly, other research has found that, compared with men, women endorse social values that promote the welfare of others (Beutel and Marini 1995) and socially compassionate social policies and moral practices that uphold marriage, the family, and organized religion (Eagly *et al.* 2004). But do these differences between women and men in general surveys hold up among more elite samples of women leaders? Although research on CEOs and board members of Swedish listed companies did replicate known sex-related differences in benevolence and universalism values (Adams and Funk 2012), considerably more research is needed on the generalizability of such findings.

Sex-related differences in leaders' values and attitudes are plausibly reflected in the behaviours of women and men in legislatures and on corporate boards. As members of legislative bodies, women are more likely than their male colleagues to advocate for changes that promote the interests of women, children, and families and that support public welfare in areas such as health care and education (for reviews, see Paxton *et al.* 2007; Wängnerud 2009). Although women are not a monolithic political block, these tendencies in general transcend political parties and nations. Similarly, a study of Indian women village leaders who gained office through a government mandate revealed that women more than men favoured policies that provided for the public good, such as bringing clean water to villages (Beaman *et al.* 2009). Moreover, the proportion of women on corporate boards in the *Fortune* 500 predicts the companies' philanthropy and charitable giving (Williams 2003). Likewise, the mandated addition of women to Norwegian corporate boards was followed by smaller workforce reductions, a change attributed to women's greater concern with the welfare of employees and their families (Matsa and Miller 2013). A related study found that women-owned private firms in the United States were less likely than firms owned by men to lay off workers during a period of financial stresses (Matsa and Miller in press).

Sex differences in personal ethics are also potentially important for leadership. Meta-analyses of studies on ethical beliefs and decision-making have shown that women are more likely than men to support ethical business practices (Franke *et al.* 1997; Borkowski and Ugras 1998; see also Kish-Gephart *et al.* 2010). Consistent with these trends, the representation of women on corporate boards related to more positive social outcomes and greater corporate responsibility, especially through companies eschewing negative, unethical business practices (Boulouta 2013). Likewise, research on mandated women village council leaders in India showed that these women were less likely to pay bribes than their male counterparts (Beaman *et al.* 2009). Finally, at the national level, larger representations of women in parliaments were associated with less corruption (Dollar *et al.* 2001; Swamy *et al.* 2001). Although some researchers have questioned whether this relation survived controls for various confounds, including the possibility that men experience more opportunities for corruption (Goetz 2007), the association between female representation in parliaments and restraint of corruption appears to be robust, including in a controlled experimental design (Rivas 2013).

In summary, although some of the research on leadership style does suggest female advantage, the effects of sex-related values, attitudes, and ethical tendencies also warrant consideration. Whether the outcomes of these other female proclivities produce female advantage would depend on the larger political context. More benevolence and universalism, for example, would fit better with the agendas of some political parties than others. Moreover, in political and business environments where corruption is endemic, ethical office holders and managers might find it difficult to make deals that benefit their constituencies or companies. Nevertheless, research on values, attitudes, and ethics suggests some differences between female and male leaders that can be advantageous for women at least in some contexts.

Female Disadvantage

Despite the advantages that women may possess in leadership style and qualities such as prosocial, benevolent values, and ethical integrity, they often encounter disadvantages that flow from scepticism about their abilities as leaders. The roots of this prejudice are apparent in the incongruity between the characteristics usually ascribed to women and the characteristics typically ascribed to leaders (Eagly and Karau 2002; see also Burgess and Borgida 1999; Heilman 2001). Specifically, people tend to believe that women excel in communal qualities such as niceness, warmth, and friendliness, whereas leaders typically excel in agentic qualities such as assertiveness, competitiveness, and ambition. Because agentic characteristics are thought to be more typical of men, these perceptions disfavour women as leaders and favour men. The incongruity between the group stereotype about women and the requirements of leader roles is the major source of the prejudice against women leaders.

Over several decades, considerable evidence has accumulated that the female gender role is less consistent with leader roles than the male gender role is. Schein (1973) initiated this research with her widely replicated *think manager, think male paradigm*. In these studies, participants rated men, women, or successful middle managers on a long list of gender-stereotypical traits. The researchers then assessed the correlational relations of leaders' traits to those of women and of men. In research implementing an *agency–communion paradigm* (Powell and Butterfield 1979), participants rated leaders on male-typical (agentic) and female-typical (communal) traits. The researchers then determined whether leaders were viewed as relatively more agentic than communal.

Also, in a *masculinity–femininity paradigm*, participants rated various managerial occupations on bipolar scales assessing masculinity versus femininity (Shinar 1975). The researchers then determined whether these occupations were on the masculine side of the androgynous midpoint of such scales. A meta-analysis by Koenig *et al.* (2011) revealed robust effects across all three of these paradigms. Leadership was perceived as decidedly masculine, especially in the minds of men. Specifically, leaders were viewed as more similar to men than women (e.g. Schein 1973), more agentic than communal (e.g. Powell and Butterfield 1979), and more masculine than feminine (e.g. Shinar 1975).

Relevant to understanding female disadvantage is a distinction that social psychologists often make about two aspects of social stereotypes: *descriptive stereotypes*, which are consensual expectations about the attributes of members of a social group, and *prescriptive stereotypes*, which are consensual expectations about the attributes that members of a social group ought to have (Fiske and Stevens 1993; Cialdini and Trost 1998). As we will explain, descriptive stereotypes create a *double standard* that reduces women's access to most leader roles, and prescriptive stereotypes create a *double bind* that complicates women's task of carrying out leader roles.

Descriptive gender stereotypes coupled with the masculine image of leadership create distrust of women's leadership abilities. Even those women who possess objectively excellent qualifications generally have to overcome others' suspicions that they are not well equipped to lead. The result is a double standard: women have to be better qualified for leadership than men (e.g. Lyness and Heilman 2006). Moreover, the descriptive aspects of the female gender role can serve as a self-fulfilling prophecy undermining women's leadership, as shown in experiments on *stereotype threat*. In this research, participants were asked to choose a leader or follower role for a group task after first viewing ads with gender stereotypical portrayals of women or non-stereotypical content. Bringing the female stereotype to mind caused the women (but not the men) to express less interest in becoming the group leader and more interest in taking a follower role (Davies *et al.* 2005). Culturally shared stereotypes about women thus seem to deter women from taking interest in becoming a leader.

Descriptive stereotypes also limit women's access to high levels of leadership by channelling women into roles that seem more communal. In particular, managerial women are more likely than their male counterparts to hold staff positions rather than line management positions. Yet, line management and its responsibility for profit and loss provide a much surer route to higher executive positions (Galinsky *et al.* 2003; Catalyst 2004). The concentration of women in staff management is one reason why women on average have less authority than men have, even when controlling for job status, education, and experience (Smith 2002).

Stereotyping also affects the complexity of the tasks that managerial women perform. Compared with male managers, female managers are less likely to gain appropriately demanding assignments, known as *developmental job experiences* (e.g. Ohlott *et al.* 1994; King *et al.* 2012). Being thought generally less competent for management, women are often assigned simpler tasks, or they end up with such work as a result of participatory task allocations (e.g. De Pater *et al.* 2009, 2010). Women therefore are more likely than men to have relatively unchallenging work that yields only limited rewards, even for outstanding performance.

Despite women's difficulties in achieving challenging assignments, they are *more* likely than men to be given highly risky assignments that often produce failures, a phenomenon known as the *glass cliff* (Ryan and Haslam 2007). Archival research on firms as well as experimental studies have found that women were more likely than men to be chosen for leadership positions in companies experiencing downturns (Ryan and Haslam 2005; see also Haslam *et al.* 2010; Gartzia *et al.* 2012). The preference for women in risky contexts derives not from the perception that women would be generally more effective in improving company performance, but from beliefs that traditional feminine traits would enable women to manage people during the crisis and, if necessary, to take the blame for failure (Ryan *et al.* 2011). The potential hazard in taking highly risky assignments—for example, a CEO position in a company that is threatened with bankruptcy—is obvious. Career success usually depends on access to tasks that are neither extremely risky nor unchallenging, but obtaining such assignments is apparently not straightforward for women.

Compounding the challenges from descriptive gender stereotypes that limit women's access to desirable leadership roles are prescriptive gender stereotypes that create conflicting demands for female leaders. Women are expected to behave communally, whereas most leader roles call for a good deal of agentic behaviour. These expectations result in a double bind in which communal female leaders may be criticized as insufficiently decisive and authoritative, whereas agentic female leaders may be criticized as insufficiently nice and considerate. Women leaders who violate the prescriptions of the leader stereotype or the female gender stereotype are disliked and experience lower evaluations of their performance (Eagly *et al.* 1992; Eagly and Karau 2002; but see Livingston *et al.* 2012 for findings on Black women). Moreover, many studies have revealed a *backlash phenomenon* whereby women are penalized for appearing to seek power (Okimoto and Brescoll 2010) and engaging in other high-status, dominant behaviours (Rudman *et al.* 2012). In addition, women, compared with men, are penalized for assertively negotiating for a higher salary for themselves (Bowles *et al.* 2007; Amanatullah and Tinsley 2012), talking a lot in public settings (Brescoll 2011), expressing angry emotions (Brescoll and Uhlmann 2008), or performing outstandingly in masculine domains (Heilman *et al.* 2004). This backlash against female agency, which researchers have repeatedly demonstrated, limits the range of behaviours that women managers can deploy to good effect.

Female leaders can alleviate the problem of the double bind to some extent by exhibiting both agentic and communal behaviour (Heilman and Okimoto 2007; Johnson *et al.* 2008; see also Eagly and Carli 2007). When women do overcome the double standard and the double bind and achieve exceptionally high levels of leadership, they are sometimes credited for having overcome those very barriers (Lyness and Heilman 2006). One study has shown, for example, that a woman described to participants as a successful CEO personally responsible for her own success was seen as more communal, agentic, and effective than a comparably successful man (Rosette and Tost 2010). Moreover, the participants perceived the woman to have relied on more feminine management tactics and to have overcome double standards more than the man, and these inferences mediated the relation between sex of leader and leader effectiveness.

THE EFFECTIVENESS OF FEMALE AND MALE LEADERS

Given the findings on female advantage and disadvantage, it would be rash to predict that women ordinarily would be judged as more effective leaders than men. Indeed, studies assessing leaders' effectiveness typically rely on participants' ratings of individual leaders, and undoubtedly such ratings may be contaminated by gender bias. Nevertheless, such ratings have some validity because leaders' success depends in part on the approval of their colleagues and subordinates. In fact, a meta-analysis of ninety-six studies comparing the effectiveness of male and female leaders in generally comparable leadership roles found no overall difference between women and men (Eagly *et al.* 1995). However, the findings showed that some contexts clearly favoured male leaders, and some favoured female leaders. Specifically, in masculine settings, particularly the military, men received higher effectiveness ratings than women; in the least masculine settings, such as in education and social services, women received somewhat higher effectiveness ratings than men (Eagly *et al.* 1995). Therefore, the overall null effect of this meta-analysis was an arbitrary consequence of the distribution of the studies into differing organizational contexts.

Leaving aside the issue of individual leaders' effectiveness, other research has explored whether groups and teams, including high-level leadership groups, perform more effectively when women are included as members. Gender-diverse groups, like groups that are diverse on other bases such as race, have the potential to outperform less diverse groups because members from a greater range of socio-demographic categories are likely to have more varied perspectives, information, and approaches to problem-solving (Page 2007). Women, for example, may have greater insight than men into certain strategic issues, especially relating to female consumers and employees. However reasonable this argument, gender diversity and other forms of demographic diversity do not consistently lead to superior group performance. The most extensive meta-analysis of the influence of diversity on group performance found that gender diversity had essentially no overall effect on objectively measured performance outcomes, although these relationships were quite heterogeneous across studies (van Dijk *et al.* in press). To gain from diversity based on demographic attributes, organizations have to leverage this resource by lessening the conflict, communication barriers, and lack of mutual respect that can develop in identity-diverse groups (Van Knippenberg *et al.* 2004; Rink and Ellemers 2009).

In other efforts, numerous researchers have explored the effects of leaders' gender diversity on the outcomes of the corporations that they lead. Specifically, interest has centred on the relation between the representation of women among corporate executives or directors and corporations' financial performance. Several studies examining *Fortune* 500 and 1000 corporations and broader samples of US companies have found that the higher the percentage of women in such positions, the better the financial outcomes (e.g. Carter *et al.* 2003; Krishnan and Park 2005; Joy *et al.* 2007). Also, a McKinsey study of large European corporations found that gender diversity was associated with better financial outcomes (Desvaux *et al.* 2007).

To the extent that the associations between proportions of women in high-level positions and firm performance have been based on relatively simple correlational analyses, they may suffer from endogeneity—that is, statistical anomalies such as reverse causation, omitted variables, selection biases, and flawed measures (Antonakis *et al.* 2010). Therefore, it is perhaps not surprising that newer research with more sophisticated controls for endogeneity has not necessarily found that female leadership promoted corporate financial success (see O'Reilly and Main 2012 for review). In particular, in a large sample of US firms, Adams and Ferriera (2009) found an overall negative average effect of the gender diversity of corporate boards when controlling for individual firm characteristics. However, this effect was moderated by how well governed firms were, as mediated by the behaviour of the directors. Specifically, female directors had better attendance at board meetings, evidently inspiring better male attendance as well, and the female directors were also especially well represented on committees involved in monitoring activities. With the greater monitoring that apparently took place in more gender-diverse boards, CEOs were held more accountable for poor performance and had greater turnover. Increased monitoring benefited the performance of firms with weak governance, but not firms that were well governed.

Another study of a similar sample of *Fortune* 1500 corporations examined gender diversity in top management teams rather than boards of directors (Dezsö and Ross 2012). The overall relation between the gender diversity of these teams and firm performance was positive, although this favourable effect was present only in firms whose strategies focused on innovation. It may be that the advantages of women's more collaborative, transformational and reward-based leadership styles are more likely to facilitate performance within management teams, rather than boards. These stylistic differences may be more helpful when companies face the complex issues involved in innovating new products and services. To realize the potential advantages of gender diversity, it may also be important that organizations foster participative managerial processes that allow all voices to be heard (Richard *et al.*, 2013).

Norwegian companies' compliance with the government-mandated 40 per cent quota for women on boards of listed corporations has presented a unique opportunity to study the effects of board gender diversity. Econometric analyses have shown a negative effect of this change on corporate profits. Ahern and Dittmar (2012) ascribed this effect to the younger age and lack of high-level managerial experience of the added female directors. In contrast, Matsa and Miller (2012*a*) ascribed the effect to the fewer workforce reductions that ensued, which increased relative labour costs and reduced short-term profits. The longer-term effects of the Norwegian quota as well as quotas imposed in other nations will allow a more complete analysis of the effects of such shifts in corporate leadership.

In summary, studies of individual managers suggest that masculine settings and highly male-dominated roles continue to be challenging for female leaders, whereas female leaders' effectiveness is more apparent in less masculine settings and roles that are less male-dominated. Other research has shown that gender diversity has mixed effects at team and firm levels. In the wake of economists' close scrutiny of corporate data, it seems that adding women leaders does not necessarily lead to increasing firm value; instead, the varying effects of gender diversity are no doubt contingent on factors such as the labour pool of qualified women, the management level of the women, the challenges that corporations face, and the prevailing economic conditions. One difficulty that precludes clear conclusions on the effects of gender diversity on firm performance based on data sets such as those of Adams and Ferriera (2009) and Dezsö and Ross (2012) is that most of the gender-diverse firms in their samples had only one woman on their top management teams or boards of directors. The token status of such women leaders could have increased the extent to which they were perceived in gender-stereotypical terms and thereby undermined their influence and authority (see Yoder 2002), weakening any potential benefits of their leadership (Torchia *et al.* 2011).

THE EFFECTS OF FEMALE LEADERSHIP ON SOCIETAL OUTCOMES

Women's leadership has broader effects on society than merely influencing organizational outcomes. Gender equality is correlated with greater economic productivity and national wealth (World Bank 2012). Obviously, expanding women's leadership, especially in the high-level positions where they are now thinly represented, increases societal gender equality merely because these positions become more integrated. In addition, the presence of female leaders can accelerate the inclusion of other women into other leadership positions. In aggregate analyses of US corporate data, higher percentages of women on boards appear to boost the hiring of female executives (Matsa and Miller 2011) and to increase the likelihood that newly appointed CEOs are women (Elsaid and Ursel 2011). In addition, organizational studies reveal that increases of women in top management correlate with increased representation of women in lower-level management (e.g. Cohen *et al.* 1998; Kurtulus and Tomaskovic-Devey 2012). Also, in a representative national US study, the inclusion of women in higher-level managerial roles was associated with a reduction in the gender wage gap (Cohen and Huffman 2007).

The reasons that female leaders foster an increase of women in leadership roles are no doubt varied. In-group biases may operate, whereby women and men favour the advancement of their own sex. General pro-equality attitudinal shifts are also possible because people have observed women adequately performing leader roles. A unique opportunity to test these hypotheses resulted when gender quotas for women leaders were randomly assigned a portion of Indian village councils. This field experiment revealed that mandated office-holding by women produced greater subsequent participation of women as candidates and office holders as well as favourable changes in attitudes towards women leaders and socially progressive changes in gender role beliefs (Beaman *et al.* 2009). Furthermore, girls' career and educational aspirations increased, the gender gap in educational attainment disappeared, and girls spent less time on domestic chores (Beaman *et al.* 2012). These effects on girls appear to reflect role modelling by female leaders. Such effects are consistent with social psychological research demonstrating that exposure to female leaders produces a reduction in automatic gender stereotyping whereby women are associated with communal traits and men with agentic traits (Dasgupta and Asgari 2004).

The Changing Context of Leadership

One of the most important findings to emerge in the last decade is clear evidence of cultural change in how people think about leadership, a shift that is consistent with the claims of many management experts (e.g. Helgesen 1995; Lipman-Blumen 2000). The meta-analysis that we noted on the masculinity of the leader stereotype thus found that, in all three of the research paradigms that have investigated this stereotype, leadership has changed to become more androgynous, apparently because leader roles are increasingly seen as requiring a good measure of culturally feminine relational skills (Koenig *et al.* 2011). This shift towards androgyny is consistent with leadership roles becoming generally more hospitable to women.

The increasing numbers of female leaders and favourable attitudes towards androgynous leadership (Carli and Eagly 2011; see also Gartzia 2010) raise questions about what women can do to attain and become successful in these roles and what organizations and societies can do to give more women opportunities to lead. Eagly and Carli (2007) argued that individual women generally gain from a blended leadership style that incorporates both agentic and communal workplace behaviours. Transformational leadership, an androgynous style, provides one type of blending of agency and communion (von Véver 2012; see also Gartzia and van Engen 2012). In this and other androgynous modes, women can address both sides of the double bind: They show that they are directive and assertive enough to be leaders but that this agency does not undermine the warmth and sociability that people expect from women. Flexibly mixing masculine and feminine behaviour in response to situational demands can help women build workplace social capital and may be a particular skill of women who are strong in the *self-monitoring* trait that favours monitoring and controlling one's own behaviour in social situations (Flynn and Ames 2006). This idea of situational flexibility of feminine and masculine behaviour harks back to the classic concept of androgyny that was especially popular in gender research of the 1970s (e.g. Bem and Lewis 1975). The advantages of androgyny emerge especially clearly in leader roles.

Now that paths to leadership are more open to women than in the past, meeting gender-related career challenges may be regarded as a problem that women should solve by themselves, but it is also a problem that should be addressed by those organizations that still obstruct women's advancement. As signs of progress, innovations that have occurred in some organizations' personnel policies have improved the retention of talented women (see Benko and Weisberg 2007). In favourable circumstances, even highly demanding leadership roles can be compatible with women's family roles (Greenhaus and Powell 2006), as research on female leaders has shown (e.g. Cheung and Halpern 2010). Because managerial roles and gender roles are in flux due to broader changes in the economy and society, building in innovations that can make organizations as welcoming to women as to men should become a priority.

Our conclusions about female advantage are generally consistent with our claims of a decade ago (Eagly and Carli 2003a). Given some degree of advantage in leadership style, female leadership offers the potential for increasing organizational effectiveness. Yet, to realize these hypothetical gains, organizations must overcome the female disadvantage inherent in cultural stereotyping of women and leadership. This change is under way as more people come to appreciate that a manager gains less from ordering others about than from establishing teams of smart, motivated collaborators who together figure out how to solve problems and get work done. Female as well as male managers can thrive in such environments.

The best news for women in the past decade is that these more androgynous ways of leading have gained cultural currency as they have become more normative and admired (Koenig *et al.* 2011). Because this shift reflects the greater complexity of the missions of most modern organizations, a return to simpler command-and-control leadership is highly unlikely. Moreover, the financial crisis and other threats should increase openings for women as leaders because gender stereotypes align women more with change and men with stability (Brown *et al.* 2011). The future is thus promising for women's greater occupancy of leader roles at all levels of organizations and governments.

NOTE

1. We use the terms *sex* and *sexes* to denote the grouping of people into female and male categories. The term *gender* refers to the meanings that societies and individuals ascribe to these female and male categories. We do not intend to use these terms to give priority to any class of causes that may underlie sex and gender effects (see Wood and Eagly 2010).

REFERENCES

Abelson, R. P. (1985). 'A Variance Explanation Paradox: When a Little is a Lot', *Psychological Bulletin*, 97: 129–33. doi:10.1037/0033-2909.97.1.129

Adams, R. B., and Ferreira, D. (2009). 'Women in the Boardroom and their Impact on Governance and Performance', *Journal of Financial Economics*, 94: 291–309. doi:10.1016/j.jfineco.2008.10.007

Adams, R. B., and Funk, P. (2012). 'Beyond the Glass Ceiling: Does Gender Matter?', *Management Science*, 58: 219–35. doi:10.1287/mnsc.1110.1452

Ahern, K., and Dittmar, A. (2012). 'The Changing of the Boards: The Impact on Firm Valuation of Mandated Female Board Representation', *Quarterly Journal of Economics*, 127: 137–97. doi:10.1093/qje/qjr049

Amanatullah, E. T., and Tinsley, C. H. (2012). 'Punishing Female Negotiators for Asserting Too Much... or Not Enough: Exploring Why Advocacy Moderates Backlash Against Assertive Female Negotiators', *Organizational Behavior and Human Decision Processes*. http://dx.doi.org/10.1016/j.obhdp.2012.03.006 Advance online publication.

Ames, D. R., and Flynn, F. J. (2007). 'What Breaks a Leader: The Curvilinear Relation between Assertiveness and Leadership', *Journal of Personality and Social Psychology*, 92: 307–24. doi:10.1037/0022-3514.92.2.307

Antonakis, J., Avolio, B. J., and Sivasubramaniam, N. (2003). 'Context and Leadership: An Examination of the Nine-Factor Full-Range Leadership Theory Using the Multifactor Leadership Questionnaire', *Leadership Quarterly*, 14: 261–95. doi:10.1016/S1048-9843(03)00030-4

Antonakis, J., Bendahan, S., Jacquart, P., and Lalive, R. (2010). 'On Making Causal Claims: A Review and Recommendations', *Leadership Quarterly*, 21: 1086–1120. doi:10.1016/j.leaqua.2010.10.010

Avolio, B. J. (2010). *Full Range Leadership Development* (2nd edn). Thousand Oaks, CA: Sage.

Ayman, R., Korabik, K., and Morris, S. (2009). 'Is Transformational Leadership Always Perceived as Effective? Male Subordinates' Devaluation of Female Transformational Leaders', *Journal of Applied Social Psychology*, 39: 852–79. doi:10.1111/j.1559-1816.2009.00463.x

Barsh, J., Cranston, S., and Lewis, G. (2009). *How Remarkable Women Lead: The Breakthrough Model for Work and Life*. New York: Crown Books.

Bass, B. M. (1998). *Transformational Leadership: Industry, Military, and Educational Impact*. Mahwah, NJ: Erlbaum.

Beaman, L., Chattopadhyay, R., Duflo, E., Pande, R., and Topalova, P. (2009). 'Powerful Women: Does Exposure Reduce Bias?', *Quarterly Journal of Economics*, 124: 1497–1540. doi:10.1162/qjec.2009.124.4.1497

Beaman, L., Duflo, E., Pande, R., and Topalova, P. (2012). 'Female Leadership Raises Aspirations and Educational Attainment for Girls: A Policy Experiment in India', *Science*, 335 (6068): 582–6. doi:10.1126/science.1212382

Bem, S. L., and Lewis, S. A. (1975). 'Sex Role Adaptability: One Consequence of Psychological Androgyny', *Journal of Personality and Social Psychology*, 31: 634–43. doi:10.1037/h0077098

Benko, C., and Weisberg, A. C. (2007). *Mass Career Customization: Aligning the Workplace with Today's Nontraditional Workforce*. Boston: Harvard Business School Press.

Beutel, A. M., and Marini, M. M. (1995). 'Gender and Values', *American Sociological Review*, 60: 436–48. doi:10.2307/2096423

Billing, Y. D., and Alvesson, M. (2000). 'Questioning the Notion of Feminine Leadership: A Critical Perspective on the Gender Labelling of Leadership', *Gender, Work and Organization*, 7: 144–57. doi:10.1111/1468-0432.00103

Borkowski, S. C., and Ugras, Y. J. (1998). 'Business Students and Ethics: A Meta-Analysis', *Journal of Business Ethics*, 17: 1117–27. doi:10.1023/A:1005748725174

Boulouta, I. (2013). 'Hidden Connections: The Link between Board Gender Diversity and Corporate Social Performance', *Journal of Business Ethics*, 113, 185–197. doi:10.1007/s10551-012-1293-7

Bowles, H. R., Babcock, L., and Lai, L. (2007). 'Social Incentives for Gender Differences in the Propensity to Initiate Negotiations: Sometimes it Does Hurt to Ask', *Organizational Behavior and Human Decision Processes*, 103: 84–103. doi:10.1016/j.obhdp.2006.09.001

Brescoll, V. L. (2011). 'Who Takes the Floor and Why? Gender, Power, and Volubility in Organizations', *Administrative Science Quarterly*, 56: 622–41. doi:10.1177/0001839212439994

Brescoll, V. L., and Uhlmann, E. L. (2008). Can an Angry Woman Get Ahead? Status Conferral, Gender, and Expression of Emotion in the Workplace', *Psychological Science*, 1: 268–75. doi:10.1111/j.1467-9280.2008.02079.x

Brown, E. R., Diekman, A. B., and Schneider, M. C. (2011). 'A Change will Do us Good: Threats Diminish Typical Preferences for Male Leaders', *Personality and Social Psychology Bulletin*, 37: 930–41. doi:10.1177/0146167211403322

Burgess, D., and Borgida, E. (1999). 'Who Women Are, Who Women Should Be: Descriptive and Prescriptive Gender Stereotyping in Sex Discrimination', *Psychology, Public Policy, and Law*, 5: 665–92. doi:10.1037/1076-8971.5.3.665

Carli, L. L., and Eagly, A. H. (2011). 'Gender and Leadership', in D. Collinson, A. Bryman, K. Grint, B. Jackson, and M. Uhl Bien (eds), *Sage Handbook of Leadership*, 269–85. London: Sage.

Carter, D. A., Simkins, B. J., and Simpson, W. G. (2003). 'Corporate Governance, Board Diversity, and Firm Value', *Financial Review*, 38: 33–53. doi:10.1111/1540-6288.00034

Catalyst. (2004). *Women and Men in U.S. Corporate Leadership: Same Workplace, Different Realities?* Retrieved from http://catalyst.org/file/74/women%20and%20men%20in%20u.s.%20corporate%20leadership%20same%20workplace,%20different%20realities.pdf

Cheung, F. M., and Halpern, D. F. (2010). 'Women at the Top: Powerful Leaders Define Success as Work + Family in a Culture of Gender', *American Psychologist*, 65: 182–93. doi:10.1037/a0017309

Cialdini, R. B., and Trost, M. R. (1998). 'Social Influence: Social Norms, Conformity and Compliance', in D. T. Gilbert, S. T. Fiske, and G. Lindzey (eds), *The Handbook of Social Psychology* (4th edn), ii. 151–92. Boston: McGraw-Hill.

Cohen, P. N., and Huffman, M. L. (2007). 'Working for the Woman? Female Managers and the Gender Wage Gap', *American Sociological Review*, 72: 681–704. doi:10.1177/000312240707200502

Cohen, L. E., Broschak, J. P., and Haveman, H. A. (1998). 'And Then there were More? The Effect of Organizational Sex Composition on the Hiring and Promotion of Managers', *American Sociological Review*, 63: 711–27. doi:10.2307/2657335

Daily Mail (2010). 'Women in Top Jobs are Viewed as "Better Leaders" than Men', 14 May. Retrieved from http://www.dailymail.co.uk/sciencetech/article-1278009/Women-jobs-viewed-better-leaders-men.html#ixzz1quPyGOPa

Dasgupta, N., and Asgari, S. (2004). 'Seeing is Believing: Exposure to Counterstereotypic Women Leaders and its Effect on the Malleability of Automatic Gender Stereotyping', *Journal of Experimental Social Psychology*, 40: 642–58. doi:10.1016/j.jesp.2004.02.003

Davies, P. G., Spencer, S. J., and Steele, C. M. (2005). 'Clearing the Air: Identity Safety Moderates the Effects of Stereotype Threat on Women's Leadership Aspirations', *Journal of Personality and Social Psychology*, 88, 276–87. doi:10.1037/0022-3514.88.2.276

De Pater, I., Van Vianen, A., Humphrey, R., Sleeth, R., Hartman, N., and Fischer, A. (2009). 'Individual Task Choice and the Division of Challenging Tasks between Men and Women', *Group and Organization Management*, 34: 563–89. doi:10.1177/1059601108331240

De Pater, I., Van Vianen, A. E. M., and Bechtoldt, M. N. (2010). 'Gender Differences in Job Challenge: A Matter of Task Allocation', *Gender, Work and Organization*, 17: 433–53. doi:10.1111/j.1468-0432.2009.00477.x

Desvaux, G., and Devillard, S. (2008). *Women Matter 2*. McKinsey & Co. Retrieved from http://www.mckinsey.com/locations/paris/home/womenmatter/pdfs/women_matter_oct2008_english.pdf

Desvaux, G., Devillard-Hoellinger, S., and Baumgarten, P. (2007). *Women Matter: Gender Diversity, a Corporate Performance Driver*. Paris: McKinsey & Co. Retrieved from http://www.mckinsey.com/locations/paris/home/womenmatter/pdfs/Women_matter_oct2007_english.pdf

Dezső, C. L., and Ross, D. G. (2012). Does Female Representation in Top Management Improve Firm Performance? A Panel Data Investigation', *Strategic Management Journal*, 33: 1072–89. doi:10.1002/smj.1955

Dollar, D., Fisman, R., and Gatti. R. (2001). 'Are Women Really the "Fairer" Sex? Corruption and Women in Government', *Journal of Economic Behavior and Organization*, 26:423–9. doi:10.1016/S0167-2681(01)00169-X

Eagly, A. H., and Carli, L. L. (2003*a*). 'The Female Leadership Advantage: An Evaluation of the Evidence', *Leadership Quarterly*, 14: 807–34. doi:10.1016/j.leaqua.2003.09.004

Eagly, A. H., and Carli, L. L. (2003*b*). 'Finding Gender Advantage and Disadvantage: Systematic Research Integration is the Solution', *Leadership Quarterly*, 14: 851–9. doi:10.1016/j.leaqua.2003.09.003

Eagly, A. H., and Carli, L. L. (2007). *Through the Labyrinth: The Truth about How Women Become Leaders*. Cambridge, MA: Harvard Business School Press.

Eagly, A. H., and Johnson, B. T. (1990). 'Gender and Leadership Style: A Meta-Analysis', *Psychological Bulletin*, 108: 233–56. doi:10.1037/0033-2909.108.2.233

Eagly, A. H., and Karau, S. J. (2002). 'Role Congruity Theory of Prejudice toward Female Leaders', *Psychological Review*, 109: 573–98. doi:10.1037/0033-295X.109.3.573

Eagly, A. H., Makhijani, M. G., and Klonsky, B. G. (1992). 'Gender and the Evaluation of Leaders: A Meta-Analysis', *Psychological Bulletin*, 111: 3–22. doi:10.1037/0033-2909.111.1.3

Eagly, A. H., Karau, S. J., and Makhijani, M. G. (1995). 'Gender and the Effectiveness of Leaders: A Meta-Analysis', *Psychological Bulletin*, 117: 125–45. doi:10.1037/0033-2909.117.1.125

Eagly, A. H., Johannesen-Schmidt, M. C., and van Engen, M. (2003). 'Transformational, Transactional, and Laissez-Faire Leadership Styles: A Meta-Analysis Comparing Women and Men', *Psychological Bulletin*, 129: 569–91. doi:10.1037/0033-2909.129.4.569

Eagly, A. H., Diekman, A. B., Johannesen-Schmidt, M. C., and Koenig, A. M. (2004). 'Gender Gaps in Sociopolitical Attitudes: A Social Psychological Analysis', *Journal of Personality and Social Psychology*, 87: 796–816. doi:10.1037/0022-3514.87.6.796

Elsaid, E., and Ursel, N. D. (2011). 'CEO Succession, Gender and Risk Taking', *Gender in Management: An International Journal*, 26: 499–512. doi:10.1108/17542411111175478

Fiske, S. T., and Stevens, L. E. (1993). 'What's So Special about Sex? Gender Stereotyping and Discrimination', in S. Oskamp and M. Costanzo (eds), *Claremont Symposium on Applied*

Social Psychology, vi. Gender Issues in Contemporary Society, 173–96. Thousand Oaks, CA: Sage.

Flynn, F. J., and Ames, D. R. (2006). 'What's Good for the Goose may Not be as Good for the Gander: The Benefits of Self-Monitoring for Men and Women in Task Groups and Dyadic Conflicts', *Journal of Applied Psychology*, 91: 272–81. doi:10.1037/0021-9010.91.2.272

Foels, R., Driskell, J. E., Mullen, B., and Salas, E. (2000). 'The Effects of Democratic Leadership on Group Member Satisfaction: Integration', *Small Group Research*, 31: 676–701. doi:10.1177/104649640003100603

Franke, G. R., Crown, D. F., and Spake, D. F. (1997). 'Gender Differences in Ethical Perceptions of Business Practices: A Social Role Theory Perspective', *Journal of Applied Psychology*, 82: 920–34. doi:10.1037//0021-9010.82.6.920

Galinsky, E., Salmond, K., Bond, J. T., Kropf, M. B., Moore, M., and Harrington, B. (2003). *Leaders in a Global Economy: A Study of Executive Women and Men*. New York: Families and Work Institute. Retrieved from http://familiesandwork.org/site/research/reports/globaltalentmgmt.pdf

Gartzia, L. (2010). 'From "Think Male" to "Think Androgynous": Implications for Gender Equality and Organizational Functioning in XXI Century Organizations', doctoral dissertation, University of the Basque Country, Spain.

Gartzia, L., and van Engen, M. (2012). 'Are (Male) Leaders "Feminine" Enough? Gendered Traits of Identity as Mediators of Sex Differences in Leadership', *Gender in Management: An International Journal*, 27: 296–314. doi:10.1108/17542411211252624

Gartzia, L., Ryan, M.K., Balluerka, N., and Aritzeta, A. (2012). 'Think Crisis—Think Female: Further Evidence', *European Journal of Work and Organizational Psychology*, 21: 603–28. doi:10.1080/1359432X.2011.591572

Gastil, J. (1994). 'A Meta-Analytic Review of the Productivity and Satisfaction of Democratic and Autocratic Leadership', *Small Group Research*, 25: 384–410. doi: 10.1177/1046496494253003

Gilligan, C. (1982). *In a Different Voice*. Cambridge, MA: Harvard University Press.

Goetz, A. M. (2007). 'Political Cleaners: Women as the New Anti-Corruption Force?', *Development and Change*, 38: 87–105. doi:10.1111/j.1467-7660.2007.00404.x

Goleman, D., Boyatzis, R. E., and McKee, A. (2002). *Primal Leadership: Realizing the Power of Emotional Intelligence*. Boston: Harvard Business School Press.

Greenhaus, J. H., and Powell, G. N. (2006). 'When Work and Family are Allies: A Theory of Work–Family Enrichment', *Academy of Management Review*, 31: 72–92. doi:10.5465/AMR.2006.19379625

Harms, P. D., and Credé, M. (2010). 'Emotional Intelligence and Transformational and Transactional Leadership: A Meta-Analysis', *Journal of Leadership and Organizational Studies*, 17: 5–17. doi:10.1177/1548051809350894

Haslam, S. A., Ryan, M. K., Kulich, C., Trojanowski, G., and Atkins, C. (2010). 'Investing with Prejudice: The Relationship between Women's Presence on Company Boards and Objective and Subjective Measures of Company Performance', *British Journal of Management*, 21: 484–97. doi:10.1111/j.1467-8551.2009.00670.x

Heilman, M. E. (2001). 'Description and Prescription: How Gender Stereotypes Prevent Women's Ascent up the Organizational Ladder', *Journal of Social Issues*, 57: 657–74. doi:10.1111/0022-4537.00234

Heilman, M. E., and Okimoto, T. G. (2007). 'Why are Women Penalized for Success at Male Tasks? The Implied Communality Deficit', *Journal of Applied Psychology*, 92: 81–92. doi:10.1037/0021-9010.92.1.81

Heilman, M. E., Wallen, A. S., Fuchs, D. and Tamkins, M. M. (2004). 'Penalties for Success: Reactions to Women Who Succeed in Male Gender-Typed Tasks', *Journal of Applied Psychology*, 89: 416–27. doi:10.1037/0021-9010.89.3.416

Helgesen, S. (1990). *The Female Advantage: Women's Ways of Leadership*. New York: Currency Doubleday.

Helgesen, S. (1995). *The Web of Inclusion: Architecture for Building Great Organizations*. New York: Doubleday.

Helgesen, S., and Johnson, J. (2010). *The Female Vision: Women's Real Power at Work*. San Francisco, CA: Berrett-Koehler.

Johnson, S., Murphy, S., Zewdie, S., and Reichard, R. (2008). 'The Strong, Sensitive Type: Effects of Gender Stereotypes and Leadership Prototypes on the Evaluation of Male and Female Leaders', *Organizational Behavior and Human Decision Processes*, 106: 39–60. doi:10.1016/j.obhdp.2007.12.002

Joseph, D. L., and Newman, D. A. (2010). 'Emotional Intelligence: An Integrative Meta-Analysis and Cascading Model', *Journal of Applied Psychology*, 95: 54–78. doi:10.1037/a0017286

Joy, L, Carter, N. M., Wagner, H. M., and Narayanan, S. (2007). *The Bottom Line: Corporate Performance and Women's Representation on Boards*. Retrieved from http://www.catalyst.org/publication/200/the-bottom-line-corporate-performance-and-womens-representation-on-boards

Judge, T. A., and Piccolo. R. F. (2004). 'Transformational and Transactional Leadership: A Meta-Analytic Test of their Relative Validity', *Journal of Applied Psychology*, 89: 901–10. doi:10.1037/0021-9010.89.5.755

Judge, T. A., Piccolo, R. F., and Ilies, R. (2004). 'The Forgotten Ones? The Validity of Consideration and Initiating Structure in Leadership Research', *Journal of Applied Psychology*, 89: 36–51. doi:10.1037/0021-9010.89.1.36

King, E. B., Botsford, W., Hebl, M. R., Kazama, S., Dawson, J. F., and Perkins, A. (2012). 'Benevolent Sexism at Work Gender Differences in the Distribution of Challenging Developmental Experiences', *Journal of Management*, 38: 1835–66. doi:10.1177/0149206310365902

Kish-Gephart, J. J., Harrison, D. A., and Treviño, L. K. (2010). 'Bad Apples, Bad Cases, and Bad Barrels: Meta-Analytic Evidence about Sources of Unethical Decisions at Work', *Journal of Applied Psychology*, 95: 1–31. doi:10.1037/a0017103

Koenig, A. M., Eagly, A. H., Mitchell, A. A., and Ristikari, T. (2011). 'Are Leader Stereotypes Masculine? A Meta-Analysis of Three Research Paradigms', *Psychological Bulletin*, 137: 616–42. doi:10.1037/a0023557

Krishnan, H. A., and Park, D. (2005). A Few Good Women—On Top Management Teams', *Journal of Business Research*, 58: 1712–20. doi:10.1016/j.jbusres.2004.09.003

Kurtulus, F. A., and Tomaskovic-Devey, D. (2012). 'Do Female Top Managers Help Women to Advance? A Panel Study Using EEO-1 Records', *Annals of the American Academy of Political and Social Science*, 639: 173–97. doi:10.1177/0002716211418445

Lipman-Blumen, J. (2000). *Connective Leadership: Managing in a Changing World*. New York: Oxford University Press.

Livingston, R. W., Rosette, A. S., and Washington, E. F. (2012). 'Can an Agentic Black Woman Get Ahead? The Impact of Race and Interpersonal Dominance on Perceptions of Female Leaders', *Psychological Science*, 23: 354–8. doi:10.1177/0956797611428079

Loden, M. (1985). *Feminine Leadership: How to Succeed in Business without Being One of the Boys*. New York: Crown.

Lowe, K. B., Kroeck, K. G., and Sivasubramaniam, N. (1996). 'Effectiveness Correlates of Transformation and Transactional Leadership: A Meta-Analytic Review of the MLQ Literature', *Leadership Quarterly*, 7: 385–425. doi:10.1016/S1048-9843(96)90027-2

Lyness, K. S., and Heilman, M. E. (2006). 'When Fit is Fundamental: Performance Evaluations and Promotions of Upper-Level Female and Male Managers', *Journal of Applied Psychology*, 91: 777–85. doi:10.1037/0021-9010.91.4.777

Matsa, D. A., and Miller, A. R. (2011). 'Chipping Away at the Glass Ceiling: Gender Spillovers in Corporate Leadership', *American Economic Review*, 101: 635–9.

Matsa, D. A., and Miller, A. R. (2013). A Female Style in Corporate Leadership? Evidence from Quotas, *American Economic Journal: Applied Economics*, 5, 136–69.

Matsa, D. A., and Miller, A. R. (in press). "Workforce Reductions at Women-Owned Businesses in the United States", *Industrial and Labor Relations Review*.

Mayer, J. D., Salovey, P., and Caruso, D. R. (2004). 'Emotional Intelligence: Theory, Findings, and Implications', *Psychological Inquiry*, 15: 197–215. doi:10.1207/s15327965pli1503_02

Melero, E. (2011). 'Are Workplaces with Many Women in Management Run Differently?', *Journal of Business Research*, 64: 385–93. doi:10.1016/j.jbusres.2010.01.009

Ohlott, P. J., Ruderman, M. N., and McCauley, C. D. (1994). 'Gender Differences in Managers' Developmental Job Experiences', *Academy of Management Journal*, 37: 46–67. doi: 10.2307/256769

Okimoto, T. G., and Brescoll, V. L. (2010). 'The Price of Power: Power Seeking and Backlash Against Female Politicians', *Personality and Social Psychology Bulletin*, 36: 923–36. doi: 10.1177/0146167210371949

O'Reilly, C. A., III, and Main, B. G. M. (2012). *Women in the Boardroom: Symbols or Substance?* Research Paper, 2098, Stanford Graduate School of Business. Retrieved from http://gsbapps.stanford.edu/researchpapers/library/RP2098.pdf

Page, S. E. (2007). *The Difference: How the Power of Diversity Creates Better Groups, Firms, Schools, and Societies*. Princeton: Princeton University Press.

Paxton, P., Kunovich, S., and Hughes, M. M. (2007). 'Gender in Politics', *Annual Review of Sociology*, 33: 263–84. doi:10.1146/annurev.soc.33.040406.131651

Powell, G. N., and Butterfield, D. A. (1979). 'The "Good Manager": Masculine or Androgynous?', *Academy of Management Journal*, 22: 395–403. doi:10.2307/255597

Powell, G. N., and Greenhaus, J. H. (2010). 'Sex, Gender, and the Work-to-Family Interface: Exploring Negative and Positive Interdependencies', *Academy of Management Journal*, 53: 513–34. doi:10.5465/AMJ.2010.51468647

Richard, O. C., Kirby, S. L., and Chadwick, K. (2013). 'The Impact of Racial and Gender Diversity in Management on Financial Performance: How Participative Strategy Making Features Can Unleash a Diversity Advantage', *The International Journal of Human Resource Management*, 24: 2571–2582. doi:10.1080/09585192.2012.744335

Rink, F., and Ellemers, N. (2009). 'Managing Diversity in Work Groups: How Identity Processes Affect Diverse Work Groups', in M. Barreto, M. K. Ryan, and M. T. Schmitt (eds), *The Glass Ceiling in the 21st Century: Understanding Barriers to Gender Equality*, 281–303. Washington, DC: American Psychological Association.

Rivas, M. F. (2013). 'An Experiment on Corruption and Gender', *Bulletin of Economic Research*, 65: 1042. doi: 10.1111/j.1467-8586.2012.00450.x

Rosener, J. B. (1990). 'Ways Women Lead', *Harvard Business Review*, 68(6): 119–25. doi:10.1007/978-90-481-9014-0_3

Rosette, A., and Tost, L. P. (2010). 'Agentic Women and Communal Leadership: How Role Prescriptions Confer Advantage to Top Women Leaders', *Journal of Applied Psychology*, 95: 221–35. doi:10.1037/a0018204

Roth, P. L., Purvis, K. L., and Bobko, P. (2012). 'A Meta-Analysis of Gender Group Differences for Measures of Job Performance in Field Studies', *Journal of Management*, 38: 719–39. doi: 10.1177/0149206310374774

Rudman, L. A., Moss-Racusin, C. A., Phelan, J. E., and Nauts, S. (2012). 'Status Incongruity and Backlash Effects: Defending the Gender Hierarchy Motivates Prejudice Against Female Leaders', *Journal of Experimental Social Psychology*, 48: 165–79. doi:10.1016/j.jesp.2011.10.008

Ryan, M. K., and Haslam, S. A. (2005). 'The Glass Cliff: Evidence that Women are Over-Represented in Precarious Leadership Positions', *British Journal of Management*, 16: 81–90. doi:10.1111/j.1467-8551.2005.00433.x

Ryan, M. K., and Haslam, S. A. (2007). 'The Glass Cliff: Exploring the Dynamics Surrounding Women's Appointment to Precarious Leadership Positions', *Academy of Management Review*, 32: 549–72. doi:10.5465/AMR.2007.24351856

Ryan, M. K., Haslam, A. S., Hersby, M. D., and Bongiorno, R. (2011). 'Think Crisis—Think Female: The Glass Cliff and Contextual Variation in the Think Manager—Think Male Stereotype', *Journal of Applied Psychology*, 96: 470–84. doi:10.1037/a0022133

Schein, V. E. (1973). 'The Relationship between Sex Role Stereotypes and Requisite Management Characteristics', *Journal of Applied Psychology*, 57: 95–100. doi:10.1037/h0037128

Schwartz, S. H., and Rubel, T. (2005). 'Sex Differences in Value Priorities: Cross-Cultural and Multimethod Studies', *Journal of Personality and Social Psychology*, 89: 1010–28. doi:10.1037/0022-3514.89.6.1010

Shinar, E. H. (1975). 'Sexual Stereotypes of Occupations', *Journal of Vocational Behavior*, 7: 99–111. doi:10.1016/0001-8791(75)90037-8

Smith, R. A. (2002). 'Race, Gender, and Authority in the Workplace: Theory and Research', *Annual Review of Sociology*, 28: 509–42. doi:10.1146/annurev.soc.28.110601.141048

Smith, C. (2009). 'No Doubts: Women are Better Managers', *New York Times*, 26 July. Retrieved from http://www.nytimes.com/2009/07/26/business/26corner.html?pagewanted=all

Swamy, A., Knack, S., Lee, Y., and Azfar, O. (2001). 'Gender and Corruption', *Journal of Development Economics*, 64: 25–55. doi:10.1016/S0304-3878(00)00123-1

Tarr-Whelan, L. (2009). *Women Lead the Way: Your Guide to Stepping up to Leadership and Changing the World*. San Francisco, CA: Berrett-Koehler.

Thompson, D. (2012). 'The Spectacular Triumph of Working Women around the World', *The Atlantic*, 7 Mar. Retrieved from http://www.theatlantic.com/business/archive/2012/03/the-spectacular-triumph-of-working-women-around-the-world/254063/

Torchia, M., Calabro, A., and Huse, M. (2011). 'Women Directors on Corporate Boards: From Tokenism to Critical Mass', *Journal of Business Ethics*, 102: 299–317. doi:10.1007/s10551-011-0815-z

van Dijk, H., van Engen, M.L., and van Knippenberg, D. (2013). Defying Conventional Wisdom: A Meta-Analytical Examination of the Differences between Demographic and Job-Related Diversity: Relationships with Performance', *Organizational Behavior and Human Decision Processes*, 86, 223–241. doi:10.1016%2Fj.obhdp.2012.06.003

van Engen, M. L. and Willemsen, T. M. (2004). 'Sex and Leadership Styles: A Meta-Analysis of Research Published in the 1990s', *Psychological Reports*, 94: 3–18. doi:10.2466/PR0.94.1.3-18 van Knippenberg, D., De Dreu, C. K. W., and Homan, A. C. (2004). 'Work Group Diversity and Group Performance: An Integrative Model and Research Agenda', *Journal of Applied Psychology*, 89: 1008–22. doi:10.1037/0021-9010.89.6.1008

Vecchio, R. P. (2002). 'Leadership and Gender Advantage', *Leadership Quarterly*, 13: 643–71. doi:10.1016/S1048-9843(02)00156-X

Vecchio, R. P. (2003). 'In Search of Gender Advantage', *Leadership Quarterly*, 14: 835–50. doi:10.1016/j.leaqua.2003.09.005

Vinkenburg, C. J., van Engen, M. L., Eagly, A. H., and Johannesen-Schmidt, M. C. (2011). 'An Exploration of Stereotypical Beliefs about Leadership Styles: Is Transformational Leadership a Route to Women's Promotion?', *Leadership Quarterly*, 22: 10–21. doi:10.1016/j.leaqua.2010.12.003

von Véver, A. (2012). 'Gender Diversity and Leadership: Which Factors are Essential for Career Success? Androgyny as a Way out of the Labyrinth of Stereotypes', doctoral dissertation, University of Munich, Germany.

Wang, G., Oh, I.-S., Courtright, S. H., and Colbert, A. E. (2011). 'Transformational Leadership and Performance across Criteria and Levels: A Meta-Analytic Review of 25 Years of Research', *Group and Organization Management*, 36: 223–70. doi:10.1177/1059601111401017

Wängnerud, L. (2009). 'Women in Parliaments: Descriptive and Substantive Representation', *Annual Review of Political Science*, 12: 51–69. doi:10.1146/annurev.polisci.11.053106.123839

Williams, R. J. (2003). 'Women on Corporate Boards of Directors and their Influence on Corporate Philanthropy', *Journal of Business Ethics*, 42: 1–10. doi:10.1023/A:1021626024014

Wood, A. H., and Eagly, A. H. (2010). 'Gender', in S. Fiske, D. Gilbert, and G. Lindzey (eds), *Handbook of Social Psychology* (5th edn), i. 629–67. New York: Wiley.

World Bank (2012). *Gender Equality and Development: World Development Report, 2012.* Washington, DC: International Bank for Reconstruction and Development, World Bank. Retrieved from http://issuu.com/world.bank.publications/docs/9780821388105

Yoder, J. D. (2002). 'Context Matters: Understanding Tokenism Processes and their Impact on Women's Work', *Psychology of Women Quarterly*, 26: 1–8.

Yukl, G. (2010). *Leadership in Organizations* (7th international edn). London: Prentice Hall.

CHAPTER 8

···

THE ROCKY CLIMB

Women's Advancement in Management

···

ISABEL METZ AND CAROL T. KULIK

*With the demographic changes and skills shortages that are now upon us,
we just cannot afford to waste the talents of any of our people—and women
account for 51 per cent of us.*

(Elizabeth Broderick, Australia's Federal Sex Discrimination
Commissioner, *Sun Herald*, 27 December 2009)

DEVELOPED countries, and the organizations operating in them, have a renewed interest in addressing the gender imbalance in leadership. Two key factors have driven this
resurgence: the shrinking labour market in developed countries and the accumulating
empirical evidence that women's participation in the workforce is positively associated
with economic growth (e.g. Goldman Sachs JBWere 2009; Hausmann *et al.* 2010). As a
result, we have witnessed the introduction of new or revised legislation and reporting
requirements that are designed to motivate organizations to take immediate and creative steps to reduce gender inequities (e.g. ASX 2009; Hausmann *et al.* 2010; Deloitte
2011). These initiatives aim to address the enduring dearth of women in leadership
(EOWA 2010; Hausmann *et al.* 2010; Soares *et al.* 2011).

Women have increased their representation at low and middle management levels
since the introduction of EEO legislation (e.g. Metz and Tharenou 1999; Cohen *et al.*
2009; Haveman and Beresford 2012), but their representation at executive and board
levels remains in single or low double digits (e.g. EOWA 2010; Hausmann *et al.* 2010;
Deloitte 2011; Soares *et al.* 2011; Catalyst 2012). For example, in 2010, women's representation was a mere 8 per cent at executive level and 8.4 per cent at board level of
the 200 largest publicly listed Australian companies (EOWA 2010). Further, women
comprised 1.8 per cent of *Financial Times* Europe 500 CEO roles (Catalyst 2011), 11 per

cent of chief executives in India (EMA Partners International 2011), and 20 per cent of all senior officials and managers in Sri Lanka (Fernando and Cohen 2011). They held only 12.5 per cent of the board seats in the United Kingdom's FSTE 100 companies (Vinnicombe *et al.* 2010), and 5 per cent of the board seats in India (Desvaux *et al.* 2010). In 2011, women held 16.1 per cent of board seats in the US Fortune 500 companies and 14.1 per cent of the Executive Officer positions (Soares *et al.* 2011). Of particular concern is the recent decline or stagnation in women's representation in leadership positions in many countries (e.g. Australia (EOWA 2010) and the US (Haveman and Beresford 2012)).

Why isn't women's advancement in management like an escalator, with a smooth upward trajectory? Why, instead, is women's advancement more like a rocky climb, with a great deal of effort expended relative to the amount of upward progress, and significant opportunities for backsliding? The quest for answers has spawned an extensive body of knowledge in the gender in organizations area. We contribute to this knowledge by reflecting on 'if and how' the barriers to women's advancement in management have changed over time. We commence this reflection with an overview of past barriers used to explain women's slow advancement, and conclude with a recommendation for change that takes into account different stages of gender equity between countries, organizations, and business units within organizations.

OVERVIEW OF PAST BARRIERS USED TO EXPLAIN WOMEN'S SLOW ADVANCEMENT IN MANAGEMENT

The literature on women's advancement in management spans several decades. This body of research offers a rich array of explanations for women's scarcity in management and puts forward recommendations for redress to both women and organizations. For example, over three decades ago, Kanter turned the spotlight on women's scarcity in management positions in organizations, and on the structural and cultural factors contributing to that scarcity, in her book *Men and Women of the Corporation* (1977*a*). One structural factor is the disproportionate number of men in management relative to women. This gender imbalance can lead to stereotyping of minority women managers that obscures women's professional capabilities and limits their work opportunities (e.g. they are more likely to be viewed as, and treated as, support staff than as managers: Kanter 1977*b*). Cultural factors include characteristics of the dominant group's culture (e.g. swearing as acceptable behaviour for men) that are used to exclude women and make it difficult for them to fully participate as employees of equal status to men (Kanter 1977*b*). Kanter's work has been widely read by practitioners and cited more than 9,000 times by scholars. Unfortunately, this rich body of knowledge has not translated into organizational change or produced tangible evidence of gender equity, particularly at high levels of organizations, as employment statistics show.

Our understanding of the reasons for women's slow progress in management is still evolving, as reflected in changes in the research focus and recommended approaches over time. For example, in 1980, Riger and Galligan found that earlier research on women's representation in management used *either* a gender-centred approach or an organization structure model to explain sex differences in attitudes to work, work values, and personal management attributes. A gender-centred model accounts for the disparity between men's and women's participation rates in management by focusing on factors internal to the individual (e.g. differences in men and women's abilities or skills). In contrast, an organization structure model emphasizes non-person factors, such as individuals' 'positions in the workplace hierarchy and their differential access to the system of workplace rewards' (Rowe and Snizek 1995: 216), as contributors to or inhibitors of organizational success. Riger and Galligan demonstrated that many person-centred explanations for women's lack of managerial advancement could be replaced by equally plausible interpretations based on women's positions within the organizational structure. However, Riger and Galligan (1980) also acknowledged that a focus on organization structure *alone* was insufficient to understand women's representation in management. Therefore, Riger and Galligan called for researchers to consider 'the interaction of both person- and situation-centred variables' (1980: 908) in explaining the scarcity of women managers.

A decade later, Fagenson (1990) echoed Riger and Galligan's (1980) call and again urged researchers to examine a more encompassing model to 'measure factors in the situation that may co-vary with the gender variable, and which may ultimately be responsible for the gender variances found' (p. 33). Specifically, Fagenson (1990) encouraged research approaches that encompassed the interaction between gender and the organizational and social environments.

By the late 1980s, scholars were responding to these calls to action. For example, Ragins and Sundstrom (1989) offered a multi-level framework for analysing the societal, organizational, interpersonal, and individual factors that might facilitate or inhibit women's advancement. Other scholars studied women's managerial career advancement from a multi-level perspective and examined interactions between person- and situation-centred variables (e.g. Metz and Tharenou 2001; Leslie and Gelfand 2008; Metz 2009). As a result, we now know that many factors can explain women's advancement in management. However, the mix and the relative importance of these factors have changed in the last forty or so years. For example, education (a person-centred variable) is no longer a direct barrier to women's advancement, as women have achieved levels of educational attainment comparable to men (Goldin 2004; England 2010; Haveman and Beresford 2012). In contrast, many organization-centred variables, including the structural barriers denounced by Kanter (1977a, 1977b), persist (e.g. Carter and Silva 2010). Thus, we focus in this chapter on those barriers to women's advancement that stand out for their persistence (e.g. stereotypes), for being new versions of their old selves (e.g. the replacement of overt forms of discrimination with covert ones), or for being new (e.g. gender fatigue).

PERSISTENT BARRIERS TO WOMEN'S
ADVANCEMENT IN MANAGEMENT

Our aim is to illustrate rather than to provide a comprehensive review of the many barriers to women's advancement that have prevailed over time. The four barriers that we focus on in this section are decision-makers' denial of gender discrimination, social gender roles, stereotypes and perceptions, and organizational culture.

Denial of Gender Discrimination

Decision-makers' denial of gender discrimination was originally based on the 'individual deficit model' (Gutek 1993). This model focused on individual deficits rather than organizational barriers: women did not advance in management because they were lacking something needed in management positions (Gutek 1993), such as leadership skills (e.g. Kaye and Scheele 1975). Organizations thus tried to 'fix' the women (Meyerson and Fletcher 2000) and many people believed that women's advancement in management was merely a matter of time (Carter and Silva 2010). As women acquired more of what they needed to advance, they would move up the organization's 'pipeline' and their representation in management would gradually increase.

The 'pipeline' argument is no longer convincing. Empirical evidence shows that women's progress up the pipeline is slower than that of their male counterparts (Metz and Harzing 2009; Carter and Silva 2010). Much of the literature ascribes women's slow climb up the organizational hierarchy to (invisible) systemic barriers, metaphorically described as 'glass ceilings' (Kanter 1977a) that block women's upward progress or 'glass walls' (Bronstein *et al.* 1993) that constrain women's occupational and professional opportunities.

Many organizations invested in policies and practices aimed at removing systemic barriers. For example, superfluous criteria that prevented women from applying for or entering specific jobs, such as being able to climb a high wall to enter the police force, have been removed (e.g. Metz and Kulik 2008). Male models of work that include long hours, spontaneous meetings, and extended time away from home, have been modified (e.g. McCracken 2000; Meyerson and Fletcher 2000). Policies to help employees combine work and family responsibilities, such as part-time work, have been introduced (e.g. French and Strachan 2007). Now that these structural barriers have been removed, any remaining discrepancies between men's and women's advancement and their representation at top levels of organizations can be, once again, attributed to individual factors, such as the 'choices' that women make between work and family (e.g. Kuperberg and Stone 2008).

In reality, however, many barriers to women's advancement in management are independent of care responsibilities (e.g. Carter and Silva 2010; Metz 2011). In particular, the argument that women 'choose' between work and family overlooks the fact that 'gender inequality is a significant disincentive to women's workforce participation' (Status of Women Minister Julie Collins, as quoted in Russell 2012). Much of the gender inequality in the workplace stems from non-individual factors, such as gender roles, stereotypes and perceptions, and organizational cultures.

Gender Roles

Social gender roles ascribe the primary child-rearing and housekeeping responsibilities to women, and the financial provider responsibilities to men. According to this social division of labour in our societies, women are the 'caregivers' and men the 'breadwinners' (Eagly 1987). As a result, women's careers differ from men's. Scholars have used metaphors such as 'cross-currents in a river' (Powell and Mainiero 1992), 'kaleidoscopes' (Mainiero and Sullivan 2005), 'labyrinths' (Eagly and Carli 2007) and 'off-ramps and on-ramps' (Hewlett 2007) to describe the non-linear complexities of women's careers. These metaphors emphasize women's need to manage the professional and caregiver roles throughout their working life by, for example, temporarily exiting the workforce or working part-time. Yet women are usually denied the opportunity to resume an interesting, challenging and promising career path after the first or second interruption (Hewlett 2007; Stone 2007; Metz 2011). Reasons for this are complex, ranging from the perseverance of the male model of work to discrimination based on gender role expectations (Mainiero and Sullivan 2005; Metz 2011). The male model of work is characterized by full-time (preferably uninterrupted) work and long work hours (Cooper and Lewis 1999; Mainiero and Sullivan 2005). The careers of most women in senior management still resemble those of their male counterparts—women have to demonstrate total commitment to their careers in order to advance (e.g. Blair-Loy 1999; Roth 2007).

Women continue to identify motherhood, or the mere possibility of it, as a barrier to advancement (Griffith and MacBride-King 1998; Hewlett 2007; Roth 2007; Metz 2011). Many decision-makers (who are often men) expect women of child-bearing age to have low commitment to work and high commitment to their existing or potential family responsibilities (Roth 2007; Hewlett et al. 2010; Hoobler et al. 2009). Based on these gender role expectations, decision-makers continue to give more developmental opportunities to men than to women (e.g. Ohlott et al. 1994; Hoobler et al. in press). Developmental opportunities lead to career success in general (Berlew and Hall 1966), and facilitate women's career advancement in particular (e.g. Mainiero 1994; Metz and Tharenou 2001). Thus, an accumulation of biased judgements against women constrains their developmental opportunities, and the resulting narrow range of work experiences places women at a pay disadvantage relative to men (Ohlott et al. 1994) and lowers their career aspirations over time (Hewlett 2007; Hoobler et al. in press). As a result, the 'choice' of some couples with children to make the male the main breadwinner and the female the principal caregiver is a rational decision reflecting the differential opportunities offered by gender roles (Gans 2008).

Organization practices may also contribute to the 'choices' women make to leave their organizations (e.g. McCracken 2000; Roth 2007; Zahidi and Ibarra 2010). For example, when women are at the cusp of senior or executive positions, they appear to 'opt out' of their careers to become 'stay-at-home' mums (e.g. Stone 2007; Kuperberg and Stone 2008). This 'choice' is disappointing, because work–family policies to help women combine work and family responsibilities were among the earliest recommendations scholars made to organizations hoping to attract and retain women (e.g. Crouter 1984). Decades later, many organizations have sound policies in place describing their work–family practices (Brown 2010; Zahidi and Ibarra 2010), but these policies and practices are often inaccessible to, or not accessed by, women (Roth 2007; Brown 2010; Cross 2010; Metz 2011). As more human resource management activities are shifted to the line (e.g. Larsen and Brewster 2003; Perry and Kulik 2008), line managers become responsible for implementing organizational work–family policies that they did not develop and may not value (Roth 2007; Rivera 2012). As a result, women in management are regularly discouraged or denied access to those policies (Cross 2010; Metz 2011), or do not access them for fear of jeopardizing their career advancement opportunities (Roth 2007; Brown 2010).

Further, as women advance, they become aware of their superiors' strong belief that management positions are strictly full-time endeavours (Metz and Tharenou 2001; Roth 2007; Stone 2007; Cross 2010; Metz 2011). As a result, women 'choose' to leave to pursue professional paths that allow for greater flexibility (McCracken 2000; Hewlett 2007; Roth 2007; Hoobler et al. 2009; Metz 2011) or to assume their ascribed role of caregivers (Stone 2007). Thus, work–family policies may attract women to organizations, but not facilitate their retention or advancement into senior positions (Brown 2010; Straub 2007).

In sum, social gender roles fuel perceptions that women are more committed to family than to work and have low career aspirations due to their family responsibilities. These perceptions can negatively influence day-to-day decisions about female staff, thus contributing to many women's unfulfilled career aspirations and, ultimately, to women's turnover. As supervisor support is important in the implementation of discrete policies and practices (Breaugh and Frye 2008), supervisors are critical players in establishing and maintaining climates supportive of women in management (Stainback et al. 2011). A climate supportive of women's advancement should at least reflect the extent to which an organization values gender equality and diversity in management, and 'therefore seeks to create and maintain diverse organizational membership' (Leslie and Gelfand 2008: 125). In contrast, an environment unsupportive of women's advancement reflects the negative stereotypes and perceptions held by supervisors (and colleagues) with regard to women as managers.

Stereotypes and Perceptions

Stereotypes and perceptions are two of the most widely recognized barriers to women's advancement (e.g. Griffith and MacBride-King 1998; Gutek 1993; Metz and Tharenou

2001; Heilman *et al.* 2004). Stereotypes and negative perceptions of women at work often underlie the unequal treatment of women and men (e.g. Powell 2005; Hoobler *et al.* 2009). Further, common stereotypes of women clash with stereotypes of managers (Eagly 1987; Schein *et al.* 1996). For example, women are expected to be submissive and nurturing (Schein 2001). In contrast, managers are expected to be male, decisive, and assertive (Schein 1973; Schein *et al.* 1996; Deal and Stevenson 1998). These perceptions have stood the test of time (Schein 1973, 2001), are consistent across countries (Schein *et al.* 1996), and are based on social role theory (Eagly 1987). As a result, discrepant stereotypes of women and managers contribute to judgements of women as unsuitable candidates for managerial positions (Eagly and Karau 2002; Hewlett *et al.* 2010; Haveman and Beresford 2012). Further, this discrepancy poses a challenge for women who want to advance, because they need to be seen as having the 'masculine' characteristics required for leadership positions without lacking the 'feminine' ones expected in a woman (Eagly and Karau 2002). Being perceived as competent (masculine) but not nice (feminine) can reduce women's opportunities for advancement (Eagly and Karau 2002; Heilman *et al.* 2004).

Stereotypes encourage organizational members to generate person-centred explanations for a woman's lack of advancement (e.g. 'we couldn't promote her because she wasn't a team player'). Person-centred explanations are popular because they place the onus on women (rather than on organizations or their decision-makers) to address the persistent scarcity of women in management. Further, such explanations are easier to address (e.g. provide interpersonal skills training for women) than non person-centred explanations (e.g. improve unsupportive organization cultures or transform decision-maker stereotypes of women and managers).

The appeal of person-centred explanations is manifest in the persistent belief that under-represented groups in management, in this case women, lack the human capital to succeed (e.g. Rivera 2012), despite a long history of empirical evidence to the contrary (e.g. Brief *et al.* 1977). Human capital can be defined as the knowledge and skills that people accumulate over time such as education, company tenure, training, work hours and work experience (Becker 1993). Human capital theory proposed that such individual factors result in increased productivity and, thus, rewards of higher status and pay (Becker 1993). However, some human capital factors affect women's advancement differently than men's. For instance, we have known for almost two decades that education level contributes less to the managerial advancement of women than of men (e.g. Simpson 1996; Tharenou and Conroy 1994), and that the number of years of work experience and company tenure have stronger effects on men's income and level than on women's (Kirchmeyer 1998; Roth 2007).

Other persistent person-centred explanations for the slow increase in women's representation in leadership include a belief that women are less committed (Kuperberg and Stone 2008) and have lower career aspirations (e.g. Hoobler *et al.* in press) than men. As previously mentioned, many of these explanations stem from the prevailing gender roles of women as primary caregivers and men as breadwinners. Yet, empirical evidence emerged more than twenty years ago showing that female and male managers were equally ambitious (Howard and Bray 1988) and committed to work (Bielby and Bielby 1989), and had similar career aspirations (Morrison *et al.* 1992). More recently, Metz and Simon (2008) found that men and women at the same management level did not differ in the hours they worked, or in their short- and long-term career aspirations. Nascent literature using non-Western samples indicate that women professionals in Eastern societies are also ambitious (e.g. Fernando and Cohen 2011). Ambition is here referred to in the traditional sense of aspiring to the top positions in one's organization or field (Howard and Bray 1988). Although ambition is a strong predictor of advancement (Howard and Bray 1988), presenting oneself as ambitious can be detrimental for women (e.g. Morrison *et al.* 1992; Phelan *et al.* 2008). The male gender stereotype places a positive value on agentic traits like ambition; the female gender stereotype does not (Rudman and Phelan 2008). As a result, ambitious women suffer a social backlash; they are less liked and are assessed as less suitable candidates than men or women whose ambitions are more consistent with gender stereotypes (Kanter 1977*b*; Phelan *et al.* 2008).

In sum, we have used several person-centred explanations for women's lack of advancement (e.g. human capital and ambition) to demonstrate how the differential treatment of men and women is strongly influenced by gender stereotypes and perceptions. Gender stereotypes and perceptions represent interpersonal barriers to women's advancement (Ragins and Sundstrom 1989). Gender stereotypes and perceptions exaggerate gender differences, and lead to performance and capability judgements that favour men over equally competent women in work settings (e.g. Roth 2007; Phelan *et al.* 2008). These interpersonal barriers to women's advancement exert a powerful influence on the behaviour of individuals within organizations, but they are difficult to change and are largely outside women's control (Ragins and Sundstrom 1989).

Organizational Culture

Organizational culture reflects employees' beliefs and judgements of how things are and should be in the organization; it 'comprises organizationally shared values, beliefs and schemas' (Lord and Maher 1991, cited in Bajdo and Dickson 2001). Despite growing evidence that gender diversity contributes to organizational (and national) performance (e.g. Goldman Sachs JBWere 2009; Desvaux et al. 2010; Hausmann *et al.* 2010), particularly in the services industry (e.g. Ali *et al.* 2011; Blum *et al.* 1994), unsupportive organization cultures continue to contribute to women's lack of advancement (Zahidi and Ibarra 2010; Stainback *et al.* 2011; Haveman and Beresford 2012), poor quality relationships between women and their supervisors and colleagues (Kanter 1977*b*; Metz 2011), and women's departure from organizations (McCracken 2000; Roth 2007; Metz 2011). 'Masculine' cultures are particularly problematic. Masculine cultures are characterized by displays of physical strength, such as working very long hours, and by gender discriminatory behaviour, such as providing preferential treatment to men (e.g. Kanter 1977*b*; Kilduff and Mehra 1996; Roth 2007) and contribute to women's lack of advancement in management (e.g. Metz and Tharenou 2001).

Fortunately, shared perceptions of cultural *practices* are more strongly associated with women's opportunities for advancement than shared perceptions of cultural *values* (Bajdo and Dickson 2001). Practices can be targeted through interventions to generate significant levels of cultural change (see Ely and Meyerson (2010) and Metz and Kulik (2008) for examples of dramatic cultural change within two masculine organizations). Interventions that change organizational practices can indirectly motivate changes in cultural values and generate faster cultural changes than interventions that try to target the underlying organizational values (Metz and Kulik 2008).

In this section, we have identified a set of barriers to women's advancement that has been constant. These barriers stem from social gender roles positioning women as caregivers and men as breadwinners and gender stereotypes that suggest women are poor fits to managerial roles. These barriers reflect person-centred explanations for women's lack of advancement in organizations—once women are dismissed as poor fits, male-dominated hierarchies and masculine cultures flourish. In the following section, we turn to a second set of barriers to women's advancement. As our societies have become more legislated and litigious (Roth 2007) some of the most egregious forms of gender discrimination have disappeared, only to be replaced by equally problematic, but more socially acceptable, alternatives. We focus here on the shift from overt to covert (or subtle) gender discrimination.

A New Version of an Old Barrier to Women's Advancement in Management

In contrast to the persistent barriers to women's advancement described in the previous section, some barriers have disappeared in response to anti-discrimination legislation. For example, we rarely see jobs advertised exclusively for men, or hear decision-makers express overt preferences for male candidates. Unfortunately, discriminatory barriers have not been fully eradicated. Instead, overt gender discrimination may have been replaced by a more covert form. However, the risk of replacing overt with covert forms of discrimination is modern sexism (Barretto and Ellemers 2005). Modern sexists acknowledge the systematic inequality in outcomes between members of different groups, while insisting that these differential outcomes do not result from systematic disadvantage (Swim *et al.* 1995). Modern sexism is expressed in explicit resistance to any 'special measures' to correct inequalities, and implicit inferences that any observed inequality must reflect a 'lack' or 'failure' among the members of the disadvantaged group. Modern sexism is difficult to detect and, thus, difficult to challenge and overcome (Swim *et al.* 1995). Modern sexism can be as detrimental to women's well-being (Barretto and Ellemers 2005) and work outcomes (Watkins *et al.* 2006) as blatant sexism.

One example of modern sexism is the propensity to appoint executive women to precarious leadership positions (Ryan and Haslam 2005; Ryan *et al.* 2007). Women are more likely to hold leadership positions in organizations that have experienced a period of below average performance or a scandal (Ryan and Haslam 2005; Brady *et al.* 2011). Appointing a woman to a senior leadership role is, on the surface, a vote of confidence in women's leadership capabilities. However, the risk of failure in these 'glass cliff' positions is disproportionately high, and when failure occurs it is more likely to be attributed to women's inability to deal with crises (Barretto and Ellemers 2005). The female leader makes a convenient scapegoat for the organization and 'when a leader is needed to later manage the crisis and turn things around, the preference for women disappears' (Ryan 2012: 15).

There are some organizational situations in which modern sexism is particularly likely to flourish. For example, business justifications can unleash the subtle prejudices harboured by modern sexists, leading to discriminatory employment behaviour. We know that modern *racism* predicts discrimination 'when a legitimate authority figure provided a business-related justification for such discrimination' (Brief *et al.* 2000: 73). Similarly, business justifications may facilitate the manifestation of modern *sexist* beliefs. In support, studies have found that the differential treatment of women in the workforce is more pronounced in bad than in good economic times (Livanos *et al.* 2009) and when an organization is undergoing change (e.g. a merger, a restructure, or downsizing; Metz 2011).

In this section, we focused on a second set of barriers to women's advancement—the barriers that have undergone a transition, making them more difficult to detect and correct. Overt forms of gender discrimination have been driven underground by legislation and social norms, only to be replaced by more subtle forms. We turn attention now to a third set of barriers. This third set represents a new problem emerging in society and only recently acknowledged in the academic literature. Women may no longer recognize gender discrimination, particularly in organizational contexts that foster an illusory perception of meritocracy.

A New Barrier to Women's Advancement in Management

One of the newest barriers to women's advancement in organizations may be also the most ironic: after such a long history of gender inequity in organizations, women may be losing both their ability to recognize gender inequity and their motivation to demand that inequity be addressed. Researchers have suggested employees are experiencing 'gender fatigue' (Kelan 2009). Employees prefer to view their workplaces as gender egalitarian, and this deeply rooted preference motivates both men and women to dismiss gender discrimination as a thing of the past. In our review, we identified as a 'persistent barrier' the evidence that *employers* deny the existence of gender discrimination. However, research now suggests that *employees* are expressing similar denials. Information communication technology workers (Kelan 2009), engineers

(Jorgenson 2002), police officers (Dick and Cassell 2002), and business school students (Kelan and Jones 2010) 'explain away' current gender inequities as isolated incidents and see it as the responsibility of individual women to overcome them. Over the last few decades, women have made substantial progress on several objective indicators—their wages have increased relative to men's and more women occupy positions of power in organizations (Spoor and Schmitt 2011). As a result, both men and women who make temporal comparisons (comparing the outcomes of historical and contemporary women) believe that gender discrimination is no longer a problem and express little interest in collective action. Only when women make intergroup comparisons (comparing the outcomes of contemporary women and contemporary men) do they acknowledge their group-based disadvantage (Spoor and Schmitt 2011). But it is exactly this collective awareness of a disadvantage that motivates organizational change (Bailyn 2003; Ely *et al.* 2011; Spoor and Schmitt 2011).

The employee preference to view workplaces as gender egalitarian has been accompanied by an organizational movement away from an 'equal opportunity' rhetoric to a 'diversity management' one. The broad 'diversity management' portfolio operating in modern organizations, with its emphasis on 'business case' arguments, may have the unintended effect of diluting organizations' attention to gender issues (Kirton and Greene 2010). Diversity management focuses on inclusion and recognizes a wide range of individual differences. This can make it more difficult for employees to identify and address specific group-based disadvantages (Kelan and Jones 2010). Further, organizations with a diversity management perspective are likely to adopt well-intentioned 'merit-based' evaluation systems that nonetheless perpetuate group-based disadvantages (Lewis 2006). These 'merit-based' systems reassure managers that non-discrimination is 'built in' to the process, so that they relax their vigilance and allow gender biases greater sway in their final judgements (Castilla and Benard 2010). The illusion that 'merit-based' systems consistently produce fair outcomes is so embedded in modern organizations that attempts to dislodge or revise them is resisted even by the groups who might benefit from such change (Krefting 2003).

THE MORE THINGS CHANGE, THE MORE THEY STAY THE SAME

As this exposition of the barriers to women's advancement in management over time has shown, women in the twenty-first century face many of the barriers to advancement their predecessors encountered approximately forty years ago (e.g. see Kanter 1977*a*, 1977*b*). This stability demonstrates that *knowing* about the barriers is not equivalent to *removing* them. Despite decades of research, there has been disappointingly little progress in changing gender stereotypes or eliminating discriminatory behaviour in organizations. To illustrate, a recent survey showed that many executives and CEOs agree that gender diversity is a performance driver, yet none considers it a top strategic priority and few have implemented strategies to increase gender diversity in their organizations (Desvaux *et al.* 2010).

We believe that the problem in overcoming person- and situation-centred barriers to women's advancement in management resides in the complex interplay between them. As a result, women need to bear some responsibility while simultaneously recognizing the situational forces that hinder their advancement. For example, women might be partly responsible for their lack of advancement, because they accept things as they are (Carnes and Radojevich-Kelley 2011). According to this reasoning, women 'choose' to abide by traditional gender roles and lower their career aspirations in response to obstacles, such as the 'glass ceiling', rather than act to overcome them. This response is not new or surprising; individuals who occupy minority positions in organizations, such as women managers, often accept things 'as they are' because of the high personal costs generated by challenging the status quo (e.g. being labelled as difficult or lacking in people skills; Kanter 1977*b*).

However, factors external to women might also be partly responsible for their lack of advancement. For example, Hoobler *et al.* (in press) found that that both male and female managers believed their female subordinates had lower career motivation than male subordinates. This finding shows that women managers can, just like their male counterparts, ascribe to traditional gender stereotypes and constrain the advancement of their female subordinates. Hoobler *et al.*'s finding is particularly worrying, because career encouragement is a predictor of women's advancement, especially at high levels of management (Metz and Tharenou 2001). If male and female superiors perceive women as having lower career motivation than men, they are less likely to encourage women than men to take on challenging work opportunities to advance their careers. Those women who hope to advance need to upwardly manage the perceptions of all superiors regardless of their gender.

These two explanations for women's slow advancement in management are not independent. When direct managers fail to encourage their female subordinates, they can lead women to doubt themselves and lose confidence in their own capabilities, thus lowering women's career aspirations relative to their male co-workers. Therefore, there is an interactive effect between women's choice and supervisory behaviour. Women interpret supervisory attitudes and act in response to those attitudes, so that their behaviour reinforces and embeds supervisory attitudes—generating a particularly problematic 'spiral'.

Many other external barriers hinder or slow women's advancement in management, particularly at high levels. For instance, women's slow progress at high levels might reflect men's fears of increased competition where the 'pickings' are few but highly coveted (Kanter 1977*a*, 1977*b*). Kanter explained how the dominant group preserves its privileged position through self-perpetuating practices. As a result, women are likely to continue to encounter group membership obstacles as members of the (female) out-group in current organizations' predominantly male hierarchies.

Social capital theory (Burt 1998; Portes 1998) posits that supportive work relationships are critical to objective measures of success (e.g. advancement, higher pay, and promotion). Supportive work relationships include mentors and powerful sponsors. Mentorship encompasses the support of a more experienced and influential member of the organization for the career development of a less experienced individual (Kram 1985). Sponsorship goes beyond mentorship; a sponsor will publicly endorse a lower level employee, enhance the employee's visibility inside and outside the organization, and 'see you to the threshold of power' (Hewlett *et al.* 2010: 6). However, women's persistent perceptions of lack of mentors, exclusion from male-dominated influential networks, and gender discrimination (e.g, Metz and Tharenou 2001; Zahidi and Ibarra 2010) indicate that most work environments remain unwelcoming of women in management. In addition, women are less likely than men to have sponsors (Hewlett 2010). Although this lack of supportive work relationships may be partly due to the difficulties of cross-gender relationships (e.g. Kram 1985; Powell and Mainiero 1992; Hewlett *et al.* 2010), it is largely attributed to persistent and damaging gender-stereotypic perceptions of women and in-group dynamics (e.g. Kanter 1977b; Roth 2007).

On the whole, the small gender equity gains we have seen coupled with large ongoing gender inequities create exactly the conditions in which gender issues in organizations are likely to 'go off the boil' and remain unaddressed. 'Gender fatigue' (Kelan 2009) may be silencing women's collective voice, so there is little incentive for organizations to engage in change. In addition, some decision-makers believe that, because their organizations have been discussing and acting on the 'gender issue' for quite some time, gender issues have been addressed and a level playing field achieved—leaving women individually responsible for any remaining disparities in their advancement compared to men's (Broadbridge and Simpson 2011). In the absence of collective demands for change, organizations are likely to continue to implement well-intentioned but largely ineffective initiatives (e.g. mentoring; Ibarra *et al.* 2010) while failing to launch more controversial but necessary ones (e.g. women-only leadership programmes (Ely *et al.* 2011) and sponsorship for talented women (Hewlett *et al.* 2010)). But change is clearly needed. Both men and women need to be aware of gender bias's subtle yet pervasive manifestations in organizations (Ely *et al.* 2011) and continue to monitor outcomes, even (or especially) those that result from ostensibly transparent and merit-based systems (Castilla and Benard 2010). Otherwise, gender inequities will become 'even more difficult, if not impossible, to address' (Kelan 2009: 206) and the climb will become rockier than ever before.

RECOMMENDATIONS FOR CHANGE

Most gains in women's representation in management, including in senior management positions, were observed during the last two or three decades of the twentieth century (Cohen *et al.* 2009; Haveman and Beresford 2012). Since then, women's representation in management has increased slowly, stalled, or even declined (e.g. Cohen *et al.* 2009; Haveman and Beresford 2012). This slow progress led academic scholars to recommend a variety of changes. Table 8.1 includes a small sample of the rich list of innovative and

Table 8.1 Examples of recommendations for change

Recommendation	Illustrative reference/s
Offer compulsory workshops to explore issues of gender in the workplace; make public commitment to change; monitor and measure change	McCracken (2000); Ibarra, Carter, and Silva, (2010)
Continuously monitor entrenched biases and experiment with possible solutions	Meyerson and Fletcher (2000)
Increase awareness of psychological drivers of prejudice towards female leaders	Eagly and Carli (2007)
Welcome women back; provide on-ramp opportunities	Mainiero and Sullivan (2005); Eagly and Carli (2007);Metz (2011)
Make managers accountable for women's turnover and advancement	Mainiero and Sullivan (2005); Ibarra et al. (2010); Metz (2011)
Open communication between managers about stereotypical assumptions about women's career aspirations and about biases in decision-making; engage in meaningful dialogue with women in times of organizational or personal change	Hoobler et al. (in press); Metz (2011)
Change the long-hours norm; establish family-friendly HR practices; encourage male participation in family-friendly benefits	Mainiero and Sullivan (2005); Eagly and Carli (2007)
Sponsor women, don't just mentor them	Hewlett et al. (2010); Ibarra et al. (2010)

useful recommendations that have been generated since the year 2000. Further, some scholars have advocated a totally different approach to give women's advancement new momentum. For example, Meyerson and Fletcher (2000) suggested a 'small wins' approach to uncovering (invisible) systemic barriers to women's advancement to executive and board levels. According to Meyerson and Fletcher, women's advancement to top-tier positions in organizations is not hindered by overt discrimination, but by biases so 'deeply embedded in organizational life' (2000: 127) that 'they're not even noticed until they are gone' (p. 128). Meyerson and Fletcher suggest that incremental changes, or 'small wins', will chip away at these entrenched biases to generate organizational change, improve overall efficiency and performance, and, thus, benefit both women and men.

However, countries, organizations, and individuals are at different stages of awareness and readiness to take on the gender equity challenge (Zahidi and Ibarra 2010). Just as individuals vary in their values and beliefs about women in management, so do organizations (as reflected by their cultures) and countries (as reflected by their political and legislative frameworks). For example, women living in the Kingdom of Saudi Arabia or in Afghanistan are still openly constrained in pursuing work interests outside the home by cultural norms, legislation, and government policies (e.g. Ahmad 2011; Icheku 2011). Even when not restricted by law and government policies, women in Eastern societies face negative attitudes towards women in the workplace and in authority positions that derive from Eastern values (Chao 2011; Fernando and Cohen 2011; Icheku 2011). These socio-cultural barriers in Eastern societies are additional to the disadvantage and discrimination in the workplace encountered by women in Western societies.

Therefore, no one approach to achieving gender equity in leadership in organizations suits all countries, organizations, or even business units within organizations. At the organizational level, approaches to enhancing gender equity need to be customized and will most likely comprise a succession of gradual steps or building blocks, and a combination of top–down and bottom–up initiatives fully supported or even spearheaded by a committed CEO and his/her executive team. Overall, we believe that the complexity surrounding women's slow advancement in management calls for a customized, step-by-step approach that might overcome 'gender fatigue' and boost 'gender equity' in organizations. Our recommended approach aims to change the 'rocky climb' to just a plain old 'climb' up the hierarchical ladder for those women who wish to advance in management.

A Customized, Step-by-Step Approach

To be truly 'customized', each organization's approach needs to start with the first step: a thorough diagnosis of where in the gender equity continuum the organization is at and why. Although there are no 'quick fixes', there might be easy-to-identify barriers, such as lack of childcare facilities and overt discrimination. Of course, easy-to-identify does not necessarily mean easy-to-address. Nevertheless, these barriers should be addressed first, because of their visibility and because possible solutions can be considered, implementation efforts can be monitored, and results can be measured.

Depending on the gender equity stage a country, organization, or business area within an organization is at, the first step may move quickly. For example, European organizations differ in their provision of work–life balance practices (Straub 2007). Scandinavian countries already have first-step policies and practices in place (e.g. reliable government funded childcare) to enable women to participate at work and achieve their career potential (Hausmann et al. 2010; Deloitte 2011). Thus, organizations operating in Scandinavian countries might only need to conduct a quick review, while those operating in other European countries might need a fuller diagnosis that identifies their position in the gender equity continuum and the reasons for that position.

Once some of the critical but easy-to-identify barriers to women's advancement have been addressed, a second step can be undertaken: the search for insidious barriers, such as biased 'merit-based' processes. One example of a biased organizational process is Meyerson and Fletcher's (2000) story of the New York advertising company that offered human-resource-type positions mostly to their talented women. These positions required high levels of the 'people' skills that women were thought to possess, but they did not help women to acquire the 'rain-making' skills they needed for further advancement. This example demonstrates why the outcomes of ostensibly fair procedures still need to be continually monitored (Castilla and Benard 2010). Further, as women comprise a small proportion of the management pool or even of the overall workforce in some fields, women's outcomes need to be regularly reviewed across organizational units so that patterns can be more visible (Bailyn 2003).

Our proposed approach is deceptively simple. We know from decades of research and practice that addressing gender equity in organizations is a very challenging endeavour; it requires sustained commitment from the CEO and top management, sometimes for many consecutive years (e.g. read about Deloitte's WIN initiative in McCracken 2000 and in Pellegrino et al. 2011). Further, although our customized, step-by-step approach incorporates some past recommendations (e.g. need the CEO's full and visible commitment to a gender-diversity programme; Desvaux et al. 2010), we emphasize the importance of taking into account differential stages of gender equity between countries, organizations, and business units within organizations. In particular, the degree of change achievable with a customized, step-by-step approach is partly facilitated (or hindered) by the legislative and regulatory context in which the organization operates. Therefore, we favour increasing legislative and regulatory pressures for gender equity in organizations in countries that trail in women's representation in leadership.

Renewed Legislative and Institutional Pressures

Organizational change is difficult to achieve (Lewin 1951). Organizations, and individuals within them, are likely to change only when there is a clear need to change (Kotter 1996). Pressures external to organizations can generate this 'need' to change. Therefore, the renewed pressure on organizations in many countries to address gender imbalances, especially at executive and board levels (Hausmann et al. 2010; Deloitte 2011) is welcome as a counterforce to 'gender fatigue'. In Australia, for example, this pressure is being felt via the Australia Security Exchange (ASX) Corporate Governance Council's expanded reporting requirements (hereafter the 'ASX requirements') announced in December 2009. According to these expanded reporting requirements, the 200 largest publicly listed companies need to report on gender diversity plans and progress at all levels after 1 January 2011 (ASX 2009). Public scrutiny of company goals and progress on gender diversity is likely to motivate these organizations to increase gender diversity, including at executive and board levels (Konrad and Linnehan 1995). Early evidence supports this rationale; the proportion of new board appointments filled with women increased from 25 per cent in 2010 to 30 per cent in 2011 in Australia (EOWA 2011).

Renewed affirmative action legislation and institutional pressures are needed to get the first cohort of women in upper management. Affirmative action legislation introduced in Western societies in the 1960s, 1970s, and 1980s helped visibly increase women's representation in low and middle management positions, but had very modest success in upper management (e.g. Metz and Tharenou 1999; Kurtulus and Tomaskovic-Devey 2012). In contrast, recent legislation on targets and quotas, and pressure by institutions such as the Australian Securities Exchange, seem to have boosted women's representation in upper management and on boards (Deloitte 2011; Zahidi and Ibarra 2010).

Increasing the percentage of female top managers through renewed affirmative action legislation and institutional pressures might not be the ideal path to change but is nevertheless welcomed, because having more female top managers seems to help women advance. For example, Kurtulus and Tomaskovic-Devey (2012) analysed EEO-1 data from the US Equal Employment Opportunity Commission covering more than 20,000 large private sector firms across all industries for a period of eight years, and found that an increase in female top managers was associated with an increase in women in middle management. This effect was most pronounced one year after the increase in female top managers.

In addition, an increase in female managers might reduce the reported experience of workplace discrimination for women (Stainback *et al.* 2011). This conclusion is consistent with what the structural feminism literature uncovered in the 1970s (e.g. Kanter 1977*a*), 1990s (e.g. Simpson 1997) and noughties (e.g. Simpson 2000). Specifically, this rich literature found that the gender mix in management was 'a critical factor defining women managers' sense of fit within the organization, the pressures they experience, and the barriers they are likely to encounter' (Simpson 2000: 15).

Unfortunately, there is a limit to what can be legislated or mandated. Past evidence indicates that legal and institutional pressure to increase women's representation in management might only be effective in organizations covered by it. The ASX's requirements do not apply to unlisted companies operating in Australia and, therefore, are likely to exercise only indirect pressure on unlisted companies to change. This indirect pressure might operate through 'mimetic processes'. Mimetic processes 'focus managers' attention on particular companies as exemplars, with the result that less successful companies attempt to imitate salient aspects of those companies' structures and practices, often with only hazy information about how these may be linked to the companies' business success' (Di Maggio 2001: 240–1). Further, unlisted companies are likely to be small or medium size. Small-to-medium size enterprises are less likely than large ones to feel institutional pressure (Blum *et al.* 1994) or to adopt formal HR practices (Bartram 2005). In addition, small organizations not bound by EEO legislation (e.g. in Australia, EEO legislation does not apply to companies with fewer than 100 employees) displayed the weakest equity indicators in Australia (Peetz *et al.* 2008) and elsewhere (e.g. Edgar 2001). Thus, unlisted and small companies are likely to show limited change, if any, in their progress on gender diversity as a result of legal and institutional pressure.

Further, governments can legislate to eliminate overt forms of discrimination and agencies (such as Security Exchanges) can require reporting on measurable outcomes from member organizations (e.g. targets for women at board of directors level; Deloitte 2011; Zahidi and Ibarra 2010), but regulatory forces cannot eliminate gender stereotypes or covert discrimination. These are the most damaging contemporary barriers to women's advancement in industrialized countries, barriers so subtle yet so strong that they are difficult to detect and eradicate (Meyerson and Fletcher 2000). These barriers exist at different levels of the organization (Eagly and Carli 2007), influence many processes and decisions affecting the careers of women managers (e.g. Phelan *et al.* 2008), and have detrimental cumulative effects on women's rate of advancement and career aspirations over time (Hewlett 2007; Hoobler *et al.* in press).

As Meyerson and Fletcher (2000: 129) explain, men 'are not to blame for the pervasive gender inequity in organizations [operating in developed Western societies] today—but neither are women'. Although both men and women can suffer from 'gender fatigue', men are still less likely than women to acknowledge that women are judged and treated differently to men in organizations (e.g. Ryan *et al.* 2007). These gender differences in the acceptance of the barriers to women's advancement are important contributors to persistent gender inequity (Prime and Moss-Racusin 2009). As men are usually the decision-makers and members of the dominant group, it is critical that they have the same understanding as women of the reasons underlying gender inequity in organizations for it to be effectively addressed (Griffith and MacBride-King 1998; McCracken 2000; Prime and Moss-Racusin 2009).

What Women Can Do

In conjunction with renewed affirmative action legislation and institutional pressures, there are some things that women can continue to do to advance. One is to be visible to one's superiors (Vinnicombe *et al.* 2000; Cross 2010). The other is to use job-focused and manager-focused strategies to enhance their chances of promotion (Singh *et al.* 2002). The literatures on impression management and the promotion process to management levels, such as to partner in law firms or executive positions, continue to show that visibility through impression management and self-promotion are necessary to break down gender stereotypes and manage perceptions (e.g. Singh *et al.* 2002; Kumra and Vinnicombe 2008). Similar strategies to advance are used by women in South Asia, where gender is arguably a more salient characteristic than in Western societies. For example, Fernando and Cohen (2011) identified eight modes of engagement Sri Lankan women use to advance their careers; two of these modes were ingratiation and networking.

Further, recent research conducted in the 'West' indicates that men still engage in behaviours aimed at enhancing their reputation more than women; examples of such behaviours are networking with decision-makers and taking credit for work achievements (e.g. Singh *et al.* 2002; Kumra and Vinnicombe 2008; Cross 2010; Hewlett *et al.* 2010). In fact, a reluctance to self-advocate was seen by women managers and CEOs across a number of countries as a top barrier to women's advancement (Desvaux *et al.* 2010). In contrast, men and women in some Eastern societies appear to be similarly inclined to engage in these career advancement strategies (Fernando and Cohen 2011). It is possible that aspects of the social context, such as the strength of gender roles and high power distance, leave women with little choice but to resort to self-promoting strategies to advance. For instance, the Sri Lankan women in Fernando and Cohen's study were very ambitious and were prepared to use various strategies, including ingratiation, adaptation, and networking, to circumvent the high social and structural barriers to their career advancement. It is also possible that such strategies are beneficial in some cultures, such as in Sri Lanka (Fernando and Cohen 2011), but are self-defeating in others, such as in the US (e.g. Kanter 1977*b*; Rudman and Phelan 2008; Hewlett *et al.* 2010). Thus, the budding research into women's advancement in Eastern countries underscores the popular saying that 'one size does not fit all'. The strategies that women can effectively use to advance can be culture and situation dependent.

Conclusion

The literature on the barriers to women's advancement is rich and spans several decades. As a result, we aimed in this chapter to illustrate rather than comprehensively examine 'if and how' some of the barriers to women's advancement in management have changed. To this end, we focused on four types of persistent barriers to women's advancement (decision-makers' denial of gender discrimination, gender roles, stereotypes and perceptions, and organizational culture), one new version of an old barrier (covert discrimination), and on one new barrier (growing gender fatigue). We conclude that the more things changed (e.g. the world of work has become less gendered) the more they stayed the same (e.g. in general women's representation at executive and board levels across countries remains low). Although no one factor explains this worldwide organizational phenomenon, societal norms clearly continue to exercise strong influence on behaviour in organizations. Time will tell if some of this social influence will diminish in importance as a result of renewed legislative and institutional pressures for organizations to address gender inequity. Part of this success will be determined by the approach each organization adopts to change. We recommend that a thorough diagnosis of the organization's position in the gender equity continuum be the first step in designing an organization-specific approach to increasing gender diversity in management. This customized approach recognizes that organizations, and the countries in which they operate, are at different stages of awareness and readiness to take on the gender equity challenge.

Acknowledgement

This research was supported by a grant from the Australian Research Council (Linkage Project 120200475). We are grateful for the support of our two Industry Partners on this grant: Aegis Diversity@Work and the Australian Senior Human Resources Roundtable.

References

Ahmad, S. Z. (2011). 'Businesswomen in the Kingdom of Saudi Arabia', *Equality, Diversity and Inclusion: An International Journal*, 30: 610–14.

Ali, M., Kulik, C. T., and Metz, I. (2011). 'The Gender Diversity–Performance Relationship in Services and Manufacturing Organizations', *International Journal of Human Resource Management*, 22: 1464–85.

ASX (2009). 'Media Release: New ASX Corporate Governance Council Recommendations on Diversity', http://www.asx.net.au/documents/about/mr_071209_asx_cgc_communique.pdf (accessed Aug. 2011).

Bailyn, L. (2003). 'Academic Careers and Gender Equity: Lessons Learned at MIT', *Gender, Work and Organization*, 10: 137–53.

Bajdo, L. M., and Dickson, M. W. (2001). 'Perceptions of Organizational Culture and Women's Advancement in Organizations', *Sex Roles*, 45(5–6): 399–414.

Barretto, M., and Ellemers, N. E. (2005). 'The Perils of Political Correctness', *Social Psychology Quarterly*, 68(1): 75–88.

Bartram, T. (2005). 'Small Firms, Big Ideas: The Adoption of Human Resource Management in Australian Small Firms', *Asia Pacific Journal of Human Resources*, 43: 137–54.

Becker, G. S. (1993). *Human Capital*. Chicago: University of Chicago Press.

Berlew, D. E., and Hall, D. T. (1966). 'The Socialization of Managers: Effects of Expectations on Performance', *Administrative Science Quarterly*, 11: 207–23.

Bielby, W. T., and Bielby, D. D. (1989). 'Family Ties: Balancing Commitments to Work and Family in Dual Earner Households', *American Sociological Review*, 4: 776–89.

Blair-Loy, M. (1999). 'Career Patterns of Executive Women in Finance: An Optimal Matching Analysis', *American Journal of Sociology*, 104: 1346–97.

Blum, T. C., Fields, D. L., and Goodman, J. S. (1994). 'Organization-Level Determinants of Women in Management', *Academy of Management Journal*, 37: 241–68.

Brady, D., Issacs, K., Reeves, M., Burroway, R., and Reynolds, M. (2011). 'Sector, Size, Stability, and Scandal', *Gender in Management: An International Journal*, 26(1): 84–105.

Breaugh, J. A., and Frye, N. K. (2008). 'Work–Family Conflict', *Journal of Business and Psychology*, 22: 345–53.

Brief, A. P., Rose, G. L., and Aldag, R. J. (1977). 'Sex Differences in Preferences for Job Attributes Revisited', *Journal of Applied Psychology*, 62: 645–6.

Brief, A. P., Dietz, J., Cohen, R. R., Pugh, S. D., and Vaslow, J. B. (2000). 'Just Doing Business', *Organizational Behavior and Human Decision Processes*, 81: 72–97.

Broadbridge, A., and Simpson, R. (2011). 'Y25 Years on: Reflecting on the Past and Looking to the Future in Gender and Management Research', *Journal of Management*, 22: 470–83.

Bronstein, P., Rothblum, E. D., and Solomon, S. E. (1993). 'Ivy Halls and Glass Walls', *New Directions for Teaching and Learning*, 53: 17–31.

Brown, L. M. (2010). 'The Relationship between Motherhood and Professional Advancement: Perceptions versus Reality', *Employee Relations*, 32: 470–94.

Burt, R. S. (1998). 'The Gender of Social Capital', *Rationality and Society*, 10(1): 5–46.

Carnes, W. J., and Radojevich-Kelley, N. (2011). 'The Effects of the Glass Ceiling on Women in the Workforce', *Review of Management Innovation and Creativity*, 4(10): 70–9.

Carter, N. M., and Silva, C. (2010). *Pipeline's Broken Promise*. New York: Catalyst.

Castilla, E. J., and Benard, S. (2010). 'The Paradox of Meritocracy in Organizations', *Administrative Science Quarterly*, 55: 543–76.

Catalyst (2011). *Current Female Heads of Companies of the Financial Times Europe 500*. New York: Catalyst. http://www.catalyst.org/publication/522/current-female-heads-of-companies-of-the-financial-times-europe-500 (accessed Mar. 2012).

Catalyst (2012). *Women in Europe*. http://www.catalyst.org/publication/285/women-in-europe (accessed: Mar. 2012).

Cohen, P. N., Huffman, M. L., and Knauer, S. (2009). 'Stalled Progress? Gender Segregation and Wage Inequality among Managers, 1980–2000', *Work and Occupations*, 36: 318–42.

Chao, Chin-Chung (2011). 'Climbing the Himalayas', *Leadership and Organization Development Journal*, 32: 720–81.

Cooper, C. L., and Lewis, S. (1999). 'Gender and the Changing Nature of Work', in G. Powell (ed.), *Handbook of Gender and Work*, 37–46. Thousand Oaks, CA: Sage.

Cross, C. (2010). 'Barriers to the Executive Suite: Evidence from Ireland', *Leadership and Organization Development Journal*, 31: 104–19.

Crouter, A. C. (1984). 'Spillover from Family to Work', *Human Relations*, 37: 425–41.

Deal, J. J., and Stevenson, M. A. (1998). 'Perceptions of Female and Male Managers in the 1990s', *Sex Roles: A Journal of Research*, 38: 287–300.

Deloitte (2011). *Women in the Boardroom: A Global Perspective*. http://www.deloitte.com/view/en_GX/global/search (accessed Mar. 2011).

Desvaux, G., Devillard, S., and Sancier-Sultan, S. (2010). *Women Matter 2010—Women at the Top of Corporations: Making it Happen*. Paris: McKinsey & Co.

Dick, P., and Cassell, C. (2002). 'Barriers to Managing Diversity in a UK Constabulary: The Role of Discourse', *Journal of Management Studies*, 39: 953–76.

Di Maggio, P. (2001). *The Twenty-First Century Firm*. Princeton: Princeton University Press.

Eagly, A. H. (1987). *Sex Differences in Social Behavior*. Hillsdale, NJ: Erlbaum.

Eagly, A. H., and Carli, L. L. (2007). 'Women and the Labyrinth of Leadership', *Harvard Business Review*, 85(9): 63–71.

Eagly, A. H., and Karau, S. J. (2002). 'Role Congruity Theory of Prejudice toward Female Leaders', *Psychological Review*, 109: 573–98.

Edgar, F. (2001). 'Equal Employment Opportunity', *New Zealand Journal of Industrial Relations*, 26: 217–26.

Ely, R. J., and Meyerson, D. E. (2010). 'An Organizational Approach to Undoing Gender', *Research in Organizational Behavior*, 30: 3–34.

Ely, R. J., Ibarra, H., and Kolb, D. M. (2011). 'Taking Gender into Account: Theory and Design for Women's Leadership Development Programs', *Academy of Management Learning and Education*, 10: 474–93.

EMA Partners International (2011). *Gender Splits*. http://www.ema-partners.com/articles/gender-split (accessed Mar. 2012).

England, P. (2010). 'The Gender Revolution', *Gender and Society*, 24: 149–66.

Equal Opportunity for Women in the Workplace Agency (EOWA) (2010). *Australian Census of Women Executive Managers and Board of Directors*. Sydney: EOWA.

EOWA. (2011). *Gender workplace statistics at a glance*. http://www.eowa.gov.au/Information_Centres/Resource_Centre/Statistics/gender%20stats%206-11_ONLINEversion.pdf (accessed July 2011).

Fagenson, E. A. (1990). 'At the Heart of Women in Management Research: Theoretical and Methodological Approaches and their Biases', *Journal of Business Ethics*, 9(1): 3–38.

Fernando, W. D. A., and Cohen, L. (2011). 'Exploring the Interplay between Gender, Organizational Context and Career', *Career Development International*, 16: 553–71.

French, E., and Strachan, G. (2007). 'Equal Opportunity Outcomes for Women in the Finance Industry in Australia', *Asia Pacific Journal of Human Resources*, 45: 314–32.

Gans, J. (2008). 'The Delicate Balance on Parental Leave', *Melbourne Review*, 4(2): 47–55.

Goldin, C. (2004). 'The Long Road to the Fast Track: Career and Family', *Annals of the American Academy*, 596: 20–35.

Goldman Sachs JBWere (2009). *Australia's Hidden Resource: The Economic Case for Increasing Female Participation*. Melbourne: Goldman Sachs JBWere.

Griffith, P. G., and MacBride-King, J. L. (1998). *Closing the Gap: Women's Advancement in Corporate and Professional Canada*. New York: Catalyst.

Gutek, B. A. (1993). 'Changing the Status of Women in Management', *Applied Psychology: An International Review*, 42: 301–11.

Hausmann, R., Tyson, L. D., and Zahidi, S. (2010). *The Global Gender Gap Report*. Geneva: World Economic Forum.

Haveman, H. A., and Beresford, L. S. (2012). 'If You're So Smart, Why Aren't You the Boss?' *Annals of the American Academy of Political and Social Science*, 639: 114–30.

Heilman, M. E., Wallen, A. S., Fuchs, D., and Tamkins, M. M. (2004). 'Penalties for Success', *Journal of Applied Psychology*, 89: 416–27.

Hewlett, S. A. (2007). *Off-Ramps and On-Ramps*. Boston: Harvard Business School Press.

Hewlett, S. A., with K. Peraino, L. Sherbin, and K. Sumberg (2010). *The Sponsor Effect*. Harvard Business Review Research Report. Boston: Harvard Business School Publishing.

Hoobler, J. M., Wayne, S. J., and Lemmon, G. (2009). 'Bosses' Perceptions of Family–Work Conflict and Women's Promotability', *Academy of Management Journal*, 52: 939–57.

Hoobler, J. M., Lemmon, G., and Wayne, S. J. (in press). 'Women's Managerial Aspirations', *Journal of Management*. doi: 10.1177/0149206311426911

Howard, A., and Bray, D. W. (1988). *Managerial Lives in Transition*. New York: Guilford Press.

Ibarra, H., Carter, N. M., and Silva, C. (2010). 'Why Men Still Get More Promotions than Women', *Harvard Business Review*, 88(9): 80–5.

Icheku, V. (2011). 'Post-Taliban Measures to Eliminate Gender Discrimination in Employment', *Equality Diversity and Inclusion: An International Journal*, 30: 563–71.

Jorgenson, J. (2002). 'Engineering Selves: Negotiating Gender and Identity in Technical Work', *Management Communication Quarterly*, 15: 350–80.

Kanter, R. M. (1977a). *Men and Women of the Corporation*. New York: Basic Books.

Kanter, R. M. (1977b). 'Some Effects of Proportions on Group Life', *American Journal of Sociology*, 82: 865–990.

Kaye, B., and Scheele, A. (1975). 'Leadership Training', *New Directions for Higher Education*, 3: 79–93.

Kelan, E. K. (2009). 'Gender Fatigue', *Canadian Journal of Administrative Sciences/Revue Canadienne des Sciences de l'Administration*, 26(3): 197–210.

Kelan, E. K., and Jones, R. D. (2010). 'Gender and the MBA', *Academy of Management Learning and Education*, 9: 26–43.

Kilduff, M., and Mehra, A. (1996) 'Hegemonic Masculinity among the Elite', in C. Cheng (ed.), *Masculinities in Organizations*, pp. 115–129. Thousand Oaks, CA: Sage.

Kirchmeyer, C. (1998). 'Determinants of Managerial Career Success: Evidence and Explanation of Male/Female Differences', *Journal of Management*, 24: 673–92.

Kirton, G., and Greene, A. (2010). 'What does Diversity Management Mean for the Gender Equality Project in the United Kingdom?', *Canadian Journal of Administrative Sciences Revue/Canadienne Des Sciences De L'Administration*, 27: 249–62.

Konrad, A. M., and Linnehan, F. (1995). 'Formalized HRM', *Academy of Management Journal*, 38: 787–820.

Kotter, J. P. (1996). *Leading Change*. Boston: Harvard Business School.

Kram, K. E. (1985). *Mentoring at Work*. Glenview, IL: Scott, Foresman & Co.

Krefting, L. A. (2003). 'Intertwined Discourses of Merit and Gender: Evidence from Academic Employment in the USA', *Gender, Work and Organization*, 10: 260–78.

Kuperberg, A., and Stone, P. (2008). 'The Media Depiction of Women Who Opt Out', *Gender and Society*, 22: 497–517.

Kumra, S., and Vinnicombe, S. (2008). 'A Study of the Promotion to Partner Process in a Professional Services Firm', *British Journal of Management*, 19: S65–S74.

Kurtulus, F. A., and Tomaskovic-Devey, D. (2012). 'Do Female Top Managers Help Women to Advance?', *Annals of the American Academy of Political and Social Science*, 639: 173–97.

Larsen, H. H., and Brewster, C. (2003). 'Line Management Responsibility for HRM: What is Happening in Europe?', *Employee Relations*, 25: 228–44.

Leslie, L. M., and Gelfand, M. J. (2008). 'The Who and When of Internal Gender Discrimination Claims', *Organisational Behavior and Human Decision Processes*, 107: 123–40.

Lewin, K. (1951). *Field Theory in Social Science*. New York: Harper & Row.

Lewis, P. (2006). 'The Quest for Invisibility', *Gender, Work and Organization*, 13: 454–69.

Livanos, I., Yalkin Ç., and Nunez, I. (2009). 'Gender Employment Discrimination: Greece and the United Kingdom', *International Journal of Manpower*, 30: 815–34.

McCracken, D. M. (2000). 'Winning the Talent War for Women', *Harvard Business Review*, 78(6): 159–60, 162, 164–7.

Mainiero, L. A. (1994). 'Getting Anointed for Advancement: The Case of Executive Women', *Academy of Management Executive*, 8(2): 53–67.

Mainiero, L. A., and Sullivan, S. E. (2005). 'Kaleidoscope Careers', *Academy of Management Executive*, 19(1): 106–23.

Metz, I. (2009). 'Firm Size, Social Factors and Women's Advancement', *Applied Psychology: An International Review*, 58: 193–213.

Metz, I. (2011). 'Women Leave Because of Family Responsibilities: Fact or Fiction?', *Asia Pacific Journal of Human Resources*, 49: 285–307.

Metz, I., and Harzing, A. W. K. (2009). 'Women in Editorial Boards of Management Journals', *Academy of Management Learning and Education*, 8: 540–57.

Metz, I., and Kulik, C. T. (2008). 'Making Public Organizations More Inclusive: A Case Study of the Victoria Police Force', *Human Resource Management*, 47: 369–87.

Metz, I., and Simon, A. (2008). 'A Focus on Gender Similarities in Work Experiences at Senior Management Levels', *Equal Opportunities International*, 27: 433–54.

Metz, I., and Tharenou, P. (1999). 'A Retrospective Analysis of Australian Women's Representation in Management in Large and Small Banks', *International Journal of Human Resource Management*, 10: 201–22.

Metz, I., and Tharenou, P. (2001). 'Women's Career Advancement', *Gender and Organisation Management*, 26: 312–42.

Meyerson, D. E., and Fletcher, J. K. (2000). 'A Modest Manifesto for Shattering the Glass Ceiling', *Harvard Business Review*, 78(1): 126–36.

Morrison, A. M., White, R. P., Van Velsor, E., and the Centre for Creative Leadership (1992). *Breaking the Glass Ceiling*. Reading, MA: Addison-Wesley.

Ohlott, P. J., Ruderman, M. N., and McCauley, C. D. (1994). 'Gender Differences in Managers' Developmental Job Experiences', *Academy of Management Journal*, 37(1): 46–67.

Peetz, D., Gardner, M., Brown, K., and Berns, S. (2008). 'Workplace Effects of Equal Employment Opportunity Legislation', *Policy Studies*, 29: 405–19.

Pellegrino, G., D'Amato, S., and Weisberg, A. (2011). *The Gender Divide: Making the Business Case for Investing in Women*. http://www.deloitte.com/genderdividend (accessed Mar. 2011).

Perry, E. L., and Kulik, C. T. (2008). 'The Devolution of HR to the Line', *International Journal of Human Resource Management*, 19: 262–73.

Phelan, J. E., Moss-Racusin, C. A., and Rudman, L. A. (2008). 'Competent and Yet Out in the Cold: Shifting Criteria for Hiring Reflect Backlash toward Agentic Women', *Psychology of Women Quarterly*, 32: 406–13.

Portes, A. (1998). 'Social Capital', *Annual Review of Sociology*, 24: 1–24.

Powell, G. N., and Mainiero, L. A. (1992). 'Cross-Currents in the River of Time: Conceptualizing the Complexities of Women's Careers', *Journal of Management*, 18: 215–37.

Powell, G. (2005). 'An Opinion: The Family-Friendly Workplace—Just an Illusion', *Industrial-Organizational Psychologist*, 42(4): 27–8.

Prime, J., and Moss-Racusin, C. A. (2009). *Engaging Men in Gender Initiatives*. New York: Catalyst. http://www.catalyst.org/publication/323/engaging--men--in--gender--initiatives--what--change--agents--need--to--know (accessed Apr. 2012).

Ragins, B. R., and Sundstrom, E. (1989). 'Gender and Power in Organisations: A Longitudinal Perspective', *Psychological Bulletin*, 105(1): 51–88.

Riger, S., and Galligan, P. (1980). 'Women in Management', *American Psychologist*, 35: 902–10.

Rivera, L. A. (2012). 'Diversity within Reach', *Annals of the American Academy of Political and Social Science*, 639: 71–90.

Roth, L. M. (2007). 'Women on Wall Street', *Academy of Management Perspectives*, 21(1): 24–35.

Rowe, R., and Snizek, W. E. (1995). 'Gender Differences in Work Values: Perpetuating the Myth', *Work and Occupations*, 22: 215–29.

Rudman, L. A., and Phelan, J. E. (2008). 'Backlash Effects for Disconfirming Gender Stereotypes in Organizations', *Research in Organizational Behavior*, 28: 61–79.

Russell, C. (2012). 'Workers to Know of Gender Issues', *The Advertiser State Edition*, 2 Mar. (accessed Mar. 2012).

Ryan, M. K. (2012). 'A Slippery Slope for Women', *The Age*, News-Opinion, 8 Mar., p. 15.

Ryan, M. K., and Haslam, S. A. (2005). 'The Glass Cliff', *British Journal of Management*, 16: 81–90. doi: 10.1111/j.1467–8551.2005.00433.x

Ryan, M. K., Haslam, S.A., and Postmes, T. (2007). 'Reactions to the Glass Cliff', *Journal of Organizational Change Management*, 20: 182–97.

Schein, V. E. (1973). 'The Relationship between Sex Role Stereotypes and Requisite Management Characteristics', *Journal of Applied Psychology*, 57: 95–100.

Schein, V. E., Mueller, R., Lituchy, T., and Liu, J. (1996). 'Think Manager—Think Male: A Global Phenomenon?', *Journal of Organizational Behavior*, 17: 33–41.

Schein, V. E. (2001). 'A Global Look at Psychological Barriers to Women's Progress in Management', *Journal of Social Issues*, 57: 675–88.

Simpson, R. (1996). 'Does an MBA Help Women?', *Gender, Work and Organization*, 3: 115–21.

Simpson, R. (1997). 'Have Times Changed?', *British Journal of Management*, 8: S121–S130.

Simpson, R. (2000). 'Gender Mix and Organisational Fit', *Women in Management Review*, 15(1): 5–19.

Singh, V., Kumra, S., and Vinnicombe, S. (2002). 'Gender and Impression Management: Playing the Promotion Game', *Journal of Business Ethics*, 37: 77–89.

Soares, R., Cobb, B., Lebow, E., *et al.* (2011). *2001 Catalyst Census: Fortune 500 Women Board Directors, Executive Officers and Top Earners*. New York: Catalyst.

Spoor, J. R., and Schmitt, M. T. (2011). '"Things Are Getting Better" Isn't Always Better', *Basic and Applied Social Psychology*, 33: 24–36.

Stainback, K., Ratliff, T., and Roscigno, V. J. (2011). 'The Organizational Context of Sex Discrimination', *Social Forces*, 894: 1165–88.

Stone, P. (2007). *Opting Out?* London: University of California Press.

Straub, C. (2007). 'A Comparative Analysis of the Use of Work–Life Balance Practices in Europe', *Women in Management Review*, 22: 289–304.

Swim, J. K., Aikin, K. J., Hall, W. S., and Hunter, B. A. (1995). 'Sexism and Racism', *Journal of Personality and Social Psychology*, 68: 199–214.

Tharenou, P., and Conroy, D. (1994). 'Men and Women Managers' Advancement', *Applied Psychology: An International Review*, 43(1): 5–31.

Vinnicombe, S., Singh, V., and Sturges, J. (2000). 'Making it to the Top in Britain', in R. J. Burke and M. Mattis (eds), *Women on Corporate Boards of Directors: International Challenges and Opportunities*, 57–74. Dordrecht: Kluwer.

Vinnicombe, S., Sealy, R., Graham, J., and Doldor, E. (2010). *The Female FTSE Board Report 2010*. UK: International Centre for Women Leaders, Cranfield School of Management.

Watkins, M. B., Kaplan, S., Brief, A., Shull, A., Dietz, J., Mansfield, M.-T., and Cohen, R. (2006). 'Does it Pay to be a Sexist?', *Journal of Vocational Behavior*, 69: 524–37.

Zahidi, S., and Ibarra, H. (2010). *The Corporate Gender Gap Report*. Geneva: World Economic Forum.

CHAPTER 9

LEADERSHIP

A Matter of Gender?

YVONNE DUE BILLING AND MATS ALVESSON

INTRODUCTION

THE theme of gender and leadership concerns a range of topics from recruitment and selection to managerial jobs to leadership styles and values, experiences in managerial positions, and to how gendered meanings associated with masculinity and femininity form leadership ideals and regulate ways of constructing, doing, and assessing leadership. The field(s) also include a variety of foci on actors, from a specific interest in females to systematic comparisons between the sexes and to an interest in cultural meanings of masculinity and femininity somewhat disconnected from individuals (and their bodies). Leadership and gender is on the whole 'biased' towards females, it fairly seldom includes a direct interest in the experiences and subjectivities of men and management (Collinson and Hearn (1996) is an exception).

The interest in gender in relation to management and organizations has come later than to other academic fields. One can of course argue that, apart from some historical exceptions—some of the great historical leaders were women—women have hardly been present on the leadership and management scene before the 1970s, and then for a time only on a small scale. This late interest was then partly due to the fact that women did not enter the professional labour market in great numbers before the 1970s and there were very few managers.

The early focus was mainly concerned about the small number of women in management and the consequence of men's dominance in management for women (e.g. Kanter 1977; Marshall 1984). Another focus was on differences between women and men in management, and if women favoured a different leadership style, sometimes referred to as feminine leadership (Helgesen 1990). What is holding women back and their similarities/differences compared to men are still addressed. A third, more recent, issue is contemporary changes on ideas and ideals of leadership, possibly involving a de-masculinization of management and whether these alleged changes not only in rhetoric but also possibly in practice are benefiting women. We are going to address these concerns from a critical position.

WOMEN AND MANAGEMENT POSITIONS

Women in management is a large and still expanding topic. Although the number of female managers has increased in many countries, mainly at the lower and middle managerial levels they are still under-represented, especially at higher levels. A gender gap still exists with regard to formal power and authority, high status, and high incomes. (See Chapters 8 and 11 in this Handbook.)

Explanations for the smaller number of women especially at the top have pointed to a number of different factors or dimensions, from differences between men and women in terms of psychological traits and/or different socialization background, work orientations, or educational/career choices (or constraints) and others point to more sociological, structural explanations (Billing and Alvesson 1994; Wilson 1998). These various factors and levels are intertwined and interact.

When addressing the small numbers of female managers, researchers tend to emphasize the lack of difference between men and women in terms of leadership abilities. On the whole, the studies of *psychological characteristics* have shown none or only minor differences between males and females, leading most commentators to suggest that psychological abilities do not account for the smaller number of women in managerial jobs (Morrison and Von Glinow 1990). As we will come back to, when the period of accounting for small numbers in and explaining under-recruitment of females to managerial jobs was over, the trend partly moved to pointing to traits and orientations indicating a female difference and a feminine advantage. The 'truth' about (no-)differences then is not stable—a phenomenon we come back to.

Women's disadvantaged position is some times attributed to less relevant education, and lacking qualified work experience: this is the *human capital* theory. As investments in education, training, and other forms of qualifying experiences are seen as the key factor behind careers, it is a bias against women if certain educations which are dominated by men are seen as the favourable ones for managerial positions.

Cultural assumptions about leadership might also work against women in leadership. The manager and leadership have traditionally been constructed in masculine terms (Schein 1973, 1975) and a so-called 'masculine ethic' is said to be part of the early image of managers.

> This 'masculine ethic' elevates the traits assumed to belong to some men to necessities for effective management: a tough-minded approach to problems; analytic abilities to abstract and plan; a capacity to set aside personal, emotional considerations in the interests of task accomplishment; and a cognitive superiority in problem-solving

and decision-making...when women tried to enter management jobs, the 'masculine ethic' was invoked as an exclusionary principle.

(Kanter 1977: 22)

These constructions are less strong or clear-cut today, a de-masculinization of leadership has started as new ideas of modern leadership have emerged. Still, there is a historical tradition and deep cultural ideas that give leadership a masculine image in most countries (Hearn and Parkin 1986/7; Collinson and Hearn 1996), which mainly is a result of men being dominant in these positions. But all jobs can be redefined.

Some authors believe that *the interests of men* prevent women from competing for privileged positions (Cockburn 1991; Lindgren 1996). Men favour men as there is a group interest in favouring oneself and members in one's network—often other men. Roper (1996: 224) uses the expression 'homosocial desire' about the preference for a single-sex group. As long as women are in a minority they may have problems with political support and getting recognized, feeling comfortable, and being promoted (Kanter 1977; Martin 1985). A critical mass is believed to be necessary for an under-represented sex to have equal opportunities along with members of a dominating social category.

According to some commentators, women who have been appointed managers seem to be more likely placed in precarious leadership positions than men. In a study of women appointed to leadership positions in top British companies, Ryan and Haslam use the metaphor 'glass cliff' to capture their findings, that

> Women were more likely than men to be placed in positions already associated with poor company performance...and female directors, thus were more likely than male directors to find themselves on a glass cliff. That is, their positions of leadership were more risky and precarious (i.e. at greater risk of being associated with failure) than those in which men found themselves.

(2007: 56)

Thus either women are seen as better suited for poorly performing companies and perhaps crisis management than men, and then it is implicit that men are not suited for dealing with crises—or women get the precarious (scapegoat) positions and can then be blamed if things do not work out well, according to Ryan and Haslam (2007: 550) who state: 'If and when that failure occurs, it is then women (rather than men) who must face the consequences and who are singled out for criticism and blame.'

However, appointing women as crisis-managers might also be seen as an attempt to capitalize on what are believed to be women's 'natural' skills and competences, whether these skills are present or not (Fletcher 2004). It can also be a reflection of women's overall weaker position in the labour market—men are more likely to get the most attractive positions and women tend to get the jobs where competition is weaker. There are, however, also indications of the opposite problems, i.e. of females *not* being placed in demanding and qualifying jobs. Women often get staff jobs, which are sometimes less challenging and have limited value compared with line positions; they may end up in less influential and prestigious fields, like HR (Brewis and Linstead 2004).

One could here switch logic and emphasize the 'positive' aspects of women's placement in specific types of managerial jobs. Getting positions in poor performing companies may give the chance to bring about significant improvements and being in HR positions could be viewed as secure. On the other hand, avoiding both extremes would seem better. A question is of course if there are forces placing females in (mainly) unfortunate career tracks and positions, although of opposing character, or if there is an active choice. More close-up empirical studies of experiences, reasoning, and social interactions of the glass 'cliffed' or 'ghettoed' female managers would be welcome and perhaps also more research investigating the extent to which women are over-represented in the glass cliffed or ghettoed tracks and how large proportions of all female managers and professionals are outside these two 'extremes' in terms of career constraints and risks.

Finally, researchers have pointed at the significance of *work–family* as the major problem preventing women from advancing (Martin 1993). Women who have a family are often less mobile, as family priorities may lead to them not to take a position if this involves longer work days, more travel, or moving geographically to a new site of employment. Lack of time is a major problem for many career and working women, because they tend to take on a double burden or double work (Davidson and Burke 2000) and this also affects how much they can socialize with colleagues, invest in social capital (Eagly and Carli 2007), and be able to prove their commitment to the organization.

The deeply culturally ingrained assumptions and expectations that women have a primary responsibility for family might in different ways influence women's attitudes and interests in careers: they might prioritize children above career (or vice versa). They may also feel ambivalence and insecurity in relation to a managerial job—or the family situation might become a stress factor in relation to a career job. There are a vast amount of studies comparing the stress levels of male and female managers, indicating that there are significant differences (Frankenheuser 1993). It appears that female managers also show more stress symptoms related to family/domestic issues than males. It is relatively common for female managers to be single and childless (Billing 1991; Frankenheuser 1993; Davidson and Burke 2000; Guillaume and Pochic 2009).

Even though the increase in the number of female managers, at least up to the middle level, indicates changes and even though most women today are less tightly committed to family work while men are increasingly, but slowly, taking more responsibility for children and housework, family concerns are still a significant obstacle for women getting managerial jobs, in particular top positions.

The extensive literature on difficulties for women attaining managerial jobs has mainly pointed to 'external' sources of problems, while women themselves are basically the same as—and thereby as good and suitable as—men for managerial jobs and the exercise of leadership. The major emphasis here has been on similarities, but some of the literature comparing men and women in leadership also indicates differences interpreted as favouring women.

STYLE OF LEADERSHIP—WOMEN COMPARED TO MEN

There are a lot of studies and arguments around women compared to men with regard to leadership. There are two competing camps, a *no-difference camp*, where the conclusion is that 'in general, comparative research indicates that there are few differences in the leadership style of female and male designated leaders' (Bartol and Martin, cited in Eagly and Johnson 1990) and a *gender-stereotypic camp*, where some crucial differences are believed to exist.

Some studies showed varieties in styles for women: some women had an open and participatory style, others favoured control in their management style (Bayes 1987; Billing 2006). Men also varied in their management styles. And comparing women and men, Kovalainen (1990) found no significant differences in a study of male and female Finnish bank managers, nor did Cliff *et al.* (2005) in a study of the organizational practices of Canadian entrepreneurs.

A common conclusion is that female and male managers may differ sometimes but mostly they do not differ (Powell 1999; Butterfield and Grinnell 1999).

As opposed to the above-mentioned studies, a number of other writers maintain that there are minor or clear differences between women and men in their management style (e.g. Loden 1986; Grant 1988; Helgesen 1990; Rosener 1990)—be it because of different socialization (Grant 1988; Rosener 1990; Lipman-Blumen 1992) and/or differences in experiences from men, for example with care responsibilities, which should then lead to different skills and a different style. Eagly and Johnson (1990) say that women 'in general' have more of the characteristics which are often said to be feminine, and on that basis there should be some spill-over, when people from this group get into management positions.

In a review of the research, Eagly and Johnson (1990) found that available (positivistic) academic studies showed another picture than the 'no-difference' one that almost all other (academic) commentators have favoured. They refer to research findings indicating that women as a group can be described as friendly, pleasant, interested in other people, expressive, and socially sensitive and this might then result in sex differences because of sex-role spill-over. In a meta-analysis of other studies they found that laboratory studies—mostly with students as research objects—typically showed sex differences in leadership style, while studies of leadership in organizations did so to a lower degree. This may indicate that structural conditions and positions influence attitudes and behaviour (Kanter 1977; Ely and Padavic 2007), something we will address more in depth below. Still, however, Eagly and Johnson (1990) found that women had a slightly more democratic leadership style than men. Eagly *et al.* (1992) also found that women who adopted stereotypically masculine styles were disliked in comparison with men.

Some research has shown that there were conditions under which men fared better than women and vice versa, 'leadership roles defined in relatively masculine terms favored male leaders and leadership roles defined in relatively feminine terms favored female leaders' (Eagly *et al.* 1995: 137). Studies indicate that sex differences are significantly correlated with the congeniality of these roles for men and women. Or to put it differently, 'women fared poorly in settings in which leadership was defined in highly masculine terms, especially in military settings. Men fared slightly worse than women in settings in which leadership was defined in less masculine terms' (Eagly *et al.* 1995: 140).

In a later contribution, Eagly and Johannesen-Schmidt (2008) emphasize the small, but important, style differences between male and female leaders. They base this on a meta-analysis of forty-five studies, comparing male and female managers, in relation to transactional, transformational, and laissez-faire leadership. They found that women managers tended to adopt a transformational style more often than men managers. The three categories are established (although far from unproblematic) distinctions.

Some use the image of a web to describe women managers' position—Ferguson (1994: 5) says that women have 'a transformational, democratic, and/or web rather than hierarchical style of leadership and more satisfied subordinates than men managers', where women see themselves as placed in the middle rather than at the top. Feminine leadership is characterized by cooperativeness, collaboration and problem-solving, based on intuition and empathy (Helgesen 1990; Rosener 1990).

A complication in drawing conclusions about gender and leadership is that the gender congeniality of leadership roles may be accompanied by different patterns. Eagly and Johnson (1990), for example, found that although male leaders were often more task oriented than females, the latter tended to be more so than males in a leadership role that was more congenial to women (e.g. head nurse). In a later study they found that females were evaluated as more task oriented than males and they conclude, 'It appears that all factors being equal, men have greater freedom than women to lead in a range of styles without encountering negative reactions' (Eagly *et al.* 1992).

There are somewhat different views on what we can conclude based on all the studies in this field. Carli and Eagly (2011: 109) state that 'it is reasonable to assume that women would lead differently than men'. Butterfield and Grinnell (1999), on the other hand, say that 'after reviewing three decades of work on the topic of gender, leadership, and managerial behavior, it appears that we have not provided conclusive answers'. So maybe, as indicated earlier, we are posing the wrong questions. Establishing general, abstract correlations between sex and leadership may be a misleading or at least a not very informative enterprise.

A complication here is of course how to determine and formulate what is a difference, i.e. when is it motivated to say that 'there is' a difference? Whether female and male managers are somewhat different or broadly similar is less a matter of reality than about what point one wants to make. It is hardly possible to make an exact quantitative statement, such as women are 2 per cent or 0.08 points more 'democratic' than men in leadership. Over time, given what is at stake in terms of women's opportunities, the 'similar to' (and as good as) men respectively 'different from' (and better than) men forms of truths have had different currency. With the expansion of female managers and possibilities to link women to recipes of good leadership fairly easy to align with femininity—and thus with women—the selling of female managers as 'different from and better than' men have become popular. This is not to say that there are no data or truth behind this type of knowledge claim, but how ambiguous phenomena 'out there' provide an input to knowledge claims is never straightforward. We can say that there has been a change in emphasis, since the 1970s, in the writings of gender and leadership, from 'women are the same and as good as men' to 'women are different from and, in some ways, better than men' with regard to management jobs and leadership. What this change of emphasis reflects or means is very hard to tell, but it would be a mistake to say that this is simply about progress in scientific rationality and now, finally, revealing the truth about women and leadership.

Problems in Investigations

One may expect robust empirical studies to come up with clear answers on issues around women and leadership. But careful consideration of difficulties indicates that one should not expect too much of research in terms of clear-cut evidence offering final, or perhaps even preliminary, truths.

Language Mirroring Reality—Not So Easy

Measuring subtle phenomena not following law-like patterns as much as bearing strong imprints of discourses, meaning, and (other) social construction processes is not so easy or necessarily a very sensible project.

From a post-structural view, one might argue that central concepts in the gender and leadership field are often used in a totalizing manner, repressing alternative understandings and drawing attention away from the local context in which they may achieve a temporary, if fluid, meaning. When Eagly and Johnson (1990: 249), for example, concluded that 'women's leadership styles were more democratic than men's' there is a rich variety of problems that can be highlighted. The idea that words (signifiers) like 'women', 'leadership', 'style', 'more', 'democratic', and 'men' stand for some objective, universal, homogeneous, robust, and easily comparable phenomena out there, mirrored in questionnaire responses or observation protocols, can be viewed as problematic. 'Leadership' and 'democracy', for example, may refer to language use, where unstable meanings cannot be lifted out of the specific context in which the act takes place and the words are used. One could also, again from a post-structuralist view, question the assumed coherence and static nature of 'leadership' and, even more so, of 'leadership style'—perhaps human actions are more processual, fragmented, varying, inconsistent, and open to alternative interpretations than these concepts and the quoted statement suggest (cf. Calás and Smircich 1991; Chia 1995; Alvesson 1996). Talking about democratic leadership may be seen as confusing, as the idea of leadership tends to contradict democracy. One may argue that leadership marks an asymmetrical relation in which the impact of the leader is far-reaching, while democracy stands for equality in terms of influence, the more of *democratic* 'leadership', the less of 'leadership'.

The statement cited indicates a crude effort to universalize across history and culture, not to say local context. The law-like nature of the statement implies that there is a fixed causal relation between sex and a 'leadership style' called 'democratic'. Also, there are different opinions among positivists. Most refute the idea of a clear difference. In the second place, much of the proof is limited to the outcomes of questionnaire filling responses. All empirical material—including laboratory studies—relies on ratings of individuals that can hardly fully avoid reflecting stereotypical cultural beliefs. Research shows that individuals are referred to in such a way that a social identity—woman, professional, Jewish, corporate employee—becomes salient, and it influences the response accordingly (Haslam 2004). So if women are aware of themselves as women when asked a question, it affects the response, which may be somewhat different than if the awareness of the self is different. Finally, the efforts to find regularities mean that the opposite—variation and inconsistencies—receives no or little attention (Alvesson 1996).

More qualitatively rich studies—in-depth interviews and ethnographies—are potentially more valuable. There are however also often problems with these. Most qualitative, experience-near studies tend to focus on general negative experiences and/or the superior moral qualities of female managers. Many studies only consider women's viewpoints, which in interviews tend to be about their own positive values, ambitions, and acting and how males and masculine norms are sources of oppression or are creating problems for females (for illustrations, see Alvesson and Billing 2009), possibly reflecting self-serving bias and the tendency for people in interviews to do moral story-telling, where the interviewee typically appears as a better person than others being portrayed (Alvesson 2011).

Talk, X-Filling Behaviour, Behaviour in Artificial and Simplified Settings, and Organizational Practice

Whether research manages to reveal 'truths' or whether gender-stereotypical ideas and expectations are guiding subjects' responses in interviews, questionnaires, and even responses when monitored in experiments (stripped of the complexities of 'real' managerial settings) is a serious issue, marginalized by most published research.

The 'women and leadership industry' is here an active force creating its own truth effects. If one learns from popular books and lectures that women lead in a particular way, female managers may adapt to that norm, in particular when asked to describe/rate one's work, and subordinates may read the behaviour of the female manager accordingly, including devaluing behaviour perceived to break with the norm of how female leaders 'are'. In this way 'knowledge' on female managers creates its own norms and 'truths', and it does not so much mirror as produce a socially constructed 'reality'. This may not impact on specific behaviours as much as on people's beliefs of what they are doing (or ought to be doing)—and thus responses in questionnaires and interviews.

The belief that females are different and have other values guiding leadership, held not just widely in society but also by many female managers, may influence their responses in interviews and questionnaires, possibly giving a misleading picture. As Cliff *et al.* (2005: 85) point out, some researchers focus on 'dissimilarities between men's and women's descriptions of their managerial orientations rather than their actual behavior, as supporting the existence of sex differences in leadership behavior'. Experimental studies, showing how management and psychology students behave in simplified and artificial settings, may also say rather little of what is happening in more complicated organizational settings. Elsesser and Lever (2011: 1571) found little or no gender bias in studies of actual bosses and they conclude: 'caution must be taken in extending laboratory results based on hypothetical bosses to actual organizational scenarios'. In the latter, there are of course also difficulties. Most research is based on interviews or questionnaires, not careful participatory observations, and these may tell us more about expectations and perceptions than actual behaviour. In the worst case empirical studies may reflect gender stereotypes more than anything else. People cannot respond accurately to interview questions and questionnaires about true beliefs, values, and ambitions abstracted from expectations and norms signalling appropriate ways of responding (Alvesson 2011).

There are also some more specific problems in sorting out how male and female managers may be compared. Managerial jobs differ tremendously. Survey studies may easily overlook that classification hides diversity—and what 'managers' really do and their 'real' social relationships to 'subordinates' may be very difficult to pinpoint—and thus compare quite different phenomena.

One should also bear in mind that, even if there should be some minor or moderate average differences between how male and female managers behave in a particular historical and cultural situation, this should not obscure that there are wide variations within the two categories: some women managers may very well be seen as autocratic and there are male managers who can be described as democratic.

This may be obvious, but is ignored by a lot of writings on the subject that try to compare 'men' and 'women' in order to establish if the groups differ or not. The search for interesting differences (or similarities) easily means an overfocus on the fairly small statistical variations and emphasizing these easily means that we come to expect some sex differences between males and females in managerial jobs. But for all practical purposes, given the great variation between people, also within the camps of males and females, one cannot predict anything in terms of leadership ideals and behaviour from the sex of a specific person. Male and female managers may exhibit any version of the entire spectrum of leadership. It is best to be quite open-minded about this.

Even if there should be some evidence behind the statement, and there are various views about this, claims such as 'it is reasonable to assume that women would lead differently than men' (Carli and Eagly 2011: 109) may in all practical contexts be unhelpful. In most settings, it may be more reasonable to assume that the sex of a specific manager does not allow any accurate predictions about his or her leading, that women is not a uniform category, and that their leading may be as diverse as men's.

REFLECTIONS ON GENDER DIFFERENCES IN LEADERSHIP: ESSENTIAL TRAITS, STRUCTURAL EFFECTS, OR THE DOINGS OF GENDER?

Many authors in the gender stereotype camp, i.e. proponents for a difference in female orientations and leadership, suggest that women managers may contribute in particular in the following important aspects: communication and cooperation, affiliation and attachment, power, and intimacy and nurturance (e.g. Grant 1988). It is argued, for example, that because women have had a lot of practice from an early age in communicating and caring for others, they are often good at it. Some authors suggest that, compared with men, women possess more flexibility, more intuition, and a greater ability to be empathetic and to create a more productive work climate (e.g. Helgesen 1990); they could exercise power in a more constructive way, mobilize human resources better, encourage creativity, and change the hierarchical structures (Rosener 1990). Lipman-Blumen (1992) talks of a 'connective leadership' in which networking and shared responsibilities are central, encouraging people to connect to others and others' goals. Eagly and Carli (2007: 68) argue that female managers tend to act through more transformational leadership (gaining trust, confidence and commitment through being inspirational) and being less laissez-faire than men, meaning that 'women's approaches are the more generally effective'.

Much of what is said is somewhat imprecise and refers to a rather idealized view of the positive contributions of women. In general, though, it makes sense to stress that women have often been socialized according to different values, norms, orientations, and psychological characteristics, which could be seen either as complementary to existing values, etc., or perhaps even as replacing some of them. However, one thing is possible socialization forms and effects, another is what is being expressed in social practice. It is not given that there are essential dispositions directly resulting in certain workplace orientations and behaviours. The 'spill-over' effect from early socialization may be weak or even non-existent.

Leadership may be more about structural conditions, here broadly defined, including organizational cultures and social norms, than psychological dispositions. Certain factors and mechanisms could neutralize such different orientations associated with possible psychological dispositions and prevent women leaders from having a significant influence on organizations in a way that is representative of broad population groups. Some would argue that the scope for action for many managers is limited, and it does not then matter whether the top boss is a man or a woman. Profit-maximization and external resource dependencies may make the sex distribution in managerial jobs of limited significance, in particular in organizations operating in a highly competitive market. Some researchers downplay the role of managers for results (cf. Pfeffer 1977; Pfeffer and Salancik 1978). The general norms and practical constraints—heavy work-load, deadlines, personnel skills amount of resources, bureaucratic regulations—on managerial behaviour may also sometimes prevent any possible effects of the sex of the manager. The job rather than the sex may make the leadership. An illustrative example is a woman who had been CEO of an advertising agency: she said in an interview that she had very little time to interact with subordinates in line with various leadership ideals, as other tasks had to be prioritized. A study of project managers indicates serious time constraints and a heavy focus on operative issues (Holmberg and Tyrstrup 2010).

Certain structural effects may, however, also, differ between men and women. Subordinates, superiors, and colleagues—and the women themselves for that matter—may evaluate female managers against the background of the traditional understanding that authority is a masculine position (see Eagly et al. 1992). According to some research, when leadership or management was carried out in an autocratic way, i.e. in a way that is stereotypically masculine, females were more strongly devalued. When leadership was exercised in a gender-congruent way, females were not devalued. But males were not devalued when engaging in 'non-masculine' leadership behaviour. In terms of difficulties for female managers, there seems to be a more restricted set of options fully acceptable for female than for male managers. As Eagly et al. (1992: 18) express it; 'they "pay a price" in terms of relatively negative evaluation if they intrude on traditionally male domains by adopting male stereotypic leadership styles or occupying male-dominated leadership positions'. It is likely that this will affect the leadership of women, adjusting leadership so that negative assessments are avoided. The leadership behaviour may thus be less a matter of psychological dispositions than an outcome of social contingencies.

Another possibility of a structural nature is that the organizational socialization process associated with management positions, at least up to now, leads to the mainstreaming of candidates, so that women's specific attributes, values, and ambitions (if there were any) are lost and gender-neutral or masculine aspects are reinforced. A related possibility is that women managers are mainly recruited in such a way that only women who do not deviate from traditionally dominating organization and leadership patterns ever attain, or aspire to, management positions. The many studies indicating that no significant differences exist in leadership style would be consistent with this opinion. Organizational cultures may mainstream people in senior positions, through workplace socialization, through selective recruitment, and through structural constraints, all reducing possible dispositional effects associated with gender. Cultural influence in organizations may also involve the production or reinforcement of gender differences, possibly interacting with dispositions associated with childhood socialization. This could be the case in conservative organizations with a strong division of labour, displaying and encouraging gender differences. One would perhaps in most organizations assume that cultural standards for good management and leadership reduce gender differences and reduce the effects of any gender-specific psychologies.

Certain structural effects may also produce specific forms of gender. Kanter (1977) emphasized the role of sex ratio in organizations: the minority tends to stand out, is attributed certain orientations, and tends to respond to these, although not necessarily in a deterministic way. Different responses are possible. Women in male-dominated groups may become more 'men-like', if there is a perceived shared male way of being, or they may emphasize their femininity or at least not deviate from what may stand out as the norm. Ely (1995) found that female lawyers in a male-dominated law firm exhibited much more feminine behaviours (in terms of dressing and flirting) than in a similar firm with more equal sex distribution. The point here is not that there is a mechanical effect of structures like recruitment patterns, job demands, and sex ratios, but that these conditions may matter as much or more than essential dispositions contingent upon gender.

A third possibility is to downplay socialization/psychology as well as institutionalized conditions/structure and emphasize the level of interaction instead. This is done by the doing gender tradition (West and Zimmerman 1987, 2009). Here people are viewed as socially accountable for their gender. With a particular body (biology) come cultural expectations that one can display gender in such a way that one meets norms and expectations for being congruent with one's biology. This does not imply fixed behaviours or strong forms of reproduction of stereotypes, but subjects in interactions doing certain things responding to the environment's monitoring of oneself in terms of gender. For example, a male kindergarten teacher may engage in more physical activities with children, express some career aspirations, or something else male (Billing 1995). A female executive may display some relational concerns in interactions with subordinates, or supplement strongly result-oriented behaviour with femininity in dress or talk about family issues. What is above addressed as a response of female managers to anticipated risks of devaluations when engaging in authoritarian forms of leadership may be seen as doing gender.

For West and Zimmerman and most of their followers, 'doing gender' is viewed as an ever-present mode of responding to a normative pressure to comply with gender stereotypes. However, other authors have pointed to radical changes in gender relations, making this less compulsory and opening up for the 'un-doing of gender', i.e. resisting the anticipations (e.g. Deutsch 2007). One could push this further and say that contemporary equal opportunity norms may also mean a 'non-doing of gender', i.e. that individuals try to avoid demonstrating what is felt to indicate gender stereotypical behaviour (Alvesson and Sandberg 2013). The compulsory nature of gender-doing may not always be central any longer, at least not in some workplaces.

The various indications on differences or similarities in leadership may thus be viewed as an outcome of men and women being psychologically (somewhat) different or (broadly) similar, affected by structural and cultural contingencies and/or an interaction effect of doing gender (or minimizing the doing of gender).

Some people may want to have a robust answer here. Friends of order may ask which of the three perspectives is the correct one and what is the major tendency in terms of gender and leadership? One could imagine the three perspectives and two major tendencies (or more specific ones) combined into $3 \times 2 = 6$ options, from fixed essential traits directing females to lead in a different way to interactionism where female managers display non-stereotypical behaviour guided by an anticipation that signs on femininity may lead to negative assessments. It is hardly likely that efforts to test the six hypothesis would lead to any conclusive answers. All research is based on assumptions and all data are impregnated by theory, making all empirical findings artefacts of theoretical ideas and specific modes of inquiries. We would argue that considering all perspectives and possible tendencies is worthwhile. Gender patterns are not so homogeneous, fixed, or one-dimensional that a dispositional, structural, or interactionist perspective can be expected to explain everything and it is very likely that different dynamics can be put into operation. Most likely there are different subjects, workplaces, interactions, organizations, and times where different perspectives and tendencies are most relevant to consider.

Also individuals may be quite multidimensional in how they think their gender makes a difference. In a study of the organizational practices of female Canadian entrepreneurs Cliff *et al.* (2005) found that the female entrepreneurs tended to claim that they had specific values and orientations making them different from male entrepreneurs but their actual behaviour as manifested in organizational structures and practices showed no difference. The exact meaning of this is not obvious—more than that people's questionnaire-filling behaviour may reflect beliefs of attitudes decoupled from behaviour. But it may indicate that certain dispositions associated with background and gendered identity influence beliefs about behaviour while structural concerns may affect their actual behaviour. From a doing gender point of view one would expect people to try to add a twist of gender-norms congruent accountability. So framing what one is doing in terms of motives, values, and objectives somewhat different from men would make the female entrepreneurs not only entrepreneurial but also sufficiently feminine to comply with cultural norms. Signalling this is 'sufficient', and the actual work can be carried out in a 'non-gendered' way.

Understanding gender and leadership calls less for a strict adherence to a specific theory and law-like pattern than an ability to consider a variety of key aspects and dynamics and a variation of tendencies and empirical outcomes (Alvesson and Billing 2009).

Here it is also important to consider changes over time. Gender is not a historical constant. The women treated for hysteria by Sigmund Freud are gone. So is, at least in many groups, the idea that 'a man's got to do what a man's got to do' and 'a woman's place is in the kitchen' (at least in most advanced, relatively equal-oriented countries). Whether historical results, say from the 1970s, are valid today or tomorrow is an open question. What would (will) happen if (or when) women had better access to higher positions or when, which some people claim (e.g. Gherardi 1995), many organizations are going through transformations involving 'demasculinization', meaning that work principles are not any longer constructed in distinctly masculine ways. It is possible that studies of gender and leadership will show other results than during times where men had close to monopoly on senior positions, leadership was defined more strongly as male, and organizational principles were less likely to be interpreted as involving 'de-masculinization'.

CULTURAL CHANGES: DE-MASCULINIZATION OF LEADERSHIP IDEALS

Leadership has traditionally been constructed with a masculine subtext and dominant views on leadership have been seen as difficult to integrate with femininity (Kanter 1977; Lipman-Blumen 1992). During recent decades there has been much talk about changes under way. There are claims about moves from more bureaucratic-technocratic modes of management to more personal-ideological or normative forms. This means that issues of a more social, subjective, and involving nature are increasingly being seen as crucial. It is quite likely that changes in management and organizational practices are grossly exaggerated in many accounts of the 'post-modern world', the 'knowledge-society', 'post-fordism', post-bureaucracy, etc. (Alvesson and Thompson 2005). There is a strong premium on portraying organizational changes being rapid, drastic, progressive, and easy to identify with. The selling of an appealing present and near future is likely to be responded to positively by naive audiences wanting good news (Alvesson 2013).

But the questionable 'truth value' of broad-brushed change talk does not prevent far-reaching claims about 'changes' being in broad circulation, creating certain effects on consciousness and willingness to adapt, at least partly, to what is viewed as new truths. Perhaps the talk about the changing nature of work, 'flexibilization', and corporate changes will make it necessary to promote some skills which are more often attributed to women than to men. For some time authors on women and leadership have connected new ideals for leadership with the natural orientations of women:

> Being a good manager...is less about competitiveness, aggression, and task orientation and more about good communication, coaching and people skills, and being intuitive and flexible, all more typically or at least stereotypically associated with women.
>
> (Cooper and Lewis 1993: 41)

Emotions are also increasingly seen as significant in organizational practice as meaning, involvement, and action to some extent replace rationality, cold calculation, and separation of decision and execution (Alvesson and Berg 1992; Fineman 1993, 2000). Themes like identity, cohesion, teams, and social integration also often point in a 'non-masculine' direction. New leadership ideals include new and non-masculine labels like post-heroic, shared, and distributed leadership (Fletcher 2004). The talk about post-heroic leadership indicates that the traditional 'heroic' leadership and the traits associated with that is surpassed by a new style, which demands traits more 'rooted in feminine-linked images and wisdom about how to "grow people" in the domestic sphere' (Fletcher 2004: 651).

This general interest in new ideas on leadership appears to have accompanied the interest in feminine leadership and/or women in management. If a more participatory, non-hierarchical, flexible, and group-oriented style of management is viewed as increasingly appropriate and formulated in feminine terms (or androgynous ones, i.e. combining what is culturally/stereotypically defined as characteristics of the two genders), then women can be marketed as carriers of suitable orientations for occupying positions as managers—network orientation, a preference for participation, etc. (Billing and Alvesson 2000). Lipman-Blumen (1992: 183), for example, believes that female leadership 'contains the seeds of connective leadership, a new integrative model of leadership more suited to the dramatically changing workplace of the twenty-first century'.

Alternatively, and minimally, the new criteria for management would at least open up for females having better access to senior positions in the organization. The incongruence between a female gender role and a management role will be gone and result in less prejudice towards women leaders (Elsesser and Lever 2011). The strong 'masculine' nature of traditional management/leadership would lose some of its appeal and the work field would be a more open terrain. The problem for women, however, is that the enactment of this style of leadership is likely to be different for women than men. Women may have a harder time distinguishing what they do as something new because it looks like they are just doing what women do (Fletcher 2004: 654). Therefore women might find it harder to be recognized as doing this 'new leadership': this will be something which is expected of women anyway and hence they may not benefit from this move from more traditional masculine models to this more 'feminine' style, according to Fletcher (2004).

More generally, there are reasons to be sceptical with regard to much of the talk about radical changes taking place in organizations leading to a large need of 'female skills' or female managers (Calás and Smircich 1993). There are, as indicated above, perhaps only superficial changes behind the rhetoric, which does not prevent some acceleration in the increase of females in low-level and middle-level managerial jobs. However, the questioning of the symbolic gender of a job 'opens' up for a new way of seeing and constructing it in perhaps less stereotypical ways. Even though leadership still probably often is given a masculine meaning, the overall picture seems to be more varied and less rigid today.

The leadership ideas and styles popular during recent years are not necessarily 'pro-women', but they accord ill with traditional ideas of the masculine character of the good manager: technocratically rational, aggressive, competitive, firm, and just. At a minimum a masculine bias is reduced. Some organizations indicate that they are actually looking for certain new values which are associated with women, such as flexibility, social skills, team orientation, etc. (e.g. Holmquist 1997). However, at the same time as there are some 'pro-female' elements in 'new' forms of leadership ideals, there are also problems, to be addressed in the subsequent two sections.

FEMININE LEADERSHIP AS IDEOLOGY AND LUBRICANT

Most formulations of women exercising leadership in different ways signal positive, humanitarian, and progressive ideals that are easy to buy into. However, the ideals expressed in interview talk and questionnaires may be rather selective, and may not put significant imprints on management practice (Cliff *et al.* 2005).

There is little in the contributions on feminine leadership or women's way of leading to seriously question corporations' (shareholders, top management's) commitment to profit, growth, and other traditional goals. Capitalism and market economy, the complexity of large-scale organizations, and other constraints may mean that any genuine female orientation may not come through very clearly in most managerial contexts, at least not as long as there are only relatively few female managers. Fierce competition between companies and domination of profit motives is not abolished by female forms of leadership. In many corporate contexts it is an open issue whether there is space, within the capitalist economy, to become really significant. Organizations are mainly about the rational exploitation of labour, cost-cutting being central and reduction of the labour force often more salient than giving priority to empathy, intuition, connectiveness, and close personal relations. This is not to deny some space for and also a certain instrumental value of the latter, but most organizations give limited space for what gender-stereotypical authors emphasize as typical for women and/or a feminine type of leadership.

There are also potential problems with leadership focusing on affiliation, nurturance, intimacy, and other relational dimensions in (most) work contexts. House and Aditay (1997) say that this may involve or lead to nepotism or problems keeping distance in decisions about wages, allocation of work tasks, promotion, or rationalizations and dismissal of people. Also some studies of feminist organizations point to problems—ambiguities, inconsistencies, anxieties, and conflict—when collectives are governed by what are claimed to be feminine ideals (Martin 1987; Ashcraft 2001).

There are two problems here. (Problems are then viewed not from a corporate point of view but from an outsider position, considering the effects of leadership ideas on consciousness and practice and what may benefit employees.) One is that the feminine leadership ideals are part of and reinforce a general management ideology portraying contemporary business and working life in idealistic, flattering ways. Inherent contradictions and often not so humanistic practices are concealed by nice-sounding formulations of leadership and the promotion of women in general as champions of appealing ways of managing. (These problems are, from a management perspective, advantages and account for some of the interest in feminine leadership and female managers.)

A second problem is that the link between a certain leadership ideal and women may mean that the latter are used in those specific situations where dealing with relations and emotions are of specific importance. Thus women could very well come to provide the necessary oil to make the machinery work better; and/or their motivational and persuasive skills could be exploited as a potential tool for carrying out unpopular rationalizations more smoothly, with women acting as mediators between the top management and the workers (Calás and Smircich 1993). Kolb (1992), for example, shows how women may be inclined to work with conflicts behind the scene, doing important work but remaining invisible and potentially preventing conflicts surfacing also in cases where airing these may be positive. The writers arguing for the special contributions that women can make relatively seldom express such points of view. Consistently, women managers seem to get the 'feminine' managerial jobs, such as staff jobs, accounting and other peripheral jobs seen in relation to the more important decision-making jobs. In these jobs women are supposed to use their so-called feminine qualities and complement male managers (Laufer 2000). Brewis and Linstead (2004: 75) call human resource management functions 'female ghettos', and say that women are dominant in these functions 'because they are widely understood to have particularly well-developed people skills, to be more intuitive, more sympathetic and more effective communicators than men'.

The two problems—of ideology and ghetto—may coexist and reinforce each other. Ideological effects may motivate a lot of rhetoric and a highly selective and limited practice. The latter may back up and provide some credibility to the former, and here the involvement of female managers—'known' for their relational skills, empathy, intuition, etc. may work as promoters of the ideology while the actual work may be to handle the dynamics and effects of corporate practices substantially quite far from what is presented as feminine and positive management and leadership practice. A corporate practice emphasizing rational and efficient use of human *resources* may be supplemented by feminine forms of leadership dealing with the *human* resources, smoothening over some awareness of the costs of the rationality and efficiency involved.

FEMININITY AND FEMININE LEADERSHIP NATURALIZING 'THE (GOOD) WOMAN': FACILITATOR OR CAGE?

There is a third problem, in addition to ideology and lubrication ones, associated with the feminine leadership, affecting those that are supposed to benefit from this concept, i.e. the female (feminine) managers themselves.

The emphasis on gender difference would mean that gender division of labour appears to some extent natural. The variety of processes tying women to primarily embracing and defining themselves through what is seen as feminine are crucial for an orientation alternative to the one of dominating masculinities (Fletcher 2004). Women typically doing feminine leadership implies that there is a normal or natural inclination for them acting in this way. This may be a support for women managers—as a set of guiding principles reducing uncertainty in how to act as leaders and/or as a discourse promoting them in their careers. But it may also function as a set of restraints, as a cage similar to roles limiting ways of acting.

As feminine leadership seems to overlap with gender-stereotypical expectations, as also noticed by Eagly and Johannesen-Schmidt (2008), it is reasonable to assume that this adopted style may be an effect of expectations and pressure from other people, most notably from their subordinates that women should act in accordance with their gender. Deviations may lead to sanctions, such as when female managers acting in authoritarian ways leads to negative assessments (and more so than facing male managers) (Eagly *et al.* 1995). There is a gender division of labour within management, there seem to be certain areas which are believed to better fit women than men; this leads to Eagly *et al.* (1995: 137) concluding that 'leadership roles defined in relatively masculine terms favored male leaders and leadership roles defined in relatively feminine terms favored female leaders'.

In another study some women managers complained about these expectations which were directed to them but not the male managers (Billing 2006). Some of these women managers however rejected these expectations and refused to change their style. One of them said, 'I am a manager, not a woman manager'. Others chose to change their style according to expectations of being more soft, better listeners, etc. 'This is what they expect from a woman', one of the women managers said.

From this study it was clear that styles changed and varied, the same manager had a different style at her former workplace, and some said that their style might change throughout the day, from some times being transactional to transformational and so on. Some of these women had books on their tables about spiritual or servant leadership and this combined with massive media covering of the female managers may influence how they think they should behave and possibly how they behave. (But as said, the link here is uncertain.) Leadership styles vary, not only between different workplaces, but also sometimes within different settings, during the day. Schnur (2008) also refers to a woman manager who felt most effective when combining masculine and feminine discursive strategies.

What is more interesting perhaps than measuring differences is then to ask under what circumstances is it possible—i.e. contingent upon the encouragement or pressure the manager faces—to act in ways that may by some be seen as masculine or feminine. The question then would be 'when is the female leader?' (cf. Alvesson and Billing 2009). This then may be a result of options and expectations sometimes giving space or putting pressure on female managers to express femininity. Whether this is an expression of a natural disposition or a compliance by a normative pressure—from people around and/or the women and leadership industry—is worth considering.

So when authors believe that women and men are socialized into being feminine or masculine, to different principles and then to different characteristics, the key question is not only if they reveal, over-emphasize, or present a misleading account for what happened in childhood (socialization) and how this may effect female managers' leadership. Truth questions are not only or mainly about mirroring reality, as the latter is uncertain and debatable, partly because it is ambiguous, complex, and changing. As Foucault puts it, the problem with knowledge is not necessarily whether it reveals or distorts the truth, the problem is that it creates it. It is the effects of talk about feminine leadership that are interesting, not so much the averages of the varieties of men and women responding to various efforts to mirror how they see themselves, which values and attitudes they have, and how they say they behave at work.

CONCLUSION

Gender and leadership has received a lot of attention. Despite all the research there are various views of the possible similarities or differences between men and women in terms of the exercise of leadership (Butterfield and Grinnell 1999; Carli and Eagly 2011). In this chapter we have argued that it is not necessarily very fruitful to seek law-like patterns in terms of gender and leadership, but we point to a variety of dynamics and relations and emphasize the need to consider different possibilities in how women tend to do leadership. Arguably, this is only partly about psychological dispositions leading to certain leadership ideals and practices.

We have also pointed to problems investigating the subject matter. Apart from all other difficulties in measuring something that perhaps is so intangible, varied, and depending on social construction processes, it is important to consider how expectations, beliefs, and normative pressures on gendering (and, sometimes, to avoid gendering) are central not only for values, identities, and behaviour, but also for how people respond to requests for reporting values, identities, and behaviour. It is perhaps naive to believe that responses to questionnaires, interviews, or experiments simply mirror gender and leadership. If males and females report different leadership values or behaviour this may say more about their inclination to signal their gender-appropriateness—doing of gender in the research situation—than reflect any possible differences in 'actual' values or behaviour, external to the specific situation. Taking this seriously would mean a downplaying of dataistic projects and up-playing of modes of reasoning for understanding a variety of aspects.

A large body of literature emphasizes feminine leadership, typically celebrating this as beneficial for organizations, employees, and female managers themselves, fitting perfectly well with the needs of the contemporary economy and working life. If we side-step the possibility of this being 'true', 'half-true', an exaggeration, ambiguous, 'false', or misleading, we can assess the value of putting forward this thesis. Of course, there are some potentially positive aspects. One is the support of progressive forms of management and leadership facilitating positive working life conditions and good social relations. Another is that it boosts the chances of respect for female managers, generally linked to feminine leadership and ideals in line with this. Given the historical and to some extent ongoing domination of men and certain forms of masculinity in the leadership field, de-masculinization can be emphasized as a balancing force.

However, the notion of feminine leadership is also problematic for different reasons. First there is not much clear-cut support for the idea that women in managerial practice operate significantly different than men in these positions, and when they do it is perhaps more a response to expectations from followers or organizational norms and guidelines for leadership. These may be much more important for the manager's behaviour than the biology of a manager. As such expectations may change over time and vary between contexts, it would be problematic to naturalize and essentialize both the difference and (fixed) expectations exercising a strong and homogeneous pressure.

Secondly, it reinforces gender stereotypes so that women and men managers conveniently are seen as fit for different management functions—as mentioned, a gender division of labour in management circles is already visible. It restricts women managers in the sense that they are believed to act and behave in a special (feminine) way to be accepted, thus perhaps they may risk being exploited as emotional labourers. Female managers may appear as lubricants in organizations mainly managed through result-orientation and exploitation of labour.

Thirdly, progressive formulations of leadership promising a much more humane, conflict-free working life, where good forms of leadership based on relationality, intuition, understanding, care, empathy, etc. will lead to effectiveness as well as satisfying work conditions may simply function as ideology. This involves legitimating organizations and management and repress awareness of contradictions between different ideals and between ideal and practice. Many popular management ideas are popular because they sound appealing. Rhetorical attraction may however lead attention away from critical thinking.

REFERENCES

Alvesson, M. (1996). 'Leadership Studies: From Procedure and Abstraction to Reflexivity and Situation', *Leadership Quarterly*, 7(4): 455–85.

Alvesson, M. (2011). *Interpreting Interviews*. London: Sage.

Alvesson, M. (2013). *The Triumph of Emptiness*. Oxford: Oxford University Press.

Alvesson, M., and Berg, P. O. (1992). *Corporate Culture and Organizational Symbolism*. Berlin and New York: de Gruyter.

Alvesson, M., and Billing, Y.D. (2009). *Understanding Gender and Organizations* (2nd edn). London: Sage.

Alvesson, M. and Sandberg, J. (2013). *Constructing Research Questions*. London: Sage.

Alvesson, M., and Thompson, P. (2005). 'Post-Bureaucracy?', in S. Ackroyd *et al.* (eds), *The Oxford Handbook of Work and Organization*, 485–507. Oxford: Oxford University Press.

Ashcraft, K. (2001) 'Organized Dissonance: Feminist Bureaucracy as Hybrid Form', *Academy of Management Journal*, 44(6): 1301–22.

Bayes, J. (1987). 'Do Female Managers in Public Bureaucracies Manage with a Different Voice?', paper presented at the 3rd international interdisciplinary congress on women. Dublin, 6–10 July.

Billing, Y. D. (1991). *Køn, karriere, familie*. Copenhagen: Juristog Økonomforbundets Forlag.

Billing, Y. D. (1995). 'A Nice Union', Working paper. Dept. of Sociology, University of Copenhagen.

Billing, Y. D. (2006). *Viljan till makt* (The Will to Power). Lund: Studentlitteratur.

Billing, Y. D., and Alvesson, M. (1994). *Gender, Managers and Organizations*. Berlin and New York: de Gruyter.

Billing, Y. D., and Alvesson, M. (2000). 'Questioning the Notion of Female Leadership: A Critical Perspective on the Gender Labelling of Leadership', *Gender, Work and Organization*, 7(3): 144–57.

Brewis, J., and Linstead, S. (2004). 'Gender and Management', in S. Linstead, L. Fulop, and S. Lilley (eds), *Management and Organization: A Critical Text*, 74–92. New York. Palgrave Macmillan.

Butterfield, D. A., and Grinnell, J. P. (1999). 'Re-viewing Gender, Leadership, and Managerial Behaviour: Do Three Decades of Research Tell us Anything?', in G. N. Powell (ed.), *Handbook of Gender and Work*, 223–239. Thousand Oaks, CA: Sage.

Calás, M., and Smircich, L. (1991). 'Voicing Seduction to Silence Leadership', *Organization Studies*, 12(4): 567–601.

Calás, M., and Smircich, L. (1993). 'Dangerous Liaisons: The "Feminine-in-Management" Meets "Globalization"', *Business Horizons* (Mar.–Apr.): 73–83.

Carli, L., and Eagly, A. (2011). 'Gender and Leadership', in A. Bryman *et al.* (eds), *Handbook of Leadership Studies*, 103–117. London: Sage.

Chia, R. (1995). 'From Modern to Postmodern Organizational Analysis', *Organization Studies*, 16(4): 579–604.

Cliff, J., Langton, N., and Aldrich, H. (2005). 'Walking the Talk? Gendered Rhetoric vs. Action in Small Firms', *Organization Studies*, 26(1): 63–91.

Cockburn, C. (1991). *In the Way of Women*. London: Macmillan.

Collinson, D., and Hearn, J. (1996). 'Breaking the Silence: On Men, Masculinities and Managements', in D. Collinson and J. Hearn (eds), *Men as Managers, Managers as Men*, 1–24. London: Sage.

Cooper, C. L., and Lewis, S. (1993). *The Workplace Revolution: Managing Today's Dual Career Families*. London: Kogan Page.

Davidson, M., and Burke, R. J. (2000). *Women in Management*. London: Sage.

Deutsch, F. M. (2007). 'Undoing Gender', *Gender and Society*, 21(1): 106–27.

Eagly, A., and Carli, L. (2007). 'Women and the Labyrinth of Leadership', *Harvard Business Review* (Sept.): 62–71.

Eagly, A., and Johannesen-Schmidt, M. A. (2008). 'Leadership Style Matters: The Small But Important Style Differences between Male and Female Leaders', in D. Bilimoria and S. K. Piderit (eds), *Handbook on Women in Business and Management*, 279–303. Northampton, MA: Edward Elgar.

Eagly, A., and Johnson, B. (1990). 'Gender and Leadership Style: A Meta-Analysis', *Psychological Bulletin*, 108(2): 233–56.

Eagly, A., Makhijani, M., and Klonsky, B. (1992). 'Gender and the Evaluation of Leaders: A Meta-Analysis', *Psychological Bulletin*, 111(1): 3–22.

Eagly, A., Karau, S. J., and Makhijani, M. G. (1995). 'Gender and the Effectivenes of Leaders: A Meta-Analysis', *Psychological Bulletin*, 117: 125–45.

Elsesser, K. M., and Lever, J. (2011). 'Does Gender Bias Against Female Leaders Persist? Quantitative and Qualitative Data from a Large-Scale Survey', *Human Relations*, 64(12): 1555–78.

Ely, R. (1995). 'The Power in Demography: Women's Social Construction of Gender Identity at Work', *Academy of Management Journal*, 38(3): 589–634.

Ely, R., and Padavic, I. (2007). 'A Feminist Analysis of Organizational Research on Sex Differences', *Academy of Managament Review*, 32: 1121–43.

Ferguson, K. (1994). 'On Bringing More Theory, More Voices and More Politics to the Study of Organization', *Organization*, 1: 81–99.

Fineman, S. (1993). 'Organizations as Emotional Arenas', in S. Fineman (ed.), *Emotions in Organizations*, 9–35. London: Sage.

Fineman, S. (2000). *Emotion in Organizations* (2nd edn). London: Sage.

Fletcher, J. (2004). 'The Paradox of Postheroic Leadership: An Essay on Gender, Power and Transformational Change', *Leadership Quarterly*, 15(5): 647–61.

Frankenheuser, M. (1993). *Kvinnligt, manligt, stressigt*. Höganäs: Bra Böcker/Nike.

Gherardi, S. (1995). *Gender, Symbolism and Organizational Cultures*. London: Sage.

Grant, J. (1988). 'Women as Managers: What can they Offer to Organizations?', *Organizational Dynamics*, 1: 56–63.

Guillaume, C., and Pochic, J. (2009). 'What would you Sacrifice? Access to Top Management and the Work–Life Balance', *Gender, Work and Organizations*, 16(1): 14–36.

Haslam, A. (2004). *Psychology of Organizations* (2nd edn). London: Sage.

Hearn, J., and Parkin, W. (1986/7). 'Women, Men and Leadership: A Critical Review of Assumptions, Practices and Change in the Industrialized Nations', *International Studies of Management and Organization*, 16: 3–4.

Helgesen, S. (1990). *The Female Advantage*. New York: Doubleday.

Holmberg, and Tyrstrup, (2010). 'Well Then—What Now? An Everyday Approach to Managerial Leadership', *Leadership*, 6(4): 353–72.

Holmquist, C. (1997). 'Den ömma bödeln', in E. Sundin (ed.), *Om makt och kön i spåren av offentliga organisationers omvandling*. SOU 83. Stockholm: Fritzes.

House, R., and Aditay, R. (1997). 'The Social Scientific Study of Leadership: Quo Vadis?', *Journal of Management*, 23(3): 409–73.

Kanter, R. M. (1977). *Men and Women of the Corporation*. New York: Basic Books.

Kolb, D. (1992). 'Women's Work: Peacemaking in Organizations', in D. Kolb and J. Bartunek (eds), *Hidden Conflict in Organizations*, 63–91. Newbury Park, CA: Sage.

Kovalainen, A. (1990). 'How do Male and Female Managers in Banking View their Work Roles and their Subordinates', *Scandinavian Journal of Management*, 6: 143–59.

Laufer, J. (2000). 'French Women Managers: A Search for Equality But Enduring Differences', in M. J. Davidson and R. J. Burke (eds), *Women in Management*, 26–39. London: Sage.

Lindgren, G. (1996). 'Broderskapets logik', *Kvinnovetenskaplig tidskrift*, 17(1): 4–14.

Lipman-Blumen, J. (1992). 'Connective Leadership: Female Leadership Styles in the 21st-Century Workplace', *Sociological Perspectives*, 35(1): 183–203.

Loden, M. (1986). *Feminine Leadership or How to Succeed in Business without Being One of the Boys*. New York: Time Books.

Marshall, J. (1984). *Women Managers: Travellers in a Male World*. Chichester: Wiley.

Martin, P. Y. (1985). 'Group Sex Composition in Work Organizations: A Structural-Normative Model', *Research in the Sociology of Organizations*, 4: 311–49.

Martin, J. (1987). 'The Black Hole. Ambiguity in Organizational Cultures', paper presented at the 3rd International Conference on Organizational Symbolism and Corporate Culture. Milan, June.

Martin, P. Y. (1993). 'Feminist Practice in Organizations: Implications for Management', in E. A. Fagenson (ed.), *Women in Management: Trends, Issues, and Challenges in Managerial Diversity*, 274–296. Thousand Oaks, CA: Sage.

Morrison, A., and Von Glinow, M. A. (1990). 'Women and Minorities in Management', *American Psychologist*, 45(2): 200–8.

Pfeffer, J. (1977). 'The Ambiguity of Leadership', *Academy of Management Review*, 2: 104–12.

Pfeffer, J., and Salancik, G. R. (1978). *The External Control of Organizations. A Resource Dependence Perspective*. New York, NY: Harper and Row.

Powell, G. (1999). 'Reflections on the Glass Ceiling', in G. Powell (ed.) *Handbook of Gender and Work*, 325–346. London: Sage.

Roper, M. (1996). 'Seduction and Succession: Circuits of Homosocial Desire in Management', in D. Collinson and J. Hearn (eds), *Men as Managers, Managers as Men*, 210–238. London: Sage.

Rosener, J. (1990). 'Ways Women Lead', *Harvard Business Review*, 68(6): 119–25.

Ryan, M., and Haslam, S. A. (2007). 'The Glass Cliff: Exploring the Dynamics Surrounding Women's Appointment to Precarious Leadership Positions', *Academy of Management Review*, 32(4): 1292–5.

Schein, V. (1973). 'The Relationship between Sex Role Stereotypes and Requisite Management Characteristics among Female Managers', *Journal of Applied Psychology*, 57: 89–105.

Schein, V. E. (1975). 'Relationships between Sex Role Stereotypes and Requisite Management Characteristics among Female Managers'. *Journal of Applied Psychology*, 60(3): 340–4.

Schnur, S. (2008). 'Surviving in a Man's World with a Sense of Humour: An Analysis of Women Leaders' Use of Humour at Work', *Leadership*, 4(3): 299–321.

West, C., and Zimmerman, D. (1987). 'Doing Gender', *Gender and Society*, 1: 125–51.

West, C. and Zimmerman, D. (2009). 'Accounting for Doing Gender', *Gender and Society*, 23(1): 112–22.

Wilson, E. (1998). 'Gendered Career Paths', *Personnel Review*, 27(5): 396–411.

CHAPTER 10

NEGATIVE INTRA-GENDER RELATIONS BETWEEN WOMEN

Friendship, Competition, and Female Misogyny

SHARON MAVIN, JANNINE WILLIAMS, AND GINA GRANDY

INTRODUCTION

BROADBRIDGE and Simpson (2011) note that key aspects of gendered management and organization may be increasingly difficult to detect, arguing for research to 'reveal' (Lewis and Simpson 2010) hidden aspects of gender and the processes of concealment within norms, practices, and values. Negative relations between women in organizations have been highlighted in different arenas since the 1960s (e.g. Goldberg 1968; Staines *et al.* 1973; Abramson 1975; Legge 1987; Nicolson 1996) but remain under-researched in management and organization studies. The following chapter offers an initial conceptual framework of women's negative intra-gender relations in organizations. The framework aims to 'reveal' some of the hidden aspects of gender and to contribute to a greater understanding of how gendered organizing contexts construct negative relations between women, and how such relations emerge through everyday organizing. In developing the framework we draw upon research from evolutionary and social psychology, sociology, management, and organization studies. Specifically we draw upon women doing gender well (in congruence with sex category), while simultaneously doing gender differently (Mavin and Grandy 2011); gendered contexts; homophily (Lazarsfeld and Merton 1954) and homosociality (Gruenfeld and Tiedens 2005); women's intra-gender competition (Campbell 2004) and processes of female misogyny (Mavin 2006a, 2006b). Our contribution focuses upon revealing hidden forms of gender in action in organizations and highlights how gendered contexts and organizing processes which impact upon women's experiences and advancement are entangled with and facilitate women's social relationships at work.

To raise women's negative intra-relations at work can be to speak the unspeakable, almost a feminist taboo, which poses risk to the speaker(s). Drawing attention to women's negative intra-gender relations in organizations also risks the reduction of the problem to individual women, rather than problematizing social relations. Negative intra-gender relations between women at work was highlighted as a challenge to women's progress by Mavin (2006a, 2006b, 2008), contributing to the maintenance of the gendered status quo and hegemonic masculinity (Connell and Messerschmidt 2005) in organizations. We have argued elsewhere (Mavin 2008; Mavin and Grandy 2012; Mavin and Williams forthcoming) that senior women in management and leadership face an oxymoron: they face expectations of positive solidarity behaviours from other women and requirements to take up the 'women in management mantle' on behalf of women in the organization, whilst in parallel they are negatively evaluated for performing masculinities, through the use of Queen Bee label (Staines *et al.* 1973; Abramson 1975). Solidarity or sisterhood behaviours (Mavin 2006a) between women are often seen as positive enablers. As numbers swell, it is suggested women are more likely to form allegiances, coalitions, and affect the culture of the organization (Kanter 1977). However, women perceived as Queen Bees are argued to disassociate themselves from their gender to survive and thrive in masculine work contexts (Derks *et al.* 2011). Individual women as Queen Bees are then positioned as 'the problem', perceived as unsupportive of other women and interpreted as attempting to hold on to power (Mavin 2008). We contend that solidarity behaviour expectations and Queen Bee evaluations are examples of women's negative intra-gender relations facilitated within gendered contexts and gendered orders. Women's experiences are complex within these gendered contexts, including the chasm in social relations with other women which requires exploration (Mavin and Williams forthcoming).

As women move into senior positions they disrupt gendered expectations and embedded gender stereotypes supporting associations of management as male, and men as managers and 'bosses', to which both men and *women* might negatively respond (Mavin 2006a, 2006b). The possibility of negative intra-relations between women can also form in horizontal as well as in vertical relationships between women at work (Gutek *et al.* 1988). These problematic relations, possibly impacted by low gender demography (Ely 1994), contribute to gendered organizations and constrain opportunities for women to be 'otherwise'. Women's intra-gender competition and processes of female misogyny (Mavin 2006a, 2006b) are further aspects of social relations between women at work, so that, contrary to gender stereotypes, women are often not friends and do not always cooperate or support each other, regardless of their hierarchical positioning. Rather women can be hostile towards women and in particular women in senior positions.

Chesler (2001: 2) contends women 'do not like, trust, respect or find their [other women] statements to be credible. To the extent that women are oppressed, we have also internalized the prevailing misogynist ideology which we uphold both in order to survive and in order to improve our own individual positions vis-à-vis all other women.' Gutek *et al.* (1988) argue that women's long history as a subordinate group has resulted in women learning to survive in a world structured by the dominant group's definitions, rules, rewards, and punishments, and therefore 'the only realistic response of many women to such overwhelming institutionally based macro-manipulation is micro-manipulation, the use of interpersonal behaviours and practices to influence, if not control the balance of power' (Lipman-Blumen 1984: 30). However, theoretical development of this argument has been limited.

Organizations have been characterised by patterns of interaction which (whether intentionally or unintentionally formed) contribute to homogeneous group structures, of which gender is one dimension (Gruenfeld and Tiedens 2005). Such constructed patterns shore up social homogeneity and hierarchical structures and are argued to contribute to organizational members' sense of security (Kanter 1977; Camussi and Leccardi 2005; Gruenfeld and Tiedens 2005). Homophily (the social process of friendship) (Lazarsfeld and Merton 1954) and homosociality (a general orientation to associate with people like oneself) (Gruenfeld and Tiedens 2005) have contributed to research investigating the gendered experiences of those in management positions through a focus upon social capital and network theory (e.g. Benschop 2009). However, a specific focus on friendship as a social process and intra-gender friendships has been lacking. In theorizing women's negative intra-gender relations, we draw upon an assumption that within work organizations and in senior positions, men experience greater opportunities for, and relationships with, others (men) and that this impacts positively on their experiences (Collinson and Hearn 2005), whilst women's workplace homophilous friendships and homosocial relations with other women are problematic and remain under-researched. Further, we integrate discussions on intra-gender competition and female misogyny (Mavin 2006*a*) to illuminate the difficulties that women may experience in accepting intra-gender differences. In turn, this highlights a greater understanding of how women negotiate organization and management within the prevailing patriarchal social order (Mavin 2006*a*, 2006*b*, 2008).

The chapter begins by outlining our understanding of gender and gendered contexts facilitating negative relations between women. This is followed by a discussion of homophily (Lazarsfeld and Merton 1954), homosociality (Gruenfeld and Tiedens 2005), and women's intra-gender competition and female misogyny (Mavin 2006*a*, 2006*b*). The conceptual framework of women's negative intra-gender relations is then discussed and summarized, to consider how negative relations between women manifest and impact on women's potential, followed by emerging questions offered to frame future research.

DOING GENDER WELL AND DIFFERENTLY

Doing gender well and differently is the first aspect for consideration within our conceptual framework, aimed to account for women's negative intra-relations in organizations.

In outlining our position on gender, we build upon current research on doing gender well, or appropriately in congruence with sex category (Mavin and Grandy 2011, 2012), and redoing or undoing gender. We contend that gender can be done well and differently through simultaneous, multiple enactments of femininity and masculinity (Mavin and Grandy 2011, 2012). In doing so, we agree with Billing (2011) who notes that gender is a fluid concept that shifts over time and place. However, we question optimistic claims that gender can ever be undone. Rather, undoing gender is really not undoing gender but redoing or doing gender differently (Messerschmidt 2009; West and Zimmerman 2009; Kelan 2010). We explicitly incorporate sex category into our understanding of doing gender, as we believe it cannot be ignored in experiences of doing gender. This does not mean that gender binaries cannot be challenged or unsettled, rather that the binary divide continues to constrain and restrict how men and women do gender.

Gender in organization studies research has progressed from essentialist perspectives which understand gender as the property of women and men manifested through ascribed individual traits, through to appreciating gender as a process. The distinction between physiological differences and social norms continues to be debated and problematized (Acker 1992), whilst intersectional studies highlight the salience of other social categories for gender relations, such as class and race (Acker 2000; Valentine 2007; Holvino 2010). For us, rather than being the property of a person, gender is always being redefined and negotiated through everyday practices and situations (Poggio 2006). Gender is a 'complex of socially guided perceptual and interactional and micropolitical activities that cast particular pursuits as expressions of masculine and feminine "natures"' (West and Zimmerman 1987: 126) and as such is a routine accomplishment (West and Zimmerman 1987).

West and Zimmerman (1987) note the distinction between gender, sex categorization, and sex. Sex is understood to be the application of biological criteria, which has been socially agreed upon. People are then placed in a sex category as sex criteria are applied to them, which is evaluated in everyday life through expectations of particular identificatory displays, indicating that one is a member of a particular sex. Such an understanding appreciates that when people do gender, they are already categorized by sex and gender is 'the activity of managing situated conduct in light of normative conceptions of attitudes and activities appropriate for one's sex category' (West and Zimmerman 2002: 5). Gender is not a possession but something achieved or accomplished through interaction, within particular social contexts, against behaviours understood to be appropriate for females or males (West and Zimmerman 1987; Messerschmidt 2009). Thus perceptions of a sex category are a facet of doing gender, as people are assessed as incumbents of a sex category. For example, as Messerschmidt (2009) argues, females who behave in ways which are considered to be masculine, may find their doing of gender is rejected as incongruent with their perceived sex category. In organizations, the use of the Queen Bee label can therefore be viewed as a sexist evaluation of senior women who perform masculinities (Mavin 2008), whereby there is a perceived incongruence between the sex category of so-called 'Queen Bees' and how they enact leadership (e.g. agentic rather than communal style).

Underpinning much doing gender research is an assumed gender binary of male/ female, masculinities/femininities (Kelan 2010). One dimension in the development of organization studies gender research has been to question the salience of binary thinking, with arguments ranging from undoing gender to destabilizing the binary (Butler 1990, 1999), although others argue that gender is done well (Mavin and Grandy 2011, 2012), differently, or redone, rather than undone (Messerschmidt 2009; West and Zimmerman 2009; Kelan 2010). Multiplicity is also argued to be one form of 'breaking' the gender binary. Linstead and Pullen (2006) argue that the gender binary can be disrupted, as gender is a social and cultural practice which is performed and practised. Performances and practices which switch positions can disrupt the binary, as embodied experiences of gender are more fluid than a binary, more akin to a rhizome (Linstead and Pullen 2006). Masculinity and femininity may therefore be co-present and simultaneous (Linstead and Pullen 2006).

While we acknowledge the social process and fluidity of gender as individual subjectivities, we cannot deny the existence of the binary divide between men and women. As Gherardi (1996) argues, the dominant symbolic order of gender is of a binary of masculinity and femininity, which should be understood as a learnt understanding of social relations (Baxter and Hughes 2004). As women's intra-gender relations are under-researched, we contend that it is impossible to analyse gender in organizations without interrogating the binary divide against which men and women as groups, and as individuals, are evaluated in organizations. The gender binary therefore cannot be ignored in theory or in practice for women in organization, as it constrains and restricts how we do gender. At the same time, we contend that an approach that takes into account efforts to do gender well and differently (Mavin and Grandy 2011, 2012) offers opportunities to recognize the fluid, contradictory, and indefinite nature of doing gender. At the heart of this approach is multiplicity, whereby women (or men) can do gender differently through simultaneous, multiple enactments of femininity and masculinity and as a result it may open up new possibilities for unsettling gender binaries over time.

Drawing from West and Zimmerman (1987) and Messerschmidt's (2009) assertion that individuals are held accountable to sex category in the doing of gender and our previous work (Mavin and Grandy 2011, 2012), we delineate the conceptualization of not simply doing gender but 'doing gender well' or appropriately in congruence with sex category and explain doing gender well in this way:

> For a woman to do gender well or appropriately, as evaluated against and accountable to her sex category, she performs expected feminine behavior through a body that is socially perceived to be female. For a man, to do gender well or appropriately, as evaluated against and accountable to his sex category, he performs expected masculine behavior, through a body that is socially perceived to be male. Thus there is congruence and balance between the perceived sex category and gender behavior, and femininity (or masculinity) is validated.

(Mavin and Grandy 2011: 3–4)

In earlier work illustrative examples were provided of how women can do gender well and differently (Mavin and Grandy 2011, 2012). Exotic dancers, for example, emphasize the sense of empowerment (independence, sexual exploration, and freedom), exploitation (objectification), temporality (means to an end, ambitious goals), professionalism (strict rules), and moral compass (faithful partner) afforded through the work (Mavin and Grandy 2011). While they do gender well, their efforts to legitimize and professionalize the work can be viewed as attempts, albeit those more aligned with masculinity, to simultaneously do gender differently. In outlining our position of doing gender well and doing gender differently, we contend that individuals can perform, either consciously or subconsciously, exaggerated expressions of femininity (or masculinity) while simultaneously performing alternative expressions of femininity or masculinity (Mavin and Grandy 2011, 2012). Further, women can also be perceived and evaluated by others as the ' "right" kind of feminine and the "wrong" kind of feminine (or masculine), even as part of simultaneous enactments of masculinity and femininity' (Mavin and Grandy 2012: 224).

Following this, our assumption is that women's negative intra-gender relations are influenced in part by women's reactions to other women when they do gender well and differently and/or are the 'wrong kind of feminine' (Mavin and Grandy 2012) according to gendered expectations and the context they are working within. Therefore while women may do gender well and differently simultaneously, thus opening up possibilities for disrupting the gender binary, this doing of gender takes place within gendered contexts and has implications for women's intra-gender relations. The doing gender well and differently aspect of the conceptual framework is shown in Figure 10.1.

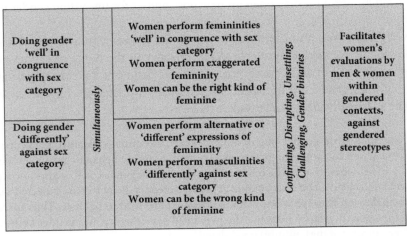

FIGURE 10.1 Doing gender well and differently

Source: Mavin and Grandy (2011, 2012).

GENDERED CONTEXTS: IMPLICATIONS FOR WOMEN'S INTER-GENDER RELATIONS

In conceptualising gendered contexts which facilitate women's negative intra-gender relations we focus upon four areas. First, we focus on the double bind experienced by women in organizations, in that they are evaluated as women and as managers or workers. Secondly, we explore the possibility of decoupling evaluations of men and women from the gender binary. However, we argue this can only be the case if everyone is able to decouple, not just some. Thirdly, we outline levels of masculine hegemony which require recognition in understanding social relations, and fourthly, we outline patriarchy as a backcloth to gendered contexts.

Research has demonstrated that work, and particularly management work, is historically and culturally associated with masculinity and men, which has contributed to establishing a gendered order (Connell 1987; Gherardi 1994) in organizations. Within this order, the ideal worker, against whom all workers are assessed, is associated with masculinity and men (Acker 1992). Men have been argued to feel comfortable with these prevailing attitudes and norms, as they perceive them as gender neutral (Simpson 1997). However, senior women in organizations face the double bind dilemma (Gherardi 1994) of expectations both of behaviour appropriate to their perceived gender role and behaviour expected of managers, the former associated with femininities, the latter with masculinities. This leads to complexity for women in negotiating organizing contexts (cf. Martin 2003; Mavin 2006a, 2006b, 2008; Eagly and Carli 2007), reflecting our position on doing gender and the challenges of gender contexts and hierarchies for women's relations.

Recent research suggests a fragmentation of gendered assumptions which contribute to a displacement of the male norm in organizing management in some contexts, for example Nordic countries (Billing 2011). Billing's (2011) position is that women managers may experience congruence with the role of manager, if they have decoupled masculinities and the male body from competencies or values associated with management. Such opportunities arise through the erosion of strong gender divisions in expectations of management or leadership (Billing 2011). However, whilst attempts may be made by some organizational members to decouple femininity and masculinity from particular behaviours and values, we contend that, unless the majority or all organizational members 'buy into' this removal of gender divisions and change their expectations of others, in interaction, then people will continue to be accountable to normative gendered expectations that draw upon sex role categorisations (Messerschmidt 2009). This reflects much doing gender research which continues to draw upon a gender binary (Kelan 2010), even when attempting to undo it (Mavin and Grandy 2011, 2012). Moreover, organizational members may also gender themselves in order to maintain a gender identity (Billing 2011). This has implications for women and senior women in organizations as their intra-gender relations are enacted within this context.

Knights and Kerfoot (2004: 446) explain that masculinity shapes 'representational knowledge' through pervasive and tacit masculine discourses which structure behaviour in organizations and which can have deleterious effects for all organizational members. Disrupting and problematizing such forms of masculinity and repressive effects for all organizational members, requires the critique of background assumptions (read masculine hegemony) which render such discourses intelligible (Knights and Kerfoot 2004). Despite the pervasiveness of the masculine hegemony, it has been argued that, in doing gender, there is space for agency (Connell and Messerschmidt 2005; Benschop 2009), in how organizational members relate to the gender socialization and patriarchy which contribute to structural restraints for members (Benschop 2009). This reflects a view that gender is constructed at micro (everyday interactions), meso (organizational), and macro (societal) level, and macro level analysis alone cannot predict what occurs or how this is achieved at the meso or micro level of activities in organizations (Billing 2011).

The concern with privileging macro level analysis at the expense of meso and micro level activities in the doing of gender also reflects a critique of patriarchy (see e.g. Walby 1989) as an explanatory framework for all gendered relations in all contexts. Critiques of patriarchy contributed to a move in research away from assuming ahistorical or universal approaches (Walby 1989), to suggest multiple hierarchies of relations between women and men. As Connell and Messerschmidt (2005) note, there are geographies of masculinities which operate and can be analysed at and between the connections of global, regional, and local levels. However, for us, patriarchy remains an important analytical category (Thornley and Thörnqvist 2009); it is useful to understand how 'gender is implicated in all social processes' (Acker 1989: 239) and in the reproduction of women's negated social positions. Patriarchy continues to be drawn upon to research different contexts in gender and organization literature (e.g. Dean 2008; Ford and Harding 2010). We contend that it is important to acknowledge the potential for multiplicity in local interactions, whilst also being cognizant of the broader social context in contributing to shaping these interactions.

Taken forward into our conceptual framework is an understanding that gendered organizational relations play out against gendered contexts and background assumptions, such as patriarchy, which contribute to maintaining assumptions of masculine hierarchical superiority (Knights and Kerfoot 2004) in organizing. Within such broader contextual influences, it is also acknowledged that the ambiguity and instability (Calás and Smircich 1992) or fluidization (Camussi and Leccardi 2005) of gender roles at work requires research into gender relations to pay careful attention to the possibilities of being otherwise. For us, women at work may draw upon agency, as well as the possibilities of being otherwise: doing gender well and differently, enacting simultaneous masculinities and femininities. However, as we outlined earlier, women experience complexities, perform gender well and differently simultaneously, but are evaluated by men and other women within gendered contexts, comprising patriarchy, masculine hegemony, the gender binary, and structures and hierarchies built upon masculine power. These gendered contexts shown in Figure 10.2, contribute to and shape women's intra-gender work relationships and work experiences.

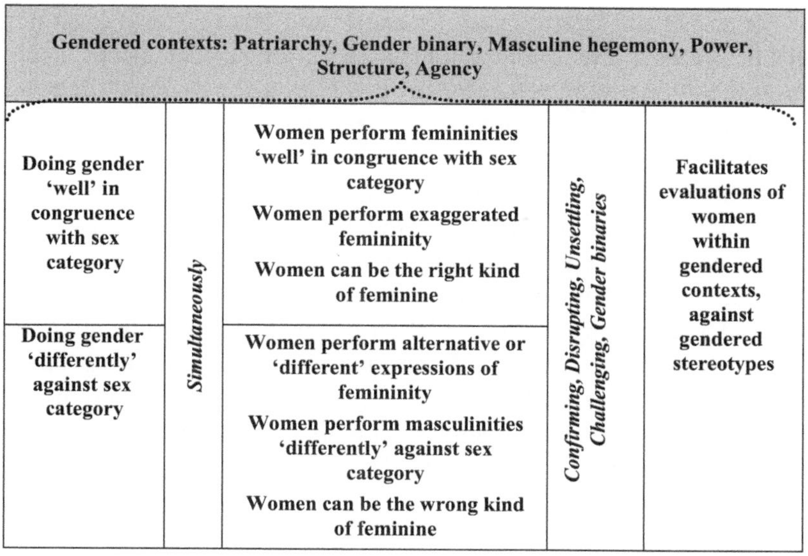

Gendered contexts: Patriarchy, Gender binary, Masculine hegemony, Power, Structure, Agency				
Doing gender 'well' in congruence with sex category	*Simultaneously*	Women perform femininities 'well' in congruence with sex category Women perform exaggerated femininity Women can be the right kind of feminine	*Confirming, Disrupting, Unsettling, Challenging, Gender binaries*	Facilitates evaluations of women within gendered contexts, against gendered stereotypes
Doing gender 'differently' against sex category		Women perform alternative or 'different' expressions of femininity Women perform masculinities 'differently' against sex category Women can be the wrong kind of feminine		

FIGURE 10.2 Gendered contexts for women's intra-gender relations

In moving to explore women's intra-gender relations within these contexts we look to the concepts of homophily and homosociality in organizations. Our assumptions are that women engage differently in these social processes, and/or that men's friendships and homosociality are more powerful and embedded within patriarchal gendered contexts.

HOMOPHILY, HOMOSOCIALITY, AND STUDIES OF GENDER

In developing an initial conceptual framework to account for women's negative intra-gender relations, we draw upon Lazarsfeld and Merton's (1954) concept of homophily, as social processes of friendship. Lazarsfeld and Merton (1954: 65) suggested that early sociological studies of friendship which emphasised 'who makes friends with whom?' were supplemented by considering the role of attitudes, values, and social status (such as race, sex, class, social standing) and the social processes which contribute to such friendship formation, alongside a concern for how friendships are maintained or disrupted. It is the '*processes* through which *social relations* interact with *cultural values* to produce diverse patterns of friendship' (Lazarsfeld and Merton 1954: 20, original emphasis) which is of interest to women's intra-gender relations. It is recognized that different levels of homophily within particular contexts and cultural values can produce functional or dysfunctional consequences which subsequently affect friendship patterns (Lazarsfeld and Merton 1954). Lazarsfeld and Merton (1954) suggest that dysfunctional or excessive levels of homophilous (friendship) relationships between men may have dysfunctional implications for an organization, for example, affecting recruitment decisions. However such intra-gender relations between women have received less attention.

When examining where, how, and when friendships form, it becomes apparent that friendship is not purely an individual or dyadic affair (Eve 2002), as there is configurational logic and strong structuring at play, whereby 'who becomes a friend seems to be determined not solely by individual attraction but above all by the potential for enriching and maintaining another relationship which is already important' (Eve 2002: 401), emphasizing the structural importance of workplace friendships and who associates with whom at work. Also, in making some friendships we distance ourselves from others, 'marking the social boundary of one's separate identity' (Eve 2002: 401). This 'marking off' a social area or boundary is part of friendship processes, for example, 'the despised colleagues which may constitute one of the main contents of the relationship via gossip, plotting, complaining, joking' (Eve 2002: 401). We argue that this 'marking the social boundary' process is also evident through expressions of women's intra-gender competitive strategies and in processes of female misogyny between women and therefore is important to explore within women's relations.

From a psychological perspective, stereotypes about same-sex friendships abound and are often contradictory (Calwell and Peplau 1982). Tiger (1969) notes that male superiority in friendship reigns, with men better able than women to form lasting bonds with same-sex partners. Donelson and Gullahorn (1977) argued that women are incapable of friendships and some women accept this view. Bell (1979), however, argued that men's friendships are superficial and lack the intimacy and emotion of women's friendships, noting that the friendships of women are more frequent, significant, and more interpersonally involved than those commonly found amongst men. Moreover, there is an issue of defining friendship and how research participants understand the term, resulting in conflicting quantitative studies of how many same-sex friendships men and women have (Calwell and Peplau 1982).

In terms of intimacy, research shows that women's friendships are affectively richer: women are more likely to have intimate confidantes and more intimate friends than men, as men have difficulty with emotional intimacy and are emotionally inexpressive, disclosing less and receiving less personal information than women (Calwell and Peplau 1982). Here the male sex role is considered as limiting emotional sharing in male–male relationships (David and Brannon 1976; Pleck 1976). Weiss and Lowenthal (1975) found that women emphasize reciprocity through help and support, while men emphasize similarity through shared experiences. This is supported by Calwell and Peplau (1982) who found women's friendships oriented towards personal sharing of information and men's friendships emphasizing joint activities, 'because the male sex role restricts men's self-disclosure to other men' (Calwell and Peplau 1982: 731). However this view, that men are unemotional, inexpressive, and impersonal, has been challenged by Keisling (2005) in his project on homosocial desire, who contends that men clearly form friendships and larger friendship groups, and must therefore manage to 'connect' with one another personally and emotionally. Keisling (2005) relates men's friendship (homophily) to male solidarity and the 'old boys club' (homosociality), as ways that men make themselves more attractive to other men (homosocial desire), arguing that these play a role in maintaining men's power as men connect with one another within a context of competition. Thus men's friendships are structurally powerful and contribute to homosociality (Gruenfeld and Tiedens 2005), as the preference to associate with people like oneself.

Homosociality is therefore a further element to consider in terms of women's intra-gender relations within gendered contexts of organizations. Homosociality has informed the broader gender literature by focusing upon men, masculinities, and male homosociality, exploring how men in management reproduce masculine hegemony (including misogynistic attitudes such as the subjugation of women) (Gregory 2009), which arguably perpetuates masculine work cultures (Bird 1996). Homosociality has been positioned as a practice (Collinson and Hearn 1996) or enactment (Worts *et al.* 2007) of masculinity, recognized to shape organizing norms and performance criteria. Homosociality is understood to be 'done' in two key ways in senior management, 'redefining competence and doing hierarchy, resulting in a preference for certain men and the exclusion of women' (Holgersson 2012: 1). This is identified as an unreflexive preference of men in organizations, so that homosociality and gender discrimination are two sides of the same coin (Holgersson 2012).

Thus men's homosociality is acknowledged to be pervasive and instrumental in maintaining power in organizations but the intimacy and emotional basis of their friendships is contested. Women's homophily in general is seen as more intimate and emotional, but their homosociality and instrumentality in work organizations is less powerful. We contend that these intra-gender relations require further research.

Wider network studies offer additional insights into how homophily has been considered. For example, McPherson *et al.*'s (2001) review of network studies, sex, and gender suggests that high levels of occupational segregation contribute to minority members' networks being more heterophilous than those of majority members, which are characterized by homophilous relations, the latter more so around friendship and support. This suggests levels of homophily are increased by majority member status (McPherson *et al.* 2001). Further research suggests that senior women in organizations may have homophilous preferences or affiliations (Cohen and Huffman 2007), as women may seek out other women to network with (Ibarra 1997). However, (numerical) constraints of available socially similar others may limit the development of such relations (Ibarra 1992). Women may also experience problems in maintaining network relations simultaneously with both men at their own level and lower status women (Ibarra 1997). Ely (1994) similarly argues demography effects relations between women, drawing a distinction between sex-integrated organizations (where there is a perception of a permeable boundary to top positions) and male-dominated organizations (a low number of women in senior positions). Beckman and Phillips's (2005) study outlines how women in senior high-status positions in law firms attract clients from women-led organizations. This suggests that at a senior level and with a high-status role, women, even when in a minority, may achieve positive intra-gender relations with similar (senior) others (Beckman and Phillips 2005), albeit with women from outside their organizations. However we question the extent to which such homophilous relations can exist between women and how they contribute to women's homosociality within gendered contexts.

In reviewing management and organizational studies research, the literature on homophily has primarily focused on gendered networks and network theory, to identify women's location, and differences between women and men, in social networks. Examples include: women's limited access to networks (Kark and Waismel-Manor 2005), women's access and contribution to networks for knowledge creation (Durbin 2011), and Benschop's (2009) call for a focus upon networking practices rather than network positions. However, network literature over-emphasizes homosociality and under-emphasises friendship, drawing less upon how Lazarsfeld and Merton (1954) conceptualize homophily as social processes of friendship. Durbin (2011), for example, suggests that homophily is shared identities or group affiliations, and Benschop (2009) similarly emphasizes interaction and socialization: both of which might be better understood as homosociality (seeking associations with those similar to self). Such narrow conceptualizations of homophily limit the possibilities of exploring women's experiences of friendship and intra-gender relations in organizations.

This conflation of homophily and homosociality reflects Gruenfeld and Tiedens's (2005) research review of organizational preferences and homogeneity and their position that consistencies have been merged between homophily, homosociality, similarity-attraction hypothesis, and in-group favouritism. Gruenfeld and Tiedens (2005) argue that this merging of theories enables an appreciation that organizing is characterized by a desire to relate to others like oneself, which subsequently shapes patterns of social relations. We argue that while this characterization of organizing is informative, important differences between the theories are lost. These differences are necessary to interrogate the under-explored forms of gender in action and how organizing processes impact upon women's relations in organizations. Homophily, as Lazarsfeld and Merton (1954) conceptualize it, emphasizes that patterns of social relations go beyond a desire to simply relate with similar others. This creates both distance and closeness between homophily and homosociality as important areas for future management and organization studies research. Historically Kanter (1977) suggested homosocial reproduction was a key mechanism in organizations through which men in senior positions secure certainty, order, and trust, by seeking association with similar others (men). This leads to the reproduction of characteristics men associate with themselves and seniority (read masculinities), as a form of social closure (Elliott and Smith 2001) which contributes to reducing uncertainty faced by those in managerial positions (Kanter 1977). While homosociality is an aspect of this process, Kanter (1977) highlights the dysfunction that occurs as an outcome of the need for trust, rather than an interest in or focus upon homophily as social processes of friendship.

To summarize discussions so far, homophily is a concept drawn upon to understand gendered relations, however this has been limited to a focus on networks not friendship, as outlined in the sociological literature. Organizational research on homosociality to date has contributed insights into how masculinities are reproduced. There is an appreciation in the literature that intra-gender relations between men, whilst they may be competitive (Connell and Messerschmidt 2005; Keisling 2005) and instrumental (Collinson and Hearn 2005), are characterized as involving degrees of cooperation, support, and friendship (Collinson and Hearn 2005), interconnected with homosocial desire (Keisling 2005). These enactments of masculinity shape organizing norms and performance criteria against which women are assessed and which contribute to the marginalization of women and their opportunities for positive intra-gender relations.

We contend that by considering homophily, as social processes of friendship, and homosociality, as a general orientation to associate with people like oneself, within the conceptual framework, we can reveal further hidden aspects of gender at work (shown in Figure 10.3). In noting men's intra-gender relations as grounded in competition and cooperation, next we consider women's intra-gender competition to further enhance our understandings of women's negative intra-gender relations.

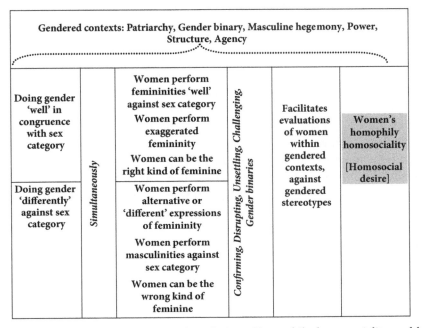

FIGURE 10.3 Women's negative intra-gender relations: Homophily, homosociality, and homosocial desire

Women's Intra-Gender Competition

Academic interest in female competition continues to grow (Campbell 2004) in diverse fields of study, yet remains largely unexplored in gender, management, and organization research. Taking an evolutionary psychology perspective, Campbell (2004) sees women's competition as an inherent part of biological status, with women less willing to escalate competition to direct aggression arising from this biology, because natural selection favoured females who avoided the danger of risking their safety or lives. Studies into men and women's competition by Tooke and Camire (1991) and Cashdan (1998) highlight how men compete with other men by exaggerating superiority, promiscuity, intensity, and popularity, while women compete with other women by alterations to their appearance, 'so that attractiveness appears to be the currency of female competition even when no mention was made of what the competition is about' (Campbell 2004: 19).

Campbell's (2004) position is that women usually compete for mates by emphasizing public qualities valued by men (beauty and sexual exclusiveness) and by using less explicit ways of denigrating rivals (through gossip and stigmatization). In this way, 'women can compete without risking their lives through acts that ostracise, stigmatize and otherwise exclude others from social interaction without risking direct physical confrontation' (Campbell 2004: 18). This indirect or relational aggression resulting from competition highlights the manipulation of social relationships through stigmatizing and exclusion (e.g. rumour and gossip) strategies which can have devastating effects upon the victim (Ahmad and Smith 1994; Simmons 2002). This indirect aggression can also be explained by gender role prescription; 'as women's direct aggression is an aberration from the female stereotype, women may seek alternative and more acceptable means of expressing competition' (Campbell 2004: 19).

Milhausen and Herold (1999) argue that gossip about sexual reputation (as a form of women's competition) is not confined to 'nice' middle-class girls. As Laidler and Hunt (2001: 668) note from research with street gangs in deprived inner-city areas, 'we find gang girls spending a great deal of energy "bitching" or casting doubt on others' reputations. This cross-cultural process operates not only as a mechanism of social control, but also of distancing and confirming one's own reputation.' This intra-gender competitive strategy uses a similar process to homophily when 'marking the social boundary' of friendship (Eve 2002) raised earlier in the chapter. As Campbell (2004: 19) argues, 'when we gossip we spread information that is damaging to the other's reputation and so diminish his or her social standing. But the act of condemnation is also an act of self-promotion: one cannot credibly accuse a rival of behaviours that one engages in oneself' (Campbell 2004: 19). These competitive strategies between women can also be seen as processes of female misogyny (Mavin 2006a, 2006b). Campbell (2004) notes that women's concern with relative attractiveness might also result from the internalization of patriarchal values, as well as from mate competition. Girls are argued to have come to ventriloquize patriarchal male attitudes about appropriate appearance and behaviour (Brown 1998), resulting in 'raging misogyny' (viewed here as female misogyny, Mavin 2006a, 2006b), as 'many women compete over things they think men value, such as looking sexy...the most dangerous outcome of this is self-hatred: girls and women disparage themselves and disassociate from other females' (Tanenbaum 2002: 47). We extend this line of thinking to organization studies with an aim to 'reveal', and to problematize, how such women's negative intra-gender relations play out in organizations.

From a sociological perspective, Connell (1987: 228) suggests women at work have experienced a different kind of regulation in the relationship with men which impacts upon their interaction with masculinities, as 'women are subjected to direct comparison with men, while being disadvantaged in the comparison from the start', through hegemonic masculinity. Hegemonic masculinity (Connell 1987), now understood as constructing a hierarchy of masculinities, where some remain more 'socially central, or more associated with authority and social power' (Connell and Messerschmidt 2005: 846), continues to shape gender relations, albeit recognizing the dynamic nature of gender relations and practices at regional and local levels (Connell and Messerschmidt 2005). Connell (1987: 228) contends that in such circumstances some femininities, constructed in a dynamic relationship with hegemonic masculinities, may reject 'withdrawal from competition as a formative personal strategy' and engage in complexities of resistance, whilst other femininities which acquiesce to hegemonic masculinities continue to be produced; they comply with hegemonic masculinity and orientate around 'accommodating the interests and desires of men' (Connell 1987: 183). The latter can be understood as 'emphasized femininity', a form of femininity performed for men, which exalts women as compliant, nurturing, and empathic (Connell 1987). According to Connell (1987) this emphasized femininity closes down the possibility of other femininities, of those ways of being which do not comply with hegemonic masculinity. This closing down of other ways of being can also shape women's responses to other women in organizations, for example, responses from women towards those women who reject emphasized femininity and engage in masculinities by 'doing competition'. Drawing upon Mavin and Grandy's (2011, 2012), doing gender well and differently simultaneously enables opportunities for women to enact emphasized femininity (doing gender well), as well as masculinity (manifested through competition), simultaneously. However, the consequences for women in this doing of gender may be more subtle and dangerous than those experienced by men. As Starr (2001) notes when considering women's competition in organizations,

> Competition between women may go deeper than professional rivalry, to include sub-conscious jealousy and competition based on age or appearance (attractiveness, weight, dress sense). This suggests that at times women may read each other's sexed bodies through men's eyes in sexual competition. At other times the perception of separation and competition is explained in work related terms through factors such as intellectual ability, professional connections, reputation, etc. Furthermore, unlike the more open forms of hostility exhibited by men, women observe that competition or opposition from women is more likely to manifest as passive resistance.

> (Starr 2001: 9)

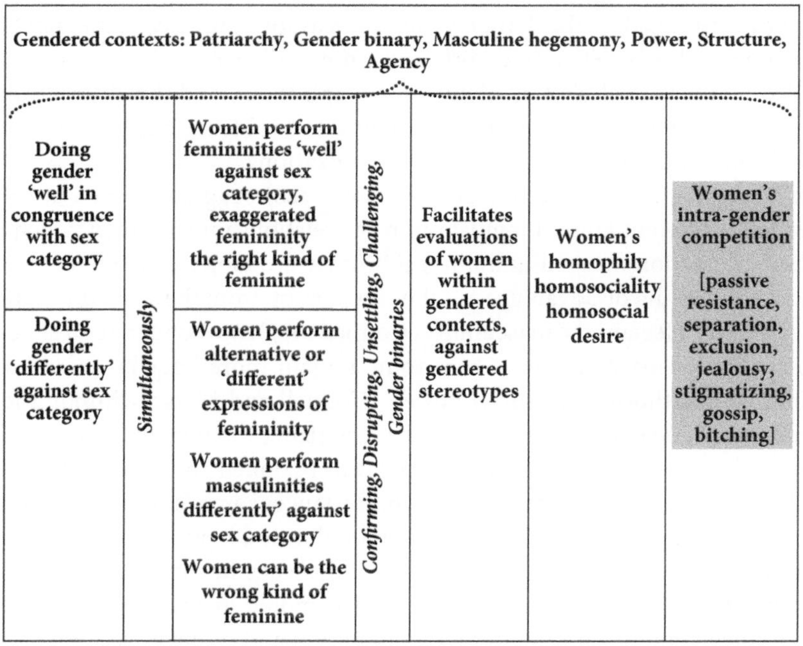

Gendered contexts: Patriarchy, Gender binary, Masculine hegemony, Power, Structure, Agency					
Doing gender 'well' in congruence with sex category	*Simultaneously*	Women perform femininities 'well' against sex category, exaggerated femininity the right kind of feminine	*Confirming, Disrupting, Unsettling, Challenging, Gender binaries*	Facilitates evaluations of women within gendered contexts, against gendered stereotypes	Women's homophily homosociality homosocial desire	Women's intra-gender competition [passive resistance, separation, exclusion, jealousy, stigmatizing, gossip, bitching]
Doing gender 'differently' against sex category		Women perform alternative or 'different' expressions of femininity Women perform masculinities 'differently' against sex category Women can be the wrong kind of feminine				

FIGURE 10.4 Women's negative intra-gender relations: Women's intra-gender competition

Whilst competition for opportunities and career advancement (as scarce resources) is a challenge in both sex-integrated and male-dominated organizations, Ely (1994) argues that gendering in male-dominated organizations contributes to women's negative assessments of other women in management and leadership positions. When there is a scarcity of women in senior positions then other women are critical of senior women's 'credentials both as women and as [law] partners: women partners not only failed to be the kind of women on whom junior women could rely for support but failed as well to be the kind of partner whose authority junior women could respect' (Ely 1994: 228). We have interpreted Ely's (1994) research as identifying negative intra-gender relations but from lower status to higher status women. However the research does not explore relations between women at different structural levels (e.g. higher status to lower, varying horizontal relations). Nor does it explain how the broader social context contributes to these relations or how conscious or unconscious women are of their intra-gender behaviours as gendered practices.

We contend that discussions of intra-gender competition between women in organizations have been almost taboo and that being a 'competitive woman' has been positioned as akin to being an 'ambitious woman': something you keep to yourself. Also, that there are unconscious and conscious competitive strategies, where women are unaware of the gendered contexts underpinning their actions and of the implications of their actions. Building upon Campbell's (2004) call to further explore the evolutionary model of female competition for mates as scarce resources within alternative sites, framed around conformity to culture-specific gender stereotypes and internalization of patriarchal values, we integrate the nature of women's intra-gender competition in organizations into the conceptual framework and question how this may manifest when the patriarchal hierarchy is disrupted by women. In doing so we extend our conceptual framework of doing gender well and differently (Mavin and Grandy 2011, 2012), organizational gendered contexts, and the nature of women's friendships and homosociality with other women, to include women's intra-gender competition (see Figure 10.4). We now turn to the concept of female misogyny (Mavin 2006a, 2006b) as the final element in our framework.

FEMALE MISOGYNY

Fotaki (2011), drawing upon Kristeva (1982), has highlighted that women occupy unstable and subordinated positions in the symbolic and are reminded of this by both men *and* women in organizations. Processes of female misogyny (Mavin 2006a, 2006b) between women in organizations are facilitated by gendered contexts and are a means by which women are reminded of their subordinate positions. Mavin (2006a, 2006b, 2008) conceptualized female misogyny from research into interactions between academic women and extended this when exploring negative behaviours and responses from women to other women in organizations and management. Female misogyny encapsulates the social processes, behaviours, and activities women engage in, consciously or unconsciously, when they subjugate, undermine, exclude, and stigmatize other women. Female misogyny can therefore be seen to encapsulate women's 'violence' towards other women. Mavin's (2006a, 2006b, 2008) argument is that as women disturb the gendered order by progressing up the managerial hierarchy, or even by showing a desire to do so, they can invoke the wrath of both men *and* women, who are enculturated to associate power and management with masculinities and men. The gendered contexts of organizations and management and the prevalence of sex-role categorizations in assessing women are argued to contribute to the backdrop of relations between women (Mavin 2008) yet they remain under-researched in management and organization literatures. Women who respond negatively towards senior women who do not meet expectations of the gender binary, for example, through solidarity behaviour, are unaware of/fail to acknowledge the complexities of the gendered organizing context and overly emphasize individual women's behaviour, or non-behaviour, as the root of the problem (Mavin 2006a, 2006b, 2008). In doing so, women also contribute to the maintenance of the 'individual woman as problem' and therefore the status quo in relation to gendered hierarchies in organizations. Research into the concept of female misogyny enables an exploration of the 'shadow side' of relations between women, constructed within gendered contexts (Mavin 2008: 573). Female misogyny, evident in social interactions between women in organizations, results from a concern for, and possible threats to, established gendered hierarchies, which become struggles over destabilization, change, and/or maintenance of the gendered status quo. Women's workplace friendships and intra-gender competition also play a part in these struggles, as complex interlocking practices and processes (Acker 2009) of women's intra-gender relations.

> Processes of female misogyny therefore emerge from the complex way in which gender order is embedded and the underlying assumptions and behaviours which socially construct and impact upon everyday experiences for women in management. The significant issue which requires further research is the way in which the privileged gendered social order evident within management, encourages and exacerbates differences between women, in order to prevent opposition in the form of successful challenges and resulting change. This is not to point to orchestrated behaviours in that many women are not conscious of their negative behaviours towards other women, rather to identify and to challenge implicit gendered assumptions which foster difference and fragmentation, which is, after all, easier to dismiss than joint action.
>
> (Mavin 2006a: 273)

Engagement in female misogyny is often an unconscious process, lacking in awareness of the gendered contexts which facilitate women's negative behaviour towards other women and of the damaging outcomes of such misogynistic behaviours. However, women's intra-gender 'violence' can be seen as horizontal and vertical aggression (Farrell 2001) in organizations. We agree with Camussi and Leccardi's (2005) argument that it is possible to see traces of misogyny in women's assessments of other women who are counter-stereotypical. Such responses are limiting for both organizations and women as they 'sanction the impossibility for women ... of constructing a different condition' (Camussi and Leccardi 2005: 135), which would allow for intra-gender differences *and* equality with men. Female misogyny (Mavin 2006a, 2006b) is an alternative perspective on relations between women in organizations which considers the gendered contexts in which these negative relations are co-constructed and a concept through which to further explore women's negative intra-gender relations in organizations.

Women's expressions of a desire for power and others' perceptions of this desire, work against women (Mavin 2006b), as such behaviours demonstrate that women are failing to live up to gendered feminine communal stereotypes associated with women generally (Okimoto and Brescoll 2010). Okimoto and Brescoll (2010) suggest that backlash responses to women who seek power are not differentiated between men and women, supporting the argument that women also respond negatively via female misogyny to women who do not meet gendered expectations (Mavin 2006a, 2006b). Rudman and Phelan (2008) suggest that women also respond negatively to other women when selecting partners for competitive tasks which have implications for their own success, and negatively to expressions of power by agentic women. They argue that this may be a form of unconscious self-oppression, so even women evaluated as competent are perceived to be 'socially unattractive' by other women (Rudman and Phelan 2008). This social unattractiveness of women to other women, constrains potential for homosocial desire (Keisling 2005) between women.

Farrell's (2001) study of relations between nurses suggests that women can act as gate-keepers to other women and such circumstances are characterised by expressions of horizontal aggression. Parks-Stamm *et al.* (2008) argue women strategically reject successful women in male-dominated roles to prevent unfavourable assessments of themselves, as successful women set a high benchmark for the assessment of other women within the organization. This argument supports a female misogyny interpretation of women's responses, as successful women are then a threat to 'the self' of other women (Parks-Stamm *et al.* 2008: 242), who can respond negatively by distancing themselves from senior women, drawing upon prescriptive norms to position successful women as unlikeable 'norm violators' (Parks-Stamm *et al.* 2008: 245). This rejection and distancing process, as female misogyny, has parallels with 'marking a boundary' in social processes of friendship (Eve 2002) and stigmatizing and exclusion (e.g. rumour and gossip) strategies of female competition (Campbell 2004). Such relational strategies and responses have significant implications for the possibility of women's positive intra-gender relations.

Our assumption is that homophilous friendships and women's positive intra-gender relations, including the ability to simultaneously compete and cooperate, particularly at different hierarchical levels, would be a positive cultural and structural enabler to women's experiences and progress, as it has been for men. However Camussi and Leccardi (2005) suggest that the very fluidization of socially sanctioned gender roles has also contributed to the levels of ambiguity and complexity in organizational members' experience. As women move away from traditional sex-role expectations and into performing multiple gender roles (care-givers and competitive careerists), 'structural ambiguity' gives way to complexity and 'fears and uncertainties', which construct space for socially shared prescriptive stereotypes to re-emerge as tools to re-establish order (Camussi and Leccardi 2005: 115). For Camussi and Leccardi (2005: 116), who draw upon Mavin and Bryans (2003) work on female misogyny, this includes intra, as well as inter, gender expectations, 'i.e. the ability to be women or men in different ways', and difficulties in accepting intra-gender differences, which result in women aligning themselves with men, in keeping other women in second place (Mavin and Bryans 2003). Integrating female misogyny into explorations of women's intra-gender relations enables a richer understanding of how women negotiate organization and management (Mavin 2006a, 2006b, 2008). This requires consideration of female misogyny as interconnected with women's intra-gender friendships, homosocial relations, and competition, constructed and reproduced within the prevailing patriarchal social order (Mavin 2006b). Therefore female misogyny offers a further contribution to our conceptual framework in enabling understanding of 'the sociocultural constraints that continue to penalize women' (Camussi and Leccardi 2005: 120). (See Figure 10.5.)

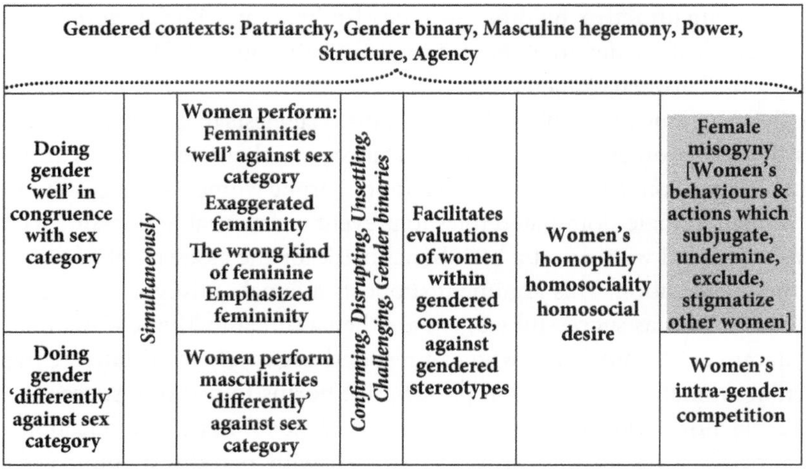

Gendered contexts: Patriarchy, Gender binary, Masculine hegemony, Power, Structure, Agency						
Doing gender 'well' in congruence with sex category	*Simultaneously*	Women perform: Femininities 'well' against sex category Exaggerated femininity The wrong kind of feminine Emphasized femininity	*Confirming, Disrupting, Unsettling, Challenging Gender binaries*	Facilitates evaluations of women within gendered contexts, against gendered stereotypes	Women's homophily homosociality homosocial desire	Female misogyny [Women's behaviours & actions which subjugate, undermine, exclude, stigmatize other women]
Doing gender 'differently' against sex category		Women perform masculinities 'differently' against sex category				Women's intra-gender competition

FIGURE 10.5 Women's negative intra-gender relations: Female misogyny

A Conceptual Framework of Women's Negative Intra-Gender Relations

We now move to summarize our discussions and highlight the research questions which emerge when exploring gender in organizations through this lens. In developing the initial conceptual framework we have given regard to the gendered contexts within which social processes and experiences take place, including agency, structure, culture, patriarchy, and hegemonic masculinity; ambiguity, instability, and fluidization of gender roles; the doing of gender well and differently simultaneously and the continued evaluation of men and women against gender binaries. These gendered contexts are surfaced as salient issues to consider as they shape, construct, and constrain women's intra-gender work relationships in organizations.

Our assumptions include women's direct comparison to men, whilst being disadvantaged from the start (Connell 1987) through hegemonic masculinity, which constructs a hierarchy of masculinities, continuing to construct gender relations at different levels (Connell and Messerschmidt 2005). In doing so we recognized Connell's (1987: 228) 'emphasized femininity' performed for men which closes down the possibility of other femininities, versus women's engagement in complexities of resistance. At the same time, our position of doing gender well and differently, enacting masculinities and femininities simultaneously (Mavin and Grandy 2011, 2012) is a process which opens up the possibilities for women to be otherwise. However, women's conscious/unconscious negative responses to other women when they engage in masculinities, do gender differently, and resist hegemonic masculinity by disrupting the gender order have been theorized here as resulting in negative intra-gender relations between women. Such negative relations contribute to the constraints around possibilities for women to be otherwise and require further research.

Our theorization of women's negative intra-gender relations has drawn upon homophily (Lazarsfeld and Merton 1954) and homosociality (Gruenfeld and Tiedens 2005), women's intra-gender competition, and female misogyny (Mavin 2006a, 2006b). If, despite complexities in relations between masculinities (Connell and Messerschmidt 2005), men's intra-gender relations can be characterized by competition, cooperation, friendship, and support, albeit potentially instrumentally (Collinson and Hearn 2005), the academy should further explore women's intra-gender relations in organizations and raise consciousness to gender contexts and resulting social relations in order to challenge prevailing gender orders.

Processes of friendship formation between women at work, and how these are constructed by gendered contexts and impact upon women's potential for homophily and homosociality, require further attention. Integrating women's intra-gender competition and female misogyny into the framework raises more questions. We have highlighted the contradictions in women's expectations of other women at work through solidarity behaviour, whilst simultaneously constructing senior women as Queen Bees (who do gender differently) and engagement in female misogyny to negatively evaluate other women. In constructing an initial conceptual framework to begin to account for women's negative intra-gender relations, a number of questions emerge to guide our future research. How does female misogyny play out when women do gender well and differently, enacting femininity and/or masculinities simultaneously? How does female misogyny undermine social processes of friendship (horizontally and vertically) between women at work and how does this restrict women's ability to engage in homosociality with women? Are women aware of their intra-gender competition and its gendered nature? How is women's intra-gender competition covert and/or overt? If female misogyny and intra-gender competition constrain women's potential to be otherwise, undermining women's solidarity, how do women develop the capacity to cooperate and compete in organizations in ways which are less damaging and more enabling than the negative relations women can operate towards each other (consciously or unconsciously)? How do women in organizations build and engage in their own homophily, homosociality, and develop alternative homosocial desire? How should the academy raise consciousness of gender contexts and resulting women's negative intra-gender relations, in order to challenge the prevailing gendered order?

A further contribution of the conceptual framework is the act of speaking the unspeakable. This in itself is a means of consciousness raising on the nature and possible impact of women's negative intra-gender relations and aims to continue the dialogue. It is critical for women to increase their gender consciousness (Martin 2003; Mavin 2006a, 2006b) and understand how gendered expectations, contexts, and order impact upon their own responses to other women (and vice versa) and to enable acceptance of intra-gender differences which have the potential to improve opportunities for, and to facilitate more positive relationships between, women in organizations.

Conclusion

In this chapter we have positioned women's negative intra-gender relations as an under-researched and hidden area of gender in organizations, worthy of exploration and problematization due to their centrality and criticality in women's experiences and progress in organizations. In reflecting back to Kanter's (1977) proposition that, as women's numbers swelled, women were more likely to form allegiances, coalitions, and affect the culture of the organization, it is time to further explore women's intra-gender relations and the shadow side of such relations, which may serve to maintain the status quo rather than challenge, disrupt, and effect change. We have endeavoured to ensure that in problematizing women's social relations at work, this is not reduced to a 'woman's problem', by acknowledging that these are a production of the gendered contexts in which they take place. We have acknowledged individual subjectivities and differences within and between individual and groups of women and men and have also drawn upon sex, sex category, and the gender binary, as we understand this remains a source of evaluation for individuals and groups.

We have theorized women's negative intra-gender relations in organizations by integrating theory in the areas of doing gender well and differently, homophily and homosociality, women's intra-gender competition and female misogyny, as complex interlocking gendered practices and processes (Acker 2009). Our contribution is a conceptual framework of women's intra-gender relations, which aims to reveal different forms of gender in action in organizations. We have extended the theoretical development of women's negative relations within gendered organizational contexts, by recognizing that they have the power to limit the potential for homosocial and homophilous relations between women. Exploring the nature of women's social relations through intra-gender competition, processes of female misogyny in organizations, and the interplay with homosociality and homophily, offers new insights into the gendered nature of organizations and how gendered organizing processes impact upon social interactions and relationships between women. Finally we have posed a number of questions to guide future research agendas and hope others will continue the dialogue.

References

Abramson, J. (1975). *The Invisible Woman: Discrimination in the Academic Profession*. London: Jossey-Bass.

Acker, J. (1989). 'The Problem with Patriarchy', *Sociology*, 23(2): 235–40.

Acker, J. (1992). 'Gendering Organizational Theory', in A. Mills and P. Tancred (eds), *Gendering Organizational Analysis*, 248–60. Newbury Park, CA: Sage.

Acker, J. (2000). 'Revisiting Class: Thinking for Gender, Race, and Organizations', *Social Politics*, 7(2): 192–214.

Acker, J. (2009). 'From Glass Ceiling to Inequality Regimes', *Sociologie du Travail*, 51(2): 199–217.

Ahmad, Y., and Smith, P. K. (1994). 'Bullying in Schools and the Issue of Sex Differences', in J. Archer (ed.), *Male Violence*, 70–83. London: Routledge.

Baxter, L., and Hughes, C. (2004). 'Tongue Sandwiches and Bagel Days: Sex, Food and Mind–Body Dualism', *Gender, Work and Organization*, 11(4): 363–80.

Beckman, C., and Phillips, D. (2005). 'Interorganizational Determinants of Promotion: Client Leadership and the Attainment of Women Attorneys', *American Sociological Review*, 20: 678–701.

Bell, C. (1979). 'Implementing Safety and Health Regulations for Women in the Workplace', *Feminist Studies*, 5(2): 286–301.

Benschop, Y. (2009). 'The Micro-Politics of Gendering in Networking', *Gender, Work and Organization*, 16(2): 217–37.

Billing, Y. D. (2011). 'Are Women in Management Victims of the Phantom of the Male Norm?', *Gender, Work and Organization*, 18(3): 298–317.

Bird, S. R. (1996). 'Welcome to the Men's Club: Homosociality and the Maintenance of Hegemonic Masculinity', *Gender and Society*, 10(2): 120–32.

Broadbridge, A., and Simpson, R. (2011). '25 Years On: Reflecting in the Past and Looking to the Future in Gender and Management Research', *British Journal of Management*, 22(3): 470–83.

Brown, B. (1998). *Unlearning Discrimination in the Early Years*. Stoke-on-Trent: Trentham.

Butler, J. (1990). *Gender Trouble: Feminism and the Subversion of Identity*. London: Routledge.

Butler, J. (1999) *Gender Trouble: Feminism and the Subversion of Identity* (2nd edn). London: Routledge.

Calás, M. B., and Smircich, L. (1992). 'Using the "F" Word: Feminist Theories and the Social Consequences of Organizational Research', in A. Mills and P. Tancred (eds), *Gendering Organizational Analysis*, 222–34. Newbury Park, CA: Sage.

Calwell, M. A., and Peplau, L. A. (1982). 'Sex Differences in Same-Sex Friendships Sex Roles', *Psychological Perspective*, 8(7): 721–32.

Campbell, A. (2004). 'Female Competition: Causes, Constraints, Content, and Context', *Journal of Sex Research*, 41(1): 16–26.

Camussi, E., and Leccardi, C. (2005). 'Stereotypes of Working Women: The Power of Expectations', *Social Science Information*, 44(1): 113–40.

Cashdan, E. (1998). 'Are Men More Competitive than Women?', *British Journal of Social Psychology*, 37(2): 213–29.

Chesler, P. (2001). *Women's Inhumanity to Woman*. New York: Thunder's Mouth Press, Nations Books.

Cohen, P. N., and Huffman, M. L. (2007). 'Working for the Woman? Female Managers and the Wage Gap', *American Sociological Review*, 72(5): 681–704.

Collinson, D. L., and Hearn, J. (1996). *Men as Managers, Managers as Men: Critical Perspectives on Men, Masculinities and Managements*. London: Sage.

Collinson, D. L., and Hearn, J. (2005). 'Men and Masculinities in Work, Organizations, and Management', in M. S. Kimmel, J. Hearn, and R. W. Connell (eds), *Handbook of Studies on Men and Masculinities*, 289–310. London: Sage.

Connell, R. W. (1987). *Gender and Power*. Sydney: Allen & Unwin.

Connell, R. W., and Messerschmidt, J. (2005). 'Hegemonic Masculinity: Rethinking the Concept', *Gender and Society*, 19/6: 829–59.

David, D. S., and Brannon, R. (1976). *The Forty-Nine Percent Majority: The Male Sex Role*. Reading, MA: Addison Wesley.

Dean, D. (2008). 'No Human Resource is an Island: Gendered, Racialized Access to Work as a Performer', *Gender, Work and Organization*, 15(2): 161–81.

Derks, B., Ellemers, N., van Laar, C., and de Groot, K. (2011). 'Do Sexist Organizational Cultures Create the Queen Bee?', *British Journal of Social Psychology*, 50(3): 519–35.

Donelson, E., and Gullahorn, J. E. (1977). *Women: A Psychological Perspective*. New York: Wiley.

Durbin, S. (2011). 'Creating Knowledge through Networks: A Gender Perspective', *Gender, Work and Organization*, 18(1): 90–112.

Eagly, A. H., and Carli, L. L. (2007). 'Women and the Labyrinth of Leadership', *Harvard Business Review*, 85(9): 62–71.

Elliott, J. R., and Smith, R. A. (2001). 'Ethnic Matching of Supervisors to Subordinate Work Groups: Findings on "Bottom–Up" Ascription and Social Closure', *Social Problems*, 48(2): 258–76.

Ely, R. J. (1994.) 'The Effects of Organizational Demographics and Social Identity on Relationships among Professional Women', *Administrative Science Quarterly*, 39: 203–38.

Eve, M. (2002). 'Is Friendship a Sociological Topic?', *European Journal of Sociology*, 43(3): 386–409.

Farrell, G. A. (2001). 'From Tall Poppies to Squashed Weeds*: Why don't Nurses Pull Together More?', *Journal of Advanced Nursing*, 35(1): 26–33.

Ford, J., and Harding, N. (2010). 'Get Back into that Kitchen, Woman: Management Conferences and the Making of the Female Professional Worker', *Gender, Work and Organization*, 17(5): 503–20.

Fotaki, M. (2011). 'The Sublime Object of Desire (for Knowledge): Sexuality at Work in Business and Management Schools in England', *British Journal of Management*, 22(1): 42–53.

Gherardi, S. (1994). 'The Gender we Think, the Gender we Do in Everyday Organizational Life', *Human Relations*, 47(6): 591–609.

Gherardi, S. (1996). 'Gendered Organisational Cultures: Narratives of Women Travellers in a Male World', *Gender Work and Organization*, 3(4): 187–201.

Goldberg, P. (1968). 'Are Women Prejudiced Against Women?' *Trans Action*, 5(5): 28–30.

Gregory, M. R. (2009). 'Inside the Locker Room: Male Homosociability in the Advertising Industry', *Gender, Work and Organization*, 16(3): 323–47.

Gruenfeld, D., and Tiedens, L. Z. (2005). 'Organizational Preferences and their Consequences', in S. T. Fiske, D. T. Gilbert, and G. Lindzey (eds), *Handbook of Social Psychology*, 1252–87. Hoboken, NJ: John Wiley & Sons.

Gutek, B. A., Stromberg, A. G., and Larwood, L. (1988). 'Women's Relationships with Women in the Workplace Vol. 1', in A. H. Stromberg, L. Larwood, and B. A. Gutek (eds), *Women and Work: An Annual Review*. London: Sage.

Holgersson, C. (2012). 'Recruiting Managing Directors: Doing Homosociality', *Gender, Work and Organization*. doi:10.1111/j.1468-0432.2012.00595.x 1-13

Holvino, E. (2010). 'Intersections: The Simultaneity of Race, Gender and Class in Organization Studies', *Gender, Work and Organization*, 17(3): 248–77.

Ibarra, H. (1992). 'Homophily and Differential Returns: Sex Differences in Network Structure and Access in an Advertising Firm', *Administrative Science Quarterly*, 37(3): 422–47.

Ibarra, H. (1997). 'Paving an Alternative Route: Gender Differences in Managerial Networks', *Social Psychology Quarterly*, 60(1): 91–102.

Kanter, R. M. (1977). *Men and Women of the Corporation*. New York: Basic Books.

Kark, R., and Waismel-Manor, R. (2005). 'Organizational Citizenship Behavior: What's Gender Got to Do with it?', *Organization*, 12(6): 889–917.

Keisling, S. B. (2005). 'Homosocial Desire in Men's Talk: Balancing and Re-creating Cultural Discourses of Masculinity', *Language in Society*, 34(5): 695–726.

Kelan, E. K. (2010). 'Gender Logic and (Un)doing Gender at Work', *Gender, Work and Organization*, 17(2): 174–94.

Knights, D., and Kerfoot, D. (2004). 'Between Representations and Subjectivity: Gender Binaries and the Politics of Organizational Transformation', *Gender, Work and Organization*, 11(4): 430–54.

Kristeva, J. (1982). *Powers of Horror: An Essay on Abjection*. New York: Columbia University Press.

Laidler, K. J., and Hunt, G. (2001). 'Accomplishing Femininity among the Girls in the Gang', *British Journal of Criminology*, 41(4): 656–78.

Lazarsfeld, P., and Merton, R. K. (1954). 'Friendship as Social Process: A Substantive and Methodological Analysis', in M. Berger, T. Abel, and C. Page (eds), *Freedom and Control in Modern Society*, 18–66. New York: Octagon Books.

Legge, K. (1987). 'Women in Personnel Management: Uphill Climb or Downhill Slide?', in A. Spencer and D. Podmore (eds), *In a Man's World: Essays on Women in Male-Dominated Professions*, 33–60. London: Tavistock.

Lewis, P., and Simpson, R. (2010). 'Introduction: Theoretical Insights into the Practises of Revealing and Concealing Gender within Organizations', in P. Lewis and R. Simpson (eds), *Revealing and Concealing Gender: Issues of Visibility in Organizations*, 1–22. Basingstoke: Palgrave Macmillan.

Linstead, S., and Pullen, A. (2006). 'Gender as Multiplicity: Desire, Displacement, Difference and Dispersion', *Human Relations*, 59(9): 1287–1310.

Lipman-Blumen, J. (1984). *Gender Roles and Power*. Englewood Cliffs, NJ: Prentice-Hall.

McPherson, M., Smith-Lovin, L. and Cook, J. M. (2001). 'Birds of a Feather: Homophily in Social Networks', *Annual Review of Sociology*, 27: 415–44.

Martin, P. Y. (2003). '"Said and Done" versus "Saying and Doing": Gendering Practices, Practicing Gender at Work', *Gender and Society*, 17(3): 342–66.

Mavin, S. (2006a). 'Venus Envy: Problematizing Solidarity Behaviour and Queen Bees', *Women in Management Review*, 21(4): 264–76.

Mavin, S. (2006b). 'Venus Envy 2: Sisterhood, Queen Bee and Female Misogyny in Management', *Women in Management Review*, 2(5): 349–64.

Mavin, S. (2008). 'Queen Bees, Wannabees and Afraid to Bees: No More Best Enemies for Women in Management', *British Journal of Management*, (1): 75–84.

Mavin, S., and Bryans, P. (2003). 'Women's Place in Organization: The Role of Female Misogyny', paper, 3rd International Gender, Work and Organization Conference, June, Keele, UK.

Mavin, S., and Grandy, G. (2011). 'Doing Gender Well and Differently in Dirty Work', *Gender, Work and Organization*. http://dx.doi.org/10.1111/j.1468-0432.2011.00567.x

Mavin, S., and Grandy, G. (2012). 'Doing Gender Well and Differently in Management', *Gender in Management: An International Journal*, 27(4): 218–31.

Mavin, S., and Williams, J. (forthcoming). 'Women's Impact on Women's Careers in Management: Queen Bees, Female Misogyny, Negative Intra-Relations and Solidarity Behaviours', in R. Burke, S. Vinnicombe, L. Moore, and S. Blake Beard (eds), *The Handbook of Research on Promoting Women's Careers*. Cheltenham: Edward Elgar.

Messerschmidt, J. (2009) '"Doing Gender": The Impact and Future of a Salient Sociological Concept', *Gender and Society*, 23(1): 85–8.

Milhausen, R. R., and Herold, E. S. (1999). 'Does the Sexual Double Standard Still Exist? Perceptions of University Women', *Journal of Sex Research*, 36(4): 361–8.

Nicolson, P. (1996). *Gender, Power and Organization: A Psychological Approach*. London: Routledge.

Okimoto, T. G., and Brescoll, V. L. (2010). 'The Price of Power: Power Seeking and Backlash Against Female Politicians', *Personality and Social Psychology Bulletin*, 36(7): 923–36.

Parks-Stamm, E. J., Heilman, M. E., and Hearns, K. A. (2008). 'Motivated to Penalize: Women's Strategic Rejection of Successful Women', *Personality and Social Psychology Bulletin*, 34(2): 237–47.

Pleck, J. H. (1976). 'Male Threat from Female Competence', *Journal of Counselling and Clinical Psychology*, 44(4): 608–13.

Poggio, B (2006). 'Outline of a Theory of Gender Practices', *Gender, Work and Organization*, 13(3): 225–33.

Rudman, L. A., and Phelan, J. E. (2008). 'Backlash Effects for Disconfirming Gender Stereotypes in Organizations', *Research in Organizational Behavior*, 28: 61–79.

Simmons, R. (2002). *Odd Girl Out: The Hidden Culture of Aggression in Girls*. New York: Harcourt Brace.

Simpson, R. (1997). 'Have Times Changed? Career Barriers and the Token Woman Manager', *British Journal of Management*, 8: 121–9.

Starr, K. (2001). 'What Makes Management Experience Different for Women? Secrets Revealed through the Structure of Cathexis', paper presented at the Rethinking Gender, Work and Organization Conference.

Staines, G., Travis, C., and Jayerante, T. E. (1973). 'The Queen Bee Syndrome', *Psychology Today*, 7(8): 55–60.

Tanenbaum, L. (2002). *Catfight: Women and Competition*. New York: Seven Stories.

Thornley, C., and Thörnqvist, C. (2009). 'Editorial: State Employment and the Gender Pay Gap', *Gender, Work and Organization*, 16(5): 529–35.

Tiger, L. (1969). *Men in Groups*. New York: Random House.

Tooke, W., and Camire, L. (1991). 'Patterns of Deception in Intersexual and Intrasexual Mating Strategies', *Ethology and Sociobiology*, 12(5): 345–64.

Valentine, G. (2007). 'Theorizing and Researching Intersectionality: A Challenge for Feminist Geography', *Professional Geographer*, 59(1): 10–21.

Walby, S. (1989). 'Theorising Patriarchy', *Sociology*, 23(2): 213–34.

Weiss, L., and Lowenthal, M. J. (1975). 'Life Course Perspectives on Friendships', in M. Thurnher, M. F. Lowenthal, and D. Chiriboga (eds), *Four Stages of Life*, 48–61. San Francisco: Jossey-Bass.

West, C., and Zimmerman, D. H. (1987). 'Doing Gender', *Gender and Society*, 1(2): 125–51.

West, C., and Zimmerman, D. H. (2002). 'Doing Gender', in C. West and S. Fenstermaker (eds), *Doing Gender, Doing Difference: Inequality Power and Institutional Change*, 3–24. New York: Routledge.

West, C., and Zimmerman, D. H. (2009). 'Accounting for Doing Gender', *Gender and Society*, 23(1): 112–22.

Worts, D., Fox, B., and McDonough, P. (2007). '"Doing Something Meaningful": Gender and Public Service during Municipal Government Restructuring', *Gender, Work and Organization*, 14(2): 162–84.

CHAPTER 11

SEX, GENDER, AND LEADERSHIP

What Do Four Decades of Research Tell Us?

GARY N. POWELL

In 1999, Tony Butterfield (my mentor) and James Grinnell reviewed the state of research on the intersection of sex, gender, and leadership in a book chapter entitled ' "Re-viewing" Gender, Leadership, and Managerial Behavior: Do Three Decades of Research Tell us Anything?' (Butterfield and Grinnell 1999). Given that research in the field that was first labelled as 'women in management' originated in the early 1970s with Schein's (1973, 1975) ground-breaking research, this review poses an analogous question, 'What do four decades of research tell us?'

In this review, as other scholars do, I distinguish between the terms of sex and gender. The term *sex* is generally used to refer to the binary categories of male and female. The term *gender* is generally used to refer to the psychosocial implications of being male or female, such as beliefs and expectations about what kinds of attitudes, values, skills, and behaviours are more appropriate for or typical of one sex than the other (Unger 1979; Archer and Lloyd 2002; Lippa 2005). Like Butterfield and Grinnell (1999: 238), I use the term *leadership* to refer to 'the behaviors and qualities of persons in formally designated leadership positions'. Thus, the study of *sex differences in leadership* examines how male and female leaders actually differ in attitudes, values, skills, behaviours, and effectiveness, whereas the study of *gender differences in leadership* focuses on how people believe that male and female leaders differ.

In addition, it is useful to distinguish between stereotypes and roles pertaining to both gender and leadership. *Gender stereotypes* represent beliefs about the psychological traits that are characteristic of members of each sex, whereas *gender roles* represent beliefs about the behaviours that are appropriate for members of each sex (Eagly *et al.* 2000; Kite *et al.* 2008; Wood and Eagly 2010). In the same vein, *leader stereotypes* represent beliefs about the psychological traits that are characteristic of leaders, whereas *leader roles* represent beliefs about the behaviours that are appropriate for leaders.

Questions about sex, gender, and leadership have always been a topic of keen public interest and often a source of debate. This topic is not simply 'hot' in the sense of being fashionable; it is also inflammatory. For example, in 2009, the *New York Times* (Room for debate 2009) conducted an online debate on the question 'Do women make better bosses?' I was one of six participants in the debate. Just the fact that the *Times* would pose such a question as worthy of debate suggests a general interest in the topic. Over 500 online comments were received from readers about the debate, many expressed in colourful language, about the general inferiority or superiority of female leaders as well as the sex composition of the debate panel, on which I was the only male. For example, the panel was alternatively disparaged as 'five females and one guy who wants to make his wife happy', not knowing that my wife has been my frequent collaborator as well (e.g. Graves and Powell 1988; Powell and Graves 2003); and 'five women and one man who was formerly the Chair of the Women in Management Division of the Academy of Management', implying that this credential provided a good basis for dismissing anything I had to say. Commenting on the responses themselves, one commentator suggested that the vitriol exhibited by male commentators demonstrated the obstacles that women face in the workplace (Room for debate 2009). Further examples of reader responses are provided in Powell (2011).

As the number and vehemence of the responses to the *New York Times* debate suggested, questions about sex differences among leaders stimulate especially heated debate. In general, some people tend to exaggerate sex differences ('alpha bias': Hare-Mustin and Maracek 1988), whereas other people tend to minimize or ignore sex differences ('beta bias'). Many people have strong beliefs about male and female similarities and differences in basic interests, abilities, attitudes and behaviours. Further, corporate leaders are given an enormous amount of attention, especially in societies that place a high value on individualism rather than collectivism such as the United States, the United Kingdom, Australia, Canada, and the Netherlands (Hofstede 2001: 215). In such societies, the success of organizations is attributed to the wisdom, values and practices of their founders or current leaders. When organizations fail to achieve expected results, their leaders are the first to be blamed. Consider the issues of sex differences and leadership together, and it is clear why so many people from all walks of life have strong opinions about what constitutes effective leadership as well as which sex, if either, is more likely to exhibit it.

Besides Butterfield and Grinnell (1999), several prior reviews of the literature on this topic have been conducted, including Terborg (1977), Bartol (1978), Riger and Galligan (1980), Nieva and Gutek (1981), Davidson and Burke (1994, 2000, 2011), Alimo-Metcalfe (2010), and Broadbridge and Simpson (2011). Each review has organized the literature somewhat differently. This review organizes the literature by examining status, preferences, stereotypes, attitudes, behaviours, and effectiveness associated with the leader role in relation to gender stereotypes and roles. First, it reviews women's status over recent decades. Second, it considers preferences for male versus female leaders in general. Third, it compares leader stereotypes with gender stereotypes and examines whether leader stereotypes have changed over time. Fourth, it reviews attitudes towards female leaders. Fifth, it investigates whether (and if so, how) female and male managers

differ in their behaviour and overall effectiveness as leaders. Finally, it considers implications for future theory, research, and practice.

Status

Table 11.1 presents statistics pertaining to US women's educational attainment and workplace status over the four-decade period between 1971 and 2010. As the table indicates,
all seven statistics were highly correlated with the year in which they were assessed
for the forty-year period. Analyses of variance by the decade in which statistics were
assessed also indicate a significant increase in each statistic across the four decades.

The educational attainment of women changed considerably across these four decades. The proportion of college degrees earned by women in all disciplines increased
between 1971 and 2010 from 43 per cent to 57 per cent at the bachelor's level and from 39
per cent to 60 per cent at the master's level. In addition, the proportion of college degrees
earned in business by women increased between 1971 and 2010 from 9 per cent to 49 per
cent at the bachelor's level and from 4 per cent to 46 per cent at the master's level. The
proportion of women earning bachelor's degrees in business increased sharply from the
1970s to the 1980s and then levelled off over the next two decades. However, the proportion of women earning master's degrees in business increased from one decade to the
next across the four decades.

The workplace role and status of women also changed considerably over the same
period of time. Women's labour force participation rate (i.e. the proportion of women
who were in the labour force) increased from 43 per cent in 1971 to 59 per cent in 2010.
The proportion of women in the labour force (i.e. the proportion of labour force participants who were female) increased from 38 per cent in 1971 to 47 per cent in 2010. Further,
although the US Department of Labor has refined its classification of which labour force
participants hold 'management positions' from time to time, the proportion of women
classified as working in management positions increased from 17 per cent in 1971 to 43
per cent in 2010. Each of these statistics increased from one decade to the next for the
first three decades examined and then levelled off over the most recent decade.

As we see, US women's status in society, as suggested by statistics on their educational
attainment and their workplace role and status, increased considerably over the past
four decades. Trends in these statistics reflect a major societal shift towards enhancement of women's academic credentials that prepare them for labour force participation
and women's increased commitment to labour force participation in general and managerial careers in particular.

However, despite these trends, women continue to be concentrated in lower managerial
levels and hold positions with less power and authority than men. The higher the organizational level, the fewer women are found, suggesting the existence of both a 'glass ceiling' that restricts women's attainment of top management positions solely because they
are women (Marshall 1984; Davidson and Cooper 1992; Powell 1999; Barreto et al. 2009)

Table 11.1 US women's status, 1971–2010

Statistic	Mean	SD	Correlation with Year	Mean Statistics by Decade				F by Decade (df = 3, 36)	Source
				1971–1980 Mean	1981–1990 Mean	1991–2000 Mean	2001–2010 Mean		
1. Proportion of women earning bachelor's degrees in all disciplines	..52	..05	..97*	..46	..51	..55	..57	1166.67*	US Dept of Education, 2011: table 283.
2. Proportion of women earning master's degrees in all disciplines	..52	..06	..98*	..44	..50	..55	..59	98.11*	US Dept of Education, 2011: table 283.
3. Proportion of women earning bachelor's degrees in business	..40	..13	..84*	..19	..44	..48	..50	84.55*	US Dept of Education, 2011: table 316.
4. Proportion of women earning master's degrees in business	..31	..13	..94*	..11	..31	..37	..43	1129.12*	US Dept of Education, 2011: table 316.
5. Women's labour force participation rate	..55	..05	..91*	..47	..55	..59	..59	93.88*	US Department of Labor, 2012: table 2.
6. Proportion of women in labour force	..44	..03	..92*	..40	..44	..46	..47	97.69*	US Dept of Labor, 2012: table 2.
7. Proportion of women in management positions	..36	..10	..90*	..21	..35	..43	..43	1140.57*	US Dept of Labor, 1972–2011: Annual table on employed persons by occupation, sex, and race for previous year.

$n = 40$ years.* $p < .001$

and a 'glass cliff' that leads women to be over-represented in top management positions that are especially precarious and difficult to retain (Ryan and Haslam 2005, 2007, 2009; Haslam and Ryan 2008). For example, although definitions of what constitutes 'top management' vary among companies, the proportion of female executive officers, typically considered as top management, is only 14.1 per cent in *Fortune* 500 corporations (Catalyst 2011*a*). This proportion has plateaued in recent years (e.g. down from 14.4 per cent the previous year), suggesting that women's progress in attaining executive positions in *Fortune 500* corporations has stalled (Catalyst 2011*b*). Although the proportion of women in chief executive positions in the US labour force as a whole is 24.2 per cent (US Department of Labor 2012: table 11), the proportion of female chief executive officers in *Fortune 500* corporations is only 4.0 per cent (Catalyst 2012). Thus, the status of women in the workplace, although greater than four decades ago, remains lower than that of men.

Preferences

Due to the increased representation of women in the managerial ranks, employees are increasingly likely to have had a female boss at some point in time. However, when people state a preference, it is more likely to be for a male boss than for a female boss. Over time, the Gallup Organization has asked people in 22 countries, 'If you were taking a new job and had your choice of a boss, would you prefer to work for a man or a woman?' Respondents could also state that the sex of their new boss would make no difference to them. All over the globe, respondents have consistently expressed a preference for a male boss (Simmons 2001).

According to the most recent poll results (Carroll 2006), twice as many Americans said that, if they were taking a new job, they would prefer a male boss (37 per cent) than those who said they would prefer a female boss (19 per cent); however, 'it makes no difference to me' was the slight favourite (44 per cent). Among men who stated a preference, 34 per cent favoured a male boss and 10 per cent a female boss. Among women who stated a preference, 40 per cent favoured a male boss and 26 per cent a female boss. In contrast, according to 1975 poll results (Simmons 2001), 63 per cent of men and 60 per cent of women preferred a male boss, whereas 4 per cent of men and 10 per cent of women preferred a female boss. Overall, while preferences for a male boss have declined over time, a male boss is still preferred over a female boss by a 2-1 margin.

There are several possible explanations for why people who state a preference tend to prefer a male boss. First, leader stereotypes that emphasize personal characteristics associated with men rather than those associated with women may account for the preference for men as leaders. Second, prejudice towards female leaders may make it difficult for women to be as effective in the leader role as men and reduce their desirability as leaders. Third, women and men may differ in their behaviours and effectiveness as leaders, with the behaviours exhibited by male leaders yielding better results than those exhibited by female leaders. The merits of these possible explanations for leader prefer-ences will be considered in the remainder of this chapter.

However, it is important to note that leader preferences differ according to the age of the person being asked. In the most recent poll (Carroll 2006), 18- to 34-year-olds were equally likely to say they preferred a male (31 per cent) or female (29 per cent) boss if they were taking a new job. Twice as many 35- to 54-year-olds preferred a male boss (38 per cent) than a female boss (19 per cent). Four times as many Americans who were 55 years old or older preferred a male boss (40 per cent) than a female boss (11 per cent). Thus, Generation Y-ers, who have had greater experience in working with women as peers in educational programmes and professional jobs, are less likely to prefer a male boss and more likely to prefer a female boss than Generation X-ers or baby boomers. These results suggest that, assuming that individuals' leader preferences do not change over their life spans, the overall sex bias in leader preferences favouring male leaders may diminish over time as new entrants to the labour force progress in their careers.

STEREOTYPES

Research on the relationships among sex, gender stereotypes, and leader stereotypes was first conducted in the 1970s. Schein (1973, 1975) compiled a list of ninety-two character-istics that people commonly believe distinguish between men and women, the basis for gender stereotypes. She then asked a sample of US middle managers to describe how well each of the characteristics fit women in general, men in general, or successful mid-dle managers in general. Schein hypothesized that, because the vast majority of manag-ers were men, the managerial job would be regarded as requiring personal attributes thought to be more characteristic of men than women. In support of her hypothesis, she found that both male and female middle managers believed that a successful middle manager possessed personal characteristics that more closely matched beliefs about the characteristics of men in general than those of women in general.

In replications of Schein's (1973, 1975) studies, women have less been inclined to view management as the domain of men. In countries with very different national cultures (e.g. the United States, United Kingdom, Germany, Japan, China, Turkey, and Sweden), both men and women believe that men are more similar to successful managers than women are, but men still endorse such beliefs to a greater extent than women do (Brenner et al. 1989; Heilman et al. 1989; Schein and Mueller 1992; Schein et al. 1996; Schein 2001; Fullagar et al. 2003). These results suggest that international beliefs about managers may be best expressed as *think manager—think male*, especially among men.

Several theoretical explanations have been offered for the think manager—think male phenomenon. For example, the *lack of fit model* (Heilman 1983, 1995) suggests that when people believe that men possess the characteristics that are best suited for the mana-gerial role in greater abundance than women, they are likely to evaluate male manag-ers more favourably than female managers, even if the managers being evaluated are exhibiting exactly the same behaviour. *Role congruity theory* (Eagly and Karau 2002) suggests that leader and gender stereotypes put female leaders at a distinct disadvantage by forcing them to deal with the perceived incongruity between the leader role and their gender role. If women conform to the female gender role, they fail to meet the requirements of the leader role. However, if women compete with men for leadership positions and conform to the leader role, they fail to meet the requirements of the female gender role, which calls for feminine niceness and deference to the authority of men (Rudman and Glick 2001).

Further, *status construction theory* (Ridgeway 1991, 2006; Webster and Hysom 1998) argues that unequal societal status is assigned to the sexes, with men granted higher status than women. Because of their weaker status position, women are required to monitor others' reactions to themselves and be responsive to interpersonal cues, leading them to specialize in interpersonally oriented traits (Aries 2006). In contrast, because of their stronger status position, men get more opportunities to initiate actions and influence decision-making, leading them to specialize in task-oriented traits. Each of these theories—lack of fit model, role congruity theory, and status construction theory—argues that the social construction of both gender and leadership exerts a powerful influence on individuals' beliefs about which sex belongs in the leader role.

Powell and Butterfield (1979, 1989, in press; Powell *et al.* 2002) have taken a different approach to the analysis of leader stereotypes in a research programme that also began in the 1970s. For four decades, they have periodically asked part-time MBA students in the United States, nearly all of whom work full-time, and undergraduate business students to describe both themselves and a 'good manager' on the Bem Sex-Role Inventory (BSRI; Bem 1974, 1981), which includes the dimensions of masculinity and femininity. *Masculinity* is defined as beliefs that people have about the extent to which they possess masculine (i.e. task-oriented, agentic) traits associated with men in gender stereotypes. *Femininity* is defined as beliefs that people have about the extent to which they possess feminine (i.e. interpersonally oriented, communal) traits that are associated with women in gender stereotypes (Eagly *et al.* 2000; Kite *et al.* 2008).

When Powell and Butterfield (1979) first collected data, the proportion of women in management positions in the United States was just beginning to rise. Based on this trend, they hypothesized that a good manager would be seen as possessing similarly high levels of masculine and feminine traits, which Bem (1981) called an 'androgynous' personal profile. However, contrary to their hypothesis, a good manager was seen as possessing predominantly masculine characteristics by a majority of respondents in all groups, including undergraduate and part-time graduate males and females. Thus, the idea of *think manager—think masculine* prevailed in these studies.

Koenig *et al.* (2011) conducted a meta-analysis of studies following Powell and Butterfield's (1979) research paradigm. According to this meta-analysis, although the proportion of respondents from different groups that describe a manager as possessing predominantly masculine characteristics has declined somewhat over time, men and women still describe a good manager in predominantly masculine terms. These results are consistent with the lack of fit model (Heilman 1983, 1995), role congruity theory

(Eagly and Karau 2002), and status construction theory (Ridgeway 1991, 2006; Webster and Hysom 1998). Overall, managerial stereotypes continue to reflect the dual notions of think manager—think masculine and think manager—think male.

Are the relationships among sex, gender stereotypes, and leader stereotypes important? The answer to this question is an emphatic 'yes'. Powell and Butterfield (in press) examined the correspondence between self-descriptions and descriptions of a 'good manager' for samples from the same two populations, undergraduate business students and part-time MBA students, over four decades. The correspondence between self- and good-manager descriptions was greater for men than women for all data combined, as well as for data collected during each decade. However, despite changes in women's status across the four decades in which data were collected, the correspondence between self- and good-manager descriptions exhibited little change over time for either women or men considered separately. Thus, men still view themselves as more similar to a good manager than women do.

These results have implications for women as they prepare to enter the labour force and once they are in it. If the correspondence between women's self- and good-manager descriptions continues to be less than that of men, women may be less likely than men to develop their managerial skills, pursue careers in the managerial ranks, or pursue careers in the *top* management ranks of organizations. For female undergraduate business students, this lack of correspondence could lead to a disinclination to pursue initial job opportunities with managerial responsibilities or place them on a career path towards positions with such responsibilities. For female part-time MBAs, who are already likely to have chosen to follow a managerial career path as indicated by their choice of degree programme, it could lead to a disinclination to pursue executive positions by 'opting out' of careers that put them on a fast track to top management ranks (Mainiero and Sullivan 2006; Still 2006).

When women decide to pursue management positions at any level, they may encounter barriers in the selection process. Given leader stereotypes that focus on masculine characteristics and males, women may be less likely than men to be hired for management positions. Moreover, once women assume leader roles, leader stereotypes act as constraints on their behaviour. Many organizations exert strong pressures on their members to conform to standards of behaviour dictated by those in power. As long as men remain in the majority in the top management ranks, the masculine leader stereotype is likely to prevail, and women throughout the organization will be expected to behave as men. A masculine stereotype of the good manager is self reinforcing and inhibits the expression of femininity by women in management positions.

In addition, the mismatch between the leader role and the female gender role constrains the advancement of female managers. When performance evaluations are conducted, women may receive lower ratings than men for similar levels of performance (Heilman 2001). Women may also be subjected to discrimination when decisions are made about promotions into higher leadership positions, resulting in a 'glass ceiling' that makes it difficult for them to rise in managerial hierarchies (Marshall 1984; Davidson and Cooper 1992; Powell 1999; Barreto et al. 2009). Thus, being competent does not ensure that a female manager will have the same amount of organizational success as her male equivalent.

Nonetheless, leader stereotypes may not apply to the actual practice of management. Stereotypes are resistant to change and do not necessarily reflect current realities. Widely held stereotypes that men are better managers and that better managers are masculine may not reflect what makes managers good. Instead, these stereotypes may reflect only that most managers have been men and that most men have been expected to live up to the masculine stereotype.

ATTITUDES

In 1965, *Harvard Business Review* published results from a survey of executives' attitudes towards women in managerial roles (Bowman *et al.* 1965). Most female executives (82 per cent) had a favourable attitude towards women in management; they believed that women should be treated as individuals rather than as a uniform group. In contrast, a large proportion of male executives (41 per cent) had an unfavourable attitude towards women in management. They believed not only that women were special but also that they had a special place, which was outside the ranks of management. Older men tended to be more accepting of women in managerial roles than were younger men. Also, men who had been superiors or peers of women managers thought more favourably of them than men who had worked only for men. Overall, few men (27 per cent) thought that they would feel comfortable working for a woman boss.

Over the next four decades, as noted earlier, the composition of the managerial ranks changed considerably. According to replications of the original *HBR* survey conducted twenty and forty years later (Sutton and Moore 1985; Carlson *et al.* 2006), male executives' attitudes about whether women belong in leader roles also changed considerably. The proportion of male executives who expressed a favourable attitude towards women in management increased from 35 per cent in 1965 to 88 per cent in 2005 (Bowman *et al.* 1965; Carlson *et al.* 2006). Similarly, the proportion who would feel comfortable working for a woman boss increased from 27 per cent in 1965 to 71 per cent in 2005. These are huge shifts in male executives' attitudes. In comparison, the proportion of female executives who expressed a favourable attitude towards women in management remained generally high, ranging from 82 per cent in 1965 to 88 per cent in 2005.

Moreover, male executives in 2005 were more positive about how female executives were being accepted in business than the women themselves. The proportion of men who thought that a woman must be exceptional to succeed in business dropped from 90 per cent in 1965 to 32 per cent in 2005, whereas this proportion dropped by a lesser amount, from 88 per cent to 69 per cent, for women (Bowman *et al.* 1965; Carlson *et al.* 2006). Similarly, the proportion of men who thought that the business community would never fully accept women executives fell from 61 per cent in 1965 to 16 per cent in 2005, whereas this proportion fell from 47 per cent to 34 per cent for women, a smaller drop-off. Thus, female executives viewed the business community's general attitude towards women as executives in a less favourable light than male executives did.

However, in studies of actual managers and their subordinates, subordinates typically express similar satisfaction with male and female managers. Subordinates do not appear to respond differently to male and female leaders for whom they have actually worked. The experience of having been supervised by a woman contributes to more positive attitudes towards women as leaders (Ezell *et al.* 1981). Being in direct contact with or proximity to women as leaders may serve to dispel stereotypes about whether women belong in leader roles. However, individuals' attitudes towards women as leaders do not become more positive with experience unless that experience itself is positive. When individuals are more *satisfied* with their interactions with women leaders, they are more positive about women in leader roles (Bhatnagar and Swamy 1995).

The focus of this section of the chapter has been on attitudes towards women, not men, as leaders. Male leaders essentially are taken for granted. Having a woman as a manager has only recently become a common experience for workers. As more people have more experience with women in leader roles, female leaders elicit less negative reactions. However, prejudices against women as leaders resulting from sexist attitudes have not disappeared, although hostility towards women as leaders is expressed less openly than in the past (Glick *et al.* 1997; Glick *et al.* 2000). Women continue to face prejudices in the leader role that men do not face (Masser and Abrams 2004; Rudman and Kilianski 2000). These prejudices make it more difficult for women to be effective as leaders.

BEHAVIOUR AND EFFECTIVENESS

To what extent do perceptions of leadership match current realities? Are men and masculine behaviours really best in leadership positions as leader stereotypes suggest? To consider these questions, I review how leadership theories have regarded the merits of stereotypically feminine or masculine behaviours. Finally, I examine research evidence on sex differences in leader behaviour and effectiveness.

Leadership Theories and Gender Stereotypes

Early theories of what leader behaviours work and do not work well were based almost entirely on studies of male managers. An early compendium of research results, *Handbook of Leadership* (Stogdill 1974), discovered few studies that examined female leaders exclusively or even included female leaders in their samples. When female managers were present in organizations being studied, they were usually excluded from the analysis because their few numbers might distort the results! It was as if female managers were less legitimate or less worthy of observation than male managers. Although management researchers no longer exclude female managers from their samples, early theories of leadership were originally developed with male managers in mind and, not surprisingly, endorsed a preponderance of masculine traits.

In recent years, transformational and transactional leadership (Bass 1985, 1998) have become the primary focus of leadership theories. Judge and Bono (2000) found a greater number of citations of transformational leadership theory (or charismatic leadership theory, as it is also called) than all other leadership theories combined. A search of the PsycINFO database by decade found that the term 'transformational leadership' appeared in the abstract of no citations from 1971 to 1980, 31 citations from 1981 to 1990, 234 citations from 1991 to 2000, and 1,041 citations from 2001 to 2010 (PsycINFO 2012).

Eagly et al. (2003) characterized transformational leadership as consisting of five dimensions: (1) idealized influence-attributes, or the display of attributes that induce followers to view the leader with pride and respect; (2) idealized influence-behaviours, or communication of a sense of values, purpose, and mission importance; (3) inspirational motivation, or optimism and excitement about the mission's importance and attainability; (4) intellectual stimulation, or encouragement of followers to question basic assumptions and consider problems and tasks from new perspectives; and (5) individualized consideration, or a focus on the development and mentoring of followers as individuals and addressing their specific needs. Of these dimensions, individualized consideration reflects the feminine gender stereotype in its developmental focus and its concern with relationships and the needs of others (Eagly and Johannesen-Schmidt 2001; Eagly et al. 2003), whereas the other dimensions seem more gender-neutral.

In contrast, transactional leadership focuses on clarifying the responsibilities of subordinates and then responding to how well subordinates execute their responsibilities (Bass et al. 1996; Bass 1998; Rafferty and Griffin 2004). It consists of two dimensions of behaviour: (1) contingent reward, or the promise and provision of suitable rewards if followers achieve their assigned objectives; and (2) management by exception, which is further divided into active and passive management by exception. Transactional leaders who engage in active management by exception monitor subordinate performance for mistakes, whereas those who engage in passive management by exception wait for subordinate difficulties to be brought to their attention before intervening. Transformational leaders may be transactional when it is necessary to achieve their goals. However, transactional leaders are seldom transformational.

Distinct from both transformational and transactional leadership is laissez-faire leadership. Laissez-faire leaders avoid taking responsibility for leadership altogether. Such leaders refrain from giving direction or making decisions and do not involve themselves in the development of their followers.

Transformational leadership has been positively associated with nurturance, a feminine trait, and negatively associated with aggression, a masculine trait (Ross and Offermann 1997). Further, Bono and Judge's (2004) meta-analysis of the relationship between transformational leadership and the Big Five personality traits (Wiggins 1996) found that it was positively related to agreeableness, which reflects the leader's warmth, kindness, gentleness and cooperativeness; agreeableness in turn is strongly linked to the feminine gender stereotype. Overall, transformational leadership appears to be more associated with the feminine gender stereotype (in its individualized consideration dimension and its positive correlations with nurturance and agreeableness) and less associated with the masculine gender stereotype (in its negative correlation with aggression) than traditional views of effective leadership that emphasize masculine traits (Bass *et al.* 1996; Eagly and Johannesen-Schmidt 2001; Koenig *et al.* 2011).

In contrast, the management by exception dimension of transactional leadership, whether active or passive, seems congruent with the masculine gender role in its focus on correcting followers' mistakes because it stresses immediate task accomplishment over long-term building of relationships and favours use of the leadership position to control others. In addition, the contingent reward dimension of transactional leadership appears to be congruent with the masculine gender role because it is primarily task-oriented.

Recall that leader stereotypes place a high value on masculine characteristics. Even though early leadership theories were developed at a time when there were far fewer women in leader roles, transformational leadership theory, which is the most popular leadership theory today, does not support these stereotypes. Thus, prevailing leadership theories do not suggest that either feminine or masculine behaviours are the key to leader effectiveness.

Sex Differences in Leader Behaviour and Effectiveness

Researchers have devoted a great deal of attention to sex differences in leader behaviour and effectiveness. In this section, I review the major research findings.

Sex differences have been examined in several types of leader behaviours. A meta-analysis of sex differences in transformational and transactional leadership found that female leaders are more transformational than their male counterparts (Eagly *et al.* 2003). Women rated higher than men on four of five dimensions of transformational leadership: idealized influence-attributes, inspirational motivation, intellectual stimulation, and individualized consideration. Women also rated higher than men on the contingent reward dimension of transactional leadership. In contrast, men rated higher than women on two dimensions of transactional leadership: active management by exception and passive management by exception. Men also rated higher than women in laissez-faire leadership.

Further evidence from meta-analysis also suggests that all of the dimensions of transformational leadership and the contingent reward dimension of transactional leadership are positively associated with leader effectiveness as reflected in individual, group, and organizational performance. In contrast, passive management by exception and laissez-faire leadership are negatively associated with leader effectiveness (Lowe *et al.* 1996). Thus, the above results suggest that women rate higher than men in behaviour that contributes to their effectiveness as leaders and lower than men in behaviour that would detract from their effectiveness.

A separate meta-analysis on sex differences in leader effectiveness found that women and men overall did not differ in their effectiveness as leaders (Eagly *et al.* 1995). Most of the studies included in this meta-analysis were conducted in organizational settings. Men were more effective than women in military settings, which are extremely male-intensive, whereas women were more effective than men in education, government, and social service settings, which are less male-intensive. Neither men nor women were more effective in business settings. Men were more effective than women when the particular leader role examined was more congruent with the male gender role and when there were a larger proportion of men as both leaders and subordinates. Further, men were more effective than women in lower-level management positions, whereas women were more effective than men in middle-level management positions. The position of middle manager is often regarded as requiring heavy use of interpersonal skills to wield influence, which would favour women according to gender stereotypes. There have not been sufficient studies of men and women in top management positions to allow a comparison of the sexes using meta-analysis.

In summary, when sex differences in leader behaviour have been found, the bulk of evidence suggests the existence of stereotypical differences. As gender stereotypes would predict, women are higher than men in the individualized consideration dimension of transformational leadership, which is associated with the feminine stereotype, and lower than men in the active and passive management by exception dimensions of transactional leadership, which are associated with the masculine stereotype. However, contrary to gender stereotypes, women are higher than men in the contingent reward dimension of transactional leadership.

Further, sex differences in leader behaviour that have been found favour women, not men. Women are higher than men in dimensions of behaviour that contribute to leader effectiveness (idealized influence-attributes, inspirational motivation, intellectual stimulation, individualized consideration, contingent reward) and lower than men in dimensions of behaviour that detract from leader effectiveness (passive management by exception, laissez-faire leadership). Reinforcing the favourability of these results for women as leaders, successful organizations are said to be shifting away from an authoritarian model of leadership and towards a more transformational and democratic model (Drucker 1988 Lawler *et al.* 1995; Hitt *et al.* 1998).

Studies that directly measure leader effectiveness, however, rate women as no more or less effective than men. Additional evidence suggests that situational factors influence whether men or women are more effective as leaders. These factors ⁱ nature of the organizational setting and leader role, the proportiᵢ and followers, and the managerial level of the position. As a result are more congenial to male leaders, whereas other leader roles are ᵢ female leaders.

Thus, the evidence clearly refutes the stereotypes that men are better leaders and that better leaders are masculine. Effective leadership today requires a combination of behaviours that are masculine (e.g. contingent reward) and feminine (e.g. individualized consideration) and the absence of other behaviours that are sex-neutral (e.g. laissez-faire leadership). Women have been found to exhibit more of behaviours that contribute to leader effectiveness than do men. However, situations differ in whether they favour women or men as leaders.

DISCUSSION

What do four decades of research tell us about the intersection of sex, gender, and leadership? Actually, they tell us a clear story. Five conclusions may be reached from this review. First, according to statistics on women's status, women have achieved higher societal status in recent decades than before, although still lower than that of men. Second, according to research on leader preferences, a male boss is still preferred over a female boss. Third, according to research on leader stereotypes, men are still believed to be better managers and better managers are still believed to be masculine. Fourth, according to research on attitudes towards women as leaders, women are still subject to hostility and prejudice when they are considered in relation to a leader role. Fifth, according to research on leader behaviour and effectiveness, female leaders exhibit a greater amount of behaviours that are positively associated with effectiveness and a lesser amount of behaviours that are negatively associated with effectiveness than male leaders. Overall, these conclusions suggest that the playing field that constitutes the managerial ranks continues to be tilted in favour of men, despite evidence suggesting that women as a group are the superior leaders.

Trends in research findings suggest a softening of the first four conclusions over time. First, women now occupy more top management positions than before, although female executives are still a distinct minority. Second, 'it makes no difference to me' has become the most endorsed response to poll questions about leader preferences, and younger respondents equally favour a male or female boss. Third, leader stereotypes that men are better managers are now endorsed more by men than by women, and the extent to which a good manager is described in predominantly masculine terms has diminished over time. Fourth, attitudes towards women as leaders become more positive among individuals who have had more satisfying experiences with female leaders, and people are increasingly having the opportunity to have such experiences. Thus, although there has been considerable stability in research findings over four decades, there is also evidence of change.

What are the implications of this review for theory, research, and practice in the next decade? The field does not suffer from a lack of theories. Indeed, four decades have yielded an abundance of theories to explain the research findings that are presented. In this chapter, in the interests of conserving space, I have cited the lack of fit model

(Heilman 1983, 1995), role congruity theory (Eagly and Karau 2002), status construction theory (Ridgeway 1991, 2006; Webster and Hysom 1998), and theories of transformational and transactional leadership (Bass 1985, 1998). However, many other theories of both gender (cf. Archer and Lloyd 2002; Lippa 2005; Wood and Eagly 2010) and leadership (cf. Avolio *et al.* 2009; Bass and Bass 2008) may be applied to explain such findings. Thus, the development of new, ground-breaking theories is not a priority.

Future research should continue to track trends in status, preferences, stereotypes, attitudes, behaviour, and effectiveness pertaining to the intersection of sex, gender, and leadership. However, new, ground-breaking research that sheds additional light on the nature of this intersection is not a priority either. We already have ample research evidence of what is going on. The most important question to be addressed in future work in this field is how to fix it. Waiting for change to happen and hoping that it will eradicate workplace disparities that place female leaders at a disadvantage is not a sufficient response.

Bem (1978: 19) said that '*behavior* should have no gender'. Ideally, to amend Bem's (1978) statement, *leader behaviour* should have no gender. The sex of individuals who hold leader roles should be of little concern. What should matter is how well individuals, male and female, respond to the demands of the particular leader role that they occupy. Evidence increasingly suggests that women tend to be better suited than men to serve as leaders in the ways required in the global economy. However, this is *not* to say that organizations should choose women for leader roles on the basis of their sex. The challenge for organizations is to take advantage of and develop the capabilities of all individuals in leader roles and then create conditions that give leaders of both sexes an equal chance to succeed (Falkenberg 1990; Yoder 2001; Hogue *et al.* 2002). No matter what the intersection of sex, gender, and leadership may look like, the goal should be to enhance the likelihood that *all* people, women and men, will be effective in leader roles. If scholars can help organizations to achieve this goal in the future, future decades of research on this topic could tell quite a different story.

References

Alimo-Metcalfe, B. (2010). 'Developments in Gender and Leadership: Introducing a New "Inclusive" Model', *Gender in Management: An International Journal*, 25: 630–9.

Archer, J., and Lloyd, B. (2002). Sex and Gender (2nd edn). Cambridge: Cambridge University Press.

Aries, E. (2006). 'Sex Differences in Interaction: A Reexamination', in K. Dindia and D. J. Canary (eds), *Sex Differences and Similarities in Communication* (2nd edn), 21–36. Mahwah, NJ: Erlbaum.

Avolio, B. J., Walumbwa, F. O., and Weber, T. J. (2009). 'Leadership: Current Theories, Research, and Future Directions', *Annual Review of Psychology*, 60: 421–49.

Barreto, M., Ryan, M. K., and Schmitt, M. T. (eds) (2009). *The Glass Ceiling in the 21st Century: Understanding Barriers to Gender Equality*. Washington, DC: American Psychological Association.

Bartol, K. M. (1978). 'The Sex Structuring of Organizations: A Search for Possible Causes', *Academy of Management Review*, 3: 805–15.

Bass, B. M. (1985). *Leadership and Performance beyond Expectations*. New York: Free Press.

Bass, B. M. (1998). *Transformational Leadership: Industry, Military, and Educational Impact*. Mahwah, NJ: Erlbaum.

Bass, B. M., and Bass, R. (2008). *The Bass Handbook of Leadership: Theory, Research, and Managerial Applications* (4th edn). New York: Free Press.

Bass, B. M., Avolio, B. J., and Atwater, L. (1996). 'The Transformational and Transactional Leadership of Men and Women', *Applied Psychology: An International Review*, 45(1): 5–34.

Bem, S. L. (1974). 'The Measurement of Psychological Androgyny', Journal of Consulting and Clinical Psychology, 42: 155–62.

Bem, S. L. (1978). 'Beyond Androgyny: Some Presumptuous Prescriptions for a Liberated Sexual Identity', in J. A. Sherman and F. L. Denmark (eds), *The Psychology of Women: Future Direction in Research*, 1–23. New York: Psychological Dimensions.

Bem, S. L. (1981). *Bem Sex-Role Inventory: Professional Manual*. Palo Alto, CA: Consulting Psychologists Press.

Bhatnagar, D., and Swamy, R. (1995). 'Attitudes toward Women as Managers: Does Interaction Make a Difference?', *Human Relations*, 48: 1285–1307.

Bono, J. E., and Judge, T. A. (2004). 'Personality and Transformational and Transactional Leadership: A Meta-Analysis', *Journal of Applied Psychology*, 89: 901–10.

Bowman, G. W., Worthy, N. B., and Greyser, S. A. (1965). 'Are Women Executives People?', *Harvard Business Review*, 4(4): 14–28, 164–78.

Brenner, O. C., Tomkiewicz, J., and Schein, V. E. (1989). 'The Relationship between Sex Role Stereotypes and Requisite Management Characteristics Revisited', *Academy of Management Journal*, 32: 662–9.

Broadbridge, A., and Simpson, R. (2011). '25 Years On: Reflecting on the Past and Looking to the Future in Gender and Management Research', *British Journal of Management*, 22: 470–83.

Butterfield, D. A., and Grinnell, J. P. (1999). 'Re-viewing Gender, Leadership, and Managerial Behavior: Do Three Decades of Research Tell us Anything?', in G. N. Powell (ed.), *Handbook of Gender and Work*, 223–38. Thousand Oaks, CA: Sage.

Carroll, J. (2006). 'Americans Prefer Male Boss to a Female Boss', *Gallup News Service*, 1 Sept. http://www.gallup.com (accessed May 2009).

Carlson, D. S., Kacmar, K. M., and Whitten, D. (2006). 'What Men Think they Know about Executive Women', *Harvard Business Review*, 84(9): 28.

Catalyst (2011a). *2011 Catalyst Census: Fortune 500 Women Executive Officers and Top Earners*. New York: Catalyst. http://www.catalyst.org (accessed Feb. 2012).

Catalyst (2011b). *No News is Bad News: Women's Leadership Still Stalled in Corporate America*. New York: Catalyst. http://www.catalyst.org (accessed Feb. 2012).

Catalyst (2012). *Women CEOs of the Fortune 1000*. New York: Catalyst. http://www.catalyst.org (accessed Aug. 2012).

Davidson, M. J., and Cooper, C. L. (1992). *Shattering the Glass Ceiling: The Woman Manager*. London: Chapman.

Davidson, M. J., and Burke, R. J. (eds) (1994). *Women in Management: Current Research Issues*. London: Chapman.

Davidson, M. J., and Burke, R. J. (eds) (2000). *Women in Management: Current Research Issues*, ii. London: Sage.

Davidson, M. J., and Burke, R. J. (eds) (2011). *Women in Management Worldwide: Progress and Prospects* (2nd edn). Aldershot: Gower.

Drucker, P. F. (1988). 'The Coming of the New Organization', *Harvard Business Review*, 88(1): 45–53.

Eagly, A. H., and Johannesen-Schmidt, M. C. (2001). 'The Leadership Styles of Women and Men', *Journal of Social Issues*, 57: 781–97.

Eagly, A. H., and Karau, S. J. (2002). 'Role Congruity Theory of Prejudice toward Female Leaders', *Psychological Review*, 109: 573–98.

Eagly, A. H., Karau, S. J., and Makhijani, M. G. (1995). 'Gender and the Effectiveness of Leaders: A Meta-Analysis', *Psychological Bulletin*, 117: 125–45.

Eagly, A. H., Wood, W., and Diekman, A. B. (2000). 'Social Role Theory of Sex Differences and Similarities: A Current Appraisal', in T. Eckes and H. M. Trautner (eds), *The Developmental Social Psychology of Gender*, 23–174. Mahwah, NJ: Erlbaum.

Eagly, A. H., Johannesen-Schmidt, M. C., and van Engen, M. L. (2003). 'Transformational, Transactional, and Laissez-Faire Leadership Styles: A Meta-Analysis Comparing Women and Men', *Psychological Bulletin*, 108: 233–56.

Ezell, H. F., Odewahn, C. A., and Sherman, J. D. (1981). 'The Effects of having been Supervised by a Woman on Perceptions of Female Managerial Competence', *Personnel Psychology*, 34: 291–9.

Falkenberg, L. (1990). 'Improving the Accuracy of Stereotypes within the Workplace', *Journal of Management*, 16: 107–18.

Fullagar, C. J., Sumer, H. C., Sverke, M., and Slick, R. (2003). 'Managerial Sex-Role Stereotyping: A Cross Cultural Analysis', *International Journal of Cross Cultural Management*, 3: 93–107.

Glick, P., Diebold, J., Bailey-Werner, B., and Zhu, L. (1997). 'The Two Faces of Adam: Ambivalent Sexism and Polarized Attitudes toward Women', *Personality and Social Psychology Bulletin*, 23: 1323–34.

Glick, P., Fiske, S. T., Mladinic, A., Saiz, J. L., Abrams, D., Masser, B., *et al.* (2000). 'Beyond Prejudice as Simple Antipathy: Hostile and Benevolent Sexism across Cultures', *Journal of Personality and Social Psychology*, 79: 763–75.

Graves, L. M., and Powell, G. N. (1988). 'An Investigation of Sex Discrimination in Recruiters' Evaluations of Actual Applicants', *Journal of Applied Psychology*, 73: 20–9.

Hare-Mustin, R. T., and Maracek, J. (1988). 'The Meaning of Difference: Gender Theory, Postmodernism, and Psychology', *American Psychologist*, 43: 455–64.

Haslam, S. A., and Ryan, M. K. (2008). 'The Road to the Glass Cliff: Differences in the Perceived Suitability of Men and Women for Leadership Positions in Succeeding and Failing Organizations', *Leadership Quarterly*, 19: 530–46.

Heilman, M. E. (1983). 'Sex Bias in Work Settings: The Lack of Fit Model', in L. L. Cummings and B. M. Staw (eds), *Research in Organizational Behavior*, v. 269–98. Greenwich, CT: JAI Press.

Heilman, M. E. (1995). 'Sex Stereotypes and their Effects in the Workplace: What we Know and What we Don't Know', *Journal of Social Behavior and Personality*, 10(6): 3–26.

Heilman, M. E. (2001). 'Description and Prescription: How Gender Stereotypes Prevent Women's Ascent up the Organizational Ladder', *Journal of Social Issues*, 57: 657–74.

Heilman, M. E., Block, C. J., Martell, R. F., and Simon, M. C. (1989). 'Has Anything Changed? Current Characterizations of Men, Women, and Managers', *Journal of Applied Psychology*, 74: 935–42.

Hitt, M. A., Keats, B. W., and DeMarie, S. M. (1998). 'Navigating in the New Competitive Landscape: Building Strategic Flexibility and Competitive Advantage in the 21st Century', *Academy of Management Executive*, 12(4): 22–42.

Hofstede, G. (2001). *Culture's Consequences: Comparing Values, Behaviors, Institutions, and Organizations across Nations* (2nd edn). Thousand Oaks, CA: Sage.

Hogue, M. B., Yoder, J. D., and Ludwig, J. (2002). 'Increasing Initial Leadership Effectiveness: Assisting Both Women and Men', *Sex Roles*, 46: 377–84.

Judge, T. A., and Bono, J. E. (2000). 'Five-Factor Model of Personality and Transformational Leadership', *Journal of Applied Psychology*, 85: 751–65.

Kite, M. E., Deaux, K., and Haines, E. L. (2008). 'Gender Stereotypes', in F. L. Denmark and M. A. Paludi (eds), *Psychology of Women: A Handbook of Issues and Theories* (2nd edn), 205–36. Westport, CT: Praeger.

Koenig, A. M., Eagly, A. H., Mitchell, A. A., and Ristikari, T. (2011). 'Are Leader Stereotypes Masculine? A Meta-Analysis of Three Research Paradigms', *Psychological Bulletin*, 137: 616–42.

Lawler, E. E., III, Mohrman, S. A., and Ledford, G. E., Jr. (1995). *Creating High Performance Organizations: Practices and Results of Employee Involvement and Total Quality Management in Fortune 1000 Companies*. San Francisco: Jossey-Bass.

Lippa, R. A. (2005). *Gender, Nature, and Nurture* (2nd edn). Mahwah, NJ: Erlbaum.

Lowe, K. B., Kroeck, K. G., and Sivasubramaniam, N. (1996). 'Effectiveness Correlates of Transformational and Transactional Leadership: A Meta-Analytic Review of the MLQ Literature', *Leadership Quarterly*, 7: 385–425.

Mainiero, L. A., and Sullivan, S. E. (2006). *The Opt-Out Revolt: Why People are Leaving Companies to Create Kaleidoscope Careers*. Mountain View, CA: Davies-Black.

Marshall, J. (1984). *Women Managers: Travellers in a Male World*. Chichester: Wiley.

Masser, B. M., and Abrams, D. (2004). 'Reinforcing the Glass Ceiling: The Consequences of Hostile Sexism for Female Managerial Candidates', *Sex Roles*, 51: 609–15.

Nieva, V. F., and Gutek, B. A. (1981). *Women in Work: A Psychological Perspective*. New York: Praeger.

Powell, G. N. (1999). 'Reflections on the Glass Ceiling: Recent Trends and Future Prospects', in G. N. Powell (ed.), *Handbook of Gender and Work*, 325–45. Thousand Oaks, CA: Sage.

Powell, G. N. (2011). 'The Gender and Leadership Wars', *Organizational Dynamics*, 40: 1–9.

Powell, G. N., and Butterfield, D. A. (1979). 'The "Good Manager": Masculine or Androgynous?', *Academy of Management Journal*, 22: 395–403.

Powell, G. N., and Butterfield, D. A. (1989). 'The "Good Manager": Did Androgyny Fare Better in the 1980s?', *Group and Organization Studies*, 14: 216–33.

Powell, G. N., and Butterfield, D. A. (In press). 'Correspondence between Self- and Good-Manager Descriptions: Examining Stability and Change over Four Decades', *Journal of Management*.

Powell, G. N., and Graves, L. M. (2003). *Women and Men in Management* (3rd edn). Thousand Oaks, CA: Sage.

Powell, G. N., Butterfield, D. A., and Parent, J. D. (2002). 'Gender and Managerial Stereotypes: Have the Times Changed?', *Journal of Management*, 28: 177–93.

PsycInfo (2012). Search of database. Retrieved 25 May 2012, from senior author's institutional subscription.

Rafferty, A. E., and Griffin, M. A. (2004). 'Dimensions of Transformational Leadership: Conceptual and Empirical Extensions', *Leadership Quarterly*, 15: 329–54.

Ridgeway, C. L. (1991). 'The Social Construction of Status Value: Gender and Other Nominal Characteristics', *Social Forces*, 70: 367–86.

Ridgeway, C. L. (2006). 'Gender as an Organizing Force in Social Relations: Implications for the Future of Inequality', in F. D. Blau, M. C. Brinton, and D. B. Grusky (eds), *The Declining Significance of Gender?*, 265–87. New York: Russell Sage Foundation.

Riger, S., and Galligan, P. (1980). 'Women in Management: An Exploration of Competing Paradigms', *American Psychologist*, 35: 902–10.

Room for debate (2009). 'Do Women Make Better Bosses?', *New York Times*, 2 Aug. http://roomfordebate.blogs.nytimes.com (accessed Aug. 2009).

Ross, S. M., and Offermann, L. R. (1997). 'Transformational Leaders: Measurement of Personality Attributes and Work Group Performance', *Personality and Social Psychology Bulletin*, 23: 1078–86.

Rudman, L. A., and Glick, P. (2001). 'Prescriptive Gender Stereotypes and Backlash toward Agentic Women', *Journal of Social Issues*, 57: 743–62.

Rudman, L. A., and Kilianski, S. E. (2000). 'Implicit and Explicit Attitudes toward Female Authority', *Personality and Social Psychology Bulletin*, 26: 1315–28.

Ryan, M. K., and Haslam, S. A. (2005). 'The Glass Cliff: Evidence that Women are Over-Represented in Precarious Leadership Positions', *British Journal of Management*, 16: 81–90.

Ryan, M. K., and Haslam, S. A. (2007). 'The Glass Cliff: Exploring the Dynamics Surrounding the Appointment of Women to Precarious Leadership Positions', *Academy of Management Review*, 32: 549–72.

Ryan, M. K., and Haslam, S. A. (2009). 'Glass Cliffs are Not So Easily Scaled: On the Precariousness of Female CEOs' Positions', *British Journal of Management*, 20: 13–16.

Schein, V. E. (1973). 'The Relationship between Sex-Role Stereotypes and Requisite Management Characteristics', *Journal of Applied Psychology*, 57: 340–4.

Schein, V. E. (1975). 'Relationships between Sex Role Stereotypes and Requisite Management Characteristics among Female Managers', *Journal of Applied Psychology*, 60: 340–4.

Schein, V. E. (2001). 'A Global Look at Psychological Barriers to Women's Progress in Management', *Journal of Social Issues*, 57: 675–88.

Schein, V. E., and Mueller, R. (1992). 'Sex Role Stereotyping and Requisite Management Characteristics: A Cross Cultural Look', *Journal of Organizational Behavior*, 13: 439–47.

Schein, V. E., Mueller, R., Lituchy, T., and Liu, J. (1996). 'Think Manager—Think Male: A Global Phenomenon?', *Journal of Organizational Behavior*, 17: 33–41.

Simmons, W. W. (2001). 'When it Comes to Choosing a Boss, Americans Still Prefer Men', *Gallup News Service*, 11 Jan. http://www.gallup.com (accessed Sept. 2001).

Still, M. C. (2006). 'The Opt-Out Revolution in the United States: Implications for Modern Organizations', *Managerial and Decision Economics*, 27: 159–71.

Stogdill, R. M. (1974). Handbook of Leadership. New York: Free Press.

Sutton, C. D., and Moore, K. K. (1985). 'Executive Women—20 Years Later', *Harvard Business Review*, 63(5): 42–66.

Terborg, J. R. (1977). 'Women in Management: A Research Review', *Journal of Applied Psychology*, 62: 647–64.

Unger, R. K. (1979). 'Toward a Redefinition of Sex and Gender', *American Psychologist*, 34: 1085–94.

US Department of Education, National Center for Education Statistics (2011). Digest of Education Statistics. http://nces.ed.gov (accessed May 2012).

US Department of Labor, Bureau of Labor Statistics (1972–2011). 'Employment and Earnings: Annual Table on Employed Persons by Occupation, Sex, and Race for Previous Year'. Archived online by HeinOnline. http://heinonline.org (accessed May 2012).

US Department of Labor, Bureau of Labor Statistics (2012). 'Labor Force Statistics from the Current Population Survey'. http://www.bls.gov/cps (accessed Aug. 2012).

Webster, M., Jr., and Hysom, S. J. (1998). 'Creating Status Characteristics', *American Sociological Review*, 63: 351–78.

Wiggins, J. S. (ed.) (1996). *The Five-Factor Model of Personality: Theoretical Perspectives*. New York: Guilford.

Wood, W., and Eagly, A. H. (2010). 'Gender', in S. T. Fiske, D. T. Gilbert, and G. Lindzey (eds), *Handbook of Social Psychology* (5th edn), i. 629–67. New York: Oxford University Press.

Yoder, J. D. (2001). 'Making Leadership Work More Effectively for Women', *Journal of Social Issues*, 57: 815–28.

CHAPTER 12

GENDERED CONSTRUCTIONS OF MERIT AND IMPRESSION MANAGEMENT WITHIN PROFESSIONAL SERVICE FIRMS

SAVITA KUMRA

INTRODUCTION

IN this chapter the concepts of merit and impression management are considered in relation to their impact on advancement decisions within professional service firms (PSFs). Of particular interest in the chapter is the gendered nature of the concepts and the implications for women's advancement. Reviewing the literature in relation to merit, we see that women are generally trustful of the concept and believe in its ability to deliver advancement processes that are objective, fair, and ability-based (Farber and Sherry 1995; Scully 1997; Kumra and Vinnicombe 2008). As a consequence, organizations proclaiming themselves as meritocratic are viewed as 'blameless', even when their senior ranks are male-dominated, as the responsibility for success or failure within the meritocratic system lies with the individual (Brennan and Naidoo 2008; Sommerlad 2012; Brink and Benschop 2012). Indeed, women are often the staunchest defenders of the status quo. For example, when the recent House of Lords Constitution Committee reported on Judicial Appointments and drew attention to the extremely low number of women in the UK judiciary, it was a female judge Lady Butler-Sloss who was among the first to call for action, but who when speaking in the House of Lords in the debate on the Second Reading of the Crime and Courts bill, strongly cautioned against any initiatives deemed unmeritocratic. In her comments, she gave 'strong support' for greater diversity in those appointed to the judiciary, with the caveat that such appointments must, of course, be 'made on merit'. She went on to say 'It will be very important that women—particularly those from ethnic minorities—who may not be able to bear the strain of the judicial process are not placed in a position where they may find themselves failing because there has been too much enthusiasm for diversity and not enough for merit' (2012).

Impression management encompasses a set of tactics, including self-promotion, managing upwards or ingratiation, and visibility (Rosenfeld *et al.* 1995). This has received increasing attention over the years as the potential to aid career advancement has become more widely recognized (Singh *et al.* 2002; Kumra and Vinnicombe 2008, 2010). However, unlike meritocratic principles, there is evidence that women are less reluctant to engage with these approaches, despite evidence that they may assist in their career advancement. The reasons most commonly cited for this reluctance is that they exemplify activities in direct opposition to meritocratic principles. They are seen to be subjective, provide an unfair advantage, and yield benefits to the individual that are not necessarily a consequence of their talent or ability.

In this chapter, I seek to explore whether merit and impression management are indeed in opposition to one another, or whether both share common characteristics. Is merit always objective and fair and is impression management always concerned with manipulation and gaining unfair advantage? The contextual background to the chapter is the professional services sector. This is chosen as there is considerable evidence to support the contention that the sector is run on meritocratic principles. At the same time, it is susceptible to the use of impression management tactics by individuals in relation to advancement decisions (Kumra and Vinnicombe 2008, 2010). This is because of the inherently performative nature of much of the work undertaken by those in the professional services and the necessity for those who are successful not only to show their value through the quality of their work but also the way in which their work is performed.

The chapter is organized as follows. In the first section the concept of meritocracy is introduced and its key characteristics assessed. Also considered are some of the inherent contradictions within the concept and the gender implications of these. The second section provides an analysis of impression management, contrasting the key characteristics of the two concepts. In the third section the nature of advancement within the professional services is discussed with particular emphasis on the performative nature of the process. In the final section gender implications are assessed and discussed.

Conceptualizing Merit

The meritocratic ideal is that positions in society should be based on the abilities and achievements of the individual rather than on characteristics such as family

they wish. In the present discussion we assess two of the most common, self-promotion and ingratiation.

Self-Promotion

Self-promotion occurs when individuals actively manage the impression others form of them such that they will be viewed as competent and knowledgeable. Self-promotion is generally a proactive process. To convince others of their competence self-promoters cannot wait in the hope that someone will notice their achievements. They need to actively say or do something. Research shows that self-promotion is quite common, particularly when exercised in relation to important audiences or on important occasions. In a key study, Gordon and Stapleton (1956) required high school students to undertake a personality test, with some students instructed that it was part of a job application and others told it was for a guidance class. The authors found higher scores were obtained when the test was for the more important job application. More recently, Ferris (1990) established an experiment in which two people gave identical performances. One, however, used impression management strategies to enhance visibility of the performance whilst the other did not. The impression management user was viewed by observers as performing at a significantly higher level than the non-user.

Ingratiation

Ingratiation is the most commonly studied impression management technique (Rosenfeld *et al.* 1995). Jones (1964) was the first to conceptualize ingratiation, viewing it as a set of inter-related acquisitive impression management tactics which aim to make the person more liked and attractive to others (Jones 1990). Whilst initially viewing ingratiation as inherently illicit (Jones 1964), organizational theorists have recently adopted a more positive view towards these attraction management tactics. They contend ingratiation is a common, and frequently effective means of obtaining organizational social influence (Ralston 1985; Ralston and Elsass 1989). Rosenfeld *et al.* (1995) argue ingratiation can lead to positive organizational benefits and as such these tactics should be welcomed in certain circumstances. Ingratiation can assist positive interpersonal relationships—increasing effective interactions both inside and outside the organization. For example, ingratiation may be of particular use to minority groups within organizations who often find themselves in the position of having to please those in the majority group holding more powerful social positions. If through ingratiatory tactics one can generate liking and feelings of goodwill, the effect of natural cognitive tendencies to stereotype and devalue those who are different may be countered (Allison and Herlocker 1994). If the ingratiator is able to achieve this positive outcome—increased liking and attraction—the ingratiator succeeds in moving from a negatively stereotyped outsider to a familiar and liked insider (Rosenfeld *et al.* 1995).

However, successful ingratiation is an extremely difficult process to manage. It is an interpersonal challenge requiring skill to avoid detection and the risk that one will be perceived as an underhand and deliberate manipulator. Failed ingratiation attempts bring their own consequences, often placing the ingratiator in a more uncomfortable impression management position than when they began, i.e. having to overcome being disliked or even mistrusted (Arkin and Shepperd 1989). Thus, the motivation incentive needs to be high for an individual to be willing to engage in ingratiatory behaviours. This is determined by how important it is for the target to like the ingratiatory, with motivation increasing the greater the dependency of the ingratiator on the target. Thus one's incentive to ingratiate with peers will generally be lower than the incentive to ingratiate with superiors (Jones and Pittman 1982).

ARE THERE GENDER DIFFERENCES IN THE USE OF IMPRESSION MANAGEMENT?

Rudman (1998) sought to determine whether women would be as likely to use self-promotion, a key impression management technique, to advance their careers and the effect this would have on their career outcomes and social acceptability. She argues that self-promotion tactics with their emphasis on internal attributions for success are particularly related to hiring and promotion decisions as they are qualities often associated with success criteria required for many occupations, such as confidence, competence, ambition, and self-motivation (Wiley and Eskilson 1985; Kacmar *et al.* 1992; Stevens and Kristof 1995). As such self-promotion may provide a means by which women can counter organizationally based gender stereotypes (Jones and Pittman 1982).

However, research clearly indicates that women behaving confidently and assertively do not have their behaviour evaluated in the same way as men adopting identical behavioural patterns (Costrich *et al.* 1975; Heilman *et al.* 1989; Butler and Geis 1990). Women with direct, task-oriented leadership styles are viewed more negatively than men (Eagly *et al.* 1992). Assertive women were less popular than men (Costrich *et al.* 1975); self-confident women scored highly on their performance evaluations but were not liked by their peers (Powers and Zuroff 1988). Thus since obtaining social influence requires not only competence but social attraction (Carli *et al.* 1995) women employing the proactive influencing styles required for self-promotion may suffer 'backlash', whereby their performance evaluations are positively affected, but achieved at the cost of social rejection (Rudman 1998).

With regard to social acceptance, research indicates self-promotion is both intuitively and normatively more acceptable for men than women (Miller *et al.* 1992). Men are culturally conditioned to take personal credit for their achievements and to compete intra-sexually for economic resources (Buss 1988). They are socialized to be individuals, to lead and to compete hierarchically for positions of power and influence, i.e. agency characteristics (Bakan 1966; Eagly 1987). This is not so for women. 'Women cooperate and men compete' is a clear and powerful message, which women learn to accept from an early age (Nelson 1978). Women's focus is often on similarity, i.e. things which bring them together. Women may also avoid self-promoting behaviours for interpersonal considerations, believing those who behave out of gender-stereotypical role risk social censure (Huston and Ashmore 1986; Deaux and Major 1987; Eagly 1987). Thus self-promoting behaviours may be rejected by women who fear they may be perceived as unfeminine, pushy, domineering, and aggressive, characteristics unlikely to recommend them for many occupations they may wish to enter (Janoff-Bulman and Wade 1996).

The position self-promoting women find themselves in has been explored by Rudman (1998). In a number of experiments looking at perceptions of a self-promoting woman, a self-promoting man, and a self-effacing man Rudman (1998) found women (though not men) saw a self-promoting woman as less competent, less socially attractive, and consequently less likely to be hired than a self-promoting man. However, when competing against a self-effacing man, the self-promoting woman was more likely to be hired. The results of this study suggest that women have to balance counter-normative behaviour in terms of its negative effect on their perceived social attraction against the career-enhancing outcomes to be gained through employment of self-promotion tactics (Kacmar et al. 1992; Wiley and Crittenden 1992; Eagly et al. 1992; Sonnert and Holton 1996).

For these reasons, extant research indicates women tend to eschew impression management as inherently subjective, manipulative, and largely self-serving. For example, in Kumra and Vinnicombe's study (2008), it was observed that women in a professional services firm, though aware of the career-enhancing benefits of impression management behaviours, chose to ignore or avoid them, believing their merit would bring them the organizational recognition they deserved for their hard work and diligence. The majority of their male colleagues had no such qualms and outperformed them in the firm's advancement processes. Closer examination of the same group of women consultants in Kumra and Vinnicombe (2010) indicated their position as 'other' was further compounded by their inability to access social capital within their firm, clearly recognized by the group as necessary to advance. A key mechanism by which this was achieved was through the acquisition of a well-placed, influential sponsor. However, here again impression management was required. Interestingly, the utilization of impression management tactics in this case was advised by female participants as a defensive strategy to dispel negative perceptions attaching to women in the firm. These issues were likeability, where the perception was that women in the firm were too task-focused and serious and it was difficult to relax or have fun around them. There was also a perception that women were not particularly ambitious compared with their male colleagues: they were not constantly discussing their achievements and contributions, asking for new challenges and demanding promotions. The final issue concerned impression management of availability. Advice was given here to avoid being perceived as too needy or high maintenance; and if unable to meet an important work commitment to be very careful about how one's absence was positioned. Also of interest was that, though female participants in the study were willing to offer advice to others on the impression management strategies they needed to adopt in order to access an influential sponsor and enhance their social capital, the majority indicated this was not behaviour they had engaged in themselves.

Another consequence accruing to women in PSFs who do not 'play the game' is not only that their merit goes unrecognized, but it is rendered invisible as it is not evidenced in expected and accepted ways. For example, Kumra (2010) found in her sample of management consultants that those most likely to wish to work flexibly were women, particularly those who had recently had a child. Interviewees expressed their disappointment in the firm's lack of commitment to its own stated and highly publicized flexible working policy, as well as the lack of perceived support and recognition of their contribution from superiors and peers. Wald (2010) explains this finding through his concept of the 'hypercompetitive' ideology. Here excellence is determined by total commitment and a willingness to provide a '24/7' round the clock service to the client. It is clear contemporary technology enables working from home and other flexible arrangements, and that clients can be contacted from home as well as from the office, but the hypercompetitive ideology does not accept this level of flexibility. The ideology requires not just attendance to the work, but physical attendance in the office as a sign of loyalty to the firm. The ideology aggravates the negative consequences accruing to those unable to comply with this requirement, rendering them disloyal and under-committed (Wald 2010). Indeed, the very act of taking advantage of such arrangements is viewed as career-limiting and those who nevertheless take up such opportunities are deemed not only under-committed and disloyal but more damagingly as individuals who either do not understand what it takes to succeed, or do understand it but are unwilling or unable to deliver. The findings can also be partially explained by the very structure of many professional services firms, which are often partnerships. The partnership structure purports to endorse a notion of a community of equals, conforming to a single construction of merit and success. In such a context, it is impossible to tolerate resentment and dissatisfaction in respect of those seeking to evidence their merit in a way that is different from the norm. The majority (for this read masculine) invokes its power to sustain a construction of justice as formal and not substantive equality (Thornton and Bagust 2007) in such a way that deviance from the benchmark precludes merit to be recognized and valued, and hence renders it invisible.

MERITOCRACY IN THE PROFESSIONAL SERVICES FIRM

The professional services industry comprises organizations that provide advice to other organizations on matters as diverse as law, accounting, management processes and practice, and strategic direction. Their defining characteristic is that the advice they give is based upon a body of complex knowledge applied to client problems. Firms typically included in professional services are accounting, law, advertising agencies, architectural practices, management and engineering consultants (Lowendahl 2000).

The principle of meritocracy lies at the heart of the professional services (Maister 1993). It is often presented as an unproblematic concept, capable of operationalization and uniquely suited to the needs and requirements of the partnership governance structure, satisfying both demand and supply-side labour requirements (Maister 1993; Malos and Campion 1995). Within this neoliberal narrative, individuals are autonomous agents free to make contracts with anyone they choose, as a consequence, both they and social institutions are presumed to be fundamentally rational. This ideal type model is generally in line with received understanding of social reproduction and expected to manifest itself most commonly in institutions explicitly committed to economic rationality. This is characterized by equating professionalism with credentialled competence and is underpinned by discourses advocating formal equality, justice, and meritocracy (Sommerlad 2012).

Promotion to partner is the position individuals aspire to for years, as it is at this point that they become not just employees of the firm but also the owners. A number of criteria have been suggested. As Swaine (1948: 24) observes: 'The choice is difficult; factors which control ultimate decisions are intangible; admittedly they are affected by the idiosyncrasies of the existing partners. Mental ability there must be, but in addition, personality, judgement, character. No pretence is made that the ultimate decisions are infallible.'

Malos and Campion (1995) view the likelihood of admission to partnership ranks as the culmination of a number of factors decided upon by existing partners after a period of extended apprenticeship/associateship with the firm. The factors necessarily include favourable business conditions to make the addition of a new partner financially viable. The individual then needs to be deemed 'suitable' for partnership—an assessment encompassing criteria such as ability to get along with clients and colleagues, to generate new business, and supervise new associates (O'Flaherty and Siow 1992). However, anecdotal evidence suggests promotion processes are often subjective or political and criteria are only partially understood and frequently inconsistently applied (Nelson 1992). Malos and Campion (1995) caution that to accurately reflect the process, biases, politics, and power differentials amongst partners and associates have to be considered. This is the case even in firms purporting to operate a more strategic approach to the management of people, as despite such processes it is likely partners are able to influence promotional outcomes through their personal sponsorship of particular associates (Sander and Williams 1992).

Temporal commitment is also key. Wald (2010) points to the rise in competition in the global market for legal services, indicating this has led to the adoption of a more explicit 'around the clock' service mentality. Terming this a 'hypercompetitive' work ethic, Wald (2010) argues the expectation is heightened that both associates and partners work and bill more hours and will be available '24/7' to serve clients' needs. Firms expect their employees to adjust to this hypercompetitive era, by displaying loyalty to the firm and its clients and reprioritizing their work–life balance. In an earlier study, Anderson-Gough *et al.* (2002) also noted that being professional entails a temporal commitment to the firm. Working long hours is the norm and the readiness to work overtime as and when required is a standard requirement (Anderson-Gough *et al.* 2002). Trainees in their study attested that they generally learnt these norms in an informal way, but in situations where norms were transgressed, seniors brought the transgression to their attention, either unofficially or through formal appraisal ratings. Most trainees, when discussing successful firm members, referred to working long hours as a key contributor to their success. By working long hours a visible commitment to getting the job done within the constraints of the project is evidenced, as is an alignment with the dominant discourse identified within the study, i.e. that of serving the client. Temporal commitment also provides evidence of a commitment to the firm's financial success.

The literature on professions has acknowledged for some time the interdependence of the production and reproduction of a profession through the enactment of particular behaviours. In Dingwall (1979) for example, learning how to behave 'correctly' and understanding key norms is viewed as crucial in presenting the appropriate image to clients and others in the profession. It is essential for a professional to know how to give the impression that the service they provide is 'special', enabling those outside the profession to place the practitioner within that special group (i.e. professionals) and themselves outside. Professional conduct requires smartness of dress, norms relating to self-conduct including timekeeping (Coffey 1994), particular ways of speaking and presenting (Goffman 1959; Harper 1989; Kumra and Vinnicombe 2008), and even the manner of handshaking and signing one's name (Coffey 1993). Professional conduct relies on a mutual construction among firms of professional solidarity. It is therefore apparent that being a meritorious professional is related to the notion of professionalism as appearance. In this view, professional conduct is relayed to new entrants through their formal appraisal procedures (cf. Covaleski *et al.* 1998) but also through informal practices. The minutiae of these practices promoting certain ways of behaving, acting, talking, looking—in short being—combine to constitute the dominant understanding of what being a professional is (Anderson-Gough *et al.* 2002; Kumra and Vinnicombe 2008).

Haynes (2012), also discussing the performative nature of 'being' a professional, assesses the embodied nature of professionalism and the way in which male and female bodies are assessed in relation to their professionalism. Drawing on interviews with female lawyers and accountants in both the UK and US, she concludes that the 'historical challenges of gendered body image and fitness to practice remain an issue in contemporary firms' (Haynes 2012: 504). The consequences are that women in professional services are required to balance distancing themselves from negatively constructed aspects of their femininity with overt displays of positively viewed masculine forms of embodiment (e.g. dress, voice, and self-presentation) if they wish to be taken seriously.

From the preceding discussion it is evident that 'becoming a professional' is much more than simply acquiring technical competence. It comprises a way of 'being'

(Anderson-Gough *et al.* 2002; Kumra and Vinnicombe 2008) and an ability to signal to seniors the possession of technical competence needed to meet clients job-related expectations, combined with the self-presentational and attitudinal orientations necessary to indicate readiness for admission to the ranks of the professional partnership. Thus to advance within professional services requires individuals to engage dual processes: that of understanding their context and making an accurate assessment of what constitutes merit; and then ensuring that through some conduit/mechanism, such as impression management; their merit is recognized, valued, and rewarded.

MERITOCRACY AND IMPRESSION MANAGEMENT

From the preceding discussion it is evident there are costs to ignoring impression management. It is also evident that reliance on meritocratic principles alone is insufficient as a career-enhancing strategy for women. In this chapter I thus propose that the concepts of meritocracy and impression management are re-examined with a view to a potential reframing. As they are currently conceptualized, it would appear that meritocracy and impression management are presented as absolute rather than (as I have asserted in this chapter) contingent concepts. The prevailing view appears to be that meritocracy takes precedence over impression management because of its inherent objectivity, fairness, and focus on individual ability. Impression management by contrast is rejected, particularly by women, because it is subjective, manipulative, and self-serving—characteristics not seen as either morally or ethically defendable when seeking to construct a professional career. However, are these concepts so diametrically opposed in practice?

From the analysis presented thus far, the answer to this question would appear to be 'no'. In Table 12.1, the key characteristics of meritocracy are presented from the literature. In the first column we see the characteristics of meritocracy listed. The second column presents the 'absolute' reading of each characteristic, i.e. the theoretical ideal. However, the third column presents the 'contingent' nature of each characteristic, i.e. recognizing that merit is socially constructed and contextually situated, in this case within the professional services. This highlights how meritocratic principles are capable of being unsettled and can be shown to operate in a manner that may be opposite to that anticipated.

It can also be argued that impression management can be considered as both an absolute, ideal type or as contingent, i.e. socially constructed and contextually situated. In Table 12.2, the key characteristics of impression management are presented from the literature. In the first column we see common characteristics of impression management listed (Rosenfeld *et al.* 1995). The second column presents the 'absolute' reading of each characteristic, i.e. the theoretical ideal. The third column presents the 'contingent' nature of each characteristic, i.e. its application in a specific context, in this case the professional services. This highlights how, rather than representing a set of subjective,

Table 12.1 Key characteristics of meritocracy—absolute and contingent view

Characteristics of Meritocracy	Absolute Characteristic	Contingent Characteristic
Objectivity	Achievement only basis upon which advancement decisions are made; ascriptive traits irrelevant (e.g. Parsons and Bales 1956; Kerr *et al.* 1960; Blau and Duncan 1967; Farber and Sherry 1995)	Subjectivity evident in advancement processes; ascriptive traits evident in decisions; definition of achievement socially constructed; context-specific (e.g. Thornton 2007; Knights and Richards 2003; Kumra 2010 Sommerlad 2012)
Fairness	Occupational positions allocated purely on the basis of individual talent and ability (e.g. Parsons and Bales 1956; Kerr *et al.* 1960; Blau and Duncan 1967; Farber and Sherry 1995)	Occupational allocation has historically based social bias embedded within it; positions awarded on socially constructed and inherently unfair basis (e.g. Williams 1991; Roithmayr 1997; Sommerlad 2012)
Reward for hard work	Individuals performing at the highest levels and achieving the best results will benefit in terms of advancement (e.g. Parsons and Bales 1956 Kerr *et al.* 1960; Blau and Duncan 1967; Farber and Sherry 1995)	Organizational processes evolve such that rewards are allocated to those displaying their merit in expected and accepted ways; those working hard but evidencing their contribution in different ways, have their merit rendered invisible (e.g. Krefting 2003; Thornton 2007; Kumra 2010; Brink and Benschop 2012)
Individual responsibility for success/failure	Individual has control over their advancement; they stand or fall on their own merit (e.g. Brennan and Naidoo 2008)	Responsibility for lack of advancement does not rest with individual alone; societal and organizational factors will contribute (Thornton 2007; Kumra 2010; Sommerlad 2012; Brink and Benschop 2012)
Equal access to career opportunities	There is equality of access to career-enhancing opportunities (e.g. Tomei 2003)	Career-enhancing opportunities allocated to those deemed deserving and evidencing their merit in expected and accepted ways (e.g. Kanter 1977; Thornton 2007; Kumra and Vinnicombe 2008, 2010)

Table 12.2 Characteristics of impression management—absolute and contingent view

Characteristics of Impression Management	Absolute Characteristic	Contingent Characteristic
Subjective	A set of inter-related tactics employed by individuals to manage the way they are viewed by others (e.g. Leary and Kowalski 1990; Rosenfeld *et al.* 1995).	Can be used to correct incorrect assumptions made through processes such as sex-role stereotyping; ensuring talent and ability are rewarded (e.g. Ferris 1990; Kumra and Vinnicombe 2008).
Manipulative	Individuals may construct an impression of themselves in terms of their abilities and contributions, which is inconsistent with their actual abilities and contributions (Leary and Kowalski 1990; Rosenfeld *et al.* 1995).	Impression management simply a set of tactics aiding day-to-day interactions; providing social scripts for the parts we each play (e.g. Goffman 1959)
Self-serving	The use of impression management benefits only the individual and has no wider societal or organizational benefits (e.g. Rosenfeld *et al.* 1995)	Impression management tactics such as ingratiation can move individuals from outsider to insider status (e.g. Alison and Herlocker 1994) and ensure decision-makers have more accurate information on which to make advancement decisions, thereby benefiting the individual and the organization (e.g. Gardner and Martinko 2007; Kumra and Vinnicombe 2008)

manipulative, and self-serving strategies utilized solely for individual benefit, impression management can be viewed as objective, seeking to align perception and reality, and of benefit to both the individual and the organization.

Analysis of Tables 12.1 and 12.2 highlights the differences between merit and impression management in their 'absolute' definitions and the similarities between them when they are considered as socially constructed and contextually situated. It is evident that merit can be articulated as subjective, unfair, and ascriptive rather than achievement based. Impression management, on the other hand, can be considered objective, not manipulative; striving to align perception and reality and as good for both the organization and the individual.

Within the professional services, the importance of 'becoming a professional' has been assessed. Extant literature indicates in this context that merit and one's professionalism entail more than simply acquiring technical competence and work-based experience. Also required is a way of 'being' (Anderson-Gough *et al.* 2002; Kumra and Vinnicombe 2008), namely, the ability to signal to seniors the possession of technical competence needed to meet clients' job-related expectations, alongside a variety of self-presentational and attitudinal orientations necessary to indicate readiness for admission to the ranks of the professional partnership. Thus advancement comprises reading the contextual environment and making an accurate assessment of what constitutes merit and what does not. It is then necessary, through an appropriate conduit/ mechanism, such as impression management, that individuals engage proactively to ensure their merit is recognized, valued, and rewarded. Having merit, however defined, is not sufficient. Merit is only valuable to the individual in terms of career capital when it is recognized, valued, and purchased by those with authority to allocate rewards and resources within the firm. It is also evident from the preceding discussion that merit does not consist solely of qualifications, skills, and work-related experiences—typically termed human capital (Becker 1964). Career capital, does indeed include human capital, but also necessary are social capital and cultural capital (Bourdieu 1986). This would require individuals to be able to read their environment and assess how each of the three forms of capital identified by Bourdieu (1986) is constructed within the professional services environment. For example, human capital would consist of qualifications, skills, and job-related experiences. Social capital would consist of equal access to influential and proactive sponsorship and key internal and external networks. Cultural capital would consist of an understanding in respect of the actual and symbolic importance of the long-hours culture, the need to engage with the discourse in respect of client service, and necessity to align self-presentation with the norms governing professional presentation. To signal to seniors that one is in possession of these key forms of merit, impression management techniques need to be employed as possession of merit without ensuring it is recognized, valued, and purchased does not guarantee career advancement. Impression management techniques useful in this respect are self-promotion, visibility, and managing upwards (Singh *et al.* 2002; Kumra and Vinnicombe 2008; Kumra and Vinnicombe 2010).

I would thus argue that a reframing of the concept of impression management is needed by women in professional services firms and its potential value as a career-enhancing mechanism should be recognized and considered in respect of the impact it can have on an individual's career. Through discussion of the concepts of merit and impression management, it is evident neither is more valuable than the other; rather they are both capable of providing sound principles upon which to base the allocation of organizational rewards and resources. However, the ability of both concepts to deliver in practice what they promise in theory is dependent upon their use or misuse by individual actors within specific contexts. Within professional services, what constitutes professional merit and hence the ability to become a partner is largely predicated on a masculine model of success. In order to meet that model of success, signals must be provided that the individual is capable of providing a level of performance similar to that currently provided by those in the partnership ranks. As Sommerlad

(2012: 2503) indicates: 'Thus the ideal worker should possess attributes already deemed appropriate not only to this construction of meritworthy professionality, but also their gender and race.' For example, a particular part of the firm may value assertiveness, but when found in a woman it may appear to be aggression and not deemed part of her merit (Silvester 1997; Sommerlad 2012). Consequently, the merit of those comprising the 'other' (e.g. women) is likely to be misrecognized or rendered invisible, leaving the cultural capital of traditional status markers unchallenged (Sommerlad 2012). Within this context, the value of the various types of capital that the 'other' brings to the firm is automatically likely to be considered as low. This means that it is even more essential that women engage with impression management behaviours, such as self-promotion (Rhodes 2011), and also that they do so in ways that are acceptable to them personally and achieve the intended outcomes, that of ensuring their merit is recognized, valued, and rewarded.

CONCLUSION

In this chapter I set out to assess the concepts of meritocracy and impression management with a view to determining whether they are as diametrically opposed as they are frequently presented in the literature. I also sought to assess why it is that in respect of career advancement, women often find engaging in impression management problematic preferring instead to rely on meritocratic principles. The contextual focus of the study is the professional services, where it is evident the meritorious professional is constructed along a masculine ideal. Those viewed as 'other', such as women, have a more difficult task in presenting their merit, as if it is not done in normatively expected and accepted ways it is likely to go unrecognized and rendered invisible (Kumra 2010; Sommerlad 2012). The chapter assesses the concepts of meritocracy and impression management and presents evidence establishing that both do not operate in practice as they are frequently presented in the literature. Thus meritocracy, far from being the guardian of objectivity, fairness, and the reward for hard work it is purported to be (e.g. Parsons and Bales 1956; Farber and Sherry 1995) often entails elements of subjectivity, unfairness, and rewards ascriptive characteristics rather than achievement (e.g. Krefting 2003; Thornton 2007; Brink and Benschop 2012). Impression management, on the other hand, rather than representing reward for utilization of subjective, manipulative, and self-serving tactics (e.g. Leary and Kowalski 1990; Rosenfeld et al. 1995); can be viewed as objective, not manipulative and good for both individuals and organizations (e.g. Goffman 1959; Ferris 1990; Alison and Herlocker 1994; Gardner and Martinko 2007).

The chapter concludes with the recognition that career capital comprises human capital, social capital, and cultural capital (Bourdieu 1986) and that to evidence possession of each of these and to signal this to seniors requires engagement with impression management activity. The processes thus work in concert with one providing the conduit (impression management) for the recognition by seniors of the other (merit). This is particularly important for women in professional services, as it is a male-dominated environment which understands and recognizes merit through a single model of success that is highly resistant to change. It is also a context within which women are deemed as 'other', with perhaps little to offer of value in respect of career capital (Kumra and Vinnicombe 2008; Sommerlad 2012). In these circumstances, the utilization of impression management strategies becomes essential (Rhodes 2011) as to fail to bring one's human, social, and cultural capital to the attention of seniors and have it rightly assessed as 'merit' can mean that merit continues to go unrecognized and is ultimately rendered invisible.

References

Allison, S. T. and Herlocker, C. E. (1994). 'Constructing Impressions in Demographically Diverse Organizational Settings: A Group Categorization Analysis', *American Behavioral Scientist*, 37: 637–652.

Anderson-Gough, F., Grey, C., and Robson, K. (2002). 'Accounting Professionals and the Accounting Profession: Linking Conduct and Context', *Accounting and Business Research*, 32(1): 41–56.

Arkin, R. M., and Shepperd, J. A. (1989). 'Self Presentation Styles in Organizations', in R. A. Giacalone and P. Rosenfeld (eds), *Impression Management in the Organization*. Hillsdale, NJ: Lawrence Erlbaum Associates.

Bakan, D. (1966). *The Duality of Human Existence*. Boston: Beacon.

Becker, G. (1964). *Human Capital*. New York: Columbia University Press.

Blau, P. M. and Duncan, O. D. (1967). *The American Occupational Structure*. New York: Wiley.

Bourdieu, P. (1986). 'The Forms of Capital', in J. G. Richardson (ed.), *Handbook of Theory and Research for the Sociology of Education*, 241–58, New York: Greenwood Press.

Brennan, J. and Naidoo, R. (2008) 'Higher Education and the Achievement (and/or Prevention) of Equity and Social Justice', *Higher Education*, 56: 287–302.

Brink, M. van den, and Benschop, Y. (2012). 'Gender Practices in the Construction of Academic Excellence: Sheep with Five Legs', *Organization*, 19(4:, 507–24.

Buss, D. M. (1988). 'The Evolution of Human Intrasexual Competition: Tactics of Male Attraction', *Journal of Personality and Social Psychology*, 54: 616–28.

Butler, D., and Geis, F. L. (1990). 'Nonverbal Affect Responses to Male and Female Leaders: Implications for Leadership Evaluations', *Journal of Personality and Social Psychology*, 58: 48–59.

Carli, L. L., LaFleur, S., and Loeber, C. C. (1995). 'Nonverbal Behaviour, Gender Influence', *Journal of Personality and Social Psychology*, 68: 1030–41.

Coffey, A. J. (1993). 'Double Entry: The Professional and Organizational Socialization of Graduate Accountants', Ph.D. thesis, University of Wales College, Cardiff.

Coffey, A. J. (1994). 'Timing is Everything: Graduate Accountants, Time and Commitment', *Sociology*, 28(4): 121–32.

Costrich, N., Feinstein, J., Kidder, L., Marecek, J., and Pascale, L. (1975). 'When Stereotypes Hurt: Three Studies of Penalties for Sex-Role Reversals', *Journal of Experimental Social Psychology*, 11: 520–30.

Covaleski, M. A., Dirsmith, M. W., and Helan, J. B. (1998). 'The Calculated and the Avowed: Techniques of Discipline and Struggles over Identity in Big 6 Public Accounting Firms', *Administrative Science Quarterly*, 43(2): 293–327.

Deaux, K., and Major, B. (1987). 'Putting Gender into Context: An Interactive Model of Gender-Related Behaviour', *Psychological Review*, 94: 369–89.

Deem, R. (2007). 'Managing a Meritocracy or an Equitable Organization? Senior Managers' and Employees' Views about Equal Opportunities Policies in UK Universities', *Journal of Education Policy*, 22(6).

Dingwall, R. (1979). 'Inequality and the National Health Service', in P. Atkinson, R. Dingwall, and A. Murcott (eds), *Prospects for the National Health*. London: Croom Helm.

Eagly, A. H. (1987). *Sex Differences in Social Behaviour: A Social-Role Interpretation*. Hillsdale, NJ: Erlbaum.

Eagly, A. H., Makhijani, M. G., and Klonsky, B. G. (1992). 'Gender and the Evaluation of Leaders: A Meta Analysis', *Psychological Bulletin*, 111: 3–22.

Farber, D. A., and Sherry, S. (1995). 'Is the Radical Critique of Merit Anti-Semitic?', *California Law Review*, 83(3): 853–84.

Ferris, G. R. (1990). 'Politics in Organizations', in R. A. Giacalone and P. Rosenfeld (eds), *Impression Management in the Organization*. Hillsdale, NJ: Lawrence Erlbaum Associates.

Gardner, W. L. (1992). 'Lessons in Organizational Dramaturgy: The Art of Impression Management', *Organizational Dynamics*, 21(1): 33–47.

Gardner, W. L. and Martinko, M. J. (1988). 'Impression Management in Organizations', *Journal of Management*, 14(2): 321–38.

Goffman, E. (1959). *The Presentation of Self in Everyday Life*. Garden City, NY: Doubleday, Anchor.

Gordon, L. V., and Stapleton, E. S. (1956). 'Fakability of a Forced-Choice Personality Test under Realistic High School Employment Conditions', *Journal of Applied Psychology*, 40: 258–62.

Greener, I. (2007). 'The Politics of Gender in the NHS: Impression Management and 'Getting Things Done'. *Gender, Work and Organizations*, 14(3): 281–289.

Harper, R. (1989). 'An Ethnography of Accountants', Ph.D. thesis, Department of Sociology, University of Manchester.

Haynes, K. (2012). 'Body Beautiful? Gender, Identity and the Body in Professional Services Firms', *Gender, Work and Organizations*, 19(5): 489–507.

Heilman, M. E., Block, C. J., Martell, R. F., and Simon, M. C. (1989). 'Has Anything Changed? Current Characterizations of Men, Women and Managers', *Journal of Applied Psychology*, 74: 935–42.

Huston, T. L., and Ashmore, R. D. (1986). 'Women and Men in Interpersonal Relationships', in R. D. Ashmore and F. K. Del Boca (eds), *The Social Psychology of Female–Male Relations: A Critical Analysis of Central Concepts*. New York: Academic Press.

Jackson, M. (2007). 'How Far Merit Selection? Social Stratification and the Labour Market'. *British Journal of Sociology*, 58(3): 367–90.

James, W. (1890). *Principles of Psychology*. New York: Holt.

Janoff-Bulman, R., and Wade, M. B. (1996). 'The Dilemma of Self-Advocacy for Women: Another Case of Blaming the Victim?', *Journal of Social and Clinical Psychology*, 15: 143–52.

Jones, E. E. (1964), *Ingratiation: A Social Psychological Analysis*. New York: Appleton-Century-Crofts.

Jones, E. E. (1990). *Interpersonal Perception*. New York: W. H. Freeman & Co.

Jones, E. E., and Pittman, T. S. (1982). 'Toward a General Theory of Strategic Self-Presentation', in J. Suls (ed.), *Psychological Perspectives on the Self*, i. Hillsdale, NJ: Lawrence Erlbaum Associates.

Kacmar, K., Delery, J. E., and Ferris, G. R. (1992). 'Differential Effectiveness of Applicant Impression Management Tactics on Employment Interview Decisions', *Journal of Applied Social Psychology*, 22: 1250–72.

Kanter, R. M. (1977). *Men and Women of the Corporation*. New York: Basic Books.

Kerr, C. J., Dunlop, F. H. and Myers, C. A. (1960). *Industrialism and Industrial Man*. New York: Oxford University Press.

Kluegel, J. R. and Smith, E. R. (1986). *Beliefs About Inequality: American's Views of What is and What Ought to Be*, New York: de Gruyter.

Knights, D., and Richards, W. (2003). 'Sex Discrimination in UK Academia', *Gender, Work and Organization*, 10(2): 213–38.

Kumra, S. (2010). 'Exploring Career 'Choices' of Work-Centred Women in a Professional Service Firm', *Gender in Management: An International Journal*, 25(3), 227–243.

Krefting, L. A. (2003). 'Intertwined Discourses of Merit and Gender: Evidence from Academic Employment in the USA', *Gender, Work and Organization*, 10(2): 260–78.

Kumra, S., and Vinnicombe, S. M. (2008). 'A Study of the Promotion to Partner Process in a Professional Services Firm: How Women are Disadvantaged', *British Journal of Management*, 19: 65–74.

Kumra, S., and Vinnicombe, S. M. (2010). 'Impressing for Success: A Gendered Analysis of a Key Social Capital Accumulation Strategy', *Gender, Work and Organization*, 17(5): 521–46.

Ladd, E. C. (1994). *The American Ideology*. Storrs, CT: Roper Center for Public Opinion Research.

Ladd, E. C. and Bowman, K. H. (1998). *Attitudes toward Economic Inequality*. Washington, DC: EI Press.

Leary, M. R., and Kowalski, R. M. (1990). 'Impression Management: A Literature Review and Two Component Model', *Psychological Bulletin*, 107: 34–47.

Lowendahl, B. R. (2000). *Strategic Management of Professional Service Firms* (2nd edn). Copenhagen: Business School Press.

Maister, D. (1993). *Managing the Professional Service Firm*. New York: Free Press.

Malos, S., and Campion, M. (1995). 'An Options-Based Model of Career Mobility in Professional Service Firms', *Academy of Management Review*, 20(3): 611–44.

McNamee, S. J. and Miller, R. K. Jr. (2004). *The Meritocracy Myth*. Lanham, MD: Rowman and Littlefield.

Miller, L. C., Cooke, L. L., Tsang, J. and Morgan, F. (1992). 'Should I brag? Nature and Impact of Positive and Boastful Disclosures for Women and Men', *Human Communication Research*, 18, 364–399.

Nelson, K. (1978). 'Modesty, Socialization in Sexual Identity: The Transition from Infant to Girl or Boy', Doctoral dissertation, University of California, Berkeley.

Nelson, R. (1992), 'Of Tournaments and Transformations: Explaining the Growth of Large Law Firms', *Wisconsin Law Review*, 38: 733–50.

O'Flaherty, B., and Siow, A. (1992). 'Up or Out Rules in the Market for Lawyers', *Journal of Labor Economics*, 32: 283–97.

Parsons, T. and Bales, R. (1956). *Family, Socialization and the Interaction Process*. London: Routledge and Kegan Paul.

Powers, T. A., and Zuroff, D. C. (1988). 'Interpersonal Consequences of Overt Self-Criticism: A Comparison with Neutral and Self-Enhancing Presentations of Self', *Journal of Personality and Social Psychology*, 54: 1054–62.

Ralston, D. A. (1985), 'Employee Ingratiation: The Role of Management', *Academy of Management Review*, 10: 477–87.

Ralston, D. A., and Elsass, P. M. (1989). 'Ingratiation and Impression Management in the organization', in R. A. Giacalone and P. Rosenfeld (eds), *Impression Management in the Organization*, 235–49, Hillsdale, NJ: Lawrence Erlbaum Associates.

Ritchie, J., and Spencer, L. (1994). 'Qualitative Data Analysis for Applied Policy Research', in A. Bryman and R. G. Burgess (eds), *Analysing Qualitative Data*, 173–94. London: Routledge.

Rhodes, D. L. (2011). 'From Platitudes to Priorities: Diversity and Gender Equity in Law Firms', *Journal of Legal Ethics*, 24: 1041–70.

Roithmayr, D. (1997). 'Deconstructing the Distinction between Bias and Merit', *California Law Review*, 85: 1449–1507.

Rosenfeld, P., Giacalone, R. A., and Riordan, C. (1995). *Impression Management in Organizations: Theory, Measurement and Practice*. London: Routledge.

Rudman, L. A. (1998). 'Self-Promotion as a Risk-Factor for Women: The Costs and Benefits of Counterstereotypical Impression Management', *Journal of Personality and Social Psychology*, 74(3): 629–45.

Sander, R. H. and Williams, E. D. (1992). 'A Little Theorizing About the Big Law Firm: Galanter, Palay, and the Economics of Growth', *Law and Social Inquiry*, 17: 391–414.

Schlenker, B. R., and Weigold, M. F. (1992). 'Interpersonal Processes Involving Impression Regulation and Management', *Annual Review of Psychology*, 43: 133–68.

Scully, M. A. (1997). 'Meritocracy', in R. E. Freeman and P. H. Werhane (eds), *Blackwell Encyclopaedic Dictionary of Business Ethics*, 413–14, Oxford: Blackwell Publishers.

Scully, M. A. (2002). 'Confronting Errors in the Meritocracy', *Organization*, 9(3): 396–401.

Sen, A. (2000) 'Merit and Justice', in Kenneth Arrow, Samuel Bowles, and Steven Durlauf (eds), *Meritocracy and Economic Inequality*. Princeton, NJ: Princeton University Press.

Silvester, J. (1997). 'Spoken Attributions and Candidate Success in Graduate Recruitment Interviews', *Journal of Occupational and Organizational Psychology*, 70: 61–73.

Singh, V., Kumra, S. and Vinnicombe, S. (2002) 'Gender and Impression Management: Playing the Promotion Game', *Journal of Business Ethics*, 37: 77–89.

Sommerlad, H. (2012). 'Minorities, Merit and Misrecognition in the Globalized Profession', *Fordham Law Review*, 80: 2481–512.

Sonnert, G., and Holton, G. (1996). 'Career Patterns of Women and Men in the Sciences', *Scientific American*, 274: 63–71.

Stevens, C. K., and Kristof, A. L. (1995). 'Making the Right Impression: A Field Study of Applicant Impression Management during Job Interviews', *Journal of Applied Psychology*, 80: 587–606.

Swaine, P. (1948). *The Cravath Firm and its Predecessors: 1819-1948*. New York: AD Press

Thornton, M. (2007). ' "Otherness" on the Bench: How Merit is Gendered', *Sydney Law Review*, 29(3): 391–413.

Thornton, M., and Bagust, J. (2007). 'The Gender Trap: Flexible Work in Corporate Legal Practice', *Osgoode Hall Law Journal*, 45(4): 773–811.

Tomei, M. (2003). 'Discrimination and Equality at Work: A Review of the Concepts', *International Labour Review*, 142(4): 401–18.

Wald, E. (2010). 'Symposium: The Economic Downturn and the Legal Profession; Glass Ceilings and Dead Ends; Professional Ideologies, Gender Stereotypes and the Future of Women Lawyers at Large Law Firms', *Fordham Law Review*, 78: 2245–84.

Wiley, M. G., and Eskilson, A. (1985). 'Speech Style, Gender Stereotypes, and Corporate Success: What if Women Talk More like Men?', *Sex Roles*, 12: 993–1007.

Wiley, M. G., and Crittenden, K. S. (1992). 'By your Attributions you shall be Known: Consequences of Attributional Accounts for Professional and Gender Identities', *Sex Roles*, 27: 259–76.

Williams, P. J. (1991). *The Alchemy of Race and Rights*. Cambridge, MA: Harvard University Press.

Young, I. M. (1990). *Justice and the Politics of Difference*. Princeton: Princeton University Press.

PART III

GENDER AND CAREERS

CHAPTER 13

GENDER AND CAREERS

Obstacles and Opportunities

VALERIE N. STREETS AND DEBRA A. MAJOR

SEVERAL metaphors have been used to describe the obstacles women encounter along their career paths, including the sticky floor, concrete wall, glass ceiling, glass cliff, and labyrinth (Hymowitz and Schellhardt 1986; Betters-Reed and Moore 1995; Bell and Nkomo 2001; Eagly and Carli 2007; Ryan and Haslam 2007). The leaky pipeline metaphor, which is depicted in Figure 13.1, has been used to describe how women and girls are disproportionately affected by obstacles encountered along the educational and career pathways (Major and Morganson 2008). In this chapter, we employ the leaky pipeline model not only to describe gendered career obstacles but also to identify opportunities for intervention in order to overcome identified barriers.

SELF-PERCEPTIONS AND CAREER ASPIRATIONS

Gendered career obstacles begin long before the formal job application process is initiated. Observation of role models, the development of self-efficacy, and internalization of gender roles operate at an early age to create individuals' self-concepts. These self-images are then compared to available information on the professional world, creating a series of expectations about one's future career identity. Because childhood and adolescence mark a critical point of identity formation, it is an ideal place to introduce intervention strategies to promote exposure to an array of professional opportunities for women.

Influences During Childhood and Adolescence

Role models, or individuals who serve as an embodiment or template of the potential success one may attain and the requisite behaviours for such achievement, play a

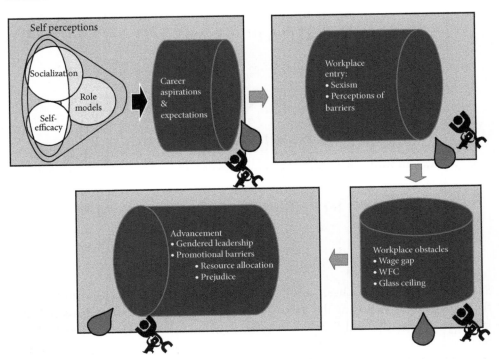

FIGURE 13.1 The leaky pipeline is marked by four phases that present disadvantages for girls and women as well as opportunities for advancement and repair

substantial role in the formation of one's professional goals and aspirations. In constructing a professional identity, individuals tend to look to similar others as a source of emulation. One critical aspect of similarity is gender, as this is a readily apparent means of comparison. This seems especially true for females, who demonstrate greater improvements in self-perceptions when using same-sex models of career success (Lockwood 2006). Furthermore, the presence of female role models within male-dominated professions, fields, and organizations undermines traditional gender stereotypes and demonstrates the possibility of excelling in a profession.

Role models serve as an early basis for career attitudes because of their impact on self-efficacy, or one's beliefs in one's own competence (Blanton 2001). Because role models personify accomplishment in a given domain, they are compared to adolescents' assessments of their own task performance, skills, traits, and abilities (Scherer *et al.* 1989). Thus, an increased presence of same-sex referents in non-traditional professions will convey the message that adolescents can develop the attributes necessary to excel in arenas outside of those within their gender norms (Barnir *et al.* 2011).

The presence of role models is especially important during childhood and early adolescent development. At this time, individuals are forming their self-concepts and ideas about their identities later in life. During this process, adolescents build prototypes of people in various professions and compare those representations with their own self-images, ultimately striving toward the prototypes that are most similar to their self-images (Eccles 1994). Certain professions are regarded as masculine or feminine in nature, with those professions emphasizing status and care-giving goals respectively. The centrality of gender in adolescents' self-concepts largely predicts gender-stereotypic career interests (Evans and Diekman 2009).

In the absence of relatable information about potential careers, girls may overlook many potential occupations that they could find desirable, as many young girls indicate that they have not considered occupations in which the prototypical worker is male (Hannover and Kessels 2004). A common example of this is the field of computer science, in which females are under-represented. Among young girls, lack of interest in technology and a preference for working with people partially explain lack of entry into the pipeline (Corneliussen 2002). However, even when female interest levels exist, girls still give considerable weight to the gendered image of a profession in their career aspirations (Rommes et al. 2007).

Career Aspirations and Expectations

The processes detailed above lay an early groundwork for career expectations and aspirations. Adolescent boys and girls alike report perceptions of most jobs as being gender segregated. Girls are especially likely to have this view, and both genders prefer 'gender-appropriate' occupations, although this preference diminishes in girls as they age (Miller and Hayward 2006). Changes in girls' preferences are not paralleled by changes in perceptions of gender segregation in occupations; girls report noticing occupational gender segregation throughout their years of schooling. Research has demonstrated that high school girls report greater anticipated career barriers relative to their male peers, including general discrimination, lack of support from colleagues, restriction of opportunity, and lack of confidence (Cardoso and Marques 2008). The anticipation of these future barriers serves as an early barrier, as evolving interests are not enough for female adolescents to feel as though male-dominated fields are accessible (Miller and Hayward 2006).

Attitudinal barriers extend through higher education, with female undergraduates anticipating a number of obstacles in their future professions. Young women expect a trade-off between their career goals and maintaining a family that is disproportionate to that of men. As a result of this juxtaposition between work and family, female students anticipate opportunities for fewer working hours and greater household responsibilities (Fetterolf and Eagly 2011). Additionally, women expect to receive less pay than men receive, even during the peak of their careers. Also, individuals pursuing female-dominated and female-typed jobs expect to receive less compensation than those pursuing male-dominated and male-typed jobs. Such findings highlight the need for increased encouragement for individuals to pursue non-stereotypic jobs, thereby diminishing gender-typing of occupations (Hogue et al. 2010).

Intervention Recommendations

The promotion of role models beginning early in education and formal occupational exploration programmes have been studied as a potential means of ameliorating gender segregation at a young age. It is important for individuals to actively encounter and relate to a multitude of role models for learning, motivation, and assistance. Observation of several role models provides the opportunity for individuals to incorporate a variety of skills, traits, and experiences into their identities (Gibson 2004). Additionally, because role models have been demonstrated to have such a strong influence on career choice, female exposure to role models in a variety of professions could be quite beneficial in helping them to make career choices. Specifically, job shadowing and small group mentoring programmes during secondary and higher education can increase exposure to and interest in non-traditional occupations. Female interest in such work is highest after interaction with successful women in related fields (Greene *et al.* 1982; Quimby and DeSantis 2006).

Career counselling has been studied as a means of strengthening females' career-related efficacy expectations. It is critical that such interventions begin by early adolescence, as decisions that influence one's professional development trajectory begin at this developmental stage. Thus, actively promoting career exploration and openly discussing occupational goals can have a profound impact on educational and occupational decisions at this stage (Jantzer *et al.* 2009). Such programmes should tackle the issues of: accomplishments in performance, vicarious experiences, emotional arousal, and verbal persuasion. Strengthening self-efficacy through these four constructs can enable young girls to realistically consider a broader array of career opportunities (Hackett and Betz 1981). Formal career exploration programmes have proven particularly effective when they incorporate career-related and self-concept building exercises via demonstrations, hands-on learning experiences, and counselling activities (Rea-Poteat and Martin 1991).

WORKPLACE ENTRY

Women's entry into the workplace is marked by two distinct obstacles: the experience of actual sexism, or gender-based discrimination, and the perception of such barriers within an organization. Both actual and perceived sexism often prevent women from obtaining desired positions

Sexism

The omnipresence of masculine norms, especially in positions of leadership, affects women in the phases of anticipation, acquisition, and actual performance of work

(Moorosi 2010). Two major forms of sexism are at work in shaping women's professional experiences. Hostile sexism, or the overarching belief that women are inferior, is linked to greater scepticism of female employees' credibility and performance levels. Benevolent sexism, or protection and chivalrous attitudes and actions towards women, is linked to a traditional gender role preference for women and men (Christopher and Wojda 2008). Recent research has demonstrated benevolent sexism to be more detrimental to women in the hiring process, as it is more subtle than hostile sexism and thereby not as easily targeted by laws and organizational policies. Such treatment by interviewers undermines the perceived abilities and intelligence of female applicants (Good and Rudman 2010).

While descriptive gender bias has historically been a problem for women attempting to enter the labour force, research has revealed its recent decline. Descriptive gender discrimination is the product of stereotypes used to make assumptions about applicants and employees on the basis of their gender (e.g. assuming women are nurturing and emotional; Glick *et al.* 1988; Luzadis *et al.* 2008). Employers much more readily demonstrate prescriptive gender bias, which emphasizes the ways in which women should or should not act. Much like benevolent sexism, this type of discrimination is less detectable and removable by policies, as employers are not overtly making assumptions about a job candidate based on his or her demographic characteristics. When employers subscribe to prescriptive gender bias, they are more likely to identify an applicant with relevant skill sets and experiences as qualified, even if his or gender does not align with the assumed nature of the job. However, these employers then stray from the predetermined objective selection criteria (e.g. educational background, work experience) and use intangible qualities to justify choosing an applicant whose gender aligns with the occupational role (Gill 2004; Luzadis *et al.* 2008).

According to Eagly's social role theory of sex differences in social behaviour (1987), expectations of each gender are influenced by the roles typically occupied by men and women on a daily basis (e.g. women assuming care-giving roles). The stereotypes that stem from this role fulfilment translate into the hiring process, as they encourage assumptions about the physical abilities, cognitive characteristics, and personality traits of men and women (Deaux and Lewis 1984). Selection decisions are often predicated upon these stereotypic assumptions of attributes; many employers report placing the most value on masculine cognitive abilities (e.g. logic, spatial reasoning; Cejka and Eagly 1999). The resultant sexism is reinforced when employers rely on subjective reactions to applicants rather than a priori selection criteria, as well as when they fail to evolve their selection criteria as jobs change, thereby maintaining traditional gender-typed job stereotypes (Guinier and Sturm 2001; Uhlmann and Cohen 2005).

The barriers of sexism do not dissipate for the women that manage to scale the hurdles of selection. Rather, the sexism that is initiated during the hiring process often marks the career progress of women. Employers tend to set lower minimum competency standards but higher ability standards for women than men, thus allowing female applicants to obtain a job but making it disproportionately difficult to receive favourable performance evaluations (Biernat and Kobrynowicz 1997). Sexism is also manifested in the allocation of work from supervisors, with women being provided with challenging assignments less often than their male colleagues, ultimately capping off their performance potential (DePater *et al.* 2010). Another notable effect of gender discrimination, as it arises from occupational stereotyping and the devaluation of the work typically performed by women, is its influence on salary allocation. Applicants are seen as more valuable to an organization when they are perceived as holding masculine characteristics or cognitive attributes, and they enjoy higher starting salaries accordingly. These pay disparities often persist over the trajectory of one's career (Alksnis *et al.* 2008).

Perceptions of Barriers

Women's perceptions of barriers in the workplace serve as an additional obstacle to workplace entry. These perceptions can limit the fields to which women apply, as perceptions of sexism serve as deterrents for entry into some areas. While men often report professional goals centred on monetary earnings and career status, women report a strong preference for helping others and working in a pleasant environment (Morgan *et al.* 2001). This further cautions women against seeking out occupations where sexism will detract from their interpersonal goals. For example, technical majors and occupations (e.g. computer science) are seen as the most sexist, consequently discouraging female participation (Fernández *et al.* 2006).

Potential barriers are often seen through a gendered lens, as gender schema influence one's perceptions. Gender schemas are cognitive frameworks for organizing learned role expectancies on the basis of one's sex (Bem 1981). Within organizational contexts, they serve as a moderator between procedural justice and both the perception of and actual opportunities available for female applicants. Gender schemas contribute to differing standards for men and women, whereas men are seen as more competent and worthy of preferential treatment unless they exhibit unfavourable performance, while women must work to prove their adequacy. As a result, hiring procedures are perceived as lacking procedural justice to the extent that they favour men and extant opportunities for women are often constrained by the gendered expectations of employers (Lemons 2003). Specifically, when an applicant's gender is salient, employers rely on gender schemas to make hiring decisions (Morrison and Von Glinow 1990). As outlined above, reliance upon gender assumptions in hiring often leads to disparities in pay allocations. Starting salary is central to the perception of fairness in the selection process (Bazerman *et al.* 1994).

Stereotype threat regularly acts as the vehicle through which perceived sexism hinders women's entry into the workforce. This phenomenon occurs when members of a minority group are aware of a negative stereotype regarding their abilities, and this awareness detracts from their actual performance in a given domain (Steele and Aronson 1995). Cognizance of these stereotypes can be triggered blatantly, with the explicit presentation of stereotypical information, or subtly, by mere awareness of one's minority status within a group (Hoyt *et al.* 2010). The presence of the stereotype works to detract from one's performance by: increasing physiological arousal, hindering self-regulation, lowering performance expectations, and reducing the cognitive resources one has available to dedicate to the desired task. This happens because the cognitive resources that would normally be devoted to performance are now spent on worry and negative thoughts related to performance. Stereotype threat is most likely to emerge when an individual is performing a task pertinent to the stereotype in question, the task is challenging, and the individual identifies with the field/skill to which the stereotype pertains (Block *et al.* 2011).

This carries two major implications for women seeking to enter the workforce. First, women's awareness of the negative stereotypes surrounding their work performance results in less female interest in male-dominated fields and organizations. Research has shown that women's career aspirations and intentions are weaker than those of men (Steele and Aronson 1995). This can partially be attributed to self-handicapping as a result of stereotype threat, in which women avoid contexts permeated by negative stereotypes to avoid failure (Keller 2002). Additionally, gendered stereotypes can lead women to perceive themselves as lacking the qualifications for a given job, therefore choosing not to apply (Kray and Shirako 2012). Second, because the activation of stereotype threat leads to actual declines in performance (Cardinu *et al.* 2005), women applying for positions in fields in which negative stereotypes exist about their performance (e.g. engineering, information technology) may have lower qualifications than do their male counterparts.

Intervention Techniques and Opportunities for Advancement

While sexism has been a persistent problem in hiring women, some simple tactics can be used to reduce its effects. Research has demonstrated that individuating information such as career history reduces gender-based discrimination in the selection process (Burke and Vinnicombe 2005). While this information has not been shown to remove employers' reliance upon gender stereotypes, proof of applicants successfully navigating gender-atypical work contexts in the past improved employers' beliefs in their abilities to perform the job in question. Women are distinctly advantaged in this case, as gender- atypical work history makes them more qualified for such work in the eyes of other organizations without detracting from their perceived abilities to perform gender-typical jobs. In contrast, men with gender-atypical work experience are perceived as less suitable for masculine jobs (Hareli *et al.* 2008).

Greater visibility of successful female career referents can improve women's career expectations and reduce their perceptions of barriers. Women have a tendency to compare themselves to women with a lower professional status than male referents, making success within a given organization seem more difficult to obtain. However, increased prominence of successful women will reduce the perceived threat of occupational barriers to women preparing to enter the workforce in that they exemplify the ability to transcend obstacles (Gibson and Lawrence 2010). This also helps to alleviate the effects of stereotype threat. By emphasizing identity security, the notion that one's social identity will not play a role in measuring job success, in recruitment and selection, employers can override stereotype threat effects (Kray and Shirako 2012).

Additionally, a stronger emphasis on multicultural organizations can greatly reduce both sexism and perceived barriers, as such organizations value a reciprocal level of cultural change for both the employee and the organization. This mutual effort bolsters social integration and identification with the organization. Women report that organizational efforts to maintain diversity and values placed on diversity are especially important in selecting a place of employment (Gilbert and Ones 1998).

WORKPLACE OBSTACLES

We focus on three obstacles women face once in the workforce: gendered pay differences, managing work and family demands, and the persistence of the glass ceiling. Although they are three distinct aspects of women's work experiences, they are best treated together, as these obstacles are synergistic in nature. For example, women who experience higher amounts of work–family conflict will be less likely to advance in the workplace and will likely suffer lower wages.

The Wage Gap

Perhaps one of the most pervasive obstacles for women at work is the gendered wage gap, which is indicative of the devaluation of women's work (Alksnis *et al.* 2008). The wage gap has been documented worldwide, though the magnitude of the pay disparity varies as a function of factors such as type of industry, occupation, and level of seniority (Kulich *et al.* 2011).

However, even within the same industries and positions, women are still disadvantaged. Research has reliably demonstrated that women are not paid directly in accordance with the outcomes of their job performance (Lips 2003). This is especially evident within organizations that use merit-based pay systems. Female leaders often find merit-based pay is predicated upon perceptions of their charisma and leadership ability, while male leaders receive merit-based pay that is based upon actual company performance (Kulich *et al.* 2007). Furthermore, merit-based pay has been demonstrated to be lower for women who exhibit equal performance scores to those of their male colleagues. The subjectivity and lack of transparency surrounding performance ratings and bonus allocations make this obstacle particularly difficult to combat (Castilla 2008).

While several mechanisms, such as lower levels of female risk-taking and confidence relative to male employees (Kulich *et al.* 2011), have been shown to sustain the wage gap, work–family conflict is believed to be a central perpetual mechanism. The gender pay gap most seriously affects mothers, a phenomenon known as 'the motherhood penalty' (Correll *et al.* 2007). This can be attributed to mothers investing less in their employment as well as managers' anticipation of divided attention from female employees with children (Lips and Lawson 2009). Additionally, the presence of work-reducing policies such as flextime and family leave have been shown to lead to lower wage levels for women by reducing their working hours (Misra *et al.* 2011).

Subscription to traditional gender roles is another instrumental force in the wage gap. Gender segregation in the labour force fuels salary disparities between male and female-typed jobs (Alksnis *et al.* 2008). Traditional gender role orientation is positively related to earnings for men and inversely so for women in the workforce. Regarding gender role orientation, women often face a self-fulfilling prophecy in which they expect and accept lower earnings than their male colleagues and seek out jobs that confirm female gender norms, which pay less than male-dominated fields (Judge and Livingston 2008).

Lastly, a lack of agency perpetuates the gap. A trade-off between likeability and promoting one's accomplishments and contributions exists for women in the workplace. By maintaining a modest demeanour regarding their job performance, women uphold their likeability (Rudman and Glick 1999). As a result of this conflict, women often refuse to be proactive in seeking raises for fear that what they may lose by defying feminine norms would not be offset by what they would hope to gain in pay (Wade 2001).

Work–Family Conflict

Women's work is largely marked by conflicting professional and familial demands. Professional women, especially those in leadership positions, often equate their success at work as striking a balance between work and family (Cheung and Halpern 2010). The difficulty in achieving such a balance is reflected by the under-representation of women in top-tier positions. In the United States, roughly half of the women who are top executives or earn a salary greater than $100,000 do not have children (Dye 2005). The struggle transcends corporate positions, as it also plagues academia. Only one-third of women who are hired in research-intensive positions ever become mothers (Hewlett 2002). Furthermore, of tenured faculty, women are twice as likely as men to divorce within twelve years of obtaining their doctorates (Mason and Goulden 2004).

This issue is so prominent that, following childbirth, women often find themselves deciding whether to 'opt in' or 'opt out' of the workforce. However, working flexible hours, part-time, or from home are all ways that mothers can 'opt in between' in an attempt at striking that balance. While 'opting in between' may seem to remedy work–family conflict, it is not without hardships; women who choose such a strategy often experience higher levels of exhaustion and perform with lower rates of success than their male colleagues (Grant-Vallone and Ensher 2011).

Within the 'sandwiched generation', women appear to be disproportionately affected by the demands of simultaneously caring for young children and elderly parents. Research shows that women with either high child-care demands or parent-care demands report significantly more work-to-family conflict than those reporting high work demands, a trend not found for men (Cullen *et al.* 2009). To meet family care-giving demands, women are more likely than their husbands to make work accom-modations, including taking time off and passing up promotions (Neal and Hammer 2007). Family responsibilities are negatively correlated with both objective and subjec-tive measures of career success as well as job satisfaction, and this relationship is medi-ated by gender, with women faring worse than men (Mayrhofer *et al.* 2008).

The Glass Ceiling

The glass ceiling is commonly understood as a set of 'artificial barriers to the advance-ment of women and minorities' (Cotter *et al.* 2001: 656) that are reflective of employer and supervisor discrimination. The ceiling is glass in that it is an invisible barrier that can-not be passed by racial and gender minorities. Cotter *et al.* (2001) identified four criteria for identifying true glass ceiling effects in the workplace. First, they are gender or racial inequalities that cannot be explained by job-relevant characteristics of the employee. Second, the experienced inequalities are amplified at higher level positions. Additionally, the glass ceiling cannot merely be understood as the proportion of women and racial minorities in senior-level positions, rather it actually precludes the advancement of such individuals. Lastly, gender or racial differences increase with career experience.

Glass ceiling effects often begin with a lack of recruitment and outreach directed towards women, coupled with a prevailing chilly climate. Such a climate is characterized by an impersonal, individualistic, and non-collaborative environment, which often coun-ters feminine behavioural norms (Foster *et al.* 1994). These effects are typically maintained by placing female employees in positions lacking a clear track to top positions and by employers' directing blame for lack of female promotions at subjective qualities of female employees (Jackson and O'Callaghan 2009). Managers tend to enable these discrepancies through their expectations of female employees. Anticipated work–family conflict com-monly leads supervisors to perceive a lack of fit between women and their occupations, particularly higher level positions (Hoobler 2007). Such expectations also lead to lower expectations of female job performance, further hindering women's progress in the work-place. However, within organizations, research has demonstrated that managers' views of women's perceived fit, rather than their actual performance outcomes, are most indicative of performance potential. As a result, promotions, performance evaluations, and delega-tions of responsibility are likely to be based more on preconceptions of women's abilities and commitment rather than their actual accomplishments (Hoobler *et al.* 2009).

Intervention Techniques and Opportunities for Advancement

Because the wage gap has largely been described as a product of women's work–life balance and disproportionate access to upper-level positions, it can be assumed that disparities in pay will be ameliorated by first addressing these obstacles. Regarding work–family conflict, potential demographic shifts in the workplace may offer some hope for working mothers, as it has been found that the percentage of women in an industry serves as a positive predictor of the amount of co-worker support and supportive work–family culture (Cook and Minnotte 2008). Furthermore, the percentage of women employed by an organization negatively predicts workers' experience of family-to-work conflict, especially for women.

Ryan and Kossek (2008) identified strategies an organization can utilize to promote a more supportive work–family environment. Most central to this framework is the role of the supervisor. By providing attention to the needs and values of individual workers, demonstrating respect and inclusion, and ensuring that the implementation of work–family policies does not adversely affect workload and social justice perceptions, supervisors can foster an organizational culture that allows employees with family commitments to thrive. Additionally, policies should be universal in nature, so as to avoid the perception of differential treatment of men and women, and negotiable to better suit the needs of each employee. Lastly, effective communication of work–family policies reveals the value placed on the issue by the organization; thus, employees are more inclined to benefit from the policies without fearing backlash or resentment from co-workers.

Pertaining to the persistence of the glass ceiling, women who have successfully obtained top-level positions can serve as mentors and examples for how to develop effective leadership and overcome stereotypes. Qualitative research on women who have effectively broken through the glass ceiling reveals a few tactics that allow women to be successful in such an endeavour (Baumgartner and Schneider 2010). First, advancing beyond the barrier must be an explicit goal to which women dedicate considerable effort. Additionally, despite gender norms, women must not be reluctant to tout their strengths and contributions to the organization. Strong communication skills and early career preparation by taking advantage of opportunities such as internships and lower-level jobs that provide diverse experience are also identified as critical components to surpassing the glass ceiling.

More generally, individuals' gender-related attitudes and beliefs have been shown to affect their reactions to gender diversity management programmes. Specifically, the centrality of one's gender identity, his or her attitudes towards affirmative action, and the acknowledgement of sexism contributed to perceived value and effectiveness of diversity management programmes. Currently, women tend to dissociate themselves from affirmative action and gender sensitivity training to avoid the stigmatizing perception of receiving preferential treatment (Heilman and Alcott 2001). Therefore, it follows that organizations must not only offer such programmes but work to manage perceptions of those efforts (Martins and Parsons 2007). Also, increased training of women in communications and technologies will increase their power to shape policy as it pertains to the workplace (Wirtenberg et al. 1981).

ADVANCEMENT

Despite the advances women have made in the workforce, they still lag behind men in terms of status, performance evaluations, and salary. By the 1980s, women matched men in their presence in the workforce, but remained excluded from most leadership positions. Many statistics suggest that gender equity is improving. For example, women compose 46 per cent of the workforce; 51.5 per cent of executive, managerial, and administrative jobs; 51 per cent of bachelor's degree recipients, and 45 per cent of advanced degree holders (BLS 2011). However, gender inequity remains at the top of work organizations, with women comprising just 3.4 per cent of the top five highest earning offices, and 14.1 per cent of CEOs (Catalyst 2012). This trend extends internationally, as women in other countries consistently earn less than their male counterparts and hold only a small proportion of management positions (Schein 2001). To better understand women's career advancement, it is important to analyse both gendered differences in leadership and the barriers that exist to the promotion of women in the workplace.

Gendered Leadership and Promotional Barriers

Many people hold stereotypical views of the ways in which each gender should demonstrate leadership behaviour. A masculine style of management is assumed to be competitive, reliant upon hierarchical authority, controlling, unemotional, and analytical. A feminine style is assumed to be cooperative, collaborative among leaders and subordinates, lower in control of the leader, intuitive, and empathetic (Loden 1985). A structural interpretation of such implied differences proposes that women's leadership style may differ from that of men because they have less access to power and fewer opportunities for advancement; therefore they behave in ways that reflect such a lack of power (Kanter 1977).

Research has pointed to a number of gender-based stereotypes that are prominent in places of employment. Perhaps most notable is the stereotype that personal characteristics typical of males are more consistent with those of effective leadership. In general, much of the prejudice against female leaders stems from this incongruity (Eagly and Karau 2002). Employees tend to be viewed through the lens of gender roles, or consensual beliefs about women and men. Furthermore, these stereotypes are based on observations of people in sex-typical social roles. As a result, communal traits, such as compassion and nurturance, are ascribed more to women, while agentic traits, such as dominance and ambition, are ascribed more to males (Cann and Siegfried 1990). Such stereotypes are deeply rooted, widely shared, and notably resistant to change (Heilman et al. 1989).

Research suggests that there may be little to no difference in the ways women and men actually lead (e.g. Feldman 1976; Hyde 2005; Miller Burke and Attridge 2011; Anderson and Hansson 2011). This is especially true of women and men who occupy the same managerial roles. Moreover, to the extent that both men and women are selected on the basis of the same set of organizational criteria, differences are likely to be minimized (Eagly and Johnson 1990).

While differences are minimal, many studies suggest that leadership styles are slightly gender stereotypic. Throughout past research, two different dichotomies of leadership style have been underscored (e.g. Powell *et al.* 1981; Helgeson 1990; Karakowsky and Siegel 1999). The first is that of task accomplishment, organizing activities to perform assigned duties; and maintenance of interpersonal relationships, attending to the morale and welfare of people in the workplace. The second dichotomy is that of an autocratic style, discouraging shared decision-making; and a democratic style, allowing subordinates to participate in decision-making. A meta-analysis by Eagly and Johnson (1990) found significant tendencies for women to be more interpersonally oriented, task-oriented, and democratic than men, thus indicating some stereotypic difference. The strongest difference was the tendency for women to be more democratic. These findings suggest that biological sex differences and sex-differentiated social experiences may shape men and women's behaviour differently, even when they occupy the same managerial roles. However, it is important to note that, in organizational studies, stereotypic sex differences were not nearly as significant as they were in lab studies, suggesting that the selection and socialization of managers minimizes gender stereotypes.

Beyond leadership style, it is also important to consider any sex differences in leadership effectiveness. Effectiveness in the workplace is commonly defined as a leader's facilitation of a group or an organization's success in meeting its goals. A meta-analysis by Eagly *et al.* (1995) found that, although women often face the initial hurdle of obtaining legitimacy, women and men do not differ in actual effectiveness. The women who make it to the top are enjoying relatively similar success in their roles as are men.

Lack of social support, insufficient training and preparation, role conflict, and masculine culture are often cited by women as barriers to their advancement in the workplace (Coronel *et al.* 2010). Differential resource allocation, such as social support and training attention, can largely be attributed to stereotyping and women's lack of fit within masculine professional contexts. Gender stereotypes play a substantial role in maintaining the division of labour between the sexes, as they limit women's upward mobility and often leave female employees with conflicting and unfavourable options (Cejka and Eagly 1999; Evans 2011). Promotional barriers can often be traced back to top executives who fail to set adequate goals for and communicate needs and expectations to female employees. Supervisors also do not hold their subordinates accountable for developing women's skills and appointing them to visible task forces and problem-solving jobs. As a result, women do not have proportionate access to typical paths of advancement (Woody and Weiss 1994).

Women who advance to leadership positions face two forms of prejudice: they have less potential to thrive because of the gender stereotypes against them, and they receive less favourable evaluations for violating their ascribed gender role. First, women face prejudice strictly on the basis of their gender and the sex-typical behaviours thought to be typical of that construct. Such prejudice restricts women from advancing in the same ways and at the same rates as men. Female leaders are often viewed in a much more negative light than are male leaders. They are perceived to be more hostile and less rational than male workers (Eagly and Karau 2002). Overall, female employees are regarded as less dominant, and consequentially receive lower evaluations and fewer promotions than male colleagues (Forsyth *et al.* 1997). Female managers often must endure a smaller salary and less availability of mentoring and informal collegial support than men (Eagly and Johnson 1990). Second, women receive particularly harsh evaluations when defying the gender roles assigned to them by stereotypes. Women, unlike men, are not regarded simply as managers; rather they are perceived in the context of their gender (Eagly *et al.* 1992). Because of this, they are expected to act in accordance with their gender stereotypes, despite the obstacles they pose in obtaining promotions. While successful leadership depends on both the ability to motivate subordinates for tasks and to meet the socio-emotional needs of the group, the task-oriented behaviours are more perceptually valued (Forsyth *et al.* 1997). As a woman adopts more masculine traits, her behaviour tends to be regarded as more extreme than that of her male counterparts. Furthermore, women occupying leadership positions experience mistakes as being considerably more detrimental than would their male equivalents. These experiences are especially true when those evaluating the leaders are male (Eagly *et al.* 1992; Heilman and Okimoto 2007; Brescoll *et al.* 2010).

Intervention Techniques and Opportunities for Advancement

The masculine construal of leadership has been decreasing over time and the incongruity between concepts of leaders and those of women are waning, possibly indicating a move towards greater acceptance of female leaders. This shift is thought to be partially attributable to changes in managerial duties in response to social and technological changes in the workplace. As careers are evolving, top–down management is giving way to democratic and interactive relationships among employees. The change is also linked to an increased presence of female leaders, which may be fostering more androgynous views of leadership practices as opposed to the stereotypical masculine beliefs (Koenig *et al.* 2011).

The existence of a female leadership advantage is beginning to be upheld by research. While male leaders tend to emphasize decision-making, women place more emphasis on employee development, communication, and vision, thereby providing them with a relational advantage in the workplace. Self-reports from employees have recently revealed a strong value placed upon fairness, honesty, development of staff, and harmony in managerial characteristics (Rosette and Tost 2010; Muchiri *et al.* 2011). Additionally, female leaders are more transformational than are male leaders, meaning that they place more emphasis on inspiring their subordinates and fostering the contributions and creativity of employees (Eagly and Carli 2003).

Based on the proposed female leadership advantage, it would follow that women interested in promotion should emphasize their own relational and motivational behaviours. Such a tactic underscores the feminine attributes associated with effective leadership without the risk of unfavourable evaluations that is associated with emphasizing more masculine traits (Vinkenburg *et al.* 2011). Additionally, the training of women for leadership positions should focus on these aspects of leadership (Eagly 2005).

CONCLUSION

Our pipeline-driven review of gender and careers indicates that women continue to encounter numerous barriers along their career pathways. However, it is also clear that there are good opportunities for patching the pipeline and repairs are being made. Although women have a role to play in capitalizing on their own opportunities, the greatest strides will be made as gender stereotypes and biases change. The process needed to mend the pipeline is currently a circular one, where attracting young girls into male-dominated careers is largely contingent upon depicting successful women in those fields. This underscores the need for further development of early intervention strategies and emphasizes the need to capitalize on the success of women who have scaled the outlined obstacles. Because barriers are prominent at all phases, intervention must be progressive in nature, attending to every leak in the career development pipeline.

REFERENCES

Alksnis, C., Desmarais, S., and Curtis, J. (2008). 'Workforce Segregation and the Gender Wage Gap: Is "Women's" Work Valued as Highly as "Men's"?', *Journal of Applied Social Psychology*, 38(6): 1416–41.

Anderson, J. A., and Hansson, P. H. (2011). 'At the End of the Road? On Differences between Women and Men in Leadership Behaviour', *Leadership and Organization Development Journal*, 32(5): e19–e26.

BarNir, A., Watson, W. E., and Hutchins, H. M. (2011). 'Mediation and Moderated Mediation in the Relationship among Role Models, Self-Efficacy, Entrepreneurial Career Intention, and Gender', *Journal of Applied Social Psychology*, 41(2): 270–97.

Baumgartner, M. S., and Schneider, D. E. (2010). 'Perceptions of Women in Management: A Thematic Analysis of Razing the Glass Ceiling', *Journal of Career Development*, 37(2): 559–76.

Bazerman, M. H., Schroth, H. A., Shah, P. P., and Diekman, K. A. (1994). 'The Inconsistent Role of Comparison Others and Procedural Justice in Reactions to Hypothetical Job Descriptions: Implications for Job Acceptance Decisions', *Organizational Behavior and Human Decision Processes*, 60(3): 326–52.

Bell, E., and Nkomo, S. (2001). *Our Separate Ways: Black and White Women and the Struggles for Professional Identity*. Boston: Harvard Business School Press.

Bem, S. L. (1981). 'Gender Schema Theory: A Cognitive Account of Sex Typing', *Psychological Review*, 88(4): 354–64.

Betters-Reed, B. L., and Moore, L. L. (1995). 'Shifting the Management Development Paradigm for Women', *Journal of Management Development*, 14: 2–24.

Biernat, M., and Kobrynowicz, D. (1997). 'Gender- and Race-Based Standards of Competence: Lower Minimum Standards But Higher Ability Standards for Devalued Groups', *Journal of Personality and Social Psychology*, 72(3): 544–57.

Blanton, H. (2001). 'Evaluating the Self in the Context of Another: The Three-Selves Model of Social Comparison Assimilation and Contrast', in G. B. Moskowitz (ed.), *Cognitive Social Psychology: The Princeton Symposium on the Legacy and Future of Social Cognition*, 72–83. Mahwah, NJ: Erlbaum.

Block, C. J., Koch, S. M., Liberman, B. E., Merriweather, T. J., and Roberson, L. (2011). 'Contending with Stereotype Threat at Work: A Model of Long-Term Responses', *Counseling Psychologist*, 39(4): 570–600.

Brescoll, V. L., Dawson, E., and Uhlmann, E. L. (2010). 'Hard Won and Easily Lost: The Fragile Status of Leaders in Gender-Stereotype-Incongruent Occupations', *Psychological Science*, 21(11): 1640–2.

Burke, R., and Vinnicombe, S. (2005). 'Advancing Women's Careers', *Career Development International*, 10(3): 165–7.

Cann, A., and Siegfried, W. D. (1990). 'Gender Stereotypes and Dimensions of Effective Leader Behavior', *Sex Roles*, 23(7–8): 413–19.

Cardinu, M., Maass, A., Rosabianca, A., and Kiesner, J. (2005). 'Why do Women Underperform under Stereotype Threat? Evidence for the Role of Negative Thinking', *Psychological Science*, 16(7): 572–8.

Cardoso, P., and Marques, J. (2008). 'Perception of Career Barriers: The Importance of Gender and Ethnic Variables', *International Journal for Educational and Vocational Guidance*, 8(1): 49–61.

Castilla, E. J. (2008). 'Gender, Race, and Meritocracy in Organizational Careers', *American Journal of Sociology*, 113(6): 1479–1526.

Catalyst (2012). *Women CEOs of the Fortune 1000*. New York: Catalyst, Mar. http://www.catalyst.org/publication/271/women-ceos-of-the-fortune-1000 (accessed February 26 2012).

Cejka, M. A., and Eagly, A. H. (1999). 'Gender-Stereotypic Images of Occupations Correspond to the Sex Segregation of Employment', *Personality and Social Psychology Bulletin*, 25(4): 413–23.

Cheung, F. M., and Halpern, D. F. (2010). 'Women at the Top: Powerful Leaders Define Success as Work + Family in a Culture of Gender', *American Psychologist*, 65(3): 182–93.

Christopher, A. N., and Wojda, M. R. (2008). 'Social Dominance Orientation, Right-Wing Authoritarianism, Sexism, and Prejudice toward Women in the Workforce', *Psychology of Women Quarterly*, 32(1): 65–73.

Cook, A., and Minnotte, K. L. (2008). 'Occupational and Industry Sex Segregation and the Work–Family Interface', *Sex Roles*, 59(11–12): 800–13.

Corneliussen, H. (2002). 'The Power of Discourse—The Freedom of Individuals: Gendered Positions in the Discourse of Computing', Department of Humanistic Informatics, University of Bergen.

Coronel, J. M., Moreno, E., and Carrasco, M. J. (2010). 'Work–Family Conflicts and the Organizational Work Culture as Barriers to Women Educational Managers', *Gender, Work and Organization*, 17(2): 219–39.

Correll, S. J., Benard, S., and Paik, I. (2007). 'Getting a Job: Is there a Motherhood Penalty?', *American Journal of Sociology*, 112(5): 1297–1338.

Cotter, D. A., Hermsen, J. M., Ovadia, S., and Vanneman, R. (2001). 'The Glass Ceiling Effect', *Social Forces*, 80(2): 655–82.

Cullen, J. C., Hammer, L. B., Neal, M. B., and Sinclair, R. R. (2009). 'Development of a Typology of Dual-Earner Couples Caring for Children and Aging Parents', *Journal of Family Issues*, 30: 458–83.

Deaux, K., and Lewis, L. L. (1984). 'The Structure of Gender Stereotypes: Interrelationships among Components and Gender Label', *Journal of Personality and Social Psychology*, 46: 991–1004.

DePater, I. E., Van Vianen, A. E. M., and Bechtoldt, M. N. (2010). 'Gender Differences in Job Challenge: A Matter of Task Allocation', *Gender, Work and Organization*, 17(4): 433–53.

Dye, J. L. (2005). *Fertility of American Women: June 2004* (Current Population Reports). US Census Bureau. http://www.census.gov/prod/2005pubs/p20-555.pdf (accessed March 1 2012).

Eagly, A. H. (1987). *Sex Differences in Social Behavior: A Social Role Interpretation*. Hillside, NJ: Erlbaum.

Eagly, A. H. (2005). 'Achieving Relational Authenticity in Leadership: Does Gender Matter?', *Leadership Quarterly*, 16(3): 459–74.

Eagly, A. H., and Carli, L. L. (2003). 'The Female Leadership Advantage: An Evaluation of the Evidence', *Leadership Quarterly*, 14(6): 807–34.

Eagly, A. H., and Carli, L. L. (2007). *Through the Labyrinth: The Truth about How Women Become Leaders*. Boston: Harvard Business School Press.

Eagly, A. H., and Johnson, B. T. (1990). 'Gender and Leadership Style: A Meta-Analysis', *Psychological Bulletin*, 108(2): 233–56.

Eagly, A. H., and Karau, S. J. (2002). 'Role Congruity Theory of Prejudice toward Female Leaders', *Psychological Review*, 109(3): 573–98.

Eagly, A. H., Karau, S. J., and Makhijani, M. G. (1995). 'Gender and the Effectiveness of Leaders: A Meta-Analysis', *Psychological Bulletin*, 117(1): 125–45.

Eagly, A. H., Makhijani, M. G., and Karau, S. J. (1992). 'Gender and the Evaluation of Leaders: A Meta-Analysis.' *Psychological Bulletin*, 111, 3–22.

Eccles, J. S. (1994). 'Understanding Women's Educational and Occupational Choices: Applying the Eccles et al. Model of Achievement-Related Choices', *Psychology of Women Quarterly*, 18: 585–609.

Evans, C. D., and Diekman, A. B. (2009). 'On Motivated Role Selection: Gender Beliefs, Distant Goals, and Career Interest', *Psychology of Women Quarterly*, 33(2): 235–49.

Evans, D. (2011). 'Room at the Top: Advancement and Equity for Women in the Business World', *National Civic Review*, 100(2): 62–4.

Feldman, D. C. (1976). 'A Contingency Theory of Socialization', *Administrative Science Quarterly*, 21: 433–52.

Fernández, M. L., Castro, Y. R., Otero, M. C., Foltz, M. L., and Lorenzo, M. G. (2006). 'Sexism, Vocational Goals, and Motivation as Predictors of Men's and Women's Career Choice', *Sex Roles*, 55: 267–72.

Fetterolf, J., and Eagly, A. (2011). 'Do Young Women Expect Gender Equality in their Future Lives? An Answer from a Possible Selves Experiment', *Sex Roles*, 65(1–2): 83–93.

Forsyth, D. R., Heiney, M. M., and Wright, S. S. (1997). 'Biases and Appraisals of Women Leaders.' *Group Dynamics: Theory, Research, and Practice*, 1: 98–103.

Foster, T. J., Foster, M. S., Flaugh, K., Kinschner, J., Locke, E., and Pidlock, N. (1994). 'An Empirical Test of Hall and Sandler's 1982 Report: Who Finds the Classroom Climate Chilly?', paper presented at the Annual Meeting of the Central States Communication Association, Oklahoma City, OK.

Gibson, D. E. (2004). 'Role Models in Career Development: New Directions for Theory and Research', *Journal of Vocational Behavior*, 65: 134–56.

Gibson, D. E., and Lawrence, B. S. (2010). 'Women's and Men's Career Referents: How Gender Composition and Comparison Level Shape Career Expectations', *Organization Science*, 21(6): 1159–75.

Gilbert, J. A., and Ones, D. S. (1998). 'Role of Informal Integration in Career Advancement: Investigations in Plural and Multicultural Organizations and Implications for Diversity Valuation', *Sex Roles*, 39(9–10): 685–704.

Gill, M. (2004). 'When Information does Not Deter Stereotyping: Prescriptive Stereotyping Can Foster Bias under Conditions that Deter Descriptive Stereotyping', *Journal of Experimental Psychology*, 40: 619–32.

Glick, P., Zion, C., and Nelson, C. (1988). 'What Mediates Sex Discrimination in Hiring Decisions?', *Journal of Personality and Social Psychology*, 55(2): 178–86.

Good, J. J., and Rudman, L. A. (2010). 'When Female Applicants Meet Sexist Interviewers: The Costs of Being a Target of Benevolent Sexism', *Sex Roles*, 62: 481–93.

Grant-Vallone, E. J., and Ensher, E. A. (2011). 'Opting in between: Strategies Used by Professional Women with Children to Balance Work and Family', *Journal of Career Development*, 38(4): 331–48.

Greene, A., Sullivan, H. J., and Beyard-Tyler, K. (1982). 'Attitudinal Effects of the Use of Role Models in Information about Sex-Typed Careers', *Journal of Educational Psychology*, 74(3): 393–8.

Guinier, L., and Sturm, S. (2001). *Who's Qualified*. Boston: Beacon Press.

Hackett, G., and Betz, N. E. (1981). 'A Self-Efficacy Approach to the Career Development of Women', *Journal of Vocational Behavior*, 18: 326–39.

Hannover, B., and Kessels, U. (2004). 'Self-to-Prototype Matching as a Strategy for Making Academic Choices: Why High School Students Do Not Like Math and Science', *Learning and Instruction*, 14: 51–67.

Hareli, S., Klang, M., and Hess, U. (2008). 'The Role of Career History in Gender Based Biases in Job Selection Decisions', *Career Development International*, 13(3): 252–69.

Heilman, M. E., Block, C. J., Martell, R. F., and Simon, M. C. (1989). 'Has Anything Changed? Current Characterizations of Men, Women, and Managers', *Journal of Applied Psychology*, 74(6): 935–42.

Heilman, M. E., and Alcott, V. B. (2001). 'What I Think You Think of Me: Women's Reactions to Being Viewed as Beneficiaries of Preferential Selection', *Journal of Applied Psychology*, 86: 574–82.

Heilman, M. E., and Okimoto, T. G. (2007). 'Why are Women Penalized for Success at Male Tasks? The Implied Community Deficit', *Journal of Applied Psychology*, 92(1): 81–92.

Helgeson, S. (1990). *The Female Advantage*. New York: Doubleday.

Hewlett, S. A. (2002). 'Executive Women and the Myth of Having it All', *Harvard Business Review*, 80: 66–73.

Hogue, M., DuBois, C. L. Z., and Fox-Cardamone, L. (2010). 'Gender Differences in Pay Expectations: The Roles of Job Intention and Self-View', *Psychology of Women Quarterly*, 34(2): 215–27.

Hoobler, J. M. (2007). 'On-Site or Out-of-Sight? Family-Friendly Childcare Provisions and the Status of Working Mothers', *Journal of Management Inquiry*, 16: 372–80.

Hoobler, J. M., Wayne, S. A., and Lemmon, G. (2009). 'Bosses' Perceptions of Family–Work Conflict and Women's Promotability: Glass Ceiling Effects', *Academy of Management Journal*, 52(5): 939–57.

Hoyt, C. I., Johnson, S. K., Murphy, S. E., and Skinnell, K. H. (2010). 'The Impact of Blatant Stereotype Activation and Group Sex-Composition on Female Leaders', *Leadership Quarterly*, 21: 716–32.

Hyde, J. S. (2005). 'The Gender Similarities Hypothesis', *American Psychologist*, 60(6): 581–92.

Hymowitz, C., and Schellhardt, T. C. (1986). 'The Glass Ceiling: Why Women Can't Seem to Break the Invisible Barrier that Blocks them from the Top Jobs', *Wall Street Journal* (24 Mar.): 4.

Jackson, J., and O'Callaghan, E. (2009). 'What do We Know about Glass Ceiling Effects? A Taxonomy and Critical Review to Inform Higher Education Research', *Research in Higher Education*, 50(5): 460–82.

Jantzer, A., Stalides, D. J., and Rottinghaus, P. J. (2009). 'An Exploration of Social Cognitive Mechanisms, Gender, and Vocational Identity among Eighth Graders', *Journal of Career Development*, 36(2): 114–38.

Judge, T. A., and Livingston, B. A. (2008). 'Is the Gap More than Gender? A Longitudinal Analysis of Gender, Gender Role Orientation, and Earnings', *Journal of Applied Psychology*, 93(5): 994–1012.

Kanter, R. M. (1977). *Men and Women of the Corporation*. New York: Basic Books.

Karakowsky, L., and Siegel, J. P. (1999). 'The Effects of Proportional Representation and Gender Orientation of the Task on Emergent Leadership Behavior in Mixed-Gender Work Groups', *Journal of Applied Psychology*, 84(4): 620–31.

Keller, J. (2002). 'Blatant Stereotype Threat and Women's Math Performance: Self-Handicapping as a Strategic Means to Cope with Obtrusive Negative Performance Expectations', *Sex Roles*, 47(3–4): 193–8.

Koenig, A. M., Eagly, A. H., Mitchell, A. A., and Ristikari, T. (2011) 'Are Leader Stereotypes Masculine? A Meta-Analysis of Three Research Paradigms', *Psychological Bulletin*, 137(4): 616–42.

Kray, L. J., and Shirako, A. (2012). 'Stereotype Threat in Organizations: An Examination of its Scope, Triggers, and Possible Interventions', in M. Inzlicht and T. Schmader (eds), *Stereotype Threat: Theory, Process, and Application*, 173–87. New York: Oxford University Press.

Kulich, C., Ryan, M. K., and Haslam, S. A. (2007). 'Where is the Romance for Women Leaders? The Effects of Gender on Leadership Attributions and Performance-Based Pay', *Applied Psychology: An International Review*, 56(4): 582–601.

Kulich, C., Trojanowski, G., Ryan, M. K., Haslam, S. A., and Renneboog, L. D. R. (2011). 'Who Gets the Carrot and Who Gets the Stick? Evidence of Gender Disparities in Executive Remuneration', *Strategic Management Journal*, 32(3): 301–21.

Lemons, M. A. (2003). 'Contextual and Cognitive Determinants of Procedural Justice Perceptions in Promotion Barriers for Women', *Sex Roles*, 49(5–6): 247–64.

Lips, H. M. (2003). 'The Gender Pay Gap: Concrete Indicator of Women's Progress toward Equality', *Analyses of Social Issues and Public Policy*, 3(1): 87–109.

Lips, H., and Lawson, K. (2009). 'Work Values, Gender, and Expectations about Work Commitment and Pay: Laying the Groundwork for the "Motherhood Penalty"?', *Sex Roles*, 61(9–10): 667–76.

Lockwood, P. (2006). ' "Someone Like Me Can Be Successful": Do College Students Need Same-Gender Role Models?', *Psychology of Women Quarterly*, 30(1): 36–46.

Loden, M. (1985). *Feminine Leadership or How to Succeed in Business without Being One of the Boys*. New York: Times Books.

Luzadis, R., Wesolowski, M., and Snavely, B. K. (2008). 'Understanding Criterion Choice in Hiring Decisions from a Prescriptive Gender Bias Perspective', *Journal of Managerial Issues*, 20(4): 468–84.

Major, D. A., and Morganson, V. J. (2008). 'Toward an Inclusive Climate IS&T Work Climate', in M. Khosrow-Pour (ed.), *Encyclopedia of Information Science and Technology*, ii. Hershey, PA (Idea Group Reference: 1899–905).

Martins, L. L., and Parsons, C. K. (2007). 'Effects of Gender Diversity Management on Perceptions of Organizational Attractiveness: The Role of Individual Differences in Attitudes and Beliefs', *Journal of Applied Psychology*, 92(3): 865–75.

Mason, M. A., and Goulden, M. (2004). 'Do Babies Matter (Part II)? Closing the Baby Gap', *Academe*, 90(6): 10–15. http://ucfamilyedge.berkeley.edu/babies%20matterII.pdf (accessed March 1 2012).

Mayrhofer, W., Meyer, M., Schiffinger, M., and Schmidt, A. (2008). 'The Influence of Family Responsibilities, Career Fields and Gender on Career Success: An Empirical Study', *Journal of Managerial Psychology*, 23(3): 292–323.

Miller, L., and Hayward, R. (2006). 'New Jobs, Old Occupational Stereotypes: Gender and Jobs in the New Economy', *Journal of Education and Work*, 19(1): 67–93.

Miller Burke, J., and Attridge, M. (2011). 'Pathways to Career and Leadership Success: Part 2— Striking Gender Similarities among $100k Professionals', *Journal of Workplace Behavioral Health*, 26(3): 207–39.

Misra, J., Budig, M., and Boeckmann, I. (2011). 'Work–Family Policies and the Effects of Children on Women's Employment Hours and Wages', *Community, Work and Family*, 14(2): 139–57.

Moorosi, P. (2010). 'South African Female Principals' Career Paths: Understanding the Gender Gap in Secondary School Management', *Educational Management Administration and Leadership*, 38(5): 547–62.

Morgan, C., Isaac, J. D., and Sansone, C. (2001). 'The Role of Interest in Understanding the Career Choices of Female and Male College Students', *Sex Roles*, 44(5–6): 295–320.

Morrison, A. M., and von Glinow, M. A. (1990). 'Women and Minorities in Management', *American Psychologist*, 45(2): 200–08.

Muchiri, M. K., Cooksey, R. W., Di Milia, L. V., and Walumbwa, F. O. (2011). 'Gender and Managerial Level Differences in Perceptions of Effective Leadership', *Leadership and Organization Development Journal*, 32(5): 462–92.

Neal, M. B., and Hammer, L. B. (2007). *Working Couples Caring for Children and Aging Parents: Effects on Work and Well-Being*. Mahwah, NJ: Lawrence Erlbaum.

Powell, G. N., Butterfield, D. A., and Mainiero, L. A. (1981). 'Sex-Role Identity and Sex as Predictors of Leadership Style', *Psychological Reports*, 49(3): 829–30.

Quimby, J. L., and DeSantis, A. M. (2006). 'The Influence of Role Models on Women's Career Choices', *Career Development Quarterly*, 54(4): 297–306.

Rea-Poteat, M. B., and Martin, P. F. (1991). 'Taking your Place: A Summer Program to Encourage Nontraditional Career Choices for Adolescent Girls', *Career Development Quarterly*, 40(2): 182.

Rommes, E., Overbeek, G., Scholte, R., Engels, R., and de Kemp, R. (2007). '"I'm Not Interested in Computers": Gender-Based Occupational Choices of Adolescents', *Information, Communication and Society*, 10(3): 299–319.

Rosette, A. S., and Tost, L. P. (2010). 'Agentic Women and Communal Leadership: How Role Prescriptions Confer Advantage to Top Women Leaders', *Journal of Applied Psychology*, 95(2): 221–35.

Rudman, L. A., and Glick, P. (1999). 'Feminized Management and Backlash toward Agentic Women: The Hidden Costs to Women of a Kinder, Gentler Image of Middle Managers', *Journal of Personality and Social Psychology*, 77(5): 1004–10.

Ryan, M. K., and Haslam, S. A. (2007). 'The Glass Cliff: Exploring the Dynamics Surrounding Women's Appointment to Precarious Leadership Positions', *Academy of Management Review*, 32: 549–72.

Ryan, A. M., and Kossek, E. E. (2008). 'Work–Life Policy Implementation: Breaking down or Creating Barriers to Inclusiveness?', *Human Resource Management*, 47(2): 295–310.

Schein, V. E. (2001). 'A Global Look at Psychological Barriers to Women's Progress in Management', *Journal of Social Issues*, 57(4): 675–88.

Scherer, R. F., Adams, J. S., Carley, S. S., and Wiebe, F. A. (1989). 'Role Model Performance Effects on Development of Entrepreneurial Career Preference', *Entrepreneurship: Theory and Practice*, 13: 53–71.

Steele, C. M., and Aronson, J. (1995). 'Stereotype Threat and the Intellectual Test Performance of African Americans', *Journal of Personality and Social Psychology*, 69(5): 797–811.

Uhlmann, E. L., and Cohen, G. L. (2005). 'Constructed Criteria: Redefining Merit to Justify Discrimination', *Psychological Science*, 16(6): 474–80.

US Department of Labor, Bureau of Labor Statistics (2011). *Women in the Labor Force: A Databook*, Dec. http://www.bls.gov/cps/wlf-databook-2011.pdf (accessed February 26 2012).

Vinkenburg, C. J., van Engen, M. L., Eagly, A. H., and Johannesen-Schmidt, M. C. (2011). 'An Exploration of Stereotypical Beliefs about Leadership Styles: Is Transformational Leadership a Route to Women's Promotion?', *Leadership Quarterly*, 22(1): 10–21.

Wade, M. E. (2001). 'Women and Salary Negotiation: The Costs of Self-Advocacy', *Psychology of Women Quarterly*, 25(1): 65–76.

Wirtenberg, J., Strausburg, G., and Alspektor, R. A. (1981). 'Educational Trends for Expanding Women's Occupational Lives', *Psychology of Women Quarterly*, 6(1): 137–59.

Woody, B., and Weiss, C. (1994). *Barriers to Workplace Advancement: The Experience of the White Female Work Force*. Washington, DC: Federal Publications, paper 130.

CHAPTER 14

THE GLASS CLIFF

Examining Why Women Occupy Leadership Positions in Precarious Circumstances

SUSANNE BRUCKMÜLLER, MICHELLE K. RYAN,
FLOOR RINK, AND S. ALEXANDER HASLAM

ALTHOUGH upper management is still a predominantly male arena (e.g. Adler 2000; Sealy and Vinnicombe 2012), women are increasingly advancing to positions of organizational leadership (e.g. Dreher 2003; Stroh *et al.* 2004; Catalyst 2009). This increasing representation of women in higher management raises important new questions about gender and leadership. For example, do the experiences of women and men holding top management positions differ, and if so, how? Under what circumstances do women achieve high-profile positions, and how do they fare?

Research has documented that the experiences of women and men in high-ranking positions differ markedly. For example, women tend to occupy positions that involve less authority, have fewer tangible rewards, and afford less opportunity for career mobility (Lyness and Thompson 1997); they also tend to receive lower remuneration than male colleagues (e.g. Blau and Kahn 2000; Kulich *et al.* 2011). Moreover, when it comes to others' perceptions of female leaders, their performance in leadership positions is placed under higher scrutiny than that of men (e.g. Eagly *et al.* 1995); many, especially men, remain sceptical about the effectiveness of female leaders (e.g. Eagly *et al.* 1992; Sczesny 2003); and subordinates tend to indicate that they prefer male over female supervisors (e.g. Simon and Landis 1989).

Taken together, while women have gained greater access to positions of organizational leadership in recent years, gender discrimination persists in higher management—although it tends to operate through increasingly subtle processes (see also Lyness and Thompson 1997; Schmitt *et al.* 2003; Agars 2004). With this is mind, the main focus of the present chapter is on a newly uncovered, and relatively subtle, barrier for women aspiring to leadership position that their male counterparts do not have to contend with: the glass cliff. We discuss the circumstances under which women are likely to be appointed to leadership positions and examine a range of possible explanations for why these tend to be more precarious than the positions of power that men obtain.

Uncovering the Glass Cliff—Archival Evidence

Archival research suggests that when women do break through the 'glass ceiling' and achieve positions of organizational leadership, they are more likely than their male colleagues to find themselves in positions that are associated with a state of crisis and thus contain an element of risk (Ryan and Haslam 2005). In an examination of the appointments of women and men to the boards of the top 100 companies listed on the London Stock Exchange (the FTSE 100), Ryan and Haslam demonstrated that, in a period of a general financial downturn, companies that appointed a woman to their board had experienced a pattern of consistently poor stockmarket performance in the months preceding the appointment, while the share price performance of a matched sample of companies who had appointed men had remained much more stable. In other words, women were more likely to achieve top management positions when a company was in crisis than when companies were doing well. In an extension of the 'glass ceiling' metaphor, Ryan and Haslam (2005, 2007) described this pattern by coining the phrase 'the glass cliff' to illustrate the precariousness of these leadership positions that women were obtaining and the associated risk of failure and criticism.

The precariousness of glass cliff positions stems from the increased risk of personal failure and because leaders of companies that perform poorly are subsequently less likely to be appointed to other leadership positions (Fama and Jensen 1983; Ferris *et al.* 2003). Such findings support the notion of the 'romance of leadership', whereby leaders rather than situational factors are blamed for poor organizational performance (Meindl *et al.* 1985). This blame may be enhanced for women in glass cliff positions. Companies in crisis are likely to attract attention from stakeholders and the media, and management decisions are likely to be placed under higher scrutiny than those made in more prosperous times (Haslam and Ryan 2008). This increased attention may be further enhanced following the appointment of a woman, because two relatively rare events—a crisis and the appointment of a female leader—co-occur (Hamilton and Gifford 1976). Indeed, experimental evidence shows that non-prototypical leaders tend to be punished more for mistakes than prototypical leaders (Brescoll *et al.* 2010; but see also Kulich *et al.* 2007).

Follow-up archival research has demonstrated that women's appointment to these precarious glass cliff positions is dependent on the way in which organizational performance is defined. For example, glass cliff positions were not in evidence when Adams *et al.* (2009) looked at men and women occupying CEO positions in US Fortune 500 companies through an examination of objective, accountancy-based measures of success, including return on equity and return on assets. However, a closer look at the nature of performance measures revealed that in FTSE 100 companies, glass cliff appointments were particularly visible when organizational success and crisis were defined in terms of subjective measures such as stockmarket performance (Haslam *et al.* 2010). This variability in findings suggests that what matters most for the glass cliff phenomenon are not the underlying financial realities of company performance, but rather *perceptions* of success or crisis.

Importantly, glass cliff appointments are not restricted to executive positions in the world's largest companies. In an archival analysis of the 2005 UK general election, Ryan *et al.* (2010) found evidence for the precariousness of women's political positions, such that, in the Conservative party, female candidates contested seats that were harder to win (i.e. seats held by an opposition party candidate with a significantly larger majority) than did male candidates. As a result, Conservative women won significantly fewer votes than their male counterparts, a tendency that was fully explained by the higher challenges faced by women.

From this research it is apparent that women and men tend to be appointed to high-profile positions under different circumstances, and that women's particular experiences are such that detrimental effects on their careers are more likely. Archival research such as this has high external validity in demonstrating *that* a glass cliff phenomenon exists. However, the observed patterns in these studies could be attributed to a number of different processes, making it quite difficult to explain *why* it exists. Most likely, there is no single factor responsible for the emergence of glass cliffs in various organizations (Ryan and Haslam 2007). So what might be important factors and processes that contribute to the glass cliff?

UNDERSTANDING THE GLASS CLIFF—EXPERIMENTAL EVIDENCE

As a first step towards answering this question, Ryan *et al.* (2007a) asked 164 participants who had read an online news story about the glass cliff for their ideas and opinions as to what might underlie the phenomenon. These participants generated a number of different explanations, ranging from relatively benign interpretations such as women's higher suitability for difficult leadership tasks to much more malign interpretations such as outright sexism or women's expendability. The explanations also ranged from those that were more intentional, such as men expressing favouritism to other men, to relatively unintentional processes such as a lack of alternative opportunities for women. These participants' speculations about the reason's underlying the glass cliff provide valuable starting points for further empirical investigation.

By now, a number of controlled scenario studies and experiments have examined the circumstances under which people are likely to prefer a woman for a leadership position, and when they are likely to prefer a man—and when and why these preferences vary with organizational performance (Ashby *et al.* 2007; Haslam and Ryan 2008; Bruckmüller and Branscombe 2010; Brown *et al.* 2011; Gartzia *et al.* 2012; Rink *et al.* 2013). In these studies, research participants typically read about a company that either performed poorly or successfully. They were then asked to select a female or male candidate to be appointed to a new leadership position, to evaluate these candidates' suitability for the position, or to evaluate the suitability of current female and/or male leaders (Haslam and Ryan 2008: study 1; Bruckmüller and Branscombe 2010; Brown *et al.* 2011; Gartzia *et al.* 2012; Rink *et al.* 2013). Although many of these studies focused on a business context they also included the selection of a candidate for a hard-to-win or easy-to-win political constituency (Ryan *et al.* 2010: study 2), of a defence lawyer for a highly criticized or a more promising legal case (Ashby *et al.* 2007), and of a youth representative for a music festival experiencing a rise or decline in popularity (Haslam and Ryan 2008: study 2). In all these studies, participants were more likely to prefer a female candidate for a leadership position in times of crisis than in times of success, where they usually favoured a male candidate. Because these studies control for a range of individual factors, such as the candidates' qualifications, they demonstrate that glass cliff appointments result, at least in part, from the judgement and decision-making processes surrounding leadership appointments. In the following, we summarize what these studies tell us about the psychological processes involved in glass cliff appointments.

The Role of Gender Stereotypes

One recurring theme that emerged when Ryan and colleagues (2007) asked participants to generate possible explanations for the glass cliff were stereotypes about the kinds of leaders that women are expected to be and the kinds of leaders that men are expected to be. Many participants suggested that women might be appointed to leadership positions during a crisis because stereotypes about women seem to match perceptions of what is needed in those situations better than stereotypes about men. For example, women's ostensibly higher ability to 'smooth things over' or their supposedly higher competence in 'crises involving other people' may make them appear particularly suitable leaders in tough times. Exploring the intuitions of these participants, a series of experimental studies have provided evidence that stereotypes about gender and leadership play an important role for the emergence of the glass cliff (Bruckmüller and Branscombe 2010; Ryan *et al.* 2011; Gratzia *et al.* in press).

Contextual Variations in Gendered Stereotypes about Leadership

Historically, leadership roles have been associated with the male gender role (Eagly 2007) and perceptions of the 'typical successful middle manager' show much higher overlap with perceptions of the 'typical man' than with perceptions of the 'typical woman' (Schein 1973). This *think-manager-think-male* (TMTM) association and the resulting perceived lack of fit between women and leadership roles (Heilman 1983) has been identified as one of the key hurdles that women must overcome to succeed in leadership (e.g. Wellington *et al.* 2003). However, perceptions of what constitutes good leadership have begun to change in recent years and stereotypically female qualities such as cooperation, communications skills, and an orientation towards teamwork are increasingly seen as important aspects of effective leadership as well. In Alice Eagly's (2007: 3) words, 'in many contexts, the Powerful Great Man model of leadership no longer holds. Good leadership is defined in terms of the qualities of a good coach or teacher.'

What are the contexts in which the traditional model of the Powerful Great Man—or the TMTM association—no longer holds? First indications come from research on charismatic or transformational leadership—a way of leading characterized by consideration, stimulation, motivation, and trust (Bass and Bass 2008) that is often associated with female leaders (e.g. Eagly *et al.* 2003) and particularly sought, and more likely to emerge, in times of crisis (e.g. Hunt 1990; Pillai 1996). Examining contextual variations in the TMTM stereotype more systematically, Ryan and colleagues (2011) demonstrated that stereotypes about typical managers in a *successful* company showed a higher overlap with the male than with the female gender stereotype (cf. Schein 1973). However, when it came to managers of *unsuccessful* companies, participants' descriptions focused on the absence of masculine traits (study 1). More importantly, when participants were asked for the attributes they considered as *desirable* for ideal leaders in successful and unsuccessful companies, no TMTM association emerged. The ideal manager for a successful company was seen as combining both stereotypically male and stereotypically female attributes, while stereotypically female attributes outweighed stereotypically male characteristics in descriptions of an ideal manager in times of crisis, indicating a 'think-crisis-think-female' association (TCTF, study 2). Thus, although leaders in general are still perceived as stereotypically male, women seem to be perceived as better suited to lead in times of crisis.

Gender Stereotypes as Causes for the Glass Cliff

Such stereotypes about gender and leadership seem like a plausible explanation for the glass cliff, and indeed, experimental studies provide direct evidence for this interpretation. Bruckmüller and Branscombe (2010: study 2) gave their participants a description of either a successful company (i.e. a company that had experienced 'a tremendous upward trend' in stockmarket value) or a company in crisis (i.e. a company that had experienced 'a tremendous downward trend'). This description was followed by the profiles of one female and one male candidate for the soon to be vacant position of CEO in this company. Participants were asked to indicate their perceptions of these two candidates in terms of pre-tested characteristics capturing the TCTF-stereotype (e.g. communication skills, willingness to cooperate, ability to encourage others) and characteristics capturing the TMTM-stereotype (e.g. independence, decisiveness, competitiveness). They then evaluated the candidates' suitability for the position and selected one of them as the new CEO. As expected, for the successful company, participants perceived the male candidate as more suitable for the leadership position, and were more likely to choose him as the new CEO, than the female candidate. For the company in crisis, participants evaluated both candidates as equally suitable and were significantly more likely to choose the female than the male candidate. Importantly, regression analyses revealed that, for the successful company, it was mostly participants' perceptions of the candidates' stereotypically male (TMTM) characteristics that predicted who they chose as new leader. When company performance was poor, however, it was mostly the perception of stereotypically female (TCTF) characteristics that predicted participants' choice. Moreover, differences in the attribution of these stereotypically female characteristics to each of the candidates partially explained the effects of company performance on participants' candidate evaluations and leader choices. In other words, participants preferred a female leader for an unsuccessful company in part because they believed that the female candidate had more abilities that are expected to matter in times of crisis (communication skills, cooperation, etc.) than the male candidate—or, similarly, because they believed that the male candidate possessed these abilities to a lower degree.

Similarly, Gartzia and colleagues (2012) presented their participants with eight different candidates for a leadership position in a company in crisis, systematically varying candidate gender and the description of these candidates with stereotypically male and stereotypically female attributes. They found that participants favoured female over male candidates, evidencing a glass cliff pattern in participants' choices, and they also favoured feminine (e.g. kind, understanding) over masculine (e.g. ambitious, competitive) candidates. These studies suggest that TCTF-notions are at least in part responsible for the glass cliff. However, they do not answer the question why stereotypically female attributes are perceived as particularly desirable in leaders in times of crisis.

Why Think-Crisis-Think-Female?

To further investigate this question, Ryan and colleagues (2011: study 3) conducted a study in which they specified what would be expected of a leader in times of crisis: (*a*) to stay in the background and endure the crisis; (*b*) to take responsibility for the inevitable failure (i.e. to act as a scapegoat); (*c*) to manage people and personnel issues through the crisis; (*d*) to be a spokesperson providing damage control; or (*e*) to take control and improve performance. They then asked participants to rate the desirability of stereotypically male and stereotypically female leader characteristics in each of these situations. Female characteristics were rated as more desirable than male characteristics in leaders who were merely expected to endure the difficult times or to take responsibility for the crisis, and the stereotypically female characteristics were rated as particularly desirable for the task of managing people through the crisis. In contrast, the female and male characteristics were rated as equally desirable in times of crisis when a leader had to act as a spokesperson or if the leader's main task was to improve performance.

Taken together, the results of this study demonstrate that one likely reason why women are selected for glass cliff positions is that they are seen to have what it takes to manage a company in times of crisis, that is, to be good people managers. However, the finding that stereotypically female traits were also associated with scapegoating and with an expectation of simply riding out the crisis, suggests that another element of glass cliff positions may be setting women up to fail. We will discuss this aspect in more detail later. First, we will summarize research that more directly connects stereotypical beliefs that women are good people managers with the glass cliff.

Two recent studies by Rink and colleagues (2013) examined the impact of the availability of social and financial resources on participants' evaluation of male and female leaders in different kinds of crisis situations. Within an organizational crisis scenario, they systematically varied the availability of financial resources and the availability of social resources, that is, whether the new leader could count on social support from shareholders and the company's board of commissioners. In contrast to previous research, participants were not asked to evaluate potential candidates for a leadership position, but were informed that the company in crisis had recently appointed a new financial director who was either female or male depending on condition. Participants were then asked to indicate how effective they anticipated this new financial director would be.

It was found that when social resources were available to a leader, participants expected the male leader to be more effective than the female leader, even though there was a crisis situation at hand. It was only in the crisis situation without social resources that the female leader was expected to be a more effective than the male leader. This enhanced preference for the female leader in a crisis where social resources were absent was due, at least in part, to the fact that participants expected her to have greater ability to establish acceptance with followers (study 1). A follow-up study demonstrated that the perception of the woman's higher suitability for leadership when social resources were unavailable was driven by the belief that she possessed more communal traits (e.g. communicative, cooperative, teamworking skills) than the male leader. That is, the ascription of the same kinds of stereotypical traits that contributed to glass cliff appointment patterns in Bruckmüller and Branscombe's (2010) study was responsible for participants' expectation that a woman would do better than a man in a leadership position without stakeholders' social support. At the same time, this study demonstrates that women are not always seen as better crisis managers than men. The male leader was evaluated more favourably than the female leader, provided that social resources were available to them. Participants then reverted back to the stereotypical TMTM association, perceiving the male leader to be more effective than the female leader because he was expected to possess more agentic traits typically associated with managerial roles (e.g. independent, decisive, competitive).

Think-Manager-Think-Male Associations and the Glass Cliff

The research outlined thus far not only suggests that stereotypes about women's communality and their presumably better 'people skills' drive perceptions that women are particularly qualified to lead in times of crisis. It also highlights the importance of considering contextual variation in TMTM associations (Ryan *et al.* 2011). Men are seen as more suitable for leadership in most situations, even in some crisis situations, because the male gender stereotype better matches with what we expect the typical manager to be like. However, these prototypical assumptions about what leaders are like and, more importantly, what they should be like, vary with context and in (certain) times of crisis they no longer hold.

Although perhaps less obvious than the importance of TCTF notions, this weakening of the TMTM-stereotype also seems to be important for the glass cliff. When Ryan and others (2011: study 2) asked their participants what leader qualities they saw as desirable in times of success versus in times of crisis, all nine stereotypically female attributes that were rated as desirable in times of success were also seen as desirable for leaders in unsuccessful companies; however, only five of the eight stereotypically male traits that were seen as desirable in times of success were also seen as desirable in a crisis. Moreover, in the study by Bruckmüller and Branscombe (2010), it was predominantly the importance of stereotypically male, agentic traits that changed with organizational performance. The attribution of these characteristics to the two candidates was highly predictive of participants' selection of leaders for a successful company and not predictive at all for their choices for a company in crisis, while the predictive power of stereotypically female, communal characteristics remained relatively stable across conditions, thus outweighing stereotypically male characteristics in times of crisis. Furthermore, it was perceptions of the male candidate's characteristics and his suitability for leadership rather than perceptions of the female candidate that changed with company performance.

Such shifting patterns of stereotypic associations lead to the prediction that both the preference for a male leader for a successful organization and the preference for a female leader for an unsuccessful organization should be particularly strong in contexts that reinforce TMTM-associations—such as in male-dominated industries, or when gender is particularly salient (Kanter 1977). The evidence for this prediction is mixed. Bruckmüller and Branscombe (2010: study 1) found that the glass cliff emerged for participants who had read about a company with a male-dominated history of leadership (presumably strengthening the TMTM-association), but not when participants had read about a company with a female-dominated history of leadership (presumably weakening these stereotypes). On the other hand, Gartzia and colleagues (2012) found that participants who had been exposed to a masculine leader as a role model were subsequently less likely to select a female candidate or a candidate with feminine attributes for a company in crisis than participants who had been exposed to a more feminine role model or no role model at all. However, since this study did not include a no-crisis comparison condition it is not possible to tease apart the influence of these role models on general preferences for male versus female leaders from potential interactions with the organizational performance context.

In summary, experimental evidence demonstrates the importance of contextual variation in the well-documented TMTM association and of stereotypes about women as better 'people managers' for glass cliff appointment patterns. These stereotypes about gender and leadership represent the most thoroughly studied factor contributing to the glass cliff phenomenon. However, there is at least initial evidence on several other possible explanations, to which we will now turn.

Perceived Need for (Signalling) Change?

The preceding discussion already hints at another factor that may contribute to the selection of women for leadership positions in times of crisis, namely, that they are simply *not men*. When an organization is performing poorly this is likely to be seen as an indication that the current approach is not working and that change from the (default) TMTM model of leadership is needed (Ryan and Haslam 2007). As one of Ryan et al.'s (2007: 119) participants put it: 'It's simply that for most companies a female CEO is something they haven't tried and so when things look bleak they start thinking what was previously "unthinkable".'

Initial support for this proposition comes from a series of studies by Brown and colleagues (2011). These authors found that conditions of threat lead to preferences for change rather than stability (study 1), and that change is implicitly associated with women, while stability is associated with men (study 2). Moreover, participants in a control condition were more likely to choose a man than a woman for a leadership position, but this tendency disappeared (study 3) or reversed (study 4) for participants who had been exposed to a threat-manipulation (via a word completion task that included threat-related words such as 'crisis', 'fear', or 'threat', or via reminding US participants of the terrorist attacks on 11 September 2001). To the extent that a severe organizational crisis causes decision members to feel threatened, these studies suggest that this psychological experience of threat might cause a desire for change and thereby make the selection of women for leadership more likely. Interestingly, Brown and others (2011) also found that preferences for a female leader under conditions of threat and for a male leader under no threat were only evident among participants who subscribed to ideologies legitimizing the current socio-political system, but not among participants with more progressive views. This suggests that, somewhat paradoxically, opting for change by selecting women under conditions of threat might actually be driven by an implicit desire to protect the status quo.

More directly relevant may be one of the studies by Bruckmüller and Branscombe (2010: study 1) mentioned above. This study found that glass cliff appointment patterns were contingent on a male-dominated history of leadership. When participants read about a successful company with a male history of leadership they chose another man as new leader; when the same company was in crisis, participants opted for a change and chose the female candidate. Importantly, this pattern disappeared but was not reversed for a company with a female-dominated history of leadership. This suggests that, although a male-dominated history of leadership may reinforce the default TMTM-association in a context of success and the perception of a female candidate as representing change in times of crisis, an organizational context of female-dominated leadership is not enough to fully eradicate widely shared stereotypes about gender and to perceive men as representing change.

A somewhat related process that is difficult to investigate via controlled experiments could be that companies deliberately appoint non-prototypical leaders in times of crisis to signal that the organization is embracing change (Lee and James 2004). Indeed, there is some evidence suggesting that, at least in Japan, poor company performance is associated with the appointment of highly visible 'outsiders', such as foreign nationals, to leadership positions (Kaplan and Minton 1994).

Sexism, in-Group Favouritism, and the Glass Cliff

Another straightforward explanation of the glass cliff, and one that was particularly common among Ryan et al.'s (2007) female participants, is that it may be a manifestation of in-group favouritism or sexism in the workplace. In-group favouritism in this context would mean that (mostly male) decision-makers select in-group members (i.e. other men) for desirable leadership positions or protect them from particularly risky positions, leaving these precarious positions for out-group members (i.e. women). In the experimental studies that have investigated leadership preferences in times of success versus crisis, female and male participants were equally likely to favour female candidates for risky glass cliffs and male candidates for more promising positions, providing no evidence of in-group favouritism (Haslam and Ryan 2008: study 1; Bruckmüller and Branscombe 2010; Brown et al. 2011; Gartzia et al. 2012; Rink et al. 2012). However, research conducted in corporate settings demonstrates that women often lack the support networks and resources that are provided to men both as they ascend the corporate ladder and once they are in leadership positions (e.g. Ibarra 1993; Tharenou et al. 1994), suggesting that dynamics of in-group favouritism might still play an important role for glass cliff appointments in real-life corporate settings.

A related, even more malign explanation of the glass cliff is to attribute it to outright sexism. Appointing women to precarious leadership positions may be one way in which decision-makers are able 'to block women's passage up the ranks' (as one participant in Ryan et al.'s 2007 study put it), while simultaneously appearing gender-fair because, superficially, they allow women to obtain leadership positions. A similar explanation is based on the notion that decision-makers may see women as more expandable than men and may therefore be more willing to appoint women to precarious positions (Ryan and Haslam 2007).

Very little research to date has looked at the role of sexism and sexist attitudes in the emergence of the glass cliff, but this research suggests that, although explaining the glass cliff by blatant sexism may be very straightforward, it may also be a bit too simplistic. First, there is no evidence that people who hold more sexist attitudes are more likely than people with more egalitarian attitudes to favour women for precarious glass cliff positions (Ashby et al. 2007). Quite to the contrary, Gartzia et al. (2012) found that participants who held more sexist attitudes were more likely to select a male or a masculine candidate for leadership in crisis than were participants with lower sexism scores. This study only provides preliminary evidence on the role of sexism for the glass cliff as it did not compare leader selection in times of crisis to leader selection for a more promising position. Nevertheless, the findings seem in contrast with the idea of blatant sexism, or the notion that decision-makers with sexist attitudes intentionally set women up to fail.

Some glass cliff studies, however, do report findings that can certainly be interpreted as the *subtle* workings of sexist assumptions. For example, stereotypically female characteristics are seen as more desirable in times of crisis when leaders are expected to act as scapegoats (Ryan *et al.* 2011: study 3), and respondents do rely on stereotypical traits to justify the selection of a female candidate for a precarious position (versus a male candidate for a promising one; Bruckmüller and Branscombe 2010; Rink *et al.* 2013). Likewise, Brown *et al.*' (2011) finding that the selection of a female candidate under conditions of threat—versus a male candidate in the control—was associated with system-legitimizing views hints at more subtle forms of sexism. Finally, in a study by Haslam and Ryan (2008: study 3) participants expected a leadership position in a context of declining company performance to be more stressful for a female candidate than for a male candidate and it was precisely this expected stressfulness that mediated the effect of company performance on evaluations of candidates. In other words, participants not only favoured a female candidate for leadership in crisis, despite their expectation that this situation would be very stressful for her, but *because* of it. While a benevolent interpretation of this pattern could be that participants were expecting the candidate to cope well with the increased stress of leading under difficult circumstances, a more malign interpretation is that participants were particularly willing to expose her to this stress.

In sum, whether or not sexism is an important factor contributing to the glass cliff largely depends on how one defines sexism. The scarce empirical evidence on this issue speaks against explicitly held sexist attitudes as a key factor, but suggests that more subtle and implicit sexist dynamics may well play an important role.

Socio-Structural Realities

In addition to the psychological processes outlined above, dynamics arising from socio-cultural realities in the corporate world might also contribute to the glass cliff. For example, risky positions are likely to have higher management turnover than more promising positions and previous research has found that women are most likely to break through the glass ceiling in companies with high turnover rates (Goodman *et al.* 2003). A relatively benevolent interpretation of such dynamics would be that the glass cliff is actually an expression of shifts towards gender equality in leadership—with companies increasingly appointing women to leadership positions and more risky positions opening up more frequently (see also Ryan *et al.* 2007). However, such appointment patterns might also indicate that women have fewer opportunities to obtain leadership positions and are afforded less choice in accepting or declining risky appointments. Indeed, one study found that glass cliff positions are seen as greater career opportunities for women than for men (Haslam and Ryan 2008: study 3), suggesting that women may expected to be less 'picky' about the leadership tasks that they take on and to see even highly precarious positions as 'golden opportunities' rather than 'poisoned chalices' (Ryan and Haslam 2007).

In sum, although the evidence for the importance of such socio-structural dynamics for the glass cliff phenomenon is scarce and rather indirect, it is certainly plausible that such processes play an important role, calling for further empirical investigation.

Do Women Actively Seek the Challenge?

The perception that risky leadership positions could offer women great career opportunities hints at another interpretation of glass cliff appointment patterns, namely, that women may not be singled out for risky positions by others, but rather that they may actively seek the challenge, looking for good opportunities to prove themselves. However, a recent study by Rink *et al.* (2012) seems to contradict this interpretation. In two studies, male and female Dutch business graduate students evaluated the attractiveness of a leadership position during an organizational crisis with either no social resources (support from employees) but sufficient financial resources, with social resources but no financial resources, or with both kinds of resources. Overall, female and male participants evaluated leadership under crisis conditions equally (un)favourable. However, this was moderated by the kind of resources available. Female participants evaluated the position lacking *social* resources as least desirable, while male participants evaluated the position without *financial* resources as least desirable. In other words, women evaluated exactly those positions as less desirable that participants in the scenario study by Rink *et al.* (2013) saw women as particularly suitable for. Although this does not directly contradict the notion that women might actively seek out particularly risky positions, it certainly makes such an interpretation appear less plausible.

Summary

Although many questions remain open for future investigation, experimental research aimed at uncovering what causes the glass cliff has provided some valuable insights. The role of gender stereotypes is certainly the most thoroughly studied contributing factor. Findings suggest that the disintegration of the default association of leadership with men in combination with stereotypical assumptions that women will be better crisis managers—better people managers and better recruiters of social resources in particular—play an important role. Women seem to be appointed to glass cliff positions because they (*a*) represent a change from the default think-manager-think-male standard and (*b*) because they are perceived as particularly skilled for bringing about the kind of change that is needed. Additional, and not entirely separable, factors that also seem to play a role are socio-structural realities and the subtle workings of sexist dynamics.

DIRECTIONS FOR FUTURE RESEARCH

An important question that remains open for future research is whether the glass cliff is a phenomenon that only affects women or whether we should expect similar appointment patterns to emerge for other groups under-represented in leadership positions. To the extent that specific gender stereotypes such as think-manager-think-male and think-crisis-think-female associations play an important role one might speculate that the glass cliff is a phenomenon specific to gender and leadership. However, considering that the default TMTM model no longer holds in times of crisis and that a perceived need for change also plays an important role for the glass cliff, it seems more likely that any candidate that represents a change from the male, white, straight, middle-class manager prototype will have better chances to be appointed in situations of crisis than under more promising circumstances. Initial research seems to support this prediction, with evidence that organizational glass cliffs occur on the basis of sexuality (Robus and Ryan 2012) and that political glass cliffs were also evident on the basis of race in the UK general elections (Kulich *et al.* in press).

However, it is difficult to conduct archival research examining actual appointment patterns of minority group members across different context simply because their numbers in high-profile positions are so low. For example, early in 2012, African Americans only comprised 1 per cent of Fortune 500 CEOs (Black Entrepreneur Profile, 2012) and there was no Black representative in the US senate (Manning and Shogan 2011). However, Cook and Glass (2013) recently analysed the appointments of National Collegiate Athletic Association (NCAA) men's basketball head coaches in the US over a thirty-year period. They found, among other things, that (*a*) non-whites were more likely than whites to be promoted to head coach for losing teams, i.e. in situations of crisis, (*b*) they held head coach positions for a significantly shorter time, and (*c*) when they were unable to generate winning records they were likely to be replace by a white 'saviour'.

This hints at another important question for future research. What happens after members of underrepresented groups have been appointed to precarious glass cliff positions? Do stakeholders acknowledge the problems that preceded their appointment or will they fall prey to the 'romance of leadership' (Meindl *et al.* 1985) and unduly blame the leader, forgetting situational factors? And is signalling change a strategy that pays off for companies in crisis? An initial study by Haslam and colleagues (2010) casts doubt on the effectiveness of such strategies. They found that, although there was no relationship with the presence of women on the boards of FTSE 100 companies and relatively objective accountancy-based measures of success, there was a relationship between women's presence on boards and more subjective, stock-based measures of company performance, suggesting that investors were more sceptical of these companies' future outlook than warranted by the underlying financial realities. This seems to support the prediction that women leaders will be placed under higher scrutiny because of their untypical status (Brescoll *et al.* 2010), particularly in times of crisis (Haslam and Ryan 2008).

Practical Implications—How to Stay Off the Glass Cliff

What does the existing research tell us about strategies that both organizations and individual women (and other minority members) who aspire to leadership positions can take to avoid glass cliff appointments? Developing specific strategies and evaluating their effectiveness is certainly an important task for future research and the advice that we can give based on the existing literature on both the glass cliff and on gender and leadership more generally is only preliminary (for a more detailed discussion, see Ryan *et al.* 2007b).

One piece of advice for individuals is to display a certain amount of scepticism and to carefully examine the opportunities offered to them. This can include investigating details about the position such as how long it has been open, what happened to the last person who took the position, and why one is seen as the right person for the job. When taking on a difficult position, it may be a good idea to make others aware of these circumstances to ensure that, in case of failure, proper acknowledgement is given of the preceding situation. Another important element is to garner social support. Professional networks and social support are, of course, important for anybody aspiring to succeed in top management positions, but they are often less easily accessible for women (Tharenou *et al.* 1994)—while ironically the expectation that they will be able to establish social support seems to be part of what causes decision-makers to appoint women to glass cliff positions (Rink *et al.* 2013).

What advice can we give to organizations to ensure that their hiring and promotion decisions are driven by candidates' actual qualifications for the position rather than stereotypes? A starting point can be to inform decision-makers about the glass cliff phenomenon and its causes (cf. Ryan *et al.* 2007b). In addition, it is advisable to overtly outline any expectations of a new leader or relevant candidate characteristics before assessing potential candidates, to reduce the influence of implicit expectations based on candidate gender and stereotypes.

Considering the importance of the TMTM-stereotype, its weakening in response to crisis situations, as well as the association of women with change (Brown *et al.* 2011) and with what is needed in times of crisis (Ryan *et al.* 2011), striving for a diverse management in general, not only in situations of crisis, seems like a promising long-term strategy. To the extent that a diverse management weakens stereotypical perceptions of a leader in this organization as male (and white, straight, middle-class, etc.), assessments of who can bring about change and turn things around in a difficult situation will less likely be driven by candidates' demographic characteristics and the associated stereotypes, and might more likely be driven by what an organization actually needs.

Conclusion

In recent years, women have been breaking through the glass ceiling in sufficient numbers to take a closer look at the situations under which they are likely to achieve top management positions. Both archival and experimental research documents that they are more likely to do so under precarious circumstances and when a position involves a higher risk of failure. The research summarized above suggests that the fact that leadership is still predominantly male—both in people's stereotypic perceptions and in actual boardrooms in the corporate world—plays an important role for these glass cliff appointments. Men are associated with stability, and women are associated with change (Brown *et al.* 2011); stereotypes that portray men as typical and desirable leaders weaken in times of crisis (Bruckmüller and Branscombe 2010; Ryan *et al.* 2011); at the same time, women are stereotypically perceived as bringing several qualities that can be desirable in a crisis—including good people-managing skills (Rink *et al.* 2013), but also including a better suitability as scapegoats or for simply enduring a difficult phase (Ryan *et al.* 2011).

Taken together, these findings suggest that the glass cliff may be one (out of several inter-related) subtle processes that continually reinforce men's prototypicality in leadership and gender stereotypes, thereby continually legitimizing and reproducing the status quo (Brown *et al.* 2011; see also Bruckmüller *et al.* 2012). This also suggests that effective strategies for combating the glass cliff as well as other subtle processes that hinder women's progress in management will be those that are aimed at increasing diversity and at reducing the strong association between men (and other privileged groups) and leadership—both in reality and in people's stereotypic perceptions. When women (and other minorities) no longer represent a remarkable change from the default standard of what we imagine a leader to be like, the glass cliff will likely crumble as well.

References

Adams, S. M., Gupta, A., and Leeth, J. D. (2009). 'Are Female Executives Over-Represented in Precarious Leadership Positions?', *British Journal of Management*, 20: 1–12.

Adler, N. (2000). 'An International Perspective on the Barriers to the Advancement of Women Managers', *Applied Psychology: An International Review*, 42: 289–300.

Agars, M. D. (2004). 'Reconsidering the Impact of Gender Stereotypes on the Advancement of Women in Organizations', *Psychology of Women Quarterly*, 28: 103–11.

Ashby, J., Ryan, M. K., and Haslam, S. A. (2007). 'Legal Work and the Glass Cliff: Evidence that Women are Preferentially Selected to Lead Problematic Cases', *William and Mary Journal of Women and the Law*, 13: 775–94.

Bass, B. M., and Bass, R. (2008). *The Bass Handbook of Leadership: Theory, Research, and Managerial Applications* (4th edn). New York: Free Press.

Black Entrepreneur Profile (2012). African American CEO's of Fortune 500 companies. http://www.blackentrepreneurprofile.com/fortune-500-ceos (accessed May 2012).

Blau, F. D., and Kahn, L. M. (2000). 'Gender Differences in Pay', *Journal of Economic Perspectives*, 14: 75–99.

Brescoll, V. L., Dawson, E., and Uhlman, E. L. (2010). 'Hard Won and Easily Lost: The Fragile Status of Leaders in Gender-Stereotype-Incongruent Occupations', *Psychological Science*, 21: 1640–2.

Brown, E. R., Diekman, A. B., and Schneider, M. C. (2011) 'A Change will Do us Good: Threats Diminish Typical Preferences for Male Leaders', *Personality and Social Psychology Bulletin*, 37: 930–41.

Bruckmüller, S., and Branscombe, N. R. (2010). 'The Glass Cliff: When and Why Women are Selected as Leaders in Crisis Contexts', *British Journal of Social Psychology*, 49: 433–51.

Bruckmüller, S., Hegarty, P., and Abele, A. E. (2012). 'Framing Gender Differences: Linguistic Normativity Affects Perceptions of Power and Gender Stereotypes', *European Journal of Social Psychology*, 42: 210–18.

Catalyst (2009). *2008 Catalyst Census of Women Corporate Officers and Top Earners of the FP500*. New York: Catalyst. http://www.catalyst.org/knowledge/2008-catalyst-census-of-womencorporate-officers-and-top-earners-of-the-fp500 (accessed July 2013).

Cook, A., and Glass, C. (2013). 'Barriers to Minority Leadership in Work Organizations?' *Social Problems*, 60: 168-87.

Dreher, G. F. (2003). 'Breaking the Glass Ceiling: The Effects of Sex-Ratios and Work-Life Programs on Female Leadership at the Top', *Human Relations*, 56: 541–62.

Eagly, A. H. (2007). 'Female Leadership Advantage and Disadvantage: Resolving the Contradictions', *Psychology of Women Quarterly*, 31: 1–12.

Eagly, A. H., Makhijani, M. G., and Klonski, B. G. (1992). 'Gender and the Evaluation of Leaders: A Meta-Analysis', *Psychological Bulletin*, 111: 3–22.

Eagly, A. H., Karau, S. J., and Makhijani, M. G. (1995). 'Gender and Leader Effectiveness: A Meta-Analysis', *Psychological Bulletin*, 117: 125–45.

Eagly, A. H., Johannesen-Schmidt, M. C., and van Engen, M. (2003). 'Transformational, Transactional, and Laissez-Faire Leadership Styles: A Meta-Analysis Comparing Women and Men', *Psychological Bulletin*, 129: 569–91.

Fama, E., and Jensen, M., (1983). 'Separation of Ownership and Control', *Journal of Law and Economics*, 26: 301–26.

Ferris, S. P., Jagannathan, M., and Pritchard, A. C. (2003). 'Too Busy to Mind the Business? Monitoring by Directors with Multiple Board Appointments', *Journal of Finance*, 58: 1087–1111.

Gartzia, L., Ryan, M. K., Balluerka, N., and Aritzeta, A. (2012). 'Think Crisis—Think Female: Further Evidence', *European Journal of Work and Organizational Psychology*, 21: 603–28.

Goodman, J. S., Fields, D. L., and Blum, T. C. (2003). 'Cracks in the Glass Ceiling: In What Kinds of Organizations do Women Make it to the Top?', *Group and Organization Management*, 28: 475–501.

Hamilton, D. L., and Gifford, R. K. (1976). 'Illusory Correlation in Intergroup Perception: A Cognitive Bias of Stereotypic Judgments', *Journal of Experimental Social Psychology*, 12: 392–407.

Haslam, S. A., and Ryan, M. K. (2008). 'The Road to the Glass Cliff: Differences in the Perceived Suitability of Men and Women for Leadership Positions in Succeeding and Failing Organizations', *Leadership Quarterly*, 19: 530–46.

Haslam, S. A., Ryan, M. K., Kulich, C., Trojanowski, G., and Atkins, C. (2010). 'Investing with Prejudice: The Relationship between Women's Presence on Company Boards and Objective and Subjective Measures of Company Performance', *British Journal of Management*, 21: 484–97.

Heilman, M. E. (1983). 'Sex Bias in Work Settings: The Lack of Fit Model', in B. Shaw and L. Cummings (eds), *Research in Organizational Behavior*, v. 269–98. Greenwich, CT: JAI Press.

Hunt, J. G. (1990). *Leadership: A New Synthesis*. Newbury Park, CA: Sage.

Ibarra, H. (1993). 'Personal Networks of Women and Minorities in Management: A Conceptual Framework', *Academy of Management Review*, 18: 56–87.

Kanter, R. M. (1977). *Men and Women of the Corporation*. New York: Basic Books.

Kaplan, S., and Minton, B. (1994). 'Appointments of Outsiders to Japanese Boards: Determinants and Implications for Managers', *Administrative Science Quarterly*, 37: 282–301.

Kulich, C., Ryan, M. K., and Haslam, S. A. (2007). 'Where is the Romance for Women Leaders? The Effect of Gender on Leadership Attributions and Performance-Based Pay', *Applied Psychology: An International Review*, 56: 582–601.

Kulich, C., Trojanowski, G., Ryan, M. K., Haslam, S. A., and Renneboog, L. D. R. (2011). 'Who Gets the Carrot and Who Gets the Stick? Evidence of Gender Disparities in Executive Remuneration', *Strategic Management Journal*, 32: 301–21.

Kulich, C., Ryan, M. K., and Haslam, S. A. (in press). 'The Political Glass Cliff: Understanding How Seat Selection Contributes to the Underperformance of Ethnic Minority Candidates', *Political Research Quarterly*. doi: 10.1177/1065912913495740

Lee, P. M., and James, E. H. (2004). 'She'-e-os: Gender Effects and Stock Price Reactions to the Announcement of Top Executive Appointments'. Darden Graduate School of Business Administration. University of Virginia Working Paper, 02-11.

Lyness, K. S., and Thompson, D. E. (1997). 'Above the Glass Ceiling? A Comparison of Matched Samples of Female and Male Executives', *Journal of Applied Psychology*, 82: 359–75.

Manning, J., and Shogan, C. (2011). *African American Members of the United States Congress: 1870–2011*. Washington, DC: CRS Report for Congress. www.crs.gov (accessed May, 2012).

Meindl, J. R., Ehrlich, S. B., and Dukerich, J. M. (1985). 'The Romance of Leadership', *Administrative Science Quarterly*, 30: 78–102.

Pillai, R. (1996). 'Crisis and the Emergence of Charismatic Leadership in Groups: An Experimental Investigation', *Journal of Applied Social Psychology*, 26: 543–62.

Rink, F., Ryan, M. K., and Stoker, J. I. (2012). 'Influence in Times of Crisis: Exploring How Social and Financial Resources Affect Men's and Women's Evaluations of Glass Cliff Positions', *Psychological Science*, 23: 1306–1313.

Rink, F., Ryan, M. K., and Stoker, J. I. (2013). 'Clarifying the Precariousness of the Glass Cliff: How Social Resources and Gender Stereotypes Affect the Evaluation of Leaders in Times of Crisis' *European Journal of Social Psychology*, 34: 381–92.

Robus, C. J., and Ryan, M. K. (2012). 'Think Crisis—Think Pink: Examining Sexuality and the Glass Cliff', manuscript in preparation, University of Exeter.

Ryan, M. K., and Haslam, S. A. (2005). 'The Glass Cliff: Evidence that Women are Over-Represented in Precarious Leadership Positions', *British Journal of Management*, 16: 81–90.

Ryan, M. K., and Haslam, S. A. (2007). 'The Glass Cliff: Exploring the Dynamics Surrounding Women's Appointment to Precarious Leadership Positions', *Academy of Management Review*, 32: 549–72.

Ryan, M. K., Haslam, S. A., and Postmes, T. (2007*a*). 'Reactions to the Glass Cliff: Gender Differences in the Explanations for the Precariousness of Women's Leadership Positions', *Journal of Organizational Change Management*, 20: 182–97.

Ryan, M. K., Haslam, S. A., Wilson-Kovacs, M. D., Hersby, M. D., and Kulich, C. (2007*b*). *Managing Diversity and the Glass Cliff: Avoiding the Dangers by Knowing the Issues*. London: Chartered Institute of Personnel and Development.

Ryan, M. K., Haslam, S. A., and Kulich, C. (2010). 'Politics and the Glass Cliff: Evidence that Women are Preferentially Selected to Contest Hard-to-Win Seats', *Psychology of Women Quarterly*, 34: 56–64.

Ryan, M. K., Haslam, S. A., Hersby, M. D., and Bongiorno, R. (2011). 'Think Crisis—Think Female: Glass Cliffs and Contextual Variation in the Think Manager—Think Male Stereotype', *Journal of Applied Psychology*, 96: 470–84.

Schein, V. E. (1973). 'The Relationship between Sex Role Stereotypes and Requisite Management Characteristics', *Journal of Applied Psychology*, 57: 95–105.

Schmitt, M. T., Ellemers, N., and Branscombe, N. R. (2003). 'Perceiving and Responding to Gender Discrimination in Organizations', in S. A. Haslam, D. Van Knippenberg, M. J. Platow, and N. Ellemers (eds), *Social Identity at Work: Developing Theory for Organizational Practice*, 277–92. Philadelphia: Psychology Press.

Sczesny, S. (2003). 'A Closer Look beneath the Surface: Various Facets of the Think-Manger-Think-Male Stereotype', *Sex Roles*, 49: 353–63.

Sealy, R., and Vinnicombe, S. (2012). *The Female FTSE Board Report*. http://www.som.cranfield.ac.uk/som/ftse (accessed Apr. 2012).

Simon, R. J., and Landis, J. M. (1989). 'Women's and Men's Attitudes about a Woman's Place and Role', *Public Opinion Quarterly*, 53: 265–76.

Stroh, L. K., Langlands, C. L., and Simpson, P. A. (2004). 'Shattering the Glass Ceiling in the New Millennium', in M. S. Stockdale and F. J. Crosby (eds), *The Psychology and Management of Workplace Diversity*, 147–67. Malden, MA: Blackwell.

Tharenou, P., Latimer, S., and Conroy, D. (1994). 'How do you Make it to the Top? An Examination of Influences on Women's and Men's Managerial Advancement', *Academy of Management Journal*, 37: 899–931.

Wellington, S., Kropf, M. B., and Gerkovich, P. (2003). 'What's Holding Women Back?', *Harvard Business Review*, 81 (June): 18–19.

CHAPTER 15

....

POWER AND RESISTANCE IN GENDER EQUALITY STRATEGIES

Comparing Quotas and Small Wins

....

YVONNE BENSCHOP AND MARIEKE VAN DEN BRINK

INTRODUCTION

....

How to change gender inequality is a topic that feminist organization scholars have been interested in for many years (Liff and Cameron 1997; Ely and Meyerson 2000b; Hearn 2000; Nentwich 2006; Deutsch 2007; Benschop *et al.* 2012). There have been successes in beating overt discrimination and gender inequalities in organizations, especially if one is willing to take a historical perspective. In the twenty-first century direct discrimination by sex and sexual harassment have been banned by laws and regulations in many Western countries. At the same time, several authors have pointed out that equal employment opportunity legislation in different countries and contexts does acknowledge social, structural, and systemic gender discrimination, but is not successfully addressing the many faces of gender inequality in organizations today (Miller *et al.* 2009; Ainsworth *et al.* 2010; Greene and Kirton 2011). Ainsworth *et al.* (2010) speak of 'a blinding lack of progress', and this assertion goes certainly beyond the Australian private sector context of their study.

One important reason for this slow pace of change is the changed nature of gender inequality at work that has gone underground (Meyerson and Fletcher 2000). Contemporary organizational life is characterized by many subtle and deeply embedded inequalities. Changing the processes and practices that reproduce gender in more opaque and subtle ways has proven to be much more difficult than changing obvious discriminatory practices. The prevailing theory about the best way to counter these subtle inequalities is the post-equity or 'small-wins' theory on social change based on the work of Karl Weick (1984). This theory contends that deeply rooted practices and beliefs can only be changed by 'a persistent campaign of incremental changes that discover and destroy the deeply embedded roots of discrimination' (Meyerson and Fletcher 2000: 128). The power of these incremental changes lies in the persistent campaign of sequential feminist experiments that target organizational processes and practices. These experiments are carefully designed so that the potential for change is maximized (Meyerson and Kolb 2000). Meyerson and Fletcher present the small wins approach as 'a powerful way of chipping away the barriers that hold women back without sparking the kind of sound and fury that scares people into resistance' (2000: 127).

This post-equity or small wins approach differs from various other approaches for organizational change. On the one hand, it differs from the popular management programmes that set out to equip (or fix) the women so that they can master the rules of the organizational power game. The post-equity approach targets organizations rather than women (Liff and Cameron 1997; Ely and Meyerson 2000a). On the other hand, it also differs from more revolutionary radical approaches to counter gender discrimination, such as equal outcome measures and gender quotas to increase the number of women in top positions (Kirton and Greene 2010). Radical interventions are powerful as they forego weakening compromises and focus on results only. Such radical interventions are also highly contested, openly resisted, and the subject of heated debates between avowed proponents and adversaries.

We focus in this chapter on the issues of power and resistance to come to a better understanding of the success or failure of different types of feminist interventions. Even though power has been the *raison d'être* of feminist organization studies, power is often the elephant in the room in the literature on organizational change towards gender equality. We see power as linked to the control of resources, structures, behaviours, agendas, ideologies, cultures and subjectivities (Kärreman and Alvesson 2009: 1118). One particularly interesting form of power in the context of organizational change towards gender equality is resistance. Resistance is a form of power that implies agency and is likely to be expressed in a multitude of mundane actions and behaviours at the workplace (Prasad and Prasad 2000: 388). Resistance is a key issue for organization and management researchers that use critical perspectives, and they have broadened the conceptualization of resistance from formal organized opposition against the exercise of power (Prasad and Prasad 2000; Kärreman and Alvesson 2009), to more informal, routinized forms of resistance in everyday practice that are inherent in the exercise of power (Thomas and Davies 2005).

We stress that changing power processes and power relations is at the heart of feminist intervention strategies, and therefore resistance is part and parcel of all such interventions. Resistance to change is typically strong when an organization's cultural norms, beliefs, attitudes, and values are the target of change efforts. This is certainly the case with projects that target gender inequalities in organizational routines (Benschop and Verloo 2006). It is thus unsurprising that studies of organizational change towards gender equality report extensive resistance, a resistance that comes in many forms and shapes. Examples include men who resist women's entry into previously masculine domains, challenges to the authority of women managers, the denial of problems with gender in the organization (Connell 2006; Martin 2006; Van den Brink 2010), requests for research or training in order to avoid action, and attempts to escape involvement in change efforts (Benschop and Verloo 2006). By bringing to the fore the role of processes of power and resistance in the different types of interventions towards gender equality in organizations, we aim to provide a balanced account of these interventions and their potential to realize the change that is needed.

This chapter is structured as follows. We begin by reviewing existing research on change towards gender equality in organizations. We summarize the different strategies and zoom in on the small wins experiments and quota regulations as contrasting intervention strategies, discussing the advantages and disadvantages of these different strategies. The analysis of the interventions leads to a discussion of the role of power and resistance in successful organizational changes towards gender equality. We conclude by summarizing our key points and suggesting questions to be explored in further research.

Changing Organizations towards Gender Equality

While an ever-growing literature has addressed questions of gender in organizations over the years and the dynamics of gender inequality at work have been documented well, much less is known about how to bring about effective gender change in organizations. The analyses of what keeps gender inequality in place outnumber the analyses of what should and could be done to change organizations towards gender equality. Nentwich presents this as a logical order when she states 'once we identify the particular ways in which certain organizational, discursive or social practices produce gender inequities, they may become potential targets for experimentation and change' (Nentwich 2006: 503). Yet, the transition from analysis to effective intervention is not so easy, as illustrated by the different strategies for organizational change that have been developed since the 1970s. Literature shows that such strategies encounter many fundamental and more practical hindrances (Walby 2005), and do not always succeed. Different conceptualizations of goals (equal opportunity, gender neutrality, or gender equality) and different diagnoses of the core problem (unequal treatment, lack of access to resources, or gendered processes of organizing) are among the fundamental problems (Benschop and Verloo 2006). The most familiar practical problems are the fragmentation of interventions, their superficiality, their poor implementation, and their addressing of women only (Benschop and Verloo 2006). Despite these problems, the strategies for gender equality continue unabated in organizations, even though the label under which they are presented may change with fashion. Below, we take a closer look at these strategies.

Reviewing the Strategies

Scholarly knowledge about organizational change towards gender equality is developed in different academic disciplines and especially in political science, gender studies, and organization studies. Bringing together the parallel developments in equality policy approaches (Jewson and Mason 1986; Kirton and Greene 2010) and feminist interventions in organizations (Ely and Meyerson 2000b; Meyerson and Kolb 2000) from these different disciplines, Benschop and Verloo (2011) come up with a two-dimensional model. The individual-structural dimension refers to the roots of social inequality in societal structures or individual action and the core target of the strategies, and the inclusion-re-evaluation-transformation dimension addresses the aspired scope of the strategies for change.

We first discuss the strategies that focus on the individual, as they are among the most popular strategies in organizations. Individual inclusion strategies are striving for equal opportunities for all individuals regardless of the social groups people belong to. The underlying principles of liberal equality strategies are universal standards of social justice, fairness, merit, and formal equal treatment for all (Jewson and Mason 1986). Policies are based on the notion of the neutral individual, as this is seen as the most efficient way to achieve a fair distribution of rewards and resources in the workplace (Kirton and Greene 2010). Those neutral individuals compete for jobs solely based on their individual merit, such as their experience and qualifications. To ensure that fair competition, all must be treated equally without regard to persons. Gender, ethnicity, age, class, sexual orientation, all such social categories should not matter, and can be ruled out by laws, rules, and regulations to ensure equal opportunities.

In the feminist organization literature, this liberal perspective is criticized for its limited conceptualization of gender as an individual characteristic. Ely and Meyerson (2000b) stress that the liberal perspective assumes that women have less skills than men to thrive in their careers and in business at large, because of sex-role socialization that allegedly produces significant differences between women and men. The strategy for change in this approach is to eliminate the differences between women and men by teaching women the political and strategic skills to 'play the game' so that women can compete with men on a par (Ely and Meyerson 2000b; Meyerson and Kolb 2000). Measures such as special management development programmes and leadership training for women are the preferred interventions in this perspective. Thus, women are the targets of these measures and they are made responsible for their own inclusion. The interventions of this approach have been popular in organizations from the early days of equal opportunity policies onwards. The programmes may have had their benefits for some: white, middle-class women who participated have entered the managerial and professional ranks. Women from different ethnicities and class backgrounds have typically not been included in the management development or leadership programmes (Betters-Reed and Moore 1995). Overall, individual inclusion strategies do not change organizations towards more gender equality as they typically target only a few of the players and leave the game and its rules intact.

Individual re-evaluation strategies are grounded in a different logic. They do not stress the neutrality of individuals nor their equality, but emphasize their differences and call for a celebration of those differences. In the managing diversity strategies that have become popular since the 1990s, the premise is that organizations can benefit from the different perspectives and experiences people from different backgrounds can bring to the workplace. These differences are individualized with the adage that all individuals have their own unique contribution to make to the organization. Differences are thus seen as positive, and diversity as something to be recognized, valued, and rewarded (Kirton and Greene 2010). Although the managing diversity approach encompasses the classic social categories of gender, ethnicity, age, class, and sexual orientation, it moves beyond those categories to include a wide array of other differences such as education, work-style, and work-orientations. The broader view on differences serves to draw attention away from social inequality and to the contribution of individuals to the effectiveness of the organization.

The currently popular managing diversity strategies are preceded by the individual re-evaluation strategy that deals with gender inequalities and call for the revaluation of the feminine (Helgesen 1990; Rosener 1990). Based on feminist standpoint theory (Harding 2004), this strategy emphasizes that the socialized differences between women and men should be recognized and that the everyday experiences and perspectives of women should be used to produce knowledge and question the practices of powerful institutions (Benschop and Verloo 2011). The disadvantage of women should be addressed by revaluing and reordering feminine skills and attributes (such as relating, nurturing, listening, emoting) relative to masculine attributes and skills (such as directing, talking, thinking, doing) (Meyerson and Kolb 2000: 562). A valuing difference strategy calls for the acknowledgement of the special contributions that feminine styles and qualities such as listening, collaborating, and communicating can bring to organizations (Fletcher 1999). Interventions in this approach include awareness raising and training to familiarize people with the differences between women's and men's skills, styles, and perspectives. Critics have pointed to the lack of proof about the effectiveness of this re-evaluation strategy and the risk that it reinforces sex stereotypes, gender segregation, and the power imbalance between women and men at work (Ridgeway 1997). The strategy has also been criticized for its particular dominant heterosexual, white, class-privileged images of femininity and masculinity (Ely and Meyerson 2000a). Again, as with the individual inclusion strategy, the individual re-evaluation strategy does not target the organizational structures and cultures that reproduce the hierarchical valuing of gender differences in the organization (Meyerson and Kolb 2000). Calling for a revaluation of contributions does little to actually achieve a different distribution of positions of power and subordination.

Table 15.1 does not contain any strategies that can be characterized as individual transformation strategies. Transformation as a goal goes beyond the individual level and addresses organizations and societies at large rather than individuals. In the same vein, there are no structural revaluation strategies included in Table 15.1. Feminist ways of organizing that disassociate from hierarchy and capitalist, masculinist ways

Table 15.1 Strategies for change

	Inclusion	Re-evaluation	Transformation
Individual	Liberal, equip the woman, create equal opportunity	Managing diversity Value difference	
Structural	Radical structural equal opportunity		Gender mainstreaming Post-equity

Source: Adapted from Benschop and Verloo (2011: 280).

of organizing (Iannello 1992) have not informed strategies for organizational change (Benschop and Verloo 2011).

We now turn to the structural dimension of the model. Structural inclusion strategies such as the radical approach to equality address the structural barriers in organizations that hinder change towards gender equality. The radical approach is labelled radical because it does not stop at ensuring equal opportunities, but wants to see equality of outcome, even as direct interventions are needed to ensure that goal (Kirton and Greene 2010). This strategy acknowledges that some social groups face systematic discrimination from dominant groups, and have to fight stereotypes and prejudice about their merits, abilities, and skills at work that seriously impede their advancement to positions of power. Furthermore, the radical approach questions the neutrality of notions such as 'skill', 'talent', and 'merit', emphasizing the socially constructed character of these notions and their effects on the organizational positions of minority groups. These questions lead to contested and political measures such as preferential selection and quotas to ensure a fair proportional representation and equal outcomes (Jewson and Mason 1986).

In the feminist organization literature, the equivalent of the radical approach is labelled 'create equal opportunity' (Meyerson and Kolb 2000). The differential opportunity structure for women and men in the workplace (Kanter 1977) is seen as the core problem as these organizational structures create sloped instead of level playing fields and hinder the managerial and professional advancement of women. To eliminate these structures, measures typically target recruitment, promotion, and evaluation procedures and practices to come with more transparent recruitment practices, more flexible work arrangements, and more variation in career paths. The structural inclusion strategies do make some changes in organizations towards greater equality, but they do not suffice to solve the problem of gender inequality in organizations. Critics note that the focus on structural barriers without systematic attention to underlying cultural norms and values related to gender, work, and organization is not enough to change organizations (Mescher *et al.* 2010). More is needed to 'disrupt the pervasive and deeply entrenched imbalance of power in the social relations between men and women' (Ely and Meyerson 2000*a*: 113). Structural inclusion strategies can be easily hijacked, covered up, or changed in direction by gender inequality practices so that sustainable organizational change is impeded (cf.Van den Brink and Benschop 2012). In the words of

Meyerson and Kolb (2000: 562), structural inclusion 'programs do not sufficiently challenge the systems of power that make them necessary in the first place'.

The final set of strategies is characterized as structural transformation strategies. The scope of change in these strategies is the most far-reaching as the goal is to transform organizational processes and routines so that they no longer reproduce gender inequality. Gender mainstreaming can be seen as an example of such a structural transformation strategy. Gender mainstreaming aims to transform organizational processes and practices by eliminating gender biases from existing routines, involving the regular actors in this transformation process (Council of Europe 1998). Gender mainstreaming is a prominent strategy in the European Union. Gender mainstreaming broadens the involvement of people responsible for gender equality in organizations from gender specialists and experts to all organizational actors involved in policy-making and implementation. It calls for the cooperation between gender experts and other actors to collectively identify the policies, practices, and routines that foster gender inequality at work. It then further develops the collaboration to design new policies and practices and to implement the change. Gender mainstreaming aims to develop new material and discursive constructions of masculinity and femininity to transform organizational systems, work practices, norms, and identities (Benschop and Verloo 2011).

Another example of a structural transformation strategy is labelled 'post-equity' or a 'non-traditional approach to gender' (Ely and Meyerson 2000b; Meyerson and Kolb 2000). In their view gender is an axis of power and a core organizing principle that shapes social structures, knowledge, and identities. This strategy bears resemblance to gender mainstreaming in its targeting of the processes and practices that reproduce gender inequalities as part of everyday organizational routines and interactions (Acker 1992). The intervention strategy proposed in the post-equity approach starts with the identification of the particular ways in which concrete organizational practices produce gender inequities. The next step is to critique the production of gender inequalities through those practices. Then these practices are targeted as sites of experimentation and change. In the final stages new narratives and stories are developed to keep the focus on the outcomes of the experiments and their contribution to change towards gender equality (Meyerson and Fletcher 2000). In contrast to all other intervention strategies, there is no clear endpoint to this approach, experimentation to change the social order goes on continuously.

The critique on the structural transformation strategies centres on two key issues. First, the transformative potential of these strategies is tamed by the need to connect to the frame of reference about gender at work of the organizational members involved in the change processes. The researchers may build on a sophisticated conception of gender as a power-based social construction of women and femininity and men and masculinity and their relationship. If gender is still perceived as an individual characteristic in the organization, this will influence the outcomes of the change projects. Second, these strategies require participation from the organization and political will and support from key figures in the organization. The strategies do not do enough to counter crucial power differences between the different parties involved, causing the need to compromise and water down transformations (Benschop and Verloo 2011).

Reviewing this spectrum of the strategies for change, we observe that processes of power and resistance play a crucial role in the success or failure to change towards greater gender equality. In the next section, we therefore zoom in on two specific strategies for structural change to further explore the importance of power and resistance to organizational change in this area. We look at the strategies that aim to change the structure of organizations rather than at the strategies that target individuals, as the latter are less contested in organizational practice precisely because they make women the central problem, not organizations. We select quotas as an exponent of structural inclusion strategies and post-equity experimentation as an exponent of structural transformation. These two approaches represent very different strategies in terms of the power processes they invoke.

Zooming in: Post-Equity Experiments versus Quotas

Post-Equity Experiments

Aim

Post-equity experiments are a strategy of evolutionary change. The core aim of this type of intervention is to continuously identify how gender inequities are produced in organizational processes, to disrupt this social order and to revise the structural, interactive, and interpretative practices in organizations accordingly (Meyerson and Fletcher 2000). There is no identifiable endpoint foreseen, the aim to change the way that work is defined, executed, and evaluated is an ongoing process (Ely and Meyerson 2000b). For the identification of gendering processes, Acker's five sets of gendering processes in organizations (1990) are used to distinguish between formal policies and procedures, informal work practices, an organization's symbols, norms, and images, everyday social interactions, and gender identities. Interventions typically consist of three components: (1) critique of the gendering processes that produce gender inequality in daily work practices, (2) experimentation with doing gender differently and improving work effectiveness, and (3) narrative generation to develop stories about the need for and success of the experiments (Meyerson and Kolb 2000). Experiments are carefully designed so that they can serve what is called the dual agenda of gender equality goals and business goals (Meyerson and Kolb 2000).

Examples

Although there is widespread common-sense acceptance about the relevance of this approach, only a few studies on post-equity experiments have been published. In the 2000 special issue of *Organization*, the Center for Gender in Organizations (CGO) research team reported about their post-equity experiments in a US global manufacturing and retail company. The company CEO and the CGO researchers shared the goal 'of creating a gender-equitable workplace' and improving work effectiveness at the same time. One experiment in the manufacturing plant concerned removing the masculine control of male supervisors over predominantly female production-line workers and creating a self-managing assembly line team. The idea was to redistribute roles and responsibilities to generate chances to value the formerly invisible work that women did. Their way to prevent problems and efforts to make the line run smoothly would be recognized and appreciated and this would also provide more opportunities for women to move ahead (Ely and Meyerson 2000a). Unfortunately, the gender equity goals got overlooked in the change process (Coleman and Rippin 2000).

Another example is the Australian work/life integration projects aiming at organizational change in one private sector organization and one public sector organization (Charlesworth and Baird 2007). The dual agenda of gender equality and organizational effectiveness was used in the collaborative research projects on work/life integration. The researchers stress the importance of gender discourses and how they are accepted in organizations: talking about 'ideal workers' and 'work/life issues' resonates better with men and women in the organization than talking about 'gender inequality' (Charlesworth and Baird 2007).

Power

We see power as linked to the control of resources, structures, behaviours, agendas, ideologies, cultures, and subjectivities (Kärreman and Alvesson 2009: 1118). An analysis of power imbues the post-equity experiments. In the words of Meyerson and Kolb (2000: 554): 'we wanted to transform work and its relations to other aspects of people's lives in ways that would fundamentally alter power relations in organizations and make them more equitable'. Interestingly, this fundamental change in power requires at least the consent and preferably the engagement of those currently in power. A political willingness at the top level is a *conditio sine-qua-non* for all strategies for gender equality and diversity management (Konrad and Linnehan 1995; Acker 2000; Williams and Clowney 2007) and the post-equity experiments are no exception to that rule. The examples above illustrate that post-equity experiments can only be conducted in organizations that are open to such interventions. The political engagement of feminists inside the organization is one way to create openness (Mergaert 2012). Emphasizing a dual agenda that links gender equality goals to business goals is another way to stimulate interest in the interventions. A dual agenda increases the urgency and legitimacy of the experiments in the eyes of managers. In a similar vein, the diversity literature increasingly emphasizes the business case of diversity, stressing how diversity in the workforce will improve competitive advantage and increase organizational effectiveness (Kirton and Greene 2010). From a critical power perspective, linking diversity and equality goals to business goals is a way to tame the transformative potential of the diversity and equality goals as they are put in the service of instrumental business goals (Zanoni and Janssens 2004; Janssens and Zanoni 2005). The documentation of the risk of losing gender to business (Coleman and Rippin 2000) suggests that, in organizational practice, the dual agenda may not be so dual after all.

In most cases, post-equity experiments are designed and implemented in close collaboration between academic researchers and organizational members. Participatory action research and collaborative interactive action research are preferred methodologies (Coleman and Rippin 2000). The collaborations are also subject to power processes. It is not always clear who initiated the projects and the collaboration, and how access to the organizations was negotiated when the academics took the initiative. Power dynamics between insiders and outsiders of the organization, between experts and non-experts, between academics and organization members and different levels of position power of various parties all impact on the success or failure of the collaborations (Benschop and Verloo 2006).

Resistance

The post-equity experiments are advocated because of their ability to prevent resistance to the changes. In post-equity experiments, there is an assumption that resistance can be played down or even avoided through close collaboration, small changes, and experiments. The experiments are presented as strategically chosen to do two things: they improve the situation for women and men, and they contribute to the efficiency and performance of the organization. This dual presentation is one important way to play down resistance. It can be questioned, however, whether that presentation is effective to avoid resistance in all the different phases in the process of change: the analysis of the current situation, the design of interventions/experiments, and their actual implementation (Mergaert 2012). The analysis phase is typically a phase in which the academic researchers take the lead as they are trained to uncover gender inequalities that are deeply embedded in daily organizational routines. They have to convey their analysis to heterogeneous organization members who are seldom familiar with feminist analyses and may resist this different take on their routines. This easily results in long discussions and reflections at the expense of practical action (Eriksson-Zetterquist and Styhre 2008). The design phase calls for close collaboration between academics and practitioners to select the experiments and agree upon the specific goals of change. There are many different ways that resistance can manifest itself in this phase, from disagreeing to the selection of experiments to missing meetings and to questioning the goals of change. Collaboration is a key element of the design of the experiments and compromising may be inevitable to keep all participants on board of the collaboration (Benschop and Verloo 2006). The implementation phase of experiments is most vulnerable to resistance. In this phase, more than discursive action is needed from more actors than the change agents. Effective strategies of resistance call into question the analyses of the problem and come back to earlier decisions, slowing down the action considerably. All in all, the post-equity experiments are complex organizational changes that entail ample opportunities for multiple actors to resist the interventions.

Gender Quotas

Aim

A quota is a (legal) tool to increase the numbers of under-represented groups in organizations or governments to correct historical under-representation. Quotas for women entail that women must constitute a certain number or percentage of the members of a body, whether it is a parliamentary assembly, a committee, or a board of directors (Dahlerup 1998). This 'equality by result' (Teigen 2000; Dahlerup 2007) does more than realize the numerical equal representation of minority groups, it also problematizes former practices of lip service without action and establishes a steady critical mass. This quota system places the burden of recruitment not on the individual woman, but on those who control the recruitment process. The appeal of quota derives in part from the failure of other gender equality initiatives in organizations to change the masculine culture in politics and organizations (Baldez 2006). Gender quotas have become an increasingly prominent solution in recent years to the under-representation of women in electoral politics (Squires 2004; Dahlerup and Freidenvall 2005; Baldez 2006; Krook 2006), but are hardly applied within business organizations. Businesses resist government control over the composition of their staff.

Examples

A well-known example is the quota law in Norway, where a law required 40 per cent of the board members in the largest companies to be of the least represented sex. The law was applied to all publicly listed companies and to state-owned and inter-municipal companies, and was later extended to all municipal companies (Armstrong and Walby 2012). The companies had five years (2003–8) to meet this requirement (Huse 2007). The quota law has led to major changes in the gender composition of corporate boards in Norway; from 6 per cent female in 2002 to 40 per cent in 2009 (Teigen 2012). The Norwegian law has sparked off a Europe-wide debate about quotas and women in leading positions. There are strong cultural differences in the acceptability of quotas, with a willingness to consider gender quota in some parts of Europe (e.g. Belgium, France, Italy), but a strongly voiced rejection in North America and other parts of Europe (UK, Sweden).

Another example of the implementation of gender quota can be found in the Dutch public sector: the police force (Benschop and Van den Brink 2009). To force more visible and rapid change, the Minister of Internal Affairs made arrangements with the corps management in 2007 to raise the number of women and ethnic minorities to 25 per cent in 2011 by marking special positions for minority groups. To accomplish this, she introduced a gender and diversity quota; 50 per cent of the new hires for vacant top positions had to be women and/or ethnic minorities. The intervention strategy comprised concrete agreements with police organization about which functions were eligible for diversity candidates. A temporary project organization was created that actively intervened in recruitment and selection procedures, by suggesting suitable candidates, advising decision-makers through all stages of the process, and monitoring the progress made. The project organization was bestowed with the power to intervene in various stages of recruitment and selection procedures, which goes against the strictly regulated and complex appointment practices. As a result, women and ethnic minorities took up half of the top appointments between 2008 and 2011, whereas earlier attempts to change the inclusiveness of the police culture did not result in a change in numbers.

Power

Power is a central element of the quota approach, as it needs top–down decision-making power to install quotas in the first place. Although the adoption of quotas in politics, organizations, and boards are often fuelled by strong feminist groups and politicians, the decision-making power at the top is crucial to pressure organization members to work according to the quota system. Without the use of power, they lack a mechanism with which to make the change happen (Hardy 1996). Quotas often come into an existence by an exercise of hierarchical power, not so much by consensus. In the case of Norway, the Minister of Economy installed the quota after an intense political and public debate. Despite enduring opponents, the minister used this political power to steer the law through parliament. Also in the Dutch police force, the Minister of Internal Affairs (who was the formal chief of the police force) forced the quota system upon her police managers despite their resistance. She argued that the long history of attempts to realize such change in the police organization failed as white men were appointed at the top time and again, and a quantitative change at the top level was never accomplished. Radical top–down interventions force the organization to take action, and to justify actions that do not lead to more diversity. The unorthodox intervention in our case features an influential project organization with the backup of the Minister of Internal Affairs who was not afraid to use her power position to enforce appointments of women and minority candidates.

Secondly, a successful implementation of the quotas highly depends on the possible sanctions for non-compliance. The ultimate sanction for companies in Norway was de-registration and hence dissolution of the company (Storvik and Teigen 2010). European countries that followed the quota initiatives from Norway varied in the legal mechanisms for the implementation and the severity of the sanctions or penalties for non-compliance. Countries with soft quotas (e.g. Finland, France, UK) are less effective. Within the Dutch police force, crucial to the power dynamics was the strong positioning of the Minister of Internal Affairs. About a year into the intervention, it seemed that few results had been achieved (15 per cent of new hires, figures from January 2009). The Minister, who had the final say over all top appointments, then decided to stall the appointment of another white man, a controversial action that made headline news. This signalled that she was serious about the change and would act upon it. After that, the first few women and minorities were appointed to top positions and the results have become significantly better since (57 per cent of new hires, figures from March 2010). Ultimately, the Minister had the final say about the appointments and she held the police organization responsible and accountable to actively search for and select women and ethnic minorities.

Third, when it comes to radical interventions such as quotas, it takes power to control the practices of recruitment and selection and enforce such a contested measure. This was not so much the case for the implementation of quotas in Norway, but in the Dutch police force, a specific project organization was equipped with ample resources in staff, money, and support to enable them to make a difference in the police organization. This project organization was mandated by the Minister to actively intervene in recruitment and selection procedures. The project organization proposed drastic and unorthodox measures that sometimes breached the rules and regulations for top appointments, but their involvement in various stages of the procedures was crucial for the final result.

Resistance

Because of its radical nature, quota interventions are widely debated (Tienari *et al.* 2009) and often meet hidden or open resistance or backlash (Holli *et al.* 2006). In the quota strategy, resistance is intrinsic, as it aims at disturbing the status quo and does not rely on consensus in the entire organization. The literature shows that privileged groups are less in favour of (radical) interventions. Privileged professional men may resist the perceived loss of control (Ashcraft 2005) over the allocation of those positions and refuse to accept their diminished chances to land such positions. And professional women may resist the quota as well for they want to be granted top positions because of their qualities, not because of their sex (Van den Brink and Stobbe forthcoming). These arguments show that positions, interests, and values are important triggers of resistance.

In Norway, there was broad political support for the quota legislation from several political parties. However, it also met strong resistance from within the Norwegian companies. The main arguments of the opponents were justice, skills, and democracy (Storvik and Teigen 2010). Quota regulations were considered to be illegitimate unequal treatment and a discrimination against men. They feared that less skilled women would take the place of experienced and skilled men. It was also claimed that the quota law would hamper owners' democratic right to recruit candidates and, in particular, interfere with the election process at the stakeholders meeting. Seven years after it was passed, the quota law is, according to Storvik and Teigen (2010), widely accepted in Norwegian politics and society.

The resistance against the quota regulations in the Dutch police force was also strong and did not resemble the subtle and covert disagreement known from many diversity management programmes (Benschop 2001). It was rather an open and quite antagonistic political opposition. Many police managers stated that they did not want diversity management dictated from above, resisted the pressure to obtain short-term results, and wanted to keep appointing new top officials through cooption as always (Benschop and Van den Brink 2009). They argued that they did not need 'blunt force' to appoint more diversity candidates and that they were perfectly capable of doing so themselves. We found different manifestations of resistance towards gender interventions, such as concerns about the challenge to quality, the decrease of men's career possibilities, and the 'violation' of meticulous and transparent procedures. These concerns can be analysed as concerns about the interventions breaching with three core values: quality, fairness, and transparency. The quota strategy triggered strong responses and opened up a political debate about what exactly constitutes quality, fairness, and transparency. This way, the quota strategy in organizations can break the ground for a much needed conflict and debate about the gendered social constructions of quality, fairness, and transparency. In an open conflict, the constructed nature of values becomes more clear and the practices of discrimination and inequalities are no longer subtle and deeply embedded, but are brought to the surface. We therefore argue that radical interventions such as gender quota provoke a resistance that can be a useful tool in organizational change strategies.

Comparing Strategies for Structural Change

In this chapter, we have discussed two structural gender equality strategies: one transformative (post-equity), and one inclusive (quota). We will describe the pros and cons of these strategies in the light of their effectiveness and impact.

Pros

Baldez (2006) argues that whether or not gender quotas are a good idea depends in part on what impact is expected from the quota regulations. Gender quotas are effective in generating a fast and significant increase in the number of women in an organization. Gender quotas provide a shock from outside that is able to change the appointment system. Hoel (2010) argues that quotas backed by legislation seem to be one of the most significant ways of effecting change. The gender balance strategies of companies alone through flexible working and mentoring and so on seem to have limited effect. Therefore, the appeal of quotas derives in part from the failure of other gender equality initiatives in organizations to change the masculine culture. Literature on political representation showed that gender quota laws can effectively break the male monopoly (Krook 2006). Furthermore, a quota strategy involves power and resistance is not only unavoidable, resistance can actually help to make gender inequality visible and question its legitimacy. By introduction of radical interventions, the resistance against those interventions helps to uncover the subtlety of gender practices. The conflict and debate that resistance generates can fuel the articulation of underlying values, and can render persistent and implicit stereotypes visible. In this way, resistance against quota can open up the possibilities for change.

In contrast, post-equity experiments are designed so that they do not come as a shock from the outside. Rather, they are the result of a collaborative effort of insiders and outsiders. They aim for a gradual transformation, but a transformation nonetheless as they bring about fundamental reforms in organizational processes. Post-equity experiments are geared towards slowly changing the system from within. A key advantage is that organization members are active participants in the process of change and have a say in the selection of which parts of the organization should be changed.

Cons

The participatory nature of the post-equity experiments has some disadvantages as well. This is one reason why the change goes so slowly, because it takes a lot of time to discuss the necessity to change gendered practices in the organization with diverse organization members. To move beyond the discussions and reflections to action is a considerable *tour de force* (Eriksson-Zetterquist and Styhre 2008).

Also, the project character of interventions makes it difficult to realize transformational change. Transformations of the post-equity experimental kind take a considerable amount of time to be effectuated and become visible. Yet, projects are inherently short-term endeavours that suggest that change can be realized in a restricted and relatively short time. Diversity managers, gender experts, and specialized consultants are needed to guide the experiments and they have to be in function long enough to see the changes through. Gender experts and diversity managers are key players in these change processes, and they have to be able to cope with frustration and slow progress (Kirton *et al.* 2007; Greene and Kirton 2009; Tatli and Özbilgin 2009).

Another disadvantage of the post-equity approach is that it depoliticizes the process of change. The post-equity strategy implies that actors in the organization are willing to learn about gender inequalities, to put aside their personal interests, positions, and values and to invest in adjusting the organizations practices. These organizational actors should endorse the problem of subtle gender inequalities, and make an active contribution to the diagnosis, dialogue, and experiments to change, even when their personal interests are breached. But how does this work in a society or organization in which there is gender fatigue (Kelan 2009) and/or gender blindness (Konrad *et al.* 2006)? In organizations where key players have fossilized norms and ideas about gender at work, these ideas are not easily changed and the 'old boys network' and images of the masculine leader, manager, professor, or police officer remain widespread, even among current business leaders (Gremmen and Benschop 2009; Van den Brink 2010). There may very well be little willingness to change in the organization, because of vested interests and privileges to be protected in the current situation. In our view it is not possible to change routines and their underlying values silently without conflict and resistance. After all, the rhetorical presentation of a dual agenda of gender quality and business performance is much easier than realizing this dual agenda (special issue *Organization* 2000).

Quotas are a much more debated intervention, with declared proponents and adversaries. The political controversy around quotas is quite strong. The core message of quotas is that the organization itself is incapable of change, and so change has to be enforced from the outside, from the government. Critics of quotas point to the limitations that quotas do nothing to change the gendered appointment systems; they only change the numbers and do no more than add women and stir. In the words of Cockburn (1989: 217) quotas 'give disadvantaged groups a boost up the ladder, while leaving the structure of that ladder and the disadvantage it entails just as before'.

Serious disadvantages of quotas are the visibility of and pressure and backlash on women that they entail. When there are very few women in an organization or profession, their visibility as women stands out so they are easily seen as representatives of their sex and much less as professionals. There is a lot of pressure on the newly appointed women to assimilate to the majority culture that sets the performance standards (Kanter 1977). Furthermore, quotas place the burden of proof on women, whose qualities can be questioned easily when quota rules apply. The suggestion that they have got their jobs unfairly by trampling over better qualified men is a symptom of the backlash that quotas entail. It is therefore not surprising that women may want to keep their distance from or openly resist quotas and rather emphasize their individual qualities for a job.

Finally, it should be noted that some (national) contexts are more open to the radical intervention of quotas than other contexts. The debates about quotas and the many varieties that have been developed in various countries illustrate that this is a feminist intervention strategy that requires a tolerance for centralized control. The opposition against quotas can be very strong, as they trigger fundamental questions about governmental control and authority over private businesses. Therefore, the power and will to see quotas through are essential of quotas' success.

This comparison between post-equity and quota strategies suggests that they can be complementary. Quotas help to increase the numbers of women in managerial positions, but if gendered processes in the organization do not change, these women might leave again. Post-equity experiments help to change these processes, and involve many organization members in the change of daily practices and underlying values. The combination of inclusive and transformational interventions leads to a synergy that is the most promising strategy for change.

CONCLUSION

We focused in this chapter on the issues of power and resistance in organizational change towards gender equality to come to a better understanding of the success or failure of different types of strategies. We reviewed the different strategies for organizational change, distinguishing between strategies that focus on the individual and strategies that focus on structures of inequalities. We zoomed in on two contrasting strategies that both aim to change the structure of organizations rather than individuals in organizations. The consensus in the gender literature is that change can only be realized when the gender inequality practices in organization structures and cultures are targeted (Acker 2006; Ely and Meyerson 2000b). Post-equity experiments emphasize the importance of gradual, incremental, even evolutionary change of gender in organizations and as such are contrasted with the radical and controversial call for quotas for top positions. In the first case, change is initiated from within the organization, whereas quotas are usually enforced by governments.

In our discussion of post-equity experiments and quotas, we analysed the processes of power and resistance that are so crucial for these strategies to change. In both approaches, the support and commitment of top management is required to initiate change towards gender equality. Ultimately, post-equity experiments aim to alter power relations in organizations, engaging those currently in power in the process by emphasizing the dual agenda of equality goals and business goals. The experiments are collaborative projects between gender researchers and organization members. The experiments have to be selected very carefully to ensure the consensus and commitment of as many organization members as possible. In contrast, quota are based on the use of hierarchical power. The core aim of quotas is to change power relations by changing the numbers in the top. Top–down decision power is needed to enforce the implementation of quotas and to sanction non-compliance. The quota strategy does not require consensus and collaboration of all organization members.

The fast change of quotas meets with more open resistance than the slow change of experiments, even though changing the numbers is more superficial than changing underlying gender practices and values. Change agents in the post-equity experiments have to do the hard work, whereas the radical interventions change the onus of proof around and make decision-makers responsible for complying or explaining their divergence from the quota. We conclude that the resistance against quotas can even be a productive tool for change towards gender equality. Seeing resistance in relation to power illuminates how the process of resistance also involves the reification and reproduction of that which is being resisted, by legitimizing and privileging it as an arena for political contest (Thomas and Davies 2005: 700). The resistance against quotas brings back political debate over organizational practices that produce the subtle inequalities that are so difficult to change. Quotas are so controversial because they are perceived as breaching the values of quality, fairness, and transparency. The exact same values of quality and fairness and to a lesser extent transparency are used both to resist and to legitimize the interventions. Resistance can be used to illuminate the conflict between the different constructions of the values and open a dialogue about these differences.

The implications for practice are that pressure from the top of the organization is sometimes needed, and that the resistance stemming from this exercise of hierarchical power does not have to be avoided. Rather, a combination of inclusive and transformative strategies seems to be the best recipe available for change towards gender equality. After all, gender and diversity change are all about a different division of power and resources, also about the power to define norms and values.

References

Acker, J. (1990). 'Hierarchies, Jobs, Bodies: A Theory of Gendered Organizations', *Gender and Society*, 4(2): 139–85.

Acker, J. (1992), 'Gendering Organizational Theory', in A. J. Mills and P. Tancred (eds), *Gendering Organizational Analysis*, 248–60. London: Sage.

Acker, J. (2000). 'Gendered Contradictions in Organizational Equity Projects', *Organization*, 7(4): 625–32.

Acker, J. (2006). 'Inequality Regimes: Gender, Class, and Race in Organizations', *Gender and Society*, 20(4): 441–64.

Ainsworth, S., Knox, A., and O'Flynn, J. (2010). '"A Blinding Lack of Progress": Management Rhetoric and Affirmative Action', *Gender, Work and Organization*, 17(6): 658–78.

Armstrong, J., and Walby, S. (2012). *Gender Quotas in Management Boards*. Brussels: European Commission (Policy Department C: Citizens' Rights and Constitutional Affairs).

Ashcraft, K. L. (2005). 'Resistance through Consent? Occupational Identity, Organizational Form, and the Maintenance of Masculinity among Commercial Airline Pilots', *Management Communication Quarterly*, 19(1): 67–90.

Baldez, L. (2006). 'The Pros and Cons of Gender Quota Laws: What Happens When You Kick Men Out and Let Women In?', *Politics and Gender*, 2(1): 102–9.

Benschop, Y. (2001), 'Gender and Organizations', *International Encyclopedia of Business and Management*, 2262–9. London: Thomson.

Benschop, Y., and Verloo, M. (2006). 'Sisyphus' Sisters: Can Gender Mainstreaming Escape the Genderedness of Organizations?', *Journal of Gender Studies*, 15: 19–33.

Benschop, Y., and Verloo, M. (2011). 'Policy, Practice and Performance: Gender Change in Organizations', in D. Knights, E. Jeanes, and P. Yancey-Martin (eds), *The Sage Handbook of Gender, Work and Organization*, 277–90. London: Sage.

Benschop, Y., and Van den Brink, M. (2009). *Teveel spelers, teveel hoepels, te weinig diversiteit: werving en selectie van kroonbenoemden bij de Nederlandse politie (Too many players, too many hoops, not enough diversity: recruitment and selection in the top of the Dutch police force)*. Nijmegen: Radboud University Nijmegen.

Benschop, Y., Helms Mills, J., Mills, A., and Tienari, J. (2012). 'Editorial: Gendering Change: The Next Step', *Gender, Work and Organization*, 19(1): 1–9.

Betters-Reed, B. L., and Moore, L. L. (1995). 'Shifting the Management Development Paradigm for Women', *Journal of Management Development*, 14(2): 24–38.

Charlesworth, S., and Baird, M. (2007). 'Getting Gender on the Agenda: The Tale of Two Organisations', *Women in Management Review*, 22(5): 391–404.

Cockburn, C. (1989). 'Equal Opportunities: The Short and Long Agenda', *Industrial Relations Journal*, 20(3): 213–25.

Coleman, G., and Rippin, A. (2000). 'Putting Feminist Theory to Work: Collaboration as a Means towards Organizational Change', *Organization*, 7(4): 573–87.

Connell, R. (2006). 'The Experience of Gender Change in Public Sector Organizations', *Gender Work and Organization*, 13: 435–52.

Council of Europe (1998). *Gender Mainstreaming: Conceptual Framework, Methodology and Presentation of Good Practices*. Strasbourg: Council of Europe.

Dahlerup, D. (1998). 'Using Quotas to Increase Women's Political Representation', in A. Karam (ed.), *Women in Parliament Beyond Numbers*, 91–106. Stockholm: IDEA.

Dahlerup, D. (2007). 'Electoral Gender Quotas: Between Equality of Opportunity and Equality of Result', *Representation*, 43(2): 73–92.

Dahlerup, D., and Freidenvall, L. (2005). 'Quotas as a "Fast Track" to Equal Representation for Women', *International Feminist Journal of Politics*, 7(1): 26–48.

Deutsch, F. M. (2007). 'Undoing Gender', *Gender and Society*, 21(1): 106–27.

Ely, R. J., and Meyerson, D. E. (2000a). 'Theories of Gender in Organizations: A New Approach to Organizational Analysis and Change', *Research in Organizational Behavior*, 22: 103–52.

Ely, R. J., and Meyerson, D. E. (2000b). 'Advancing Gender Equity in Organizations: The Challenge and Importance of Maintaining a Gender Narrative', *Organization*, 7(4): 589–608.

Eriksson-Zetterquist, U., and Styhre, A. (2008). 'Overcoming the Glass Barriers: Reflection and Action in the "Women to the Top" Programme', *Gender, Work and Organization*, 15(2): 133–60.

Fletcher, J. (1999). *Disappearing Acts: Gender, Power, and Relational Practice at Work*. Cambridge, MA: MIT Press.

Greene, A. M., and Kirton, G. (2009). *Diversity Management in the UK: Organizational and Stakeholder Experiences*. Abingdon: Routledge.

Greene, A., and Kirton, G. (2011). 'Diversity Management Meets Downsizing: The Case of a Government Department', *Employee Relations*, 33(1): 22–39.

Gremmen, I., and Benschop, Y. (2009). 'Walking the Tightrope: Constructing Gender and Professional Identities in Account Management', *Journal of Management and Organization*, 15(5): 596–610.

Harding, S. G. (ed.) (2004), *The Feminist Standpoint Theory Reader: Intellectual and Political Controversies*. New York: Routledge.

Hardy, C. (1996). 'Understanding Power: Bringing about Strategic Change', *British Journal of Management*, 7: S3–S16.

Hearn, J. (2000). 'On the Complexitiy of Feminist Intervention in Organizations', *Organization*, 7(4): 609–624.

Helgesen, S. (1990). *The Female Advantage: Women's Ways of Leading*, New York: Doubleday.

Hoel, M. (2010), 'The Quota Story: Five Years of Change in Norway', in S. Vinnicombe, V. Singh, R. J. Burke, D. Bilimoria, M. Huse (eds), 79–87 *Women on Corporate Boards of Directors: International Research and Practice*. Cheltenham: Edward Elgar.

Holli, A. M., Luhtakallio, E., and Raevaara, E. (2006). 'Quota Trouble: Talking about Gender Quotas in Finnish Local Politics', *International Feminist Journal of Politics*, 8(2): 169–93.

Huse, M. (2007). *Boards, Governance and Value Creation: The Human Side of Corporate Governance*. Cambridge: Cambridge University Press.

Iannello, K. P. (1992). *Decisions without Hierarchy: Feminist Interventions in Organization Theory and Practice*. London: Routledge.

Janssens, M., and Zanoni, P. (2005). 'Many Diversities for Many Services: Theorizing Diversity (Management) in Service Companies', *Human Relations*, 58(3): 311–40.

Jewson, N., and Mason, D. (1986). 'The Theory and Practice of Equal Opportunities Policies: Liberal and Radical Approaches', *Sociological Review*, 34(2): 307–34.

Kanter, R. M. (1977). *Men and Women of the Corporation*. New York: Basic Books.

Kärreman, Dan, and Alvesson, Mats (2009). 'Resisting Resistance: Counter-Resistance, Consent and Compliance in a Consultancy Firm', *Human Relations*, 62(8): 1115–44.

Kelan, E. (2009). 'Gender Fatigue: The Ideological Dilemma of Gender Neutrality and Discrimination in Organizations', *Canadian Journal of Administrative Sciences/Revue Canadienne des Sciences de l'Administration*, 26(3): 197–210.

Kirton, G., and Greene, A. M. (2010). *The Dynamics of Managing Diversity: A Critical Approach.* Oxford: Elsevier Butterworth-Heinemann.

Kirton, G., Greene, A. M., and Dean, D. (2007). 'British Diversity Professionals as Change Agents: Radicals, Tempered Radicals or Liberal Reformers?', *International Journal of Human Resource Management*, 18(11): 1979–94.

Konrad, A. M., and Linnehan, F. (1995). 'Race and Sex Differences in Line Managers' Reactions to Equal Employment Opportunity and Affirmative Action Interventions', *Group and Organization Management*, 20: 409–39.

Konrad, A. M., Prasad, P., and Pringle, J. K. (2006). *Handbook of Workplace Diversity.* London: Sage.

Krook, M. L. (2006). 'Gender Quotas, Norms, and Politics', *Politics and Gender*, 2(1): 110–18.

Liff, S., and Cameron, I. (1997). 'Changing Equality Cultures to Move beyond "Women's Problems"', *Gender, Work and Organization*, 4(1): 35–46.

Martin, P. Y. (2006). 'Practising Gender at Work: Further Thoughts on Reflexivity', *Gender, Work and Organization*, 13(3): 254–76.

Mergaert, L. (2012). 'The Reality of Gender Mainstreaming Implementation: The Case of the EU Research Policy', Ph.D. thesis, Radboud University Nijmegen.

Mescher, S., Benschop, Y., and Doorewaard, H. (2010). 'Representations of Work–Life Balance Support', *Human Relations*, 63(1): 21–39.

Meyerson, D. E., and Fletcher, J. (2000). 'A Modest Manifesto for Shattering the Glass Ceiling', *Harvard Business Review*, 78(1): 126–36.

Meyerson, D. E., and Kolb, D. M. (2000). 'Moving out of the 'Armchair': Developing a Framework to Bridge the Gap between Feminist Theory and Practice'. *Organization*, 7(4): 553–71.

Miller, G. E., Mills, Albert J., and Mills, J. Helms (2009). 'Introduction: Gender and Diversity at Work: Changing Theories, Changing Organizations', *Canadian Journal of Administrative Sciences/Revue Canadienne des Sciences de l'Administration*, 26(3): 173–5.

Nentwich, J. (2006). 'Changing Gender: The Discursive Construction of Equal Opportunities', *Gender, Work and Organization*, 13(6): 499–521.

Prasad, P., and Prasad, A. (2000). 'Stretching the Iron Cage: The Constitution and Implications of Routine Workplace Resistance', *Organization Science*, 11(4): 387–403.

Ridgeway, C. L. (1997). 'Interaction and the Conservation of Gender Inequality: Considering Employment', *American Sociological Review*, 62(2): 218–35.

Rosener, J. B. (1990). 'Ways Women Lead', *Harvard Business Review*, (Nov.–Dec.): 119–25.

Squires, J. (2004). 'Gender Quotas: Comparative and Contextual Analyses', *European Political Science*, 3(3): 51–8.

Storvik, A., and Teigen, M. (2010). *Women on Board: The Norwegian Experience.* Berlin: Friedrich Ebert Stiftung, International Policy Analysis.

Tatli, A., and Özbilgin, M. F. (2009). 'Understanding Diversity Managers' Role in Organizational Change: Towards a Conceptual Framework', *Canadian Journal of Administrative Sciences/ Revue Canadienne des Sciences de l'Administration*, 26(3): 244–58.

Teigen, M. (2000). 'The Affirmative Action Controversy', *NORA/Nordic Journal of Feminist and Gender Research*, 8(2): 63–77.

Teigen, M. (2012). 'Gender Quotas on Corporate Boards: On the Diffusion of a Distinct National Policy Reform', in *Firms, Boards and Gender Quotas: Comparative Perspectives = Comparative Social Research*, 29: 115–46.

Thomas, R., and Davies, A. (2005). 'Theorizing the Micro-Politics of Resistance: New Public Management and Managerial Identities in the UK Public Services', *Organization Studies*, 26(5): 683–706.

Tienari, J., Holgersson, C., Meriläinen, S., and Höök, P. (2009). 'Gender, Management and Market Discourse: The Case of Gender Quotas in the Swedish and Finnish Media', *Gender, Work and Organization*, 16(4): 501–21.

Van den Brink, M. (2010) *Behind the Scenes of Sciences: Gender Practices in Recruitment and Selection for Professors in the Netherlands*. Amsterdam: Pallas Publications.

Van den Brink, M., and Benschop, Y. (2012). 'Gender Practices in the Construction of Academic Excellence: Sheep with Five Legs', *Organization*, 19(4): 507–24.

Van den Brink, M., and Stobbe, L. (forthcoming). 'The Support Paradox: Overcoming Dilemmas in Gender Equality Programs', *Scandinavian Journal of Management*.

Walby, S. (2005). 'Gender Mainstreaming: Productive Tensions in Theory and Practice', *Social Policy*, 12(3): 321–43.

Weick, K. E. (1984). 'Small Wins: Redefining the Scale of Social Problems', *American Psychologist*, 39(1): 40–9.

Williams, D. A., and Clowney, C. (2007). 'Strategic Planning for Diversity and Organizational Change: A Primer for Higher-Education Leadership', *Effective Practices for Academic Leaders*, 2(3): 1–16.

Zanoni, P., and Janssens, M. (2004). 'Deconstructing Difference: The Rhetoric of Human Resource Managers' Diversity Discourses', *Organization Studies*, 25(1): 55–74.

....

CHAPTER 16

SEXUAL HARASSMENT IN THE WORKPLACE

....

SANDRA L. FIELDEN AND CARIANNE HUNT

THE emergence of the term 'sexual harassment' (SH) can be traced back to the mid-1970s in North America, although, in the UK, the first successful case when SH was argued to be a form of sex discrimination was in 1986, under the Employment Protection Act (Hodges Aeberhard 2001). However, SH has been somewhat ignored over recent years, with much of the academic literature focusing specifically on bullying. Precise quantification of workplace SH is problematic as there appears to be a lack of consensus regarding the definition of SH, particularly when examining the behaviours and the circumstances in which SH occurs (Bimrose 2004). SH can be defined as:

> Unwanted conduct of a sexual nature, or other conduct based on sex affecting the dignity of women and men at work which include physical verbal and non verbal conduct. This conduct must be done with the purpose of, or have the effect of, violating your dignity, or of creating an intimidating, hostile, degrading, humiliating or offensive environment for you.
>
> (Equality and Human Rights Council 2012)

This chapter explores the current literature relating to SH in the workplace and provides a model which can be adapted for use in organizations to help overcome the problems of SH.

SCALE OF HARASSMENT

....

From a review of the literature there does not appear to be one clearly agreed incident survey of SH, which makes it difficult to ascertain the scale of SH in UK organizations. This is compounded by the links between SH and bullying, particularly when considering the importance of power (Wilson and Thompson 2001). In order to address this,

Hearn and Parkin (2005) suggest that SH, bullying, and physical violence can all be seen in terms of organizational violation. Prevalence estimates diverge markedly according to methodological protocols such as sample size and diversity; whether the surveys targeted random samples from the community or a specific industry or sector; whether SH was operationalized according to a legal or behavioural definition; and the retrospective time-frame specified to participants (McDonald 2012). American estimates indicate that 40–75 per cent of women and 13–31 per cent of men have experienced workplace SH (Aggarwal and Gupta 2000). In a meta-analytic review of seventy-four national European studies in eleven member states, Timmerman and Bajema (1999) stated that between 17 and 81 per cent of employed women reported experiencing some form of SH in the workplace. Asking respondents directly whether or not they have experienced SH according to legally defined objective measures leads to substantially lower estimates than studies using perceptual measures where behaviours believed to constitute SH are listed (Australian Human Rights Commission (AHRC) 2008). In a meta-analytic study conducted in the US Illies *et al.* (2003), reported that rates of SH by respondents' own definitions were less than half the number of reports of potentially harassing incidents believed by researchers to constitute SH.

Lockwood (2008) found that in UK industrial tribunals (1995–2005) 90 per cent reported experiencing verbal abuse, 30 per cent (within sample) also non-verbal, 60 per cent physical harassment (duration ranging from three days to eight years duration). Telephone (including text) harassment has also been found to be a particularly problem for call centre employees (50 per cent females vs. 2 per cent males) (Sczesny and Stahlberg 2000). Furthermore, the last decade has seen an increasing number of complaints about internet SH, including emails, social networking sites, etc. (Khoo and Senn 2004). In reference to IT and SH Tyler (2002: 195) commented: 'the internet seems to have created a new way of doing old things'.

ANTECEDENTS OF SH

Historically, research centred on SH has tended to focus on the victim and the perpetrator's behaviour, including the psychological profile of a harasser (Jansma 2000). A more holistic approach to SH needs to be taken, as the majority of published research has paid little attention to the specific social contexts in which SH takes place (Welsh 1999), however there is now increasing understanding that harassment will, and does, take various forms depending on different contexts (Dellinger and Williams 2002). It is evident that organizational culture can play a key role when examining the various risk factors, causes, and how SH occurs (Dougherty and Smythe 2004; Handy 2006). For example, if employees feel that it is not being tackled then they may believe that such behaviour is tolerated and even condoned, which in turn can lead to a culture of SH. Studies have also shown that SH and others forms of harassment and victimization can be more prevalent in certain work situations, for example, in jobs where there is an unequal sex ratio; where there are large power differentials between women and men; during periods of job insecurity; or when a new supervisor or manager is appointed (Keily and Henbest 2001; Kohlman 2004). Studies have also shown that leadership styles can have an affect on the level of SH experienced in organizations, for example, an authoritarian style where there is limited consultation with staff; and a laissez-faire style where management fails to lead or intervene in workplace behaviour (O'Moore 2000; Di Martino *et al.* 2003).

Building on Gruber's concept of *double dominance*, de Haas and Timmerman (2010) found that the nature of male-dominated work environments mediated the relationship between numerical male dominance and SH. This and other research has shown that SH is more problematic in blue-collar male-dominated settings such as fire-fighting, where jobs are typically highly physical and where cultural norms associated with sexual bravado and posturing and the denigration of female behaviour are sanctioned, compared to white-collar male-dominated occupations such as accounting (Chamberlain *et al.* 2008; de Haas and Timmerman 2010). Other research by Handy (2006) in New Zealand and Timmerman and Bajema (1999), who reviewed European studies, has also shown that organizational norms and cultures, such as the level of sensitivity to the problem of balancing work and personal obligations, and the extent to which the culture is employee rather than job-oriented, are more important in predicting the frequency of SH incidents than organizational sex ratios. Indeed, despite high rates of SH in male-dominated workplaces, international research suggests that SH is by no means confined to these environments, but occurs in a wide range of organizational settings (e.g. Ellis *et al.* 1991; McCabe and Hardman 2005).

Individual and situational factors can also be combined to increase the propensity of repeated sexually harassing behaviours (Pryor *et al.* 1995). For example, SH is increasingly likely to be apparent in highly sexualized settings and work environments. Person factors can include the personal characteristics which can contribute to the likelihood of an individual to sexually harass, e.g. sexual harassers can often have reputations for exhibiting sexually exploitative behaviour (Perry 1983). Individual factors can also have an impact on the likelihood of SH occurring. These are socio-demographic factors such as sex, age, and marital status; personality characteristics; specific types of behaviour; and an individual's employment characteristics such as length of tenure and level of responsibility (Chappell and Di Martino 2000; Kohlman 2004). Societal characteristics can also affect the level of SH experienced in countries. High levels of violent crime, dramatic economic change, and rapid social change can be seen as predictors of increasing violence in the workplace (Sheehan 1999; Chappell and Di Martino 2000; Di Martino 2003).

SEXUALITY, POWER, AND ORGANIZATIONAL VIOLATION

SH can be viewed as a result of men's dominance in society and men exercising power over women (Sedley and Benn 1982). Thus, it is important to explore how heterosexuality is embedded within language which is used in organizations; this can be seen when looking at the language often used in business, such as penetrating markets (Hearn and Parkin 1995; Collinson and Hearn 1996). Hearn and Parkin (2005) believe that taking a non-gendered approach when examining organizational violation will have many limitations: (1) the majority of victims of SH are women, (2) violence by men towards men tends to be related to the construction of men and masculinity, and (3) even when an organization does not appear to be dominated by men and could even have an equal gender ratio of senior management, masculine norms can still be apparent and overriding. This can be illustrated when looking at the profession of midwifery. This is overwhelmingly a female profession, where women care for women, but is still controlled by the male medical discourse. Another example would be primary school education, the majority of teachers being female, whereas the headteachers tend to be male (Hearn and Parkin 2005).

There are also assumptions about what management is, i.e. 'strong', 'masculine' environments are seen as being positive and desirable (Collinson 1988; Einarsen and Raknes 1997). Therefore, the gender of the management does not appear to be relevant, because the majority of organizations tend to have a masculine form of management. Hearn and Parkin (2005) believe that hierarchical and managerial power is a central theme when analysing organization violations. As men remain in positions of power and tend to dominate management structures, they have increased opportunities to exercise their power in a negative manner when compared with women.

While it is evident that there are links between SH and power Wilson and Thompson (2001) believe that sexuality, power, and organizations are too complex to highlight specific forms and types of harassment within dimensions of power. They use Luke's (1986: 9) model to offer an analysis of SH in organizations:

- The one-dimensional view relates to the interests of individuals being revealed by political behaviour in decision-making. In this case the organizational hierarchy creates the power, which is used within the organization. In these structures men are typically in positions of power and women are not.
- The two-dimensional perspective is based on the assumption that power is exercised over others by the control of agendas and the organization will ultimately dictate what is seen as normal behaviour, e.g. SH is normalized within the organization.
- The three-dimensional view 'allows that power may operate to shape and modify desires and beliefs in a manner contrary to people's interests' (Luke 1986: 9). Here SH is seen to have a broader power base, where men exercise power over women and sexuality is just one factor. Collinson and Collinson (1989: 107) state that 'men's sexuality and organizational power are inextricably linked'.

Hearn and Parkin (2005) believe that, as organizational violation increases, the more it is likely to be taken for granted, and the result is that certain behaviours will become normalized in the organization. Creating and managing violation-free organizations is a difficult task to achieve; however, Ishmael and Alemro (1999: 47) state that the creation of a 'positive work environment' can be attained by auditing the organization's culture. Key to this audit and assessment is the issue of management and leadership styles. A violation-free organization will ultimately allow and empower individuals to speak out about violation and ensure that management deals with it explicitly. This can be further enhanced by the use of independent agencies, which are placed outside the organizational hierarchy. Such agencies could help to deal with problematic situations and prevent individuals from having to take a complaint through the formal organizational structure, which may have actually enabled and accepted the harassment (Collier 1995). There is an increasing need for expertise in organizational violation. Hearn and Parkin (2005) believe that this may be partly achieved by education in school, and in organizations, management, and professions.

Non Face-to-Face Forms of SH

SH does not only take place in the physical presence of the victim and harasser, with harassers increasingly using technology to harass from a distance. With the ever increasing number of individuals using the internet, it is not surprising that there has also been a general increase in unsolicited email, including emails which can be perceived as inappropriate or harassing (Khoo and Senn 2004). Tyler (2002: 195) states that 'the internet seems to have created a new way of doing old things', but unfortunately there is limited research so far on people's responses to email harassment. Specifically for SH there is a need to focus on 'immediate responses' (Woodzicka and LaFrance 2001: 19), so we can understand the experience of SH as it occurs. Khoo and Senn's (2004) study found that women perceived email messages with sexual content to be more offensive than men did. There were also marked differences in the response to a sexual proposition, with women finding it to be extremely offensive while men appeared to rate this as somewhat enjoyable. Soewita and Kleiner (2000) suggest that companies can do a range of things to protect their employees. First, they can adopt a clear policy on the actual use of company email and the internet prior to allowing employee access to the internet. The organization's written policy should also make this clear.

SH over the telephone is also a problem which needs to be addressed, particularly when considering the increasing number of employees who work in call centres. Whilst research examining SH over the telephone in the workplace is scarce, there have been a few studies examining SH over the telephone in private settings. For example, a study based on a sample of a hundred German students ($n = 49$ females and 51 males), found that 50 per cent of women and 2 per cent of men had at some point experienced SH over the telephone (Sczesny 1997). The most serious form of SH over the telephone contained 'groaning' in 55 per cent of cases, 'sexual advances' in 49 per cent and 'silence' in 26 per cent of cases (Sczesny 1997: 158).

THE VICTIM

When examining the individual antecedents it is necessary to examine literature pertaining to the perpetrator and the victim. It is clear that individual factors can influence the incidence and process of SH in the workplace and some patterns are emerging (Di Martino *et al.* 2003). A study by Kohlman (2004) found that for women age, education, race, and marital status appeared to be more significant than their occupational position, whereas occupational position was a particularly important factor for men. Further, women in lower level or in typically 'male' occupations were particularly likely to report SH, whereas men in higher status occupations or who were employed in positions which had a high percentage of female staff, were more likely to report it. Another survey by the UK Ministry of Defence/EHRC (Rutherford *et al.* 2006) found that younger women were more likely to have experienced unwelcome sexual behaviours. Seventy seven per cent of women under 23 had experienced behaviours, compared to 44 per cent of women in their forties. Those in lower ranks were also more likely to find themselves a victim of SH (20 per cent) (Rutherford *et al.* 2006). Wearing a uniform can actually have an effect in two ways, either as a deterrent or as an antecedent, depending on the overall attitude towards people in uniform (Chappell and Di Martino 2000).

Short-term contracts can also lead to increased SH because the individual may have less employment rights than employees in long-term positions. Equally, research has shown that there appears to be a higher risk of bullying within the public sector when compared with the private sector. To explain this some have argued that is the use of tenure, 'a job for life', which leaves employees with less room for flexibility and mobility, therefore fewer leave the organization after experiencing conflict (Zapf *et al.* 2003). A study of SH cases which had been taken to industrial tribunals, reported that, in the majority of cases, the victim had been working for their present employer for less than one year, suggesting that staff with shorter tenure may be at particular risk (Equal Opportunities Commission 2002). Coyne *et al.* (2000) also found that victims tended to be less assertive and competitive when compared with a control group, although it could be argued that this was an effect of the SH they had experienced.

Although it is overwhelmingly women who experience SH, it is not exclusive to women; men also experience such harassment. A Department of Trade and Industry survey found that two-fifths of UK employees who said they had been sexually harassed were men (Grainger and Fitzner 2005). Furthermore, a report compiled for the European Commission (2001), although based on a very small sample, found that 51 per cent of male healthcare workers reported being subject to SH. A survey of Australians (Human Rights and Equal Opportunity Commission 2003) found that: 28 per cent of adults (41 per cent female and 14 per cent male) had experienced SH and that two-thirds of these incidents were in the workplace (22 per cent in the past year).

SHORT- AND LONG-TERM EFFECTS OF SH

SH can have a negative effect both in the short and long term. The effect on the victim and the organization can include illness, humiliation, anger, loss of self-confidence and decreased job satisfaction, psychological distress, and damage to business performance and costing the organization money, particularly with regard to compensation and negative impact on employee turnover—SH can be seen as predictor of negative mood and self-esteem which can ultimately lead to employees leaving the organization (Sczesny and Stahlberg 2000; Chartered Institute of Personnel and Development 2005; Ballard *et al.* 2006). As the frequency of SH increases, it is increasingly likely to be seen as an unwelcome and harassing behaviour and victims will tend to believe that the events will continue to recur. Thus, experiences of SH can have a detrimental effect, causing work-related negative moods, which may have a negative impact on employee turnover.

According to the EOC (2005), SH also damages business performance and is costly in terms of compensation for personal injury. Studies have shown that, when women are disrespected in organizations, this ultimately has an impact on all employees, rather than simply the women who are involved (Miner-Rubino and Cortina 2004). Employees who witness SH may conclude that the organization does not care about the workforce and this may result in negative assumptions regarding organizational norms and behaviours, specifically relating to fairness and justice (Lamertz 2002). When employees perceive their organization to be permissive of SH, they tend to be less satisfied with their jobs and their physical health. There does not appear to be one clearly agreed method of coping and reacting to SH, with different studies advocating either a confrontational or passive coping strategy (Knapp *et al.* 1997; Mann and Guadagno 1999; Sczesny and Stahlberg 2000; Sigal *et al.* 2003; Stockdale 2005).

REPORTING SH

The problem of reporting SH may be caused by organizations finding it difficult to monitor, particularly when complaints are made informally. When examining the issue of filing a complaint, it is evident in many cases that the victims only route to report the SH may be through the hierarchical system which has actually perpetuated the SH. In addition, victims may have concerns regarding the press, facing the perpetrator, and the outcomes of the proceedings. All of these factors may prevent an individual from filing a complaint (Earnshaw and Davidson 1994). Setting an informal and formal route for dealing with harassment can be seen as a way of combating many of the problems faced by the victims (EOC 2002). The Ministry of Defence/EOC survey (2006) found that only 5 per cent of survey respondents who had suffered a particularly upsetting experience actually made a formal written complaint. Reasons for not filing a complaint were similar to why respondents were not willing to tell anyone about their experience, i.e. wanting to handle the situation themselves (67 per cent), fear of being labelled a troublemaker (39 per cent), fear of the complaint having a bad effect on their career (35 per cent), feeling they would not be believed (19 per cent), and 39 per cent felt that nothing would be done about it (Rutherford *et al.* 2006).

LEGAL AND ADMINISTRATIVE RESPONSES TO SH

There are a variety of legal responses which have been put into practice to address the problem of SH. For example, the European Commission code of practice: protecting the dignity of women and men at work (1991); the Sex Discrimination Act (1975); the EOC's code of practice on sex discrimination in employment explains the statutory requirements of the Act; the Health and Safety at Work Act (1974); the European Equal Treatment Directive has extended the definition of sex discrimination to cover any act which leads to intimidation or degradation. *Adapting to Change in Work and Society: A New Community Strategy on Health and Safety 2002–2006'* (European Commission 2002), stresses the need to adapt the legal framework to cover the emerging psychosocial risks. In addition, the European Commission's Advisory Committee on Safety, Hygiene and Health Protection at Work (2001), in its *Opinion on Violence at the Workplace,* calls for the Commission to issue guidelines in this area. The Commission recommendation on the protection of the dignity of women and men at work is also particularly useful. Whilst the code of practice is not legally binding it does offer advice which is consistent with current human resource management best practices.

POLICIES

When taking a preventive management perspective, it is essential to implement policies and initiatives early on. Primary interventions or preventive measures aim to address the root cause, thus preventing a problem from developing (Quick 1999). At this point the organization and the employees have certain characteristics which have the potential of merging together to create an unhealthy organization, whereby SH is embedded in the culture of the organization. Bell *et al.* (2002: 162) believe that there are organizational actions that can be implemented which can help to prevent this from happening. SH policies can be introduced to educate potential harassers and potential victims by providing employees with a clear statement of the types of conduct and behaviour which may constitute harassment, and to make it clear that harassment is not tolerated within the organization (Thomas 2004).

A strong zero tolerance perspective towards SH is an important factor. A formal SH policy can help to set behavioural guidelines which potential harassers may be deterred by and may encourage potential victims of SH to report the harassment (Gruber and Smith 1995). It is essential that the zero tolerance perspective is communicated to, and understood by, all employees. However, a recent survey by the Ministry of Defence/EHRC (2006: 44) found that respondents had some concerns with a zero tolerance policy; there were 'fears that too draconian approach would lead to political correctness and people treading too carefully'. Respondents also stated that a strict zero tolerance approach might lead to increased numbers of formal complaints and that might not be the best route to take. Consequently, Bell *et al.* (2002) and Thomas (2004) advocate the importance of taking a consultative approach when designing and implementing SH policies and procedures. Wilson and Thompson (2001) concur with the view of taking a bottom–up approach and engaging multiple stakeholders, for example, employee groups and trade unions. They assert that trying to change the situation is not simply about relying on anti-discrimination law; it requires 'assault upon established practices and privileges already in place in organisations' (Wilson and Thompson 2001: 76).

Training

Training can be an effective method to employ at the primary intervention stage, as it can be used to inform staff of SH and can help to equip individuals with the necessary skills to deal with SH if it occurs (York *et al.* 1997). In general, the studies which have been conducted to examine the effectiveness of training programmes show that they can be effective in tackling SH (Antecol and Cobb-Clark 2003). In addition, there appears to be an absence of research examining tertiary-level interventions. Tertiary interventions are the rehabilitative procedures which are provided for the victim and, in some cases, the perpetrator (Di Martino 2003).

Educational workshops have been found to be an effective method of addressing the issue of SH in organizations. A study by Barak (1994) found that educational workshops had primary and secondary effects, for example the knowledge of their existence, the sharing of experiences by participants at work following the completion of the workshop, and the participants' behaviour as a model, which other women workers could use, all had a positive effect on SH in the organizations. Workshops and training have also been found to have a significant impact, specifically on male subjects, as it increases awareness and sensitivity regarding the issue of SH (Beauvais 1986). One of the few studies to examine the effectiveness of this has been conducted by Newman *et al.* (2003). The study found that on average men and older workers were more likely to perceive training as effective, when compared with women and younger workers. There also appeared to be a tendency for individuals with higher levels of education at relatively higher grades and those who were divorced to perceive the training as less effective. Whilst this information is important to understand the differences in perception of workers, it is necessary to delve deeper into this area and examine the reasons why certain groups of individuals felt that SH training was effective or ineffective.

TERTIARY INTERVENTIONS

There is absence of empirical research examining tertiary interventions in SH programmes. Once SH has occurred, the primary concern should be for the victim and rehabilitative procedures should be examined in order to ensure that the victims' lives are returned to normal as quickly and effectively as possible (Di Martino *et al.* 2003). Di Martino *et al.* (2003) illustrate a case study where a sexual harassment and mobbing advice centre was established in the University of Vienna, Austria, to help support victims of SH (Bukowska and Schnepf 2001). The main aim of the centre was to be a refuge for victims and it was operated by a female psychotherapist and a social worker, with interventions focused mainly on debriefing and psychosocial counselling, legal advice, and support. The service provided:

- crisis intervention, which means clarifying the situation and providing the necessary support, information on possible measures within the university or externally, and referring to external long-term counsel for as long as required;
- support in the decision making process about any legal action to be taken;
- case specific legal advice after clarification of the situation.

To date 100 consultations have been conducted and the services concludes that effects of the interventions are promising, although there is no firm data to support these claims (European Foundation for the Improvement of Living and Working Conditions 2004).

BEST PRACTICE MODEL

Based on all of the available literature a best practice model (Figures 16.1 and 16.2) has been designed which incorporates primary, secondary, and tertiary interventions. The model emphasizes the importance of taking a consultative and participatory approach to SH and the importance of monitoring and evaluation, advocating a proactive rather than a reactive strategy to SH policies and procedures. The best practice model highlights the various stages of intervention.

The primary intervention stage refers to prevention, activities which the organization can implement in order to prevent SH from occurring, for example, implementing an effective policy and procedure. The secondary intervention stage refers to how the organization will respond when faced with SH for example, ensuring that an effective complaints procedure is in place, and finally the tertiary intervention stage refers to the follow-up procedure which the organization can implement which will help to deal with problems once SH has occured, for example, ensuring that effective rehabilitation is provided. At each of these stages it is essential that the organization conducts thorough monitoring and evaluation to determine whether these procedures are effective. It is also essential that a consultative and participatory approach is taken at each stage.

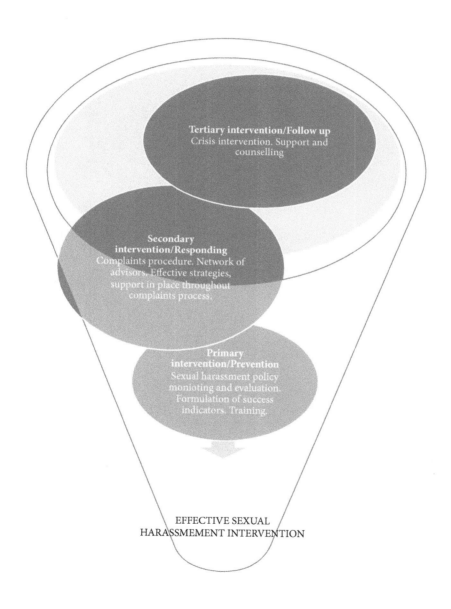

Tertiary intervention/Follow up
Crisis intervention. Support and counselling

Secondary intervention/Responding
Complaints procedure. Network of advisors. Effective strategies, support in place throughout complaints process.

Primary intervention/Prevention
Sexual harassment policy monioting and evaluation. Formulation of success indicators. Training.

EFFECTIVE SEXUAL HARASSMEMENT INTERVENTION

FIGURE 16.1   Sexual harassment intervention model (Continued)

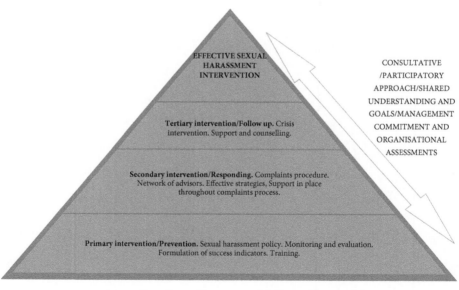

FIGURE 16.2   Sexual harassment intervention model

Conclusion

SH is increasingly being seen as a management and leadership problem which needs to be addressed. Studies have suggested that men's sexuality and organizational power are linked and that SH can be viewed as a result of men's dominance in society (Einarsen and Raknes 1997; Hearn and Parkin 2005). Hearn and Parkin (2005) suggest that SH, bullying, and physical violence can all be seen in terms of organizational violation. Societal characteristics can also affect the level of SH experienced in countries. High levels of violent crime, dramatic economic change, and rapid social change can be seen as predictors of increasing violence in the workplace (Chappell and Di Martino 2000; Di Martino 2003).

Understanding individuals' and collective perceptions of SH is crucial, particularly because individuals define SH differently and a victim may have problems with labelling sexually harassing behaviours. Victims may not in fact define their experience as SH, which can distort incidence rates (Lee 2001). Even when a victim defines their experience as SH, they may face further problems when filing a complaint. Whilst there are a number of legal and administrative responses to SH in the workplace, one must consider whether such responses are being implemented in practice. The victims' only route to report the SH may be through the hierarchical system which has actually perpetuated

Table 16.1 The way forward for SH research

What is already known from the literature?	What is needed?
• Sexual harassment definitions	• Empirical research examining
• Organizational violation, power and sexualilty	• Incidence of sexual harassment
• Forms of the Sexual harassment	• Leadership styles and their impact on Sexual harassment incident rate
• Groups at risk of Sexual harassment	
• Antecedents of Sexual harassment (individual/organizational/societal)	• Ethnicity and sexual harassment
	• Same sex sexual harassment
• Short and long term effects on the individual and organization	• Disability and sexual harassment
	• Sexual harassment via electronic media
• Difficulties facing individuals who wish to file a complaint	• Attitudes towards sexual harassment and the impact on the individual of filling complaint
• Legal responses	• Effectiveness on sexual harassment programmes and training
• Examples of best practice procedures	
	• Comparative review of pubic and private sector policies and procedures
	• The gap between policy and practice

the SH. In addition, victims may have concerns regarding the press, facing the perpetrator, and the outcomes of the proceedings. All of these factors may prevent an individual from filing a complaint (Table 16.1).

Gender-based discrimination and violence continue to be persistent and multifaceted, with discrimination on multiple grounds becoming the rule rather than the exception (European Foundation for the Improvement of Living and Working Conditions, 2004). Non-discrimination and non-violence are at the core of recovery and action policies. The global financial crisis has resulted in women potentially facing increased precariousness of work and reductions in social protection. Because of the recession women may tend to accept jobs below their education and qualification levels and may create conditions whereby women have fewer possibilities to protect themselves from harassment and violence at work as they may fear losing their job if they report it (Triems 2012). Therefore it is essential that organizations have effective policies and procedures in order to ensure that they are implementing effective SH strategies.

Moving forward, an examination of all forms of SH is imperative to understanding the extent of this phenomenon, for example, sexual harassment via text, email, and social networking sites. An understanding of the perpetrators of SH is also required if interventions are to be comprehensive in addressing individual behaviour and organizational culture. Many of those working in this area have reservations about the profiling of perpetrators, but without such knowledge the picture of SH is incomplete. This is a crime like any other and needs to be treated as such.

ACKNOWLEDGEMENT

This chapter has been adapted from research co-funded by the Equality and Human Rights Commission and Manchester Business School.

REFERENCES

Aggarwal, A., and Gupta, M. (2000). *Sexual Harassment in the Workplace* (3rd edn). Vancouver, BC: Butterworths.

Antecol, H., and Cobb-Clark, D. (2003). 'Does Sexual Harassment Training Change Attitudes? A View from the Federal Level', *Social Science Quarterly*, 84(4): 826–42.

Australian Human Rights Commission (2008). *Sexual Harassment: Serious Business. Results of the 2008 Sexual Harassment National Telephone Survey*. Sydney: AHRC.

Ballard, T. J., Romito, P., Lauria, L., Vigiliano, V., Caldora, M., Mazzanti, C., and Verdecchia, A. (2006). 'Self Perceived Health and Mental Health among Women Flight Attendants', *Occupational and Environmental Medicine*, 63: 33–38.

Barak, A. (1994). 'A Cognitive-Behavioural Educational Workshop to Combat Sexual Harassment in the Workplace', *Journal of Counseling and Development*, 72(6): 595–603.

Beauvais, K. (1986). 'Workshops to Combat Sexual Harassment: A Case Study of Changing Attitudes', *Signs: Journal of Women in Culture and Society*, 12(1): 130–45.

Bell, M. P., Quick, J. C., and Cycyota, C. S. (2002). 'Assessment and Prevention of Sexual Harassment of Employees: An Applied Guide to Creating Health Organizations', *International Journal of Selection and Assessment*, 10(1–2): 160–7.

Bimrose, J. (2004). 'Sexual Harassment in the Workplace: An Ethical Dilemma for Career Guidance Practice?', *British Journal of Guidance and Counselling*, 32(1): 109–21.

Bukowska, S., and Schnepf, E. (2001). *Beratungsstelle Sexuelle Belästigung und Mobbing*. Vienna: University of Vienna. www.univie.ac.at/women/index.htm

Chamberlain, L., Crowley, M., Tope, D., and Hodson, R. (2008). 'Sexual Harassment in Organizational Context', *Work and Occupations*, 35: 262–95.

Chappell, D., and Di Martino, V. (2000) *Violence at Work* (2nd edn). Geneva: ILO.

Chartered Institute of Personnel and Development (CIPD) (2005). *Harassment at Work: Diversity and Equality*. London: CIPD.

Collier, R. (1995). *Combating Sexual Harassment in the Workplace*. Buckingham and Philadelphia: Open University Press.

Collinson, D. L. (1988). 'Engineering Humor: Masculinity, Joking and Conflict in Shop Floor Relationships', *Organization Studies*, 9: 181–99.

Collinson, D. L., and Collinson, M. (1989). 'Sexuality in the Workplace: The Domination of Men's Sexuality', in J. Hearn, D. L. Sheppard, P. Tancred-Sheriff, and G. Burrell. (eds) *The Sexuality of Organization*, 91–109 London: Sage.

Collinson, D. L., and Hearn, J. (eds) (1996). *Men as Managers, Managers as Men: Critical Perspectives on Men, Masculinities, and Management*. London: Sage.

Coyne, I., Seigne, E., and Randall, P. (2000). 'Predicting Workplace Victim Status from Personality', *European Journal of Work and Organisational Psychology*, 9: 335–49.

Deadrick, D. L., Bruce McAfee, R., and Champagne, P. J. (1996). 'Preventing Workplace Harassment: An Organizational Change Perspective', *Journal of Organizational Change Management*, 9(2): 66–75.

De Haas, S., and Timmerman, G. (2010). 'Sexual Harassment in the Context of Double Male Dominance', *European Journal of Work and Organizational Psychology*, 19: 717–34.

Dellingfer, K., and Williams, C.L. (2002). 'The Locker Room and the Dorm Room: Workplace Norms and the Boundaries of Sexual Harassment in Magazine Editing', *Social Problems*, 49(2): 242–57.

Di Martino, V. (2003). *Relationship between Stress and Violence in the Health Sector*. Geneva: ILO.

Di Martino, V., Hoel, H., and Cooper, C. L. (2003). *Preventing Violence and Harassment in the Workplace*. European Foundation for the Improvement of Living and Working Conditions. Luxembourg: Office for Official Publications of the European Communities.

Dougherty, D. S., and Smythe, M. J. (2004). 'Sensemaking, Organizational Culture, and Sexual Harassment', *Journal of Applied Communication Research*, 32(4): 293–317.

Earnshaw, J., and Davidson, M. J. (1994). 'Remedying Sexual Harassment via Industrial Tribunal Claims: An Investigation of the Legal and Psychosocial Process', *Personnel Review*, 23(8): 3–6.

Einarsen, S., and Raknes, B. I. (1997). 'Harassment in the Workplace and the Victimization of Men', *Violence and Victims*, 12: 247–63.

Ellis, S., Barak, A., and Pinto, A. (1991). 'Moderating Effect of Personal Cognitions on Experiences and Perceived Sexual Harassment of Women at the Workplace', *Journal of Applied Social Psychology*, 21: 1320–37.

Equal Opportunities Commission (EOC) (2002). *Policy Statement: Analysis of Sexual Harassment Tribunal Cases*. London: EOC. http://www.eoc.org.uk/Default.aspx?page=14998

Equal Opportunities Commission (EOC) (2005). *Dealing with Sexual Harassment*. London: EOC. http://www.eoc.org.uk/Default.aspx?page=15398andlang=e

Equality and Human Rights Commission (EHRC) (2012). *Sexual Harassment—Your Rights*. http://www.equalityhumanrights.com/advice-and-guidance/your-rights/gender/sex-discrimination-your-rights-at-work/sexual-harassment/sexual-harassment-your-rights

European Commission (2002). *Adapting to Change in Work and Society: A New Community Strategy on Health and Safety at Work 2002–2006.* 11.03.2002 COM 118 final. Brussels: European Commission.

European Commission, Advisory Committee on Safety, Hygiene and Health Protection at Work (2001). *Opinion on Violence at the Workplace*. Opinion adopted 29 Nov. Brussels: European Commission.

European Foundation for the Improvement of Living and Working Conditions (2004). 'Progressive Initiatives to Combat Harassment: Three Case Examples. Accessed 28 Aug. 2013. http://www.eurofound.europa.eu/emcc/content/source/eu04010a.htm

Grainger, H., and Fitzner, G. (2005). *Department of Trade and Industry 14: Fair Treatment at Work Survey*. London: DTI.

Gruber, J. E., and Smith, M. E. (1995). 'Women's Responses to Sexual Harassment: A Multivariate Analysis', *Basic and Applied Social Psychology*, 17: 543–62.

Handy, J. (2006). 'Sexual Harassment in Small-Town New Zealand: A Qualitative Study of Three Contrasting Organizations', *Gender Work and Organization*, 13(1): 1–24.

Hearn, J., and Parkin, W. (1995). *'Sex' at 'Work': The Power and Paradox of Organization Sexuality* (rev edn). Hemel Hempstead: Prentice Hall/St Martin's/Harvester Wheatsheaf.

Hearn, J., and Parkin, W. (2005). 'Recognition Processes in Sexual Harassment, Bullying and Violence at Work: The Move to Organizational Violations', in J. E. Gruber and P. Morgan (eds), *In the Company of Men: Male Dominance and Sexual Harassment*, 92–116. Boston: Northeastern University Press.

Hodges Aeberhard, J. (2001). 'Sexual Harassment in Employment: Recent Judicial and Arbitral Trends', in M. F. Loutfi (ed.), *Women, Gender and Work*, 503–40. Geneva: ILO.

Hoel, H., Einarsen S, Zapf D, Vartia M. (2003). 'Empirical Findings of Bullying in the Workplace', in S. Einarsen, H. Hoel, D. Zapf and C. L. Cooper (eds), *Bullying and Emotional Abuse in the Workplace: International Perspectives in Research and Practice*, 103–26. London: Taylor & Francis.

Human Rights and Equal Opportunity Commission (HREOC) (2003). *Sexual Harassment: A Bad Business Review of Sexual Harassment in Employment Complaints 2002*. Sydney: HREOC.

Illies, R., Hauserman, N., Schwochau, S., and Stibal, J. (2003). 'Reported Incidence Rates of Work-Related Sexual Harassment in the US: Using Meta-Analysis to Explain Reported Rate Disparities', *Personnel Psychology*, 56: 607–18.

Ishmael, A., and Alemro, B. (1999). *Harassment, Bullying and Violence at Work*. London: Industrial Society.

Jansma, L. L. (2000). 'Sexual Harassment Research: Integration, Reformulation, and Implications for Mitigation Efforts', in M. E. Roloff (ed.), *Communication Yearbook 23*, 163–225. Thousand Oaks, CA: Sage.

Keily, J., and Henbest, A. (2001). 'Sexual Harassment at Work: Experiences from an Oil Refinery', *Women in Management Review*, 15(2): 65–80.

Khoo, P. N., and Senn, C. Y. (2004). 'Not Wanted in the Inbox! Evaluations of Unsolicited and Harassing Emails', *Psychology of Women Quarterly*, 28: 204–14.

Knapp, D. E., Faley, R. H., Ekeberg, S. E., and Dubois, C. L. Z. (1997). 'Determinants of Target Responses to Sexual Harassment: A Conceptual Framework', *Academy of Management Review*, 22: 687–729.

Kohlman, M. H. (2004). 'Person or Position? The Demographics of Sexual Harassment in the Workplace', *Equal Opportunities International*, 23(3–5): 143–62.

Lamertz, K. (2002). 'The Social Construction of Fairness: Social Influence and Sense Making in Organizations', *Journal of Organizational Behavior*, 23: 19–37.

Lee, D. (2001). 'He didn't Sexually Harass Me, as in Harassed for Sex... He Was Just Horrible: Women's Definitions of Unwanted Sexual Conduct at Work', *Women's Studies International Forum*, 24(1): 25–38.

Lockwood, G. (2008). 'A Legal Analysis of Sexual Harassment Employment Tribunal Cases 1995–2005', *International Journal of Law and Management*, 50(6): 316–32.

Luke, S. (1986). *Power: A Radical View*. New York: New York Press.

McCabe, M., and Hardman, L. (2005). 'Attitudes and Perceptions of Workers to Sexual Harassment', *Journal of Social Psychology*, 145: 719–40.

McDonald, P. (2012). 'Workplace Sexual Harassment 30 Years on: A Review of the Literature', *International Journal of Management Reviews*, 14: 1–17.

Mann, A. I., and Guadagno, R. E. (1999). 'Perceptions of Sexual Harassment Victims as a Function of Labeling and Reporting', *Sex Roles*, 41: 921–40.

Miner-Rubino, K., and Cortina, L. M. (2004). 'Working in a Context of Hostility toward Women: Implications for Employees' Well-Being', *Journal of Occupational Health Psychology*, 9(2): 107–22.

Newman, M. A., Jackson, R. A., and Baker, D. D. (2003). 'Sexual Harassment in the Federal Workplace', *Public Administration Review*, 63(4, July/Aug.): 472–83.

O'Moore, M. (2000). *Bullying at Work in Ireland: A National Study*. Dublin: Anti Bullying Centre.

Perry, S. (1983). 'Sexual Harassment in the Campuses: Deciding Where to Draw the Line', *Chronicle of Higher Education*, (26 Mar.): 21–2.

Pryor, J. B., Giedd, J. L., and Williams, K. B. (1995). 'A Social Psychological Model for Predicting Sexual Harassment', *Journal of Social Issues*, 51: 69–84.

Quick, C. (1999). 'Organizational Health Psychology: Historical Roots and Future Directions', *Health Psychology*, 18: 82–8.

Rutherford, S., Schneider, R., and Walmsley, A. (2006). *Agreement on Preventing and Dealing Effectively with Sexual Harassment: Quantitative and Qualitative Research into Sexual Harassment in the Armed Forces*. London: Ministry of Defence/Equal Opportunities Commission.

Sczensky, S. (1997). 'Sexuelle Belaestigung am Telefon', *Eine sozialpsychologie*, 30: 151–64.

Sczesny, S., and Stahlberg, D. (2000). 'Sexual Harassment over the Telephone: Occupational Risk at Call Centres', *Work and Stress*, 14(2): 121–36.

Sedley, A., and Benn, M. (1982). *Sexual Harassment at Work*. London: NCLL Rights for Women Unit.

Sheehan, M. (1999). Workplace Bullying: Responding with Some Emotional Intelligence. *International Journal of Manpower*, 20: 57–69.

Sigal, J., Braden-Maguire, J., Patt, I., Goodrich, C., and Perrino, C. S. (2003). 'Effects of Type of Coping Responses, Setting, and Social Context on Reactions to Sexual Harassment', *Sex Roles*, (Feb.): 157–67.

Soewita, S., and Kleiner, B. H. (2000). 'How to Monitor Electronic Mail to Discover Sexual Harassment', *Equal Opportunities International*, 19(6–7): 45–7.

Stockdale, M. S. (2005). 'The Sexual Harassment of Men: Articulating the Approach-Rejection Theory of Sexual Harassment', in J. E. Gruber and P. Morgan (eds), *In the Company of Men: Male Dominance and Sexual Harassment*, 117–142. Boston: Northeastern University Press.

Thomas, A. N. (2004). 'Politics, Policies and Practice: Assessing the Impact of Sexual Harassment Policies in UL Universities', *British Journal of Sociology of Education*, 25(2): 143–60.

Timmerman, G., and Bajema, C. (1999). 'Incidence and Methodology in Sexual Harassment Research in Northwest Europe', *Women's Studies International Forum*, 22: 673–81.

Triems, B. (2012). *Preventing Violence Against Women at Work: The EU Must Act Now! Equality in the Workplace: Zero Tolerance on Gender-Based Discrimination and Harassment in Europe*. Brussels: EWL, 6 Mar.

Tyler, T. R. (2002). 'Is the Internet Changing Social Life? It Seems the More Things Change, the More they Stay the Same', *Journal of Social Issues*, 58: 195–205.

Welsh, S. (1999). 'Gender and Sexual Harassment', *Annual Review Sociology*, 25: 169–90.

Wilson, F., and Thompson, P. (2001). 'Sexual Harassment as an Exercise of Power', *Gender Work and Organization*, 18(1): 61–83.

Woodzicka, J. A., and LaFrance, M. (2001). 'Real versus Imagined Gender Harassment', *Journal of Social Issues*, 57: 15–30.

York, K. M., Barclay, L. A., and Zajack, A. B. (1997). 'Preventing Sexual Harassment: The Effect of Multiple Training Methods', *Employee Responsibilities and Rights Journal*, 10(4): 277–89.

Zapf, D., Einarsen, S., Hoel, H., and Vartia, M. (2003). 'Empirical Findings on Bullying', in S. Einarsen, H. Hoel, D. Zapf, and C. L. Cooper (eds), *Workplace Bullying and Emotional Abuse in the Workplace: International Perspectives in Research and Practice*, 103–126. London and New York: Taylor & Francis.

CHAPTER 17

....

ORGANIZATIONAL CULTURE, WORK INVESTMENTS, AND THE CAREERS OF MEN

Disadvantages to Women?

....

RONALD J. BURKE

WOMEN WORKING IN MEN'S CULTURES

....

THREE major hypotheses have been proposed to explain the failure of women to achieve executive-level positions in organizations (Fagenson 1990). The first emphasizes ways in which women are different from men (the men are from Mars and women are from Venus school of thought (Gray 1992)) and hold women's deficiencies as the reason for their lack of progress. These differences include attitudes, behaviours, traits, socialization practices and experiences, lack of job training, and not enough time in the career pipeline. Research support for this hypothesis is limited; there are generally few or no differences between managerial and professional women and men in their traits, abilities, education, motivation, and ambition (Powell 2011).

A second hypothesis emphasizes bias and discrimination of the majority towards the minority (men against women). Men hold negative stereotypes and negative expectations and biases against women, reflected in the worldwide notion of 'think manager—think male' (Schein 1994, 1975). Individuals have consistently rated 'good' managers as masculine (Powell and Butterfield 1989), thus there is some support for this hypothesis.

The third hypothesis focuses on structural, systemic, and cultural discrimination based on policies and practices affecting the treatment of women and ultimately their career advancement (Morrison 1992; Marshall 1995). These policies and practices influence women's access to opportunities and power, resulting in the current situation in senior management. They produce a lack of mentors and sponsors, tokenism, reduced development opportunities, and few visible and challenging assignments for women (Morrison 1992; Catalyst 1998; Wittenberg-Cox and Maitland 2008).

This should not be surprising since managerial and professional women live and work in a larger patriarchal society in which men have greater access to privilege, power, and wealth. This chapter considers ways that organizational culture and organizational values advantage men and disadvantage women.

Male-created and male-dominated organizational cultures are uncomfortable for women and impede women's career progress. Cultures of male advantage and privilege protect men's interests and power (Acker 1990; Smith and Cockburn 2009). Masculine values in the structure, culture, and practices of organizations are reflected in the gendered nature of power relations, the importance of paid work as an important source of masculine identity, status, and power, and the provider role for men (Hearn and Morgan 1990). Organizational cultures obviously reflect gender relations in the wider society (Kimmel 2009), and men's resistance to giving up power and privilege (Cockburn 1991). As a result, changing organizational cultures has proven difficult, reflected in the small numbers of women currently at senior organizational levels (Barreto *et al.* 2009).

Working life in organizations and organizations themselves were created by men for men (Marshall 1984). Women are merely travellers in this male world. Traditional organizational views assumed that workers were male, with women in lower paying support roles, mirroring their positions in society at large. Culture serves to include some and exclude others (Rogers 1988; McDowell 1997). Rutherford (2011, 2002, 2001) has written extensively on the difficulties women face as they move into male cultures. Workplace cultures are male-dominated and masculine, they are better suited then for men than for women (Wacjman 1998). As such they tend to marginalize and exclude women, in both obvious and subtle ways. Women are less comfortable in these cultures and have more difficulties advancing their careers in them.

Rutherford defines culture as follows: 'the attitudes, values, beliefs and patterns of behavior of organizational members. It is expressed in the management style, work ideologies (what is and isn't work), language and communication, physical artifacts, informal socializing and temporal structuring of work, and the gender awareness and expression of sexuality' (Rutherford 2011: 28). Several of these concepts serve as barriers to women; management is masculine.

Women have more responsibility for 'second shift' work, are more likely to have career gaps for having and raising children, are more likely to work part-time at some point in their careers, have to re-enter the workplace following career gaps, have less time for socializing and informal networking after work, are less likely to be 'visible' to the organization, and are more likely to receive sexual advances. Time seems to be a more available resource for men than for women even though they work longer hours and are more likely to be married, have children, and family commitments.

Cultural Obstacles Facing Women in Management

There are obviously more men than women in senior management jobs and many more men than women seeking to obtain these jobs. It is difficult to identify 'objective' criteria for success in these jobs and 'objective' credentials such as education appear to not be all that important in many cases or common to both women and men. These male decision-makers are likely to use gender-related factors and criteria in their selection and promotion decisions. Decision-making criteria at this level are generally unstructured (use of gut feel) and rarely scrutinized (were they unbiased?).

Women themselves may not go after top-level jobs because of their family responsibilities and a preference for not making extreme personal sacrifices (Hewlett 2002a, 2002b). A double standard still exists on the child-care front. Women have always taken time off work to raise young children and take time off if a child is ill. When men now take time off to raise a child or take care of a sick child they are championed in the press.

Women appear to be less interested in pursuing both business education and managerial jobs because of what Kennedy and Kray (2012) term a 'taboo trade-off'. Women more than men associated business with immorality and questionable ethics and values, women more than men become morally outraged at compromised ethics and values, and women were less interested than men in making as much money as possible using questionable tactics.

Women's Contributions and Behaviours are Viewed Differently than Men's

There is considerable evidence that the contributions and behaviours of women are viewed differently than those of men. Bergeron (2012) studied organizational citizenship behaviours (OCBs) engaged in by both women and men. OCBs were of two types: communal and agentic. The first type were stereotypically seen to be engaged in by women and the second type engaged in by men. Women engaged in the same number of agentic OCBs, engaged in more communal than agentic OCBs, yet received fewer career benefits from engaging in their OCBs than did men.

In addition, managerial and professional women who violate role stereotype expectations tend to get punished. Agentic women are less likely to be hired, promoted, or liked (Rudman and Glick 2001).

Finally, business is increasingly being undertaken globally across borders. In large organizations, obtaining experience in another country is very important to further career advancement, yet only a very small increase in the number of women undertaking international assignments has been observed (Altman and Shortland, 2008).

Executive Men and Women See the Issues Facing Women Differently

Bringing about cultural change is also difficult due to the fact that male CEOs and female managers and professionals disagree on the major barriers facing women (Conference Board of Canada 1997; Ragins *et al.* 1998). Key barriers noted by male CEOs were: lack of general management or line experience, not having enough job experience (a shorter time in the pipeline), male stereotypes and preconceptions, an incompatible corporate culture, and exclusion from informal networks. Their first two were very commonly held. Senior women noted male stereotypes and preconceptions, exclusion from informal networks, lack of general management or line experience, an inhospitable corporate culture, and not having enough job experience (length of tine in the pipeline). One can classify these barriers into 'fix the women' or 'fix the culture' categories. Male CEOs endorsed more 'fix the women' barriers and senior women endorsed more 'fix the culture' barriers. A Canadian study (Conference Board of Canada 1997) came to very similar conclusions and differences, with the additional barrier that more senior women than male CEOs believed that commitment to family responsibilities and having children hindered women's advancement.

Work Investments: The Long Work Hours Culture

Long work hours and excessive work investment includes actual hours worked, overtime hours, 'face' time, workload, work intensity, and work addiction. Although the focus will be on the effects of these on individuals, effects are also present on families, workplaces, and innocent bystanders (Taylor 2003). This section reviews the literature on antecedents and consequences of working hours, work intensity, and work addiction particularly among managers and professionals. The dependent variables associated with these include health-related illnesses, injuries, sleep patterns, fatigue, heart rate and hormone level changes, as well as several work/non-work life balance issues.

There is evidence that managers in industrialized countries are working more hours now than previously (Greenhouse 2001; Schor 1991, 2003; Bunting 2004). Working long hours may in fact be a prerequisite for achieving senior leadership positions (Wallace 1997; Jacobs and Gerson 1998). In addition, organizational downsizing efforts have increased the workloads of survivors (*Globe and Mail* 2011). The globalization of business may mean that managers and professionals communicate with others across several time zones, thus requiring both early starts and late finishes. The globalization of business also increases travel requirements to other countries and increases the prospects of international assignments, sometimes harder for women and men to undertake as well.

The importance of focusing on work hours and work investment is multifaceted. First, a large number of employees are unhappy about the number of hours they work (Jacobs and Gerson 1998; Clarkberg and Moen 2001; Dembe 2005; MacInnes 2005). Second, the amount of time demanded by work is an obvious and important way in which work affects other parts of one's life (Galinsky 1999; Shields 1999). Third, work hours are a widely studied structural output of employment (Adam 1993). Fourth, the study of work hours and well-being outcomes has produced both inconsistent and complex results (Barnett 1998).

New technology and job flexibility have facilitated working from home or outside the office, which has contributed to an increase in hours worked (Golden and Figart 2000). We have become a twenty-four-hour society (Kreitzman 1999). With more women now working, workers are increasingly married to other workers, making the family work week longer. In addition, smaller families also allow couples to work more.

People have expected an association of long working hours and adverse well-being consequences for well over 100 years. These concerns were first raised during the Industrial Revolution in 1830 and in later years, and efforts were made to legislate limits in working hours to ten hours per day at that time (Golden 2006).

Hewlett and Luce (2006) examined 'extreme jobs', jobs in which incumbents worked 70 hours a week or more, were high wage earners, and had jobs having at least five characteristics of work intensity (e.g. unpredictable flow of work, fast-paced work under tight deadlines, inordinate scope of responsibility that amounts to more than one job, work-related events outside regular work hours, availability to clients 24/7, and a large amount of travel).

They carried out surveys of high wage earners in the US and high earning managers and professionals working in large multinational organizations. In addition, they conducted focus groups and interviews to better understand the motivations for, and the effects of, working in these jobs. Their two surveys of high-earning managers and professionals revealed four characteristics that created the most intensity and pressure: unpredictability (cited by 91 per cent of respondents), fast pace with tight deadlines (86 per cent), work-related events outside business hours (66 per cent), and 24/7 client demands (61 per cent).

They concluded that managers and professionals were now working harder than ever. Of the extreme job holders, 48 per cent said they were now working an average of 16.6 hours per week more than they did five years ago. And 42 per cent took ten or fewer vacation days per year (less then they were entitled to) and 55 per cent indicated that they had to cancel vacation plans regularly. Forty-four per cent of their respondents felt the pace of their work was extreme.

But extreme job holders (66 per cent in the US sample and 76 per cent in the global sample) said that they loved their jobs. Several reasons were given for working these long hours: the job was stimulating and challenging (over 85 per cent), working with high-quality colleagues (almost 50 per cent), high pay (almost 50 per cent), recognition (almost 40 per cent), and power and status (almost 25 per cent). In addition, increased competitive pressures, improved communication technologies, downsizings and restructurings resulting in few higher level jobs, flattened organizational hierarchies, and values changes in the broader society supportive of 'extremes' also played a role (Wiehert 2002).

Individuals holding 'extreme' jobs, however, had to let some things go. These included: home maintenance (about 70 per cent), relationships with their children (almost 50 per cent), relationships with their spouses/partners (over 45 per cent), and an unsatisfying sex life (over 40 per cent).

Extreme jobs were more likely to be held by men than women (17 per cent versus 4 per cent in the US sample; 30 per cent versus 15 per cent in the global sample). Women were more likely to be in jobs with high demands but working fewer hours. Women were not afraid of jobs having high levels of responsibility. Both women and men in extreme jobs indicated 'difficulties' with their children (acting up, eating too much junk food, under-achieving in school, too little adult supervision).

Most women (57 per cent) in extreme jobs did want to continue to be working as hard in five years (US sample), 48 per cent of men felt the same way. Most women (57 per cent) in extreme jobs did not want to be working as hard in five years. The numbers were higher in the global survey: women, 80 per cent and men, 55 per cent; women, 5 per cent and men, 17 per cent.

Work Hours and their Effects

A variety of outcome measures have been examined in connection with working long hours (van der Hulst 2003). Many studies of long work hours have been conducted in Japan where 'karoshi', sudden death due to long hours and insufficient sleep, was first observed. The Japanese coined this term to refer to deaths of individuals from over-work or working long work hours, and have actually defined such deaths (Kanai 2006). Several hypotheses have been advanced to explain the relationship between long work hours and adverse health outcomes. Working long hours affects the cardiovascular system through chronic exposure to increases in blood pressure and heart rate (Buell and Breslow 1960; Uehata 1991; Iwasaki *et al.* 1998). Working long hours produces sleep deprivation and lack of recovery leading to chronic fatigue, poor health-related behaviours, and ill health(Defoe *et al.* 2001; Ala-Mursjula *et al.* 2002; Liu and Tanaka 2002). Working long hours makes it more difficult to recover from job demands and the stress of long work hours. Finally, working long hours has been associated with more errors and accidents (Schuster and Rhodes 1985; Nachreiner *et al.* 2000; Gander *et al.* 2005).

More specifically, the literature suggests that long hours are associated with adverse health effects and increased safety risk, poor psychological health, excessive fatigue, burnout, more work–family conflict, worrying, and irritability (see Burke (2006) for a review).

FACE TIME OR PRESENTEEISM

Some of the hours spent at work are unnecessary. Let's take 'face' time or presenteeism. We undertook a project with an international public accounting firm to determine why so few women achieved partnership and why valued and qualified women were leaving the firm. The emphasis on billable hours meant staff worked long hours. We also found that staff would remain in their offices, visible, until the managing partner went home. Then staff left, though they could have left earlier. We heard of a staff member leaving an extra jacket on his chair with his office lights on while he attended a performance at a nearby theatre with his wife, then returning to his office to reclaim his jacket and turn out the lights.

LONG WORK HOURS DISADVANTAGE WOMEN

The fact that organizations expect or require women and men to work long hours does not constitute gender discrimination, at least not in a legal sense. The fact that long work hours may be a requirement for career advancement disadvantages women. Women are unwilling or unable to work as many hours as men (Hochschild 1997, 1990; Hewlett and Luce 2006). Organizations today need to become an employer of choice; the need to attract, develop, and retain talent of all employees. The long work hours, culture makes it difficult to fully utilize some of the population. There are men who prefer not to work long hours and some women who are comfortable with working very long hours.

WORK INTENSITY

Besides working longer hours, jobs themselves have also become more intense. Work intensity is generally conceptualized as an effort-related activity (Green 2001, 2004*a*, 2004*b*). In this regard, it is very similar to the 'work effort' concept discussed by Green (2001: 56) as: 'the rate of physical and/or mental input to work tasks performed during the working day... in part, effort is inversely linked to the "porosity" of the working day, meaning those gaps between tasks during which the body or mind rests'. Obviously, it would be difficult to measure such effort objectively; it can only be determined through self-reports, or extraordinarily well-controlled laboratory experiments. Burchell and Fagan (2004) used the 'speed of work' to mean work intensity, and reported that Europeans were working more intensely (in 2000 compared to 1991). Green (2001) focused on 'effort change' (respondents were asked to compare their current jobs with that of five years previously on items that included 'how fast you work' and 'the effort you have to put into your job'), and 'work effort' ('How much effort do you put into your job beyond what is required?' and 'My job requires that I work very hard'). He found that work effort had increased in Britain. While these are good starting points for conceptualizing work intensity, they measure only certain aspects of it. There is no research that attempts to capture a more extensive list of attributes. Others, in different samples, using different measures and time-frames, have also shown the increasing intensification of work (Worrall and Cooper 1999; Adams et al. 2000; Green 2001).

Increases in work effort or intensity are represented in employees having less idle time (less time between tasks), having to work harder now, needing more skills (multi-skilling), greater use of performance goals and appraisals, use of total quality management and just-in-time processes, needing to work faster, having more deadlines, and having more responsibility (George 1997; Green and McIntosh 2001; White et al. 2003).

WORKAHOLISM AND WORK ADDICTION

Oates (1971) defined a workaholic as 'a person whose need for work has become so excessive that it creates noticeable disturbance or interference with his bodily health, personal happiness, and interpersonal relationships, and with his smooth social functioning' (Oates 1971). Others (see Killinger 1991; Porter 1996; Robinson 1998) also define workaholism in negative terms. Most writers use the terms excessive work, workaholism, and work addiction interchangeably. Men are more likely than women to describe themselves as workaholics, often with pride.

Types of Workaholics

Some researchers have proposed the existence of different types of workaholic behaviour patterns, each having potentially different antecedents and associations with job performance, work, and life outcomes (Scott et al. 1997). Spence and Robbins (1992) propose three workaholic profiles based on their workaholic triad notion. The workaholic triad consists of work involvement, feeling driven to work because of inner pressures, and work enjoyment. Work Addicts (WAs) score high on work involvement, high on feeling driven to work, and low on work enjoyment. Work Enthusiasts (WE) score high on work involvement, low on feeling driven to work, and high on work enjoyment. Enthusiastic Addicts (EA) score high on all three workaholism components.

Non-workaholic profiles also exist (e.g. men and women scoring low on all three work-aholic triad components) and workaholic types worked significantly more hours per week than did the non-workaholic types.

We compared the personal demographics, job behaviours, work outcomes, extra-work outcomes, and psychological health of the three types of workaholics proposed by Spence and Robbins (1992) in a series of studies.

Job Behaviours

Burke (1999a) considered these relationships in a large sample of Canadian MBA graduates. First, there were no differences between WAs and WEs on hours worked per week or extra-hours worked per week; workaholism types worked significantly more hours and extra-hours per week than did non-workaholism types. WAs reported higher levels of work stress, more perfectionism, and greater unwillingness or difficulty in delegating than one or both of the other workaholism types.

Work Outcomes

Burke (1999b) compared levels of work and career satisfaction and success among the workaholism profiles. WAs scored lower than WEs and EAs on job satisfaction, career satisfaction, and future career prospects, and higher than WEs on intent to quit.

Workaholism Types and Flow at Work

Csikszentmihalyi (1990) uses the term 'flow' to refer to times when individuals feel in control of their actions and masters of their own destinies. Optimal experiences commonly result from hard work and meeting challenges head on. Burke and Matthiesen (2004) found that journalists scoring higher on work enjoyment and lower on feeling driven to work because of internal needs indicated higher levels of flow or optimal experience at work. In this same study, Burke and Matthiesen found that WEs and EAs indicated higher levels of flow than did WAs.

Psychological Well-Being

Burke (1999c) compared the three workaholism types on three indicators of psychological and physical well-being in a sample of 530 employed women and men MBA graduates. WAs had more psychosomatic symptoms than both WEs and EAs and poorer physical and emotional well-being than did WEs.

Extra-Work Satisfactions and Family Functioning

Burke (1999*d*) compared the three workaholism types on three aspects of life or extra-work satisfaction: family, relationship, and community. The comparisons of the workaholism types on the three measures of life or extra-work satisfactions provided moderate support for the hypothesized relationships. WAs reported less satisfaction on all three extra-work satisfaction measures than did WEs and less satisfaction on one (family) than did EAs.

Are Women Specifically Disadvantaged?

Here are a sample of some research findings relevant to long work hours, work addiction, and the experiences of men and women.

Workaholism or Work Addiction

Burke (1999*e*), in a study of 530 male and female Canadian MBA graduates from the same large university, found that although males and females differed on several personal demographic and work situation characteristics (men were at higher organizational levels, earned more money, were more likely to be married and to have children, and worked more hours), men and women scored similarly on three workaholism components (work involvement, joy in work, and feeling driven to work), and women reported higher levels of particular workaholic job behaviours (e.g. perfectionism and job stress) likely to be associated with lower levels of satisfaction and emotional health.

In a study of male and female Australian psychologists, Burgess *et al.* (2006) again found considerable gender differences on personal and work situation characteristics (males more likely to have children, to earn more income, to have fewer career interruptions, and to have worked part-time), but males also scored higher on two of the three workaholism components (work involvement, feeling driven to work). Female psychologists scored higher again on perfectionism and job stress. Female psychologists scoring higher on feeling driven to work also indicated higher levels of workaholic job behaviours, less job satisfaction and greater intent to quit, and more emotional exhaustion and psychosomatic symptoms. Burke *et al.* (2006) reported similar findings in a study of Australian female managers and professionals in early career.

Comparing the three workaholic types among female Australian psychologists (Work Addicts, Work Enthusiasts, and Enthusiastic Addicts) showed, consistent with findings from samples of adult working males, that Work Addicts reported less satisfaction with their jobs and careers, and higher levels of psychological distress (Burke *et al.* 2004).

Extreme Jobs

Burke and Fiksenbaum (2009*a*) compared work and well-being outcomes among males and females in the Canadian sample of MBA graduates in 'extreme jobs' in which they worked fifty-six hours a week or more. A higher percentage of men than women held 'extreme jobs'. Men in 'extreme jobs' were older, more likely married and to have children. There were relatively few differences on workaholism components, job behaviours, work outcomes, extra-work outcomes, and psychological well-being. Females did however indicate more psychosomatic symptoms, more friend satisfaction, and tended to indicate higher levels of perfectionism and job stress. Holding extreme jobs did not seem to be a problem for married men with children.

Burke and Fiksenbaum (2009*b*), using females from the Canadian sample of MBA graduates, then compared work and well-being outcomes for women in 'extreme jobs' working fifty-six hours a week or more with women working fifty-five or fewer hours per week. Women in 'extreme jobs' were less likely to be parents, and earned higher salaries. Women in 'extreme jobs' reported benefits such as higher levels of job satisfaction, future career prospects and salary; the costs included higher levels of job stress and psychosomatic symptoms and lower levels of family satisfaction and emotional health. Having children seemed to be a problem for women with children.

Workaholism

Burgess *et al.* (2006) compared male and female psychologists in Australia on workaholism components, workaholic behaviours, and work and psychological well-being outcomes. Women scored higher than on two of the three workaholism components (work involvement, feeling driven to work because of inner pressures). Women worked fewer extra-hours but reported higher levels of job stress, perfectionism, and tended to report higher levels of difficulty in delegating. Males and females reported similar levels on work outcomes and emotional and physical health, with females reporting more psychosomatic symptoms.

Burke and Matthiesen (2009) considered gender differences among Norwegian journalists. Female journalists reported higher levels of feeling driven to work because of inner pressures, and indicated higher levels of both negative affect and exhaustion and lower feelings of personal efficacy.

Burke (1999*e*) compared male and female MBA graduates on measures of workaholism components, workaholic job behaviours, and work and well-being outcomes. Males and females in this sample indicated similar levels of the three workaholism components, but females scored higher on a workaholic job behaviour and a work outcome (perfectionism, job stress) found to be associated with diminished well-being.

Two other studies considered gender differences and workaholism, its antecedents, and outcomes. First, Spence and Robbins (1992) found, in a US sample of social workers holding academic positions, that females scored higher on feeling driven to work, job stress, job involvement, time commitment, and health complaints. Second, Doerfler and Kammer (1986) considered the relationship of levels of workaholism with both gender and sex role orientation (masculine, feminine, androgynous) in a sample that included attorneys, physicians, and psychologists. The number of respondents classified as workaholic were similar across the two genders, and the three occupational groups. Interestingly, a majority of single workaholics were female, and female workaholics reported more masculine and androgynous characteristics than feminine characteristics.

Conclusions from two other related bodies of research should also be mentioned. First, there is considerable evidence that female work addicts, in the Spence and Robbins (1992) classification, report more negative outcomes than do male work addicts (Burgess *et al.* 2006). Second, as more women enter professional and managerial positions, the role of gender and gender differences is likely to diminish. Aziz and Cunningham (2008), in a study of male and female managers and professionals, found similar levels of workaholism and job stress in the two groups and that gender did not moderate the workaholism and job stress or workaholism and work–life imbalance relationships.

ORGANIZATIONAL CULTURE AND WORKAHOLISM

Ng *et al.* (2007) suggest that industry and organizational culture are likely associated with levels of workaholism. For example, law firms are noted for long work hours and pressures for billable hours (Wallace 1997). In addition, they suggest that masculine cultures are more likely to be associated with workaholism—masculine cultures are seen as competitive, power hungry, task oriented, and fearful of failure, while feminine cultures are seen as displaying modesty and showing concern for relationships (Newman and Nollen 1996; Detert *et al.* 2000). Workaholism would be more common in masculine cultures and workaholics would perform better in masculine cultures.

Johnstone and Johnston (2005) found that workaholism was higher among employees working in the business sector than in the social service sector and that cultures higher on work pressure were associated with higher levels of feeling driven to work because of inner pressures.

Burke (2001), in a study of 530 MBA graduates from the same Canadian university, reported that organizational values less supportive of work–personal life balance were significantly higher in workaholic types than in non-workaholic types, using the Spence and Robbins (1992) classifications.

WORKING AT HOME

There are at least two dimensions to the hours women and men spend working at home. One dimension involves telework, in which women and men put in their typical day but do it at home rather than going into the office. A second dimension is going into the office and taking work home with you, which is then undertaken in the evening or on weekends. A recent study undertaken by the US Bureau of Labor Statistics showed that 9 per cent of non-farm-workers brought some of their work home on a typical work-day. Most of those who did so were male and married. The main reason for taking work home was to finish work tasks or to catch up. Women were less able or willing to do this.

UNPAID OVERTIME

Some employee groups in North America have filed class action suits against their employers (e.g. in the banking sector) for pay for working hours of unpaid overtime. These employees are paid on a weekly rate for a given number of hours. Unfortunately many of these employees are 'coerced' by their managers to work more hours in order to complete their work, hours for which they receive no pay.

Individuals, families, organizations, and societies need to be concerned about the effects of long work hours, work intensity, and work addiction. These have been found to diminish employee well-being and health, levels of family functioning and organizational performance, and increase social welfare and health care costs borne by society (Kirkcaldy et al. 2009). Healthy organizations and societies need healthy individuals and families; coming to grips with long work hours in intense and demanding jobs, and work addiction is an important start.

BOTH WOMEN AND MEN BENEFIT

An important question then becomes whether women and men benefit from more supportive organizational cultures. The answer appears to be 'yes' (Burke 2010). Burke et al. (2005), in a study of female and male psychologists in Australia, reported that both women and men benefited from organizational cultures more supportive of work–personal life integration (less job stress, more joy in work, lower intention to quit, more job and career satisfaction, fewer psychosomatic symptoms, and more positive emotional and physical well-being). There were more independent and significant correlates of organizational cultures supportive of work–personal life integration among women than men however. Organizational values more supportive of work–personal life integration were found to be associated with a greater number of favourable work and well-being outcomes among female than male psychologists (Burke et al. 2005).

Earlier, Burke (2000), in a study of 283 male MBA graduates from the same large Canadian university, reported that respondents indicating organizational values more supportive of work–personal life integration or balance also reported working fewer hours, fewer extra-hours, less job stress, greater joy in work, lower intentions to quit, greater job, career and life satisfaction, fewer psychosomatic symptoms, and more positive emotional and physical well-being,

Conclusions

Men are advantaged in the following ways in contemporary organizations because of gendered workplace cultures reflecting perceptions that exist in the wider society.

- They receive privileges for living in patriarchal societies and working in male-dominated masculine organizational cultures-think manager, think male.
- They are comfortable in and fit well with these cultures. Men have self-identities consistent with the managerial and leadership role, socialization, education, and training (business school education) that reinforces these privileges.
- They have the ability to work long hours in intense jobs though having home and family connections.
- They have access to household support.
- They have access to mentoring, sponsorship, and networking activities both at work and after work hours.

But men may endure some costs such as job stress, diminished psychological and physical well-being, work–family role conflict, and limited contributions to home and family functions (Burke and Nelson 1998; Burke 2002). Corporate masculinity has several kinds of dysfunctions (Maier 1991) such as limited managerial effectiveness, overwork, and limited views of careers success.

Having It All

The article by Anne-Marie Slaughter (2012) in the July/August issue of the *Atlantic*, the appointment of a pregnant Marissa Mayer as president of Yahoo, and the statement of Cheryl Sandberg, a mother and CEO of Facebook came together to raise the issue of why women can't have it all. Slaughter resigned from a high-level policy adviser position with US Secretary of State Hillary Clinton and returned to her professorship at Princeton because she was unable to spend enough time with her family. People wondered whether a pregnant and about to be mother could handle her CEO job (apparently Mayer will take one week off after the birth of her child), and though Sandberg made it home for family dinners, she worked several hours each night following these dinners. Slaughter said that neither women nor men can have it all under the current work and business expectations: the long work hours culture. Long hours, face time, endless meetings make it difficult for women. Even elite women, with lots of control and resources, cannot have satisfying home and family lives. Men can't have it all either.

Long work hours interferes with their home and family relationships and leisure. Some women may be disadvantaged in their careers by these work demands and some men may be disadvantaged in their personal and home lives by these work demands. A worthwhile agenda would be to address these demands in ways that keep organizations productive while satisfying the needs of both women and men—a win-win all around.

Implications

The research evidence shows that managerial and professional women continue to make only very slow progress in career advancement and in obtaining corporate board directorships (Davidson and Burke 2011; Vinnicombe *et al.* 2012). We are still seeing 'firsts' by women in several areas of endeavour (e. g, first woman to head a particular company, first woman to receive a Nobel prize, first woman to become a gondolier in Venice). Yet women are increasingly obtaining the necessary education, work experience, and time in the pipeline in preparation for these more senior roles.

There are some encouraging signs that things may be changing in the near future. Several factors have recently come together to support women's career development and advancement. First, women are increasingly becoming a significant economic force in many countries (Wittenberg-Cox and Maitland 2008; Silverstein and Sayre 2009). Second, demographic changes have resulted in shortages of skilled workers and produced a war for talent (Michaels *et al.* 2001). Organizations and societies can no longer afford to waste the skills and talents of half their populations. Attracting, retaining, and advancing qualified women has now become a business issue, not a women's issue (Catalyst 1998; Wittenberg-Cox 2010).

Organizations then need to create work environments that work for both women and men. Barnett and Hall (2001) suggest that organizations noted for fewer working hours may in fact have a significant recruiting advantage. Fortunately there is considerable guidance on how this can be realized (Catalyst 1998; Wittenberg-Cox and Maitland 2008) and rich descriptions of successful company efforts (Munck 2001; Bailyn 2002; Spinks and Tombari 2002; Lee *et al.* 2002; Weil and Kivland 2005). Individuals can also reduce their working hours through more efficient time utilization and priority-setting initiatives (Philipson 2002; Song 2006; Burwell and Chen 2008).

Let us now consider organizational initiatives targeted at the work–family interface, a major barrier for women and a topic of increasing interest to men. Historically men went to work while women stayed home to shoulder responsibilities for home and family functioning. In addition, work and family were seen as 'separate worlds'. Things have changed. More women are now in the workforce and work and family are no longer seen as 'separate worlds'. Events and experiences at work influence home and family; events and experiences at home and family influence work, though to a lesser extent (Byron 2005: Michel *et al.* 2011). Managerial and professional women report higher levels of work–family and family–work interference than do men.

As an example, a female acquisitions editor at a major UK publisher indicated to me that I would be receiving a contract from her assistant as she had to leave the office for a day or two to take care of her young child who was ill and needed to be removed from her child care centre. Did her husband take any time off?

Given the important role played by organizational culture in women's satisfaction and career progress and the respect and recognition they receive, women should place the fit of organizational culture with their values, needs, and skills before salary and job titles in selecting employers.

Organizations have attempted to meet this challenge by implementing policies and programmes that address work–family conflict and support work–family integration. Examples would include: onsite childcare resources, flexible working hours, telecommuting, compressed work weeks, reduced workloads, job sharing arrangements, use of performance evaluation tools that focus on results rather than hours worked. Rapoport *et al.* (2002) describe organizational interventions that identify workplace policies and practices that limit performance and interfere with work–family integration, that when addressed improve organizational performance and work–family integration simultaneously.

Organizational change efforts in the past have fallen short since many followed a piecemeal or limited approach. To be successful, organizational change efforts must be undertaken at three levels simultaneously (Rutherford 2011): *organizational*—policies, practices, and structures; *interpersonal*—the ways people behave towards each other; and *personal*—an examination of one's long-held attitudes and beliefs, and aspects of one's self-identity.

Change, to be successful, needs the commitment of leadership. Most organizations do not yet see the advancement of women as a key business issue. Change begins with an audit of the present organizational situation. Rutherford (2011: 4) proposes three broad questions as starters. 'What kind of an organization do we want to be? What is stopping us being that? What do we need to do to get there?'

It is important that more men become more supportive of these organizational changes and the business case goals of supporting and advancing the careers of women. Men hold the majority of leadership roles, control the organizational agenda, manage the allocation of scarce organizational resources, such as time, energy, and money, and are the source of resistance and backlash. What can men do to reduce male privilege? First, they must acknowledge it and become accountable. However, common male responses maintain the status quo (Maier 1994). These include: male privilege doesn't exist; male privilege exists but it hasn't helped me; male privilege alone cannot explain the central roles in play in the world; male privilege exists, but the system is thousands of years old and cannot be changed; I will make efforts to help women, but I won't work to lessen men's privilege; and male privilege exists and I will work to lessen it as it is unfair. The last view represents men who get it.

Thus changing men, masculinities, and gendered organizations is a necessary step in bringing about positive change. And there will be benefits for men for undertaking this journey (Burke 2002).

Acknowledgement

Preparation of this chapter was supported in part by York University.

References

Acker, J. (1990). 'Hierarchies, Jobs and Bodies: A Theory of Gendered Organizations', *Gender and Society*, 4: 139–58.

Adam, B. (1993). 'Within and beyond the Time Economy of Employment Relations: Conceptual Issues Pertinent to Research on Time and Work', *Social Science Information*, 32: 163–84.

Adams, A., Chase, J., Arber, S., and Band, S. (2000). 'Skill Mix Changes and Work Intensification in Nursing', *Work, Employment and Society*, 14: 541–55.

Ala-Mursjula, L., Vahtera., J., Kivimaki, M., Kevin, M. V., and Penttij, J. (2002). 'Employee Control over Working Times: Associations with Subjective Health and Sickness Absences', *Journal of Epidemiology and Community Health*, 56: 272–8.

Altman, Y., and Shortland, S. (2008). Women in International Assignments: Taking Stock—A 25 Year Review', *Human Resource Management*, 42: 199–216.

Aziz, S., and Cunningham, J. (2008). 'Workaholism, Work Stress, Work-Life Imbalance: Exploring Gender's Role', *Gender in Management: An International Journal*, 23: 553–66.

Bailyn, L. (2002). 'Time in Organizations: Constraints on, and Possibilities for Gender Equity in the Workplace', in R. J. Burke and D. L. Nelson (eds), *Advancing Women's Careers*, 262–72. Oxford: Blackwell Publishers.

Barnett, R. C. (1998). 'Towards a Review and Reconceptualization of the Work/Family Literature', *Genetic Social and General Psychology Monograph*, 124: 125–82.

Barnett, R. C., and Hall, D. T. (2001). 'How to Use Reduced Hours to Win the War for Talent', *Organizational Dynamics*, 29: 192–210.

Barreto, M., Ryan, M. K., and Schmitt, M. T. (2009). *The Glass Ceiling in the 21st Century: Understanding Barriers to Gender Equality.* Washington, DC: American Psychological Association.

Bergeron, D. (2012). 'The Stability of Organizational Citizenship Behavior over Time: Women as Good Soldiers', paper presented at the Annual Meeting of the Academy of Management, Boston, Aug.

Buell, P., and Breslow, L., (1960). 'Mortality from Coronary Heart Disease in Californian Men Who Work Long Hours', *Journal of Chronic Disease*, 11: 615–26.

Bunting, M. (2004). *Willing Slaves.* London: Harper & Collins.

Burchell, B., and Fagan, C. (2004). 'Gender and the Intensification of Work: Evidence from the European Working Conditions Survey', *Eastern Economic Journal*, 30: 627–42.

Burgess, Z., Burke, R. J., and Oberklaid, F. (2006). 'Workaholism among Australian Psychologists: Gender Differences', *Equal Opportunities International*, 25: 48–59.

Burke, R. J. (1999a). 'Workaholism in Organizations: Measurement Validation and Replication', *International Journal of Stress Management*, 6: 45–55.

Burke, R. J. (1999b). 'Are Workaholics Job Satisfied and Successful in their Careers?', *Career Development International*, 26: 149–58.

Burke, R. J. (1999c). 'Workaholism in Organizations: Psychological and Physical Well-Being Consequences', *Stress Medicine*, 16: 11–16.

Burke, R. J. (1999d). 'Workaholism and Extra-Work Satisfactions', *International Journal of Organizational Analysis*, 7: 352–64.

Burke, R. J. (1999e). 'Workaholism in Organizations: Gender Differences', *Sex Roles*, 41: 333–45.

Burke, R. J. (2000). 'Do Managerial Men Benefit from Organizational Values Supporting Work–Personal Life Balance?', *Gender in Management: An International Journal*, 25: 91–9.

Burke, R. J. (2001). 'Workaholism in Organizations: The Role of Organizational Values', *Personnel Review*, 30: 637–45.

Burke, R. J. (2002). 'Men, Masculinities, and Health', in D. L. Nelson and R. J. Burke (eds), *Gender, Work Stress and Health*, 35–54. Washington, DC: American Psychological Association.

Burke, R. J. (2006) *Research Companion to Working Time and Work Addiction.* Cheltenham: Edward Elgar.

Burke, R. J. (2010). 'Managers, Balance, and Fulfilling Lives', *Gender in Management: An International Journal*, 25: 86–90.

Burke, R. J., and Fiksenbaum, L. (2009a). 'Managerial and Professional Women in "Extreme Jobs": Benefits and Costs', *Equal Opportunities International*, 28: 432–42.

Burke, R. J., and Fiksenbaum, L. (2009b). 'Are Managerial Women in "Extreme Jobs" Disadvantaged?', *Gender in Management: An International Journal*, 24: 5–13.

Burke, R. J. and Matthiesen, S. (2004). 'Workaholism among Norwegian Journalists: Antecedents and Consequences', *Stress and Health*, 20: 301–8.

Burke, R. J., and Matthiesen, S. B. (2009). 'Workaholism among Norwegian Journalists: Gender Differences', *Equal Opportunities International*, 28: 452–64.

Burke, R. J., and Nelson, D. L. (1998). 'Organizational Men: Masculinity and its Discontents', in D. L. Cooper and I. T. Robertson (eds), *International Review of Industrial and Organizational Psychology*, xiii. 225–72. Chichester: John Wiley.

Burke, R. J., Oberklaid, F., and Burgess, Z (2004). 'Workaholism among Australian Women Psychologists: Antecedents and Consequences', *International Journal of Management*, 21: 263–77.

Burke, R. J., Oberklaid, F., and Burgess, Z. (2005). 'Organizational Values, Job Experiences and Satisfactions among Female and Male Psychologists', *Community, Work and Family*, 8: 53–68.

Burke, R. J., Burgess, Z., and Fallon, B. (2006). 'Workaholism among Australian Female Managers and Professionals: Job Behaviors, Satisfactions and Psychological Health', *Equal Opportunities International*, 25: 200–13.

Burwell, R. M, and Chen, C. P. (2008). 'Positive Psychology for Work–Life Balance: A New Approach', in R. J. Burke and C. L. Cooper (eds), *The Long Work Hours Culture: Causes, Consequences and Choices*, 295–313. Bingley: Emerald.

Byron, K. (2005). 'A Meta-Analytic Review of Work–Family Conflict and its Antecedents', *Journal of Vocational Behavior*, 67: 169–98.

Catalyst (1998). *Advancing Women in Business: The Catalyst Guide.* San Francisco: Jossey-Bass.

Clarkberg, M., and Moen, P. (2001). 'The Time Squeeze: Is the Increase in Working Time Due to Employer Demands or Employee Preferences?', *American Behavioral Scientist*, 44: 1115–36.

Cockburn, C. (1991). *In the Way of Women: Men's Resistance to Sex Equality in Organizations.* Basingstoke: Macmillan.

Conference Board of Canada (1997). *Closing the Gap: Women's Advancement in Corporate and Professional Canada.* Ottawa: Conference Board of Canada.

Csikszentmihalyi, M. (1990). *Flow: The Psychology of Optimal Experience.* New York: Harper Collins.

Davidson, M. J., and Burke, R. J. (2011). *Women in Management Worldwide: Progress and Prospects.* Aldershot: Gower Publishing.

Defoe, D. M., Power, M. L., Holzman, G. B., Carpentieri, A., and Schulkin, J. (2001). 'Long Hours and Little Sleep: Work Schedules of Residents in Obstetrics and Gynecology', *Obstetrics and Gynecology,* 97: 1015–18.

Dembe, A. E. (2005). 'Long Working Hours: The Scientific Bases for Concern', *Perspectives on Work,* (Winter): 20–2.

Detert, J. R., Schroeder, R. G., and Mauriel, J. J. (2000). 'A Framework for Linking Culture and Improvement Initiatives in Organizations', *Academy of Management Review,* 25: 850–63.

Doerfler, M. C., and Kammer, P. P. (1986). 'Workaholism: Sex and Sex Role Stereotyping among Female Professionals', *Sex Roles,* 14: 551–60.

Fagenson, A. E. (1990). 'At the Heart of Women in Management Research: Theoretical and Methodological Approaches and their Biases', *Journal of Business Ethics,* 9: 267–74.

Galinsky, E. (1999). *Ask the Children: What America's Children Really Think about Working Parents.* New York: William Morrow.

Gander, P. H., Merry, A., Millar, M. M., and Weller, J. (2000). 'Hours of Work and Fatigue-Related Error: A Survey of New Zealand Anaesthetists', *Anaesthetic and Intensive Care,* 28: 178–83.

George, D. (1997). 'Working Longer Hours: Pressure from the Boss or Pressure from the Marketers?', *Review of Social Economy,* 60: 33–65.

Globe and Mail (2011). 'Heavier Workloads from Layoffs Still Not Easing: Poll', (12 Mar.): B19.

Golden, L., and Figart, D. M. (2000), *Work Time: International Trends, Theory and Policy Perspectives.* London: Routledge.

Golden, L. (2006). 'How Long? The Historical Economic and Cultural Factors behind: Working Hours and Overwork', in R. J. Burke (ed.), *Research Companion to Working Hours and Work Addiction,* 36–57. Cheltenham: Edward Elgar.

Gray, J. (1992). *Men are from Mars: Women are from Venus.* New York: Harper Collins.

Green, F. (2004a). Why has Work Effort Become More Intense?', *Industrial Relations* (Oct.), 709–741.

Green, F. (2004b). Work Intensification, Discretion, and the Decline in Well-Being at Work', *Eastern Economic Journal,* 30: 615–25.

Green, F. (2001). 'It's been a Hard Day's Night: The Concentration and Intensification of Work in Late Twentieth-Century Britain', *British Journal of Industrial Relations,* 39: 53–80.

Green, F., and McIntosh, S. (2001). 'The Intensification of Work in Europe', *Labour Economics,* (May): 291–308.

Greenhouse, S. (2001). 'Report Shows Americans have More "Labor Days"', *New York Times,* (1 Sept.): A6.

Hearn, J., and Morgan, D. (1990). *Men, Masculinities and Social Theory.* London: Unwin Hyman.

Hewlett, S. A. (2002a). 'Executive Women and the Myth of Having it All', *Harvard Business Review,* 80: 66–73.

Hewlett, S. A. (2002b). *Creating a Life: Professional Women and the Quest for Children.* New York: Hyperion.

Hewlett, S. A., and Luce, C. B. (2006). 'Extreme Jobs: The Dangerous Allure of the 70-Hour Work Week', *Harvard Business Review,* (Dec.): 49–59.

Hochschild, A. R. (1990). 'The Second Shift: Working, Parents and the Revolution at Home'. New York: Avon Books.

Hochschild, A. R. (1997). The Time Bind: When Work Becomes Home and Home Becomes Work. New York: Metropolitan Books.

Iwasaki, K., Sasaki, T., Oka, T., and Hisanaga, N. (1998). 'Effect of Working Hours on Biological Functions Related to Cardiovascular System among Salesmen in a Machinery Manufacturing Company', Industrial Health, 36: 361–7.

Jacobs, J. A., and Gerson, K. (1998). 'Who are the Overworked Americans?', Review of Social Economy, 56: 442–59.

Johnstone, A., and Johnston, L. (2005). 'The Relationship between Organizational Climate, Occupational Type and Workaholism', New Zealand Journal of Psychology, 34: 181–8.

Kanai, A. (2006). 'Economic and Employment Conditions, Karoshi Work to Death and the Trend of Studies on Workaholism in Japan', in R. J. Burke (ed.), Research Companion to Working Time and Work Addiction, 158–71. Cheltenham: Edward Elgar.

Kennedy, J. A., and Kray, L. J. (2012). 'Who is Willing to Sacrifice Values for Money and Social Status? Gender Differences in Reactions to Taboo Trade-Offs', paper presented at the Annual Meeting of the Academy of Management, Boston, Aug.

Killinger, B. (1991). Workaholics: The Respectable Addicts. New York: Simon & Schuster.

Kimmel, M. (2009). The Gendered Society. Oxford: Oxford University Press.

Kirkcaldy, B., Furnham, A., and Shephard, R. (2009). 'The Impact of Working Hours and Working Patterns on Physical and Psychological Health', in S. Cartwright and C. L. Cooper (eds), The Oxford Handbook of Organizational Well-Being, 303–35. Oxford: Oxford Publishing.

Kreitzman, L. (1999). The 24-Hour Society. London: Profile Books.

Lee, M. D., Engler, L., and Wright, L. (2002). 'Exploring the Boundaries in Professional Careers: Reduced Load Work Arrangements in Law, Medicine, and Accounting', in R. J. Burke and D. L. Nelson (eds), Advancing Women's Careers, 174–205. Oxford: Blackwell Publishers.

Liu, Y., and Tanaka, H., The Fukuoka Heart Study Group (2002). 'Overtime Work, Insufficient Sleep, and Risk of Non-Fatal Acute Myocardial Infarction in Japanese Men', Occupational Environmental Medicine, 59(7): 447–51.

Loomis, D. (2005). 'Long Work Hours and Occupational Injuries: New Evidence on Upstream Causes', Occupational and Environmental Medicine, 62: 585.

McDowell, L. (1997). Capital Culture: Gender at Work in the City. Oxford: Blackwell.

MacInnes, J. (2005). 'Work–Life Balance and the Demands for Reduction in Working Hours: Evidence from the British Social Attitudes Survey 2002', British Journal of Industrial Relations, 43: 273–95.

Maier, M. (1991). 'The Dysfunctions of "Corporate Masculinity"; Gender and Diversity Issues in Organizational Development', Journal of Management in Practice, 8: 49–63.

Maier, M. (1994). 'Save the Males: Reflections on White Male Privileges in Organizations', paper presented at the New York State Political Science Association 48th Annual Meeting, Albany, NY, Apr.

Marshall, J. (1984). Women Managers: Travelers in a Male World. Chichester: John Wiley.

Marshall, J. (1995). 'Working at Senior Management and Board Levels: Some of the Issues for Women', Women in Management Review, 10: 21–5.

Michaels, E., Handfield-Jones, H. J, and Axelrod, B. (2001). The War for Talent. Boston: Harvard Business School Press.

Michel, J. S. Kotrba, L. M., Mitchelson, J. K., Clark, M. A., and Baltes, B. B. (2011). 'Antecedents of Work–Family Conflict: A Meta-Analytic Review', *Journal of Organizational Behavior*, 32: 689–725.

Morrison, A. M. (1992). *The New Leaders*. San Francisco: Jossey-Bass.

Munck, B. (2001). 'Changing a Culture of Face Time', *Harvard Business Review*, (Nov.): 167–72.

Nachreiner, F., Akkermann, S., and Haenecke, K. (2000). 'Fatal Accident Risk as a Function of Hours into Work', in S. Hornberger, P. Knauth, G. Costa, and S. Folkard (eds), *Shift Work in the 21st Century*, 19–24. Frankfurt: Peter Lang.

Newman, K. L., and Nollen, S. D. (1996). 'Culture and Congruence: The Fit between Management Practices and National Culture', *Journal of International Business Studies*, 4: 753–79.

Ng, T. W. H., Sorensen, K., and Feldman, D. C. (2007). 'Dimensions, Antecedents, and Consequences of Workaholism: A Conceptual Integration and Extension', *Journal of Organizational Behavior*, 28: 111–36.

Oates, W. (1971). 'Confessions of a Workaholic: The Facts about Work Addiction', *New York: World of Work, Environment and Health*, 24: 43–8.

Philipson, J. (2002). *Married to the Job: Why we Live to Work and What we can Do about it*. New York: Simon & Schuster.

Porter, G. (1996). 'Organizational Impact of Workaholism: Suggestions for Researching the Negative Outcomes of Excessive Work', *Journal of Occupational Health Psychology*, 1: 70–84.

Powell, G. N. (2011). *Women and Men in Management* (4th edn). Thousand Oaks, CA: Sage.

Powell, G. N., and Butterfield, D. A. (1989). 'The "Good Manager": Did Androgyny Fare Better in the 1980s?', *Group and Organization Studies*, 14: 216–33.

Ragins, B. R., Townsend, B., and Mattis, M. C. (1998). 'Gender Gap in the Executive Suite: CEOs and Female Executives Report on Breaking the Glass Ceiling', *Academy of Management Executive*, 12: 28–42.

Rapoport, R., Bailyn, L, Fletcher, J. K., and Pruitt, B. H. (2002). *Beyond Work–Family Balance: Advancing Gender Equity and Workplace Performance*. San Francisco: Jossey-Bass.

Robinson, B. E. (1998). *Chained to the Desk: A Guidebook for Workaholics, their Partners and Children and the Clinicians Who Treat them*. New York: New York University Press.

Rogers, B. (1988). *Men Only: An Investigation into Men's Organizations*. London: Pandora.

Rudman, L. A., and Glick, P. (2001). 'Proscriptive Gender Stereotypes and Backlash toward Agentic Women', *Journal of Social Issues*, 57: 743–62.

Rutherford, S. (2001). 'Are you Going Home Already? The Long Hours Culture, Women Managers and the Patriarchal Closure', *Time and Society*, 10: 259–76.

Rutherford, S. (2002). 'Organizational Culture, Women Managers and Exclusion', *Women in Management Review*, 8: 127–39.

Rutherford, S. (2011). *Women's Work, Men's Cultures: Overcoming Resistance and Changing Organizational Cultures*. Basingstoke: Palgrave Macmillan.

Schein, V. E. (1975). 'Relationships between Sex Role Stereotypes and Requisite Management Characteristics among Female Managers', *Journal of Applied Psychology*, 60: 340–4.

Schein, V. E. (1994). 'Managerial Sex Typing: A Persistent and Pervasive Barrier to Women's Opportunities', in M. J. Davidson and R. J. Burke (eds), *Women in Management*, 41–53. London: Routledge.

Schor, J. B. (1991). *The Overworked American*, New York: Basic Books.

Schor, J. B. (2003). 'The (Even More) Overworked American', in J. deGraaf (ed.), *Take Back your Time*, 6–11. San Francisco: Berrett-Koehler.

Schuster, M., and Rhodes, S. (1985). 'The Impact of Overtime Work on Industrial Accident Rates', *Industrial Relations*, 24: 234–46.

Scott, K. S., Moore, K. S., and Miceli, M. P. (1997). 'An Exploration of the Meaning and Consequences of Workaholism', *Human Relations*, 50: 287–314.

Shields, M., (1999). 'Long Working Hours and Health', *Health Reports*, 11: 33–48.

Silverstein, M. J., and Sayre, K. (2009). 'The Female Economy: Companies Ignore Women, "the Largest Market Opportunity in the World"', *Harvard Business Review*, (Sept.): 46–53.

Slaughter, A.-M. (2012). 'Why Women Still Can't Have it All', *Atlantic*, (July/Aug.): 85–102.

Smith, S. N., and Cockburn, T. (2009). 'Cultural Sexism in the UK Airline Industry', *Gender in Management: An International Journal*, 24: 32–45.

Song, M. (2006). *The Hamster Revolution: How you Manage your E-mail Before it Manages you*. San Francisco: Berrett-Koehler.

Spence, J. T., and Robbins, A. S. (1992). 'Workaholism: Definition, Measurement, and Preliminary Results', *Journal of Personality Assessment*, 58: 160–78.

Spinks, N. L., and Tombari, N. (2002). 'Flexible Work Arrangements: A Successful Strategy for the Advancement of Women at the Royal Bank Financial Group', in R. J. Burke and D. L. Nelson (eds), *Advancing Women's Careers*, 220–40. Oxford: Blackwell Publishers.

Taylor, B. (2003). *What Kids Really Want that Money Can't Buy*. New York: Warner Books.

Uehata, T. (1991). 'Long Working Hours and Occupational Stress-Related Cardiovascular Attacks among Middle Aged Workers in Japan', *Journal of Human Ergonomics*, 20: 147–53.

van der Hulst, M. (2003). 'Long Work Hours and Health', *Scandinavian Journal of Work, Environment and Health*, 2: 171–88.

Vinnicombe, S. E., Burke, R. J., Moore, L., and Blake-Beard, S. (2012). *Handbook of Research on Supporting Women's Career Advancement*. Cheltenham: Edward Elgar.

Wacjman, J. (1998). *Managing Like a Man*. Cambridge: Polity Press.

Wallace, J. E. (1997). 'It's about Time: A Study of Hours Worked and Work Spillover among Law Firm Lawyers', *Journal of Vocational Behavior*, 50: 227–48.

Weil, P. A., and Kivland, C. (2005). 'Work–Life Balance Practices in Healthcare Organizations: A 2003 Status Report', in R. J. Burke and M. C. Mattis (eds), *Supporting Women's Career Advancement: Challenges and Opportunities*, 210–41. Cheltenham: Edward Elgar.

White, M., Haill, S., McGovern, P., Mills, C., and Smeaton, D. (2003). '"High-Performance": Management Practices, Working Hours and Work–Life Balance', *British Journal of Industrial Relations*, 41: 175–95.

Wiehert, I. C. (2002). 'Job Insecurity and Work Intensification: The Effects on Health and Well-Being', in B. J. Burchell, D. Lapido, and F. Wilkinson (eds), *Job Security and Work Intensification*, 57–74. London: Routledge.

Wittenberg-Cox, A., and Maitland, A. (2008). *Why Women Mean Business: Understanding the Emergence of our Next Economic Revolution*. San Francisco: Jossey-Bass.

Wittenberg-Cox, A. (2010). *How Women Mean Business*. Chichester: John Wiley.

Worrall, L., and Cooper, C. L. (1999). *Quality of Work Life Survey*. London: Institute of Management.

CHAPTER 18

CHALLENGING GENDER BOUNDARIES

Pressures and Constraints on Women in Non-Traditional Occupations

BARBARA BAGILHOLE

Sex segregation has a history as old as the labour force itself.

(Reskin and Roos 1990: p. ix)

INTRODUCTION

The pervasive nature of labour-market gender segregation, both horizontally into different occupations and vertically with women generally concentrated at the bottom of occupational hierarchies if in the same job as men, is internationally well documented. The extent of segregation in the UK is illustrated by evidence that 60 per cent of women are employed in just ten out of seventy-seven occupations, and highly concentrated in 'the five Cs'; caring, cashiering, catering, cleaning, and clerical (HMSO 2005: 6). However, this chapter is about women who have broken into non-traditional, male-dominated work. The overall aim of the chapter is to examine and analyse common pressures and constraints these women experience. These are particularly identified around three areas: the continuing over-representation of women in domestic responsibilities and its impact on their perceived professionalism; structural barriers in the way the occupations are performed; and finally hostile cultural environments within the workplace.

It will use four exemplars of non-traditional occupations where the author has conducted qualitative empirical research; civil service management, academia, construction engineering, and the priesthood in the Church of England. The civil service was feminized but men retained the management of it; academia has been identified as one of the last bastions of male power; construction is the most male occupation after coal mining; and the Church of England refused to ordain women as priests until the 1990s.

The focus of research on the segregation of the labour market has changed from the measurement of occupational segregation to a case study methodology, trying to explain occupation segregation by determining the factors behind it (Plantenga and Rubery 1999). As Gonas and Lehto (1999: 25) argue:

> The persistence of gender segregation is bound up to a large extent with internal processes in organizations.... To reach some kind of understanding of the forces that shape and reshape segregation patterns, researchers have to go beyond quantitative statistical analysis. They have to probe deeper into organizational cultures...and reveal the silent, hidden processes at work behind the official ideologies of eo [equal opportunities] and equality.

What do we Mean by Non-Traditional Occupations?

For the purposes of this chapter non-traditional employment is taken to denote any occupation which is traditionally undertaken by men. Of course the terms 'non-traditional' and 'traditional' with regard to gendered work remain culturally and historically specific. For example, it is still relatively unusual to find women engaged in the construction industry in Europe. However, this is not so in other parts of the world, particularly in poorer countries, such as India, where particular social, cultural, and economic conditions affect women's employment activity (Bagilhole 2000). Importantly, such differences in the allocation of work are proof of the fact that work roles are not necessarily assigned on the basis of biological or physical attributes.

Governments, agencies, institutions, and employers in different countries have increasingly had to reach a contextually agreed definition in relation to anti-discrimination legislation, or the funding of special training and employment schemes for the under-represented sex. The most widely used cut-off point has emerged, if one sex is represented by roughly less than a third of those involved in an occupation, it is designated as non-traditional for the under-represented sex.

There have always been women who have 'stepped outside the stereotype' (Whittock 2000: 7). However, apart from the two world wars when there was a general, if often grudging recognition, of women's participation in non-traditional forms of work, such activity has been 'hidden from history' (Rowbotham 1974: 1). Therefore, in an effort to recover what Connell refers to as 'a marginalised form of femininity' (1993: 188), this chapter introduces the reader to women's experiences in four non-traditional occupations.

Why Study Women in Non-Traditional Occupations?

Women and paid work are increasingly seen as a popular and fascinating topic both worldwide and across disciplines. This reflects the continuing marked rise in women's involvement in labour markets in all countries, and governments' continual desire for their increased participation in order to utilize their talents and skills to the full. Therefore, as women make inroads into non-traditional occupations, this becomes an important topic. More women have been gaining the qualifications to embark on careers formerly dominated by men, and more women have been stepping onto the lower rungs of such careers.

Common themes appear to emerge when women enter non-traditional occupations where they seem to have a fairly consistent experience. Often the women who enter fields predominantly occupied by men are highly qualified and less likely to have partners or be married and, if married, less likely to have children than their male colleagues. If they do have partners and families, they continue to bear the major workload and responsibility for housework and carework. They generally do less well than male colleagues in the recruitment, selection, and promotion processes of their occupation, and they report the existence of a 'glass ceiling' affecting professional advancement. They are paid less than the men and cluster in lower-status specialties within their professions. Finally, they suffer from reported incidences of sexual harassment and discrimination. Women in male-dominated occupations demonstrate the dynamics of exclusion and marginalization. As Spencer and Podmore (1987: 1) point out: 'The severity of the damaging effects of this for women varies between the different professions discussed, but in each case it is clear that women encounter considerable difficulties in their careers as a result of their "deviant" gender'.

It is too simplistic to argue that women pose a threat just by their presence in most organizations. Rather, it is by their claims to equality. Women entering non-traditional occupations equivalently (or better) qualified for professional roles than their male colleagues are seen as a threat. 'Somehow when integration supports men's super-ordinate status there is no objection. Men cheerfully permit women into their workplaces when they are secretaries to their manager roles, and nurses to their doctor roles' (Epstein 1997: 206). On the other hand: 'Women who in their career choices, ambitions and behaviour, seem to challenge gender stereotypes will be perceived as odd, as different (or even deficient) in essential aspects of femininity, caring and relatedness. They will be admired by a few but they will mostly be criticised, by other women as well as men, in their attempts to break new ground' (Evetts 2000: 60).

No Change at Home?

> At the moment women fight for personal improvement with one arm tied behind their back.
>
> (civil service manager)

Women continue to combine wider and heavier responsibilities to their families and home with their paid work demands. This can be conceptualized as 'spillover theory', where there is a reciprocal relationship between paid work and home and family life (Caligiuri and Cascio 1998, cited in Linehan and Walsh 2000). Women are expected to be available for unpaid work and are therefore viewed as deviant in the labour market.

Cockburn (1991) uses the idea of women being 'defined in domesticity'. In this way, men represent women as a problem in the workplace, and 'what is being problematised is women's relation to the domestic sphere' (Cockburn 1991: 76). Whether women have children or not, they are believed to bring 'problems' associated with their reproductive role into the workplace. Their commitment to the male-dominated public sphere, especially non-traditional occupations, is thus constantly open to scrutiny, treated with suspicion, and detracts from the notion that they can be committed professionals. As Cockburn (1991: 76) points out; 'Even if the woman in question is celibate or childless she is seen and represented as one of the maternal sex'.

Women civil service managers divined this assumption in their managers' attitudes, even though it is outlawed by equal opportunities policies and therefore mostly covert.

> *When managers were writing annual reports, they used to ask women if they're having babies or not. They've stopped asking, but attitudes still carry on in their minds. They've not changed just because they don't ask the question.*

Therefore the trend of 'economising' on domestic demands is most pronounced in women in managerial careers and non-traditional professional occupations. As Liff and Ward (2001: 32) show, the old adage 'think manager, think male' has changed with the increasing numbers of women managers, to the new image of 'think female manager, think childless superwoman'.

Looking at the four occupations in this chapter, civil service managers, academics, and engineers have fewer partners and even fewer children compared to other groups of women and their male colleagues. Many women have given up the idea of having husbands/partners and families in order to be successful. The only exceptions are the women priests. The majority are, or have been, married and most of these have children. The women themselves acknowledge that this was because they had a long hard struggle for their right to be ordained, and therefore got on with fulfilment through their family 'careers'.

In the civil service, the encouragement given to some of these women managers by gaining promotion contributes to, and in some cases provokes, their decision not to have children so as not to inhibit their careers.

> *I've made the decision now not to have any children. If my husband had earned more and I'd earned less, we probably would have conformed to a normal pattern.*

Again relatively few of the academics are married or have a partner, and only a third have children. It would appear that women can succeed in higher education, but that they do so at a price.

> *There is a difficulty in combining the conventional role of a woman, family, children, with the expectations of a successful academic. Successful women academics have to demonstrate that they are more committed to the task in hand, which involves no domestic distractions.*

Construction work and particularly site-based roles in nationally based divisions are seen as demanding and time-consuming and to impinge on social activities and family responsibilities.

> I would think twice about getting married because they would see it as a signal that my career was over.

Structural Barriers

> It's very hard to get in with the 'old boys network' when you're not old and you're not a boy!
>
> (Construction engineer)

Institutional structures impede women's success in non-traditional work. This is not to deny women's agency, but this approach is set in opposition to others that cite women's choice as the key element in patterns of work (Hakim 1995). In contrast, it is suggested here that any choices that women make are shaped and to a large extent determined by the structural environment at work. Therefore, 'it is the context which constrains and limits women's achievements, not women themselves or their career strategies' (Pascall et al. 2000: 66).

In construction engineering an initial period of site-based work experience is a prerequisite for eventual progression to senior positions, because it offers significant levels of responsibility at relatively junior levels, which facilitates rapid vertical advancement in careers. However, women are encouraged by senior male managers to enter supporting roles, which tend to be office-based, removed from the production function and widely acknowledged not to afford as many opportunities for rapid advancement.

The accepted and dominant pattern of a career is still built around continuous lifetime service, working hard at all times, making oneself available at all times, working long hours, and with no career breaks. This pressure for a continuous career is seen in three of the non-traditional occupations studied here. Academics are under pressure to ensure a continuous publishing record. Time-consuming family responsibilities distract and divert women from research activities and the traditional research-based career path, often normalized and formalized through the process of performance appraisal (Bagilhole and White 2011). In the civil service, having children holds women back in their career, so that male peers overtake them. In construction, women largely find it impossible to deal with the responsibilities of a family and stay within the profession. The women clergy are the exceptional case, with many having had children before becoming priests, as discussed previously. Any consequences of this for their careers will only be measurable in the future.

However, it is important to point out that, even when women conform to this male, continuous career model, they still encounter problems. Bagilhole and White (2011) find that women academic leaders whose employment patterns are similar to men face problems when trying to progress through organizational career routes and paths. This leads them to conclude that processes within work organizations rather than within families are probably the most crucial influences on women's careers.

RECRUITMENT

The first obstacle that women face with male-dominated occupations is making it through the recruitment process, be it formal or informal. The four occupations under scrutiny in this book are prime examples of Lipman-Blumen's (1976) 'homosocial' institutions, therefore, the rules pertaining to appointment are male-driven and are evaluated according to male standards. Greed (1997) attributes women's under-representation in construction to men blocking the entrance of people and ideas that are seen as different and/or unsettling.

Informal recruitment processes are recognized as one of the major areas where discrimination and prejudice can creep into an organization. The women priests feel very strongly that there is not fair and open recruitment in the Church of England, because of the secrecy and informality of the system.

> No—it's word of mouth. The old boy network prevails and is very powerful. Most jobs are not advertised. Criteria are not disclosed, decisions are made secretly.

Recruitment in the construction industry is often informal through personal contacts (Druker and White 1996). The effectiveness of networking for informal contacts is gender-specific, with many women discussing difficulties they have experienced.

> Recruitment works very informally. I think it's a personality industry and so you are recruited by people you know.

The women construction engineering professionals generally find they experience more rigorous recruitment and selection techniques than their male colleagues. In interviews, they are subjected to an in-depth examination of their commitment and professional competence.

> There is no consistency. Some people get a real grilling, others especially the men seem to waltz in without so much as an informal chat.

There is also a connection between the depth and rigour of the recruitment process and women's ages. Women in their late twenties and early thirties had encountered particular difficulties and many feel that this stems from assumptions that they would have families at this age.

> As the interview went on they began to ask me some very pointed questions about my future plans with regards to having children and getting married. I tried to fend them

off as much as possible, but it was clear that they would not back down until they had an answer on my future plans . . . needless to say that I didn't want children.

CAREER DEVELOPMENT

Career development begins with the induction process of new staff into the organization, which is problematic for women. Women construction professionals complain of inadequate supervision and assistance during this stage.

It's a bit embarrassing when you start on site because you don't want to ask people for help every five minutes. It's easier for men because if they make a mistake everyone on the site isn't going to pick up on it.

For women academics their probationary period is a crucial time for establishment within the profession. Over half of the women academics report problems with the process, involving insufficient and inappropriate advice and support, and confusion over the regulations. Over two-thirds of the established women report that they had problems during probation.

I was extremely unhappy for my first two years. Everyone was very helpful if I went and asked, if you didn't ask you were left to get on with it. I had an enormous teaching load with no time to establish my research. I was not protected by my probationary adviser.

The vast majority of women academics report that they had received no career advice. This is particularly detrimental for women because, in a minority, they lack role models, informal networks, 'sponsors', or mentors.

No, not in a definite way, casually in conversations picked up things. I'm only told to publish more, but I'm given no support or encouragement.

Appraisal processes are seen as the principal mechanism for monitoring performance, assessing training needs, and allocating staff development opportunities. In academia, given the lack of seniority of women, both women and men will more often be appraised by men. For most of the women academics in this study appraisal is a negative experience because they feel they have different perceptions of their job to their male appraisers.

I tried to push forward the fact that, what women do cannot be measured in the same way; caring for students. Measured against male successes it doesn't seem to count as much. Women are involved more with explaining what they're doing and why, because they can't hide behind male authority.

In the construction industry, women feel that appraisal is used by some male managers as a vehicle to undermine women's careers. They give women lower appraisal scores, make unfair assessments of women's training needs, or use appraisal assessments as a way to restrict women's intra-organizational mobility. Also, as women are more likely than their male colleagues to be in support positions, they feel that their performance cannot be fairly assessed as they had less tangible outputs.

> *It's fine if you're a project manager and you have control of your project…but if you're doing business development…how on earth can you appraise what I'm doing?*

Selection and Promotion

There is evidence of the persistence of certain work practices relating to promotion, which are discriminatory towards women (e.g. see Bagilhole and White 2011). In the American clergy, Nesbitt (1997: 596) concludes that by maintaining these discriminatory mechanisms 'men are unlikely to surrender their dominance of religious leadership'. The position in the UK is likely to be even more stark with the controversy about women becoming bishops. The majority of the women priests feel that there is not equity between women and men priests in terms of selection and promotion.

> *No—and it's not a 'glass' ceiling more a lead roof! Women are generally passed over for 'plum' parishes. Sometimes a job is said to be too big for a woman.*

The women civil service managers recognize the phenomenon of women's greater likelihood of leaving the civil service, having contributed very efficient work at a low level. They see this as extremely useful to the working of the system.

> *If women stayed in the Civil Service things would be more equal. Pressures of numbers would force the promotion issue. It's easy for the Civil Service at the moment, when so many women leave and drift off. Therefore, the Civil Service doesn't have to face the problem of their promotion.*

Civil service promotion boards are likely to be male-dominated because they are in a majority in the top grades. One woman found her experiences at promotion boards changed her mind about wanting promotion.

> *I've been for three promotion boards and failed them all. I was offered a fourth, but I turned it down because I'd had enough. I was a career girl but now I'm not particularly ambitious. I find the boards horrifying.*

There is a high level of ambition in the group of women academics; the vast majority want promotion.

> *I hope to rise up the hierarchy of the department and the university. I'm ambitious. I ultimately want a chair. I'll be very disappointed if I don't make senior lectureship.*

However, in contrast to their high aspirations, their expectations are much lower. Only a minority have positive views about their promotion chances. These assessments dovetail with the view of the Association of University Teachers (AUT 1992: 13) 'that the combination of secrecy, subjectivity and amateurism which surround too many promotion procedures, is lethal to the objectives of equal opportunities'.

> *I try to look at it positively but looking at the statistics it's not very hopeful. There is an unofficial agenda. If you publish enough papers you'll get on, even if you fail at teaching students. It helps if you're male. Being a woman is a disadvantage.*

WORKING CONDITIONS

Physical working conditions in certain non-traditional occupations disproportionately impact on women to their disadvantage. The women construction professionals identify sites with no female toilets, and safety clothes and gear that are the wrong size as problems.

> *I find toilet facilities on site are horrendous. You're sitting on the toilet with holes through the ceiling and no lock on the door, and that's the case on nearly every site I've been on.*

The long working hours culture is prevalent in the UK. The construction industry is a prime example because economic considerations usually require the shortest possible construction period, so sites work long hours. Most people and their families find these characteristics difficult to cope with, but they disproportionately affect women. The long hours' culture is seen by many women as being ingrained within the work practices of the industry.

> *When I was working on the tunnel, I was a down there at seven and rarely go out before eight, six days so I saw virtually no daylight for 18 months. That's the way it is though, and it will never change. Either you accept it and work with it or leave, it's as simple as that.*

This is also a problem in terms of the clergy. Charlton (1997: 608) finds 'lack of privacy, lack of differentiation between workplace, community, and home, and emotional labour intensified by blurred boundaries' are often cited by women as reasons for leaving ministry. The hours demanded of them disproportionately impact on women priests.

I have to do home and work. It's complicated by the fact that work is at home too.

GEOGRAPHICAL MOBILITY

The requirement for geographical mobility, either as a condition of doing the job, or for promotion, is a particular problem for women with family responsibilities. For this reason, many academic women report that they have decided not to have a partner. Women priests find it difficult to get parishes geographically compatible with their families, and women civil service managers have given up promotion because of the requirement to move.

The construction industry has excessive mobility requirements. Construction projects are transitory. Construction companies arrive on site, build, and leave for the next project. The project site may be far distant from the headquarters of the construction company, and in the case of major international construction projects, it may well be in another country. Therefore, a significant problem concerns the geographical allocation of staff, which is seen as being carried out with little regard for employees' personal needs. In an attempt to remain geographically stable, several women travel long distances to work.

The hours aren't too bad but I have to travel so far to get here. It adds four hours to my day. It's my choice, and I know it could be worse, but in the end it's worth it to maintain my career, and my relationship.

Women perceive that there are hidden agendas in the allocation of staff and it is felt that companies think that they get more work output from employees working away from home.

I think it's the uncertainty that gets to you, particularly if you have kids. I mean this job finishes in two months and we don't know where we are going next. We have heard on the grapevine that a new job is starting in Leeds, but if I get sent there I don't know what I'll do about child care.

The possible requirement for geographical mobility for promotion is also identified by the women civil service managers. The mobility requirement creates problems for women with family responsibilities because of the necessity of considering their family's willingness or ability to move. Many married women had turned down promotion because of the difficulty with mobility.

It's a problem. I'm waiting for a posting. If I'm offered London I can't move because of my husband's job.

Hostile Reactions: Male Culture

It's a man's world, and if you don't f***ing like it, don't f***ing come in to it.

(Construction engineer)

This section examines the cultural environments of male-dominated occupations that prevent women achieving 'like men'. Work in this area has produced some recurring themes that are all particularly prevalent in male-dominated occupations. They basically relate to men's exclusionary behaviour, where men tend to share information predominantly with other men, recruit in their own image, ostracise Don't understand this – no change and undermine women, and generally act to perpetuate ways of working and forms of interaction with which they feel comfortable. 'Gender cultures' in non-traditional occupations are hierarchical, patriarchal, sex-segregated, sexually divided, sex-stereotyped, sex-discriminatory, sexualized, sexist, misogynist, resistant to change, and encompass gendered power structures. Women in male-dominated occupations become tired of managing gender relations and a highly gendered culture. When women enter occupations that were previously seen as the preserve of men, they encounter hostility, both overt and covert opposition, and resistance to their success.

Recruiting in One's Own Likeness

The 'gendered power system' is another concept used in studying women as a minority in male-dominated workplaces. Powerful individuals (mostly men) can influence or even define values within an organizational context by influencing the recruitment and promotion of those whom they consider to be similar to themselves (Bagilhole and White 2011). Liff and Ward (2001) find that women in banking have problems presenting themselves as plausible candidates for promotion since the dominant organizational model of those that will succeed is strongly sex-typed male. Morgan and Knights (1991) find a rationale amongst male managers as to why women are not good sales representatives, which includes the idea that women will disrupt the 'esprit de corps' based on male gender identity and will not fit comfortably within the ethos and culture of the sales force.

Kanter (1977) in her classic study of managers in a large US corporation finds this culture of homogeneity. Men like working with those they are most sure of, that is, other men. Lipmen-Blumen's Should be Lipman-Blumen's (1976) 'homosocial theory of sex roles' also offers a useful conceptualization of the problem, where men are dominant in sex-segregated institutions and act to exclude women from participation. Academia, the construction industry, civil service management, and the Church of England are prime examples of homosocial institutions, run by men who reward those most like themselves and exclude and marginalize women.

This preference for men similar to themselves is confirmed in the study of women construction engineering professionals

> *A lot of the directors are very fierce men, they like to see that quality in the people following them up. But this is something, which is unlikely to be found in most of the women managers. . . . if I see a potential conflictual situation when I'm negotiating a payment to a sub-contractor, I would rather work a way around the problem. My line manager would rather go in shouting and hoping to force them into submission.*

Men's Networks: Men's Club

Organizations are political cultural systems that promote competition and cooperation simultaneously, and form arenas for the manifestation of the power and interests of their members and only those who understand this organizational power and politics are likely to progress (Kvande and Rasmussen 1994). An essential way of obtaining this understanding is being included in powerful networks, from which women are largely excluded in male-dominated occupations. In universities, women tell of being outsiders in homosocial male territories and the existence of an influential 'old boy's club' (Bagilhole and White 2011). Greed (2000: 183) in looking at the construction industry utilizes the concept of men's 'closure' against women entering or remaining within the industry: 'This is worked out on a daily basis at an interpersonal level. Some people are made to feel awkward, unwelcome and "wrong", others are welcomed into the subculture, and made to feel comfortable and part of the team'.

This woman academic explains the problem for her.

> *Going over to a group of men at a conference and joining in their conversation is difficult. When I start talking to a man, it never occurs to me that he might think I'm making a pass at him, but it has happened and it is embarrassing. I try to be professional with them and not be aware of this, but it's a problem*

This exclusion means that women find it very difficult to accrue the necessary resources to perform valued professional activities. Even when women try to network themselves this is discouraged. The following is an example from the construction industry:

> *A few women got together to attempt to arrange some kind of informal networking meeting, but we were told by the director to stop. It was perceived as a threat.*

Male workers are well aware that important business and contacts are made at social and sporting events. However, many women academics comment on their exclusion from their colleagues' social life.

> *You're barred from meeting them outside the university because they're involved in sport, bars, things I can't participate in.*

In construction, the women's workplace isolation is exacerbated by their deliberate exclusion from social events, which are an effective way of networking within organizations.

I am about 400 miles from home here, and I sometimes feel quite lonely, but not once have I been invited for a drink after work. They even ask each other if they are going for a drink in front of me.

SOCIALIZATION INTO THE PROFESSION

Included in the resources needed for successful occupational performance is the 'indeterminate' knowledge of professional and organizational life. The essential performance skills of a profession are never explicitly taught, but are communicated via male networks. Thus the problem for women is not their human capital: women's investment in technical knowledge and expertise. Rather, the problem is professional contact, style, and legitimacy (Sagebiel and Dahmen 2006)

Another side of socialization into most jobs is the experience gained from doing particular tasks and having particular responsibilities, which is very important for promotion. In construction, women find themselves competing unsuccessfully for the necessary career experience.

I've asked for the opportunity to do some international work but there's a guy there now who does it all. He knows that if I get this experience it will help me get my promotion and that's why he won't let it go, and he is so in with the boss that he is unlikely to make him share it.

Also well over half of the women civil service managers feel that the experience needed for promotion is reserved for their male colleagues.

Men get the jobs which are important for promotion. They try to give most of the men at least a chance at them. This isn't true of the women.

In construction, some women find that they are allocated low-level tasks, or that they are overloaded to the extent that their failure to complete tasks leads colleagues to question their competence.

My site manager asked me if I would temporarily look after the site safety inductions... that was five months ago and I'm still doing it. There are so many other tasks that would better utilize my skills and experience, and I keep telling him that I'm finding it soul destroying, but it's the way that he sees the role for women on site.

WOMEN GIVEN DIFFICULT JOBS

Maddock (1999: 168) finds that women are made vulnerable by being given difficult management roles, and 'generally held in low esteem until they proved otherwise'. Haslem and Ryan (2008) identified the concept of a 'glass cliff' in academia where women only achieve promotion into leadership roles that men do not want. The following woman civil service manager gives her experience of this.

> When I was first in my grade I had to doubly prove myself, especially with the nastier side of the job, for example, disciplining staff. It was as if I was on trial. I didn't realize this until later.

Also, in construction, male project staff are encouraged by managers to test women's commitment and competence.

> I have found that my project manager quite often encourages other managers to test me out a bit, see how I'll take it. They have a dig at me, make a rude remark, try to undermine decisions and things like that.

The stress on women in male-dominated occupations to perform well for herself, for her colleagues, and as a role model is immense. Several priests feel that being a woman marks them out as different and risky, and their visibility opens them up to being judged for their mistakes.

> It marks one out as being different. There is a sense in which we can't be an individual because all women priests are judged by our mistakes.

ROLE MODELS AND MENTORS

Male networks also provide role models and mentors, whose sponsorship enhances men's self-esteem, self-confidence, and careers. Less than a quarter of the women academics have a mentor, although they do feel a need for them and their benefits are appreciated. Success in the academic market place requires a high level of educational attainment, but moving through the system of rewards and status requires knowing colleagues who can provide guidance, support, and advocacy to the apprentice. Male colleagues' greater access to these benefits is perceived.

> They come off the walls at them, don't they? I've no access to the grapevine. They intertwine well and it helps their work.

In construction, most women have no female colleagues working with them on a daily basis, and, if they try to break down gender barriers in an attempt to find a mentor, they experience unfortunate consequences.

I wish that I had someone I could get on really well with informally that would listen to me and get to know me and give me a bit of advice really. It's just some recognition that the way I am doing things is OK and that I am going in the right direction. But I don't know any senior women and the men either don't seem interested, or think I'm after something else!

CHILLY CLIMATE

For women to perform effectively in non-traditional roles, they must form good relationships with their male colleagues, which they can find difficult. Kanter (1977) finds that women entering male-dominated occupations experience 'boundary heightening', where men exaggerate the differences between themselves and the tokens and treat them as outsiders. These effects of tokenism can actually increase as women increase in number, as they present a greater threat to men. Thus men are found to be culturally active in creating an environment in which 'women don't flourish' (Kanter 1977). This can be oblique and subtle. Lorber (1994) identifies the 'Salieri phenomenon' named after the composer who allegedly sabotaged Mozart's career. 'Salieri's damning with faint praise is one way women are undermined by their male colleagues and bosses, often without being aware of it' (Lorber 1994: 237). Also, men generate a masculine technological or managerial culture in and around their work that can make women feel, without being told so in so many words, 'you are out of place here' (Cockburn 1991: 65). This is identified in many studies as the 'chill factor'.

Many women priests have experienced hostility or open opposition to their priesthood. In fact, many report serious and upsetting negative reactions from male colleagues.

I have had rudeness and downright brutish behaviour from male colleagues. I had to be removed from the parish because I was bullied. I have been bawled at down the phone by male clergy who cannot bear the idea of women priests. There are still constant battles—or at least bitter pills to be swallowed.

Nearly half of the women academics report negative relationships with male colleagues.

It's more like non-communication rather than different treatment. The department is male-dominated. They are quite standoffish. They don't tell me the things I need to know. They are pleasant but underneath the knife is going in. You can't do the job properly if you're not given information.

Also, they report feelings of intrusion.

> *They don't know how to react and respond to me. They are boys together and I don't fit in with that. As a woman, it's hard to penetrate that. The language in the university is male. I am made to feel that I am in men's territory. You are always aware of being a woman in this university.*

In construction, women's visibility as minorities in the workplace has led to considerable resentment from their male colleagues when they managed to gain promotion.

> *There was real deep resentment at the fact I had got promoted as a woman. A lot of people thought that they deserved promotion rather than me and lots of nasty rumours went around about why I had got promoted when they hadn't. . . . It made my life very difficult.*

Threat to Masculinity

Men's masculinity is closely related to and constructed through their paid work. Therefore, the question of whether notions of traditional masculinity are challenged or transformed for men when women enter their occupation is raised. Men may be threatened by competent women, whose presence may result in a demystifying of their achievements, and a blurring of the boundaries between masculine and feminine identities. Men resent the intrusion of women, on the grounds that it suppresses normal conversation, and takes away the pleasure of working in an all-male environment.

In construction, men certainly resent women entering their environment.

> *We had been arguing about something to do with work, and no matter how hard I tried I could not get him to compromise. It's because they are frightened, that's all it is, it's fright that their closed world may be changing.*

A woman priest gives an example of unease about and resistance to the numbers of women increasing:

> *There is real anxiety about all women teams running parishes. But there's no problem with all male teams! When a team ministry of eight appointed a third woman, there was a frisson of anxiety—'the women are taking over'.*

Sexual Harassment

Many studies have explored the perpetuation of male domination through the way masculine sexuality is incorporated into organizations via sexual harassment (SH). Studies in the US and UK reveal that from 42 per cent to as many as 88 per cent of women workers experience SH in the workplace, at some time in their lives (Ragins and Scandura 1995). SH is a major issue, particularly for women who enter non-traditional occupations. They will often be more at risk of SH and more aggressive SH than women in traditional forms of female employment (Bagilhole and Woodward 1995). SH is especially pervasive in male-dominated jobs, since it serves as a means for male workers to reassert dominance and control over women colleagues who otherwise would be their equals. 'Emergent and potentially powerful women must be cut down to size by sexual means' (Cockburn 1991: 141).

Also gender and sexuality are closely intertwined in women's experiences of the workplace. Heavily male-dominated work environments offer a site where sexuality is brought into even sharper focus. Women are constantly reminded of their female embodiment and women are often viewed in terms of their sexual availability to men, and may be cast as a lesbian is they are not available (McDowell 1997). Cockburn (1991: 196) observed that lesbianism can be 'used as a category with which to control heterosexual women'.

As a woman civil service manager explains:

> *Unless you are extraordinarily ambitious, the sexism and sexual harassment soon puts you in your place.*

Virtually all the priests have experienced SH.

> *I get mild sexual harassment from colleagues. It's minor, irritating, but very hurtful.*

Also, some have experienced very serious incidents of violence from men in the public as a consequence of their occupation.

> *I've had problems with a stalker, and an incident with indecent exposure including going to a magistrates court case and experiencing an abusive defence lawyer.*

Just over a quarter of the women academics have experienced SH. This ranges across personal remarks, sexual touching, and sexual advances.

> *A senior member of the department is always very 'friendly' he puts his arm around you, his hand on your knee. I don't think he means anything by it but he doesn't do it to his male colleagues. I take evasive action, which is ridiculous.*

Also, a very serious incident is reported, which could be categorized as sexual assault, and includes actions that involve physical force and require resistance.

> *At one Christmas party I was backed into a corner by a male colleague and kept there. He said, 'you're the most frightening female I've ever met'. Other male colleagues joined in saying things like 'I'm not surprised he's frightened of you'. It was a ludicrous situation.*

These incidences occur predominantly with male staff, but also behaviour from some male students, especially in the male-dominated disciplines in engineering and science can be difficult.

> *I stopped wearing short skirts following comments about my legs. I have to give as good as I get, sexual innuendo. I can handle it but some female staff find it more difficult.*

Importantly, a further substantial number of the women academics say they have not experienced any SH. However, they go on to describe incidents which can be classified as such, when they are asked questions about any comments, experiences, or incidents that they feel are linked to being a woman academic, and about their relationship with male colleagues.

> *There are a few members of staff who have a reputation for touching, but they reserve it for students and secretaries. I've made it clear that I won't stand for it. When I first arrived a couple pursued me but they soon cooled off.*

This discourse of personal control implies that they could or should be responsible for men's behaviour, and shaped some women's definitions of what does and does not constitute SH. There is a reluctance to recognize themselves as victims of SH, which reflects their identity as professional women.

In construction, overt SH is found to be common.

> *There were a few guys at my last job that used to get me down. They had these rude calendars, I asked them twice if they would take them down but they refused . . . the language, it's horrendous, the crudeness is completely unnecessary. I have spoken to my line manager. Swear words and that usually don't worry me, it's the crudeness about sexual things. I find that offensive but they still do it in front of me, and they seem to think that it's quite funny too.*

Again, there is a strong sense that the women feel they should be able to handle it.

> *It was difficult on site because they used to bring their nudey calendars in and leave copies of 'Penthouse' on my desk. The way I used to handle it was to turn over the calendar on the first day of every month and say 'right boys what do you reckon?' It was all very 'cor blimey your tits are looking nice today', . . . it's no good as a woman coming in thinking that you are going to change the world, it's never going to happen, you should accept the industry for what it is, and learn to cope with it.*

Despite the evidence from these studies, which shows that severe SH is experienced by many of these women, very few of the incidents are officially reported. These women either feel that, as professional women, they should be able to cope and deal with the situation, or they use avoidance tactics for fear of retaliation. Collinson and Collinson (1996) find women's responses to SH (resistance, integration, indifference, distancing, and denial) all prove to be self-defeating in the asymmetrical gendered power relations and fail to deal effectively with SH. Also, even when women do report and complain about SH very few are satisfied with the outcomes. This example comes from the construction industry.

> *Things were getting very messy on site. This foreman had been harassing me daily for months, so in the end I actually put in an official complaint to my senior engineer. He told me that he had spoken to him, and then told me to let it lie. It was like I got told to keep my mouth shut. I was told that, if the contracts manager got to hear about it, that he would sack the person straight away, but then no one would have anything to do with me again.*

A common feature of women in non-traditional occupations is that they tend to keep a 'low profile' and are reluctant to admit that discrimination has affected them. They experience tension between balancing resistance, on the one hand, with the need to protect professional interests on the other. Male cultures have the effect of silencing women, either because women silence themselves, or because when they attempt to articulate their thinking they are met with disbelief or non-understanding (Maddock 1999).

The following is an example of a woman in construction being silenced.

> *There is a guy on this site who thinks that women shouldn't be allowed in the industry. He says that I have done a good man out of a job. I just stay out of his way. Women that confront that kind of behaviour only make a rod for their own back. I think that you just have to accept that as a woman in this industry you are going to have problems.*

On the other hand, the few women who confront such behaviour attract additional criticism. In male-dominated occupations, there is a 'dominance of 'gender narratives' that ridicule radical and challenging women' (Maddock 1999: 7).

Conclusion

The pressures and constraints that women in non-traditional occupations experience can be identified and examined around three major areas; the continuing over-representation of women in domestic responsibilities and its impact on their perceived professionalism; structural barriers in the way the occupation is performed; and finally hostile cultural environments within the workplace. Within these areas both formal and informal processes mitigate against the equality of women.

There certainly seems to be a spillover in the two-way relationship between unpaid and paid work for women, with the real and assumed impact of unpaid work on paid work having the greatest impact. Assumptions are made about women's 'domesticity' and how this constrains their performance at work. There is an assumption of women's homogeneity, and the primacy they will all give to family responsibilities, even if they do not have one. The mirror image to this is the assumed primacy men will give to their paid work, and thus their families and family responsibilities remain invisible.

It must be acknowledged that there is a very real effect on women of taking on a disproportionate amount of unpaid work. The spillover theory shows how work affects home and home affects work. However, this two-way process is not gender neutral. Women's unpaid work impinges on their paid work limiting their chances, whereas men's paid work impinges on their desire or ability to take on more unpaid work. The roles of wife, or partner, and mother carry their own ideologies for women and also practical problems, which act negatively on their careers, ambitions, and aspirations. Women's perception of the reality of often being required to 'choose' between paid work or a family has led to demographic changes. More women, especially young women, are choosing their careers. The civil service managers, academics, and the engineers have all 'economized' on domestic demands. They have relatively few partners and children, compared to other groups of women and their male colleagues. The exception to this pattern is the women clergy, but they formed their families before they had won the struggle to be ordained, and acknowledge this as a reason for their relatively bigger families.

Structural barriers within the workplace also disadvantage women in non-traditional occupations. We can see that these institutional structures limit women's success from the recruitment processes, both informal and formal, through career development, selection and promotion procedures, and finally into their day-to-day working conditions. These include physical facilities, particularly in the construction industry, the long working hours culture, and importantly the requirement for geographical mobility. These structural constraints are not the only reasons for women's relative lack of success in hierarchical terms in these occupations, but they are very important contributory factors.

Clearly, consistent negative themes emerge in the cultural environment of non-traditional occupations that the women in these 'male territories' have to use energy to deal with. This leaves them less time than their male colleagues to concentrate on their careers. The overall culture is one that is hostile to women's presence and consists of both overt and covert resistance by men. This begins with issues around the identified 'homosociability' of men who work in male-dominated occupations. They prefer to select men like themselves and use the dominance of the male career model to assist. Within the occupations, there are well-established, powerful men's networks that provide men with the necessary tools for their trade and exclude women. Men are thus provided with resources to do the job, including the necessary 'indeterminant' skills and experience, role models and mentors.

In contrast, women are given the less useful jobs in terms of career enhancement, and tested out in difficult tasks and roles. Women's relationships with their male colleagues create a 'chilly climate' in which they find it hard to thrive. Women's entry can provoke emotional feelings of loss of masculinity and fear of loss of occupational prestige for the men. Some men try to regain their perceived loss of status by reasserting their authority through the insidious method of sexual harassment. Women in the studies have not found any effective solutions to this harassment, and do not appear to be assisted by the organizations involved. As professional women, they feel the need to cope themselves and to control the men's behaviour, or they use avoidance tactics for fear of predictable retaliation.

Given the accumulation of these persistent pressures and constraints for women in non-traditional occupations, without greater numbers of women in these occupations, alongside effective and supportive equal opportunities strategies, these women will continue to underachieve.

REFERENCES

AUT (1992). *Sex Discrimination in Universities: Report of an Academic Pay Audit Carried out by the Association of University Teachers' Research Department.* London: AUT.

Bagilhole, B., and Woodward H. (1995). 'An Occupational Hazard Warning: Academic Life Can Seriously Damage your Health. An Investigation of Sexual Harassment of Women Academics in a UK University', *British Journal of Sociology of Education,* 16(1): 37–51.

Bagilhole, B. (2000). 'Through the Upgrading of Skills to Empowerment: A Collaborative Project between East and West for the Career Promotion of Women in India', paper presented at International Women's Conference, 'Women's Status, Vision and Reality, Bridging East and West', New Delhi, 27 Feb.–3 Mar.

Bagilhole, Barbara, and White, Kate (eds) (2011). *Gender, Power and Management: A Cross-Cultural Analysis of Higher Education.* Houndmills: Palgrave Macmillan.

Charlton, J. (1997). 'Clergywomen of the Pioneer Generation: A Longitudinal Study', *Journal for the Scientific Study of Religion,* 36(4): 599–613.

Cockburn, C. (1991). *In the Way of Women.* London: Macmillan.

Collinson, M., and Collinson, D. L. (1996). 'It's Only Dick: The Sexual Harassment of Women Managers in Insurance Sales', *Work, Employment and Society,* 10(1): 29–56.

Connell, R. W. (1993). *Gender and Power.* Cambridge: Polity Press.

Druker, J., and White, G. (1996). *Managing People in Construction.* London: Institute of Personnel and Development.

Epstein, C. F. (1997). 'The Myths and Justifications of Sex Segregation in Higher Education: VMI and the Citadel', *Duke Journal of Gender Law and Policy,* 4: 185–210.

Evetts, J. (2000). 'Analysing Change in Women's Careers: Culture, Structure and Action Dimensions', *Gender, Work and Organization,* 7(1): 57–67.

Gonas, L., and Lehto, A. (1999) 'Segregation of the Labour Market', in European Commission, *Women and Work: Equality between Women and Men,* 55–65. Luxembourg: Office of the Official Publications for the European Communities.

Greed, C. (1997). 'Cultural Change in Construction', in Proceedings 13th Annual ARCOM Conference, Kings College Cambridge, (1 Sept.): 11–21.

Greed, C. (2000). 'Women in the Construction Professions: Achieving Critical Mass', *Gender, Work and Organization,* 7(3): 181–96.

Hakim, C. (1995). 'Five Feminist Myths about Women's Employment', *British Journal of Sociology,* 46(3): 429–55.

Haslem, S. A., and Ryan, M. K. (2008). 'The Road to the Glass Cliff: Differences in the Perceived Suitability of Men and Women for Leadership Positions in Succeeding and Failing Organizations', *Leadership Quarterly,* 19: 530–48.

HMSO (2005). *Jobs for the Girls: The Effect of Occupational Segregation on the Gender Pay Gap. House of Commons Trade and Industry Committee Report of Session 2004–5*. London: The Stationery Office.

Kanter, Rosabeth Moss (1977). *Men and Women of the Corporation*. New York: Basic.

Kvande, E., and Rasmussen, B. (1994). 'Men in Male-Dominated Organizations and their Encounter with Women Intruders', *Scandinavian Journal of Management*, 10(2): 163–74.

Liff, Sonia, and Ward, Kate (2001). 'Distorted Views through the Glass Ceiling: The Construction of Women's Understandings of Promotion and Senior Management Positions', *Gender, Work and Organization*, 8(1): 19–36.

Linehan, Margaret, and Walsh, James S. (2000). 'Work–Family Conflict and the Senior Female International Manager', *British Journal of Management*, 11: 49–58.

Lipman-Blumen, J. (1976). 'Towards a Homosocial Theory of Sex Roles: An Explanation of the Sex-Segregation of Social Institutions', *Signs*, 3: 15–22.

Lorber, Judith (1994). *Paradoxes of Gender*. New Haven: Yale University Press.

McDowell, L. (1997). *Capital Culture, Gender at Work in the City*. Oxford: Blackwell.

Maddock, S. (1999). *Challenging Women: Gender, Culture and Organization*. London: Sage.

Morgan, G., and Knights, D. (1991). 'Gendering Jobs: Corporate Strategy, Managerial Control and the Dynamics of Job Segregation', *Work, Employment and Society*, 5(2): 181–200.

Nesbitt, Paula (1997). 'Clergy Feminization: Controlled Labor or Transformative Change?', *Journal for the Scientific Study of Religion*, 36(4): 585–98.

Pascall, Gillian, Parker, Susan, and Evetts, Julia (2000). 'Women in Banking Careers: A Science of Muddling Through?', *Journal of Gender Studies*, 9(1): 63–73.

Plantenga, J., and Rubery, J. (1999). 'Introduction and Summary of Main Results', in European Commission, *Women and Work: Equality between Women and Men*, 1–10. Luxembourg: Office for Official Publication of the European Communities.

Ragins, B. R., and Scandura, T. A. (1995). 'Antecedents and Work-Related Correlates of Reported Sexual Harassment: An Empirical Investigation of Competing Hypotheses', *Sex Roles*, 32: 429–55.

Reskin, Barbara F., and Roos, Patricia A. (eds) (1990). *Job Queues, Gender Queues: Explaining Women's Inroads into Male Occupations*. Philadelphia: Temple University Press.

Rowbotham, Sheila (1974). *Hidden from History*. London: Pluto.

Sagebiel, F., and Dahmen, J. (2006). 'Masculinities in Organizational Cultures in Engineering Education in Europe: Results of the European Union Project WomEng', *European Journal of Engineering Education*, 31: 14–23.

Spencer, A., and Podmore, D. (1987). 'Introduction', in A. Spencer and D. Podmore (eds), *In a Man's World: Essays on Women in Male-Dominated Profession*, 1–13. London: Tavistock Publications.

Whittock, Margaret (2000). *Feminising the Masculine? Women in Non-Traditional Employment*. Aldershot: Ashgate.

PART IV

....

MASCULINITIES IN ORGANIZATIONS

....

CHAPTER 19

CONTEXTUALIZING MEN, MASCULINITIES, LEADERSHIP, AND MANAGEMENT

*Gender/Intersectionalities,
Local/Transnational, Embodied/Virtual,
Theory/Practice*

JEFF HEARN

INTRODUCTION

IN most workplaces, industries and countries men continue to dominate leadership and management. It is typically with the managerial function that organizational power, decision-making, authority, and leadership formally reside. In that sense, management and managerial leadership are among the primary public domain institutional forms of capitalist, and even non-capitalist organizations, and what is sometimes called patriarchy. In turn, mainstream organizations can often be understood as 'men's organizations', places of 'men's organizing', full of unnamed and usually non-gender-conscious 'men's groups'.

Yet, the taken-for-grantedness of most men's leadership is rife. Men's domination of organizational leadership and management is still not an explicit, gendered topic of concern in mainstream social science, and not even in most critical social science. Most mainstream studies of leadership and management do not seem to notice that they are often talking mainly about men and masculinities: they generally do not gender men. On the other hand, there are now many studies that have made explicit the gendering of men and masculinities in workplaces. Even so, within Critical Studies of Men and Masculinities (CSMM) many researches have underplayed the relevance of work, organizations, leadership, and management for understanding men, men's practices, and men's power. It may be that in the effort to see men and masculinities *differently* within these critical studies the more obvious associations of men and paid, organizational work, and their leadership and management, have been diminished.

In this chapter, I focus mainly on leadership, though connections with management are also considered. This emphasis is partly because there has been more critical attention to the gendering of men and masculinities in management than the more specific area of leadership. Following discussion of, first, organizations, leadership, and management, and, second, gendering and non-gendering, I reflect on my own interest in this area, as a prelude to examining recent developments in CSMM, and then three neglected aspects or absences: gender and intersectionalities; localization and transnationalization; and embodiment and virtualization. The chapter concludes with remarks on the relations of theory and practice.

ORGANIZATIONS, LEADERSHIP, AND MANAGEMENT

Though in some contexts the two words—leadership and management—are used almost interchangeably, or almost so, in English these terms are usually rather clearly distinguishable. Management refers to both those people, managers, named, often formally, as such for managing and organizing the organization, and the wider process of managing, which may involve people, technologies, systems, and social processes in organizations more generally. Leadership is usually used as a more specific term to indicate acts and processes of leading, that is, showing, explicitly or implicitly, some direction and initiative. It is also sometimes used to mean those formally given the task of leading, as in 'the leadership'—even if they, like some managers, do not necessarily show leadership, in the sense of direction or initiative. However, these different emphases between leadership and management are somewhat differently constructed in different languages.[1] Thus any given fixed difference or relation between management and leadership cannot be assumed.

Organizations are sites of men's power and masculinities, and workplace issues such as organizational control, decision-making, remuneration, and culture often reflect and reinforce masculine material-discursive practices. Many ways of being men are formed and constructed in work processes of control, collaboration, innovation, competition, conformity, resistance, homosociality, and contradiction. As managers, corporate leaders, entrepreneurs, innovators, owners, board members, supervisors, team leaders, administrators, trade unionists, workers, and the unemployed, men have been prominent in the formation, development, and change of organizations. Emphasizing paid work as central in many men's identity, status, and power, feminist organizational studies have critiqued how 'most organisations are saturated with masculine values' (Burton 1991: 3).

Looked at in a global context, men still dominate, both numerically and politically, much management and leadership, at least in formal terms, in most social spheres, and especially so at the highest levels. There is growing literature on different aspects of men and masculinities in management (Collinson and Hearn 1996). The many other issues around men, masculinities, leadership, and management include: historical relations of men and management in reproducing patriarchies (Hearn 1992); relations of bureaucracy, men, and masculinities (Bologh 1990); historical transformations in managerial masculinities (Roper 1994); management–labour relations seen as interrelations of masculinities; managerial identity formation processes (Kerfoot and Knights 1993, 1998); masculine models, stereotypes, and symbols in management; men managers' discrimination against women; and the possibility of men's non-oppressive, even pro-feminist, management and leadership. Men, especially in mixed working situations, like other 'members of dominant and status identity groups typically display more aggressive nonverbal behaviors, speak more often, interrupt others more often, state more commands and have more opportunity to influence' (Merrill-Sands *et al.* 2003: 334).[2]

While various masculinities shape leadership and managerial practices, leadership and managerial practices in turn impact on specific masculinities. Routinely, pervasive dominant managerial masculinities and what I call here 'leader masculinities' take the form of different workplace control practices. The diversity of leader masculinities is shaped by different forms and sites of work, varying by, for example, industry, class, and organizational type. In effect there are many ways in which the authority and status of managers can signify 'men', and vice versa.

Non-Gendering and Gendering

As noted, the notions of leadership and management can at times be ambiguous in referring to both embodied leaders and managers, formal and informal, and broader more embedded organizational and social processes, In the case of leadership practice and research, it is perhaps not so surprising that the construction of both leadership and leaders tends to be highly individualistic (Fletcher 2004), as exemplified in the 'great man' theory of leadership (and history)—often listing, through US eyes, such men as Thomas Edison, Benjamin Franklin, and Henry Ford. Accordingly, leadership practice and theory is pervaded by what one might call the *ideology* of leading-ness... *and there shall be leaders!* Critiquing this, Amanda Sinclair (2005: 388) has written:

> The bulk of books [on leadership] are righteous and banal, journal articles offer tediously empirical tests of little consequence. Much writing colludes with the lionization of leadership as a normative performance. Research behaves as if leadership was

gendered and disembodied. The infatuation with transformational and inspiring leadership offers little consolation in its tired references to vision and charisma.

At the same time, much mainstream leadership research has been strongly located within organizational psychology, and also often somewhat separate from organizational and management research, with its stronger framing from institutional and kindred theories. Many psychological models of leadership have promoted parallels between conventional qualities of the male sex role and masculinity, and (the assumption of) masculinist qualities of good or successful leadership, including the will to lead itself (see Hearn 1989a; Calás and Smircich 1993; Ford 2010).

An important way to understand both leadership and management practice, and leadership and management research, in their contemporary and dominant forms, is to see them as specifically historical phenomena. Many models of leadership have strong historical antecedents. This is no clearer than in the frequent associations of leadership and militarism in many cultural traditions. The revered 'good leader' in many societies is he, and occasionally she, who has led the nation, be it in war, as with Churchill, or in national struggle, as with Mandela. Wartime and militarism have also figured prominently in the promotion of research on leadership, especially in terms of selecting (male) leaders under wartime pressures, for example, in the Second World War through the work of the US Navy research on group dynamics, and the UK War Office Selection Boards (WOSBs) and Tavistock Institute.

Despite such obvious genderings, in theory and practice, dominant forms and processes of leadership have long been seen as supposedly either 'gender-neutral' or implicitly and unremarkably men's/male. Much of what men do in and as leadership is *not* seen as related to gender at all. Much of men's leadership practices, in work, negotiations, persuasion, networking, lobbying, pressurizing, and so on is not seen as gendered. They are generally done, perceived, and felt as (if they were) 'normal'; they 'just happen'! Men's leadership practices may often easily be coincide with what is considered and counts as the usual, even the official, way of doing things, as made clear by Patricia Yancey Martin (2001) in her studies of organizational decision-making. Such practices of men are ordinary, mundane; women's are often noteworthy—or worse.

The historical impetus for seeing, analysing, and changing leadership through the gender lens has come mainly from women and the critical assessment of women's position in organizations and management. Gendering leadership involves attending to both gender and sexuality, and also recognizing the relations of women to leadership; stopping the invisibility of women's leadership and management; promoting (in both senses) women's leadership; and problematizing 'leadership' and asking 'whose leadership?' But gendering leadership also means gendering men and leadership, and how different men, masculinities, and men's practices relate both to each other and to leadership. This means addressing the relationality of gender and leadership—not leadership as a separate reified 'thing'! Discourses on leadership are still typically understood to involve 'core elements of masculinity', thus sustaining asymmetrical gender relations (Ford 2006).

A Personal-Political Interlude

My own interest in leadership goes back to teaching and researching on gender and organizations from the late 1970s, and realizing, working with Wendy Parkin, the lacuna around leadership even within gendered studies of work (Hearn and Parkin 1987, 1988). Then, in 1988 I was asked by David Butcher to edit a special issue of *Leadership and Organization Development Journal* on men, masculinities, and leadership. At the time, few gender scholars were explicitly addressing issues of men and leadership, and few leadership scholars were explicitly gendering men. Over the next year a journal issue of five papers was assembled. The issue was accepted and ready for publishing, but by then David had stopped being journal editor and another editor was in place. The final manuscript was forwarded to him, and then I had a very unpleasant phone call: he thought it dealt with 'moral' rather than 'scientific' questions; and he wouldn't publish it. As he was now the editor, he had control of it, and simply *he didn't like it*. I explained this was contrary to the agreement with the previous editor, and whilst I was of course interested in his views, that was not the question in terms of the professional and academic (moral?) matters in hand. The call ended badly! I spoke to David and, thanks to him and the publisher, a solution was found. The publishing house had various other journals in its stable, and another, *Equal Opportunities International*, simply accepted and published the whole issue complete and unamended.[3] So this was how I came to edit in 1989 a special issue of *Equal Opportunities International* on the theme of 'Men, Masculinities and Leadership' (Hearn 1989b).

In the special issue a number of key themes were highlighted including: relations of men and leadership within patriarchy; styles of 'masculine' leadership, and followers' perceptions of masculinity and leadership; men's organizational cultures; the predominance of men-/masculine-coded language and images in leadership, for example, the use of military or sporting metaphors; masculine psychodynamics in leadership; and the relation of theory and practice (Hearn, 1989a). In addition, another broad issue was how some well-established models of leadership can be interpreted as forms of men's leadership following Weber's (1964) distinctions between traditional authority, charismatic authority, and bureaucratic authority: These three forms can in turn be translated to men's leadership embedded in cultural domination, such as symbolic father-figures or patriarchal dynasties; grounded in individual qualities and appearances; and based in rationalistic organization. Many of these points still apply.

So, what has changed over the last twenty-five years? To what extent has the relationship of men and leadership, in practice and theory, changed (if at all?)? How are we to go beyond the obvious links of men, masculinities, and leadership? I see two sets of developments with strong implications for understanding men, masculinities, and leadership: first, the major growth of CSMM; and, second, major social changes, for example, globalization, and the spread of information and communication technologies (ICTs), and associated conceptualizations, around, for example, intersectionality, human–machine interfaces. Such contextualizing aspects now seem much more important than twenty-five years ago. I will now address these two broad developments in turn.

Critical Studies on Men and Masculinities

The considerable growth of CSMM over recent decades has a number of implications for how men and leadership might now be seen and understood. Some of the main features of CSMM can be summarized as:

- a *specific,* rather than an implicit or incidental, *focus* on the topic of men and masculinities;
- taking account of *feminist, gay, and other critical gender scholarship*;
- attending to the *explicit gendering* of men and masculinities;
- understanding men and masculinities as *socially constructed, produced, and reproduced* rather than as just 'naturally' one way or another;
- seeing men and masculinities as *variable and changing* across time (history) and space (culture), within societies, and through life courses and biographies;
- emphasizing men's relations, albeit differentially, to *gendered power*;
- spanning both *the material and the discursive* in analysis;
- interrogating the *intersections of gender with other social divisions* in the construction of men and masculinities (Connell *et al.* 2005: 3).

CSMM can be summarised as *historical, cultural, relational, materialist, deconstructive, anti-essentialist* studies on men. One of the most important set of influences on this body of work has come from what may be called 'masculinities theory', especially that as propounded and developed by Raewyn Connell and colleagues (Carrigan *et al.* 1985; Connell 1995). My own synthesis of some of the main features of this approach is:

- critique of sex role theory;
- the use of a power-laden concept of masculinities;
- emphasis on men's unequal relations to men, as well as men's relations to women;
- attention to the implications of gay scholarship and sexual hierarchies;
- distinguishing between hegemonic, complicit, subordinated, and marginalized (and sometimes other) masculinities;
- highlighti4ng of contradictions, and at times resistance(s);
- the analysis of the institutional/social, interpersonal and intrapsychic (psychodynamics) aspects of masculinities; and
- explorations of transformations and social change.

While Gramscian theory has been a major force framing this theory, other influences include socialist feminism, critique of gender categoricalism, pluralism, intersectionality, practice theory, structuration theory, and psychodynamics. This may explain both its diverse appeal and its variable interpretation (Hearn 2004). In *Masculinities*, Connell (1995) discusses hegemonic masculinity in terms of the link with analysis of economic class relations through the operation of cultural dynamics, and notes hegemonic masculinity is open to challenge and possible change. In that book hegemonic masculinity is defined as:

> the configuration of gender practice which embodies the currently accepted answer to the problem of legitimacy of patriarchy, which guarantees (or is taken to guarantee) the dominant position of men and the subordination of women.
>
> (Connell 1995: 77)

At times, hegemony, domination, complicity, subordination and marginalization are referred to as social processes; at other times, hegemonic masculinity is described as a 'configuration of gender practice'. Whilst both of these approaches are not easily compatible with seeing hegemonic masculinity simply as a specific *type* of masculinity, others have often used the term in this way. Vigorous debates fill this work with many different strands and emphases.

So what has been learnt from CSMM, and what are the implications of such studies for debates on men and leadership? How do this approach and its critiques translate into the question of (men's) leadership? Perhaps a first implication is the problematization of the well-established sex differences and sex role approaches to leadership. In many such studies, men are (found to be) more likely to adopt 'transactional' leadership styles, exchanging rewards or punishments for performance, and using power from their organizational position/formal authority, and less likely to adopt person-oriented, transformational leadership (Eagly and Johannesen-Schmidt 2001). On the other hand, in some studies little difference is found (Boulgarides 1984) as part of gender convergence, perhaps through common organizational experiences; and in some situations some women managers and leaders may be more achievement-oriented than both men and other women (Donnell and Hall 1980). Importantly, in most sex differences and sex role approaches there is little sense of the political and social contingency of men, masculinities, and leadership. Moreover, finding sex differences (or not) does not explain gender power relations; for that, there is a need to look to wider networks, relations, structures, as emphasized in CSMM.

Despite well recognized methodological problems with sex/gender differences and sex role approaches (Eichler 1980; Carrigan *et al.* 1985), they continue to live on, especially in public, managerial, and media discourses. They are especially popular in some popular business, governmental, and international debates. They tend to reduce the problem of gender to variations in behaviour, not power and resources. Sex differences and sex role approaches are not only about styles of doing leadership, but also gendered constructions of leadership. For example, men are reported as less likely than women to construe leadership in transformational terms, and to describe their own leadership as transformational, and moreover may be less likely to be described by others as transformational. Rather they are more likely to be described as 'laissez-faire' or in terms of management-by-exception (Alimo-Metcalfe and Alban-Metcalfe 2005). This raises a deep and complex question around what it means to label leadership (or other activities) as masculine or feminine, male or female. There is an iterative and self-reproducing gendered process here by which labelling leadership as such in turn solidifies what the masculine and the feminine might mean (Eichler 1980; Clatterbaugh 1998).

A second set of implications concerns how different versions of men's leadership link with different masculinities. At least some forms of men's leadership can be interpreted within the frame of hegemonic masculinity. This is most obvious in relation to those models of (men's) leadership defined in terms of individualism, toughness, heroism, and decisiveness. However, in doing this, there are dangers of returning to a modified (male sex) role theory; there is a need to move beyond analysis of men and leadership in terms of leadership styles.

At the same time, masculinities theory and theories of hegemonic masculinity have been subject to an increasing range of critiques. First, there is often a lack of clarity in the very concept of masculinity. What does 'masculinity' mean? Does hegemonic masculinity refer to cultural representations, everyday practices, or institutional structures? Can it be reduced to fixed set of practices? How does masculinity, hegemonic or otherwise, address the complexities of everyday inculcation and resistance? Second, there are detailed empirical studies of how men behave and men talk about themselves that complicate or contradict some of this theory (e.g. Wetherell and Edley 1999). Third, there are more general theoretical critiques: historical, poststructuralist, postcolonial, Gramscian (e.g. Howson 2006).[4]

These various critiques raise further sets of implications for the analysis of the relations of men and leadership. One problem is that the concept of hegemonic masculinity is often used very loosely and variably, sometimes as a form of masculinity, sometimes as more general social processes. In this context, it is unclear whether it is leadership through force or leadership through control of resources or by example or by persuasion that would be hegemonic (or not). In considering the relationship of hegemonic masculinity (and other masculinities) to leadership, one may ask: how do various *dominant/dominating* forms of leadership interconnect with each other? How different forms of men's leadership might link with each other is difficult to specify. In one sense it may appear to suggest that men's dominant, dominating, or heroic leadership is hegemonic. But we may question: are dominant images of men's leadership part of or illustrative of hegemonic masculinity or not? Not all versions of men's dominant or dominating leadership rely on heroism. To put this differently: is men's heroism in leadership or men's heroic leadership really so hegemonic in many contemporary organizations? Indeed, why is heroism 'hegemonic' at all? So is there a case for seeing tendencies in men's contemporary leadership towards the post-heroic? Is post-heroic leadership, emphasizing the development of subordinates' capabilities, becoming hegemonic, especially within distributed, project, and network organizing? Thus, it is important to go beyond what appear obvious connections.

Relating hegemonic leadership to heroism is just one aspect—it is possible to change men's heroic leadership and for men to continue dominating organizations and management. There are many different ways of being a 'successful', or dominant, man leader. To draw on an earlier analysis of men and management (Collinson and Hearn 1994), some versions of men's leadership are strong on detail, and might be considered by some to be formal, even boring and pedantic: *bureaucratic* (cf. Weber 1964). In contrast, some are entrepreneurial in style: *entrepreneurialism*; others are jolly and chummy: *informalism*; and some, perhaps many, are self-serving: *careerism*. Thus, it is important not to see men's leadership or indeed 'masculine leadership' as one thing. It should not be reified. Pronounced variations in men's leadership masculinities are also apparent in the public political sphere, with such dramatically different and more, or indeed less, 'successful' styles as those of Berlusconi, Blair, Brown, and the Bushes (Messerschmidt 2010). Indeed, much of men's leadership is mundane, and not so dramatic. Moreover, some forms of men's leadership might exemplify complicit or even subordinated masculinity.

Some of these critiques also move focus from masculinity or specific masculinities to men. Masculinities may change, as in the examples above, but the power of men less so. Indeed the very 'flexibility' of masculinity and elusiveness of patriarchy is part of men's power in leadership and management. To address this means going beyond the hegemony of hegemonic masculinity (as indeed often used outside of its Gramscian frame) to examine the hegemony of men: the simultaneous construction of men, individually, collectively, and as a social category, and men's hegemonic domination in the gender order (Hearn 2004).

And yet despite all these important questions around the relations of men, masculinities, and leadership, they still rarely figure in most contemporary scientific debates. For example, the progressive Sage journal *Leadership* has since its inception in 2005 produced 31 issues and 157 scientific papers, of which ten are broadly on women/gender, and nine briefly mention women/gender. Only one of these explicitly addresses in any detail the gendering of 'masculine leadership', as a point of relation to a psychosocial analysis of one woman's leadership (Ford 2010). This is even though the journal presents itself as:

> the leading scholarly journal in the field of researching leadership studies, at the cutting edge of the theory and practice of leadership and organization. The main emphasis in the journal is on interdisciplinary, diverse and critical analyses of leadership processes in contemporary organizations. *Leadership* encourages new ways of researching and conceptualizing leadership.[5]

The continuing dominance of men in leadership and leadership 'roles' appears 'normal' or 'natural' and largely escapes critical analysis or commentary, even in progressive forums.

PUTTING 'MEN, LEADERSHIP AND MANAGEMENT' IN CONTEXT

The second set of issues which now seem more important than twenty-five years ago concern social change and social contextualizations. Key themes here are globalization, transnationalization, intersectionality, embodiment, and virtuality. With these issues in mind, there is the need to go beyond the obvious in analysing men, masculinities, and leadership. Instead, 'men and leadership' needs to placed into several complex contexts, and within them be spoken of beyond binaries.

Gender/Intersectionalities

The first element of contextualization that now seems much more significant concerns the relations of gender and intersectionalities: the intersections of multiple social divisions, of gender with age, class, ethnicity, and sexuality, amongst other social divisions and differences. This applies in terms of specifying such categories as 'middle-aged men leaders', 'white men leaders', 'middle class men leaders', 'heterosexual men leaders', as well as more novel notions such as 'father managers' (Hearn and Niemistö 2012). These intersections are now more obvious than twenty-five years ago. Managers and leaders can no longer be seen or analysed as simply and only men or women.

Let us go back a step. Initially, much critical research on men and masculinities in organizations tended to concentrate on those in subordinate positions, particularly working-class and manual workers, and to some extent their relations with men and women in middle-class positions (Cockburn 1983; Collinson 1992; Morgan 2005). In due course, further emphasis has been developed on other differences in class, status, hierarchy, function, and occupation, and on sexuality, age, disabilities, and other divisions. Those in subordinate positions are often made 'other'; they are initially recognized as socially constructed in an explicit relation to the social division in question (Hearn and Collinson 2006). This has been so for black and minority ethnic men. There is a large explicitly focused research and policy literature on such men compared with that which is explicitly focused on ethnic majority or white men. In many ways this might suggest a turn from the more visible hegemonic to the less visible complicit leader masculinity.

As with heterosexual masculinities, whiteness and white masculinities are characteristically left invisible, taken-for-granted, assumed but unstated, as absent but present privilege. Future research needs to analyse and critique white leader masculinities, and their intersections, such as age, class, nationality, religion—in effect, to deconstruct the dominant (Hearn 1996). Postcolonialism in its many forms challenges unified white Western male positionality. Accordingly, categories of white men and WHAM (white, heterosexual able-bodied men) need to be part of analysis (Hearn and Collinson 1994). Alongside this argument, black, minority ethnic, and non-white masculinities also need to be critically examined; oppressions recur in multiple ways.

The intersection of racial and ethnic disadvantage and diversity with gender and masculinities produces complex dynamics. Men's experience of ethnic and racial subordination may contradict their gender advantage relative to women of similar ethnicity or racialization (McGuire and Reskin 1993; Edmondson Bell and Nkomo 2001). Men's experience of social subordination may challenge forms of masculinity that assume established claims of privilege and advantage (Eveline 1994; Pease 2010). Similar contradictions may persist in workplaces in the relations of minority ethnic men and black men to white and ethnic majority women, who may indeed be supervisors, managers, or leaders. At the same time, there are gradually growing numbers of ethnic minority men, black men, and men of colour entering supervision and management, especially lower and middle management in some sectors and countries. In the UK, for example, men of Chinese and Indian ethnic origin are more likely than ethnic white men to be in managerial and professional positions (EHRC 2011: 427).

In many parts of the world, such as in most of Africa and Asia, it is non-white men who lead governments, if less so business organizations. At times, and in some locations, ethnic or racialized transformations of men leaders and leadership are relatively swift, as in South Africa, albeit after long struggle. Often changes in such leadership are more immediate amongst political leaders than amongst business leaders. All such relative marginalizations by ethnicity and racialization may be compounded by class, nationality, language, religion, and other divisions and oppressions. The issue of gender/intersectionalities is in turn closely linked to the relations of the local/transnational.

Local/Transnational

Much, probably most, research on leadership has been national in its purview or context. Even critical gender scholarship has been characterized by 'methodological nationalism' (Chernilo 2006). Many studies of leadership still have a US, Anglophone, or Western bias, and are highly normative in orientation. Questions of globalization, transnationalization, and postcolonialism are becoming increasingly important in contemporary debate. These major social processes cannot be ignored in developing more complex analyses of men and leadership. Men's leadership and leader masculinities exist within, and constitute, various forms of culture: team/group cultures, work cultures, leadership, organizational and management cultures, national cultures, regional cultures, and transnational cultures. Such various cultures are contested and complex, and frequently hybrid, rather than pre-existing fixed contexts in which men work and live. Leader masculinities may also be shaped by national and regional cultures.

With greater transnationalization of organizations, dominant constructions of men and masculinities are likely to impact even more widely on forms of transnational leader masculinity. This is clear from even cursory consideration of international organizations, international human resource management (IHRM), or international politics. There are an immense number of issues around men and masculinities to be explored in the field of international management and leadership, such as men's homosocial relations, mergers and acquisitions, and supply chains, even if this is very far from obvious in mainstream IHRM (Hearn *et al.* 2012).

Raewyn Connell (1998) has posited a form of 'transnational business masculinity' that may be increasingly hegemonic and directly connected to patterns of world trade and communication dominated by the West and the global 'North', as opposed to the 'South' (also see Connell and Wood 2005). This masculinity is marked by egocentrism, precarious and conditional forms of loyalty to employers, and a declining sense of responsibility. It differs from traditional bourgeois masculinity in its libertarian sexuality and tendency to commodify relations with women. This pattern, however, represents only one of several versions of senior managerial men's practices. It is premature to see this as a general form, and indeed detailed empirical work has shown considerable variation in different national and transnational locations in how corporate leaders live their lives; for example, some appear to adopt much more conventional marriage-type social relations and lifestyles (Hearn *et al.* 2008), the form of which vary between different national contexts (Reis 2004). Furthermore, leaders of multinational corporations can still remain intensely national in identification (Tienari *et al.* 2010), thus creating possibilities for hybrid forms of local/transnational leadership.

In seeking to make sense of these overarching yet variable transnational patterns I have found it useful to develop the concept of trans(national)patriarchies or simply transpatriarchies (Hearn 2009). Men's transnational patriarchal leadership occurs in business, militarism, international relations, and transgovernmental institutions, in relation to transport, energy, environment, and ICTs. Transpatriarchies encompass the gendered transnational or multinational corporation (Hearn and Louvrier 2011), within and between which men routinely lead, organize, and manage. Many large multinational enterprises are very much men's arenas, with clear structural gendered hierarchies, even with clear attempts by some companies to recruit more women into management, especially at middle levels. Such transnational organizations provide definite hierarchical spaces for men and women, often wholly or largely taken-for-granted. (Transnational) taken-for-granted gendered hierarchy is usually a major aspect of men's leadership in large multinational corporations. While the extent to which this is made explicit or problematic varies, there are various ways of referring to the domination of managerial work, such as 'top management', 'the management team', and 'headquarters'. Here, men managers', and the relatively few women top managers', power is maintained partly through commonalities. Yet transnational management also differentiates men, between managers and non-managers, and between different types of leaders and managers. In these modes of organizing, men's leadership can be seen as forms of gendered/intersectional local/transnational practice.

A further aspect of gendering of the gendered multinational corporation, as organizational transpatriarchies, is men's relations to the local domestic home and the transnational work. Complex interrelations may develop between personal, marriage, and family relations, gendered careers, transnational managerial leadership, and transnational mobility. In a recent Finnish study of forty senior managers in large transnational companies, women and men senior managers inhabited very different domestic social worlds. Men's leadership was usually founded on very traditional marriage-type relations, unlike the women managers (Hearn *et al.* 2008, 2009). The majority of the men lived in very traditional social worlds; of twenty men managers interviewed, one had a woman partner in a comparable management career; half lived with 'housewives'; all were fathers, and with twice as many children as the equivalent women managers. This small-scale study fits with wider social patterns in some countries whereby better off, even higher educated, men tend to have more children, while better off, and even higher educated, women tend to have less children. In this situation, the notion of 'father managers' may be useful to make sense of such men's lives and leadership (Hearn and Niemistö 2012). Transnationalization can further accentuate such gendered processes.

Of twenty women managers interviewed, half had men partners in comparable high-level management careers, often reported as relatively supportive and companionate husbands. There were more traditional expatriate assignments for men than women managers. This appeared relevant to the promotion of varied, yet core, business-oriented careers for men. If the men took practical responsibilities within the family, they were often reported as 'helpers' rather than active adults responsible for everyday life. Such 'duties' were usually minor tasks connected to 'quality time', rather than routine, tiring, or monotonous everyday family obligations. Of the twelve men who had worked abroad (their own words), ten 'did nothing at home', one 'something', one was unclear. A typical interview extract follows:

Q: Did your wife work or stay at home while you were abroad?
A: She was at home for 10 years. When we returned to [Nordic country] and then later to Finland she was ready to start work again... She has the same education as I have. *For us as a family, it was an excellent solution that she could stay at home and take care of the children and home.* (My emphasis)

With growing moves towards globalization and transnationalization, the amount of transnational leadership and management in their various forms are likely to increase, with the associated complex local/transnational lives and leadership. Just as the neologism 'glocalization' has been coined to address how globalization can itself be local in form and facilitate localization (Robertson 1995), so the local/transnational demands a parallel concept, perhaps suggesting 'trocal', and thus men's 'trocal' leadership and management.

Embodied/Virtual

A third matter that now deserves closer attention is the relation of embodiment and virtuality, not least through the spread of the use of ICTs in doing, experiencing, and being affected by leadership. There is a growing amount of research on both men's embodiment (Hearn 2012) and virtual leadership (Boje and Rhodes 2005; Zimmerman *et al.* 2008), but little on men's embodied/virtual leadership practices.

An embodied perspective on men's leadership fits well with the notions of homosociality in organizations (men's greater valuation of men, and preference for men and men's company, rather than women and women's company), and cultural cloning (the tendency to reproduce more of the same leaders, by gender, ethnicity, organizational culture, tradition, and so on) (Essed and Goldberg 2002). This may, for example, be enacted through the valuation of sport in developing and maintaining organizational leadership (Knoppers 2011), as in Obama's high-profile all-male basketball and golf sessions.[6]

As Lipman-Blumen (1976) put it, 'men are attracted to, stimulated by, and interested in other men'; '[m]en can and commonly do seek satisfaction for most of their needs from other men'. Homosociality also involves power between men: information between men; emotional 'socio-erotic' charge between men; and even dispensability of individual men, but affirmation of men more generally. An example of foregrounding the body in leadership was Roper's (1996) study of homosocial desire, admiration, emulation between men in management, emphasizing the importance of dress, clothing, hair, style, and pose. As one manager put it, 'he was a joy to watch'. This is in contrast to the so-called 'missing body' in leadership research (Ropo and Sauer 2008; also see Kerfoot and Knights 1996; Sinclair 1998; Jeanes *et al.* 2011: s. 2). Such ('heterosexual') homosociality can sometimes go hand in hand with heterosexism, and homophobia as a consequence of underlying homoerotic desire (Kimmel 1993). Additionally, perhaps confusingly for some, 'more feminine', personalistic leadership by men towards and between men might sometimes be (reinterpreted as) a form of male bonding.

Such embodied approaches to men's leadership are themselves complicated by relations, sometimes contradictions, with virtual leadership practices, and the paradoxical play of the embodiment of the virtual, and the virtuality of embodiment. The spread of ICTs has brought many qualitative changes in social life: time/space compression, instantaneousness, asynchronicity, reproducibility of images, creation of virtual bodies, blurring of 'real' and 'representational', wireless portability, globalized connectivity, and personalization. These 'smart technologies' (phones, networks, grids) have numerous implications not only for 'smart economic growth',[7] but also 'smart masculinities' and 'smart leadership'. Quintessentially 'smart' technological leader masculinity has been that personified by the founder and former CEO of Apple, Steve Jobs, combining managerial leadership, capitalist accumulation, personal charisma, technological aesthetics, virtuality, visuality, and simplicity of soundbites—in particular, *buy it!* (Gallo 2010; Isaacson 2011; Sharma and Grant 2011, for non-gendered accounts). We might see in such leadership performances a form of leadership as embodied love of technology, a 'natural', 'harmonious' prosthetics of men/masculinity, making a leadership transcending the embodied/virtual. The real and the virtual may coincide.

This is not only a matter for world leaders in the heady world of ICT capitalism. Some while ago I witnessed a slightly different kind of 'smart' leaderly performance in a keynote conference presentation by a senior science leader. Given without notes or overheads, he referred to his very few research sources but without naming any of their authors or referencing any whatsoever—the message was simply *trust me*—whilst at the same time repeatedly urging 'smart' decision-making. The result was a strange embodied/disembodied mixture of male preacher, consultant, and avatar. These new uses of the word 'smart' seems to connote cleverness and ease, along with success and positivity (Collinson 2012), without conflict, resistance, mess, or social divisions, and all presumably with the subtle, behind-the-scenes facilitation of ICTs. In the world of soundbites, such performative style appears an increasingly important feature of leadership, in both academia and business.

Another arena for the intersections of embodiment and virtuality, and a fast-changing one, is what might be called geek leadership. The male nerd can be both asocial, apparently disembodied, and antithetic to dominant masculine leadership, a serving follower and non-leader, sometimes non-masculinized, even feminized, and yet also represent new forms of male embodied techno-homosociality (Bell 2013; see Glen 2003; Nugent 2008). Changing forms of men's embodied/virtual leadership are developing in online blogging, webfora, chatrooms, online meetings, interactive online education, user-generated content, anonymous action, and so on (e.g. Kendall 2002; Olson 2012), with their potential for homosociality, competition, 'total' control by the online 'host', as well as gender switching and female masculinity.

The neglect of men's embodied/virtual leadership in research is perhaps surprising, as the theme of human-machine (or 'man-machine') interaction is a well-worn track within organization studies, at least from 1950s studies of socio-technical systems (Trist and Bamforth 1951), and subsequently Actor Network Theory, and Science and Technology Studies. These latter perspectives have been taken up in critical work on men, masculinities, technology, and work, notably in studies of machinic masculinities and technomasculinities (Mellström 2004; Poster 2013). There is here the basis for far-reaching embodied/virtual analyses of men, masculinities, leadership, and management from the perspective of ICTs and other technologies. This is likely to be an important area for future research on leadership, and men and leadership.

CONCLUDING REMARKS: ON THEORY/PRACTICE

Finally, there is the irresistible relation of theory and practice. This was an important aspect of the 1989 *EOI* special issue. More complex issues of praxis in hegemony, in both work organizations and research practice, may appear. There is debate on what is to count as theory, and the gendering of men in theory. Similarly, we may ask: what is to count as practice? There is widespread support for transformational leadership, but the question remains for men—transforming to and for what? For (pro)feminist leadership, this means men supporting women; men changing men against sexism, patriarchy, oppression, pay inequity; challenging what gender and sexuality seem; changing the *quantity* of men leaders, the *quality* of men's leadership, and the *content* of men's leadership—a possible virtuous circle!

To analyse and change the relations of men and leadership in theory and practice involves addressing resistances, responsibilities, and reaching out to other men, for example, in gender equality policy and research practice. Men's *resistance* comes for many reasons: patriarchal practices, sexism, maintenance of power, complicity in current arrangements, preference for men's company. *Responsibilities* of men for involvement in gender equality range across work, home, sexuality, violence, education, health, sport and so on. *Reaching out* refers to contact with men, individually and collectively, who may be uninterested, even hostile, around these issues, and taking these questions into men's arenas—management, trade unions, workgroups, and elsewhere. Men leaders need to look critically at themselves, not only as workers or managers, but also through gendered/intersectional eyes. Within these arenas there are many 'gender-non-conscious' and some 'gender conscious' forms of organizing and organization. Men's 'gender consciousness' in leadership varies from reproducing and advancing men's privilege to occasionally opposing such privilege to emphasizing men's differences from each other (cf. Messner 1997; Egeberg Holmgren and Hearn 2008).

In making sense of these complex processes, there is a need to deconstruct leadership, management, and indeed men and gender. Mainstream leadership and management, typically presented as 'gender-neutral', remain structurally predominantly forms of men's practices. Seeing 'men and leadership' as a specific arena of activity is rare. The 'Man problem' remains obscure(d), partly because so much leadership is about men, yet is not recognized as such. To move beyond some of the more obvious links of leadership, men, and masculinities, there is a need to attend to both debates within CSMM, and the major contextualizing questions of gender/intersectionalities, the local/transnational, and the embodied/virtual—and beyond the binaries so easily suggested. Increasingly, men's leadership is likely to be, can become, and can be seen as simultaneously gendered/intersectional, local/transnational, embodied/virtual, theory/practice.

NOTES

1. E.g. 'In Finnish, the word often used corresponding to "management" is a generic term "johtaminen", referring to both leading and managing; however, the exact meaning of the term compared to the English term remains somewhat obscure. A corresponding more accurate translation of "management" into Finnish is reached by referring to "management of issues", "asioiden johtaminen". "Leadership" in Finnish is quite unproblematically translated as "johtajuus", also often referring to the leading of people ...'. (Husu *et al.* 2011: p. vi).

2. Citing Sessa and Jackson 1995; Canney Davison and Ward 1999.

3. The issue comprised an introductory article (Hearn 1989*a*), and further papers by Peter Bluckert, Brian McAndrew, Sean Ruth, and Michael Simmons, all of whom were involved in leadership development.

4. In partial response Connell and Messerschmidt (2005) have reviewed some, but only some, of the critiques of the hegemonic masculinity framework. They argue that too simple a model of global gender dominance should be rejected, and instead suggest reformulations of hegemonic masculinity towards more holistic understanding of gender hierarchy; geography of masculinities; social embodiment; and dynamics of masculinities, including contestation and democratization.

5. There have been conspicuous missed opportunities, including several articles on men leaders, such as that titled 'Leadership in Crisis: "Events, My Dear Boy, Events" ' (Mabey

and Morrell 2011), and another on the 2007–9 Scottish banking crisis that is all about men, without noting this (Kerr and Robinson 2011).

6. I am grateful to Michele Gregory for bringing this to my attention.

7. Significantly, one of the pillars of the European Commission's Horizon 2020 strategy is 'smart economic growth' (EC 2010).

References

Alimo-Metcalfe, B., and Alban-Metcalfe, J. (2005). 'Leadership: Time for a New Direction?', *Leadership*, 1(1): 51–71.

Bell, D. (2013). 'Geek Myths: Technologies, Masculinities, Globalizations', in J. Hearn, M. Blagojević and K. Harrison (eds), *Rethinking Transnational Men: Beyond, Between and Within Nations*, 76–90. New York: Routledge.

Boje, D. M., and Rhodes, C. (2005). 'The Virtual Leader Construct: The Mass Mediatization and Simulation of Transformational Leadership', *Leadership*, 1(4): 407–28.

Bologh, R. W. (1990). *Love or Greatness? Max Weber and Masculine Thinking—A Feminist Inquiry*. London: Unwin Hyman.

Boulgarides, J. D. (1984). 'A Comparison of Male and Female Business: Managers', *Journal of Leadership and Organization Development*, 5(5): 27–31.

Burton, C. (1991). *The Promise and the Price*. Sydney: Allen & Unwin.

Calás, L., and Smircich, L. (1993). 'Dangerous Liaisons: The Feminine-in-Management Meets Globalization', *Business Horizons*, (Mar.–Apr.): 73–83.

Canney Davison, S., and Ward, K. (1999). *Leading International Teams*. London: McGraw-Hill.

Carrigan, T., Connell, R., and Lee, J. (1985). 'Towards a New Sociology of Masculinity', *Theory and Society*, 14(5): 551–604.

Chernilo, D. (2006). 'Social Theory's Methodological Nationalism: Myth and Reality', *European Journal of Social Theory*, 9(1): 5–22.

Clatterbaugh, K. (1998). 'What is Problematic about Masculinities?', *Men and Masculinities*, 1(1): 24–45.

Cockburn, C. (1983). *Brothers: Male Dominance and Technological Change*. London: Pluto.

Collinson, D. L. (1992). *Managing the Shopfloor: Subjectivity, Masculinity and Workplace Culture*. Berlin: Walter de Gruyter.

Collinson, D. L. (2012). 'Prozac Leadership and the Limits of Positive Thinking', *Leadership*, 8(2): 87–107.

Collinson, D. L., and Hearn, J. (1994). 'Naming Men as Men: Implications for Work, Organizations and Management', *Gender, Work and Organization*, 1(1): 2–22.

Collinson, D. L., and Hearn, J. (eds) (1996). *Men as Managers, Managers as Men: Critical Perspectives on Men, Masculinities and Managements*. London: Sage.

Connell, R. (1995). *Masculinities*. Cambridge: Polity.

Connell, R.W. (1998). 'Globalization and Masculinities', *Men and Masculinities*, 1(1): 3–23.

Connell, R., Hearn, J., and Kimmel, M. (2005). 'Introduction', in M. Kimmel, J. Hearn, and R. Connell (eds), *Handbook of Studies on Men and Masculinities*, 1–12. Thousand Oaks, CA: Sage.

Connell, R., and Messerschmidt, J. W. (2005). 'Hegemonic Masculinity: Rethinking the Concept', *Gender and Society*, 19(6): 829–59.

Connell, R., and Wood, J. (2005). 'Globalization and Business Masculinities', *Men and Masculinities*, 7(4): 347–64.

Donnell, S. M., and Hall, J. (1980). 'Men and Women as Managers: A Significant Case of No Significant Differences', *Organizational Dynamics*, 8(4): 60–77.

Eagly, A. H., and Johannesen-Schmidt, M. C. (2001). 'The Leadership Styles of Women and Men', *Journal of Social Issues*, 57(4): 781–97.

Edmondson Bell, E. L., and Nkomo, S. M. (2001). *Our Separate Ways: Black and White Women and the Struggle for Professional Identity*. Boston: Harvard Business School Press.

Egeberg Holmgren, L., and Hearn, J. (2008). 'Framing "Men in Feminism": Theoretical Locations, Local Contexts and Practical Passings in Men's Gender-Conscious Positionings on Gender Equality and Feminism', *Journal of Gender Studies*, 18(4): 403–18.

EHRC (2011). *Equality, Human Rights and Good Relations in 2010: The First Triennial Review*. London: Equality and Human Rights Commission.

Eichler, M. (1980). *The Double Standard*. London: Croom Helm.

Essed, P., and Goldberg, D. T. (2002). 'Cloning Cultures: The Social Injustices of Sameness', *Ethnic and Racial Studies*, 25(6): 1066–82.

European Commission (2010). *Communication from the Commission EUROPE 2020: A Strategy for Smart, Sustainable and Inclusive Growth*. COM(2010) 2020 final. Brussels: EC.

Eveline, J. (1994). 'The Politics of Advantage', *Australian Feminist Studies*, 9(Autumn): 129–54.

Fletcher, J. K. (2004). 'The Paradox of Postheroic Leadership: An Essay on Gender, Power, and Transformational Change', *Leadership Quarterly*, 15(5): 647–61.

Ford, J. (2006). 'Discourses of Leadership: Gender, Identity and Contradiction in a UK Public Sector Organization', *Leadership*, 2(1): 77–99.

Ford, J. (2010). 'Studying Leadership Critically: A Psychosocial Lens on Leadership Identities', *Leadership*, 6(1): 47–65.

Gallo, C. (2010). *The Presentation Secrets of Steve Jobs*. New York: McGraw-Hill.

Glen, P. (2003). *Leading Geeks: How to Manage and Lead People Who Deliver Technology*. San Francisco: Jossey-Bass.

Hearn, J. (1989a). 'Leading Questions for Men: Men's Leadership, Feminist Challenges and Men's Responses', *Equal Opportunities International*, 8(1): 3–11.

Hearn, J. (ed.) (1989b). Special issue 'Men, Masculinities and Leadership: Changing Patterns and New Initiatives'. *Equal Opportunities International*, 8(1).

Hearn, J. (1992). *Men in the Public Eye: The Construction and Deconstruction of Public Men and Public Patriarchies*. London: Routledge.

Hearn, J. (1996). 'Deconstructing the Dominant: Making the One(s) the Other(s)', *Organization: The Interdisciplinary Journal of Organization, Theory and Society*, 3(4): 611–26.

Hearn, J. (2004). 'From Hegemonic Masculinity to the Hegemony of Men', *Feminist Theory*, 5(1): 49–72.

Hearn, J. (2009). 'Patriarchies, Transpatriarchies and Intersectionalities', in E. Oleksy (ed), *Intimate Citizenships: Gender, Sexualities, Politics*, 177–92. London: Routledge.

Hearn, J. (2012). 'Male Bodies, Masculine Bodies, Men's Bodies: The Need for a Concept of Gex', in B. S. Turner (ed), *Handbook of the Body*, 305–18. London: Routledge.

Hearn, J., and Collinson, D. L. (1994). 'Theorizing Unities and Differences between Men and between Masculinities', in H. Brod and M. Kaufman (eds), *Theorizing Masculinities*, 148–62. Newbury Park, CA: Sage.

Hearn, J., and Collinson, D. L. (2006). 'Men, Masculinities and Workplace Diversity/ Diversion: Power, Intersections and Contradictions', in A. Konrad, P. Prasad, and J. Pringle (eds), *Handbook of Workplace Diversity*, 299–322. London: Sage.

Hearn, J., and Louvrier, J. (2011). 'The Gendered Intersectional Corporation and Diversity Management', in S. Gröschl (ed.), *Diversity in the Workplace: Multi-disciplinary and International Perspectives*, 133–46. Aldershot: Gower.

Hearn, J., and Niemistö, C. (2012). 'Men, Managers, Fathers and Home–Work Relations: National Context, Organisational Policies, and Individual Lives', in P. McDonald and E. Jeanes (eds), *Men's Wage and Family Work*, 95–113. London: Routledge.

Hearn, J., and Parkin, W. (1987). 'Frauen, Männer und Führung' (Women, Men and Leadership), in A. Kieser, G. Reber, and R. Wundurer (eds), *Handwörterbuch der Führung (The Encyclopaedia of Leadership)*, 326–40. Stuttgart: C. E. Poeschel.

Hearn, J., and Parkin, W. (1988). 'Women, Men and Leadership: A Critical Review of Assumptions, Practices and Change in the Industrialized Nations', in N. Adler and D. Izraeli (eds), *Women in Management Worldwide*, 17–40. New York: M. E. Sharpe.

Hearn, J., Jyrkinen, M., Piekkari, R., and Oinonen, E. (2008). '"Women Home and Away": Transnational Managerial Work and Gender Relations', *Journal of Business Ethics*, 83(1): 41–54.

Hearn, J., Piekkari, R., and Jyrkinen, M. (2009). *Managers Talk about Gender: What Managers in Large Transnational Corporations Say about Gender Policies, Structures and Practices*. Helsinki: Edita.

Hearn, J., Metcalfe, B. D., and Piekkari, R. (2012). 'Gender, Intersectionality and International Human Resource Management', in G. Ståhl, I. Björkman, and S. Morris (eds), *Handbook of Research on International Human Resource Management*, 509–31. Cheltenham: Edward Elgar.

Howson, R. (2006). *Challenging Hegemonic Masculinity*. London: Routledge.

Husu, L., Hearn, J., Lämsä, A.-M., and Vanhala, S. (2011). 'Introduction—Johdanto', in L. Husu, J. Hearn, A.-M. Lämsä, and S. Vanhala (eds), *Women, Leadership and Management*, pp. v–viii. Helsinki: Edita.

Isaacson, W. (2011). *Steve Jobs*. New York: Simon & Schuster.

Jeanes, E. L., Knights, D., and Yancey Martin, P. (eds) (2011). *Handbook of Gender, Work and Organization*. Chichester: Wiley.

Kendall, L. (2002). *Hanging Out in the Virtual Pub: Masculinities and Relationships Online*. Berkeley, CA: University of California Press.

Kerfoot, D., and Knights, D. (1993). 'Management, Masculinity and Manipulation: From Paternalism to Corporate Strategy in Financial Services in Britain', *Journal of Management Studies*, 30(4): 659–79.

Kerfoot, D., and Knights, D. (1996). '"The Best is Yet to Come?" The Quest for Embodiment in Managerial Work', in D. L. Collinson and J. Hearn (eds), *Men as Managers, Managers as Men: Critical Perspectives on Men, Masculinities and Managements*, 78–98. London: Sage.

Kerfoot, D., and Knights, D. (1998). 'Managing Masculinity in Contemporary Organizational Life: A "Man"agerial Project', *Organization*, 5(1): 7–26.

Kerr, R., and Robinson, S. (2011). 'Leadership as an Elite Field: Scottish Banking Leaders and the Crisis of 2007–2009', *Leadership*, 7(2): 151–73.

Kimmel, M. (1993). 'Masculinity as Homophobia: Fear, Shame, and Silence in the Construction of Gender Identity', in H. Brod and M. Kaufman (eds), *Theorizing Masculinities*, 119–41. Newbury Park, CA, and London: Sage.

Knoppers, A. (2011). 'Giving Meaning to Sport Involvement in Managerial Work', *Gender, Work and Organization*, 18(S1): e1–e22.

Lipman-Blumen, J. (1976). 'Toward a Homosocial Theory of Sex Roles: An Explanation of the Sex Segregation of Social Institutions', in M. Blaxall and B. Reagan (eds), *Women and the Workplace: The Implications of Occupational Segregation*, 15–31. Chicago: University of Chicago Press.

Mabey, C., and Morrell, K. (2011). 'Leadership in Crisis: "Events, My Dear Boy, Events"', *Leadership*, 7(2): 105–17.

McGuire, G., and Reskin, B.F. (1993). 'Authority Hierarchies at Work: The Impacts of Race and Sex', *Gender and Society*, 7(4): 487–507.

Martin, P. Y. (2001). '"Mobilizing Masculinities": Women's Experiences of Men at Work', *Organizations*, 8(4): 587–618.

Mellström, U. (2004). 'Machines and Masculine Subjectivity: Technology as an Integral Part of Men's Life Experiences', *Men and Masculinities*, 6(4): 368–82.

Merrill-Sands, D., Holvino, E., and Cummings, J. (2003). 'Working with Diversity: A Focus on Global Organizations', in R. Ely, E. Foldy, and M. Scully (eds), *Reader in Gender, Work and Organization*, 327–42. Oxford and New York: Blackwell.

Messerschmidt, J. W. (2010). *Hegemonic Masculinities and Camouflaged Politics: On the Bush Dynasty and its War against Iraq*. Boulder, CO: Paradigm.

Messner, M. (1997). *Politics of Masculinities*. Thousand Oaks, CA: Sage.

Morgan, D. H. J. (2005). 'Class and Masculinity', in M. Kimmel, J. Hearn, and R. W. Connell (eds), *Handbook of Studies on Men and Masculinities*, 165–77. Thousand Oaks, CA: Sage.

Nugent, B. (2008). *American Nerd: The Story of my People*. New York: Scribner.

Olson, P. (2012). *We are Anonymous*. New York: Little, Brown.

Pease, B. (2010). *Undoing Privilege*. London: Zed.

Poster, W. (2013). 'Subversions of Techno-Masculinity: Indian ICT Professionals in the Global Economy', in J. Hearn, M. Blagojević, and K. Harrison (eds), *Rethinking Transnational Men: Beyond, Between and Within Nations*, 113–33. New York: Routledge.

Reis, C. (2004). *Men Managers in a European Multinational Company*. Mering-Munich: Rainer Humpp Verlag.

Robertson, R. (1995). 'Glocalization: Time-Space and Homogeneity-Heterogeneity', in M. Featherstone, S. Lash, and R. Robertson (eds), *Global Modernities*, 25–44. London: Sage.

Roper, M. (1994). *Masculinity and the British Organization Man since 1945*. Oxford: Oxford University Press.

Roper, M. (1996). '"Seduction and Succession": Circuits of Homosocial Desire in Management', in D. L. Collinson and J. Hearn (eds), *Men as Managers, Managers as Men: Critical Perspectives on Men, Masculinities and Managements*, 210–26. London: Sage.

Ropo, A., and Sauer, E. (2008). 'Corporeal Leaders', in D. Barry and H. Hansen (eds), *New Approaches in Management and Organization*, 469–78. London: Sage.

Sessa, V., and Jackson, S. (1995). 'Diversity in Decision-Making Teams: All Differences are Not Created Equal', in M. Chemers, S. Oskamp, and M. Costanzo (eds), *Diversity in Organizations: New Perspectives for a Changing Workplace*, 133–56. London: Sage.

Sharma, A., and Grant, D. (2011). 'Narrative, Drama and Charismatic Leadership: The Case of Apple's Steve Jobs'. *Leadership*, 7(1): 3–26.

Sinclair, A. (1998). *Doing Leadership Differently: Gender, Power, and Sexuality in a Changing Business Culture*. Victoria, Australia: Melbourne University Press.

Sinclair, A. (2005). 'Body Possibilities in Leadership'. *Leadership*, 1(4): 387–406.

Tienari, J., Vaara, E., and Merilainen, S. (2010). 'Becoming an International Man: Top Manager Masculinities in the Making of a Multinational Corporation', *Equality, Diversity and Inclusion*, 29(1): 38–52.

Trist, E. L., and Bamforth, K. W. (1951). 'Some Social and Psychological Consequences of the Longwall Method of Coal-Getting', *Human Relations*, 4: 3–38.

Weber, M. (1964). *The Theory of Economic and Social Organization*. London: Routledge & Kegan Paul.

Wetherell, M., and Edley, N. (1999). 'Negotiating Hegemonic Masculinity: Imaginary Positions and Psycho-Discursive Practices', *Feminism and Psychology*, 9(3): 335–56.

Zimmerman, P., Wit, A., and Gill, R. (2008). 'The Relative Importance of Leadership Behaviours in Virtual and Face-to-Face Communication Settings', *Leadership*, 4(3): 321–37.

......

CHAPTER 20

MASCULINITIES IN MANAGEMENT

Hidden, Invisible, and Persistent

......

STEPHEN M. WHITEHEAD

INTRODUCTION

......

THIS chapter examines an ongoing phenomenon of management, that is, the masculinist paradigm which envelops it and the ways in which masculinities persist within organizations while remaining hidden and invisible, not least to those men managers who practise them daily. This phenomenon is not just of interest to those critically examining gender in organizations, but to all concerned with how organizations are controlled, directed, and managed. The issue of masculinities in management surfaces, I suggest, in the actions, attitudes, and ethics of employees and leaders in both the public or private spheres and in local and global fields. It connects directly with the ethical principles, people skills, work practices, and leadership abilities of not only influential and publicly visible senior managers such bankers, politicians, headteachers, PLC directors, and entrepreneurs, but also junior and aspiring managers everywhere. Recognizing such, it can be persuasively argued that the work and leadership style of leaders and managers has a direct and profound influence on the very culture of an organization and therefore ultimately impacts on employee attitudes, relationships, and even their identities (Knights and Willmott 1999; Clegg *et al.* 2005; Holmes 2006; Jackson and Carter 2007). Given the ubiquitous character of (work) organizations and their (in)direct influence on families, communities, and societies, the linearity between management masculinities in action and the social fabric itself becomes apparent.

In analysing the otherwise hidden, invisible, and persistent character of masculinities in management, this chapter draws on those feminist poststructuralist insights and theories (Butler 1990) which recognize masculinity as plural, contingent, and discursively enabled and ontologized (Petersen 1998; Whitehead 2002). The aims are first to highlight

the relationship between a prevailing 'masculinist' (Brittan 1989) organizati
and management behaviour, and, secondly, to show how this work culture i
tained through managers taking up those gender discourses which refle
ways of being. As the chapter reveals, this process not only reinforces a manager's subj-
tive relationship to work, it has the capacity to validate her/his very sense of self.

The chapter is structured as follows. The first section examines global changes in
women's representation in management over recent decades and considers the emergence of new work cultures in multinational corporations especially. The second section explores the relationship between men and masculinities, in particular the key
theoretical perspectives that have contributed to this relatively new field of critical
sociological enquiry. Section three moves on to look at how masculinities can come to
influence and be replicated within organizational cultures and management practices.
One particularly dominant management practice in contemporary organizations is
performativity and section four considers its gender character and its capacity to reinforce masculine-management subjectivities. The following three sections of the chapter
examine the conditions by which masculinities in management become and are, respectively, hidden, invisible, and persistent.

WOMEN IN MANAGEMENT AND ORGANIZATIONS

In the early 1990s, when the first substantial academic research was undertaken into
the relationship between men, masculinities, and management (see e.g. Kerfoot
and Knights 1993; Collinson and Hearn 1993) the representation of women in senior
management and directorship positions was low; typically, between 2 and 5 per cent
in the UK, USA, and Australia. The 1995 National Management Survey (Institute of
Management 1995) actually found a fall in the number of women middle managers
from 10.2 per cent in 1993, to 9.8 per cent in 1994 (Collinson and Hearn 1996). These
statistics were, for the most part, indicative of the relative absence of women in middle and senior management worldwide and across public and private sectors (see also
Mills and Tancred 1992 Marshall 1995; Itzin and Newman 1995; Ledwith and Colgan
1996; Whitehead and Moodley 1999). Indeed, as early as 1977, Rosabeth Moss Kanter
was speculating on the impact on individual organizational cultures should the number
of women managers ever reach a magical 15 per cent. Similarly, terms such 'glass ceiling' (Davidson and Cooper 1992) and 'organizational man' (Roper 1994) were highlighting both the intransigence of men's dominance of management and their subjective and
existential relation to it.

Two decades later and the picture is rather different. At the time of writing, recently
published research into women in senior management positions worldwide reveals
wide disparities globally, but overall a significant numerical rise since the early 1990s.
For example, in the UK women now represent 23 per cent of all senior managers, a similar percentage to Brazil, Chile, and many EU countries. In South Africa, 28 per cent

of senior managers are women, in Italy it is 36 per cent, and Poland 31 per cent (Grant Thornton 2012), In the EU, 40 per cent of company directors are women. Already in the UK, young women professionals are now, for the first time, earning more than their male counterparts, a trend that is also apparent in North America (ONS 2010). In Canada, women now represent 36 per cent of the legal profession (30 per cent in the USA) and are increasing their representation in medicine, chemistry, business, and engineering (Kay and Gorman 2008). In 2012 the number of women CEOs in the Fortune 500 companies reached a record twenty, with women now leading multinational corporations such as PepsiCo, Hewlett Packard, Western Union, Sierbert, Kraft, DuPont, Yahoo, IBM, WellPoint, Xerox, Avon, Sempra Energy, and Pearson (Whitehead *et al.* 2013).

In Asia, changes in women's representation in management is equally if not more marked. In Hong Kong women now occupy 30 per cent of management positions (up from 15 per cent in 1992) and 35 per cent of senior management posts. The percentage of women in senior management posts in China is 34 per cent, Taiwan 30 per cent, Singapore 30 per cent, Philippines 35 per cent, and Malaysia 31 per cent. Globally, Thailand and Russia have the highest number of women in senior management at 40 per cent and 45 per cent, respectively, while in the Philippines women have the highest global representation of women in all management positions at 56 per cent (Grant Thornton 2012; Whitehead *et al.* 2013).

Globally, women currently hold 21 per cent of senior management positions. While this global average is subject to periodic fluctuation, the evidence is overwhelming that women, since the early 1990s, have at least doubled their representation in management, an upward trend that continues (Grant Thornton 2012).

Increases in the number of women in management at all levels would suggest corresponding changes in the work culture of organizations, perhaps to something more recognizable as 'feminine work practices', e.g. those which have a high emotional intelligence factor (see, for discussion, Fletcher 1999). Indeed, there are indicators this is beginning to happen. For example, internationally recognizable companies such as Google, Mercedez Benz (USA), Microsoft, Goldman Sachs, Adobe, American Express, Dreamworld, PricewaterhouseCoopers, Marriott International, Mattel, KPMG, Capital One, and Cisco consistently make it into the 'Top 100 Companies to Work For' (cnn.com 2012). In making this list these and similar corporations are offered, and are keen to present themselves, as examples of organizations which have succeeded not least by adopting more egalitarian, perhaps even collegiate, and certainly less hierarchical work systems and practices. These include, for example, a 'softer' human resources management approach, stronger ethical values, employee diversity, family-friendly work policies, non-discriminatory policies on sexual orientation, a high percentage of women and ethnic minorities across the organization, an emphasis on teamworking and individual initiative where challenging the 'old ways of doing things' is encouraged. Alongside global corporations such as Semco, these 'top 100' companies are also being presented as indicative of a much to be encouraged 'postmodern turn' in organizational culture and leadership styles (Clegg *et al.* 2005; also Child and Rodrigues 2002).

At the same time, while such companies demonstrate a welcome maturity and self-awareness in their management styles and culture, they remain a distinct minority. There is little evidence of such changes emerging wholesale across the public and private sectors. Indeed it could be argued that many public sector organizations, for example educational institutions, are actually reversing the trend towards more collegiate work systems (see Dent and Whitehead 2001; Bottery 2004) and directly or indirectly sustaining the men/masculinity/management relationship.

The fact that, globally, 34 per cent of businesses still have no women in senior management and that women are markedly absent from the higher echelons of banking and finance has led to increasing speculation that today's economic crisis would not have occurred, or at least been so severe, if more women were in the boardrooms of major financial institutions:

> If Lehman Brothers had been 'Lehman Sisters', today's economic crises would look quite different.
> Although most women do not want to be in the City, if they had been there, we wouldn't have had the same problem.

> (Quotes from Grant Thornton 2012: 6)

As many gender theorists have observed, there is a direct correspondence between rigid, hierarchical, and authoritarian management/leadership styles; a 'boy's own' work culture which serves to reinforce men's presence as leaders; macho managerialist behaviours; and dominant expressions of masculinity (Cockburn 1983; Collinson *et al.* 1990; Kerfoot and Knights 1993; Cheng 1996; Kerfoot and Whitehead 1998; Hearn and Collinson 2001).

To understand better the processes of gender acculturation at work it is necessary to first clarify the term 'masculinities' and appreciate how they become replicated and reinforced within the binary of male–female within organizations.

MEN AND MASCULINITIES

The subject of masculinities has now acquired a credible if not impressive academic history. Arising out of first- and second-wave feminist theorizing of the late 1960s and 1970s (Tong 1994; Beasley 2005), the critical study of men and masculinities has become a substantial sociological topic in its own right, offering important and valuable insights into men's practices in all social settings, such as their emotional labours, ideologies, power formations, sexualities, social and economic materialities, and processes of identity work (Hearn 1992; Connell 2000; Whitehead and Barrett 2001; Haywood and Mac an Ghaill 2003; Whitehead 2006a). The subject continues to develop and expand its theoretical horizons and in line with this has increasingly been linked to feminist poststructuralist concepts and perspectives (Whitehead 2002).

One of the main strengths of feminist poststructuralist theory is that it provides a persuasive understanding of identity, highlighting the ways in which femininities and masculinities interact, resist, and connect to the ongoing subjective identity work of the discursively situated individual (Foucault 1998). That is, by acknowledging the Foucauldian insight that no identity or sense of self is sovereign nor constituted outside of discourse, feminist poststructuralism exposes the contingent, fluid, and multiple aspects of being and becoming a woman or man (see Butler 1990; Whitehead 2002; Weedon 2003; Salih and Butler 2004).

In respect of men and masculinities it is possible, then, to see the discursive relationship between dominant social assumptions about how men should behave and the persistence of masculinist values and practices in many social sites. Once we take the premise that all identity is socially mediated through the actions of dominant and subordinate discourses, then it is possible to see how masculinities, rather than being predictably fixed to the sex category 'man', are more complex ways of being which attach to males and men through the social codes, cultures, and beliefs operating in any particular social and cultural location. In this way, a masculinist culture is replicated and reinforced—not through the inevitable biology of preordained gender action, but through the persistently dominant ways of thinking about men and males. Such social action is framed within a gender binary that, while attempting to distinguish between male and female, men and women, also provides the means by which masculinities and femininities exist in a mutual and self-sustaining association (Connell 2000; Whitehead 2002; Whitehead *et al.* 2013).

Males do not exhibit and 'know' masculinity at some predetermined and biological level, but come to learn what it means to be male and man in their particular social setting. What are being learned or assimilated here are actually discourses that signal both 'appropriate and 'inappropriate' masculine behaviours, with additional variables such as sexuality, ethnicity, age, religion, social positioning and contributing to the individual male identity that is performed. This is a complex, dynamic process of intersectionality (Mann and Huffman 2005), whereby identity and potential points of oppression configure the space within which an individual expresses and experiences their sense of self in relation to others. The subsequent validation of gender behaviour can be termed 'identity work', though this is not work that is done knowingly and cognitively by the individual in all instances. Rather, it is a contingent process whereby (an individual's) identity is never fully attained but always in a state of being and becoming (Deleuze and Guattari 1988; Whitehead 2002). It is a highly subjective and intensely unstable process, not least because no social setting remains constant and given. Below I look more closely at the model of organizational culture being adopted here, but central to it is the notion of contingency and fluidity within a discursive constituency.

In short, there is no true and pure identity of man, merely a complex dynamic of interacting, reinforcing, sometimes conflicting, discourses of identity, all of which have some association with the category of male and with given social understandings of maleness. Man (and woman) are, in essence then, discursive subjects seeking ontological validation of their (gender and sex) identity, along with other aspects of their being.

Importantly for this chapter, a very powerful social arena for such ontological validation to occur in is work organizations, especially management and leadership (see Knights and Willmott 1999).

A further question that requires addressing when critically examining men and masculinities is the extent to which men as a gender category operate as a hegemonic group that subordinates women and 'others'. The work of Connell (2000) is especially significant here because it provides theoretically rigorous insights into male power, oppression, and gender advantage. Utilizing his concept of 'hegemonic masculinity' Connell argues that men's material power arises largely through their continued association with a way of being a man which privileges rationality, homosociability, competition, machismo, and heterosexuality (see also Kimmel 2000). Consequently, many researchers have connected hegemonic masculinity with a masculine work culture showing how particular dominant practices of men forge an organizational culture and value system which marginalizes, if not excludes, women and gay men (see e.g. 1996; Barrett 2001; Prokos and Padavic 2006; Hinojosa 2010).

While hegemonic masculinity is a persuasive concept, it offers us little understanding of resistance, male diversity, nor the ways in which otherwise marginalized identities can become powerful. Moreover, in respect of women in management and leadership, hegemonic masculinity fails to explain how women who break through the hegemonically reinforced 'glass ceiling' often fail to change or challenge the masculinist work culture surrounding them (for a detailed critique of hegemonic masculinity, see Speer 2001; Whitehead 2002; Connell and Messerschmidt 2005).

The work of Judith Butler provides a way of seeing how identity becomes politicized, not through cognitive hegemonic processes but through the subject's desire for ontological grounding. Butler's concept of 'performativity' highlights how discourses of gender signification are repeated, habitually and over time by the subject (individual) (Butler 1990, 1993). This is a performative process that serves to render such discourses 'real' and fixed, not least in the mind of the subject, but also reinforces them visibly, in organizations and in society. Individualization is 'created' but out of gender signifying practices which are already politicized within the social domain prior to them being taken up by the subject. Performativity is especially helpful in understanding how gender, in particular masculinities, remains persistent, hidden, and also invisible within management. Likewise performativity offers us insights into how women can come to do masculinities (Halberstam 1998). For, as I discuss below, we should not assume men and masculinities to be in an exclusive symbiotic relationship; women can do masculinities too.

To summarize, men and masculinities have a self-sustaining but not predetermined relationship. Men, as discursive subjects, can acquire a high level of ontological validation as men through performatively taking up the gender signifying practices of maleness that are considered 'valid' if not essential male behaviours in a particular social setting. The agency of the individual resides in their ability and willingness to take on such practices or reject them. What individual men will do, when faced with powerful dominant discourses that act to reinforce dominant masculine behaviour and marginalize others, cannot be predicted.

As I now go on to discuss, the reality is that management is no longer the sole province of men. Women are managers and leaders too. In which case we need to look more closely at the character of the managerial space and the masculinist practices that envelop and in many cases, substantiate it.

The Managerial Space: A Masculine Culture

In order to grasp the complex dynamics that serve to reinforce both management identities and a gendered managerial work culture, we first need to be clear about the very idea of organizations. In this chapter I am taking two related views of organizational culture. The first is to see organization not as an entity but as a discursive process without conclusion or settlement. In being ultimately and only ever discursive, the organization can be recognized as being inevitably dynamic, fluid, and contingent. The managements that arise within any organization are similarly unstable and ontologically vulnerable, not least to the vagaries of external and often unforeseen and uncontrollable factors. The second view is to recognize the process of organization as having a strong conceptual aspect, in that meanings, languages, and symbols are attached to it that serve to render symbolic association, discipline members, and encourage loyalty, commitment, motivation, and belonging (see, for discussion, Jackson and Carter 2007; Knights and Willmott 2012; also Holmes 2006). So all organizations exist within a contested space between the ongoing uncertain dynamics which arise from the process, and the prevailing meanings which individual members attach to this process and which assist in reinforcing their identity to the organization itself.

A further condition to management is that it arises from a gendered history that has long privileged man as the 'natural' manager and leader (Collinson and Hearn 1996). Many dominant practices and values of management continue to serve and reinforce this gendered association—e.g. instrumental rational behaviour; individualized competitive attitudes; paternalism; ruthlessness and violence; control and unquestioned authority; punitive audit and accountability; sexism; aggressive leadership; presenteeism; authoritarian hierarchical structures; managed intimacy; hard expressions of human resource management (Cheng 1996; Kerfoot and Whitehead 1998; Fletcher 1999; Kerfoot 2001; Hearn and Parkin 2001). These characteristics of the managerial space may be more or less apparent in particular organizations. They are not confined to the private sector but are also apparent in the public sector, for example, through expressions of 'new' public management and performative work cultures (Dent and Whitehead 2001; Barry *et al.* 2003).

This is not to say that all managers and leaders engage with these work practices willingly and readily. As is apparent, many do resist while also seeking to create working environments that are collegiate, supportive, emotionally enhancing, balanced, considerate, and yet also highly productive and creative. However, we should not assume that such 'feminine' characteristics are the inevitable province of women managers, nor the outcome of having more women managers in the organization. Gender operates in a contested space, especially with those men who feel their masculinity to be exposed and vulnerable. For example, in those professions and work sites that are female-dominated, men will seek to reproduce and maintain masculinities that might otherwise be undermined by a dominant female culture (see e.g. Simpson 2004).

The managerial space is accordingly a gendered political arena, made possible through the actions and behaviours of those individuals who occupy it. Those individuals, whether men or women, gay, straight, or bisexual, are themselves discursive subjects whose desire for ontological validation supersedes their desire for managerial status, though the two aspects of self are clearly connected and potentially self-supporting (Knights and Willmott 1999).

Management and leadership can therefore provide a powerful ontological validation for the discursive subject. As my own research shows (Whitehead 2006b), management offers assumptions of power, control, and status; it suggests success, attainment, advantage, and reinforced association with the organizational rhetoric. To be a manager, particularly a senior manager or leader, is to be placed high up the corporate ladder, elevated to a position which carries with it strong existential and ontological reinforcements for the discursive subject (also Whitehead 2001a, 2003).

However, at the same time, management is always unpredictable. Organizations change, as do people. The movement of people into and out of an organization has a powerful effect on the organizational process. People bring with them certain dominant discourses when they enter an organization and they may well remove such discourses, or weaken them, when they leave. This may be the case at senior management level where there are fewer people and where managers often try hard to appear convincing in their management and leadership role, a state of being which requires a significant physical, emotional, and ontological effort (Fineman 1993; Casey 1995; Parker 2000).

We have then, at these levels, a relatively small pool of individuals who have strong subjective associations with being a manager and an ontological and existential investment in being successful and therefore legitimized and validated in such a role. I would argue that this desire for managerial identity validation is no less strong in women than it is in men. Indeed, it may be stronger given the fact that management has long been considered a male arena which few women can comfortably enter. Those women who do make it to the highest levels or management do so in spite of, not because of, male attitudes and the masculinist culture that tends to prevail (see Whitehead (2001a) for discussion). As I discuss below, this process of attainment against the 'gender grain' of management can elicit in many women a strong awareness of the gender reality within which they work, i.e. its masculinist culture.

MASCULINIST CULTURES AND PERFORMATIVITY

By masculinist culture I am referring to those dominant managerial practices, together with their underpinning rationale, that reinforce male identity—in particular competition, individualization, emotional distancing, and strongly assertive and aggressive behaviour. An example of such a practice would be Lyotard's (1984) concept of performativity, i.e. the obsession with audit and performance measurement in all aspects of an organization's activities. This practice becomes legitimized as 'scientific', objective, and therefore an impartial but necessary regular assessment of an individual's achievements in order to achieve the larger aims of the organization (Dent and Whitehead 2001; Ball 2003; Blackmore 2005).

Performativity is often deployed in tandem with 'hard' systems of human resource management not least in order to function as a mechanism for disciplining managers and employees and intensifying labour (Legge 2005). In their more benign and supportive formations, performative measures to encourage 'success' and 'attainment' can be advantageous for managers and employees in that they assist the clarification and achievement of organizational goals. However, the most common type of performativity is one where targets are imposed on individuals without negotiation, continually ratcheted up once attained, and often operating in an atmosphere of uncertainty alongside individual appraisal and audit (Dent and Whitehead 2001; Ball 2003).

Performativity is a discourse about how to manage and control, but with evident material and emotional consequences. It has the potential to reduce motivation and commitment; increase employees' anxieties and stress; encourage 'silo' mentalities within teams of workers (e.g. competition and a lack of collaboration between teams); 'validate' aggressive leadership styles; and instil a culture of presenteeism especially within management. As such, performativity not only fits well with the sorts of behaviours often exhibited by senior men managers but, as research is now revealing, has the potential to reinforce men managers' very sense of themselves as men and as masculine (Whalley 2011).

Numerous discourses about 'how to be a manager' can exist within a given managerial space, though performativity has become a particularly common and dominant one. Another dominant discourse is the notion that managers/leaders are powerful and should constantly express confidence about their aims, direction, and vision for the organization. Being a leader-manager suggests the individual has both the power and opportunity to bring about change, exercise control over others, and, not least, themselves. Again, this discourse highlights the connection between management and masculinity where the manager is both powerful, rational, and potent enough to remain 'in control' whatever circumstances might arise. However, management is a very uncertain and insecure space. Managers can feel isolated, stressed, highly anxious, and desperately weak and insecure. They can feel emotional and out of control of events. They can certainly feel threatened by change and by events over which they can have little control but which have the power to diminish their organizational status and therefore threaten their future career (see e.g. Hodgson 2005).

It is a paradox and irony of management that managers must work hard at presenting themselves as convincingly masculine—e.g. competitive, controlling, authoritarian, individualistic, rational, instrumental, and totally committed to the organization above all else—while remaining deeply uncertain, often anxious and highly insecure. In this respect, management serves to expose the deep paradox and unresolved tension within those individuals, especially men but also women, who draw too heavily on masculine ways of being in order to validate their ontological identity. In other words, dominant masculinity (both in and outside the management arena) can require a performance of self that necessitates physical and emotional commitment to a singular aim, lacks empathy, mitigates against intimate relations, becomes emotionally dysfunctional over time, and can never be fully resolved for the simple reason that it constantly requires accomplishment in order to appear momentarily valid and convincing, not least to the individual performing it.

HIDDEN MASCULINITIES

As I have discussed, masculinities can emerge in management through practices that reinforce men's association with management and organizational power. At the same time, women managers are also required to engage in such practices as evidence of their 'capability' to perform as 'effective' managers and leaders. So managerialist practices such as performativity and harder forms of HRM, are not just required of men managers: *they are required of all managers regardless of their gender*. These practices usually exist prior to the manager entering the organization, becoming embedded in the organizational culture. As I go on to explore in this section, masculinities get 'hidden' in organizations precisely because what is being performed as a dominant managerial practice appears gender neutral, not least because all managers are expected to impose it, when in fact it draws heavily on masculinist ways of being.

Taking a highly common if not legitimate management practice such as audit, and analysing it through the lens of performativity, allows us to see more clearly how so-called 'scientific' and objective work practices are never entirely neutral but are invested through discourses about organizational 'reality'. These discourses have a gendered dimension in that they are powerfully and historically associated with men. This association with men is through those ways of being which define (male) leadership. Likewise to talk of 'powerful leaders' is invariably to conjure up male, not female, images. The rhetoric of the strong, assertive, confident, and unchallengeable leader or manager is enveloped in discourses which allow little space for femininity, and certainly do little to encourage traditional expressions of it in managers.

In this way masculinities become hidden in management. We can scrutinize management practices such as audit, accountability, recruitment, HRM, and appraisal. We can research the varied leadership styles of individuals. And we can subject managerialism, performativity, presenteeism, silo working, power and resistance, and work cultures to critical examination. But to do so without recognizing their gender sub-text is to fail to see how these practices are gendered ways of being, not only privileging masculinities, but encouraging and reinforcing an organizational culture which is inherently masculinist in its values and actions. Masculinities get concealed in management precisely because of this inability to connect the discourses of management with the discourses of masculinity.

A second reason why masculinities get hidden in management is because we too readily associate them with men and fail to see that women do masculinities also. As Chatterjee (2007) argues, 'why do we (feminist theorists) insist on, or at least act like we believe in, an inherently natural bond between masculinity and biological male bodies?' (quoted in Whitehead *et al.* 2013: 287). Chatterjee is alerting us to the fact that women do masculinities and that we need to overcome what Halberstam (1998: 15) describes as the 'general disbelief in female masculinity...a collective failure of imagination [to recognize] that female-born people have been making convincing and powerful assaults on the coherence of a male masculinity for well over a hundred years'.

As I discuss in relation to the concepts of 'post-modern men' and 'post-modern women' (Whitehead *et al.* 2013), the gender binary reinforces the link between men and masculinities and women and femininities, even while there is increasing evidence all around us of women and men operating effectively and confidently across both spectrums at different times in their lives and in diverse social settings. This is almost certainly the case in contemporary management where increasing numbers of women are effectively performing the masculinities which envelop and define 'good' management practice across countless organizational sites and worldwide. Thus, masculine performances of management are not the exclusive province of men. To assume so is to merely reinforce the gender binary and fail to recognize that both women and men are ultimately discursive subjects taking up, knowingly or not, those gender signifying practices at their disposal. This performative action is not defined or predetermined by a person's sex, but highly influenced by the cultural and social codes within which they are operating as discursive subjects in search of, ultimately, ontological validation.

Here, management and certainly senior management provide a powerful existential and ontological setting for both men and for women (Whitehead 2001a). As Gherardi (1995) notes, women moving into highly masculinist work sites will be constantly reminded of their femininity in ways in which men will not be constantly reminded of their masculinity. This can create in women managers a 'schizogenic' relationship to management whereby they are expected to retain an inherent and 'natural' femininity, while also practising masculine codes of management as effectively, or more so, than men managers.

Yet despite the obvious and long-standing gender and social barriers to women succeeding as leaders and senior managers, increasing numbers are now reaching the highest levels in organizations. They can be seen on this basis to be operating successfully in masculinist arenas. We should not assume, therefore, that women managers will, by virtue of being women, automatically challenge masculinist management cultures. Some may challenge, others may readily encourage them—just as not all men will readily enact masculinist management practices and instead try to resist them. The point is, in feminist poststructuralist terms, the discourses of management provide both men and women with the opportunity for ontological validation as gendered subjects. We should not assume only one gender is taking up these discourses.

To summarize, masculinities get hidden in management for two reasons. First, there is a failure to recognize that dominant managerial practices are framed within codes of behaviour that have historically been associated with male ways of being. The second reason masculinities get hidden in management is because we can fail to recognize that many women managers are now effectively performing masculinities not just as managerial practices, but as ways for themselves to achieve validation as women. What we are seeing here is a female masculine subject achieving validation of her particular expression of femininity and doing so quite effectively within the masculine/managerial space.

INVISIBLE MASCULINITIES

A recurring and long-standing theme within feminist theorizing has been the issue of what Middleton (1992) described as 'men's blocked reflexivity', i.e. their apparent lack of self-awareness as to their own gender. Roberts (1984) and Segal (1990) also referred to 'obliviousness' in men when it comes to their lack of awareness of their gendered attitudes, prejudices, and motivations. Pro-feminist theorist, Michael Kimmel, has remarked that in his opinion 'men do not know they have a gender' (Kimmel, cited in Middleton, 1992: 11). Indeed, one of the main motivations for the 'consciousness-raising' men's groups of the early 1970s, and the subsequent wider social critique of traditional male sex roles by feminists and other social commentators, was to alert men to the ways in which their everyday behaviours and assumptions about 'being a man' can be profoundly damaging to both themselves and others (Pleck 1981).

The concern was, and still is, can men change? Moreover, can they do so having reflexively and critically engaged with their own history as males and men, their relationships to others, their sexual identities, and their gendered practices? This question is not merely rhetorical but cuts to the point made by many (e.g. Giddens 1991) which is that, while we are seeing profound and dramatic changes in women and (post)modern expressions of femininity, corresponding and supportive changes in men are less marked and certainly less uniform.

Of course, as feminists and pro-feminists have long noted there are very powerful reasons for men to resist self-critique and subsequent change to a more enlightened and equitable way of being a man (Morgan 1992). Those men who have been raised in a patriarchal culture, one that has historically privileged males over females, may well be reluctant if not strongly resistant to a perceived loss of gendered power which comes with any rebalancing of the gender status quo. This resistance can be interpreted as some men's strategic response to the emergence of women in positions of power and authority; they recognize the threat to their own gender status and seek to actively challenge it. There are examples of this around the world today, not least in the actions and rhetoric of certain radical religious groups.

However, arguably the more insidious and persistent barrier to gender equality, in both the public and private spheres, is less the overt resistance by men (which is increasingly challenged in law and society) and more the absolute inability of men to recognize their own practices of gender identity as being a contributing factor. In other words, men may verbally articulate support for gender equality, and in fact have a strong desire for such, yet at the same time be unable to make the leap to becoming more self-aware and reflexive as to how their own masculine identity remains founded on the very gendered assumptions and practices which have longed sustained women's marginalization.

This brings us back to the point made by Kimmel, which is that most men appear unable to recognize they have a gender. That being so, how are they then able to move towards a state of gender equality with women if they cannot make a connection between their own masculine identity and their everyday practices as men?

Research has shown how women become reminded of if not confronted with the social, cultural, and material conditions arising from their sex and gender identity through both the actions of men and their experiences not least in work organizations (e.g. Gherardi 1995). That is, being and becoming a woman (leader) entails an almost inevitable process of self-awareness as to women's social relationship to men within organizations. While this may or may not engender an 'epistemic community' (Assister 1996) and subsequent 'feminist standpoint' (Smith 1988) it certainly helps women become aware as to how their gender influences their experiences as a manager and leader, and this is especially the case when they move into organizations that are dominated by men and which operate within a masculinist culture.

My own research (Whitehead 2001b) clearly revealed the discrepancies that exist between most women managers' awareness of their femininity as a major contributing factor to their experiences in the organization, and men managers' awareness as to the same. Interviews I undertook through the 1990s and subsequently ten years later with women senior managers in UK further education showed an almost universal recognition amongst them of the gendered conditions prevailing in their respective work cultures. While these women had achieved high levels of seniority in educational management, they had been constantly reminded of their gender through the actions and behaviours of male managers. These behaviours ranged from comments about the women managers' attractiveness and speculation about their sex lives to outright intimidation, threatening behaviour, and misogynistic comments about women's ability to lead and manage in organizations. One woman vice principal of a large further education college in the north of England was typical of many in that she worked under a male principal who was under a lot of stress not least through his own inability to delegate. Instead of dealing with people in an even-handed way he used his female VP as an outlet for his frustrations and anger. Consequently, middle managers would be wary of approaching him. The woman vice principal therefore unwillingly assumed the position of 'buffer' between a concerned staff and the principal. As she put it to me:

> It's like I have to mother him (the Principal) and protect him from a lot of what's going on in the college. He just cannot cope and when he gets stressed he loses his temper and flies off at anyone close by, which is usually me. The staff don't like to approach him themselves so they come to me. If there is a problem I end up dealing with it, trying to smooth it all out so it doesn't come to the attention of the Principal. I know the Principal partly does this on purpose—he likes having a woman VP in the

organization, someone who can be maternalistic and soften his harder edges. I do stand up to him but it takes a lot of emotional energy to continually do so.

While bullying in organizations is not exclusive to male managers, this is an explicit example of how gender identity becomes entwined in the expectations and actions of managers and leaders. The 'harder' forms of people management can quickly be rendered invisible in terms of their gender origin, not least because they get interpreted as men just acting out 'naturally' macho and aggressive leadership styles. Men are excused such behaviour on the grounds that it is 'what men leaders do'. Women managers, by contrast, are always under a gender gaze that can subject them to sexist comments if they themselves act out macho performances. It is much more acceptable for them to be maternalistic in style. This highlights the 'schizogenic' situations to which women in masculinist organizations can be exposed, that is, on one hand they are expected to conform to some larger feminine model of behaviour that suggests 'natural' maternal, supportive, and empathetic behaviour, yet on the other hand they are expected to replicate a more masculine style of leadership and management in order to be judged 'credible leaders'.

There is extensive research undertaken over many decades which confirms how women managers come to acquire a heightened gender subjectivity as a consequence of their experiences climbing the 'greasy pole' of management (e.g. Itzin and Newman 1995; Marshall 1995; Whitehead and Moodley 1999; Kerfoot et al. 2000; Miller 2004; Prokos and Padavic 2006; Pacholok 2009; Bird and Rhoton 2011). This is not just in respect of having to overcome a 'glass ceiling' in their career journey, but is also indicative of a more fundamental level of self-awareness as to the relationship between being a woman and their experiences in management and organizations.

These experiences are symptomatic of the gendered divide which operates in many organizations where women managers are required to straddle both feminine and masculine behaviours if they are to progress up the management ladder, while men managers can remain locked in an state of apparent invisibility as to their gendered behaviour. The extent of such invisibility in management was exposed in research I undertook into men education managers during the late 1990s (Whitehead 2002). I interviewed nineteen men managers for a larger research project, asking each manager the following question: 'Do you think your experiences as a manager have in any way been affected by you being a man? If so, how?'.

I was looking to assess to what extent the men managers were able to reflect on the fact that being a man might itself facilitate their rise up the management ladder, influence their relationships with colleagues and staff, affect the way they responded to situations, present challenges and opportunities to their career pathway. However, the responses I received merely confirmed the invisibility of masculinities in management, at least to these men managers:

Response 1: 'Mmm... interesting; can you expand on that to help me?'
Response 2: 'I don't know... I really don't know how I could answer that.'
Response 3: 'I don't know... I've just been lucky really.'

Response 4: 'That is difficult. Difficult to give you a realistic answer...I'm not a woman.'

Response 5: 'Mmm...[pause]...mmm...another question to ask my wife.'

Response 6: 'I don't know. I can't answer that.'

Response 7: 'Don't know; most of my appointments have been women.'

Response 8: 'No idea...I can't answer that question.'

Response 9: 'I've never thought about it. I don't treat women differently.'

Response 10: 'I'm not aware of it, not conscious of it...a difficult question.'

Response 11: 'I don't think so...you can't know.'

Response 12: 'Oh God! I don't know...I've always worked tried to treat men and women similarly.'

Response 13: 'No, I'm lucky, I've always worked with men.'

Response 14: 'Difficult to answer that; I've never been a woman.'

Response 15: 'Yes, but I'm not sure where I go with that.'

Response 16: 'Yes, but...'

Response 17: 'I'm not sure; it must have been...yes it's made a difference.'

Response 18: 'Yes, but I tend to be a systems man.'

Response 19: ' Yes, because I work in a male-orientated environment.'

(Taken from Whitehead 2001*b*: 361–2)

Out of nineteen male senior education managers, all working in arguably one of the most important professional spaces for empowering individuals, regardless of their class, age, race, ethnicity, sexuality, and gender, only one is able to make a direct and confident link between his gender as a man and his professional experiences. It is not that these men are necessarily sexist or anti-equal opportunities. Indeed, my research showed a very clear desire in these managers to be empowering, supportive, and enabling of all those under their remit, both students and staff. What is patently missing is the reflexivity, their awareness of themselves as gendered subjects. Some of these men managers were instrumental in writing and developing equal opportunities policies in their institution and were politically committed to such. But at the same time they lacked any appreciation of their own gender, their own masculine behaviour, their own masculine subjectivity as men and as managers. Gender was a word, perhaps even a political space, they had learned to be sensitive of as educationalists—but they didn't recognize their own gender.

How, then, can we interpret these very different responses from women and men to their understanding of how gender impacts on their experiences of being a manager? Masculine behaviour not only exists in management; in many instances it comes to define management and leadership. The relationship between men and management has a historical precedence and therefore already exists prior to any individual entering the management domain. But individuals have identities and those identities are gendered in some form or another. For women to enter management is to expose them to expectations, actions, languages, and behaviours which range from simplistic gendered stereotyping to outright bullying and misogynistic behaviour by men. For men to enter management is merely to locate them in a work arena that has come to be largely defined by male practices and men's presence.

The stark contrast between most women managers' experiences of organizational life and most men managers' experiences of organizational life reveals the gender under-current at work. Research shows that women are much more likely to become sensitive, alert, and reflexive of this than men. The men can, indeed, continue to be the 'invisible gendered subject' in management not least because, for the most part, they are not constantly forced to reflect on how their gender is shaped through work.

PERSISTENT MASCULINITIES

As I have discussed, there have been, over the recent decades, profound if not dramatic changes in terms of the numerical presence of women in middle and senior management. Yet the same time, and with few notable exceptions, masculinist attitudes and behaviours tend to prevail in organizations. While they are more visible in, say, the uniformed services than many other professional sites (see e.g. Prokos and Padavic 2006; Hinojosa 2010) it can be argued that the gender paradigm within which work organizations flourish remains persistently masculinist. Consequently, management continues to privilege and reward those managers (men or women) who are prepared to replicate such values and styles of leadership. Inevitably, such values and styles will mitigate strongly against ways of working, managing, and leading in organizations which are more empathetic, balanced, gender neutral, indeed challenging of hegemonic masculine ways of being.

The question then arises as to why damaging/marginalizing expressions of masculinities persist in management, given several decades of feminist critique, more women in management/leadership, and most organizations recognizing, at least on the surface, equal opportunities. My own research and studies into men and masculinities over several decades suggests the following;

1. The gender binary remains a core factor in sustaining both masculine and feminine distinctions, to some extent throughout all societies. This division of being, reinforced through language, stereotype, and culture is crucially sustained in work organizations, even those that are active in challenging such stereotypes. The sexual division of labour in organizations framed largely through the public/private divide still posits women into caring and supportive roles and men into positions of authority and leadership. In short, the gender binary continues to provide the dualism by which the vast majority of identities and gender subjectivities become realized, so it should be no surprise that they are replicated in work organizations also.

2. The historical relationship of men to management/leadership positions continues to provide the template by which managers and leaders are recruited, measured, and identified. Any manager, regardless of their sex, gender, or sexuality,

is operating in an arena that has come, over time, to be defined through male ways of being. So the performance and identity work which leader/managers must enact in order to appear 'credible' draws heavily on masculine values. For example, it might be the case that an overtly camp transsexual one day becomes President of the United States, Prime Minister of Russia, or CEO of General Motors, though that seems highly improbable at time of writing.

3. Management, and especially leadership, is potentially highly ontologically seductive (Whitehead 2002). That is, the very act of being a leader-manager requires a sense of grounded identity, a way of being which presents itself as assured, confident, and determined. Managers/leaders cannot afford to present ambiguity and insecurity. Indeed, the very discourses of being a leader-manager are bound up in notions of direction and vision. The language of leadership must present authority to be convincing to those it seeks to persuade and control. To therefore offer a way of being a leader-manager which is soft, yielding, passive, empathetic, and stereotypically feminine would not only be very difficult to perform but would be unlikely to be accepted by those who are its intended recipients, not least the shareholders of public limited companies.

4. Changes in management practices over the past few decades have come to reinforce masculine ways of being a leader-manager, not lessen them. The most obvious example is in the public sector where the introduction of market forces, neo-liberalist ideology, and performative appraisal systems all serve to emphasize a harder-edged competitiveness, removed rationality, fixation on outcomes over processes, audit and accountability systems, and instrumental intimacy/emotionality. Consequently, professional identity itself becomes increasingly invested in a masculinist value system that is highly difficult to challenge at the individual level. To be 'successful' as a leader-manager under these conditions requires a commitment to the 'bottom-line' that is unswerving. Management and leaders may well speak of the organization as a 'family' and the workforce as the most valuable resource available to it, but the reality is that the individual is a readily disposable unit.

The persistence of masculinities in management is, then, reinforced through the historical relationship of men/masculinities to leadership, the embedded practices of contemporary management across both the public and privates spheres, and in the seductive gendered discourses reflected in notions of the 'strong, powerful, focused and determined leader'. Behind these factors lies the largely unchanging gender binary, the very ways in which societies are framed around a male and female dualism.

Conclusion

This chapter has, by exposing the invisible and hidden masculine character of contemporary management, shown how a masculinist paradigm persists in organizations despite the numerical advancement of women into management.

As gender theorists now widely accept, there are different types and expressions of masculinity in the social arena. Indeed I would argue these differences are becoming more apparent and acute. We can see, for example, postmodern expressions in that ways of being a man do not always reflect traditional discourses of macho-masculinity but appear to celebrate a more confident and blended feminine/masculine performance (Anderson 2011; Whitehead *et al.* 2013). Similarly, there is also increasing evidence that ways of being a leader are changing, reflecting less the modern authoritarian (male) leader and more the postmodern collegiate and facilitatory approach to management and leadership (Clegg *et al.* 2005). Alongside this there is the ongoing tussle in organizations and managements between the 'hard and soft' forms of human resource management (Legge 2005) and the requirements on workers and management to operate not so much in fixed hierarchical structures but as self-autonomous and responsible teams enabled through a more sensitive and situational leadership approach (Bottery 2004; Clegg *et al.* 2005).

These changes are to be welcomed and, combined with the numerical advancement of women into senior management positions worldwide, could herald a real challenge to the male/masculinist paradigm that has historically defined practices and notions of leadership and management and reinforced men's symbiotic relationship to authority.

My concern with masculinities in management is, then, not so much that they exist at all, but the often damaging ways of being a manager/leader they can privilege and reinforce and the organizational cultures they subsequently come to inform. My second concern is that women managers, if they are to have career progression, become compelled into ways of behaving that can be potentially existentially, emotionally, and organizationally problematic for them. That is, they are required to operate some sort of balanced duality identity work; being both women (engaging a feminine discourse) and managers (engaging a masculine discourse). They then get judged and assessed on what is in effect their ability to maintain a continuous balancing act between femininity and masculinity. Any slippage leads to censure or ridicule.

Ideally, we should arrive at a place where we are no more concerned at having masculinities in management than we are at having femininities in management. What will remain of concern to feminist researchers are the discourses that are configuring management/leadership, their gendered origin, the (damaging) practices they can encourage, and the extent such discourses (as language and practice) openly or covertly serve to marginalize particular social groups and individuals, especially women and LGBT persons.

For the moment, management, leadership, and organizational cultures remain, to a greater rather than lesser extent, located in a discursive arena which privileges both men and/or ways of working which are toxically masculinist. By being masculinist as described, they require of their subjects a commitment to practices and organizational relationships that are neither collegiate in approach nor empowering of the individual.

This compels us to a recognition which is that it is most important to make the distinction *between a numerical gender balance in management and a management/organizational work culture which is gender balanced.* The two are not the same and certainly it is apparent that women in management do not, in themselves, inevitably challenge masculinities in management. From a feminist poststructuralist perspective this is hardly surprising. For to assume that women managers are, by virtue of being female, unable to perform masculinities would be untenable given what we now recognize to be the contingent, performative, and discursive character of gender identity and feminine and masculine subjectivities.

That said, as I have clearly argued in this chapter and shown through the research available, women managers do operate, for the most part, in a different discursive space to that of most men managers. Their sense of gender subjectivity is influenced to some extent by their experiences of being and becoming a leader-manager. Given the masculinist paradigm within which work organizations operate then one can appreciate that for many women while they may well come to replicate dominant expressions of masculinity as leaders themselves, they do so in a way which is not so seamless and unreflective as it is for most men leader-managers.

Therefore, one obvious way forward in terms of promoting more gender-neutral and sensitive behaviour in managers and leaders is to assist men managers especially to reflect on their masculine identity, their gendered assumptions, their priorities as men, and their subjective masculine relationship to power and authority. In this way we might bridge the current gap between the gendered subjectivities of women managers and the gendered subjectivities of men managers, thereby encouraging sensitivity and self-awareness, and positive gender and cultural transformations both in organizations and society generally.

REFERENCES

Anderson, E. (2011). *Inclusive Masculinity: The Changing Nature of Masculinities*. London: Routledge.

Assister, A. (1996). *Enlightened Women: Modernist Feminism in a Post-Modern Age*. London: Routledge.

Ball, S. J. (2003). 'The Teacher's Soul and the Terrors of Performativity', *Journal of Education Policy*, 18(2): 215–28.

Barrett, F. J. (2001). 'The Organizational Construction of Hegemonic Masculinity: The Case of the US Navy', in S. M. Whitehead and F. J. Barrett (eds), *The Masculinities Reader*, 77–99. Cambridge: Polity.

Barry, J, Dent, M., and O'Neill, M. (eds) (2003). *Gender, Professionalism and Managerial Change: An International Perspective*. London: Routledge.

Beasley, C. (2005). *Gender and Sexuality: Critical Theories, Critical Thinkers*. London: Sage.

Bird, S. R., and Rhoton, L. A. (2011). 'Women Professionals' Gender Strategies: Negotiating Gendered Organizational Barriers', in E. L. Jeanes, D. Knights, and P. Y. Martin (eds), *Handbook of Gender, Work and Organization*, 245–62 New York: Wiley.

Blackmore, J. (2005). 'The Emperor has No Clothes: Professionalism, Performativity and Educational Leadership in High-Risk Postmodern Times', in J. Collard and C. Reynolds (eds), *Leadership, Gender, and Culture in Education*, 173–194. Buckingham: Open University Press.

Bottery, M. (2004). *The Challenges of Educational Leadership*. London: Paul Chapman.

Brittan, A. (1989). *Masculinity and Power*. Oxford: Basil Blackwell.

Butler, J. (1990). *Gender Trouble: Feminism and the Subversion of Identity*. New York: Routledge.

Butler, J. (1993). *Bodies that Matter: On the Discursive Limits of Sex*. London: Routledge.

Casey, C. (1995). *Work, Self and Society: After Industrialism*. London: Routledge.

Chatterjee, R. (2007). 'Charlotte Dacre's Nymphomaniacs and Demon Lovers: Teaching Female Masculinities', in B. Knights (ed.) *Masculinities in Text and Teaching*, 75–89. London: Palgrave Macmillan.

Cheng, C. (ed.) (1996). *Masculinities in Organizations*. Thousand Oaks, CA: Sage.

Child, J., and Rodrigues, S. B. (2002). 'Corporate Governance and New Organizational Forms: The Problem of Double and Multiple Agency', Academy of Management Meeting, Denver, Aug.

Clegg, S., Kornberger, M., and Pitsis, T. (2005). *Managing and Organizations*. London: Sage.

cnn.com (2012) '100 Best Companies To Work For'. http://money.cnn.com/magazines/fortune/best-companies/2012/full_list (accessed 1 Oct. 2012).

Cockburn, C. (1983). *Brothers: Male Dominance and Technological Change*. London: Pluto Press.

Collinson, D. L. and Hearn, J. (1996). *Men as Managers, Managers as Men*. London: Sage.

Collinson, D., and Hearn, J. (2001 [1993]). 'Naming Men as Men: Implications for Work, Organization and Management', in S. M. Whitehead and F. J. Barrett (eds), *The Masculinities Reader*, 144–69. Cambridge: Polity.

Collinson, D., Collinson, M., and Knights, D. (1990). *Managing to Discriminate*. New York: Routledge.

Connell, R. W. (2000). *Masculinities*. Cambridge: Polity Press.

Connell, R. W. and Messerschmidt, J. W. (2005). 'Hegemonic Masculinity: Rethinking the Concept', *Gender and Society*, 19(6): 829–59.

Davidson, M., and Cooper, C. L. (1992). *Shattering the Glass Ceiling: The Woman Manager*. London: Paul Chapman.

Deleuze, G., and Guattari, F. (1980 [1988]). *A Thousand Plateaus*, tr. B. Massumi. London: Athlone.

Dent, M., and Whitehead, S. (eds) (2001). *Managing Professional Identities: Knowledge, Performativity and the 'New' Professional*. London: Routledge.

Fineman, S. (ed.) (1993). *Emotion in Organization*. London: Sage.

Fletcher, J. K. (1999). *Disappearing Acts: Gender, Power and Relational Practice*. Cambridge, MA: MIT Press.

Foucault, M. (1998). 'Technologies of the Self', in, L. H. Martin, J. Gutman, and P. Hutton (eds), *Technologies of the Self: A Seminar with Michel Foucault*, 16–49. Boston: University of Massachusetts Press.

Gherardi, S. (1995). *Gender, Symbolism and Organizational Cultures*. London: Sage.

Giddens, A. (1991). *Modernity and Self-Identity: Self and Society in the Late Modern Age*. Stanford, CA: Stanford University Press.

Grant Thornton (2012). *International Business Report 2012*. http://www.internationalbusiness-report.com/Press-room/2012_women.asp (accessed 21 Sep. 2012).

Halberstam, J. (1998). *Female Masculinity*. Durham, NC: Duke University Press.

Haywood, C., and Mac an Ghaill, M. (2003). *Men and Masculinities: Theory, Research and Social Practice*. Buckingham: Open University Press.

Hearn, J. (1992). *Men in the Public Eye*. London: Routledge.

Hearn, J., and Parkin, W. (2001). *Gender, Sexuality and Violence in Organizations*. London: Sage.

Hinojosa, R. (2010) 'Doing Hegemony: Military, Men, and Constructing a Hegemonic Masculinity', *Journal of Men's Studies*, 18(2): 179–96.

Hodgson, D. (2005). 'Putting on a Professional Performance: Performativity, Subversion and Project Management', *Organization*, 12(1): 51–68.

Holmes, J. (2006). *Gendered Discourse in the Workplace: Constructing Gender Identity through Workplace Discourse*. Oxford: Blackwell.

Institute of Management (1995). *National Management Salary Survey*. Kingston upon Thames: Institute of Management.

Itzin, C., and Newman, J. (eds) (1995). *Gender, Culture and Organizational Change*. London: Routledge.

Jackson, N., and Carter, P. (2007). *Rethinking Organisational Behaviour: A Poststructuralist Framework*. London: Prentice Hall.

Kanter, R. M. (1977). *Men and Women of the Corporation*. New York: Basic Books.

Kay, F., and Gorman, E. (2008). 'Women in the Legal Profession', *Annual Review of Law and Social Sciences*, 4: 299–332.

Kerfoot, D. (2001). 'The Organization of Intimacy: Managerialism, Masculinity and the Masculine Subject', in S. M. Whitehead and F. J. Barrett (eds), *The Masculinities Reader*, 233–52. Cambridge: Polity.

Kerfoot, D., and Knights, D. (2006 [1993]). 'Management, Masculinity and Manipulation: From Paternalism to Corporate Strategy in Financial Services in Britain', in S. M. Whitehead (ed.), *Men and Masculinities: Critical Concepts in Sociology*, ii. 7–27. London: Routledge.

Kerfoot, D., and Whitehead, S. (1998). ' "Boy's Own" Stuff: Masculinity and the Management of Further Education', *Sociological Review*, 46(3): 436–57.

Kerfoot, D., Prichard, C., and Whitehead, S. (eds) (2000). 'Special Issue: (En)Gendering Management: Work, Organisation and Further Education', *Journal of Further and Higher Education*, 24(2).

Kimmel, M. S. (2000). *The Gendered Society*. Oxford: Oxford University Press.

Knights, D., and Willmott, H. (1999). *Management Lives: Power and Identity in Work Organizations*. London: Sage.

Knights, D., and Willmott, H. (2012). *Introducing Organizational Behaviour and Management*. London: Thomson Learning.

Ledwith, S., and Colgan, F. (eds) (1996). *Women in Organisations*. London: Palgrave Macmillan.

Legge, K. (2005). *Human Resource Management: Rhetorics and Realities*. New York: Palgrave Macmillan.

Lyotard, J.-F. (1984). *The Postmodern Condition: A Report on Knowledge*. Manchester: Manchester University Press.

Mann, S. A., and Huffman, D. J. (2005). 'The Decentering of Second Wave Feminism and the Rise of the Third Wave', *Science and Society*, 69(1): 56–91.

Marshall, J. (1995). *Women Managers Moving On: Exploring Life and Career Choices*. London: Routledge.

Middleton, P. (1992). *The Inward Gaze: Masculinity and Subjectivity in Modern Culture*. London: Routledge.

Miller, G. E. (2004). 'Frontier Masculinity in the Oil Industry: The Experience of Women Engineers', *Gender, Work and Organization*, 11(1): 47–73.

Mills, A. J., and Tancred, P. (eds) (1992.) *Gendering Organizational Analysis*. London: Sage.

Morgan, D. (1992). *Discovering Men*. London: Routledge.

Office of National Statistics (2010). *Annual Survey of Hours and Earnings, 2010*. London: ONS.

Pacholok, S. (2009). 'Gendered Strategies of Self: Navigating Hierarchy and Contesting Masculinities', *Gender, Work and Organization*, 16(4): 471–500.

Parker, M. (2000). *Organizational Culture and Identity*. London: Sage.

Petersen, A. (1998). *Unmasking the Masculine: 'Men' and 'Identity' in a Sceptical Age*. London: Sage.

Pleck, J. H. (1981). *The Myth of Masculinity*. Cambridge, MA: MIT Press.

Prokos, A., and Padavic, I. (2006). ' "There Oughta Be a Law Against Bitches": Masculinity Lessons in Police Academy Training', in S. M. Whitehead (ed.), *Men and Masculinities: Critical Concepts in Sociology*, ii. 75–95. London: Routledge.

Roberts, Y. (1984). *Man Enough: Men of Thirty-Five Speak Out*. London: Chatto.

Roper, M. (1994). *British Organizational Man since 1945*. Oxford: Oxford University Press.

Salih, S., and Butler, J. (eds) (2004). *The Judith Butler Reader*. Malden, MA: Blackwell Publishing.

Segal, L. (1990). *Slow Motion: Changing Masculinities, Changing Men*. London: Virago.

Simpson, R. (2004). 'Masculinity at Work: The Experiences of Men in Female Dominated Occupations', *Work, Employment and Society*, 18(2): 349–68.

Smith, D. E. (1988). *The Everyday World as Problematic: A Feminist Sociology*. Milton Keynes: Open University Press.

Speer, S. A. (2001). 'Participant's Orientations, Ideology and the Ontological Status of Hegemonic Masculinity: A Rejoinder to Nigel Edley', *Feminism and Psychology*, 11(1): 141–4.

Tong, R. (1994). *Feminist Thought*. London: Routledge.

Weedon, C. (2003). *Feminist Practice and Poststructuralist Theory*. Oxford: Blackwell.

Whalley, M. A. (2011). 'Legitimising Educational Management Identity: Seductive Discourses of Professionalism, Masculinity and Performativity', doctoral thesis. Keele University, May.

Whitehead, S. (2001a). 'Woman as Manager: A Seductive Ontology', *Gender, Work and Organization*, 8(1): 84–107.

Whitehead, S. (2001b). 'Man: The Invisible Gendered Subject?', in S. M. Whitehead and F. J. Barrett (eds), *The Masculinities Reader*, 351–69. Cambridge: Polity.

Whitehead, S. M. (2002). *Men and Masculinities: Key Themes and New Directions*. Cambridge: Polity.

Whitehead, S. (2003). 'Identifying the Professional Manager: Masculinity, Professionalism and the Search for Legitimacy', in J. Barry, M. Dent, and M. O'Neill (eds), *Gender, Professionalism, and Managerial Change: An International Perspective*, 85–103. London: Macmillan.

Whitehead, S. M. (2006a). *Men and Masculinities: Critical Concepts in Sociology*, i–v. London: Routledge.

Whitehead, S. M. (2006b). 'Contingent Masculinites: Disruptions to "Man" Agerialist Identity', in S. M. Whitehead (ed.), *Men and Masculinities: Critical Concepts in sociology*, ii. 51–74. London: Routledge.

Whitehead, S. M., and Barrett, F. J. (eds) (2001). *The Masculinities Reader*. Cambridge: Polity.

Whitehead, S., and Moodley, R. (eds) (1999). *Transforming Managers: Gendering Change in the Public Sector*. London: UCL Press.

Whitehead, S. M., Talahite, A., and Moodley, R. (2013). *Gender and Identity: Key Themes And New Directions*. Oxford: Oxford University Press.

CHAPTER 21

MASCULINITY AND SEXUALITY AT WORK

Incorporating Gay and Bisexual Men's Perspectives

NICK RUMENS

INTRODUCTION

As the title of this *Handbook* indicates, the study of gender and organization has often been accorded priority by scholars over other aspects of individual difference, including sexuality. This has led not only to an important and varied feminist literature on the gender of organization and work, but also a critical scholarship on men and masculinities in the workplace. One motive for studying men and masculinity has been to expose a familiar but, paradoxically, often unacknowledged link between men, masculinity, and organization. In other words, it might be 'obvious' that men's identities, men's practices, and masculinities are inclined to dominate organizations but, as scholars associated with critical studies on masculinity and organizations aver (Collinson and Hearn 1994; Cheng 1996; Whitehead 2001; Hearn 2002), this connection has not always been established and, at times, intentionally avoided. There has been a sizeable boom in the number of studies that have 'named men as men', problematizing the relationship between men, masculinity, and organization, particularly the conflation of men with organizational power. The same literature advocates speaking about masculinity not in the singular but in the plural. But if there are indeed so many types of men and masculinities, why should men who identify as 'gay' and 'bi' (bisexual) be routinely overlooked within these debates?

This chapter contends that the critical organizational scholarship on men and masculinities is heteronormative in its tendency to normalize white, middle-class, heterosexual, able-bodied gender norms. If it is to avoid accusations of complicity in reproducing heteronormative analyses of men and masculinities, the field itself must acknowledge and address its omissions. One aim of this chapter is to name some of those missing men and masculinities by drawing on the sexuality of organization scholarship, especially the segment that examines sexual minorities in the workplace. While accounts of gay men are more visible in this research, in-depth analyses of bi men's perspectives and experiences are difficult to identify. Indeed, the sexuality of organization literature is not without its gaps, of which the study of bisexualities is one stark example. As such, another aim of this chapter is to engender dialogues between organizational scholars interested in men and masculinities and those scholars involved in examining gay and bi sexualities at work. It is hoped that researchers interested in genders and sexualities of organization may become better equipped to respond the question of 'which men?' when investigating how gender and sexuality influence the work lives of all men and women.

Undertaking such an endeavour, the first part of this chapter reviews some of the theoretical and empirical contributions in the critical literature on men and masculinities in organizations. Likewise, in the next section I review the sexualities of organization literature, paying close attention to the coverage of gay and bi-identified men, and how gender influences the construction of gay and bi sexualities within organizational settings. In the remainder of the chapter, I bring to the fore two examples of different men's perspectives—gay and bi men—in order to enrich research on masculinities and sexualities at work. Part of this discussion introduces queer theory to provide conceptual inspiration and support for maintaining scholarly efforts to account for the dynamics between men, masculinities, and organizations in ways that transcend gender and sexual binaries. Throughout the sections on gay and bi sexualities I indicate underdeveloped areas of research which warrant further investigation.

MEN AND MASCULINITIES IN ORGANIZATION

As mentioned above, one significant development within the field of work and organization is the growth in scholarship on gender in general and the critical studies of masculinity in particular. It is within the latter body of literature that a critical account of men, masculinity, and organization has emerged over the last two decades or so (Collinson and Hearn 1994; Cheng 1996; Kerfoot and Whitehead 2000; Hearn 2002), often with the aim of illuminating and problematizing how organizational cultures are numerically and symbolically gendered in terms of masculinity. Sociological research by Carrigan *et al.* (1987) and Connell (1995) has been particularly influential for coining and advancing the term 'hegemonic masculinity', which fosters an understanding of masculinity that recognizes dominant definitions of being masculine are organized by social institutions, such as work organizations. Thus hegemonic masculinity regards male dominance as an exercise in gendered power relations, enabling researchers to cultivate analyses of how some men may 'reproduce the social relations that generate their dominance' (Carrigan *et al.* 1987: 179). As noted by pro-feminist scholars (Whitehead 2002), the influence of hegemonic masculinity has been profound within the field of critical gender research, but later theoretical developments have sought to explain differently the dynamics between men, masculinities, gendered power, and organization.

As one prominent example, poststructuralist theories have enabled gender critical organizational scholars to import into their analyses a discursive understanding of gender and power. More than most, poststructuralist theories have opened up possibilities for claiming there is no stable, immanent, and universal relationship between men and masculinity, just as there is no intrinsic relationship between women and femininity. It is no coincidence that scholars influenced by poststructuralism often refer to the term 'masculinity' in the plural—*masculinities*—to signify the diversity in how masculinity may be variously formed, expressed, and attributed meaning in specific contexts at different moments in time (Kerfoot and Knights 1993, 1998; Whitehead 2002). As such, analysing masculinities in the plural rejects masculinity (and femininity) as a fixed essential property of an individual and underlines the fluidity and contextual contingency of masculinities. In some strains of poststructuralist theorizing, masculinity is not just taken to be constructed, but also 'performative'. Judith Butler's (1990) *Gender Trouble*, which is heavily indebted to feminism and poststructuralism, is a well-thumbed text for scholars seeking a performative notion of gender. In this frame, gender is said to be continuously (re)constructed through processes of signification, enabling gender critical scholars to understand how the performance of organizational masculinities is constrained by dominant gender norms. At the same time, a Butlerian approach accounts for the discursive opportunities that might be available for individuals to enact gender identities in ways that transcend designated boundaries and norms. This chapter broadly subscribes to a Butlerian conceptualization of gender, not least because it can guard against relaxing into essentialist accounts of gender that reproduce restrictive and harmful gender binaries.

Poststructuralism has provided the conceptual scaffolding for some of the most frequently cited accounts of masculinity, work, and organization (e.g. Kerfoot and Knights 1993, 1998; Collinson and Hearn 1994; Kerfoot and Whitehead 2000), and it is remarkable how few of these discuss the multiplicity of organizational men and masculinities in relation to those who identify as gay and bisexual. This is a depressing state of affairs, illustrated vividly in research on the relationship between men, masculinity, and management. Over the last few decades or so, this literature has made an important contribution to the study of men within the world of management, especially as, in the first sentence of the opening paragraph of one landmark text, *Men as Managers, Managers as Men* (1996: 1), Collinson and Hearn assert that 'most managers in most organizations in most countries are men'. The taken-for-granted nature of the relationship between men and management is the subject of critical examination in this text and others that have subsequently followed. None the less, despite the intense scrutiny this topic has attracted over the last twenty-five years or so (Broadbridge and Simpson 2011), it is striking that gay and bisexual men barely figure in these analyses. On the one hand, I doubt if the paucity of accounts on gay and bisexual men as managers betrays a serious aversion among scholars to researching the connections between gay and bisexual men, masculinities and management. On the other hand, the overwhelming focus displayed in this scholarship on heterosexual (and white, middle-class, able-bodied) men exposes the implicit heterosexism that has been imported into much of this scholarship, which also pervades the broader literature on men, masculinities, and organization (Steinman 2011).

While there are challenging methodological issues that make members of these 'hard to find' groups of men 'difficult' subjects to recruit as research participants (Ward and Winstanley 2004), these are not insurmountable obstacles. Put another way, not enough scholars have taken strident steps to engage with these groups of men within (non)managerial contexts, which is problematic given that so many critical gender commentators advocate dismantling the gender binary in the light of growing concerns about the normalization of white, middle-class, gender norms in the field of management and in other work contexts (Ashcraft 2009). The examination of non-conforming men in that respect is vital, of which many but crucially not all gay and bisexual men might qualify as illustrative examples. Disappointingly, the scarcity of literature on this topic reveals the limit of ambition among some scholars committed to undermining sexual and gender binaries. For insights into the employment experiences of gay and bi-identified men, it is necessary to examine the scholarship on the sexuality of organization.

SEXUALITIES OF ORGANIZATION

The study of sexuality in organization is an important body of literature, animated by early seminal texts including Hearn *et al.'s* (1989) *Sexuality of Organisation* and Hearn and Parkin's (1987) *Sex at Work*. These texts marked a radical departure from understanding sexuality (and gender) as something that could be imported into and potentially contaminate 'neutral' places of work towards understanding how sexuality and organization exist in a mutually constitutive relationship. Put differently, sexuality shapes organizational forms just as organizations can shape sexualities, in terms of how they are organized, expressed, and assigned meaning. Researchers on the sexuality of organization have come at this dynamic from different and competing theoretical directions; a review of these theories falls outside the scope of this chapter. However, one approach, which this chapter subscribes to, suggests organizations are constituted in and through discourses that relate to, among other things, gender *and* sexuality (Halford and Leonard 2006; Tyler and Cohen 2010). As such, organizations are, in part, shaped by restrictive, hierarchical binaries that help us to make sense of complex and 'messy' matters such as gender (masculine/feminine) and sexuality (heterosexuality/homosexuality).

The question that then arises is how to theorize the relationship between sexuality and gender in the workplace. There are too many possibilities here to sketch out—for a useful overview of the gender–sexuality dynamic more generally, see Richardson (2007). However, for Judith Pringle (2008), who draws on the gender management literature to illustrate her complaint, separating the concepts of gender and sexuality is deeply problematic. For one thing it neglects the myriad ways heterosexuality is normalized within the construct of gender, obscuring our understanding of how gender is informed by assumptions and displays of heterosexuality. This compels Pringle to utilize interview data gathered from lesbian managers in New Zealand in order to articulate this relationship as 'heterogender'. Here heterosexuality is viewed as an influential but often concealed organizing principle of gender. For others, gender and sexuality are seen to be interlinked but analytically separate, with some organizational researchers embracing the work of Judith Butler (1990, 1993, 2004) to that end. In an effort to carry through the full potential of Butler's arguments within organization studies, scholars have asserted that it is the recitation of sexual and gender norms which partly affords the subject recognition in organizational life (Borgerson 2005; Hancock and Tyler 2007; Tyler and Cohen 2010). This is to say that performances of sexuality and gender are 'intelligible' when they comply with sexual and gender norms that form part of an epistemological schema that privileges heterosexuality and masculinity, or what Butler (1990) calls the 'heterosexual matrix'. Performances of sexuality and gender that deviate from established hetero-norms are troublesome for they risk becoming 'unintelligible' and, potentially, 'queer' (Butler 2004). This directs a critical gaze towards 'queer possibilities' for performing gender and sexuality at work, surfaced when we mobilize queer theories to study the dynamic between organizational masculinities and sexualities: a deconstructive and destabilizing enterprise that I discuss later in this chapter. The exhortation to consider 'queer' performances of gender and sexuality within organizational environments marked by heteronormativity has a direct bearing on those who might be positioned as 'non-conformers' in that sense: gay and bi-identified men are two important examples.

GAY AND BISEXUAL MEN IN THE WORKPLACE

Organizational research on gay and bi-identified men is typically found in the literature on the employment experiences of sexual minorities. Emerging in the late 1970s, studies on employment discrimination on the grounds of sexual orientation have dominated the field, documenting numerous 'misery stories' about the multifarious practices and inimical effects of workplace heterosexism and homophobia (Levine 1979; Hall 1989; Woods and Lucas 1993; Ward and Winstanley 2003; Bowring and Brewis 2009). Related to this literature, studies on identity disclosure and identity management issues published since the 1980s have also increased our knowledge about dilemmas for interaction and identity building in the workplace. This body of work typically examines the antecedents and outcomes of identity disclosure (or 'coming out' as it has been termed in popular parlance) and the different strategies employed by members of sexual minority groups to 'manage' sexual identities within organizational life (Woods and Lucas 1993; Button 2004; Clair et al. 2005; DeJordy 2008). As lesbian, gay, bisexual, and transsexual (LGBT) people have increasingly gained rights and recognition in the public sphere

(Colgan *et al.* 2007; Giuffre *et al.* 2008; Colgan 2011), studies have emerged over the last three decades or so that have moved the sexualities of organization literature forward in at least two regards. First, recent studies remind us that, although many LGBT employees are able to understand and normalize sexual orientation at work in positive and meaningful ways, in the process new restrictions emerge imposing limits on recognition and visibility in the workplace (Williams *et al.* 2009; Rumens and Kerfoot 2009). One key message from much of this research is that while 'pro-gay/lesbian' organizations may acknowledge sexual diversity and engender feelings of 'inclusiveness', LGB sexualities are subject to particular processes of organizing that open and foreclose opportunities for constructing positive and 'open' identities. Studies by Williams *et al.* (2009) and others (Giuffre *et al.* 2008; Rumens and Kerfoot 2009) show that some LGB employees who are able to openly participate in organizational life still find themselves confined by colleagues to harmful stereotypes about how LGB people are expected to look, act, and work. Williams *et al.* (2009) coin the term 'gay-friendly closet' to describe the 'forced' choices many LGB employees make between acceptance and visibility within 'inclusive' places of work.

Second, following the above work, there is a growing concern that not all sexual minorities have received equal coverage or been studied with necessary sensitivity in regard to axes of differences that relate to gender, among others. For instance, gay men and lesbians figure centrally in many studies, while bisexuals and trans people remain on the periphery of many organizational research agendas (Green *et al.* 2011; Thanem 2011). For some researchers, crucial gender differences are obscured when the employment experiences of gay men are conflated with lesbians. For example, Wright (2011) rebukes Ward and Winstanley (2006) for not differentiating sufficiently the gender dimension in their study of study on gay men and lesbians in the UK fire services. Such criticism reminds us that we should not presume or prejudge the employment experiences of men and women at a group level, as countless individuals make different choices about how they wish to live as LGBT people at and outside work.

So, with the above in mind, what can we draw from the sexuality of organization literature for advancing the study of sexualities and masculinities in the context of work organizations? One pertinent observation is that, while the organizational literature on men and masculinities directs attention to how men are expected to work as 'men', it follows that we must also examine how and why men are expected to be *particular kinds of men*. This is a challenge not yet convincingly met in a great deal of extant research, as I have suggested above. By not being sufficiently critical in how we might locate specific men in relation to sexuality, binary analyses of men and women remain intact. As noted by gender critical scholars (Alvesson and Billing 2009; Ashcraft 2009), this is partly explained by the propensity for masculinities (and femininities) to take precedence over other facets of difference which are treated as secondary. Although the sexuality of organization literature has addressed this to some extent, plenty of scope exists for notions of masculinities and sexualities to be profoundly ruptured. The remainder of this chapter pursues this theme more pointedly, in relation to gay and bi-identified men.

GAY MALE SEXUALITIES

Although there is not a dire shortage of organizational research on gay men in the workplace, there are gaping holes in current research agendas. For example, studies on the diversity of gay masculinities within different work locales are uncommon, despite the predilection among gender critical scholars for theorizing masculinity in terms of multiplicity. While sexualities scholars often agree that gender shapes the construction of gay male sexualities, and vice versa, organizational research on sexual minorities which include gay men seldom fully accounts for the influence of gender when examining, for example, the disclosure and management of sexual identities at work (Button 2004; Ward and Winstanley 2005). In some studies there is not a clear sense of how gay men differ in terms of gender or, indeed, age, class, ethnicity, race, and (dis)ability, and how these formations of difference might shape the motives and outcomes of adopting specific identity disclosure and management strategies. One way to avoid treating gay men as an undifferentiated group and thereby enrich the organizational literature on masculinities is to burrow more deeply into the realm of gender and its intersection with gay sexualities. In so doing, we may challenge dualistic modes of thinking that endorse gender in a binary formation. Facets of the critical scholarship on masculinities and sexualities can inspire organization studies scholars in that endeavour.

One general point of consensus to emerge from these bodies of literature is that gay men frequently pose a threat to hegemonic definitions of masculinity that valorize characteristics such as assertiveness, rationality, conquest, and control (Kimmel 1994; Pascoe 2007). This is due partly to the antagonistic relationship between gay masculinities and traditional masculinities within a dichotomous sex/gender system of signification. Both rely on the other to define themselves in terms of what they are not, but heterosexuality sustains the differences that reproduce the dominance of masculinity. Homophobia has an important policing role to play here. This is illustrated vibrantly in research about how gay men manage gender and sexuality identities in male-dominated work contexts that have a significant influence in shaping wider cultural discourses of masculinity, such as the armed forces (Kaplan and Ben-Ari 2000). As an organizing principle of masculinities, homophobia regulates relations between men, helping them to determine which of them measure up and fall short of idealized standards of masculinity that invoke what it is to be a 'real man'. Significantly, this configuration of masculinities is structured within a binary formation, with the assumption that gay men are more likely to come under the category of those men who 'fail' at masculinity: men who can never be 'real men'. Crucially, however, the relationship between hegemonic masculinities and gay men is not as static and straightforward as this binary arrangement might have us believe.

Organizational researchers stand to benefit from investigating how organizations can open up and foreclose opportunities for gay men to construct understandings of masculinity that are less vulnerable to being understood as marginal, failed, or Othered. Such insights may be derived from sociological and queer scholarship on the variability of gay masculinities: from gay skins, 1970s clones, faeries, bears, leathermen to gay chavs (Hennen 2008; Brewis and Jack 2010). Not all these gay masculinities coexist peacefully, as gay men battle to perform and attribute particular meanings to certain masculinities in specific contexts at different moments in time. The delineation of different 'types' of gay men and masculinities has, especially in popular culture, suffered from crude caricaturing, but there is no reason to doubt that gay men might perform masculinities in diverse ways, as specific work situations enable and constrain such gender constructions and performances. One important avenue of inquiry is how gay sexualities and masculinities mingle, creating opportunities for their construction in terms of empowerment and pleasure (Hennen 2008), terms rarely associated with the study of gay men within heteronormative work contexts.

It is worth expanding on this, albeit briefly. For instance, research on gay men's workplace friendships shows how they can afford the individuals involved in them 'safe' relational contexts for inventing new ways of relating and identifying that resist the discursive pull of restrictive gender and sexual categories (Rumens 2008, 2010, 2011a). Such modes of organizing informally at work can be politically edged, undermining dominant binary discourses on gender and sexuality, producing a transformative effect by reshaping the gendered contours of some work environments (Rumens 2011a). Workplace friendships that involve men who identify differently sexually potentially represent a fresh empirical focal point for studying the power differentials between different men. For instance, some of the heterosexual men in Rumens (2010) experienced less pressure to perform constructions of traditional heteronormative masculinity in workplace friendships with gay men, revealing how gay sexualities may provide an effective way of dismantling how men perform gender at work. Indeed, some recent TV shows such as Bravo's *Queer Eye for the Straight Guy* (US) demonstrate there may be a thing or two gay men can teach heterosexual men about the merits of consumer masculinity, which rejects aspects of traditional heterosexual masculinity (Hart 2004). Acknowledging that some men might wish to draw on particular gay men for lessons in performing masculinity differently, researchers must also recognize the variability in the organizational contexts in which these gender performances are constituted rather than contained.

A great deal of organizational research on sexual minorities has been conducted in public sector organizations, largely because of a 'progressive' track record many claim to demonstrate in sexual orientation equality work (Carabine and Monro 2004; Colgan and Wright 2011). While research on sexual minorities in private sector companies continues to address the asymmetry in sector analysis (Colgan 2011), there is a limited engagement with organizations run by or set up to provide services and products to LGBT people. Extant research in this area examines the possibilities for gay men to perform gender and sexuality differently in work environments within gay nightclubs, bars, and leisure parks (Deverell 2001; Orzechowicz 2010; Andersson 2011). In some of these settings, gay rather than heterosexual masculinities may be dominant, privileged, energising, constructed as a source of 'professionalism', and experienced as pleasurable.

These work sites may be characterized by organizational cultures that accept and celebrate gay sexualities. In others, gay men may outnumber heterosexuals, furnishing gay men with opportunities to openly discuss partners, home life, engage in flirting and sexual banter, and develop intimacies with other men (Orzechowicz 2010; Rumens 2010). Whether the dominance of gay men and masculinities in such work contexts destabilizes the heterosexual/homosexual binary is questionable, requiring further empirical research. For example, Orzechowicz's (2010) study of a gay male-dominated parade department within a US leisure park suggests such work cultures can simultaneously fracture and reproduce gender and sexual binaries. Given the numerical dominance of gay men within one department, employees were assumed to be 'gay until proven to be straight', conditioning possibilities for certain gay masculinities to be experienced as pleasurable and empowering. At the same time, the assumption of homosexuality sustained the heterosexual/homosexual binary by, for example, excluding and marginalizing alternative sexualities relating to lesbians and bisexuals. Such studies are a novel twist on existing research that investigates how 'minority' employees are perceived by 'majority' employees. Indeed, research conducted along these lines could enrich the literature on men who enter female-dominated occupations and women who enter male-dominated occupations, which often do not take into full account the influence of sexual orientation (Lupton 2000; Simpson 2004).

In such work arenas, where conceptualizations of femininity and masculinity are highly visible and susceptible to contestation, new insights may be cultivated into how men are (dis)advantaged by their gendered 'token' status. For example, in male-dominated occupations such as the police services, some openly gay men may find themselves gendered by colleagues as naturally more 'feminine' than heterosexual men, and thus corralled into departments (e.g. sexual abuse and rape) where the display of feminine characteristics is deemed to be important (Rumens and Broomfield 2012). Recent research on gay male masculinities within academic women's studies problematizes facile stereotypes about, for example, female-dominated work contexts being the 'natural province' of gay men (Murphy 2011). One challenge confronted by gay men in female-dominated feminist work contexts is how to legitimate their presence without reasserting male dominance or retreating into the closet by playing down their sexuality and overlooking their disruptive potential to the gender binary. This, to my mind, helps to undermine the blithe assumptions made about gay men being 'naturally' closer to femininity and more comfortable in the company of women. For some gay men this stereotype may not be such a bad thing (Rumens 2012), but foregrounding the nuances in gay men's perspectives on how men manage masculinity at work will expand the stream of research that is committed to exposing the diversity within gender groups.

Another take on this is how gender non-conformity may no longer be a strong indicator of homosexuality (Richardson 2004). Such debates need to be understood against a social and cultural context in which processes of gay and lesbian normalization involve aligning homosexuality with idealized constructions of heterosexuality. Some sociology researchers contend that the meanings attached to gender non-conformity appear to have altered (Seidman 2002). Historically, 'real' homosexuals were typically those who were gender non-conformers—those who overidentified with femininity (Chauncey 1994). However, contemporary 'Western' constructions of 'normal' gay and lesbian identities often appear indistinguishable from heterosexual identities, giving rise to questions about whether the heterosexual/homosexual binary is becoming blurred or being dissolved. For some this is a powerful indicator of how many gay men and lesbians conform to hetero-gender norms willingly for a number of reasons, not the least of them being to seek acceptance, respectability, and be understood as 'normal'. Notably, gay men's bodies are also implicated in this gendered process of normalization, as some studies expose the concerns voiced by gay men about not wishing to appear 'too gay' at work, through excessive body sculpting, hyper-muscularity, or being seen as narcissistic through a love of fashion (Rumens and Kerfoot 2009; Rumens 2011b). Accountability to such gender norms is likely to vary within and between work organizations, variously shaping how gay men's bodies are discursively constructed, signposting an overlooked but significant line of future inquiry.

For some critical commentators, in order for some gay men to identify as 'normal', other (gay) sexualities must be positioned as the Other. For example, some gay men may borrow from hegemonic formations of dominant heterosexual masculinities to distance themselves discursively from those gay men considered 'effeminate' (Clarkson 2008). Effeminate gay men may find themselves attracting scorn not only from heterosexual men who endorse hegemonic heterosexual masculinities predicated on conquest, dominance, and control, but also from other gay men who subscribe to the same masculine characteristics. In the workplace, some gay men may be strategic in how they position themselves in relation to those men who are labelled 'effeminate'. Research shows how effeminacy in gay men can be negatively read as a sign of 'incompetence', 'unprofessionalism', and flaunting their sexuality in the face of colleagues (Rumens and Kerfoot 2009). Such is the ferocity of this distaste of the feminine displayed by some gay men that 'effeminophobia' rather than 'homophobia' has been used to label this form of gender regulation among men (Richardson 2009). For others, gay masculinities that draw on idealized constructions of heterosexual masculinities are contingent on both a high level of effeminophobia and homophobia, in order for gay men to slam gender performances that 'flaunt' gay sexuality. For Clarkson (2008), the apparent rejection of the association between effeminacy and homosexuality signals the potential to destabilize heteronormative, dichotomous discourses on gender and sexuality. From another perspective, effeminophobic behaviour bolsters heteronormative constructions of masculinity through marginalizing certain gay men and women. Such workplace issues merit future research but further analyses are also needed that can build on existing insights into how gay masculinities framed as 'effeminate' can, in some work locations, subordinate gay masculinities infused with traditional heterosexual masculinities, disrupting heteronormative gender binaries (Orzechowicz 2010).

In the efforts to complicate organizational masculinities, especially in the context of organizational heteronormativity, the focus on gay men represents one promising avenue of inquiry that may guard against treating the category of 'men' as an undifferentiated group. Still, there are potential pitfalls to be wary of. As Ashcraft (2009) cautions, the challenge confronting critical gender scholars is not to stretch gender or sexual binaries by simply proliferating differences within a dualistic frame. One effect of this is the proliferation of a multiplicity of genders, particularly in terms of identities, but within the same dualistic frame of masculinity and femininity. Still, there are other theoretical and empirical possibilities for scholars within the field of organizational masculinities and sexualities to rupture binary formations, as discussed in the next section.

Queering Organizational Masculinities and Sexualities: Male Bisexualities

In regard to male (and female) bisexualities in organizational settings, there is little research to go on. Some organization studies scholars have touched on the subject when discussing how to move beyond binary thinking in gender relations (Linstead and Pullen 2006). The promise bisexualities appear to hold in that respect relate to the way bisexuals are frequently considered to be fluid and dynamic in their sexuality. Indeed, the literature on bisexualities more generally is largely conceptual, typically addressing definitional issues (Steinman 2011). Empirical accounts that solely focus on bisexuality as a lived experience have grown appreciably over the last ten years or so (Elia and Eliason 2012), revealing the opportunities and limitations of the conceptual fluidity routinely allotted to bisexuality (discussed below), although these rarely extend into the workplace. As suggested previously, within the organization literature on sexual minorities, bisexuals tend to figure in small numbers, often outnumbered in study samples by gay men and lesbians. In sum, the coverage of bisexualities as a serious organizational topic of study is highly variable. Before considering some of the options available to researchers for incorporating bisexualities into existing debates on organizational masculinities and sexualities at work, it is useful to begin by noting how bisexualities might be theorized. For insight here we must venture beyond the organization studies literature and tap into the well-established social science and humanities scholarship on sexualities.

One feature of the sexualities literature is that there has been much interest in using queer theory as a lens through which to examine and often applaud the disruptive potential of bisexualities for shattering restrictive sexual and gender binaries (Burrill 2009; Callis 2009; Erickson-Schroth and Mitchell 2009; Feldman 2009; Steinman 2011). This marks a welcome point of department for queer theory analyses, which have traditionally focused on homosexuality. This is not altogether surprising, given queer theory's suspicion of and relentless resistance to neatly defined categories imbued with a sense of fixity, whether they relate to sexuality and gender (Butler 1990, 2004) or relationships and intimacies (Warner 1999). Nonetheless, for some commentators, queer theory has become synonymous with gay men and gay male sexualities that have excluded the concerns of (lesbian) women (Jeffreys 2003) and bisexuals (Storr 1999). Some of these criticisms are valid, but the appeal of queer theories for scholars interested in analysing bisexualities remains strong. It is worth elaborating on this.

Queer theory is famously difficult to define but its definitional indeterminacy makes queer theoretically attractive for questioning taken-for-granted assumptions that underpin everyday life (Jagose 1996). For many theorists associated with queer theory, sexuality and gender are viewed as discursive effects, thereby denaturalizing sexual and gender categories as fixed, natural, and predictable (Butler 1990, 2004; Sedgwick 1990). The crisp focus on sexuality and gender partly distinguishes queer theory from poststructuralism more generally, which had a huge bearing on its germination within the humanities departments of universities during the 1990s. In destabilizing sexuality and gender, queer theorists expose the constraining power of normative frameworks, as experienced by many LGBT people who must ward off violence (e.g. homophobia, bi-negativity) done by restrictive norms, in order to openly participate in a heteronormative society (Butler 1990). While queer theory is adept at underscoring the oppressive effects of heteronormativity, it also allows us to see how lives and identities can be constructed differently, at some distance from restrictive norms (Butler 2004). Here then queer theorizations of bisexualities may be seen to bear fruit, because they nourish a politics of transgression and subversion based on strategies such as deconstructing unified sexual categories (e.g. heterosexual/homosexual) understood in hierarchical and binary terms. From a queer perspective, when shoehorned into a binary schema that posits individuals as either heterosexual or homosexual, bi-identified people may find themselves struggling to mobilize relevant discursive resources for articulating their experiences as gendered and sexual subjects (Bereket and Brayton 2008).

One benefit of this approach to theorizing bisexualities is that it resists constructing bisexuality as 'immature' when compared to heterosexuality and homosexuality (Erickson-Shroth and Mitchell 2009). This is conceptually empowering but the realities of openly participating in and outside work as 'bisexual' reveals the harsh opprobrium and discrimination bi-identified individuals may attract (Eliason 2000; Klesse 2011). Defying rigid binary formations has its costs, as bisexuality and bisexuals are dogged by cruel stereotypes that emphasize internal conflict, instability, and immaturity in comparison to heterosexuality and homosexuality, which are assumed to be stable and 'mature' sexual categories. As such, compelling research on bisexualities exposes the multifarious practices of bi-negativity that feed the misapprehension that bisexuality is not a sexual orientation and that bisexuals are merely people who cannot make up their minds when it comes to deciding between two objects of desire (Klesse 2011). Such crude assumptions reinforce a restrictive gender binary by apportioning to bisexuality an excessive fluidity and indeterminacy which reinforces it as a non-sexuality. Queer theories may be seen to reject this mode of understanding bisexuality, viewing it as a subversive option in disassembling a binary system that privileges heterosexuality but depends on homosexuality as a subordinated Other (Horncastle 2008; Callis 2009; Erickson-Shroth and Mitchell 2009).

Turning to the particular issues that relate to theorizing and living male bisexualities, one general observation is pertinent here. As mentioned earlier, research on bisexualities is a relatively new phenomenon, but it is one that has attracted increasing attention among sexualities scholars. For instance, the *Journal of Bisexuality*, just over a decade old, exemplifies some of the achievements in the critical enterprise of scholars to firmly position bisexualities on research agendas. Despite these progressive developments within the field of bisexualities, it is striking that studies on male bisexualities are relatively uncommon. As Steinman (2011) notes, in regard the content of *Journal of Bisexuality* over the last ten years, there is appreciably less research on and produced by bi-identified men than research about and conducted by bi women. A number of reasons may be put forward to explain this discrepancy (e.g. bi men appear less willing to articulate their experiences), but one consequence of this asymmetry is that we do not know enough about the influence of gender in the construction of bisexualities. Addressing the deficit of research on bi men, scholars are encouraged to establish links with the critical literature on masculinities. One advantage of yoking together the critical scholarship on masculinities and research on bisexualities is deeper theoretical and empirical insights into how male bisexualities are gendered in specific contexts, and to what effect.

Taking the workplace as one important example, preliminary research published in this area reveals some startling but, arguably, not wholly surprising findings. Green *et al.* (2011) conducted an international survey of the experience of bisexual people in the workplace, which indicates that being 'out' as bi at work is linked to a higher quality of work life, especially when employers are committed to developing policies and organizational practice that targets both sexual orientation and gender identity issues. At the same time, being out is a risky enterprise due to the pervasiveness of bi-negativity. Many study respondents reported that gay and heterosexual colleagues misunderstood bisexuality, found they were not accepted as 'legitimate' members of LGBT employee resource groups, and perceived by colleagues as 'untrustworthy', 'unreliable and/or indecisive'. Many survey participants felt this had a damaging effect on their career advancement prospects. A similar picture is painted by Chamberlain's (2009) research conducted through the UK LGB charity Stonewall. It is not altogether remarkable that, when faced with the prospect of having to 'explain' their behaviour and bi identity to others, bi men and women are dissuaded from disclosing to colleagues. From this research, organizational studies scholars can formulate an initial picture of contemporary organizational life for bisexual men and women. However, in the near absence of published studies that directly focus on bi-men's experiences of the workplace, the lacuna in current knowledge about how masculinities and male bisexualities intersect remains cavernous. One consequence of this omission is the sketchy current understanding of how bi men may construct and experience specific workplace encounters and relationships in terms of gender. The salience of this line of inquiry is underscored by the critical masculinities literature which shows how much is at stake for men who struggle to construct a seemingly coherent gender identity (Kimmel 1994).

As examples of men who risk being read as gender incoherent, bi men may encounter 'gender troubles' in the workplace that relate to the epistemological violence that is done when individuals do not fit sexual categories (Butler 2004). One particular issue relates to the difficulties men face in managing relations with other men in ways that are pleasurable without incurring accusations of homosexuality (Nardi 1992). This leads Steinman (2011) and others (Eliason 2000) to suggest that bi men might find it extremely challenging, perhaps more so than bi women, to construct positive gendered identities and subjectivities in the workplace, especially within environments where traditional notions of masculinity hold firm. Issues like these are likely to be of great importance to critical researchers interested in organizational masculinities, especially in occupations dominated by certain types of men and women (Lupton 2000; Simpson 2004). The significance of this point is underlined by the various ways bi men might be positioned. For example, heterosexuals who elide bisexuality with promiscuity are likely to construct bi men as sexual perverts and a public health risk, especially in terms of spreading sexual diseases (Steinman 2011). For some gay men (and lesbians), the perceived sexual greediness and promiscuity of bi men threatens to thwart discursive efforts to normalize homosexuality in relation to idealized versions of heterosexuality, in terms of monogamy. From another viewpoint, for those gay men (and lesbians) committed to understanding homosexuality as an *essential* part of themselves, male bisexualities raise uncomfortable and inconvenient questions about the constructed nature of sexuality. Of course female bisexualities may inspire similar concerns and anxieties, but the dearth of comparative analyses on gender differences between bi men and women does very little to turn present speculation into empirically informed insight.

In sum, bisexualities can be attacked on a number of (un)expected fronts. Therefore, it important to remember that while epistemological transgressions can be transformative and exhilarating for some bi men, for others it is exactly the opposite (McLean 2008). Not all bi (and gay) men have an appetite for binary bashing. Some bi men may wish to evade persecution and discrimination by not disclosing their bi identity to colleagues, while others may actively court 'gender trouble' in the workplace, by refusing to be constrained by organizational gender and sexual norms (Rumens 2012). What is more, many bisexual men and women may want to infuse their bisexual identities with a sense of stability rather than fluidity, creating a discrete and fixed 'third' sexual category. This might well create a new dynamic, one characterized by an abandonment of extremes and absolutes but, as Linstead and Pullen (2006) rightly argue, this might be difficult to realize for those individuals who have a substantial investment in one end of the binary. How these discursive tensions and their material effects are played out within organizations, amidst the welter of organizational gender and sexual norms that influence what is accepted as '(ab)normal', merits future research. Returning then to the notion of queering organizational masculinities, the matter of bisexualities is central to such an endeavour. Attentiveness to queer possibilities for performing gender and sexuality at work, as they might be enacted by bi-identified men, may seek to reduce the marginalizing effects of current heteronormative research agendas and promote a more explicitly critical organizational literature on masculinities and sexualities. This is not to overlook the apprehensions of those for whom queer theory has favoured the concerns of gay men over women and bisexuals. Rather, this is to suggest that queering organizational masculinities needs to strike roots in the critical masculinities literature and sexualities scholarship more deeply, generating a sharper critical conscious about how bisexualities might constitute a new aperture for moving beyond binary thinking in gender and sexuality (Linstead and Pullen 2006). Part of this investigation entails examining whether organizational bisexualities, in the way they are constructed, sustained, embodied, and experienced generates fluidity of the kind that dissolves entrenched binary boundaries. Organizational scholars have their work cut out here, and queer theories are one conceptual resource of many for supporting these investigations.

CONCLUDING REMARKS

I have suggested in this chapter that the study of men and masculinities in the workplace has demonstrated limited engagement with men who identify as gay and bi. Before outlining a horizon of different possibilities in respect to gay and bi men's perspectives and experiences, two broad segments of scholarship have been sketched out: (1) the critical gender literature which focuses on organizational masculinities; (2) the sexuality of organization literature. While applauding some of the scholarly contributions to naming men as men in organizations, I have also argued that this scholarship tends to normalize heterosexuality which unhelpfully maintains dualistic modes of understanding sexuality and gender. Gay and bi-identified men do not regularly figure in the critical masculinities literature and there appears to be low commitment among scholars for positioning bisexualities as a focus for empirical investigation in the organizational sexuality literature. The task of developing a richer organizational literature on gender, to which this *Handbook* is deeply committed, is vital and it must account fully for how gender is shaped by a range of sexualities. Otherwise we perpetuate the tendency to prioritize gender over other aspects of difference such as sexuality, thereby inadvertently treating sexuality and gender as detached conceptual outposts. Gay and bi men's sexualities are two examples presented here to challenge the heteronormativity of the gender binary and enrich the discussion of gender within the sexualities of organization scholarship. However, other examples that deserve scholarly attention include the sexualities and genders that relate to trans people, bisexual women, pansexual, and metrosexual men and women.

I am optimistic about the type of fruit such scholarly endeavours might produce. Critical gender scholars on organizational masculinities benefit from incorporating gay masculinities into organization analyses as a way of resisting the normalization of heterosexuality in gender analyses and problematizing dualistic understandings of dominant/subordinate masculinities. Scholars in the sexuality of organization literature can derive deeper empirical insights into bisexualities, which suffer from the effects of marginalization within the organizational literature on sexual minorities. Building bridges between these two broad literatures is essential, not least for developing theoretical analyses of how gender and sexuality are mutually influencing within organization contexts. Thus one aim of this chapter has been to engender alternative ways of understanding men, masculinities and sexualities in the workplace. And I would cite queer theory as one but crucially not the only way forward here, especially given its limited impact on the field of organization studies so far (Rumens 2013). Queer theory has its detractors and limitations, but there are exciting opportunities for queer theories to inform empirical inquiry aimed at exploring the gender and sexuality of organizations, especially as numerous scholars working within these fields routinely clamour for the dismantling of gender and sexual binaries. Articulating fresh perspectives on this matter, like those expressed by bisexual men, helps us to search for the 'queer existences' and 'queer possibilities' that inform alternative ways of accounting for the gender and sexuality of organization.

REFERENCES

Alvesson, M. and Billing, Y. D. (2009). *Understanding Gender and Organizations*. London: Sage.

Andersson, J. (2011). 'Vauxhall's Post-Industrial Pleasure Gardens: "Death Wish" and Hedonism in 21st-Century London', *Urban Studies*, 48(1): 85–100.

Ashcraft, K. L. (2009). 'Gender and Diversity: Other Ways to "Make a Difference"', in M. Alvesson, T. Bridgman, and H. Willmott (eds), *The Oxford Handbook of Critical Management Studies*, 304–27. Oxford: Oxford University Press.

Bereket, T., and Brayton, J. (2008). '"Bi" No Means: Bisexuality and the Influence of Binarism on Identity', *Journal of Bisexuality*, 8(1–2): 51–61.

Borgerson, J. (2005). 'Judith Butler: On Organizing Subjectivities', *Sociological Review*, 53(1): 63–79.

Bowring, M., and Brewis, J. (2009). 'Truth and Consequences: Managing Lesbian and Gay Identity in the Canadian Workplace', *Equal Opportunities International*, 28(5): 361–77.

Brewis, J., and Jack, G. (2010). 'Consuming Chavs: The Ambiguous Politics of Gay Chavinism', *Sociology*, 44(2): 251–68.

Broadbridge, A., and Simpson, R. (2011). '25 Years On: Reflecting on the Past and Looking to the Future in Gender and Management Research', *British Journal of Management*, 22(3): 470–83.

Burrill, K. G. (2009). 'Queering Bisexuality', *Journal of Bisexuality*, 9(3–4): 491–9.

Butler, J. (1990). *Gender Trouble: Feminism and the Subversion of Identity*. London: Routledge.

Butler, J. (1993). *Bodies that Matter: On the Discursive Limits of 'Sex'*. New York: Routledge.

Butler, J. (2004). *Undoing Gender*. London: Routledge.

Button, S. (2004). 'Identity Management Strategies Used by Gay and Lesbian Employees: A Quantitative Investigation', *Group and Organization Management*, 29(4): 470–94.

Callis, A. S. (2009). 'Playing with Butler and Foucault: Bisexuality and Queer Theory', *Journal of Bisexuality*, 9(3–4): 213–33.

Carabine, J., and Monro, S. (2004). 'Lesbian and Gay Politics and Participation in New Labour's Britain', *Social Politics*, 11(2): 312–27.

Carrigan, T., Connell, B., and Lee, J. (1987). 'Hard and Heavy: Toward a New Sociology of Masculinity', in M. Kaufman (ed.), *Beyond Patriarchy: Essays by Men on Pleasure, Power, and Change*, 139–92. Toronto: Oxford University Press.

Chamberlain, B. (2009). *Bisexual People in the Workplace: Practical Advice for Employers*. London: Stonewall.

Chauncey, G. (1994). *Gay New York: Gender, Urban Culture, and the Makings of the Gay Male World, 1890–1940*. New York: Basic Books.

Cheng, C. (ed.) (1996). *Masculinities in Organizations*. Thousand Oaks, CA: Sage.

Clair, J., Beatty, J., and MacLean, T. (2005). 'Out of Sight But Not Out of Mind: Managing Invisible Social Identities in the Workplace', *Academy of Management Review*, 30(1): 78–95.

Clarkson, J. (2008). 'The Limitations of the Discourse of Norms: Gay Visibility and Degrees of Transgression', *Journal of Communication Inquiry*, 32(4): 368–82.

Colgan, F. (2011). 'Equality, Diversity and Corporate Responsibility: Sexual Orientation and Diversity Management in the UK Private Sector', *Equality, Diversity and Inclusion: An International Journal*, 30(8): 719–34.

Colgan, F., and Wright, T. (2011). 'Lesbian, Gay and Bisexual Equality in a Modernizing Public Sector 1997–2010: Opportunities and Threats', *Gender, Work and Organization*, 18(5): 548–70.

Colgan, F., Creegan, C., McKearney, A., and Wright, T. (2007). 'Equality and Diversity Policies and Practices at Work: Lesbian, Gay, and Bisexual Workers', *Equal Opportunities International*, 26(6): 590–609.

Collinson, D. L., and Hearn, J. (1994). 'Naming Men as Men: Implications for Work, Organization and Management', *Gender, Work and Organization*, 1(1): 2–22.

Collinson, D. L. and Hearn, J. (eds) (1996). *Men as Managers, Managers as Men*. London: Sage.

Connell, R. W. (1995). *Masculinities*. Berkeley, CA: University of California Press.

DeJordy, R. (2008). 'Just Passing Through: Stigma, Passing, and Identity Decoupling in the Workplace', *Group and Organization Management*, 33(5): 504–31.

Deverell, K. (2001). *Sex, Work and Professionalism: Working in HIV/AIDS*. London: Routledge.

Elia, J. P., and Eliason, M. (2012). 'A Decade of the *Journal of Bisexuality*: Some Notes on Content and Future Directions', *Journal of Bisexuality*, 12(1): 4–12.

Eliason, M. (2000). 'Bi-negativity', *Journal of Bisexuality*, 1(2–3): 137–54.

Erickson-Schroth, L., and Mitchell, J. (2009). 'Queering Queer Theory, or Why Bisexuality Matters', *Journal of Bisexuality*, 9(3–4): 297–315.

Feldman, S. (2009). 'Reclaiming Sexual Difference: What Queer Theory Can't Tell us about Sexuality', *Journal of Bisexuality*, 9(3–4): 259–78.

Giuffre, P., Dellinger, K., and Williams, C. L. (2008). 'No Retribution for Being Gay? Inequality in Gay-Friendly Workplaces', *Sociological Spectrum*, 28(3): 254–77.

Green, H. B., Payne, N. R., and Green, J. (2011). 'Working Bi: Preliminary Findings from a Survey on Workplace Experiences of Bisexual People', *Journal of Bisexuality*, 11(2–3): 300–16.

Halford, S., and Leonard, P. (2006). *Negotiating Gendered Identities at Work*. Basingstoke: Palgrave Macmillan.

Hall, M. (1989). 'Private Experiences in the Public Domain: Lesbians in Organizations', in J. Hearn, D. L. Sheppard, P. Tancred-Sheriff, and G. Burrell (eds), *The Sexuality of Organization*, 125–38. London: Sage.

Hancock, P., and Tyler, M. (2007). 'Un/doing Gender and the Aesthetics of Organizational Performance', *Gender, Work and Organization*, 14(1): 512–33.

Hart, K. R. (2004). 'We're Here, We're Queer—and We're Better than You: The Representational Superiority of Gay Men to Heterosexuals on *Queer Eye for the Straight Guy*', *Journal of Men's Studies*, 12(3): 241–53.

Hearn, J. (2002). 'Alternative Conceptualizations and Theoretical Perspectives on Identities and Organizational Cultures: A Personal Review of Research on Men in Organizations', in I. Aaltio and A. J. Mills (eds), *Gender, Identity and the Culture of Organizations*, 39–56. London: Routledge.

Hearn, J., and Parkin, W. (1987). *Sex at Work: The Power and Paradox of Organisation*. Brighton: Wheatsheaf.

Hearn, J., Sheppard, D. L., Tancred-Sheriff, P., and Burrell, G. (eds) (1989). *The Sexuality of Organization*. London: Sage.

Hennen, P. (2008). *Fairies, Bears and Leathermen: Men in Community Queering the Masculine*. Chicago: University of Chicago Press.

Horncastle, J. (2008). 'Queer Bisexuality: Perceptions of Bisexual Existence, Distinctions, and Challenges', *Journal of Bisexuality*, 8(1–2): 25–49.

Jagose, A. (1996). *Queer Theory: An Introduction*. Melbourne: Melbourne University Press.

Jeffreys, S. (2003). *Unpacking Queer Politics: A Lesbian Feminist Perspective*. Cambridge: Polity Press.

Kaplan, D., and Ben-Ari, E. (2000). 'Brothers and Others in Arms: Managing Gay Identity in Combat Units of the Israeli Army', *Journal of Contemporary Ethnography*, 29(4): 396–432.

Kerfoot, D., and Knights, D. (1993). 'Management, Masculinity and Manipulation: From Paternalism to Corporate Strategy in Financial Services in Britain', *Journal of Management Studies*, 30(4): 659–79.

Kerfoot, D., and Knights, D. (1998) 'Managing Masculinity in Contemporary Organizational Life: A "Man"agerial Project', *Organization*, 5(1): 7–26.

Kerfoot, D., and Whitehead, S. (2000). 'Keeping All the Balls in the Air: Further Education and the Masculine/Managerial Subject', *Journal of Further and Higher Education*, 24(2): 183–202.

Kimmel, M. S. (1994) 'Masculinity as Homophobia: Fear, Shame and Silence in the Construction of Gender Identity', in H. Brod and M. Kaufman (eds), *Theorizing Masculinities*, 119–41. Thousand Oaks, CA: Sage.

Klesse, C. (2011). 'Shady Characters, Untrustworthy Partners, and Promiscuous Sluts: Creating Bisexual Intimacies in the Face of Heteronormativity and Biphobia', *Journal of Bisexuality*, 11(2–3): 227–44.

Levine, M. P. (1979). 'Employment Discrimination Against Gay Men', *International Review of Modern Sociology*, 9: 151–63.

Linstead, S., and Pullen, A. (2006). 'Gender as Multiplicity: Desire, Displacement, Difference and Dispersion', *Human Relations*, 59(9): 1287–310.

Lupton, B. (2000). 'Maintaining Masculinity: Men Who Do "Women's Work"', *British Journal of Management*, 11(3): 33–48.

McLean, K. (2008). 'Inside, Outside, Nowhere: Bisexual Men and Women in the Gay and Lesbian Community', *Journal of Bisexuality*, 8(1–2): 63–80.

Murphy, M. J. (2011). 'You'll Never be More of a Man: Gay Male Masculinities in Academic Women's Studies', *Men and Masculinities*, 14(2): 173–89.

Nardi, P. M. (1992). 'Sex, Friendship, and Gender Roles among Gay Men', in P. M. Nardi (ed.), *Men's Friendships*, 173–85. Newbury Park, CA: Sage.

Orzechowicz, D. (2010). 'Fierce Bitches on Tranny Lane: Gender, Sexuality, Culture, and the Closet in Theme Park Parades', in C. L. Williams and K. Dellinger (eds), *Gender and Sexuality in the Workplace*, 227–52. Research in the Sociology of Work, 20. Bingley: Emerald Group.

Pascoe, C. J. (2007). *Dude, You're a Fag: Masculinity and Sexuality in High School*. Berkeley, CA: University of California Press.

Pringle, J. K. (2008). 'Gender in Management: Theorizing Gender as Heterogender', *British Journal of Management*, 19(1): 110–19.

Richardson, D. (2004). 'Locating Sexualities: From Here to Normality', *Sexualities*, 7(4): 391–411.

Richardson, D. (2007). 'Patterned Fluidities: (Re)Imagining the Relationship between Gender and Sexuality', *Sociology*, 41(3): 457–74.

Richardson, N. (2009). 'Effeminophobia, Misogyny and Queer Friendship: The Cultural Themes of Channel 4's *Playing It Straight*', *Sexualities*, 12(4): 525–44.

Rumens, N. (2008). 'Working at Intimacy: Gay Men's Workplace Friendships', *Gender, Work and Organization*, 15(1): 10–30.

Rumens, N. (2010). 'Workplace Friendships between Men: Gay Men's Perspectives and Experiences', *Human Relations*, 63(10): 1541–62.

Rumens, N. (2011a). *Queer Company: The Role and Meaning of Friendship in Gay Men's Work Lives*. Aldershot: Ashgate.

Rumens, N. (2011b). 'Minority Support: Friendship and the Career Experiences of Gay and Lesbian Managers', *Equality, Diversity and Inclusion: An International Journal*, 30(6): 444–62.

Rumens, N. (2012). 'Queering Cross-Sex Friendships: An Analysis of Gay and Bi Men's Workplace Friendships with Heterosexual Women', *Human Relations*, 65(8): 955–78.

Rumens, N. (2013). 'Organization Studies: Not Nearly Queer Enough', in Y. Taylor and M. Addison (eds), Queer Presences and Absences, 241–259. Basingstoke: Palgrave Macmillan.

Rumens, N., and Broomfield, J. (2012). 'Gay Men in the Police: Identity Disclosure and Management Issues', *Human Resource Management Journal*, 22(3): 283–98.

Rumens, N., and Kerfoot, D. (2009). 'Gay Men at Work: (Re)constructing the Self as Professional', *Human Relations*, 62(5): 763–86.

Sedgwick, E. K. (1990). *The Epistemology of the Closet*. Berkeley, CA: University of California Press.

Seidman, S. (2002). *Beyond the Closet: The Transformation of Gay and Lesbian Life*. New York: Routledge.

Simpson, R. (2004). 'Masculinity at Work: The Experiences of Men in Female Dominated Occupations', *Work, Employment and Society*, 18(2): 349–68.

Steinman, E. (2011). 'Revisiting the Invisibility of (Male) Bisexuality: Grounding (Queer) Theory, Centering Bisexual Absences and Examining Masculinities', *Journal of Bisexuality*, 11(4): 399–411.

Storr, M. (1999). 'Postmodern Bisexuality', *Sexualities*, 2(3): 309–25.

Thanem, T. (2011). 'Embodying Transgender in Studies of Gender, Work, and Organization', in E. Jeanes, D. Knights, and P. Y. Martin (eds), *Handbook of Gender, Work, and Organization*, 191–204. Oxford: Wiley/Blackwell.

Tyler, M., and Cohen, L. (2010). 'Spaces that Matter: Gender Performativity and Organizational Space', *Organization Studies*, 31(2): 175–98.

Ward, J., and Winstanley, D. (2003). 'The Absent Present: Negative Space within Discourse and the Construction of Minority Sexual Identity in the Workplace', *Human Relations*, 56(10): 1255–80.

Ward, J., and Winstanley, D. (2004). 'Sexuality and the City: Exploring the Experience of Minority Sexual Identity through Storytelling', *Culture and Organization*, 10(3): 219–36.

Ward, J., and Winstanley, D. (2005). 'Coming out at Work: Performativity and the Recognition and Renegotiation of Identity', *Sociological Review*, 53(3): 447–75.

Ward, J., and Winstanley, D. (2006). 'Watching the Watch: The UK Fire Service and its Impact on Sexual Minorities in the Workplace', *Gender, Work and Organization*, 13(2): 193–219.

Warner, M. (1999). *The Trouble with Normal: Sex, Politics, and the Ethics of Queer Life*. New York: Free Press.

Whitehead, S. (2001). 'Woman as Manager: A Seductive Ontology', *Gender, Work and Organization*, 8(1): 84–107.

Whitehead, S. M. (2002). *Men and Masculinities*. Oxford: Blackwell.

Williams, C. L., Giuffre, P. A., and Dellinger, K. (2009). 'The Gay-Friendly Closet', *Sexuality Research and Social Policy*, 6(1): 29–45.

Woods, J. D., and Lucas, J. H. (1993). *The Corporate Closet: The Professional Lives of Gay Men in America*. New York: Free Press.

Wright, T. (2011). 'A "Lesbian Advantage"? Analysing the Intersections of Gender, Sexuality and Class in Male Dominated Work', *Equality, Diversity and Inclusion: An International Journal*, 30(8): 686–701.

CHAPTER 22

....

DOING GENDER
DIFFERENTLY

Men in Caring Occupations

....

RUTH SIMPSON

INTRODUCTION

....

IN this chapter I pull together the findings from a number of studies of men in non-traditional occupations, including my own research in Australia and the UK, to explore how men 'do gender' in 'feminized' work, i.e. occupations that are traditionally held by women and which are notable for requiring skills and attributes (e.g. sensitivity, service, nurturance, care) that society associates with femininity (Hochschild 1983; Heilman 1997). The chapter draws on a perspective that sees gender not as the property of the individual or as a simple, unambiguous category, but as 'situated doing' (West and Zimmerman 2002) or performance (Butler 1990). This focuses on how gender differentiated practices take place in the light of normative and localized conceptions of what it means to be a woman or a man (Moloney and Fenstermaker 2002).

The context of 'feminized' work is a powerful one for exploring masculinities and how they are constructed, resisted, and maintained. While in other contexts masculinity is often taken for granted as the normative case and therefore 'opaque' to analysis (Collinson and Hearn 1994), in feminized work, as Morgan (1992) points out, conceptualizations of masculinity are 'on the line', highly visible and vulnerable to challenge. Men in these contexts 'stand out' and their gender is rendered visible: they are seen as gendered subjects, as *men* working in gender-atypical roles. Equally, men face challenges in a non-traditional career that can help surface particular gender dynamics. Being both a man and a nurse for example involves occupying two contradictory subject positions in the 'discourse on work' (Fletcher 2003), based on 'masculine' attributes of independence and detachment as well as 'feminine' notions of nurturance and care.

It is through these contradictions that new knowledge can be gained, with potential, as Connell (2000) argues, to understand and challenge the gender order.

In a general context, occupational 'sex typing' and the labelling of occupations as 'masculine' or 'feminine' mean that jobs carry assumptions concerning their suitability, in terms of the skills and attributes required, for women or men (Acker 1990; Williams 1995; Fletcher 2003). Different meanings attached to these skills and attributes can lead to a greater value placed on those that are associated with men and an undervaluation of those traditionally performed or seen to be possessed by women (often translating into differentials in status and pay). The persistence of these associations can be seen in the high levels of gender segregation by occupational groups. In the UK, for example, 81 per cent of care, leisure, and service workers as well as 77 per cent of administrative and sec-retarial staff are women, while 92 per cent of skilled trades, 89 per cent of machine oper-atives, and 68 per cent of managers and senior officials are men (ONS 2011). Further, women represent just 6 per cent of engineering, 13 per cent of ICT, and 14 per cent of architectural posts (EHRC 2011). In fact, a report by the EOC (2007) has indicted that over 60 per cent of occupations are performed mainly by men or by women. Despite this, there has been a trend for men and women to move into gender-atypical areas. For example, men now account for 9.6 per cent of all nurses in Australia (AIHW 2011), while the figure in the UK is currently 12 per cent (ONS 2011). In fact, latest data from the Teaching Agency (2012) shows more men are becoming primary school teachers, with the number of male trainee primary teachers having increased by more than 50 per cent in the last four years.

These figures suggest that, despite the persistence of gender-based occupational seg-regation referred to above, a small but growing number of men are 'crossing over' into female-dominated occupations—perhaps attracted by a desire for a professional sta-tus (Lupton 2006), for more 'meaningful' work (Simpson 2009), or because of declin-ing employment options elsewhere. Greater opportunities for promotion may present themselves as key motivating factors for some groups of men, as the privileges attached to their gender translate into a 'glass escalator' (Williams 1993) in terms of career suc-cess. In primary schools, for example, many men have been attracted to the profession because of a targeted recruitment, reflecting a strategic shift in education policy, that has placed emphasis on the benefits through role modelling for male pupils as well as on prospects for career progression. Similarly, as Lupton (2006) found, men often seek a professional status that may not be open to them in more 'masculine' occupations. They may also be motivated by a desire for personal fulfilment and the opportunity to develop the 'affective' domain of their lives (Galbraith 1992; Simpson 2005). However, whatever the motivation to enter a non-traditional career, men often face challenges as they nego-tiate the potential mismatch between the (feminine) nature of the job and a gendered (masculine) identity.

These challenges relate to probable sacrifices in terms of pay and status in a non-traditional occupation as well as to a possible questioning, on the part of colleagues, managers, peers, of men's motives, of their 'masculinity', and hence of their suitability for the job (Bradley 1993; Williams 1993; Lupton 1999). As Hochschild (1983) pointed out, work involving service and care may call for special abilities that only women are seen to possess. This can create problems for men who call into question their competence and suitability if they assert a traditional masculinity and yet who invite challenges to their masculinity, as well as their sexuality, if they adopt a more feminine approach. In the context of primary school teaching, men have been found to be in a double bind: their presumed masculine interests in sport, assumptions of leadership potential and ability to act as 'role models' for boys give them an initial hiring advantage but these same characteristics can alienate them from female staff (Williams 1993). This raises issues about how male workers reconcile the feminine nature of their work with the demands of a hegemonically masculine gender regime.

Work in the area suggests that men engage in various practices to reconcile the two and that these often involve a distancing from the feminine. One strategy may be a 'careerist' one, i.e. to move into management or supervisory positions (Williams 1995); to identify with more powerful male groups, such as hospital doctors or headteachers (Simpson 2009); to emphasize the male and downplay the female elements of the job by moving into what may be seen as more 'masculine' specialisms (Williams 1993). In nursing, for example, men often gravitate towards mental health, with historic (and 'masculine') links to custodialism, or accident and emergency, seen as more 'adrenalin charged' than general nursing care (Williams 1995; Squires 1995). Other work suggests that men create a sense of comfort with their occupational role through a process of 'naming and reframing'. Piper and Collamer (1991) in the context of librarianship found that men often renamed the job (e.g. information scientist, information manager) to redress the weak image of the occupation and chose 'masculine' specialisms that involved new technology within its domain. Such strategies suggest a tension for men in non-traditional roles between the 'feminine' nature of the job and dominant discourses of masculinity—and the need to maintain a separation from what can be defined as female in order to restore a comfortable sense of self.

More recent work has moved away from a singular focus on separation strategies to explore how men move between and draw on both the masculine and feminine domains (Pullen and Simpson 2009; Simpson 2009). This has a focus on multiplicities of difference (Pullen and Linstead 2005) and how these may 'play out' in organizational contexts. Men in non-traditional occupations may draw on the privileges bestowed by gender to support a masculine identity but at the same time may experience and draw on 'multiple otherness' and forms of femininity. As Bruni and Gherardi (2002) found in a different context, lines of division may be sometimes deployed and sometimes resisted so that difference is both activated and dismantled. This suggests a less stable orientation to gender as a doing or performance, discussed below. Drawing on recent work, including an extensive study I conducted on men in non-traditional occupations in Australia and the UK, the chapter explores different strategies of compliance and resistance; how bodies are 'marked' in a feminized context and the ways bodies are deployed in gender performances; how 'doing emotions' are implicated in these performances and how spaces are drawn upon as both gendered and gendering. These themes help to highlight the complexities and uncertainties of 'doing masculinity' at work. We start with an overview of gender as practice and performance.

GENDER AS PRACTICE AND PERFORMANCE

The doing or practising of gender is associated with two key strands of thought that draw on symbolic interactionism, whereby difference and the gender binary are actively produced as part of the work of gender in everyday interactions (West and Zimmerman 2002; Fenstermaker and West 2002) and poststructuralism (Butler 1990, 1993) that conceives of gender as a 'performance' produced through discourse (see Kelan 2010 for a helpful overview). In terms of the former, gender is seen to be created through social interaction, i.e. through gesture, speech and body language, so that gender is an 'interactional achievement' or accomplishment that is produced according to normative conceptions of what it means to be a woman or a man. Here the focus is on how gender is 'done' in terms of the ways in which the gendered order, in the forms of a stable gender binary, is reproduced and maintained. Failure to do gender according to normative expectations, as West and Zimmerman (2002) point out, will call individuals (their character, motives, predispositions) to account. Individuals are thus 'accountable' for appropriate performances of gender in that they are subject to the regulatory force of what it means to be a woman or a man. This has clear implications for the individual performances of men in 'feminized' roles.

For poststructuralists, the gender binary is produced through discourse, i.e. signs, labels, expressions and rhetoric that help to construct meaning and which accordingly shape our thinking, attitudes, and behaviour, as well as our sense of self. Meanings about gender as well as sexuality are framed by rhetorical strategies about the 'appropriate' placing and behaviour of men and women where, as Butler (1990, 1993) maintains, these behaviours are shaped by and viewed in relation to heterosexual norms. Thus, articulations of dominant gender regimes and associated practices of gender are brought into being by the reiterative and citational practices of individuals within them (Kelan 2010)—a process Butler (1993) refers to as 'performativity' (Butler 1993). Individuals create themselves as gendered subjects through citing gendered discourses—so that discourse 'produces the effects that it names' (Butler 1993: 2, cited in Kelan 2010: 181).

Doing or practising gender accordingly involves creating and recreating difference in specific interactional and institutional contexts—either through interaction (West and Zimmerman 2002) or through discursive effects (Butler 1990, 1993). These provide a 'repertoire of practices' (Martin 2003) concerning the doing of gender at work. In the institutional context of teaching, for example, men 'do' masculinity in that they are often called upon to be the disciplinarian or to take on difficult or challenging groups (Simpson 2004, 2009)—a form of difference and 'special contribution' that can be discursively mobilized by men to support a desired identity. However, as Deutsch (2007) argues, the concept of 'doing' may leave little room for the incorporation of the challenge to and dismantling of difference. Recent work (e.g. Butler 2004: Pullen and Knights 2007) has therefore focused on the 'undoing' of gender, defined by Deutsch as those social interactions and associated discourses that reduce, dismantle, or challenge gender difference. In this respect, men's decision to enter a non-traditional career can be seen as a form of resistance to dominant conceptions of masculinity. Gender performances in these contexts are likely to render the salience of undoing gender more visible and alert us to the complications of doing and the undoing of difference. In other words, gender can be seen to be done and undone in action and interaction—produced through discourse (Kerfoot and Knights 1998), symbolism (Gherardi 1995), or performances (Butler 1993).

DOING MASCULINITY IN NON-TRADITIONAL WORK

In so far as gender is a meaning that is produced through interaction with others and through discourse, issues of complexity, ambiguity, and fluidity become central themes. From this perspective, work on masculinity (e.g. Kerfoot and Knights 1993, 1998; Connell 2000) has focused on the fragmented, insecure and uncertain nature of gender identity (Kerfoot and Knights 1998; Collinson 2003; Pullen 2006). Such work has explored the dynamic nature of masculinity, how it is constructed and reconstructed, how it is experienced at a subjective level and how multiple masculinities exist in relation to the dominant (hegemonic) form.

This has highlighted the insecure nature of masculinity as men construct themselves as gendered subjects. Part of this insecurity may emanate from an erosion of stable reference points as a source of identity (e.g. birth, religion, class) and the rise of paid employment, with all its uncertainties, as the source of a valued sense of self (Collinson 2003). For Kimmel (1994) insecurity relates to the perceived failure to live up to a masculine hegemonic ideal. As Connell (2000) argues, while discourses of hegemonic masculinity are not always and everywhere the same and while different masculinities exist in definite (hierarchical, exclusionary) relations with each other, there is in most situations some form of hegemonic masculinity—the most honoured or desired—which can be expressed and represented in everyday interactions. Accordingly, despite differences and variations, successful performance of masculinity is, more often than not, equated with being heterosexual, successful, capable, reliable, and in control (Collinson and Hearn 1994; Alvesson 1998; Connell 2000; Knights and McCabe 2001). From Kimmel (1994), failure to ahieve this 'ideal' in all its manifestations is thus a constant source of anxiety and insecurity for most men.

As Kerfoot and Knights (1998) argue, the masculine desire for control (over self, over others, over environment) needs to be set in the context of this ontological insecurity. In this respect, competitive masculinity is one way to achieve a sense of certainty and stability in an uncertain world—but its foundation is shaky in that it is based on an identity that is continuously threatened as a consequence of the 'failure, potential or otherwise, to maintain control' (Kerfoot and Knights 1998: 9). This supports the need for identity work as men negotiate and renegotiate their conditions of insecurity as an ever incomplete process. Given that the cultural resources available for identity management will vary from organization to organization, as well as in different cultural contexts, 'doing masculinity' will be multiplicitous, fluid as well as multidimensional. Organizations also 'do' gender and have codes for performing gender identities. They are thus home to a variety of masculinities (Collinson and Hearn 1994), with implications for how the dynamics of different masculinities—some prioritized, others suppressed—inform men's gender identities at work.

Managing masculinity in non-traditional occupations is likely to throw up particular challenges for men as they seek ways to reconcile the work they do with an identity they can accept. Here, discourses of service and care can collide with dominant conceptions of masculinity that are arraigned around rationality, detachment and profit (Ross-Smith and Kornberger 2004). Organizational interactions and practices may reflect non-masculine ways of working and men may find such femininity 'unlivable', in that its constituent elements (nurturance, passivity, service) have been defined and rejected by masculine identities. The insecurities referred to above as integral to masculinity may accordingly become exacerbated in non-traditional roles. As Williams (1993) argues, men who 'cross over' upset the gendered assumptions embedded in such work so they are suspected of not being 'real' men—generating insecurity in relation to their positioning against the hegemonic ideal. This is supported by Lupton (1999) who suggests that men in non-traditional roles fear feminization and stigmatization—fears which are particularly acute under the gaze of other men. Associations with homosexuality for example can clash with dominant discourses of heterosexual masculinity with sometimes painful implications for men in these roles (Lupton 1999). Men must therefore manage their identity under specific conditions of insecurity.

GENDER RESISTANCE IN NON-TRADITIONAL WORK

At a micro-political level, resistance occurs when individuals challenge or refuse to accept subjectivities and identities defined by dominant discourses. Resistance from this perspective concerns individual struggles over meanings and subjectivities rather than on specific behaviours or acts. As Davies and Thomas (2004) point out, resistance arises at 'points of contestation and contradictions within discursive fields, presenting spaces for alternative meanings and subjectivities, and new forms of practice' (Davies and Thomas 2004: 105). Individuals thus strive to subvert and evade the identities imposed on them by dominant discourses and associated attempts at classification.

Men in non-traditional occupations, as we have seen, may struggle to align discourses of masculinity with their (feminine) occupational identity and may resist the categorizations and subjectivities that are imposed. The resources open to individuals are varied. Overt resistance can be seen in the hostility often expressed towards more higher status men (Simpson 2004, 2009)—as male nurses, for example, through antagonistic interactions, resist the subordinated identity impressed by their proximity to privileged masculinity in the workplace. Resistance is thus endemic to how men do gender in these contexts.

As we have seen, one form of resistance to the implications of non-masculine associations is to create and maintain distance from the feminine (Lupton 1999; Simpson 2004, 2009). Thus, male cabin crew often focus on the safety and security aspects of the job, overriding associations with deferential service and customer care (Simpson 2009), discussed further below. Men also engage in compensatory gendered practices so as to 'restore' a dominating position. This includes, as discussed above, 'careerism' whereby men aspire to (masculine) management or supervisory posts and away from (feminine) day-to-day professional practice (Heikes 1992; Williams 1995) and/or identification with more powerful male groups, such as male hospital doctors or headteachers (Floge and Merrill 1986; Simpson 2004, 2009). Blatant sexism in these contexts (Cross and Bagilhole 2002) is another way in which traditional masculinity can be emphasized.

Resistance can take more subtle forms. It can be activated through humour when employees make fun of customers or management and in so doing recapture a sense of dignity in what may be seen as a powerless or demeaning role. Cabin crew, for example, encounter asymmetric relations of deference in their interactions with passengers—where those interactions are driven by discourses of consumer sovereignty and the need to please (Abbott and Tyler 1998; Simpson 2009). They also have limited resources from which to draw, such as those around professional care or vocationalism, to create a compensatory and more valued identity. Crew often made fun of passengers through subtle changes in intonation (e.g. an exaggerated yes sir, no sir). Through 'twists of meaning' (Kondo 1990) they conveyed ironic service as a practice of subversion.

As Davies and Thomas (2004) argue, individuals can practise resistance by taking up alternative identities—embracing an Other status and using it to their advantage. Some cabin crew would act out a flamboyant and parodic homosexuality to cause discomfort in the flight deck—where male pilots were seen to embody normative and heterosexual masculinity and to be the source of a conventionally gender-based 'gaze' (Simpson 2009). Male crew thus reified and celebrated alterity—playing on difference to unsettle the mainstream.

Some men deliberately invoke gender and acknowledge the power of masculinity and some of its conventional understandings. Men can reconstruct a different masculinity through feminine practices that involve resistance to normative conceptions of gender: engaging in traditionally 'feminine' behaviours ('female talk' in the staffroom; performing domestic chores during offduty moments), valuing the special status as 'New Men'. Similarly, men can present themselves as special in their reflexive capacity to be critical of masculine norms and values, for example, adopting a gentle tone while listening carefully and sympathetically to elderly patients. Men are thus able to create a satisfying identity that builds on yet also subverts traditional notions of gender. They recognize the power of masculinity but reject it in such a way that can also reinforce its status. In support of this, as Davies and Thomas (2004) argue, 'to resist something is also to reify it', privileging it, legitimising and reproducing 'the very subject position that is being denied' (Davies and Thomas 2004: 115). Men may therefore challenge masculinity through discourses and practices of denial but its purchase is dependent on masculinity's continuing, hegemonic influence and presence. The challenge is supportive of the very thing that is resisted and refused.

This points to the complex forms of resistance and the diverse ways in which individuals respond to dissonance between subjectivity and the context in which he or she moves—where that context may mean a subject position on offer that is disconfirming to a sense of self. Men therefore engage in practices that create distance from as well as embrace the feminine while also, by mobilizing special capacities and qualities, calling into question masculine norms and values. Much of these practices and performances have an embodied dimension—an aspect of non-traditional work that is discussed below.

MASCULINE BODY PERFORMANCES

As Connell (2000) points out, bodies matter in that biological differences between men and women play an important part in determining what is seen to be 'masculine' and 'feminine' work. In other words, gendered norms and expectations are 'written on' the sexed bodies of men and women with implications for what is seen as 'gender appropriate' occupations. From Butler's (1993) work, bodies carry meanings that help authenticate particular gender performances which then take on the semblance of 'natural' dispositions. Men and women accordingly 'do' gender partly through their bodies and these performances become sedimented as essentially gendered and universal. Norms of deference or of nurturance for example are often inscribed onto 'softer' female bodies (Trethewey 1999) while professional norms (e.g. of competence, expertise) and meanings attached to leadership roles are associated with the bodies of men. Men and women then come to be seen as 'naturally' predisposed to these roles and activities.

In a similar vein, women are often seen as more 'embodied' than men. As Hassard *et al.* (2000) point out, women are linked to nature through bodily associations of fluidity, flux, fecundity, and passion and associated with the private sphere where 'natural' bodily, emotional (and hence lower) functions are seen to occur. Men by contrast signify the 'organized body' that has meanings of dryness, solidity, containment—a body that is disciplined and controlled and the standard against women and their bodies are judged to be problematic. In short, men in general are disembodied in their divorce from bodily considerations—standing for universal personhood, organization, and rationality.

In non-traditional occupations, drawing on essential notions of femininity, the masculine body is largely at odds with this social definition (e.g. as rational, detached, disembodied) making it more difficult for masculinity to 'take hold' in these contexts. This potential mismatch is exacerbated, arguably, by the primacy afforded to body characteristics (gender, age, personality) in the production of service and care—where the embodied performances of the worker, such as facial expression and bodily demeanour, can be seen to be partly constitutive of the product. As Leidner (1991) has argued, these performances (and ultimately the nature and quality of the service produced) draw on looks, voices, emotion, personalities, and facial and body displays. Meanings attached to men's bodies are therefore likely to be strongly implicated in how service and care are produced and less evident in other occupations, with implications for men in how such roles and scripted are performed.

Here, research suggests that men's bodies are often coopted in a non-traditional role—as men are expected to undertake physically demanding work and to act as disciplinarian in the organization. Milligan (2001) found male nurses are routinely expected to deal with physically aggressive patients; Sargent (2001) that male primary school teachers are given the more challenging classes, while Simpson (2009) found that male cabin crew are often sent to deal with difficult passengers. In all these cases there is explicit recognition of the physicality of the masculine body so that men are deployed as 'boundary setters' (Forseth 2005) in aggressive or challenging situations where they signify an authoritative, reassuring, and regulatory presence. This is irrespective of men's actual body size or personal disposition. Some men may be small in stature or non-confrontational in disposition—resentful of and resisting the pressure to take on these roles—suggesting that 'body capital' such as body build and physical characteristics such as strength, despite their individual variability, can become a 'currency' of masculinity in more general terms (Evans 2004).

Other men can welcome these meanings in that they enable the uptake and performance of a 'protector' role in the work context. This is presented in a traditional way as masculine chivalry and a concern for the welfare of women (Pullen and Simpson 2009), thereby creating a distinctive space for the practice of masculinity through meanings associated with heroic qualities of endurance and strength. In some instances, however, these meanings are out of place and men must manage their bodies, for example reducing semblance of body size and volume of voice, in order to present a non-threatening, caring self (e.g. sitting, rather than standing, at the end of the bed). Here men spoke of the need to overcome the inherent disadvantages of the male body in some practices of care. Therefore, in primary schools while men are called upon to be the disciplinarian or to act as masculine role model to boys (e.g. taking an active role in sport), they must at the same time conform to stereotypically feminine qualities in order to work effectively with young children—including bodily displays of 'maternal care' (Vogt 2002). This can create a 'double bind' (Sargent 2001): if men enact a traditional masculinity, demanded in their role modelling for boys and anticipated by many female colleagues and managers, they reinforce common assumptions concerning the appropriate behaviour of men and reinforce ideas of their often supposed ineptness. If they do teaching in the same way as women (e.g. as maternal care), their masculinity and their motivations are called into question.

From the above, while men may be valued for their bodies—assigned for example to work that demands physical strength or discipline—they are also 'marked' as different from the female norm where body size and other masculine features sit uncomfortably with the unmarked bodies of women. As Dahl (2005) notes, men can be seen as intruders creating disorder in a system over which women claim jurisdiction. Male teachers and nurses are thus often fast-tracked into more 'body congruent' specialties and levels of hierarchy and away from the classroom or the ward (Sargent 2001; Evans 2004). One nurse commented in Simpson's (2009) study:

> There was this constant undercurrent that a. you shouldn't be here, either because you're a bloke or b. because you're too intelligent and you should be a doctor.

Male bodies can thus be 'matter out of place' in a non-traditional context. Men's sexuality and masculinity can be seen as undesirable, potentially dangerous and disruptive in some of the work they do. Here, the presence of men in a non-masculine role can call into question and challenge the heterosexual norm in the workplace—with implications for the meanings attached to men's bodies in these roles. Thus, irrespective of individual sexuality, men are routinely labelled 'gay'. Equally, while women's touch is seen as harmless or non-threatening, men's touch is sexualized (Evans 2002), leading to 'cautious care-giving' as men manage the line between relationality and propriety (Sargent 2001). In the context of teaching the sexualization of men's touch is often seen to be synonymous with the potential for child molestation, with often painful implications for men. Although they want to have the same level of close interaction with the children that female colleagues enjoy, they must also distance themselves in order to avoid the very real dangers of suspicion. As Sargent (2001) notes, referring to the double bind above, they then reinforce the very behaviours (e.g. of stoic, distant men) which suggest that men are unsuited to primary teaching in the first place.

From the above, we can see how we do masculinity and femininity through our bodies, drawing them into displays of appropriate (or inappropriate) gender behaviour. At the same time, bodies are inscribed with meanings, beyond our control, which have implications for how we do gender and for our sense of self. As Butler (2004) has pointed out, we therefore both do and have gender done to us. The lived body, i.e. the ways in which it is represented and used in specific contexts, is an amalgam of active and passive, of doing and being done to, of signifying and signified which together strongly influence how gender is performed in specific contexts. Men in non-traditional roles can therefore draw on traditional body ascriptions to present a particular self, coopting these meanings to produce a desired effect; they manage their bodies to redefine meanings associated with, for example, body size seen as inappropriate for some forms of care and they negotiate the implications of the marking of their bodies in day-to-day practices and performances.

Performing Emotions, Service, and Masculinity

Both emotions and service are involved to greater and lesser extents in men's non-traditional occupations. This may include professional 'caring' roles as in nursing, teaching, social work or 'para-professional' roles including front-line service workers (e.g. cabin crew, hotel workers, call centre operators, retail assistants) engaged in interactive work. Both categories have cultural connections with femininity through associations with service and care or through the domestic nature of some of the tasks involved (Tyler and Abbott 1998; Adib and Guerrier 2003). Such work is therefore often constructed as a 'natural' part of doing femininity (Hall 1993; Adkins 2001) through embodied performances and gendered scripts. This has implications for men as they do or undo masculinity in non-traditional roles—roles that are embedded with gendered norms and expectations. For example, as MacDonald and Sirianni (1996) argue, men are less likely to embrace some service demands, particularly those involving deference, because they do not fit their notion of gender-appropriate behaviour—drawing on other voices and repertoires (e.g. of authority, of rational expert) to side-step the practices and identity implications of such work.

In the context of teaching and nursing, there are several implications for men in terms of how they perform service and care. Some emotions are not seen as suitable for men: open displays of sorrow for example can be seen to be out of place in the workplace and a source of embarrassment even if the circumstances are painful—supporting Parkin's (1993) claim that an ethos of control over emotions is seen as part of being professional. Thus, male nurses have been found to mobilise a 'cool headed detachment' (Pullen and Simpson 2009) in their descriptions of how they care, creating a difference from women who are positioned deficiently as overly emotional—so that they fail to comply with what men see as a professional stance. Further, men are more likely than women to present their emotional labour—work that alters the ways individuals manage emotions to make them appropriate with an expected organizational goal (Hochschild 1983)—as special and distinctive (Simpson 2009; Pullen and Simpson 2009). Here, high standards of care are often seen as a 'natural' part of femininity and hence undervalued and invisible when undertaken by women—(Taylor and Tyler, 2000; Williams 2003). One outcome is that emotional labour performed by men is more likely to be presented—and received—as 'philanthropic' (Bolton 2005), i.e. as a particular type of 'gift exchange' that goes beyond the call of duty. Relatedly, men can combine philanthropic with maverick performances as they present themselves as 'going the extra mile'—circumventing bureaucracy and mobilizing resources for those in their care. This is highlighted in the following example:

> What do I mean by care? Well, for me it's being able to do the best for the patient, and if that means breaking rules then so be it…so you've got to understand it from

a patient's point of view…if a patient is obviously dying, he wants a cigarette, there's no point refusing him a cigarette, I would take him outside whereas my female colleagues would say 'no you're not allowed to smoke'. You try and give him a cigarette when they're going to die anyway. They're at the end of the road and what he needs most in this case is a cigarette, he wanted a cigarette. Now, with the women he'll get a hard and fast, 'this is what's good for you' and that's it, that's the way…I don't feel that way. You tell me what's good for you, what you want—and I'll do it…

(cited in Pullen and Simpson 2009: 571)

Presenting women as 'rule bound' and cautious allows the uptake of a more satisfying oppositional identity around independence and non-conformity. At the same time, as Pullen and Simpson (2009) found, men can promote femininity, commenting and placing value on their feminine side. Here men dis-identify with normalized conceptions of masculinity, presenting themselves in terms of 'New Manhood' as they celebrate their 'feminine' dispositions to nurture and care. Activities around emotional labour may accordingly provide men with repertoires of 'doing' gender that, by drawing on displays of risk-taking, emotional control, and distinctive capacity, have positive identity implications (Pierce 1995; Brannen 2005; Lewis and Simpson 2007). As Swan (2008) has argued, the mobilization of feminine emotions can be a source of power for men. Men restore status and privilege (threatened, for example, by the feminization of work practices) through a number of 'appropriating moves'—including the cooption of the feminine, the performance of care as a masculine rationality and detachment, and the mobilisation of New Manhood to enhance the value and specificity of their caring skills.

PERFORMING GENDER IN SPACE

As Halford and Leonard (2006) argue, place, space, and time in organizations combine with more generic resources such as those afforded by gender and occupation to offer multiple and competing resources for the construction of working identities. Individuals operate and interact with and in space—through gesture and bodily movements and influenced by norms of engagement in these specific contexts. These spatial contexts are not gender neutral but offer resources that are both gendered and gendering (Massey 1994, 2005). Environmental 'variables' such as access to space and opportunities for mobility as well as meanings attached to space can come together to strengthen gender power relations. The primary school classroom, for example, where teachers spend most of their day with the same group of children, can be seen to be a feminine space (diminutive, confined, nurturing)—supported by a maternal logic with implications for men who work within that domain. The presence of men, as we have seen, may alter the nature of the space so that it becomes less symbolic, for example, of 'safe' maternal care. Similarly, male nurse speak of simple presence ('just standing there') to contain behaviour of unruly patients—as their bodies create meanings around authority and discipline in the public space of a hospital waiting room. At the same time, as we have seen, men can be 'marked' through occupancy of a feminine space (as non-masculine, as potentially disruptive) with implications for professional practice and identity.

Cabin crew work provides an interesting example of how gendered spaces, both supportive and incorporative of power relations, can be drawn upon to assist in a performance and subjectivity of 'male' and 'female'. Here, the flight deck and the aircraft cabin can be seen as masculine and feminine spaces respectively (Simpson 2009). As Mills (1995) has illustrated, the highly technologized space of the flight deck is underscored with meanings around rationality, danger and expertise that have core connections with discourses of (heterosexual) masculinity. Militaristic uniforms worn by the (mostly male) occupants are symbolic resources that further enhance the masculinity of this space. By contrast, the cabin, or main body of the aircraft, can be defined as a feminine space. It is in this arena that consumption, service, and the trivia of entertainment occur—culturally associated with femininity. Further, this space has sexual connotations in that it continues to be the site for the mobilization by airlines of a subordinated heterosexual femininity (Hochschild 1983; Williams 2003) based on perceptions of sexual availability and norms of deferential service. When men enter this space as crew they too become associated with femininity (Hochschild 1983) and with a denigrated homosexuality.

These gendered meanings attached to spaces are not stable however but can shift with the practices occurring within them—such as the nature and extent of physical movement and the satisfaction of service demands. For example, while movement and walking are often associated with a (masculine) purpose and authority (e.g. as crew undertake safety and security checks in the aisle), movement in response to the demands of others ('running around' after the passengers) can be seen to be part of deferential and hence devalued 'feminine' service. In terms of the former, as Simpson (2009) found, male crew often draw on (masculine) discourses of safety and security, positioning themselves as a source of strength and security against the less 'reassuring' presence of female crew—masculine attributes that have arguably been given greater purchase since the events of 9/11. However, once safety checks and take-off are complete, the appearance of the trolley (as feminine artefact) in the aisle signifies a shift in power relations as consumer sovereignty and service demands take hold. The aisle is thus transformed from a space of authority and expertise to one associated with devalued domesticity and deferential service—impressing painfully on male crew (many referred to the uncomfortable 'trolley dolly' image of their role). Space is thus complex and unstable—its meanings shifting with the activities and practices contained—both influencing and influenced by gendered identity processes and gendered performances. When men enter a feminine space, meanings attached to that space may alter as we have seen above. At the same time, male bodies (as well as the spaces they occupy) become 'marked' as different from the heterosexual norm.

In the latter respect, in a 'reversal of gaze' and through 'movements of exhibition' as male crew perform their role within the highly visible space of the aisle, attention is drawn to men's Otherness with often contradictory consequences. On the one hand, assumptions of homosexuality, irrespective of individual sexual preferences of male crew, routinely insinuate into on-board service interactions through customer attitudes and behaviours. The power saturated 'gaze' is fundamentally patriarchal and heterosexual. On the other hand, men can gain pleasure and pride from aesthetic displays. As Warhurst and Nickson (2007) found in the context of service organizations generally, the need for airlines to present a particular professional 'face' is associated with aesthetic codes of body management and with corporeal dispositions that express friendly and helpful service. In contrast to women where the 'panoptic gaze' is often seen to be experienced as oppressive and patriarchal (Tyler and Abbott 1998; Taylor and Tyler 2000), and assisted by celebrations of physical strength and ascriptions of proficiency and autonomy, men often gain pleasure and pride from their bodies (an one male crew commented of his visibility at work: 'I love it, I'm like that, bring it on!'). Male crew, therefore, may celebrate their bodies as the bearers of an aesthetic ideal as well as a claimed source of strength and security in the day-to-day operations of the cabin.

We have seen how men draw on strategies of resistance, for example, to a subordinated identity in a non-traditional role. Spaces are implicated in different ways in these processes and behaviours. Both men and women use the galley as an 'off-stage' region out of sight of customers for mutual support. Here emotions can be expressed without the danger of violating organizational display norms (Morris and Feldman 1996). Crew let off steam, can refer disparagingly to passengers, and indulge in sexual innuendo, gossip, and flirtatious banter. In more subtle ways, and drawing on the greater licence for humour often afforded to men, resistance can be practised through ironic recognition of the dominant meanings attached to space (e.g. of deference and sexual availability). In the following quote a male crew member playfully contrasted an acceptance of a subordinated identity within the service encounter of the aisle with a more authoritative, expert role:

> So you're a trolley dolly because they see you up and down the aisles with the coffee or tea and I used to laugh and make a joke, I'd say you can't afford me so you might as well have the tea or coffee or whatever but when you collapse in the aisle with a heart attack, this trolley dolly has got to know what to do...

Male crew also engaged in status levelling tactics associated with the masculine space of the flight deck. As suggested earlier, the proximity of this space, culturally associated with a superior and heterosexual masculinity, may serve to underscore the femininity of the meanings and material practices within the cabin. The antagonism displayed towards the flight deck and its occupants (perceived as arrogant, bullying, and homophobic) may be testimony to levels of discomfort experienced as status differences impress subordinated identities on male crew. One crew briefly reflected on his negative feelings towards pilots: 'I'm only a steward and you're flight deck... it's a male thing I suppose'. Putting pilots in their place ('winding them up') was a preoccupation of male crew—achieved through potentially embarrassing banter (e.g. through excessive, flamboyant displays of homosexuality and camp), withholding food, and refusing a servile role. As Lupton (1999) argues, while women are frequently subordinated in organizations, the take-up of an inferiorized identity can be more difficult for men. This may be particularly the case when men are forced to confront a privileged masculine space in the workplace.

We have seen how a 'masculine' authoritative space created through discourses and activities of safety and security can be easily disrupted and subverted by feminine activities of service and care—creating tensions for men as they manage such contestations of meaning. Equally, the presence of male bodies can convert the cabin and aisle as sites of female heterosexuality, encouraged by past and current promotional practices suggestive of female service and availability (Tyler and Abbott 1998; Tyler and Taylor, 1998; Williams 2003), into a space saturated with and marked by homosexual meanings. Men accordingly move within and draw on space in ways that support dominant conceptions of (deferential) service and, within the feminized space of the cabin, strive in different ways to reinvigorate traditional notions of masculinity. Taken together, this highlights how the performance of gender is deeply contextual—not in the sense of context as a 'backdrop' to gender expression, but as an active resource in the 'doing' of gender at work.

CONCLUSION

This chapter has explored some of the ways in which men, in the context of Australia and the UK, 'do' gender in a non-traditional occupational context. We have charted the challenges men face in a non-traditional (e.g. service and/or caring) role and have positioned much of the existing literature on men's experiences and practices as oriented towards a 'distancing' from the feminine. Through our focus on bodies and embodiment, the gendering of service and care and the significance of gendered spaces, we have considered some of the diverse ways in which men manage gender in these contexts. These may encompass distancing strategies and reaffirmations of masculinity—such as the ways in which masculinity (rationality, detachment) can be mobilized in performances of service and care, as well how men value (and can be valued for) their bodies and for the meanings masculine bodies convey. A focus on distancing however can hide more complex and often contradictory processes—as men negotiate the marking of their bodies as 'out of place' in ways that then support essentialized notions of men's non-suitability for the job; how masculinity can be invoked in strategies of 'new manhood' that seek to reject its status while, paradoxically, reifying its normativity and, relatedly, how men can restore status and privilege by colonizing the feminine through practices and discourses of distinctiveness.

Equally, we have seen how men may work to manage their masculinity (voice, demeanour) and their embodied performances to comply with more appropriate and non-assertive notions of care; how men must manage the implications of a gaze that is founded on conventional notions of masculinity, femininity and sexuality. Alterity can be a challenging, painful and sometimes pleasurable experience—it can confer subordinated identities or support rewarding and less conventional selves. In the latter respect, men comfortably engage with behaviour and activities culturally coded feminine in day-to-day practices that define their occupation. In other words, while some aspects of occupational identity may be resisted and challenged, others will be simultaneously embraced. Some elements of gender relations may be questioned, others may become further entrenched, and still others quietly transformed in day-to-day activities of service and care.

As Deutsch (2007) argues, the concept of gender 'doing' leaves little room for dismantling gender difference or diminishing gender inequity and oppression. While 'doing' gender can theoretically capture both conformity and resistance to gendered norms, the focus on 'accountability', as a regulatory (and institutionally based) mechanism of gender assessment, tilts the balance towards conformity. The single conceptualization of 'doing' may not therefore adequately capture the dismantling of and resistance to difference. By capturing the social interactions and associated discourses can reduce or challenge difference, 'undoing' goes some way to address these weakness. Undoing alerts us to the ways in which difference is drawn upon, activated, and denied to dismantle or disrupt traditional notions of gender. In this respect, the analysis of the experiences of men in gender-atypical occupations allows a dramatization of difference which may be rendered more visible than in other more conventional contexts. This chapter has highlighted how, by entering a non-traditional occupation, men simultaneously 'do' and 'undo' gender in different ways and with different levels of contestation and challenge—acting to reinforce as well as to destabilize gender in its stereotypical forms.

REFERENCES

Acker, J. (1990). 'Hierarchies, Jobs, Bodies: A Theory of Gendered Organization', *Gender and Society*, 4: 139–58.

Adib, A., and Guerrier, Y. (2003). 'The Interlocking of Gender with Nationality, Race, Ethnicity and Class: The Narratives of Women in Hotel Work', *Gender Work and Organization*, 10(4): 413–32.

Adkins, L. (2001). 'Cultural Feminisation: Money, Sex and Power for Women', *Signs*, 26(3): 669–95.

AIHW (2011). *Nursing and Midwifery Labour Force 2009*. Sydney: AIHW.

Alvesson, M. (1998). 'Gender Relations and Identity at Work: A Case Study of Masculinities and Femininities in an Advertising Agency', *Human Relations*, 51(8): 969–1005.

Bolton, S. (2005). *Emotion Management in the Workplace*. London: Palgrave.

Bradley, H. (1993). 'Across the Great Divide', in C. Williams (ed.), *Doing Women's Work: Men in Non-Traditional Occupations*, 10–28. London: Sage.

Bruni, A., and Gherardi, S. (2002). 'Omega's Story: The Heterogeneous Engineering of a Gendered Professional Self', in M. Dent and S. Whitehead (eds), *Managing Professional Identities: Knowledge Performativity and the New Professional*, 174–200. London: Routledge.

Butler, J. (1990). *Gender Trouble: Feminism and the Subversion of Identity*. London: Routledge.

Butler, J. (1993). *Bodies that Matter: On the Discursive Limits of Sex*. London: Routledge.

Butler, J. (2004). *Undoing Gender*, London: Routledge.

Connell, R. (1995). *Masculinities*. Berkeley, CA: University of California Press.

Collinson, D., and Hearn, J. (1994). 'Naming Men as Men: Implications for Work, Organization and Management', *Gender Work and Organization*, 1(1): 2–22.

Collinson, D. (2003). 'Identities and Insecurities: Selves at Work', *Organization*, 10(3): 527–48.

Connell, R. (2000). *The Men and the Boys*. Cambridge: Polity Press.

Cross, S., and Bagilhole, B. (2002) 'Girl's Jobs for the Boys? Men, Masculinity and Non-Traditional Occupations', *Gender Work and Organization*, 9(2): 204–26.

Dahl, R. (2005). 'Men Bodies and Nursing', in D. Morgan, B. Brandth and E. Kvande (eds), *Gender Bodies and Work*, 127–138. Aldershot: Ashgate.

Davies, A., and Thomas, R. (2004). 'Gendering Resistance in the Public Services', in R. Thomas, A. Mills, and J. Helms Mills (eds), *Identity Politics at Work: Resisting Gender, Gendering Resistance*, 105–122. London: Routledge

Deutsch, F. (2007). 'Undoing Gender', *Gender and Society*, 21(1): 106–27.

EHRC (2011). *How Fair is Britain? Equality, Human Rights and Good Relations in 2010*. The First Triennial Review. London: EHRC.

EOC (2007). *Facts about Men and Women in Great Britain*, Equal Opportunities Commission. UK: Manchester.

Evans, J. (2002). 'Cautious Caregivers: Gender Stereotypes and the Sexualisation of Men Nurses' Touch', *Journal of Advanced Nursing*, 40(4): 441–8.

Evans, J. (2004). 'Bodies Matter: Men Masculinity and the Gendered Division of Labour in Nursing', *Journal of Occupational Science*, 11(1): 14–22.

Fenstermaker, S., and West, C. (2002). *Doing Gender, Doing Difference*. London: Routledge.

Fletcher, J. (2003). 'The Greatly Exaggerated Demise of Heroic Leadership: Gender Power and the Myth of the Female Advantage', in R. J. Ely, E. Foldy, and M. A. Scully (eds), *Reader in Gender Work and Organization*, 204–10. Victoria, Australia: Blackwell.

Forseth, U. (2005). 'Gender Matters? Exploring How Gender is Negotiated in Service Encounters', *Gender Work and Organization*, 12(5): 440–59.

Galbraith, M. (1992). 'Understanding the Career Choices of Men in Elementary Education', *Journal of Educational Research*, 85(4): 246–253.

Gherardi, S. (1995). *Gender, Symbolism and Organizational Culture*. London: Sage.

Halford, S. and Leonard, P. (2006). *Negotiating Gendered Identities at Work: Place, Space and Time*. Basingstoke: Palgrave Macmillan.

Hall, E. (1993). 'Smiling, Deferring and Flirting: Doing Gender by Giving Good Service', *Work and Occupations*, 20(4): 452–471.

Hassard, J., Holliday, R. and Willmott, H. (2000). *Body and Organization*. London: Sage.

Hearn, J. (1994). 'Changing Men and Changing Management: Social Change, Social Research and Social Action', in M. Davidson and R. Burke (eds), *Women in Management: Current Research Issues*, 192–212. London: Paul Chapman.

Heikes, J. (1992). 'When Men are in the Minority: The Case of Men in Nursing', *Sociological Quarterly*, 32(3): 389–401.

Heilman, M. (1997). 'Sex Discrimination and the Affirmative Action Remedy: The Role of Sex Stereotypes', *Journal of Business Ethics*, 16(9): 877–99.

Hochschild, A. (1983). *The Managed Heart: Commercialisation of Feeling*. Berkeley, CA: University of California Press.

Kelan, E. (2010). 'Gender Logic and (Un)doing Gender at Work', *Gender Work and Organization*, 17(2): 174–94

Kerfoot, D. and Knights, D. (1993). 'Management Masculinity and Manipulation: From Paternalism to Corporate Strategy in Financial Services in Britain', *Journal of Management Studies*, 30(4): 659–77.

Kerfoot, D., and Knights, D. (1998). 'Managing Masculinity in Contemporary Organizational Life: A Man(agerial) Project', *Organization*, 5(1): 7–26.

Kimmel, M. (1994). 'Masculinity as Homophobia: Fear Shame and Silence in the Construction of Gender Identity', in H. Brod and M. Kaufman (eds), *Theorising Masculinities*, 119–41. London: Sage.

Knights, D., and McCabe, D. (2001). 'A Different World: Shifting Masculinities in the Transition to Call Centres', *Organization*, 8(4): 619–45.

Kondo, D. (1990). *Crafting Selves: Power, Gender and Discourses of Identity in a Japanese Workplace*. Chicago: University of Chicago Press.

Leidner, R. (1991). 'Serving Hamburgers and Selling Insurance: Gender Work and Identity in Interactive Service Jobs', *Gender and Society*, 5(2): 154–77.

Lewis, P., and Simpson, R. (eds) (2007). *Gendering Emotion in Organizations*. London: Palgrave.

Lupton, B. (1999). 'Maintaining Masculinity: Men who do Women's Work', *British Journal of Management*, 11: S33–S48.

Lupton, B. (2006). 'Explaining Men's Entry into Female Concentrated Occupations: Issues of Masculinity and Class', *Gender Work and Organization*, 13(2): 103.

Macdonald, C., and Sirianni, C. (1996). *Working in the Service Society*. Philadelphia: Temple University Press.

Moloney, M., and Fenstermaker, S. (2002). 'Performance and Accomplishment: Reconciling Feminist Conceptions of Gender', in S. Fenstermaker and C. West (eds), *Doing Gender, Doing Difference*, 189–204. London: Routledge.

Martin, P. Y. (2003). '"Said and Done" versus "Saying and Doing": Gendering Practices, Practicing Gender at Work', *Gender and Society*, 17(3): 342–66.

Massey, D. (1994). *Space, Place and Gender*. London: Methuen

Massey, D. (2005). *For Space*. London: Sage.

Milligan, F. (2001). 'The Concept of Care in Male Nurse Work: An Ontological Hermeneutic Study in Acute Hospitals', *Journal of Advanced Nursing*, 35(1): 7–16.

Mills, A. (1995). 'Cockpits, Hangars, Boys and Galleys: Corporate Masculinities and the Development of British Airways', *Gender Work and Organization*, 5(3): 172–88.

Morgan, D. (1992). *Discovering Men*. London: Routledge.

ONS (2009). 'Presentation of the Gender Pay Gap', ONS Position Paper. London: ONS.

ONS (2011). *Labour Market Statistics*. London: ONS.

Parkin, W. (1993). 'The Public and the Private: Gender Sexuality and Emotion', in S. Fineman (ed.), *Emotion in Organizations*, 167–89. London: Sage.

Pierce, J. L. (1995). *Gender Trials: Emotional Lives in Contemporary Law Firms*. Berkeley, CA: University of California Press.

Piper, P., and Collamer, B. (1991). 'Male Librarians: Men in a Feminized Profession', *Journal of Academic Librarianship*, 27(5): 406–11.

Pullen, A. (2006). *Managing Identity*. London: Palgrave.

Pullen, A., and Knights, D. (2007). 'Undoing Gender: Organizing and Disorganizing Performance', *Gender Work and Organization*, 14(6): 505–11.

Pullen, A., and Linstead, S. (2005). *Organization and Identity*. London: Routledge.

Pullen, A., and Simpson, R. (2009). 'Managing Difference in Feminized Work: Men Otherness and Social Practice', *Human Relations*, 62(4): 561–87.

Ross-Smith, A., and Kornberger, M. (2004). 'Gendered Rationality: A Genealogical Exploration of the Philosophical and Sociological Conceptions of Rationality, Masculinity and Organization', *Gender Work and Organization*, 11(3): 280–305.

Sargent, P. (2001). *Real Men or Real Teachers? Contradictions in the Lives of Male Elementary School Teachers*. Harriman, TN: Men's Studies Press.

Simpson, R. (2004). 'Masculinity at Work: The Experiences of Men in Female Dominated Occupations', *Work, Employment and Society*, 18(2): 349–68.

Simpson, R. (2005). 'Men in Non-Traditional Occupations: Career Entry, Career Orientation and Experience of Role Strain', *Gender Work and Organization*, 12(4): 363–80.

Simpson, R. (2009). *Men in Caring Occupations: Doing Gender Differently*. Basingstoke: Palgrave.

Squires, T. (1995). 'Men in Nursing', *RN Journal*, 58(7): 26–8.

Swan, E. (2008). '"You Make me Feel Like a Woman": Therapeutic Cultures and the Contagion of Femininity', *Gender Work and Organization*, 15(1): 88–107.

Taylor, S., and Tyler, M. (2000). 'Emotional Labour and Sexual Difference in the Airline Industry', *Work, Employment and Society*, 14(1): 77–95.

Teaching Agency (2012). London: Department of Education. http://www.education.gov.uk/inthenews/inthenews/a00211812/record-numbers-of-men-teaching-in-primary-schools

Trethewey, A. (1999). 'Disciplined Bodies: Women's Embodied Identities at Work', *Organization Studies*, 20(3): 423–50.

Tyler, M., and Abbott, P. (1998). 'Chocs Away: Weight Watching in the Contemporary Airline Industry', *Sociology*, 32(3): 433–50.

Tyler, M., and Taylor, S. (1998). 'The Exchange of Aesthetics: Women's Work and "The Gift"', *Gender Work and Organization*, 5(3): 165–71.

Vogt, F. (2002). 'A Caring Teacher: Explorations into Primary School Teachers' Professional Identity and Ethic of Care', *Gender and Education*, 14(3): 251–64.

Warhurst, C., and Nickson, D. (2007). 'Employee Experience of Aesthetic Labour in Retail and Hospitality', *Work Employment and Society*, 21(1): 103–20.

West, C., and Zimmerman, D. (2002). 'Doing Gender', in S. Fenstermaker and C. West (eds), *Doing Gender, Doing Difference*, 121–51 London: Routledge.

Williams, C. (ed.) (1993). *Doing Women's Work: Men in Non-Traditional Occupations*. London: Sage.

Williams, C. (1995). *Still in a Man's World: Men Who Do Women's Work*. London: University of California Press.

Williams, C. (2003). 'Sky Service: The Demands of Emotional Labour in the Airline Industry', *Gender Work and Organization*, 10(5): 513–50.

CHAPTER 23

MASCULINITY IN THE FINANCIAL SECTOR

DAVID KNIGHTS AND MARIA TULLBERG

INTRODUCTION

MAINSTREAM or malestream writing generally denies or denigrates the significance of gender in the study of management and organizational practice. By contrast, gender studies and liberal feminist studies have focused on women and their experience of, and location within (patriarchal) organizational structures and their differential status in the paid labour market (Ely *et al.* 2003). One result of this neglect of men, and in particular, of discourses of masculinity as a core problematic in management and organization is that they 'remain taken for granted, hidden and unexamined' (Collinson and Hearn 1994: 3). We follow this literature in seeking to theorize discourses of masculinity and management in organizations but with the limited focus of contributing to an understanding of the global financial crisis of 2008 and its aftermath. Our target of exploring discourses of management and masculinity with reference primarily to the financial sector is not at the expense of the sphere of work and organization and the gender dimension more generally, since it can be argued that these different spheres are impossible to understand independently of one another. However, because of the particularly aggressive macho competitiveness and preoccupation with financial reward (Lewis 2010), the financial sector renders more visible the discourses of masculinity which otherwise remain hidden and unexamined. It is for this reason if no other that it is an important site for examining and studying masculine discourses and practices. This would be so regardless of the devastating social and political damage incurred by the excesses that eventually helped to produce the global financial crisis but the latter provides even further justification for addressing the problems of masculinity in this sector.

The analysis of gender, whether under the auspices of gender theory, feminist research, women's studies, the study of men and masculinity, literary criticism, psychoanalysis, philosophy, equal opportunity, or labour-market research has, over recent years, been proliferating within academia, education, the media, and politics (Butler 1990, 1993, 2004a; Knights and Kerfoot 2004; Connell and Messerschmidt 2005; Young 2005; Dolan 2011). It is also beginning to be taken seriously within the academic study of management and organizations as a number of contemporary publications testify (Ferguson 1984; Butler 1990, 2004a; Collinson et al. 1990; Acker 1992; Mills and Tancred 1992; Collinson and Hearn 1996; Connell and Messerschmidt 2005; Young 2005; Dolan 2011; Jeanes et al. 2011).

Sex discrimination, inequality of opportunity, sex stereotyping, gendered job segregation, and sexual harassment have been central concerns of both students of gender at work, and the anti-discrimination and equal opportunity agencies. However, there has been a comparatively limited literature that attempts an integration of the issues from a perspective that focuses on management and masculinity (cf. Kerfoot and Knights 1992, 1993, 1996; Collinson and Hearn 1994, 1996; Aaltio-Marjosola and Lehtinen 1998; Wahl 2001; Mörck and Tullberg 2005). It is also the case that some of them are only peripherally engaged in theorizing the relationship between masculinity and management.

Within Western economies, a tough, macho-masculinity in the management or mismanagement of organizations (Knights and Tullberg 2012) has been stimulated and legitimized by neo-liberal political philosophies that emphasize market principles, forceful competition, and self-sufficiency. As yet, this shift to a more aggressively masculine, political, and managerial economy has gone comparatively unchallenged, even when it can be seen to have contributed significantly to the events that unfolded in the global financial crisis of 2008 and the Eurozone crisis of 2011 (cf. Knights and Tullberg 2012).

The financial sector is particularly important because 'it exerts more and more influence over more and more people's everyday life' (Blomberg 2009: 204) especially as discourses and practices of financialization effect a reduction of social relations to financially instrumental transactions (Dembinski 2009) that displace the very trade they traditionally were designed to service (Knights 2009; Lewis 2010).[1] It is this crisis that we are drawing on to illustrate and analyse masculinity in management and business in this chapter. While we believe discourses of masculinity are important for studying many aspects of contemporary society, their contribution to the global financial crisis and its negative impact on almost everyone cannot be ignored. We will not argue that masculine discourses have single-handedly generated the crisis, for there are multiplicities of contributing conditions to which we will make passing reference in this chapter. However, our interest in focusing upon discourses of masculinity is based on their almost total neglect in the voluminous academic and media discussions of the crisis.[2]

In this chapter we first review the gender literature, and in particular discourses of masculinity in the context of management and organization, since it is central to our argument that the very concept of management reflects and reproduces modes of discourse and behaviour that we identify as masculine.[3] Our approach to this literature is selective as we specifically support a view of discourses of gender and masculinity as socially constructed and reproduced through routine and recurrent performances. Our central concern is to provide the analytical context for an examination of the global financial crisis in terms of what we argue are dominant discourses of masculinity within management. This focus on discourses of masculinity is unique in terms of understanding management, trading, and institutional practices in the financial sector. Of course, we acknowledge that numerous diverse conditions are responsible for the events that culminated in major international banks and insurance companies either passing into administration or accepting a bailout from taxpayers through their national governments. Consequently, we document some of the most frequently reported media accounts of the crisis before turning to an analysis of what has been almost completely neglected or ignored—the contribution of discourses of masculinity.

Discourses on Management and Masculinity

The discussion in this chapter is based on a view of gender as continually constructed and sustained through different performative acts. Gender is not there, to be played out as a set of role expectations; it has to be performed in everyday life and work is one important arena for this process of *doing gender* (West and Zimmerman 1987; Butler 1990) and in particular of doing masculinity (Game 1990; Clough 1992; Knights 1997). Other scholars developed this perspective further and have shown how the allocation of labour, of wages, and power positions contribute to the social construction of gender (Acker 1992; Gheradi 1994; Liff and Ward 2001; Smithson and Stokoe 2005). Following Butler (1993/7, 2004*b*) we eschew any sense of an essential gender identity that lies behind the performance; gender is all drag, a stylized repetition of acts. The gender binary has to be continually repeated and reproduced and its effects are to sustain the dominance of heterosexuality, but the performance is not simply 'within the grasp or control of the individual as it is social and cultural' (Butler 1997, quoted in Emig and Rowland 2010: 7). Nor is it the subject that performs; it is the performance that constitutes the subject. The performance includes several elements, where some are more important than others in the construction of gender and gender identity. To become a mother is more performative in the construction of femininity than to be a professor but it also depends on the actions as a mother. Not all mothers are seen as 'the right kind of woman'. Each single element implies different meanings in diverse contexts; it is the interplay, which constitutes the final result. Also, the performance has to be interpreted and confirmed by others. Some glimpses from a study on masculinity constructions among senior managers and board members in a Swedish bank (Tullberg and Mörck 2012) provide an example:

> An organization in crisis needs to recover and to restore confidence among customer and the public. Previous board members and top management are changed and at the AGM the new chairman of the board is presented to the shareholders. He

introduces himself as a very experienced man (he is more than 60 years of age) and he describes his new position as similar 'to doing military service' which is compulsory for young men in Sweden. The same day he is interviewed on the TV news and one part of the interview is performed in a gym where he is training with weights, and there are close-ups of the muscles on his arms. Military service is a task only for the 'right kind' of men—women and disabled men are excluded. To do military service is also a community service—not anything that you do for financial gain or to boost your career. To be bodily strong and fit through regular training, and refusing to change any routines in response to a request for an interview from a journalist signals control, both mentally and physically. For many of the people in the TV audience and generally in the media he was confirmed as a man of the 'right kind' for this important management position. Some maybe instead interpreted this performance as absurd or ridiculous—that a man near retirement is drawing a parallel between himself and a young draftee and so obviously that he needed to demonstrate his physical abilities by being filmed in the gym.

The concepts of *performance* and *performativity* relate to the perspective of understanding social life in terms of theatrical metaphors (Schechner 2002). Consistent with this metaphor the society is the stage, and as such is set up for different kinds of performances. In various times and places, there are a range of conditions that invite but also preclude certain kinds performances and, in writing about gender, a diversity of scripts along the masculinity/femininity continuum are available within current discourse. The setting of the scene is produced in interplay between the actors, sometimes with serious conflicts and difficult negotiations, which become evident in times of change. Examples of such conflicts are when different measures are introduced to increase the number of women on company boards or to encourage more men to use the rights to parental leave. The familiar performative elements in the construction of gender become threatened and many, both men and women, protest loudly largely because they see that traditional constructions of masculinity and femininity are being threatened. Women often want to remain primary with respect to childcare since not to do so threatens their (socially mediated) construction of what it is to be a mother. Men, however, hardly ever challenge this primacy but secure their identity more exclusively through work. Similarly, the 'breadwinner' image of masculinity is one that is preserved by both men and women as men are supposed to privilege work and women domestic relations. In this context, it is interesting to see that, despite a long period over which the of the association of breadwinner with men has been discredited, a well-known popular author has claimed that men are hard-wired to protect and provide and to lose his breadwinner status would be equivalent to having his penis cut off (BBC 2012).

Our focus on management is because it is a scene for discourses of masculinity. Masculinity is frequently a presupposition underlying the production of most representations of management but it remains tacit and unspoken. Clearly it is not something that can be observed or pointed to in a concrete sense, for masculinity is like many other cultural phenomena (e.g. patriotism, misogyny) only visible indirectly through behaviour and discourse. Only for purposes of analysis do we attribute the label masculinity to these behaviours and discourses and to the subjectivity that is its medium and out-come. This is not to suggest that what is eventually labelled masculine is absent from conversation or behaviour prior to it being attributed as such. Masculinity, no less than femininity, is seen as manifest in a multiplicity of forms and not as a 'fixed' entity. Also within the financial sector there are obvious differences among male financial opera-tors, i.e. between the patriarchs from the boardroom and the macho dealers from the floor (McDowell 1997). Accepting the term masculinity as problematic, in that there are clearly diverse masculinities across racial and ethnic difference within and between countries (Gilmore 1990) and across time and location. Subsequent work on men and masculinity has sought to escape the confines of dualistic gender essentialism. Moreover, beyond the obvious plurality of masculinities, the differing experiences of masculinity within the lifespan of individuals forces reconsideration of whether masculinity as an all-embracing descriptor for the behaviours of men is of any significant value.

The work of Connell (2005/1995) resonated with that begun by Brittan (1989), argu-ing that the failure to recognize masculine identities as plural could be found in the hitherto unacknowledged understanding that masculinity had been conceived as an internally undifferentiated category. Connell's contention was that the failure to recog-nize the complexities and differences amongst men had resulted in a skewed analysis of social relations and in a politics of the sexes wherein, as a consequence of this theo-retical slippage, all men were pitted against all women. Whilst there are clearly multi-ple masculinities, culturally and historically, what remained at issue in the discussion and development of the literature on masculinity was the shared characteristics of these behaviours. Following the attempts of Carrigan et al. (1985) to theorize men's (domi-nant) behaviour in terms of masculinity, the term 'hegemonic masculinity' achieved prominence. We are aware of the criticism of this concept because of its 'universalism', 'inherent rationalism', and neglect of the body and its emotions (Siedler 2009: 10) and that it has many of the same limitations as did the concept of patriarchy as 'too mono-lithic, ethnocentric, and one dimensional' (Beechey 1979; Rowbotham 1979, quoted in Hearn and Morrell 2012: 5). However, we are limiting our use of the concept largely to Western cultures of management as implicated in the US and European driven financial crisis of 2008. Also in the form of homosocial bonding as contributing to the excessive bonuses in banking, we believe the masculine hegemony does incorporate an emotional and bodily dimension, albeit not one conducive to self-transformation independently of external intervention.

What we would want to describe as discourses of hegemonic masculinity are what characterize most business and indeed non-commercial organizations. While in their own terms tacit and non-explicit, discourses of masculinity nonetheless prevail to struc-ture and sustain behaviour of certain sorts. It is ordinarily behaviour that is technically rational, performance oriented, highly instrumental, devoid of intimacy yet preoccu-pied with identity, and driven by rarely reflected upon corporate or bureaucratic goals that are presumed inviolable. These masculine discourses thereby have the effect of con-stituting both managers and employees as subjects that secure their sense of identity, meaning, and reality through the rational, efficient, and singularly uncritical pursuit of goals and objectives handed down from above. Conditioned by this privileged and per-vasive form of masculinity, the modern manager in the private sector is ritually engaged in coordinating and controlling others in pursuit of the instrumental goals of perfor-mance, productivity, and profit. In the public or voluntary sector, the same rigorous pursuits continue by adopting various substitutes for the market measure of profit accu-mulation such as efficiency, 'best value', best practice, etc.

It is appropriate to embellish our earlier sketch of the genesis of management and masculinity with not so much the genesis of management *per se* but of the way it has evolved, and continues to be sustained, predominantly as a masculine enterprise. This is the case, we would argue, even when partly due to equal opportunity policies but increasingly because of the feminization of many large corporations, more women are becoming managers. Most accounts of the history of modern management indicate that, although women were very often to be found in factory employment or the mines, virtu-ally no women occupied managerial positions during the Industrial Revolution (Pollard 1965; Anthony 1977). In the early development of managerial work, then, men and mas-culinity were a dominant feature. Even today, there remain few promotions of women into senior managerial positions above what has euphemistically been described as the 'glass ceiling' (Davidson and Cooper 1992) or the 'glass cliff' (Ryan *et al.* 2007). But even if women were able to secure managerial positions, this would not be sufficient to negate the description of management as a masculine enterprise. Wajcman (1998) showed how women managers had to adopt very much the same behaviour as men in the same posi-tion. For, as we go on to argue, masculinity is neither exclusive to men nor exhaustive of their discursive being; nonetheless, contemporary organizations create, sustain, and reproduce masculine modes of behaviour that privilege heterosexual men and discrimi-nate against women and those with non-conventional sexual preferences or identities.

These discourses are abstract and highly instrumental with respect to controlling their objects, thus sustaining a mode of relating to externalities that is self-estranged and wholly disembodied (Williams and Bendelow 1998; Knights and Thanem 2006). Rarely does masculinity embrace the world or even itself with a sense of wonderment, pleasure, or engagement, for it labours ceaselessly under a struggle to control and pos-sess the objects of its desire, whilst at the same time self-deceptively presuming itself to be free of desire. That is to say, the desire is so buried beneath a series of rationalities and rationalizations as to be virtually invisible to its agents. Yet it may be suggested that the hidden agenda behind masculine discursive struggles for control is a desire to produce a stable world in which the identity of certain individuals can be rendered safe and secure (Game 1990; Clough 1992). For example, these authors attribute the tendency for men to produce 'grand theory' in the social sciences to this struggle for order and control.

Where this masculine preoccupation with ordering the world reaches its culmination in modern management is in its attraction to strategic discourse. As 'Big Boy Caprice' in the film *Dick Tracy* puts it very neatly "A man is not a man without a plan"' (Kerfoot and Knights 1996: 82). Of all the managerialist concepts peddled by consultants and gurus, strategy has probably had a more universal impact on modern management than has any other (Whittington 1993). In fact Hoskin and Macve (1988) treat strategy[4] as synonymous with the genesis of modern management in the sense that it was the point when exact records and calculable decisions rendered the future amenable to the precise and predictable controls of a strategic plan. While Hoskin does not focus on the gendered nature of these technologies of control, it is clear that strategic management coincides closely with those contemporary forms of masculinity that turn everything into an object of conquest and control (Seidler 1989). It involves a disembodied and emotionally estranged conception of reason (Kerfoot and Knights 1993) that attaches itself to a ceaseless pursuit of strategic goals, the attainment of which confirms the promise, though rarely the reality, of a secure masculine identity. In this sense 'the best is yet to come' (Kerfoot and Knights 1996) is the illusion of perfect security, contentment, and control that pushes masculinity towards ever more heroic and self-sacrificing struggles.

Corporate capitalism is both the vehicle for the expression of this masculinity and a major driving force in its continued discursive domination within management and organizations more generally. For in elevating competitive success and the ethic of accumulation or 'possessive individualism' (MacPherson 1962) as the reason for existence, the capitalist corporation provides a legitimate outlet for masculine preoccupations with conquest and control. But as we have already indicated, it also reproduces the conditions that sustain the dominance of masculine discourses. First, the corporation stimulates the pursuit of abstract, instrumental objectives, such as output, profit, and growth, above all else and this reflects and reinforces a disembodied and estranged relationship between masculine subjects and their worlds, including that of other subjects. Yet second and simultaneously, it generates the conditions of competition and uncertainty where such strategic pursuits continually leave individuals feeling threatened and insecure. Since, as we have intimated, masculine projects are often a disguised or disavowed desire to order the world for purposes of establishing a secure identity, these insecurities only serve to reinforce the masculine drive for success. In short, the very pursuit of security through achievement, conquest, success, and wealth generates precisely those self-same conditions of insecurity that make the pursuit compulsive and unending, thus closing the 'vicious circle' and denying masculinity the possibility of escaping from itself. Of course, in the sense that masculinity knows no other mode of being and denigrates feminine, gay, lesbian, or alternative lifestyle discourses that fail to sustain this way of relating to the world, it cannot even contemplate any necessity for such an escape.

In management as in most male sites to be a man is to 'not be a woman', which is constantly demonstrated in everyday life, for example by the frequency of invectives against men's sexist references to the feminine body; their resistance to wearing clothes that in any way resemble feminine attire or to engage in what is seen as feminine interests. The 2000 movie about *Billy Elliot*, who preferred ballet to boxing and breakdance illustrates very clearly the dilemma a boy gets into when not following the right script. Scholars have also described the two-sided problem to get more men into what is seen as feminine occupations like nursing and pre-school teaching (Williams 1995; Sargent 2000; Nordberg 2002; Ekstrand 2005). There are difficulties in attracting men to, and problems for men working in, feminized occupations, especially in proving that that they are 'real men'. One could argue that this dilemma is the same for women who break gender norms but there are some important differences. In general, the feminine scripts are more open and less constraining in regard to dress: girls can wear trousers and play like boys but for heterosexual boys and men the fear of being seen as effeminate or 'gay' means that they rarely wear women's clothes. Recent androgynous fashions worn by 'metrosexuals'—young men with nail varnish, eyeliner, and interesting haircuts—most publicly represented by David Beckham wearing a sarong, confirm rather than contradict this view. Only a very masculine man, preferably a successful sportsman, visible and public heterosexual (or even better, a notorious womanizer) can use female attributes without losing his position of masculinity within conventional male hierarchies.

In research on work organization and management it is well documented that women are discriminated against to the point of almost being excluded from male-dominated occupations (Cockburn 1983; Wahl 1995) and that men in both male and female professions are often favoured to enjoy a faster career progress. We can speculate about different accounts for understanding this state of affairs but there is little doubt that men exercise more power than women largely because they disproportionately occupy senior positions in organizational hierarchies (Kanter 1977; Ferguson 1984). But men also feel the need to exclude women because they represent a potential threat to their performance of masculinity and this is partly why women are under-represented in senior management despite a long history of equal opportunity practice. Men retain their privileged status in the performance of masculinity and in maintaining some level of distance from aspects of femininity and sometimes from women through homosocial relations. In work situations where women form a significant minority, the performative aspects of masculinity or not-to-be-a-woman are threatened so that it is no longer a job or an arena for 'real men'.

The sacrifices that are made in transforming everything into an object of control and conquest are especially visible in the field of management where leisure, personal life, family, and even physical and mental health are often subordinated to the greater goal of conquest, mastery, and competitive material and symbolic success. These sacrifices are made largely in the name of the organization and its hierarchical structure but in the service of the masculine self. Of course, the desire to be in control and always winning or succeeding in competitive conquests manifests itself in many different ways. Within business and management, at least two of these manifestations are illustrative of masculine conquest, competition, and control. First is when it takes the form of predatory corporate mergers and acquisitions despite evidence that they are often unprofitable.[5]

A merger may often have more to do with glory-seeking than business strategy. The executive ego, which is boosted by buying the competition, is a major force in M&A, especially when combined with the influences from the bankers, lawyers, and other assorted advisers who can earn big fees from clients engaged in mergers. Most CEOs get to where they are because they want to be the biggest and the best, and many top executives get a big bonus for merger deals, no matter what happens to the share price later (McClure 2012).

These examples clearly demonstrate how masculine conquest, competition, and control dominate business in general, but we now turn to examine the global financial crisis suggesting that it can be understood more clearly through an analysis of discourses of masculinity and more specifically, how money makes the man.

MONEY MAKES THE MAN

Before focusing on how money makes an important contribution to the construction and reconstruction of discourses and identities of masculinity, we need to provide a brief summary of the financial stage upon which this drama and its turning into a crisis takes place.[6] The finance sector generated the conditions of crisis in the first decade of the twenty-first century in the area of domestic credit provision where a range of mortages with different risk profiles were securitized; that is, transformed into tradeable bonds for an immediate profit rather than awaiting returns at maturity. The packaging of these securities into bundles (credit debt obligations) that were able to secure triple A ratings as assets only to prove toxic once the bubble burst, and creating insurance products called credit default swaps that insured against any default exacerbated an already non-sustainable market in credit. The ratings and the insurance against risk generated a moral hazard whereby the overprovision of credit created and sustained a boom of such proportions that it had inevitably to bust. Of course, it is much more complicated than this and indeed the proliferation of financial instruments that fuelled it were often not understood by many that traded them. However, for our purposes it is only necessary to understand that even in less candyfloss times as prior to the crisis, those working in this sector operate principally from a position of making as much money as possible on the basis that they never expect such bonanzas to last. In other words, their very actions anticipate the crash but as a self-fulfilling prophecy since they behave in precisely those ways that make it inevitable.

As indicated earlier, we adopt a perspective on gender as socially constructed as individuals interact with others where there is an anticipation or expectation of the social confirmation of self (Mead 1934; Blumer 1969). Of course, because it is impossible to predict let alone control this social confirmation not least because of shifting desires and values, the self or identity is under continuous reconstruction and change. The concepts of performance and performativity that were introduced earlier relate to the perspective of understanding social life in terms of theatrical metaphors (Schechner 2002; Bial 2004) and these enable us to identify what a person has to do, to think, to desire, to show (or to pretend), in order to become the 'right' kind of man (or woman) in the eyes of others, as well as themselves. There is a considerable amount of effort required to produce and reproduce gender relations as they are presently constituted since it is a 'historical process, not a self-producing system' (Connell and Messerschmidt 2005: 844).

According to Butler, it is not the subject that performs; it is the performance that constitute the subject (1993/1997) and different cultures and subcultures provide individuals with distinct scripts for these acts. In so far as subjects can only secure social confirmation when they perform the scripts and comply with the expectations of others and of the wider culture they inhabit, performativity is clearly foundational to social life and existence. However, there is a sense in which this perspective appears wholly deterministic whereas there are a multiplicity of dramas and scripts from which to choose and the transformation of individuals into subjects (Foucault 1982) that secure a sense of identity through engaging in the practices that the exercise of power invokes (Knights 1992, 2002) is variable. Power is never exhaustive and individuals are selective in their pursuit of specific identities, subscribing to some but resisting or refusing other exercises of power in accordance with particular aesthetics, ethics, meanings, talents, and values, even though these are themselves clearly cultural products.

In this chapter we have focused on the discourses of hegemonic masculinity (Connell 2005/1995) in contemporary management and, in particular, the financial sector where they could be seen to manifest some of their most extreme forms. There are certain instrumental elements in the performance of different masculinities that we want to illustrate here. These elements are not unique to this context but, in combination with the access that the finance sector has to economic power, we argue can have a devastating impact on our economic and social well-being, as has been the case since the global financial crisis of 2008.

Without seeking to grade these different elements, we begin with the notion of competitiveness that everywhere but especially in the finance sector frequently takes the form of individuals seeking to win, be visibly successful through securing the highest income, status, and power. The continued furore over bankers' bonuses is a visible manifestation of this preoccupation with winning and being among the most highly paid of executives but this is supported by a range of remuneration committees whose members invariably benefit from being part of an elite club that is often characterized by homosocial male bonding. It has been reported in the UK that the majority of non-executive directors who sit on remuneration committees come from business and in particular financial intermediation (High Pay Centre 2012: 4). For example, former lead executives represent 46 per cent of those sitting on remuneration committees of FTSE 100 companies and of the 366 non-executive directors who sit on remuneration committees only thirty-seven (10 per cent) are from outside business or financial intermediation. Furthermore, 45 per cent of the companies in this survey had all-male remuneration committees and only 16 per cent of the total memberships of these committees were women (High Pay Centre 2012: 4). This evidence supports our claim that homosocial bonding in remuneration committees is effective in protecting privileges and high pay and bonuses in business (Knights and Tullberg 2012).

Obviously these are well-known performative elements in several masculine discourses even though the concepts of competitiveness and success may imply different kinds of performance, not necessarily concerned exclusively with economic reward. Masculine competitive performance is a major preoccupation of, for example, politicians (Knights 2004), sports people (Dashper 2011), and even criminals (Britton 2011). For criminals, 'being masculine is screwing the most girls, who could fight the best...who could hold the most beer, smoke the most weed' (Britton 2011: 50). Among the business elite, success is equated with financial power and above all it is manifested in growth, profit, or income. As an owner or manager of a company you could become visibly, economically successful in different ways: the company grows, takes over other businesses, becomes a market leader, and so on. The owner doesn't have to exhibit great personal wealth as there are many other ways to be a winner. This doesn't apply to most managers in the financial sector where corporate and institutional rather than individual, personal ownership is dominant; the majority of managers are employees thus executing power on behalf of a range of different institutional or private investors. Although it is not routine, executive managers as well as board members can in principle easily be replaced at short notice. Combined with the basic insecurities surrounding masculine identities, this fear of how the income, power, and status of executive managers are precarious became especially intense during the credit boom of the early twenty-first century. This was because, based as it was on a candyfloss structure of escalating credit, 'deep down' most of the beneficiaries even without knowledge of economic cycles must have felt that the boom had a limited lifespan. Indeed, as one retired senior public official remarked at a recent conference on financial services, 'managers with such high incomes believe it cannot go on for much longer so they grab all they can before the Armageddon' (Bristol Business School 2010).

We argue that this masculine insecurity and sense of precariousness had a major impact on the financial sector and the crisis it perpetrated in 2008. First, as we have argued, the sense of the boom and bonanza being too good to be true stimulated those at the heart of it to push for as much as possible while it still lasted—hence the astronomical bonuses in the sector. This may be seen as part of the explanation for what appears to be excessive risk-taking, greed, and irrational hubris in the sector. Of course, politicians who believed in finance as a salvation for Western economies in global manufacturing decline and academics who constructed models justifying the risk-taking, the proliferation of new financial instruments, and the bonus culture supported this. There was a neo-liberal consensus about the 'free' market and the benefits of new financial instruments among Western politicians and within academia, the Black-Scholes model provided mathematical justification for risk calculations[7] and finance economists developed principal-agency theory to justify stock options and bonuses as incentives for corporate managers.

Despite the horrendous consequences of the global financial crisis in terms of economic recession, austerity programmes, and threats to human well-being, executives in the financial sector continue to be awarded excessive bonuses, only marginally restrained by public and political pressure.[8] For example, executive management in general and particularly those in the financial sector seek to have the highest salaries or achieve the biggest bonuses regardless of corporate performance. Between 2008 and 2012 there have been frequent scandals concerning the continued payment of large bonuses to executives and traders in the financial sector against a background of falling profits and returns to shareholders, resulting in what journalists were describing in 2012 as a Spring Shareholder uprising. While there have been some high-profile shareholder revolts about executive pay (e.g. Citigroup, Barclays, and Aviva) 'of 200 or so proxy votes completed this year, only four received a majority negative vote on executive compensation' so the challenge to excessive remuneration is less than is claimed in the media (Waggoner and Krantz 2012). Of course, there is good reason for this state of affairs because institutional investment companies are the owners of most corporations and their managers enjoy similar remuneration privileges and so have a vested interest in the high pay culture that dominates the corporate boardroom. Also as we argue below, they are part of the homosocial elite from which remuneration committees recruit their members (High Pay Centre 2012). Moreover, transparency programmes may well create a greater visibility of the remuneration packages with the intention of restraining excessive pay and bonuses but it generally has the opposite effect since masculine discourse reinforces a competitive pursuit of the highest pay and everybody joins in the chase. Thus transparancy has the unintended consequence of undermining its own objectives. The public represented by the media and politicians have been incensed when the banks that had to be bailed out by taxpayers continue to reward their executives excessively not only in terms of salary and bonuses but also through pension benefits and agreements on parachutes on retirement or dismissal regardless of success or failure.

In several countries the politicians have introduced or tried to introduce rules to limit the bonuses but, for example in Sweden, the bank managers and brokers then raised the fixed salary in order to achieve the same level of reward (*Dagens Industri* 2012). Defenders of the bonus system argue that removing them would not only result in losing talented executives to competitors but also a displacement of bonuses into higher fixed salaries, thus removing the incentive element. Yet, without denying the importance of economic insecurity and individual greed, albeit simply a reflection of elite and capitalist values, we argue that there are also other driving forces behind the rapidly rising curve of income among the (white, male) business elite (Knights and Tullberg 2012) closely connected with the masculine discourse in this circle of men. For senior executives, high income continues to be the most obvious sign of importance, position, and success in comparison with others. To be among the best paid makes one a winner. In addition, a winner never has to provide any further rationale to explain their advantages, as competitive success is its own justification. Thus no explanations or reasons are ever given in the media, annual reports, or other texts justifying the basis for the high income. Instead, what are published are comparisons with other companies or managers and sometimes listings of the best paid CEOs. These revelations simply fuel the high income/bonus culture within executive management.

While insecurity is inescapable, having its roots in ontological, existential, and social conditions of human life (Knights and Willmott 1999), it is exacerbated by discourses of both management and masculinity. Moreover in Western society, discourses of masculinity and management are mutually reinforcing. That is to say, managers and the culture that surrounds them tends to reflect and reproduce a range of masculine discourses while the very subscription to masculine discourses invites a management mentality of control, competition, and conquest (Kerfoot and Knights 1996). Masculinity is a performative element in the construction of management and leadership, which is one reason why it is so much harder for women to secure access to these positions (Wahl 2001) and why successful women in management are often depicted as being more masculine in their assertiveness, determination, and drive. It may be expected that, as the context becomes even more insecure, the relationship between management and masculinity will become even more important to maintain and protect. There is little doubt that the financial sector is a very insecure field, and this is not simply because of the general insecurity around management work (Knights and Willmott 1999).[9] The insecurity of management in general is exacerbated within the financial sector by the intangible, risky, and unpredictable nature of its products and services (Morgan and Sturdy 2000) and the speed and intensity of the circulation of money and decisions surrounding it (Lewis 2010). Success, as well as failure, is highly visible and in the direct spotlight of the media.

Having the solidarity of group membership reduces vulnerability for the individual, as this can offer some feelings of safety and protection if often they may prove somewhat illusory. In a business context group bonding and networking can be a resource when seeking new positions in cases of dismissal or resignation, a source of privileged information and knowledge, access to resources, and other advantages of value in business transactions. The groups become very important and is protected by strong, often homosocial group norms. Classic theory on group dynamics informs us that the more threatened a group is, the stronger the norms become and the more outsiders are stereotyped as the threat to internal solidarity. For this reason internal traitors or deviant members are heavily policed and punished since they can be seen as an even greater threat to group solidarity (Tajfel 1970; Tajfel and Turner 1986). One is in-the-group if following the norms, or out of the group and without protection if one is questioning the dominant values or set of rules in the group or is outside the group and therefore an out-group subject to stereotyping.

Several scholars have discussed homosociality among men, how men seek out other men for friendship,[10] and prefer to employ and work with men, rather than women. The phenomena has been connected with asymmetries in the numbers and power of men in most organizations (Kanter 1977); with male domination and that men and male interests and activities are perceived to be more valuable in society (Roper 1996; Bird 1996; Holgersson 2004. Clearly, homosociality is prevalent in the financial sector and can be seen as one of the reasons why men seek mutual dependence in one another, thus facilitating a closing of ranks as has been evident in relation to the public hostility to excessive pay and bonuses since the financial crisis (Knights and Tullberg 2012). Homosocial male bonding increases material and symbolic security and provides a means of confirming a form of mutual masculinity.

Earlier research has shown how the strong norms among the business elite, as well as among different occupational groups in the financial sector, are embodied in the uniform style of dressing, in the language, and in the codes of behaviour (McDowell 1997; Blomberg 2009; Tullberg and Mörck 2012). The documentary film *Inside Job* (2010) clearly illustrate the economic short-term and anti-social values, the attitudes, and behaviour in the financial sector but also how people are loyal to operations and decisions that they are actually questioning from an individual ethical point of view. To be an in-group member means you have to join in the dominant masculine discourse, not reveal any information about doubtful operations, keep closed ranks, and last but not least, never challenge or question any kind of remunerations however excessive they may appear. This also applies to the minority of women who are holding senior positions in the financial sector (Wajcman 1998; Pullen and Knights 2007) for in climbing the 'greasy pole' they have been forced to subscribe to discourses of masculinity and they get generally rewarded if they can be seen as one of the 'lads' and thereby not a threat to masculine norms.

Earlier crises, scandals, and crimes led to a demand for greater transparency and new rules and standards were set for company reports that included the remuneration of board members and top executive management. The boards of the main stockmarket-listed companies then formed remuneration committees to set the levels of pay and their work is separately reported in the AGMs and in the Annual Reports. In the written report the figures are shown, but verbally in the AGM meetings no money is mentioned. What's reported there is the work process, most often in terms of number of meetings and the basic principles of how to distribute the remuneration according to the relation between fixed and variable parts. No justification of the levels of remuneration is provided and the construction of the bonus system or how it is connected to different requirements or targets is never explained and intimate questions about the remunerations are rejected by the chairman (Mörck and Tullberg 2005).

SUMMARY AND CONCLUSION

In this chapter we have discussed the link between the discourses of masculinity in contemporary management and the financial crisis/crises recently experienced in Western countries. Our aim has not been to displace all other accounts of the global financial crisis for clearly there is a proliferation of conditions that would need to be examined. To mention just a few, new financial instruments such as collateral debt obligations and credit default swaps, the neo-liberal or Washington political consensus, economic deregulation, bonus mania, inadequate or light touch regulation, corporate, government, and individual over-indebtedness all help to account for the crisis. We have simply focused here on understanding the financial crisis and its aftermath through a neglected gender perspective on masculinity that complements but also challenges some of the conventional wisdom in this field.

Mainstream writing on management generally discounts or dismisses the significance of gender as a concept in understanding management and organizational practices. Still, while remaining tacit and unspoken, masculinity is a presupposition underlying the production of many representations of management. Clearly it is not something that can be observed or pointed to in a concrete manner for masculinity is like many other cultural phenomena (e.g. patriotism, misogyny) only visible indirectly through behaviour and discourse. Also it is only for purposes of analysis that we attribute the label masculinity to different behaviours and discourses and to the subjectivity that is their medium and outcome. Masculinity, no less than femininity, is seen as manifest in a multiplicity of forms and not as a 'fixed' entity and also within the financial sector there are obvious differences among male financial actors (McDowell 1997). Masculine discourses do not reduce managers to a single unitary set of subjects; however, they do appear to encourage particular interests and preoccupations that coalesce around pursuing high salaries and bonuses, both of which can be seen to have fuelled and fired the crisis.

While acknowledging the various critiques of hegemonic masculinity (Connell 2005/1995) that claim it is over-rational and disembodied (Seidler 2007; Hearn and Morrell 2012), we believe it is still of relevance for the financial sector. This is because it is in the financial sector and not just on the trading floor where some of the most extreme manifestations of unadulterated masculine discourses are performed and perpetuated. Certain elements, such as conquest, competitiveness, and visible success, have been emphasized that are not unique to this context, but in combination with the centrality of its role in the modern economy and thereby its access to power, the finance sector has had a devastating impact on our economic and social well-being.

The discussion has been based on a view of gender as continually constructed and sustained through different performative acts, where masculinities as well as femininities are constructed in everyday life and where work is an important arena for this process of *doing gender* (Butler 1990). The concepts of *performance* and *performativity* relate to the perspective of understanding social life in terms of theatrical metaphors (Schechner 2002). Consistent with these metaphors, masculinities are staged not only as entertaining dramas but, as we have sought to indicate, ones that have very real consequences for us all that are far from theatrical.

Management is a scene for discourses of masculinity that prevent too many women invading the protected space where homosocial constructions and masculine performances reproduce power and privilege for men. It keeps the feminine at a safe distance so as to minimize the threat of the Other to masculine discourses and their performative outcomes, for the greatest fear is to fail to be a 'real man'.

It may be expected that the relationship between management and masculinity will become even more important to maintain and protect, the more insecure the context. There is little doubt that the financial sector is a very insecure stage, and this is not simply because of the general insecurity around management work (Knights and Willmott 1999). The insecurity of management in general is exacerbated within the financial sector by the intangible, risky, and unpredictable nature of its products and services (Morgan and Sturdy 2000) and the speed and intensity of the circulation of money and decisions surrounding it (Lewis 2010). Success, as well as failure, is highly visible and in the direct spotlight of the media.

Earlier crises, scandals, and crimes led to a demand for greater transparency and new rules and standards were set for company reports that included the remuneration of board members and top executive management. We argue that these attempts to restrain pay generally have had the opposite effect. The greater visibility of remuneration involves all senior executives chasing the highest levels of pay, for that is a sign of their manliness and a demonstration of their masculine sense of being first past the post.

NOTES

1. E.g. at the height of the bubble in 2007 hedge funds were trading at several trillion dollars, equal to the gross national product of the whole globe.
2. The literature both academic and popular as well as the media coverage has become too extensive to document. However, the following academic books were consulted and none of them provide any analysis based on gender or masculinity: Harvey 2010; Keen 2010; Mellor 2010; Rajan 2010; Stiglitz 2010; Rogoff 2009; Roubini and Mihm 2011. There have been some limited contributions from feminists in the media where the crisis was attributed to gender imbalances in business and management but no sustained analysis of masculinity as attempted in this chapter.
3. Despite emphasizing the multiplicity and diversity of masculine identities or masculinities, the singular term masculinity or masculine identity is also often used to convey certain of its ideal typical features. Among these are such features as disembodiment, instrumental rationality, and a preoccupation with control that are commonly, though variably, produced and reproduced in all masculine discourses.
4. As is often the case, by far the best definition of strategy derives not from the management literature but from philosophy. De Certeau (1984: p. xix) argues that strategy is that calculus of force relationships that becomes possible when a subject of power can be isolated from an 'environment'.
5. Studies reveal that between 40 and 80% of mergers and acquisitions prove to be disappointing. The reason is that their value on the stockmarket deteriorates (http://finance.mapsofworld.com/merger-acquisition/failure.html).
6. Ironically, in the 1980s there was a regular advert on UK TV for domestic insurance arguing that 'we never turn a drama into a crisis' because of the speed and efficiency of settling the claim for damage and losses.
7. While winning a Noble prize for their efforts, the model developed by Fischer Black and Myron Scholes had already been discredited long before the financial crisis, as its authors lost money as fast as they had made it by applying the model to their investments. The model was not wrong in providing a method for assessing the value of a financial derivative but it was then used to turn derivatives into commodities traded in their own right—the Midas Formula for turning everything into gold. The herd instinct did the rest in terms of sending the global economy over the precipice (Stewart 2012).
8. This restraint has been most marked in the UK where a £1million bonus to the CEO of one of the nationalized banks was withdrawn only after political pressure from the opposition in parliament.
9. Ironically novels are a better source than academic publications for evidence of the insecurity among managers (e.g. Joseph Heller, *Something Happened*. New York: Alfred A. Knopf, 1974; Tom Wolfe, *The Bonfire of the Vanities*, New York: Farrar Straus Giroux, 1987; David Lodge, *Nice Work*, Harmondsworth: Penguin, 1989).
10. A recent piece of research found that telephone usage was highly gendered. Women spent more time communicating with their children and other close relatives whereas men were

more likely to call their male friends (Smoreda and Licoppe 2010: 244). Males are more profound texters than females in almost all the categories of users (Iqbal 2010: 517).

References

Aaltio-Marjosola, B., and Lehtinen, J. (1998). 'Male Managers as Fathers? Contrasting Management, Fatherhood, and Masculinity', *Human Relations*, 51: 121

Acker, J. (1992). 'Gendering Organizational Theory', in A. Mills and P. Tancred (eds), *Gendering Organizational Analysis*, 248–60. London: Sage.

Anthony, P. D. (1977). *The Ideology of Work*. London: Tavistock.

BBC Radio 4 (2012). A special edition devoted to the role of men on *Woman's Hour*, 22 May.

Beechey, V. (1979). 'On Patriarchy', *Feminist Review*, 3: 66–82.

Bial, H. (ed.) (2004). *The Performance Studies Reader*. London: Routledge.

Bird, S. R. (1996). 'Welcome to the Men's Club: Homosociality and the Maintenance of Hegemonic Masculinity', *Gender and Society*, 10(2): 120–32.

Blomberg, J. (2009). 'Gendering Finance: Masculinities and Hierarchies at the Stockholm Stock Exchange', *Organization*, 16: 203.

Blumer, H. (1969). *Symbolic Interactionism: Perspective and Method*. Englewood Cliffs, NJ: Prentice Hall.

Bristol Business School (2010). 'Governance and Accountability in the Financial Services Sector', Centre for Global Finance and Bristol Centre for Leadership and Organizational Ethics, University of West of England, 12 Mar.

Brittan, A. (1989). *Masculinity and Power*. Oxford: Blackwell.

Britton, D. M. (2011). *The Gender of Crime*. Plymouth: Rowman & Littlefield.

Butler, J. (1990). *Gender Trouble: Feminism and the Subversion of Identity*. London: Routledge.

Butler, J. (1993/1997). *Bodies that Matter: On the Discursive Limits of 'Sex'*. New York: Routledge.

Butler, J. (2004a). *Undoing Gender*. New York: Routledge.

Butler, J. (2004b). 'Performative Acts and Genre Constitutions: An Essay in Phenomenology and Feminist Theory', in H. Bial (ed.), *The Performance Studies Reader*, 154–6. London and New York: Routledge.

Carrigan, T., Connell, R. W., and Lee, J. (1985). 'Towards a New Sociology of Masculinity', *Theory and Society*, 14(5): 551–604.

Clough, P. T. (1992). *The End(s) of Ethnography*. London: Sage.

Cockburn, C. (1983). *Brothers: Male Dominance and Technological Change*. London: Pluto Press.

Collinson, D., and Hearn, J. (1994). 'Naming Men as Men: Implications for Work Organization and Management', *Gender, Work and Organization*, 1: 2–22.

Collinson, D., and Hearn, J. (eds) (1996). *Masculinity and Management*. London: Sage.

Collinson D., Knights, D., and Collinson, M. (1990). *Managing to Discriminate*. London: Routledge.

Connell, R. (2005/1995). *Masculinities*. Cambridge: Polity Press.

Connell, R. W., and Messerschmidt, James W. (2005). 'Hegemonic Masculinity: Rethinking the Concept', *Gender and Society*, 19(6): 829–59.

Dagens Industri (2012). 'Slut på bonusfest, ... men de fasta lönerna höjs' (End of the bonuses party, ... but fixed wages increased), Stockholm, 21 Feb.

Dashper, K. (2011). 'Gender in Sport: An Analysis of Equestrian Sports', doctorate, Keele University, UK.

Davidson, M. J., and Cooper, C. L. (1992). *Shattering the Glass Ceiling: The Women Manager*. London: Paul Chapman Publishing.

De Certeau, M (1984). *The Practice of Everyday Life*. New Haven: Yale University Press.

Dembinski, P. H. (2009). *Finance: Servant or Deceiver, Observatoire de la Finance*, tr. K. Cook. Basingstoke: Palgrave Macmillan.

Dolan, A. (2011). ' "You Can't Ask for a Dubonnet and Lemonade!": Working Class Masculinity and Men's Health Practices', *Sociology of Health and Illness*, 33(4): 586.

Ekstrand P. (2005). *Tarzan and Jane*. Uppsala: Uppsala University.

Emig, R., and Rowland, A. (eds) (2010). *Performing Masculinity*. Basingstoke: Palgrave Macmillan.

Ferguson, K. E. (1984). *The Feminist Case Against Bureaucracy*. Philadelphia: Temple University Press.

Foucault, M. (1982). 'The Subject and Power', in H. L. Dreyfus and P. Rabinow (eds), *Beyond Structuralism and Hermeneutics*, 208–26. Brighton: Harvester Press.

Game, A. (1990). *Undoing the Social*. London: Routledge.

Gheradi, S. (1994). 'The Gender we Think, the Gender we Do in our Everyday Organizational Lives', *Human Relations*, 47(6): 591–610.

Gilmore, David C. (1990). *Manhood in the Making: Cultural Concepts of Masculinity*. New Haven: Yale University Press.

Hearn, J., and Morrell, R. (2012). 'Reviewing Hegemonic Masculinities and Men in Sweden and South Africa', *Men and Masculinities*, 15(1): 3–10.

High Pay Centre (2012). *The New Closed Shop: Who's Deciding on Pay? The Make up of Remuneration Committees*. London: High Pay Centre. www.highpaycentre.org (accessed 5 June 2012).

Holgersson, C. (2004). *The Recruitment of Managing Directors: A Study of Homosociality*. Stockholm: Stockholm School of Economics.

Hoskin, K. W., and Macve, R. H (1988). 'The Genesis of Accountability: The West Point Connections', *Accounting, Organisations and Society*, 13(1): 37–73.

Jeanes, E., Knights, D., and Yancey Martin P. (eds) (2011). *A Handbook on Gender, Work and Organization*. London and New York: Wiley.

Kanter, R. Moss (1977). *Men and Women of the Corporation*. New York: Basic Books.

Kerfoot, D., and Knights, D. (1992). 'Planning for Personnel? HRM Reconsidered', *Journal of Management Studies*, 29(5): 651–68.

Kerfoot, D., and Knights, D. (1993). 'Management, Manipulation and Masculinity: From Paternalism to Corporate Strategy in Financial Services', *Journal of Management Studies*, 30(4): 659–77.

Kerfoot, D., and Knights, D. (1996). ' "The Best is Yet to Come?" Searching for Embodiment in Management', in D. Collinson and J. Hearn (eds), *Masculinity and Management*, 78–98. London: Sage.

Knights, D. (1992). 'Changing Spaces: The Disruptive Impact of a New Epistemological Location for the Study of Management', *Academy of Management Review*, 17(3): 514–36.

Knights, D. (1997). 'Organization Theory in the Age of Deconstruction: Dualism, Gender and Postmodernism Revisited', *Organization Studies*, 18(1): 1–19.

Knights, D. (2002). 'Writing Organization Analysis into Foucault', *Organization*, 9(4): 575–93.

Knights, D. (2004). 'Passing the Time in Pastimes, Professionalism and Politics: Reflecting on the Ethics and Epistemology of Time Studies', *Time and Society*, 15(3): 251–6.

Knights, D. (2009). 'The Political Economy of Regulation in Financial Services: Discursive Reflections', presented at the Political Economy, Financialisation and Discourse Theory Conference, Cardiff University Business School, May.

Knights, D., and Kerfoot, D. (2004). 'Between Representations and Subjectivity: Gender Binaries and the Politics of Organizational Transformation', *Gender, Work and Organization*, 11(4): 430–54.

Knights, D., and Thanem, T. (2006). 'Embodying Emotional Labour', in David Morgan, Berit Brandth, and Elin Kvande (eds), *Gender, Bodies and Work*, 31–43. Aldershot: Ashgate.

Knights, D., and Tullberg, M. (2012). 'Managing Masculinity/Mismanaging the Corporation', *Organization*, 19(4) pp. 385–404.

Knights, D., and Willmott, H. (1999). *Management Lives: Power and Identity in Work Organisations*. London: Sage.

Lewis, Michael (2010). *The Big Short: Inside the Doomsday Machine*. New York: W. W. Norton & Co.

Liff, S., and Ward, K. (2001). 'Distorted Views through the Glass Ceiling: The Construction of Women's Understanding of Promotion and Senior Management Positions', *Gender, Work and Organization*, 8(1): 19.

McDowell, L. (1997). *Capital Culture. Gender at Work in the City*. Oxford: Blackwell.

McClure, B (2012). 'Mergers and Acquisitions: Why They Can Fail', Investopedia. http://www.investopedia.com/university/mergers/mergers5.asp#axzz1tuW9Kn9U (accessed 5 June 2012).

MacPherson, C. B. (1962). *The Political Theory of Possessive Individualism*. New York: Oxford University Press.

Mills, A., and Tancred, P. (1992). *Gendering Organizational Analysis*. London: Sage.

Mead, G. H. (1934). *Mind, Self and Society*. Chicago: University of Chicago Press.

Mörck, M., and Tullberg, M. (2005). 'Bolagsstämman: En homosocial ritual', *Kulturella Perspektiv*, 2: 33–42.

Morgan, G., and Sturdy, A. (2000). *Beyond Organizational Change: Structure, Discourse and Power in UK Financial Services*. London: Palgrave Macmillan.

Nordberg, M. (2002). 'Constructing Masculinity in Women's Worlds: Men Working as Pre-School Teachers and Hairdressers', *Nordic Journal of Feminist and Gender Research (NORA)*, 10(1): 26–37.

Pollard S (1965). *The Genesis of Modern Management: A Study of the Industrial Revolution in Great Britain*. Cambridge, MA: Harvard University Press.

Pullen, A., and Knights, D. (2007). 'Editorial: Undoing Gender: Organizing and Disorganizing Performance', *Gender, Work and Organization*, 14(6): pp. 505–11.

Roper, M. (1996). 'Seduction and Succession: Circuits of Homosocial Desire in Management', in D. Collinson and J. Hearn (eds), *Men as Managers: Managers as Men*, pp. 77–91. London: Sage.

Ryan, M. K., Haslam, S. A., and Postmes, T. (2007). 'Reactions to the Glass Cliff: Gender Differences in the Explanations for the Precariousness of Women's Leadership Positions', *Journal of Organizational Change Management*, 20(2): 182–97.

Sargent, P. (2000). 'Real Men or Real Teachers? Contradictions in the Lives of Men Elementary Teachers', *Men and Masculinities*, 2(4): 410–33.

Schechner, R. (2002). *Performance Studies: An Introduction*. London: Routledge.

Seidler, V. J. (1989). *Rediscovering Masculinity*. London: Routledge.

Seidler, V. J. (2007). 'Masculinities, Bodies and Emotional Life', *Men and Masculinities*, 10(1): 9–21.

Smithson, J., and Stokoe E. H. (2005). 'Discourses of Work–Life Balance: Negotiating "Genderblind" Terms in Organizations', *Gender, Work and Organization*, 12(2): 147.

Stewart, I. (2012). 'The Mathematical Equation that Caused the Banks to Crash', *The Observer*, 12 Feb. 2012. http://www.theguardian.com/science/2012/feb/12/black-scholes-equation-credit-crunch (accessed 5 June 2012).

Tajfel, H. (1970). 'Experiments in Intergroup Discrimination', *Scientific American*, 223: 96–102.

Tajfel, H., and Turner, J. C. (1986). 'The Social Identity Theory of Intergroup Behavior', in S. Worchel and W. G. Austin (eds), *Psychology of Intergroup Relations*, 25–34. Chicago: Nelson-Hall.

Tullberg, M., and Mörck, M. (2012). 'The Business Suit and the Performance of Masculinity', unpublished manuscript, University of Gothenburg, School of Business, Economics and Law.

Waggoner, J., and Krantz, M. (2012). 'Analysis: Shareholder Uprisings on Executive Pay Limited, So Far', *USA Today*, 20 Apr. http://www.calgaryherald.com/business/business/6492427/story.html#ixzz1tucvJAHc (accessed 5/6/12).

Wajcman, J. (1998). *Managing like a Man: Women and Men in Corporate Management*. University Park, PA: Pennsylvania University Press.

Wahl, A. (ed.) (1995). *Men's Perceptions of Women and Management*. Stockholm: Ministry of Health and Social Affairs.

Wahl, A. (2001). 'From Lack to Surplus', in S.-E. Sjöstrand, J. Sandberg, and M. Thyrstrup (eds), *Invisible Management*, 63–82. Mitcham: Thomson Learning.

West, C., and Zimmerman, D. H. (1987). 'Doing Gender', *Gender and Society*, 1(2): 125–51.

Whittington, R. (1993). *What is Strategy—and does it Matter?* London: Routledge.

Williams, C. L. (1995). 'The Glass-Escalator: Hidden Advantages for Men in the Female Professions', in M. Kimmel and M. Messner (eds), *Men's Lives*, 78–89. Boston: Allyn and Bacon.

Williams, Simon J., and Bendelow, G. (1998). *The Lived Body: Sociological Themes, Embodied Issues*. London: Routledge.

Young, I. M. (2005). *On Female Body Experience: Throwing like a Girl and Other Essays*. Oxford: Oxford University Press.

CHAPTER 24

MASCULINITIES IN MULTINATIONALS

JANNE TIENARI AND ALEXEI KOVESHNIKOV

INTRODUCTION

> Global managers have exceptionally open minds...[they] push the limits
> of the culture...[and] sort through the debris of cultural excuses and find
> opportunities to innovate.
>
> (Gupta and Govindarajan 2002: 116)

'GLOBAL managers' find their playing field in multinationals. Business is increasingly carried out across national boundaries and companies with operations in several countries reign supreme, often with ownership distributed across nations. The term multinational corporation prevails, although some scholars and practitioners prefer to talk about multinational enterprises. The management of these organizations is portrayed as particularly complex, with a need to embrace global integration and overcome cultural challenges in fostering innovation, effectiveness, and performance (Perlmutter 1969; Bartlett and Ghoshal 1992; Gupta and Govindarajan 2002). The message is that managing multinationals requires exceptional people who are capable of heroic deeds.

Management in multinationals is infused with specific masculinities as the top echelons of these organizations are taken up by a particular type of man and the ways in which they operate routinely exclude others from positions of influence. A masculinist image of the 'ideal' corporate executive is reproduced (Connell and Wood 2005; Tienari et al. 2010). This affects recruitment, reward, and promotion practices in the multinational as specific forms of masculinity are valorized, supported, and celebrated while others are silenced, marginalized, and excluded. Nevertheless, the extant literature in the field of international business and management remains blind to these questions.

This is unfortunate, as the importance of multinationals transcends their organizational boundaries in the contemporary global economy. Encouraging and rewarding specific forms of masculinity reproduces particular assumptions about family, father-hood, and motherhood (Calás and Smircich 1993; Hearn *et al.* 2008). More generally, multinationals are social spaces where transnational elites operate. Their privileged key people draw from and reproduce a global system of inclusion and exclusion, which serves to perpetuate inequality within and across societies (Bauman 1998; Acker 2004).

In this chapter, we revisit influential texts on the management of multinationals. These widely read and cited texts represent a genre of academic writing where the focus is on individuals who thrive in the multinational context. This genre has remained popular throughout recent decades. We argue that the texts and the idealized image of managers that they represent are based on problematic assumptions. They (re)create meanings that go far beyond the global openness and cultural sensitivity which they claim to advocate. We scrutinize these assumptions and disrupt the masculine image of managing multinationals.

MANAGING MULTINATIONALS

Burgeoning research interest in multinational corporations reflects the growing power of these organizations (Seno-Alday 2010). In the global economy, multinationals account for a large part of international trade. They demand favourable conditions for their operations and they 'vote' in national elections (Vaaler 2008). Through mergers, acquisitions and direct investments, the ownership base of an increasing number of for-merly domestically contolled companies becomes multinational. Foreign ownership, in turn, tends to change management practices in these organizations (Tainio *et al.* 2003). Through outsourcing and offshoring decisions multinationals further impact upon the restructuring of national economies and societies.

The ways in which multinationals operate have been studied from a variety of theo-retical perspectives. The bulk of this research has been functionalist in orientation, dis-playing what is known as the economic perspective (Buckley and Casson 1976), while Hofstede's (1980) idea of culture as the 'programming of the mind' offered a template for advancing a culturalist perspective on multinationals. The evolutionary perspective, in turn, has conceptualized the multinational as a differentiated network (Nohria and Ghoshal 1997) and pointed towards the importance of socialization processes, man-agement of shared values, and normative integration in addressing cross-cultural and cross-divisional conflicts in multinationals. Multinationals have also been subjected to comparative institutional (Kostova and Zaheer 1999) and national business systems (Whitley *et al.* 2001) analyses. These institutionalist perspectives emphasize how institu-tions in both home and host environments affect the operations of multinationals.

We selected three well-known texts by influential scholars to represent state-of-the-art theorizing on the management of multinationals. The texts span more than three decades. First, as early as the 1960s, Howard Perlmutter pondered the degrees of multi-nationality and identified three different attitudes held by executives toward building a multinational enterprise. He termed them ethnocentrism, polycentrism, and geocentrism (Perlmutter 1969). He presented geocentrism—the ability of an organization to seek the most competent individuals, regardless of nationality, to solve problems anywhere in the world—as the most advanced form of managing multinationals. Perlmutter cemented the idea that the management of multinationals requires specific qualities that only outstanding individuals possess.

Second, Christopher Bartlett and Sumantra Ghoshal (1992) advocated the notion of transnational managers who are attentive to the local specifics of national markets, but are simultaneously able to maintain a global view. They argued that transnationals require three kinds of specialists (business, country, and functional managers) as well as senior executives who nurture the specialists and coordinate their efforts. Third, and finally, Anil Gupta and Vijay Govindarajan (2002) suggested that creating a global mindset is one of the central ingredients required for business in the multinational context. Building intelligence in observing and interpreting the dynamic world, they argued, is paramount in managing multinationals. A global mindset allows executives to be open towards diversity across cultures and markets and to benefit from it.

We suggest that these three texts represent an influential genre of academic writing that develops particular assumptions about what managing multinationals means. They are concerned with individuals and enthusiastic about the opportunities offered by the global marketplace. They seek out insightful ways to benefit from these opportunities. We suggest that, in so doing, the texts (re)construct an idealized image of managing multinationals and define particular personal traits and capabilities required for such management. The underlying assumptions in the texts are particularly intriguing from the point of view of men and masculinities.

REVISITING MANAGEMENT TEXTS

Drawing insights from critical studies of men as well as feminist theory, we reread the articles by Perlmutter (1969), Bartlett and Ghoshal (1992), and Gupta and Govindarajan (2002). This is done in three moves. First, we search for explicit references to men and women in the texts. We locate passages where they could have been referred to but, for some reason, were not. In this way, we uncover their taken-for-granted assumptions about men and women as managers and discern qualities that are attached to the exceptional people who are needed to run multinationals. We use Collinson and Hearn's (1994) idea of multiple masculinities (authoritarianism, paternalism, entrepreneurialism, informalism, and careerism) as a sense-making device, and consider how the image of the 'ideal' executive in the multinational is (re)constructed through meanings related to men and masculinities.

Second, we focus on what the texts assume about the lives of the executives valorized by them. Feminist research argues that, according to the dominant discourse on multinationals, women are not able to travel as much as is required of executives due to their

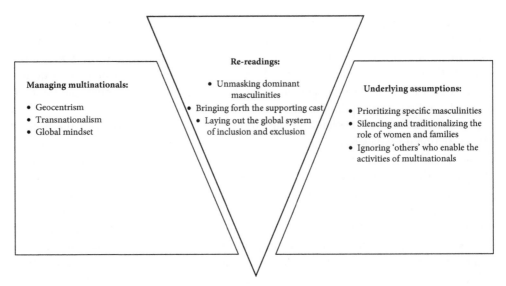

FIGURE 24.1 Revisiting management texts

family obligations, and are thus deemed unable to meet the challenges of the global playing field (Calás and Smircich 1993). Executive management in multinationals remains in
practice reserved for men whose wives and domestic help take care of the home (Calás
and Smircich 1993; Reis 2004; Hearn *et al.* 2008). We consider how descriptions of the
'ideal' executive's qualities and tasks are related to assumptions about their family status
and arrangements. These assumptions are hidden in the texts, and unmasking them is
an analytical exercise.

Third, and in similar manner, we consider the kind of operational context (re)con-
structed by the texts for the 'ideal' executives. From a critical perspective, multinationals
have been conceived of as a platform on which new transnational business elites operate (Beaverstock 2002) and where others are systematically disadvantaged and excluded
(Bauman 1998). Hence it has been argued that multinationals act as spaces where abun-
dant societal inequality originates. In other words, assumptions about management in
multinationals have repercussions far beyond the boundaries of these organizations
(Connell 1998; Acker 2004). We read the three exemplary texts against this backdrop
and consider how they describe the playing field of the exceptional individuals valorized
by them (see Figure 24.1).

GEOCENTRIC MEN

Howard Perlmutter (1969) innocently talks about the 'best men' when he discusses
the building blocks of sound management in multinationals. Perlmutter mentions

the staffing of managerial ranks as a key challenge in these organizations: 'Whenever replacements for … men are sought, it is the practice, if not the policy, to "look next to you at the head office" and pick someone (usually a home-country national) you know and trust' (1969: 9–10). The apparent problem, then, is that men tend to favour their fellow nationals when they make choices for key positions. Perlmutter suggests that this problem can be directly confronted in what he calls the geocentric organization where the recruitment of senior executives is based on merit and initiative: 'Within legal and political limits, [geocentric organizations] seek the best men, regardless of nationality, to solve the company's problems anywhere in the world' (p. 13).

In a section titled 'A Geocentric Man—?' Perlmutter concludes that 'the geocentric enterprise depends on having an adequate supply of men who are geocentrically oriented' (p. 17). This is not easy because 'building trust between persons of different nationality is a central obstacle' (p. 18). Perlmutter nevertheless finishes the article on a positive note, envisioning that 'if we are to judge men … "by the kind of world they are trying to build," the senior executives engaged in building the geocentric enterprise could well be the most important social architects of the last third of the twentieth century' (p. 18). In a nutshell, Perlmutter argues for a decidedly meritocratic management system populated by individuals free of national sentiments. This would enable the multinational organization to reach its full potential.

Perlmutter's article makes no reference to homosociality. Back in 1969, even those international business and management scholars who would have been interested in these questions lacked the vocabulary to discuss homosocial reproduction and its effects (Kanter 1977). Nevertheless, 'picking someone … you know and trust' clearly suggests that relations between men of the same nationality in the multinational context is a relevant basis for homosociality. However, Perlmutter's alternative meritocratic system remains available in a similar way to men only or, more precisely, to a specific type of man.

Geocentric men must be able to withstand considerable pressure. They need to be constantly available and willing to move wherever they are needed. To accomplish this, geocentric men need a supporting cast at home. Pressure is transferred to their families. When explicitly discussing the costs and risks of the geocentric orientation, Perlmutter lists a number of political and social repercussions such as excessive distribution of power and problems with personnel and with executives re-entering their home organization. He only subsequently mentions 'the human stress' of a geocentric career, such as major adjustments expected from the managers' family. Perlmutter goes on to optimistically conclude that the payoffs of the geocentric orientation are, in any case, much greater both for the company and the manager than any costs that it involves on a personal level.

On this basis, particular forms of masculinity can be read in Perlmutter's text. The notion of 'best men' reconstructs the masculinity of *careerism*. The 'best men' are committed to their organization's goals, willing to work long hours, travel extensively, and move overseas when necessary. Their lives—as well as the lives of their family members—are guided by their organizational commitments and career aspirations. These men also embrace the masculinity of *informalism*. Being geocentric in orientation, they need to be able to build a network of similarly oriented men whom 'they know and trust.' Thus, a form of homosociality based on national identification (which is presumably prevalent among ethnocentric managers), is replaced by homosociality between geocentrically minded men who are not preoccupied with national interests and sentiments, but who build their networks out of 'the best men of different nations' (p. 17).

Finally, the masculinity of *entrepreneurialism* is pertinent to geocentric men. Perlmutter states that these men 'erect promises of a greater universal sharing of wealth and a consequent control of the explosive centrifugal tendencies of our evolving world community' (p. 18). They take their multinational towards its full potential by being motivated to work for worldwide objectives. Being open to new business ideas that can surface anywhere in the world to serve present and future customers, and constantly seeking new possibilities to raise money, to build a plant, or to conduct R&D constitute the essence of geocentric men. Only with an entrepreneurial mindset can a multinational become genuinely geocentric.

Howard Perlmutter, professor at the Wharton School at the University of Pennsylvania, became a world-renowned authority on globalization issues. 'He always had a gift for prediction', the Wharton website tells us in 2012. Perlmutter's work on *The Multinational Firm and the Future* (1972) 'accurately forecast viability and legitimacy issues for multinational corporations'. Perlmutter's ideas on developing the geocentric organization where the 'best men' prevail stuck. When we fast-forward to the early 1990s, this rhetoric turns to considerations of transnationality. The essence of management in the multinational is, however, still about the ability of exceptional individuals to see the big picture and to work for the whole rather than its constituent parts. This is exemplified by Christopher Bartlett and Sumantra Ghoshal's influential ideas on transnational managers.

Transnational Managers

In an article titled 'What is a Global Manager?', Bartlett and Ghoshal (1992) offer inspiring stories of great leaders in multinationals, testifying to their extraordinary skills. Although no explicit references are made to either men or women, all the examples cited are male executives. The text leads with the heroic tale of Mr Leif Johansson, a visionary Swede who is said to have transformed the household appliance division of Electrolux into a globally integrated and strategically agile business unit that was able 'to capture valuable global-scale efficiencies while reaping the benefits of a flexible response to national market fragmentation' (p. 103). Johansson represents the transnational business manager in Bartlett and Ghoshal's model.

Three other stories of successful men follow. Mr Howard Gottlieb of NEC (the transnational country manager) and Mr Wahib Zaki of Procter & Gamble (the transnational functional manager) are introduced as two leaders who were able to turn their respective units into 'hothouses of entrepreneurship and innovation', thereby changing the role of these units from mere implementers of corporate strategies into important contributors. Finally, the transnational corporate manager—the leader, talent scout, and developer who keeps the multinational together—is given the face of Mr Floris Maljers, the Dutch co-chairman of Unilever. We are told that Mr Maljers has made the recruitment, training, and development of promising executives a top corporate priority and, by so doing, developed a strong pool of specialized senior managers in the organization.

A specific type of man prevails in Bartlett and Ghoshal's text, a man very similar to Perlmutter's. He is a visionary who, in a world 'riven by ideology, religion, and mistrust…more fragmented, more at odds, than at any time since, arguably, World War II' (p. 101), figures out how to run global operations in an integrated way, innovatively, efficiently and profitably. This man is not 'an elite jet-setter' or 'a big-picture overseer' (although such an interpretation could be made on the basis of the text), but instead possesses 'the skills, knowledge, and sophistication to operate in a more tightly linked and less classically hierarchical network' (p. 102) that consists of transnational business, country, regional, and functional managers. The focus of this man's job resolves around 'the intense interchanges and subtle negotiations' within the network (p. 108).

Following the categorization by Collinson and Hearn (1994), specific masculinities are embraced in the representation of transnational managers in Bartlett and Ghoshal's text. First, the masculinity of *careerism* can be discerned. The implicitly present hierarchical division between business, country, and functional managers suggests that, by fulfilling the specific requirements and roles imposed on them by the transnational multinational, managers are able to move ever higher in the corporate hierarchy. In no way, the authors claim, are any of these positions meant to be 'a warehouse for corporate misfits' or 'a graveyard for managerial has-beens'. All are important and, if managers perform their functions well, they can be catapulted at any moment 'into a much more central role' (p. 106).

Second, the very essence of the transnational multinational presupposes that transnational managers develop and function in a tight and dispersed network that includes managers from different hierarchical levels and geographic regions. Channelling important information between different units effectively and in a timely manner is presented as one of its most critical features. Bartlett and Ghoshal suggest that one way for transnational business managers to coordinate these information flows is through informal communication. Similarly, functional managers need to develop their informal networks to channel information concerning opportunities that pop up around the world. Hence, the vitality of informal networks points towards the importance of the masculinity of *informalism* for managers in Bartlett and Ghoshal's transnational organization.

Third, the masculinity of *entrepreneurialism* is embraced. For instance, transnational functional managers must continuously 'scan their local environment for new developments…and champion innovations with transnational applications' (p. 108). Country managers, in turn, are to develop their units into 'hothouses of entrepreneurship and innovation' and business managers are to be 'the architects' who initiate and lead the debate on how to achieve an efficient distribution of assets and resources. This seeks to optimize robustness and flexibility in the long term rather than to minimize short-term costs. Efforts to achieve this presuppose a particular entrepreneurial masculinity that is continuously creative and innovative.

In all, then, it can be seen that, despite the gap of some twenty-three years between the texts, the forms of masculinity embraced by managers in Bartlett and Ghoshal's transnational multinational are notably similar to those represented by Perlmutter's geocentric men. Both images of 'ideal' executives rely heavily on their informal networks, both need to be continuously innovative and entrepreneurial, and, by so doing, both strive to move up the corporate ladder. As in the case of the geocentric men, we can assume that transnational managers rely on support from their families, although in fact Bartlett and Ghoshal do not touch upon this issue at all. The role of managers' families in enabling the transnational multinational remains silenced and unrecognized. Managers manage, without responsibilities at home.

By the early 1990s, Christopher Bartlett, professor at Harvard Business School, and Sumantra Ghoshal, professor at London Business School, were already seminal figures in the field of international business and management. Their *Transnational Management* had become a hugely influential textbook, which continues to be used in universities across the world. The search for the Holy Grail of managing multinationals continued. Anil Gupta and Vijay Govindarajan's (2002) work is illustrative of this.

GLOBAL MINDSET MANAGERS

Gupta and Govindarajan (2002: 116) begin their article by quoting the (ex-)CEO of ABB, Mr Percy Barnevik: 'global managers have exceptionally open minds...push the limits of the culture...[and] sort through the debris of cultural excuses and find opportunities to innovate'. Gupta and Govindarajan introduce a particular type of manager blessed with a 'global mindset'. These individuals combine 'an openness to diversity and an awareness of it across cultures and markets with a propensity and ability to synthesize across this diversity'. As one of the exemplary managers in the article puts it, a global mindset is about the ability to ambitiously combine 'Chinese costs with Japanese quality, European design, and American marketing'.

No direct references to men or women are made, but all but one of the examples in the texts deal with men. Gupta and Govindarajan state that to achieve a global mindset takes continuous cultivation. They call it 'the quest', as every multinational should be constantly searching for managers with a global mindset. Gupta and Govindarajan propose a number of managerial characteristics that the global mindset is connected to. These are (1) curiosity about the world and commitment to becoming smarter about how the world works; (2) the ability to explicitly and self-consciously articulate current mindsets; (3) openness to diversity and novelty; and (4) and the discipline required to develop an integrated perspective that weaves together diverse strands of knowledge about cultures and markets.

Multinationals need to value 'global experience' as the highest merit on the basis of which managers are promoted to senior executive levels. Managers, in turn, should be motivated to actively interact with other global mindset people from other cultures and markets and not be 'trapped in one's own mental web'. To illustrate this, Gupta and Govindarajan narrate the heroic stories of Mr Douglas Daft (Coca-Cola Company) and Mr Andreas Renschler (DaimlerChrysler), who are presented as culturally sensitive and knowledgeable global mindset men; they are 'always away from home'. We are told that both have managed to steer their multinationals successfully towards a 'think global, act local' approach.

Gupta and Govindarajan suggest that there are a number of ways through which global mindset managers can cultivate knowledge regarding diverse cultures and markets. First, they need to obtain formal education, which can take the form of self-study courses, university-based education, in-company seminars, and management education programmes. Second, these managers must actively take part in the cross-border endeavours of the multinational and in job rotations between regions, divisions, and functions. This will facilitate development of their interpersonal and social ties with other managers based in different locations. Finally, and most importantly, global mindset managers need to possess immersion experiences in foreign cultures and eagerly take on expatriate assignments.

Similarly to Perlmutter's geocentric multinationals and Bartlett and Ghoshal's transnationals, global mindset multinationals are committed to using 'merit'—the most valuable of which is global experience—rather than nationality as the prime driver for rewards and career mobility. High value is put on the ability of managers to move overseas. The aim is, again, to develop individuals who can rise above national sentiments to see and comprehend the 'big picture' and to integrate various local opinions, views, and knowledge. Managers are to be assisted in developing interpersonal and social networks to help them to stay open to other cultures and markets.

Global mindset managers are depicted as representing forms of masculinity similar to those embraced by geocentric men and transnational managers. First, the important role of social and interpersonal ties across the world points toward the masculinity of *informalism*. Second, the constant urge of global mindset managers to increase their global experience, which is perceived to constitute the highest merit, indicates that the masculinity of *careerism* is privileged. Finally, the overarching idea behind cultivating global mindsets for multinationals to combine 'Chinese costs with Japanese quality, European design, and American marketing' points towards privileging the masculinity of *entrepreneurialism,* which prioritizes performance levels, budget targets, and profits, and makes economic efficiency the prime criterion of success for global mindset managers.

Taking on expatriate assignments is acknowledged to be very expensive for the company as well as for the individual, due to the increasing preponderance of dual-career marriages. Curiously, after acknowledging this, Gupta and Govindarajan go on to suggest that 'companies need to target expatriate assignments toward high-potential managers...to ensure that their stay abroad fosters cultural learning rather than cultural isolation' (p. 122). A dual-career marriage, then, might reduce the likelihood that a manager would be identified as having high potential.

Anil K. Gupta is Professor of Global Strategy and Entrepreneurship at the Smith School of Business, the University of Maryland at College Park. Vijay Govindarajan is Professor of International Business at Tuck School of Business at Dartmouth College. Both are regarded as leading experts on strategy and globalization, and they continue to be active in the circuit of global business thinkers and speakers.

DISRUPTING THE IMAGE OF THE 'IDEAL' MANAGER

The three influential texts revisited above (re)construct an idealized image of management and managers in multinationals. They are written by men about men for men, as management texts tend to be (Collinson and Hearn 1994). Not all men qualify, however. A particular type of man prevails. The texts carry assumptions not only about preferred masculinities, but also about family, fatherhood, and motherhood. Descriptions of the right qualities of multinational executives (needed for the tasks envisioned) promote, first, masculinities such as careerism, informalism, and entrepreneurialism that are presented as 'normal' traits of ambitious individuals. The qualities remain unquestioned and alternatives are not considered. Second, the texts reinforce and render 'natural' a traditional male breadwinner family model where women are destined to stay at home. This is assumed rather than explicitly addressed and argued for, let alone questioned. Overall, the construed image is uncompromising, and it has not changed to any significant extent since the 1960s.

While mainstream international business and management research remains blind to these questions, critical studies have focused on family-related assumptions that underlie career-making in multinationals. In their study of managerial work and gender relations in the multinational, for example, Hearn et al. (2008) bring to the fore that someone else must take care of his home while the executive develops his exceptional qualities and climbs the organizational hierarchy. Hearn et al. (2008) focus on Finland-based multinationals, and highlight the important and often decisive role of domestic arrangements for the development and success of managerial careers. Their study shows that, constrained by their obligations to fulfil a traditional woman's role of being primarily responsible for children and the household, female managers cannot in practice engage to the same extent as male managers in transnational managerial work, which includes foreign assignments and constant travelling. At the same time, the support of the (woman) spouse and family are seen as crucial for the (man) expatriate's adjustment to the new cultural environment and his performance overseas. To provide that support, working women have to interrupt their own careers in order to follow their husbands' career moves abroad.

In her study of a European multinational, Cristina Reis (2004) similarly finds that the families of male managers are curiously important for how they are perceived by themselves and others. Being married (to a woman) is regarded as crucial for men as it is considered to be a sign of stability in life—and, thus, a prerequisite for good job performance as a manager in the multinational context. As in the study by Hearn and colleagues, the managers' wives studied by Reis are expected to adjust their employment to fit their husband's career aspirations. It becomes clear in both studies that married men are able to take advantage of their traditional family arrangements in living up to the stereotypical expectations for management in a multinational. A masculinized model of management *and* a conservative gender pattern persists in these organizations (Wajcman 1999).

While Perlmutter (1969), Bartlett and Ghoshal (1992), and Gupta and Govindarajan (2002) theorize on managers in multinationals as if they lived their lives in a vacuum, Wajcman (1999), Reis (2004), and Hearn *et al.* (2008) highlight the problematic assumptions behind representations of exceptional men and their success. The argument is that what managers in multinationals are expected to do at work, on the one hand, and how they organize their private life, on the other, are two sides of the same coin, although this remains unsaid. Multinationals have significant gendered consequences in the societies where they operate. It has been argued, for example, that the gender egalitarian tradition in the Nordic countries such as Finland, which is based on a double-earner family model, is being eroded by the norms and practices of multinationals (Tienari *et al.* 2005, 2010; Hearn *et al.* 2008). The traditional male-breadwinner model is making a triumphant comeback in Finnish society in and through multinationals.

Third, it is relevant to consider how texts such as the three examples revisited above (re)construct a particular context for the 'ideal' executives to operate in. In developing their qualities and jumping from one job assignment to the next, they rely not only on their partners at home, but on the work of many others. Marta Calás and Linda Smircich (2011) revisit Hearn *et al.*'s (2008) study of Finnish-based multinationals and note that, by focusing on highly skilled managers in a wealthy European country, Hearn *et al.* make 'various others disappear from view' (p. 422). Calás and Smircich ask who are the nannies taking care of the children of Finnish expatriates and where do they come from? The point is that it does not suffice for us to theorize on the operations of multinationals from the point of view of managers that comprise global elites. The 'various others' who enable the local activities of multinationals should also be included in the analyses to acknowledge that any transnational activity occurs 'in transnational social fields populated by people [with] interconnections and linkages in various national and transnational networks' (p. 423). Reis (2004) offers a similar argument. In the rare cases where the managers' wives in her study had demanding full-time jobs, they relied on less privileged women to do the 'unspoken work' in the home.

While exceptional individuals use multinationals as springboards for their careers, others experience multinationals in a fundamentally different way, providing them with the necessary infrastructure to be constantly and comfortably on the move. These people are expected to stay put. When they migrate, they are treated as unwanted aliens. Zygmunt Bauman (1998) speaks of tourists and vagabonds. Managers in multinationals are like tourists; they are welcome everywhere and pampered by others. Vagabonds, in contrast, are only welcome in the West when they are needed as nannies and the like, in jobs that the locals are reluctant to take up.

The argument in Calás and Smircich's (2011) criticism of Hearn *et al*'s (2008) focus on gender relations in Finnish families relates to this dynamic in the global economy. Echoing privileged views, Western management academics—mainstream and critical alike—tend to inadvertently valorize the tourists and disregard the vagabonds. We reproduce the idea that multinational organizations are primarily about exceptional individuals with the right qualities for management, and our criticism merely extends to highlighting gender-based inequalities in relation to white Western women.

The invisibility of others in the literature sustains crucial gender, race, and class-based disparities in globalization processes. The fact remains unchallenged that the ways in which multinationals function works to the detriment of poor women of colour around the world (Mohanty 2004; Berry and Bell 2012). Decisions made by managers embodying the right masculinities in multinationals have consequences that are shrugged off as inevitable economic forces or disembodied social trends (Acker 2004). However, there is nothing inevitable about these decisions. Intersecting markers of difference such as gender and ethnicity serve to legitimize practices that produce inequalities in the division of labour on a global scale (Calás *et al.* 2010).

Transnational Business Masculinity and its Discontents

Those men who are able to live up to the right masculinities in multinationals and who have a supporting cast at home that enables them to move from one job assignment to the next, gain entry into mobile elites. Multinationals become social spaces where aspiring elite members earn their spurs in what unfold as transnational social networks. Fiona Moore (2004) studied what she calls the social maps of such people. According to Moore, they do not belong to a 'global culture' that exists with little references to the local—rather, they possess complex connections to various global and local groups. The focal point in their lives is not a specific organization. In contrast, they are ambitious individuals who are concerned about their own personal development and career. They are likely to be less preoccupied with the long-term success of a particular multinational than with increasing their own opportunities to gain the right capabilities and credentials and to acquire the right contacts to move on and take up new challenges elsewhere. They are likely to jump from one multinational to another several times during the course of their career—and return favours by recruiting and promoting other individuals of the same persuasion (Gee *et al.* 1996).

In this way, Raewyn Connell (1998, 2001) argues, a specific type of hegemonic masculinity has found a home in multinationals operating in global markets. The ideal executive is a self-assertive and energetic individual who is constantly mobile and available for work. Connell notes that the masculinity thus cultivated and celebrated—somewhat ironically—motivates managers to become more egocentric, career oriented, and less organizationally committed. Transnational social networks that spread around 'global cities' facilitate such a mindset and lifestyle (Beaverstock 2002; Faulconbridge *et al.* 2009), which reinforces a sense of short-termism and a focus on maximizing one's personal gains (Gee *et al.* 1996). The dominant forms of masculinities that we identified in the three texts above can thus be argued to transcend the boundaries of specific multinationals. They are portable, and they become the currency found in transnational social networks and nurtured by them.

The other side of the coin is a highly insecure and stressful work environment. In order to be considered part of elite networks, managers in multinationals need to constantly prove their worth. The right masculinities and lifestyles must be continuously demonstrated. The pace is hectic, every new challenge is a decisive one, and successes in the past do not matter much. At the same time, Connell and Wood describe management in multinationals as driven by an intense multidirectional peer scrunity. Despite encouraging egocentrism, the managers' work is increasingly collective in the sense that it is characterized by mutual control and comparison; a particular form of homosociality, perhaps. All this leads Connell and Wood (2005) to ask the question: 'What holds people in high-stress, unhealthy, insecure jobs?' Their answer is a simple and persuasive one: 'Lots of money is part of the answer . . . Also significant is the access to power' (p. 358).

Connell and Wood (2005) argue that collective scrunity among managers in multinationals is a powerful countercurrent to the competitive individualism of contemporary business ideology. 'An intense and stressful labor process creates multiple linkages among managers and subjects them to mutual scrutiny, a force for gender conservatism' (p. 347). This is reflected in managers' conformity of appearance. Not only complying to the right dress code, but managing one's own body as 'an entity' becomes a significant part of the job. Appearing to be fit and energetic is crucial. Amanda Sinclair (2011: 126) argues that 'there is a danger that [this] new body awareness is recruited to the project of becoming a "corporate athlete": a leader who can work harder and longer towards ends that are exploitative'. Again, our attention is drawn to masculinities and male bodies. They are inescapable, no matter how one views management and managers in multinationals.

Knights and Tullberg (2012) take the explanatory power of masculinities in the global economy perhaps the furthest in their analysis of the financial crisis that shocked the Western world in 2008. Knights and Tullberg suggest a link between managing masculinity and mismanaging the corporation. They argue that not only individual material and symbolic self-interest played a role in the events. Masculine fragilities conditioned the excesses that led to the crisis. Again, we can see how masculinities and masculine discourses within the global business elite transcend the boundaries of individual organizations and show their systemic nature.

Finally, Tienari *et al.* (2010) take a different perspective to look at masculinities in multinationals. They observe the prevalence of transnational business masculinity that is shared by managers of different nationalities who 'live in a space where in exchange for handsome material rewards they are expected to offer their body and mind to the rhythm and spatial challenges [of multinationals]' (p. 41). However, Tienari and colleagues are convinced that the national identification of managers persists as a form of masculinity that provides a counterbalance to the pressures in their mobile transnational work. National identification offers 'protection' in the 'crazy life' (p. 47) of the weary men in their hectic jobs. This comforting side of ethnocentrism is homosociality among male compatriots and nurturing the possibility of returning to a job back home. This is silenced in management texts' enthusiastic celebrations of geocentrism, transnationalism, and global mindset. But it is likely to lurk beneath the surface.

CONCLUDING REMARKS

In this chapter, we have revisited influential texts on the management of multinational corporations from the viewpoint of men and masculinities. We have disrupted the masculine image that the texts (re)construct. We have suggested that the 'ideal' executive in the multinational reflects the gendered nature of today's global capitalism, and that it has repercussions for men, women, families, and gender relations across societies. The implications of managing multinationals in specific ways by a particular type of masculine individual spill over organizational boundaries. Multinationals transfer Western knowledge, which is deemed globally superior. They dictate what types of individuals are best suited to turn that knowledge into profit and value. The image of the Western white male business executive, embracing the masculinities of informalism, careerism, and entrepreneurialism, is reproduced. This takes the form of a transnational business masculinity that is superficially sensitive to difference and diversity, but remains egocentric and exploitative of others (Connell and Wood 2005).

However, multinationals remain complex organizations. A spectrum of gender patterns across contexts must be recognized in the increasingly complex global business environment (Connell and Wood 2005). Different socio-cultural traditions of displaying masculinity jockey for position in multinationals and transnational business masculinity is challenged by prevalent masculinities in different national and local contexts (Tienari et al. 2010). Contrary to what Perlmutter (1969) and others after him have envisioned, national identification is not left behind when one makes a career in a multinational. Transnational business masculinity and national identification are likely to be in constant tension in these organizations.

One possible direction for future research on masculinities in multinationals would deal with the ways in which transnational business masculinity is challenged when it is imposed on the non-Western affiliates of Western multinationals. Even more importantly, considering the increasing global activity of multinationals originating from emerging markets such as China, India, and Russia, the hegemonic Western masculinity may be contested by hegemonic masculinities prominent in those contexts. The result is a meeting of Western business masculinity and those originating elsewhere. As the economic and geopolitical domination of Western multinationals is increasingly challenged, we are likely to see changes in the forms of masculinities that are idealized among corporate managers and management scholars.

When we study these 'new' masculinities we should not lose sight of the wider repercussions of the dominance of particular types of men in multinationals. We must not forget about the various others who provide the infrastructure for the exceptional individuals who get all the credit. While we are all traversed by multi- or transnational social and economic conditions fuelled by powerful ideological and political apparatuses, we articulate our experiences from the particular vantage point of our particular place (Calás *et al.* 2010). Analyses of transnational processes must be nuanced by paying attention to local specificities as globalized capitalism takes various forms in different nation-states and local settings.

REFERENCES

Acker, J. (2004). 'Gender, Capitalism, and Globalization', *Critical Sociology*, 30(1): 17–42.

Bartlett, C., and Ghoshal, S. (1992). 'What is a Global Manager?', *Harvard Business Review*, 70(5): 101–8.

Bauman, Z. (1998). *Globalization: The Human Consequences*. New York: Columbia University Press.

Beaverstock, J. V. (2002). 'Transnational Elites in Global Cites: British Expatriates in Singapore's Financial District', *Geoforum*, 33(4): 525–38.

Berry, D. P., and Bell, M. P. (2012). '"Expatriates": Gender, Race and Class Distinctions in International Management', *Gender, Work and Organization*, 19(1): 10–28.

Buckley, P. J., and Casson, M. C. (1976). *The Future of the Multinational Enterprise*. London: Holmes & Meier.

Calás, M. B., and Smircich, L. (1993). 'Dangerous Liaisons: The Feminine in Management Meets Globalization', *Business Horizons*, (Apr.): 73–83.

Calás, M. B., and Smircich, L. (2011). 'In the Back and Forth of Transmigration: Rethinking Organization Studies in a Transnational Key', in E. Jeanes, D. Knights, and P. Yancey Martin (eds), *Handbook of Gender, Work and Organization*, 411–28. Chichester: John Wiley & Sons.

Calás, M. B., Smircich, L., Tienari, J., and Ellehave, C. F. (2010). 'Observing Globalized Capitalism: Gender and Ethnicity as an Entry Point', *Gender, Work and Organization*, 17(3): 243–7.

Collinson, D., and Hearn, J. (1994). 'Naming Men as Men: Implications for Work, Organization and Management', *Gender, Work and Organization*, 1(1): 2–22.

Connell, R. W. (1998). 'Masculinities and Globalization', *Men and Masculinities*, 1(1): 3–23.

Connell, R. W. (2001). 'Understanding Men: Gender Sociology and the New International Research on Masculinities', *Social Thought and Research*, 24: 13–31.

Connell, R. W., and Wood, J. (2005). 'Globalization and Business Masculinities', *Men and Masculinities*, 7: 347–64.

Faulconbridge, J. R., Beaverstock, J. V., Hall, S., and Hewitson, A. (2009). 'The "War for Talent": The Gatekeeper Role of Executive Search Firms in Elite Labour Markets', *Geoforum*, 40: 800–8.

Gee, J. P., Hull, G., and Lankshear, C. (1996). *The New Work Order: Behind the Language of the New Capitalism*. Sydney: Allen & Unwin.

Gupta, A. K., and Govindarajan, V. (2002). 'Cultivating a Global Mindset', *Academy of Management Executive*, 16(1): 116–26.

Hearn, J., Jyrkinen, M., Piekkari, R., and Oinonen, E. (2008). '"Women Home and Away": Transnational Managerial Work and Gender Relations', *Journal of Business Ethics*, 83: 41–54.

Hofstede, G. (1980). *Culture's Consequences: International Differences in Work-Related Values*. London: Sage.

Kanter, R. M. (1977). *Men and Women of the Corporation*. New York: Basic Books.

Knights, D., and Tullberg, M. (2012). 'Managing Masculinity/Mismanaging the Corporation', *Organization*, 19(4): 385–404.

Kostova, T., and Zaheer. S. (1999). 'Organizational Legitimacy under Conditions of Complexity: The Case of the Multinational Enterprise', *Academy of Management Review*, 24(1): 64–81.

Mohanty, C. T. (2004). *Feminism without Borders* (4th printing). Durham, NC, and London: Duke University Press.

Moore, F. (2004). 'Tales of the Global City: German Expatriate Employees, Globalisation and Social Mapping', *Anthropology Matters Journal*, 6(1): 1–12.

Nohria, N., and Ghoshal, S. (1997). 'Differentiated Fit and Shared Values: Alternatives for Managing Headquarters–Subsidiary Relations', *Strategic Management Journal*, 15(6): 491–502.

Perlmutter, H. V. (1969). 'The Tortuous Evolution of the Multinational Corporation', *Columbia Journal of World Business*, 4: 9–18.

Perlmutter, H. V. (1972). 'The Multinational Firm and the Future', *Annals of the American Academy of Political and Social Science*, 403(1): 139–52.

Reis, C. (2004). *Men Working as Managers in a European Multinational Company*. Mering, Germany: Rainer Hampp Verlag GmbH.

Seno-Alday, S. (2010). 'International Business Thought: A 50-Year Footprint', *Journal of International Management*, 16: 16–31.

Sinclair, A. (2011) 'Leading with Body', in E. L. Jeanes, D. Knights, and P. Yancey Martin (eds), *Handbook of Gender, Work and Organization*, 117–30. Chichester: John Wiley & Sons.

Tainio, R., Huolman, M., Pulkkinen M., Ali-Yrkkö, J., and Ylä-Anttila, P. (2003). 'Global Investors Meet Local Managers: Shareholder Value in the Finnish Context', in M.-L. Djelic and S. Quack (eds), *Globalization and Institutions: Redefining the Rules of the Economic Game*, 37–56. Cheltenham: Edward Elgar.

Tienari, J., Søderberg, A.-M., Holgersson, C., and Vaara, E. (2005). 'Gender and National Identity Constructions in the Cross-Border Merger Context', *Gender, Work and Organization*, 12(3): 217–41.

Tienari, J., Vaara, E., and Meriläinen, S. (2010). 'Becoming an International Man: Top Manager Masculinities in the Making of a Multinational Corporation', *Equality, Diversity and Inclusion: An International Journal*, 29(1): 38–52.

Vaaler, P. M. (2008). 'How do MNCs Vote in Developing Country Elections?', *Academy of Management Journal*, 51(1): 21–43.

Wajcman, J. (1999). *Managing like a Man: Women and Men in Corporate Management*. Sydney: Allen & Unwin.

Whitley, R., Morgan, G., and Kristensen, P. H. (2001). *The Multinational Firm: Organising Across Institutional and National Divides*. Oxford: Oxford University Press.

Index